T0180762

Lecture Notes in Artificial Intelligence   12976

Subseries of Lecture Notes in Computer Science

More information about this subseries at http://www.springer.com/series/1244

Nuria Oliver · Fernando Pérez-Cruz ·
Stefan Kramer · Jesse Read ·
Jose A. Lozano (Eds.)

Machine Learning and Knowledge Discovery in Databases

Research Track

European Conference, ECML PKDD 2021
Bilbao, Spain, September 13–17, 2021
Proceedings, Part II

 Springer

Editors
Nuria Oliver (iD)
ELLIS - The European Laboratory
for Learning and Intelligent Systems
Alicante, Spain

Avda Universidad, San Vicente del Raspeig
Alicante, Spain

Vodafone Institute for Society
and Communications
Berlin, Germany

Data-Pop Alliance
New York, USA

Stefan Kramer
Johannes Gutenberg University of Mainz
Mainz, Germany

Jose A. Lozano (iD)
Basque Center for Applied Mathematics
Bilbao, Spain

Fernando Pérez-Cruz (iD)
ETHZ and EPFL
Zürich, Switzerland

Jesse Read (iD)
École Polytechnique
Palaiseau, France

ISSN 0302-9743   ISSN 1611-3349 (electronic)
Lecture Notes in Artificial Intelligence
ISBN 978-3-030-86519-1   ISBN 978-3-030-86520-7 (eBook)
https://doi.org/10.1007/978-3-030-86520-7

LNCS Sublibrary: SL7 – Artificial Intelligence

This Springer imprint is published by the registered company Springer Nature Switzerland AG
The registered company address is: Gewerbestrasse 11, 6330 Cham, Switzerland

Preface

This edition of the European Conference on Machine Learning and Principles and Practice of Knowledge Discovery in Databases (ECML PKDD 2021) has still been affected by the COVID-19 pandemic. Unfortunately it had to be held online and we could only meet each other virtually. However, the experience gained in the previous edition joined to the knowledge collected from other virtual conferences allowed us to provide an attractive and engaging agenda.

ECML PKDD is an annual conference that provides an international forum for the latest research in all areas related to machine learning and knowledge discovery in databases, including innovative applications. It is the leading European machine learning and data mining conference and builds upon a very successful series of ECML PKDD conferences. Scheduled to take place in Bilbao, Spain, ECML PKDD 2021 was held fully virtually, during September 13–17, 2021. The conference attracted over 1000 participants from all over the world. More generally, the conference received substantial attention from industry through sponsorship, participation, and also the industry track.

The main conference program consisted of presentations of 210 accepted conference papers, 40 papers accepted in the journal track and 4 keynote talks: Jie Tang (Tsinghua University), Susan Athey (Stanford University), Joaquin Quiñonero Candela (Facebook), and Marta Kwiatkowska (University of Oxford). In addition, there were 22 workshops, 8 tutorials, 2 combined workshop-tutorials, the PhD forum, and the discovery challenge. Papers presented during the three main conference days were organized in three different tracks:

- Research Track: research or methodology papers from all areas in machine learning, knowledge discovery, and data mining.
- Applied Data Science Track: papers on novel applications of machine learning, data mining, and knowledge discovery to solve real-world use cases, thereby bridging the gap between practice and current theory.
- Journal Track: papers that were published in special issues of the Springer journals Machine Learning and Data Mining and Knowledge Discovery.

We received a similar number of submissions to last year with 685 and 220 submissions for the Research and Applied Data Science Tracks respectively. We accepted 146 (21%) and 64 (29%) of these. In addition, there were 40 papers from the Journal Track. All in all, the high-quality submissions allowed us to put together an exceptionally rich and exciting program.

The Awards Committee selected research papers that were considered to be of exceptional quality and worthy of special recognition:

- Best (Student) Machine Learning Paper Award: Reparameterized Sampling for Generative Adversarial Networks, by Yifei Wang, Yisen Wang, Jiansheng Yang and Zhouchen Lin.

- First Runner-up (Student) Machine Learning Paper Award: "Continual Learning with Dual Regularizations", by Xuejun Han and Yuhong Guo.
- Best Applied Data Science Paper Award: "Open Data Science to fight COVID-19: Winning the 500k XPRIZE Pandemic Response Challenge", by Miguel Angel Lozano, Oscar Garibo, Eloy Piñol, Miguel Rebollo, Kristina Polotskaya, Miguel Angel Garcia-March, J. Alberto Conejero, Francisco Escolano and Nuria Oliver.
- Best Student Data Mining Paper Award: "Conditional Neural Relational Inference for Interacting Systems", by Joao Candido Ramos, Lionel Blondé, Stéphane Armand and Alexandros Kalousis.
- Test of Time Award for highest-impact paper from ECML PKDD 2011: "Influence and Passivity in Social Media", by Daniel M. Romero, Wojciech Galuba, Sitaram Asur and Bernardo A. Huberman.

We would like to wholeheartedly thank all participants, authors, Program Committee members, area chairs, session chairs, volunteers, co-organizers, and organizers of workshops and tutorials for their contributions that helped make ECML PKDD 2021 a great success. We would also like to thank the ECML PKDD Steering Committee and all sponsors.

September 2021

Jose A. Lozano
Nuria Oliver
Fernando Pérez-Cruz
Stefan Kramer
Jesse Read
Yuxiao Dong
Nicolas Kourtellis
Barbara Hammer

Organization

General Chair

Jose A. Lozano   Basque Center for Applied Mathematics, Spain

Research Track Program Chairs

Nuria Oliver   Vodafone Institute for Society and Communications,
       Germany, and Data-Pop Alliance, USA
Fernando Pérez-Cruz   Swiss Data Science Center, Switzerland
Stefan Kramer   Johannes Gutenberg Universität Mainz, Germany
Jesse Read   École Polytechnique, France

Applied Data Science Track Program Chairs

Yuxiao Dong   Facebook AI, Seattle, USA
Nicolas Kourtellis   Telefonica Research, Barcelona, Spain
Barbara Hammer   Bielefeld University, Germany

Journal Track Chairs

Sergio Escalera   Universitat de Barcelona, Spain
Heike Trautmann   University of Münster, Germany
Annalisa Appice   Università degli Studi di Bari, Italy
Jose A. Gámez   Universidad de Castilla-La Mancha, Spain

Discovery Challenge Chairs

Paula Brito   Universidade do Porto, Portugal
Dino Ienco   Université Montpellier, France

Workshop and Tutorial Chairs

Alipio Jorge   Universidade do Porto, Portugal
Yun Sing Koh   University of Auckland, New Zealand

Industrial Track Chairs

Miguel Veganzones   Sherpa.ia, Portugal
Sabri Skhiri   EURA NOVA, Belgium

Award Chairs

Myra Spiliopoulou   Otto-von-Guericke-University Magdeburg, Germany
João Gama   University of Porto, Portugal

PhD Forum Chairs

Jeronimo Hernandez   University of Barcelona, Spain
Zahra Ahmadi   Johannes Gutenberg Universität Mainz, Germany

Production, Publicity, and Public Relations Chairs

Sophie Burkhardt   Johannes Gutenberg Universität Mainz, Germany
Julia Sidorova   Universidad Complutense de Madrid, Spain

Local Chairs

Iñaki Inza   University of the Basque Country, Spain
Alexander Mendiburu   University of the Basque Country, Spain
Santiago Mazuelas   Basque Center for Applied Mathematics, Spain
Aritz Pèrez   Basque Center for Applied Mathematics, Spain
Borja Calvo   University of the Basque Country, Spain

Proceedings Chair

Tania Cerquitelli   Politecnico di Torino, Italy

Sponsorship Chair

Santiago Mazuelas   Basque Center for Applied Mathematics, Spain

Web Chairs

Olatz Hernandez   Basque Center for Applied Mathematics, Spain
  Aretxabaleta
Estíbaliz Gutièrrez   Basque Center for Applied Mathematics, Spain

ECML PKDD Steering Committee

Andrea Passerini   University of Trento, Italy
Francesco Bonchi   ISI Foundation, Italy
Albert Bifet   Télécom ParisTech, France
Sašo Džeroski   Jožef Stefan Institute, Slovenia
Katharina Morik   TU Dortmund, Germany
Arno Siebes   Utrecht University, The Netherlands
Siegfried Nijssen   Université Catholique de Louvain, Belgium

Luís Moreira-Matias   Finiata GmbH, Germany
Alessandra Sala   Shutterstock, Ireland
Georgiana Ifrim   University College Dublin, Ireland
Thomas Gärtner   University of Nottingham, UK
Neil Hurley   University College Dublin, Ireland
Michele Berlingerio   IBM Research, Ireland
Elisa Fromont   Université de Rennes, France
Arno Knobbe   Universiteit Leiden, The Netherlands
Ulf Brefeld   Leuphana Universität Lüneburg, Germany
Andreas Hotho   Julius-Maximilians-Universität Würzburg, Germany
Ira Assent   Aarhus University, Denmark
Kristian Kersting   TU Darmstadt University, Germany
Jefrey Lijffijt   Ghent University, Belgium
Isabel Valera   Saarland University, Germany

Program Committee

Guest Editorial Board, Journal Track

Richard Allmendinger   University of Manchester
Marie Anastacio   Leiden University
Ana Paula Appel   IBM Research Brazil
Dennis Assenmacher   University of Münster
Ira Assent   Aarhus University
Martin Atzmueller   Osnabrueck University
Jaume Bacardit   Newcastle University
Anthony Bagnall   University of East Anglia
Mitra Baratchi   University of Twente
Srikanta Bedathur   IIT Delhi
Alessio Benavoli   CSIS
Viktor Bengs   Paderborn University
Massimo Bilancia   University of Bari "Aldo Moro"
Klemens Böhm   Karlsruhe Institute of Technology
Veronica Bolon Canedo   Universidade da Coruna
Ilaria Bordino   UniCredit R&D
Jakob Bossek   University of Adelaide
Ulf Brefeld   Leuphana Universität Luneburg
Michelangelo Ceci   Universita degli Studi di Bari "Aldo Moro"
Loïc Cerf   Universidade Federal de Minas Gerais
Victor Manuel Cerqueira   University of Porto
Laetitia Chapel   IRISA
Silvia Chiusano   Politecnico di Torino
Roberto Corizzo   American University, Washington D.C.
Marco de Gemmis   Università degli Studi di Bari "Aldo Moro"
Sébastien Destercke   Università degli Studi di Bari "Aldo Moro"
Shridhar Devamane   Visvesvaraya Technological University

Carlotta Domeniconi	George Mason University
Wouter Duivesteijn	Eindhoven University of Technology
Tapio Elomaa	Tampere University of Technology
Hugo Jair Escalante	INAOE
Nicola Fanizzi	Università degli Studi di Bari "Aldo Moro"
Stefano Ferilli	Università degli Studi di Bari "Aldo Moro"
Pedro Ferreira	Universidade de Lisboa
Cesar Ferri	Valencia Polytechnic University
Julia Flores	University of Castilla-La Mancha
Germain Forestier	Université de Haute Alsace
Marco Frasca	University of Milan
Ricardo J. G. B. Campello	University of Newcastle
Esther Galbrun	University of Eastern Finland
João Gama	University of Porto
Paolo Garza	Politecnico di Torino
Pascal Germain	Université Laval
Fabian Gieseke	University of Münster
Josif Grabocka	University of Hildesheim
Gianluigi Greco	University of Calabria
Riccardo Guidotti	University of Pisa
Francesco Gullo	UniCredit
Stephan Günnemann	Technical University of Munich
Tias Guns	Vrije Universiteit Brussel
Antonella Guzzo	University of Calabria
Alexander Hagg	Hochschule Bonn-Rhein-Sieg University of Applied Sciences
Jin-Kao Hao	University of Angers
Daniel Hernández-Lobato	Universidad Autónoma de Madrid
Jose Hernández-Orallo	Universitat Politècnica de València
Martin Holena	Institute of Computer Science, Academy of Sciences of the Czech Republic
Jaakko Hollmén	Aalto University
Dino Ienco	IRSTEA
Georgiana Ifrim	University College Dublin
Felix Iglesias	TU Wien
Angelo Impedovo	University of Bari "Aldo Moro"
Mahdi Jalili	RMIT University
Nathalie Japkowicz	University of Ottawa
Szymon Jaroszewicz	Institute of Computer Science, Polish Academy of Sciences
Michael Kamp	Monash University
Mehdi Kaytoue	Infologic
Pascal Kerschke	University of Münster
Dragi Kocev	Jozef Stefan Institute
Lars Kotthoff	University of Wyoming
Tipaluck Krityakierne	University of Bern

Peer Kröger	Ludwig Maximilian University of Munich
Meelis Kull	University of Tartu
Michel Lang	TU Dortmund University
Helge Langseth	Norwegian University of Science and Technology
Oswald Lanz	FBK
Mark Last	Ben-Gurion University of the Negev
Kangwook Lee	University of Wisconsin-Madison
Jurica Levatic	IRB Barcelona
Thomar Liebig	TU Dortmund
Hsuan-Tien Lin	National Taiwan University
Marius Lindauer	Leibniz University Hannover
Marco Lippi	University of Modena and Reggio Emilia
Corrado Loglisci	Università degli Studi di Bari
Manuel Lopez-Ibanez	University of Malaga
Nuno Lourenço	University of Coimbra
Claudio Lucchese	Ca' Foscari University of Venice
Brian Mac Namee	University College Dublin
Gjorgji Madjarov	Ss. Cyril and Methodius University
Davide Maiorca	University of Cagliari
Giuseppe Manco	ICAR-CNR
Elena Marchiori	Radboud University
Elio Masciari	Università di Napoli Federico II
Andres R. Masegosa	Norwegian University of Science and Technology
Ernestina Menasalvas	Universidad Politécnica de Madrid
Rosa Meo	University of Torino
Paolo Mignone	University of Bari "Aldo Moro"
Anna Monreale	University of Pisa
Giovanni Montana	University of Warwick
Grègoire Montavon	TU Berlin
Katharina Morik	TU Dortmund
Animesh Mukherjee	Indian Institute of Technology, Kharagpur
Amedeo Napoli	LORIA Nancy
Frank Naumann	University of Adelaide
Thomas Dyhre	Aalborg University
Bruno Ordozgoiti	Aalto University
Rita P. Ribeiro	University of Porto
Pance Panov	Jozef Stefan Institute
Apostolos Papadopoulos	Aristotle University of Thessaloniki
Panagiotis Papapetrou	Stockholm University
Andrea Passerini	University of Trento
Mykola Pechenizkiy	Eindhoven University of Technology
Charlotte Pelletier	Université Bretagne Sud
Ruggero G. Pensa	University of Torino
Nico Piatkowski	TU Dortmund
Dario Piga	IDSIA Dalle Molle Institute for Artificial Intelligence Research - USI/SUPSI

Gianvito Pio	Università degli Studi di Bari "Aldo Moro"
Marc Plantevit	LIRIS - Université Claude Bernard Lyon 1
Marius Popescu	University of Bucharest
Raphael Prager	University of Münster
Mike Preuss	Universiteit Leiden
Jose M. Puerta	Universidad de Castilla-La Mancha
Kai Puolamäki	University of Helsinki
Chedy Raïssi	Inria
Jan Ramon	Inria
Matteo Riondato	Amherst College
Thomas A. Runkler	Siemens Corporate Technology
Antonio Salmerón	University of Almería
Joerg Sander	University of Alberta
Roberto Santana	University of the Basque Country
Michael Schaub	RWTH Aachen
Lars Schmidt-Thieme	University of Hildesheim
Santiago Segui	Universitat de Barcelona
Thomas Seidl	Ludwig-Maximilians-Universitaet Muenchen
Moritz Seiler	University of Münster
Shinichi Shirakawa	Yokohama National University
Jim Smith	University of the West of England
Carlos Soares	University of Porto
Gerasimos Spanakis	Maastricht University
Giancarlo Sperlì	University of Naples Federico II
Myra Spiliopoulou	Otto-von-Guericke-University Magdeburg
Giovanni Stilo	Università degli Studi dell'Aquila
Catalin Stoean	University of Craiova
Mahito Sugiyama	National Institute of Informatics
Nikolaj Tatti	University of Helsinki
Alexandre Termier	Université de Rennes 1
Kevin Tierney	Bielefeld University
Luis Torgo	University of Porto
Roberto Trasarti	CNR Pisa
Sébastien Treguer	Inria
Leonardo Trujillo	Instituto Tecnológico de Tijuana
Ivor Tsang	University of Technology Sydney
Grigorios Tsoumakas	Aristotle University of Thessaloniki
Steffen Udluft	Siemens
Arnaud Vandaele	Université de Mons
Matthijs van Leeuwen	Leiden University
Celine Vens	KU Leuven Kulak
Herna Viktor	University of Ottawa
Marco Virgolin	Centrum Wiskunde & Informatica
Jordi Vitrià	Universitat de Barcelona
Christel Vrain	LIFO – University of Orléans
Jilles Vreeken	Helmholtz Center for Information Security

Willem Waegeman	Ghent University
David Walker	University of Plymouth
Hao Wang	Leiden University
Elizabeth F. Wanner	CEFET
Tu Wei-Wei	4paradigm
Pascal Welke	University of Bonn
Marcel Wever	Paderborn University
Man Leung Wong	Lingnan University
Stefan Wrobel	Fraunhofer IAIS, University of Bonn
Zheng Ying	Inria
Guoxian Yu	Shandong University
Xiang Zhang	Harvard University
Ye Zhu	Deakin University
Arthur Zimek	University of Southern Denmark
Albrecht Zimmermann	Université Caen Normandie
Marinka Zitnik	Harvard University

Area Chairs, Research Track

Fabrizio Angiulli	University of Calabria
Ricardo Baeza-Yates	Universitat Pompeu Fabra
Roberto Bayardo	Google
Bettina Berendt	Katholieke Universiteit Leuven
Philipp Berens	University of Tübingen
Michael Berthold	University of Konstanz
Hendrik Blockeel	Katholieke Universiteit Leuven
Juergen Branke	University of Warwick
Ulf Brefeld	Leuphana University Lüneburg
Toon Calders	Universiteit Antwerpen
Michelangelo Ceci	Università degli Studi di Bari "Aldo Moro"
Duen Horng Chau	Georgia Institute of Technology
Nicolas Courty	Université Bretagne Sud, IRISA Research Institute Computer and Systems Aléatoires
Bruno Cremilleux	Université de Caen Normandie
Philippe Cudre-Mauroux	University of Fribourg
James Cussens	University of Bristol
Jesse Davis	Katholieke Universiteit Leuven
Bob Durrant	University of Waikato
Tapio Elomaa	Tampere University
Johannes Fürnkranz	Johannes Kepler University Linz
Eibe Frank	University of Waikato
Elisa Fromont	Université de Rennes 1
Stephan Günnemann	Technical University of Munich
Patrick Gallinari	LIP6 - University of Paris
Joao Gama	University of Porto
Przemyslaw Grabowicz	University of Massachusetts, Amherst

Eyke Hüllermeier	Paderborn University
Allan Hanbury	Vienna University of Technology
Daniel Hernández-Lobato	Universidad Autónoma de Madrid
José Hernández-Orallo	Universitat Politècnica de València
Andreas Hotho	University of Wuerzburg
Inaki Inza	University of the Basque Country
Marius Kloft	TU Kaiserslautern
Arno Knobbe	Universiteit Leiden
Lars Kotthoff	University of Wyoming
Danica Kragic	KTH Royal Institute of Technology
Sébastien Lefèvre	Université Bretagne Sud
Bruno Lepri	FBK-Irst
Patrick Loiseau	Inria and Ecole Polytechnique
Jorg Lucke	University of Oldenburg
Fragkiskos Malliaros	Paris-Saclay University, CentraleSupelec, and Inria
Giuseppe Manco	ICAR-CNR
Dunja Mladenic	Jozef Stefan Institute
Katharina Morik	TU Dortmund
Sriraam Natarajan	Indiana University Bloomington
Siegfried Nijssen	Université catholique de Louvain
Andrea Passerini	University of Trento
Mykola Pechenizkiy	Eindhoven University of Technology
Jaakko Peltonen	Aalto University and University of Tampere
Marian-Andrei Rizoiu	University of Technology Sydney
Céline Robardet	INSA Lyon
Maja Rudolph	Bosch
Lars Schmidt-Thieme	University of Hildesheim
Thomas Seidl	Ludwig-Maximilians-Universität München
Arno Siebes	Utrecht University
Myra Spiliopoulou	Otto-von-Guericke-University Magdeburg
Yizhou Sun	University of California, Los Angeles
Einoshin Suzuki	Kyushu University
Jie Tang	Tsinghua University
Ke Tang	Southern University of Science and Technology
Marc Tommasi	University of Lille
Isabel Valera	Saarland University
Celine Vens	KU Leuven Kulak
Christel Vrain	LIFO - University of Orléans
Jilles Vreeken	Helmholtz Center for Information Security
Willem Waegeman	Ghent University
Stefan Wrobel	Fraunhofer IAIS, University of Bonn
Min-Ling Zhang	Southeast University

Area Chairs, Applied Data Science Track

Francesco Calabrese	Vodafone
Michelangelo Ceci	Università degli Studi di Bari "Aldo Moro"
Gianmarco De Francisci Morales	ISI Foundation
Tom Diethe	Amazon
Johannes Frünkranz	Johannes Kepler University Linz
Han Fang	Facebook
Faisal Farooq	Qatar Computing Research Institute
Rayid Ghani	Carnegie Mellon Univiersity
Francesco Gullo	UniCredit
Xiangnan He	University of Science and Technology of China
Georgiana Ifrim	University College Dublin
Thorsten Jungeblut	Bielefeld University of Applied Sciences
John A. Lee	Université catholique de Louvain
Ilias Leontiadis	Samsung AI
Viktor Losing	Honda Research Institute Europe
Yin Lou	Ant Group
Gabor Melli	Sony PlayStation
Luis Moreira-Matias	University of Porto
Nicolò Navarin	University of Padova
Benjamin Paaßen	German Research Center for Artificial Intelligence
Kitsuchart Pasupa	King Mongkut's Institute of Technology Ladkrabang
Mykola Pechenizkiy	Eindhoven University of Technology
Julien Perez	Naver Labs Europe
Fabio Pinelli	IMT Lucca
Zhaochun Ren	Shandong University
Sascha Saralajew	Porsche AG
Fabrizio Silvestri	Facebook
Sinong Wang	Facebook AI
Xing Xie	Microsoft Research Asia
Jian Xu	Citadel
Jing Zhang	Renmin University of China

Program Committee Members, Research Track

Hanno Ackermann	Leibniz University Hannover
Linara Adilova	Fraunhofer IAIS
Zahra Ahmadi	Johannes Gutenberg University
Cuneyt Gurcan Akcora	University of Manitoba
Omer Deniz Akyildiz	University of Warwick
Carlos M. Alaíz Gudín	Universidad Autónoma de Madrid
Mohamed Alami	Ecole Polytechnique
Chehbourne Abdullah Alchihabi	Carleton University
Pegah Alizadeh	University of Caen Normandy

Reem Alotaibi — King Abdulaziz University
Massih-Reza Amini — Université Grenoble Alpes
Shin Ando — Tokyo University of Science
Thiago Andrade — INESC TEC
Kimon Antonakopoulos — Inria
Alessandro Antonucci — IDSIA
Muhammad Umer Anwaar — Technical University of Munich
Eva Armengol — IIIA-SIC
Dennis Assenmacher — University of Münster
Matthias Aßenmacher — Ludwig-Maximilians-Universität München
Martin Atzmueller — Osnabrueck University
Behrouz Babaki — Polytechnique Montreal
Rohit Babbar — Aalto University
Elena Baralis — Politecnico di Torino
Mitra Baratchi — University of Twente
Christian Bauckhage — University of Bonn, Fraunhofer IAIS
Martin Becker — University of Würzburg
Jessa Bekker — Katholieke Universiteit Leuven
Colin Bellinger — National Research Council of Canada
Khalid Benabdeslem — LIRIS Laboratory, Claude Bernard University Lyon I
Diana Benavides-Prado — Auckland University of Technology
Anes Bendimerad — LIRIS
Christoph Bergmeir — University of Granada
Alexander Binder — UiO
Aleksandar Bojchevski — Technical University of Munich
Ahcène Boubekki — UiT Arctic University of Norway
Paula Branco — EECS University of Ottawa
Tanya Braun — University of Lübeck
Katharina Breininger — Friedrich-Alexander-Universität Erlangen Nürnberg
Wieland Brendel — University of Tübingen
John Burden — University of Cambridge
Sophie Burkhardt — TU Kaiserslautern
Sebastian Buschjäger — TU Dortmund
Borja Calvo — University of the Basque Country
Stephane Canu — LITIS, INSA de Rouen
Cornelia Caragea — University of Illinois at Chicago
Paula Carroll — University College Dublin
Giuseppe Casalicchio — Ludwig Maximilian University of Munich
Bogdan Cautis — Paris-Saclay University
Rémy Cazabet — Université de Lyon
Josu Ceberio — University of the Basque Country
Peggy Cellier — IRISA/INSA Rennes
Mattia Cerrato — Università degli Studi di Torino
Ricardo Cerri — Federal University of Sao Carlos
Alessandra Cervone — Amazon
Ayman Chaouki — Institut Mines-Télécom

Paco Charte	Universidad de Jaén
Rita Chattopadhyay	Intel Corporation
Vaggos Chatziafratis	Stanford University
Tianyi Chen	Zhejiang University City College
Yuzhou Chen	Southern Methodist University
Yiu-Ming Cheung	Hong Kong Baptist University
Anshuman Chhabra	University of California, Davis
Ting-Wu Chin	Carnegie Mellon University
Oana Cocarascu	King's College London
Lidia Contreras-Ochando	Universitat Politècnica de València
Roberto Corizzo	American University
Anna Helena Reali Costa	Universidade de São Paulo
Fabrizio Costa	University of Exeter
Gustavo De Assis Costa	Instituto Federal de Educação, Ciência e Tecnologia de Goiás
Bertrand Cuissart	GREYC
Thi-Bich-Hanh Dao	University of Orleans
Mayukh Das	Microsoft Research Lab
Padraig Davidson	Universität Würzburg
Paul Davidsson	Malmö University
Gwendoline De Bie	ENS
Tijl De Bie	Ghent University
Andre de Carvalho	Universidade de São Paulo
Orphée De Clercq	Ghent University
Alper Demir	İzmir University of Economics
Nicola Di Mauro	Università degli Studi di Bari "Aldo Moro"
Yao-Xiang Ding	Nanjing University
Carola Doerr	Sorbonne University
Boxiang Dong	Montclair State University
Ruihai Dong	University College Dublin
Xin Du	Eindhoven University of Technology
Stefan Duffner	LIRIS
Wouter Duivesteijn	Eindhoven University of Technology
Audrey Durand	McGill University
Inês Dutra	University of Porto
Saso Dzeroski	Jozef Stefan Institute
Hamid Eghbalzadeh	Johannes Kepler University
Dominik Endres	University of Marburg
Roberto Esposito	Università degli Studi di Torino
Samuel G. Fadel	Universidade Estadual de Campinas
Xiuyi Fan	Imperial College London
Hadi Fanaee-T.	Halmstad University
Elaine Faria	Federal University of Uberlandia
Fabio Fassetti	University of Calabria
Kilian Fatras	Inria
Ad Feelders	Utrecht University

Songhe Feng   Beijing Jiaotong University
Àngela Fernández-Pascual   Universidad Autónoma de Madrid
Daniel Fernández-Sánchez   Universidad Autónoma de Madrid
Sofia Fernandes   University of Aveiro
Cesar Ferri   Universitat Politécnica de Valéncia
Rémi Flamary   École Polytechnique
Michael Flynn   University of East Anglia
Germain Forestier   Université de Haute Alsace
Kary Främling   Umeå University
Benoît Frénay   Université de Namur
Vincent Francois   University of Amsterdam
Emilia Gómez   Joint Research Centre - European Commission
Luis Galárraga   Inria
Esther Galbrun   University of Eastern Finland
Claudio Gallicchio   University of Pisa
Jochen Garcke   University of Bonn
Clément Gautrais   KU Leuven
Yulia Gel   University of Texas at Dallas and University
                                 of Waterloo
Pierre Geurts   University of Liège
Amirata Ghorbani   Stanford University
Heitor Murilo Gomes   University of Waikato
Chen Gong   Shanghai Jiao Tong University
Bedartha Goswami   University of Tübingen
Henry Gouk   University of Edinburgh
James Goulding   University of Nottingham
Antoine Gourru   Université Lumière Lyon 2
Massimo Guarascio   ICAR-CNR
Riccardo Guidotti   University of Pisa
Ekta Gujral   University of California, Riverside
Francesco Gullo   UniCredit
Tias Guns   Vrije Universiteit Brussel
Thomas Guyet   Institut Agro, IRISA
Tom Hanika   University of Kassel
Valentin Hartmann   Ecole Polytechnique Fédérale de Lausanne
Marwan Hassani   Eindhoven University of Technology
Jukka Heikkonen   University of Turku
Fredrik Heintz   Linköping University
Sibylle Hess   TU Eindhoven
Jaakko Hollmén   Aalto University
Tamas Horvath   University of Bonn, Fraunhofer IAIS
Binbin Hu   Ant Group
Hong Huang   UGoe
Georgiana Ifrim   University College Dublin
Angelo Impedovo   Università degli studi di Bari "Aldo Moro"

Nathalie Japkowicz	American University
Szymon Jaroszewicz	Institute of Computer Science, Polish Academy of Sciences
Saumya Jetley	Inria
Binbin Jia	Southeast University
Xiuyi Jia	School of Computer Science and Technology, Nanjing University of Science and Technology
Yuheng Jia	City University of Hong Kong
Siyang Jiang	National Taiwan University
Priyadarshini Kumari	IIT Bombay
Ata Kaban	University of Birmingham
Tomasz Kajdanowicz	Wroclaw University of Technology
Vana Kalogeraki	Athens University of Economics and Business
Toshihiro Kamishima	National Institute of Advanced Industrial Science and Technology
Michael Kamp	Monash University
Bo Kang	Ghent University
Dimitrios Karapiperis	Hellenic Open University
Panagiotis Karras	Aarhus University
George Karypis	University of Minnesota
Mark Keane	University College Dublin
Kristian Kersting	TU Darmstadt
Masahiro Kimura	Ryukoku University
Jiri Klema	Czech Technical University
Dragi Kocev	Jozef Stefan Institute
Masahiro Kohjima	NTT
Lukasz Korycki	Virginia Commonwealth University
Peer Kröger	Ludwig Maximilian University of Münich
Anna Krause	University of Würzburg
Bartosz Krawczyk	Virginia Commonwealth University
Georg Krempl	Utrecht University
Meelis Kull	University of Tartu
Vladimir Kuzmanovski	Aalto University
Ariel Kwiatkowski	Ecole Polytechnique
Emanuele La Malfa	University of Oxford
Beatriz López	University of Girona
Preethi Lahoti	Aalto University
Ichraf Lahouli	Euranova
Niklas Lavesson	Jönköping University
Aonghus Lawlor	University College Dublin
Jeongmin Lee	University of Pittsburgh
Daniel Lemire	LICEF Research Center and Université du Québec
Florian Lemmerich	University of Passau
Elisabeth Lex	Graz University of Technology
Jiani Li	Vanderbilt University
Rui Li	Inspur Group
Wentong Liao	Lebniz University Hannover

Jiayin Lin	University of Wollongong
Rudolf Lioutikov	UT Austin
Marco Lippi	University of Modena and Reggio Emilia
Suzanne Little	Dublin City University
Shengcai Liu	University of Science and Technology of China
Shenghua Liu	Institute of Computing Technology, Chinese Academy of Sciences
Philipp Liznerski	Technische Universität Kaiserslautern
Corrado Loglisci	Università degli Studi di Bari "Aldo Moro"
Ting Long	Shanghai Jiaotong University
Tsai-Ching Lu	HRL Laboratories
Yunpu Ma	Siemens AG
Zichen Ma	The Chinese University of Hong Kong
Sara Madeira	Universidade de Lisboa
Simona Maggio	Dataiku
Sara Magliacane	IBM
Sebastian Mair	Leuphana University Lüneburg
Lorenzo Malandri	University of Milan Bicocca
Donato Malerba	Università degli Studi di Bari "Aldo Moro"
Pekka Malo	Aalto University
Robin Manhaeve	KU Leuven
Silviu Maniu	Université Paris-Sud
Giuseppe Marra	KU Leuven
Fernando Martínez-Plumed	Joint Research Centre - European Commission
Alexander Marx	Max Plank Institue for Informatics and Saarland University
Florent Masseglia	Inria
Tetsu Matsukawa	Kyushu University
Wolfgang Mayer	University of South Australia
Santiago Mazuelas	Basque center for Applied Mathematics
Stefano Melacci	University of Siena
Ernestina Menasalvas	Universidad Politécnica de Madrid
Rosa Meo	Università degli Studi di Torino
Alberto Maria Metelli	Politecnico di Milano
Saskia Metzler	Max Planck Institute for Informatics
Alessio Micheli	University of Pisa
Paolo Mignone	Università degli studi di Bari "Aldo Moro"
Matej Mihelčić	University of Zagreb
Decebal Constantin Mocanu	University of Twente
Nuno Moniz	INESC TEC and University of Porto
Carlos Monserrat	Universitat Politécnica de Valéncia
Corrado Monti	ISI Foundation
Jacob Montiel	University of Waikato
Ahmadreza Mosallanezhad	Arizona State University
Tanmoy Mukherjee	University of Tennessee
Martin Mundt	Goethe University

Mohamed Nadif	Université de Paris
Omer Nagar	Bar Ilan University
Felipe Kenji Nakano	Katholieke Universiteit Leuven
Mirco Nanni	KDD-Lab ISTI-CNR Pisa
Apurva Narayan	University of Waterloo
Nicolò Navarin	University of Padova
Benjamin Negrevergne	Paris Dauphine University
Hurley Neil	University College Dublin
Stefan Neumann	University of Vienna
Ngoc-Tri Ngo	The University of Danang - University of Science and Technology
Dai Nguyen	Monash University
Eirini Ntoutsi	Free University Berlin
Andrea Nuernberger	Otto-von-Guericke-Universität Magdeburg
Pablo Olmos	University Carlos III
James O'Neill	University of Liverpool
Barry O'Sullivan	University College Cork
Rita P. Ribeiro	University of Porto
Aritz Pèrez	Basque Center for Applied Mathematics
Joao Palotti	Qatar Computing Research Institute
Guansong Pang	University of Adelaide
Pance Panov	Jozef Stefan Institute
Evangelos Papalexakis	University of California, Riverside
Haekyu Park	Georgia Institute of Technology
Sudipta Paul	Umeå University
Yulong Pei	Eindhoven University of Technology
Charlotte Pelletier	Université Bretagne Sud
Ruggero G. Pensa	University of Torino
Bryan Perozzi	Google
Nathanael Perraudin	ETH Zurich
Lukas Pfahler	TU Dortmund
Bastian Pfeifer	Medical University of Graz
Nico Piatkowski	TU Dortmund
Robert Pienta	Georgia Institute of Technology
Fábio Pinto	Faculdade de Economia do Porto
Gianvito Pio	University of Bari "Aldo Moro"
Giuseppe Pirrò	Sapienza University of Rome
Claudia Plant	University of Vienna
Marc Plantevit	LIRIS - Universitè Claude Bernard Lyon 1
Amit Portnoy	Ben Gurion University
Melanie Pradier	Harvard University
Paul Prasse	University of Potsdam
Philippe Preux	Inria, LIFL, Universitè de Lille
Ricardo Prudencio	Federal University of Pernambuco
Zhou Qifei	Peking University
Erik Quaeghebeur	TU Eindhoven

Tahrima Rahman	University of Texas at Dallas
Herilalaina Rakotoarison	Inria
Alexander Rakowski	Hasso Plattner Institute
María José Ramírez	Universitat Politècnica de València
Visvanathan Ramesh	Goethe University
Jan Ramon	Inria
Huzefa Rangwala	George Mason University
Aleksandra Rashkovska	Jožef Stefan Institute
Joe Redshaw	University of Nottingham
Matthias Renz	Christian-Albrechts-Universität zu Kiel
Matteo Riondato	Amherst College
Ettore Ritacco	ICAR-CNR
Mateus Riva	Télécom ParisTech
Antonio Rivera	Universidad Politécnica de Madrid
Marko Robnik-Sikonja	University of Ljubljana
Simon Rodriguez Santana	Institute of Mathematical Sciences (ICMAT-CSIC)
Mohammad Rostami	University of Southern California
Céline Rouveirol	Laboratoire LIPN-UMR CNRS
Jože Rožanec	Jožef Stefan Institute
Peter Rubbens	Flanders Marine Institute
David Ruegamer	LMU Munich
Salvatore Ruggieri	Università di Pisa
Francisco Ruiz	DeepMind
Anne Sabourin	Télécom ParisTech
Tapio Salakoski	University of Turku
Pablo Sanchez-Martin	Max Planck Institute for Intelligent Systems
Emanuele Sansone	KU Leuven
Yucel Saygin	Sabanci University
Patrick Schäfer	Humboldt Universität zu Berlin
Pierre Schaus	UCLouvain
Ute Schmid	University of Bamberg
Sebastian Schmoll	Ludwig Maximilian University of Munich
Marc Schoenauer	Inria
Matthias Schubert	Ludwig Maximilian University of Munich
Marian Scuturici	LIRIS-INSA de Lyon
Junming Shao	University of Science and Technology of China
Manali Sharma	Samsung Semiconductor Inc.
Abdul Saboor Sheikh	Zalando Research
Jacquelyn Shelton	Hong Kong Polytechnic University
Feihong Shen	Jilin University
Gavin Smith	University of Nottingham
Kma Solaiman	Purdue University
Arnaud Soulet	Université François Rabelais Tours
Alessandro Sperduti	University of Padua
Giovanni Stilo	Università degli Studi dell'Aquila
Michiel Stock	Ghent University

Lech Szymanski	University of Otago
Shazia Tabassum	University of Porto
Andrea Tagarelli	University of Calabria
Acar Tamersoy	NortonLifeLock Research Group
Chang Wei Tan	Monash University
Sasu Tarkoma	University of Helsinki
Bouadi Tassadit	IRISA-Université Rennes 1
Nikolaj Tatti	University of Helsinki
Maryam Tavakol	Eindhoven University of Technology
Pooya Tavallali	University of California, Los Angeles
Maguelonne Teisseire	Irstea - UMR Tetis
Alexandre Termier	Université de Rennes 1
Stefano Teso	University of Trento
Janek Thomas	Fraunhofer Institute for Integrated Circuits IIS
Alessandro Tibo	Aalborg University
Sofia Triantafillou	University of Pittsburgh
Grigorios Tsoumakas	Aristotle University of Thessaloniki
Peter van der Putten	LIACS, Leiden University and Pegasystems
Elia Van Wolputte	KU Leuven
Robert A. Vandermeulen	Technische Universität Berlin
Fabio Vandin	University of Padova
Filipe Veiga	Massachusetts Institute of Technology
Bruno Veloso	Universidade Portucalense and LIAAD - INESC TEC
Sebastián Ventura	University of Cordoba
Rosana Veroneze	UNICAMP
Herna Viktor	University of Ottawa
João Vinagre	INESC TEC
Huaiyu Wan	Beijing Jiaotong University
Beilun Wang	Southeast University
Hu Wang	University of Adelaide
Lun Wang	University of California, Berkeley
Yu Wang	Peking University
Zijie J. Wang	Georgia Tech
Tong Wei	Nanjing University
Pascal Welke	University of Bonn
Joerg Wicker	University of Auckland
Moritz Wolter	University of Bonn
Ning Xu	Southeast University
Akihiro Yamaguchi	Toshiba Corporation
Haitian Yang	Institute of Information Engineering, Chinese Academy of Sciences
Yang Yang	Nanjing University
Zhuang Yang	Sun Yat-sen University
Helen Yannakoudakis	King's College London
Heng Yao	Tongji University
Han-Jia Ye	Nanjing University

Kristina Yordanova	University of Rostock
Tetsuya Yoshida	Nara Women's University
Guoxian Yu	Shandong University, China
Sha Yuan	Tsinghua University
Valentina Zantedeschi	INSA Lyon
Albin Zehe	University of Würzburg
Bob Zhang	University of Macau
Teng Zhang	Huazhong University of Science and Technology
Liang Zhao	University of São Paulo
Bingxin Zhou	University of Sydney
Kenny Zhu	Shanghai Jiao Tong University
Yanqiao Zhu	Institute of Automation, Chinese Academy of Sciences
Arthur Zimek	University of Southern Denmark
Albrecht Zimmermann	Université Caen Normandie
Indre Zliobaite	University of Helsinki
Markus Zopf	NEC Labs Europe

Program Committee Members, Applied Data Science Track

Mahdi Abolghasemi	Monash University
Evrim Acar	Simula Research Lab
Deepak Ajwani	University College Dublin
Pegah Alizadeh	University of Caen Normandy
Jean-Marc Andreoli	Naver Labs Europe
Giorgio Angelotti	ISAE Supaero
Stefanos Antaris	KTH Royal Institute of Technology
Xiang Ao	Institute of Computing Technology, Chinese Academy of Sciences
Yusuf Arslan	University of Luxembourg
Cristian Axenie	Huawei European Research Center
Hanane Azzag	Université Sorbonne Paris Nord
Pedro Baiz	Imperial College London
Idir Benouaret	CNRS, Université Grenoble Alpes
Laurent Besacier	Laboratoire d'Informatique de Grenoble
Antonio Bevilacqua	Insight Centre for Data Analytics
Adrien Bibal	University of Namur
Wu Bin	Zhengzhou University
Patrick Blöbaum	Amazon
Pavel Blinov	Sber Artificial Intelligence Laboratory
Ludovico Boratto	University of Cagliari
Stefano Bortoli	Huawei Technologies Duesseldorf
Zekun Cai	University of Tokyo
Nicolas Carrara	University of Toronto
John Cartlidge	University of Bristol
Oded Cats	Delft University of Technology
Tania Cerquitelli	Politecnico di Torino

Prithwish Chakraborty	IBM
Rita Chattopadhyay	Intel Corp.
Keru Chen	GrabTaxi Pte Ltd.
Liang Chen	Sun Yat-sen University
Zhiyong Cheng	Shandong Artificial Intelligence Institute
Silvia Chiusano	Politecnico di Torino
Minqi Chong	Citadel
Jeremie Clos	University of Nottingham
J. Albert Conejero Casares	Universitat Politécnica de Vaécia
Evan Crothers	University of Ottawa
Henggang Cui	Uber ATG
Tiago Cunha	University of Porto
Padraig Cunningham	University College Dublin
Eustache Diemert	CRITEO Research
Nat Dilokthanakul	Vidyasirimedhi Institute of Science and Technology
Daizong Ding	Fudan University
Kaize Ding	ASU
Michele Donini	Amazon
Lukas Ewecker	Porsche AG
Zipei Fan	University of Tokyo
Bojing Feng	National Laboratory of Pattern Recognition, Institute of Automation, Chinese Academy of Science
Flavio Figueiredo	Universidade Federal de Minas Gerais
Blaz Fortuna	Qlector d.o.o.
Zuohui Fu	Rutgers University
Fabio Fumarola	University of Bari "Aldo Moro"
Chen Gao	Tsinghua University
Luis Garcia	University of Brasília
Cinmayii Garillos-Manliguez	University of the Philippines Mindanao
Kiran Garimella	Aalto University
Etienne Goffinet	Laboratoire LIPN-UMR CNRS
Michael Granitzer	University of Passau
Xinyu Guan	Xi'an Jiaotong University
Thomas Guyet	Institut Agro, IRISA
Massinissa Hamidi	Laboratoire LIPN-UMR CNRS
Junheng Hao	University of California, Los Angeles
Martina Hasenjaeger	Honda Research Institute Europe GmbH
Lars Holdijk	University of Amsterdam
Chao Huang	University of Notre Dame
Guanjie Huang	Penn State University
Hong Huang	UGoe
Yiran Huang	TECO
Madiha Ijaz	IBM
Roberto Interdonato	CIRAD - UMR TETIS
Omid Isfahani Alamdari	University of Pisa

Guillaume Jacquet	JRC
Nathalie Japkowicz	American University
Shaoxiong Ji	Aalto University
Nan Jiang	Purdue University
Renhe Jiang	University of Tokyo
Song Jiang	University of California, Los Angeles
Adan Jose-Garcia	University of Exeter
Jihed Khiari	Johannes Kepler Universität
Hyunju Kim	KAIST
Tomas Kliegr	University of Economics
Yun Sing Koh	University of Auckland
Pawan Kumar	IIIT, Hyderabad
Chandresh Kumar Maurya	CSE, IIT Indore
Thach Le Nguyen	The Insight Centre for Data Analytics
Mustapha Lebbah	Université Paris 13, LIPN-CNRS
Dongman Lee	Korea Advanced Institute of Science and Technology
Rui Li	Sony
Xiaoting Li	Pennsylvania State University
Zeyu Li	University of California, Los Angeles
Defu Lian	University of Science and Technology of China
Jiayin Lin	University of Wollongong
Jason Lines	University of East Anglia
Bowen Liu	Stanford University
Pedro Henrique Luz de Araujo	University of Brasilia
Fenglong Ma	Pennsylvania State University
Brian Mac Namee	University College Dublin
Manchit Madan	Myntra
Ajay Mahimkar	AT&T Labs
Domenico Mandaglio	Università della Calabria
Koji Maruhashi	Fujitsu Laboratories Ltd.
Sarah Masud	LCS2, IIIT-D
Eric Meissner	University of Cambridge
João Mendes-Moreira	INESC TEC
Chuan Meng	Shandong University
Fabio Mercorio	University of Milano-Bicocca
Angela Meyer	Bern University of Applied Sciences
Congcong Miao	Tsinghua University
Stéphane Moreau	Université de Sherbrooke
Koyel Mukherjee	IBM Research India
Fabricio Murai	Universidade Federal de Minas Gerais
Taichi Murayama	NAIST
Philip Nadler	Imperial College London
Franco Maria Nardini	ISTI-CNR
Ngoc-Tri Ngo	The University of Danang - University of Science and Technology

Anna Nguyen	Karlsruhe Institute of Technology
Hao Niu	KDDI Research, Inc.
Inna Novalija	Jožef Stefan Institute
Tsuyosh Okita	Kyushu Institute of Technology
Aoma Osmani	LIPN-UMR CNRS 7030, Université Paris 13
Latifa Oukhellou	IFSTTAR
Andrei Paleyes	University of Cambridge
Chanyoung Park	KAIST
Juan Manuel Parrilla Gutierrez	University of Glasgow
Luca Pasa	Università degli Studi Di Padova
Pedro Pereira Rodrigues	University of Porto
Miquel Perelló-Nieto	University of Bristol
Beatrice Perez	Dartmouth College
Alan Perotti	ISI Foundation
Mirko Polato	University of Padua
Giovanni Ponti	ENEA
Nicolas Posocco	Eura Nova
Cedric Pradalier	GeorgiaTech Lorraine
Giulia Preti	ISI Foundation
A. A. A. Qahtan	Utrecht University
Chuan Qin	University of Science and Technology of China
Dimitrios Rafailidis	University of Thessaly
Cyril Ray	Arts ct Metiers Institute of Technology, Ecole Navale, IRENav
Wolfgang Reif	University of Augsburg
Kit Rodolfa	Carnegie Mellon University
Christophe Rodrigues	Pôle Universitaire Léonard de Vinci
Natali Ruchansky	Netflix
Hajer Salem	AUDENSIEL
Parinya Sanguansat	Panyapiwat Institute of Management
Atul Saroop	Amazon
Alexander Schiendorfer	Technische Hochschule Ingolstadt
Peter Schlicht	Volkswagen
Jens Schreiber	University of Kassel
Alexander Schulz	Bielefeld University
Andrea Schwung	FH SWF
Edoardo Serra	Boise State University
Lorenzo Severini	UniCredit
Ammar Shaker	Paderborn University
Jiaming Shen	University of Illinois at Urbana-Champaign
Rongye Shi	Columbia University
Wang Siyu	Southwestern University of Finance and Economics
Hao Song	University of Bristol
Francesca Spezzano	Boise State University
Simon Stieber	University of Augsburg

Sponsors

Google **ASML**

MEMBER OF BASQUE RESEARCH
& TECHNOLOGY ALLIANCE

Contents – Part II

Graphs and Networks

Generative Models

Non-exhaustive Learning Using Gaussian Mixture Generative Adversarial Networks

Jun Zhuang$^{(\boxtimes)}$ and Mohammad Al Hasan$^{(\boxtimes)}$

Indiana University-Purdue University Indianapolis, Indianapolis, IN 46202, USA
junz@iu.edu, alhasan@iupui.edu

Abstract. Supervised learning, while deployed in real-life scenarios, often encounters instances of unknown classes. Conventional algorithms for training a supervised learning model do not provide an option to detect such instances, so they miss-classify such instances with 100% probability. Open Set Recognition (OSR) and Non-Exhaustive Learning (NEL) are potential solutions to overcome this problem. Most existing methods of OSR first classify members of existing classes and then identify instances of new classes. However, many of the existing methods of OSR only makes a binary decision, i.e., they only identify the existence of the unknown class. Hence, such methods cannot distinguish test instances belonging to incremental unseen classes. On the other hand, the majority of NEL methods often make a parametric assumption over the data distribution, which either fail to return good results, due to the reason that real-life complex datasets may not follow a well-known data distribution. In this paper, we propose a new online non-exhaustive learning model, namely, Non-Exhaustive Gaussian Mixture Generative Adversarial Networks (NE-GM-GAN) to address these issues. Our proposed model synthesizes Gaussian mixture based latent representation over a deep generative model, such as GAN, for incremental detection of instances of emerging classes in the test data. Extensive experimental results on several benchmark datasets show that NE-GM-GAN significantly outperforms the state-of-the-art methods in detecting instances of novel classes in streaming data.

Keywords: Open set recognition · Non-exhaustive learning

1 Introduction

Numerous machine learning models are supervised, relying substantially on labeled datasets. In such datasets, the labels of training instances enable a supervised model to learn the correlation between the labels and the patterns in the features, thus helping the model to achieve the desired performance in different kinds of classification or recognition tasks. However, many realistic machine learning problems originate in non-stationary environments where instances of unseen classes may emerge naturally. The presence of such instances weakens the robustness of conventional machine learning algorithms, as these algorithms

© Springer Nature Switzerland AG 2021
N. Oliver et al. (Eds.): ECML PKDD 2021, LNAI 12976, pp. 3–18, 2021.
https://doi.org/10.1007/978-3-030-86520-7_1

do not account for the instances from unknown classes, either in the train or the test environments. To overcome this challenge, a series of related research activities has become popular in recent years; examples include anomaly detection (AD) [13,15,27,34], few-shot learning (FSL) [12,25], zero-shot learning (ZSL) [21,29], open set recognition (OSR) and open-world classification (OWC) [1,2,4,5,8,11,14,17,19,20,23,26,30,31]. Collectively, each of these works belongs to one of the four different categories [6], differing on the kind of instances observed by the model during train and test. If L refers to labeling and I refers to self-information (e.g., semantic information in image dataset), the categories C can be denoted as the Cartesian product of two sets L and I, as shown below:

$$C = L \times I = \{(l,i) \; : \; l \in L \; \& \; i \in I\}, \tag{1}$$

both L and I have two elements: known (K) and unknown (U). Thus, there are four categories in C: (K, K), (K, U), (U, K), (U, U). For example, (U, U) refers to the learning problem in which instances belonging to unknown classes having no self-information.

Conventional supervised learning task belongs to the first category, as for such a task all instances in train and test datasets belong to (K, K). The anomaly detection (AD) task, a.k.a. one-class classification or outlier detection, detects a few (U, U) instances from the majority of (K, K) instances; for AD, the (U, U) instances may only (but not necessary) exist in the test set. FSL and ZSL are employed to identify (U, K) instances in the test set. The main difference between FSL and ZSL is that the training set of FSL contains a limited number of (U, K) instances while for the case of ZSL, the number of (U, K) instances in the train set is zero. In other words, ZSL identifies (U, K) instances in the test set only by associating (K, K) instances with (U, K) instances through self-information. Finally, works belonging to open set recognition (OSR) identify (U, U) instances in the test set. These works are the most challenging; unlike AD, whose objective is to detect only one class (outlier), OSR handles both (K, K) and (U, U) in the test set. Similar to OSR, OWC also incrementally learns the new classes and rejects the unseen class. Nevertheless, most existing methods of OSR or OWC do not distinguish the test instances among incremental unseen classes, which is more close to the realistic scenario. The scope of our work falls in the OSR category which only deals with (K, K) and (U, U) instances. In Table 1, we present a summary of the discussion of this paragraph.

Some works belonging to OSR have also been referred as Non-Exhaustive Learning (NEL). The term, Non-Exhaustive, means that the training data does not have instances of all classes that may be expected in the test data. The majority of early research works of NEL employ Bayesian methods with Gaussian mixture model (GMM) or infinite Gaussian mixture model (IGMM) [24,33]. However, these works suffer from some limitations; for instance, they assume that the data distribution in each class follows a mixture of Gaussian, which may not be true in many realistic datasets. Also, in the case of GMM, its ability to recognize unknown classes depends on the number of initial clusters that it uses. IGMM can mitigate this restriction by allowing cluster count to grow on the

Table 1. The background of related tasks (conv. for conventional method)

Tasks	Training set	Testing set	GOAL
Conv.	(K, K)	(K, K)	Supervised learning with (K, K)
AD	(K, K) w./wo. outliers	(K, K) w. outliers	Detect outliers
FSL	(K, K) w. limited (U, K)	(U, K)	Identify (U, K) in test set
ZSL	(K, K) w. self-info.	(U, K)	Identify (U, K) in test set
OSR	(K, K)	(K, K) & (U, U)	Distinguish (U, U) from (K, K)
NEL	(K, K)	(K, K) & (U, U)	Incrementally learn (U, U)

fly, but the inference mechanism of IGMM is time-consuming, no matter what kind of sampling method it uses for inferring the probabilities of the posterior distribution.

To address these issues, in this work we propose a new non-exhaustive learning model, Non-exhaustive Gaussian mixture Generative Adversarial Networks (NE-GM-GAN), which synthesizes the Bayesian method and deep learning technique. Comparing to the existing methods for OSR, our proposed method has several advantages: First, NE-GM-GAN takes multi-modal prior as input to better fit the real data distribution; Second, NE-GM-GAN can deal with class-imbalance problem with end-to-end offline training; Finally, NE-GM-GAN can achieve accurate and robust online detection on large sparse dataset while avoiding noisy distraction. Extensive experiments demonstrate that our proposed model has superior performance over competing methods on benchmark datasets. The contribution of this paper can be summarized as follows:

- We propose a new model for non-exhaustive learning, namely NE-GM-GAN, which can detect novel classes in online test data accurately and defy the class-imbalance problem effectively.
- NE-GM-GAN integrates Bayesian inference with the distance-based and the threshold-based method to estimate the number of emerging classes in the test data. It also devises a novel scoring method to distinguish the UCs (unknown classes) from KCs (known classes).
- Extensive experiments on four datasets (3 real and 1 synthetic) demonstrate that our model is superior to existing methods for accurate and robust online detection of emerging classes in streaming data.

2 Related Work

Anomaly detection (AD) basically can be divided into two categories, conventional methods, and deep learning techniques. Majority of conventional methods widely focus on distance-based approaches [15,28], reconstruction-based approaches [9], and unsupervised clustering. Deep learning techniques usually include autoencoder and GAN. An autoencoder identifies the outlier instances through reconstruction loss [34]. GAN has also been used as another means for computing reconstruction loss and then identifying anomalies [27,32]. In our app-roach, we use bi-directional GAN (BiGAN) with multi-modal prior distribution to improve the performance of UCs extraction.

AD mainly detects one class of anomalies whereas realistic data often con-tains multiple UCs. OSR is the right technique that solves this kind of problem. According to [6], OSR models are categorized into two types, discriminative and generative. The first type includes SVM-based methods [26] and distance-based method [1,2]. A collection of recent OSR works venture towards the generative direction [4,11,19,31]. A subset of OSR methods, named NEL, mainly employ Bayesian methods, such as infinite Gaussian mixture model (IGMM) [24] to learn the UCs. For example, Zhang et al. [33] use a non-parametric Bayesian framework with different posterior sampling strategies, such as one sweep Gibbs sampling, for detecting novel classes in online name disambiguation. However, IGMM-type methods can only handle small datasets that follow Gaussian distri-bution. To address this issue, we propose a novel algorithm that can achieve high accuracy on the large sparse dataset, which does not necessarily follow Gaussian distribution.

3   Background

Generative Adversarial Networks (GAN). Vanilla GAN [7] consists of two key components, a generator $\mathcal{G}$, and a discriminator $\mathcal{D}$. Given a prior distribution Z as input, $\mathcal{G}$ maps an instance $\mathbf{z} \sim Z$ from the latent space to the data space as $\mathcal{G}(\mathbf{z})$. On the other hand, $\mathcal{D}$ attempts to distinguish a data instance $\mathbf{x}$ from a synthetic instance $\mathcal{G}(\mathbf{z})$, generated by $\mathcal{G}$. We use the terminology $p_Z(\mathbf{z})$ to denote that $\mathbf{z}$ is a sampled instance from the distribution Z. The training process is set up as if $\mathcal{G}$ and $\mathcal{D}$ are playing a zero-sum game, a.k.a. minimax game; $\mathcal{G}$ tries to generate the synthetic instances that are as close as possible to actual data instances; on the other hand, $\mathcal{D}$ is responsible for distinguishing the real instances from the synthetic instances. In the end, GAN converges when both $\mathcal{G}$ and $\mathcal{D}$ reach a Nash equilibrium; at that stage, $\mathcal{G}$ learns the data distribution and is able to generate data instances that are very close to the actual data instances. The objective function of GAN can be written as follows:

$$\min_{\mathcal{G}} \max_{\mathcal{D}} V(\mathcal{D}, \mathcal{G}) = \mathbb{E}_{\mathbf{x} \sim X}[\log \mathcal{D}(\mathbf{x})] + \mathbb{E}_{\mathbf{z} \sim Z}[\log(1 - \mathcal{D}(\mathcal{G}(\mathbf{z})))] \tag{2}$$

where X is the distribution of $\mathbf{x}$ and Z is the distribution from which $\mathcal{G}$ samples.

Bidirectional Generative Adversarial Networks (BiGAN). Besides train-ing a generator $\mathcal{G}$, BiGAN [10] also trains an encoder $\mathcal{E}$, that maps real instances $\mathbf{x}$ into latent feature space $\mathcal{E}(\mathbf{x})$. Its discriminator $\mathcal{D}$ takes both $\mathbf{x}$ and $p_Z(\mathbf{z})$ as input in order to match the joint distribution $p_{\mathcal{G}}(\mathbf{x}, \mathbf{z})$ and $p_{\mathcal{E}}(\mathbf{x}, \mathbf{z})$. The objective function of BiGAN can be written as follows:

$$\min_{\mathcal{G}, \mathcal{E}} \max_{\mathcal{D}} V(\mathcal{D}, \mathcal{E}, \mathcal{G}) = \mathbb{E}_{\mathbf{x} \sim X}[\log \mathcal{D}(\mathbf{x}, \mathcal{E}(\mathbf{x}))] + \mathbb{E}_{\mathbf{z} \sim Z}[\log(1 - \mathcal{D}(\mathcal{G}(\mathbf{z}), \mathbf{z}))] \tag{3}$$

The objective function achieves the global minimum if and only if the distribution of both generator and encoder matches, i.e., $p_{\mathcal{G}}(\mathbf{x}, \mathbf{z}) = p_{\mathcal{E}}(\mathbf{x}, \mathbf{z})$.

4 Methodology

In this paper, we propose a novel model, Non-Exhaustive Gaussian Mixture Generative Adversarial Networks (NE-GM-GAN) for online non-exhaustive learning. The whole process is displayed in Fig. 1. Given a training set X_{train} with k_0 KCs, in the training step (offline), the proposed NE-GM-GAN employs a bidirectional GAN to train its encoder $\mathcal{E}$ and generator $\mathcal{G}$, by matching the joint distribution of encoder (X, Z) with the same of the generator. Note that the prior distribution Z of $\mathcal{G}$ is a multi-modal Gaussian (shown as Gaussian clusters on the top-middle part of the figure). After training, the generator and encoder of the GAN can take $\mathbf{z}$ and $\mathbf{x}$ as input and generate $\mathcal{G}(\mathbf{z})$ and $\mathcal{E}(\mathbf{x})$ as output, respectively.

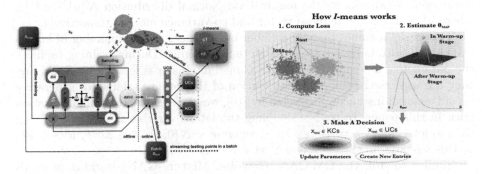

Fig. 1. The model architecture of NE-GM-GAN (left-hand side) and the workflow of I-means in Algorithm (2) (right-hand side)

The test step (online) shown on the right side of the model architecture and it is run on a batch of input instances, X_{test}. For all data instance from a batch (say, $\mathbf{x}$ is one such instance), NE-GM-GAN computes the $UCS(\mathbf{x})$ (unknown class score) of all instances in that batch; UCS score is derived from the reconstruction loss $L_{rec} = |\mathbf{x} - \mathcal{G}(\mathcal{E}(\mathbf{x}))|$. Using this score, the instances of a batch are partitioned into two groups: KCs and UCs. Elements in KCs belong to the known class, whereas the elements in UCs are potential UC instances. Using instances of UCs group, the model estimates the number of emerging class, k_{new}. After estimation, the model updates the prior of the $\mathcal{G}$ by adding the number of new classes k_{new} to k_0 as shown in the top right part of the model architecture. The GMM is then retrained for clustering both KCs and UCs. At this stage, the online test process for one test batch is finished.

In subsequent discussion, $X_{train} \in \mathbb{R}^{r \times d}$ is considered to be training data, containing r data instances, each of which is represented as a d-dimensional vector. K is the total number of known classes in X_{train}. X_{test} is test data that may contain instances of KCs and also instances of UCs. The dimensionality of latent space is denoted by p.

Offline Training: Computing Multi-modal Prior Distribution
In the vanilla form, generators of both GAN and BiGAN has a unimodal distribution as prior; in other words, the random variables $p_Z(\mathbf{z})$ is an instance from a unimodal distribution. Enlightened by [16], in this paper, we consider a multi-modal distribution as prior since this prior can better fit the real-life distribution of multi-class datasets. Thus,

$$p_Z(\mathbf{z}) = \sum_{k=1}^{K} \alpha^{\{k\}} \cdot p_k(\mathbf{z}) \qquad (4)$$

We assume that the number of initial clusters in the Gaussian distribution matches with the number of known classes (K) in X_{train}. $\alpha^{\{k\}}$ is the mixing parameter, $p_k(\mathbf{z})$ denotes the multivariate Normal distribution $\mathcal{N}(u^{\{k\}}, \Sigma^{\{k\}})$, where $u^{\{k\}}$ and $\Sigma^{\{k\}}$ are mean vector and co-variance matrix, respectively.

The model assumes that the number of instances and the number of known classes in the training set are given at the beginning. During training (offline), the parameters $u^{\{k\}}$ and $\Sigma^{\{k\}}$ of each Gaussian cluster is learned by GMM and they are used as the sampling distribution of the latent variable for generating the adversarial instances. Suggested by [16], we also use the re-parameterization trick in this paper. Instead of sampling the latent variable $\mathbf{z} \sim N(u^{\{k\}}, \Sigma^{\{k\}})$, the model samples $\mathbf{z} = A^{\{k\}}\epsilon + u^{\{k\}}$, where $\epsilon \sim N(0, I)$, $A \in \mathbb{R}^{p \times p}$, $u^{\{k\}} \in \mathbb{R}^p$. In this scenario, $u(\mathbf{z}) = u^{\{k\}}$ and $\Sigma(\mathbf{z}) = A^{\{k\}}A^{\{k\}T}$.

Similar to [10], the GM-GAN (Gaussian Mixture-GAN) learning proceeds as follows. The model takes sampled instance $\mathbf{z}$, sampled from the Gaussian multi-modal prior and a real instances $\mathbf{x}$ as input. Generator $\mathcal{G}$ attempts to map this sampled $p_Z(\mathbf{z})$ to data space as $\mathcal{G}(\mathbf{z})$. Encoder $\mathcal{E}$ maps real instances $\mathbf{x}$ into latent feature space as $\mathcal{E}(\mathbf{x})$. Discriminator $\mathcal{D}$ takes both $p_Z(\mathbf{z})$ and $\mathbf{x}$ as input for matching their joint distributions. After the model converges, theoretically, $\mathcal{G}(\mathbf{z}) \sim \mathbf{x}$ and $\mathcal{E}(\mathbf{x}) \sim p_Z(\mathbf{z})$. Note that NE-GM-GAN encodes X_{train} for offline training. To do so, GMM takes encoded X_{train} as input and then generates encoded u and Σ.

Extracting Potential Unknown Class
UC extraction of NE-GM-GAN is an online process that works on unlabeled data. During online detection, the model assumes that the test instance $\mathbf{x}$ is coming in a batch of the test set $X_{test} \in \mathbb{R}^{b \times d}$, where b is batch size and d is the dimension of feature space. Unlike [10], whose purpose is to generate the fake images as real as possible, our model aims at extracting the UC as accurately as possible. More specifically, our model generates the reconstructed instance $\mathcal{G}(\mathcal{E}(\mathbf{x}))$ at first and then computes the reconstruction loss between $\mathbf{x}$ and $\mathcal{G}(\mathcal{E}(\mathbf{x}))$. This step returns a size-b 1-D vector, consisting of reconstruction losses of the b points in the current batch, which is defined below:

$$L_{rec} = \|\mathbf{x} - \mathcal{G}(\mathcal{E}(\mathbf{x}))\| \qquad (5)$$

To distinguish the UC from KC in each test batch, we propose a metric, unknown class score, in short, UCS; the larger the score for an instance, the more likely that the instance belongs to an unknown class. To compute UCS of a test instance $\mathbf{x}$, NE-GM-GAN first computes, for each KC (out of K KCs), a baseline reconstruction loss, which is equal to the median of reconstruction losses of all train objects belonging to that known class. Then, UCS of $\mathbf{x}$ is equal to the minimum of the differences between $\mathbf{x}$'s reconstruction loss and each of the K baseline reconstruction losses. The pseudo-code of UCS computation is shown in Algorithm 1.

The intuition of UCS function is that GAN models instances of KCs with smaller reconstruction loss than the instances of UCs, but different known classes may have different baseline reconstruction loss, so we want an unknown class's reconstruction loss larger than the worst loss among all the KCs. This mechanism is inspired by [32]. Nevertheless, unlike [32], which assumes the prior as unimodal distribution and the UC must be far away from KC, our approach considers a multi-modal prior. After computing the UCS, the model extracts the potential UC from KC with a given threshold. For online detection, the threshold for the first test batch is empirically given whereas subsequent thresholds are decided by the percentage of UCs from previous test batches. Note that, the UCs objects may belong to multiple classes, but the model has no knowledge yet about the number of classes.

Algorithm 1: UCS for multi-modal prior

Input: Matrix $X_{train} \in \mathbb{R}^{r \times d}$ and $X_{test} \in \mathbb{R}^{b \times d}$
1 Compute $L_{test}(x_{test})$ with Eq. (5);
2 **for** $i \leftarrow 1$ **to** b **do**
3   **for** $k \leftarrow 1$ **to** K **do**
4   Compute $L_{train}(x_{train})^{\{k\}}$ with Eq. (5);
5   Select the median of $L_{train}(x_{train})^{\{k\}}$;
6          $UCS(x_{test})^{\{k\}} = \left| L_{test}(x_{test})^{\{i\}} - L_{train}(x_{train})^{\{k\}}_{median} \right|$;
7   **end**
8      $UCS^{\{i\}}_{min} = \min \left(UCS(x_{test})^{\{1\}}, ..., UCS(x_{test})^{\{K\}} \right)$;
9 **end**
10 $UCS = [UCS^{\{1\}}_{min}, ..., UCS^{\{b\}}_{min}]$;
11 **return** *Vector* $UCS \in \mathbb{R}^{b \times 1}$

Estimating the Number of Emerging Class

The previous extraction only extracts potential UCs. In practice, a small number of anomalous KC instances may be selected as UC instances. So, we use a subsequent step that distinctly identifies instances of unknown classes together with the number of UC and their parameters (mean, and covariance matrix of each of the UCs). We name this step as Infinite Means (I-means); the name reflects the fact that the number of unknown classes can increase as large as needed based on the test instances. Using I-means, a test instance is assigned to a new class if it is positioned far from the mean of all the KCs, and discovered novel classes prior to seeing that instance. To achieve this, for i-th test instance $x_{test}^{\{i\}}$, as shown in Eq. (6), I-means computes the distance $L_\mu^{\{k\}}$ between $x_{test}^{\{i\}}$ and the mean vector $\mu^{\{k\}}$ for the k-th KC and then selects the minimum of these values as $loss_{min}$ in Eq. (7).

$$L_\mu^{\{k\}} = \|x_{test}^{\{i\}} - \mu^{\{k\}}\|, \forall k \in [1..K] \tag{6}$$

$$loss_{min} = \min\left(L_\mu^{\{1\}}, L_\mu^{\{2\}}, ..., L_\mu^{\{K\}}\right), \; idx = \arg\min\left(L_\mu^{\{1\}}, L_\mu^{\{2\}}, ..., L_\mu^{\{K\}}\right) \tag{7}$$

A small value of $loss_{min}$ indicates that $x_{test}^{\{i\}}$ may potentially be a member of class idx; on the other hand, a large value $loss_{min}$ indicates that $x_{test}^{\{i\}}$ possibly belongs to a UC. To make the determination, we use a Bayesian approach, which dynamically adjusts the probability that a test point that is closest to cluster idx's mean vector belongs to cluster idx or not. The process is described below.

For a test instance, $x_{test}^{\{i\}}$ for which $idx = k$, the binary decision whether the instance belongs to k-th existing cluster or an emerging cluster follows Bernoulli distribution with parameter θ_k, which is modeled by using a Beta prior with parameter α_k, and β_k, where $\alpha_k, \beta_k \geq 1$ and $\theta_k = \frac{\alpha_k}{\alpha_k + \beta_k}$. The value of α_k and β_k are updated using Bayes rule. Based on the Bayes' theorem, the posterior distribution $p(\theta_k | x_{test}^{\{i\}})$, where $\theta_k \in [0, 1]$, is proportional to the prior distribution $p(\theta_k)$ multiplied by the likelihood function $p(x_{test}^{\{i\}} | \theta_k)$:

$$p(\theta_k | x_{test}^{\{i\}}) \propto p(x_{test}^{\{i\}} | \theta_k) \cdot p(\theta_k) \tag{8}$$

The posterior $p(\theta_k | x_{test}^{\{i\}})$ in Eq. (8) can be re-written as following:

$$
\begin{aligned}
p(\theta_k | x_{test}^{\{i\}}) &\propto \theta_k^{\alpha_{k0}}(1 - \theta_k)^{\beta_{k0}} \cdot \theta_k^{\alpha_k - 1}(1 - \theta_k)^{\beta_k - 1} \\
&= \theta_k^{\alpha_{k0} + \alpha_k - 1} \cdot (1 - \theta_k)^{\beta_{k0} + \beta_k - 1} \\
&= beta(\theta_k | \alpha_{k0} + \alpha_k, \beta_{k0} + \beta_k)
\end{aligned}
\tag{9}
$$

As the test instances are coming in streaming fashion, for any subsequent test instance for which $idx = k$, the posterior $p(\theta_k | x_{test}^{\{i\}})$ will act as prior for the next update. For the very first iteration, α_{k0} and β_{k0} are shape parameters of beta prior, which we learn in a warm-up stage. In the warm-up stage, we apply the three-sigma rule to compute the beta priors, α_{k0}, and β_{k0}. Each test point in the warm-up stage, for which $idx = k$, contributes a count of 1 to α_{k0} if the point is further than 3 standard deviation away from the mean, otherwise it contributes a count of 1 to β_{k0}. After the warm-up stage, we employ the Maximum-A-Posteriori (MAP) estimation to obtain the θ_{MAP_k} at which the posterior $p(\theta_k | x_{test}^{\{i\}})$ reaches its maximum value. According to the property of beta distribution, the θ_{MAP_k} is most likely to occur at the mean of posterior $p(\theta_k | x_{test}^{\{i\}})$. Thus, we can estimate the θ_{MAP_k} by:

$$\theta_{MAP_k} = \arg\max_{\theta_k} p(\theta_k | x_{test}^{\{i\}}) = \frac{\alpha_{k0} + \alpha_k}{\alpha_{k0} + \alpha_k + \beta_{k0} + \beta_k} \tag{10}$$

After estimating the θ_{MAP_k} by Eq. (10), I-means makes a cluster membership decision for each $x_{test}^{\{i\}}$ based on θ_{MAP_k}. This decision simulates the Bernoulli process, i.e., among the test instances which are close to the k-th cluster, approximately θ_{MAP_k} fraction of those will belong to the emerging cluster, whereas the remaining $(1 - \theta_{MAP_k})$ fractions of such instances will belongs to the k-th cluster. After each decision, corresponding parameters will be updated. If $x_{test}^{\{i\}}$ is clustered as a member of $KC^{\{k\}}$, we update the parameters $\mu_k^{\{i\}} \in \mathbb{R}^{1 \times d}$, $\sigma_k^{\{i\}} \in \mathbb{R}^{d \times d}$ of the k-th cluster by Eq. (11) and Eq. (12), respectively. The shape parameter β_k is increased by 1. Otherwise, if $x_{test}^{\{i\}}$ is considered as a member of UC, the shape parameter α_k, k_{new} are increased by 1, and the mean and covariance matrix of this new class are initialized by assigning current $x_{test}^{\{i\}}$ as new mean vector and creating a zero vector with the same shape of $x_{test}^{\{i\}}$ as new standard deviation vector.

$$\mu_k^{\{i\}} = \mu_k^{\{i-1\}} + \frac{x_{test}^{\{i\}} - \mu_k^{\{i-1\}}}{i} \tag{11}$$

$$v_k^{\{i\}} = v_k^{\{i-1\}} + \left(x_{test}^{\{i\}} - \mu_k^{\{i-1\}}\right)\left(x_{test}^{\{i\}} - \mu_k^{\{i\}}\right), \ \sigma_k^{\{i\}} = \sqrt{\frac{v_k^{\{i\}}}{(i-1)}} \tag{12}$$

The entire process of this paragraph is summarized I-means in Algorithm 2.

Table 2. Statistics of datasets (#Inst. denotes the number of instances; #F. denotes to the number of features after one-hot embedding or dropping for network intrusion dataset; #C. denotes to the number of classes.)

Dataset	#Inst.	#F.	#C.	Selected UCs
KDD99	494,021	121	23	Neptune, normal, back, satan, ipsweep, portsweep, warezclient, teardrop
NSL-KDD	148,517	121	40	Neptune, satan, ipsweep, smurf, portsweep, nmap, back, guess_passwd
UNSW-NB15	175,341	169	10	Generic, exploits, fuzzers, DoS, reconnaissance, analysis, backdoor, shellcode
Synthetic	100,300	121	16	No. 3, 4, 5, 6, 7, 8, 9, 10

5   Experiments

In this section, we show experimental results for validating the superior performance of our proposed NE-GM-GAN over different competing methods for multiple capabilities. Firstly, we compare the performance of potential UCs extraction. Furthermore, we compare the estimation of the number of distinct unknown classes. Finally, we show some experimental results for studying the effect of user-defined parameters on the algorithm's performance.

Algorithm 2: INFINITE Means (*I*-means)

Input: Testing batch $X_{test} \in \mathbb{R}^{b \times d}$, mean matrix, co-variance matrix

1 **for** *all* $x^{\{i\}} \in X_{test}$ **do**
2   **for** *all* $\mu^{\{k\}} \in M$ **do**
3   Compute $L_\mu^{\{k\}}$ by Eq. (6);
4   **end**
5   Get the index, idx, of minimum loss by Eq. (7);
6   **if warm-up stage then**
7   Select beta prior α_{idx0} and β_{idx0} based on Three-sigma Rule;
8   **end**
9   **else**
10   Estimate the θ_{MAP_k} by Eq. (10);
11   **end**
12   **if Uniform** $(0, 1) \leq \theta_{MAP_k}$ **then**
13   Update corresponding μ and σ by Eq. (11) and Eq. (12);
14        $\beta_{idx} \leftarrow \beta_{idx} + 1$;
15   **end**
16   **else**
17        $\alpha_{idx} \leftarrow \alpha_{idx} + 1$;
18        $k_{new} \leftarrow k_{new} + 1$;
19   **end**
20 **end**
21 **return** *The number of new emerging clusters* k_{new}

Dataset. We evaluate NE-GM-GAN on four datasets. Three of the datasets are real-life network intrusion datasets and the remaining one is a synthetic dataset. The network intrusion is very common for non-exhaustive classification because attackers constantly update their attack methods, so the classification model must adapt to novel class scenarios. The datasets are: (1) KDD Cup 1999 network intrusion dataset (**KDD99**), which contains 494,021 instances and 41 features with 23 different classes. One of the class represents "Normal" activity and the rest 22 represent various network attacks; (2) NSL-KDD dataset (**NSL-KDD**) [3], which is also a network intrusion dataset built by filtering some records from KDD99; (3) UNSW-NB15 dataset (**UNSW-NB15**) [18], which hybridizes real normal network activities with synthetic attack; (4) Synthetic dataset (**Synthetic**), which contains non-isotropic Gaussian clusters. Many of the features in the intrusion datasets are categorical or binary, so we employ one-hot embedding for such features. We also drop some columns which are redundant or whose values are almost zero or missing along the column. After that, we select eight of the classes as unknown classes (UCs) for each dataset. The test set is constructed from two parts. The first part is randomly sampled 20% of KCs instances and the second part is all the instances of the UCs. Rest 80% of KC instances are left for the training set. In the synthetic dataset, noises are injected into Gaussian clusters, each cluster representing a class. The injected noise is homocentric to the corresponding normal class but with a larger variance. The detailed statistics of the datasets are provided in Table 2.

Competing Methods. The performance of UCs extraction is evaluated with three competing methods, AnoGAN [27], DAGMM [34], and ALAD [32]. AnoGAN is the first GAN-based model for UC detection. Similarly, ALAD is another GAN-based model, which uses reconstructed errors to determine the UC. In contrast, DAGMM implements the autoencoder for the same task instead. The experimental setting follows [32] for this experiment. On the other hand, the capability of estimating the number of new emerging classes is compared against two competing methods, X-means [22], and IGMM [24,33]. X-means is a classical distance-based algorithm that can efficiently search the data space without knowing the initial number of clusters. On the contrary, IGMM is a Bayesian mixture model which uses the Dirichlet process prior and Gibbs sampler to efficiently identify new emerging entities. This experiment uses one sweep Gibbs sampler for IGMM [33]. For IGMM, we select the tunable parameters as following; $h = 10$, $m = h + 100$, $\kappa = 100$ and $\alpha = 100$, which is identical to the parameter values in [33]. Both models can return the number of online classes as NE-GM-GAN does, so they are selected as competing methods.

Evaluation Metrics. We use an external clustering evaluation metric, such as F1-score, to evaluate the performance of UCs extraction. For evaluating the prediction of the number of UCs (a regression task), we propose a new metric, Symmetrical R-squared (S-R^2). To obtain this, the root mean square error ($RMSE$) for both NE-GM-GAN and a competing method are computed and plugged into Eq. 13. S-$R^2 \in [-1, 1]$ gets more close to 1 if NE-GM-GAN defeats the competing method. On the contrary, its value will become more close -1. S-R^2 is exactly equal to 1 when the proposed model gets perfect prediction while the competing method doesn't. S-R^2 is zero when both methods have similar performance. The motivation to propose a new metric rather than using R-squared (R^2) is that R^2 would be less distinctive if two methods get much worse predictions because of using mean square error (MSE) inside. Besides, baseline sometimes achieves better performance, but R^2 cannot reflect this scenario as its range is from negative infinity to positive one.

$$S\text{-}R^2 = \begin{cases} 1 - \dfrac{RMSE_m}{RMSE_{bl}}, & RMSE_m < RMSE_{bl} \\ \dfrac{RMSE_{bl}}{RMSE_m} - 1, & RMSE_m > RMSE_{bl} \end{cases} \tag{13}$$

where $RMSE_m$ and $RMSE_{bl}$ denote the $RMSE$ of our model and baseline model, respectively.

The Capability of Unknown Class Extraction. In Table 3, we show the F1-score values of NE-GM-GAN and the competing methods for detecting the unknown class instances (the best results are shown in bold font). The result is computed by running each model 10 times and then taking the average. Out of the four datasets, NE-GM-GAN has the best performance in three with a healthy margin over the second-best method. In the largest dataset, our model received a

0.99 F1-score, a very good performance considering the fact that unknown class instances are assembled from 8 different classes. Only in the NSL-KDD dataset, NE-GM-GAN came out as the second-best. The performance of the other three models is mixed without a clear winner. One observation is that all the methods perform better on the larger dataset (KDD99).

To understand NE-GM-GAN's performance further, we perform an ablation study by switching the prior, as shown in Table 4. As we can see Gaussian multi-modal prior used in NE-GM-GAN is better suited than Unimodal prior generally used in traditional GAN. For all datasets multi-modal prior has 1% to 2% better F-score. A possible reason is that multi-modal prior is more closer to the real distribution of the training data.

Table 3. The F1-score of four models for UCs extraction

Data	NE-GM-GAN	AnoGAN	DAGMM	ALAD
KDD99	**0.99**	0.87	0.97	0.94
NSL-KDD	0.75	0.68	**0.79**	0.73
UNSW-NB15	**0.57**	0.49	0.53	0.51
Synthetic	**0.74**	0.51	0.70	0.56

Table 4. F1 score from our proposed model by using different prior

Prior	KDD99	NSL-KDD	UNSW-NB15	Synthetic
Unimodal	0.98	0.74	0.55	0.72
Multi-modal	**0.99**	**0.75**	**0.57**	**0.74**

The Estimation of the Number of New Classes. In this experiment, we compare NE-GM-GAN against two competing methods on all four datasets. To extend the scope of experiments, we vary the number of unknown classes from 2 to 6 by choosing all possible combinations of UCs and build multiple copies of one dataset and report performance results over all those copies. The motivation of using a combination of different UCs is to validate the robustness of the methods against varying numbers of UC counts. The result is shown in Table 5 using S-R^2 metric discussed earlier. The result close to 1 (the majority of the values in the table are between 0.85 and 0.95) means NE-GM-GAN substantially outperforms the competing methods. We argue that both competing methods assume that data distribution in each class follows mixture of Gaussian and thus fail to achieve good performance on realistic datasets. In only one dataset (Synthetic), X-means was able to obtain identical performance as ours' method, as both methods have the perfect prediction.

The same results are also shown in Fig. 2 as bar charts. In this Figure, y-axis is the number of predicted clusters, and each group of bars denotes the number of actual clusters for different methods. As we can see, NE-GM-GAN's prediction is very close to the actual prediction, whereas the results of the completing methods are way-off, except for the X-means method on the synthetic dataset. These experimental results demonstrate that our NE-GM-GAN outperforms the competing methods in terms of accuracy and robustness.

Table 5. The $S\text{-}R^2$ between NE-GM-GAN and baselines on 4 datasets (We denote "UCs" as the number of unknown classes in this table)

Datasets	Methods	UCs = 2	UCs = 3	UCs = 4	UCs = 5	UCs = 6
KDD99	X-means	0.8301	0.8805	0.8628	0.9105	0.8812
	IGMM	0.9528	0.8991	0.8908	0.9303	0.9248
NSL-KDD	X-means	0.8892	0.8604	0.9539	0.9228	0.9184
	IGMM	0.8771	0.8647	0.9517	0.9285	0.9238
UNSW-NB15	X-means	0.8892	0.8604	0.9539	0.9228	0.9184
	IGMM	0.8771	0.8647	0.9517	0.9285	0.9238
Synthetic	X-means	0	0	0	0	0
	IGMM	1	1	1	1	1

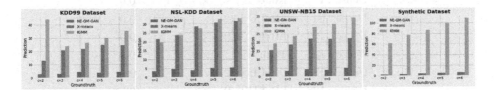

Fig. 2. Comparison on the estimation of new emerging class among three methods

Study of User-Defined Parameters. We perform a few experiments to justify some of our parameter design choices. For instance, to build the initial beta priors we used three-sigma rule. In Table 6, we present the percentage of instances of points that falls within the three standard deviations of the mean. The four columns correspond to the four datasets. As can be seen in the third row of the table, for all datasets, almost 100% of the points falls within the three standard deviations away from the mean. So, the priors selected in the warm-up stage based on three-sigma rule can sufficiently distinguish the UCs from the known class instances.

We also show unknown class prediction results over different values of WS (epochs of the warm-up stage) for different (between 2 to 6) unknown class that counts for all datasets. In Fig. 3, each curve represent a specific UC count. As can be seen, the prediction of the unknown class gets better with a larger number of WS. In most cases, the prediction converges when the number of epochs in the warm-up stage (WS) reaches 200 or above. In all our experiments, we select the WS value 200 for all datasets.

Table 6. Test of three-sigma rule (%)

Range	KDD99	NSL-KDD	UNSW-NB15	Synthetic
$\mu \pm 1\sigma$	94.37	61.17	58.67	56.64
$\mu \pm 2\sigma$	99.58	99.87	99.92	99.81
$\mu \pm 3\sigma$	99.60	100.00	100.00	100.00

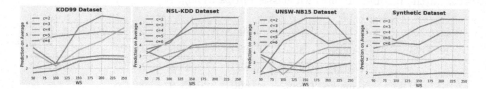

Fig. 3. Investigation on the number of epochs in the warm-up stage (WS) for I-means on four datasets

Table 7. Model architectures

	Layers	Units	Activation	Batch norm.	Dropout
$\mathcal{E}(\mathbf{x})$	Dense	64	LReLU (0.2)	×	0.0
	Dense	1	None	×	0.0
$\mathcal{G}(\mathbf{z})$	Dense	64	LReLU (0.2)	×	0.0
	Dense	128	LReLU(0.2)	×	0.0
	Dense	121	Tanh	×	0.0
$\mathcal{D}(\mathbf{x}, \mathbf{z})$	Dense	128	LReLU (0.2)	✓	0.5
	Dense	128	LReLU(0.2)	✓	0.5
	Dense	1	Sigmoid	×	0.0

Reproducibility of the Work. The model is implemented using Python 3.6.9 and Keras 2.2.4. For optimization, Adam is used with $\alpha = 10^{-5}$ and $\beta = 0.5$; mini-batch size is 50, latent dimension is 32, and the number of training epochs equal to 1000. The source code is available at https://github.com/junzhuang-code/NEGMGAN. The details of the BiGAN model architecture is given in Table 7.

6 Conclusion

In this paper, we propose a new online non-exhaustive model, Non-Exhaustive Gaussian Mixture Generative Adversarial Network (NE-GM-GAN), that synthesizes Bayesian method and deep learning technique for incremental learning the new emerging classes. NE-GM-GAN consists of three main components: (1) Gaussian mixture clustering generating multi-modal prior and re-clusters both KCs and UCs for parameter updating. (2) Bidirectional adversarial learning reconstructs the loss for extracting imbalanced UCs from KCs in an online testing batch. (3) A novel algorithm, I-means, estimates the number of new emerging classes for incremental learning the UCs on large sparse datasets. Experimental results illustrate that NE-GM-GAN significantly outperforms the competing methods for online detection across several benchmark datasets.

Acknowledgments. This research is partially supported by National Science Foundation with grant number IIS-1909916.

References

1. Bendale, A., Boult, T.: Towards open world recognition. arXiv preprint arXiv:1412.5687 (2014)
2. Bendale, A., Boult, T.: Towards open set deep networks. arXiv preprint arXiv:1511.06233 (2015)
3. Dhanabal, L., Shantharajah, S.: A study on NSL-KDD dataset for intrusion detection system based on classification algorithms. Int. J. Adv. Res. Comput. Commun. Eng. **4**(6), 446–452 (2015)
4. Ge, Z., Demyanov, S., Chen, Z., Garnavi, R.: Generative OpenMax for multi-class open set classification. arXiv preprint arXiv:1707.07418 (2017)
5. Geng, C., Chen, S.: Collective decision for open set recognition. IEEE Trans. Knowl. Data Eng. (2020)
6. Geng, C., Huang, S.J., Chen, S.: Recent advances in open set recognition: a survey. arXiv preprint arXiv:1811.08581 (2018)
7. Goodfellow, I.J., et al.: Generative adversarial networks. arXiv preprint arXiv:1406.2661 (2014)
8. Hassen, M., Chan, P.K.: Learning a neural-network-based representation for open set recognition. In: Proceedings of the 2020 SIAM International Conference on Data Mining. SIAM (2020)
9. Hubert, M., Rousseeuw, P.J., Vanden Branden, K.: ROBPCA: a new approach to robust principal component analysis. Technometrics **47**(1), 64–79 (2005)
10. Donahue, J., Krähenbühl, P., Darrell, T.: Adversarial feature learning. arXiv preprint arXiv:1605.09782 (2017)
11. Jo, I., Kim, J., Kang, H., Kim, Y.D., Choi, S.: Open set recognition by regularising classifier with fake data generated by generative adversarial networks. In: 2018 IEEE International Conference on Acoustics, Speech and Signal Processing (ICASSP). IEEE (2018)
12. Koch, G., Zemel, R., Salakhutdinov, R.: Siamese neural networks for one-shot image recognition. In: ICML Deep Learning Workshop. Lille (2015)
13. Liu, F.T., Ting, K.M., Zhou, Z.H.: Isolation forest. In: 2008 Eighth IEEE International Conference on Data Mining, pp. 413–422. IEEE (2008)
14. Liu, Z., Miao, Z., Zhan, X., Wang, J., Gong, B., Yu, S.X.: Large-scale long-tailed recognition in an open world. In: Proceedings of the IEEE/CVF Conference on Computer Vision and Pattern Recognition, pp. 2537–2546 (2019)
15. Manevitz, L.M., Yousef, M.: One-class SVMs for document classification. J. Mach. Learn. Res. **2**(Dec), 139–154 (2001)
16. Matan Ben-Yosef, D.W.: Gaussian mixture generative adversarial networks for diverse datasets, and the unsupervised clustering of images. arXiv preprint arXiv:1808.10356 (2018)

17. Mensink, T., Verbeek, J., Perronnin, F., Csurka, G., et al.: Distance-based image classification: generalizing to new classes at near zero cost. IEEE Trans. Pattern Anal. Mach. Intell. **35**, 2624–2637 (2013)
18. Moustafa, N., Slay, J.: UNSW-NB15: a comprehensive data set for network intrusion detection systems (UNSW-NB15 network data set). In: 2015 Military Communications and Information Systems Conference (MilCIS), pp. 1–6. IEEE (2015)
19. Neal, L., Olson, M., Fern, X., Wong, W.K., Li, F.: Open set learning with counterfactual images. In: Proceedings of the European Conference on Computer Vision (ECCV), pp. 613–628 (2018)
20. Oza, P., Patel, V.M.: C2AE: class conditioned auto-encoder for open-set recognition. In: Proceedings of the IEEE/CVF Conference on Computer Vision and Pattern Recognition (2019)
21. Palatucci, M., Pomerleau, D., Hinton, G.E., Mitchell, T.M.: Zero-shot learning with semantic output codes. In: Advances in Neural Information Processing Systems, pp. 1410–1418 (2009)
22. Pelleg, D., Moore, A.W., et al.: X-means: extending k-means with efficient estimation of the number of clusters. In: ICML, vol. 1, pp. 727–734 (2000)
23. Perera, P., et al.: Generative-discriminative feature representations for open-set recognition. In: Proceedings of the IEEE/CVF Conference on Computer Vision and Pattern Recognition (2020)
24. Rasmussen, C.E.: The infinite gaussian mixture model. In: NIPS (2000)
25. Ravi, S., Larochelle, H.: Optimization as a model for few-shot learning (2016)
26. Scheirer, W.J., Jain, L.P., Boult, T.E.: Probability models for open set recognition. IEEE Trans. Pattern Anal. Mach. Intell. **36**(11), 2317–2324 (2014)
27. Schlegl, T., Seebök, P., Waldstein, S.M., Schmidt-Erfurth, U., Langs, G.: Unsupervised anomaly detection with generative adversarial networks to guide marker discovery. arXiv preprint arXiv:1703.05921 (2017)
28. Schölkopf, B., Williamson, R.C., Smola, A.J., Shawe-Taylor, J., Platt, J.C.: Support vector method for novelty detection. In: Advances in Neural Information Processing Systems, pp. 582–588 (2000)
29. Socher, R., Ganjoo, M., Manning, C.D., Ng, A.: Zero-shot learning through cross-modal transfer. In: Advances in Neural Information Processing Systems (2013)
30. Wang, Y., et al.: Iterative learning with open-set noisy labels. In: Proceedings of the IEEE Conference on Computer Vision and Pattern Recognition (2018)
31. Yang, Y., Hou, C., Lang, Y., Guan, D., Huang, D., Xu, J.: Open-set human activity recognition based on micro-doppler signatures. Pattern Recogn. **85**, 60–69 (2019)
32. Zenati, H., Romain, M., Foo, C.S., Lecouat, B., Chandrasekhar, V.: Adversarially learned anomaly detection. In: 2018 IEEE International Conference on Data Mining (ICDM). IEEE (2018)
33. Zhang, B., Dundar, M., Hasan, M.A.: Bayesian non-exhaustive classification a case study: online name disambiguation using temporal record streams. arXiv preprint arXiv:1607.05746 (2016)
34. Zong, B., et al.: Deep autoencoding Gaussian mixture model for unsupervised anomaly detection (2018)

Unsupervised Learning of Joint Embeddings for Node Representation and Community Detection

Rayyan Ahmad Khan[1,3]($\boxtimes$), Muhammad Umer Anwaar[1,3], Omran Kaddah[3], Zhiwei Han[2], and Martin Kleinsteuber[1,3]

[1] Mercateo AG, Munich, Germany
[2] Fortiss GmbH, Munich, Germany
[3] Technical University of Munich, Munich, Germany
rayyan.khan@tum.de

Abstract. In graph analysis community detection and node representation learning are two highly correlated tasks. In this work, we propose an efficient generative model called **J-ENC** for learning **J**oint **E**mbedding for **N**ode representation and **C**ommunity detection. **J-ENC** learns a community-aware node representation, i.e., learning of the node embeddings are constrained in such a way that connected nodes are not only "closer" to each other but also share similar community assignments. This joint learning framework leverages community-aware node embeddings for better performance on these tasks: node classification, overlapping community detection and non-overlapping community detection. We demonstrate on several graph datasets that **J-ENC** effectively outperforms many competitive baselines on these tasks. Furthermore, we show that **J-ENC** not only has quite robust performance with varying hyperparameters but also is computationally efficient than its competitors.

1 Introduction

Graphs are flexible data structures that model complex relationships among entities, i.e. data points as nodes and the relations between nodes via edges. One important task in graph analysis is community detection, where the objective is to cluster nodes into multiple groups (communities). Each community is a set of densely connected nodes. The communities can be overlapping or non-overlapping, depending on whether they share some nodes or not. Several algorithmic [1,5] and probabilistic approaches [9,20,34,38] to community detection have been proposed. Another fundamental task in graph analysis is learning

---

R. A. Khan, M. U. Anwaar, O. Kaddah and Z. Han—Equal contribution.

Electronic supplementary material The online version of this chapter (https://doi.org/10.1007/978-3-030-86520-7_2) contains supplementary material, which is available to authorized users.

N. Oliver et al. (Eds.): ECML PKDD 2021, LNAI 12976, pp. 19–35, 2021.
https://doi.org/10.1007/978-3-030-86520-7_2

the node embeddings. These embeddings can then be used for downstream tasks like graph visualization [8,28,33,34] and classification [3,27].

In the literature, these tasks are usually treated separately. Although the standard graph embedding methods capture the basic connectivity, the learning of the node embeddings is independent of community detection. For instance, a simple approach can be to get the node embeddings via DeepWalk [23] and get community assignments for each node by using k-means or Gaussian mixture model. Looking from the other perspective, methods like Bigclam [36], that focus on finding the community structure in the dataset, perform poorly for node-representation tasks e.g. node classification. This motivates us to study the approaches that jointly learn community-aware node embeddings.

Recently several approaches, like CNRL [30], ComE [4], vGraph [26] etc., have been proposed to learn the node embeddings and detect communities simultaneously in a unified framework. Several studies have shown that community detection is improved by incorporating the node representation in the learning process [3,18]. The intuition is that the global structure of graphs learned during community detection can provide useful context for node embeddings and vice versa.

The joint learning methods (CNRL, ComE and vGraph) learn two embeddings for each node. One node embedding is used for the node representation task. The second node embedding is the "context" embedding of the node which aids in community detection. As CNRL and ComE are based on Skip-Gram [22] and DeepWalk [23], they inherit "context" embedding from it for learning the neighbourhood information of the node. vGraph also requires two node embeddings for parameterizing two different distributions. In contrast, we propose learning a single community-aware node representation which is directly used for both tasks.

In this paper, we propose an efficient generative model called **J-ENC** for jointly learning both community detection and node representation. The underlying intuition behind **J-ENC** is that every node can be a member of one or more communities. However, the node embeddings should be learned in such a way that connected nodes are "closer" to each other than unconnected nodes. Moreover, connected nodes should have similar community assignments. Formally, we assume that for i-th node, the node embeddings z_i are generated from a prior distribution $p(z)$. Given z_i, the community assignments c_i are sampled from $p(c_i|z_i)$, which is parameterized by node and community embeddings. In order to generate an edge (i, j), we sample another node embedding z_j from $p(z)$ and respective community assignment c_j from $p(c_j|z_j)$. Afterwards, the node embeddings and the respective community assignments of node pairs are fed to a decoder. The decoder ensures that embeddings of both the nodes and the communities of connected nodes share high similarity. This enables learning such node embeddings that are useful for both community detection and node representation tasks.

We validate the effectiveness of our approach on several real-world graph datasets. In Sect. 4, we show empirically that **J-ENC** is able to outperform the baseline methods including the direct competitors on all three tasks i.e. node classification, overlapping community detection and non-overlapping community detection. Furthermore, we compare the computational cost of training different algorithms. **J-ENC** is up to 40x more time-efficient than its competitors. We also conduct hyperparameter sensitivity analysis which demonstrates the robustness of our approach. Our main contributions are summarized below:

- We propose an efficient generative model called **J-ENC** for joint community detection and node representation learning.
- We adopt a novel approach and argue that a single node embedding is sufficient for learning both the representation of the node itself and its context.
- Training **J-ENC** is extremely time-efficient in comparison to its competitors.

2   Related Work

2.1   Community Detection

Early community detection algorithms are inspired from clustering algorithms [35]. For instance, spectral clustering [29] is applied to the graph Laplacian matrix for extracting the communities. Similarly, several matrix factorization based methods have been proposed to tackle the community detection problem. For example, Bigclam [36] treats the problem as a non-negative matrix factorization (NMF) task. Another method CESNA [38] extends Bigclam by modelling the interaction between the network structure and the node attributes. Some generative models, like vGraph [26], Circles [20] etc., have also been proposed to detect communities in a graph.

2.2   Node Representation Learning

Many successful algorithms which learn node representation in an unsupervised way are based on random walk objectives [10,11,23]. Some known issues with random-walk based methods (e.g. DeepWalk, node2vec etc.) are: (1) They sacrifice the structural information of the graph by putting over-emphasis on the proximity information [24] and (2) great dependence of the performance on hyperparameters (walk-length, number of hops etc.) [10,23]. Some interesting GCN based approaches include graph autoencoders e.g. GAE and VGAE [17] and DGI [32].

2.3   Joint Community Detection and Node Representation Learning

In the literature, several attempts have been made to tackle both these tasks in a single framework. Most of these methods propose an alternate optimization process, i.e. learn node embeddings and improve community assignments with them and vice versa [4,30]. Some approaches (CNRL [30], ComE [4]) are inspired from random walk, thus they inherit the issues discussed above. Others, like GEMSEC [25], are limited to the detection of non-overlapping communities. Some generative models like CommunityGAN [13] and vGraph [26] also jointly learn community assignments and node embeddings. CNRL, ComE and vGraph require learning two embeddings for each node for simultaneously tackling the two tasks. Unlike them, **J-ENC** learns a single community-aware node representation which is directly used for both tasks.

It is pertinent to highlight that although both vGraph and **J-ENC** adopt a variational approach but the underlying models are quite different. vGraph assumes that each node can be represented as a mixture of multiple communities and is described by a multinomial distribution over communities, whereas **J-ENC** models the node embedding by a single distribution. For a given node, vGraph, first draws a community assignment and then a connected neighbor node is generated based on the assignment. Whereas, **J-ENC** draws the node embedding from prior distribution and then community assignment is conditioned on a single node only. In simple terms, vGraph also needs edge information in the generative process whereas **J-ENC** does not require it. **J-ENC** relies on the decoder to ensure that embeddings of the connected nodes and their communities share high similarity with each other.

3   Methodology

3.1   Problem Formulation

Suppose an undirected graph $\mathcal{G} = (\mathcal{V}, \mathcal{E})$ with the adjacency matrix $\boldsymbol{A} \in \mathbb{R}^{N \times N}$ and a matrix $\boldsymbol{X} \in \mathbb{R}^{N \times F}$ of F-dimensional node features, N being the number of nodes. Given K as the number of communities, we aim to jointly learn the node embeddings and the community embeddings following a variational approach such that:

- One or more communities can be assigned to every node.
- The node embeddings can be used for both community detection and node classification.

3.2   Variational Model

Generative Model: Let us denote the latent node embedding and community assignment for i-th node by the random variables $z_i \in \mathbb{R}^d$ and c_i respectively. The generative model is given by:

$$p(\boldsymbol{A}) = \int \sum_c p(\boldsymbol{Z}, \boldsymbol{c}, \boldsymbol{A}) d\boldsymbol{Z}, \tag{1}$$

where $\boldsymbol{c} = [c_1, c_2, \cdots, c_N]$ and the matrix $\boldsymbol{Z} = [z_1, z_2, \cdots, z_N]$ stacks the node embeddings. The joint distribution in (1) is mathematically expressed as

$$p(\boldsymbol{Z}, \boldsymbol{c}, \boldsymbol{A}) = p(\boldsymbol{Z}) \, p_\theta(\boldsymbol{c}|\boldsymbol{Z}) \, p_\theta(\boldsymbol{A}|\boldsymbol{c}, \boldsymbol{Z}), \tag{2}$$

where θ denotes the model parameters. Let us denote elements of $\boldsymbol{A}$ by a_{ij}. Following existing approaches [14,17], we consider z_i to be *i.i.d* random variables.

Furthermore, assuming $c_i|z_i$ to be *i.i.d* random variables, the joint distributions in (2) can be factorized as

$$p(\boldsymbol{Z}) = \prod_{i=1}^{N} p(\boldsymbol{z}_i) \tag{3}$$

$$p_\theta(\boldsymbol{c}|\boldsymbol{Z}) = \prod_{i=1}^{N} p_\theta(c_i|\boldsymbol{z}_i) \tag{4}$$

$$p_\theta(\boldsymbol{A}|\boldsymbol{c}, \boldsymbol{Z}) = \prod_{i,j} p_\theta(a_{ij}|c_i, c_j, \boldsymbol{z}_i, \boldsymbol{z}_j), \tag{5}$$

where Eq. (5) assumes that the *edge decoder* $p_\theta(a_{ij}|c_i, c_j, \boldsymbol{z}_i, \boldsymbol{z}_j)$ depends only on $c_i, c_j, \boldsymbol{z}_i$ and $\boldsymbol{z}_j$.

Inference Model: We aim to learn the model parameters θ such that $\log(p_\theta(\boldsymbol{A}))$ is maximized. In order to ensure computational tractability, we introduce the approximate posterior

$$q_\phi(\boldsymbol{Z}, \boldsymbol{c}|\mathcal{I}) = \prod_i q_\phi(\boldsymbol{z}_i, c_i|\mathcal{I}) \tag{6}$$

$$= \prod_i q_\phi(\boldsymbol{z}_i|\mathcal{I})q_\phi(c_i|\boldsymbol{z}_i, \mathcal{I}), \tag{7}$$

where $\mathcal{I} = (\boldsymbol{A}, \boldsymbol{X})$ if node features are available, otherwise $\mathcal{I} = \boldsymbol{A}$. We maximize the corresponding ELBO bound (for derivation, refer to the supplementary material), given by

$$
\begin{aligned}
\mathcal{L}_{\text{ELBO}} \approx &- \sum_{i=1}^{N} D_{KL}\Big(q_\phi(\boldsymbol{z}_i|\mathcal{I}) \parallel p(\boldsymbol{z}_i) \Big) \\
&- \sum_{i=1}^{N} \frac{1}{M} \sum_{m=1}^{M} D_{KL}\Big(q_\phi(c_i|\boldsymbol{z}_i^{(m)}, \mathcal{I}) \parallel p_\theta(c_i|\boldsymbol{z}_i^{(m)}) \Big) \\
&+ \sum_{(i,j)\in\mathcal{E}} \mathbb{E}_{(\boldsymbol{z}_i, \boldsymbol{z}_j, c_i, c_j)\sim q_\phi(\boldsymbol{z}_i, \boldsymbol{z}_j, c_i, c_j|\mathcal{I})} \Big\{ \log\Big(p_\theta(a_{ij}|c_i, c_j, \boldsymbol{z}_i, \boldsymbol{z}_j) \Big) \Big\},
\end{aligned}
\tag{8}
$$

where $D_{KL}(.\|.)$ represents the KL-divergence between two distributions. The distribution $q_\phi(\boldsymbol{z}_i, \boldsymbol{z}_j, c_i, c_j|\mathcal{I})$ in the third term of Eq. (8) is factorized into two conditionally independent distributions i.e.

$$q_\phi(\boldsymbol{z}_i, \boldsymbol{z}_j, c_i, c_j|\mathcal{I}) = q_\phi(\boldsymbol{z}_i, c_i|\mathcal{I})q_\phi(\boldsymbol{z}_j, c_j|\mathcal{I}). \tag{9}$$

3.3 Design Choices

In Eq. (3), $p(z_i)$ is chosen to be the standard gaussian distribution for all i. The corresponding approximate posterior $q_\phi(z_i|\mathcal{I})$ in Eq. (7), used as node embeddings encoder, is given by

$$q_\phi(z_i|\mathcal{I}) = \mathcal{N}\big(\boldsymbol{\mu}_i(\mathcal{I}), \mathrm{diag}(\boldsymbol{\sigma}^2{}_i(\mathcal{I}))\big). \qquad (10)$$

The parameters of $q_\phi(z_i|\mathcal{I})$ can be learnt by any encoder network e.g. graph convolutional network [16], graph attention network [31], GraphSAGE [11] or even two matrices to learn $\boldsymbol{\mu}_i(\mathcal{I})$ and $\mathrm{diag}(\boldsymbol{\sigma}^2{}_i(\mathcal{I}))$. Samples are then generated using reparametrization trick [6].

For parameterizing $p_\theta(c_i|z_i)$ in Eq. (4), we introduce community embeddings $\{\boldsymbol{g}_1, \cdots, \boldsymbol{g}_K\}$; $\boldsymbol{g}_k \in \mathbb{R}^d$. The distribution $p_\theta(c_i|z_i)$ is then modelled as the softmax of dot products of z_i with $\boldsymbol{g}_k$, i.e.

$$p_\theta(c_i = k|z_i) = \frac{\exp(<z_i, \boldsymbol{g}_k>)}{\sum\limits_{\ell=1}^{K} \exp(<z_i, \boldsymbol{g}_\ell>)}. \qquad (11)$$

The corresponding approximate posterior $q_\phi(c_i = k|z_i, \mathcal{I})$ in Eq. (7) is affected by the node embedding z_i as well as the neighborhood. To design this, our intuition is to consider the similarity of $\boldsymbol{g}_k$ with the embedding z_i as well as with the embeddings of the neighbors of the i-th node. The overall similarity with neighbors is mathematically formulated as the average of the dot products of their embeddings. Afterwards, a hyperparameter α is introduced to control the bias between the effect of z_i and the set $\mathcal{N}_i$ of the neighbors of the i-th node. Finally, a softmax is applied, i.e.

$$q_\phi(c_i = k|z_i, \mathcal{I}) = \mathrm{softmax}\Big(\alpha < z_i, \boldsymbol{g}_k >$$
$$+ (1-\alpha)\frac{1}{|\mathcal{N}_i|} \sum_{j \in \mathcal{N}_i} < z_j, \boldsymbol{g}_k > \Big). \qquad (12)$$

Hence, Eq. (12) ensures that graph structure information is employed to learn community assignments instead of relying on an extraneous node embedding as done in [4, 26]. Finally, the choice of edge decoder in Eq. (5) is motivated by the intuition that the nodes connected by edges have a high probability of belonging to the same community and vice versa. Therefore we model the edge decoder as:

$$p_\theta(a_{ij}|c_i = \ell, c_j = m, z_i, z_j) = \frac{\sigma(<z_i, \boldsymbol{g}_m>) + \sigma(<z_j, \boldsymbol{g}_\ell>)}{2}. \qquad (13)$$

For better reconstructing the edges, Eq. (13) makes use of the community embeddings, node embeddings and community assignment information simultaneously. This helps in learning better node representations by leveraging the global information about the graph structure via community detection. On the other hand, this also forces the community assignment information to exploit the local graph structure via node embeddings and edge information.

3.4   Practical Aspects

The third term in Eq. (8) is estimated in practice using the samples generated by the approximate posterior. This term is equivalent to the negative of binary cross-entropy (BCE) loss between observed edges and reconstructed edges. Since community assignment follows a categorical distribution, we use Gumbel-softmax [12] for backpropagation of the gradients. As for the second term of Eq. (8), it is also enough to set $M = 1$, i.e. use only one sample per input node.

For inference, non-overlapping community assignment can be obtained for i-th node as

$$C_i = \underset{k \in \{1, \cdots, K\}}{\arg\max} \; q_\phi(c_i = k | \mathbf{z}_i, \mathcal{I}). \qquad (14)$$

To get overlapping community assignments for i-th node, we can threshold its weighted probability vector at ϵ, a hyperparameter, as follows

$$C_i = \left\{ k \; \middle| \; \frac{q_\phi(c_i = k | \mathbf{z}_i, \mathcal{I})}{\max_\ell q_\phi(c_i = \ell | \mathbf{z}_i, \mathcal{I})} \geq \epsilon \right\}, \quad \epsilon \in [0, 1]. \qquad (15)$$

3.5   Complexity

Computation of dot products for all combinations of node and community embeddings takes $\mathcal{O}(NKd)$ time. Solving Eq. (12) further requires calculation of mean of dot products over the neighborhood for every node, which takes $\mathcal{O}(|\mathcal{E}|K)$ computations overall as we traverse every edge for every community. Finally, we need softmax over all communities for every node in Eq. (11) and Eq. (12) which takes $\mathcal{O}(NK)$ time. Equation (13) takes $\mathcal{O}(|\mathcal{E}|)$ time for all edges as we have already calculated the dot products. As a result, the overall complexity becomes $\mathcal{O}(|\mathcal{E}|K + NKd)$. This complexity is quite low compared to other algorithms designed to achieve similar goals [4,39].

4   Experiments

4.1   Synthetic Example

We start with a synthetic dataset, consisting of 3 communities with 5 points per community. This dataset is actually a random partition graph generated by the python package networkx. The encoder simply consists of two matrices that give $\boldsymbol{\mu}_i(\mathcal{I})$ and $\mathrm{diag}(\boldsymbol{\sigma}^2{}_i(\mathcal{I}))$. The results of the community assignments discovered by **J-ENC** are given in Fig. 1, where the node sizes are reciprocal to the confidence of **J-ENC** in the community assignments. We choose 3 communities for demonstration because the probabilistic community assignments in such case can be thought of as rgb values for coloring the nodes. It can be seen that **J-ENC** discovers the correct community structure. However, the two bigger nodes in the center can be assigned to more than one communities as **J-ENC** is not very confident in case of these nodes. This is evident from the colors that are a mix of red, green and blue. We now proceed to the experiments on real-world datasets.

4.2   Datasets

We have selected 18 different datasets ranging from 270 to 126,842 edges. For non-overlapping community detection and node classification, we use 5 the citation datasets [2,40]. The remaining datasets [20,37], used for overlapping community detection, are taken from SNAP repository [19]. Following [26], we take 5 biggest ground truth communities for youtube, amazon and dblp. Moreover, we also analyse the case of large number of communities. For this purpose, we prepare two subsets of amazon dataset by randomly selecting 500 and 1000 communities from 2000 smallest communities in the amazon dataset (Table 1).

Fig. 1. Visualization of community assignments discovered by **J-ENC** in the synthetic dataset of 15 points divided in three communities.

Table 1. Every dataset has $|\mathcal{V}|$ nodes, $|\mathcal{E}|$ edges, K communities and $|F|$ features. $|F| = $ N/A means that either the features were missing or not used.

Dataset	$|\mathcal{V}|$	$|\mathcal{E}|$	K	$|F|$	Overlap
CiteSeer	3327	9104	6	3703	N
CiteSeer-full	4230	10674	6	602	N
Cora	2708	10556	7	1433	N
Cora-ML	2995	16316	7	2879	N
Cora-full	19793	126842	70	8710	N
fb0	333	2519	24	N/A	Y
fb107	1034	26749	9	N/A	Y
fb1684	786	14024	17	N/A	Y
fb1912	747	30025	46	N/A	Y
fb3437	534	4813	32	N/A	Y
fb348	224	3192	14	N/A	Y
fb414	150	1693	7	N/A	Y
fb698	61	270	13	N/A	Y
Youtube	5346	24121	5	N/A	Y
Amazon	794	2109	5	N/A	Y
Amazon500	1113	3496	500	N/A	Y
Amazon1000	1540	4488	1000	N/A	Y
Dblp	24493	89063	5	N/A	Y

4.3   Baselines

For overlapping community detection, we compare with the following competitive baselines: **MNMF** [34] learns community membership distribution by using joint non-negative matrix factorization with modularity based regularization. **BIGCLAM** [36] also formulates community detection as a non-negative matrix factorization (NMF) task. It simultaneously optimizes the model likelihood of observed links and learns the latent factors which represent community affiliations of nodes. **CESNA** [38] extends BIGCLAM by statistically modelling the interaction between the network structure and the node attributes. **Circles** [20] introduces a generative model for community detection in ego-networks by learning node similarity metrics for every community. **SVI** [9] formulates membership of nodes in multiple communities by a Bayesian model of networks.

vGraph [26] simultaneously learns node embeddings and community assignments by modelling the nodes as being generated from a mixture of communities. **vGraph+**, a variant further incorporates regularization to weigh local connectivity. **ComE** [4] jointly learns community and node embeddings by using gaussian mixture model formulation. **CNRL** [30] enhances the random walk sequences (generated by DeepWalk, node2vec etc.) to jointly learn community and node embeddings. **CommunityGAN** (ComGAN)is a generative adversarial model for learning node embeddings such that the entries of the embedding vector of each node refer to the membership strength of the node to different communities. Lastly, we compare the results with the communities obtained by applying k-means to the learned embeddings of **DGI** [32].

For non-overlapping community detection and node classification, in addition to MNMF, DGI, CNRL, CommunityGAN, vGraph and ComE, we compare **J-ENC** with the following baselines: **DeepWalk** [23] makes use of SkipGram [22] and truncated random walks on network to learn node embeddings. **LINE** [27] learns node embeddings while attempting to preserve first and second order proximities of nodes. **Node2Vec** [10] learns the embeddings using biased random walk while aiming to preserve network neighborhoods of nodes. **Graph Autoencoder (GAE)** [17] extends the idea of autoencoders to graph datasets. We also include its variational counterpart i.e. **VGAE**. **GEMSEC** is a sequence sampling-based learning model which aims to jointly learn the node embeddings and clustering assignments.

4.4 Settings

For overlapping community detection, we learn mean and log-variance matrices of 16-dimensional node embeddings. We set $\alpha = 0.9$ and $\epsilon = 0.3$ in all our experiments. Following [17], we first pre-train a variational graph autoencoder. We perform gradient descent with Adam optimizer [15] and learning rate $= 0.01$. Community assignments are obtained using Eq. (15). For the baselines, we employ the results reported by [26]. For evaluating the performance, we use *F1-score* and *Jaccard similarity*.

For non-overlapping community detection, since the default implementations of most the baselines use 128 dimensional embeddings, for we use $d = 128$ for fair comparison. Equation (14) is used for community assignments. For vGraph, we use the code provided by the authors. We employ *normalized mutual information (NMI)* and *adjusted random index (ARI)* as evaluation metrics.

For node classification, we follow the training split used in various previous works [16,32,40], i.e. 20 nodes per class for training. We train logistic regression using LIBLINEAR [7] solver as our classifier and report the evaluation results on rest of the nodes. For the algorithms that do not use node features, we train the classifier by appending the raw node features with the learnt embeddings. For evaluation, we use *F1-macro* and *F1-micro* scores.

All the reported results are the average over five runs. Further implementation details can be found in: https://github.com/RayyanRiaz/gnn_comm_det.

4.5   Discussion of Results

Tables 2 and 3 summarize the results of the performance comparison for the overlapping community detection tasks.

Table 2. F1 scores (%) for overlapping communities. Best and second best values are bold and underlined respectively.

Dataset	MNMF	Bigclam	CESNA	Circles	SVI	vGraph	vGraph+	ComE	CNRL	ComGan	DGI	J-ENC
fb0	14.4	29.5	28.1	28.6	28.1	24.4	26.1	31.1	11.5	**35.0**	27.4	_34.7_
fb107	12.6	39.3	37.3	24.7	26.9	28.2	31.8	39.7	20.2	_47.5_	35.8	**59.7**
fb1684	12.2	50.4	51.2	28.9	35.9	42.3	43.8	_52.9_	38.5	47.6	42.8	**56.4**
fb1912	14.9	34.9	34.7	26.2	28.0	25.8	_37.5_	28.7	8.0	35.6	32.6	**45.8**
fb3437	13.7	19.9	20.1	10.1	15.4	20.9	22.7	21.3	3.9	_39.3_	19.7	**50.2**
fb348	20.0	49.6	53.8	51.8	46.1	55.4	53.1	46.2	34.1	_55.8_	54.7	**58.2**
fb414	22.1	58.9	60.1	48.4	38.9	64.7	_66.9_	55.3	25.3	43.9	56.9	**69.6**
fb698	26.6	54.2	58.7	35.2	40.3	54.0	_59.5_	45.8	16.4	58.2	52.2	**64.0**
Youtube	59.9	43.7	38.4	36.0	41.4	50.7	52.2	_65.5_	51.4	43.6	47.8	**67.3**
Amazon	38.2	46.4	46.8	53.3	47.3	53.3	53.2	50.1	_53.5_	51.4	44.7	**58.1**
Amazon500	30.1	52.2	57.3	46.2	41.9	_61.2_	60.4	59.8	38.4	59.3	33.8	**67.6**
Amazon1000	19.3	28.6	30.8	25.9	21.6	_54.3_	47.3	50.3	27.1	52.7	37.7	**60.5**
Dblp	21.8	23.6	35.9	36.2	33.7	39.3	39.9	_47.1_	46.8	34.9	44.0	**53.9**

Table 3. Jaccard scores (%) for overlapping communities. Best and second best values are bold and underlined respectively.

Dataset	MNMF	Bigclam	CESNA	Circles	SVI	vGraph	vGraph+	ComE	CNRL	ComGan	DGI	J-ENC
fb0	08.0	18.5	17.3	18.6	17.6	14.6	15.9	19.5	06.8	_24.1_	16.8	**24.7**
fb107	06.9	27.5	27.0	15.5	17.2	18.3	21.7	28.7	11.9	_38.5_	25.3	**46.8**
fb1684	06.6	38.0	38.7	18.7	24.7	29.2	32.7	_40.3_	25.8	37.9	38.8	**42.5**
fb1912	08.4	24.1	23.9	16.7	20.1	18.6	_28.0_	18.5	04.6	13.5	22.5	**37.3**
fb3437	07.7	11.5	11.7	05.5	09.0	12.0	13.3	12.5	02.0	_33.4_	11.6	**36.2**
fb348	11.3	35.9	40.0	39.3	33.6	_41.0_	40.5	34.4	21.7	23.2	_41.8_	**43.5**
fb414	12.8	47.1	47.3	34.2	29.3	51.8	_55.9_	42.2	15.4	53.6	46.4	**58.4**
fb698	16.0	41.9	45.9	22.6	30.0	43.6	_47.7_	33.8	09.6	46.9	42.1	**50.4**
Youtube	46.7	29.3	24.2	22.1	28.7	34.3	34.8	_52.5_	35.5	44.0	32.7	**53.3**
Amazon	25.2	35.1	35.0	36.7	36.4	36.9	36.9	34.6	_38.7_	38.0	29.1	**41.9**
Amazon500	20.8	51.2	53.8	47.2	45.0	59.1	_59.6_	58.4	41.1	57.3	23.3	**64.9**
Amazon1000	20.3	26.8	28.9	24.9	23.6	_54.3_	49.7	52.0	26.9	54.1	23.2	**57.1**
Dblp	20.9	13.8	22.3	23.3	20.9	25.0	25.1	27.9	_32.8_	25.0	29.2	**37.3**

Table 4. Non-overlapping community detection results. Best and second best values are bold and underlined respectively.

Alg.	NMI(%)					ARI(%)				
	CiteSeer	CiteSeer-full	Cora	Cora-ML	Cora-full	CiteSeer	CiteSeer-full	Cora	Cora-ML	Cora-full
MNMF	14.1	09.4	19.7	37.8	42.0	02.6	00.4	02.9	24.1	06.1
DeepWalk	08.8	15.4	39.7	43.2	48.5	09.5	16.4	31.2	33.9	22.5
LINE	08.7	13.0	32.8	42.3	40.3	03.3	03.7	14.9	32.7	11.7
Node2Vec	14.9	22.3	39.7	39.6	48.1	08.1	10.5	25.8	27.9	18.8
GAE	17.4	55.1	39.7	<u>48.3</u>	48.3	14.1	50.6	29.3	41.8	18.3
VGAE	16.3	48.4	40.8	<u>48.3</u>	47.0	10.1	40.6	34.7	<u>42.5</u>	17.9
DGI	<u>37.8</u>	<u>56.7</u>	<u>50.1</u>	46.2	39.9	**38.1**	<u>50.8</u>	<u>44.7</u>	42.1	12.1
GEMSEC	11.8	11.1	27.4	18.1	10.0	00.6	01.0	04.8	01.0	00.2
CNRL	13.6	23.3	39.4	42.9	47.7	12.8	20.2	31.9	32.5	<u>22.9</u>
ComGAN	03.2	16.2	05.7	11.5	15.0	01.2	04.9	03.2	06.7	00.6
vGraph	09.0	07.6	26.4	29.8	41.7	05.1	04.2	12.7	21.6	14.9
ComE	18.8	32.8	39.6	47.6	<u>51.2</u>	13.8	20.9	34.2	37.2	19.7
J-ENC	**38.5**	**59.0**	**52.7**	**56.3**	**55.2**	<u>35.2</u>	**60.3**	**45.1**	**49.8**	**28.8**

Table 5. Node classification results. Best and second best values are bold and underlined respectively.

Alg.	F1-macro (%)					F1-micro (%)				
	CiteSeer	CiteSeer-full	Cora	Cora-ML	Cora-full	CiteSeer	CiteSeer-full	Cora	Cora-ML	Cora-full
MNMF	57.4	68.6	60.9	64.2	30.4	60.8	68.1	62.7	64.2	32.9
DeepWalk	49.0	56.6	69.7	75.8	41.7	52.0	57.3	70.2	75.6	<u>48.3</u>
LINE	55.0	60.2	68.0	75.3	39.4	57.7	60.0	68.3	74.6	42.1
Node2Vec	55.2	61.0	71.3	78.4	<u>42.3</u>	57.8	61.5	71.4	78.6	48.1
GAE	57.9	<u>79.9</u>	71.2	76.5	36.6	61.6	<u>79.6</u>	73.5	77.6	41.8
VGAE	59.1	74.4	70.4	75.2	32.4	62.2	74.4	72.0	76.4	37.7
DGI	<u>62.6</u>	**82.1**	71.1	72.6	16.5	<u>67.9</u>	**81.8**	73.3	75.4	21.1
GEMSEC	37.5	53.3	60.3	70.6	35.8	39.4	53.5	59.4	72.5	38.9
CNRL	50.0	58.0	70.4	77.8	41.3	53.2	57.9	70.4	78.4	45.9
ComGAN	55.9	65.7	56.6	62.5	27.7	59.1	64.9	58.5	62.8	29.4
vGraph	30.8	28.5	44.7	59.8	33.4	32.1	28.5	44.6	62.3	37.6
ComE	59.6	69.9	<u>71.6</u>	<u>78.5</u>	42.2	63.1	70.2	<u>74.2</u>	<u>79.5</u>	47.8
J-ENC	**64.8**	76.8	**73.1**	**80.2**	**43.1**	**68.2**	77.0	**75.6**	**82.0**	**49.6**

First, we note that our proposed method **J-ENC** outperforms the competitors on all datasets in terms of Jaccard score as well as F1-score, with the dataset (*fb0*) being the only exception where **J-ENC** is the second best. These results demonstrate the capability of **J-ENC** to learn multiple community assignments quite well and hence reinforces our intuition behind the design of Eq. (12).

Second, we observe that there is no consistent performing algorithm among the competitive methods. That is, excluding **J-ENC** , the best performance is achieved by vGraph/vGraph+ on 5, ComGAN on 4 and ComE on 3 out of 13 datasets in terms of F1-score. A a similar trend can be seen in Jaccard Similarity. It is worth noting that all the methods, which achieve the second-best performance, are solving the task of community detection and node representation learning jointly.

Third, we observe that vGraph+ results are generally better than vGraph. This is because vGraph+ incorporates a regularization term in the loss function which is based on Jaccard coefficients of connected nodes as edge weights. However, it should be noted that this prepossessing step is computationally expensive for densely connected graphs.

Table 4 shows the results on non-overlapping community detection. First, we observe that MNMF, DeepWalk, LINE and Node2Vec provide a good baseline for the task. However, these methods are not able to achieve comparable performance on any dataset relative to the frameworks that treat the two tasks jointly. Second, **J-ENC** consistently outperforms all the competitors in NMI and ARI metrics, except for *CiteSeer* where it achieves second best ARI. Third, we observe that GCN based models i.e. GAE, VGAE and DGI show competitive performance. That is, they achieve second best performance in all the datasets except *CiteSeer*. In particular, DGI achieves second best NMI results in 3 out of 5 datasets and 2 out of 5 datasets in terms of ARI. Nonetheless, DGI results are not very competitive in Table 2 and Table 3, showing that while DGI can be a good choice for learning node embeddings for attributed graphs with non-overlapping communities, it is not the best option for non-attributed graphs or overlapping communities.

The results for node classification are presented in Table 5. **J-ENC** achieves best F1-micro and F1-macro scores on 4 out of 5 datasets. We also observe that GCN based models i.e. GAE, VGAE and DGI show competitive performance, following the trend in results of Table 4. Furthermore, we note that the node classification results of CommunityGan (ComGAN) are quite poor. We think a potential reason behind it is that the node embeddings are constrained to have same dimensions as the number of communities. Hence, different components of the learned node embeddings simply represent the membership strengths of nodes for different communities. The linear classifiers may find it difficult to separate such vectors.

4.6   Hyperparameter Sensitivity

We study the dependence of **J-ENC** on ϵ and α by evaluating on four datasets of different sizes: $fb698(N = 61)$, $fb1912(N = 747)$, $amazon1000(N=1540)$ and $youtube(N = 5346)$.

Effect of ϵ: We sweep for $\epsilon = \{0.1, 0.2, \cdots, 0.9\}$. For demonstrating effect of α, we fix $\epsilon = 0.3$ and sweep for $\alpha = \{0.1, 0.2, \cdots, 0.9\}$. The average results of five runs for ϵ and α are given in Fig. 2a and Fig. 2b respectively. Overall **J-ENC** is quite robust to the change in the values of ϵ and α. In case of ϵ, we see a general trend of decrease in performance when the threshold ϵ is set quite high e.g. $\epsilon > 0.7$. This is because the datasets contain overlapping communities and a very high ϵ will cause the algorithm to give only the most probable community assignment instead of potentially providing multiple communities per node. However, for a large part of sweep space, the results are almost consistent.

Effect of α: When ϵ is fixed and α is changed, the results are mostly consistent except when α is set to a low value. Equation (12) shows that in such a case the node itself is almost neglected and **J-ENC** tends to assign communities based upon neighborhood only, which may cause a decrease in the performance. This effect is most visible in *amazon1000* dataset because it has only 1.54 points on average per community. This implies a decent chance for neighbours of a point of being in different communities. Thus, sole dependence on the neighbors will most likely result in poor results.

4.7   Training Time

Now we compare the training times of different algorithms in Fig. 3. As some of the baselines are more resource intensive than others, we select aws instance type `g4dn.4xlarge` for fair comparison of training times. For vGraph, we train for 1000 iterations and for **J-ENC** for 1500 iterations. For all other algorithms

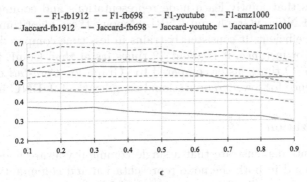

(a) Effect of ϵ. Overall a slight decrease in scores can be observed after $\epsilon = 0.7$ mark.

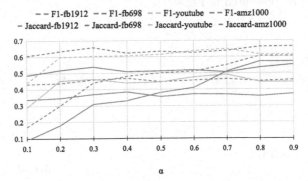

(b) Effect of α. The scores generally tend to decrease for small values of α.

Fig. 2. Effect of hyperparameters on the performance. F1 and Jaccard scores are in solid and dashed lines respectively.

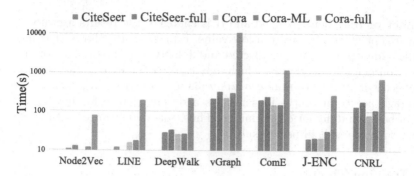

Fig. 3. Comparison of running times of different algorithms. We can see that **J-ENC** outperforms the direct competitors. The time on y-axis is in log scale.

we use the default parameters as used in Sect. 4.4. We observe that the methods that simply output the node embeddings take relatively less time compared to the algorithms that jointly learn node representations and community assignments e.g. **J-ENC** , vGraph and CNRL. Among these algorithms **J-ENC** is the most time efficient. It consistently trains in less time compared to its direct competitors. For instance, it is about 12 times faster than ComE for *CiteSeer-full* and about 40 times faster compared to vGraph for *Cora-full* dataset. This provides evidence for lower computational complexity of **J-ENC** in Sect. 3.5.

4.8 Visualization

Our experiments demonstrate that a single community-aware node embedding is sufficient to aid in both the node representation and community assignment tasks. This is also qualitatively demonstrated by graph visualizations of node embeddings (obtained via t-SNE [21]) and inferred communities for two datasets, fb107 and fb3437, presented in Fig. 4.

(a) fb107

(b) fb3437

Fig. 4. Graph visualization with community assignments (better viewed in color)

5   Conclusion

We propose a scalable generative method **J-ENC** to simultaneously perform community detection and node representation learning. Our novel approach learns a single community-aware node embedding for both the representation of the node and its context. **J-ENC** is scalable due to its low complexity, i.e. $\mathcal{O}(|\mathcal{E}|K + NKd)$. The experiments on several graph datasets show that **J-ENC** consistently outperforms all the competitive baselines on node classification, overlapping community detection and non-overlapping community detection tasks. Moreover, training the **J-ENC** is highly time-efficient than its competitors.

References

1. Ahn, Y.Y., Bagrow, J.P., Lehmann, S.: Link communities reveal multiscale complexity in networks. Nature **466**(7307), 761–764 (2010)
2. Bojchevski, A., Günnemann, S.: Deep Gaussian embedding of graphs: unsupervised inductive learning via ranking. arXiv preprint arXiv:1707.03815 (2017)
3. Cao, S., Lu, W., Xu, Q.: GraRep: learning graph representations with global structural information. In: Proceedings of the 24th ACM International on Conference on Information and Knowledge Management (CIKM 2015), pp. 891–900. Association for Computing Machinery, New York (2015). https://doi.org/10.1145/2806416.2806512
4. Cavallari, S., Zheng, V.W., Cai, H., Chang, K.C.C., Cambria, E.: Learning community embedding with community detection and node embedding on graphs. In: Proceedings of the 2017 ACM on Conference on Information and Knowledge Management, pp. 377–386 (2017)
5. Derényi, I., Palla, G., Vicsek, T.: Clique percolation in random networks. Phys. Rev. Lett. **94**(16), 160202 (2005)
6. Doersch, C.: Tutorial on variational autoencoders. arXiv preprint arXiv:1606.05908 (2016)
7. Fan, R.E., Chang, K.W., Hsieh, C.J., Wang, X.R., Lin, C.J.: LIBLINEAR: a library for large linear classification. J. Mach. Learn. Res. **9**(Aug), 1871–1874 (2008)
8. Gao, S., Denoyer, L., Gallinari, P.: Temporal link prediction by integrating content and structure information. In: Proceedings of the 20th ACM International Conference on Information and Knowledge Management, pp. 1169–1174 (2011)
9. Gopalan, P.K., Blei, D.M.: Efficient discovery of overlapping communities in massive networks. Proc. Natl. Acad. Sci. **110**(36), 14534–14539 (2013)
10. Grover, A., Leskovec, J.: node2vec: scalable feature learning for networks. In: Proceedings of the 22nd ACM SIGKDD International Conference on Knowledge Discovery and Data Mining, pp. 855–864 (2016)
11. Hamilton, W., Ying, Z., Leskovec, J.: Inductive representation learning on large graphs. In: Advances in Neural Information Processing Systems, pp. 1024–1034 (2017)
12. Jang, E., Gu, S., Poole, B.: Categorical reparameterization with Gumbel-Softmax. arXiv preprint arXiv:1611.01144 (2016)
13. Jia, Y., Zhang, Q., Zhang, W., Wang, X.: CommunityGAN: community detection with generative adversarial nets. In: The World Wide Web Conference, pp. 784–794 (2019)

14. Khan, R.A., Anwaar, M.U., Kleinsteuber, M.: Epitomic variational graph autoencoder (2020)
15. Kingma, D.P., Ba, J.: Adam: a method for stochastic optimization. arXiv preprint arXiv:1412.6980 (2014)
16. Kipf, T.N., Welling, M.: Semi-supervised classification with graph convolutional networks. arXiv preprint arXiv:1609.02907 (2016)
17. Kipf, T.N., Welling, M.: Variational graph auto-encoders. arXiv preprint arXiv:1611.07308 (2016)
18. Kozdoba, M., Mannor, S.: Community detection via measure space embedding. In: Proceedings of the 28th International Conference on Neural Information Processing Systems (NIPS 2015), vol. 2, pp. 2890–2898. MIT Press, Cambridge (2015)
19. Leskovec, J., Krevl, A.: SNAP datasets: stanford large network dataset collection, June 2014. http://snap.stanford.edu/data
20. Leskovec, J., Mcauley, J.J.: Learning to discover social circles in ego networks. In: Advances in Neural Information Processing Systems, pp. 539–547 (2012)
21. Maaten, L.V.D., Hinton, G.: Visualizing data using t-SNE. J. Mach. Learn. Res. 9(Nov), 2579–2605 (2008)
22. Mikolov, T., Chen, K., Corrado, G., Dean, J.: Efficient estimation of word representations in vector space. arXiv preprint arXiv:1301.3781 (2013)
23. Perozzi, B., Al-Rfou, R., Skiena, S.: DeepWalk: online learning of social representations. In: Proceedings of the 20th ACM SIGKDD International Conference on Knowledge Discovery and Data Mining, pp. 701–710 (2014)
24. Ribeiro, L.F., Saverese, P.H., Figueiredo, D.R.: struc2vec: learning node representations from structural identity. In: Proceedings of the 23rd ACM SIGKDD International Conference on Knowledge Discovery and Data Mining, pp. 385–394 (2017)
25. Rozemberczki, B., Davies, R., Sarkar, R., Sutton, C.: GEMSEC: graph embedding with self clustering. In: Proceedings of the 2019 IEEE/ACM International Conference on Advances in Social Networks Analysis and Mining, pp. 65–72 (2019)
26. Sun, F.Y., Qu, M., Hoffmann, J., Huang, C.W., Tang, J.: vGraph: a generative model for joint community detection and node representation learning. In: Advances in Neural Information Processing Systems, pp. 514–524 (2019)
27. Tang, J., Qu, M., Wang, M., Zhang, M., Yan, J., Mei, Q.: LINE: large-scale information network embedding. In: Proceedings of the 24th International Conference on World Wide Web, pp. 1067–1077 (2015)
28. Tang, J., Aggarwal, C., Liu, H.: Node classification in signed social networks. In: Proceedings of the 2016 SIAM International Conference on Data Mining, pp. 54–62. SIAM (2016)
29. Tang, L., Liu, H.: Leveraging social media networks for classification. Data Min. Knowl. Discov. 23(3), 447–478 (2011). https://doi.org/10.1007/s10618-010-0210-x
30. Tu, C., et al.: A unified framework for community detection and network representation learning. IEEE Trans. Knowl. Data Eng. 31(6), 1051–1065 (2018)
31. Veličković, P., Cucurull, G., Casanova, A., Romero, A., Lio, P., Bengio, Y.: Graph attention networks. arXiv preprint arXiv:1710.10903 (2017)
32. Velickovic, P., Fedus, W., Hamilton, W.L., Liò, P., Bengio, Y., Hjelm, R.D.: Deep graph infomax (2019)
33. Wang, D., Cui, P., Zhu, W.: Structural deep network embedding. In: Proceedings of the 22nd ACM SIGKDD International Conference on Knowledge Discovery and Data Mining, pp. 1225–1234 (2016)

34. Wang, X., Cui, P., Wang, J., Pei, J., Zhu, W., Yang, S.: Community preserving network embedding. In: AAAI, vol. 17, pp. 203–209 (2017)
35. Xie, J., Kelley, S., Szymanski, B.K.: Overlapping community detection in networks: the state-of-the-art and comparative study. ACM Comput. Surv. (CSUR) **45**(4), 1–35 (2013)
36. Yang, J., Leskovec, J.: Overlapping community detection at scale: a nonnegative matrix factorization approach. In: Proceedings of the sixth ACM International Conference on Web Search and Data Mining, pp. 587–596 (2013)
37. Yang, J., Leskovec, J.: Defining and evaluating network communities based on ground-truth. Knowl. Inf. Syst. **42**(1), 181–213 (2013). https://doi.org/10.1007/s10115-013-0693-z
38. Yang, J., McAuley, J., Leskovec, J.: Community detection in networks with node attributes. In: 2013 IEEE 13th International Conference on Data Mining, pp. 1151–1156. IEEE (2013)
39. Yang, L., Cao, X., He, D., Wang, C., Wang, X., Zhang, W.: Modularity based community detection with deep learning. IJCAI **16**, 2252–2258 (2016)
40. Yang, Z., Cohen, W., Salakhudinov, R.: Revisiting semi-supervised learning with graph embeddings. In: International Conference on Machine Learning, pp. 40–48. PMLR (2016)

GraphAnoGAN: Detecting Anomalous Snapshots from Attributed Graphs

Siddharth Bhatia[1(✉)], Yiwei Wang[1], Bryan Hooi[1], and Tanmoy Chakraborty[2]

[1] National University of Singapore, Singapore, Singapore
{siddharth,y-wang,bhooi}@comp.nus.edu.sg
[2] IIIT-Delhi, Delhi, India
tanmoy@iiitd.ac.in

Abstract. Finding anomalous snapshots from a graph has garnered huge attention recently. Existing studies address the problem using shallow learning mechanisms such as subspace selection, ego-network, or community analysis. These models do not take into account the multifaceted interactions between the structure and attributes in the network. In this paper, we propose GraphAnoGAN, an anomalous snapshot ranking framework, which consists of two core components – generative and discriminative models. Specifically, the generative model learns to approximate the distribution of anomalous samples from the candidate set of graph snapshots, and the discriminative model detects whether the sampled snapshot is from the ground-truth or not. Experiments on 4 real-world networks show that GraphAnoGAN outperforms 6 baselines with a significant margin (28.29% and 22.01% higher precision and recall, respectively compared to the best baseline, averaged across all datasets).

Keywords: Anomaly detection · Graph snapshot · Generative adversarial network

1 Introduction

Anomaly detection on graphs is a well-researched problem and plays a critical role in cybersecurity, especially network security [13]. Majority of the proposed approaches focus on anomalous nodes [2,24,27,34], anomalous edges [18,38,43], community structures [44], or sudden surprising changes in graphs [8,10,14].

However, we focus our attention on *detecting anomalous snapshots from attributed graphs*. This problem is motivated by the following cybersecurity threats: (a) fraudulent customers controlling the sentiment (customers operate in a way that they can not be tracked individually), (b) hackers targeting the network (attacks such as DDOS, phishing), (c) black-market syndicates in online social media [17], and (d) camouflaged financial transactions.

Detecting anomalous snapshots in a graph has received little attention; SPOT-LIGHT [19] is one of them. However, SPOTLIGHT does not take into account the

© Springer Nature Switzerland AG 2021
N. Oliver et al. (Eds.): ECML PKDD 2021, LNAI 12976, pp. 36–51, 2021.
https://doi.org/10.1007/978-3-030-86520-7_3

patterns being formed in the graph even if there is no outburst of edges. Moreover, it tends to ignore the node features as well. On the other hand, convolutional architectures nicely capture the complex interactions between the structure and the attributes, taking data sparsity and non-linearity into account.

Therefore, we propose **GraphAnoGAN**, a generative adversarial network (GAN) based framework that takes advantage of its structure in the following two ways: (i) the generative model learns to find the anomalous snapshots via the signals from the discriminative model and the global graph topology; (ii) the discriminative model achieves the improved classification of snapshots by modeling the data provided by the generative model and ground-truth. To the best of our knowledge, GraphAnoGAN is the first GAN-based method for detecting anomalous snapshots in graphs.

Our convolutional architecture plays an integral role in choosing the required feature descriptors, to be utilized in identifying anomalous snapshots. Moreover, we demonstrate that GraphAnoGAN is able to learn complex structures and typical patterns required in such problem settings.

Finally, we evaluate the performance of GraphAnoGAN on 4 datasets and compare it with 6 state-of-the-art baselines. Experimental results show that GraphAnoGAN outperforms all the baselines by a significant margin – it achieves 65.75% precision (*resp.* 66.5% recall) on average across all the datasets, which is 28.29% (*resp.* 22.01%) higher than the best baseline.

Reproducibility: Our code and datasets are publicly available at https://github.com/LCS2-IIITD/GraphAnoGAN-ECMLPKDD21.

2   Related Work

Readers are encouraged to go through [3] and [31] for extensive surveys on graph-based and GAN-based anomaly detection. Traditional methods for anomaly detection can be (i) reconstruction-based: PCA [25], kernel PCA [22]; (ii) clustering-based: GMM [49], KDE [1]; (iii) one class classification-based: OC-SVM [40]. More recently, deep learning based methods for anomaly detection have been popular. These methods include (i) energy-based: DSEBM [48], MEG [29]; (ii) autoencoder-based: DAGMM [50]; and (iii) GAN-based: AnoGAN [39], Ganomaly [4], FenceGAN [32], MemGAN [47], TAnoGAN [7,23], and ExGAN [11]. There has been work on attributed graphs as well [5,6,9,33,36,37,42]. However, these methods do not detect graph anomalies.

Graph-Based GAN Frameworks: GraphGAN [45] has two components – Generator which tries to model the true connectivity distribution for all the vertices, and Discriminator which detects whether the sampled vertex is from the ground-truth or generated by the Generator. NetGAN [12] uses a novel LSTM architecture and generates graphs via random walks. Generator in such models tries to form the whole graph which is computationally challenging and not scalable. Instead of generating the whole graph, our proposed Generator in GraphAnoGAN learns to retrieve anomalous snapshots from a pool via signals from the Discriminator.

Graph-Based Anomaly Detection: These methods can be divided into four categories. (i) Using community or ego-network analysis to spot the anomaly. AMEN [35] detects anomalous neighborhoods in attributed graphs. A neighborhood is considered normal if it is internally well-connected and nodes are similar to each other on a specific attribute subspace as well as externally well separated from the nodes at the boundary. SPOTLIGHT [19] used a randomized sketching-based approach, where an anomalous snapshot is placed at a larger distance from the normal snapshot. (ii) Utilizing aggregated reconstruction error of structure and attribute. DOMINANT [15] has a GCN and autoencoder network, where an anomaly is reported if aggregated error breaches the threshold. (iii) Using residuals of attribute information and its coherence with graph information. ANOMALOUS [34] is a joint framework to conduct attribute selection and anomaly detection simultaneously based on cut matrix decomposition [30] and residual analysis. (iv) Performing anomaly detection on edge streams. SSKN [38] takes neighbors of a node, and historic edges into account to classify an edge. We consider all of these as baselines.

Since we focus on attributed graphs, the abnormality is determined jointly by mutual interactions of nodes (i.e., topological structure) and their features (i.e., node attributes). As shown in Table 1, GraphAnoGAN satisfies all the four aspects – it takes into account *node attributes*; it *classifies graph snapshots*; it can be *generalized* for weighted/unweighted and directed/undirected graphs; and it considers *structures/patterns* exhibited by anomalies.

Table 1. Comparison of GraphAnoGAN with baseline approaches.

	SPOTLIGHT	AMEN	ANOMALOUS	DOMINANT	SSKN	GraphAnoGAN
Node attribute		✓	✓	✓		✓
Snapshot anomaly	✓	✓				✓
Generalizable	✓	✓	✓	✓	✓	✓
Structure/pattern						✓

3   Problem Definition

Let $g = \{g_1, \cdots, g_T\}$ be T different snapshots[1] of an attributed graph $G = \{V, E, X\}$ with $|V| = n$ nodes, $|E| = m$ edges, and $|X| = d$ node attributes. Each snapshot $g_t = \{V_t, E_t, X_t\}$ contains $|V_t| = n_t$ nodes, $|E_t| = m_t$ edges, and each node in V_t is associated with a d-dimensional attribute vector X_t. A and A_t indicate the adjacency matrices of G and g_t, respectively. Each graph snapshot g_t is associated with a label, $y_t \in \mathcal{Y}$, where $\mathcal{Y} \in \{0, 1\}$ (0 represents normal and 1 represents anomalous snapshot). Our goal is to detect anomalous snapshots by

---

[1] A 'snapshot' of a graph is used as a general term and can refer to any subgraph of the graph, e.g., a particular area of the graph, a temporal snapshot of the graph, an egonet of a node, etc.

leveraging node attributes, structure, and complex interactions between different information modalities. The aim is to learn a model which could utilise the above information, analyse and identify snapshots that are anomalous in behaviour.

The problem of detecting anomalous graph snapshots from attributed graphs is as follows: Given a set of snapshots $\{g_1, \cdots, g_T\}$ from a graph $G = \{V, E, X\}$ with node attributes, analyze the structure and attributes of every snapshot, and return the top K anomalous snapshots.

4   Proposed Algorithm

We introduce GraphAnoGAN to detect anomalous snapshots of a given graph. GraphAnoGAN captures complex structures and unusual patterns to rank the snapshots according to the extent of their anomalous behavior. GraphAnoGAN follows a typical GAN architecture. There are two components: a Generator and a Classifier/Discriminator. The Generator will select those snapshots from the candidate pool which it deems to be anomalous (similar to the ground-truth), and therefore, fools the Discriminator; whereas the Discriminator will distinguish between the ground-truth and the generated snapshots. Essence of our architecture is that the Discriminator will try to classify the graph snapshots, and in doing so, learn the representation of anomalous/normal snapshots. On the other hand, the Generator will learn to find a list of anomalous snapshots from the candidate set. Figure 1 depicts the schematic architecture of GraphAnoGAN.

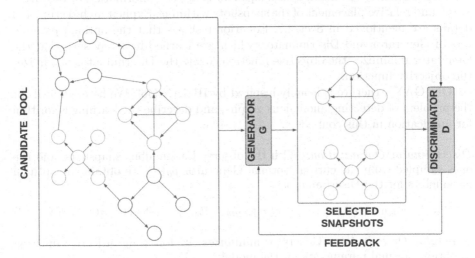

Fig. 1. Illustration of GraphAnoGAN. The aim of Generator is to report K anomalous snapshots from a pool of samples. Discriminator is fed with samples, of which it identifies if the sample belongs to the ground-truth or is produced by Generator.

4.1   GAN Modeling

Given the set of T candidate snapshots $\{g_1, \cdots, g_T\}$ of G, we want to detect k anomalous snapshots where $k \ll T$. We unify two different types of models (i.e., Generator and Discriminator) through a minimax game. The formulation is described below [46]:

$$J = \min_\theta \max_\phi \left(\mathbb{E}_{g_t \sim p_{\mathrm{true}}(g)}[\log D(g_t)] + \mathbb{E}_{g_t \sim p_\theta(g)}[\log(1 - D(g_t))] \right) \quad (1)$$

Here we represent parameters for Generator and Discriminator as θ and ϕ, respectively. $p_{true}(g)$ represents the distribution of anomalous snapshots in the ground-truth. Generator and Discriminator are written as $p_\theta(g)$ and $f_\phi(g_t)$, respectively. Discriminator score, D captures the probability of the snapshot being sampled from the ground-truth, calculated using sigmoid function represented below:

$$D(g_t) = \sigma(f_\phi(g_t)) = \frac{\exp(f_\phi(g_t))}{1 + \exp(f_\phi(g_t))} \quad (2)$$

Discriminator $f_\phi(g_t)$ is trained on the labeled snapshots and instances received from the Generator. The objective of Discriminator is to maximize the log-likelihood of correctly distinguishing anomalous snapshots from the ones provided by Generator.

Generator $p_\theta(g)$ tries to generate (or select) anomalous snapshots from the candidate pool, i.e., approximates $p_{true}(g)$ as much as possible. Generator learns the distribution of the anomalous snapshots using the information of the entire graph and relative placement of the snapshots within it. Explicit architecture and details are mentioned in Sect. 4.2. Equation 1 shows that the optimal parameters of Generator and Discriminator, which are learned in a way such that the Generator minimizes the objective function, while the Discriminator maximizes the objective function.

Our GAN structure is heavily inspired by IRGAN [46]. We here adopt it for the problem of detecting anomalous graphs, and describe the learning algorithm for application in this context.

Discriminator Optimization: With the observed anomalous snapshots, and the ones sampled from the current optimal Generator $p_\theta(g)$, we obtain the optimal parameters for the Discriminator:

$$\phi^* = \arg\max_\phi (\mathbb{E}_{g_t \sim p_{\mathrm{true}}(g)}[\log(\sigma(f_\phi(g_t))] + \mathbb{E}_{g_t \sim p_{\theta*}(g)}[\log(1 - \sigma(f_\phi(g_t))]) \quad (3)$$

Generator Optimization: Generator minimizes the following objective function to obtain optimal parameters for the model:

$$\theta^* = \arg\min_\theta \mathbb{E}_{g_t \sim p_{\mathrm{true}}(g)}[\log \sigma(f_\phi(g_t))] + \mathbb{E}_{g_t \sim p_\theta(g)}[\log(1 - \sigma(f_\phi(g_t)))] \quad (4)$$

Taking reference of Eq. 2, Eq. 4 can be rewritten as:

$$\theta^* = \arg\min_\theta \mathbb{E}_{g_t \sim p_\theta(g)} [\log \frac{\exp(f_\phi(g_t))}{(1 + \exp(f_\phi(g_t)))^2}]$$

$$\simeq \arg\max_\theta \underbrace{\mathbb{E}_{g_t \sim p_\theta(g)} [\log(1 + \exp(f_\phi(g_t)))]}_{\text{denoted as } J} \tag{5}$$

We keep $f_\phi(g)$ fixed. Note that we can not employ gradient descent to solve the problem as g is discrete. We approach the problem using policy gradient based reinforcement learning [46] as follows:

$$\nabla_\theta J = \nabla_\theta \mathbb{E}_{g_t \sim p_\theta(g)} [\log(1 + \exp(f_\phi(g_t)))]$$

$$= \sum_{i=1}^{T} \nabla_\theta p_\theta(g_i) \log(1 + \exp(f_\phi(g_i)))$$

$$= \sum_{i=1}^{T} p_\theta(g_i) \nabla_\theta \log p_\theta(g_i) \log(1 + \exp(f_\phi(g_i))) \tag{6}$$

$$\simeq \frac{1}{K} \sum_{k=1}^{K} \nabla_\theta \log p_\theta(g_k) \log(1 + \exp(f_\phi(g_k)))$$

where we perform a sampling approximation in the last step of Eq. 6 in which g_k is the kth snapshot sampled from the output obtained from the Generator, i.e., $p_\theta(g)$. With reinforcement learning, the term $\log(1 + \exp(f_\phi(g_k)))$ acts as a reward for the policy $p_\theta(g)$ taking an action g_k.

4.2 Architecture

The convolutional architecture used in GAN comprises the following components: a graph convolutional layer, a DegPool layer, 1D convolutional layer and a fully connected layer. We discuss individual components below:

Graph Convolution Layers: We use Graph Convolutional Network (GCN) [26]. The forward convolution operation used layer-wise is represented below:

$$Z_{l+1} = \sigma(\hat{D}_t^{-\frac{1}{2}} \hat{A}_t \hat{D}_t^{-\frac{1}{2}} Z_l W_l)$$

Z_l and Z_{l+1} represent the input and output at layer l, respectively. Z_0 is initialised with X_t for the graph snapshot g_t. A_t depicts the adjacency matrix for the specified snapshot. D_t is the diagonal matrix corresponding to A_t, used to normalise in order to scale down the factor introduced by A_t. $\hat{A}_t$ is represented by $A_t + I$, where I is the identity matrix. W_l is the trainable weights corresponding to the layer, and $\sigma(.)$ signifies the activation function (ReLU in our case).

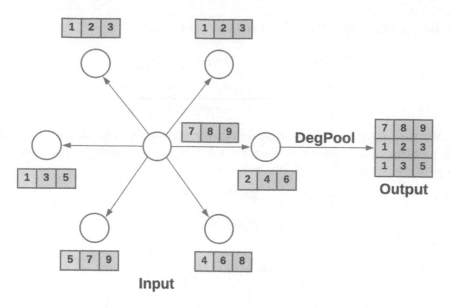

Fig. 2. Visualization of DegPool layer.

DegPool Layer: Before feeding the convolved input to a 1D convolutional layer, we want its size to be consistent, and here DegPool plays an important role. The principle of the DegPool layer is to sort the feature descriptors according to the degree of vertices. We sort the vertices in decreasing order of their outdegree. Vertices with the same outdegree are sorted according to the convolutional output. In this layer, the input is a concatenated tensor $\mathbf{Z}^{1:h}$ of size $n \times \sum_1^h c_t$, where h represents graph convolution layers, each row corresponds to a vertex feature descriptor, each column corresponds to a feature channel, and c_t represents the number of output channels at layer l. For vertices having the same degree, we keep on preceding the channel until the tie is broken. The output of DegPool is a $k \times \sum_1^h c_t$ tensor, where k is a user-defined integer. We use the degree as the first order of sorting since the nodes emitting denser edges are more probable of being part of anomalous snapshots.

To make size of the output consistent, DegPool evicts or extends the output tensor, makes the number of vertices from n to k. This is done to feed consistent and equal tensors to 1D convolution. If $n > k$, extra $n - k$ vertices are evicted; whereas if $n < k$, the output is extended by appending zeros. Figure 2 visualises the layer when $n = 7$ and $k = 3$. The numbers in rectangles, attributed to each vertex represent the convolution input to DegPool layer.

Model: The architecture of Generator and that of Discriminator are similar to each other, the difference being the adjacency matrix utilized. **Generator** forms a graph by combining edges over all timesteps and then applies a convolution over the entire graph. Before feeding to the DegPool layer, the global structure is used which allows overall placement and broader structural details to be taken into account. DegPool only considers the vertices present in the snapshot, and it is followed by a 1D convolutional and fully connected layer. It helps the Generator to effectively model features and learn the distribution using complete graph information. **Discriminator** takes the snapshot into consideration and takes only features and structure as input.

Discriminator-Only Architecture vs. GAN: One may argue – *What is the requirement of Generator if we consider only Discriminator which can use certain activation function (i.e., softmax) and predict top K anomalous snapshots from the pool?* Although Discriminator can be trained to choose top anomalous snapshots, it requires a massive amount of labeled data to generate representations of the snapshots. On unlabeled data, the Discriminator may not be able to mine the signals and representations required. We have observed that Generators are able to successfully learn the distribution of data (i.e., node attribute and structure), and thus act as an important component in our model. In GraphAnoGAN, Generator utilizes the complete graph information, as opposed to the Discriminator. The snapshot placement in the whole graph is of utmost importance and plays a crucial role in determining its state. Generator and Discriminator help each other through minimax game and learn through signals received from each other. Comparative analysis in Table 3 shows that Discriminator-only model (henceforth, `Discriminator`) is not as effective as GraphAnoGAN.

4.3   Training Procedure

How do the Generator and Discriminator train each other? Consider Discriminator to be an obstacle, which restricts non-anomalous samples passing through. Generator aims to misguide Discriminator by pushing instances through the obstacle, while the obstacle tries to allow only anomalous samples to pass through. Generator learns to push positive but unobserved samples (which have not passed through the obstacle yet), and Discriminator learns to allow only anomalous samples to pass through. Figure 3 visually represents the training procedure. Convergence is obtained when positive (anomalous) and negative (normal) snapshots are separated. Since the unobserved positive examples are linked to the observed positive examples, eventually they should be able to pass through the obstacle, and (unobserved) negative samples should settle.

In a normal two-player game, Generator and Discriminator have their own loss functions and they try to minimize them. However, in the current scenario, Generator tries to select top anomalous snapshots, and the Discriminator identifies whether the output is from the ground-truth or from the Generator. Generator ultimately learns to identify snapshots that represent top anomalous examples.

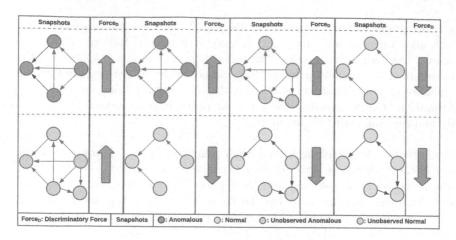

Fig. 3. Visualization of the training procedure.

Time Complexity: The time complexity of GraphAnoGAN is similar to other GANs [21]. Complexity of training each GAN iteration is $O(CT)$, where T represents the number of iterations, and C is the complexity of convolution filters, where the time taken by filter l is $O(|E|dc_l)$ [26].

5 Datasets

We utilise four attributed graphs, namely ACM, BLOGC, DARPA, and ENRON – we inject anomalies synthetically in first two graphs; for the remaining two, the ground-truth anomalies are already annotated. Table 2 shows the statistics of the graphs.

Table 2. Statistics of the datasets used in our experiments.

Dataset	# nodes	# edges	# attributes
ACM	16,484	71,980	8,337
BLOGC	5,196	171,743	8,189
DARPA	32,000	4,500,000	24
ENRON	13,533	176,987	20

ACM: This graph is constructed using citations among papers published before 2016. Attributes are obtained by applying bag-of-words on the content of paper [16].

Table 3. Performance (precision@K, recall@K) of the competing methods on different datasets. First (second) row corresponding to each model indicates precision (recall). Best value in bold.

Method	ACM				BlogC				DARPA				Enron			
	50	100	150	200	50	100	150	200	50	100	150	200	50	100	150	200
SPOTLIGHT	0.56	0.52	0.49	0.42	0.57	0.53	0.46	0.43	0.55	0.48	0.45	0.49	0.58	0.51	0.46	0.44
	0.36	0.46	0.51	0.58	0.26	0.34	0.43	0.54	0.33	0.43	0.48	0.56	0.39	0.45	0.52	0.56
AMEN	0.61	0.56	0.51	0.49	0.63	0.59	0.49	0.44	0.60	0.55	0.53	0.51	0.62	0.56	0.52	0.48
	0.37	0.49	0.54	0.60	0.34	0.42	0.51	0.62	0.36	0.48	0.57	0.61	0.42	0.49	0.56	0.59
ANOMALOUS	0.51	0.47	0.40	0.37	0.53	0.49	0.43	0.40	0.50	0.45	0.37	0.32	0.53	0.48	0.44	0.35
	0.35	0.42	0.49	0.52	0.28	0.37	0.45	0.50	0.41	0.45	0.48	0.51	0.31	0.36	0.45	0.52
DOMINANT	0.49	0.43	0.39	0.32	0.50	0.45	0.40	0.37	0.47	0.39	0.35	0.31	0.48	0.42	0.36	0.30
	0.33	0.40	0.46	0.51	0.31	0.40	0.46	0.49	0.37	0.44	0.49	0.50	0.35	0.41	0.42	0.49
SSKN	0.46	0.41	0.35	0.29	0.41	0.43	0.39	0.30	0.45	0.37	0.32	0.28	0.44	0.38	0.31	0.27
	0.28	0.39	0.45	0.49	0.26	0.35	0.41	0.45	0.31	0.37	0.41	0.46	0.29	0.33	0.38	0.45
Discriminator	0.64	0.60	0.56	0.53	0.67	0.62	0.55	0.52	0.65	0.60	0.58	0.55	0.68	0.60	0.56	0.52
	0.36	0.48	0.55	0.61	0.35	0.45	0.55	0.63	0.40	0.50	0.58	0.65	0.43	0.50	0.58	0.61
GraphAnoGAN	**0.74**	**0.70**	**0.65**	**0.61**	**0.76**	**0.71**	**0.66**	**0.59**	**0.74**	**0.70**	**0.68**	**0.66**	**0.75**	**0.69**	**0.64**	**0.58**
	0.42	**0.59**	**0.69**	**0.74**	**0.44**	**0.56**	**0.66**	**0.73**	**0.43**	**0.55**	**0.64**	**0.77**	**0.49**	**0.58**	**0.67**	**0.72**

BlogC: Blogcatalog is an online website that is designed to share blogs, articles, and content. Users act as nodes, and 'follow' relationships are used to draw edges. Attributes are obtained from the content of users' blogs [15]. We extract T snapshots from each graph by randomly sampling P vertices and taking their ego networks [35]. Since the ground-truth anomalies are not annotated in both these graphs, we inject anomalies. Initially, normal snapshots, i.e., having low conductance cuts [20] are chosen. These snapshots are considered to have the lowest anomaly factor. Of the chosen set, we add structural and attribute anomaly. The former is injected by forming a clique in the network, while the latter is injected by sampling random nodes, and replacing their features with the node having maximum dissimilarity in the network.

DARPA: This graph is composed of known graph attacks, e.g., portsweep, ipsweep, etc. Each communication is a directed edge, and attribute set constitutes duration, numFailedLogins, etc. [41]. We obtain 1463 snapshots of the graph by aggregating edges on an hourly basis. The snapshot is considered as anomalous if it contains at least 50 edges [19]. **Enron**: The dataset is used to form a graph having $50k$ relationships, having emails over a three year period among 151 employees of the company. Each email is a directed edge, and the attributes include average content length, average number of recipients, etc. We create graph snapshots on a per-day basis and obtain a stream of 1139 snapshots. The anomalies are labeled by verifying with the major events of the scandal[2] [19].

---

[2] http://www.agsm.edu.au/bobm/teaching/BE/Enron/timeline.html.

6   Experiments

6.1   Baselines

We consider 6 baselines for comparison – first two (AMEN and SPOTLIGHT) focus on detecting snapshots of a graph as anomalies; the next two (ANOMALOUSand DOMINANT) focus on anomaly detection of nodes on attributed graphs; while the fifth one (SSKN) focuses on detecting edges as anomalies (see Sect. 2). The last one (`Discriminator`) is the model which uses only Discriminator to detect anomalous snapshots (as discussed in Sect. 4.2).

AMEN considers egonets of the constituent nodes in the snapshot and then takes geometric mean of their anomaly scores to classify the snapshot. For ANOMA-LOUSand DOMINANT, an anomalous snapshot is determined using geometric mean of the per-node anomaly scores. An anomaly score is assigned to the snapshot according to the following formula, $anomalyScore(g_t) = -\log \ell(g_t)$, where $\ell(g_t) = \prod_{v \in V_t} \ell(v)$. For SSKN, the likelihood of a snapshot $g_t(V_t, E_t)$ is computed as the geometric mean of the per-edge likelihoods $\ell(g_t) = \left(\prod_{e \in E_t} \ell(e)^w\right)^{1/W}$, where W and w represent the total edge weight and the edge weight of e, respectively. A graph is less anomalous if it is more likely, i.e., $anomalyScore\,(g_t) = -\log \ell(g_t)$.

6.2   Comparative Evaluation

For a fair evaluation, we keep the same test set for all the competing methods; we use 10-fold cross-validation; for each fold, the same test set is used for all competing methods to measure the performance. The experiment is then performed 10 times, and the average performance is reported.

GraphAnoGAN has five graph convolutional layers with 64, 64, 64, 32, 1 output channels, respectively. The convolutional layer corresponding to DegPool has one channel and ensures that k in DegPool constitutes at least 70% nodes of the snapshot. It is followed by two 1-D convolutional layers consisting of 32 and 16 output channels. Finally, the dense layer is composed of 32 nodes. It is followed by softmax for Generator, while sigmoid is used in the Discriminator. Softmax allows to sort the vertices and pick top K; whereas sigmoid helps for binary classification. A dropout layer with a dropout rate of 0.3 is used after the dense layer. The nonlinear function ReLU is used in the GCN layers. We observe that after 100 epochs, the Generator becomes stable. Similar behavior is observed for Discriminator after 60 epochs.

Comparison: Table 3 shows the comparative analysis of the competing methods. In general, GraphAnoGAN outperforms all baselines.[3] AMEN turns out to be the best baseline across all the datasets. However, GraphAnoGAN beats AMEN with a relative improvement of (21–27)%, (20–34)%, (11–21)%, and (20–23)% in precision for ACM, BLOGC, DARPA and ENRON, respectively. Similar

---

[3] All the improvements in the results are significant at $p < 0.05$ with paired t-test.

results are obtained in recall where GraphAnoGAN beats AMEN with a relative improvement of (10–23)%, (20–34)%, (11–21)%, and (20–23)% for ACM, BLOGC, DARPA and ENRON, respectively.

6.3 Side-by-Side Diagnostics

We further dive deeper to analyze why GraphAnoGAN performs better than state-of-the-art baselines. Figure 4 shows a detailed comparison of the competing methods at every timestep to detect anomalous snapshots on DARPA. A time period between 0–200 is used as the training period for the models, and the remaining period is used for testing.

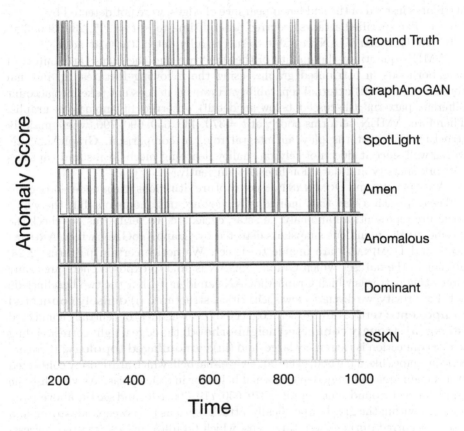

Fig. 4. Anomaly examples identified by GraphAnoGAN and baselines on the graph carved out of DARPA at different timesteps.

SPOTLIGHT captures the situation well when edges (dis)appear in a large amount. As it does not take node attributes into account, it does not work well in the given setting. Buffer overflow, rootkit, or ipsweep which depend upon

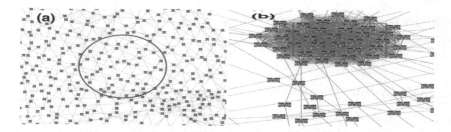

Fig. 5. Graph projection of DARPA dataset, drawn at two timestamps: (a) graph having anomalous attributes, and (b) graph having anomalous structure.

attributes instead of the sudden appearance of edges, were not detected by SPOT-LIGHT. The specified attacks occurred around timesteps, $t = 205, 230, 300, 320, 450, 620$. GraphAnoGAN was able to successfully detect these attacks.

AMEN quantifies the quality of the structure and the focus (attributes) of neighborhoods in attributed graphs, using the following features: (i) internal consistency and (ii) external separability. However, it does not take into account different patterns (discussed below in detail) observed in anomalous graphs. Therefore, AMEN performs poorly at $t = 370, 430, 650, 690, 790$, where majorly structure and patterns play an integral role. In comparison, GraphAnoGAN works well since it does not rely on shallow learning mechanisms and captures network sparsity and data non-linearity efficiently.

ANOMALOUS and DOMINANT do not capture situations when nodes have camouflaged in such a way that individually they are not anomalous but the whole structure represents an anomaly, i.e., star, cycle, etc. To verify this, we check the ground-truth anomalous snapshots detected by GraphAnoGAN which ANOMALOUS and DOMINANT are unable to detect. We notice certain patterns [2,28] present in the dataset which GraphAnoGAN is able to expose. There are many interesting patterns which GraphAnoGAN can identify, but the two baselines do not. For brevity, we discuss a few such critical structures: (i) **densely connected components:** vertices are densely connected (near-cliques) or centrally connected (stars); (ii) **strongly connected neighborhood:** the edge weight corresponding to one connection is extremely large, and (iii) **camouflaged topology:** there are certain shapes like a barbell, cycle, and wheel-barbell, which are densely connected at a certain area but may depict normal behavior in other areas. We visualise the graph formed around timesteps, $t = 370, 650, 710, 790, 840$, and see the above given patterns within the graph. Specifically, attacks such as portsweep, udpstorm, and mscan occurred during these timesteps, which GraphAnoGANdetected successfully (see Fig. 4).

Figure 5 shows the two detected anomalies by GraphAnoGAN from DARPA at two different timestamps. Specifically, Fig. 5(a) denotes the anomaly detected using attributes present in the dataset when the structure did not behave abnormally. Figure 5(b) demonstrates the capability of GraphAnoGAN to capture anomalous patterns, near-clique in the given example. In a nutshell, GraphAnoGAN can help us find the anomalous snapshot of different patterns.

SSKN performed poorly in detecting attacks, specifically neptune, which majorly appear in the DARPA network. SSKN considers the following three aspects to detect a snapshot as normal: (i) snapshots with edges present before, (ii) nodes densely connected, and (iii) nodes sharing neighbors. These aspects work fine on the graph evolving slowly, which does not hold for most of the real-world networks; thus, it performs poorly here.

7   Conclusion

In this paper, we addressed the problem of detecting anomalous snapshots from a given graph. We proposed GraphAnoGAN, a GAN-based framework, that utilizes both the structure and the attribute while predicting whether a snapshot is anomalous. We demonstrate how GraphAnoGAN is able to learn typical patterns and complex structures. Extensive experiments on 4 datasets showed the improvement of GraphAnoGAN compared to 6 other baseline methods. Future work could examine ways to carve out anomalous snapshots from graphs and analysis in a temporal setting.

Acknowledgments. We would like to thank Deepak Thukral and Alex Beutal for the active discussions. The project was partially supported by the Early Career Research Award (DST) and the Ramanujan Fellowship.

References

1. Aggarwal, C.C.: Outlier analysis. In: Data Mining, pp. 237–263. Springer, Cham (2015). https://doi.org/10.1007/978-3-319-14142-8_8
2. Akoglu, L., McGlohon, M., Faloutsos, C.: Oddball: spotting anomalies in weighted graphs. In: Zaki, M.J., Yu, J.X., Ravindran, B., Pudi, V. (eds.) PAKDD 2010. LNCS (LNAI), vol. 6119, pp. 410–421. Springer, Heidelberg (2010). https://doi.org/10.1007/978-3-642-13672-6_40
3. Akoglu, L., Tong, H., Koutra, D.: Graph based anomaly detection and description: a survey. Data Min. Knowl. Disc. **29**(3), 626–688 (2014). https://doi.org/10.1007/s10618-014-0365-y
4. Akcay, S., Atapour-Abarghouei, A., Breckon, T.P.: GANomaly: semi-supervised anomaly detection via adversarial training. In: Jawahar, C.V., Li, H., Mori, G., Schindler, K. (eds.) ACCV 2018. LNCS, vol. 11363, pp. 622–637. Springer, Cham (2019). https://doi.org/10.1007/978-3-030-20893-6_39
5. Atzmüller, M., Soldano, H., Santini, G., Bouthinon, D.: MinerLSD: efficient mining of local patterns on attributed networks. Appl. Netw. Sci. **4**, 1–33 (2019)
6. Baroni, A., Conte, A., Patrignani, M., Ruggieri, S.: Efficiently clustering very large attributed graphs. In: 2017 IEEE/ACM International Conference on Advances in Social Networks Analysis and Mining (ASONAM), pp. 369–376 (2017)
7. Bashar, M.A., Nayak, R.: TAnoGAN: time series anomaly detection with generative adversarial networks. In: SSCI (2020)
8. Belth, C., Zheng, X., et al.: Mining persistent activity in continually evolving networks. In: KDD (2020)

9. Bendimerad, A., Mel, A., Lijffijt, J., Plantevit, M., Robardet, C., De Bie, T.: SIAS-miner: mining subjectively interesting attributed subgraphs. Data Min. Knowl. Disc. **34**(2), 355–393 (2019). https://doi.org/10.1007/s10618-019-00664-w
10. Bhatia, S., Hooi, B., Yoon, M., Shin, K., Faloutsos, C.: MIDAS: microcluster-based detector of anomalies in edge streams. In: AAAI (2020)
11. Bhatia, S., Jain, A., Hooi, B.: ExGAN: adversarial generation of extreme samples. In: AAAI Conference on Artificial Intelligence (AAAI) (2021)
12. Bojchevski, A., Shchur, O., et al.: NetGAN: generating graphs via random walks. In: ICML (2018)
13. Borghesi, A., Bartolini, A., et al.: Anomaly detection using autoencoders in high performance computing systems. In: AAAI (2019)
14. Chang, Y.Y., Li, P., et al.: F-FADE: frequency factorization for anomaly detection in edge streams. In: WSDM (2021)
15. Ding, K., Li, J., Bhanushali, R., Liu, H.: Deep anomaly detection on attributed networks. In: SDM (2019)
16. Ding, K., Li, J., Liu, H.: Interactive anomaly detection on attributed networks. In: WSDM (2019)
17. Dutta, H.S., Chakraborty, T.: Blackmarket-driven collusion on online media: a survey. arXiv preprint arXiv:2008.13102 (2020)
18. Eswaran, D., Faloutsos, C.: SedanSpot: detecting anomalies in edge streams. In: ICDM (2018)
19. Eswaran, D., Faloutsos, C., Guha, S., Mishra, N.: Spotlight: detecting anomalies in streaming graphs. In: KDD (2018)
20. Gleich, D.F., Seshadhri, C.: Vertex neighborhoods, low conductance cuts, and good seeds for local community methods. In: KDD (2012)
21. Goodfellow, I.: NIPS tutorial: generative adversarial networks (2016)
22. Günter, S., Schraudolph, N.N., Vishwanathan, S.: Fast iterative kernel principal component analysis. J. Mach. Learn. Res. (2007)
23. Han, X., Chen, X., Liu, L.: GAN ensemble for anomaly detection. arXiv: 2012.07988 (2020)
24. Jiang, M., Cui, P., Beutel, A., Faloutsos, C., Yang, S.: Catching synchronized behaviors in large networks: a graph mining approach. In: TKDD (2016)
25. Jolliffe, I.T.: Principal components in regression analysis. In: Principal Component Analysis. Springer Series in Statistics. Springer, New York (1986). https://doi.org/10.1007/978-1-4757-1904-8_8
26. Kipf, T.N., Welling, M.: Semi-supervised classification with graph convolutional networks. In: ICLR (2017)
27. Kleinberg, J.M.: Authoritative sources in a hyperlinked environment. JACM (1999)
28. Koutra, D., Vogelstein, J.T., Faloutsos, C.: DELTACON: a principled massive-graph similarity function. In: SDM (2013)
29. Kumar, R., Goyal, A., Courville, A.C., Bengio, Y.: Maximum entropy generators for energy-based models. arXiv:1901.08508 (2019)
30. Mahoney, M.W., Drineas, P.: CUR matrix decompositions for improved data analysis. PNAS **106**(3), 697–702 (2009)
31. Mattia, F.D., Galeone, P., Simoni, M., Ghelfi, E.: A survey on GANs for anomaly detection. arXiv:abs/1906.11632 (2019)
32. Ngo, C., et al.: Fence GAN: towards better anomaly detection. arXiv preprint arXiv:1904.01209 (2019)
33. Pei, Y., Huang, T., van Ipenburg, W., Pechenizkiy, M.: ResGCN: attention-based deep residual modeling for anomaly detection on attributed networks. arXiv:abs/2009.14738 (2020)

34. Peng, Z., Luo, M., Li, J., Liu, H., Zheng, Q.: ANOMALOUS: a joint modeling approach for anomaly detection on attributed networks. In: IJCAI (2018)
35. Perozzi, B., Akoglu, L.: Scalable anomaly ranking of attributed neighborhoods. In: SDM (2016)
36. Perozzi, B., Akoglu, L.: Discovering communities and anomalies in attributed graphs. ACM Trans. Knowl. Disc. Data (TKDD) **12**, 1–40 (2018)
37. Pienta, R., Tamersoy, A., Tong, H., Chau, D.H.: MAGE: matching approximate patterns in richly-attributed graphs. In: 2014 IEEE International Conference on Big Data (Big Data), pp. 585–590 (2014)
38. Ranshous, S., Harenberg, S., Sharma, K., Samatova, N.F.: A scalable approach for outlier detection in edge streams using sketch-based approximations. In: SDM (2016)
39. Schlegl, T., Seeböck, P., Waldstein, S.M., Schmidt-Erfurth, U., Langs, G.: Unsupervised anomaly detection with generative adversarial networks to guide marker discovery. In: Niethammer, M., et al. (eds.) IPMI 2017. LNCS, vol. 10265, pp. 146–157. Springer, Cham (2017). https://doi.org/10.1007/978-3-319-59050-9_12
40. Schölkopf, B., Platt, J.C., Shawe-Taylor, J., Smola, A., Williamson, R.: Estimating the support of a high-dimensional distribution. Neural Comput. **13**(7), 1443–1471 (2001)
41. Song, J., Takakura, H., Okabe, Y.: Description of Kyoto University benchmark data (2006)
42. Soundarajan, S., et al.: Generating graph snapshots from streaming edge data. In: Proceedings of the 25th International Conference Companion on World Wide Web (2016)
43. Sricharan, K., Das, K.: Localizing anomalous changes in time-evolving graphs. In: SIGMOD (2014)
44. Sun, J., Faloutsos, C., Papadimitriou, S., Yu, P.S.: GraphScope: parameter-free mining of large time-evolving graphs. In: SIGKDD (2007)
45. Wang, H., Wang, J., et al.: GraphGAN: graph representation learning with generative adversarial nets. In: AAAI (2018)
46. Wang, J., et al.: IRGAN: a minimax game for unifying generative and discriminative information retrieval models. In: SIGIR (2017)
47. Yang, Z., Zhang, T., Bozchalooi, I.S., Darve, E.F.: Memory augmented generative adversarial networks for anomaly detection. arXiv:abs/2002.02669 (2020)
48. Zhai, S., Cheng, Y., et al.: Deep structured energy based models for anomaly detection. In: ICML (2016)
49. Zimek, A., Schubert, E., Kriegel, H.P.: A survey on unsupervised outlier detection in high-dimensional numerical data. Stat. Anal. Data Min.: ASA Data Sci. J. **5**, 363–387 (2012)
50. Zong, B., Song, Q., et al.: Deep autoencoding gaussian mixture model for unsupervised anomaly detection. In: ICLR (2018)

The Bures Metric for Generative Adversarial Networks

Hannes De Meulemeester[1](✉) ⓘ, Joachim Schreurs[1] ⓘ, Michaël Fanuel[2] ⓘ,
Bart De Moor[1] ⓘ, and Johan A. K. Suykens[1] ⓘ

[1] ESAT-STADIUS, KU Leuven, Kasteelpark Arenberg 10, 3001 Leuven, Belgium
{hannes.demeulemeester,joachim.schreurs,bart.demoor,
johan.suykens}@kuleuven.be
[2] Univ. Lille, CNRS, Centrale Lille, UMR 9189 - CRIStAL, 59000 Lille, France
michael.fanuel@univ-lille.fr

Abstract. Generative Adversarial Networks (GANs) are performant generative methods yielding high-quality samples. However, under certain circumstances, the training of GANs can lead to mode collapse or mode dropping. To address this problem, we use the last layer of the discriminator as a feature map to study the distribution of the real and the fake data. During training, we propose to match the real batch diversity to the fake batch diversity by using the Bures distance between covariance matrices in this feature space. The computation of the Bures distance can be conveniently done in either feature space or kernel space in terms of the covariance and kernel matrix respectively. We observe that diversity matching reduces mode collapse substantially and has a positive effect on sample quality. On the practical side, a very simple training procedure is proposed and assessed on several data sets.

Keywords: Generative Adversarial Networks · Mode collapse · Optimal transport

1 Introduction

In several machine learning applications, data is assumed to be sampled from an implicit probability distribution. The estimation of this empirical implicit distribution is often intractable, especially in high dimensions. To tackle this issue, generative models are trained to provide an algorithmic procedure for sampling from this unknown distribution. Popular approaches are Variational Auto-Encoders proposed by [16], Generating Flow models by [32] and Generative Adversarial

---

M. Fanuel—Most of this work was done when MF was at KU Leuven.

Electronic supplementary material The online version of this chapter (https://doi.org/10.1007/978-3-030-86520-7_4) contains supplementary material, which is available to authorized users.

N. Oliver et al. (Eds.): ECML PKDD 2021, LNAI 12976, pp. 52–66, 2021.
https://doi.org/10.1007/978-3-030-86520-7_4

Networks (GANs) initially developed by [10]. The latter are particularly successful approaches to produce high quality samples, especially in the case of natural
images, though their training is notoriously difficult. The vanilla GAN consists
of two networks: a generator and a discriminator. The generator maps random
noise, usually drawn from a multivariate normal, to fake data in input space. The
discriminator estimates the likelihood ratio of the generator network to the data
distribution. It often happens that a GAN generates samples only from a few of
the many modes of the distribution. This phenomenon is called 'mode collapse'.

Contribution. We propose BuresGAN: a generative adversarial network which
has the objective function of a vanilla GAN complemented by an additional term,
given by the squared Bures distance between the covariance matrix of real and
fake batches in a latent space. This loss function promotes a matching of fake and
real data in a feature space $\mathbb{R}^f$, so that mode collapse is reduced. Conveniently,
the Bures distance also admits both a feature space and kernel based expression.
Contrary to other related approaches such as in [4] or [35], the architecture of
the GAN is unchanged, only the objective is modified. We empirically show that
the proposed method is robust when it comes to the choice of architecture and
does not require an additional fine architecture search. Finally, an extra asset of
BuresGAN is that it yields competitive or improved IS and FID scores compared
with the state of the art on CIFAR-10 and STL-10 using a ResNet architecture.

Related Works. The Bures distance is closely related to the Fréchet distance [6]
which is a 2-Wasserstein distance between multivariate normal distributions.
Namely, the Fréchet distance between multivariate normals of equal means is
the Bures distance between their covariance matrices. The Bures distance is also
equivalent to the exact expression for the 2-Wasserstein distance between two
elliptically contoured distributions with the same mean as shown in [8] and [28].
Noticeably, it is also related to the Fréchet Inception Distance score (FID), which
is a popular manner to assess the quality of generative models. This score uses
the Fréchet distance between real and generated samples in the feature space of
a pre-trained inception network as it is explained in [33] and [12].

There exist numerous works aiming to improve training efficiency of generative networks. For mode collapse evaluation, we compare BuresGAN to the
most closely related works. GDPP-GAN [7] and VEEGAN [35] also try to enforce
diversity in 'latent' space. GDPP-GAN matches the eigenvectors and eigenvalues of the real and fake diversity kernel. In VEEGAN, an additional recon-
structor network is introduced to map the true data distribution to Gaussian
random noise. In a similar way, architectures with two discriminators are analysed by [26], while MADGAN [9] uses multiple discriminators and generators.
A different approach is taken by Unrolled-GAN [23] which updates the generator with respect to the unrolled optimization of the discriminator. This allows
training to be adjusted between using the optimal discriminator in the generator's objective, which is ideal but infeasible in practice. Wasserstein GANs [1, 11]
leverage the 1-Wasserstein distance to match the real and generated data distributions. In MDGAN [4], a regularization is added to the objective function, so

that the generator can take advantage of another similarity metric with a more predictable behavior. This idea is combined with a penalization of the missing modes. Some other recent approaches to reducing mode collapse are variations of WGAN [37]. Entropic regularization has been also proposed in PresGAN [5], while metric embeddings were used in the paper introducing BourGAN [38]. A simple packing procedure which significantly reduces mode collapse was proposed in PacGAN [19] that we also consider hereafter in our comparisons.

2 Method

A GAN consists of a discriminator $D : \mathbb{R}^d \rightarrow \mathbb{R}$ and a generator $G : \mathbb{R}^\ell \rightarrow \mathbb{R}^d$ which are typically defined by neural networks and parametrized by real vectors. The value $D(\boldsymbol{x})$ gives the probability that $\boldsymbol{x}$ comes from the empirical distribution, while the generator G maps a point $\boldsymbol{z}$ in the latent space $\mathbb{R}^\ell$ to a point in the input space $\mathbb{R}^d$. The training of a GAN consists in solving

$$\min_G \max_D \mathbb{E}_{\boldsymbol{x} \sim p_d}[\log D(\boldsymbol{x})] + \mathbb{E}_{\boldsymbol{x} \sim p_g}[\log(1 - D(\boldsymbol{x}))], \tag{1}$$

by alternating two phases of training. In (1), the expectation in the first term is over the empirical data distribution p_d, while the expectation in the second term is over the generated data distribution p_g, implicitly given by the mapping by G of the latent prior distribution $\mathcal{N}(0, \mathbb{I}_\ell)$. It is common to define and minimize the discriminator loss by

$$V_D = -\mathbb{E}_{\boldsymbol{x} \sim p_d}[\log D(\boldsymbol{x})] - \mathbb{E}_{\boldsymbol{x} \sim p_g}[\log(1 - D(\boldsymbol{x}))]. \tag{2}$$

In practice, it is proposed in [10] to minimize generator loss

$$V_G = -\mathbb{E}_{\boldsymbol{z} \sim \mathcal{N}(0, \mathbb{I}_\ell)}[\log D(G(\boldsymbol{z}))], \tag{3}$$

rather than the second term of (1), for an improved training efficiency.

Matching Real and Fake Data Covariance. To prevent mode collapse, we encourage the generator to sample fake data of similar diversity to real data. This is achieved by matching the sample covariance matrices of real and fake data respectively. Covariance matching and similar ideas were explored for GANs in [25] and [7]. In order to compare covariance matrices, we propose to use the squared Bures distance between positive semi-definite $\ell \times \ell$ matrices [2], i.e.,

$$\mathcal{B}(A, B)^2 = \min_{U \in O(\ell)} \|A^{1/2} - B^{1/2}U\|_F^2$$
$$= \text{Tr}(A + B - 2(A^{\frac{1}{2}} B A^{\frac{1}{2}})^{\frac{1}{2}}).$$

Being a Riemannian metric on the manifold of positive semi-definite matrices [21], the Bures metric is adequate to compare covariance matrices. The covariances are defined in a feature space associated to the discriminator. Namely, let $D(\boldsymbol{x}) = \sigma(\boldsymbol{w}^\top \phi(\boldsymbol{x}))$, where $\boldsymbol{w}$ is the weight vector of the last dense layer and σ is the sigmoid function. The last layer of the discriminator, denoted by $\phi(x) \in \mathbb{R}^{d_\phi}$, naturally defines a feature map. We use the normalization $\bar{\phi}(x) = \phi(x)/\|\phi(x)\|_2$, after the centering of $\phi(x)$. Then, we define a covariance matrix as follows: $C(p) = \mathbb{E}_{x \sim p}[\bar{\phi}(x)\bar{\phi}(x)^\top]$. For simplicity, we denote the real data and generated data covariance matrices by $C_d = C(p_d)$ and $C_g = C(p_g)$, respectively. Our proposal is to replace the generator loss by $V_G + \lambda \mathcal{B}(C_d, C_g)^2$. The value $\lambda = 1$ was found to yield good results in the studied data sets. Two specific training algorithms are proposed. Algorithm 1 deals with the squared Bures distance as an additive term to the generator loss, while an alternating training is discussed in Supplementary Material (SM) and does not introduce an extra parameter. The training described in Algorithm 1 is analogous to the training of GDPP GAN, although the additional generator loss is rather different. The computational advantage of the Bures distance is that it admits two expressions which can be evaluated numerically in a stable way. Namely, there is no need to calculate a gradient update through an eigendecomposition.

Algorithm 1. BuresGAN

Sample a real and fake batch
Update G by minimizing $V_G + \lambda \mathcal{B}(\hat{C}_r, \hat{C}_g)^2$
Update D by maximizing $-V_D$;

Feature Space Expression. In the training procedure, real $x_i^{(d)}$ and fake data $x_i^{(g)}$ with $1 \le i \le b$ are sampled respectively from the empirical distribution and the mapping of the normal distribution $\mathcal{N}(0, \mathbb{I}_\ell)$ by the generator. Consider the case where the batch size b is larger than the feature space dimension. Let the embedding of the batches in feature space be $\Phi_\alpha = [\phi(x_1^{(\alpha)}), \dots, \phi(x_b^{(\alpha)})]^\top \in \mathbb{R}^{b \times d_\phi}$ with $\alpha = d, g$. The covariance matrix of one batch in feature space[1] is $\hat{C} = \bar{\Phi}^\top \bar{\Phi}$, where $\bar{\Phi}$ is the ℓ_2-normalized centered feature map of the batch. Numerical instabilities can be avoided by adding a small number, e.g. 1e−14, to the diagonal elements of the covariance matrices, so that, in practice, we only deal with strictly positive definite matrices. From the computational perspective, an interesting alternative expression for the Bures distance is given by

$$\mathcal{B}(C_d, C_g)^2 = \text{Tr}(C_d + C_g - 2(C_g C_d)^{\frac{1}{2}}), \qquad (4)$$

whose computation requires only one matrix square root. This identity can be obtained from Lemma 1. Note that an analogous result is proved in [27].

Lemma 1. *Let A and B be symmetric positive semidefinite matrices of the same size and let $B = Y^\top Y$. Then, we have: (i) AB is diagonalizable with nonnegative eigenvalues, and (ii) $\text{Tr}((AB)^{\frac{1}{2}}) = \text{Tr}((Y A Y^\top)^{\frac{1}{2}})$.*

---

[1] For simplicity, we omit the normalization by $\frac{1}{b-1}$ in front of the covariance matrix.

Proof. (i) is a consequence of Corollary 2.3 in [13]. (ii) We now follow [27]. Thanks to (i), we have $AB = PDP^{-1}$ where D is a nonnegative diagonal and the columns of P contain the right eigenvectors of AB. Therefore, $\mathrm{Tr}((AB)^{1/2}) = \mathrm{Tr}(D^{1/2})$. Then, $YAY^\top$ is clearly diagonalizable. Let us show that it shares its nonzero eigenvalues with AB. a) We have $ABP = PD$, so that, by multiplying of the left by Y, it holds that $(YAY^\top)YP = YPD$. b) Similarly, suppose that we have the eigenvalue decomposition $YAY^\top Q = Q\Lambda$. Then, we have $BAY^\top Q = Y^\top Q\Lambda$ with $B = Y^\top Y$. This means that the non-zero eigenvalues of $YAY^\top$ are also eigenvalues of BA. Since A and B are symmetric, this completes the proof.

Kernel Based Expression. Alternatively, if the feature space dimension f is larger than the batch size b, it is more efficient to compute $\mathcal{B}(\hat{C}_d, \hat{C}_g)$ thanks to $b \times b$ kernel matrices: $K_d = \bar{\Phi}_d \bar{\Phi}_d^\top$, $K_g = \bar{\Phi}_g \bar{\Phi}_g^\top$ and $K_{dg} = \bar{\Phi}_d \bar{\Phi}_g^\top$. Then, we have the kernel based expression

$$\mathcal{B}(\hat{C}_d, \hat{C}_g)^2 = \mathrm{Tr}\left(K_d + K_g - 2\left(K_{dg}K_{dg}^\top\right)^{\frac{1}{2}}\right), \tag{5}$$

which allows to calculate the Bures distance between covariance matrices by computing a matrix square root of a $b \times b$ matrix. This is a consequence of Lemma 2.

Lemma 2. *The matrices $X^\top XY^\top Y$ and $YX^\top XY^\top$ are diagonalizable with nonnegative eigenvalues and share the same non-zero eigenvalues.*

Proof. The result follows from Lemma 1 and its proof, where $A = X^\top X$ and $B = Y^\top Y$.

Connection with Wasserstein GAN and Integral Probability Metrics. The Bures distance is proportional to the 2-Wasserstein distance $\mathcal{W}_2$ between two ellipically contoured distributions with the same mean [8]. For instance, in the case of multivariate normal distributions, we have

$$\mathcal{B}(A, B)^2 = \min_{\pi} \mathbb{E}_{(X,Y) \sim \pi} \|X - Y\|_2^2 \text{ s.t. } \begin{cases} X \sim \mathcal{N}(0, A) \\ Y \sim \mathcal{N}(0, B), \end{cases}$$

where the minimization is over the joint distributions π. More precisely, in this paper, we make the approximation that the implicit distribution of the real and generated data in the feature space $\mathbb{R}^{d_\phi}$ (associated to $\phi(\boldsymbol{x})$) are ellipti-cally contoured with the same mean. Under different assumptions, the Generative Moment Matching Networks [18,31] work in the same spirit, but use a different approach to match covariance matrices. On the contrary, WGAN uses the Kantorovich dual formula for the 1-Wasserstein distance: $\mathcal{W}_1(\alpha, \beta) = \sup_{f \in \mathrm{Lip}} \int f \mathrm{d}(\alpha - \beta)$, where α, β are signed measures. Generalizations of such integral formulae are called integral probability metrics (see for instance [3]). Here, f is the discriminator, so that the maximization over Lipschitz functions f plays the role of the maximization over discriminator parameters in the min-max game of (1). Then, in the training procedure, this maximization alternates with a minimization over the generator parameters.

We can now discuss the connection with Wasserstein GAN. Coming back to the definition of BuresGAN, we can now explain that the 2-Wasserstein distance provides an upper bound on an integral probability metric. Then, if we assume that the densities are elliptically contoured distributions in feature space, the use of the Bures distance to calculate W_2 allows to spare the maximization over the discriminator parameters – and this motivates why the optimization of $\mathcal{B}$ only influences updates of the generator in Algorithm 1. Going more into detail, the 2-Wasserstein distance between two probability densities (w.r.t. the same measure) is equivalent to a Sobolev dual norm, which can be interpreted as an integral probability metric. Indeed, let the Sobolev semi-norm $\|f\|_{H^1} = (\int \|\nabla f(x)\|^2 dx)^{1/2}$. Then, its dual norm over signed measures is defined as $\|\nu\|_{H^{-1}} = \sup_{\|f\|_{H^1} \leq 1} \int f d\nu$. It is then shown in [29] and [28] that there exist two positive constants c_1 and c_2 such that

$$c_1 \|\alpha - \beta\|_{H^{-1}} \leq W_2(\alpha, \beta) \leq c_2 \|\alpha - \beta\|_{H^{-1}}.$$

Hence, the 2-Wasserstein distance gives an upper bound on an integral probability metric.

Algorithmic Details. The matrix square root in (4) and (5) is obtained thanks to the Newton-Schultz algorithm which is inversion free and can be efficiently calculated on GPUs since it involves only matrix products. In practice, we found 15 iterations of this algorithm to be sufficient for the small scale data sets, while 20 iterations were used for the ResNet examples. A small regularization term $1e-14$ is added to the matrix diagonal for stability. The latent prior distribution is $\mathcal{N}(0, \mathbb{I}_\ell)$ with $\ell = 100$ and the parameter in Algorithm 1 is always set to $\lambda = 1$. In the tables hereafter, we indicate the largest scores in bold, although we invite the reader to also consider the standard deviation.

3 Empirical Evaluation of Mode Collapse

BuresGAN's performances on synthetic data, artificial and real images are compared with the standard DCGAN [33], WGAN-GP, MDGAN, Unrolled GAN, VEEGAN, GDPP and PacGAN. We want to emphasize that the purpose of this experiment is not to challenge these baselines, but to report the improvement obtained by adding the Bures metric to the objective function. It would be straightforward to add the Bures loss to other GAN variants, as well as most GAN architectures, and we would expect an improvement in mode coverage and generation quality. In the experiments, we notice that adding the Bures loss to the vanilla GAN already significantly improves the results.

A low dimensional feature space ($d_\phi = 128$) is used for the synthetic data so that the feature space formula in (4) is used, while the dual formula in (5) is used for the image data sets (Stacked MNIST, CIFAR-10, CIFAR-100 and STL-10) for which the feature space is larger than the batch size. The architectures used for the image data sets are based on DCGAN [30], while results using ResNets are given in Sect. 4. All images are scaled in between -1 and 1 before running the algorithms. Additional information on the architectures and data sets is given in SM. The hyperparameters of other methods are typically chosen as suggested in the authors' reference implementation. The number of unrolling steps in Unrolled GAN is chosen to be 5. For MDGAN, both versions are implemented but only MDGAN-v2 gives interesting results. The first version, which corresponds to the mode regularizer, has hyperparameters $\lambda_1 = 0.2$ and $\lambda_2 = 0.4$, for the second version, which corresponds to manifold diffusion training for regularized GANs, has $\lambda = 10^{-2}$. WGAN-GP uses $\lambda = 10.0$ and $n_{\text{critic}} = 5$. All models are trained using Adam [15] with $\beta_1 = 0.5$, $\beta_2 = 0.999$ and learning rate 10^{-3} for both the generator and discriminator. Unless stated otherwise, the batch size is 256. Examples of random generations of all the GANs are given in SM. Notice that in this section we report the results achieved only at the end of the training.

3.1   Artificial Data

Synthetic. RING is a mixture of eight two-dimensional isotropic Gaussians in the plane with means $2.5 \times (\cos((2\pi/8)i), \sin((2\pi/8)i))$ and std 0.01 for $1 \leq i \leq 8$. GRID is a mixture of 25 two-dimensional isotropic normals in the plane with means separated by 2 and with standard deviation 0.05. All models have the same architecture, with $\ell = 256$ following [7], and are trained for 25k iterations. The evaluation is done by sampling 3k points from the generator network. A sample is counted as high quality if it is within 3 standard deviations of the nearest mode. The experiments are repeated 10 times for all models and their performance is compared in Table 1.

BuresGAN consistently captures all the modes and produces the highest quality samples. The training progress of the BuresGAN is shown on Fig. 1, where we observe that all the modes early on in the training procedure, afterwards improving the quality. The training progress of the other GAN models listed in Table 1 is given in SM. Although BuresGAN training times are larger than most other methods for this low dimensional example, we show in SM that BuresGAN scales better with the input data dimension and architecture complexity.

Stacked MNIST. The Stacked MNIST data set is specifically constructed to contain 1000 known modes. This is done by stacking three digits, sampled uniformly at random from the original MNIST data set, each in a different channel. BuresGAN is compared to the other methods and are trained for 25k iterations. For the performance evaluation, we follow [23] and use the following metrics: the number of captured modes measures mode collapse and the KL divergence, which also measures sample quality. The mode of each generated image is identified by using a standard MNIST classifier which is trained up to 98.43% accuracy on the validation set (see Supplementary Material), and classifies each channel of the fake sample. The same classifier is used to count the number of captured modes. The metrics are calculated based on 10k generated images for all the models. Generated samples from BuresGAN are given in Fig. 2. As it was observed by previous papers such as in [19] and [38], even the vanilla GAN can achieve an excellent mode coverage for certain architectures.

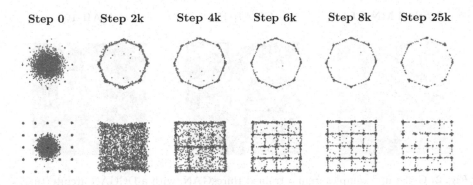

Fig. 1. Figure accompanying Table 1, the progress of BuresGAN on the synthetic examples. Each column shows 3k samples from the training of the generator in blue and 3k samples from the true distribution in green (Color figure online).

Table 1. Experiments on the synthetic data sets. Average (std) over 10 runs. All the models are trained for 25k iterations.

	Grid (25 modes)		Ring (8 modes)	
	Nb modes	% in 3σ	Nb modes	% in 3σ
GAN	22.9 (4)	76 (13)	7.4 (2)	76 (25)
WGAN-GP	24.9 (0.3)	77 (10)	7.1 (1)	9 (5)
MDGAN-v2	**25** (0)	68 (11)	5 (3)	20 (15)
Unrolled	19.7 (1)	78 (19)	**8**(0)	77 (18)
VEEGAN	**25** (0)	67 (3)	8 (0)	29 (5)
GDPP	20.5 (5)	79 (23)	7.5 (0.8)	73 (25)
PacGAN2	23.6 (4)	65 (28)	8 (0)	81 (15)
BuresGAN (ours)	**25** (0)	**82** (1)	8 (0)	**82**(4)

To study mode collapse on this data set, we performed extensive experiments for multiple discriminator layers and multiple batch sizes. In the main section, the results for a 3 layer discriminator are reported in Table 2. Additional experiments can be found in SM. An analogous experiment as in VEEGAN [35] with a challenging architecture including 4 convolutional layers for both the generator and discriminator was also included for completeness; see Table 3. Since different authors, such as in the PacGAN's paper, use slightly different setups, we also report in SM the specifications of the different settings. We want to emphasize the interests of this simulation is to compare GANs in the same consistent setting, while the results may vary from those reported in e.g. in [35] since some details might differ.

Interestingly, for most models, an improvement is observed in the quality of the images – KL divergence – and in terms of mode collapse – number of modes attained – as the size of the batch increases. For the same batch size, architecture

Stacked MNIST   **CIFAR-10**   **CIFAR-100**

Fig. 2. Generated samples from a trained BuresGAN, with a DCGAN architecture.

Table 2. KL-divergence between the generated distribution and true distribution for an architecture with 3 conv. layers for the Stacked MNIST dataset. The number of counted modes assesses mode collapse. The results are obtained after 25k iterations and we report the average(std) over 10 runs.

		Nb modes (↑)			KL div. (↓)		
	Batch size	64	128	256	64	128	256
3 conv. layers	GAN	993.3(3.1)	995.4(1.7)	**998.3**(1.2)	0.28(0.02)	**0.24**(0.02)	0.21(0.02)
	WGAN-GP	980.2 (57)	838.3 (219)	785.1 (389)	**0.26** (0.34)	1.05 (1)	1.6 (2.4)
	MDGAN-v1	233.8 (250)	204.0 (202)	215.5 (213)	5.0(1.6)	4.9 (1.3)	5.0 (1.2)
	MDGAN-v2	299.9 (457)	300.4 (457)	200.0 (398)	4.8(3.0)	4.7 (3.0)	5.5 (2.6)
	UnrolledGAN	934.7 (107)	874.1 (290)	884.9 (290)	0.72(0.51)	0.98 (1.46)	0.90 (1.4)
	VEEGAN	974.2 (10.3)	687.9 (447)	395.6 (466)	0.33(0.05)	2.04 (2.61)	3.52 (2.64)
	GDPP	894.2 (298)	897.1 (299)	997.5 (1.4)	0.92(1.92)	0.88 (1.93)	**0.20** (0.02)
	PacGAN2	989.8 (4.0)	993.3 (4.8)	897.7 (299)	0.33(0.02)	0.29 (0.04)	0.87 (1.94)
	BuresGAN (ours)	**993.5**(2.7)	**996.3** (1.6)	997.1 (2.4)	0.29 (0.02)	0.25 (0.02)	0.23 (0.01)

and iterations, the image quality is improved by BuresGAN, which consistently performs well over all settings. The other methods show a higher variability over the different experiments. MDGANv2, VEEGAN, GDPP and WGAN-GP often have an excellent single run performance. However, when increasing the number of discriminator layers, the training of these models has a tendency to collapse more often as indicated by the large standard deviation. Vanilla GAN is one of the best performing models in the variant with 3 layers.

We observe in Table 3 for the additional experiment that vanilla GAN and GDPP collapse for this architecture. WGAN-GP yields the best result and is followed by BuresGAN. However, as indicated in Table 2, WGAN-GP is sensitive to the choice of architecture and hyperparameters and its training time is also longer as it can be seen from the corresponding timings table in SM. More generally, these results depend heavily on the precise architecture choice and to a lesser extent on the batch size. These experiments further confirm the finding that most GAN models, including the standard version, can learn all modes with careful and sufficient architecture tuning [19, 22]. Finally, it can be concluded that BuresGAN performs well for all settings, showing that it is robust when it comes to batch size and architecture.

Table 3. Stacked MNIST experiment for an architecture with 4 conv. layers. All the models are trained for 25k iterations with a batch size of 64, a learning rate of 2×10^{-4} for Adam and a normal latent distribution. The evaluation is over 10k samples and we report the average(std) over 10 runs.

		Nb modes (↑)	KL div. (↓)
4 conv. layers	GAN	21.6 (25.8)	5.10 (0.83)
	WGAN-GP	**999.7** (0.6)	**0.11** (0.006)
	MDGAN-v2	729.5 (297.9)	1.76 (1.65)
	UnrolledGAN	24.3 (23.61)	4.96 (0.68)
	VEEGAN	816.1 (269.6)	1.33 (1.46)
	GDPP	33.3 (39.4)	4.92 (0.80)
	PacGAN2	972.4 (12.0)	0.45 (0.06)
	BuresGAN (ours)	989.9 (4.7)	0.38 (0.06)

3.2   Real Images

Metrics. Image quality is assessed thanks to the Inception Score (IS), Fréchet Inception Distance (FID) and Sliced Wasserstein Distance (SWD). The latter was also used in [7] and [14] to evaluate image quality as well as mode-collapse. In a word, SWD evaluates the multiscale statistical similarity between distributions of local image patches drawn from Laplacian pyramids. A small Wasserstein distance indicates that the distribution of the patches is similar, thus real and fake images appear similar in both appearance and variation at this spatial resolution. The metrics are calculated based on 10k generated images for all the models.

CIFAR Data Sets. We evaluate the GANs on the $32 \times 32 \times 3$ CIFAR data sets, for which all models are trained for 100k iterations with a convolutional architecture. In Table 4, the best performance is observed for BuresGAN in terms of image quality, measured by FID and Inception Score, and in terms of mode collapse, measured by SWD. We also notice that UnrolledGAN, VEEGAN and WGAN-GP have difficulty converging to a satisfactory result for this architecture. This is contrast to the 'simpler' synthetic data and the Stacked MNIST data set, where the models attain a performance comparable to BuresGAN. Also, for this architecture and number of training iterations, MDGAN did not converge to a meaningful result in our simulations. In [1], WGAN-GP achieves a very good performance on CIFAR-10 with a ResNet architecture which is considerably more complicated than the DCGAN used here. Therefore, results with a Resnet architecture are reported in Sect. 4.

Table 4. Generation quality on CIFAR-10, CIFAR-100 and STL-10 with DCGAN architecture. For the CIFAR images, Average(std) over 10 runs and 100k iterations for each. For the STL-10 images, Average(std) over 5 runs and 150k iterations for each. For improving readability, SWD score was multiplied by 100. The symbol '∼' indicates that no meaningful result could be obtained for these parameters.

	CIFAR-10			CIFAR-100			STL-10		
	IS (↑)	FID (↓)	SWD (↓)	IS (↑)	FID (↓)	SWD (↓)	IS (↑)	FID (↓)	SWD (↓)
GAN	5.67 (0.22)	59 (8.5)	3.7 (0.9)	5.2 (1.1)	91.7 (66)	7.8 (4.9)	2.9 (1.8)	237 (54)	12.3 (4.1)
WGAN-GP	2.01 (0.47)	291 (87)	8.3 (1.9)	1.2 (0.5)	283 (113)	9.7 (2.5)	∼	∼	∼
UnrolledGAN	3.1 (0.6)	148 (42)	9.0 (5)	3.2 (0.7)	172.9 (40)	13.1 (9.2)	∼	∼	∼
VEEGAN	2.5 (0.6)	198 (33.5)	12.0 (3)	2.8 (0.7)	177.2 (27)	12.8 (3.9)	∼	∼	∼
GDPP	5.76 (0.27)	62.1 (5.5)	4.1 (1.1)	5.9 (0.2)	65.0 (8)	4.4 (1.9)	3.3 (2.2)	232 (84)	8.2 (4.0)
PacGAN2	5.51 (0.18)	60.61 (5.9)	2.95 (1)	5.6 (0.1)	59.9 (5.2)	4.0 (1.8)	4.7 (1.5)	161 (36)	8.1 (4.3)
BuresGAN (ours)	**6.34** (0.17)	**43.7** (0.9)	**2.1** (0.6)	**6.5** (0.1)	**47.2** (1.2)	**2.1** (1.0)	**7.6** (0.3)	**109** (7)	**2.3** (0.3)

CIFAR-100 data set consists of 100 different classes and is therefore more diverse. Compared to the original CIFAR-10 data set, the performance of the studied GANs remains almost the same. An exception is vanilla GAN, which shows a higher presence of mode collapse as measured by SWD.

Fig. 3. Images generated by BuresGAN with a ResNet architecture for CIFAR-10 (left) and STL-10 (right) data sets. The STL-10 samples are full-sized 96 × 96 images.

STL-10. The STL-10 data set includes higher resolution images of size $96 \times 96 \times 3$.

The best performing models from previous experiments are trained for 150k iterations. Samples of generated images from BuresGAN are given on Fig. 3. The metrics are calculated based on 5k generated images for all the models. Compared to the previous data sets, GDPP and vanilla GAN are rarely able to generate high quality images on the higher resolution STL-10 data set. Only BuresGANs are capable of consistently generating high quality images as well as preventing mode collapse, for the same architecture.

Timings. The computing times for these data sets are in SM. For the same number of iterations, BuresGAN training time is comparable to WGAN-GP training for the simple data in Table 1. For more complicated architectures, BuresGAN scales better and the training time was observed to be significantly shorter with respect to WGAN-GP and several other methods.

4 High Quality Generation Using a ResNet Architecture

Table 5. Best achieved IS and FID, using a ResNet architecture. Results with an asterisk are quoted from their respective papers (std in parenthesis). BuresGAN results were obtained after $300k$ iterations and averaged over 3 runs. The result indicated with † are taken from [37]. For all the methods, the STL-10 images are rescaled to $48 \times 48 \times 3$ in contrast with Table 4.

	CIFAR-10		STL-10	
	IS (↑)	FID (↓)	IS (↑)	FID (↓)
WGAN-GP ResNet [11]*	7.86 (0.07)	18.8†	/	/
InfoMax-GAN [17]*	8.08(0.08)	17.14 (0.20)	8.54 (0.12)	37.49 (0.05)
SN-GAN ResNet [24]*	8.22 (0.05)	21.7(0.21)	9.10 (0.04)	40.1 (0.50)
ProgressiveGAN [14]*	8.80 (0.05)	/	/	/
CR-GAN [39]*	8.4	14.56	/	/
NCSN [34]*	**8.87** (0.12)	25.32	/	/
Improving MMD GAN [36]*	8.29	16.21	9.34	37.63
WGAN-div [37]*	/	18.1†	/	/
BuresGAN ResNet (Ours)	8.81(0.08)	**12.91** (0.40)	**9.67** (0.19)	**31.42** (1.01)

As noted by [20], a fair comparison should involve GANs with the same architecture, and this is why we restricted in our paper to a classical DCGAN architecture. It is natural to question the performance of BuresGAN with a ResNet architecture. Hence, we trained BuresGAN on the CIFAR-10 and STL-10 data sets by using the ResNet architecture taken from [11]. In this section, the STL-10 images are rescaled to a resolution of $48 \times 48 \times 3$ according the procedure described in [17,24,36], so that the comparison of IS and FID scores with other works is meaningful. Note that BuresGAN has no parameters to tune, except for the hyperparameters of the optimizers.

The results are displayed in Table 5, where the scores of state-of-the-art unconditional GAN models with a ResNet architecture are also reported. In contrast with Sect. 3, we report here the best performance achieved at any time during the training, averaged over several runs. To the best of our knowledge, our method achieves a new state of the art inception score on STL-10 and is within a standard deviation of state of the art on CIFAR-10 using a ResNet architecture.

The FID score achieved by BuresGAN is nonetheless smaller than the reported FID scores for GANs using a ResNet architecture. A visual inspection of the generated images in Fig. 3 shows that the high inception score is warranted, the samples are clear, diverse and often recognizable. BuresGAN also performs well on the full-sized STL-10 data set where an inception score of 11.11 ± 0.19 and an FID of 50.9 ± 0.13 is achieved (average and std over 3 runs).

5 Conclusion

In this work, we discussed an additional term based on the Bures distance which promotes a matching of the distribution of the generated and real data in a feature space $\mathbb{R}^{d_\phi}$. The Bures distance admits both a feature space and kernel based expression, which makes the proposed model time and data efficient when compared to state of the art models. Our experiments show that the proposed methods are capable of reducing mode collapse and, on the real data sets, achieve a clear improvement of sample quality without parameter tuning and without the need for regularization such as a gradient penalty. Moreover, the proposed GAN shows a stable performance over different architectures, data sets and hyperparameters.

Acknowledgments. EU: The research leading to these results has received funding from the European Research Council under the European Union's Horizon 2020 research and innovation program/ERC Advanced Grants (787960, 885682). This paper reflects only the authors' views and the Union is not liable for any use that may be made of the contained information. Research Council KUL: Optimization frameworks for deep kernel machines C14/18/068, projects C16/15/059, C3/19/053, C24/18/022, C3/20/117), Industrial Research Fund (Fellowships 13-0260, IOF/16/004) and several Leuven Research and Development bilateral industrial projects; Flemish Government: FWO: projects: GOA4917N (Deep Restricted Kernel Machines: Methods and Foundations), EOS Project no G0F6718N (SeLMA), SBO project S005319N, Infrastructure project I013218N, TBM Project T001919N; PhD Grants (SB/1SA1319N, SB/1S93918, SB/1S1319N), EWI: the Flanders AI Research Program. VLAIO: Baekeland PhD (HBC.20192204) and Innovation mandate (HBC.2019.2209), CoT project 2018.018. Other funding: Foundation 'Kom op tegen Kanker', CM (Christelijke Mutualiteit). Ford KU Leuven Research Alliance Project KUL0076 (Stability analysis and performance improvement of deep reinforcement learning algorithms).

References

1. Arjovsky, M., Chintala, S., Bottou, L.: Wasserstein generative adversarial networks. In: Proceedings of the 34th International Conference on Machine Learning (ICML) (2017)
2. Bhatia, R., Jain, T., Lim, Y.: On the Bures-Wasserstein distance between positive definite matrices. Expositiones Mathematicae **37**(2), 165–191 (2019)
3. Binkowski, M., Sutherland, D.J., Arbel, M., Gretton, A.: Demystifying MMD GANs. In: Proceedings of the International Conference on Learning Representations (ICLR) (2018)

4. Che, T., Li, Y., Jacob, A.P., Bengio, Y., Li, W.: Mode regularized generative adversarial networks. In: Proceedings of the International Conference on Learning Representations (ICLR) (2017)
5. Dieng, A.B., Ruiz, F.J.R., Blei, D.M., Titsias, M.K.: Prescribed generative adversarial networks. arxiv:1910.04302 (2020)
6. Dowson, D., Landau, B.: The Fréchet distance between multivariate normal distributions. J. Multivar. Anal. 12(3), 450–455 (1982)
7. Elfeki, M., Couprie, C., Riviere, M., Elhoseiny, M.: GDPP: learning diverse generations using determinantal point processes. In: Proceedings of the 36th International Conference on Machine Learning (ICML) (2019)
8. Gelbrich, M.: On a formula for the L2 Wasserstein metric between measures on Euclidean and Hilbert spaces. Mathematische Nachrichten 147(1), 185–203 (1990)
9. Ghosh, A., Kulharia, V., Namboodiri, V.P., Torr, P.H., Dokania, P.K.: Multi-agent diverse generative adversarial networks. In: Proceedings of the IEEE Conference on Computer Vision and Pattern Recognition (CVPR) (2018)
10. Goodfellow, I., et al.: Generative adversarial nets. In: Advances in Neural Information Processing Systems, vol. 27 (2014)
11. Gulrajani, I., Ahmed, F., Arjovsky, M., Dumoulin, V., Courville, A.C.: Improved training of Wasserstein GANs. In: Advances in Neural Information Processing Systems, vol. 31 (2017)
12. Heusel, M., Ramsauer, H., Unterthiner, T., Nessler, B., Hochreiter, S.: GANs trained by a two time-scale update rule converge to a local nash equilibrium. In: Advances in Neural Information Processing Systems, vol. 30 (2017)
13. Hong, Y., Horn, R.A.: The Jordan canonical form of a product of a Hermitian and a positive semidefinite matrix. Linear Algebra Appl. 147, 373–386 (1991)
14. Karras, T., Aila, T., Laine, S., Lehtinen, J.: Progressive growing of GANs for improved quality, stability, and variation. In: Proceedings of the International Conference on Learning Representations (ICLR) (2017)
15. Kingma, D.P., Ba, J.: Adam: a Method for Stochastic Optimization. In: Proceedings of the International Conference on Learning Representations (ICLR) (2015)
16. Kingma, D.P., Welling, M.: Auto-encoding variational Bayes. In: Proceedings of the International Conference on Learning Representations (ICLR) (2014)
17. Lee, K.S., Tran, N.T., Cheung, N.M.: InfoMax-GAN: mutual information maximization for improved adversarial image generation. In: NeurIPS 2019 Workshop on Information Theory and Machine Learning (2019)
18. Li, C.L., Chang, W.C., Cheng, Y., Yang, Y., Poczos, B.: MMD GAN: towards deeper understanding of moment matching network. In: Advances in Neural Information Processing Systems, vol. 30 (2017)
19. Lin, Z., Khetan, A., Fanti, G., Oh, S.: PacGAN: the power of two samples in generative adversarial networks. In: Advances in Neural Information Processing Systems, vol. 31 (2018)
20. Lucic, M., Kurach, K., Michalski, M., Gelly, S., Bousquet, O.: Are GANs created equal? A large-scale study. In: Advances in Neural Information Processing Systems, pp. 700–709 (2018)
21. Massart, E., Absil, P.A.: Quotient geometry with simple geodesics for the manifold of fixed-rank positive-semidefinite matrices. SIAM J. Matrix Anal. Appl. 41(1), 171–198 (2020)
22. Mescheder, L., Nowozin, S., Geiger, A.: The numerics of GANs. In: Proceedings of the 31st International Conference on Neural Information Processing Systems, p. 1823–1833 (2017)

23. Metz, L., Poole, B., Pfau, D., Sohl-Dickstein, J.: Unrolled generative adversarial networks. In: Proceedings of the International Conference on Learning Representations (ICLR) (2017)
24. Miyato, T., Kataoka, T., Koyama, M., Yoshida, Y.: Spectral normalization for generative adversarial networks. In: Proceedings of the International Conference on Learning Representations (ICLR) (2018)
25. Mroueh, Y., Sercu, T., Goel, V.: McGan: mean and covariance feature matching GAN. In: Proceedings of the 34th International Conference on Machine Learning (ICML) (2017)
26. Nguyen, T., Le, T., Vu, H., Phung, D.: Dual discriminator generative adversarial nets. In: Advances in Neural Information Processing Systems, vol. 30 (2017)
27. Oh, J.H., Pouryahya, M., Iyer, A., Apte, A.P., Deasy, J.O., Tannenbaum, A.: A novel kernel Wasserstein distance on Gaussian measures: an application of identifying dental artifacts in head and neck computed tomography. Comput. Biol. Med. **120**, 103731 (2020)
28. Peyré, G., Cuturi, M., et al.: Computational optimal transport. Found. Trends Mach. Learn. **11**(5–6), 355–607 (2019)
29. Peyre, R.: Comparison between W_2 distance and H^{-1} norm, and localization of Wasserstein distance. ESAIM: COCV **24**(4), 1489–1501 (2018)
30. Radford, A., Metz, L., Chintala, S.: Unsupervised representation learning with deep convolutional generative adversarial networks. In: Proceedings of the International Conference on Learning Representations (ICLR) (2016)
31. Ren, Y., Zhu, J., Li, J., Luo, Y.: Conditional Generative Moment-Matching Networks. In: Advances in Neural Information Processing Systems, vol. 29 (2016)
32. Rezende, D.J., Mohamed, S.: Variational inference with normalizing flows. In: Proceedings of the 32th International Conference on Machine Learning (ICML) (2015)
33. Salimans, T., Goodfellow, I., Zaremba, W., Cheung, V., Radford, A., Chen, X.: Improved techniques for training GANs. In: Advances in Neural Information Processing Systems, vol. 29 (2016)
34. Song, Y., Ermon, S.: Generative modeling by estimating gradients of the data distribution. In: Advances in Neural Information Processing Systems, pp. 11918–11930 (2019)
35. Srivastava, A., Valkov, L., Russell, C., Gutmann, M.U., Sutton, C.: VEEGAN: reducing mode collapse in GANs using implicit variational learning. In: Advances in Neural Information Processing Systems, vol. 30 (2017)
36. Wang, W., Sun, Y., Halgamuge, S.: Improving MMD-GAN training with repulsive loss function. In: Proceedings of the International Conference on Learning Representations (ICLR) (2019)
37. Wu, J., Huang, Z., Thoma, J., Acharya, D., Van Gool, L.: Wasserstein divergence for GANs. In: Ferrari, V., Hebert, M., Sminchisescu, C., Weiss, Y. (eds.) ECCV 2018. LNCS, vol. 11209, pp. 673–688. Springer, Cham (2018). https://doi.org/10.1007/978-3-030-01228-1_40
38. Xiao, C., Zhong, P., Zheng, C.: BourGAN: generative networks with metric embeddings. In: Advances in Neural Information Processing Systems, vol. 32 (2018)
39. Zhang, H., Zhang, Z., Odena, A., Lee, H.: Consistency regularization for generative adversarial networks. In: Proceedings of the International Conference on Learning Representations (ICLR) (2020)

Generative Max-Mahalanobis Classifiers for Image Classification, Generation and More

Xiulong Yang, Hui Ye, Yang Ye, Xiang Li, and Shihao Ji[✉]

Department of Computer Science, Georgia State University, Atlanta, Georgia
{xyang22,hye2,yye10,xli62,sji}@gsu.edu

Abstract. Joint Energy-based Model (JEM) of [11] shows that a standard softmax classifier can be reinterpreted as an energy-based model (EBM) for the joint distribution $p(x, y)$; the resulting model can be optimized to improve calibration, robustness and out-of-distribution detection, while generating samples rivaling the quality of recent GAN-based approaches. However, the softmax classifier that JEM exploits is inherently discriminative and its latent feature space is not well formulated as probabilistic distributions, which may hinder its potential for image generation and incur training instability. We hypothesize that generative classifiers, such as Linear Discriminant Analysis (LDA), might be more suitable for image generation since generative classifiers model the data generation process explicitly. This paper therefore investigates an LDA classifier for image classification and generation. In particular, the Max-Mahalanobis Classifier (MMC) [30], a special case of LDA, fits our goal very well. We show that our Generative MMC (GMMC) can be trained discriminatively, generatively or jointly for image classification and generation. Extensive experiments on multiple datasets show that GMMC achieves state-of-the-art discriminative and generative performances, while outperforming JEM in calibration, adversarial robustness and out-of-distribution detection by a significant margin. Our source code is available at https://github.com/sndnyang/GMMC.

Keywords: Energy-based models · Generative models · Max-Mahalanobis Classifier

1 Introduction

Over the past few years, deep neural networks (DNNs) have achieved state-of-the-art performance on a wide range of learning tasks, such as image classification, object detection, segmentation and image captioning [14,18]. All of these breakthroughs, however, are achieved in the framework of discriminative models, which are known to be exposed to several critical issues, such as adversarial examples [10], calibration of uncertainty [13] and out-of-distribution detection [15]. Prior works have shown that generative training is beneficial to these

© Springer Nature Switzerland AG 2021
N. Oliver et al. (Eds.): ECML PKDD 2021, LNAI 12976, pp. 67–83, 2021.
https://doi.org/10.1007/978-3-030-86520-7_5

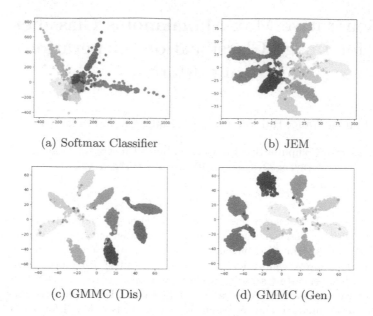

(a) Softmax Classifier   (b) JEM

(c) GMMC (Dis)   (d) GMMC (Gen)

Fig. 1. t-SNE visualization of the latent feature spaces learned by different models trained on CIFAR10.

models and can alleviate some of these issues at certain levels [5,7]. Yet, most recent research on generative models focus primarily on qualitative sample quality [3,33,36], and the discriminative performances of state-of-the-art generative models are still far behind discriminative ones [2,6,8].

Recently, there is a flurry of interest in closing the performance gap between generative models and discriminative models [1,8,9,11]. Among them, IGEBM [8] and JEM [11] are the two most representative ones, which reinterpret CNN classifiers as the energy-based models (EBMs) for image generation. Since the CNN classifier is the only trained model, which has a high compositionality, it is possible that a single trained CNN model may encompass the generative capabilities into the discriminative model without sacrificing its discriminative power. Their works realize the potential of EBMs in hybrid modeling and achieve improved performances on discriminative and generative tasks. Specifically, JEM [11] reinterprets the standard softmax classifier as an EBM and achieves impressive performances in image classification and generation simultaneously, and ignites a series of follow-up works [9,12,41].

However, the softmax classifier that JEM exploits is inherently discriminative, which may hinder its potential in image generation. To investigate this, we visualize the latent feature spaces learnt by a standard softmax classifier and by JEM through t-SNE [21] in Figs. 1(a) and 1(b), respectively. Apparently, the feature space of the softmax classifier has been improved significantly by JEM as manifested by higher inter-class separability and intra-class compactness. However, JEM's latent space is not well formulated as probabilistic distributions, which may limit its generative performance and incur training instability as observed in [11]. We hypothesize that generative classifiers (e.g., LDA) might be more suitable for image classification and generation. This is because generative classifiers model the data generation process explicitly with probabilistic distributions, such as mixture of Gaussians, which aligns well with the generative process of image synthesis. Therefore, in this paper we investigate an LDA classifier for image classification and generation. In particular, the Max-Mahalanobis Classifier (MMC) [30], a special case of LDA, fits our goal very well since MMC formulates the latent feature space explicitly as the Max-Mahalanobis distribution [29]. Distinct to [30], we show that MMC can be trained discriminatively, generatively or jointly as an EBM. We term our algorithm Generative MMC (GMMC) given that it is a hybrid model for image classification and generation, while the original MMC [30] is only for classification.

As a comparison, Figs. 1(c) and 1(d) illustrate the latent feature spaces of GMMC optimized with discriminative training and generative training (to be discussed in Sect. 3), respectively. It can be observed that the latent feature spaces of GMMC are improved even further over that of JEM's with higher inter-class separability and intra-class compactness. Furthermore, the explicit generative modeling of GMMC leads to many auxiliary benefits, such as adversarial robustness, calibration of uncertainty and out-of-distribution detection, which will be demonstrated in our experiments. Our main contributions can be summarized as follows:

1. We introduce GMMC, a hybrid model for image classification and generation. As an alternative to the softmax classifier utilized in JEM, GMMC has a well-formulated latent feature distribution, which fits well with the generative process of image synthesis.
2. We show that GMMC can be trained discriminatively, generatively or jointly with reduced complexity and improved stability as compared to JEM.
3. Our model matches or outperforms prior state-of-the-art hybrid models on multiple discriminative and generative tasks, including image classification, image synthesis, calibration of uncertainty, out-of-distribution detection and adversarial robustness.

2   Background and Related Work

2.1   Energy-Based Models

Energy-based models (EBMs) [20] define an energy function that assigns low energy values to samples drawn from data distribution and high values otherwise, such that any probability density $p_\theta(x)$ can be expressed via a Boltzmann distribution as

$$p_\theta(x) = \exp\left(-E_\theta(x)\right)/Z(\theta), \qquad (1)$$

where $E_\theta(x)$ is an energy function that maps each input $x \in \mathbb{R}^D$ to a scalar, and $Z(\theta)$ is the normalizing constant (also known as the partition function) such that $p_\theta(x)$ is a valid density function.

The key challenge of training EBMs lies in estimating the partition function $Z(\theta)$, which is notoriously intractable. The standard maximum likelihood estimation of parameters θ is not straightforward, and a number of sampling-based approaches have been proposed to approximate it effectively. Specifically, the gradient of the log-likelihood of a single sample x w.r.t. θ can be expressed as

$$\frac{\partial \log p_\theta(x)}{\partial \theta} = \mathbb{E}_{p_\theta(x')} \frac{\partial E_\theta(x')}{\partial \theta} - \frac{\partial E_\theta(x)}{\partial \theta},\qquad(2)$$

where the expectation is over model distribution $p_\theta(x')$, sampling from which is challenging due to the intractable $Z(\theta)$. Therefore, MCMC and Gibbs sampling [17] have been proposed previously to estimate the expectation efficiently. To speed up the mixing for effective sampling, recently Stochastic Gradient Langevin Dynamics (SGLD) [38] has been used to train EBMs by exploiting the gradient information [8,11,28]. Specifically, to sample from $p_\theta(x)$, SGLD follows

$$x_0 \sim p_0(x),\qquad x_{i+1} = x_i - \frac{\alpha}{2}\frac{\partial E_\theta(x_i)}{\partial x_i} + \alpha\epsilon,\ \ \epsilon \sim \mathcal{N}(0,1),\qquad(3)$$

where $p_0(x)$ is typically a uniform distribution over $[-1,1]$, whose samples are refined via noisy gradient decent with step-size α, which should be decayed following a polynomial schedule.

Besides JEM [11] that we discussed in the introduction, [39] is an earlier work that derives a generative CNN model from the commonly used discriminative CNN by treating it as an EBM, where the authors factorize the loss function $\log p(x|y)$ as an EBM. Following-up works, such as [8,27], scale the training of EBMs to high-dimensional data using SGLD. However, all of these previous methods define $p(x|y)$ or $p(x)$ as an EBM, while our GMMC defines an EBM on $p(x,y)$ by following a mixture of Gaussian distribution, which simplifies the maximum likelihood estimation and achieves improved performances in many discriminative and generative tasks.

2.2   Alternatives to the Softmax Classifier

Softmax classifier has been widely used in state-of-the-art models for discriminative tasks due to its simplicity and efficiency. However, softmax classifier is known particularly vulnerable to adversarial attacks because the latent feature space induced by softmax classifier is typically not well separated (as shown in Fig. 1(a)). Some recent works propose to use generative classifiers to better formulate the latent space distributions in order to improve its robustness to adversarial examples. For example, Wan et al. [37] propose to model the latent feature space as mixture of Gaussians and encourages stronger intra-class compactness and larger inter-class separability by introducing large margins between classes. Different from [37], Pang et al. [29] pre-design the centroids based on the Max-Mahalanobis distribution (MMD), other than learning them from data. The authors prove that if the latent feature space distributes as an MMD, the LDA classifier will have the best robustness to adversarial examples. Taking advantage of the benefits of MMD, Pang et al. [30] further propose a max-Mahalanobis center regression loss, which induces much denser feature regions and improves the robustness of trained models. Compared with softmax classifier, all these works can generate better latent feature spaces to improve the robustness of models for the task of classification. Our GMMC is built on the basic framework of MMC, but we reinterpret MMC as an EBM for image classification and generation. Moreover, we show that the generative training of MMC can further improve calibration, adversarial robustness and out-of-distribution detection.

3   Methodology

We assume a Linear Discriminant Analysis (LDA) classifier is defined as: $\phi(x)$, $\mu = \{\mu_y, y = 1, 2, \cdots, C\}$ and $\pi = \{\pi_y = \frac{1}{C}, y = 1, 2, \cdots, C\}$ for C-class classification, where $\phi(x) \in \mathbb{R}^d$ is the feature representation of x extracted by a CNN, parameterized by ϕ[1], and $\mu_y \in \mathbb{R}^d$ is the mean of a Gaussians distribution with a diagonal covariance matrix $\gamma^2 I$, i.e., $p_\theta(\phi(x)|y) = \mathcal{N}(\mu_y, \gamma^2 I)$. Therefore, we can parameterize LDA by $\theta = \{\phi, \mu\}$[2]. Instead of using this regular LDA classifier, in this paper the max-Mahalanobis classifier (MMC) [30], a special case of LDA, is considered. Different from the LDA modeling above, in MMC $\mu = \{\mu_y, y = 1, 2, \cdots, C\}$ is pre-designed to induce compact feature representations for model robustness. We found that the MMC modeling fits our goal better than the regular LDA classifier due to its improved training stability and boosted adversarial robustness. Therefore, in the following we focus on the MMC modeling for image classification and generation. As such, the learnable parameters of MMC reduce to $\theta = \{\phi\}$, and the pseudo-code of calculating pre-designed μ can be found in Algorithm 1 of the supplementary material[3]. Figure 2 provides an overview of the training and test of our GMMC algorithm, with the details discussed below.

Instead of maximizing $p_\theta(y|x)$ as in standard softmax classifier, following JEM [11] we maximize the joint distribution $p_\theta(x, y)$, which follows a mixture of Gaussians distribution in GMMC. To optimize $\log p_\theta(x, y)$, we can consider three different approaches.

3.1   Approach 1: Discriminative Training

According to the MMC modeling above, the joint distribution $p_\theta(x, y)$ can be expressed as

$$p_\theta(x, y) = p(y)p_\theta(x|y) \propto \frac{1}{C}(2\pi\gamma^2)^{-d/2}\exp(-\frac{1}{2\gamma^2}||\phi(x) - \mu_y||_2^2)$$

$$= \frac{\exp(-\frac{1}{2\gamma^2}||\phi(x) - \mu_y||_2^2)}{Z(\theta)} = \frac{\exp(-E_\theta(x, y))}{Z(\theta)} \qquad (4)$$

---

[1] To avoid notational clutter in later derivations, we use ϕ to denote a CNN feature extractor and its parameter. But the meaning of ϕ is clear given the context.

[2] We can treat γ as a tunable hyperparameter or we can estimate it by post-processing. In this work, we take the latter approach as discussed in Sect. 3.1.

[3] Supplementary material: https://arxiv.org/abs/2101.00122.

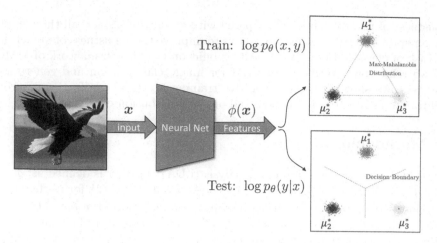

Fig. 2. Overview of GMMC for training and test, where the model can be trained discriminatively, generatively or jointly. $\{\boldsymbol{\mu}_1^*, \boldsymbol{\mu}_2^*, \cdots, \boldsymbol{\mu}_C^*\}$ are pre-designed according to MMD [29]. Only $\boldsymbol{\theta} = \{\boldsymbol{\phi}\}$ is learned from data.

where we define $E_{\boldsymbol{\theta}}(\boldsymbol{x}, y) = \frac{1}{2\gamma^2}\|\boldsymbol{\phi}(\boldsymbol{x}) - \boldsymbol{\mu}_y\|_2^2$, and $Z(\boldsymbol{\theta}) = \int \exp(-E_{\boldsymbol{\theta}}(\boldsymbol{x}, y))d\boldsymbol{x}dy$, which is an intractable partition function. To avoid the expense of evaluating the partition function, we follow Mnih and Teh [23] and approximate $Z(\boldsymbol{\theta})$ as a constant (e.g., $Z(\boldsymbol{\theta}) = 1$). This turns out to be an effective approximation for neural networks with lots of parameters as it encourages the model to have "self-normalized" outputs. With this approximation, the log of the joint distribution can be simplified as

$$\log p_{\boldsymbol{\theta}}(\boldsymbol{x}, y) = -\frac{1}{2\gamma^2}\|\boldsymbol{\phi}(\boldsymbol{x}) - \boldsymbol{\mu}_y\|_2^2 - \log Z(\boldsymbol{\theta})$$
$$\approx -E_{\boldsymbol{\theta}}(\boldsymbol{x}, y) + \text{constant}. \tag{5}$$

To optimize the parameters $\boldsymbol{\theta}$, we can simply compute gradient of Eq. 5 w.r.t. $\boldsymbol{\theta}$, and update the parameters by stochastic gradient descent (SGD) [31]. Note that γ is a constant in Eq. 5, its effect can be absorbed into the learning rate when optimizing Eq. 5 via SGD. After convergence, we can estimate $\gamma^2 = \frac{1}{d}\left(\frac{1}{N}\sum_{i=1}^{N}\|\boldsymbol{\phi}(\boldsymbol{x}_i) - \boldsymbol{\mu}_i\|_2^2\right)$ from training set by using optimized $\boldsymbol{\phi}$ and pre-designed $\boldsymbol{\mu}$.

Note that Eq. 5 boils down to the same objective that MMC [30] proposes, i.e. the center regression loss. While MMC reaches to this objective from the perspective of inducing compact feature representations for model robustness, we arrive at this objective by simply following the principle of maximum likelihood estimation of model parameter $\boldsymbol{\theta}$ of joint density $p_{\boldsymbol{\theta}}(\boldsymbol{x}, y)$ with the "self-normalization" approximation [23].

3.2   Approach 2: Generative Training

Comparing Eq. 4 with the definition of EBM (1), we can also treat the joint density

$$p_\theta(\boldsymbol{x}, y) = \frac{\exp(-\frac{1}{2\gamma^2}\|\phi(\boldsymbol{x}) - \boldsymbol{\mu}_y\|_2^2)}{Z(\theta)} \tag{6}$$

as an EBM with $E_\theta(\boldsymbol{x}, y) = \frac{1}{2\gamma^2}\|\phi(\boldsymbol{x}) - \boldsymbol{\mu}_y\|_2^2$ defined as an energy function of $(\boldsymbol{x}, y)$.

Following the maximum likelihood training of EBM (2), to optimize Eq. 6, we can compute its gradient w.r.t. $\boldsymbol{\theta}$ as

$$\frac{\partial \log p_\theta(\boldsymbol{x}, y)}{\partial \theta} = \beta \mathbb{E}_{p_\theta(\boldsymbol{x}', y')} \frac{\partial E_\theta(\boldsymbol{x}', y')}{\partial \theta} - \frac{\partial E_\theta(\boldsymbol{x}, y)}{\partial \theta}, \tag{7}$$

where the expectation is over the joint density $p_\theta(\boldsymbol{x}', y')$, sampling from which is challenging due to the intractable $Z(\boldsymbol{\theta})$, and β is a hyperparameter that balances the contributions from the two terms. Different value of β has a significant impact to the performance. From our experiments, we find that $\beta = 0.5$ works very well in all our cases. Therefore, we set $\beta = 0.5$ as the default value. Notably, the two terms of our GMMC (Eq. 7) are both defined on the same energy function $E_\theta(\boldsymbol{x}, y)$, while the two terms of JEM [11] are defined on $p(\boldsymbol{x})$ and $p(y|\boldsymbol{x})$, respectively, which are computationally more expensive and might incur training instability as we will discuss in Sect. 3.3.

The tricky part is how to generate samples $(\boldsymbol{x}', y') \sim p_\theta(\boldsymbol{x}', y')$ to estimate the first term of Eq. 7. We can follow the mixture of Gaussians assumption of MMC. That is, $p_\theta(\boldsymbol{x}', y') = p(y')p_\theta(\boldsymbol{x}'|y')$: (1) sample $y' \sim p(y') = \frac{1}{C}$, and then (2) sample $\boldsymbol{x}' \sim p_\theta(\boldsymbol{x}'|y') \propto \mathcal{N}(\boldsymbol{\mu}_{y'}, \gamma^2 \boldsymbol{I})$. To sample $\boldsymbol{x}'$, again we can consider two choices.

(1) Staged Sampling. We can first sample $\boldsymbol{z}_{\boldsymbol{x}'} \sim \mathcal{N}(\boldsymbol{\mu}_{y'}, \gamma^2 \boldsymbol{I})$, and then find an $\boldsymbol{x}'$ to minimize $E_\theta(\boldsymbol{x}') = \frac{1}{2\gamma^2}\|\phi(\boldsymbol{x}') - \boldsymbol{z}_{\boldsymbol{x}'}\|_2^2$. This can be achieved by

$$\boldsymbol{x}'_0 \sim p_0(\boldsymbol{x}'), \qquad \boldsymbol{x}'_{t+1} = \boldsymbol{x}'_t - \alpha \frac{\partial E_\theta(\boldsymbol{x}'_t)}{\partial \boldsymbol{x}'_t}, \tag{8}$$

where $p_0(\boldsymbol{x})$ is typically a uniform distribution over $[-1, 1]$. Note that this is similar to SGLD (see Eq. 3) but without a noisy term. Thus, the training could be more stable. In addition, the function $E_\theta(\boldsymbol{x}') = \frac{1}{2\gamma^2}\|\phi(\boldsymbol{x}') - \boldsymbol{z}_{\boldsymbol{x}'}\|_2^2$ is just an L_2 regression loss (not an LogSumExp function as used in JEM [11]). This may lead to additional numerical stability.

(2) Noise Injected Sampling. We can first sample $\boldsymbol{z} \sim \mathcal{N}(\boldsymbol{0}, \boldsymbol{I})$, then by the reparameterization trick we have $\boldsymbol{z}_{\boldsymbol{x}'} = \gamma \boldsymbol{z} + \boldsymbol{\mu}_{y'}$. Finally, we can find an $\boldsymbol{x}'$ to minimize

$$E_\theta(x') = \frac{1}{2\gamma^2}||\phi(x') - z_{x'}||_2^2 = \frac{1}{2\gamma^2}||\phi(x') - \mu_{y'} - \gamma z||_2^2$$

$$= \frac{1}{2\gamma^2}||\phi(x') - \mu_{y'}||_2^2 + \frac{1}{2}||z||_2^2 - \frac{1}{\gamma^2} <\phi(x') - \mu_{y'}, \gamma z>$$

$$= E_\theta(x', y') + \frac{1}{2}||z||_2^2 - \frac{1}{\gamma} <\phi(x') - \mu_{y'}, z> . \tag{9}$$

This can be achieved by

$$x'_0 \sim p_0(x'),$$

$$x'_{t+1} = x'_t - \alpha\frac{\partial E_\theta(x'_t)}{\partial x'_t} = x'_t - \alpha\frac{\partial E_\theta(x'_t, y')}{\partial x'_t} + \alpha\frac{1}{\gamma} < \frac{\partial\phi(x'_t)}{\partial x'_t}, z >, \tag{10}$$

where we sample a different z at each iteration. As a result, Eq. 10 is an analogy of SGLD (see Eq. 3). The difference is instead of using an unit Gaussian noise $\epsilon \sim \mathcal{N}(0, 1)$ as in SGLD, a gradient-modulated noise (the 3rd term) is applied.

Algorithm 1. Generative training of GMMC: Given model parameter $\theta = \{\phi\}$, step-size α, replay buffer B, number of steps τ, reinitialization frequency ρ

1: **while** not converged **do**
2:      Sample x and y from dataset $\mathcal{D}$
3:      Sample $(x'_0, y') \sim B$ with probability $1 - \rho$, else $x'_0 \sim \mathcal{U}(-1, 1), y' \sim p(y') = \frac{1}{C}$
4:      Sample $z_{x'} \sim \mathcal{N}(\mu_{y'}, \gamma^2 I)$ if staged sampling
5:      **for** $t \in [1, 2, \cdots, \tau]$ **do**
6:          Sample $z \sim \mathcal{N}(0, I), z_{x'} = \mu_{y'} + \gamma z$ if noise injected sampling
7:          $x'_t = x'_{t-1} - \alpha\frac{\partial E_\theta(x'_{t-1})}{\partial x'_{t-1}}$ (Eq. 8)
8:      **end for**
9:      Calculate gradient with Eq. 7 from (x, y) and (x'_τ, y') for model update
10:     Add / replace updated (x'_τ, y') back to B
11: **end while**

Algorithm 1 provides the pseudo-code of the generative training of GMMC, which follows a similar design of IGEBM [8] and JEM [11] with a replay buffer B. For brevity, only one real sample $(x, y) \sim \mathcal{D}$ and one generated sample $(x', y') \sim p_\theta(x', y')$ are used to optimize the parameter θ. It is straightforward to generalize the pseudo-code above to a mini-batch training, which is used in our experiments. Compared to JEM, GMMC needs no additional calculation of $p_\theta(y|x)$ and thus has reduced computational complexity.

3.3 Approach 3: Joint Training

Comparing Eq. 5 and Eq. 7, we note that the gradient of Eq. 5 is just the second term of Eq. 7. Hence, we can use (Approach 1) discriminative training to pretrain θ, and then finetune θ by (Approach 2) generative training. The transition between the two can be achieved by scaling up β from 0 to a predefined value (e.g., 0.5). Similar joint training strategy can be applied to train JEM as well. However, from our experiments we note that this joint training of JEM is extremely unstable. We hypothesize that this is likely because the two terms of JEM are defined on $p(\boldsymbol{x})$ and $p(y|\boldsymbol{x})$, respectively, while the two terms of our GMMC (Eq. 7) are defined on the same energy function $E_\theta(\boldsymbol{x}, y)$. Hence, the learned model parameters from the two training stages are more compatible in GMMC than in JEM. We will demonstrate the training issues of JEM and GMMC when we present results.

3.4 GMMC for Inference

After training with one of the three approaches discussed above, we get the optimized GMMC parameters $\boldsymbol{\theta} = \{\boldsymbol{\phi}\}$, the pre-designed $\boldsymbol{\mu}$ and the estimated $\gamma^2 = \frac{1}{d}\left(\frac{1}{N}\sum_{i=1}^{N}||\boldsymbol{\phi}(\boldsymbol{x}_i) - \boldsymbol{\mu}_i||_2^2\right)$ from training set. We can then calculate class probabilities for classification

$$p_\theta(y|\boldsymbol{x}) = \frac{\exp(-\frac{1}{2\gamma^2}||\boldsymbol{\phi}(\boldsymbol{x}) - \boldsymbol{\mu}_y||_2^2)}{\sum_{y'}\exp(-\frac{1}{2\gamma^2}||\boldsymbol{\phi}(\boldsymbol{x}) - \boldsymbol{\mu}_{y'}||_2^2)}. \tag{11}$$

4 Experiments

We evaluate the performance of GMMC on multiple discriminative and generative tasks, including image classification, image generation, calibration of uncertainty, out-of-distribution detection and adversarial robustness. Since GMMC is inspired largely by JEM [11], for a fair comparison, our experiments closely follow the settings provided in the source code of JEM[4]. All our experiments are performed with PyTorch on Nvidia RTX GPUs. Due to page limit, details of the experimental setup are relegated to the supplementary material.

4.1 Hybrid Modeling

We train GMMC on three benchmark datasets: CIFAR10, CIFAR100 [18] and SVHN [26], and compare it to the state-of-the-art hybrid models, as well as standalone generative and discriminative models. Following the settings of JEM, we use the Wide-ResNet [40] as the backbone CNN model for JEM and GMMC. To evaluate the quality of generated images, we adopt Inception Score (IS) [32] and Fréchet Inception Distance (FID) [16] as the evaluation metrics.

---

[4] https://github.com/wgrathwohl/JEM.

Table 1. Hybrid modeling results on CIFAR10.

Class	Model	Acc % ↑	IS ↑	FID ↓
Hybrid	Residual Flow	70.3	3.60	46.4
	Glow	67.6	3.92	48.9
	IGEBM	49.1	8.30	37.9
	JEM	92.9	**8.76**	38.4
	GMMC (Ours)	**94.08**	7.24	**37.0**
Disc.	WRN w/BN	95.8	N/A	N/A
	WRN w/o BN	93.6	N/A	N/A
	GMMC (Dis)	94.3	N/A	N/A
Gen.	SNGAN	N/A	8.59	25.5
	NCSN	N/A	8.91	25.3

Table 2. Test accuracy (%) on SVHN and CIFAR100.

Model	SVHN	CIFAR100
Softmax	96.6	72.6
JEM	96.7	72.2
GMMC (Dis)	97.1	**75.4**
GMMC (Gen)	**97.2**	73.9

The results on CIFAR10, CIFAR100 and SVHN are shown in Table 1 and Table 2, respectively. It can be observed that GMMC outperforms the state-of-the-art hybrid models in terms of accuracy (94.08%) and FID score (37.0), while being slightly worse in IS score. Since no IS and FID scores are commonly reported on SVHN and CIFAR100, we present the classification accuracies and generated samples on these two benchmarks. Our GMMC models achieve 97.2% and 73.9% accuracy on SVHN and CIFAR100, respectively, outperforming the softmax classifier and JEM by notable margins. Example images generated by GMMC for CIFAR10 are shown in Fig. 3. Additional GMMC generated images for CIFAR100 and SVHN can be found in the supplementary material.

4.2 Calibration

While modern deep models have grown more accurate in the past few years, recent researches have shown that their predictions could be over-confident [13].

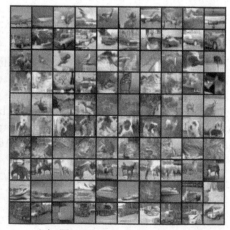

(a) Unconditional Samples   (b) Class-conditional Samples

Fig. 3. Generated CIFAR10 Samples.

Outputting an incorrect but confident decision can have catastrophic consequences. Hence, calibration of uncertainty for DNNs is a critical research topic. Here, the confidence is defined as $\max_y p(y|\boldsymbol{x})$ which is used to decide when to output a prediction. A well-calibrated, but less accurate model can be considerably more useful than a more accurate but less-calibrated model.

We train GMMC on the CIFAR10 dataset, and compare its Expected Calibration Error (ECE) score [13] to that of the standard softmax classifier and JEM. Results are shown in Fig. 4 with additional results on SVHN and CIFAR100 provided in the supplementary material. We find that the model trained by GMMC (Gen) achieves a much smaller ECE (1.33% vs. 4.18%), demonstrating GMMC's predictions are better calibrated than the competing methods.

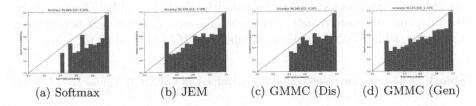

(a) Softmax   (b) JEM   (c) GMMC (Dis)   (d) GMMC (Gen)

Fig. 4. Calibration results on CIFAR10. The smaller ECE is, the better.

4.3 Out-Of-Distribution Detection

The OOD detection is a binary classification problem, which outputs a score $s_\theta(\boldsymbol{x}) \in \mathbb{R}$ for a given query $\boldsymbol{x}$. The model should be able to assign lower scores to OOD examples than to in-distribution examples, such that it can be used to distinguish two sets of examples. Following the settings of JEM [11], we use the Area Under the Receiver-Operating Curve (AUROC) [15] to evaluate the performance of OOD detection. In our experiments, three score functions are considered: the input density $p_\theta(\boldsymbol{x})$ [24], the predictive distribution $p_\theta(y|\boldsymbol{x})$ [15], and the approximate mass $\|\frac{\partial \log p_\theta(\boldsymbol{x})}{\partial \boldsymbol{x}}\|$ [11].

(1) Input Density. A natural choice of $s_\theta(\boldsymbol{x})$ is the input density $p_\theta(\boldsymbol{x})$. For OOD detection, intuitively we consider examples with low $p(\boldsymbol{x})$ to be OOD. Quantitative results can be found in Table 3 (top). The corresponding distributions of scores are visualized in Table 4. The GMMC model assigns higher likelihoods to in-distribution data than to the OOD data, outperforming all the other models by a significant margin.

(2) Approximate Mass. Recent work of [25] has found that likelihood may not be enough for OOD detection in high-dimensional space. Real samples from a distribution form the area of high probability *mass*. But a point may have a high density while the surrounding area has a very low density, which indicates the density can change rapidly around it and that point is likely not a sample from the real data distribution. Thus, the norm of gradient of the log-density will be large compared to examples in the area *mass*. Based on this reasoning, Grathwohl et al. propose a new OOD score: $\theta(x) = -\|\frac{\partial \log p_\theta(x)}{\partial x}\|_2$. Adopting this score function, we find that our model still outperforms the other competing methods (JEM and IGEBM), as shown in Table 3 (bottom).

(3) Predictive Distribution. Another useful OOD score is the maximum probability from a classifier's predictive distribution: $s_\theta(x) = \max_y p_\theta(y|x)$. Hence, OOD performance using this score is highly correlated with a model's classification accuracy. The results can be found in Table 3 (middle). Interestingly, with this score function, there is no clear winner over four different benchmarks consistently, while GMMC performs similarly to JEM in most of the cases.

Table 3. OOD detection results. Models are trained on CIFAR10. Values are AUROC.

$s_\theta(x)$	Model	SVHN	CIFAR10 Interp	CIFAR100	CelebA
$\log p_\theta(x)$	Unconditional Glow	.05	.51	.55	.57
	Class-Conditional Glow	.07	.45	.51	.53
	IGEBM	.63	.70	.50	.70
	JEM	.67	.65	.67	.75
	GMMC (Gen)	**.84**	**.75**	**.84**	**.86**
$\max_y p_\theta(y|x)$	Wide-ResNet	**.93**	**.77**	.85	.62
	Class-Conditional Glow	.64	.61	.65	.54
	IGEBM	.43	.69	.54	.69
	JEM	.89	.75	**.87**	**.79**
	GMMC (Gen)	.84	.72	.81	.31
$\|\frac{\partial \log p_\theta(x)}{\partial x}\|$	Unconditional Glow	**.95**	.27	.46	.29
	Class-Conditional Glow	.47	.01	.52	.59
	IGEBM	.84	.65	.55	.66
	JEM	.83	.78	.82	.79
	GMMC (Gen)	.88	**.79**	**.85**	**.87**

Table 4. Histograms of $\log_\theta p(x)$ for OOD detection. Green corresponds to in-distribution dataset, while red corresponds to OOD dataset.

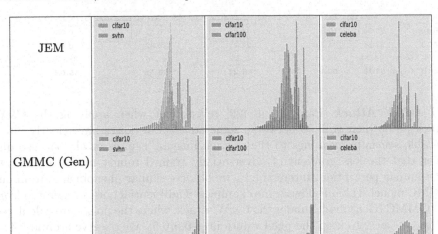

In summary, among all three different OOD score functions, GMMC outperforms the competing methods by a notable margin with two of them, while being largely on par with the rest one. The improved performance of GMMC on OOD detection is likely due to its *explicit* generative modeling of $p_\theta(x, y)$, which improves the evaluation of $p_\theta(x)$ over other methods.

4.4 Robustness

DNNs have demonstrated remarkable success in solving complex prediction tasks. However, recent works [10,19,34] have shown that they are particularly vulnerable to adversarial examples, which are in the form of small perturbations to inputs but can lead DNNs to predict incorrect outputs. Adversarial examples are commonly generated through an iterative optimization procedure, which resembles the iterative sampling procedure of SGLD in Eq. 3. GMMC (and JEM) further utilizes sampled data along with real data for model training (see Eq. 7). This again shares some similarity with adversarial training [10], which has been proved to be the most effective method for adversarial defense. In this section, we show that GMMC achieves considerable robustness compared to other methods thanks to the MMC modeling [30] and its generative training.

(1) PGD Attack. We run the white-box PGD attack [22] on the models trained by standard softmax, JEM and GMMC. We use the same number of steps (40) with different ϵ's as the settings of JEM for PGD. Table 5 reports the test accuracies of different methods. It can be observed that GMMC achieves much higher accuracies than standard softmax and JEM under all different attack strengths. The superior defense performance of GMMC (Dis) over JEM mainly attributes to the MMC modeling [30], while GMMC (Gen) further improves its robustness over GMMC (Dis) due to the generative training.

Table 5. Classification accuracies when models are under L_∞ PGD attack with different ϵ's. All models are trained on CIFAR10.

Model	Clean (%)	PGD-40 $\epsilon = 4/255$	PGD-40 $\epsilon = 8/255$	PGD-40 $\epsilon = 16/255$	PGD-40 $\epsilon = 32/255$
Softmax	93.56	19.05	4.95	0.57	0.06
JEM	92.83	34.39	21.23	5.82	0.50
GMMC (Dis)	**94.34**	44.83	42.22	41.76	38.69
GMMC (Gen)	94.13	**56.29**	**56.27**	**56.13**	**55.02**

(2) C&W Attack. Pang et al. [29] reveal that when applying the C&W attack [4] on their trained networks some adversarial noises have clearly interpretable semantic meanings to the original images. Tsipras et al. [35] also discover that the loss gradient of adversarially trained robust model aligns well with human perception. Interestingly, we observe similar phenomenon from our GMMC model. Table 6 shows some examples of adversarial noises generated from the GMMC (Gen) model under the C&W attack, where the noises are calculated as $(x_{adv} - x)/2$ to keep the pixel values in $[-0.5, 0.5]$. We observe around 5% of adversarial noises have clearly interpretable semantic meanings to their original images. These interpretable adversarial noises indicate that GMMC (Gen) can learn robust features such that the adversarial examples found by the C&W attack have to weaken the features of the original images as a whole, rather than generating salt-and-pepper like perturbations as for models of lower robustness.

To have a quantitative measure of model robustness under C&W attack, we apply the C&W attack to the models trained by the standard softmax, JEM and GMMC. In terms of classification accuracy, all of them achieve almost 100% error rate under the C&W attack. However, as shown in Table 7, the adversarial noises to attack the GMMC (Gen) model have a much larger L_2 norm than that of adversarial noises for other models. This indicates that to attack GMMC (Gen), the C&W attack has to add much stronger noises in order to successfully evade the network. In addition, GMMC (Dis) achieves a similar robustness as JEM under the C&W attack, while GMMC (Gen) achieves an improved robustness over GMMC (Dis) likely due to the generative training of GMMC.

Table 6. Example adversarial noises generated from the GMMC (Gen) model under C&W attack on CIFAR10.

Table 7. L_2 norms of adversarial perturbations under C&W attack on CIFAR10.

Model	Untarget iter $= 100$	Target iter $= 1000$
Softmax	0.205	0.331
JEM	0.514	0.905
GMMC (Dis)	0.564	0.784
GMMC (Gen)	**1.686**	**1.546**

4.5 Training Stability

Compared to JEM, GMMC has a well-formulated latent feature distribution, which fits well with the generative process of image synthesis. One advantage we observed from our experiments is that GMMC alleviates most of the instability issues of JEM. Empirically, we find that JEM can train less than 60 epochs before running into numerical issues, while GMMC can run 150 epochs smoothly without any numerical issues in most of our experiments.

4.6 Joint Training

Finally, we compare the joint training of JEM and GMMC on CIFAR10. The results show that joint training of GMMC is quite stable in most of our experiments, while JEM experiences substantial numerical instability issues. However, the quality of generated images from joint training of GMMC is not as good as generative training of GMMC from scratch. Due to page limit, details of the comparison are relegated to the supplementary material.

5 Conclusion and Future Work

In this paper, we propose GMMC by reinterpreting the max-Mahalanobis classifier [30] as an EBM. Compared to the standard softmax classifier utilized in JEM, GMMC models the latent feature space explicitly as the max-Mahalanobis distribution, which aligns well with the generative process of image synthesis. We show that GMMC can be trained discriminative, generatively or jointly with reduced complexity and improved stability compared to JEM. Extensive experiments on the benchmark datasets demonstrate that GMMC can achieve state-of-the-art discriminative and generative performances, and improve calibration, out-of-distribution detection and adversarial robustness.

As for future work, we plan to investigate the GMMC models trained by different methods: discriminative vs. generative. We are interested in the differences between the features learned by different methods. We also plan to investigate the joint training of GMMC to improve the quality of generated images further because joint training speeds up the learning of GMMC significantly and can scale up GMMC to large-scale benchmarks, such as ImageNet.

Acknowledgment. We would like to thank the anonymous reviewers for their comments and suggestions, which helped improve the quality of this paper. We would also gratefully acknowledge the support of VMware Inc. for its university research fund to this research.

References

1. Ardizzone, L., Mackowiak, R., Rother, C., Köthe, U.: Training normalizing flows with the information bottleneck for competitive generative classification. In: Neural Information Processing Systems (NeurIPS) (2020)

2. Behrmann, J., Grathwohl, W., Chen, R.T., Duvenaud, D., Jacobsen, J.H.: Invertible residual networks. arXiv preprint arXiv:1811.00995 (2018)
3. Brock, A., Donahue, J., Simonyan, K.: Large scale GAN training for high fidelity natural image synthesis. In: ICLR (2019)
4. Carlini, N., Wagner, D.: Towards evaluating the robustness of neural networks. In: IEEE Symposium on Security and Privacy (S&P) (2017)
5. Chapelle, O., Scholkopf, B., Zien, A.: Semi-supervised learning. IEEE Trans. Neural Netw. **20**(3), 542–542 (2009)
6. Chen, R.T., Behrmann, J., Duvenaud, D., Jacobsen, J.H.: Residual flows for invertible generative modeling. arXiv preprint arXiv:1906.02735 (2019)
7. Dempster, A.P., Laird, N.M., Rubin, D.B.: Maximum likelihood from incomplete data via the EM algorithm. J. Roy. Stat. Soc.: Ser. B (Methodol.) **39**(1), 1–22 (1977)
8. Du, Y., Mordatch, I.: Implicit generation and generalization in energy-based models. arXiv preprint arXiv:1903.08689 (2019)
9. Gao, R., Nijkamp, E., Kingma, D.P., Xu, Z., Dai, A.M., Wu, Y.N.: Flow contrastive estimation of energy-based models. In: IEEE/CVF Conference on Computer Vision and Pattern Recognition (CVPR) (2020)
10. Goodfellow, I., Shlens, J., Szegedy, C.: Explaining and harnessing adversarial examples. In: International Conference on Learning Representations (ICLR) (2015)
11. Grathwohl, W., Wang, K.C., Jacobsen, J.H., Duvenaud, D., Norouzi, M., Swersky, K.: Your classifier is secretly an energy based model and you should treat it like one. In: International Conference on Learning Representations (ICLR) (2020)
12. Grathwohl, W., Wang, K.C., Jacobsen, J.H., Duvenaud, D., Zemel, R.: Learning the stein discrepancy for training and evaluating energy-based models without sampling. In: Proceedings of the 37th International Conference on Machine Learning (ICML) (2020)
13. Guo, C., Pleiss, G., Sun, Y., Weinberger, K.Q.: On calibration of modern neural networks. In: Proceedings of the 34th International Conference on Machine Learning, vol. 70, pp. 1321–1330. JMLR. org (2017)
14. He, K., Zhang, X., Ren, S., Sun, J.: Deep residual learning for image recognition. In: IEEE Conference on Computer Vision and Pattern Recognition (CVPR) (2016)
15. Hendrycks, D., Gimpel, K.: A baseline for detecting misclassified and out-of-distribution examples in neural networks. In: International Conference on Learning Representations (ICLR) (2016)
16. Heusel, M., Ramsauer, H., Unterthiner, T., Nessler, B., Hochreiter, S.: GANs trained by a two time-scale update rule converge to a local nash equilibrium. In: Advances in Neural Information Processing Systems, pp. 6626–6637 (2017)
17. Hinton, G.E.: Training products of experts by minimizing contrastive divergence. Neural Comput. **14**(8), 1771–1800 (2002)
18. Krizhevsky, A., Hinton, G.: Learning multiple layers of features from tiny images. Technical report, Citeseer (2009)
19. Kurakin, A., Goodfellow, I., Bengio, S.: Adversarial machine learning at scale. In: International Conference on Learning Representations (ICLR) (2017)
20. LeCun, Y., Chopra, S., Hadsell, R., Ranzato, M., Huang, F.: A tutorial on energy-based learning. Predicting Structured Data **1**(0) (2006)
21. Maaten, L.V.D., Hinton, G.: Visualizing data using t-SNE. J. Mach. Learn. Res. (JMLR) **9**(Nov), 2579–2605 (2008)
22. Madry, A., Makelov, A., Schmidt, L., Tsipras, D., Vladu, A.: Towards deep learning models resistant to adversarial attacks. In: ICLR (2018)

23. Mnih, A., Teh, Y.W.: A fast and simple algorithm for training neural probabilistic language models. In: International Conference on Machine Learning (2012)
24. Nalisnick, E., Matsukawa, A., Teh, Y.W., Gorur, D., Lakshminarayanan, B.: Do deep generative models know what they don't know? arXiv preprint arXiv:1810.09136 (2018)
25. Nalisnick, E., Matsukawa, A., Teh, Y.W., Lakshminarayanan, B.: Detecting out-of-distribution inputs to deep generative models using a test for typicality. arXiv preprint arXiv:1906.02994 (2019)
26. Netzer, Y., Wang, T., Coates, A., Bissacco, A., Wu, B., Ng, A.Y.: Reading digits in natural images with unsupervised feature learning (2011)
27. Nijkamp, E., Hill, M., Han, T., Zhu, S.C., Wu, Y.N.: On the anatomy of MCMC-based maximum likelihood learning of energy-based models. arXiv preprint arXiv:1903.12370 (2019)
28. Nijkamp, E., Zhu, S.C., Wu, Y.N.: On learning non-convergent short-run MCMC toward energy-based model. arXiv preprint arXiv:1904.09770 (2019)
29. Pang, T., Du, C., Zhu, J.: Max-mahalanobis linear discriminant analysis networks. In: ICML (2018)
30. Pang, T., Xu, K., Dong, Y., Du, C., Chen, N., Zhu, J.: Rethinking softmax cross-entropy loss for adversarial robustness. In: ICLR (2020)
31. Robbins, H., Monro, S.: A stochastic approximation method. Ann. Math. Stat. 400–407 (1951)
32. Salimans, T., Goodfellow, I., Zaremba, W., Cheung, V., Radford, A., Chen, X.: Improved techniques for training GANs. In: NeurIPS (2016)
33. Santurkar, S., Ilyas, A., Tsipras, D., Engstrom, L., Tran, B., Madry, A.: Image synthesis with a single (robust) classifier. In: Advances in Neural Information Processing Systems (2019)
34. Szegedy, C., et al.: Intriguing properties of neural networks. In: International Conference on Learning Representations (ICLR) (2014)
35. Tsipras, D., Santurkar, S., Engstrom, L., Turner, A., Madry, A.: Robustness may be at odds with accuracy. arXiv preprint arXiv:1805.12152 (2018)
36. Vahdat, A., Kautz, J.: NVAE: a deep hierarchical variational autoencoder. In: Neural Information Processing Systems (NeurIPS) (2020)
37. Wan, W., Zhong, Y., Li, T., Chen, J.: Rethinking feature distribution for loss functions in image classification. In: Proceedings of the IEEE Conference on Computer Vision and Pattern Recognition (CVPR), pp. 9117–9126 (2018)
38. Welling, M., Teh, Y.W.: Bayesian learning via stochastic gradient langevin dynamics. In: ICML, pp. 681–688 (2011)
39. Xie, J., Lu, Y., Zhu, S.C., Wu, Y.: A theory of generative convnet. In: International Conference on Machine Learning, pp. 2635–2644 (2016)
40. Zagoruyko, S., Komodakis, N.: Wide residual networks. In: The British Machine Vision Conference (BMVC) (2016)
41. Zhao, S., Jacobsen, J.H., Grathwohl, W.: Joint energy-based models for semi-supervised classification. In: ICML 2020 Workshop on Uncertainty and Robustness in Deep Learning (2020)

Gaussian Process Encoders: VAEs with Reliable Latent-Space Uncertainty

Judith Bütepage$^{(\boxtimes)}$ ⓘ, Lucas Maystre ⓘ, and Mounia Lalmas

Spotify, Stockholm, Sweden
{judithb,lucasm,mounia}@spotify.com

Abstract. Variational autoencoders are a versatile class of deep latent variable models. They learn expressive latent representations of high dimensional data. However, the latent variance is not a reliable estimate of how uncertain the model is about a given input point. We address this issue by introducing a sparse Gaussian process encoder. The Gaussian process leads to more reliable uncertainty estimates in the latent space. We investigate the implications of replacing the neural network encoder with a Gaussian process in light of recent research. We then demonstrate how the Gaussian Process encoder generates reliable uncertainty estimates while maintaining good likelihood estimates on a range of anomaly detection problems. Finally, we investigate the sensitivity to noise in the training data and show how an appropriate choice of Gaussian process kernel can lead to automatic relevance determination.

Keywords: Variational Autoencoder · Uncertainty estimation · Anomaly detection · Gaussian process

1 Introduction

Generative models can represent a joint probability distribution over observed and latent variables. Modern generative models often combine the representational power of deep neural networks with the structured representations encoded by probabilistic graphical models [12]. One popular class of deep latent variable models are Variational Autoencoders (VAEs) [14,22]. VAEs generate samples of the data distribution by transforming a sample from a simple noise distribution, the prior, into an output distribution in data space with the help of a neural network (NN), the decoder network. To determine the latent variable distribution for a given data point, an encoder network, representing the approximate posterior, is used to determine the form of the latent variable of each data point. VAEs are trained using the Evidence Lower BOund (ELBO), which regularizes the data likelihood under the approximate posterior with the Kullback-Leibler divergence (KL) between the approximate posterior and a prior distribution.

While this inference scheme usually works well for the mean parameter of the latent variable, it often fails to learn an informative variance parameter for each data point [3]. In many cases, the latent variance fails to correlate with how

© Springer Nature Switzerland AG 2021
N. Oliver et al. (Eds.): ECML PKDD 2021, LNAI 12976, pp. 84–99, 2021.
https://doi.org/10.1007/978-3-030-86520-7_6

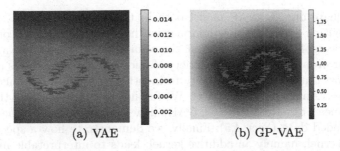

Fig. 1. The latent variance of a (a) VAE encoder and (b) GP-VAE encoder trained on the two-moon dataset and evaluated over a grid of points around the training data points. The red dots visualize the latent mean of the training data. (Color figure online)

uncertain the model is about the input. An example of this is shown in Fig. 1(a), which depicts the latent variance estimates of a VAE trained on the two-moon dataset [15]. Contrary to expectation, the uncertainty is decaying the further away from the training data we evaluate.

This behavior becomes problematic when one relies on the estimates of the latent uncertainty. For example in reinforcement learning, when sampling in the latent space of a temporal VAE to predict the next observation given an uncertain input, it is important to sample a variety of possible futures [7]. Note that the problem of modelling latent uncertainty is different to modelling an accurate variance in the data space using the decoder as discussed in e.g. [24]. The output uncertainty of a VAE centers around a single data point, while a high latent variance produces a larger variety of samples. Another example that requires reliable estimates of the latent uncertainty is anomaly or out-of-distribution (OOD) detection using generative models. As demonstrated in [9,17] the likelihood distribution of a VAE cannot reliably detect OOD data. Variance estimates in the latent space can be an alternative.

In this work, we extend VAEs to enable reliable latent uncertainty estimates for in-distribution (ID) and OOD data. We replace the neural network encoder traditionally used in VAEs with a Gaussian process (GP) encoder (Sect. 2.1). We refer to this model as a GP-VAE. Our formulation considers the GP encoder as a drop-in replacement of neural networks. With this, we retain the versatility of VAEs while gaining several advantages: reliable uncertainty estimates, reduced overfitting, and increased robustness to noise. For scalability, we parameterize the GP by using a small number of inducing points, similar to sparse variational GPs [26]. This enables us to learn a compact mapping from the data space to the latent space.

The GP encoder learns to represent data points in the latent space based on their similarity, by using a kernel function. It produces principled uncertainty estimates as shown for the moon dataset in Fig. 1(b). In contrast to the standard VAE in Fig. 1(a), the latent variance of the GP-VAE increases with the distance to the latent means of the training data.

We evaluate our GP-VAE model both in terms of how well it fits the ID data distribution as well as how informative the latent variance is compared to baselines (Sects. 4.1 and 4.2). To test the reliability of the latent variance, we evaluate our proposed model on a variety of anomaly detection tasks (Sect. 4.3). We also show that the model's ability to identify OOD data is robust to OOD noise in the training data (Sect. 4.4). We further demonstrate how the reliable latent variance estimates of the GP-VAE allow for meaningful synthesized variants of encoded data (Sect. 4.5). Finally, we demonstrate how a specific choice of the GP kernel, namely an additive kernel, leads to interpretable models and allows us to identify which input features are important to distinguish between ID and OOD data (Sect. 4.6).

1.1 Contributions

Our contributions are threefold: a) we introduce a Gaussian process encoder for VAEs that infers reliable uncertainty estimates in the latent space of a VAE; b) we derive a scalable inference scheme for the GP encoder using a set of inducing points; and finally c) we describe how to use additive kernels to create interpretable models that can be used to identify features that distinguish ID from OOD data.

2 Background

We begin by introducing the general ideas behind variational autoencoders. We then discuss why the common choice of a neural network encoder leads to poor latent uncertainty estimates. In Sect. 3.2 we relate back to this section and discuss the implications of replacing the NN with a GP encoder after having introduced the GP encoder formally in Sect. 3.1.

2.1 Variational Autoencoder

Let $\{x_1, \ldots, x_N\}$ be a set of N data points, where $x_n \in \mathbf{R}^K$, and $X \in \mathbf{R}^{N \times K}$ be the collection of data points as a stacked matrix. We construct a generative model of the data with parameters θ that maximizes the data log-likelihood $\log p_\theta(X)$. We follow the same generative model as assumed for VAEs, i.e., that each data point x is generated independently and is conditioned on a latent variable $z \in \mathbf{R}^D$, where typically $D \ll K$.

$$p_\theta(X) = \prod_n \int p_\theta(x_n \mid z)p(z)dz. \tag{1}$$

As commonly assumed for VAEs, we assume an i.i.d. zero-mean and spherical Gaussian prior for the latent variables $p(z) = \mathcal{N}(z \mid 0, \sigma^2 I)$, and model the conditional likelihood of the data by using $p_\theta(x \mid z) = \mathcal{N}[\mu_\theta(z), \sigma_\theta^2(z)]$, where $\mu_\theta(\cdot)$ and $\sigma_\theta^2(\cdot)$ are feed-forward neural networks. The choice of the likelihood distribution depends on the dataset and does not need to be Gaussian. When generating the data, we first sample a latent variable from the prior $p(z)$ and subsequently sample a data point using the decoder $p_\theta(x \mid z)$.

Directly maximizing Eq. 1 (or the log thereof) is intractable. Therefore, following the derivation of VAEs, we use Jensen's inequality to derive the evidence lower-bound

$$\log p_\theta(X) \geq \sum_n \{ \mathbf{E}_{q(z|x_n)} [\log p_\theta(x_n \mid z)] - \mathrm{KL}[q(z \mid x_n)\|p(z)]\}, \qquad (2)$$

where $q(z \mid x)$ is an auxiliary distribution that approximates the posterior distribution $p(z \mid x)$ [14]. This variational distribution is commonly chosen to be a Gaussian with diagonal covariance where the mean and covariance matrix are functions of the input data point:

$$q(z \mid x) = \mathcal{N}\left\{ \boldsymbol{\mu}_\phi(x), \mathrm{diag}[\boldsymbol{\sigma}_\phi^2(x)]\right\}. \qquad (3)$$

These functions are paramterized by ϕ. As such, we can think of the functions $\boldsymbol{\mu}_\phi(\cdot)$ and $\boldsymbol{\sigma}_\phi^2(\cdot)$ as *encoding* the data point into the latent space. In the literature on VAEs, these functions are usually chosen to be neural networks, as illustrated in Fig. 3(a).

Training a VAE entails maximizing Eq. (2) over θ and ϕ. In contrast to Eq. (1), this can be done efficiently, provided that the encoder and decoder are differentiable. We refer to [13] for more background on VAEs.

2.2 Latent Variance Estimates of NN

Neural Network encoders can exhibit different learning behaviors when optimizing the ELBO. For example, one common phenomenon occurs when the KL divergence in the ELBO is too strong in the early stages of training, which then forces the approximate posterior to be equal to the prior. This can impact either all latent dimensions, called posterior collapse [8], or a subset of latent dimensions, called the dying units problem [31].

A second behavior discussed by [3] is that the ELBO pushes both the encoder and decoder variance to zero to achieve minimal reconstruction errors. Consider the KL divergence part of the ELBO in Equation (2). In the case of Gaussian latent variables with a standard normal prior we have $\mathrm{KL}[q(z \mid x_n)\|p(z)] = \frac{1}{2}\sum_{d=1}^{D}[\sigma_{d,\phi}^2(x) + \mu_{d,\phi}(x)^2 - 1 - \log(\sigma_{d,\phi}(x)^2)]$. The part of the KL concerned with the latent variance rarely dominates the ELBO even when the latent variance values are relatively small. Therefore, a NN encoder can neglect that part of the KL divergence and set the latent variance to very small values. This allows the model to focus on maximizing the expected log likelihood while simultaneously minimizing the distance of the latent means and the prior mean as dictated by the KL, i.e. $\mu_{d,\phi}(x)^2$. We can view this vanishing latent variance as a form of overfitting. This behavior is illustrated in Fig. 2(a), which depicts 250 encoded mean and variance values of a VAE trained on the FashionMnist dataset [30]. The latent mean is clustered around the prior mean, namely zero, while the latent variance is too small to be visible.

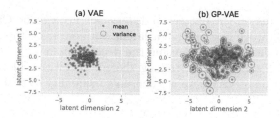

Fig. 2. Latent space mean (blue) and variance (red) of 250 data points of the Fashion-Mnist dataset. (a) The latent variance of the VAE is too small to be visible. (b) The GP-VAE spreads its probability mass to accommodate larger latent variance estimates. (Color figure online)

2.3 Mismatch Between the Prior and Approximate Posterior

The preference of the VAE to minimize the reconstruction error can lead to inaccurate inference, where the approximate posterior aggregated over all training data points is no longer equal to the prior. When this is the case, sampling from the prior will not reconstruct the entire data distribution. The authors of [3] propose a two-stage approach that consists of a standard VAE and a second generative model trained to emulate samples from the aggregated approximate posterior. Instead of a two-stage solution, it has also been proposed to approximate the aggregated posterior by a mixture of variational posteriors with pseudo-inputs during training [27].

An alternative solution to the misalignment between prior and the aggregated approximate posterior is proposed by [32] who suggest to add an additional loss term, namely the mutual information between the data and the latent variables. In their experiments, they remove the KL term from the ELBO altogether and implement the mutual information as the Maximum Mean Discrepancy (MMD). They show that the aggregated approximate posterior of their model, the InfoVAE, is closer to the prior compared to the VAE's. However, this is only attributable to the latent mean, not the latent variance which converges to zero. Since the MMD is only concerned with samples from the latent distribution and not the latent variance parameters, the InfoVAE is free to set the latent variance essentially to zero.

3   Methodology

We suggest to use the predictive equations of a Gaussian process to parametrize the encoder. At a high level, we define the encoder as the posterior distribution of a Gaussian process, given a small number of pseudo-observations that we treat as parameters. The variance in the latent space is thus estimated in a principled way, by using Bayes' rule in a well-defined probabilistic model. In contrast to using a neural network encoder, using a GP encoder leads to inductive biases that prevent the variance from vanishing.

In this section we describe how we replace a neural network encoder with a sparse Gaussian process. In short, we learn a number of representative data

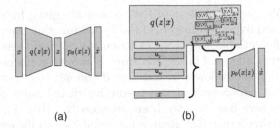

(a)   (b)

Fig. 3. Model structure: **(a)** The usual structure of a VAE with neural networks as encoder and decoder. **(b)** Our proposed VAE with a sparse GP as encoder and a neural network as decoder.

points, so called inducing points, and their corresponding value in the latent space. This allows us to employ a Gaussian process to infer the latent distribution for ID data points, which in turn can be decoded by a neural network into the data space.

3.1   Gaussian Process Encoder

In contrast to the common formulation of VAEs, we propose using a sparse Gaussian process encoder instead of a neural network, as depicted in Fig. 3(b). We can view the encoder as the output of an auxiliary Gaussian process recognition model $\hat{z}(x)$ that maps from the data space to the latent space. We assume, for simplicity, that the D dimensions of the multivariate GP's output are a priori identically and independently distributed, i.e.,

$$\hat{z}(x) \sim \mathcal{GP}[0, k(x, x')I], \tag{4}$$

where $k(x, x')$ is a positive-definite kernel. We assume that the correlation between data points can be captured by a set of M pseudo-inputs or inducing points $u_1, \ldots, u_M \in \mathbf{R}^K$. Suppose that, for each inducing point $m = 1, \ldots, M$, we observe that, at input u_m, the GP has value $y_m \in \mathbf{R}^D$ with Gaussian measurement noise $v_m \in \mathbf{R}^D_{>0}$. Then, we can formalize the distribution of $\hat{z}$ at a new input x as the predictive distribution of a GP.

To this end, we stack these vectors into matrices $Y, V \in \mathbf{R}^{M \times D}$ and denote by y^d and v^d the dth column of the respective matrix. Finally, let $K = [k(u_m, u_{m'})]$ be the $M \times M$ matrix obtained by evaluating the kernel at each pair of inducing points.

Given these, we define the dth dimension of the encoder (Eq. (3)) by

$$\mu_{d,\phi}(x) = k^\top \left[K + \operatorname{diag}(v^d) \right]^{-1} y^d,$$
$$\sigma^2_{d,\phi}(x) = k(x, x) - k^\top \left[K + \operatorname{diag}(v^d) \right]^{-1} k, \tag{5}$$

where $k = [k(x, U)]$. As in Eq. (3), we therefore have can formalize the encoder as $q(z \mid x) = \mathcal{N}(\mu_\phi(x), \operatorname{diag}[\sigma^2_\phi(x)])$, where $\mu_\phi(x)$ and $\sigma^2_\phi(x)$ are the stacked outputs of Eq. (5). The parameters of the encoder are given by $\phi = \{U, Y, V\}$, to which we can add hyperparameters of the kernel function. Since the functions in Eq. (5) are differentiable with respect to ϕ, we can use them to replace the neural networks typically used in the literature. The VAE is trained as previously, by optimizing the evidence lower bound over θ and ϕ.

Optimizing the parameters of the predictive distribution of a sparse GP directly as we do here has recently been proposed by [10]. The authors show that Parametric Predictive GP Regression exhibits significantly better calibrated uncertainties than e.g. Fully Independent Training Conditional approximations [10,25]

Aggregated Approximated Posterior. As discussed in Sect. 2.3, to sample from VAEs effectively, we need to make use of the aggregated approximated posterior, when it is not equal to the prior. Our formulation of a sparse VAE lends itself to represent the aggregated approximated posterior using the inducing points as

$$\hat{q}(z) = \frac{1}{M} \sum_m q(z \mid u_m) = \frac{1}{M} \sum_m \mathcal{N}(y_m, \operatorname{diag}(v_m)). \tag{6}$$

This formulation is similar to the aggregated approximated posterior using pseudo-inputs proposed by [27]. Contrary to a neural network encoder, the sparse GP learns these pseudo-inputs in form of the inducing points automatically.

3.2 The Implications of a Gaussian Process Encoder

As discussed in Sect. 2.2, neural networks can fail to learn reliable uncertainty estimates. The latent variance of a GP encoder on the other hand reflects how close a data point is to the training data or inducing points. It is therefore constraint in its ability to set the latent variance to zero. To accommodate reliable latent variance estimates while minimizing the reconstruction error, the GP-VAE needs to minimize both the latent mean and the latent variance part of the KL. The loss connected to the variance will be smaller, allowing the KL loss connected to the mean to be larger, i.e. the latent mean spreads more than in the case of a VAE. As shown in Fig. 2(b), the GP-VAE trained on the FashionMnist dataset spreads the probability mass of the approximate posterior more than encouraged by the prior, while maintaining reliable latent variance estimates. Thus, to reconstruct the data distribution we cannot sample from the prior, falling back to the problem discussed in Sect. 2.3. In contrast to a two-stage solution however, the use of inducing points in the GP encoder allows us to formulate an aggregated approximate posterior without the need of fitting a separate model (see Sect. 3.1).

3.3 Out-of-Distribution Detection

Given a test dataset $\{x_1^*, \ldots, x_{N_t}^*\}$, we want to evaluate whether the latent variance can be used to determine whether a data point has been drawn from the same distribution as the training data or is a OOD data point. To this end, we average over the D values of the diagonal of covariance matrix $\boldsymbol{\sigma}_n^* \doteq \frac{1}{D}\sum_{d=1}^{D}\sigma_\phi^2(\boldsymbol{x}_n^*)^d$ of the Gaussian approximate posterior $q(\boldsymbol{z} \mid \boldsymbol{x}_n^*)$ in Eq. (5). These values can be interpreted as a measure of how uncertain the model is about the data point and should therefore distinguish between ID and OOD test data. In the experiments, we report the area under the Receiver Operating Characteristics roc and Precision-Recall curve pr for $\{\boldsymbol{\sigma}_1^*, \ldots, \boldsymbol{\sigma}_{N_t}^*\}$. To use the values for OOD detection in practice, one needs to determine an appropriate threshold, e.g. with a small number of labeled data points.

4   Experiments

We evaluate the GP-VAE both in terms of how well it models the training data distribution as well as how reliable the estimates of the latent variance are compared to other models. We evaluate our approach on a number of OOD detection datasets (Sects. 4.1, 4.2 and 4.3). To test for the robustness of the latent variance, we introduce OOD examples into the training ID data and analyze how this noise impacts OOD detection on the test data (Sect. 4.4). We also demonstrate that informative uncertainty estimates can be used to generate a diverse set of data samples (Sect. 4.5). Finally, we describe how the choice of an additive kernel function leads to an interpretable model that allows identifying the input features contributing to OOD detection (Sect. 4.6).

Datasets. To compare the ability of our model to detect OOD data samples, we run extensive experiments on the Outlier Detection DataSets (ODDS) Library [20]. This library contains a number of datasets from different domains. We test our approach on 20 datasets in the ODDS library. Each dataset consists of ID data points and OOD data points. We create an ID training set by randomly selecting half of the ID data points and test on the remaining ID data points and the OOD data points. The exact specifics for the datasets are described in the Appendix. We also test our approach on an image dataset, namely the FashionMnist (FM) [30] and Mnist (M) [16] dataset, as done in [17].

Model Specifics. We compare our GP-VAE model to baselines trained under the same conditions as ours. The two generative models that we compare to are the standard VAE and the InfoVAE [32]. The encoder and decoder neural networks of all VAE-based models consist of two fully connected layers and one latent variable layer. As a kernel function for the GP-VAE, we use the squared exponential kernel. More specifics on the model architecture and training protocols can be found in the Appendix. Additionally, we train a recent supervised approach based on deviation networks (Dev-Net) [19] on all datasets as a state-of-the-art baseline for OOD detection. We evaluate the models after training the VAE-based models for 200 epochs and the Dev-Net for 50 epochs and repeat the experiment with five different seeds.

4.1   Log Likelihood

We start by comparing the ability to model the ID data distribution by computing the log likelihood for all test datasets. We approximate the log likelihood with the help of importance sampling by

$$\log(p(\boldsymbol{x}_n)) = \log\left(\mathbf{E}_{q(z|\boldsymbol{x}_n)}\left[p_\theta(\boldsymbol{x}_n \mid z)\frac{p(z)}{q(z \mid \boldsymbol{x}_n)}\right]\right). \tag{7}$$

As shown in the first three columns in Table 1, the GP-VAE performs similar to the VAE and InfoVAE in terms of likelihood estimates. The small latent variance values of the VAE and InfoVAE result in high denominator values in the above equation, which can impact their likelihood estimate negatively. The standard deviations across runs and reconstruction errors of all three models for the OOD datasets are presented in the Appendix.

4.2   Uncertainty in the Latent Space

We introduced the GP encoder as a means to reliably express uncertainty about unfamiliar data points. A model should be more uncertain about data points that have low similarity to the training data compared to more typical examples. Thus, we expect a positive correlation between the average distance of a data point to all training data points and its latent variance. To test how well the different VAE-based models follow this behavior, we train an VAE, InfoVAE and GP-VAE on all 20 datasets selected from the ODDS. We then compute the average euclidean distance of each ID and OOD test data point to all data points in the ID training dataset and infer the latent variance values using the specific encoders. Finally we compute the Pearson correlation coefficient (PCC) between the distances and the latent variances. As shown in Fig. 4, both the VAE and InfoVAE fail to capture the similarity of a data point to the training data as the average correlation is PCC $= -0.069$ and PCC $= -0.0019$ respectively. In contrast, the GP-VAE correlates positively with an average of PCC $= 0.48$.

4.3   Benchmarking OOD Detection

To test whether the latent variance reliably indicates if a data point is ID or OOD, we use the latent variance values for OOD detection. We repeat model

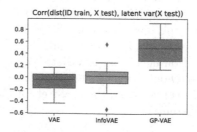

Fig. 4. The Pearson correlation coefficient between the latent variance of each test point (X test) and the average distance between X test and the all data points in the training dataset (ID train).

Table 1. OOD detection performance of the GP-VAE, a standard VAE, a InfoVAE and deviation network models on the OODS datasets and FashionMnist vs Mnist. We present both roc and pr. Bold values indicate the best performing OOD detection mechanism.

Metric	log(p(x))			roc				pr			
	VAE	InfoVAE	GP-VAE	VAE	InfoVAE	GP-VAE	Dev-Net	VAE	InfoVAE	GP-VAE	Dev-Net
annthyroid	-47.44	-49.96	**-8.73**	0.64	0.36	**0.94**	0.68	0.32	0.16	**0.67**	0.28
arrhythmia	-306.86	-307.26	**-257.96**	0.62	0.61	**0.81**	0.50	0.55	0.53	**0.69**	0.26
breastw	**-12.73**	-13.06	-13.86	0.44	0.66	**0.99**	0.94	0.62	0.76	**0.99**	0.92
cardio	-28.10	-28.86	**-24.89**	0.31	0.27	**0.98**	0.75	0.13	0.15	**0.93**	0.55
glass	-12.79	**-12.02**	-15.59	0.39	0.61	0.84	**0.86**	0.12	0.17	0.31	**0.32**
ionosphere	-68.74	-75.78	**-36.69**	0.79	0.79	**0.96**	0.81	0.86	0.84	**0.97**	0.80
letter	-98.08	-99.03	**-49.47**	0.60	0.66	**0.85**	0.58	0.25	0.31	**0.49**	0.16
lympho	-37.13	-38.10	**-18.10**	0.55	0.60	**0.89**	0.66	0.34	0.36	**0.66**	0.31
mnist	-132.50	-132.60	**-122.48**	0.59	0.56	**0.96**	0.67	0.40	0.34	**0.86**	0.39
musk	-207.80	-208.09	**-182.41**	0.01	0.01	**1.00**	0.84	0.03	0.03	**1.00**	0.63
optdigits	-92.59	-93.71	**-68.35**	0.61	0.63	0.98	**0.99**	0.09	0.11	0.66	**0.95**
pendigits	**-21.45**	-23.60	-23.22	0.35	0.38	**1.00**	0.93	0.04	0.08	**0.98**	0.79
pima	**-7.53**	-8.12	-13.69	0.47	0.41	**0.70**	0.55	0.52	0.48	**0.69**	0.55
satellite	-167.27	-197.56	**-59.73**	0.48	0.49	0.80	0.63	0.53	0.53	**0.86**	0.61
satimage	-155.36	-153.43	**-59.08**	0.30	0.09	**1.00**	0.82	0.02	0.01	**0.98**	0.58
shuttle	**-10.97**	-11.28	-24.38	0.92	0.85	**1.00**	0.97	0.74	0.69	**0.98**	0.95
thyroid	-23.83	-16.54	**-9.18**	0.55	0.71	**0.98**	0.90	0.28	0.49	**0.80**	0.69
vertebral	-9.14	**-8.43**	-9.84	0.38	0.53	0.59	**0.66**	0.20	0.30	0.27	**0.33**
wbc	**-33.07**	-34.25	-45.16	0.29	0.02	**0.97**	0.78	0.25	0.06	**0.84**	0.52
wine	-14.72	**-14.35**	-22.24	0.22	0.34	**0.97**	0.77	0.10	0.22	**0.84**	0.53
FashionMnist	-790.66	-805.52	**-753.12**	0.13	0.23	0.93	**0.97**	0.33	0.37	0.89	**0.97**

training after randomly splitting the ID data into train and test set with five random seeds and report the average performance. The standard deviations across runs are depicted in the Appendix. Compared to supervised OOD methods, we do not actively train the VAE-based models to detect OOD data.

The area under the Receiver Operating Characteristics *roc* and Precision-Recall curve *pr* values are presented in column 4–11 of Table 1. The bold values indicate the best performing models. A general observation is that the GP-VAE outperforms the other VAE-based models on all datasets. This indicates that the GP-VAE produces more reliable latent variance estimates than the standard VAE and InfoVAE. The GP-VAE also outperforms the supervised Dev-Net on 17 out of the 20 OOD datasets. This suggests that our approach is well suited for OOD detection while not requiring labeled training data.

FashionMnist vs Mnist. We train each model on the training data of FashionMnist (FM) and test on the test data of FashionMnist and Mnist (M). As before, we compare our approach to a standard VAE, the InfoVAE and Dev-Net. The *roc* and *pr* values are listed in the last row of Table 1. We see that the latent variance of the standard VAE and InfoVAE have no discriminative power. To understand this discrepancy, we depict the VAE's and GP-VAE's latent uncertainty values over the FashionMnist training and testing data and the Mnist testing data in Fig. 5. While the OOD data (Mnist) has lower latent uncertainty values than the ID data in the case of the standard VAE, the GP-VAE assigns higher latent uncertainty values to the OOD data compared to the ID data. The extreme behavior of the VAE might be explained by similar arguments as

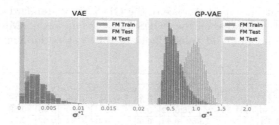

Fig. 5. Histograms of latent uncertainties σ^{*1} of the first latent dimension generated by the GP-VAE and the VAE models trained on the FM dataset.

brought forward in [21]; from a statistical viewpoint the Mnist density lies within the FashionMnist density with lower variance of pixel values. The NN encoder of the VAE might be influenced by these low-level statistics and therefore underestimate the uncertainty over OOD data points.

4.4   OOD Polution of the Training Data

While we investigated whether the latent uncertainty estimates of the GP-VAE can reliably distinguish between ID and OOD data in Sect. 4.3, in this section we look at how stable the latent uncertainty estimates are in the presence of OOD noise in the ID training data. We therefore analyze how the GP-VAE reacts to different levels of data pollution and compare the behavior to standard VAEs. To this end, we take a look at the Breast Cancer Wisconsin (Original) dataset (breastw) in the ODDS library. The dataset consists of 444 ID data points (split into 222 training and 222 testing points) and 239 OOD data points. Except for introducing OOD noise into the training data, we keep all other settings the same as described in Sect. 4.3.

We compare the behavior of the GP-VAE to the VAE in Fig. 6 by removing 0, 10, 50, 100 and 200 OOD samples from the testing data and adding these to the ID training data. This data split is performed randomly, with ten different seeds over which we average the performance in terms of OOD detection (measured in *roc* and *pr*). The standard deviation between trials is visualized by the

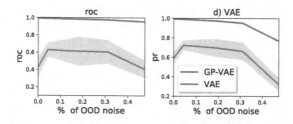

Fig. 6. Performance of the GP-VAE compared to a VAE under the influence of noise in the training data. We present *roc* (left) and *pr* values (right) average and variance over ten random seeds. The between-run variance of the GP-VAE is very small.

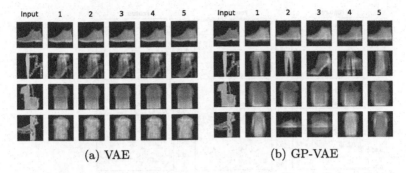

Fig. 7. Decoded samples from the encoded latent variables for the Input image (left row). The top row shows an intact image of the FM dataset. The three bottom rows show OOD images.

colored shadows in Fig. 6. We see that the GP-VAE is less susceptible to noise. In addition, the performance of the VAE is random seed dependent, which causes the large variations between trials while the across-trial variance of the GP-VAE is too small to be visible in the Figure. This implies that the GP-VAE's reaction to noise is more stable across different noise samples.

4.5   Synthesizing Variants of Input Data

As described in the introduction, VAEs are used to generate different variants of an encoded data point. We would expect the model to generate a larger variety when it is uncertain about the input. To demonstrate how the GP-VAE compares to a VAE in that regard, we use the FM dataset. We generate OOD data points by concatenating two halves of different FM images as shown in the three bottom rows of the *Input* column in Fig. 7. We then sample variants of these inputs as shown in columns 1–5 in Fig. 7. For comparison, we also sample around an ID image from the FM dataset in the top row. We can see that the VAE fails to sample a larger variety for OOD data as the latent variance is not reliably expressing uncertainty. In contrast, the GP-VAE samples a larger variety and even generates samples from both fashion items that constitute the input image, e.g. a high heel or trousers in the second row.

4.6   Interpretable Kernels

One advantage of GPs is that we are free to choose the kernel function. Building on [5], we use additive kernels of the form $k(\boldsymbol{x}, \boldsymbol{x}') = \sum_d k^d(\boldsymbol{x}^d, \boldsymbol{x}'^d)$, that is, summing over separate kernels for each feature. This gives rise to interpretable models and automatic relevance determination as each feature dimension is treated independently. We use the same approach to determine which features distinguish ID and OOD data on the *thyroid* dataset. We choose a squared exponential kernel for each input feature dimension and keep the remaining settings

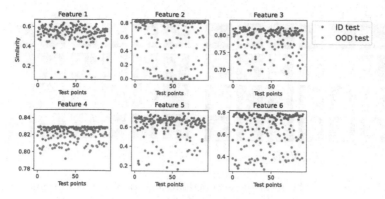

Fig. 8. The similarity$^d = \sum_j k^d(\boldsymbol{x}^d, \hat{\boldsymbol{x}}_j^d)$ score for ID and OOD test data points of the *thyroid* dataset.

the same. To determine how close a data point is to the feature representation learned by the model, we compute the similarity between dimension d of point $\boldsymbol{x}$ and the inducing points $similarity^d = \sum_m k^d(\boldsymbol{x}^d, \boldsymbol{u}_m^d)$. We hypothesize that feature dimensions that are important for OOD detection, meaning in which ID and OOD data points differ, will also differ in terms of $similarity^d$.

We analyze whether this hypothesis holds for the *thyroid* dataset, which consists of six continuous features describing hormone levels. To determine the importance of each feature to detect OOD, we train a random forest, a decision tree and logistic regression to classify ID and OOD data in a supervised fashion. We normalize the feature importance values determined by these models and average over the three models (see the Appendix for all values). This gives us the feature importance values of $\boldsymbol{x}^1 = 0.01$, $\boldsymbol{x}^2 = 0.39$, $\boldsymbol{x}^3 = 0.11$, $\boldsymbol{x}^4 = 0.19$, $\boldsymbol{x}^5 = 0.04$, $\boldsymbol{x}^6 = 0.26$, which indicates that the hormone features 2 and 6 are important for OOD detection.

When inspecting the similarity scores in Fig. 8, it becomes apparent that these features also exhibit abnormal behavior as determined by the additive kernel. The similarity score of the OOD data is lower than for the ID data for these features. To quantify these findings, we fit a Gaussian distribution to the ID similarity values of each feature and compute the log likelihood of the OOD similarity values under each Gaussian. This gives us the feature importance values of $ll^1 = 83.53$, $ll^2 = -6151.94$, $ll^3 = 120.36$, $ll^4 = 68.84$, $ll^5 = 49.52$, $ll^6 = -488.53$, where smaller values indicate higher importance.

By determining whether a set of test points behaves differently in the similarity space of each dimension, we gain a better understanding of which features contribute to OOD detection. This knowledge can be used to make subsequent decisions, e.g. when building a rule-based anomaly detection system.

5   Related Work

A number of works have identified the problem of small, uninformative latent variance values and proposed different ways to overcome them. One solution is to set the latent variance equal to the neural network output plus a small constant value [18]. In a similar manner, the encoded latent variance can be removed from the model and replaced by a constant that is treated as a hyperparameter [1]. While these approaches achieve the effect that one can sample different variants of an encoded data point, the variance is not reliably representing how uncertain the model is over a given data point. In contrast, our proposed GP encoder learns to express meaningful latent variance estimates for each data point. Another solution to the latent variance problem is to add an additional weighted constraint to the ELBO that keeps the latent variance values from vanishing [23]. However, this additional objective does not ensure reliable uncertainty estimates and comes with the additional burden of having to determine how to optimally weight the additional constraint in the loss function.

As our approach builds on Gaussian processes, we discuss different approaches combining Gaussian processes and VAEs proposed in recent years.

Deep Gaussian Processes have been employed as VAEs, replacing all neural network layers in the decoder network by Gaussian Processes [4] and using neural networks to infer the variational parameters of these GPs. We propose instead to replace the encoder by a GP to infer accurate uncertainty estimates while keeping the neural network structure of the decoder.

Similar to our work, variational Gaussian processes [28] warp samples from a simple distribution with a GP to model the approximate posterior, which in turn is regularized by an auxiliary distribution. Our approach relies on inducing points in the data space and therefore does not require additional auxiliary distributions.

To model correlations in the latent space, several works have introduced a GP prior reflecting structural correlations in the data, see e.g. [2,11]. It replaces the common choice of an i.i.d. standard normal prior. Since the encoder, or approximate posterior, is still driven by a neural network, obtaining useful latent-space variance estimates remains problematic. In contrast, our work keeps the i.i.d. standard normal prior but replaces the neural network encoder with a sparse GP approximate posterior. It is possible to obtain a model similar to (but distinct from) ours with a structured prior, as discussed in the Appendix.

Finally, [6] employ Gaussian processes for time-series imputation by modeling temporal dependencies in the latent space with a GP. However, the encoding and decoding is performed by neural networks. In contrast, we propose to substitute the encoder with a GP.

6   Conclusion

In this work, we introduced a Gaussian process encoder with sparse inducing points for Variational autoencoders. The combination of Gaussian processes and neural networks, as proposed in this paper, merges the advantages of GPs, such as the ability to encode structure through a kernel function and reliable uncertainty estimates, with the advantages coming with neural networks, such as efficient representation learning and scalability. Our experiments show that GP-VAEs have additional advantages over standard VAEs, such as robustness to noise and the freedom to choose the kernel function.

One disadvantage of GPs compared to neural networks is their limited capability of representing structured, high dimensional data such as images. The kernel function constraints the generalization capability of GPs to local metrics, such as the euclidean distance. However, novel kernel functions, such as convolutional kernels, can be used to operate even in image spaces [29]. In future work we plan to extend our approach to include such priors as well as temporal dynamics and data from multiple sources.

References

1. Braithwaite, D.T., Kleijn, W.B.: Bounded information rate variational autoencoders. In: KDD Deep Learning Day (2018)
2. Casale, F.P., Dalca, A., Saglietti, L., Listgarten, J., Fusi, N.: Gaussian process prior variational autoencoders. In: Advances in Neural Information Processing Systems, pp. 10369–10380 (2018)
3. Dai, B., Wipf, D.: Diagnosing and enhancing VAE models. In: International Conference on Learning Representations (2018)
4. Dai, Z., Damianou, A.C., González, J., Lawrence, N.D.: Variational auto-encoded deep gaussian processes. In: ICLR (Poster) (2016)
5. Duvenaud, D.: Automatic model construction with Gaussian processes. Ph.D. thesis, University of Cambridge (2014)
6. Fortuin, V., Rätsch, G., Mandt, S.: Multivariate time series imputation with variational autoencoders. In: 23rd International Conference on Artificial Intelligence and Statistics (AISTATS) (2020)
7. Ha, D., Schmidhuber, J.: Recurrent world models facilitate policy evolution. In: Advances in Neural Information Processing Systems, pp. 2450–2462 (2018)
8. He, J., Spokoyny, D., Neubig, G., Berg-Kirkpatrick, T.: Lagging inference networks and posterior collapse in variational autoencoders. In: International Conference on Learning Representations (2018)
9. Hendrycks, D., Mazeika, M., Dietterich, T.G.: Deep anomaly detection with outlier exposure. In: 7th International Conference on Learning Representations, ICLR 2019, New Orleans, LA, USA, 6–9 May 2019. OpenReview.net (2019)
10. Jankowiak, M., Pleiss, G., Gardner, J.: Parametric gaussian process regressors. In: International Conference on Machine Learning, pp. 4702–4712. PMLR (2020)
11. Jazbec, M., Ashman, M., Fortuin, V., Pearce, M., Mandt, S., Rätsch, G.: Scalable gaussian process variational autoencoders. In: International Conference on Artificial Intelligence and Statistics, vol. 130, pp. 3511–3519. PMLR (2021)
12. Johnson, M.J., Duvenaud, D.K., Wiltschko, A., Adams, R.P., Datta, S.R.: Composing graphical models with neural networks for structured representations and fast inference. In: Advances in Neural Information Processing Systems, pp. 2946–2954 (2016)

13. Kingma, D.P., Welling, M.: An introduction to variational autoencoders. Found. Trends® Mach. Learn. **12**(4), 307–392 (2019)
14. Kingma, D.P., Welling, M.: Auto-encoding variational Bayes. In: 2nd International Conference on Learning Representations, ICLR 2014 (2014)
15. scikit learn: two moons dataset. In: https://scikit-learn.org/stable/modules/generated/sklearn.datasets.make_moons.html, scikit-learn dataset make_moons (2021). Accessed 2021
16. LeCun, Y., Cortes, C., Burges, C.: MNIST handwritten digit database. ATT Labs [Online]. http://yann.lecun.com/exdb/mnist 2 (2010)
17. Nalisnick, E.T., Matsukawa, A., Teh, Y.W., Görür, D., Lakshminarayanan, B.: Do deep generative models know what they don't know? In: 7th International Conference on Learning Representations, ICLR 2019, New Orleans, LA, USA, 6–9 May 2019. OpenReview.net (2019)
18. Nash, C., Williams, C.K.: The shape variational autoencoder: a deep generative model of part-segmented 3D objects. In: Computer Graphics Forum. vol. 36, pp. 1–12. Wiley Online Library (2017)
19. Pang, G., Shen, C., van den Hengel, A.: Deep anomaly detection with deviation networks. In: Proceedings of the 25th ACM SIGKDD International Conference on Knowledge Discovery & Data Mining, pp. 353–362 (2019)
20. Rayana, S.: Odds library. In: Stony Brook University, Department of Computer Sciences (2016)
21. Ren, J., et al.: Likelihood ratios for out-of-distribution detection. In: Advances in Neural Information Processing Systems, pp. 14680–14691 (2019)
22. Rezende, D.J., Mohamed, S., Wierstra, D.: Stochastic backpropagation and approximate inference in deep generative models. In: International Conference on Machine Learning (2014)
23. Rubenstein, P., Schölkopf, B., Tolstikhin, I.: Learning disentangled representations with Wasserstein auto-encoders. In: International Conference on Learning Representations (ICLR 2018) Workshops (2018)
24. Skafte, N., Jørgensen, M., Hauberg, S.: Reliable training and estimation of variance networks. In: Advances in Neural Information Processing Systems, pp. 6326–6336 (2019)
25. Snelson, E., Ghahramani, Z.: Local and global sparse gaussian process approximations. In: Artificial Intelligence and Statistics, pp. 524–531 (2007)
26. Titsias, M.: Variational learning of inducing variables in sparse Gaussian processes. In: Artificial Intelligence and Statistics, pp. 567–574 (2009)
27. Tomczak, J., Welling, M.: VAE with a VampPrior. In: International Conference on Artificial Intelligence and Statistics, pp. 1214–1223 (2018)
28. Tran, D., Ranganath, R., Blei, D.M.: The variational Gaussian process. In: 4th International Conference on Learning Representations, ICLR 2016 (2016)
29. Van der Wilk, M., Rasmussen, C.E., Hensman, J.: Convolutional gaussian processes. In: Advances in Neural Information Processing Systems, pp. 2849–2858 (2017)
30. Xiao, H., Rasul, K., Vollgraf, R.: Fashion-MNIST: a novel image dataset for benchmarking machine learning algorithms (2017)
31. Zhang, C., Bütepage, J., Kjellström, H., Mandt, S.: Advances in variational inference. IEEE Trans. Pattern Anal. Mach. Intell. **41**(8), 2008–2026 (2018)
32. Zhao, S., Song, J., Ermon, S.: Infovae: balancing learning and inference in variational autoencoders. Proc. AAAI Conf. Artif. Intell. **33**, 5885–5892 (2019)

Variational Hyper-encoding Networks

Phuoc Nguyen[1]($\boxtimes$), Truyen Tran[1], Sunil Gupta[1], Santu Rana[1],
Hieu-Chi Dam[2], and Svetha Venkatesh[1]

[1] A2I2, Deakin University, Geelong, Australia
{phuoc.nguyen,truyen.tran,sunil.gupta,santu.rana,
svetha.venkatesh}@deakin.edu.au
[2] Japan Advanced Institute of Science and Technology, Nomi, Japan
dam@jaist.ac.jp

Abstract. We propose a framework called HyperVAE for encoding distributions of distributions. When a target distribution is modeled by a VAE, its neural network parameters are sampled from a distribution in the model space modeled by a hyper-level VAE. We propose a variational inference framework to implicitly encode the parameter distributions into a low dimensional Gaussian distribution. Given a target distribution, we predict the posterior distribution of the latent code, then use a matrix-network decoder to generate a posterior distribution for the parameters. HyperVAE can encode the target parameters in full in contrast to common hyper-networks practices, which generate only the scale and bias vectors to modify the target-network parameters. Thus HyperVAE preserves information about the model for each task in the latent space. We derive the training objective for HyperVAE using the minimum description length (MDL) principle to reduce the complexity of HyperVAE. We evaluate HyperVAE in density estimation tasks, outlier detection and discovery of novel design classes, demonstrating its efficacy.

Keywords: Deep generative models · Meta-learning · Hyper networks

1 Introduction

Humans can extract meta knowledge across tasks such that when presented with an unseen task they can use this meta knowledge, adapt it to the new context and quickly solve the new task. Recent advance in meta-learning [5,8] shows that it is possible to learn a single model such that when presented with a new task, it can quickly adapt to the new distribution and accurately classify unseen test points. Since meta-learning algorithms are designed for few-shot or one-shot learning where labeled data exists, it faces challenges *when there is none*[1] to assist backpropagation when testing.

Hyper-networks [9] can generate the weights for a target network given a set of embedding vectors of those weights. Due to its generative advantage, it can be used to generate a distribution of parameters for a target network [9,12]. In

---

[1] This is not the same as zero-shot learning where label description is available.

© Springer Nature Switzerland AG 2021
N. Oliver et al. (Eds.): ECML PKDD 2021, LNAI 12976, pp. 100–115, 2021.
https://doi.org/10.1007/978-3-030-86520-7_7

practice, due to the high dimensional parameter space, it only generates scaling factors and biases for the target network. This poses a problem that the weight embedding vectors only encode partial information about the target task, and thus are not guaranteed to perform well on unseen tasks.

On the other hand, variational autoencoders (VAEs) [11, 19] is a class of deep generative models that can model complex distributions. A major attractive feature of VAEs is that we can draw from simple, low-dimensional distributions (such as isotropic Gaussians), and the model will generate high-dimensional data instantly without going through expensive procedures like those in the classic MCMC. This suggests VAEs can be highly useful for high dimensional design exploration [7]. In this work, we lift this idea to one more abstraction level, that is, using a *hyper VAE to generate VAE models*. While the VAEs work at the individual design level, the *hyper VAE* works at the class level. This permits far more flexibility in exploration, because not only we can explore designs within a class, we can explore multiple classes. The main insight here is that the model parameters can also be treated as a design in a model design space. Hence, we can generate the model parameters using another VAE given some latent low-dimensional variable.

We propose HyperVAE, a novel class of VAEs, as a powerful deep generative model to learn to generate the parameters of VAE networks for modeling the distribution of different tasks. The versatility of the HyperVAE to produce VAE models allows it to be applied for a variety of problems where model flexibility is required, including density estimation, outlier detection, and novelty seeking. For the latter, since HyperVAE enforces a smooth transition in the model family, interpolating in this space will enable us to extrapolate to models of new tasks which are *close* to trained tasks. Thus as global search techniques can guide the generation of latent spaces of VAEs, search enables HyperVAE to produce novel classes of discovery. We use Bayesian Optimization (BO) [20], to search in the low dimensional encoding space of VAE. Once a low dimensional design is suggested, we can decode it to the corresponding high dimensional design.

We demonstrate the ability of HyperVAE on three tasks: density estimation, robust outlier detection and discovery of unseen design classes. Our main contributions and results are: (i) Development of a hyper-encoding framework, guided through MDL; (ii) Construction of a versatile HyperVAE model that can tackle density estimation tasks and outlier detection; and (iii) Demonstration of novel designs produced from our model coupled with BO.

2 Variational Autoencoder (VAE)

Let x denote an $\mathcal{X}$-value random variable associated with a $\mathcal{Z}$-value random variable z through a joint distribution $p(x, z)$. We consider a parametric family $\mathcal{P}$ of generative models $p(x, z; \theta)$ factorized as a conditional $p(x|z; \theta)$ and a simple prior $p(z)$, usually chosen as $\mathcal{N}(0, I)$. Maximum likelihood estimate (MLE) of $\theta \in \Theta$, where Θ is the parameter space, over the marginal $\log p(x; \theta) = \log \int p(x, z; \theta) dz$ is intractable, thus requiring alternatives such as expectation-maximization and variational inference. VAE is an amortized variational inference that jointly learns the generative model $p(x|z; \theta)$ and the variational inference model $q(z|x; \theta)^2$. Its ELBO objective,

$$\mathcal{L}(x, p, q; \theta) = \mathbb{E}_{q(z|x;\theta)} \log p(x|z; \theta) - D_{KL}\left(q(z|x; \theta)\|p(z)\right) \qquad (1)$$

lower-bounds the marginal log-likelihood $\log p(x; \theta)$. In practice, Monte Carlo estimate of the ELBO's gradient is used to update θ. The form of q and p in Eq. 1 makes an encoder and a decoder, hence the name auto-encoder [11].

3   Variational Hyper-encoding Networks

We assume a setting where there is a sequence of datasets (or tasks) and model parameters $\{(D_t, \theta_t)\}_t$ a sender wish to transmit to a receiver using a minimal combined code length.

3.1   Hyper-auto-encoding Problem

Given a set of T distributions $\{D_t\}_{t=1}^T$ called tasks, each containing samples $x \sim p_{D_t}(x)$, our problem is first fitting each parametric model $p(x; \theta_t)$, parameterized by $\theta_t \in \Theta$, to each D_t:

$$\hat{\theta}_t = \underset{\theta \in \Theta}{\mathrm{argmax}}\, p(D_t; \theta) \qquad (2)$$

then fitting a parametric model $p(\theta; \gamma)$, parameterized by $\gamma \in \Gamma$ to the set $\{\hat{\theta}_t\}_{t=1}^T$. However, there are major drawbacks to this approach. First, the number of tasks may be insufficient to fit a large enough number of θ_t for fitting $p(\theta; \gamma)$. Second, although we may resample D_t and refit θ_t to create more samples, it is computationally expensive. A more practical approach is to jointly learn the distribution of θ and D.

3.2   Hyper-encoding Problem

Our problem is to learn the joint distribution $p(\theta, D; \gamma)$ for some parameters γ^3.

HyperVAE. We propose a framework for this problem called *HyperVAE* as depicted in Fig. 1. The main insight here is that the *VAE model parameters* $\theta \in \Theta$ can also be treated as a normal input in the parameter space Θ. Hence, we can generate the model parameters θ using *another VAE at the hyper level* whose generative process is $p_\gamma(\theta|u)$ for some low-dimensional latent variable $u \sim p(u) \equiv \mathcal{N}(0, I)$, the prior distribution defined over the latent manifold $\mathcal{U}$ of Θ. The joint distribution $p(\theta, D; \gamma)$ can be expressed as the marginal over the latent representation u:

$$p(\theta, D) = \int p(\theta, D|u)p(u)\mathrm{d}u = \int p(D|\theta)p(\theta|u)p(u)\mathrm{d}u \qquad (3)$$

---

[2] We use $\theta = (\theta_p, \theta_q)$ to denote the set of parameters for p and q.

[3] We assume a Dirac delta distribution for γ, i.e. a point estimate, in this study.

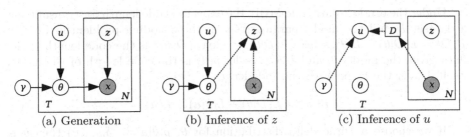

(a) Generation   (b) Inference of z (c) Inference of u

Fig. 1. HyperVAE networks, $D = \{x_n\}_{n=1}^N$.

Generation of a random data point x is as follows, c.f., Fig. 1(a):

$$u_t \sim N(0, I)$$
$$\theta_t \sim p_\gamma(\theta \mid u_t)$$
$$z \sim N(0, I)$$
$$x \sim p_{\theta_t}(x \mid z)$$

Inference of z given x and θ, Fig. 1(b), is approximated by a Gaussian distribution, $q(z|x, \theta) = \mathcal{N}(z|\mu_\theta(x), \sigma_\theta^2(x))$, where μ_θ and σ_θ^2 are neural networks generating the mean and variance parameter vectors.

Inference of u is shown in Fig. 1(c). We also assume a Gaussian posterior distribution $q(u|D, \theta) = \mathcal{N}(u|\mu(d_t), \sigma^2(d_t))$ parameterized by neural networks $\mu(.)$ and $\sigma^2(.)$. Since θ can be trained on D thus depending on D, we can approximate this dependency implicitly using the inference network itself, thus $q(u|D, \theta) \approx q(u|D)$. The next problem is that since D_t is a set, $q(u|D_t)$ is a function of set, which is an interesting problem on its own. Here we use a simple method to summarize D_t into a vector,

$$d_t = s(D_t) \tag{4}$$

and this turns $q(u|D_t)$ into $q(u|d_t)$. For example, $s(\cdot)$ can be a mean function, a random draw from the set, or a description of the set. In this study, we simply choose a random draw x from the set D_t.

3.3   Minimum Description Length

It is well-known that variational inference is equivalent to the Minimum Description Length (MDL) principle [10]. In this section, we use MDL to compute the total code length of the model and data misfits. From the total code length, we show that a shorter code length and a simpler model can be achieved by redesigning the distribution of the model space. We used a Dirac delta distribution centered at $\mu(u)$ for θ given each latent code u, $p(\theta|u) = \delta_{\mu(u)}(\theta)$ parameterized by a neural network $\mu(u)$ for each latent code u. This results in an implicit distribution for θ represented by the compound distribution $p(\theta) = \int \delta_{\mu(u)}(\theta)p(u)\mathrm{d}u$.

Under the crude 2-part code MDL, the expected code length for transmitting a dataset D and the model parameters θ in the encoding problem in Eq. 2 is $L(D) = L(D|\theta) + L(\theta)$, where $L(D|\theta) = -\log p(D; \theta)^4$ is the code length of the data given the model θ, and $L(\theta) = -\log p(\theta)$ is the code length of the model itself. Under the HyperVAE the code length of D is:

$$L(D) = L(D|\theta) + L(\theta|u) + L(u) \tag{5}$$

If we choose a Dirac delta distribution for θ, $p(\theta|u) = \delta_{\mu(u)}(\theta)$ then θ is deterministic from u and we can eliminate the code length $L(\theta|u)$, thus making the total code length shorter:

$$L(D) = L(D|\theta(u)) + L(u) \tag{6}$$

Additionally, bits-back coding can recover the additional information in the entropy of the variational posterior distribution $q(u|D)$, thus this information should be subtracted from the total code length [10,22]. The total expected code length is then:

$$
\begin{aligned}
L(D) &= \mathbb{E}\left[L\left(D|\theta(u)\right) - \log p(u) + \log q(u|D)\right] \\
&= \mathbb{E}\left[L\left(D|\theta(u)\right)\right] + D_{KL}\left(q(u|D)\|p(u)\right)
\end{aligned}
\tag{7}
$$

where the expectation is taken over the posterior distribution $q(u|D)$. The description length of a dataset $D = \{x_i\}_{i=1}^{|D|}$ is the summation of the description length of every data point:

$$
\begin{aligned}
L\left(D|\theta(u)\right) &= \sum_{i=1}^{|D|} L\left(x_i|\theta(u)\right) \\
&= \sum_{i=1}^{|D|} \left(\mathbb{E}_{q_\theta(z|x_i)} L\left(x_i|z, \theta\right) + D_{KL}\left(q_\theta(z|x_i)\|p(z)\right)\right)
\end{aligned}
\tag{8}
$$

where we ignored the dependence of θ on u to avoid clutter. We train the Hyper-VAE parameters by minimizing the description length in Eq. 7. In our experiment, we scale down this objective by multiplying it by $1/|D|$ to have a similar scale as a normal VAE's objective. The training objective for HyperVAE is then:

$$L(D) = \frac{1}{|D|} \left[L\left(D|\theta(u)\right) + D_{KL}\left(q(u|D)\|p(u)\right)\right] \tag{9}$$

Mini-batches as Tasks. In practice, the number of tasks is too small to adequately train the hyper-parameters γ. Here we simulate tasks using data mini-batches in the typical stochastic gradient learning. That is, each mini-batch is treated as a task. To qualify as a task, each mini-batch needs to come from the same class. For example, for handwritten digits, the class is the digit label.

---

[4] We abused the notation and use p to denote both a density and a probability mass function. Bits-back coding is applicable to continuous distributions [10].

3.4   Compact Hyper-decoder Architecture

Since neural networks weights are matrices that are highly structured and often overparameterized, we found that a more efficient method is to use a matrix generation network [3] for generating the weights. More concretely, a matrix hyper-layer receives an input matrix H and computes a weight matrix W as $W = \sigma(UHV + B)$, where U, V, B are parameters. As an example, if H is a 1D matrix of size 400×1 and a target weight W of size 400×400, a matrix-layer will require 176 thousand parameters, a 3 order of magnitude reduction from 64.16 million parameters of the standard fully-connected hyper-layer. This compactness allows for complex decoder architecture for generating the target network, unlike hyper-networks methods which rely on a linear layer of an embedding vector for each target-network layer.

3.5   Applications

We can use the HyperVAE framework for density estimation, outlier detection and novelty discovery. In the following, we use *HyperVAE* to denote the whole VAE-of-VAEs framework, *hyper VAE* for the hyper level VAE, and *main VAE* for the VAE of each target task.

Density Estimation. After training, HyperVAE can be used to estimate the density of a new dataset/task. Let D_t is the new task data. We first infer the posterior distribution $q(u|D_t) \approx q(u|d_t) = \mathcal{N}(u|\mu(d_t), \sigma^2(d_t))$, where d_t is a summary of D_t, Eq. 4, which we choose as random in this study. Next we select the mean of this posterior distribution and decode it into θ using $p_\gamma(\theta|u)$. We use this θ to create the main VAE for D_t then use importance sampling to estimate the density of $x \in D_t$ as follows:

$$p(x) = \mathbb{E}_{q(z|x,\theta)} \frac{p(x|z)p(z)}{q(z|x,\theta)} \approx \frac{1}{N} \sum_{i=1}^{N} \frac{p(x|z_i)p(z_i)}{q(z_i|x,\theta)}$$

where N is a chosen number of importance samples, $\{z_i\}_{i=1}^N$ are samples from the proposal distribution $q(z|x,\theta)$ to reduce the variance of the density estimate, and $p(z_i)/q(z_i|x,\theta)$ is the multiplicative adjustment to compensate for sampling from $q(z|x,\theta)$ instead of $p(z)$.

Outlier Detection. Similar to the density estimation application above, we first encode a test vector x_t into a latent distribution $q(u|x_t)$ then decode its mean vector into θ_t to create a VAE model. We then use the description length of x_t, c.f. Eq. 8, under this VAE as the outlier score. Our assumption is that outliers are unseen to the trained model, thus incompressible under this model and should have longer description lengths.

Novelty Discovery. HyperVAE provides an extra dimension for exploring the model space Θ in addition to exploring the design space $\mathcal{X}$. Once trained, the network can guide exploration of new VAE models for new tasks with certain similarity to the trained tasks.

Given no prior information, we can freely draw models $\theta(u)$ from $u \sim p(u)$ and designs $x \sim p_\theta(x|z)$ with $z \sim p_{\theta(u)}(z)$ and search for the desired x^* satisfying some property $F(x^*)$. An intuitive approach is to employ a global search technique such as Bayesian Optimization (BO) in both the model latent space of u and in the data latent space of z. However searching for both $u \in \mathcal{U}$ and $z \in \mathcal{Z}$ is expensive due to the combined number of dimensions can be very high. Furthermore, reducing the latent dimension would affect the capacity of VAE. To overcome this major challenge, we use BO for optimizing the z space and replace the search in u space by an iterative search heuristic. The workflow starts with an initial exemplar x_0^* which can be completely uninformative (e.g., an empty image for digits or a random design), or properly guided (e.g., from the best choice thus far in the database, or from what is found by VAE+BO itself). The search process for the optimal design at step $t = 1, 2, ..., T$ is as follows:

$$u_t \sim q(u \mid d_{t-1}); \qquad \theta_t = g_\gamma(u_t);$$
$$z^* \leftarrow \text{BO}(g_{\theta_t}(z)); \qquad x_t^* \leftarrow g(z^*). \qquad (10)$$

where $d_{t-1} \leftarrow x_{t-1}^*$.

The optimization step in the z space maximizes a function $\max_x F(x) = \max_z F \circ g_{\theta_t}(z)$ for a fixed generator θ_t. Let z_t^* and thus $x_t^* = g_{\theta_{t-1}}(z_{t-1}^*)$ be the solution found at step t. The generator parameter in the subsequent step is set as $\theta_t \leftarrow \theta(\mu(x_t^*))$ where μ is the posterior mean. Thus the HyperVAE step transforms the objective function with respect to z by shifting θ.

4   Experiments

We evaluate HyperVAE on three tasks: density estimation, robust outlier detection, and novel discovery.

4.1   Data Sets

We use four datasets: MNIST handwritten digits, Omniglot handwritten characters, Fashion MNIST, and Aluminium Alloys datasets. The MNIST contains 60,000 training and 10,000 test examples of 10 classes ranging from 0 to 9. The Omniglot contains 24,345 training and 8,070 test examples. The Fashion MNIST dataset contains the same number of training and test examples as well as the number of classes. In these three datasets, the images are statically binarized to have pixel values in $\{0, 1\}$.

The Alloys dataset (https://tinyurl.com/tmah538), previously studied in [16], consists of 15,000 aluminium alloys. Aluminium alloy is a combination of about 85% aluminium and other elements.

   Phase diagram contains important characteristics of alloys, representing variations between the states of compounds at different temperatures and pressures. They also contain thermodynamic properties of the phases. In this experiment, a phase diagram is coded as a 2D matrix, in which each cell is the prevalence of a phase at a particular temperature.

4.2   Model Settings

We use a similar architecture for the encoder and decoder of all VAE in all datasets. The encoder has 2 convolution layers with 32 and 64 filters of size 3×3, stride 2, followed by one dense layer with 100 hidden units, then two parallel dense layers to output the mean and log variance of $q(z|x; \theta)$. The decoder architecture exactly reverses that of the encoder to map from z to x, with transposed convolution layers in place of convolution layers, and outputs the Bernoulli mean of $p(x|z; \theta)$. For the alloys dataset, the convolution layers are replaced by matrix layers with size 200×200, as in [4]. We also use a similar architecture for HyperVAE in all datasets. The encoder uses the same architecture as the VAE's encoder. The decoder use a dense layer with 100 hidden units, followed by L parallel matrix layers generating the weights, biases, and filters of the main VAE network, resembling the parameter θ. The input to the matrix layer is reshaped into size 20×20. All layers except the last layer use RELU activation. The z-dimension and u-dimension is 10 for all datasets. We used Adam optimizer with parameters $\beta_1 = 0.9$, $\beta_2 = 0.999$, learning rate $\eta = 0.0003$, minibatches of size 100, and ran for 10000 iterations or when the models converge.

4.3   Model Behavior

We study whether the HyperVAE learns a meaningful latent representation and data distribution for the MNIST and Omniglot datasets. We use negative log-likelihood (NLL) and $D_{KL}(q(z|x)\|p(z))$ as measures. NLL is calculated using importance sampling with 1024 samples.

Table 1. Negative log-likelihood $(-LL)$, and $D_{KL}(q(z|x)\|p(z))$ (KL). Smaller values are better.

		VAE	MetaVAE	HyperVAE
MNIST	$-LL$	99.4	93.0	**88.2**
	KL	18.8	15.5	18.5
Omniglot	$-LL$	111.4	128.1	**105.5**
	KL	17.1	**13.2**	18.1
Fashion MNIST	$-LL$	237.7	232.7	**231.8**
	KL	14.5	**13.7**	13.9

   Table 1 compares the performance of VAE, MetaVAE, and HyperVAE. As shown, HyperVAE has better NLL on the three datasets. MetaVAE has smallest

KL, which is due to it has a separate and fixed generator for each task. HyperVAE has slightly smaller KL on the MNIST and Fashion MNIST dataset than VAE. Note that better log-likelihoods can be achieved by increasing the number of latent dimensions, e.g. $\dim(z) = 50$, instead of $\dim(z) = 10$ in this experiment.

Table 2. Number of parameters (rounded to thousands).

	VAE	MetaVAE	HyperVAE
Inference	445	445	445
Generative	445	$445 \times \#\text{task}$	445
Total	890	$445 + 445 \times \#\text{task}$	890

Table 2 compares the number of parameters between networks. Note that while MetaVAE shares the same inference network for all tasks, it needs a separate generative network for each task. For HyperVAE, the trainable parameters are from the hyper level VAE, whereas the main VAE network of each task obtains its parameters by sampling from the HyperVAE network. Therefore, for the comparison we only count the number of trainable parameters, which is what eventually saved to disk. The real parameters for the target networks will be generated on-the-fly given a target task. Thus, it will take extra generation time for each task, c.f. Table 3.

Table 3. Time measured in milliseconds for a batch of 100 inputs.

	Generation	Inference	Total time
VAE	0.12	0.12	0.24
MetaVAE	0.12	0.12	0.24
HyperVAE	0.12 (x)	0.12 (z)	1.11
	0.75 (θ)	0.12 (u)	

Overfitting. VAE is trained on the combined dataset therefore it is less affected by overfitting due to high variance in the data. Whereas MetaVAE is more susceptible to overfitting when the number of examples in the target task is small, which is the case for Omniglot dataset, c.f. Table 1. While the training of MetaVAE's encoder is amortized across all datasets, the training of its decoder is task-specific. As a result, when a (new) task has a small number of examples, the low variance data causes overfitting to this task's decoder. Therefore MetaVAE is not suitable for transfer learning to new/unseen tasks. HyperVAE can avoid overfitting by taking a Bayesian approach.

HyperVAE Complexity. The algorithmic complexity of HyperVAE is about double that of VAE, since it is a VAE of VAE, plus the extra generation time of the parameters. Specifically, it runs the VAE at the hyper level to sample a weight parameter θ, then it runs the VAE to reconstruct a set of inputs given this parameter θ. Due to the difference in matrix sizes of different layers in the target network, we generate each weight matrix and bias vector at a time, resulting in $O(D)$ time[5] with D being the depth of the target network. Therefore, the time complexity of the hyper generation network is $O(L_{\text{hyper}} + L_{\text{VAE}} + D)$, where L_{hyper} and L_{VAE} are the number of layers of the hyper and the primary generation networks respectively, and we assumed the average hidden size of the layers is a constant. However, more efficient methods is also possible. For example, inspired from [9], we can reshape matrices into batches of blocks of the same size, then stacking along the batch dimension in to a large 3D tensor. Then, we can use a matrix network to generate this tensor in $O(1)$ time[6] whence the time complex will be $O(L_{\text{hyper}} + L_{\text{VAE}})$. We leave this implementation for future investigation. Table 3 shows the wall-clock time comparison between methods on a Tesla P100 GPU.

Table 4. Outlier detection on MNIST. AUC: Area Under ROC Curve, FPR: False Positive Rate, FNR: False Negative Rate, KL: KL divergence, -EL: mean negative loglikelihood and KL.

		VAE	MetaVAE		HyperVAE
			KL	-EL	
MNIST	AUC	93.0	54.7	52.2	**95.3**
	FPR	16.3	47.5	49.4	**15.6**
	FNR	15.5	45.0	50.5	**8.0**
Omniglot	AUC	98.3	87.3	97.5	**98.7**
	FPR	5.5	18.8	7.2	**4.9**
	FNR	6.4	20.9	9.0	**5.9**
Fashion MNIST	AUC	74.6	58.2	56.8	**76.8**
	FPR	**33.5**	44.1	45.8	33.6
	FNR	32.0	43.5	44.5	**28.7**

4.4   Robust Outlier Detection

Next, we study HyperVAE model for outlier detection tasks. We use three datasets: MNIST, Omniglot, and Fashion MNIST to create three outlier detection experiments. For each experiment, we select one dataset as the normal class and 20% random samples from another dataset as outliers. All methods are trained on only normal data. VAE and HyperVAE use the negative log-likelihood and KL for calculating the outlier score, which is equivalent to the negative ELBO. MetaVAE does not have a generative network for new data. Therefore we use two scoring

---

[5] We assumed a matrix multiplication takes $O(1)$ time in GPU.
[6] Batched matrix multiplication can be paralleled in GPU.

methods: (1) KL divergence only, and (2) mean negative log-likelihood and KL, using all trained generative networks. For training MetaVAE and HyperVAE, we define the task as before, i.e. the data in each task have a similar class label.

Table 4 compares the performance of all methods on the three datasets. Overall, HyperVAE has better AUCs compared to VAE and MetaVAE. The MetaVAE has the lowest AUC. This could be due to the use of a discrete set of generative networks for each task, making it unable to handle new, unlabeled data.

While the false positive rates of VAE and HyperVAE models are similar, the false negative rates for HyperVAE are lower than that of VAE. This is because HyperVAE was trained across tasks, thus it has a better support between tasks.

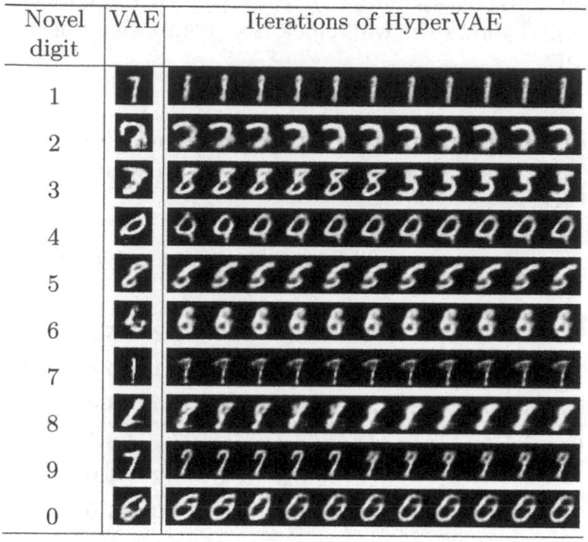

Fig. 2. Best digits found at iterative steps in searching for a new class of digits, corresponding to the performance curves in Fig. 3.

4.5   Novelty Discovery

We demonstrate the effectiveness of HyperVAE+BO for finding realistic designs close to an ideal design, which lies outside known design classes. The performance measure is how close we get to the given ideal design, as measured in cosine distance for simplicity.

In each of the following two experiments, the BO objective is to search for a novel unseen design x^*, an unseen digit or alloy, by maximizing a Cosine distance $F(x^*)$. The maximum number of BO iterations is set to 300 and the search space is $[-5, 5]$ for each z and u dimension.

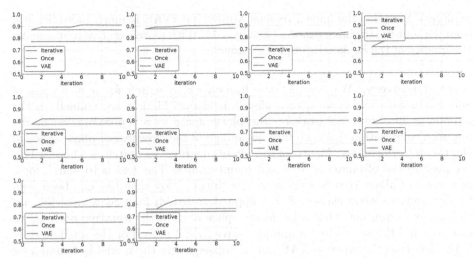

(a) Searching for unseen MNIST digits $\{1, \ldots, 9, 0\}$, from left to right, top to bottom.

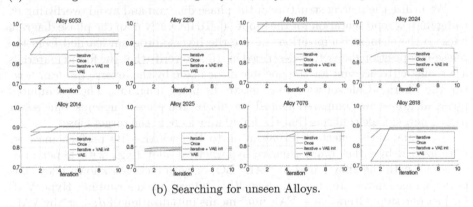

(b) Searching for unseen Alloys.

Fig. 3. Searching for: (a) unseen digits, and (b) unseen alloys designs. Cosine distance between target and best found vs iterations. Best viewed in color. (Color figure online)

Digit Discovery. This experiment illustrates the capability of HyperVAE+BO in novel exploration on MNIST. For each experiment, one digit is held out. We used nine digit classes for training and tested the model ability to search for high quality digits of the remaining unseen digit class. BO is applied to search for new digits that are similar to a given new exemplar in the z-space.

In the iterative process, an empty image $d_1 = \mathbf{0}$ is given at the first step, and subsequently updated as $d_t = x^*_{t-1}$. After each step t we set $u_t = \mu(x^*_t)$. The quality curves are presented Fig. 3(a). Examples of discovery process are listed in Fig. 2. The figures show that VAE has a very limited capability to support exploration outside the known regions, while HyperVAE is much more flexible, even without the iterative process (#Step = 1). With more iterative refinements, the quality of the explored samples improves.

Alloy Discovery. We now use the framework to search for a new class of alloys. For each experiment, one alloy is held out. Models are trained on the remaining 29 alloys. We work on the phase space as a representation of the material composition space, to take advantage of the closeness of phase space to the target performance. We treat the phase diagrams as matrices whose values are proportions of phases at different temperatures. The goal is to search for a new class of alloys that is similar to the "ideal" alloy that has not been seen in any previous alloy classes. BO is applied to search for new alloys that are similar to a given new ideal alloy in the space of z. In the iterative process, we can initialize the search by an uninformative model $u_1 = \mathbf{0}$ or the one found by VAE+BO (the "Iterative + VAE init"). Subsequently the model is updated by setting $d_t = x_{t-1}^*$. The u variable is set to $u_t = \mu(x_t^*)$ after each step t.

We utilize the matrix structure of the phase diagram and avoid overfitting by using matrix representation for the input [4]. To inversely map the phase diagram back to the element composition, we use the inverse program learned from the phase-composition dataset, as described in [16]. To verify that the found materials are realistic (to account for the possible error made by the inverse program), we run the Thermo-Calc software to generate the phase diagrams. These computed phase diagrams are compared against the discovered phase diagrams. The result from Thermo-Calc confirms that the found alloys are in the target class.

To examine the effect of initialization to HyperVAE+BO performance, we initialized it by either uninformative hypothetical alloy (e.g., with hyper prior of zeros), with the alloy found by VAE+BO, or with a chosen known alloy. The performance curves are shown in Fig. 3(b). "Once" means running HyperVAE for just one step. "Iterative + VAE init" means initialization of $d_1 = x^*$ by VAE. It shows: (a) For a majority of cases, HyperVAE+BO initialized uninformatively could find a better solution than VAE+BO, and (b) initializing HyperVAE+BO with solution found by VAE+BO boosts the performance further, sometimes a lot more. This suggests that care must be taken for initializing HyperVAE+BO.

We examine the results of the ten most difficult to find alloy targets, i.e. the alloys whose distance to their nearest alloy are largest, in descending order of difficulty. Table 5 shows the element composition errors of found alloys. The results show that most alloys are found to be in the target class and all found alloys are close to the boundaries of their targets (at ±20%). Table 5 also shows that the Thermo-Calc phase calculation agrees with the predicted phase, i.e. small errors. The alloy 6951 and 6463 have the smallest errors compared to others.

Table 5. Column A - Element composition errors of found alloys (the composition is predicted by the method in [16]). The found alloys are expected to reside within ±20% relative error to the target alloy to stay within its class. **Column B** - Verification of phase in Thermo-Calc simulator, where the phase error is calculated as the mean error relative to the maximum proportion of each phase. The errors of the best method are reported. Alloys are ranked by their relative distance to the nearest neighbor, in decreasing order.

Alloy	A - Element error (%)	B - Phase error (%)
6053	**11.3**	3.0
2219	26.6	3.0
6951	**20.0**	0.4
2024	31.5	3.5
2014	**18.5**	1.4
2025	**4.4**	2.9
7076	31.8	2.5
2618	23.9	3.6

5   Related Work

Our method can be considered as a lossless compression strategy where the HyperVAE compresses a family of networks that parameterize the parameters of distributions across datasets. The total code length of both the model and data misfits are minimized using HyperVAE, thus help it generalize to unseen data. This is in contrast to the lossy compression strategy [1] where local information of images are freely decoded independent of the compressed information.

The HyperVAE shares some insight with the recent MetaVAE [2], but this is different from ours in the target and modeling, where the latent z is factored into data latent variable z and the task latent variable u. HyperVAE is related to Bayesian VAE, where the model is also a random variable generated from some hyper-prior. There has been some work on priors of VAE [13,21], but using VAE as a prior for VAE is new.

HyperGAN [18] is a recent attempt to generate the parameters of model for classification. This framework generates all parameters from a single low dimension Gaussian noise vector. Bayesian neural networks (BNN) in [23] also use GAN framework for generating network parameters θ that looks real similar to one drawn from BNN trained with stochastic gradient Langevin dynamics. However, GAN is not very successful for exploration but more for generating realistic samples.

Continual learning are gaining ground in recent years. Variational continual learning [15], for example, solves catastrophic forgetting problems in supervised learning, but it still needs a set of prototype points for old tasks. [17] tackles this problem in unsupervised tasks and also does task inference as ours, however our settings and approaches are different. Meta-learning frameworks for classification and regression [6,14,24] is another direction where the purpose is to learn agnostic models that can quickly adapt to a new task.

6   Conclusion

We proposed a new method called HyperVAE for encoding a family of neural network models into a simple distribution of latent representations. A neural network instance sampled from this family is capable of modeling the end task in which the family is trained on. Furthermore, by explicitly training the variational hyper-encoder network over a complex distribution of tasks, the hyper-network learns the smooth manifold of the family encoded in the posterior distribution of the family. This enables the model to extrapolate to new tasks close to trained tasks, and to transfer common factors of variation across tasks. In the handwritten digit example, the transferable factors may include writing styles, font face and size. It can be thought of as expanding the support of the distribution of trained model, thus is useful for downstream tasks such as searching for a data distribution close to existing ones and reducing the false positive error in outlier detection.

Acknowledgments. This research was partially funded by the Australian Government through the Australian Research Council (ARC). Prof Venkatesh is the recipient of an ARC Australian Laureate Fellowship (FL170100006).

References

1. Chen, X., et al.: Variational lossy autoencoder. arXiv preprint arXiv:1611.02731 (2016)
2. Choi, K., Wu, M., Goodman, N., Ermon, S.: Meta-amortized variational inference and learning. arXiv preprint arXiv:1902.01950 (2019)
3. Do, K., Tran, T., Venkatesh, S.: Matrix-centric neural networks. arXiv preprint arXiv:1703.01454 (2017)
4. Do, K., Tran, T., Venkatesh, S.: Learning deep matrix representations. arXiv preprint arXiv:1703.01454 (2018)
5. Finn, C., Abbeel, P., Levine, S.: Model-agnostic meta-learning for fast adaptation of deep networks. In: Proceedings of the 34th International Conference on Machine Learning, vol. 70, pp. 1126–1135 (2017). JMLR.org
6. Finn, C., Levine, S.: Meta-learning and universality: deep representations and gradient descent can approximate any learning algorithm. In: ICLR (2018)
7. Gómez-Bombarelli, R., et al.: Automatic chemical design using a data-driven continuous representation of molecules. ACS Central Sci. 4(2), 268–276 (2018)
8. Grant, E., Finn, C., Levine, S., Darrell, T., Griffiths, T.: Recasting gradient-based meta-learning as hierarchical Byes. arXiv preprint arXiv:1801.08930 (2018)
9. Ha, D., Dai, A., Le, Q.V.: Hypernetworks. arXiv preprint arXiv:1609.09106 (2016)
10. Hinton, G., Van Camp, D.: Keeping neural networks simple by minimizing the description length of the weights. In: Proceedings of the 6th Annual ACM Conference on Computational Learning Theory. Citeseer (1993)

11. Kingma, D.P., Welling, M.: Auto-encoding variational Bayes. arXiv preprint arXiv:1312.6114 (2013)
12. Krueger, D., Huang, C.-W., Islam, R., Turner, R., Lacoste, A., Courville, A.: Bayesian hypernetworks. arXiv preprint arXiv:1710.04759 (2017)
13. Le, H., Tran, T., Nguyen, T., Venkatesh, S.: Variational memory encoder-decoder. In: NeurIPS (2018)
14. Mishra, N., Rohaninejad, M., Chen, X., Abbeel, P.: A simple neural attentive meta-learner. In: ICLR 2018 (2018)
15. Nguyen, C., Li, Y., Bui, T.D., Turner, R.E.: Variational continual learning. In: ICLR (2018)
16. Nguyen, P., Tran, T., Gupta, S., Rana, S., Barnett, M., Venkatesh, S.: Incomplete conditional density estimation for fast materials discovery. In: Proceedings of the 2019 SIAM International Conference on Data Mining, pp. 549–557. SIAM (2019)
17. Rao, D., Visin, F., Rusu, A., Pascanu, R., Teh, Y.W., Hadsell, R.: Continual unsupervised representation learning. In: Advances in Neural Information Processing Systems, pp. 7645–7655 (2019)
18. Ratzlaff, N., Fuxin, L.: HyperGAN: a generative model for diverse. In: Performant Neural Networks, ICML (2019)
19. Rezende, D.J., Mohamed, S., Wierstra, D.: Stochastic backpropagation and approximate inference in deep generative models. arXiv preprint arXiv:1401.4082 (2014)
20. Shahriari, B., Swersky, K., Wang, Z., Adams, R.P., De Freitas, N.: Taking the human out of the loop: a review of Bayesian optimization. Proc. IEEE **104**(1), 148–175 (2016)
21. Tomczak, J., Welling, M.: VAE with a VampPrior. arXiv preprint arXiv:1705.07120 (2017)
22. Townsend, J., Bird, T., Barber, D.: Practical lossless compression with latent variables using bits back coding. arXiv preprint arXiv:1901.04866 (2019)
23. Wang, K.-C., Vicol, P., Lucas, J., Gu, L., Grosse, R., Zemel, R.: Adversarial distillation of Bayesian neural network posteriors. In: International Conference on Machine Learning, pp. 5177–5186 (2018)
24. Yoon, J., Kim, T., Dia, O., Kim, S., Bengio, Y., Ahn, S.: Bayesian model-agnostic meta-learning. In: Advances in Neural Information Processing Systems, pp. 7332–7342 (2018)

Principled Interpolation in Normalizing Flows

Samuel G. Fadel[1,2]([✉]) [ID], Sebastian Mair[2] [ID], Ricardo da S. Torres[3] [ID],
and Ulf Brefeld[2]

[1] University of Campinas, Campinas, Brazil
samuel.fadel@ic.unicamp.br
[2] Leuphana University, Lüneburg, Germany
{mair,brefeld}@leuphana.de
[3] Norwegian University of Science and Technology, Ålesund, Norway
ricardo.torres@ntnu.no

Abstract. Generative models based on normalizing flows are very successful in modeling complex data distributions using simpler ones. However, straightforward linear interpolations show unexpected side effects, as interpolation paths lie outside the area where samples are observed. This is caused by the standard choice of Gaussian base distributions and can be seen in the norms of the interpolated samples as they are outside the data manifold. This observation suggests that changing the way of interpolating should generally result in better interpolations, but it is not clear how to do that in an unambiguous way. In this paper, we solve this issue by enforcing a specific manifold and, hence, change the base distribution, to allow for a principled way of interpolation. Specifically, we use the Dirichlet and von Mises-Fisher base distributions on the probability simplex and the hypersphere, respectively. Our experimental results show superior performance in terms of bits per dimension, Fréchet Inception Distance (FID), and Kernel Inception Distance (KID) scores for interpolation, while maintaining the generative performance.

Keywords: Generative modeling · Density estimation · Normalizing flows

1 Introduction

Learning high-dimensional densities is a common task in unsupervised learning. Normalizing flows [9, 26, 28, 31, 32] provide a framework for transforming complex distributions into simple ones: a chain of L parametrized bijective functions $\mathbf{f} = \mathbf{f}_1 \circ \mathbf{f}_2 \circ \cdots \circ \mathbf{f}_L$ converts data into another representation that follows a given base distribution. The likelihood of the data can then be expressed as the

---

S. G. Fadel and S. Mair—Equal contribution.

Electronic supplementary material The online version of this chapter (https://doi.org/10.1007/978-3-030-86520-7_8) contains supplementary material, which is available to authorized users.

N. Oliver et al. (Eds.): ECML PKDD 2021, LNAI 12976, pp. 116–131, 2021.
https://doi.org/10.1007/978-3-030-86520-7_8

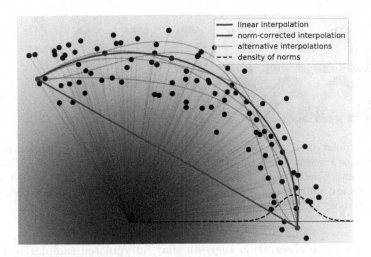

Fig. 1. Illustration of different interpolation paths of points from a high-dimensional Gaussian. The figure also shows that, in high dimensions, points are not concentrated at the origin.

likelihood of the base distribution and the determinants of the Jacobians of the transformations $\mathbf{f}_i$. In contrast to generative adversarial networks (GANs) [12], the likelihood of the data can be directly optimized, leading to a straightforward training procedure. Moreover, unlike other approaches, such as variational autoencoders (VAEs) [17,18], there is no reconstruction error since all functions $\mathbf{f}_i$ within this chain are bijections.

In flow-based generative models, data are generated by drawing samples from a base distribution, where the latter is usually given by a simple distribution, such as a standard Gaussian [23]. The Gaussian samples are then mapped to real data using the chain $\mathbf{f}$. A prevalent operation is to linearly interpolate samples and consider the interpolation path in data space. In generative modeling, interpolations are frequently used to evaluate the quality of the learned model and to demonstrate that the model generalizes beyond what was seen in the training data [25].

The consequences for interpolation, however, are not immediately apparent for Gaussian base distributions. Figure 1 shows a linear interpolation (*lerp*) of high-dimensional samples from a Gaussian. The squared Euclidean norms of the samples follow a χ_d^2-distribution as indicated by the dashed black line. Data points have an expected squared Euclidean norm of length d, where d is the dimensionality. This implies that there is almost no point around the origin. As seen in the figure, the norms of a linear interpolation path (green line) of two samples drop significantly and lie in a low-density area w.r.t. the distribution of the norms (dashed black line) [33].

Instead of a linear interpolation (green line), an interpolation that preserves the norm distribution of interpolants is clearly preferable (blue and red lines): the blue and red interpolation paths stay in the data manifold and do not enter low-

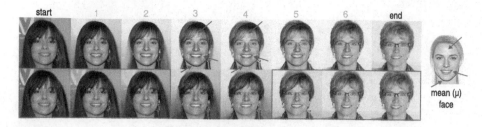

Fig. 2. Interpolation of samples from CelebA. *Top*: a linear interpolation path. The central images resemble features of the mean face as annotated in red. *Bottom*: an alternative interpolation path using a norm-correction. Note that the first and last three images are almost identical as annotated in blue. *Right*: decoded expectation of base distribution, i.e., the mean face. (Color figure online)

density areas. The observation suggests that interpolated samples with norms in a specific range should generally result in better interpolations. This can be achieved, i.e., by shrinking the variance of the density or norms (dashed lines), which yields a subspace or manifold with a fixed norm.

In this paper, we propose a framework that respects the norm of the samples and allows for a principled interpolation, addressing the issues mentioned above. We study base distributions on supports that have a fixed norm. Specifically, we consider unit p-norm spheres for $p \in \{1, 2\}$, leading to the Dirichlet ($p = 1$) and the von Mises-Fisher ($p = 2$) distributions, respectively. The conceptual change naturally implies technical difficulties that arise with restricting the support of the base distribution to the simplex or the unit hypersphere. We thus need to identify appropriate bijective transformations into unit p-norm spheres.

The next sections are organized as follows. In Sect. 2, we propose a simple heuristic to the problem and discuss its problems before we introduce normalizing flows in Sect. 3. Section 4 contains the main contribution, a framework for normalizing flows onto unit p-norm spheres. Empirical results are presented in Sect. 5, and related work is discussed in Sect. 6. Section 7 provides our conclusions.

2   An Intuitive Solution

The blue path in Fig. 1 is obtained by a norm correction of the linear interpolation via also interpolating the norms. Mathematically, that is

$$\gamma(\lambda) = \underbrace{((1 - \lambda)\mathbf{z}_a + \lambda\mathbf{z}_b)}_{\text{linear interpolation}} \cdot \underbrace{\frac{(1 - \lambda)\|\mathbf{z}_a\|_2 + \lambda\|\mathbf{z}_b\|_2}{\|(1 - \lambda)\mathbf{z}_a + \lambda\mathbf{z}_b\|_2}}_{\text{norm correction}}, \tag{1}$$

for endpoints $\mathbf{z}_a, \mathbf{z}_b$ and $\lambda \in [0, 1]$. We refer to this approach as norm-corrected linear interpolation (*nclerp*). However, the depicted red lines also stay within the manifold, hence it remains unclear how a unique interpolation path can be obtained.

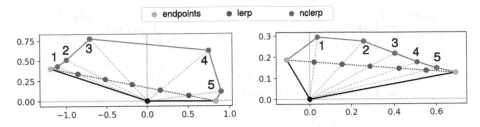

Fig. 3. Two examples showing the issues caused by a norm-corrected linear interpolation (*nclerp*).

Figure 2 depicts two interpolation paths for faces taken from CelebA [14] created using Glow [19], a state-of-the-art flow-based model that uses a standard Gaussian as base distribution. The leftmost and rightmost faces of the paths are real data, while the other ones are computed interpolants. The face on the right depicts the so-called *mean face*, which is given by the mean of the Gaussian base distribution and is trivially computed by decoding the origin of the space. The top row shows a linear interpolation similar to the green line in Fig. 1. The interpolation path is close to the origin, and the interpolants consequentially resemble features of the mean face, such as the nose, mouth, chin, and forehead shine, which neither of the women have. We highlighted those features in red in Fig. 2.

By contrast, the bottom row of Fig. 2 shows the norm-corrected interpolation sequence (as the blue line in Fig. 1): the background transition is smooth and not affected by the white of the mean face, and also subtleties like the shadow of the chin in the left face smoothly disappears in the transition. The norm correction clearly leads to a better transition from one image to the other. However, the simple heuristic in Eq. (1) causes another problem: the path after norm correction is no longer equally spaced when λ values are equally spaced in $[0, 1]$. Implications of this can be seen in blue in the bottom row of Fig. 2, where the first and last three faces are almost identical. We provide additional examples in the supplementary material.

In Fig. 3, we illustrate two examples, comparing a linear interpolation (*lerp*) and a norm-corrected linear interpolation (*nclerp*) between points from a high-dimensional Gaussian (green points). For equally-spaced λ values in $[0, 1]$, a linear interpolation yields an equally-spaced interpolation path (red line). Evidently, the norm-corrected interpolation (blue line) keeps the norms of interpolants within the range observed in data.

However, the interpolants are no longer evenly spaced along the interpolation path. Hence, control over the interpolation mixing is lost. This problem is more pronounced on the left example, where points 1 through 3 are closer to the start point, while points 4 and 5 are closer to the endpoint. Consequently, evaluations such as Fréchet Inception Distance (FID) scores [13] will be affected. Such scores are computed by comparing two sets of samples, in this case, the real data and interpolated data. As those points will be clearly more similar to the endpoints, which are samples from the data set itself, the scores are in favor of the norm-corrected interpolation.

Fig. 4. An example of a one-dimensional normalizing flow.

3 Normalizing Flows

Let $\mathcal{X} = \{\mathbf{x}_1, \ldots, \mathbf{x}_n\} \subset \mathbb{R}^d$ be instances drawn from an unknown distribution $p_x(\mathbf{x})$. The goal is to learn an accurate model of $p_x(\mathbf{x})$. Let $\mathbf{f}^{(\theta)} : \mathbb{R}^d \to \mathbb{R}^d$ be a bijective function parametrized by θ. Introducing $\mathbf{z} = \mathbf{f}^{(\theta)}(\mathbf{x})$ and using the change of variable theorem allows us to express the unknown $p_x(\mathbf{x})$ by a (simpler) distribution $p_z(\mathbf{z})$, defined on $\mathbf{z} \in \mathbb{R}^d$, given by

$$p_x(\mathbf{x}) = p_z\left(\mathbf{f}^{(\theta)}(\mathbf{x})\right)\left|\det J_{\mathbf{f}}^{(\theta)}(\mathbf{x})\right|,$$

where $J_{\mathbf{f}}^{(\theta)}(\mathbf{x})$ is the Jacobian matrix of the bijective transformation $\mathbf{f}^{(\theta)}$. We denote $p_z(\mathbf{z})$ as the base distribution and drop the subscript of the distribution p whenever it is clear from the context.

Representing $\mathbf{f}^{(\theta)}$ as a chain of L parametrized bijective functions, i.e., $\mathbf{f}^{(\theta)} = \mathbf{f}_L^{(\theta)} \circ \mathbf{f}_{L-1}^{(\theta)} \circ \cdots \circ \mathbf{f}_1^{(\theta)}$, creates a *normalizing flow* that maps observations $\mathbf{x}$ into representations $\mathbf{z}$ that are governed by the base distribution $p(\mathbf{z})$. Let $\mathbf{z}_0 = \mathbf{x}$ be the input data point and $\mathbf{z}_L = \mathbf{z}$ be the corresponding output of the chain, where every intermediate variable is given by $\mathbf{z}_i = \mathbf{f}_i(\mathbf{z}_{i-1})$, where $i = 1, \ldots, L$. A one-dimensional example is depicted in Fig. 4. The data log-likelihood can then be expressed as the log-likelihood of the base distribution and the log-determinant of the Jacobians of each transformation as

$$\log p(\mathbf{x}) = \log p(\mathbf{z}) + \sum_{i=1}^{L} \log \left|\det J_{\mathbf{f}_i^{(\theta)}}(\mathbf{z}_{i-1})\right|.$$

Flow-based generative models can be categorized by how the Jacobian structure of each transformation $\mathbf{f}_i$ is designed since computing its determinant is crucial to its computational efficiency. The Jacobians either have a lower triangular structure, such as autoregressive flows [20], or a structured sparsity, such as coupling layers in RealNVP [10] and Glow [19]. Transformations with free-form Jacobians allow much higher expressibility by replacing the computation of the determinant with another estimator for the log-density [6]. For more information regarding flow-based generative models, we refer the reader to [23].

In the remainder, we simplify the notation by dropping the superscript θ from $\mathbf{f}$. We also note that a normalizing flow defines a generative process. To create a new sample $\mathbf{x}$, we first sample $\mathbf{z}$ from the base distribution $p(\mathbf{z})$ and then transform $\mathbf{z}$ into $\mathbf{x}$ using the inverse chain of transformations $\mathbf{f}^{-1}$.

4 Base Distributions on p-Norm Spheres

Motivated by earlier observations illustrated in Figs. 1 and 2, we intend to reduce ambiguity by shrinking the variance of the norms of data. We achieve this by considering base distributions on restricted subspaces. More specifically, we focus on unit p-norm spheres defined by

$$\mathbb{S}_p^d = \left\{ \mathbf{z} \in \mathbb{R}^{d+1} \ \middle| \ \|\mathbf{z}\|_p^p = \sum_{j=1}^{d+1} |z_j|^p = 1 \right\}. \tag{2}$$

We distinguish two choices of p and discuss the challenges and desirable properties that ensue from their use. We consider $p \in \{1, 2\}$ as those allow us to use well-known distributions, namely the Dirichlet distribution for $p = 1$ and the von Mises-Fisher distribution for $p = 2$.

4.1 The Case p = 1

For $p = 1$, the Dirichlet distribution defined on the standard simplex Δ^d is a natural candidate. Its probability density function is given by

$$p(\mathbf{s}) = \frac{1}{Z(\alpha)} \prod_{k=1}^{d+1} s_k^{\alpha_k - 1},$$

$$\text{with} \quad Z(\alpha) = \frac{\prod_{k=1}^{d+1} \Gamma(\alpha_k)}{\Gamma\left(\sum_{k=1}^{d+1} \alpha_k\right)},$$

where Γ is the gamma function and $\alpha_k > 0$ are the parameters. In order to make use of it, we also need to impose a non-negativity constraint in addition to Eq. (2).

Let $\mathbf{z} \in \mathbb{R}^d$ be an unconstrained variable. The function $\phi : \mathbb{R}^d \to (0, 1)^d$ transforms $\mathbf{z}$ into a representation $\mathbf{s}$ by first transforming each dimension z_k into intermediate values v_k with

$$v_k = \sigma\left(z_k - \log\left(d + 1 - k\right)\right)$$

which are used to write $\mathbf{s}$ as

$$s_k = \left(1 - \sum_{l=1}^{k-1} s_l\right) \cdot v_k,$$

where $\sigma(\cdot)$ denotes the sigmoid function. We note a few details of this transformation. First, a property of ϕ is that $0 < \sum_{k=1}^{d} s_k < 1$. Therefore, a point in Δ^d can be obtained with an implicit additional coordinate $s_{d+1} = 1 - \sum_{k=1}^{d} s_k$. Second, the difference in dimensionality does not pose a problem for computing its Jacobian as ϕ establishes a bijection within $\mathbb{R}^d$ while the mapping to Δ^d is given implicitly. Third, ϕ maps $\mathbf{z} = \mathbf{0}$ to the center of the simplex $\mathbf{s} = (d+1)^{-1}\mathbf{1}$. Fourth, since $\mathbf{s}$ consists of solely positive numbers which sum up to one, numerical problems may arise for high-dimensional settings. We elaborate on this issue in Sect. 5.

The Jacobian J_ϕ has a lower triangular structure and solely consists of non-negative entries. Hence, the log-determinant of this transformation can be efficiently computed in $\mathcal{O}(d)$ time via

$$\log|\det J_\phi| = \sum_{k=1}^{d} \log\left(v_k\left(1 - v_k\right)\right)$$

$$+ \log\left(1 - \sum_{l=1}^{k-1} s_l\right).$$

The inverse transformation $\phi^{-1} : \mathbb{R}^d \to \mathbb{R}^d$ is given by

$$z_k = \sigma^{-1}\left(\frac{s_k}{1 - \sum_{l=1}^{k-1} s_l}\right) + \log(d + 1 - k).$$

The interpolation of two points $\mathbf{a}, \mathbf{b} \in \Delta^d$ within the unit simplex is straightforward. A linear interpolation $(1 - \lambda)\mathbf{a} + \lambda\mathbf{b}$ using $\lambda \in [0, 1]$ is guaranteed to stay within the simplex by definition.

4.2   The Case p = 2

For $p = 2$, data points lie on the surface of a d-dimensional hypersphere. The von Mises-Fisher (vMF) distribution, defined on $\mathbb{S}_2^d$, is frequently used in directional statistics. It is parameterized by a *mean direction* $\boldsymbol{\mu} \in \mathbb{S}_2^d$ and a *concentration* $\kappa \geq 0$, with a probability density function given by

$$p(\mathbf{s}) = C_{d+1}(\kappa) \exp(\kappa\boldsymbol{\mu}^\top \mathbf{s}),$$

$$\text{with} \quad C_\nu(\kappa) = \frac{\kappa^{\nu/2-1}}{(2\pi)^{\nu/2} I_{\nu/2-1}(\kappa)},$$

where I_w denotes the modified Bessel function of the first kind at order w.

Again, let $\mathbf{z} \in \mathbb{R}^d$ be an unconstrained variable. We employ a stereographic projection, for both its invertibility and its Jacobian, whose log-determinant can be efficiently computed. The transformation $\psi : \mathbb{R}^d \to \mathbb{S}_2^d$ maps a point $\mathbf{z} \in \mathbb{R}^d$ to a point $\mathbf{s} \in \mathbb{S}_2^d \subset \mathbb{R}^{d+1}$ on the hypersphere via

$$\psi(\mathbf{z}) = \mathbf{s} = \begin{bmatrix} \mathbf{z}\rho_{\mathbf{z}} \\ 1 - \rho_{\mathbf{z}} \end{bmatrix}, \quad \text{with} \quad \rho_{\mathbf{z}} = \frac{2}{1 + \|\mathbf{z}\|^2}.$$

The transformation ψ, which has no additional parameters, ensures that its image is on the unit hypersphere, allowing the use of a vMF distribution to model $p(\mathbf{s})$. Two points in $\mathbb{S}_2^d$ are of special interest, namely the *south pole* and the *north pole*, where the last coordinate of $\mathbf{s}$ is either -1 or 1, respectively. By construction, the transformation is symmetric around zero and sends $\mathbf{z} = \mathbf{0}$ to the *south pole*, which we choose as the mean direction $\boldsymbol{\mu}$. Furthermore, it is bijective up to an open neighborhood around the *north pole*, as $\rho_{\mathbf{z}} \to 0$ whenever $\|\mathbf{z}\|^2 \to \infty$. For this reason, we avoid choosing a uniform distribution on the hypersphere, which is obtained for $\kappa = 0$. Figure 5 shows an example.

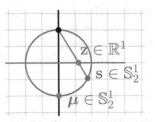

Fig. 5. A stereographic projection mapping $\mathbf{z} \in \mathbb{R}^1$ to $\mathbf{s} \in \mathbb{S}_2^1$ using the north pole depicted as a black dot. The mean direction $\boldsymbol{\mu} \in \mathbb{S}_2^1$ is shown in orange.

Contrary to the previous case, the log-determinant of J_ψ alone is not enough to accommodate the density change when transforming from $\mathbb{R}^d$ to $\mathbb{S}_2^d$ [11]. The correct density ratio change is scaled by $\sqrt{\det J_\psi^\top J_\psi}$ instead, whose logarithm can be computed in $\mathcal{O}(d)$ time as

$$\log \sqrt{\det J_\psi^\top(\mathbf{z}) J_\psi(\mathbf{z})} = d \log \frac{2}{1 + \|\mathbf{z}\|^2} = d \log \rho_{\mathbf{z}},$$

with $\rho_{\mathbf{z}}$ given as stated above. The inverse function $\psi^{-1} : \mathbb{S}_2^d \subset \mathbb{R}^{d+1} \to \mathbb{R}^d$ is

$$\psi^{-1}(\mathbf{s}) = \mathbf{z} = \frac{[\mathbf{s}]_{1:d}}{1 - [\mathbf{s}]_{d+1}},$$

where $[\mathbf{s}]_{1:d}$ denotes the first d coordinates of $\mathbf{s}$ and $[\mathbf{s}]_{d+1}$ is the $(d+1)$-th coordinate of $\mathbf{s}$.

To interpolate points on the hypersphere, a spherical linear interpolation (*slerp*) [30] can be utilized. It is defined as follows. Let $\mathbf{s}_a$ and $\mathbf{s}_b$ be two unit vectors and $\omega = \cos^{-1}(\mathbf{s}_a^\top \mathbf{s}_b)$ be the angle between them. The interpolation path is then given by

$$\gamma(\lambda) = \frac{\sin((1-\lambda)\omega)}{\sin(\omega)} \mathbf{s}_a + \frac{\sin(\lambda\omega)}{\sin(\omega)} \mathbf{s}_b, \quad \text{for } \lambda \in [0, 1].$$

5   Experiments

We now evaluate the restriction of a normalizing flow to a unit p-norm sphere and compare them to a Gaussian base distribution. As we focus on a principled way of interpolating in flow-based generative models, we employ a fixed architecture per data set instead of aiming to achieve state-of-the-art density estimation. We use Glow [19] as the flow architecture for the experiments in the remainder of this section. However, our approach is not limited to Glow, and the transformations and changes in the base distribution can also be used in other architectures. We also do not compare against other architectures as our contribution is a change of the base distribution, allowing for better interpolations.

5.1 Performance Metrics and Setup

Performance is measured in terms of *bits per dimension (BPD)*, calculated using $\log_2 p(\mathbf{x})$ divided by d; *Fréchet Inception Distance (FID) scores*, which have been shown to correlate highly with human judgment of visual quality [13]; and *Kernel Inception Distance (KID) scores* [3]. KID is similar to FID as it is based on Inception scores [29]. While the FID first fits a Gaussian distribution on the scores of a reference set and a set of interest and then compares the two distributions, the KID score is non-parametric, i.e., it does not assume any distribution and compares the Inception scores based on Maximum Mean Discrepancy (MMD). We follow previous work [3] and employ a polynomial kernel with degree three for our evaluations.

We measure bits per dimension on the test set and on interpolated samples. FID and KID scores are evaluated on generated and interpolated samples and then compared to a reference set, which is the training data. When generating data, we draw as many samples from the base distribution as we have for training. For interpolation, we focus on interpolation within classes and adopt regular linear interpolation for Gaussian-distributed samples, while using a spherical linear interpolation on the sphere for vMF-distributed samples. In this operation, we sample $n/5$ pairs of images from the training set and generate five equally spaced interpolated data instances per pair, resulting in n new images. From those interpolation paths, we only use the generated points and not the points which are part of training data. Hence, we are only considering previously unseen data.

We also compare against the norm-corrected linear interpolation (*nclerp*) defined in Eq. (1). Note that a linearly spaced interpolation path is no longer linearly spaced after norm correction. The resulting interpolation paths are composed of images located closer to the endpoints and thus bias the evaluation. We include the results nevertheless for completeness.

The reported metrics are averages over three independent runs and include standard errors. The code is written in PyTorch [24]. All experiments run on an Intel Xeon CPU with 256GB of RAM using an NVIDIA V100 GPU.

5.2 Data

In our experiments, we utilize **MNIST** [22], **Kuzushiji-MNIST** [7], and **Fashion-MNIST** [34], which contain gray-scale images of handwritten digits, Hiragana symbols, and images of Zalando articles, respectively. All MNIST data sets consist of 60,000 training and 10,000 test images of size 28×28. In addition, we evaluate on **CIFAR10** [21], which contains natural images from ten classes. The data set has 50,000 training and 10,000 test images of size 32×32.

Table 1. Results for generative modeling averaged over three independent runs including standard errors.

	Base dist.	Test	Sample	
		BPD	FID	KID
MNIST	Gaussian	1.59 ± 0.06	$\mathbf{34.53} \pm 0.83$	$\mathbf{0.033} \pm 0.001$
	vMF $\kappa = 1d$	$\mathbf{1.46} \pm 0.07$	40.07 ± 2.46	0.037 ± 0.001
	vMF $\kappa = 1.5d$	1.54 ± 0.09	40.39 ± 1.40	0.036 ± 0.001
	vMF $\kappa = 2d$	1.82 ± 0.08	39.82 ± 0.26	0.038 ± 0.001
	Dirichlet $\alpha = 2$	1.76 ± 0.12	40.08 ± 0.72	0.039 ± 0.001
K-MNIST	Gaussian	2.58 ± 0.11	35.34 ± 0.76	0.041 ± 0.001
	vMF $\kappa = 1d$	2.63 ± 0.06	36.63 ± 0.37	0.041 ± 0.001
	vMF $\kappa = 1.5d$	$\mathbf{2.48} \pm 0.06$	$\mathbf{35.00} \pm 0.61$	$\mathbf{0.040} \pm 0.001$
	vMF $\kappa = 2d$	2.51 ± 0.04	36.45 ± 0.42	0.041 ± 0.001
	Dirichlet $\alpha = 2$	2.50 ± 0.05	35.54 ± 0.39	$\mathbf{0.040} \pm 0.001$
F-MNIST	Gaussian	3.24 ± 0.04	66.64 ± 1.29	0.064 ± 0.003
	vMF $\kappa = 1d$	$\mathbf{3.16} \pm 0.03$	$\mathbf{60.45} \pm 3.34$	$\mathbf{0.055} \pm 0.005$
	vMF $\kappa = 1.5d$	3.30 ± 0.07	61.89 ± 1.29	0.056 ± 0.002
	vMF $\kappa = 2d$	3.22 ± 0.06	60.60 ± 3.47	$\mathbf{0.055} \pm 0.004$
CIFAR10	Gaussian	3.52 ± 0.01	71.34 ± 0.45	$\mathbf{0.066} \pm 0.001$
	vMF $\kappa = 1d$	3.43 ± 0.00	71.07 ± 0.78	0.069 ± 0.001
	vMF $\kappa = 1.5d$	$\mathbf{3.42} \pm 0.00$	$\mathbf{70.58} \pm 0.40$	0.068 ± 0.001
	vMF $\kappa = 2d$	$\mathbf{3.42} \pm 0.01$	71.00 ± 0.28	0.068 ± 0.001

5.3 Architecture

We employ the Adam optimizer [16] with a learning rate of 10^{-3}, clip gradients at 50, and use linear learning rate warm-up for the first ten epochs. Models were trained on MNIST data and CIFAR10 using mini-batches of size 256 and 128, respectively. All models are trained for 100 epochs without early stopping. We keep all architectures as close as possible to Glow, with the following deviations. For MNIST data, we use random channel permutations instead of invertible 1×1 convolutions. The number of filters in the convolutions of the affine coupling layers is 128. In Glow terms, we employ $L = 2$ levels of $K = 16$ steps each. For CIFAR10, our models have $L = 3$ levels of $K = 24$ steps each, while the affine coupling layers have convolutions with 512 filters. The architecture is kept the same across base distributions, except for the additional parameterless transformations to the restricted subspaces introduced in Sect. 4.

When comparing base distributions, we consider the following hyperparameters. For the vMF distribution, we use concentration values for which the partition function is finite. For consistency, the values we use are the same multiples of the data dimensionality d for each data set. The concentration values for the Dirichlet distribution are set to $\alpha = 2$, which refers to $2 \cdot 1^{d+1} \in \mathbb{R}^{d+1}$.

Table 2. Results for interpolation averaged over three independent runs including standard errors. Interpolations are in-class only and use five intermediate points; *lerp* refers to a linear interpolation; *nclerp* refers to the norm-corrected linear interpolation (Sect. 1) and *slerp* refers to the spherical interpolation.

	Base dist.	Type	BPD	FID	KID
MNIST	Gaussian	lerp	1.33 ± 0.05	5.10 ± 0.14	0.003 ± 0.000
	Gaussian	nclerp	1.44 ± 0.06	5.12 ± 0.30	0.003 ± 0.000
	vMF $\kappa = 1d$	slerp	$\mathbf{1.31} \pm 0.09$	$\mathbf{3.84} \pm 0.36$	$\mathbf{0.002} \pm 0.000$
	vMF $\kappa = 1.5d$	slerp	1.40 ± 0.10	4.22 ± 0.12	$\mathbf{0.002} \pm 0.000$
	vMF $\kappa = 2d$	slerp	1.63 ± 0.10	4.45 ± 0.06	$\mathbf{0.002} \pm 0.000$
	Dirichlet $\alpha = 2$	lerp	1.61 ± 0.10	5.81 ± 0.36	0.004 ± 0.001
K-MNIST	Gaussian	lerp	1.91 ± 0.17	19.71 ± 1.59	0.021 ± 0.002
	Gaussian	nclerp	2.15 ± 0.15	17.60 ± 1.48	0.020 ± 0.002
	vMF $\kappa = 1d$	slerp	2.08 ± 0.15	17.93 ± 3.72	0.020 ± 0.004
	vMF $\kappa = 1.5d$	slerp	$\mathbf{1.80} \pm 0.07$	22.72 ± 2.65	0.025 ± 0.003
	vMF $\kappa = 2d$	slerp	2.03 ± 0.14	$\mathbf{14.54} \pm 2.51$	$\mathbf{0.016} \pm 0.003$
	Dirichlet $\alpha = 2$	lerp	1.81 ± 0.04	24.09 ± 2.35	0.026 ± 0.003
F-MNIST	Gaussian	lerp	2.84 ± 0.10	13.06 ± 0.62	0.007 ± 0.001
	Gaussian	nclerp	2.93 ± 0.03	7.80 ± 0.13	0.004 ± 0.000
	vMF $\kappa = 1d$	slerp	$\mathbf{2.66} \pm 0.03$	$\mathbf{12.16} \pm 0.13$	$\mathbf{0.006} \pm 0.000$
	vMF $\kappa = 1.5d$	slerp	2.84 ± 0.07	12.19 ± 1.07	$\mathbf{0.006} \pm 0.001$
	vMF $\kappa = 2d$	slerp	2.70 ± 0.05	15.11 ± 0.85	0.008 ± 0.001
CIFAR10	Gaussian	lerp	2.64 ± 0.06	58.63 ± 1.26	0.053 ± 0.001
	Gaussian	nclerp	3.32 ± 0.01	14.29 ± 0.16	0.010 ± 0.000
	vMF $\kappa = 1d$	slerp	2.78 ± 0.05	$\mathbf{51.08} \pm 0.37$	0.010 ± 0.000
	vMF $\kappa = 1.5d$	slerp	2.66 ± 0.05	55.23 ± 5.14	0.047 ± 0.005
	vMF $\kappa = 2d$	slerp	$\mathbf{2.58} \pm 0.08$	52.65 ± 3.34	0.044 ± 0.004

5.4   Quantitative Results

We first evaluate the generative modeling aspects of all competitors. Table 1 summarizes the results in terms of bits per dimension on test data and FID and KID scores on generated samples for all data sets. Experiments with the Dirichlet base distribution were not successful on all data sets. The restrictions imposed to enable the use of the distribution demand a high numerical precision since every image on the simplex is represented as a non-negative vector that sums up to one. Consequently, we only report results on MNIST and Kuzushiji-MNIST. Using the vMF as a base distribution clearly outperforms the Gaussian in terms of bits per dimension on test data. As seen in the FID and KID scores, we perform competitive compared to the Gaussian for generating new data. Hence, the generative aspects of the proposed approach are either better or on par with the default choice of a Gaussian. Note that lower bits per dimension on test data and lower FID/KID scores on generated data might be obtained with more sophisticated models.

start   1   2   3   4   5   end

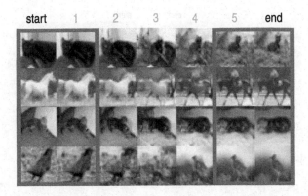

Fig. 6. Five interpolation paths of the norm-corrected linear interpolation (*nclerp*) depicting the problem of almost repeated endpoints (highlighted in blue) and thus a biased evaluation on CIFAR10. (Color figure online)

We now evaluate the quality of interpolation paths generated via various approaches. Table 2 shows the results in terms of bits per dimension, FID, and KID scores for all data sets. The experiments confirm our hypothesis that an interpolation on a fixed-norm space yields better results as measured in bits per dimension, FID, and KID scores. The norm-corrected interpolation yields better FID and KID scores for Fashion-MNIST and CIFAR10. However, this heuristic produces interpolation paths that are biased towards the endpoints (cf. Fig. 3) and hence are naturally closer to observed data, thus yield better FID and KID scores. This is depicted in Fig. 6 where the first and last interpolant is very close to real data. More results on general interpolations within classes and across classes are provided in the supplementary material.

5.5   Qualitative Results

Figure 7 displays interpolation paths with five interpolants of four pairs of data from CIFAR10, created using the same architecture trained on different base distributions. We pick the best-performing model on BPD on test data from the multiple training runs for each base distribution. We visually compare a linear interpolation using a Gaussian base distribution against a spherical linear interpolation using a vMF base distribution with different concentration values. Naturally, the images in the center show the difference and the effects resulting from the choice of base distribution and, hence, the interpolation procedure.

Overall, the linear interpolation with a Gaussian tends to show mainly darker objects on brighter background (almost black and white images) in the middle of the interpolation path. This is not the case for the spherical interpolations using a vMF base distribution. Specifically, in the second example showing dogs, the checkerboard background of the left endpoint smoothly fades out for the vMF ($\kappa = 2d$) model while the Gaussian shows an almost white background. A similar effect happens in the last pair of images, highlighting the weaknesses of

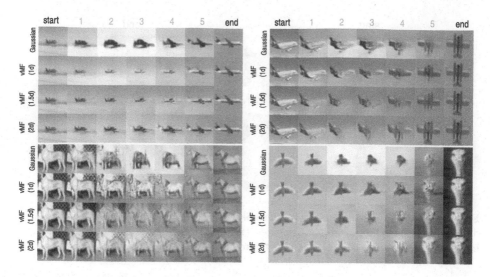

Fig. 7. Interpolation paths of four pairs of data from CIFAR10 using different models.

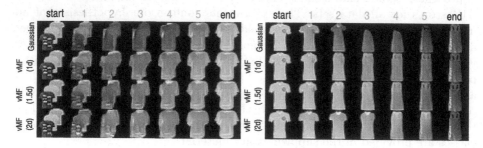

Fig. 8. Interpolation paths of two pairs from Fashion-MNIST using different models.

a linear interpolation once again. By contrast, the vMF models generate images where those effects are either less prominent or non-existent, suggesting a path that strictly follows the data manifold. We provide more interpolation paths on CIFAR10 in the supplementary material.

Figure 8 depicts interpolation paths with five interpolants on two pairs of data from Fashion-MNIST. In both cases, the Gaussian model produces suboptimal images with visible color changes, which is not consistent with the endpoints. Furthermore, there is visible deformation of the clothing items.

6   Related Work

Interpolations are commonplace in generative modeling, being particularly useful for evaluating them. Spherical linear interpolations [30] are also proposed [33] to circumvent the problems depicted in Fig. 1 in GANs and VAEs. However, as the Gaussian is kept as a base distribution, the difference in norms causes problems similar to the norm corrected approach. The problem of interpolation is also investigated for GANs [1]. Specifically, they show that the quality of the generated images in the interpolation path improves when attempting to match the distribution of norms between interpolants and the GAN prior. The problem with the distribution mismatch while interpolating is also studied in [15].

Simultaneously learning a manifold and corresponding normalizing flow on it is also possible [5]. By contrast, in this paper, we employ a prescribed manifold, i.e., a p-norm sphere, on which the interpolation can be done in a principled way. Using a vMF distribution as a prior of VAEs is also used to encourage the model to learn better latent representations on data with hyperspherical structure [8, 35]. While results show improvements over a Gaussian prior, properties of our interest, such as interpolation, are not addressed.

Employing normalizing flows on non-Euclidean spaces, such as the hypersphere, was first proposed by [11]. They introduce a mapping for doing normalizing flows on hyperspherical data. The main difference from our setting is that the data is already on a sphere and is moved to $\mathbb{R}^d$, an unrestricted space, performing the entire flow in there instead, before moving back to the sphere. This avoids defining a flow on the sphere, which is studied in [27] for tori and spheres. Besides, normalizing flows on hyperbolic spaces are beneficial for graph-structured data [4].

A geometric analysis of autoencoders, showing that they learn latent spaces, which can be characterized by a Riemannian metric, is provided by [2]. With this, interpolations follow a geodesic path under this metric, leading to higher quality interpolations. Compared to our contribution, these approaches do not change the standard priors but propose alternative ways to interpolate samples. In contrast, we propose an orthogonal approach by changing the base distribution and imposing constraints on the representation in our training procedure. Consequently, standard interpolation procedures, such as the spherical linear interpolation, can be used in a principled way.

7   Conclusion

This paper highlighted the limitations of linear interpolation in flow-based generative models using a Gaussian base distribution. As a remedy, we proposed to focus on base representations with a fixed norm where the interpolation naturally overcomes those limitations and introduced normalizing flows onto unit p-norm spheres. Specifically, we showed for the cases $p \in \{1, 2\}$ that we could operate on the unit simplex and unit hypersphere, respectively. We introduced a computationally efficient way of using a Dirichlet distribution as a base distribution for the case of $p = 1$ and leveraged a von Mises-Fisher distribution using a stereographic projection onto a hypersphere for the case $p = 2$. Although the former suffered from numerical instabilities in a few experiments, our experimental results showed superior performance in terms of bits per dimension on test data and FID and KID scores on interpolation paths that resulted in natural transitions from one image to another. This was also confirmed by visually comparing interpolation paths on CIFAR10 and Fashion-MNIST.

Acknowledgements. This research was financed in part by the Coordenação de Aperfeiçoamento de Pessoal de Nível Superior (CAPES), Brazil, Finance Code 001 and by FAPESP (grants 2017/24005-2 and 2018/19350-5).

References

1. Agustsson, E., Sage, A., Timofte, R., Gool, L.V.: Optimal transport maps for distribution preserving operations on latent spaces of generative models. In: International Conference on Learning Representations (2019)
2. Arvanitidis, G., Hansen, L.K., Hauberg, S.: Latent space oddity: on the curvature of deep generative models. In: International Conference on Learning Representations (2018)
3. Bińkowski, M., Sutherland, D.J., Arbel, M., Gretton, A.: Demystifying MMD GANs. In: International Conference on Learning Representations (2018)
4. Bose, J., Smofsky, A., Liao, R., Panangaden, P., Hamilton, W.: Latent variable modelling with hyperbolic normalizing flows. In: International Conference on Machine Learning, pp. 1045–1055. PMLR (2020)
5. Brehmer, J., Cranmer, K.: Flows for simultaneous manifold learning and density estimation. In: Advances in Neural Information Processing Systems, vol. 33 (2020)
6. Chen, T.Q., Behrmann, J., Duvenaud, D.K., Jacobsen, J.H.: Residual flows for invertible generative modeling. In: Advances in Neural Information Processing Systems, pp. 9913–9923 (2019)
7. Clanuwat, T., Bober-Irizar, M., Kitamoto, A., Lamb, A., Yamamoto, K., Ha, D.: Deep learning for classical Japanese literature. CoRR abs/1812.01718 (2018)
8. Davidson, T.R., Falorsi, L., De Cao, N., Kipf, T., Tomczak, J.M.: Hyperspherical variational auto-encoders. In: Proceedings of the Thirty-Fourth Conference on Uncertainty in Artificial Intelligence, pp. 856–865 (2018)
9. Dinh, L., Krueger, D., Bengio, Y.: NICE: non-linear independent components estimation. In: 3rd International Conference on Learning Representations, ICLR 2015, San Diego, CA, USA, 7–9 May 2015, Workshop Track Proceedings (2015)
10. Dinh, L., Shol-Dickstein, J., Bengio, S.: Density estimation using Real NVP. In: International Conference on Learning Representations (2017)
11. Gemici, M.C., Rezende, D., Mohamed, S.: Normalizing flows on Riemannian manifolds. CoRR abs/1611.02304 (2016)
12. Goodfellow, I., et al.: Generative adversarial nets. In: Advances in Neural Information Processing Systems, pp. 2672–2680 (2014)
13. Heusel, M., Ramsauer, H., Unterthiner, T., Nessler, B., Hochreiter, S.: GANs trained by a two time-scale update rule converge to a local nash equilibrium. In: Advances in Neural Information Processing Systems, pp. 6626–6637 (2017)
14. Karras, T., Aila, T., Laine, S., Lehtinen, J.: Progressive growing of GANs for improved quality, stability, and variation. In: International Conference on Learning Representations (2018)
15. Kilcher, Y., Lucchi, A., Hofmann, T.: Semantic interpolation in implicit models. In: International Conference on Learning Representations (2018)

16. Kingma, D.P., Ba, J.: Adam: A method for stochastic optimization. In: 3rd International Conference on Learning Representations, ICLR 2015, San Diego, CA, USA, 7–9 May 2015, Conference Track Proceedings (2015)
17. Kingma, D.P., Welling, M.: Auto-encoding variational Bayes. In: 2nd International Conference on Learning Representations, ICLR 2014, Banff, AB, Canada, 14–16 April 2014, Conference Track Proceedings (2014)
18. Kingma, D.P., Welling, M., et al.: An introduction to variational autoencoders. Found. Trends® Mach. Learn. **12**(4), 307–392 (2019)
19. Kingma, D.P., Dhariwal, P.: Glow: generative flow with invertible 1 x 1 convolutions. In: Advances in Neural Information Processing Systems, pp. 10215–10224 (2018)
20. Kingma, D.P., Salimans, T., Jozefowicz, R., Chen, X., Sutskever, I., Welling, M.: Improved variational inference with inverse autoregressive flow. In: Advances in Neural Information Processing Systems, pp. 4743–4751 (2016)
21. Krizhevsky, A., Hinton, G.: Learning multiple layers of features from tiny images. Technical report (2009)
22. LeCun, Y., Cortes, C., Burges, C.: Mnist handwritten digit database. ATT Labs [Online]. http://yann.lecun.com/exdb/mnist, February 2010
23. Papamakarios, G., Nalisnick, E., Rezende, D.J., Mohamed, S., Lakshminarayanan, B.: Normalizing flows for probabilistic modeling and inference. J. Mach. Learn. Res. **22**(57), 1–64 (2021)
24. Paszke, A., et al.: Pytorch: an imperative style, high-performance deep learning library. In: Advances in Neural Information Processing Systems, pp. 8024–8035 (2019)
25. Radford, A., Metz, L., Chintala, S.: Unsupervised representation learning with deep convolutional generative adversarial networks. In: International Conference on Learning Representations, ICLR 2016 (2016)
26. Rezende, D., Mohamed, S.: Variational inference with normalizing flows. In: International Conference on Machine Learning, pp. 1530–1538. PMLR (2015)
27. Rezende, D.J., et al.: Normalizing flows on tori and spheres. In: International Conference on Machine Learning, pp. 8083–8092 (2020)
28. Rippel, O., Adams, R.P.: High-dimensional probability estimation with deep density models. CoRR abs/1302.5125 (2013)
29. Salimans, T., Goodfellow, I., Zaremba, W., Cheung, V., Radford, A., Chen, X.: Improved techniques for training GANs. In: Advances in Neural Information Processing Systems, pp. 2234–2242 (2016)
30. Shoemake, K.: Animating rotation with quaternion curves. In: Proceedings of the 12th Annual Conference on Computer Graphics and Interactive Techniques, pp. 245–254 (1985)
31. Tabak, E.G., Turner, C.V.: A family of nonparametric density estimation algorithms. Commun. Pure Appl. Math. **66**(2), 145–164 (2013)
32. Tabak, E.G., Vanden-Eijnden, E.: Density estimation by dual ascent of the log-likelihood. Commun. Pure Appl. Math. **8**(1), 217–233 (2010)
33. White, T.: Sampling generative networks. CoRR abs/1609.04468 (2016)
34. Xiao, H., Rasul, K., Vollgraf, R.: Fashion-MNIST: a novel image dataset for benchmarking machine learning algorithms. CoRR abs/1708.07747 (2017)
35. Xu, J., Durrett, G.: Spherical latent spaces for stable variational autoencoders. In: Proceedings of the 2018 Conference on Empirical Methods in Natural Language Processing, pp. 4503–4513 (2018)

CycleGAN Through the Lens
of (Dynamical) Optimal Transport

Emmanuel de Bézenac[1](✉), Ibrahim Ayed[1,2](✉), and Patrick Gallinari[1,3]

[1] LIP6, Sorbonne Université, Paris, France
{emmanuel.de-bezenac,ibrahim.ayed,patrick.gallinari}@lip6.fr
[2] Theresis Lab, Thales, Paris, France
[3] Criteo AI Lab, Criteo, Paris, France

Abstract. Unsupervised Domain Translation (UDT) is the problem of finding a meaningful correspondence between two given domains, without explicit pairings between elements of the domains. Following the seminal CycleGAN model, variants and extensions have been used successfully for a wide range of applications. However, although there have been some attempts, they remain poorly understood, and lack theoretical guarantees. In this work, we explore the implicit biases present in current approaches and demonstrate where and why they fail. By expliciting these biases, we show that UDT can be reframed as an Optimal Transport (OT) problem. Using the dynamical formulation of Optimal Transport, this allows us to improve the CycleGAN model into a simple and practical formulation which comes with theoretical guarantees and added robustness. Finally, we show how our improved model behaves on the CelebA dataset in a standard then in a more challenging setting, thus paving the way for new applications of UDT. Supplementary material is available at https://arxiv.org/pdf/1906.01292.

Keywords: Deep learning · Optimal transport · Generative models

1 Introduction

Given pairs of elements from two different domains, *domain translation* consists in learning a mapping from one domain to another, linking paired elements together. A wide range of problems can be formulated as translation, including image-to-image [16], video-to-video [25], image captioning [30], natural language translation [2], etc. However, obtaining paired examples is often difficult and for this reason has motivated a growing interest towards the more general *unpaired* or *unsupervised* setting where only samples from both domains are available without pairing. A seminal and influential work for solving Unsupervised Domain Translation (UDT) has been the CycleGAN model [31]. It has spurred many variants and extensions leading to impressive results in several application domains [7,8,12,18,28].

N. Oliver et al. (Eds.): ECML PKDD 2021, LNAI 12976, pp. 132–147, 2021.
https://doi.org/10.1007/978-3-030-86520-7_9

More formally, Unsupervised Domain Translation (UDT) is the problem of finding, for any element a of a domain $\mathcal{A}$, its best representative b in another given domain $\mathcal{B}$. Both domains are generally provided in the form of a finite number of samples and we will model them here as absolutely continuous probability measures, respectively α and β. We will make the additional hypothesis that both domain are compact in $\mathcal{R}^d$, with regular boundaries. CycleGAN-like models can then be framed as follows: Given samples from the two probability measures α and β, learn transformations T and S that map one distribution onto the other, while being each other's mutual inverse. This problem thus involves minimizing the following loss:

$$\mathcal{L}(T, S, \mathcal{A}, \mathcal{B}) = \mathcal{L}_{\mathrm{gan}}(T, S, \mathcal{A}, \mathcal{B}) + \mathcal{L}_{\mathrm{cyc}}(T, S, \mathcal{A}, \mathcal{B}) \tag{1}$$

where $\mathcal{L}_{\mathrm{gan}}$ ensures, at optimality, that[1]

$$T_\sharp \alpha = \beta \quad \text{and} \quad S_\sharp \beta = \alpha$$

while $\mathcal{L}_{\mathrm{cyc}}$ ensures cycle-consistency, namely that both transformations are mutual inverses.

Despite its popularity and empirical successes, there is no clear understanding on why CycleGAN is so effective. As shown in [14,21,27], the kernel or null space of the CycleGAN loss, *i.e.* the set of couples (T, S) such that $\mathcal{L}(T, S, \mathcal{A}, \mathcal{B}) = 0$, is not reduced to a singleton except in trivial cases and is often infinite in most cases of interest. By studying the kernel of the loss, [21] show more precisely that elements of the null space as well as solutions obtained through the extended version of the loss, where the loss is regularized so that the transformations are close to the identity function, can lead to arbitrarily undesirable solutions of UDT. Thus, there is a discrepancy between what the loss of CycleGAN-like models captures and their practical usefulness. [14] postulate that obtained solutions are of minimal *complexity*, a notion related in their work to the minimal number of neural layers necessary to represent a function, and conjecture that mappings of minimal complexity represent a small subset of the CycleGAN's loss kernel. Although their definition of complexity is not satisfying and they do not explain why these solutions would correspond to satisfactory ones, this intuition is a valuable one and we build upon it in this work.

More generally, this paper attempts to explain empirically and theoretically why and in which conditions CycleGAN works, and proposing a framework which opens the way for more robust and more flexible CycleGAN models. More precisely,

- We assess the desiderata ensuring satisfactory results for UDT and conduct an empirical analysis of CycleGAN which shows a systematic implicit bias towards low energy transformations, i.e. transformations that displace the inputs as little as possible, and that this bias not only explains its success, but predicts where it fails.

---

[1] The *push-forward measure* $f_\sharp \rho$ is defined as $f_\sharp \rho(B) = \rho(f^{-1}(B))$, for any measurable set B. Said otherwise, we need T to map α to β and S does the reverse.

- Building on this idea, we reformulate the general problem of UDT as an Optimal Transport (OT) problem, thus allowing us to use results from OT theory. This ensures the well-posedness of the problem and regularity of the solution. We are also able to solve problems where CycleGAN methods fail.
- We illustrate our findings by proposing a simple instance of the formulation and conducting illustrative experiments. Using the dynamical formulation of OT, our model is more robust, allows for smooth interpolations and halves the number of necessary parameters by providing an inverse mapping for free after training.

2   Desiderata for UDT and Analysis of CycleGAN

Here, we characterize qualitatively then quantitatively how a good UDT model should behave and show that CycleGAN-like models tend to compute low-energy transformations.

2.1   What Should Be the Properties of a UDT Solution?

Qualitatively, good solutions of a UDT problem are the ones which translate an input a from $\mathcal{A}$ to $\mathcal{B}$ while still conserving as much as possible the characteristics of a, and conversely from $\mathcal{B}$ to $\mathcal{A}$. The CycleGAN seminal paper tries to enforce this through the cycle-consistency loss but, as discussed above and in previous papers, this loss is null for any invertible mapping T by taking the couple (T, T^{-1}), without necessarily conserving any characteristics across domains. In other words, this loss doesn't really add any constraints on the mapping and infinitely many undesirable can still be theoretically recovered by the model.

This intuition has already been formulated in [14] in the notion of "semantics preserving mappings". The authors, recognizing that preserving semantics is a vague notion, propose to measure it through the minimal number of layers necessary for neural networks to represent the transformation. However, while we think that it provides a useful step forward in understanding UDT, such a formulation has several shortcomings: There is no reason why complexity should always be measured as the number of layers of a non-residual NN [29] and it is not even clear whether such a minimal number is always finite for relevant transformations; This notion doesn't provide theoretical insights on how and why CycleGAN performs so well in practice or why it seems to work well even with very deep networks; Crucially, there are no guarantees regarding the uniqueness of minimal complexity mappings.

It is also interesting to consider the extended loss for CycleGAN introduced in the original paper [31] as a regularization forcing T and S to be close from the identity mapping. While, as shown theoretically and empirically in [21], adding this regularization doesn't prevent undesirable mappings to be reached by the model, the fact that it was necessary to further constrain the objective for certain tasks in this way shows that it can be helpful to have transformations which do not transform inputs too much. This is coherent with the view of [14]. We aim to extend both approaches in a more adaptive, robust and theoretically grounded formalism.

Generalizing those discussions, in our view, there are two main important desirable features in UDT models as used in many practical settings:

- The mapping T (and, symmetrically S) should be constrained to be as conservative as is possible, in the sense that they should be as close to the identity as is possible.
- The mappings T and S should also be regular. Indeed, in the case of image-to-image translation from paintings to photographs for example, if we take two paintings a, a' representing nearly the same scene then we would want the corresponding photos $T(a), T(a')$ to be similar as well. This property would mean that T and S are endowed with some functional regularity, at least a form of continuity.

While the first feature extends the points of view already discussed in previous works, the second one is novel, up to our knowledge. It seems difficult to enforce directly the regularity of the estimated mappings but we show in the following that our approach seamlessly satisfies both properties.

2.2   CycleGAN Is Biased Towards Low Energy Transformations

In practice, the success of CycleGAN models is made possible by the presence of inductive biases that constrain the set of solutions and that are imposed through the combination of the choices made for SGD-based methods, networks architectures, weight parameterization and initialization. In order to develop a better understanding and identify implicit biases, we have conducted an exploratory analysis to characterize the influence of CycleGAN hyperparameters. Our main finding is that the initialization gain σ, *i.e.* the standard deviation of the weights of the residual network (along with a fixed small learning rate), has the most substantial and consistent impact, among all the hyperparameters, on the retrieved mappings. These findings are illustrated in the following experiments.

2D Toy Example. Figure 1 shows the effect of changing the gain from a small value, $\sigma = 0.01$, to a higher one, $\sigma = 1$ when learning to map one circular distribution to another. This changes the obtained mapping from a simple translation aligning the two distributions with a minimum displacement to a more disorderly one. In other words, it seems that higher initialization gains lead to higher *energy* mappings. Further quantifying the effect of initialization gain on the retrieved mappings, we use a natural characterisation of *disorder/complexity* of a mapping: the average distance between a sample x from α and its image $T(x)$. Using the squared Euclidean distance, this corresponds to the *kinetic energy* of the displacement and can be written as $\int_{\mathcal{R}^d} \|x - T(x)\|_2^2 \, d\alpha(x)$. This quantity is also the *quadratic transport cost* used in Optimal Transport [22]. Using the mapping minimizing this cost as a reference, we see, on the left of Fig. 2, that the larger σ becomes, the further CycleGAN's mapping (blue curve) is from it. The

$\sigma = 0.01$ $\sigma = 1$

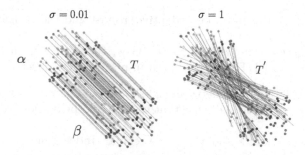

Fig. 1. Pairings between domains obtained with CycleGAN. Both domains correspond to uniform distributions on a 2d-sphere with shifted centers. Small initialization values lead to simple and ordered mappings (Left), whereas larger ones yield complex and disordered ones (Right). Colors highlight original pairing between domains, before shifting.

right curve confirms this finding: As σ grows, so does the transport cost of the trained CycleGAN. For both experiments, the variance across runs increases i.e. the model yields very different mappings across runs, corroborating the ill-posed nature of CycleGAN's optimization problem.

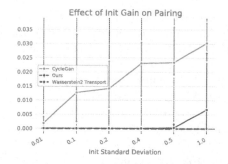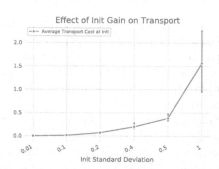

Fig. 2. Left: L^2 distance to the Optimal Transport mapping "*Wasserstein 2 Transport*" as a function of the initialization gain (domains are illustrated in Fig. 1). "Ours" refers to the model presented subsequently. Right: Transport cost of the CycleGAN mapping as a function of initialization gain. Metrics are averaged across 5 runs, and the standard deviation is plotted. (Color figure online)

High-Dimensional Analysis. We also conducted a similar analysis with high-dimensional distributions of images on the CelebA dataset. While in this case calculating the exact OT map is intractable, we can visualize samples obtained with the CycleGAN mapping for different values of σ. The task is male to female transformation where one wants to keep as many characteristics as possible from the original image in the generated one. The result in Fig. 5 confirms the low-dimensional findings: while in all cases the distributions have been successfully aligned, as all males are transformed into females, the mappings initialized with low σ values perform a minimal transformation of the input while high σ values produce unwanted changes in the features (hair color, face, skin color,...). This is corroborated by measuring the transportation cost incurred by CycleGAN which goes from 0.15 for $\sigma = 0.01$ to 9.7 for $\sigma = 1.5$, showing that this behaviour is linked with *high transport costs*.

In summary, for common UDT tasks where the input is to be preserved as much as possible, successful CycleGAN models tend to consistently converge to low energetic mappings and this bias is induced by a small initialization gain. However, the CycleGAN model doesn't give any explicit control over this bias, thus warranting a blind hyper-parameter / architecture search for each new task. In the following section, we use OT to define a class of explicitly controllable models with theoretical guarantees.

3   UDT as Optimal Transport

Using Optimal Transport theory, this section formalizes the findings of the previous one.

3.1   A (Dynamical) OT Model for UDT

Let us consider the classical Monge problem formulation for OT:

$$\min_{T} \quad \mathcal{C}(T) = \int_{\mathcal{R}^d} c(x, T(x)) \, d\alpha(x)$$
$$\text{s.t.} \quad T_{\sharp}\alpha = \beta \tag{2}$$

with the *ground cost* being defined as $c(x, y) = h(x - y)$ with h strictly convex.

Using OT as a way to solve UDT seems very natural as, for most applications, the user's criteria are about preserving input features as much as possible: this is precisely what is given by the OT mapping, its associated cost defining which features are to be preserved. Our idea is that any solution of the Monge problem would be a good candidate for a UDT forward mapping.

Moreover, for a wide range of costs, *e.g.* cost of the form[2] $c(x, y) = \|x - y\|^p$ for $p > 1$, there exists a dynamical point of view of OT equivalent to the Monge formulation[3], similar in intuition and formulation to the equations of fluid dynamics. The general idea is to produce T by using a velocity field v which gradually transports particles from α to β. The OT map can then be recovered from a path of minimal length, with v solving the optimization problem:

---

[2] A larger family of costs can be considered at the expense of some technicalities, see [13].

[3] Which was pioneered in [5] and for which a detailed modern presentation is given in chapters 4 and 5 of [22].

$$
\begin{aligned}
\min_{v} \quad & \mathcal{C}^{\mathrm{dyn}}(v) = \int_0^1 \|v_t\|_{L^p((\phi_t^{\cdot})_{\sharp}\alpha)}^p \, \mathrm{d}t \\
\text{s.t} \quad & \partial_t \phi_t^x = v_t(\phi_t^x) \\
& \phi_0^{\cdot} = \mathrm{id}_{\mathcal{A}} \\
& (\phi_1^{\cdot})_{\sharp}\alpha = \beta
\end{aligned}
\tag{3}
$$

where the function $\phi_t^{\cdot} : \mathcal{A} \to \mathcal{R}^d$, induced by the vector field v, is the transport map at time t. This problem can be treated as a continuous-time optimal control problem, and can thus be solved using standard techniques [22].

We then have, using results from Optimal Transport theory:

Proposition 1 (Existence, Uniqueness and Interpolation). *With the hypothesis already made for α, β and c, (2) admits a unique minimum realized with an invertible map $T^\star$.*

Moreover, for $p > 1$, when $c(x,y) = \|x - y\|^p$, (3) also admits a unique minimal vector field $v^\star$. In addition, we have that the corresponding curve $(\phi_t^{\cdot})_{\sharp}\alpha$ interpolates geodesically between α and β in $\mathbb{W}_p$.

Finally, we have that $T^\star = \phi_1^{\cdot}$ and we recover the transport cost in the static Monge formulation, i.e. $\mathcal{C}^{dyn}(v^\star) = \mathcal{C}(T^\star)$.

Proof. α, β and c verify the hypothesis for [22, Theorem 1.17] which gives existence and uniqueness of the OT map $T^\star$. Its invertibility is justified by Remark 1.20 of the same reference.

Taking $\mu_t = (\phi_t^{\cdot})_{\sharp}\alpha$, we have that (μ_t, v_t) solves the continuity equation:

$$
\partial_t \mu_t + \nabla \cdot (\mu_t v_t) = 0
$$

(3) then becomes a problem of finding the curve of minimal length in $\mathbb{W}_p$ between α and β. This space being a geodesic one, such a curve always exists and is unique. This also justifies the equivalence of (2) and (3) as well as the fact that $T^\star = \phi_1^{\cdot}$. A more rigorous justification is given in chapters 4 and 5 of [22]. $\square$

Here, $\mathbb{W}_p$ is the metric space of absolutely continuous probability measures with finite p-th moment where the distance between two measures μ, ν is defined as the p-th root of the OT cost between them.

The key claim of this work, which is supported by the experiments conducted in Sect. 2.2, is the following one: The OT map T for the quadratic cost behaves very similarly to the solution of UDT approximated by CycleGAN-like models when they behave correctly.

3.2 Regularity of OT Maps

Let us recall the definition of Hölder continuity: A function $f : \mathcal{X} \to \mathcal{Y}$ is said to be η-Hölder continuous if:

$$
\forall x, y \ \|f(x) - f(y)\| \leq M \|x - y\|^{\eta}
$$

for $\eta \in]0, 1]$. Moreover, the space of functions whose k-th derivative is η-Hölder continuous is denoted by $C^{k,\eta}(\mathcal{X}, \mathcal{Y})$.

Using the same notation as above and recent results obtained for OT maps [20], we have the following:

Proposition 2 (Regularity). $T^\star$ *is everywhere differentiable, except on a set of null α measure.*

Additionally, if $T^\star$ does not have singularities, there exists $\eta > 0$ and A, respectively B, relatively closed in $\mathcal{A}$, respectively $\mathcal{B}$, of null Lebesgue measure, such that $T^\star$ is η-Hölder continuous from $\mathcal{A} \setminus A$ to $\mathcal{B} \setminus B$.

Moreover, if the densities of α and β are $C^{k,\eta}$, then $T^\star \in C^{k+1,\eta}(\mathcal{A} \setminus A, \mathcal{B} \setminus B)$.

This notion of regularity is exactly the one that one wants for UDT as the regularity of the mappings has to be linked to that of their underlying domains. Here, the recovered map is even one degree more regular than the domains themselves.

Moreover, the fact that regularity excludes a negligible set of points of the domains is also coherent with what we should expect: In the transported domains, there can be points which are close but nevertheless represent elements from different classes and thus should be transported far from each other. For example, in image-to-image translation between photographs and paintings, two images with the same background can represent different objects and thus be translated into very different paintings. Thus, this regularity result supports our claim for the transport cost to be the right measure of "complexity" for UDT mappings.

3.3   Computing the Inverse

Consider the optimal vector field of (3) and the following system of differential equations, for all $x \in \mathcal{B}$:

$$\begin{cases} \partial_t \psi_t^x = -v_t^\star(\psi_t^x) \\ \psi_0^x = x \end{cases} \tag{4}$$

Then we have the following:

Proposition 3. *The solution curve $(\psi_t)_t$ of (4) geodesically interpolates between β and α. In particular, $S^\star = \psi_1^\cdot$ is the inverse of $T^\star$, verifies $S_\sharp^\star \beta = \alpha$ and is the OT map between β and α.*

Proof. Let us consider $\nu_t = (\psi_t^\cdot)_\sharp \beta$. Then $(\nu_t, -v_t)$ solves the continuity equation. On the other hand, by a direct calculation and taking previous notations, we have that:

$$\frac{\mathrm{d}}{\mathrm{d}t} \int f \mathrm{d}\mu_{1-t} = -\int \nabla f(x) \cdot v_t(x) \mathrm{d}\mu_{1-t}$$

for any C^1 test function f which means that $(\mu_{1-t}, -v_t)$ also solve the continuity equation with the same initial condition β. This means, by uniqueness, that we have $\nu_t = \mu_{1-t}$ which proves the result. □

This result shows that $(T^\star, S^\star)$ does indeed solve the UDT problem and is in the null space of the CycleGAN loss. Moreover, in order to compute $S^\star$, there is no need to parametrize it nor to solve a difficult optimization problem. It is only necessary to discretize the associated differential equation which is of the same nature as the one for the forward mapping, meaning that the same scheme can be used.

4 A Residual Instantiation from Dynamical OT

This section proposes an instantiation of our model which closely follows the CycleGAN implementation and experiments are conducted to compare both on the CelebA dataset.

4.1 Linking the Dynamical Formulation with CycleGAN

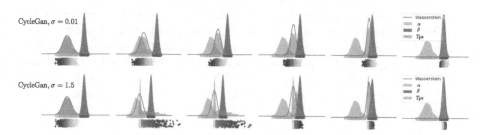

Fig. 3. Visualization of the hidden layers of CycleGAN when mapping the yellow gaussian distribution to the green one with different initializations: As shown by the colored points representing samples under the histograms, when σ increases, the mapping goes from a simple translation (Top) to a more complicated mapping (Bottom), thus inducing an increase in transport cost. (Color figure online)

Let us show that CycleGAN corresponds to a specific implementation of our dynamic formulation with the added transport minimization.

Discretization. If v_k corresponds to the residual for layer k of the residual block defined by $\phi_k^x = \phi_{k-1}^x + v_k(\phi_{k-1}^x)$, then, taking the continuous time limit, one recovers the differential equation $\partial_t \phi_t^x = v_t(\phi_t^x)$ [26] which appears as a constraint in Eq. 3. Thus, if we discretize the forward equation in (3) using an Euler numerical scheme, we recover the forward map in the CycleGAN architecture[4].

In CycleGAN, the first boundary condition $\phi_0 = \mathrm{id}$ is satisfied by construction, while the second $(\phi_1^\cdot)_\sharp \alpha = \beta$ is enforced with the GAN loss.

---

[4] Other schemes could be used, which would lead to other architectures, and could arguably be more suited for stability reasons but this is beyond the scope of this work.

Thus we recover CycleGAN as a particular implementation of this model when there is no transport cost minimization. We actually construct our instantiation in a similar fashion in order to have meaningful comparisons: The differential equations are discretized using an Euler scheme and boundary conditions are enforced using an iterative penalization of the GAN loss. More involved schemes can be used here such as any suitable parametrized solver [6].

The fully discretized optimization problem is then the following:

$$\min_{\theta} \quad \mathcal{C}_d(\theta) = \sum_{k=1}^{K} \sum_{x \in \text{Data}_\alpha} \|v^{\theta_k}(\phi_k^x)\|_p^p$$

$$\text{s.t} \quad \forall x, \ \forall k, \ \phi_{k+1}^x = \phi_k^x + \Delta t \, v^{\theta_k}(\phi_k^x)$$

$$\phi_0^{\cdot} = \text{id}, \quad (\phi_1^{\cdot})_\sharp \alpha = \beta \tag{5}$$

Let us also notice that using small initialization gains for the network (See 2.2) tends to bias the $\|v^\theta\|$s to small values, linking latent trajectories of residual networks with minimal length ones as in Fig. 3. It remains to be proven that this fact is indeed stable after training via gradient descent and we consider this to be an interesting problem to analyze in the future.

Enforcing Boundary Conditions. The constraint $(\phi_1^{\cdot})_\sharp \alpha = \beta$ ensuring that input domain α maps to the target domain β isn't straightforward to implement. We do so by optimizing an iterative Lagrangian relaxation associated to (5), introducing a measure of discrepancy D between output and target domains:

$$\min_{\theta} \quad \mathcal{C}_d(\theta) + \frac{1}{\lambda_i} D((\phi_1^{\cdot})_\sharp \alpha, \beta) \tag{6}$$

where the sequence of Lagrange multipliers $(\lambda_i)_i$ converges linearly to 0 during optimization. At the limit, as the sequence of multipliers converges to 0, the constraint is satisfied.

Each λ_i induces an optimization problem which is solved using stochastic gradient based techniques. As in most approaches for UDT, D may be implemented using generative adversarial networks, or any other appropriate measure of discrepancy between measures, such as kernel distances. Moreover, in order to stabilize the adversarial training which enforces boundary conditions for both our model and CycleGAN, we use an auto-encoder to a lower dimensional latent space. This limits the sharpness of output images but produces consistent and reproducible results, thus allowing meaningful comparisons which is the objective here.

Algorithm. Training is done only for the forward equation and the reverse is obtained by iterating $y_{k-1} = y_k - \Delta t \, v^{\theta_k}(y_k)$, starting from a sample y_K from β, as Sect. 3.3 allows to. Algorithm 1 gives all necessary details of the procedure.

Algorithm 1. Training procedure

Input: Dataset of unpaired images (I_A, I_B) sampled from (α, β),

initial coefficient λ_0, decay parameter d, initial parameters θ, minimal penalization ϵ

Pretrain Encoder E and decoder D

Make dataset of encodings $(x = E(I_A), y = E(I_B))$

for $i = 1, \ldots, M$ **do**

  Randomly sample a mini-batch of x, y

  Solve forward equation $\phi_{k+1}^x = \phi_k^x + \Delta t\, v^{\theta_k}(\phi_k^x)$, starting from $\phi_0^x = x$

  Estimate loss $\mathcal{L} = \mathcal{C}_d(\theta) + \frac{1}{\lambda_i} D((\phi_1)_\# \alpha, \beta)$ on mini-batch

  Compute gradient $\frac{d\mathcal{L}}{d\theta}$ backpropagating through forward equation

  Update θ in the steepest descent direction

  $\lambda_{i+1} \leftarrow \max(\lambda_i - d, \epsilon)$

end for

Output: Learned parameters θ.

Architectures. Implementation is performed via DCGAN and ResNet architectures as described below. For the Encoder, we use a standard DCGAN architecture[5], augmenting it with 2 self-attention layers, mapping the images to a fixed, 128 dimensional latent vector. For the Decoder, we use residual up-convolutions, also augmented with 2 self-attention layers. We use 9 temporal steps, corresponding to as many residual blocks which consist of a linear layer, batch normalization, a non-linearity, and a final linear layer. The discrepancy D is implemented using generative adversarial networks with the discriminator being a simple MLP architecture of depth 3, consisting of linear layers with spectral normalization, and LeakyReLU($p = 0.2$).

Moreover, in the experiments below, our dataset is the CelebA dataset, resizing images to 128×128 pixels, without any additional transformation. The initial coefficient is $\lambda_0 = 1$, and the decay factor is set depending on the number of total iterations M, so as to be ϵ on the final iteration. Throughout all the experiments, we use the Adam optimizer with $\beta_1 = 0.5$ and $\beta_2 = 0.999$.

4.2   A Typical UDT Task

Taking the CelebA dataset, we consider the male to female task where the objective is to change the gender of the input image while keeping other characteristics of the image unchanged as much as possible.

Figure 4 illustrates how our model works for Male to Female translation (forward) and back (reverse) on the CelebA dataset, displaying intermediate images as the input distribution gradually transforms into the target distribution. **No**

---

[5] https://github.com/pytorch/examples/tree/master/dcgan

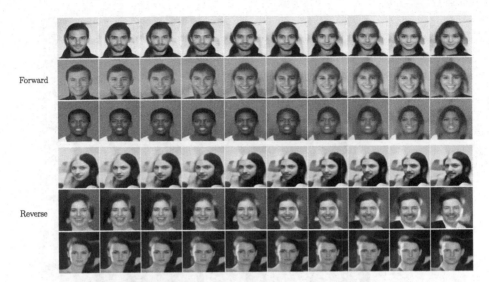

Fig. 4. Male to Female translation (top) and the inverse (bottom). Intermediate images are the interpolations provided by the network's intermediate layers. The reverse mapping is not trained.

cycle-consistency is being explicitly enforced here and the **reverse is not directly parametrized nor trained** but still performs well. The model changes relevant high-level attributes when progressively aligning the distributions but doesn't change non-relevant features (hair or skin color, background,...) which is coherent to what is expected for an optimal map w.r.t. an attractive cost function (here the squared Euclidean one). All the experiments conducted in this work with our proposed OT framework have been implemented using this dynamical formulation.

Figure 5 shows that for a low initialization gain, both our method and Cycle-GAN give satisfying and similar solutions. When changing the value of this hyper-parameter, the CycleGAN mapping becomes unstable, producing outputs very different from the inputs.

The non-uniqueness of the solution of CycleGAN's optimization problem is highlighted here by the multiple mappings found for different initializations. It is also worth noting that, for CycleGAN, using a large σ made convergence of the optimization harder. As already observed before, the chaotic behavior of the CycleGAN model correlates with an increase in the transport cost of the obtained mappings. This validates the L^2 OT bias of CycleGAN, showing that this model only works as an implicit OT mapping for a quadratic cost given a certain architecture, initialized and trained in a certain way. For this example, the prior induced by the quadratic transport cost is the right one and correctly captures the geometry of the task, as one wants to preserve as much as possible the characteristics of the input. By explicitly enforcing optimality w.r.t. the

Input

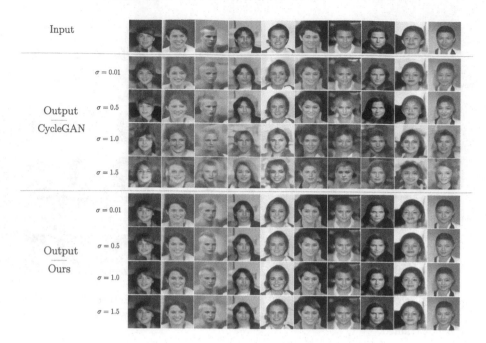

CycleGAN

Ours

Fig. 5. Each column associates one input image to its outputs for different models: CycleGAN and our model with different initial gain parameters. We have ensured convergence of all models to the same fit to the target distribution.

quadratic cost, the model becomes robust to changes in the initialization as the OT problem admits a unique solution for this cost.

4.3 Imbalanced CelebA task

Here, we tackle the case of a corrupted dataset where **structural bias** is present in the target domain, which can be an important use case of UDT when fairness of the datasets is an issue [11]: samples from the target dataset are systematically corrupted. We consider a subset of the CelebA dataset, where domains correspond, respectively, to female faces with **black hair** which are **non-smiling** for α, and **smiling** for β. However, we only have access to biased samples from β, where female faces have **blond** instead of **black** hair.

In Fig. 6, we report results with CycleGAN and our approach with the quadratic cost: the hair color is modified along with the *smile* feature, and black-haired non smiling faces are mapped to blond smiling ones as should be expected from both. This highlights a particular case where CycleGAN's implicit bias fails.

Using our presented formulation, we are able to solve this task by changing the cost function: We use a non-standard cost which is more suited to the geometry of the problem:

$$c(x, y) = \|H(x) - H(y)\|_2^2 \tag{7}$$

Input	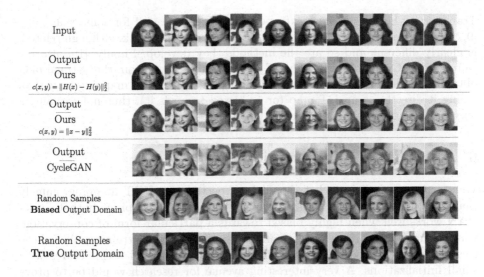
Output Ours $c(x,y) = \|H(x) - H(y)\|_2^2$	
Output Ours $c(x,y) = \|x - y\|_2^2$	
Output CycleGAN	
Random Samples **Biased** Output Domain	
Random Samples **True** Output Domain	

Fig. 6. Results for imbalanced CelebA task. We wish to map faces that have the **Non-Smiling** and **Black Hair** attributes to **Smiling, Black-Hair** faces, while only accessing **Smiling, Blond Hair** faces for the target domain.

where $H(I)$ is a histogram function of the image I. More precisely, H is computed as a soft histogram over the colors of the image of 20 bins, using a Gaussian kernel with $\sigma = 0.05$ for the smoothing. This cost allows to take into account the texture of the image, thus helping to find an OT map which preserves hair color in this case and re-balances the dataset as needed.

This task is an example of a case where a simple cost may help achieve non-trivial results when appropriate information is injected into it. In other words, by using prior knowledge on the corruption of the dataset, a cost function can be tailored to correct it.

More generally, it is not difficult to prove that a cost can almost always be designed to find the right solution for a given task between two distributions α and β among the infinity of candidates in the kernel of CycleGAN's loss.

5 Related Work

As discussed before, our work is motivated by the observations of works such as [3,14] which have linked well-behaved UDT models with a notion of simplicity which we tried here to frame in a more rigorous and more useful formulation, making it task dependant. Moreover, similarly to us, [4,14] show that learning a one-sided mapping is possible but do not directly obtain the inverse mapping as we do. Others have tried a hybrid approach between paired and unpaired translation [24], which still doesn't solve the problem of ill-posedness as there generally still are infinitely many possible mappings. Also similar to us, [15] uses a progressive interpolation. In the domain adaptation field, using Optimal

Transport to help a classifier extrapolate has been around for some years, e.g. [9,10] use a transport cost to align two distributions. The task, although related, is clearly different and so are the methods they develop. Finally, [19] also try to regularize CycleGAN through OT but use barycenters from the optimal plan obtained in the discrete, static setting in order to guide the mapping instead of seeing it directly as an OT map (or as biased towards it), thus not explaining why CycleGAN works in practice.

6 Discussion and Conclusion

We start by formalizing what should be expected of a UDT mapping, namely that it should be conservative and regular. We then show empirically that Cycle-GAN works well when highly biased towards a particular form of conservation, which is unexpected as this is not enforced explicitly during training. We believe this is in particular due to gradient descent using residual architectures with small initializations. A very interesting avenue for research would be to prove this theoretically, potentially making use of recent developments in the implicit regularization effect of gradient descent [1,17,23].

We believe the proposed OT formulation is particularly adapted to UDT and allows us to leverage the plethora of theoretical and practical tools developed in this community. Typically, we were able to guarantee not only the existence and uniqueness of the solution, but also provide fine grained regularity results for the solution map. Moreover, we have also adapted practical algorithms from OT and have made analogies between residual networks and Dynamical OT which have resulted in an improved UDT model which is more robust and can be useful in settings where CycleGAN's biases fail.

References

1. Arora, S., Cohen, N., Hu, W., Luo, Y.: Implicit regularization in deep matrix factorization (2019)
2. Bahdanau, D., Cho, K., Bengio, Y.: Neural machine translation by jointly learning to align and translate (2015). http://arxiv.org/abs/1409.0473
3. Benaim, S., Galanti, T., Wolf, L.: Estimating the success of unsupervised image to image translation. In: Ferrari, V., Hebert, M., Sminchisescu, C., Weiss, Y. (eds.) ECCV 2018. LNCS, vol. 11209, pp. 222–238. Springer, Cham (2018). https://doi.org/10.1007/978-3-030-01228-1_14
4. Benaim, S., Wolf, L.: One-sided unsupervised domain mapping (2017)
5. Benamou, J., Brenier, Y.: A computational fluid mechanics solution to the Monge-Kantorovich mass transfer problem. Numerische Mathematik **84**, 375–393 (2000). https://doi.org/10.1007/s002110050002
6. Chen, T.Q., Rubanova, Y., Bettencourt, J., Duvenaud, D.K.: Neural ordinary differential equations. CoRR abs/1806.07366 (2018)
7. Choi, Y., Choi, M., Kim, M., Ha, J., Kim, S., Choo, J.: StarGAN: unified generative adversarial networks for multi-domain image-to-image translation (2018)

8. Chung, Y., Weng, W., Tong, S., Glass, J.: Unsupervised cross-modal alignment of speech and text embedding spaces (2018)
9. Courty, N., Flamary, R., Tuia, D., Rakotomamonjy, A.: Optimal transport for domain adaptation. CoRR abs/1507.00504 (2015)
10. Damodaran, B.B., Kellenberger, B., Flamary, R., Tuia, D., Courty, N.: DeepJDOT: deep joint distribution optimal transport for unsupervised domain adaptation (2018)
11. Dwork, C., Hardt, M., Pitassi, T., Reingold, O., Zemel, R.S.: Fairness through awareness (2012)
12. Felix, R., Kumar, V.B.G., Reid, I., Carneiro, G.: Multi-modal cycle-consistent generalized zero-shot learning (2018)
13. Figalli, A.: Optimal Transportation and Action-Minimizing Measures. Tesi 8, Ed. Della Normale, Pisa (2008). http://www.sudoc.fr/128156937
14. Galanti, T., Wolf, L., Benaim, S.: The role of minimal complexity functions in unsupervised learning of semantic mappings (2018)
15. Gong, R., Li, W., Chen, Y., Gool, L.V.: DLOW: domain flow for adaptation and generalization (2019)
16. Isola, P., Zhu, J.Y., Zhou, T., Efros, A.A.: Image-to-image translation with conditional adversarial networks. arXiv (2016)
17. Ji, Z., Telgarsky, M.: Gradient descent aligns the layers of deep linear networks (2019)
18. Lample, G., Conneau, A., Denoyer, L., Ranzato, M.: Unsupervised machine translation using monolingual corpora only (2018)
19. Lu, G., Zhou, Z., Song, Y., Ren, K., Yu, Y.: Guiding the one-to-one mapping in cycleGAN via optimal transport (2018)
20. Ma, X., Trudinger, N., Wang, X.: Regularity of potential functions of the optimal transportation problem. Archive for Rational Mechanics and Analysis (2005)
21. Moriakov, N., Adler, J., Teuwen, J.: Kernel of cycleGAN as a principle homogeneous space (2020)
22. Santambrogio, F.: Optimal Transport for Applied Mathematicians: Calculus of Variations, PDEs and Modeling. Springer, New York (2015). https://doi.org/10.1007/978-3-319-20828-2
23. Soudry, D., Hoffer, E., Nacson, M.S., Srebro, N.: The implicit bias of gradient descent on separable data. J. Mach. Learn. Res. 19(1), 2822–2878 (2018)
24. Tripathy, S., Kannala, J., Rahtu, E.: Learning image-to-image translation using paired and unpaired training samples (2018)
25. Wang, T., et al.: Video-to-video synthesis. CoRR abs/1808.06601 (2018). http://arxiv.org/abs/1808.06601
26. Weinan, E.: A proposal on machine learning via dynamical systems. Commun. Math. Stat. 5, 1–11 (2017)
27. Yang, C., Kim, T., Wang, R., Peng, H., Kuo, C.J.: ESTHER: extremely simple image translation through self-regularization (2018)
28. Yuan, Y., Liu, S., Zhang, J., Zhang, Y., Dong, C., Lin, L.: Unsupervised image super-resolution using cycle-in-cycle generative adversarial networks (2018)
29. Zhang, C., Bengio, S., Hardt, M., Recht, B., Vinyals, O.: Understanding deep learning requires rethinking generalization (2017)
30. Zhang, H., et al.: StackGAN: text to photo-realistic image synthesis with stacked generative adversarial networks (2016)
31. Zhu, J., Park, T., Isola, P., Efros, A.A.: Unpaired image-to-image translation using cycle-consistent adversarial networks. CoRR abs/1703.10593 (2017). http://arxiv.org/abs/1703.10593

Decoupling Sparsity and Smoothness in Dirichlet Belief Networks

Yaqiong Li[1], Xuhui Fan[2(✉)], Ling Chen[1], Bin Li[3], and Scott A. Sisson[2]

[1] Australian Artificial Intelligence Institute, University of Technology,
Sydney, Australia
`yaqiong.li@student.uts.edu.au, ling.chen@uts.edu.au`
[2] UNSW Data Science Hub, and School of Mathematics and Statistics,
University of New South Wales, Sydney, Australia
`{xuhui.fan,scott.sisson}@unsw.edu.au`
[3] Shanghai Key Laboratory of IIP, School of Computer Science, Fudan University,
Shanghai, China
`libin@fudan.edu.cn`

Abstract. The Dirichlet Belief Network (DirBN) has been proposed as a promising deep generative model that uses Dirichlet distributions to form layer-wise connections and thereby construct a multi-stochastic layered deep architecture. However, the DirBN cannot simultaneously achieve both *sparsity*, whereby the generated latent distributions place weights on a subset of components, and *smoothness*, which requires that the posterior distribution should not be dominated by the data. To address this limitation we introduce the sparse and smooth Dirichlet Belief Network (ssDirBN) which can achieve both sparsity and smoothness simultaneously, thereby increasing modelling flexibility over the DirBN. This gain is achieved by introducing binary variables to indicate whether each entity's latent distribution at each layer uses a particular component. As a result, each latent distribution may use only a subset of components in each layer, and smoothness is enforced on this subset. Extra efforts on modifying the models are also made to fix the issues which is caused by introducing these binary variables. Extensive experimental results on real-world data show significant performance improvements of ssDirBN over state-of-the-art models in terms of both enhanced model predictions and reduced model complexity.

Keywords: Dirichlet belief networks · Markov chain Monte Carlo · Sparsity

1 Introduction

The Dirichlet Belief Network (DirBN) [20] was recently proposed as a promising deep probabilistic framework for learning *interpretable* hierarchical latent distributions for entities (or objects). To date, DirBN has been successfully implemented in two application areas: (1) topic structure learning [20], where the entities represent topics and the entities' latent distributions describe the

© Springer Nature Switzerland AG 2021
N. Oliver et al. (Eds.): ECML PKDD 2021, LNAI 12976, pp. 148–163, 2021.
https://doi.org/10.1007/978-3-030-86520-7_10

topic's vocabulary distributions; and (2) relational modelling [4,5,12], where the entities represent individuals and the latent distributions characterise an individual's membership distribution over community structures. By constructing a deep architecture for the latent distributions, the DirBN can effectively model high-order dependency between topic-vocabulary distributions (for topic models) and individual's membership distributions (for relational models).

$\alpha = (0.1, 0.1, 0.1)$ $\alpha = (0.1, 0.1, 1.0)$ $\alpha = (1, 1, 10)$

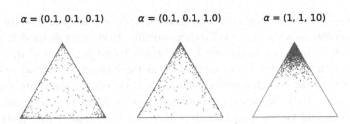

Fig. 1. Small Dirichlet concentration parameters generate sparse latent distributions. 1500 samples (red dots) generated from a 3-dimensional Dirichlet distribution are shown on the 2-dimensional unit simplex, $x_1 + x_2 + x_3 = 1$, with different concentration parameters α. When α is small (left and middle panels), most samples reside on vertices or edges, placing most mass on one or two dimensions. When α is not small (right panel), most samples lie inside the triangle, placing mass on all three dimensions. (Color figure online)

While promising, DirBN currently has some structural limitations which reduce their modelling flexibility. One limitation is that the length of each entity's latent distribution is restricted to be the same over all entities and all layers. As constructed, this restriction can reveal inadequate modelling flexibility in the DirBN when entities are related to different subsets of components in different layers (e.g. when individuals belong to different communities for different layers in the relational modelling setting). Since the latent distributions are linearly scaled (by Gamma-distributed variables) when being propagated into each subsequent layer, the resulting changes in the latent distributions can be too slow to adequately model the rapid changes inherent in the data.

A second limitation of DirBN is that it is unable to achieve the desirable properties of sparsity and smoothness simultaneously. Sparsity is achieved when the generated latent distributions place large weight on a subset of components – for the Dirichlet distribution, this occurs when the concentration parameters approach zero (see Fig. 1). In this case, however, the resulting posterior distribution over the latent distributions would be less smooth across layers as the empirical counts will then dominate the posterior distribution. Typically (though not exclusively) the posterior distribution is expected to be smooth, so as to reduce sensitivity to rarely-occurring latent distribution components.

In order to resolve these issues, we propose a sparse and smooth Dirichlet Belief Network (ssDirBN), which introduces binary variables into the layer-wise connections of the DirBN. In particular, each binary variable $b_{ik}^{(l)}$, which is generated by a Bernoulli distribution with entity-specific parameter, determines whether entity i's latent distribution at layer l uses component k ($b_{ik}^{(l)} = 1$) or not ($b_{ik}^{(l)} = 0$). Under this representation, sparsity is achieved through the Bernoulli distribution that generates the binary variable $b_{ik}^{(l)}$, flexibly permitting the latent distributions of different layers to solely focus on different subsets of components. Smoothness can then be enforced over those components with non-zero $b_{ik}^{(l)}$ through the Dirichlet concentration parameters. In this manner sparsity and smoothness are decoupled, and the benefits of both may be simultaneously obtained.

To ensure latent distributions to be defined appropriately and enable efficient posterior inference, we make two further modifications on the model: (1) fixing $b_{iK}^{(l)} = 1$, so that the last component K is certain to be propagated into the next layer, which can guarantee latent distributions be defined on at least one component; (2) letting those components with $b_{ik}^{(l)} = 0$ be propagated into component K, which can satisfy the specific condition (specified in the last paragraph of Sect. 2) for efficient Gibbs sampling algorithms on the membership distributions.

We explore the effectiveness of the ssDirBN in context of relational models, which use multi-stochastic layered latent distributions to model individual's membership distributions over communities. In this setting, the ssDirBN permits individuals to belong to different subsets of communities within different layers, and can thereby obviate placing unnecessary small probability masses on unrelated communities. Our experimental results on real-world data show significant performance improvements of ssDirBN over DirBN and other state-of-the-art models, in terms of reduced model complexity and improved link prediction performance. Similar to DirBN that can be considered as a self-contained module [20], the ssDirBN can be flexibly combined with alternative emission models and be implemented in these applications such as topic data, collaborative filtering data, etc.

2   Preliminary Knowledge

We first give a brief review on the DirBN model, where we use N to denote the number of entities (or number of topics in topic modeling) in each layer, K to denote the number of components in the entity's latent distributions and L to denote the number of layers. In general, DirBN assumes each entity has latent distributions $\pi_i^{(l-1)}$ at layer $l-1$ and uses Dirichlet distributions to generate entities' latent distribution at layer l, with the concentration parameters being the linear sum of entities' latent distributions at layer $l-1$. Within the relational modelling setting, $\{\pi_i^{(l)}\}_{l=1}^{L}$ represent entity i's membership distributions over K communities at L layers. The generative process of propagating the membership distributions $\{\pi_j^{(l-1)}\}_j$ to $\pi_i^{(l)}$ at layer l can be briefed as follows:

1. $\beta_{ji}^{(l)} \sim \text{Gam}(c_j, d), \forall i, j = 1, \dots, N$
2. $\pi_i^{(l)} \sim \text{Dir}(\sum_j \beta_{ji}^{(l)} \pi_j^{(l-1)}), \forall i = 1, \dots, N, l = 1, \dots, L$

where $\text{Gam}(c, d)$ is the Gamma distribution with mean c/d and variance c/d^2. $\beta_{ji}^{(l)}$ represents the information propagation coefficient from entity j at layer $l-1$

to entity i in at layer l, c_j, d are the hyper-parameters. After generating entity i's membership distribution at layer L, [4] uses counting vectors $\boldsymbol{m}_i^{(L)}$, which is sampled from Multinomial distribution with $\boldsymbol{\pi}_i^{(L)}$ as event probabilities, and community compatibility matrix to form probability function for generating the entity-wise relations.

DirBN is mostly inferred through Markov chain Monte Carlo (MCMC) methods. To enable efficient Gibbs sampling algorithm for DirBN, we note that the probability density function of $\boldsymbol{\pi}_i^{(l)}$ is written as:

$$P(\boldsymbol{\pi}_i^{(l)}|-) = \frac{\Gamma(\sum_k \sum_j \beta_{ji}^{(l)} \pi_{jk}^{(l-1)})}{\prod_k \Gamma(\sum_j \beta_{ji}^{(l)} \pi_{jk}^{(l-1)})} \prod_k (\pi_{ik}^{(l)})^{\sum_j \beta_{ji}^{(l)} \pi_{jk}^{(l-1)} - 1} \tag{1}$$

where $(-)$ refers to the set of conditional variables related to $\boldsymbol{\pi}_i^{(l)}$ and $\Gamma(\cdot)$ is the Gamma function. As $\{\pi_{jk}^{(l-1)}\}_j$ appear in the Gamma function, the prior and posterior distributions of $\boldsymbol{\pi}_i^{(l-1)}$ are not conjugate and it is difficult to implement efficient Gibbs sampling for $\boldsymbol{\pi}_i^{(l-1)}$. A strategy of first upward propagating latent counts and then downward sampling variables has been developed in [20] to address this issue, which is detailed below.

Upward Propagating Latent Counts. W.l.o.g., we assume the observation at layer l is the counts $\boldsymbol{m}_i^{(l)}$, which is obtained through Multinomial distribution with $\boldsymbol{\pi}_i^{(l)}$ as event probabilities. We may first integrate $\boldsymbol{\pi}_i^{(l)}$ out and obtain the likelihood term of the latent counts $\boldsymbol{m}_i^{(l)}$ as:

$$P(\{m_{ik}^{(l)}\}_k | \{\pi_{jk}^{(l-1)}\}_{j,k}) \propto \prod_k \frac{\Gamma(\sum_j \beta_{ji}^{(l)} \pi_{jk}^{(l-1)} + m_{ik}^{(l)})}{\Gamma(\sum_j \beta_{ji}^{(l)} \pi_{jk}^{(l)})} \tag{2}$$

The r.h.s. in Eq. (2) can be augmented through a random counts $y_{ik}^{(l)}$ from the Chinese Restaurant Table (CRT) distribution (i.e. $y_{ik}^{(l)} \sim \text{CRT}(m_{ik}^{(l)}, \sum_j \beta_{ji}^{(l)} \pi_{jk}^{(l)})$) as:

$$P(\{y_{ik}^{(l)}\}_k, \{m_{ik}^{(l)}\}_k | \{\pi_{ik}^{(l-1)}\}_k) \propto \prod_k [(\sum_j \beta_{ji}^{(l)} \pi_{jk}^{(l-1)})^{y_{ik}^{(l)}} (m_{ik}^{(l)} - y_{ik}^{(l)})!]$$

By further distributing the 'derived' count $y_{ik}^{(l)}$ into the entities at layer $l-1$ through a Multinomial distribution as: $(h_{1ik}^{(l)}, \ldots, h_{Nik}^{(l)}) \sim \text{Multi}(y_{ik}^{(l)}; \frac{\{\pi_{jk}^{(l-1)} B_{ji}^{(l-1)}\}_j}{\sum_j \beta_{ji}^{(l)} \pi_{jk}^{(l)}})$ and the terms associated with $\{\boldsymbol{\pi}_i^{(l-1)}\}_i$ are abstracted as:

$$P(\{h_{jik}^{(l-1)}\}_{j,k} | \{\pi_{jk}^{(l-1)}\}_k) \propto \prod_k (\pi_{jk}^{(l-1)})^{\sum_j h_{jik}^{(l-1)}}$$

The latent counts $\boldsymbol{m}_i^{(l-1)} = (\sum_j h_{ji1}^{(l-1)}, \ldots, \sum_j h_{jiK}^{(l-1)})$ can be regarded as a random draw from a Multinomial distribution, with $\boldsymbol{\pi}_i^{(l-1)}$ as event probabilities.

Downward Sampling Variables. After the counts are propagated to entity i at each layer l as $m_{ik}^{(l)}$, the posterior distribution of $\pi_i^{(l)}$ follows as:

$$\pi_i^{(l)} \sim \mathrm{Dir}(\sum_j \beta_{ji}^{(l)} \pi_j^{(l-1)} + m_i^{(l-1)})$$

3 Sparse and Smooth Dirichlet Belief Networks

3.1 Generative Process

ssDirBN aims at enabling each entity's latent distribution at each layer to be defined on individual subsets of components and thus simultaneously obtain the benefits of sparsity and smoothness of the Dirichlet distribution. Given N entities' latent distributions $\{\pi_i^{(l-1)}\}_{i=1}^N$ at layer $l-1$, we use the following method to generate entity i's latent distribution $\pi_i^{(l)}$ at layer l:

1. $\beta_{ji}^{(l)} \sim \mathrm{Gam}(c_j, d), \forall i, j = 1, \ldots, N$
2. $\omega_i^{(l)} \sim \mathrm{Beta}(\gamma, s), b_{iK}^{(l)} = 1, \{b_{ik}^{(l)}\}_{k=1}^{K-1} \sim \mathrm{Bernoulli}(\omega_i^{(l)})$
3. $\pi_i^{(l)} \sim \mathrm{Dir}\left(\sum_j \beta_{ji}^{(l)}(b_i^{(l)} \cdot \pi_j^{(l-1)} + e_K \cdot (\sum_k \pi_{jk}^{(l-1)} \delta_{b_{ik}^{(l)}=0}))\right), \forall i = 1, \ldots, N$

where $e_K = [0, \ldots, 0, 1]$ is a K-length vector with the last element 1 and 0 elsewhere; the two multiplication operations $\cdot$ in step 3 represents element-wise vector multiplication and scalar and vector multiplication respectively; c_j, d, γ, s are the hyper-parameters and we set Gamma priors on them as: $c_j \sim \mathrm{Gam}(c_0/K, d_0), d \sim \mathrm{Gam}(e_0, f_0), \gamma, s \sim \mathrm{Gam}(g_0, h_0)$.

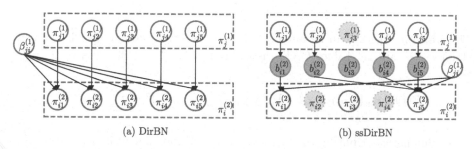

(a) DirBN   (b) ssDirBN

Fig. 2. Example visualisations on propagating latent distributions of $\pi_j^{(1)}$ at layer 1 to $\pi_i^{(2)}$ at layer 2. In DirBN (left panel), all the components in $\pi_j^{(1)}$ are one-to-one propagated to their corresponding components in $\pi_i^{(2)}$. In ssDirBN (right panel), binary variables are inserted between $\pi_j^{(1)}$ and $\pi_i^{(2)}$. Red entities of $b_{ik}^{(2)}$ represent "$b_{ik}^{(2)} = 0$" and in this case, $\pi_{jk}^{(1)}$ will be propagated to $\pi_{i5}^{(2)}$, which is the last component of $\pi_i^{(2)}$. Green $b_{ik}^{(2)}$ entities represent that $b_{ik}^{(2)} = 1$ and in this case, $\pi_{jk}^{(1)}$ will be propagated to $\pi_{ik}^{(2)}$. Grey dotted entities represent non-existing components. Since $\pi_{j3}^{(1)}$ is not existing, $\pi_j^{(1)}$ can not propagate component 3 to $\pi_i^{(2)}$ (other entities with existing component 3 will propagate to it.). (Color figure online)

In this generative process, step 1 generates layer l's entity-wise information propagation coefficient $\beta_{ji}^{(l)}$, which is same as in DirBN and represents the information propagation coefficient from entity j's latent distribution at layer $l-1$ to that of entity i at layer l.

Step 2 generates a component including variable $\omega_i^{(l)}$ and a subsequent K-length binary vector $\boldsymbol{b}_i^{(l)} \in \{0,1\}^{1 \times K}$ for entity i at layer l. When larger values of $\omega_i^{(l)}$ encourage more "1" entries in $\boldsymbol{b}_i^{(l)}$, the case of $b_{ik}^{(l)} = 1$ denotes that entity i's latent distribution at layer l includes component k and vice versa. That is, $\boldsymbol{b}_i$ specifies a small simplex through its "1" entries. An exception is the component K, for which we fix $b_{iK}^{(l)}$ as $b_{iK}^{(l)} = 1$ to make sure $\pi_i^{(l)}$ is well defined ($\pi_i^{(l)}$ will be problematic if $b_{i,k}^{(l)} = 0$ for $k = 1, \dots, K$). $\boldsymbol{b}_i^{(l)}$ determines the subset of components for entity i's latent distribution at layer l, i.e., $|\pi_i^{(l)}| = \sum_k b_{ik}^{(l)}$.

Step 3 also uses Dirichlet distribution to generate $\pi_i^{(l)}$ at layer l. The related concentration parameter is a linear sum of the entities' components weight at layer $l-1$. For component k, its weight propagation would be proceeded differently based on the value of $b_{ik}^{(l)}$: (1) when $b_{ik}^{(l)} = 1$, $\pi_{jk}^{(l-1)}$ will be added to component k of the concentration parameters; (2) when $b_{ik}^{(l)} = 0$, $\pi_{jk}^{(l-1)}$ will be added to component K of the concentration parameters. Figure 2 shows an example to visualise the propagation of latent distributions in DirBN and ssDirBN respectively.

Our ssDirBN has the following advantages when compared to DirBN:

- **Flexible subsets of components.** By introducing the binary variables, the latent distributions for different entities at different layers can be defined on different subsets of components. The model complexity is reduced as fewer components are involved. Also, the flexible usage of components may help each latent distribution focus on closely related components without assigning weights to unrelated ones.
- **Flexible weight ratios between components.** Recall that the ratios of component weights are unchanged when propagated from the current layer to the next layer. Thus, the concentration parameters of Dirichlet distributions should also follow these ratios generally. In ssDirBN, since we allow some components to be non-existing, the ratios of components can thus change greatly during the layer-wise connections, which enhances the representation capability of the model.
- **Decoupling sparsity and smoothness.** The sparsity in our ssDirBN is controlled by the Bernoulli parameter ω_i for entity i, with component k retained only if $b_{ik}^{(l)} = 1$. The smoothing effect is placed on the remaining components and controlled through the linear coefficient $\beta_{ji}^{(l)}$. In this way, the sparsity and smoothness are decoupled and we can obtain the benefits of both properties at the same time.

3.2   Necessity of Fixing $b_{iK}^{(l)} = 1$

Fixing $b_{iK}^{(l)} = 1$ $(\forall i, l)$ and making component k propagate to component K when $b_{ik}^{(l)} = 0$ $(\forall k)$ are key steps to guarantee the feasibility of upward count propagation method in ssDirBN. Recall that we have $\sum_j \beta_{ji}^{(l)} = \sum_k \sum_j \beta_{ji} \pi_{jk}^{(l)}$ to ensure that generated counts follow a Multinomial distribution. If we directly introduce the binary variable $b_{ik}^{(l)}$, which makes $b_{ik}^{(l)} \sim \text{Bernoulli}(\omega_i)(\forall k)$, we have

$$P(\{h_{jik}^{(l)}\}_j | \{\pi_{jk}^{(l-1)}\}_k) \propto [q_i^{(l)}]^{\sum_k \sum_j \beta_{ji} b_{ik}^{(l)} \pi_{jk}^{(l)}} \prod_k (\pi_{jk}^{(l-1)})^{h_{jik}^{(l)}}$$

That is, the counts of $\{h_{jik}^{(l)}\}_j$ *cannot* form a Multinomial distribution. However, we can still obtain $\sum_j \beta_{ji}^{(l)} = \sum_k \sum_j \beta_{ji} b_{jk}^{(l)} \pi_{jk}^{(l)}$ in ssDirBN, which enables the upward count propagation.

4   Related Work

In addition to the DirBN variants mentioned in the introduction, ssDirBN is also closely related to Gamma Belief Networks (GBN) [22], which is another multi-stochastic layered deep generative model. Instead of Dirichlet distributions, GBN used Gamma distributions to propagate scalar variables between layers and was the first to develop upward latent counts propagation and downward variable sampling method for model inference. Applications of GBN and the related inference technique have been observed in natural language modelling [7], Dynamic Systems [8,13,14,17] and even variational auto-encoder methods [18]. GBN does not enjoy the unique sparsity property of Dirichlet distribution and cannot be used to model latent distributions.

The basic idea of our ssDirBN is inspired by the sparse topic models (sparseTM) [2,16]. Compared with our ssDirBN, sparseTM places binary variables for all the components of the Dirichlet distribution. However, as a shallow model, sparseTM cannot model the complex entity-wise dependencies. Our usage of binary variables may also be similar to the techniques of Bayesian-dropout [6], which uses binary variables to decide whether or not to propagate the corresponding neuron to the next layer. In ssDirBN, we propagate the "neuron" to the last component when the binary variable equals to 0, rather than directly discarding it.

5   ssDirBN for Relational Modelling

We apply our ssDirBN in the setting of relational modelling, which focuses on constructing multi-stochastic layered membership distributions over communities for entities. The detail generative process is expressed as follows:

1. $\boldsymbol{\pi}_i^{(1)} \sim \text{Dirichlet}(\boldsymbol{\alpha})$;
2. For $l = 2, \ldots, L$

$$- \beta_{ji}^{(l-1)} \begin{cases} \sim \text{Gam}(c_i, d), \ j \in \{j : R_{ji} = 1\} \cup \{i\}; \\ = 0, \qquad \text{otherwise}; \end{cases}$$

$$- \omega_i^{(l)} \sim \text{Beta}(\gamma, s), b_{iK}^{(l)} = 1, \{b_{ik}^{(l)}\}_{k=1}^{K-1} \sim \text{Bernoulli}(\omega_i^{(l)})$$

$$- \boldsymbol{\pi}_i^{(l)} \sim \text{Dir}\left(\sum_j \beta_{ji}^{(l)} (\boldsymbol{b}_i^{(l)} \cdot \boldsymbol{\pi}_j^{(l-1)} + \boldsymbol{e}_K \cdot (\sum_k \pi_{jk}^{(l-1)} \delta_{b_{ik}^{(l)} = 0}))\right)$$

3. $M_i \sim \text{Poisson}(M), (X_{i1}, \ldots, X_{iK}) \sim \text{Multi}(M_i; \pi_{i1}^{(L)}, \ldots, \pi_{iK}^{(L)});$
4. $\Lambda_{k_1 k_2} \sim \text{Gam}(k_\Lambda, \frac{1}{\theta_\Lambda});$
5. $Z_{ij,k_1 k_2} \sim \text{Poisson}(X_{ik_1} \Lambda_{k_1 k_2} X_{jk_2});$
6. $R_{ij} = \mathbf{1}(\sum_{k_1,k_2} Z_{ij,k_1 k_2} > 0).$

In this generative process, $\boldsymbol{\alpha}$ in line 1 represents the concentration parameter in the Dirichlet distribution in generate all the entities' latent distribution at layer 1; line 2 represents our proposed ssDirBN structure, which can generate sparse and smooth latent distributions at layer L; line 3 generates latent count variable $(X_{i1}, \ldots, X_{iK})$ for entity i's latent distribution $\boldsymbol{\pi}_i^{(L)}$ at layer L; $\Lambda_{k_1 k_2}$ in line 4 is a community compatibility parameter such that a larger value of $\Lambda_{k_1 k_2}$ indicates a larger possibility of generating the links between community k_1 and community k_2; and $Z_{ij,k_1 k_2,t}$ is a community-to-community latent integer for each relation R_{ij}.

It is noted that, through the Multinomial distributions with $\boldsymbol{\pi}_i^{(L)}$ as event probabilities, $\boldsymbol{X}_i$ can be regarded as an estimator of $\boldsymbol{\pi}_i^{(L)}$. Since the sum also follows a Poisson distribution as $M_i \sim \text{Poisson}(M)$, according to the Poisson-Multinomial equivalence, each X_{ik} is equivalently distributed as $X_{i,k} \sim \text{Poisson}(M\pi_{ik}^{(L)})$. Therefore, both the prior distribution for generating X_{ik} and the likelihood based on X_{ik} are Poisson distributions. We may form feasible categorical distribution on its posterior inference. This trick is inspired by the recent advances in data augmentation and marginalisation techniques [4], which allows us to implement posterior sampling for X_{ik} efficiently.

The counts $\boldsymbol{X}_i$ lead to the generation of the $K \times K$ integer matrix $Z_{ij,k_1 k_2}$. Based on the Bernoulli-Poisson link function [3,21], the observed R_{ij} is mapped to the latent Poisson count random variable matrix $\boldsymbol{C}_{ij}$. It is shown in [4] that $\{C_{ij,k_1 k_2}\}_{k_1,k_2} = 0$ if $R_{ij} = 0$. That is, only the non-zero links are involved during the inference for $\boldsymbol{C}_{ij,k_1 k_2}$, which largely reduces the computational complexity, especially for large and sparse dynamic relational data. That is, since we use the Poisson-Bernoulli likelihood in modeling the relations, the computational cost of our model scales to the number of positive links.

5.1   Inference

We adopt the Markov chain Monte Carlo (MCMC) algorithm to iteratively sample the random variables from their posterior distributions. The latent conditional distributions of random variables we are approximating are: $\{\boldsymbol{\pi}_i^{(l)}\}_{i,l}$, which involves upward propagating the counting variable to each layer and downward sampling the variables of $\boldsymbol{\pi}$, the binary variable $b_{ik}^{(l)}$; $\boldsymbol{\alpha}$, which is the concentration parameters in generating the latent distributions at layer 1; scaling parameter M, which controls the latent count variables at layer L; $\{\Lambda_{k_1 k_2}\}_{k_1, k_2}$, which denotes the compatibility values between community k_1 and community k_2; $\{Z_{ij, k_1 k_2}\}_{i, j, k_1, k_2}$, which represents the latent integer variable for the relation from entity i to entity j from community k_1 to community k_2.

Upward Propagating Latent Counts. Using similar techniques of DirBN, we first upward propagate entity i's latent counts $\boldsymbol{m}_i^{(l)}$ at layer l to all the entities at layer $l-1$ via the following steps:

– sample 'derived latent counts' $y_{ik}^{(l)}$ and Beta random variable $q_i^{(l)}$ as

$$y_{ik}^{(t)} \sim \mathrm{CRT}(m_{ik}^{(s)}, \sum_j \beta_{ji} \pi_{jk}^{(l-1)}), q_i^{(s)} \sim \mathrm{Beta}(\sum_j \beta_{ji}^{(l)}, \sum_k m_{ik}^{(s)}),$$

$$y_{iK}^{(t)} \sim \mathrm{CRT}(m_{iK}^{(s)}, \sum_j \beta_{ji} \pi_{jK}^{(l-1)} + \sum_{j,k} \beta_{ji} \pi_{jk}^{(l-1)} \delta_{b_{ik}=0});$$

– distribute count $y_{ik}^{(l)}$ to the entities at layer $l-1$ as follows:

$$\{h_{ijk}^{(l-1)}\}_j \sim \mathrm{Multi}(y_{ik}^{(l)}; \frac{\{\pi_{jk}^{(l-1)} \beta_{ji}^{(l)}\}_j}{\sum_j \beta_{ji} \pi_{jk}^{(l)}}), \qquad \text{if } k = 1, \ldots, K-1;$$

$$\{\{h_{ijK}^{(l-1)}\}_j, \{h_{ijk}^{(l-1)}\}_{(i,k):b_{ik}^{(l)}=0}\} \sim \mathrm{Multi}(y_{iK}^{(l)}; \frac{\{\{\pi_{jK}^{(l-1)} \beta_{ji}^{(l)}\}_j, \{\pi_{jk}^{(l-1)} \beta_{ji}^{(l)}\}_{(j,k):b_{jk}^{(l)}=0}\}}{\sum_j \beta_{ji} \pi_{jK}^{(l)} + \sum_{(j,k):b_{jk}^{(l)}=0} \pi_{jk}^{(l-1)} \beta_{ji}^{(l)}}), \text{otherwise.}$$

– collect all the latent counts propagated to entity i at layer $l-1$ as $m_{ik}^{(l-1)} = \sum_j h_{jik}^{(l-1)}$.

Downward Sampling $\boldsymbol{\pi}_i^{(l)}$. Given the upward propagated latent counts $\boldsymbol{m}_i^{(l)}$, entity i's latent distribution at layer l can be sampled as:

$$\boldsymbol{\pi}_i^{(l)} \sim \mathrm{Dir}(\sum_j \beta_{ji}^{(l)} (\boldsymbol{b}_i^{(l)} \cdot \boldsymbol{\pi}_j^{(l-1)} + \boldsymbol{e}_K \cdot (\sum_k \pi_{jk}^{(l-1)} \delta_{b_{ik}^{(l)}=0})) + \boldsymbol{m}_i^{(l)})$$

Sampling $b_{ik}^{(l)}$. We integrate out $\omega_i^{(l)}$ to sample $b_{ik}^{(l)}$. Since new values of $b_{ik}^{(l)}$ will lead to new $\boldsymbol{\pi}_i^{(l)}$ (the length of $\boldsymbol{\pi}_i^{(l)}$ is different), we also need to generate new value for $\boldsymbol{\pi}_i^{(l)}$ when calculating the acceptance ratio. Given the current binary value of $b_{ik}^{(l)}$, we use the generative process to generate a new latent distribution $\boldsymbol{\pi}_i^{(l,*)}$ based on the opposite value of $b_{ik}^{(l)}$ and then accept the opposite value of $b_{ik}^{(l)}$ and $\boldsymbol{\pi}_i^{(l,*)}$ with a ratio of $\min(1, \alpha)$, where α is defined as:

$$\alpha = \frac{P(b_{ik}^{(l,*)}|b_{i,/k}^{(l)}, \gamma, s)}{P(b_{ik}^{(l)}|b_{i,/k}^{(l)}, \gamma, s)} \cdot \frac{P(\{\boldsymbol{\pi}_j^{(l-1)}, \boldsymbol{\pi}_j^{(l+1)}\}_j | \boldsymbol{\pi}_i^{(l,*)}, -)}{P(\{\boldsymbol{\pi}_j^{(l-1)}, \boldsymbol{\pi}_j^{(l+1)}\}_j | \boldsymbol{\pi}_i^{(l)}, -)} \tag{3}$$

Sampling $\{\beta_{ji}^{(l)}\}_{j,i,l}$. For $j \in \{j : R_{ji} \neq \{i\}\}$, the prior for $\beta_{ji}^{(l)}$ is $\mathrm{Gam}(c_i, d)$, the posterior distribution is

$$\beta_{ji}^{(l)} \sim \mathrm{Gam}(c_j + \sum_k h_{jik}^{(l)}, d - \log q_i^{(l)}) \tag{4}$$

Sampling $\{X_{ik}\}_{i,k}$: We have $M_i \sim \text{Poisson}(M)$,

$$(X_{i1}, \ldots, X_{iK}) \sim \text{Multi}(M_i; \pi_{i1}^{(L)}, \ldots, \pi_{iK}^{(L)}) \overset{d}{=} X_{ik} \sim \text{Poisson}(M\pi_{ik}^{(L)}), \forall k.$$

Both the prior distribution for generating X_{ik} and the likelihood parametrised by X_{ik} are Poisson distributions. The full conditional distribution of X_{ik} (assuming $z_{ii,..} = 0, \forall i$) is then

$$P(X_{ik}|M, \boldsymbol{\pi}, \boldsymbol{\Lambda}, \boldsymbol{Z}) \propto \frac{\left[M\pi_{ik}^{(L)} e^{-\Sigma_{j\neq i, k_2} X_{jk_2}(\Lambda_{kk_2}+\Lambda_{k_2 k})}\right]^{X_{ik}}}{X_{ik}!} (X_{ik})^{\Sigma_{j_1, k_2} Z_{ij_1, kk_2} + \Sigma_{j_2, k_1} Z_{j_2 i, k_1 k}}.$$

$$(5)$$

This follows the form of Touchard polynomials, where $1 = \frac{1}{e^x T_n(x)} \sum_{k=0}^{\infty} \frac{x^k k^n}{k!}$ with $T_n(x) = \sum_{k=0}^{n} \{{n \atop k}\} x^k$ and where $\{{n \atop k}\}$ is the Stirling number of the second kind. A draw from (5) is then available by comparing a Uniform$(0,1)$ random variable to the cumulative sum of $\{\frac{1}{e^x T_n(x)} \cdot \frac{x^k k^n}{k!}\}_k$.

Sampling $\{Z_{ij,k_1 k_2}\}_{i,j,k_1,k_2}$. We first sample $Z_{ij,..}$ from a Poisson distribution with positive support:

$$Z_{ij,..} \sim \text{Poisson}_+(\sum_{k_1,k_2} X_{ik_1} X_{jk_2} \Lambda_{k_1 k_2}), \text{where } Z_{ij,..} = 1, 2, 3, \ldots \qquad (6)$$

Then, $\{Z_{ij,k_1 k_2}\}_{k_1,k_2}$ can be obtained through the Multinomial distribution as:

$$(\{Z_{ij,k_1 k_2}\}_{k_1,k_2}) \sim \text{Multinomial}\left(Z_{ij,..}; \left\{\frac{X_{ik_1} X_{jk_2} \Lambda_{k_1 k_2}}{\sum_{k_1,k_2} X_{ik_1} X_{jk_2} \Lambda_{k_1 k_2}}\right\}_{k_1,k_2}\right) \qquad (7)$$

Sampling $\{\Lambda_{k_1 k_2}\}_{k_1,k_2}$. For $\Lambda_{k_1 k_2}$'s posterior distribution, we get

$$P(\Lambda_{k_1 k_2}|-) \propto \exp\left(-\Lambda_{k_1 k_2}(\sum_{i,j} X_{ik_1} X_{jk_2})\right) \Lambda_{k_1 k_2}^{\Sigma_{i,j} Z_{ij,k_1 k_2}} \cdot \exp\left(-\Lambda_{k_1 k_2} \theta_\Lambda\right) \Lambda^{k_\Lambda - 1}$$

$$(8)$$

Thus, we get

$$\Lambda_{k_1 k_2} \sim \text{Gam}\left(\sum_{i,j} Z_{ij,k_1 k_2} + k_\Lambda, \frac{1}{\theta_\Lambda + \sum_{i,j} X_{ik_1} X_{jk_2}}\right) \qquad (9)$$

Sampling M. Given Gamma distribution $\text{Gam}(k_M, \theta_M)$ as the prior distribution for M, M's posterior distribution is:

$$P(M|-) = M^{k_M - 1} \exp(-\theta_M M) \prod_{i,k} \left(\exp(-M\pi_{ik}^{(L)})\right) M^{\Sigma_{i,k} X_{ik}} \qquad (10)$$

Thus, we sample M from:

$$M \sim \text{Gam}\left(k_M + \sum_{i,k} X_{ik}, \frac{1}{\theta_M + N}\right) \tag{11}$$

Sampling α. Similarly, given Gamma distribution $\text{Gam}(k_\alpha, \theta_\alpha)$ as the prior distribution for α, α's posterior distribution is

$$\alpha \sim \text{Gam}(k_\alpha + \sum_{i,k} h_{i\alpha k}^{(1)}, \frac{1}{\theta_\alpha - \sum_i \log q_i^{(1)}}) \tag{12}$$

Computational Complexity. It is noted that our ssDirBN for relational modeling does not increase the computational scalability of the DirBN for relational modeling. We have changed the counts allocation in the detailed process of latent counts propagation, however, the computational complexity remained the same in our ssDirBN and DirBN. It is easy to see that the complexity of sampling the binary variable b also scales to the number of positive relations. Thus, the computational complexity of our ssDirBN for relational modeling is the same as that of DirBN.

Table 1. Dataset information. N is the number of entities, N_E is the number of positive links.

Dataset	N	N_E	Dataset	N	N_E	Dataset	N	N_E	Dataset	N	N_E
Citeer	3 312	4 715	Cora	2 708	5 429	Pubmed	2 000	17 522	PPI	4 000	105 775

5.2   Experimental Results

Dataset Information. We apply our ssDirBN in the following four real-world sparse datasets: three standard citation networks (*Citeer, Cora, Pubmed* [15] and one protein-to-protein interaction network (*PPI*) [23]. The number of entities N and the number of relations N_E for these datasets are provided in Table 1. In the citation datasets, entities correspond to documents and edges represent citation links, whereas in the protein-to-protein dataset, we model the protein-to-protein interactions [9]. We do not include the feature information of entities in the experiments. Instead, we use an identity matrix I_N as the feature matrix when these information are needed in specific models.

Evaluation Criteria. We primarily focus on link prediction and use this to evaluate model performance. We use AUC (Area Under ROC Curve) and Average Precision value on test relational data as the two comparison criteria. The AUC value represents the probability that the algorithm will rank a randomly chosen existing-link higher than a randomly chosen non-existing link. Therefore, the higher the AUC value, the better the predictive performance. Each reported criteria value is the mean of 10 replicate analyses.

Experimental Settings. For hyper-parameters, we let $c_0 = d_0 = e_0 = f_0 = g_0 = h_0 = 0.1$ for all the datasets. The re-sampling of hyper-parameters is specified in the similar way as that of [4]. Each run uses 2 000 MCMC iterations with the first 1 000 samples discarded as burn-in and the mean value of the second 1 000 samples' performance score is used report the required score. Unless specified, reported AUC values are obtained using 90% (per row) of the relational data as training data and the remaining 10% as test data. After testing various scenarios for different settings of L and K, we set the number of layers $L = 3$ and the number of communities $K = 10$ for all the testing cases as the performance can be obtained in the balance of excellence performance scores and fast running time.

Comparison Methods: Several Bayesian methods for relational data and two Graph Auto-Encoder models are used for comparison: the Mixed-Membership Stochastic Blockmodel [1], the Hierarchical Latent Feature Relational Model (HLFM) [10], the Node Attribute Relational Model (NARM) [19], the Hierarchical Gamma Process-Edge Partition Model (HGP-EPM) [21], graph autoencoders (GAE) and variational graph autoencoders (VGAE) [11]. The NARM, HGP-EPM, GAE and VGAE methods are executed using their respective implementations from the authors, under their default settings. The MMSB and HLFM are implemented to the best of our abilities and we set the number of layers and length of latent binary representation in HLFM as same as those in ssDirBN. For the GAE and VGAE, the AUC and Precision values are calculated based on the pairwise similarities between the entity representations.

Performance of ssDirBN for Different Values of K, L. Figure 3 shows the link prediction performance of ssDirBN for relational modeling on the cases of $K = 5, 10, 15, 20, 30$ and $L = 2, 3, 4, 5$. In terms of the number of communities K, we find that the performance of $K = 10$ is significantly better than the one of $K = 5$ and slightly worse than those of $K = 15, 20, 30$. The performance of $L = 3$ has similar behaviours as it is much better than that of $L = 2$, which is likely because the insufficient deep structure, and slightly worse than those of $L = 4, 5$, which may possibly due to that $L = 3$ is deep enough. The behaviours of AUC and Precision are quite consistent in forming these conclusions.

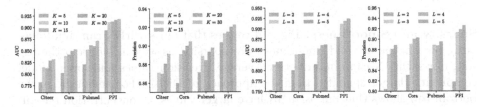

Fig. 3. Link prediction performance (AUC and Precision) of ssDirBN for relational modelling on different values of K and L.

Table 2. Link prediction performance comparison. It is noted that we do not use the entities' feature information and only use the binary relational data for each dataset.

AUC (mean and standard deviation)			
Model	Citeer	Cora	Pubmed	PPI
MMSB	0.690 ± 0.004	0.743 ± 0.007	0.774 ± 0.005	0.801 ± 0.003
NARM	0.759 ± 0.003	0.809 ± 0.003	0.808 ± 0.004	0.821 ± 0.002
HGP-EPM	0.763 ± 0.003	0.810 ± 0.003	0.803 ± 0.006	0.834 ± 0.004
HLFM	0.781 ± 0.010	0.829 ± 0.005	0.829 ± 0.005	0.856 ± 0.010
GAE	0.789 ± 0.004	0.846 ± 0.006	0.822 ± 0.004	0.874 ± 0.009
VGAE	0.790 ± 0.003	0.849 ± 0.004	0.826 ± 0.002	0.880 ± 0.007
DirBN	0.779 ± 0.004	0.832 ± 0.008	0.845 ± 0.008	0.892 ± 0.007
ssDirBN	$\mathbf{0.815} \pm 0.007$	$\mathbf{0.839} \pm 0.007$	$\mathbf{0.853} \pm 0.004$	$\mathbf{0.912} \pm 0.002$
Average precision (mean and standard deviation)			
Model	Citeer	Cora	Pubmed	PPI
MMSB	0.661 ± 0.004	0.704 ± 0.005	0.742 ± 0.004	0.823 ± 0.003
NARM	0.781 ± 0.004	0.831 ± 0.004	0.771 ± 0.005	0.844 ± 0.002
HGP-EPM	0.776 ± 0.002	0.840 ± 0.003	0.786 ± 0.006	0.864 ± 0.004
HLFM	0.793 ± 0.004	0.842 ± 0.003	0.802 ± 0.003	0.883 ± 0.008
GAE	0.839 ± 0.004	0.884 ± 0.007	0.846 ± 0.004	0.889 ± 0.003
VGAE	0.846 ± 0.003	0.889 ± 0.004	0.850 ± 0.003	0.882 ± 0.004
DirBN	0.819 ± 0.004	0.875 ± 0.03	0.860 ± 0.007	0.884 ± 0.002
ssDirBN	$\mathbf{0.871} \pm 0.007$	$\mathbf{0.891} \pm 0.003$	$\mathbf{0.889} \pm 0.006$	$\mathbf{0.913} \pm 0.005$

Link Prediction Performance. Table 2 displays the AUC and Average Precision values over the testing relational data. As we can see, our ssDirBN performs the best among all these comparison methods. The performance of deep hierarchical models (i.e., VGAE, HLFM, DirBN, ssDirBN) are usually better than the shallow models, which verifies the advantages of deep structures. The competitive performance of our ssDirBN, as well as the one of DirBN, over GAE and VGAE verifies the promising aspects of using Dirichlet distributions to construct the deep generative models.

Sparsity and Smoothness. Figure 4 verifies that our ssDirBN can simultaneously obtain sparsity and smoothness. For sparsity, as we can see from the top row, our ssDirBN can obtain better Average Precision values and lower model complexity (larger sparsity) than the approach of DirBN. For smoothness, the bottom row shows that our ssDirBN have larger values of concentration parameters in generating the membership distributions than that of DirBN. Given the same output counts, larger concentration parameters will lead to smoother posterior distributions. Thus, the posterior distributions of $\pi_i^{(l)}$ in our ssDirBN would be smoother than that in DirBN.

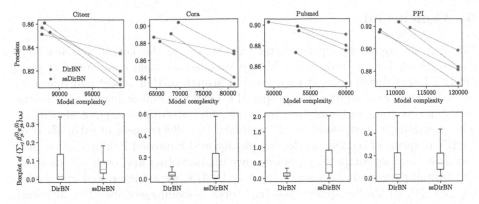

Fig. 4. Top row: model complexity versus Precision value for the datasets of *Citeer, Cora, Pubmed, PPI*. We define 'model complexity' as the number of community membership values for all entities. In DirBN, it is calculated as NKL, whereas in ssDirBN, it is calculated as: $\sum_{i,k,l} \delta_{b_{ik}^{(l)}=1}$. Each dotted line represents that its connected red and blue dots are evaluated on the same training and testing dataset. Bottom row: boxplot of concentration parameters $\{\sum_j \beta_{ji}^{(l)} \pi_{jk}^{(l)}\}_{i,k,l}$ in generating membership distributions for DirBN and ssDirBN on *Citeer, Cora, Pubmed, PPI*.

Visualizations on Membership Distributions. Figure 5 displays example membership distributions over the first 50 entities in the Citeer data for DirBN and ssDirBN. For DirBN, the membership distributions become more dominated by a few communities from layer 1 to layer 3. However, the components' weight ratio does not change too much. For ssDirBN, the membership distributions clearly show more changes across different layers. The representation capability of our ssDirBN can be thus enhanced through the larger changes of membership distributions across different layers.

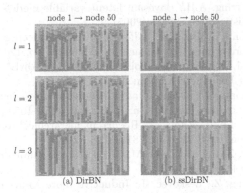

Fig. 5. Visualizations of the membership distributions ($\{\pi_{i=1:50}^{(l)}\}_{l=1}^3$) on the *Citeer* data set for DirBN (left) and ssDirBN (right). The panels in the top, middle and bottom row represent the membership distributions at layer $l = 1, 2, 3$. In each panel, columns represent entities. Colors represent communities (with $K = 10$) and the length of color occupations in one column represents the membership value of the particular community for that entity. (Color figure online)

6   Conclusion

We have decoupled the sparsity and smoothness in the Dirichlet Belief Networks (DirBN) by introducing a binary variable for each component of each entity's latent distribution at each layer. Through further model and inference modifications, we guarantee the proposed ssDirBN is well defined for the latent distributions and can be inferred by using efficient Gibbs sampling algorithm. The promising experimental results validate the effectiveness of ssDirBN over DirBN in terms of reduced model complexity and improved link prediction performance, and its competitive performance against other approaches. Given the substantial performance improvement over DirBN, we are interested in combining ssDirBN with other applications (e.g. topic modelling, collaborative filtering) in the future.

Acknowledgments. Yaqiong Li is a recipient of UTS Research Excellence Scholarship. Xuhui Fan and Scott A. Sisson are supported by the Australian Research Council (ARC) through the Australian Centre of Excellence in Mathematical and Statistical Frontiers (ACEMS, CE140100049), and Scott A. Sisson through the ARC Future Fellow Scheme (FT170100079). Bin Li is supported in part by STCSM Project (20511100400), Shanghai Municipal Science and Technology Major Projects (2018SHZDZX01, 2021SHZDZX0103), and the Program for Professor of Special Appointment (Eastern Scholar) at Shanghai Institutions of Higher Learning.

References

1. Airoldi, E.M., Blei, D.M., Fienberg, S.E., Xing, E.P.: Mixed membership stochastic block models. J. Mach. Learn. Res. **9**, 1981–2014 (2008)
2. Burkhardt, S., Kramer, S.: Decoupling sparsity and smoothness in the Dirichlet variational autoencoder topic model. J. Mach. Learn. Res. **20**(131), 1–27 (2019)
3. Dunson, D.B., Herring, A.H.: Bayesian latent variable models for mixed discrete outcomes. Biostatistics **6**(1), 11–25 (2005)
4. Fan, X., Li, B., Li, C., Sisson, S., Chen, L.: Scalable deep generative relational model with high-order node dependence. In: NeurIPS, pp. 12637–12647 (2019)
5. Fan, X., Li, B., Li, Y., Sisson, S.: Poisson-randomised DirBN: large mutation is needed in dirichlet belief networks. In: ICML (2021)
6. Gal, Y., Ghahramani, Z.: Dropout as a Bayesian approximation: representing model uncertainty in deep learning. In: ICML, pp. 1050–1059 (2016)
7. Guo, D., Chen, B., Lu, R., Zhou, M.: Recurrent hierarchical topic-guided RNN for language generation. In: ICML, pp. 3810–3821 (2020)
8. Guo, D., Chen, B., Zhang, H., Zhou, M.: Deep Poisson gamma dynamical systems. In: NeurIPS, pp. 8442–8452 (2018)
9. Hamilton, W., Ying, Z., Leskovec, J.: Inductive representation learning on large graphs. In: NIPS, pp. 1024–1034 (2017)
10. Hu, C., Rai, P., Carin, L.: Deep generative models for relational data with side information. In: ICML, pp. 1578–1586 (2017)
11. Kipf, T.N., Welling, M.: Variational graph auto-encoders. arXiv preprint arXiv:1611.07308 (2016)

12. Li, Y., Fan, X., Chen, L., Li, B., Yu, Z., Sisson, S.A.: Recurrent Dirichlet belief networks for interpretable dynamic relational data modelling. In: IJCAI, pp. 2470–2476 (2020)
13. Schein, A., Linderman, S., Zhou, M., Blei, D., Wallach, H.: Poisson-randomized gamma dynamical systems. In: NeurIPS, pp. 782–793 (2019)
14. Schein, A., Wallach, H., Zhou, M.: Poisson-gamma dynamical systems. In: NIPS, pp. 5005–5013 (2016)
15. Sen, P., Namata, G., Bilgic, M., Getoor, L., Galligher, B., Eliassi-Rad, T.: Collective classification in network data. AI Mag. **29**, 93 (2008)
16. Wang, C., Blei, D.: Decoupling sparsity and smoothness in the discrete hierarchical Dirichlet process. In: NIPS, pp. 1982–1989. Curran Associates Inc. (2009)
17. Yang, S., Koeppl, H.: A Poisson gamma probabilistic model for latent node-group memberships in dynamic networks. In: AAAI (2018)
18. Zhang, H., Chen, B., Guo, D., Zhou, M.: WHAI: Weibull hybrid autoencoding inference for deep topic modeling. In: International Conference on Learning Representations (2018)
19. Zhao, H., Du, L., Buntine, W.: Leveraging node attributes for incomplete relational data. In: ICML, pp. 4072–4081 (2017)
20. Zhao, H., Du, L., Buntine, W., Zhou, M.: Dirichlet belief networks for topic structure learning. In: NeurIPS, pp. 7955–7966 (2018)
21. Zhou, M.: Infinite edge partition models for overlapping community detection and link prediction. In: AISTATS, pp. 1135–1143 (2015)
22. Zhou, M., Cong, Y., Chen, B.: Augmentable gamma belief networks. J. Mach. Learn. Res. **17**(163), 1–44 (2016)
23. Zitnik, M., Leskove, J.: Predicting multicellular function through multi-layer tissue networks. Bioinformatics **33**, i190–i198 (2017)

Algorithms and Learning Theory

Self-bounding Majority Vote Learning Algorithms by the Direct Minimization of a Tight PAC-Bayesian C-Bound

Paul Viallard[1(✉)], Pascal Germain[2], Amaury Habrard[1], and Emilie Morvant[1]

[1] Univ Lyon, UJM-Saint-Etienne, CNRS, Institut d Optique Graduate School,
Laboratoire Hubert Curien UMR 5516, 42023 Saint-Etienne, France
{paul.viallard,amaury.habrard,emilie.morvant}@univ-st-etienne.fr
[2] Département d'informatique et de génie logiciel, Université Laval, Québec, Canada
pascal.germain@ift.ulaval.ca

Abstract. In the PAC-Bayesian literature, the C-Bound refers to an insightful relation between the risk of a majority vote classifier (under the zero-one loss) and the first two moments of its margin (*i.e.*, the expected margin and the voters' diversity). Until now, learning algorithms developed in this framework minimize the empirical version of the C-Bound, instead of explicit PAC-Bayesian generalization bounds. In this paper, by directly optimizing PAC-Bayesian guarantees on the C-Bound, we derive self-bounding majority vote learning algorithms. Moreover, our algorithms based on gradient descent are scalable and lead to accurate predictors paired with non-vacuous guarantees.

Keywords: Majority vote · PAC-Bayesian · Self-bounding algorithm

1 Introduction

In machine learning, ensemble methods [10] aim to combine hypotheses to make predictive models more robust and accurate. A weighted majority vote learning procedure is an ensemble method for classification where each voter/hypothesis is assigned a weight (*i.e.*, its influence in the final voting). Among the famous majority vote methods, we can cite Boosting [13], Bagging [5], or Random Forest [6]. Interestingly, most of the kernel-based classifiers, like Support Vector Machines [3,7], can be seen as a majority vote of kernel functions. Understanding when and why weighted majority votes perform better than a single hypothesis is challenging. To study the generalization abilities of such majority votes, the PAC-Bayesian framework [25,34] offers powerful tools to obtain Probably Approximately Correct (PAC) generalization bounds. Motivated by the fact that PAC-Bayesian analyses can lead to tight bounds (*e.g.*, [28]), developing algorithms to minimize such bounds is an important direction (*e.g.*, [11,14,15,24]).

Electronic supplementary material The online version of this chapter (https://doi.org/10.1007/978-3-030-86520-7_11) contains supplementary material, which is available to authorized users.

© Springer Nature Switzerland AG 2021
N. Oliver et al. (Eds.): ECML PKDD 2021, LNAI 12976, pp. 167–183, 2021.
https://doi.org/10.1007/978-3-030-86520-7_11

We focus on a class of PAC-Bayesian algorithms minimizing an upper bound on the majority vote's risk called the C-Bound[1] in the PAC-Bayesian literature [20]. This bound has the advantage of involving the majority vote's margin and its second statistical moment, *i.e.*, the diversity of the voters. Indeed, these elements are important when one learns a combination [10,19]: A good majority vote is made up of voters that are "accurate enough" and "sufficiently diverse". Various algorithms have been proposed to minimize the C-Bound: MINCQ [31], P-MINCQ [2], CQBOOST [32], or CB-BOOST [1]. Despite being empirically efficient, and justified by theoretical analyses based on the C-Bound, all these methods minimize *only* the empirical C-Bound and not directly a PAC-Bayesian generalization bound on the C-Bound. This can lead to vacuous generalization bound values and thus to poor risk certificates.

In this paper, we cover three different PAC-Bayesian viewpoints on generalization bounds for the C-Bound [20,26,33]. Starting from these three views, we derive three algorithms to optimize generalization bounds on the C-Bound. By doing so, we achieve *self-bounding algorithms* [12]: the predictor returned by the learner comes with a statistically valid risk upper bound. Importantly, our algorithms rely on fast gradient descent procedures. As far as we know, this is the first work that proposes both efficient algorithms for C-Bound optimization and non-trivial risk bound values.

The paper is organized as follows. Section 2 introduces the setting. Section 3 recalls the PAC-Bayes bounds on which we build our results. Our self-bounding algorithms leading to non-vacuous PAC-Bayesian bounds are described in Sect. 4. We provide experiments in Sect. 5, and conclude in Sect. 6.

2   Majority Vote Learning

2.1   Notations and Setting

We stand in the context of learning a weighted majority vote for binary classification. Let $\mathcal{X} \subseteq \mathbb{R}^d$ be a d-dimensional input space, and $\mathcal{Y} = \{-1, +1\}$ be the label space. We assume an unknown data distribution $\mathcal{D}$ on $\mathcal{X} \times \mathcal{Y}$, we denote by $\mathcal{D}_{\mathcal{X}}$ the marginal distribution on $\mathcal{X}$. A learning algorithm is provided with a learning sample $\mathcal{S} = \{(\mathbf{x}_i, y_i)\}_{i=1}^m$ where each example $(\mathbf{x}_i, y_i)$ is drawn *i.i.d.* from $\mathcal{D}$, we denote by $\mathcal{S} \sim \mathcal{D}^m$ the random draw of such a sample. Given $\mathcal{H}$ a hypothesis set constituted by so-called *voters* $h : \mathcal{X} \rightarrow \mathcal{Y}$, and $\mathcal{S}$, the learner aims to find a weighted combination of the voters from $\mathcal{H}$; the weights are modeled by a distribution on $\mathcal{H}$. To learn such a combination in the PAC-Bayesian framework, we assume a *prior* distribution $\mathcal{P}$ on $\mathcal{H}$, and—after the observation of $\mathcal{S}$—we learn a *posterior* distribution $\mathcal{Q}$ on $\mathcal{H}$. More precisely, we aim to learn a well-performing classifier that is expressed as a $\mathcal{Q}$-*weighted majority vote* $\mathrm{MV}_{\mathcal{Q}}$ defined as

$$\forall \mathbf{x} \in \mathcal{X}, \quad \mathrm{MV}_{\mathcal{Q}}(\mathbf{x}) \triangleq \mathrm{sign}\left(\mathop{\mathbb{E}}_{h \sim \mathcal{Q}} h(\mathbf{x})\right) = \mathrm{sign}\left(\sum_{h \in \mathcal{H}} \mathcal{Q}(h)h(\mathbf{x})\right).$$

---

[1] The C-Bound was introduced by Breiman in the context of Random Forest [6].

We thus want to learn $\mathrm{MV}_\mathcal{Q}$ that commits as few errors as possible on unseen data from $\mathcal{D}$, *i.e.*, that leads to a low true risk $r_\mathcal{D}^{\mathrm{MV}}(\mathcal{Q})$ under the 0-1-loss defined as

$$r_\mathcal{D}^{\mathrm{MV}}(\mathcal{Q}) \triangleq \mathop{\mathbb{E}}_{(\mathbf{x},y)\sim\mathcal{D}} \mathbf{I}\Big[\mathrm{MV}_\mathcal{Q}(\mathbf{x}) \neq y\Big], \quad \text{where } \mathbf{I}[a] = \begin{cases} 1 \text{ if the assertion } a \text{ is true,} \\ 0 \text{ otherwise.} \end{cases}$$

2.2 Gibbs Risk, Joint Error and C-Bound

Since $\mathcal{D}$ is unknown, a common way to try to minimize $r_\mathcal{D}^{\mathrm{MV}}(\mathcal{Q})$ is the minimization of its empirical counterpart $r_\mathcal{S}^{\mathrm{MV}}(\mathcal{Q}) = \frac{1}{m}\sum_{i=1}^{m} \mathbf{I}[\mathrm{MV}_\mathcal{Q}(\mathbf{x}_i)\neq y_i]$ computed on the learning sample $\mathcal{S}$ through the Empirical Risk Minimization principle. However, learning the weights by the direct minimization of $r_\mathcal{S}^{\mathrm{MV}}(\mathcal{Q})$ does not necessarily lead to a low true risk. One solution consists then in looking for precise estimators or generalization bounds of the true risk $r_\mathcal{D}^{\mathrm{MV}}(\mathcal{Q})$ to minimize them. In the PAC-Bayesian theory, a well-known estimator of the true risk $r_\mathcal{D}^{\mathrm{MV}}(\mathcal{Q})$ is the **Gibbs risk** defined as the $\mathcal{Q}$-average risk of the voters as

$$r_\mathcal{D}(\mathcal{Q}) = \mathop{\mathbb{E}}_{h\sim\mathcal{Q}} \mathop{\mathbb{E}}_{(\mathbf{x},y)\sim\mathcal{D}} \mathbf{I}[h(\mathbf{x}) \neq y].$$

Its empirical counterpart is defined as $r_\mathcal{S}(\mathcal{Q}) = \frac{1}{m}\sum_{i=1}^{m} \mathbb{E}_{h\sim\mathcal{Q}}\mathbf{I}[h(\mathbf{x}_i) \neq y_i]$. However, in ensemble methods where one wants to combine voters efficiently, the Gibbs risk appears to be an unfair estimator since it does not take into account the fact that a combination of voters has to compensate for the individual errors. This is highlighted by the decomposition of $r_\mathcal{D}(\mathcal{Q})$ in Eq. (1) (due to Lacasse *et al.* [20]) into the expected **disagreement** and the expected **joint error**, respectively defined by

$$d_\mathcal{D}(\mathcal{Q}) = \mathop{\mathbb{E}}_{h_1\sim\mathcal{Q}} \mathop{\mathbb{E}}_{h_2\sim\mathcal{Q}} \mathop{\mathbb{E}}_{\mathbf{x}\sim\mathcal{D}_\mathcal{X}} \mathbf{I}\big[h_1(\mathbf{x}) \neq h_2(\mathbf{x})\big],$$

$$\text{and} \quad e_\mathcal{D}(\mathcal{Q}) = \mathop{\mathbb{E}}_{h_1\sim\mathcal{Q}} \mathop{\mathbb{E}}_{h_2\sim\mathcal{Q}} \mathop{\mathbb{E}}_{(\mathbf{x},y)\sim\mathcal{D}} \mathbf{I}\big[h_1(\mathbf{x}) \neq y\big]\mathbf{I}\big[h_2(\mathbf{x}) \neq y\big].$$

Indeed, an increase of the voter's diversity, captured by the disagreement $d_\mathcal{D}(\mathcal{Q})$, have a negative impact on the Gibbs risk, as

$$r_\mathcal{D}(\mathcal{Q}) = e_\mathcal{D}(\mathcal{Q}) + \tfrac{1}{2}d_\mathcal{D}(\mathcal{Q}). \tag{1}$$

Despite this unfavorable behavior, many PAC-Bayesian results deal only with the Gibbs risks, thanks to a straightforward upper bound of the majority vote's risk which consists in upper-bounding it by twice the Gibbs risk [21], *i.e.*,

$$r_\mathcal{D}^{\mathrm{MV}}(\mathcal{Q}) \leq 2\,r_\mathcal{D}(\mathcal{Q}) = 2e_\mathcal{D}(\mathcal{Q}) + d_\mathcal{D}(\mathcal{Q}). \tag{2}$$

This bound is tight only when the Gibbs risk is low (*e.g.*, when voters with large weights perform well individually [14, 21]). Recently, Masegosa *et al.* [24] propose to deal directly with the joint error as

$$r_\mathcal{D}^{\mathrm{MV}}(\mathcal{Q}) \leq 4e_\mathcal{D}(\mathcal{Q}) = 2r_\mathcal{D}(\mathcal{Q}) + 2e_\mathcal{D}(\mathcal{Q}) - d_\mathcal{D}(\mathcal{Q}). \tag{3}$$

Equation (3) is tighter than Eq. (2) if $e_{\mathcal{D}}(\mathcal{Q}) \leq \frac{1}{2} d_{\mathcal{D}}(\mathcal{Q}) \Leftrightarrow r_{\mathcal{D}}(\mathcal{Q}) \leq d_{\mathcal{D}}(\mathcal{Q})$; This captures the fact that the voters need to be sufficiently diverse and commit errors on different points. However, when the joint error $e_{\mathcal{D}}(\mathcal{Q})$ exceeds $\frac{1}{4}$, the bound exceeds 1 and is uninformative. Another bound—known as the C-Bound in the PAC-Bayes literature [20]—has been introduced to capture this trade-off between the Gibbs risk $r_{\mathcal{D}}(\mathcal{Q})$ and the disagreement $d_{\mathcal{D}}(\mathcal{Q})$, and is recalled in the following theorem.

Theorem 1 (C-Bound). *For any distribution $\mathcal{D}$ on $\mathcal{X} \times \mathcal{Y}$, for any voters set $\mathcal{H}$, for any distribution $\mathcal{Q}$ on $\mathcal{H}$, if $r_{\mathcal{D}}(\mathcal{Q}) < \frac{1}{2} \iff 2e_{\mathcal{D}}(\mathcal{Q}) + d_{\mathcal{D}}(\mathcal{Q}) < 1$, we have*

$$r_{\mathcal{D}}^{MV}(\mathcal{Q}) \leq 1 - \frac{(1 - 2r_{\mathcal{D}}(\mathcal{Q}))^2}{1 - 2d_{\mathcal{D}}(\mathcal{Q})} \triangleq C_{\mathcal{D}}(\mathcal{Q})$$

$$= 1 - \frac{\left(1 - [2e_{\mathcal{D}}(\mathcal{Q}) + d_{\mathcal{D}}(\mathcal{Q})]\right)^2}{1 - 2d_{\mathcal{D}}(\mathcal{Q})}.$$

*The **empirical C-Bound** is denoted by $C_{\mathcal{S}}(\mathcal{Q})$ where the empirical disagreement is defined by $d_{\mathcal{S}}(\mathcal{Q}) = \frac{1}{m} \sum_{i=1}^{m} \mathbb{E}_{h_1 \sim \mathcal{Q}} \mathbb{E}_{h_2 \sim \mathcal{Q}} \mathbf{I}[h_1(\mathbf{x}_i) \neq h_2(\mathbf{x}_i)]$, and the empirical joint error is defined by $e_{\mathcal{S}}(\mathcal{Q}) = \frac{1}{m} \sum_{i=1}^{m} \mathbb{E}_{h_1 \sim \mathcal{Q}} \mathbb{E}_{h_2 \sim \mathcal{Q}} \mathbf{I}[h_1(\mathbf{x}_i) \neq y_i] \mathbf{I}[h_2(\mathbf{x}_i) \neq y_i]$.*

As Eq. (3), the C-Bound is tighter than Eq. (2) when $r_{\mathcal{D}}(\mathcal{Q}) \leq d_{\mathcal{D}}(\mathcal{Q})$ and looks for a good trade-off between individual risks and disagreement. The main interest of the C-bound compared to Eq. (3) is that when $e_{\mathcal{D}}(\mathcal{Q})$ is close to $\frac{1}{4}$, the C-Bound can be close to 0 depending on the value of the disagreement $d_{\mathcal{D}}(\mathcal{Q})$: the C-bound is then more precise. Moreover, it is important to notice that the C-Bound is always tighter than Eq. (3) and tighter than Eq. (2) when $r_{\mathcal{D}}(\mathcal{Q}) \leq d_{\mathcal{D}}(\mathcal{Q})$. We summarize the relationships between Eqs. (2), (3) and $C_{\mathcal{D}}(\mathcal{Q})$ in the next theorem.

Theorem 2 (From Germain et al. [32] and Masegosa et al. [24]). *For any distribution $\mathcal{D}$ on $\mathcal{X} \times \mathcal{Y}$, for any voters set $\mathcal{H}$, for any distribution $\mathcal{Q}$ on $\mathcal{H}$, if $r_{\mathcal{D}}(\mathcal{Q}) < \frac{1}{2}$, we have*

(i) $C_{\mathcal{D}}(\mathcal{Q}) \leq 4e_{\mathcal{D}}(\mathcal{Q}) \leq 2r_{\mathcal{D}}(\mathcal{Q})$, if $r_{\mathcal{D}}(\mathcal{Q}) \leq d_{\mathcal{D}}(\mathcal{Q})$,
(ii) $2r_{\mathcal{D}}(\mathcal{Q}) \leq C_{\mathcal{D}}(\mathcal{Q}) \leq 4e_{\mathcal{D}}(\mathcal{Q})$, otherwise.

In this paper, we focus on the minimization of PAC-Bayesian generalization bounds on the C-Bound to get a low-risk majority vote. In Sect. 3, we recall such PAC-Bayesian bounds that have been introduced in the literature.

2.3 Related Works

Previous algorithms have been developed to minimize the *empirical* C-Bound $C_{\mathcal{S}}(\mathcal{Q})$. Roy *et al.* [31] first proposed MinCq where this minimization is expressed as a quadratic problem. MinCq considers a specific voters' set to regularize the minimization process. One drawback of MINCQ is that the optimization problem is not scalable to large datasets. Lately, Bauvin *et al.* [1] proposed CB-BOOST that minimizes $C_S(Q)$ in a greedy procedure with the advantage to be more scalable while obtaining sparser majority vote. However, since both MINCQ and CB-BOOST minimize the empirical $C_S(Q)$, the PAC-Bayesian generalization bound associated with their learned majority vote predictors can be vacuous. Note that CB-BOOST has been proposed to improve another algorithm called CQBOOST [32].

When it comes to deriving a learning algorithm that directly minimizes a PAC-Bayesian bound, it is mentioned in the literature that optimizing a PAC-Bayesian bound on the C-bound is not trivial [22,24]. This underlines the need of other majority vote learning algorithms based on the C-Bound, which motivates our contributions of Sect. 4.

3 PAC-Bayesian C-Bounds

We recall now three PAC-Bayesian generalization bounds on the C-Bound referred hereafter as the **PAC-Bayesian C-Bounds**. Considering these three approaches has the interest to offer a large coverage of the PAC-Bayesian C-bound literature. Our contribution, described in Sect. 4, consists in deriving a self-bounding algorithm for each of these PAC-Bayesian C-Bounds. This shows that the PAC-Bayesian C-Bound offers various ways to learn majority votes that might have been overlooked until now.

3.1 An Intuitive Bound—McAllester's View

We recall the most intuitive and interpretable PAC-Bayesian C-Bound [32]. It consists in upper-bounding separately the Gibbs risk $r_D(Q)$ and the disagreement $d_D(Q)$ with the usual PAC-Bayesian bound of McAllester [26] that bounds the deviation between true and empirical values with the Euclidean distance.

Theorem 3 (PAC-Bayesian C-Bound of Roy et al. [32]). *For any distribution D on $X \times Y$, for any prior distribution P on H, for any $\delta > 0$, we have*

$$\Pr_{S \sim D^m}\left(\forall Q \text{ on } \mathcal{H}, \; C_D(Q) \leq \underbrace{1 - \frac{\left(1 - 2\min\left[\frac{1}{2}, r_S(Q) + \sqrt{\frac{1}{2}\psi_r(Q)}\right]\right)^2}{1 - 2\max\left[0, d_S(Q) - \sqrt{\frac{1}{2}\psi_d(Q)}\right]}}_{C_S^M(Q)}\right) \geq 1 - 2\delta,$$

(4)

with $\psi_r(Q) = \frac{1}{m}\left[\text{KL}(Q\|P) + \ln\frac{2\sqrt{m}}{\delta}\right]$, and $\psi_d(Q) = \frac{1}{m}\left[2\,\text{KL}(Q\|P) + \ln\frac{2\sqrt{m}}{\delta}\right]$, and $\text{KL}(Q\|P) = \mathbb{E}_{h \sim Q} \ln\frac{Q(h)}{P(h)}$ is the KL-divergence between Q and P.

While there is no algorithm that directly minimizes Eq. (4), this kind of interpretable bound can be seen as a justification of the optimization of $r_S(Q)$ and

$d_S(Q)$ in the empirical C-Bound such as for MINCQ [31] or CB-BOOST [1]. In Sect. 4.1, we derive a first algorithm to directly minimize it.

However, this PAC-Bayesian C-Bound can have a severe disadvantage with a small m and a Gibbs risk close to $\frac{1}{2}$: even for a $KL(Q\|P)$ close to 0 and a low empirical C-Bound, the value of the PAC-Bayesian C-Bound will be close to 1. To overcome this drawback, one solution is to follow another PAC-Bayesian point of view, the one proposed by Seeger [33] that compares the true and empirical values through $kl(a\|b) = a \log \left[\frac{a}{b}\right] + (1-a) \log \left[\frac{1-a}{1-b}\right]$, knowing that $|a-b| \leq \sqrt{\frac{1}{2}kl(a\|b)}$ (Pinsker's inequality).

In the next two subsections, we recall such bounds. The first one in Theorem 4 involves the risk and the disagreement, while the second one in Theorem 5 simultaneously bounds the joint error and the disagreement.

3.2   A Tighter Bound—Seeger's View

The PAC-Bayesian generalization bounds based on the Seeger's approach [33] are known to produce tighter bounds [15]. As for Theorem 3, the result below bounds independently the Gibbs risk $r_D(Q)$ and the disagreement $d_D(Q)$.

Theorem 4 (PAC-Bayesian C-Bound (PAC-Bound 1) of Germain et al. [15]). *Under the same assumptions and notations as Theorem 3, we have*

$$\Pr_{S\sim D^m}\left(\forall Q \text{ on } \mathcal{H}, \ C_D(Q) \leq 1 - \underbrace{\frac{\left(1-2\min\left[\frac{1}{2}, \overline{kl}\left(r_S(Q) \mid \psi_r(Q)\right)\right]\right)^2}{1-2\max\left[0, \underline{kl}\left(d_S(Q) \mid \psi_d(Q)\right)\right]}}_{C_S^S(Q)}\right) \geq 1-2\delta,$$

$$(5)$$

with $\overline{kl}(q|\psi) = \max\{p \in (0,1)|kl(q\|p) \leq \psi\}$, *and* $\underline{kl}(q|\psi) = \min\{p \in (0,1)|kl(q\|p) \leq \psi\}$.

The form of this bound makes the optimization a challenging task: the functions $\overline{kl}$ and $\underline{kl}$ do not benefit from closed-form solutions. However, we see in Sect. 3.2 that the optimization of $\overline{kl}$ and $\underline{kl}$ can be done by the bisection method [30], leading to an easy-to-solve algorithm to optimize this PAC-Bayesian C-Bound.

3.3   Another Tighter Bound–Lacasse's View

The last theorem on which we build our contributions is described below. Proposed initially by Lacasse *et al.* [20], its interest is that it simultaneously bounds the joint error and the disagreement (as explained by Germain *et al.* [15]). Here, to compute the bound, we need to find the worst C-Bound value that can be obtained with a couple of joint error and disagreement denoted by (e, d) belonging to the set $A_S(Q)$ that is defined by

$$A_S(Q) = \left\{(e, d) \ \middle| \ kl\left(e_S(Q), d_S(Q)\|e, d\right) \leq \kappa(Q)\right\},$$

Algorithm 1. Minimization of Equation (4) by GD

Given: learning sample $\mathcal{S}$, prior distribution $\mathcal{P}$ on $\mathcal{H}$, the objective function $G_{\mathcal{S}}^{\text{M}}(\mathcal{Q})$

Update function[2] UPDATE-$\mathcal{Q}$
Hyperparameters: number of iterations T
function MINIMIZE-$\mathcal{Q}$
    $\mathcal{Q} \leftarrow \mathcal{P}$
    for $t \leftarrow 1$ to T **do** $\mathcal{Q} \leftarrow$UPDATE-$\mathcal{Q}(G_{\mathcal{S}}^{\text{M}}(\mathcal{Q}))$
    return $\mathcal{Q}$

---

where $\kappa(\mathcal{Q}) = \frac{1}{m}\left[2\text{KL}(\mathcal{Q}\|\mathcal{P}) + \ln\frac{2\sqrt{m}+m}{\delta}\right]$,

and $\text{kl}(q_1,q_2\|p_1,p_2) = q_1\ln\frac{q_1}{p_1} + q_2\ln\frac{q_2}{p_2} + (1-q_1-q_2)\ln\frac{1-q_1-q_2}{1-p_1-p_2}$.

The set $A_{\mathcal{S}}(\mathcal{Q})$ can actually contain some pairs not achievable by any $\mathcal{D}$, it can then be restricted to the valid subset $\widetilde{A}_{\mathcal{S}}(\mathcal{Q})$ defined in the theorem below.

Theorem 5 (PAC-Bayesian C-Bound (PAC-Bound 2) of Germain et al. [15]). *Under the same assumptions as Theorem 3, we have*

$$\Pr_{\mathcal{S}\sim\mathcal{D}^m}\left(\forall\mathcal{Q}\text{ on }\mathcal{H},\ C_{\mathcal{D}}(\mathcal{Q}) \leq \sup_{(e,d)\in\widetilde{A}_{\mathcal{S}}(\mathcal{Q})}\left[1 - \frac{(1-(2e+d))^2}{1-2d}\right]\right) \geq 1-\delta,$$

where $\widetilde{A}_{\mathcal{S}}(\mathcal{Q}) = \left\{(e,d)\in A_{\mathcal{S}}(\mathcal{Q}) \mid d \leq 2\sqrt{e}-2e,\ 2e+d < 1\right\}$.

Optimizing this bound *w.r.t.* $\mathcal{Q}$ can be challenging, since it boils down to optimize indirectly the set $\widetilde{A}_{\mathcal{S}}(\mathcal{Q})$. Hence, a direct optimization by gradient descent is not possible. In Sect. 4.3 we derive an approximation easier to optimize.

4 Self-bounding Algorithms for PAC-Bayesian C-Bounds

In this section, we present our contribution that consists in proposing three self-bounding algorithms to directly minimize the PAC-Bayesian C-Bounds.

4.1 Algorithm Based on McAllester's View

We derive in Algorithm 1 a method to directly minimize the PAC-Bayesian C-Bound of Theorem 3 by Gradient Descent (GD). An important aspect of the optimization is that if $r_{\mathcal{S}}(\mathcal{Q})+\sqrt{\frac{1}{2}\psi_r(\mathcal{Q})} \geq \frac{1}{2}$, the gradient of the numerator in $C_{\mathcal{S}}^{\text{M}}(\mathcal{Q})$ with respect to $\mathcal{Q}$ is 0 which makes the optimization impossible. Hence,

---

[2] UPDATE-$\mathcal{Q}$ is a generic update function, *i.e.*, it can be for example a standard update of GD or the update of another algorithm like Adam [18] or COCOB [27].

we aim at minimizing the following constraint optimization problem:

$$\min_{\mathcal{Q}} \underbrace{\left[1 - \frac{\left(1 - 2\min\left[\frac{1}{2}, r_{\mathcal{S}}(\mathcal{Q}) + \sqrt{\frac{1}{2}\psi_r(\mathcal{Q})}\right]\right)^2}{1 - 2\max\left[0, d_{\mathcal{S}}(\mathcal{Q}) - \sqrt{\frac{1}{2}\psi_d(\mathcal{Q})}\right]} \right]}_{C_{\mathcal{S}}^{\mathrm{M}}(\mathcal{Q})} \quad \text{s.t} \quad r_{\mathcal{S}}(\mathcal{Q}) + \sqrt{\tfrac{1}{2}\psi_r(\mathcal{Q})} \le \tfrac{1}{2}.$$

From this formulation, we deduce a non-constrained optimization problem: $\min_{\mathcal{Q}}\left[C_{\mathcal{S}}^{\mathrm{M}}(\mathcal{Q}) + \mathbf{B}(r_{\mathcal{S}}(\mathcal{Q}) + \sqrt{\frac{1}{2}\psi_r(\mathcal{Q})} - \frac{1}{2})\right]$, where $\mathbf{B}$ is the barrier function defined as $\mathbf{B}(a) = 0$ if $a \le 0$ and $\mathbf{B}(a) = +\infty$ otherwise. Due to the nature of $\mathbf{B}$, this problem is not suitable for optimization: the objective function will be infinite when $a > 0$. To tackle this drawback, we replace $\mathbf{B}$ by the approximation introduced by Kervadec *et al.* [17] called the log-barrier extension and defined as

$$\mathbf{B}_\lambda(a) = \begin{cases} -\frac{1}{\lambda}\ln(-a), & \text{if } a \le -\frac{1}{\lambda^2}, \\ \lambda a - \frac{1}{\lambda}\ln(\frac{1}{\lambda^2}) + \frac{1}{\lambda}, & \text{otherwise.} \end{cases}$$

In fact, $\mathbf{B}_\lambda$ tends to $\mathbf{B}$ when λ tends to $+\infty$. Compared to the standard log-barrier[3], the function $\mathbf{B}_\lambda$ is differentiable even when the constraint is not satisfied, *i.e.*, when $a > 0$. By taking into account the constraint $r_{\mathcal{S}}(\mathcal{Q}) + \sqrt{\frac{1}{2}\psi_r(\mathcal{Q})} \le \frac{1}{2}$, we solve by GD with Algorithm 1 the following problem:

$$\min_{\mathcal{Q}} G_{\mathcal{S}}^{\mathrm{M}}(\mathcal{Q}) = \min_{\mathcal{Q}} C_{\mathcal{S}}^{\mathrm{M}}(\mathcal{Q}) + \mathbf{B}_\lambda\left(r_{\mathcal{S}}(\mathcal{Q}) + \sqrt{\tfrac{1}{2}\psi_r(\mathcal{Q})} - \tfrac{1}{2}\right).$$

For a given λ, the optimizer will thus find a solution with a good trade-off between minimizing $C_{\mathcal{S}}^{\mathrm{M}}(\mathcal{Q})$ and the log-barrier extension function $\mathbf{B}_\lambda$. As we show in the experiments, minimizing the McAllester-based bound does not lead to the tightest bound. Indeed, as mentioned in Sect. 3, such bound is looser than Seeger-based bounds, and leads to a looser PAC-Bayesian C-Bound.

4.2 Algorithm Based on Seeger's View

In order to obtain better generalization guarantees, we should optimize the Seeger-based C-bound of Theorem 4. In the same way as in the previous section, we seek at minimizing the following optimization problem:

$$\min_{\mathcal{Q}} \underbrace{\left[1 - \frac{\left(1 - 2\min\left[\frac{1}{2}, \overline{\mathrm{kl}}\left(r_{\mathcal{S}}(\mathcal{Q}) \mid \psi_r(\mathcal{Q})\right)\right]\right)^2}{1 - 2\max\left[0, \underline{\mathrm{kl}}\left(d_{\mathcal{S}}(\mathcal{Q}) \mid \psi_d(\mathcal{Q})\right)\right]} \right]}_{C_{\mathcal{S}}^{\mathrm{S}}(\mathcal{Q})} \quad \text{s.t} \quad \overline{\mathrm{kl}}\left(r_{\mathcal{S}}(\mathcal{Q}) \mid \psi_r(\mathcal{Q})\right) \le \tfrac{1}{2},$$

---

[3] The reader can refer to [4] for an introduction of interior-point methods.

with $\overline{\mathrm{kl}}(q|\psi) = \max\{p \in (0,1)|\mathrm{kl}(q\|p) \leq \psi\}$, and $\underline{\mathrm{kl}}(q|\psi) = \min\{p \in (0,1)|\mathrm{kl}(q\|p) \leq \psi\}$. For the same reasons as for deriving Algorithm 1, we propose to solve by GD:

$$\min_{\mathcal{Q}} G_{\mathcal{S}}^{\mathrm{S}}(\mathcal{Q}) = \min_{\mathcal{Q}} C_{\mathcal{S}}^{\mathrm{S}}(\mathcal{Q}) + \mathbf{B}_{\lambda}\left(\overline{\mathrm{kl}}\left(r_{\mathcal{S}}(\mathcal{Q}) \mid \psi_r(\mathcal{Q})\right) - \tfrac{1}{2}\right).$$

The main challenge to optimize it is to evaluate $\overline{\mathrm{kl}}$ or $\underline{\mathrm{kl}}$ and to compute their derivatives. To do so, we follow the bisection method to calculate $\overline{\mathrm{kl}}$ and $\underline{\mathrm{kl}}$ proposed by Reeb et al. [30]. This method is summarized in the functions COMPUTE-$\overline{\mathrm{kl}}(q|\psi)$ and COMPUTE-$\underline{\mathrm{kl}}(q|\psi)$ of Algorithm 2, and consists in refining iteratively an interval $[p_{\min}, p_{\max}]$ with $p \in [p_{\min}, p_{\max}]$ such that $\mathrm{kl}(q\|p) = \psi$. For the sake of completeness, we provide the derivatives of $\underline{\mathrm{kl}}$ and $\overline{\mathrm{kl}}$ with respect to q and ψ, that are:

$$\frac{\partial k(q|\psi)}{\partial q} = \frac{\ln\frac{1-q}{1-k(q|\psi)} - \ln\frac{q}{k(q|\psi)}}{\frac{1-q}{1-k(q|\psi)} - \frac{q}{k(q|\psi)}}, \text{ and } \frac{\partial k(q|\psi)}{\partial \psi} = \frac{1}{\frac{1-q}{1-k(q|\psi)} - \frac{q}{k(q|\psi)}}, \quad (6)$$

with k is either $\underline{\mathrm{kl}}$ or $\overline{\mathrm{kl}}$. To compute the derivatives with respect to the posterior $\mathcal{Q}$, we use the chain rule for differentiation with a deep learning framework (such as PyTorch [29]). The global algorithm is summarized in Algorithm 2.

Algorithm 2. Minimization of Equation (4) by GD

Given: learning sample $\mathcal{S}$, prior distribution $\mathcal{P}$ on $\mathcal{H}$, the objective function $G_{\mathcal{S}}^{\mathrm{M}}(\mathcal{Q})$

Update function UPDATE-$\mathcal{Q}$
Hyperparameters: number of iterations T
function MINIMIZE-$\mathcal{Q}$
    $\mathcal{Q} \leftarrow \mathcal{P}$
    for $t \leftarrow 1$ to T **do**
        Compute $G_{\mathcal{S}}^{\mathrm{S}}(\mathcal{Q})$ using COMPUTE-$\overline{\mathrm{kl}}(q|\psi)$ and COMPUTE-$\underline{\mathrm{kl}}(q|\psi)$
        $\mathcal{Q} \leftarrow$ UPDATE-$\mathcal{Q}(G_{\mathcal{S}}^{\mathrm{S}}(\mathcal{Q}))$ (thanks to the derivatives in Equation (6))
    return $\mathcal{Q}$

Hyperparameters: tolerance ϵ, maximal number of iterations $T_{\max}$
function COMPUTE-$\overline{\mathrm{kl}}(q|\psi)$ (RESP. COMPUTE-$\underline{\mathrm{kl}}(q|\psi)$)
    $p_{\max} \leftarrow 1$ and $p_{\min} \leftarrow q$ (resp. $p_{\max} \leftarrow q$ and $p_{\min} \leftarrow 0$)
    for $t \leftarrow 1$ to $T_{\max}$ **do**
        $p = \tfrac{1}{2}[p_{\min} + p_{\max}]$
        if $\mathrm{kl}(q\|p) = \psi$ or $(p_{\min} - p_{\max}) < \epsilon$ **then return** p
        if $\mathrm{kl}(q\|p) > \psi$ **then** $p_{\max} = p$ (resp. $p_{\min} = p$)
        if $\mathrm{kl}(q\|p) < \psi$ **then** $p_{\min} = p$ (resp. $p_{\max} = p$)
    return p

4.3 Algorithm Based on Lacasse's View

Theorem 5 jointly upper-bounds the joint error $e_{\mathcal{D}}(\mathcal{Q})$ and the disagreement $d_{\mathcal{D}}(\mathcal{Q})$; But as pointed out in Sect. 3.3 its optimization can be hard. To ease its manipulation, we derive below a C-Bound resulting of a reformulation of the constraints involved in the set $\tilde{A}_{\mathcal{S}}(\mathcal{Q}) = \{(e,d) \in A_{\mathcal{S}}(\mathcal{Q}) \mid d \leq 2\sqrt{e} - 2e, 2e + d < 1\}$.

Theorem 6. *Under the same assumptions as Theorem 3, we have*

$$\mathop{P}_{\mathcal{S} \sim \mathcal{D}^m} \left(C_{\mathcal{D}}(\mathcal{Q}) \leq \sup_{(e,d) \in \widehat{A}_{\mathcal{S}}(\mathcal{Q})} \underbrace{\left[1 - \frac{[1-(2e+d)]^2}{1-2d} \right]}_{C^{\mathrm{L}}(e,d)} \right) \geq 1-\delta, \qquad (7)$$

where $\widehat{A}_{\mathcal{S}}(\mathcal{Q}) = \left\{ (e,d) \in A_{\mathcal{S}}(\mathcal{Q}) \mid d \leq 2\sqrt{\min\left(e, \tfrac{1}{4}\right)} - 2e, \ d < \tfrac{1}{2} \right\}$,

and $A_{\mathcal{S}}(\mathcal{Q}) = \{(e,d) \mid \mathrm{kl}\left(e_{\mathcal{S}}(\mathcal{Q}), d_{\mathcal{S}}(\mathcal{Q}) \| e, d\right) \leq \kappa(\mathcal{Q})\}$, *with* $\kappa(\mathcal{Q}) = \frac{2\mathrm{KL}(\mathcal{Q}\|\mathcal{P}) + \ln \frac{2\sqrt{m}+m}{\delta}}{m}$.

Proof. Beforehand, we explain how we fixed the constraints involved in $\widehat{A}_{\mathcal{S}}(\mathcal{Q})$. We add to $A_{\mathcal{S}}(\mathcal{Q})$ three constraints: $d \leq 2\sqrt{e} - 2e$ (from Prop. 9 of [15]), $d \leq 1 - 2e$, and $d < \tfrac{1}{2}$. We remark that when $e \leq \tfrac{1}{4}$, we have $2\sqrt{e} - 2e \leq 1 - 2e$. Then, we merge $d \leq 2\sqrt{e} - 2e$ and $d \leq 1 - 2e$ into $d \leq 2\sqrt{\min\left(e, \tfrac{1}{4}\right)} - 2e$. Indeed, we have

$$d \leq 2\sqrt{\min(e, \tfrac{1}{4})} - 2e \iff \begin{cases} d \leq 2\sqrt{e} - 2e & \text{if } e \leq \tfrac{1}{4}, \\ d < 1 - 2e & \text{if } e \geq \tfrac{1}{4}. \end{cases}$$

We prove now that under the constraints involved in $\widehat{A}_{\mathcal{S}}(\mathcal{Q})$, we still have a valid bound on $C_{\mathcal{D}}(\mathcal{Q})$. To do so, we consider two cases.

Case 1: If for all $(e,d) \in \widehat{A}_{\mathcal{S}}(\mathcal{Q})$ we have $2e + d < 1$.

In this case $(e_{\mathcal{D}}(\mathcal{Q}), d_{\mathcal{D}}(\mathcal{Q})) \in \widehat{A}_{\mathcal{S}}(\mathcal{Q})$, then we have $2e_{\mathcal{D}}(\mathcal{Q}) + d_{\mathcal{D}}(\mathcal{Q}) < 1$ and Theorem 1 holds. We have $C_{\mathcal{D}}(\mathcal{Q}) = 1 - \frac{[1-(2e_{\mathcal{D}}(\mathcal{Q})+d_{\mathcal{D}}(\mathcal{Q}))]^2}{1-2d_{\mathcal{D}}(\mathcal{Q})} \leq \sup_{(e,d) \in \widehat{A}_{\mathcal{S}}(\mathcal{Q})} C^{\mathrm{L}}(e,d)$.

Case 2: If there exists $(e,d) \in \widehat{A}_{\mathcal{S}}(\mathcal{Q})$ such that $2e + d = 1$.

We have $\sup_{(e,d) \in \widehat{A}_{\mathcal{S}}(\mathcal{Q})} C^{\mathrm{L}}(e,d) = 1$ that is a valid bound on $C_{\mathcal{D}}(\mathcal{Q})$. ☐

Theorem 6 suggests then the following constrained optimization problem:

$$\min_{\mathcal{Q}} \left\{ \sup_{(e,d) \in [0,\frac{1}{2}]^2} \left(1 - \frac{[1-(2e+d)]^2}{1-2d} \right) \text{ s.t. } (e,d) \in \widehat{A}_{\mathcal{S}}(\mathcal{Q}) \right\} \text{ s.t. } 2e_{\mathcal{S}}(\mathcal{Q}) + d_{\mathcal{S}}(\mathcal{Q}) \leq 1,$$

with $\widehat{A}_S(Q) = \{(e,d) | d \leq 2\sqrt{\min(e,\frac{1}{4})} - 2e,\ d < \frac{1}{2},\ \mathrm{kl}(e_S(Q),d_S(Q)\|e,d) \leq \kappa(Q)\}$. Actually, we can rewrite this constrained optimization problem into an unconstrained one using the barrier function. We obtain

$$\min_Q \left\{ \max_{(e,d)\in[0,\frac{1}{2}]^2} \left(C^{\mathrm{L}}(e,d) - \mathbf{B}\left[d - 2\sqrt{\min(e,\tfrac{1}{4})} - 2e\right] - \mathbf{B}\left[d - \tfrac{1}{2}\right] \right. \right.$$

$$\left. \left. - \mathbf{B}\left[\mathrm{kl}(e_S(Q),d_S(Q)\|e,d) - \kappa(Q)\right] \right) + \mathbf{B}\left[2e_S(Q) + d_S(Q) - 1\right] \right\},\quad (8)$$

where $C^{\mathrm{L}}(e,d) = 1 - \frac{(1-(2e+d))^2}{1-2d}$ if $d < \frac{1}{2}$, and $C^{\mathrm{L}}(e,d) = 1$ otherwise. However, this problem cannot be optimized directly by GD. In this case, we have a min-max optimization problem, *i.e.*, for each descent step we need to find the couple (e,d) that maximizes the $C^{\mathrm{L}}(e,d)$ given the three constraints that define $\widehat{A}_S(Q)$ before updating the posterior distribution Q.

First, to derive our optimization procedure, we focus on the inner maximization problem when $e_S(Q)$ and $d_S(Q)$ are fixed in order to find the optimal (e,d). However, the function $C^{\mathrm{L}}(e,d)$ we aim at maximizing is not concave for all $(e,d) \in \mathbb{R}^2$, implying that the implementation of its maximization can be hard[4]. Fortunately, $C^{\mathrm{L}}(e,d)$ is quasi-concave [15] for $(e,d) \in [0,1] \times [0,\frac{1}{2}]$. Then by definition of quasi-concavity, we have:

$$\forall \alpha \in [0,1], \quad \left\{ (e,d) \mid 1 - \frac{[1-(2e+d)]^2}{1-2d} \geq 1 - \alpha \right\}$$

$$\iff \forall \alpha \in [0,1], \quad \left\{ (e,d) \mid \alpha(1-2d) - [1-(2e+d)]^2 \geq 0 \right\}.$$

Hence, for any fixed $\alpha \in [0,1]$ we can look for (e,d) that maximizes $C^{\mathrm{L}}(e,d)$ and respects the constraints involved in $\widehat{A}_S(Q)$. This is equivalent to solve the following problem for a given $\alpha \in [0,1]$:

$$\max_{(e,d)\in[0,\frac{1}{2}]^2} \quad \alpha(1-2d) - [1-(2e+d)]^2$$

$$\text{s.t. } d \leq 2\sqrt{\min(e,\tfrac{1}{4})} - 2e \qquad \text{and} \qquad \mathrm{kl}(e_S(Q),d_S(Q)\|e,d) \leq \kappa(Q). \quad (9)$$

In fact, we aim at finding $\alpha \in [0,1]$ such that the maximization of Eq. (9) leads to $1-\alpha$ equal to the largest value of $C^{\mathrm{L}}(e,d)$ under the constraints. To do so, we make use of the "Bisection method for quasi-convex optimization" [4] that is summarized in MAXIMIZE-*e*-*d* in Algorithm 3. We denote by (e^*,d^*) the solution of Eq. (9). It remains then to solve the outer minimization problem that becomes:

$$\min_Q \left\{ \mathbf{B}\left[2e_S(Q) + d_S(Q) - 1\right] - \mathbf{B}\left[\mathrm{kl}(e_S(Q),d_S(Q)\|e^*,d^*) - \kappa(Q)\right] \right\}.$$

---

[4] For example, when using CVXPY [9], that uses Disciplined Convex Programming (DCP [16]), the maximization of a non-concave function is not possible.

Algorithm 3. Minimization of Equation (7) by GD

Given: learning sample $\mathcal{S}$, prior $\mathcal{P}$ on $\mathcal{H}$, the objective function $G_{\mathcal{S}}^{e^*,d^*}(\mathcal{Q})$

Update function UPDATE-$\mathcal{Q}$
Hyperparameters: number of iterations T
function MINIMIZE-$\mathcal{Q}$
    $\mathcal{Q} \leftarrow \mathcal{P}$
    for $t \leftarrow 1$ to T **do**
        $(e^*, d^*) \leftarrow$ MAXIMIZE-e-d$(e_{\mathcal{S}}(\mathcal{Q}), d_{\mathcal{S}}(\mathcal{Q}))$
        $\mathcal{Q} \leftarrow$ UPDATE-$\mathcal{Q}(G_{\mathcal{S}}^{e^*,d^*}(\mathcal{Q}))$
    return $\mathcal{Q}$

Given: learning sample $\mathcal{S}$, joint error $e_{\mathcal{S}}(\mathcal{Q})$, disagreement $d_{\mathcal{S}}(\mathcal{Q})$
Hyperparameters: tolerance ϵ
function MAXIMIZE-e-d$(e_{\mathcal{S}}(\mathcal{Q}), d_{\mathcal{S}}(\mathcal{Q}))$
    $\alpha_{\min} = 0$ and $\alpha_{\max} = 1$
    while $\alpha_{\max} - \alpha_{\min} > \epsilon$ **do**
        $\alpha = \frac{1}{2}(\alpha_{\min} + \alpha_{\max})$
        $(e, d) \leftarrow$ Solve Equation (9)
        if $C^{\mathrm{L}}(e, d) \geq 1-\alpha$ **then** $\alpha_{\max} \leftarrow \alpha$ **else** $\alpha_{\min} \leftarrow \alpha$
    return (e, d)

Since the barrier function $\mathbf{B}$ is not suitable for optimization, we approximate this problem by replacing $\mathbf{B}$ by the log-barrier extension $\mathbf{B}_\lambda$, *i.e.*, we have

$$\min_{\mathcal{Q}} \ G_{\mathcal{S}}^{e^*,d^*}(\mathcal{Q}) \ = \ \min_{\mathcal{Q}} \big\{ \ \mathbf{B}_\lambda \left[2e_{\mathcal{S}}(\mathcal{Q})+d_{\mathcal{S}}(\mathcal{Q})-1\right] \\ -\mathbf{B}_\lambda \left[\mathrm{kl}\left(e_{\mathcal{S}}(\mathcal{Q}), d_{\mathcal{S}}(\mathcal{Q})\|e^*, d^*\right) - \kappa(\mathcal{Q})\right] \ \big\}.$$

The global method is summarized in Algorithm 3. As a side note, we mention that the classic Danskin Theorem [8] used in min-max optimization theory is not applicable in our case since our objective function is not differentiable for all $(e, d) \in [0, \frac{1}{2}]^2$. We discuss this point in Supplemental.

5 Experimental Evaluation

5.1 Empirical Setting

Our experiments[5] have a two-fold objective: *(i)* assessing the guarantees given by the associated PAC-Bayesian bounds, and *(ii)* comparing the performance of the different C-bound based algorithms in terms of risk optimization. To achieve this objective, we compare the three algorithms proposed in this paper to the following state-of-the-art PAC-Bayesian methods for majority vote learning:

---

[5] Experiments are done with PyTorch [29] and CVXPY [9]. The source code is available at https://github.com/paulviallard/ECML21-PB-CBound.

Table 1. Comparison of the true risks "r_T^{MV}" and bound values "Bnd" obtained for each algorithm. "Bnd" is the value of the bound that is optimized, excepted for MinCq and CB-Boost for which we report the bound obtained with Theorem 6 instantiated with the majority vote learned. Results in **bold** are the couple (r_T^{MV},Bnd) associated to **the lowest risk value**. Italic and underlined results are the couple (r_T^{MV},Bnd) associated respectively to *the lowest bound* value and the second lowest bound values.

	Algorithm 1		Algorithm 2		Algorithm 3		CB-Boost		MinCq		MasEgosa		2R	
	r_T^{MV}	Bnd	r_T^{MV}	Bnd	r_T^{MV}	Bnd	r_T^{MV}	Bnd	r_T^{MV}	Bnd	r_T^{MV}	Bnd	r_T^{MV}	Bnd
letter:AvsB	.009	.323	.018	.114	**.000**	**.085**	.000	.104	.009	.451	.004	.070	.018	.056
letter:DvsO	**.013**	.469	.018	.298	.018	.205	.022	.224	.022	.999	.018	.185	.044	.174
letter:OvsQ	.017	.489	.017	.332	**.009**	**.229**	.017	.249	.039	1	.013	.210	.030	.201
credit	.141	.912	.141	.874	.129	.816	.144	.855	**.126**	**.929**	.132	.869	.150	.651
glass	.047	.904	.047	.832	.056	.798	**.037**	**.911**	.056	.999	.056	.903	.047	.566
heart	.250	.976	.264	.962	.250	.955	.270	.981	.270	1	**.243**	**1.19**	.250	.787
tictactoe	.063	.815	.084	.750	**.056**	**.610**	.063	.649	.071	.782	.058	.580	.152	.511
usvotes	.041	.741	.046	.584	.037	.508	.037	.590	.046	.985	**.032**	**.490**	.060	.342
wdbc	.060	.725	.053	.603	.032	.523	**.025**	**.591**	.039	.992	.035	.513	.063	.362
mnist:1vs7	.006	.161	.005	.061	**.005**	**.038**	.005	.040	.015	.994	.006	.034	.006	.043
mnist:4vs9	.017	.238	.016	.167	.016	.110	.016	.113	.046	.960	.016	.106	.063	.148
mnist:5vs6	.011	.210	.011	.124	.011	.078	.011	.081	.035	.999	**.011**	**.073**	.036	.109
fash:COvsSH	**.108**	**.462**	.109	.433	.110	.366	.110	.371	.185	.894	.111	.358	.146	.409
fash:SAvsBO	.018	.217	.018	.134	.019	.094	.019	.097	.034	1	.018	.087	.020	.114
fash:TOvsPU	.029	.245	.029	.165	**.029**	**.133**	.030	.136	.045	.809	.030	.125	.051	.123
adult	.163	.532	.163	.514	**.163**	**.492**	.163	.495	.204	1	**.163**	**.492**	.200	.413
Mean	.062	.526	.065	.434	.059	.378	.061	.405	.078	.925	.059	.393	.083	.313

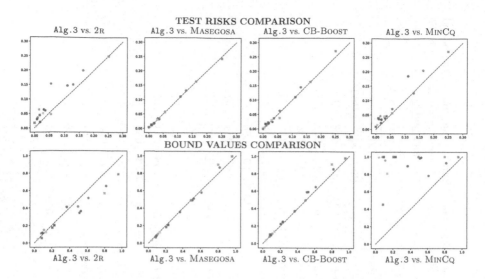

Fig. 1. Pairwise comparisons of the test risks (first line) and the bounds (second line) between Algorithm 3 and the baseline algorithms. Algorithm 3 is represented on the x-axis, while the y-axis is used for the other approaches. Each dataset corresponds to a point in the plot and a point above the diagonal indicates that Algorithm 3 is better.

- MINCQ [31] and CB-BOOST [1] that are based on the minimization of the empirical C-Bound. For comparison purposes and since MINCQ and CB-BOOST do not explicitly minimize a PAC-Bayesian bound, we report the bound values of Theorem 6 instantiated with the models learned;
- The algorithm proposed by Masegosa *et al.* [24] that optimizes a PAC-Bayesian bound on $r_{\mathcal{D}}^{MV}(\mathcal{Q}) \leq 4e_{\mathcal{D}}(\mathcal{Q})$ (see Theorem 9 of [24]);
- An algorithm[6], denoted by 2R, to optimize a PAC-Bayesian bound based only on the Gibbs risk [21]: $r_{\mathcal{D}}^{MV}(\mathcal{Q}) \leq 2r_{\mathcal{D}}(\mathcal{Q}) \leq 2\overline{\mathrm{kl}}(r_{\mathcal{S}}(\mathcal{Q})|\psi_r(\mathcal{Q}))$.

We follow a general setting similar to the one of Masegosa *et al.* [24]. The prior distribution $\mathcal{P}$ on $\mathcal{H}$ is set as the uniform distribution, and the voters in $\mathcal{H}$ are decision trees: 100 trees are learned with 50% of the training data (the remaining part serves to learn the posterior $\mathcal{Q}$). More precisely, for each tree $\sqrt{d}$ features of the d-dimensional input space are selected, and the trees are learned by using the Gini criterion until the leaves are pure.

In this experiment, we consider 16 classic datasets[7] that we split into a train set $\mathcal{S}$ and a test set $\mathcal{T}$. We report for each algorithm in Table 1, the test risks (on $\mathcal{T}$) and the bound values (on $\mathcal{S}$, such that the bounds hold with prob. at least 95%). The parameters of the algorithms are selected as follows. **1) For** Masegosa's algorithm we kept the default parameters [24]. **2) For** all the other bounds minimization algorithms, we set $T = 2,000$ iterations for all the datasets

---

[6] The algorithm 2R is similar to Algorithm 2, but without the numerator of the C-Bound (*i.e.*, the disagreement). More details are given in the Supplemental.

[7] An overview of the datasets is presented in the Supplemental.

except for adult, fash and mnist where $T = 200$. We fix the objective functions with $\lambda = 100$, and we use COCOB-Backprop optimizer [27] as UPDATE-$\mathcal{Q}$ (its parameter remains the default one). For Algorithm 3, we fix the tolerance $\epsilon = .01$, *resp.* $\epsilon = 10^{-9}$, to compute $\underline{kl}$, *resp.* $\overline{kl}$. Furthermore, the maximal number of iterations T_{max} in MAXIMIZE-e-d is set to $1,000$. **3) For** MINCQ, we select the margin parameter among 20 values uniformly distributed in $[0, \frac{1}{2}]$ by 3-fold cross validation. Since this algorithm is not scalable due to its high time complexity, we reduce the training set size to $m = 400$ when learning with MINCQ on the large datasets: adult, fash and mnist (MINCQ is still competitive with less data on this datasets). For CB-BOUND which is based on a Boosting approach, we fix the maximal number of boosting iterations to 200.

5.2   Analysis of the Results

Beforehand, we compare only our three self-bounding algorithms. From Table 1, as expected we observe that Algorithm 1 based on the McAllester's bound (that is more interpretable but less tight) provides the worst bound. Algorithm 3 always provides tighter bounds than Algorithms 1 and 2, and except for letter:DvsO, fash:COvsSH, and fash:SAvsBO Algorithm 3 leads to the lowest test risks. We believe that Algorithm 3 based on the Lacasse's bound provides lower bounds than Algorithm 2 based on the Seeger's bound because the Lacasse's approach bounds simultaneously the joint error and the disagreement. Algorithm 3 appears then to be the best algorithm among our three self-bounding algorithms that minimize a PAC-Bayesian C-Bound.

In the following we focus then on comparing our best contribution represented by Algorithm 3 to the baselines; Fig. 1 summarizes this comparison.

First, 2R gives the lowest bounds among all the algorithms, but at the price of the largest risks. This clearly illustrates the limitation of considering *only* the Gibbs risk as an estimator of the majority vote risk: As discussed in Sect. 2.2, the Gibbs risk is an unfair estimator since an increase of the diversity between the voters can have a negative impact on the Gibbs risk.

Second, compared to Masegosa's approach, the results are comparable: Algorithm 3 tends to provide tighter bounds, and similar performances that lie in the same order of magnitude, as illustrated in Table 1. This behavior was expected since minimizing the bound of Masegosa [24] or the PAC-Bayesian C-Bound boils down to minimize a trade-off between the risk and the disagreement.

Third, compared to empirical C-bound minimization algorithms, we see that Algorithm 3 outputs better results than CB-BOOST and MINCQ for which the difference is significative and the bounds are close to 1 (*i.e.*, non-informative). Optimizing the risk bounds tend then to provide better guarantees that justify that optimizing the empirical C-bound is often too optimistic.

Overall, from these experiments, our Algorithm 3 is the one that provides the best trade-off between having good performances in terms of risk optimization and ensuring good theoretical guarantees with informative bounds.

6   Conclusion and Future Work

In this paper, we present new learning algorithms driven by the minimization of PAC-Bayesian generalization bounds based on the C-Bound. More precisely, we propose to solve three optimization problems, each one derived from an existing PAC-Bayesian bound. Our methods belong to the class of *self-bounding* learning algorithms: The learned predictor comes with a tight and statistically valid risk upper bound. Our experimental evaluation has confirmed the quality of the learned predictor and the tightness of the bounds with respect to state-of-the-art methods minimizing the C-Bound.

As future work, we would like to study extensions of this work to provide meaningful bounds for learning (deep) neural networks. In particular, an interesting perspective would be to adapt the C-Bound to control the diversity and the weights in a neural network.

Acknowledgements. This work was supported by the French Project APRIORI ANR-18-CE23-0015. Moreover, Pascal Germain is supported by the NSERC Discovery grant RGPIN-2020-07223 and the Canada CIFAR AI Chair Program. The authors thank Rémi Emonet for insightful discussions.

References

1. Bauvin, B., Capponi, C., Roy, J., Laviolette, F.: Fast greedy C-bound minimization with guarantees. Mach. Learn. **109**(9), 1945–1986 (2020). https://doi.org/10.1007/s10994-020-05902-7
2. Bellet, A., Habrard, A., Morvant, E., Sebban, M.: Learning a priori constrained weighted majority votes. Mach. Learn. **97**(1–2), 129–154 (2014). https://doi.org/10.1007/s10994-014-5462-z
3. Boser, B., Guyon, I., Vapnik, V.: A training algorithm for optimal margin classifiers. In: COLT (1992)
4. Boyd, S., Vandenberghe, L.: Convex Optimization. Cambridge University Press, Cambridge (2004)
5. Breiman, L.: Bagging predictors. Mach. Learn. **24**(2), 123–140 (1996). https://doi.org/10.1007/BF00058655
6. Breiman, L.: Random forests. Mach. Learn. **45**(1), 5–32 (2001). https://doi.org/10.1023/A:1010933404324
7. Cortes, C., Vapnik, V.: Support-vector networks. Mach. Learn. **20**(3), 273–297 (1995). https://doi.org/10.1007/BF00994018
8. Danskin, J.: The theory of max-min, with applications. SIAM J. Appl. Math. **14**(4), 641–664 (1966)
9. Diamond, S., Boyd, S.: CVXPY: a python-embedded modeling language for convex optimization. J. Mach. Learn. Res. **17**(1), 2909–2913 (2016)
10. Dietterich, T.G.: Ensemble methods in machine learning. In: Kittler, J., Roli, F. (eds.) MCS 2000. LNCS, vol. 1857, pp. 1–15. Springer, Heidelberg (2000). https://doi.org/10.1007/3-540-45014-9_1
11. Dziugaite, G.K., Roy, D.: Computing nonvacuous generalization bounds for deep (stochastic) neural networks with many more parameters than training data. In: UAI (2017)

12. Freund, Y.: Self bounding learning algorithms. In: COLT (1998)
13. Freund, Y., Schapire, R.: Experiments with a new boosting algorithm. In: ICML (1996)
14. Germain, P., Lacasse, A., Laviolette, F., Marchand, M.: PAC-Bayesian learning of linear classifiers. In: ICML (2009)
15. Germain, P., Lacasse, A., Laviolette, F., Marchand, M., Roy, J.: Risk bounds for the majority vote: from a PAC-Bayesian analysis to a learning algorithm. J. Mach. Learn. Res. (2015)
16. Grant, M., Boyd, S., Ye, Y.: Disciplined convex programming. In: Liberti, L., Maculan, N. (eds.) Global Optimization, pp. 155–210. Springer, Boston (2006). https://doi.org/10.1007/0-387-30528-9_7
17. Kervadec, H., Dolz, J., Yuan, J., Desrosiers, C., Granger, E., Ayed, I.B.: Constrained Deep Networks: Lagrangian Optimization via Log-Barrier Extensions. CoRR abs/1904.04205 (2019)
18. Kingma, D., Ba, J.: Adam: a method for stochastic optimization. In: ICLR (2015)
19. Kuncheva, L.: Combining Pattern Classifiers: Methods and Algorithms. Wiley, Hoboken (2014)
20. Lacasse, A., Laviolette, F., Marchand, M., Germain, P., Usunier, N.: PAC-Bayes bounds for the risk of the majority vote and the variance of the gibbs classifier. In: NIPS (2006)
21. Langford, J., Shawe-Taylor, J.: PAC-Bayes & margins. In: NIPS (2002)
22. Lorenzen, S.S., Igel, C., Seldin, Y.: On PAC-Bayesian bounds for random forests. Mach. Learn. 108(8), 1503–1522 (2019). https://doi.org/10.1007/s10994-019-05803-4
23. Madry, A., Makelov, A., Schmidt, L., Tsipras, D., Vladu, A.: Towards deep learning models resistant to adversarial attacks. In: ICLR (2018)
24. Masegosa, A., Lorenzen, S.S., Igel, C., Seldin, Y.: Second order PAC-Bayesian bounds for the weighted majority vote. In: NeurIPS (2020)
25. McAllester, D.: Some PAC-Bayesian theorems. Mach. Learn. 37(3), 355–363 (1999). https://doi.org/10.1023/A:1007618624809
26. McAllester, D.: PAC-Bayesian stochastic model selection. Mach. Learn. 51(1), 5–21 (2003). https://doi.org/10.1023/A:1021840411064
27. Orabona, F., Tommasi, T.: Training deep networks without learning rates through coin betting. In: NIPS (2017)
28. Parrado-Hernández, E., Ambroladze, A., Shawe-Taylor, J., Sun, S.: PAC-Bayes bounds with data dependent priors. J. Mach. Learn. Res. 13(1), 3507–3531 (2012)
29. Paszke, A., et al.: PyTorch: an imperative style, high-performance deep learning library. In: NeurIPS (2019)
30. Reeb, D., Doerr, A., Gerwinn, S., Rakitsch, B.: Learning Gaussian processes by minimizing PAC-Bayesian generalization bounds. In: NeurIPS (2018)
31. Roy, J., Laviolette, F., Marchand, M.: From PAC-Bayes bounds to quadratic programs for majority votes. In: ICML (2011)
32. Roy, J., Marchand, M., Laviolette, F.: A column generation bound minimization approach with PAC-Bayesian generalization guarantees. In: AISTATS (2016)
33. Seeger, M.: PAC-Bayesian generalisation error bounds for gaussian process classification. J. Mach. Learn. Res. 3, 233–269 (2002)
34. Shawe-Taylor, J., Williamson, R.: A PAC analysis of a Bayesian estimator. In: COLT (1997)

Midpoint Regularization: From High Uncertainty Training Labels to Conservative Classification Decisions

Hongyu Guo$^{(\boxtimes)}$

National Research Council Canada, 1200 Montreal Road,
Ottawa, ON K1A 0R6, Canada
hongyu.guo@nrc-cnrc.gc.ca

Abstract. Label Smoothing (LS) improves model generalization through penalizing models from generating overconfident output distributions. For each training sample the LS strategy smooths the one-hot encoded training signal by distributing its distribution mass over the non-ground truth classes. We extend this technique by considering example pairs, coined PLS. PLS first creates midpoint samples by averaging random sample pairs and then learns a smoothing distribution during training for each of these midpoint samples, resulting in midpoints with high uncertainty labels for training. We empirically show that PLS significantly outperforms LS, achieving up to 30% of relative classification error reduction. We also visualize that PLS produces very low winning softmax scores for both in and out of distribution samples.

Keywords: Label smoothing · Model regularization · Mixup

1 Introduction

Label Smoothing (LS) is a commonly used output distribution regularization technique to improve the generalization performance of deep learning models [4, 13,15,20,21,23,30]. Instead of training with data associated with one-hot labels, models with label smoothing are trained on samples with soft targets, where each target is a weighted mixture of the ground truth one-hot label with the uniform distribution of the classes. This penalizes overconfident output distributions, resulting in improved model generalization [16,18,19,26,29].

When smoothing the one-hot training signal, existing LS methods, however, only consider the distance between the only gold label and the non-ground truth targets. This motivates our Pairwise Label Smoothing (PLS) strategy, which takes a pair of samples as input. In a nutshell, the PLS first creates a midpoint sample by averaging the inputs and labels of a sample pair, and then distributes the distribution mass of the two ground truth targets of the new midpoint sample over its non-ground truth classes. Smoothing with a pair of ground truth labels enables PLS to preserve the relative distance between the two truth labels while

© Springer Nature Switzerland AG 2021
N. Oliver et al. (Eds.): ECML PKDD 2021, LNAI 12976, pp. 184–199, 2021.
https://doi.org/10.1007/978-3-030-86520-7_12

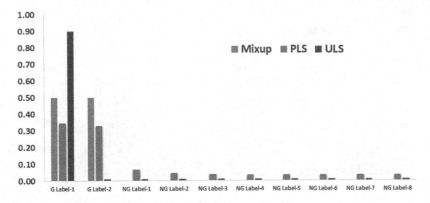

Fig. 1. Average distribution mass (Y-axis) of PLS and Mixup (from sample pairs with different labels) and ULS on Cifar10's 10 classes (X-axis). PLS' class-wise training signals are spread out, with gold labels probabilities close to 0.35, which has larger uncertainty than that of 0.5 and 0.9 from Mixup and ULS, receptivity.

being able to further soften that between the truth labels and the other class targets. Also, unlike current LS methods, which typically require to find a global smoothing distribution mass through cross-validation search, PLS automatically learns the distribution mass for each input pair during training. Hence, it effectively eliminates the turning efforts for searching the right level of smoothing strength for different applications.

Also, smoothing with a pair of labels empowers PLS to be trained with low distribution mass (i.e., high uncertainty) for the ground truth targets. Figure 1 depicts the average values (over all the midpoint samples created from sample pairs with different labels) of the 10 target classes in Cifar10 used by PLS for training. These ground truth targets are smaller than 0.35, which represents much larger uncertain than the 0.5 and 0.9 in Mixup and the uniform label smoothing (denoted as ULS), respectively. Owing to such high uncertainty ground truth target training signals, PLS produces extremely low winning softmax scores during testing, resulting in very conservative classification decisions. Figure 2 depicts the histograms of the winning softmax scores of Mixup, ULS and PLS on the Cifar10 validation data. Unlike Mixup and ULS, PLS produced very conservative softmax scores, with extremely sparse distribution in the high confidence region. Such conservative classifications are produced not only on in-distribution data but also on out-of-distribution samples.

We empirically show that PLS significantly outperforms LS and Mixup [28], with up to 30% of relative classification error reduction. We also visualize that PLS produces extremely conservative predictions in testing time, with many of its winning predicted softmax scores slightly over 0.5 for both in and out of distribution samples.

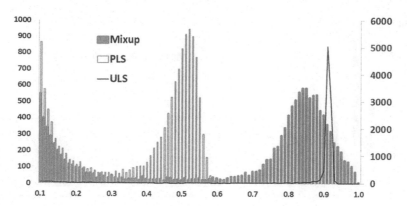

Fig. 2. Histograms of the winning softmax scores on Cifar10 validation data, generated by ULS (right Y-axis) and by Mixup and PLS (left Y-axis) where the X-axis depicts the softmax score. The PLS produces very low softmax winning scores; the differences versus the ULS and Mixup are striking: the PLS models are extremely sparse in the high confidence region at the right end.

2   Related Work

Label smoothing has shown to provide consistent model generalization gains across many tasks [15, 16, 18, 19, 21, 24]. Unlike the above methods which apply label smoothing to each single input, our smoothing strategy leverages a pair of inputs for label smoothing. Furthermore, unlike existing LS approaches which deploy one static and uniform smoothing distribution for all the training samples, our strategy automatically learns a dynamic distribution mass for each sample during training.

Mixup methods [2, 8–10, 22, 27, 28] create *a large number* of synthetic samples with various features from a sample pair, through interpolating the pair's both features and labels with mixing coefficients randomly sampling between [0,1]. In contrast, our approach creates *only the midpoint* sample of a sample pair (equivalent to Mixup with a fixed mixing ratio of 0.5). More importantly, our method adds a label smoothing component on the midpoint samples. That is, we adaptively learn a distribution mass to smooth the pair of ground truth labels of each midpoint sample, with the aims of deploying high uncertainty training targets (much higher than that of Mixup) for output distribution regularization. Such high uncertain training labels result in PLS producing very conservative classification decisions and superior accuracy than Mixup.

The generation of the smoothing distribution in our method is also related to self-distillation [1, 6, 12, 17, 25]. However, these approaches treat the final predictions as target labels for a new round of training, and the teacher and student architectures are identical [17]. In our method, the classification and the smoothing distribution generator have different network architectures, and the training targets for the classification are a mixture of the outputs of the two networks.

3   Label Smoothing over Midpoint Samples

Our proposed method PLS leverages a pair of samples, randomly selected from the training data set, to conduct label smoothing. It first creates a midpoint sample for each sample pair, and then adaptively learns a smoothing distribution for each midpoint sample. These midpoint samples are then used for training to regularize the learning of the networks.

3.1   Preliminaries

We consider a standard classification setting with a given training data set $(X;Y)$ associated with K candidate classes $\{1, 2, \cdots, K\}$. For an example x_i from the training dataset $(X;Y)$, we denote the ground truth distribution q over the labels as $q(y|x_i)$ ($\sum_{y=1}^{K} q(y|x_i) = 1$). Also, we denote a neural network model to be trained as f_θ (parameterized with θ), and it produces a conditional label distribution over the K classes as $p_\theta(y|x_i)$:

$$p_\theta(y|x_i) = \frac{\exp(z_y)}{\sum_{k=1}^{K} \exp(z_{y_k})}, \tag{1}$$

with $\sum_{y=1}^{K} p_\theta(y|x_i) = 1$, and z is noted as the logit of the model f_θ. The logits are generated with two steps: the model f_θ first constructs the m-dimensional input embedding $S_i \in R^m$ for the given input x_i, and then passes it through a linear fully connected layer $W_l \in R^{K \times m}$:

$$S_i = f_\theta(x_i), z = W_l S_i. \tag{2}$$

During learning, the model f_θ is trained to optimize the parameter θ using the n examples from $(X;Y)$ by minimizing the cross-entropy loss:

$$\ell = -\sum_{i=1}^{n} H_i(q, p_\theta). \tag{3}$$

Instead of using one-hot encoded vector for each example x_i in $(X;Y)$, label smoothing (LS) adds a smoothed label distribution (i.e., the prior distribution) $u(y|x_i)$ to each example x_i, forming a new target label, namely soft label:

$$q'(y|x_i) = (1 - \alpha)q(y|x_i) + \alpha u(y|x_i), \tag{4}$$

where hyper-parameter α is a weight factor ($\alpha \in [0,1]$) needed to be tuned to indicate the smoothing strength for the one-hot label. This modification results in a new loss:

$$\ell' = -\sum_{i=1}^{n} \left[(1 - \alpha)H_i(q, p_\theta) + \alpha H_i(u, p_\theta) \right]. \tag{5}$$

Usually, the $u(y|x_i)$ is a uniform distribution, independent of data x_i, as $u(y|x_i) = 1/K$, and hyper-parameter α is tuned with cross-validation.

3.2   Midpoint Generation

For a sample, denoted as x_i, from the provided training set $(X; Y)$ for training, PLS first randomly selects[1] another training sample x_j. For the pair of samples $(x_i; y_i)$ and $(x_j; y_j)$, where x is the input and y the one-hot encoding of the corresponding class, PLS then generates a synthetic sample through element-wisely averaging both the input features and the labels, respectively, as follows:

$$x_{ij} = (x_i + x_j)/2, \tag{6}$$

$$q(y|x_{ij}) = (y_i + y_j)/2. \tag{7}$$

These synthetic samples can be considered as the *midpoint samples* of the original sample pairs. It is also worth noting that, the midpoint samples here are equivalent to fixing the linear interpolation mixing ratios as 0.5 in the Mixup [28] method. Doing so, for the midpoint sample x_{ij} we have the ground truth distribution q over the labels as $q(y|x_{ij})$ ($\sum_{y=1}^{K} q(y|x_{ij}) = 1$). The newly resulting midpoint x_{ij} will then be used for label smoothing (will be discussed in detail in Sect. 3.3) before feeding into the networks for training. In other words, the predicted logits as defined in Eq. 2 is computed by first generating the m-dimensional input embedding $S_{ij} \in R^m$ for x_{ij} and then passing through the fully connected linear layer to construct the logit z:

$$S_{ij} = f_\theta(x_{ij}), \tag{8}$$

$$z = W_l S_{ij}. \tag{9}$$

Hence, the predicted conditional label distribution over the K classes produced by the networks is as follows:

$$p_\theta(y|x_{ij}) = \frac{\exp(z_y)}{\sum_{k=1}^{K} \exp(z_{y_k})}. \tag{10}$$

3.3   Learning Smoothing Distribution for Midpoints

PLS leverages a learned distribution, which depends on the input x, to dynamically generate the smoothing distribution mass for midpoint samples to distribute their ground truth target distribution to the non-target classes. To this end, the PLS implements this by adding a fully connected layer to the network f_θ. That is, the f_θ produces two projections from the *penultimate* layer representations of the network: one for the logits as the original network (Eq. 9), and another for generating the smoothing distribution as follows.

In specific, an additional fully connected layer $W_t \in R^{K \times m}$ is added to the original networks f_θ to produce the smoothing distribution over the

---

[1] For efficiency purpose, we implement this by randomly selecting a sample from the same mini-batch during training.

K classification classes. That is, for the given input image x_{ij}, its smoothing distributions over the K classification targets, denoted as $u'_\theta(y|x_{ij})$, are computed as follows:

$$u'_\theta(y|x_{ij}) = \frac{\exp(v_y)}{\sum_{k=1}^{K} \exp(v_{y_k})}, \tag{11}$$

$$v = \sigma(W_t S_{ij}), \tag{12}$$

where σ denotes the Sigmoid function, and S_{ij} is the same input embedding as that in Eq. 9. In other words, the two predictions (i.e., Eqs. 9 and 12) share the same networks except the last fully connected layer. That is, the only difference between PLS and the original networks is the added fully connected layer W_t. The added Sigmoid function here aims to squash the smoothing distributions learned for different targets to the same range of $[0, 1]$.

After having the smoothing distributions $u'_\theta(y|x_{ij})$, PLS then uses them to smooth the ground truth labels $q(y|x_{ij})$ as described in Eq. 7, with average:

$$q'(y|x_{ij}) = (q(y|x_{ij}) + u'_\theta(y|x_{ij}))/2. \tag{13}$$

The loss function of PLS thus becomes the follows:

$$\ell' = -\sum_{i=1}^{n} \Big[0.5 \cdot H_i(q(y|x_{ij}), p_\theta(y|x_{ij})) + 0.5 \cdot H_i(u'_\theta(y|x_{ij}), p_\theta(y|x_{ij})) \Big]. \tag{14}$$

Coefficient 0.5 here helps prevent the network from over-distributing its label distribution to the non-ground truth targets. Over-smoothing degrades the model performance, as will be shown in the experiments. Also, 0.5 here causes the resulting training signals to have high uncertainty regarding the ground truth targets. Such high uncertainty helps train the models to make conservative predictions.

3.4 Optimization

For training, PLS minimizes, with gradient descent on mini-batch, the loss ℓ'. One more issue needs to be addressed for the training. That is, the midpoint samples used for training, namely $(x_{ij}; y)$, may lack information on the original training samples $(x_i; y)$ due to the average operation in midpoint generation. To compensate this, we alternatively feed inputs to the networks with either a mini-batch from the original inputs, i.e., x_i or x_{ii}, or a mini-batch from the midpoint samples, i.e., x_{ij}. Note that, when training with the former, the networks still need to learn to assign the smoothing distribution $u'_\theta(y|x_{ii})$ to form the soft targets $q'(y|x_{ii})$ for the sample x_{ii}. As will be shown in the experiment section, this training strategy is important to PLS' regularization effect.

4 Experiments

We first show our method's superior accuracy, and then visualize its high uncertainty training labels and conservative classifications.

Table 1. Error rate (%) of the testing methods with PreAct ResNet-18 [11] as baseline. We report mean scores over 5 runs with standard deviations (denoted ±). Best results are in **Bold**.

Methods	MNIST	Fashion	SVHN	Cifar10	Cifar100	Tiny-ImageNet
PreAct ResNet-18	0.62 ± 0.05	4.78 ± 0.19	3.64 ± 0.42	5.19 ± 0.30	24.19 ± 1.27	39.71 ± 0.08
ULS-0.1	0.63 ± 0.02	4.81 ± 0.07	3.20 ± 0.06	4.95 ± 0.15	21.62 ± 0.29	38.85 ± 0.56
ULS-0.2	0.62 ± 0.02	4.57 ± 0.05	3.14 ± 0.11	4.89 ± 0.11	21.51 ± 0.25	38.54 ± 0.32
ULS-0.3	0.60 ± 0.01	4.60 ± 0.06	3.12 ± 0.03	5.02 ± 0.12	21.64 ± 0.27	38.32 ± 0.37
Mixup	0.56 ± 0.01	4.18 ± 0.02	3.37 ± 0.49	3.88 ± 0.32	21.10 ± 0.21	38.06 ± 0.29
Mixup-ULS0.1	0.53 ± 0.02	4.13 ± 0.10	2.96 ± 0.39	4.00 ± 0.17	21.51 ± 0.51	37.23 ± 0.48
Mixup-ULS0.2	0.53 ± 0.03	4.18 ± 0.09	3.02 ± 0.42	3.95 ± 0.13	21.41 ± 0.55	38.21 ± 0.38
Mixup-ULS0.3	0.50 ± 0.02	4.15 ± 0.06	2.88 ± 0.31	4.06 ± 0.04	20.94 ± 0.49	38.93 ± 0.43
PLS	**0.47 ± 0.03**	**3.96 ± 0.05**	**2.68 ± 0.09**	**3.63 ± 0.10**	**19.14 ± 0.20**	**35.26 ± 0.10**

Table 2. Error rate (%) of the testing methods with ResNet-50 [11] as baseline. We report mean scores over 5 runs with standard deviations (denoted ±). Best results are in **Bold**.

Methods	MNIST	Fashion	SVHN	Cifar10	Cifar100	Tiny-ImageNet
ResNet-50	0.61 ± 0.05	4.55 ± 0.14	3.22 ± 0.05	4.83 ± 0.30	23.10 ± 0.62	35.67 ± 0.50
ULS-0.1	0.63 ± 0.02	4.58 ± 0.16	2.98 ± 0.02	4.98 ± 0.25	23.90 ± 0.99	35.02 ± 0.39
ULS-0.2	0.62 ± 0.03	4.52 ± 0.04	3.08 ± 0.03	5.00 ± 0.35	23.88 ± 0.73	36.19 ± 0.66
ULS-0.3	0.65 ± 0.03	4.51 ± 0.15	3.04 ± 0.07	5.16 ± 0.16	23.17 ± 0.50	36.14 ± 0.06
Mixup	0.57 ± 0.03	4.31 ± 0.05	2.85 ± 0.07	4.29 ± 0.28	19.48 ± 0.48	32.36 ± 0.53
Mixup-ULS0.1	0.60 ± 0.04	4.28 ± 0.12	2.90 ± 0.09	4.02 ± 0.27	21.58 ± 0.86	32.11 ± 0.09
Mixup-ULS0.2	0.58 ± 0.02	4.33 ± 0.09	2.89 ± 0.07	4.09 ± 0.10	20.87 ± 0.51	32.81 ± 0.48
Mixup-ULS0.3	0.57 ± 0.04	4.29 ± 0.11	2.84 ± 0.10	4.19 ± 0.18	21.64 ± 0.41	33.94 ± 0.56
PLS	**0.51 ± 0.02**	**4.15 ± 0.09**	**2.36 ± 0.03**	**3.60 ± 0.18**	**18.65 ± 1.08**	**30.73 ± 0.20**

4.1 Datasets, Baselines, and Settings

We use six benchmark image classification tasks. **MNIST** is a digit (1–10) recognition dataset with 60,000 training and 10,000 test 28 × 28-dimensional gray-level images. **Fashion** is an image recognition dataset with the same scale as MNIST, containing 10 classes of fashion product pictures. **SVHN** is the Google street view house numbers recognition data set. It has 73,257 digits, 32 × 32 color images for training, 26,032 for testing, and 531,131 additional, easier samples. Following literature, we did not use the additional images. **Cifar10** is an image classification task with 10 classes, 50,000 training and 10,000 test samples. **Cifar100** is similar to Cifar10 but with 100 classes and 600 images each. **Tiny-ImageNet** [5] has 200 classes, each with 500 training and 50 test 64 × 64 × 3 images.

We conduct experiments using the popular benchmarking networks PreAct ResNet-18 and ResNet-50 [11]. We compare with the label smoothing methods [16,18] (denoted as ULS) with various smoothing coefficients (i.e., α in Eq. 5), where ULS-0.1, ULS-0.2, and ULS-0.3 denote the smoothing coefficient of 0.1, 0.2, and 0.3, respectively. We also compare with the input-pair based data augmentation method Mixup [28]. We further compare with methods that stacking the ULS on top of Mixup, denoted Mixup-ULS0.1, Mixup-ULS0.2, and Mixup-ULS0.3.

For Mixup, we use the authors' code at[2] and the uniformly selected mixing coefficients between [0,1]. For PreAct ResNet-18 and ResNet-50, we use the implementation from Facebook[3]. For PLS, the added fully connected layer is the same as the last fully connected layer of the baseline network with a Sigmoid function on the top. All models are trained using mini-batched (128 examples) backprop, with the *exact settings* as in the Facebook codes, for 400 epochs. Each reported value (accuracy or error rate) is the mean of five runs on a NVIDIA GTX TitanX GPU with 12 GB memory.

4.2 Predictive Accuracy

The error rates obtained by ULS, Mixup, Mixup-ULS, and PLS using ResNet-18 as baseline on the six test datasets are presented in Table 1. The results with ResNet-50 as baselines are provided in Table 2.

Table 1 shows that PLS outperforms, in terms of predictive error, the ResNet-18, the label smoothing models (ULS-0.1, ULS-0.2, ULS-0.3), Mixup, stacking ULS on top of Mixup (Mixup-ULS0.1, Mixup-ULS0.2, Mixup-ULS0.3) on all the six datasets. For example, the relative improvement of PLS over ResNet-18 on Cifar10 and MNIST are over 30% and 24%, respectively. When considering PLS and the average error obtained by the three ULS models, on both Cifar10 and MNIST, the relative improvement is over 23%. It is also promising to see that on Tiny-ImageNet, PLS reduced the absolute error rates over Mixup, Mixup-ULS, and ULS from about 38% to 35%.

For the cases with ResNet-50 as baselines, Table 2 indicates that similar error reductions are obtained by PLS. Again, on all the testing datasets, PLS outperforms all the comparison baselines. For example, for Cifar10, the relative improvement achieved by PLS over ResNet-50 and the average of the three label smoothing strategies (i.e. ULS) are 25.47% and 28.67%, respectively. For Cifar100 and Tiny-ImageNet, PLS reduced the absolute error rates of ULS from 23% and 36% to 18.65% and 30.73% respectively.

An important observation here is that, stacking label smoothing on top of Mixup (i.e., Mixup-ULS) did not improve, or even degraded, Mixup's accuracy. For example, for Cifar10, Cifar100, and Tiny-ImageNet (the last three columns in Tables 1 and 2), Mixup-ULS obtained similar or slightly higher error rate than Mixup. The reason here is that Mixup creates samples with soft labels through linearly interpolating between [0, 1], which is a form of label smoothing regularization [3]. Consequently, stacking another label smoothing regularizer on top of Mixup can easily mess up the soft training targets, resulting in underfitting. Promisingly, PLS was able to improve over Mixup and Mixup-ULS. For example, on Tiny-ImageNet PLS outperformed Mixup and Mixup-ULS by reducing the error from 38.06% to 35.26% and from 32.36% to 30.73%, respectively, when using ResNet-18 and ResNet-50. Similar error reduction can be observed on Cifar10 and Cifar100.

---

[2] https://github.com/facebookresearch/mixup-cifar10.
[3] https://github.com/facebookresearch/mixup-cifar10/models/.

4.3   Ablation Studies

Impact of Learned Distribution and Original Samples. We evaluate the impact of the key components in PLS using ResNet-18 and ResNet-50 on Cifar100. Results are in Table 3. The key components include 1) removing the learned smoothing distribution mass in Eq. 13, 2) excluding the original training inputs as discussed in the method section, and 3) replacing the learned smoothing distribution mass with uniform distribution with weight coefficients of 0.1, 0.2, and 0.3 (denoted UD-0.1, UD-0.2 and UD-0.3).

The error rates obtained in Table 3 show that, both the learned smoothing distribution and the original training samples are critical for PLS. In particular, when excluding the original samples from training, the predictive error of PLS dramatically increased from about 19% to nearly 24% for both ResNet-18 and ResNet-50. The reason, as discussed in the method section, is that, without the original training samples, the networks are trained with midpoint samples only, thus may lack information on the validation samples with one-hot labels.

Table 3. Error rates (%) on Cifar100 by PLS while varying its key components: no learned distribution, no original samples, replacing learned smoothing distribution with uniform distribution.

PLS	ResNet-18	ResNet-50
	19.14	18.65
—— no learned distribution	21.06	19.35
—— no original samples	23.84	24.42
—— UD 0.1	19.50	18.91
—— UD 0.2	19.25	18.81
—— UD 0.3	19.31	18.89

Also, Table 3 indicates that, replacing the learned smoothing distribution in PLS with manually tuned Uniform distribution (i.e., UD) obtained slightly larger errors. This indicates that PLS is able to learn the distribution to smooth the two target labels in the midpoint samples, resulting in superior accuracy and excluding the need for the coefficient search for different applications.

Learning Smoothing Coefficient. We also conducted experiments of learning the smoothing coefficient, instead of learning the smoothing distribution, for each midpoint example. That is, we replace Eq. 12 with $\sigma(W_t S_{ij})$ where W_t is $R^{1 \times m}$ instead of $R^{K \times m}$. Table 4 presents the error rates (%) obtained using PreAct ResNet-18 and ResNet-50 on Cifar100 and Cifar10. These results suggest that learning to predict smoothing distribution significantly better than predicting the smoothing coefficient in PLS.

Re-weight Smoothing Strength. PLS distributes half of a midpoint sample's ground truth distribution mass over the non-ground truth targets (Eq. 13). We

Table 4. Error rates (%) on Cifar100 and Cifar10 obtained by PLS while predicting the smoothing coefficient (pred. coeff.) instead of learning the smoothing distribution (PLS) for each sample pair.

PLS	ResNet-18 pred. coeff./PLS	ResNet-50 pred. coeff./PLS
Cifar100	21.60/19.14	19.21/18.65
Cifar10	4.67/3.63	4.06/3.60

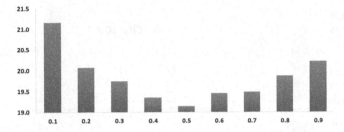

Fig. 3. Error rates (%, Y-axis) on Cifar100 obtained by varying the weight factor in PLS from 0.1 to 0.9 (X-axis).

here evaluate the impact of different weight factors between the ground truth and non-ground truth targets, by varying it from 0.1 to 0.9 (0.5 equals to the average used by PLS). The results obtained by PLS using ResNet-18 on Cifar100 are in Fig. 3. The error rates in Fig. 3 suggest that average used by midpoints in PLS provides better accuracy than other weighting ratios. This is expected as discussed in Sect. 3.3.

4.4 Uncertainty Label and Conservative Classification

This section will show that PLS utilizes high uncertainty labels from midpoint samples for training and produces very conservative classification decisions in testing.

High Uncertainty Labels in Training. We visualize, in Fig. 4, the soft target labels used for training by PLS with ResNet-18 on Cifar100 (top) and Cifar10 (bottom). Figure 4 depicts the soft label values of the training samples for both the ground truth targets (in green) and their top 5 largest non-ground truth classes (in orange). The figure presents the average values over all the training samples resulting from sample pairs with two different one-hot true labels. Here, X-axis depicts the training targets, and Y-axis the corresponding distribution mass.

Results in Fig. 4 indicate that, PLS uses much smaller target values for the ground truth labels during training, when compared to the one-hot representation label used by the baselines and the soft targets used by label smoothing ULS and Mixup. For example, the largest ground truth targets for PLS are around 0.25 and 0.35 (green bars), respectively, for Cifar100 and Cifar10. These values are much smaller than 1.0 used by the baselines and 0.9 and 0.5 used by ULS-0.1 and Mixup respectively. Consequently, in PLS, the distance between the ground truth targets and the non-ground truth targets are much smaller than that in the baseline, ULS-0.1 and Mixup. In addition, the distance between the two ground truth targets in PLS (green bars) is very small, when compared to the distance between the ground truth and non-ground truth target values (green vs. orange bars).

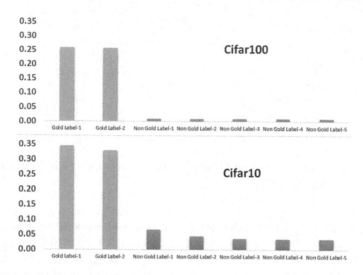

Fig. 4. Average distribution of ground truth targets (green bars) and the top 5 largest non-ground truth targets (orange bars) used by PLS in training with ResNet-18 on Cifar100 (top) and Cifar10 (bottom). X-axis is the classes and Y-axis the distribution mass. (Color figure online)

These results indicate that the training samples in PLS have smoother training targets across classes, and those training signals are far from 1.0 (thus with much higher uncertainty), which in turn impacts how PLS makes its classification decisions as will be discussed next.

Low Predicted Softmax Scores in Testing. One effect of the high uncertainty training signals (far from 1.0) as discussed above was reflected on the model's prediction scores made in test time. Figure 5 visualizes the predicted winning softmax scores made by ResNet-18 (top row), ULS-0.1 (second row), Mixup (third row) and PLS (bottom row) on all the 10K test data samples in Cifar100 (left column) and Cifar10 (right column). To have better visualization, we have removed the predictive scores less than 0.1 for all the methods since all models obtained similar results for confidences smaller than 0.1.

For Cifar100, results on the left of Fig. 5 indicate that ResNet-18 produced very confident predictions (top-left subfigure), namely skewing large mass of its predicted softmax scores on 1.0 (i.e., 100% confidence). On the other hand,

ULS and Mixup were able to decrease its prediction confidence at test time (middle rows/left). They spread their predictive confidences to the two ends, namely moving most of the predicted scores into two bins [0.1–0.2] and [0.8–1.0]. In contrast, PLS produced very conservative predicted softmax scores, by distributing many of its predicted scores to the middle, namely 0.5, and with sparse distribution for scores larger than 0.7 (bottom left subfigure).

For Cifar10, results on the right of Fig. 5 again indicate that ResNet-18 (top) produced very confident predictions, putting large mass of its predicted softmax scores near 1.0. For ULS and Mixup (middle rows), the predicted softmax scores were also distributed near the 1.0, but they are much less than that of ResNet-18. In contrast, PLS (bottom right) again generated very conservative predicted softmax scores. Most of them distributes near the middle of the softmax score range, namely 0.5, with a very few larger than 0.6.

These results suggest that, resulting from high uncertainty training signals across classes, PLS becomes extremely conservative when generating predicted scores in test time, producing low winning softmax scores for classification.

Impact on Model Calibration. The conservative predictions generated by PLS improve the model's accuracy, but how such low softmax scores affect the calibration of the output probabilities? In this section, we report the Expected Calibration Error (ECE) [7] obtained by the baseline PreAct ResNet-18, ULS-0.1, Mixup and PLS on the test set with 15 bins as used in [18] for both Cifar100 and Cifar10.

Results in Fig. 6 indicate that ULS (dark curve) is able to reduce the miscalibration error ECE on the Cifar100 data set (left subfigure), but for the Cifar10 dataset (right subfigure), ULS has larger ECE error after 100 epochs of training than the baseline model. Also, Mixup has higher ECE error than ULS on both Cifar100 and Cifar10. However, the ECE errors obtained by the PLS methods for both the Cifar100 and Cifar10 are much larger than the baseline, ULS and Mixup. Note that, although the authors in [7] state that the Batch Normalization (BN) strategy [14] also increases the miscalibration ECE errors for unknown reasons, we doubt that the PLS will have the same reason as the BN approach. This is because the main characteristic of the PLS model is that it produces extremely conservative winning softmax scores which is not the case for the BN strategy. We here suspect that the high ECE score of the PLS method may be caused by the fact that ECE is an evenly spaced binning metrics but the PLS produces sparse dispersion of the softmax scores across the range.

To verify the above hypothesis, we further investigate the Temperature Scaling (TS) method [7], which enables redistributing the distribution dispersion after training with no impact on the testing accuracy. During testing, TS multiplies the logits by a scalar before applying the softmax operator. We apply this TS technique to PLS, and present the results in Fig. 6, depicting by red curve in the left and right subfigures for the Cifar100 and Cifar10, respectively. The TS factors was 0.5 and 0.2 respectively for Cifar100 and Cifar10, which were found by a search with 10% of the training data. Figure 6 indicates that TS can signif-

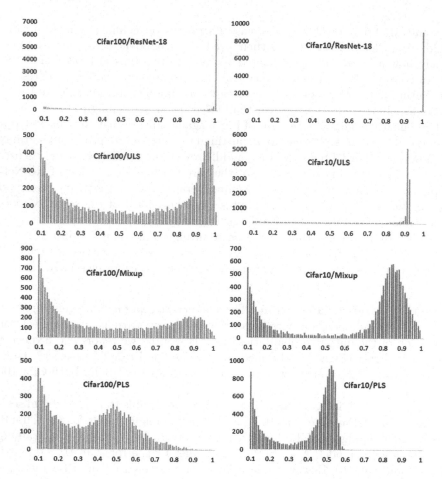

Fig. 5. Histograms of predicted softmax scores on the validation data by ResNet-18 (top row), ULS-0.1 (second row), Mixup (third row), and PLS (bottom row) on Cifar100 (left) and Cifar10 (right).

icantly improve the calibration of PLS for both cases. The ECE errors obtained by PLS-TS for both Cifar100 and Cifar10 (red curves) are lower than that of the baseline, ULS, and Mixup.

4.5 Testing on Out-of-Distribution Data

Section 4.4 shows that PLS produces very conservative classification decisions for in-distribution samples. We here explore the effect of PLS on out-of-distribution data.

In this experiment, we first train a ResNet-18 or ResNet-50 network (denoted as vanilla models) with in-distribution data (using either Cifar10 or Cifar100) and then let the trained networks to predict on the testing samples from the

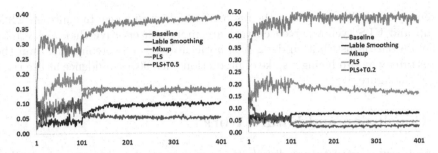

Fig. 6. ECE curves on the validation data of Cifar100 (left) and Cifar10 (right) from ResNet-18, ULS-0.1, Mixup, PLS and PLS with TS. X-axis is the training epoch and Y-axis the ECE value. (Color figure online)

SVHN dataset (i.e., out-of-distribution samples). We compare our method with the vanilla baseline (ResNet-18 or ResNet-50), Mixup, and ULS-0.1. The winning predicted softmax scores on the SVHN testing samples are presented in Fig. 7, where the top and bottom rows depict the results for ResNet-18 and ResNet-50, respectively.

Figure 7 shows that, PLS again produces low winning predicted softmax scores (namely less confident) against the samples from the SVHN dataset when training with either Cifar10 or Cifar100 data, when compared to Mixup, ULS, and the vanilla models. For example, when being trained with Cifar10 and tested on the SVHN data with ResNet-18 (the top-left subfigure), PLS produced (orange curve) a spike of score distribution near the midpoint 0.5, with

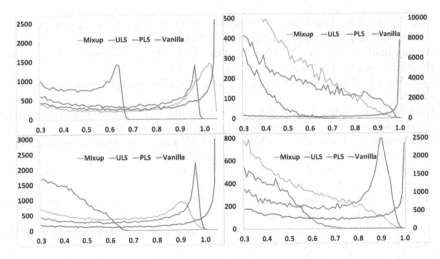

Fig. 7. Winning softmax scores produced by PLS, Mixup, ULS, and baseline models, when being trained with Cifar10 (left) and Cifar100 (right) and then testing on SVHN with PreAct ResNet-18 (top) and ResNet-50 (bottom). (Color figure online)

extremely spare distribution in the regions of high confidence. In contrast, while Mixup and ULS (yellow/green curves) are more conservative than the vanilla model on out-of-distribution data, it is noticeably more overconfident than the PLS strategy by producing a spike of prediction with 90% confidence at the right end of the figure.

5   Conclusion and Future Work

We proposed a novel output distribution regularization technique, coined PLS, which learns smoothing distribution for midpoint samples that average random sample pairs. We empirically showed PLS' superior accuracy over label smoothing and Mixup models. We visualized the high uncertainty training labels of the midpoint samples, which cause PLS to produce very low winning softmax scores for unseen in and out of distribution samples.

Our studies here suggest some interesting directions for future investigation. For example, what are the other benefits arising from high uncertainty training labels and conservative classification? Would such uncertain predictions help with beam search or ranking process for some downstream applications? Does the proposed method benefit from mixing more than two samples for smoothing? Another interesting research direction would be providing theoretical explanation on why the method works considering that the synthetic images are not realistic. We are also interested in applying our strategy to other domains beyond image.

References

1. Ahn, S., Hu, S.X., Damianou, A., Lawrence, N.D., Dai, Z.: Variational information distillation for knowledge transfer. arXiv (2019)
2. Archambault, G.P., Mao, Y., Guo, H., Zhang, R.: Mixup as directional adversarial training, vol. abs/1906.06875 (2019)
3. Carratino, L., Cissé, M., Jenatton, R., Vert, J.P.: On mixup regularization. arXiv (2020)
4. Chorowski, J., Jaitly, N.: Towards better decoding and language model integration in sequence to sequence models. In: INTERSPEECH (2016)
5. Chrabaszcz, P., Loshchilov, I., Hutter, F.: A downsampled variant of imagenet as an alternative to the CIFAR datasets. arXiv (2017)
6. Furlanello, T., Lipton, Z.C., Tschannen, M., Itti, L., Anandkumar, A.: Born again neural networks. In: ICML (2018)
7. Guo, C., Pleiss, G., Sun, Y., Weinberger, K.Q.: On calibration of modern neural networks. In: ICML 2017. JMLR.org (2017)
8. Guo, H.: Nonlinear mixup: out-of-manifold data augmentation for text classification. In: AAAI, pp. 4044–4051 (2020)
9. Guo, H., Mao, Y., Zhang, R.: Augmenting data with mixup for sentence classification: an empirical study, vol. abs/1905.08941 (2019)
10. Guo, H., Mao, Y., Zhang, R.: Mixup as locally linear out-of-manifold regularization. In: AAAI, pp. 3714–3722 (2019)

11. He, K., Zhang, X., Ren, S., Sun, J.: Identity mappings in deep residual networks. In: Leibe, B., Matas, J., Sebe, N., Welling, M. (eds.) ECCV 2016. LNCS, vol. 9908, pp. 630–645. Springer, Cham (2016). https://doi.org/10.1007/978-3-319-46493-0_38
12. Hinton, G., Vinyals, O., Dean, J.: Distilling the knowledge in a neural network. arXiv (2015)
13. Huang, Y., et al.: GPipe: efficient training of giant neural networks using pipeline parallelism. In: NeurIPS (2019)
14. Ioffe, S.: Batch renormalization: towards reducing minibatch dependence in batch-normalized models. In: Guyon, I., et al. (eds.) NeurIPS, pp. 1945–1953 (2017)
15. Li, W., Dasarathy, G., Berisha, V.: Regularization via structural label smoothing. In: AISTAT (2020)
16. Lukasik, M., Bhojanapalli, S., Menon, A.K., Kumar, S.: Does label smoothing mitigate label noise? In: ICML (2020)
17. Mobahi, H., Farajtabar, M., Bartlett, P.L.: Self-distillation amplifies regularization in hilbert space. arXiv (2020)
18. Müller, R., Kornblith, S., Hinton, G.E.: When does label smoothing help? In: NIPS (2019)
19. Pereyra, G., Tucker, G., Chorowski, J., Kaiser, L., Hinton, G.E.: Regularizing neural networks by penalizing confident output distributions. In: ICLR Workshop (2017)
20. Real, E., Aggarwal, A., Huang, Y., Le, Q.V.: Regularized evolution for image classifier architecture search. In: AAAI (2019)
21. Szegedy, C., Vanhoucke, V., Ioffe, S., Shlens, J., Wojna, Z.: Rethinking the inception architecture for computer vision. In: 2016 IEEE Conference on Computer Vision and Pattern Recognition (CVPR), pp. 2818–2826 (2016)
22. Tokozume, Y., Ushiku, Y., Harada, T.: Learning from between-class examples for deep sound recognition. In: ICLR (2018)
23. Vaswani, A., et al.: Attention is all you need. arXiv abs/1706.03762 (2017)
24. Xie, L., Wang, J., Wei, Z., Wang, M., Tian, Q.: Disturblabel: regularizing CNN on the loss layer. In: 2016 IEEE Conference on Computer Vision and Pattern Recognition (CVPR), pp. 4753–4762 (2016)
25. Yang, C., Xie, L., Qiao, S., Yuille, A.L.: Training deep neural networks in generations: a more tolerant teacher educates better students. In: AAAI, pp. 5628–5635. AAAI Press (2019)
26. Yuan, L., Tay, F.E.H., Li, G., Wang, T., Feng, J.: Revisit knowledge distillation: a teacher-free framework. arXiv (2019)
27. Yun, S., Han, D., Chun, S., Oh, S.J., Yoo, Y., Choe, J.: Cutmix: regularization strategy to train strong classifiers with localizable features. In: ICCV, pp. 6022–6031. IEEE (2019)
28. Zhang, H., Cissé, M., Dauphin, Y.N., Lopez-Paz, D.: Mixup: beyond empirical risk minimization. In: ICLR (2018)
29. Zhu, Z., et al.: Viewpoint-aware loss with angular regularization for person re-identification. In: AAAI, pp. 13114–13121 (2020)
30. Zoph, B., Vasudevan, V., Shlens, J., Le, Q.V.: Learning transferable architectures for scalable image recognition. In: 2018 IEEE/CVF Conference on Computer Vision and Pattern Recognition, pp. 8697–8710 (2018)

Learning Weakly Convex Sets
in Metric Spaces

Eike Stadtländer[1], Tamás Horváth[1,2,3]([☒]), and Stefan Wrobel[1,2,3]

[1] Department of Computer Science, University of Bonn, Bonn, Germany
{stadtlaender,horvath,wrobel}@cs.uni-bonn.de
[2] Fraunhofer IAIS, Schloss Birlinghoven, Sankt Augustin, Germany
[3] Fraunhofer Center for Machine Learning, Sankt Augustin, Germany

Abstract. We introduce the notion of weak convexity in metric spaces, a generalization of ordinary convexity commonly used in machine learning. It is shown that weakly convex sets can be characterized by a closure operator and have a unique decomposition into a set of pairwise disjoint connected blocks. We give two generic efficient algorithms, an extensional and an intensional one for learning weakly convex concepts and study their formal properties. Our experimental results concerning vertex classification clearly demonstrate the excellent predictive performance of the extensional algorithm. Two non-trivial applications of the intensional algorithm to polynomial PAC-learnability are presented. The first one deals with learning k-convex Boolean functions, which are already known to be efficiently PAC-learnable. It is shown how to derive this positive result in a fairly easy way by the generic intensional algorithm. The second one is concerned with the Euclidean space equipped with the Manhattan distance. For this metric space, weakly convex sets form a union of pairwise disjoint axis-aligned hyperrectangles. We show that a weakly convex set that is consistent with a set of examples and contains a minimum number of hyperrectangles can be found in polynomial time. In contrast, this problem is known to be NP-complete if the hyperrectangles may be overlapping.

Keywords: Abstract convexity · Concept learning · Vertex classification

1 Introduction

Several results in the theory of machine learning are concerned with concept classes defined by various forms of *convexity* (e.g., polygons formed by the intersection of a bounded number of half-spaces [2], conjunctions [14], or geodesic convexity in graphs [11]). In a broad sense, convex sets constitute *contiguous* subsets of the domain. This property can, however, be a drawback when the target concept cannot be represented by a single convex set. To overcome this problem, we relax the notion of convexity by introducing that of *weak convexity* for *metric spaces*. More precisely, a subset A of a metric space is *weakly convex* if

N. Oliver et al. (Eds.): ECML PKDD 2021, LNAI 12976, pp. 200–216, 2021.
https://doi.org/10.1007/978-3-030-86520-7_13

for all $x, y \in A$ and for all points z in the ground set, z belongs to A whenever x and y are *near* to each other and the three points satisfy the triangle inequality with *equality*. This definition has been inspired by the following *relaxation* of convexity in the Hamming metric space [6]: A Boolean function is k-*convex* for some positive integer k if for all true points x and y having a Hamming distance of at most k, all points on all shortest paths between x and y are also true. Our definition of weak convexity generalizes this notion to *arbitrary* metric spaces.

We present some properties of weakly convex sets of a metric space. In particular, we show that they form a *convexity space* [15] and hence, a *closure system*. Furthermore, they give rise to a *unique* decomposition into a set of "connected" blocks that have a pairwise minimum distance. We also study two scenarios for *learning* weakly convex sets. The first one considers the case that the metric space is *finite* and weakly convex sets are given *extensionally*. For this setting we define a *preclosure* operator and show that weakly convex sets can be characterized by a *closure* operator defined by the fixed points of the iterative applications of this preclosure operator. This characterization gives rise to an *efficient* algorithm computing the weakly convex hull for any set of points. We then prove that a weakly convex set that is *consistent* with a set of examples and has the *smallest* number of blocks can be found in polynomial time. This result makes use of the unique decomposition of weakly convex sets. As a proof of concept, we *experimentally* demonstrate on graph vertex classification that a remarkable accuracy can be obtained already with a relatively small training data set.

The second scenario deals with the case that the metric spaces are *not* necessarily finite and that weakly convex sets are given *intensionally*, using some compact representation. We present a simple generic algorithm, which iteratively "merges" weakly convex connected blocks and give sufficient conditions for the efficiency of a more sophisticated version of this naïve algorithm. Similarly to the extensional setting, we prove that a weakly convex set consistent with a set of examples and containing a *minimum* number of blocks can be found in *polynomial* time if certain conditions are fulfilled. We also present two non-trivial applications of this general result to polynomial PAC-learnability [14]. The first one deals with learning k-*convex Boolean functions*, for which there already exists a positive PAC result [6]. We still consider this problem because we show that the same result can be obtained in a very simple way by our intensional learning algorithm. Furthermore, our general purpose algorithm calculates the k-convex Boolean function for a set of examples in the same asymptotic time complexity as the domain specific one in [6]. The second application deals with the metric space defined by $\mathbb{R}^d$ endowed with the Manhattan (or L_1) distance. Weakly convex sets for this case are the union of a set of pairwise disjoint axis-aligned closed *hyperrectangles*. Using our general learning algorithm, we prove in a very simple way that the concept class formed by weakly convex sets containing at most k hyperrectangles is *polynomially* PAC-learnable. To underline the strength and utility of our approach, we note that the consistent hypothesis finding problem for the related problem that the hyperrectangles are not required to be pairwise disjoint is NP-complete, even for $d = 2$ (see, e.g., [1]).

Related Work. To the best of our knowledge, our notion of weak convexity in metric spaces is *new*. As mentioned above, it has been inspired by the definition of *k-convex Boolean functions* introduced in [6]. In fact, our notion generalizes that for k-convex Boolean functions to a broad class of metric spaces, including infinite ones as well. We also mention the somewhat related, but fundamentally distinct notion of *α-hulls* (see, e.g., [5]). They are defined as the intersection of enclosing generalized disks, but only for *finite* subsets of $\mathbb{R}^2$ and $\mathbb{R}^3$. Furthermore, it is known that the α-hull operator is *not* idempotent [8]. In contrast, our notion results in an abstract convexity structure in the sense of [15] and has therefore a corresponding closure operator defined for *arbitrary* subsets of a broad class of metric spaces. Last, but not least, even though our definitions resemble those of the density-based clustering approach [7], DBSCAN clusters are generally *not* weakly convex, except for very specific parameter values.

Outline. The rest of the paper is organized as follows. We collect the necessary notions and fix the notation in Sect. 2. In Sect. 3 we define weakly convex sets in metric spaces and prove some of their basic properties. Sections 4 and 5 are devoted to learning weakly convex sets in the extensional and intensional problem settings. Finally, we conclude in Sect. 6 and mention some problems for future work. Most of the proofs are omitted for space limitations. They can be found in [13].

2   Preliminaries

In this section we collect the necessary notions and fix the notation. For any $n \in \mathbb{N}$, $[n]$ denotes the set $\{1, 2, \ldots, n\}$. The family of all finite subsets of a set X is denoted by $[X]^{<\omega}$. A *metric space* is a pair (X, D), where X is a set and D is a metric on X (i.e., (i) $\mathrm{D}(x,y) = 0$ iff $x = y$, (ii) $\mathrm{D}(x,y) = \mathrm{D}(y,x)$, and (iii) $\mathrm{D}(x,y) \leq \mathrm{D}(x,z) + \mathrm{D}(z,y)$ for all $x, y, z \in X$).

A *closure system* over some ground set X is a pair $(X, \mathcal{C})$ with $\mathcal{C} \subseteq 2^X$ such that $\mathcal{C}$ is closed under arbitrary intersection, where 2^X denotes the *power set* of X. We require that $X \in \mathcal{C}$. The elements of $\mathcal{C}$ are called *closed sets*. One of the elementary properties of closure systems is that they can be characterized in terms of closure operators (see, e.g., [3]). More precisely, a function $\rho : 2^X \to 2^X$ is a *closure operator* if it satisfies the following properties for all $A, B \subseteq X$: (i) $A \subseteq \rho(A)$ (*extensivity*), (ii) $\rho(A) \subseteq \rho(B)$ whenever $A \subseteq B$ (*monotonicity*), and (iii) $\rho(\rho(A)) = \rho(A)$ (*idempotency*). If ρ is extensive and monotone, but not necessarily idempotent, then it is a *preclosure operator*. The fixed points of a closure operator are called *closed sets* and the set system $(X, \mathcal{C}_\rho)$ with $\mathcal{C}_\rho = \{A \subseteq X : \rho(A) = A\}$ is always a closure system. Conversely, for any closure system $(X, \mathcal{C})$, the function $\rho : 2^X \to 2^X$ with $\rho : A \mapsto \bigcap \{C \in \mathcal{C} : A \subseteq C\}$ is a closure operator satisfying $\mathcal{C} = \{\rho(A) : A \subseteq X\}$. Finally, a *convexity space* [15] over a set X is a closure system $(X, \mathcal{C})$ such that (i) $\emptyset, X \in \mathcal{C}$ and (ii) $\mathcal{C}$ is closed under *nested unions* (i.e., $\bigcup \mathcal{D} \in \mathcal{C}$ for any $\mathcal{D} \subseteq \mathcal{C}$ that is totally ordered w.r.t. set inclusion).

Our notion of *weak convexity* defined in the next section is inspired by that of *k-convexity* introduced in the seminal paper by Ekin, Hammer, and Kogan [6].

More precisely, consider the metric space $(\mathbb{H}_d, D_H)$, where $\mathbb{H}_d = \{0,1\}^d$ is the d-dimensional *Hamming cube* and D_H is the Hamming distance. A subset X of $\mathbb{H}_d$ is k-*convex* for an integer $k \geq 1$ if for all $x, y \in X$ with $D_H(x,y) \leq k$ and for all $z \in \mathbb{H}_d$, $z \in X$ whenever the triangle inequality holds with equality (i.e., $D_H(x,y) = D_H(x,z) + D_H(z,y)$).

An *(undirected) graph* is a pair $G = (V, E)$, where V is a finite set of vertices and $E \subseteq \{e \subseteq V : |e| = 2\}$ is a set of edges. An edge $\{x, y\}$ will sometimes be denoted by xy. A graph $G' = (V', E')$ is a *subgraph* of G if $V' \subseteq V$ and $E' \subseteq E$. A *path* is a graph $P = (V_P, E_P)$ with $V_P = \{v_1, \ldots, v_n\}$ and $E_P = \{v_i v_{i+1} : i \in [n-1]\}$. The *length* of a path P is the number of edges it contains. A graph is *connected* if all pairs of its vertices are connected by a path. If two vertices of a graph G are connected by a path, we define their *geodesic distance* by the length of a shortest path connecting them. Note that it is a metric on the set of vertices for connected graphs. A subset $X \subseteq V$ is called *(geodesically) convex* in a graph $G = (V, E)$ if for all $u, v \in V$ and for all shortest paths $P = (V_P, E_P)$ connecting u and v, we have $V_P \subseteq X$.

For the standard definitions of *concepts, concept classes, VC-dimension,* and *polynomial PAC-learnability* from computational learning theory, the reader is referred to some standard text book about learning theory (see, e.g., [9]). Let $\mathcal{C}$ be a concept class over some domain X. The k-*fold union* of $\mathcal{C}$ for some $k \geq 1$ integer is defined by $\mathcal{C}_\cup^k = \{C_1 \cup \ldots \cup C_k : C_i \in \mathcal{C} \text{ for all } i \in [k]\}$. Note that the definition does not require the C_is to be pairwise different. The following problem is central to concept learning:

Problem 1 (The Consistency Problem). *Given a concept class* $\mathcal{C} \subseteq 2^X$ *over some domain* X *and disjoint sets* $E^+, E^- \subseteq X$ *of examples, return a concept* $C \in \mathcal{C}$ *that is consistent with* E^+ *and* E^-, *i.e.,* $E^+ \subseteq C$ *and* $E^- \cap C = \emptyset$ *if such a concept exists; o/w return* "No".

In order to prove polynomial PAC-learnability, we will use the following results from computational learning theory [2].

Theorem 2. *Let* $\mathcal{C} \subseteq 2^X$ *be a concept class over some domain* X *with VC-dimension* $d > 0$.

(i) $\mathcal{C}$ is polynomially PAC-learnable if d is bounded by a polynomial of its parameters and Problem 1 can be solved in polynomial time in the parameters.
(ii) For all $k \geq 1$, the VC-dimension of $\mathcal{C}_\cup^k$ is at most $2dk \log(3k)$.

3 Weak Convexity in Metric Spaces

In this section we relax the notion of convexity defined for Euclidean spaces to *weak convexity* in *metric spaces* and discuss some basic formal properties of weakly convex sets. The main result of this section is formulated in Theorem 4. It states that weakly convex sets have a *unique* decomposition into a set of weakly convex "connected" blocks that have a pairwise minimum distance from each other. To define weak convexity, recall that a subset $A \subseteq \mathbb{R}^d$ is *convex* if

$$D_2(x, z) + D_2(z, y) = D_2(x, y) \text{ implies } z \in A \tag{1}$$

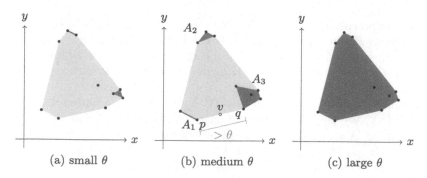

(a) small θ (b) medium θ (c) large θ

Fig. 1. Illustration of weakly convex sets in the Euclidean plane $\mathbb{R}^2$. (Color figure online)

for all $x, y \in A$ and for all $z \in \mathbb{R}^d$, where D_2 is the Euclidean distance. Our notion of weak convexity in metric spaces incorporates a relaxation of (1), motivated by the fact that convex sets defined by (1) are always "contiguous" and cannot therefore capture well-separated regions of the domain. We address this problem by adapting the idea of *k-convexity* over Hamming metric spaces [6] to *arbitrary* ones. Analogously to [6], we do not require (1) to hold for all points x and y, but only for such pairs which have a distance of at most a user-specified threshold. In other words, while ordinary convexity is based on a *global* condition resulting in a single "contiguous" region, our notion of weak convexity relies on a *local* one, resulting in potentially several isolated regions, where the spread of locality is controlled by the above mentioned user-specified threshold. This consideration yields the following formal definition of *weakly convex* sets in *metric spaces*:

Definition 3. Let (X, D) be a metric space and $\theta \geq 0$. A set $A \subseteq X$ is θ-convex (or simply, *weakly convex*) if for all $x, y \in A$ and $z \in X$ it holds that $z \in A$ whenever $D(x, y) \leq \theta$ and $z \in \triangle_=(x, y)$, where

$$\triangle_=(x, y) = \{z \in X : D(x, z) + D(z, y) = D(x, y)\} \ . \tag{2}$$

Notice that (2) does not require $x \neq y$. In particular, $\triangle_=(x, x) = \{x\}$ for all $x \in X$. The family of all weakly convex sets is denoted by $\mathcal{C}_{\theta,D}$; we omit D if it is clear from the context. It always holds that $\mathcal{C}_{0,D} = 2^X$.

In order to illustrate the notion of weak convexity, consider the finite set of points $A \subseteq \mathbb{R}^2$ depicted by filled dots in Fig. 1b. The strongly (i.e., ordinary) convex hull of A is indicated by the gray area. In contrast, the $\subseteq$-smallest θ-convex set containing A for some suitable $\theta \geq 0$ is drawn in red. The most obvious difference is that there are three separated regions A_1, A_2, and A_3, instead of a single contiguous area. In other words, weakly convex sets need not be connected despite that strongly convex sets in $\mathbb{R}^2$ do. This is a consequence of considering only such pairs for membership witnesses that have a distance of at most θ (see, also, Fig. 1a and 1c). For example, the points p and q in Fig. 1b have a distance strictly greater than θ, implying that they do not witness the membership of the point v. Notice that in the same way as strongly convex sets, (parts of) weakly convex sets may degenerate. While A_2 and A_3 are regions with strictly positive area, A_1 is just a segment. We may even have isolated points as shown in Fig. 1a.

Despite this unconventional behavior of weakly convex sets, $(X, \mathcal{C}_\theta)$ forms a *convexity space*. To see this, note that $\emptyset, X \in \mathcal{C}_\theta$. Furthermore, $\mathcal{C}_\theta$ is stable for arbitrary intersections and nested unions. Indeed, if $\mathcal{F} \subseteq \mathcal{C}_\theta$ is a family of θ-convex sets, $x, y \in \bigcap \mathcal{F}$ with $D(x, y) \leq \theta$ then $\triangle_=(x, y) \subseteq F$ for all $F \in \mathcal{F}$ implying that $\bigcap \mathcal{F}$ is θ-convex. If, in addition, $\mathcal{F}$ is totally ordered by inclusion and $x, y \in \bigcup \mathcal{F}$ with $D(x, y) \leq \theta$ then there are $F_x, F_y \in \mathcal{F}$, say $F_x \subseteq F_y$, such that $x \in F_x$ and $y \in F_y$. Then, according to (2), $\triangle_=(x, y) \subseteq F_y$ implying that $\bigcup \mathcal{F}$ is θ-convex. Hence, $\mathcal{C}_\theta$ is a convexity space as claimed.

Since $\mathcal{C}_\theta$ is stable for arbitrary intersections, it has an associated *closure operator* $\rho_\theta : 2^X \to 2^X$ with $A \mapsto \bigcap \{C \in \mathcal{C}_\theta : A \subseteq C\}$ for all $A \subseteq X$. That is, ρ_θ maps a set A to the $\subseteq$-smallest θ-convex set containing A. It is called the *weakly convex hull operator* and its fixed points (i.e., the ρ_θ-closed sets) form exactly $\mathcal{C}_\theta$. Moreover, ρ_θ is *domain finite* [15], i.e., $\rho_\theta(A) = \bigcup \{\rho_\theta(F) : F \subseteq [A]^{<\omega}\}$.

3.1   Some Basic Properties of Weakly Convex Sets

We now present some basic properties of weakly convex sets that make this kind of closed sets interesting for machine learning from a practical as well as from a theoretical viewpoint. As already mentioned, weakly convex sets need *not* be contiguous (cf. Fig. 1), in contrast to, for instance, ordinary convex sets in the Euclidean space. Instead, one can observe regions that are separated from each other. This is again due to the fact that the notion of weak convexity utilizes a distance threshold θ. As a consequence, *separate* regions may arise with a pairwise distance of at least θ. In Theorem 4 below, which is one of our main technical results for this work, we formally state this property of weakly convex sets. We note that this result generalizes that stated in [6, Proposition 3.2] for the Hamming metric space to arbitrary metric spaces.

We first introduce some necessary notions. Let $\mathcal{M} = (X, D)$ be a metric space, $\theta \geq 0$, and $A \subseteq X$. Two points $a, b \in A$ are θ-*connected* w.r.t. A, denoted $a \sim_{\theta, A} b$, if there is a finite sequence $a = p_1, p_2, \ldots, p_r = b \in A$ such that $D(p_i, p_{i+1}) \leq \theta$ for all $i \in [r-1]$. A is θ-*connected* if $a \sim_{\theta, A} b$ for all $a, b \in A$. Note that $\sim_{\theta, A}$ is an equivalence relation on A; its equivalence classes, denoted by $[a]_{\sim_{\theta, A}} = \{b \in A : a \sim_{\theta, A} b\}$ for all $a \in A$, will be referred to as θ-*connected components*.

Theorem 4. *Let (X, D) be a metric space and $\theta \geq 0$. Then $A \subseteq X$ is θ-convex iff there is a uniquely defined family of non-empty sets $(A_i \subseteq A)_{i \in I}$ for some index set I satisfying the following conditions:*

(i) $A = \bigcup_{i \in I} A_i$,
(ii) A_i is θ-convex for all $i \in I$,
(iii) A_i is θ-connected for all $i \in I$,
(iv) for all $i, j \in I$ with $i \neq j$, $D(a, b) > \theta$ for all $a \in A_i, b \in A_j$.

In what follows, the family $(A_i)_{i \in I}$ satisfying conditions (i)–(iv) in Theorem 4 will be referred to as the θ-*decomposition* of the θ-convex set A. Furthermore, the sets A_i in the θ-decomposition of A will be called θ-*blocks* or simply, *blocks*. The theorem above tells us that weakly convex sets can be partitioned *uniquely* into a family of non-empty blocks in a way that the distance between each pair of such weakly convex components is at least θ. The uniqueness of the θ-decomposition in Theorem 4 gives rise to a naïve algorithm for computing the weakly convex hull of a finite set intensionally (cf. Algorithm 2 in Sect. 5). The idea is to start with the singletons and enforce conditions (i)–(iv) by repeatedly merging invalid pairs of blocks. However, that requires the strict inequality in condition (iv) not only to hold for pairs of points, but also between blocks. Corollary 5 below is concerned with metric spaces in which this property holds.

Corollary 5. *Let $\mathcal{M} = (X, D)$ be a metric space, $\theta \geq 0$, $A \subseteq X$, and $(A_i)_{i \in I}$ the θ-decomposition of $\rho_\theta(A)$.*

(i) We have $|I| \leq |A|$. In particular, if A is finite, then I is finite.

(ii) If $\mathcal{M}$ is complete and A_i and A_j are (topologically) closed for some $i \neq j$ then $D(A_i, A_j) = \inf\limits_{a \in A_i, b \in A_j} D(a, b) > \theta$.

Accordingly, Corollary 5 motivates the following definition of *well-behaved* metric spaces. A metric space $\mathcal{M} = (X, D)$ is *compatible* with the convexity space $(X, \mathcal{C}_\theta)$ if $\rho_\theta(A)$ is (topologically) closed for all $A \in [X]^{<\omega}$ [15]. If, in addition, $\mathcal{M}$ is complete, we call it *well-behaved*.

Finally, we claim that the weakly convex hull operator is monotone w.r.t. θ. This property will be utilized by our consistent hypothesis finding algorithms.

Proposition 6. *Let (X, D) be a metric space and $0 \leq \theta \leq \theta'$. Then for all $A \subseteq X$, (i) $\rho_\theta(A) \subseteq \rho_{\theta'}(A)$ and (ii) for all $x, y \in \rho_\theta(A)$, x, y are in the same θ'-block of the θ'-decomposition of $\rho_{\theta'}(A)$ if they are in the same θ-block of the θ-decomposition of $\rho_\theta(A)$.*

4   Learning in the Extensional Problem Setting

In this section we consider the case that the underlying metric space is *finite* and weakly convex sets are represented *extensionally*, e.g. because they have *no* (natural) compact representation. Examples of this scenario include, among others, the case that the metric space is given by the set of vertices of a graph together with some distance on vertices. To formulate some basic properties of ρ_θ introduced in Sect. 3, we define a preclosure operator $\hat{\rho}_\theta$ over X. More precisely, let $\mathcal{M} = (X, D)$ be a finite metric space and $\theta \geq 0$. For all $x, y \in X$, let $W_\theta(x, y) = \triangle_=(x, y)$ if $D(x, y) \leq \theta$; o/w $W_\theta(x, y) = \emptyset$. Finally, define the function $\hat{\rho}_\theta : 2^X \to 2^X$ by $\hat{\rho}_\theta(A) = \bigcup\limits_{x, y \in A} W_\theta(x, y)$ for all $A \subseteq X$.

Lemma 7. *The function $\hat{\rho}_\theta$ over $\mathcal{M}$ is a preclosure operator.*

Algorithm 1. EXTENSIONAL WEAKLY CONVEX HULL $\hat{\rho}_\theta$

Require: finite metric space (X, D) and $S_x = \{(x', \mathrm{D}(x, x')) : x' \in X \setminus \{x\}\}$ sorted in increasing order in the second component, for all $x \in X$
Input: $A \subseteq X$ and $\theta \geq 0$
Output: θ-decomposition of $\rho_\theta(A)$

1: $C, E \leftarrow \emptyset$, queue $Q \leftarrow A$
2: mark all elements in A
3: **while** $Q \neq \emptyset$ **do**
4:      $x \leftarrow \mathrm{DEQUEUE}(Q)$, $C \leftarrow C \cup \{x\}$
5:      **for all** $y \in N_\theta(x) \cap C$ **do**
6:          $E \leftarrow E \cup \{xy\}$
7:          **for all** $z \in N_{\mathrm{D}(x,y)}(x) \cap N_{\mathrm{D}(x,y)}(y)$ **do**
8:              **if** z is unmarked and $z \in \triangle_=(x, y)$ **then**
9:                  mark z, $\mathrm{ENQUEUE}(Q, z)$
10: **return** $\mathcal{D} = \{V(Z) : Z$ is a connected component of $G_\theta = (C, E)\}$

Let $\hat{\rho}_\theta^0(A) = A$ and $\hat{\rho}_\theta^{i+1}(A) = \hat{\rho}_\theta(\hat{\rho}_\theta^i(A))$ for all $i \in \mathbb{N}$ and $A \subseteq X$. Since $\hat{\rho}_\theta$ is monotone by Lemma 7 and X is finite, for all $A \subseteq X$ there exists a positive integer $\gamma(A)$ such that $\hat{\rho}_\theta^{\gamma(A)}(A) = \hat{\rho}_\theta^{\gamma(A)+1}(A)$, implying

$$\hat{\rho}_\theta^{\gamma(A)}(A) = \hat{\rho}_\theta^{\gamma(A)+l}(A) \tag{3}$$

for all $l \geq 0$. Furthermore, $\Gamma = \max\{\gamma(A) : A \subseteq X\} < \infty$. In the theorem below we claim that $\hat{\rho}_\theta^\Gamma$ yields exactly ρ_θ.

Theorem 8. *Let (X, D) be a finite metric space and $\theta \geq 0$. Then*

(i) $\rho : 2^X \to 2^X$ with $\rho(A) = \hat{\rho}_\theta^\Gamma(A)$ for all $A \subseteq X$ is a closure operator and (ii) for all $A \subseteq X$, $\rho_\theta(A) = A$ iff $\rho(A) = A$.

We now consider the problem of computing weakly convex sets for the case that the metric space is *finite* and weakly convex sets are represented *extensionally*. More precisely, we are interested in the following problem setting:

Problem 9 (The Extensional Weakly Convex Hull Problem). Given a finite metric space $\mathcal{M} = (X, \mathrm{D})$ with $|X| = n$, a set $A \subseteq X$, and a threshold $\theta \geq 0$, compute the θ-decomposition $A_1, \ldots, A_\ell$ of $\rho_\theta(A)$, where the A_is are given extensionally.

The algorithm solving Problem 9 is given in Algorithm 1. Its input consists of a set $A \subseteq X$ for some finite metric space (X, D) and a non-negative real number θ. The algorithm assumes that the pairwise distances for (X, D) are given explicitly and that each element $x \in X$ is associated with a sorted sequence S_x of pairs $(x', \mathrm{D}(x, x'))$, for all $x' \in X \setminus \{x\}$. We assume that these sequences are calculated and stored once in a preprocessing step. The reason behind this assumption is that in order to solve a related consistency problem defined below, Algorithm 1 will be called with different values of θ. For any $\delta \geq 0$, these sequences allow the δ-neighborhood $N_\delta(x) = \{y \in X : \mathrm{D}(x,y) \leq \delta\}$ of a point $x \in X$ to be calculated in time $\mathcal{O}(|N_\delta(x)|)$ for all $\delta \geq 0$ (cf. lines 5 and 7 of Algorithm 1).

Algorithm 1 maintains three variables. In particular, one can show in the proof of the theorem below, the set $\rho_\theta(A)$ is calculated in C. All elements of C are added first to the queue Q, which is initialized with A (cf. line 1). The elements of Q are processed one by one (cf. lines 4–9). In particular, for the element x of Q considered in line 4, we move x from Q to C (line 4) and take all elements y in the θ-neighborhood of x that have already been added to $C \subseteq \rho_\theta(A)$ (line 5). In variable E we maintain the set of edges of the θ-neighborhood graph over C, i.e., two elements of C are connected by an edge iff their distance is at most θ. As x is a new element in C, in line 6 we connect it with all y considered in line 5. In lines 7–9 we take all $z \in W_\theta(x,y)$ that have not yet been considered, mark z, and add it to the queue Q. Regarding line 7, note that the triangle inequality implies that if $z \in W_\theta(x,y)$ then $\mathrm{D}(x,z), \mathrm{D}(y,z) \leq \mathrm{D}(x,y)$. Finally, after we have processed all elements that have been added to Q, in line 10 we calculate the connected components of the θ-neighborhood graph $G_\theta = (C, E)$ and return the family formed by the sets of vertices of the connected components.

Theorem 10. *Algorithm 1 is correct and solves Problem 9 in $\mathcal{O}(nd^2)$ time, where d is the degree of the θ-neighborhood graph $G_\theta = (C, E)$.*

In Sect. 4.1 we will be concerned with an application scenario of the following *consistent hypothesis finding (CHF)* problem:

Problem 11 (The CHF Problem for Extensional Weakly Convex Hulls). Given a finite metric space $\mathcal{M} = (X, \mathrm{D})$ with $|X| = n$, disjoint sets $E^+, E^- \subseteq X$ of positive and negative examples, and an integer $k > 0$, return "YES" and the θ-decomposition of a θ-convex set consistent with E^+ and E^- that consists of at most k blocks, if it exists for some θ; o/w the answer "NO".

Remark 12. Note that if Problem 11 can be solved in polynomial time then, as k cannot be greater than $|E^+|$ by (i) of Corollary 5, a consistent hypothesis with the *smallest* number of blocks can be found in polynomial time. It always exists, as $\rho_0(E^+) = E^+$ and $E^+ \cap E^- = \emptyset$ by assumption.

Theorem 13. *Problem 11 can be solved in $\mathcal{O}(T_P(\mathcal{M}) + n^3 \log n)$ time, where $T_P(\mathcal{M})$ is the time complexity of computing all pairwise distances for X.*

4.1 Application Scenario: Vertex Classification

As a proof of concept, in this section we *empirically* demonstrate the learnability of weakly convex concepts over *graphs*. More precisely, we consider the metric space $\mathcal{M} = (V, \mathrm{D}_g)$ for some undirected graph $G = (V, E)$, where D_g is the *geodesic* distance on V. In the learning setting, V is partitioned into V^+ and V^-, such that V^+ is θ-convex for some $\theta \geq 0$. The *target concept* V^+ as well as θ are *unknown* to the learning algorithm. The problem we investigate empirically

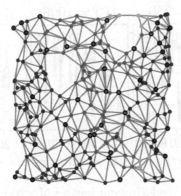

•	true neg.
•	true pos.
·	false neg.
○	training set

Fig. 2. Example of a graph with 250 vertices and 40 training examples. (Color figure online)

is to approximate V^+ given some *small* labeled set $E = E^+ \cup E^-$ of positive and negative examples.

We solve this learning task by computing the hypothesis $C = \rho_{\theta'}(E^+)$ for the greatest $\theta' \leq \max_{u,v \in V} D(u,v)$ that is consistent with E. Such a θ' always exists (cf. Remark 12). Furthermore, C is computed by performing a binary search for θ' (see [13] for the details). To measure the predictive performance, we use *accuracy* (i.e., number of correctly classified vertices in $V \backslash E$ over $|V \backslash E|$) and compare it to the *baseline* $\max\{|V^+|/|V|, |V^-|/|V|\}$ defined by majority vote. We stress that the only purpose of these experiments is to empirically demonstrate that weakly convex concepts can be learned with a remarkable accuracy, *without* utilizing any domain specific properties and with using only a *few* training examples. An adaptation of our approach to the domain specific problem of learning on graphs and a rigorous empirical comparison of its predictive performance with state-of-the-art problem specific algorithms goes far beyond the scope of this paper.

We generated 50 random graphs for $|V| = 100, 250, 1000$, and 2500 for the experiments as follows: According to Proposition 6, the diameter of a graph is an upper bound on the parameter θ. In order to provide a diverse set of target concepts and possible values for θ, we generated random graphs based on *Delaunay triangulations* [4] as follows: After choosing the respective number of nodes $V \subset [0,1]^2$ uniformly at random, we have computed the Delaunay triangulation. We then connected two nodes in V by an (undirected) edge iff they co-occur in at least one simplex of the triangulation. We considered the two cases that the edges are unweighted or they are weighted with the Euclidean distance between their endpoints. However, the resulting graph often contains a small number of very long edges (in terms of the Euclidean distance), especially near the "outline" of the chosen point set. Since such edges reduce the graph's diameter substantially, we removed the longest 5% of the edges, i.e., those that are *not* contained in the 95th percentile w.r.t. the Euclidean distance of their endpoints.

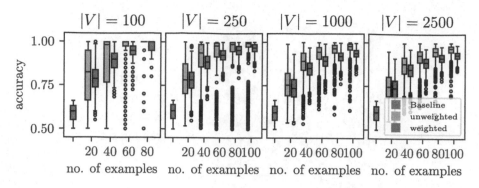

Fig. 3. Results for Delaunay-based graphs for varying number of vertices ($|V|$). (Color figure online)

For each graph $G = (V, E)$ in the resulting dataset, we have generated random partitionings (V^+, V^-) of V in a way that V^+ and V^- are balanced (i.e., $|V^+| \approx |V^-|$) and V^+ is θ-convex. We note that for all random partitionings obtained, V^+ was not strongly (i.e., ordinary) convex. The training examples E^+ and E^- have been sampled uniformly at random from V^+ and V^-, respectively, such that $|E^+| \approx |E^-|$. The number of training examples (i.e., $|E^+ \cup E^-|$) was varied over $20, 40, 60, 80, 100$. This overall procedure generates $5,000$ learning tasks (50 graphs $\times$ 20 random target concepts $\times$ 5 training set sizes), for each graph size $|V| = 100, 250, 1000, 2500$. In Fig. 2 we give an example graph with $|V| = 250$, together with the node prediction using 40 training examples. The training examples are marked with black outline and the predictions are encoded by colors. In particular, dark red corresponds to true positive, dark blue to true negative, and light red to false negative nodes. In the example we have no false positive node, which was the case for most graphs.

Figure 3 shows the accuracy (y-axes) of the baseline (blue box plots) and our learner (orange box plots for unweighted and red ones for weighted edges) grouped by the number of provided examples (x-axes) and the graph sizes $|V|$. In all cases, our learner outperforms the baseline significantly by noting that for $|V| = 100$, the high accuracy results obtained from 60 training examples are less interesting. For $|V| > 100$ it is remarkable that the learner does *not* require much more examples with increasing graph size. For example, for graphs with $2,500$ vertices, already 80 examples are sufficient to achieve an average accuracy of 0.94 for unweighted graphs. Notice that the baseline is in all cases very close to 0.6. This is due to our construction of the target concepts: We chose θ maximal such that $2|V^+| < |V|$. Therefore, in almost all cases there are about 10% less positive nodes than negative. We have tested the generated weakly convex sets for strong convexity: almost all of them were *not* strongly convex. In summary, our experimental results clearly show that a remarkable predictive accuracy can be obtained already with relatively small training sets with our generic approach, without utilizing any domain specific knowledge.

Algorithm 2. INTENSIONAL WEAKLY CONVEX HULL (NAÏVE)

Require: well-behaved metric space $\mathcal{M} = (X, \mathrm{D})$ and representation scheme μ for $\mathcal{M}$
Input: $A \in [X]^{<\omega}$ and $\theta \geq 0$
Output: $\mu(\theta, A_1), \ldots, \mu(\theta, A_\ell)$, where $A_1, \ldots, A_\ell$ is the θ-decomposition of $\rho_\theta(A)$

1: $\mathcal{D} \leftarrow \{\mu(\theta, \{x\}) : x \in A\}$
2: **while** $\exists B_i, B_j \in \mathcal{D}$ such that $B_i \neq B_j$ and $\overline{\mathrm{D}}(B_i, B_j) \leq \theta$ **do**
3: $\quad \mathcal{D} \leftarrow (\mathcal{D} \setminus \{B_i, B_j\}) \cup \{\mathrm{MERGE}(\theta, A, B_i, B_j)\}$
4: **return** $\mathcal{D}$

---

5   The Intensional Problem Setting

In this section we consider the *intensional* problem setting, that is, the scenario that weakly convex sets have some compact representation. In contrast to the extensional case, the metric spaces in this section are allowed to be *infinite*. They are, however, required to be *well-behaved* (see Sect. 3 for the definition). To formulate the problem setting considered in this section, we introduce the following notion for a metric space $\mathcal{M} = (X, \mathrm{D})$: A *representation scheme* for $\mathcal{M}$ is a function $\mu : \mathbb{R}_{\geq 0} \times [X]^{<\omega} \to \{0, 1\}^*$ satisfying $\mu(\theta, A) = \mu(\theta, B)$ iff $\rho_\theta(A) = \rho_\theta(B)$ for all $A, B \in [X]^{<\omega}$ and $\theta \geq 0$. In other words, μ returns some representation of $\rho_\theta(A)$ for all finite subsets $A \subseteq X$. Note that $\rho_\theta(A)$ can be infinite. Analogously to Problem 9, we are interested in the following computational problem:

Problem 14 (The Intensional Weakly Convex Hull Problem).   Given a well-behaved metric space $\mathcal{M} = (X, \mathrm{D})$, a representation scheme μ for $\mathcal{M}$, a set $A \subseteq [X]^{<\omega}$ with $|A| = m$, and $\theta \geq 0$, compute $\mu(\theta, A)$.

For page limitation, we give a very simple naïve algorithm for Problem 14 (see Algorithm 2), by noting that it is not optimal. It assumes a well-behaved metric space $\mathcal{M} = (X, \mathrm{D})$ and some representation scheme μ for $\mathcal{M}$. The input to the algorithm consists of a finite subset $A \subseteq X$ and a distance threshold $\theta \geq 0$. Its output is the set $\{\mu(\theta, A_1,), \ldots, \mu(\theta, A_\ell)\}$ of binary strings representing the blocks $A_1, \ldots, A_\ell$ in the θ-decomposition of $\rho_\theta(A)$. The algorithm first initializes the variable $\mathcal{D}$ with the set of the representations of $\rho_\theta(\{x\}) = \{x\}$ for all $x \in A$ (cf. line 1). It then iteratively selects two different blocks $B_i, B_j \in \mathcal{D}$ such that $\overline{\mathrm{D}}(B_i, B_j) = \min_{x \in B_i, y \in B_j} \mathrm{D}(x, y) \geq \theta$. If there are no such B_i and B_j, then it terminates by returning $\mathcal{D}$; o/w it updates $\mathcal{D}$ by removing B_i, B_j and adding their merge defined by $\mathrm{MERGE}(\theta, A, B_i, B_j) = \mu(\theta, (\mathrm{ext}(B_i) \cup \mathrm{ext}(B_j)) \cap A)$ if $\overline{\mathrm{D}}(B_i, B_j) \leq \theta$; o/w $\mathrm{MERGE}(\theta, A, B_i, B_j) = \bot$, where $\mathrm{ext}(B_i), \mathrm{ext}(B_j)$ denote the extensions of B_i, B_j, respectively. The proof of the proposition below follows by induction on $|\mathcal{D}|$ from Theorem 4 and Corollary 5.

Proposition 15. *Algorithm 2 is correct.*

Let T_S, T_D, and T_M denote the time complexity of computing $\mu(\theta, \{x\})$, the distance between B_i and B_j, and the merge of B_i and B_j, respectively, for any

$x \in X$ and θ-blocks B_i and B_j. One can easily check that the time complexity of Algorithm 2 is $\mathcal{O}(mT_S + m^3 T_D + mT_M)$. Using a more sophisticated version of Algorithm 2 not presented for space limitations, we have the following improved result.

Theorem 16. *Problem 14 can be solved in time $\mathcal{O}(mT_S + m^2 T_D + mT_M)$.*

We consider the consistency problem also for the intensional scenario.

Problem 17 (The Consistency Problem for Intensional Weakly Convex Hulls). Given a well-behaved metric space $\mathcal{M} = (X, \mathrm{D})$, representation scheme μ for $\mathcal{M}$, disjoint finite sets $E^+, E^- \subseteq X$ of labeled examples with $|E^+ \cup E^-| = m$, and $k > 0$, return "YES" and the representations of the blocks in the θ-decomposition of a θ-convex set that is consistent with E^+ and E^- and has at most k blocks, if such a decomposition exists for some $\theta > 0$; o/w the answer "No".

Note that Remark 12 applies also to the problem above. Using the same idea as for the solution of Problem 11 (i.e., to decide whether a desired θ exists, we perform a binary search on the sorted set of pairwise distances between the elements in A), we have the following result on the above problem:

Theorem 18. *Problem 17 can be solved in $\mathcal{O}((mT_S + m^2 T_D + mT_M) \log m)$ time.*

In Sects. 5.1 and 5.2 below we present two non-trivial applications of Theorem 18 to polynomial PAC-learnability.

5.1   Learning Weakly Convex Boolean Functions

As a first application of Theorem 18, we show that the concept class formed by *weakly convex Boolean functions* is efficiently PAC-learnable. This result is not new, it has been obtained with a *domain specific* algorithm in [6]. Still, we present it as an application because, as we show below, we can derive it in a very simple way by using Theorem 18. Furthermore, our *general purpose* algorithm solving Problem 14 has the same asymptotic complexity on this problem as the *domain specific* one in [6].

Consider the metric space $\mathcal{M}_H = (\mathbb{H}_n, \mathrm{D}_H)$ for some $n \in \mathbb{N}$ (see Sect. 2). Clearly, the finiteness of $\mathbb{H}_n$ implies that $\mathcal{M}_H$ is well-behaved for all $\theta \geq 0$. A Boolean function $f : \{0, 1\}^n \to \{0, 1\}$ $(n \in \mathbb{N})$ is θ-convex for some $\theta \geq 0$ if for all $x, y, z \in \mathbb{H}_n$, $f(z) = 1$ whenever $f(x) = f(y) = 1$, $\mathrm{D}_H(x, y) \leq \theta$, and $z \in \triangle_=(x, y)$. Note that for $\mathcal{M}_H$ it suffices to consider the values in $[n]$ for θ.

Throughout this section we will use the following notation: L_n denotes the set $\{x_1, \neg x_1, \ldots, x_n, \neg x_n\}$ of Boolean literals. A *term* T is a conjunction of literals from L_n; T is sometimes regarded as the set of literals it contains. A *conflict* between two terms T_i and T_j over L_n is an integer $p \in [n]$ such that $x_p \in T_i$ and $\neg x_p \in T_j$ or vice versa. Finally, for a Boolean function f, $\mathrm{ext}(f)$ denotes the *extension* of f (i.e., $\mathrm{ext}(f) = \{x \in \mathbb{H}_n : f(\mathrm{x}) = 1\}$). We will use the following auxiliary result:

Lemma 19. *Let $A \subseteq \mathbb{H}_n$ be θ-convex and θ-connected for some $\theta \geq 2$. Then A is convex (i.e., n-convex) and can be represented by a term T over L_n.*

For any $n > 0$, the concept class $\mathcal{B}_{n,k} \subseteq 2^{\mathbb{H}_n}$ is defined as follows: For all $A \subseteq \mathbb{H}_n$, $A \in \mathcal{B}_{n,k}$ iff A is θ-convex for some $\theta \geq 0$ and its θ-decomposition has at most k blocks. Theorem 4 and Lemma 19 together imply that any such weakly convex set A can be represented *uniquely* by a k-term DNF F such that the extensions of the terms in F represent precisely the blocks in the θ-decomposition of A. Since the blocks are non-empty, no term contains a variable and its negation. This gives rise to the following definition of the representation scheme μ for $\mathcal{M}_H$: For all $S \subseteq \mathbb{H}_n$, define $\mu_H : \mathbb{R}_{\geq 0} \times \mathbb{H}_n \to \{0,1\}^*$ by $\mu_R(\theta, S) = F$, where F is the *unique* DNF representation of $\rho_\theta(S)$, if $\rho_\theta(S)$ is θ-connected; o/w by $\perp$.

Lemma 20. *Problem 17 can be solved in $\mathcal{O}(nm^2 \log m)$ time for $\mathcal{M}_H$.*

Proof. Let μ in Problem 17 be defined by μ_H. We show that T_S, T_D, T_M in Theorem 18 are all in $\mathcal{O}(n)$. For T_S, the claim follows from $\mu_H(\theta, \{x\}) = \bigwedge_i l_i$, where $l_i = x_i$ if $x_i = 1$; o/w $l_i = \neg x_i$. Let T_i and T_j be terms over L_n. Their distance $\overline{D}_H(T_i, T_j)$ is equal to the number of conflicts between T_i and T_j, implying $T_D \in \mathcal{O}(n)$. Finally, if $\overline{D}_H(T_i, T_j) \leq \theta$ then $\text{MERGE}(T_i, T_j)$ is the term T with literals $T_i \cap T_j$. Thus, $T_{\text{MERGE}} = \mathcal{O}(n)$. The claim then follows by Theorem 18. $\square$

Theorem 21. *For all $d, k \geq 0$, $\mathcal{B}_{n,k}$ is polynomially PAC-learnable.*

Proof. Since $\mathcal{B}_{n,k} \subseteq (\mathcal{B}_{n,1})^k_\cup$, VC-dim$(\mathcal{B}_{n,k}) \leq$ VC-dim$((\mathcal{B}_{n,1})^k_\cup) \leq 4nk \log(3k)$ by VC-dim$(\mathcal{B}_{n,1}) \leq 2n$ and by (ii) of Theorem 2. Hence, the VC-dimension of $\mathcal{B}_{n,k}$ is polynomial in n and k. Furthermore, by Lemma 20, the consistency problem for $\mathcal{B}_{n,k}$ can be solved in time polynomial in n, k, and $m = |E^+ \cup E^-|$. The theorem then follows by (i) of Theorem 2. $\square$

Note that if the extensions of the terms in the DNF are *not* required to be pairwise disjoint then, in contrast to the positive result in Theorem 21, k-term DNF formulas are *not* polynomially PAC-learnable for any $k \geq 2$ if P $\neq$ RP [10]. In [6] it is shown that the class of θ-convex Boolean functions is not polynomially PAC-learnable for $\theta > n/2 - 1$. The reason is that the number of terms having a pairwise distance greater than $n/2 - 1$ can be exponential in n. Notice that the number of terms in $\mathcal{B}_{n,k}$ is bounded by the *parameter* k in our problem setting. Finally we note that the time complexity of the *domain specific* algorithm in [6] that solves Problem 14 for $\mathbb{H}_n$ is $\mathcal{O}(m^2 n)$, which is the same as that of the sophisticated version of our *general purpose* Algorithm 2 in [13].

5.2   Learning Weakly Convex Axis-Aligned Hyperrectangles

Our second application of Theorem 18 is concerned with polynomial PAC-learnability of weakly convex sets in $\mathcal{M}_R = (\mathbb{R}^d, D_1)$, where D_1 is the *Manhattan* (or L_1) distance, i.e., $D_1(x, y) = \sum_i |x_i - y_i|$ for all $x = (x_1, \ldots, x_d)$ and $y = (y_1, \ldots, y_d) \in \mathbb{R}^d$. Note that $\mathcal{M}_R$ can be regarded as a generalization of $\mathcal{M}_H$ considered in the previos section, as D_1 becomes equal to D_H over the domain $\mathbb{H}_d \subseteq \mathbb{R}^d$. Clearly, $\mathcal{M}_R$ is complete. Furthermore, for all $x, y, z \in \mathbb{R}^d$, $D_1(x, z) + D_1(z, y) = D_1(x, y)$ iff z belongs to the smallest axis-aligned (topologically) closed hyperrectangle in $\mathbb{R}^d$ that contains x and y. This implies that all axis-aligned closed hyperrectangles are θ-convex for all $\theta > 0$ and $\rho_\theta(A)$ is closed for all finite subsets $A \subset \mathbb{R}^d$.

All concepts in the concept class $\mathcal{R}_{d,k}$ considered in this section are defined by the union of at most k pairwise disjoint axis-aligned closed hyperrectangles in $\mathbb{R}^d$, for some $d, k > 0$. More precisely, for all $R \subseteq \mathbb{R}^d$, $R \in \mathcal{R}_{d,k}$ iff R is θ-convex for some $\theta > 0$ with respect to $\mathcal{M}_R$ and the θ-decomposition of R consists of at most k blocks (i.e., axis-aligned closed hyperrectangles). For all finite sets $S \in [\mathbb{R}^d]^{<\omega}$, define $\mu_R : \mathbb{R}_{\geq 0} \times \mathbb{R}_{\geq 0} \times [\mathbb{R}^d]^{<\omega} \to \{0,1\}^*$ by $\mu_R(\theta, S) = (S_{\min}, S_{\max})$ if $\rho_\theta(S)$ is θ-connected; o/w by $\perp$,[1] where $S_{\min}$ (resp. $S_{\max}$) denotes the componentwise minimum (resp. maximum) of the points in S.

Lemma 22. *Problem 17 can be solved in $\mathcal{O}(dm^2 \log m)$ time for $\mathcal{M} = (\mathbb{R}^d, D_1)$.*

Proof. Let μ in Problem 17 be defined by μ_R. We prove the claim by showing that $T_S, T_D, T_{\mathrm{MERGE}}$ in Theorem 18 are all in $\mathcal{O}(d)$. In particular, $T_S \in \mathcal{O}(d)$ follows from $\mu_R(\theta, \{x\}) = (x, x)$. Let B_i (resp. B_j) be an axis-aligned closed hyperrectangle, $u = \min B_i$, and $v = \max B_i$ (resp. $x = \min B_j$ and $y = \max B_j$). We have $T_D \in \mathcal{O}(d)$ by the fact that $\overline{D}_1(B_i, B_j) = \sum_{i=1}^d D_1'([u_i, v_i], [x_i, y_i])$, where $D_1'([u_i, v_i], [x_i, y_i]) = \min\{|x_i - v_i|, |u_i - y_i|\}$ if $[u_i, v_i] \cap [x_i, y_i] \neq \emptyset$; o/w $D_1'([u_i, v_i], [x_i, y_i]) = 0$. Finally, if $\overline{D}(B_i, B_j) \leq \theta$ then $\mathrm{MERGE}(B_i, B_j)$ is the smallest axis-aligned closed hyperrectangle containing $\min\{u, x\}$ and $\max\{v, y\}$, implying $T_{\mathrm{MERGE}} = \mathcal{O}(d)$. The claim then follows by Theorem 18. □

Theorem 23. *For all $d, k \geq 0$, $\mathcal{R}_{d,k}$ is polynomially PAC-learnable.*

Proof. Since $\mathcal{R}_{d,k} \subseteq (\mathcal{R}_{d,1})_\cup^k$, VC-dim$(\mathcal{R}_{d,k}) \leq$ VC-dim$((\mathcal{R}_{d,1})_\cup^k) \leq 4dk \log(3k)$ by VC-dim$(\mathcal{R}_{d,1}) = 2d$ and by (ii) of Theorem 2. Hence, the VC-dimension of $\mathcal{R}_{d,k}$ is polynomial in d and k. Furthermore, by Lemma 22, the consistency problem for $\mathcal{R}_{d,k}$ can be solved in time polynomial in d, k, and $|E^+ \cup E^-|$. Thus, the theorem follows by (i) of Theorem 2. □

While Lemma 22 implies that a consistent hypothesis that has the smallest number of pairwise disjoint axis-aligned d-dimensional closed hyperrectangles can be found in polynomial time for all $d \geq 1$, this problem becomes NP-complete even for $d = 2$, if disjointness is not required (see, e.g., [1]).

6 Concluding Remarks

The theoretical and experimental results of this paper demonstrate the usefulness of weakly convex sets for machine learning. While our focus in this paper was

---

[1] We assume that real numbers are represented in $\mathcal{O}(1)$ space up to a certain precision.

solely on applications to *machine learning*, weakly convex sets seem to be useful for *data mining* applications (e.g., itemset mining, subgroup discovery) as well.

The notion of weak convexity can be uninteresting for certain metric spaces. For example, for finite subspaces of $(\mathbb{R}^d, D_2)$, $\triangle_=(x, y) = \{x, y\}$ holds almost surely for all points x and y. To overcome this problem, in the long version of this paper we allow the triangle inequality to hold up to some tolerance ε instead of equality. All results of Sect. 3 can be generalized to this relaxed definition.

There are several interesting questions for further research. Note, for example, that Algorithm 2 is very similar to *single linkage clustering*, raising the following question: Can the time complexity in Theorem 18 be further improved by using techniques (e.g., in [12]) designed for single linkage clustering algorithms?

Acknowledgments. This work was partly supported by the Ministry of Education and Research of Germany (BMBF) under project ML2R (grant number 01/S18038C) and by the Deutsche Forschungsgemeinschaft (DFG, German Research Foundation) under Germany's Excellence Strategy - EXC 2070 - 390732324.

References

1. Bereg, S., Cabello, S., Díaz-Báñez, J., Pérez-Lantero, P., Seara, C., Ventura, I.: The class cover problem with boxes. Comput. Geom. **45**(7), 294–304 (2012)
2. Blumer, A., Ehrenfeucht, A., Haussler, D., Warmuth, M.K.: Learnability and the Vapnik-Chervonenkis dimension. J. ACM **36**(4), 929–965 (1989)
3. Davey, B.A., Priestley, H.A.: Introduction to Lattices and Order, 2nd edn. Cambridge University Press, Cambridge (2002)
4. Delaunay, B.N.: Sur la sphère vide. Bull. Acad. Sci. URSS **6**, 793–800 (1934)
5. Edelsbrunner, H., Mücke, E.P.: Three-dimensional alpha shapes. ACM Trans. Graph. **13**(1), 43–72 (1994)
6. Ekin, O., Hammer, P.L., Kogan, A.: Convexity and logical analysis of data. Theoret. Comput. Sci. **244**(1), 95–116 (2000)
7. Ester, M., Kriegel, H.P., Sander, J., Xu, X., et al.: A density-based algorithm for discovering clusters in large spatial databases with noise. In: Proceedings of the Second International Conference on Knowledge Discovery and Data Mining (KDD), vol. 96, pp. 226–231. AAAI Press (1996)
8. Hemmer, M., Portaneri, C., Alliez, P.: Alpha Hulls. Research report, November 2020
9. Kearns, M.J., Vazirani, U.V.: An Introduction to Computational Learning Theory. MIT Press, Cambridge (1994)
10. Pitt, L., Valiant, L.G.: Computational limitations on learning from examples. J. ACM **35**(4), 965–984 (1988)
11. Seiffarth, F., Horváth, T., Wrobel, S.: Maximal closed set and half-space separations in finite closure systems. In: Brefeld, U., Fromont, E., Hotho, A., Knobbe, A., Maathuis, M., Robardet, C. (eds.) ECML PKDD 2019. LNCS (LNAI), vol. 11906, pp. 21–37. Springer, Cham (2020). https://doi.org/10.1007/978-3-030-46150-8_2
12. Sibson, R.: SLINK: an optimally efficient algorithm for the single-link cluster method. Comput. J. **16**(1), 30–34 (1973)

13. Stadtländer, E., Horváth, T., Wrobel, S.: Learning weakly convex sets in metric spaces. CoRR abs/2105.06251 (2021). https://arxiv.org/abs/2105.06251
14. Valiant, L.G.: A theory of the learnable. Commun. ACM **27**(11), 1134–1142 (1984)
15. van de Vel, M.: Theory of Convex Structures, North-Holland Mathematical Library, vol. 50. Elsevier (1993)

Disparity Between Batches as a Signal for Early Stopping

Mahsa Forouzesh$^{(\boxtimes)}$ and Patrick Thiran

Ecole Polytechnique Fédérale de Lausanne, Lausanne, Switzerland
{mahsa.forouzesh,patrick.thiran}@epfl.ch

Abstract. We propose a metric for evaluating the generalization ability of deep neural networks trained with mini-batch gradient descent. Our metric, called *gradient disparity*, is the ℓ_2 norm distance between the gradient vectors of two mini-batches drawn from the training set. It is derived from a probabilistic upper bound on the difference between the classification errors over a given mini-batch, when the network is trained on this mini-batch and when the network is trained on another mini-batch of points sampled from the same dataset. We empirically show that gradient disparity is a very promising early-stopping criterion (i) when data is limited, as it uses all the samples for training and (ii) when available data has noisy labels, as it signals overfitting better than the validation data. Furthermore, we show in a wide range of experimental settings that gradient disparity is strongly related to the generalization error between the training and test sets, and that it is also very informative about the level of label noise.

Keywords: Early stopping · Generalization · Gradient alignment · Overfitting · Neural networks · Limited datasets · Noisy labels

1 Introduction

Early-stopping using a separate validation set is one of the most popular techniques used to avoid under/over fitting deep neural networks trained with iterative methods, such as gradient descent [1–3]. The optimization is stopped when the performance of the model on a validation set starts to diverge from its performance on the training set. Early stopping requires an accurately labeled validation set, separated from the training set, to act as an unbiased proxy on the unseen test error. Obtaining such a reliable validation set can be expensive in many real-world applications as data collection is a time-consuming process that might require domain expertise. Furthermore, deep learning is becoming popular in applications for which there is simply not enough available data [4,5]. Finally, inexperienced label collectors, complex tasks (e.g., distinguishing a guinea pig from a hamster), and corrupted labels due for instance to adversarial attacks result in datasets that contain noisy labels [6]. Deep neural networks have the unfortunate ability to overfit to such small and/or noisy labeled datasets, an

© Springer Nature Switzerland AG 2021
N. Oliver et al. (Eds.): ECML PKDD 2021, LNAI 12976, pp. 217–232, 2021.
https://doi.org/10.1007/978-3-030-86520-7_14

issue that cannot be completely solved by popular regularization techniques [7]. A signal of overfitting during training is therefore particularly useful, if it does *not* need a separate, accurately labeled validation set, which is the purpose of this paper.

Let S_1 and S_2 be two mini-batches of points sampled from the available (training) dataset. Suppose that S_1 is selected for an iteration (step) of the mini-batch gradient descent (SGD), at the end of which the parameter vector is updated to w_1. The average loss over S_1 (denoted by $L_{S_1}(h_{w_1})$) is in principle reduced, given a sufficiently small learning rate. The average loss $L_{S_2}(h_{w_1})$ over the other mini-batch S_2 is not as likely to be reduced. It is more likely to remain larger than the loss $L_{S_2}(h_{w_2})$ computed over S_2, if it was S_2 instead of S_1 that had been selected for this iteration. The difference $\mathcal{R}_2 = L_{S_2}(h_{w_1}) - L_{S_2}(h_{w_2})$ is the penalty that we pay for choosing S_1 over S_2 (and similarly, $\mathcal{R}_1$ is the penalty that we would pay for choosing S_2 over S_1). $\mathcal{R}_2$ is illustrated in

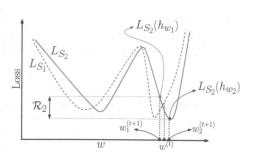

Fig. 1. An illustration of the penalty term $\mathcal{R}_2$, where the y-axis is the loss, and the x-axis indicates the parameter of the model. L_{S_1} and L_{S_2} are the average losses over mini-batches S_1 and S_2, respectively. $w^{(t)}$ is the parameter at iteration t and $w_i^{(t+1)}$ is the parameter at iteration $t+1$ if batch S_i was selected for the update step at iteration t, with $i \in \{1, 2\}$.

Fig. 1 for a hypothetical non-convex loss as a function of a one dimensional parameter w. The expected penalty measures how much, in an iteration, a model updated on one batch (S_1) is able to generalize on average to another batch (S_2) from the dataset. Hence, we call $\mathcal{R}$ the *generalization penalty*.

We establish a probabilistic upper bound on the sum of the expected penalties $\mathbb{E}[\mathcal{R}_1] + \mathbb{E}[\mathcal{R}_2]$ by adapting the PAC-Bayesian framework [8–10], given a pair of mini-batches S_1 and S_2 sampled from the dataset (Theorem 1). Interestingly, under some mild assumptions, this upper bound is essentially a simple expression driven by $\|g_1 - g_2\|_2$, where g_1 and g_2 are the gradient vectors over the two mini-batches S_1 and S_2, respectively. We call it *gradient disparity*: it measures how much a small gradient step on one mini-batch negatively affects the performance on the other one.

We propose gradient disparity as an effective early stopping criterion, because of its computational tractability that makes it simple to use during the course of training, and because of its strong link with generalization error, as evidenced in the experiments that we run on state-of-the-art configurations. Gradient disparity is particularly well suited when the available dataset has limited labeled data, because it does not require splitting the available dataset into training and validation sets: all the available data can be used during training, unlike for instance k-fold cross-validation. We observe that using gradient disparity,

Table 1. The test loss and area under the receiver operating characteristic curve (AUC score) of the MRNet dataset [11] when using 5-fold cross-validation (5-fold CV) and gradient disparity (GD) as early stopping criteria for detecting the presence of abnormally, ACL tears, and meniscal tears from the sagittal plane MRI scans. The corresponding curves during training are shown in Fig. 15 in the full version [12] (see Appendix F.3 in the full version [12] for more details). The results of early stopping are given, both when the metric (GD or validation loss) has increased for 5 epochs from the beginning of training and between parenthesis when the metric has increased for 5 consecutive epochs. Using GD outperforms 5-fold CV with either choice of the early stopping threshold. The standard deviations are obtained from 5 runs.

Task	Method	Test loss	Test AUC score (in percentage)
Abnormal	5-fold CV	$0.284_{\pm 0.016}(0.307_{\pm 0.057})$	$71.016_{\pm 3.66}(87.44_{\pm 1.35})$
	GD	$\mathbf{0.274}_{\pm 0.004}(\mathbf{0.275}_{\pm 0.053})$	$\mathbf{72.67}_{\pm 3.85}(\mathbf{88.12}_{\pm 0.35})$
ACL	5-fold CV	$0.973_{\pm 0.111}(1.246_{\pm 0.142})$	$79.80_{\pm 1.23}(89.32_{\pm 1.47})$
	GD	$\mathbf{0.842}_{\pm 0.101}(\mathbf{1.136}_{\pm 0.121})$	$\mathbf{81.81}_{\pm 1.64}(\mathbf{91.52}_{\pm 0.09})$
Meniscal	5-fold CV	$0.758_{\pm 0.04}(1.163_{\pm 0.127})$	$73.53_{\pm 1.30}(72.14_{\pm 0.74})$
	GD	$\mathbf{0.726}_{\pm 0.019}(\mathbf{1.14}_{\pm 0.323})$	$\mathbf{74.08}_{\pm 0.79}(\mathbf{73.80}_{\pm 0.24})$

instead of an unbiased validation set, results in a predictive improvement of at least 1% for classification tasks with limited and very costly available data, such as the MRNet dataset, which is a small size image-classification dataset used for detecting knee injuries (Table 1).

Moreover, we find that gradient disparity is a more accurate early stopping criterion than validation loss when the available dataset contains noisy labels. Gradient disparity reflects the label noise level quite well throughout the training process, especially at early stages of training. Finally, we observe that gradient disparity has a strong positive correlation with the test error across experimental settings that differ in training set size, batch size, and network width.

2 Related Work

The coherent gradient hypothesis [13] states that the gradient is stronger in directions where similar examples exist and towards which the parameter update is biased. He and Su [14] study the local elasticity phenomenon, which measures how the prediction over one sample changes, as the network is updated on another sample. Motivated by [14], reference [15] proposes generalization upper bounds using locally elastic stability. The generalization penalty introduced in our work measures how the prediction over one sample (batch) changes when the network is updated on the same sample, instead of being updated on another sample.

Finding a practical metric that completely captures the generalization properties of deep neural networks, and in particular indicates the level of label noise and decreases with the size of the training set, is still an active research direction

[16–19]. Recently, there have been a few studies that propose similarity between gradients as a generalization metric. The benefit of tracking generalization by measuring the similarity between gradient vectors is its tractability during training, and the dispensable access to unseen data. Sankararaman et al. [20] propose gradient confusion, which is a bound on the inner product of two gradient vectors, and shows that the larger the gradient confusion is, the slower the convergence is. Gradient interference (when the gradient inner product is negative) has been studied in multi-task learning, reinforcement learning and temporal difference learning [21–23]. Yin et al. [24] study the relation between gradient diversity, which measures the dissimilarity between gradient vectors, and the convergence performance of distributed SGD algorithms. Fort et al. [25] propose a metric called stiffness, which is the cosine similarity between two gradient vectors, and shows empirically that it is related to generalization. Fu et al. [26] study the cosine similarity between two gradient vectors for natural language processing tasks. Reference [27] measures the alignment between the gradient vectors within the same class (denoted by Ω_c) , and studies the relation between Ω_c and generalization as the scale of initialization (the variance of the probability distribution the network parameters are initially drawn from) is increased. These metrics are usually not meant to be used as early stopping criteria, and indeed in Table 2 and Table 12 in the appendix of the full version [12], we observe that none of them consistently outperforms k-fold cross-validation.

Another interesting line of work is the study of the variance of gradients in deep learning settings. Negrea et al. [28] derive mutual information generalization error bounds for stochastic gradient Langevin dynamics (SGLD) as a function of the sum (over the iterations) of square gradient incoherences, which is closely related to the variance of gradients. Two-sample gradient incoherences also appear in [29], which are taken between a training sample and a "ghost" sample that is not used during training and therefore taken from a validation set (unlike gradient disparity). The upper bounds in [28,29] are cumulative bounds that increase with the number of iterations and are not intended to be used as early stopping criteria. As shown in Appendix H (in the full version [12]), gradient disparity can be used as an early stopping criterion not only for SGD with additive noise (such as SGLD), but also other adaptive optimizers. Reference [30] shows that the variance of gradients is a decreasing function of the batch size. However, reference [31] hypothesizes that gradient variance counter-intuitively increases with the batch size, by studying the effect of the learning rate on the variance of gradients, which is consistent with our results on convolutional neural networks in Sect. 6. References [30,31] mention the connection between variance of gradients and generalization as promising future directions. Our study shows that variance of gradients used as an early stopping criterion outperforms k-fold cross-validation (see Table 12 of the full version [12]).

Liu et al. [32] propose a relation between gradient signal-to-noise ratio (SNR), called GSNR, and the one-step generalization error, with the assumption that both the training and test sets are large. Mahsereci et al. [33] also study gradient SNR and propose an early stopping criterion called evidence-based criterion (EB) that eliminates the need for a held-out validation set. Reference [34] proposes an early stopping criterion based on the signal-to-noise ratio figure, which is further studied in [35], a study that shows the average test error achieved by standard early stopping is lower than the one obtained by this criterion. Zhang et al. [36] empirically show that the variance term in the bias-variance decomposition of the loss function dominates the variations of the test loss, and hence propose optimization variance (OV) as an early stopping criterion.

Summary of Comparison to Related Work. In Table 2 and Appendix I of the full version [12], we compare gradient disparity (GD) to EB, GSNR, gradient inner product, sign of the gradient inner product, variance of gradients, cosine similarity, Ω_c, and OV. We observe that the only metrics that consistently outperform k-fold cross-validation as early stopping criteria across various settings (see Table 12 in the full version [12]), and that reflect well the label noise level (see in Figs. 26 and 27 of the full version [12] that metrics such as EB and $\text{sign}(g_i \cdot g_j)$ do not correctly detect the label noise level), are gradient disparity and variance of gradients. The two are analytically very close as discussed in Appendix I.2 of the full version [12]. However, we observe that the correlation between gradient disparity and the test loss is in general larger than the correlation between variance of gradients and the test loss (see Table 13 in the full version [12]).

Table 2. The test error (TE) and test loss (TL) achieved by using various metrics as early stopping criteria for an AlexNet trained on the MNIST dataset with 50% random labels. See Table 12 in the appendix of the full version [12] for further details and experiments.

	Min	GD/Var	EB	GSNR	$g_i \cdot g_j$	$\text{sign}(g_i \cdot g_j)$	$\cos(g_i \cdot g_j)$	Ω_c	OV	k-fold	No ES
TE	13.76	**16.66**	24.63	35.68	37.92	24.63	35.68	29.40	34.36	17.86	25.72
TL	0.75	1.08	**0.86**	1.68	1.82	**0.86**	1.68	1.46	1.65	1.09	0.91

3   Generalization Penalty

Consider a classification task with input $x \in \mathcal{X} := \mathbb{R}^n$ and ground truth label $y \in \{1, 2, \cdots, k\}$, where k is the number of classes. Let $h_w \in \mathcal{H} : \mathcal{X} \to \mathcal{Y} := \mathbb{R}^k$ be a predictor (classifier) parameterized by the parameter vector $w \in \mathbb{R}^d$, and $l(\cdot, \cdot)$ be the 0–1 loss function $l(h_w(x), y) = \mathbb{1}[h_w(x)[y] < \max_{j \neq y} h_w(x)[j]]$ for all $h_w \in \mathcal{H}$ and $(x, y) \in \mathcal{X} \times \{1, 2, \cdots, k\}$. The expected loss and the empirical loss over the training set S of size m are respectively defined as

$$L(h_w) = \mathbb{E}_{(x,y) \sim D}[l(h_w(x), y)], \tag{1}$$

and

$$L_S(h_w) = \frac{1}{m} \sum_{i=1}^{m} l(h_w(x_i), y_i), \tag{2}$$

where D is the probability distribution of the data points and $S = \{(x_i, y_i)\}^m$ is a collection of m i.i.d. samples drawn from D. Similar to the notation used in [16], distributions on the hypotheses space $\mathcal{H}$ are simply distributions on the underlying parameterization. With some abuse of notation, ∇L_{S_i} refers to the gradient with respect to the surrogate differentiable loss function, which in our experiments is cross entropy[1].

In a mini-batch gradient descent (SGD) setting, let mini-batches S_1 and S_2 have sizes m_1 and m_2, respectively, with $m_1 + m_2 \leq m$. Let $w = w^{(t)}$ be the parameter vector at the beginning of an iteration t. If S_1 is selected for the next iteration, w gets updated to $w_1 = w^{(t+1)}$ with

$$w_1 = w - \gamma \nabla L_{S_1}(h_w), \tag{3}$$

where γ is the learning rate. The generalization penalty $\mathcal{R}_2$ is defined as the gap between the loss over S_2, $L_{S_2}(h_{w_1})$, and its target value, $L_{S_2}(h_{w_2})$, at the end of iteration t.

When selecting S_1 for the parameter update, Eq. (3) makes a step towards learning the input-output relations of mini-batch S_1. If this negatively affects the performance on mini-batch S_2, $\mathcal{R}_2$ will be large; the model is learning the data structures that are unique to S_1 and that do not appear in S_2. Because S_1 and S_2 are mini-batches of points sampled from the same distribution D, they have data structures in common. If, throughout the learning process, we consistently observe that, in each update step, the model learns structures unique to only one mini-batch, then it is very likely that the model is memorizing the labels instead of learning the common data-structures. This is captured by the generalization penalty $\mathcal{R}$.

We adapt the PAC-Bayesian framework [8,9] to account for the trajectory of the learning algorithm; For each learning iteration t we define a prior, and two possible posteriors depending on the choice of the mini-batch selection. Let $w \sim P$ follow a prior distribution P, which is a $\mathcal{F}_t$-measurable function, where $\mathcal{F}_t$ denotes the filtration of the available information at the beginning of iteration t. Let h_{w_1}, h_{w_2} be the two learned single predictors, at the end of iteration t, from S_1 and S_2, respectively. In this framework, for $i \in \{1,2\}$, each predictor h_{w_i} is randomized and becomes h_{ν_i} with $\nu_i = w_i + u_i$, where u_i is a random variable whose distribution might depend on S_i. Let Q_i be the distribution of ν_i, which is a distribution over the predictor space $\mathcal{H}$ that depends on S_i via w_i and possibly u_i. Let $\mathcal{G}_i$ be a σ-field such that $\sigma(S_i) \cup \mathcal{F}_t \subset \mathcal{G}_i$ and such that the

---

[1] We have also studied networks trained with the mean square error in Appendix E.3 of the full version [12], and we observe that there is a strong positive correlation between the test error/loss and gradient disparity for this choice of the surrogate loss function as well (see Fig. 11 of the full version [12]).

posterior distribution Q_i is $\mathcal{G}_i$-measurable for $i \in \{1, 2\}$. We further assume that the random variable $\nu_1 \sim Q_1$ is statistically independent from the draw of the mini-batch S_2 and, vice versa, that $\nu_2 \sim Q_2$ is independent from the batch $S_1{}^2$, i.e., $\mathcal{G}_1 \perp\!\!\!\perp \sigma(S_2)$ and $\mathcal{G}_2 \perp\!\!\!\perp \sigma(S_1)$.

Theorem 1. *For any $\delta \in (0, 1]$, with probability at least $1 - \delta$ over the sampling of sets S_1 and S_2, the sum of the expected penalties conditional on S_1, and S_2, respectively, satisfies*

$$\mathbb{E}\left[\mathcal{R}_1\right] + \mathbb{E}\left[\mathcal{R}_2\right] \leq \sqrt{\frac{2KL(Q_2||Q_1) + 2\ln\frac{2m_2}{\delta}}{m_2 - 2}} + \sqrt{\frac{2KL(Q_1||Q_2) + 2\ln\frac{2m_1}{\delta}}{m_1 - 2}}.$$

$$(4)$$

In this paper, the goal is to get a signal of overfitting that indicates at the beginning of each iteration t whether to stop or to continue training. This signal should track the performance of the model at the end of iteration t by investigating its evolution over all the possible outcomes of the batch sampling process during this iteration. For simplicity, we consider two possible outcomes: either mini-batch S_1 or mini-batch S_2 is chosen for this iteration (we later in the next section extend to more pairs of mini-batches). If we were to use bounds such as the ones in [10,38] for one iteration at a time, the generalization error at the end of that iteration can be bounded by a function of either $KL(Q_1||P)$ or $KL(Q_2||P)$, depending on the selected mini-batch. Therefore, as each of the two mini-batches is equally likely to be sampled, we should track $KL(Q_1||P)$ and $KL(Q_2||P)$ for a signal of overfitting at the end of the iteration, which requires in turn access to the three distributions P, Q_1 and Q_2. In contrast, the upper bound on the generalization penalty given in Theorem 1 only requires the two distributions Q_1 and Q_2, which is a first step towards a simpler metric since, loosely speaking, the symmetry between the random choices for S_1 and S_2 should carry over these two distributions, leading us to assume the random perturbations u_1 and u_2 to be identically distributed. If furthermore we assume them to be Gaussian, then we show in the next section that $KL(Q_2||Q_1)$ and $KL(Q_1||Q_2)$ are equal and boil down to a very tractable generalization metric, which we call gradient disparity.

4 Gradient Disparity

In Sect. 3, the randomness modeled by the additional perturbation u_i, conditioned on the current mini-batch S_i, comes from (i) the parameter vector at the beginning of the iteration w, which itself comes from the random parameter initialization and the stochasticity of the parameter updates until that iteration, and (ii) the gradient vector ∇L_{S_i} (simply denoted by g_i), which may also be

---

[2] Mini-batches S_1 and S_2 are drawn without replacement, and the random selection of indices of mini-batches S_1 and S_2 is independent from the dataset S. Hence, similarly to [28,37], we have $\sigma(S_1) \perp\!\!\!\perp \sigma(S_2)$.

random because of the possible additional randomness in the network structure due for instance to dropout [39]. A common assumption made in the literature is that the random perturbation u_i follows a normal distribution [38,40]. The upper bound in Theorem 1 takes a particularly simple form if we assume that for $i \in \{1,2\}$, u_i are zero mean i.i.d. normal variables ($u_i \sim \mathcal{N}(0, \sigma^2 I)$), and that w_i is fixed, as in the setting of [16].

As $w_i = w - \gamma g_i$ for $i \in \{1,2\}$, the KL-divergence between $Q_1 = \mathcal{N}(w_1, \sigma^2 I)$ and $Q_2 = \mathcal{N}(w_2, \sigma^2 I)$ (Lemma 1 in Appendix B of the full version [12]) is simply

$$\mathrm{KL}(Q_1 \| Q_2) = \frac{1}{2} \frac{\gamma^2}{\sigma^2} \| g_1 - g_2 \|_2^2 = \mathrm{KL}(Q_2 \| Q_1), \tag{5}$$

which shows that, keeping a constant step size γ and assuming the same variance for the random perturbations σ^2 in all the steps of the training, the bound in Theorem 1 is driven by $\| g_1 - g_2 \|_2$. This indicates that the smaller the ℓ_2 distance between gradient vectors is, the lower the upper bound on the generalization penalty is, and therefore the closer the performance of a model trained on one mini-batch is to a model trained on another mini-batch.

For two mini-batches of points S_i and S_j, with respective gradient vectors g_i and g_j, we define the *gradient disparity* (GD) between S_i and S_j as

$$\mathcal{D}_{i,j} = \| g_i - g_j \|_2 . \tag{6}$$

To compute $\mathcal{D}_{i,j}$, a first option is to sample S_i from the training set and S_j from the held-out validation set, which we refer to as the "train-val" setting, following [25]. The generalization penalty $\mathcal{R}_j$ in this setting measures how much, during the course of an iteration, a model updated on a training set is able to generalize to a validation set, making the resulting ("train-val") gradient disparity $\mathcal{D}_{i,j}$ a natural candidate for tracking overfitting. But it requires access to a validation set to sample S_j, which we want to avoid. The second option is to sample both S_i and S_j from the training set, as proposed in this paper, to yield now a value of $\mathcal{D}_{i,j}$ that we could call "train-train" gradient disparity (GD) by analogy. Importantly, we observe a strong positive correlation between the two types of gradient disparities ($\rho = 0.957$) in Fig. 2. Therefore, we can expect that both of them do (almost) equally well in detecting overfitting, with the advantage that the latter does not require to set data aside, contrary to the former. We will therefore consider GD when both batches are sampled from the training set and evaluate it in this paper.

To track the upper bound of the generalization penalty for more pairs of batches, we can compute an average gradient disparity over B batches, which requires all the B gradient vectors at each iteration, which is computationally expensive if B is large. We approximate it by computing GD over only a much smaller subset of the batches, of size $s \ll B$,

$$\overline{\mathcal{D}} = \sum_{i=1}^{s} \sum_{j=1, j \neq i}^{s} \frac{\mathcal{D}_{i,j}}{s(s-1)}.$$

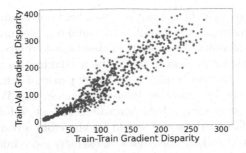

Fig. 2. "Train-val" gradient disparity versus "train-train" gradient disparity for 220 experimental settings that vary in architecture, dataset, training set size, label noise level and initial random seed. Pearson's correlation coefficient is $\rho = 0.957$.

In our experiments, $s = 5$; we observe that such a small subset is already sufficient (see Appendix E.2 of the full version for an experimental comparison of different values of s).

Consider two training iterations t_1 and t_2 where $t_1 \ll t_2$. At earlier stages of the training (iteration t_1), the parameter vector $(w^{(t_1)})$ is likely to be located in a steep region of the training loss landscape, where the gradient vector of training batches, g_i, and the training loss $L_{S_i}(h_{w^{(t_1)}})$ take large values. At later stages of training (iteration t_2), the parameter vector $(w^{(t_2)})$ is more likely in a flatter region of the training loss landscape where g_i and $L_{S_i}(h_{w^{(t_2)}})$ take small values. To compensate for this scale mismatch when comparing the distance between gradient vectors at different stages of training, we re-scale the loss values within each batch before computing $\overline{\mathcal{D}}$ (see Appendix E.1 in the full version [12] for more details). Note that this re-scaling is only done for the purpose of using GD as a metric, and therefore does not have any effect on the training process itself.

We focus on the vanilla SGD optimizer. In Appendix H of the full version [12], we extend the analysis to other stochastic optimization algorithms: SGD with momentum, Adagrad, Adadelta, and Adam. In all these optimizers, we observe that GD (Eq. (6)) appears in $\mathrm{KL}(Q_1\|Q_2)$ with other factors that depend on a decaying average of past gradient vectors. Experimental results support the use of GD as an early stopping metric also for these popular optimizers (see Fig. 25 in Appendix H of the full version [12]). For vanilla SGD optimizer, we also provide an alternative and simpler derivation leading to gradient disparity from the linearization of the loss function in Appendix D of the full version [12].

5   Early Stopping Criterion

In the presence of *large* amounts of *reliable* data, it is affordable to split the available dataset into a training and a validation set and to perform early stopping by evaluating the performance of the model on the held-out validation set. However, if the dataset is *limited*, this approach makes an inefficient use of the data because the model never learns the information that is still present in the validation set. Moreover, if the dataset is *noisy*, held-out validation might poorly estimate the performance of the model as the validation set might contain a high percentage of noisy samples. To avoid these issues, k-fold cross-validation [41] is a solution that makes an efficient usage of the available data while providing an unbiased estimate of the performance, at the expense of a high computational overhead and of a possibly underestimated variance [42]. While each of its k rounds is itself a setting with a held-out validation set, k-fold cross-validation (as opposed to held-out validation) would be therefore advantageous to use in the presence of limited and/or noisy data. It extracts more information from the dataset as it uses all the data samples for both training and validation, and it is less dependent on how the data is split into training and validation sets.

Table 3. The test loss and accuracy when using gradient disparity (GD) and k-fold cross-validation (CV) (k =5) as early stopping criteria when the available dataset is limited: (top) VGG-13 trained on 1.28 k samples of the CIFAR-10 dataset, and (bottom) AlexNet trained on 256 samples of the MNIST dataset. The corresponding curves during training are presented in Fig. 13 of the full version [12]. The results below are obtained by stopping the optimization when the metric (either validation loss or GD) has increased for 5 epochs from the beginning of training.

Setting	Method	Test loss	Test accuracy
CIFAR-10, VGG-13	5-fold CV	$1.846_{\pm 0.016}$	$35.982_{\pm 0.393}$
	GD	$\mathbf{1.793}_{\pm 0.016}$	$\mathbf{36.96}_{\pm 0.861}$
MNIST, AlexNet	5-fold CV	$1.123_{\pm 0.25}$	$62.62_{\pm 6.36}$
	GD	$\mathbf{0.656}_{\pm 0.080}$	$\mathbf{79.12}_{\pm 3.04}$

The baseline to beat is therefore k-fold cross-validation (CV). We compare gradient disparity to CV in the two target settings: (i) when the available dataset is limited and (ii) when the available dataset has corrupted labels. Medical applications are one of the practical examples of setting (i), where datasets are costly because they require the collection of patient data, and the medical staff's expertise to label the data. An example of such an application is the MRNet dataset [11], which contains a limited number of MRI scans to study the presence of abnormally, ACL tears and meniscal tears in knee injuries. This dataset is by nature limited and we use the entire available data for both early stopping methods GD and k-fold CV. In addition, to further simulate setting (i), we use small subsets of three image classification benchmark datasets: MNIST, CIFAR-10 and CIFAR-100. Performing early stopping in the presence of label noise (setting (ii)) is also practically very important, because it has been empirically observed that deep neural networks trained on noisy datasets overfit to noisy labeled samples at later stages of training. A good early stopping signal can therefore prevent such an overfitting [43–45]. To simulate setting (ii), we use a corrupted version of these image classification benchmark datasets, where for a fraction of the samples (the amount of noise), we choose the labels at random.

(i) We observe that using gradient disparity instead of a validation loss in k-fold CV results in an improvement of more than 1% (on average over all three tasks) in the test AUC score of the MRNet dataset, and therefore adds a correct detection for more than one patient for each task (see Table 1). Furthermore, we observe that gradient disparity performs better than k-fold CV as an early stopping criterion for image-classification benchmark datasets as well (see Table 3). A plausible explanation for the better peformance of GD over k-fold CV is that, although CV uses the entire set of samples over the k rounds for both training and validation, it trains the model only on a $(1 - 1/k)$ portion of the dataset in each individual round. In contrast, GD allows to train the model over the entire dataset in a single run, which therefore results in a better performance on the final unseen (test) data when data is limited. For more experimental results refer to Table 10 and Figs. 13 and 15 in Appendix F of the full version [12].

(ii) We observe that gradient disparity performs better than k-fold cross-validation as an early stopping criterion when data is noisy (see Table 4). When the labels of the available data are noisy, the validation set is no longer a reliable estimate of the test set. Nevertheless, and although it is computed over the noisy training set, gradient disparity reflects the performance on the test set quite well[3] For more experimental results refer to Table 11 and Fig. 14 in Appendix F of the full version [12].

Quite surprisingly, we observe that GD performs better in terms of accuracy than an extension of k-fold CV, which we call k^+-fold CV, which uses the entire dataset for training with the early stopping signal found by k-fold CV (see Table 4, where $k = 10$ for these settings). More precisely, k^+-fold CV is done in 3 steps: (1) perform k-fold CV, (2) compute the stopping epoch by tracking the validation loss found in step (1), and (3) retrain the model on the entire dataset and stop at the epoch obtained in step (2). k^+-fold CV uses therefore $k + 1$ rounds because of step (3), thus one more round than k-fold CV, but unlike k-fold CV (and similarly to GD), k^+-fold CV produces models that are trained on the entire dataset. It is therefore interesting to note that using GD still outperforms k^+-fold CV in terms of accuracy (although not in terms of loss).

The metrics used as early stopping criteria, whether they are the validation loss or gradient disparity, are measured on signals that are subject to random fluctuations. As a result, they rely on a pre-defined threshold p (sometimes called *patience* by practitioners) that sets the number of iterations during which the metric increases before the algorithm is stopped. We use two popular thresholds: (t1) the first one is to stop the algorithm when the metric (GD or validation loss) has increased for $p = 5$ (possibly non consecutive) epochs from the beginning of training, and (t2) the second is the same as (t1) but the $p = 5$ epochs must be consecutive. Both GD and k-fold CV might be sensitive to the choice of (t1) or (t2), or even to the value of p itself. It is therefore important to study the sensitivity of an early stopping metric to the choice of the threshold p, which

---

[3] See for example Fig. 14 (left column) in the full version [12] where the validation loss fails to estimate the test loss, but where GD (Fig. 14 (middle left column)) does signal overfitting correctly.

Table 4. The test loss and accuracy when using gradient disparity (GD) and k-fold cross-validation (CV) ($k = 10$) as early stopping criteria when the available dataset is noisy: 50% of the available data has random labels. The corresponding curves during training are shown in Fig. 14 in the full version [12]. The results below are obtained by stopping the optimization when the metric (either validation loss or GD) has increased for 5 epochs from the beginning of training. The last row in each setting, which we call 10^+-fold CV, refers to the test loss and accuracy reached at the epoch suggested by 10-fold CV, for a network trained on the entire set. Notice that for the CIFAR-100 experiments (the top rows), for computational reasons, the models are trained on only 1.28 k samples of the dataset which explains the very low test accuracy for this experiment. However, for the MNIST experiments (the bottom rows), the models are trained on the entire dataset, and we observe rather high test accuracies.

Setting	Method	Test loss	Test accuracy
CIFAR-100, ResNet-18	10-fold CV	$5.023_{\pm 0.083}$	$1.59_{\pm 0.15}$ (top-5: $6.47_{\pm 0.52}$)
	GD	$\mathbf{4.463}_{\pm 0.038}$	$\mathbf{3.68}_{\pm 0.52}$ (top-5: $\mathbf{15.22}_{\pm 1.24}$)
	10^+-fold CV	$4.964_{\pm 0.057}$	$1.68_{\pm 0.24}$ (top-5: $7.05_{\pm 0.71}$)
MNIST, AlexNet	10-fold CV	$0.656_{\pm 0.034}$	$97.28_{\pm 0.20}$
	GD	$0.654_{\pm 0.031}$	$\mathbf{97.32}_{\pm 0.27}$
	10^+-fold CV	$\mathbf{0.639}_{\pm 0.029}$	$97.31_{\pm 0.15}$

is done in Appendix F.1 of the full version [12] for both GD and k-fold CV for ten different values of $p \in \{1, \cdots, 10\}$ and the two thresholds (t1) and (t2). We observe that GD always gives similar or higher test accuracy than k-fold CV for all 20 possible thresholds (see Fig. 3 of the full version [12]). More importantly, GD is much more robust to the choice of the early stopping threshold (see Table 5).

Table 5. Sensitivity of each method to the choice of the early stopping threshold. The sensitivity is computed from the reported values of Tables 7 and 8 according to Eq. 14 in the appendix of the full version [12].

Method	Sensitivity of the test accuracy	Sensitivity of the test loss
GD	**0.916**	**0.886**
CV	1.613	1.019

When data is abundant and clean, the validation loss is affordable and trustworthy to use as an early stopping signal. GD does also correctly signal overfitting in this case (see for example Fig. 4 in the full version [12]). However, when data is limited and/or noisy (which is also when early stopping is particularly important), we observe that the validation loss is costly and unreliable to use as an early stopping signal. In contrast, in these settings, GD does not cost a separate held-out validation set and is a reliable signal of overfitting even in the presence of label noise. In practice, the label noise level of a given dataset is in general not known a priori and we do not know whether the size of the dataset is large enough to afford sacrificing a subset for validation. We often do not know whether we are in the former setting, with abundant and clean data, or in the later setting, with limited and/or noisy data. It is therefore important to have a good early stopping criterion that works for both settings. Unlike the validation loss, GD is such a signal.

6 Discussion and Final Remarks

We propose gradient disparity (GD), as a simple to compute early stopping criterion that is particularly well-suited when the dataset is limited and/or noisy. Beyond indicating the early stopping time, GD is well aligned with factors that contribute to improve or degrade the generalization performance of a model, which have an often strikingly similar effect on the value of GD as well. We briefly discuss in this section some of these observations that further validate the use of GD as an effective early stopping criterion; more details are provided in the appendix.

Label Noise Level. We observe that GD reflects well the label noise level throughout the training process, even at early stages of training, where the generalization gap fails to do so (see Figs. 5, 17, 21, and 24 in Appendix G of the full version [12]).

Training Set Size. We observe that GD, similarly to the test error, decreases with training set size, unlike many previous metrics as shown by [17,18]. Moreover, we observe that applying data augmentation decreases the values of both GD and the test error (see Figs. 6 and 22 in the full version [12]).

Batch Size. We observe that both the test error and GD increase with batch size. This observation is counter-intuitive because one might expect that gradient vectors get more similar when they are averaged over a larger batch. GD matches the ranking of test errors for different networks, trained with different batch sizes, as long as the batch sizes are not too large (see Fig. 23 in the full version [12]).

Width. We observe that both the test error and GD (normalized with respect to the number of parameters) decrease with the network width for ResNet, VGG and fully connected neural networks (see Figs. 8 and 20 in Appendix G of the full version [12]).

Gradient disparity belongs to the same class of metrics based on the similarity between two gradient vectors [20,25–27,31]. A common drawback of all these metrics is that they are not informative when the gradient vectors are very small. In practice however, we observe (see for instance Fig. 18 in the appendix of the full version [12]) that the time at which the test and training losses start to diverge, which is the time when overfitting kicks in, does not only coincide with the time at which gradient disparity increases, but also occurs much before the training loss becomes infinitesimal. This drawback is therefore unlikely to cause a problem for gradient disparity when it is used as an early stopping criterion. Nevertheless, as a future direction, it would be interesting to explore this further especially for scenarios such as epoch-wise double-descent [46].

References

1. Prechelt, L.: Early stopping - but when? In: Orr, G.B., Müller, K.-R. (eds.) Neural Networks: Tricks of the Trade. LNCS, vol. 1524, pp. 55–69. Springer, Heidelberg (1998). https://doi.org/10.1007/3-540-49430-8_3
2. Yao, Y., Rosasco, L., Caponnetto, A.: On early stopping in gradient descent learning. Constr. Approximation **26**(2), 289–315 (2007)
3. Gu, J., et al.: Recent advances in convolutional neural networks. Pattern Recogn. **77**, 354–377 (2018)
4. Roh, Y., Heo, G., Whang, S.E.: A survey on data collection for machine learning: a big data-ai integration perspective. IEEE Transactions on Knowledge and Data Engineering (2019)
5. Ipeirotis, P.G., Provost, F., Wang, J.: Quality management on amazon mechanical turk. In: Proceedings of the ACM SIGKDD Workshop on Human Computation, pp. 64–67 (2010)
6. Frénay, B., Verleysen, M.: Classification in the presence of label noise: a survey. IEEE Trans. Neural Netw. Learn. Syst. **25**(5), 845–869 (2013)
7. Zhang, C., Bengio, S., Hardt, M., Recht, B., Vinyals, O.: Understanding deep learning requires rethinking generalization. arXiv preprint arXiv:1611.03530 (2016)
8. McAllester, D.A.: Pac-bayesian model averaging. In: Proceedings of the Twelfth Annual Conference on Computational Learning Theory, pp. 164–170 (1999)
9. McAllester, D.A.: Some pac-bayesian theorems. Mach. Learn. **37**(3), 355–363 (1999)
10. McAllester, D.: Simplified PAC-Bayesian margin bounds. In: Schölkopf, B., Warmuth, M.K. (eds.) COLT-Kernel 2003. LNCS (LNAI), vol. 2777, pp. 203–215. Springer, Heidelberg (2003). https://doi.org/10.1007/978-3-540-45167-9_16
11. Bien, N., et al.: Deep-learning-assisted diagnosis for knee magnetic resonance imaging: development and retrospective validation of mrnet. PLoS Med. **15**(11), e1002699 (2018)
12. Forouzesh, M., Thiran, P.: Disparity between batches as a signal for early stopping. arXiv preprint arXiv:2107.06665 (2021)
13. Chatterjee, S.: Coherent gradients: an approach to understanding generalization in gradient descent-based optimization. In: International Conference on Learning Representations (2020). https://openreview.net/forum?id=ryeFY0EFwS
14. He, H., Su, W.: The local elasticity of neural networks. In: International Conference on Learning Representations (2020). https://openreview.net/forum?id=HJxMYANtPH
15. Deng, Z., He, H., Su, W.J.: Toward better generalization bounds with locally elastic stability. arXiv preprint arXiv:2010.13988 (2020)

16. Dziugaite, G.K., Roy, D.M.: Computing nonvacuous generalization bounds for deep (stochastic) neural networks with many more parameters than training data. arXiv preprint arXiv:1703.11008 (2017)
17. Neyshabur, B., Bhojanapalli, S., McAllester, D., Srebro, N.: Exploring generalization in deep learning. In: Advances in Neural Information Processing Systems, pp. 5947–5956 (2017)
18. Nagarajan, V., Kolter, J.Z.: Uniform convergence may be unable to explain generalization in deep learning. In: Advances in Neural Information Processing Systems, pp. 11611–11622 (2019)
19. Chatterji, N., Neyshabur, B., Sedghi, H.: The intriguing role of module criticality in the generalization of deep networks. In: International Conference on Learning Representations (2020). https://openreview.net/forum?id=S1e4jkSKvB
20. Sankararaman, K.A., De, S., Xu, Z., Huang, W.R., Goldstein, T.: The impact of neural network overparameterization on gradient confusion and stochastic gradient descent. arXiv preprint arXiv:1904.06963 (2019)
21. Riemer, M., et al.: Learning to learn without forgetting by maximizing transfer and minimizing interference. arXiv preprint arXiv:1810.11910 (2018)
22. Liu, V., Yao, H., White, M.: Toward understanding catastrophic interference in value-based reinforcement learning. Optimization Foundations for Reinforcement Learning Workshop at NeurIPS (2019)
23. Bengio, E., Pineau, J., Precup, D.: Interference and generalization in temporal difference learning. arXiv preprint arXiv:2003.06350 (2020)
24. Yin, D., Pananjady, A., Lam, M., Papailiopoulos, D., Ramchandran, K., Bartlett, P.: Gradient diversity: a key ingredient for scalable distributed learning. arXiv preprint arXiv:1706.05699 (2017)
25. Fort, S., Nowak, P.K., Jastrzebski, S., Narayanan, S.: Stiffness: A new perspective on generalization in neural networks. arXiv preprint arXiv:1901.09491 (2019)
26. Fu, J., Liu, P., Zhang, Q., Huang, X.: Rethinking generalization of neural models: A named entity recognition case study. arXiv preprint arXiv:2001.03844 (2020)
27. Mehta, H., Cutkosky, A., Neyshabur, B.: Extreme memorization via scale of initialization. arXiv preprint arXiv:2008.13363 (2020)
28. Negrea, J., Haghifam, M., Dziugaite, G.K., Khisti, A., Roy, D.M.: Information-theoretic generalization bounds for sgld via data-dependent estimates. In: Advances in Neural Information Processing Systems, pp. 11013–11023 (2019)
29. Haghifam, M., Negrea, J., Khisti, A., Roy, D.M., Dziugaite, G.K.: Sharpened generalization bounds based on conditional mutual information and an application to noisy, iterative algorithms. arXiv preprint arXiv:2004.12983 (2020)
30. Qian, X., Klabjan, D.: The impact of the mini-batch size on the variance of gradients in stochastic gradient descent. arXiv preprint arXiv:2004.13146 (2020)
31. Jastrzebski, S., et al.: The break-even point on optimization trajectories of deep neural networks. arXiv preprint arXiv:2002.09572 (2020)
32. Liu, J., Bai, Y., Jiang, G., Chen, T., Wang, H.: Understanding why neural networks generalize well through gsnr of parameters. In: International Conference on Learning Representations (2020). https://openreview.net/forum?id=HyevIJStwH
33. Mahsereci, M., Balles, L., Lassner, C., Hennig, P.: Early stopping without a validation set. arXiv preprint arXiv:1703.09580 (2017)
34. Liu, Y., Starzyk, J.A., Zhu, Z.: Optimized approximation algorithm in neural networks without overfitting. IEEE Trans. Neural Netw. 19(6), 983–995 (2008)
35. Piotrowski, A.P., Napiorkowski, J.J.: A comparison of methods to avoid overfitting in neural networks training in the case of catchment runoff modelling. J. Hydrol. 476, 97–111 (2013)

36. Zhang, X., Wu, D., Xiong, H., Dai, B.: Optimization variance: Exploring generalization properties of dnns (2021). https://openreview.net/forum?id=ZAfeFYKUek5
37. Dziugaite, G.K., Hsu, K., Gharbieh, W., Roy, D.M.: On the role of data in pac-bayes bounds. arXiv preprint arXiv:2006.10929 (2020)
38. Neyshabur, B., Bhojanapalli, S., Srebro, N.: A pac-bayesian approach to spectrally-normalized margin bounds for neural networks. arXiv preprint arXiv:1707.09564 (2017)
39. Srivastava, N., Hinton, G., Krizhevsky, A., Sutskever, I., Salakhutdinov, R.: Dropout: a simple way to prevent neural networks from overfitting. J. Mach. Learn. Res. **15**(1), 1929–1958 (2014)
40. Bellido, I., Fiesler, E.: Do backpropagation trained neural networks have normal weight distributions? In: Gielen, S., Kappen, B. (eds.) ICANN 1993, pp. 772–775. Springer, London (1993). https://doi.org/10.1007/978-1-4471-2063-6_214
41. Stone, M.: Cross-validatory choice and assessment of statistical predictions. J. R. Stat. Soc. Ser. B (Methodol.) **36**(2), 111–133 (1974)
42. Bengio, Y., Grandvalet, Y.: No unbiased estimator of the variance of k-fold cross-validation. J. Mach. Learn. Res. **5**(9), 1089–1105 (2004)
43. Li, M., Soltanolkotabi, M., Oymak, S.: Gradient descent with early stopping is provably robust to label noise for overparameterized neural networks. In: International Conference on Artificial Intelligence and Statistics, pp. 4313–4324. PMLR (2020)
44. Song, H., Kim, M., Park, D., Lee, J.G.: Prestopping: How does early stopping help generalization against label noise? (2020). https://openreview.net/forum?id=BklSwn4tDH
45. Xia, X., et al.: Robust early-learning: hindering the memorization of noisy labels. In: International Conference on Learning Representations (2021). https://openreview.net/forum?id=Eql5b1_hTE4
46. Heckel, R., Yilmaz, F.F.: Early stopping in deep networks: Double descent and how to eliminate it. arXiv preprint arXiv:2007.10099 (2020)

Learning from Noisy Similar and Dissimilar Data

Soham Dan[1]([⊠]), Han Bao[2,3], and Masashi Sugiyama[2,3]

[1] University of Pennsylvania, Philadelphia, USA
sohamdan@seas.upenn.edu
[2] RIKEN Center for Advanced Intelligence Project, Tokyo, Japan
tsutsumi@ms.k.u-tokyo.ac.jp, sugi@k.u-tokyo.ac.jp
[3] The University of Tokyo, Tokyo, Japan

Abstract. With the widespread use of machine learning for classification, it becomes increasingly important to be able to use weaker kinds of supervision for tasks in which it is hard to obtain standard labeled data. One such kind of supervision is provided *pairwise* in the form of Similar (S) pairs (if two examples belong to the same class) and Dissimilar (D) pairs (if two examples belong to different classes). This kind of supervision is realistic in privacy-sensitive domains. Although the basic version of this problem has been studied recently, it is still unclear how to learn from such supervision under *label noise*, which is very common when the supervision is, for instance, crowd-sourced. In this paper, we close this gap and demonstrate how to learn a classifier from noisy S and D labeled pairs. We perform a detailed investigation of this problem under two realistic noise models and propose two algorithms to learn from noisy SD data. We also show important connections between learning from such pairwise supervision data and learning from ordinary class-labeled data. Finally, we perform experiments on synthetic and real-world datasets and show our noise-informed algorithms outperform existing baselines in learning from noisy pairwise data.

Keywords: Classification · Pairwise supervision · Noisy supervision

1 Introduction

In the standard supervised learning framework, a classifier is trained with labeled data points, which are usually collected through human annotation. While collecting labeled data points is the traditional way to apply supervised classification, *pairwise comparison* is often more appealing for human decision making [10], where annotators are requested to compare two instances and give relative relationships between them; e.g., which instance has stronger stimulus,

---

S. Dan—Work done during an internship at RIKEN-AIP.

Electronic supplementary material The online version of this chapter (https://doi.org/10.1007/978-3-030-86520-7_15) contains supplementary material, which is available to authorized users.

N. Oliver et al. (Eds.): ECML PKDD 2021, LNAI 12976, pp. 233–249, 2021.
https://doi.org/10.1007/978-3-030-86520-7_15

whether two instances belong to the same category, and so on. This is partly because (a) decision makers tend to be subjective at directly choosing a single hypothesis,[1] and (b) decision makers are often biased about picking an opinion.[2]

This relative ease of making pairwise comparisons over direct point-wise labeling has inspired several successful large-scale annotation frameworks, for example, crowd-clustering [11, 26]. There are broadly two ways to incorporate pairwise comparisons for identifying the latent classes of data:

(1) Semi-Supervised Clustering based methods [5]:, which utilizes pairwise super-vision indicating whether two instances belong to the same cluster or not (known as must-link and cannot-link constraints), guiding clustering as deci-sion makers desire. This class of methods suffer from dataset-dependent assumptions.

(2) Empirical Risk Minimization (ERM) based methods: [1, 23] which trains an inductive classifier from the pairwise comparisons thereby establishing a con-nection to standard supervised learning in the ERM framework. These meth-ods outperform semi-supervised clustering based methods because it does not make similar assumptions as the latter. In this paper, our primary focus is on the second class of methods, aiming to learn inductive classifiers from pair-wise data, and we shall see in Sect. 5 they outperform the first class of meth-ods empirically. It is important to note that both methods assume that the pairwise comparisons are noise-free, i.e., instances marked similar are indeed from the same class.

While learning from pairwise comparisons has been highly successful [8, 10, 13, 14, 21], it is sensitive to the quality of the annotations. Large-scale frameworks like crowd-clustering are especially prone to noisy annotations [26]. Existing techniques to learn an inductive classifier from noisy labels [12, 15, 19, 20] are inapplicable in this setting, since they only work on pointwise, class-labeled data. Moreover, there lacks a systematic characterization of the kinds of noise that might arise in pairwise-annotated data. We aim to bridge this gap by char-acterizing the two unique types of errors that arise in this setting, as depicted in Fig. 1. The first error results from *pairing corruption*: some pairs of instances are hard to identify whether they belong to the same category or not. The sec-ond error is from *labeling corruption*: labels of some instances are intrinsically ambiguous and thus, subsequent pairwise comparison is also affected. Each of these situations give rise to a specific noise model for pairwise supervision.

In this paper, we thoroughly investigate classification with noisy pairwise supervision, where the noise is present in pairwise comparison and follows either pairing corruption or labeling corruption, and provide two distinct strategies to

---

[1] [24] has studied a relationship between relative comparison and a single hypothesis on stimuli, which is known as the law of comparative judgement.

[2] This bias is known as social desirability bias [9]; questionees are unconsciously led to a socially desirable opinion when they are asked to reveal their opinions in a direct way. Such a tendency is observed especially in answering their sensitive matters such as criminal records.

Pairing Corruption Noise Model   **Labeling Corruption Noise Model**

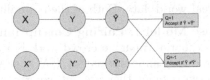

Fig. 1. The two noise models. Q indicates whether the pair is similar ($Q = 1$) or dissimilar ($Q = -1$)

deal with this problem. In the first strategy, we introduce a corrected loss function, which induces an unbiased estimator of the classification risk in the presence of noise for pairwise data. Subsequently, a classifier can be obtained through the minimization of the corrected loss. This extends previous approaches [19, 20] to the pairwise setting. The second strategy is motivated from the insight that the Bayes classifier of the classification risk under the noise-free distribution corresponds to that of the weighted risk under the noisy distribution. This extends cost-sensitive classification [7, 22] to the pairwise setting. Each of these strategies can handle both the pairwise noise models.

We make the following contributions in this paper:

- We provide two distinct, realistic data generating scenarios for SD data: the pairing corruption noise model and the labeling corruption noise model.
- We provide two algorithms based on loss correction and weighted classification which can be applied to either data generating scenario mentioned.
- We theoretically analyze performance bounds of our algorithms and provide two new performance bounds for the noise-free SD learning problem [23].
- We perform extensive experiments on various datasets to show that the proposed algorithms work well in practice and outperform existing methods.

2   Problem Setup

Let $\mathcal{X}$ denote the instance space, $\mathcal{Y} = \{+1, -1\}$ the label space and $\mathcal{Z}$ the underlying distribution over $(\mathcal{X}, \mathcal{Y})$. We want to perform well with respect to $\mathcal{Z}$, i.e., the test data is drawn from $\mathcal{Z}$ and we want to minimize the risk of f (a real valued decision function) w.r.t. the 0-1 loss:

$$R_{\mathcal{Z}}(f) = \mathbb{E}_{(x,y) \sim \mathcal{Z}}[\mathbb{1}_{\{\text{sign}(f(x)) \neq y\}}], \tag{1}$$

where $\mathbb{1}_{\{\cdot\}}$ denotes the indicator function, $f \in \mathcal{F}$ and $\mathcal{F} \subseteq \mathbb{R}^{\mathcal{X}}$ is a hypothesis class. However, we assume that due to the domain constraints we are unable to procure direct class-labeled data from $\mathcal{Z}$ and only have access to pairwise supervision—whether a pair of instances (X, X') is from the same class ($Y = Y'$) or from different classes ($Y \neq Y'$). There is a latent variable Q dictating whether the pair is similar ($Q = 1$) or dissimilar ($Q = -1$). Our training dataset, $\mathcal{D} = \{X_i, X_i', Q_i\}_{i=1}^{n}$, consists of n noisy similar or dissimilar pairs which are generated from one of the following noise models that reflect what we may expect in real-world data. Both of them are illustrated in Fig. 1.

Noise Model 1: Pairing Corruption: This model is motivated by the following scenario—imagine crowd-workers are given a pool of instances drawn from $\mathcal{X}$ and they annotate pairs (X, X') as $Q = 1$ if they believe $Y = Y'$ and $Q = -1$ otherwise. Since they are not experts, they often make mistakes in this process and sometimes assigns $Q = -1$ when it should have been $Q = 1$ and vice versa. Formally, this model comprises of the following steps:

1. Two samples $(X, Y),(X', Y')$ are drawn from the underlying distribution $\mathcal{Z}$.
2. If $Y = Y'$, this pair of samples is labeled as similar ($Q = 1$) with probability $1 - \rho_S$ and if $Y \neq Y'$ this pair is labeled as dissimilar ($Q = -1$) with probability $1 - \rho_D$, where ρ_S and ρ_D are the noise rate for similar (S) and dissimilar (D) data respectively and $0 \leq \rho_S, \rho_D \leq 1$.

In this noise model, the S and D samples are drawn from mixtures of the true S and D distributions: $P(Q = 1|Y = Y') = 1 - \rho_S$ and $P(Q = -1|Y \neq Y') = 1 - \rho_D$. We assume that $\rho_S + \rho_D < 1$, without which it is impossible to learn a classifier.

Noise Model 2: Labeling Corruption: Consider that we are dealing with a privacy sensitive domain where responders do not want to reveal their individual labels. In such cases, responders may intentionally reveal wrong labels. Hence, the pointwise labels that we obtain are intrinsically noisy. A moderator converts the pointwise data (X, Y) to pairwise data $(X, X', Q = \pm 1)$, to preserve privacy. Formally, there are the following steps in this noise model:

1. Two samples $(X, Y),(X', Y')$ are drawn from the underlying distribution $\mathcal{Z}$.
2. Then the labels are flipped with probability $\rho_\pm$ (this is class conditioned: if a sample originally has label $+1$ it is flipped to label -1 with probability ρ_+ and respectively, ρ_- for the other case). Formally, this can be expressed as $P(\tilde{Y} = +1|Y = -1) = \rho_-$ and $P(\tilde{Y} = -1|Y = +1) = \rho_+$.
3. Then in the next step the similar or dissimilar labels are assigned (there is assumed to be no noise in this step since the moderator is an expert).

Thus, $P(Q = 1|\tilde{Y} = \tilde{Y}') = 1$ and $P(Q = -1|\tilde{Y} \neq \tilde{Y}') = 1$. Again, in this noise model we assume $\rho_+ + \rho_- < 1$. In the following, we derive the conditional density functions separately for the noise-free, pairing noise and labeling noise scenarios and show how one can view the pairwise samples as pointwise samples.

Conditional Density Functions in the Noise-Free Case: We consider the one-sample case of the problem of learning from noisy SD labels, which means only one training dataset of SD samples are drawn (as opposed to the two sample case of separate S and D data) from a joint distribution $P(x, x', Q)$. Let us consider the simpler, noise-free scenario first. We can write the joint distribution as:

$$P(x, x', Q) = P(x, x'|Q = 1)P(Q = 1) + P(x, x'|Q = -1)P(Q = -1).$$

If there were no noise corruption, S data would comprise of two positive instances or two negative instances and D data would comprise of one positive instance and one negative instance. Accordingly, we obtain the following conditional distributions for the S and D pairs as follows:

$$P(x, x'|Q = 1) = P(x, x'|y = y' = 1 \lor y = y' = -1),$$
$$= \frac{\pi^2 P_+(x) P_+(x') + (1 - \pi)^2 P_-(x) P_-(x')}{\pi^2 + (1 - \pi)^2},$$
$$P(x, x'|Q = -1) = P(x, x'|y = 1, y' = -1 \lor y = -1, y' = 1),$$
$$= \frac{\pi(1 - \pi)P_+(x)P_-(x') + \pi(1 - \pi)P_-(x)P_+(x')}{2\pi(1 - \pi)}. \tag{2}$$

where $\pi = P(Y = +1)$ denotes the class prior, $P_+(X) = P(X|Y = +1)$ and $P_-(X) = P(X|Y = -1)$, $P(Q = 1) = \pi^2 + (1 - \pi)^2$ and $P(Q = -1) = 2\pi(1-\pi)$. One can further marginalize out x' and get the densities in terms of a single data point x. This view is important for our subsequent analysis and has been previously used in [1]. The implication is that we can now treat the SD data as pointwise data from the Similar (S) and Dissimilar (D) classes.

$$P_S(x) = P(x|Q = 1) = \frac{\pi^2 P_+(x) + (1 - \pi)^2 P_-(x)}{\pi^2 + (1 - \pi)^2},$$
$$P_D(x) = P(x|Q = -1) = \frac{P_+(x) + P_-(x)}{2}. \tag{3}$$

Conditional Density Functions Under Pairwise Noise: The above expressions are derived under the noise-free assumption. Now, we present the expression for the densities, $\tilde{P}_S(x)$ and $\tilde{P}_D(x)$, under each of the noise models presented above. The derivations follow from the graphical model in Fig. 1 and can be found in the Appendix 1.

For the pairing corruption model we get:

$$\tilde{P}_S(x) = (1 - \rho_S)P_S(x) + \rho_D P_D(x), \quad \tilde{P}_D(x) = \rho_S P_S(x) + (1 - \rho_D)P_D(x), \tag{4}$$

and for the labeling corruption model we get,

$$\tilde{P}_S(x) = \frac{(\pi(1 - \rho_+)P_+(x) + (1 - \pi)\rho_- P_-(x))\tilde{\pi}}{\tilde{\pi}^2 + (1 - \tilde{\pi})^2}$$
$$+ \frac{((1 - \pi)(1 - \rho_-)P_-(x) + \pi\rho_+ P_+(x))(1 - \tilde{\pi})}{\tilde{\pi}^2 + (1 - \tilde{\pi})^2},$$
$$\tilde{P}_D(x) = \frac{(\pi\rho_+ P_+(x) + (1 - \pi)\rho_- P_-(x))\tilde{\pi}}{2}$$
$$+ \frac{(\pi(1 - \rho_+)P_-(x) + (1 - \pi)(1 - \rho_-)P_+(x))(1 - \tilde{\pi})}{2}, \tag{5}$$

where $\tilde{\pi} = \pi(1 - \rho_+) + (1 - \pi)\rho_-$.

Let $\mathcal{Z}_Q$ denote the distribution over $(\mathcal{X}, Q)$ where $P(X|Q = 1) = \tilde{P}_S(X)$ and $P(X|Q = -1) = \tilde{P}_D(X)$. Given the marginalized representations in (4) and (5), we can now think of the training data of n pairwise instances (drawn from either of the noise models) to be equivalent to a training data of $2n$ pointwise instances drawn from $\mathcal{Z}_Q$, i.e., $\mathcal{D} \triangleq \{X_i, X_i', Q_i\}_{i=1}^n \equiv \mathcal{D}' \triangleq \{X_i, Q_i\}_{i=1}^{2n} \sim \mathcal{Z}_Q$.

3   Loss Correction Approach

In this section, we present the first of our two proposed algorithms for learning from noisy pairwise data. In this method, we derive an unbiased estimator of the classification risk (Eq. 1) on noisy SD data, by modifying the standard loss function for binary classification. We assume that the noise rates (ρ_S and ρ_D for the pairing corruption model and $\rho_\pm$ for the labeling corruption model) are available beforehand, which are then used to obtain an unbiased estimator of the classification risk. [20] studies a technique for loss correction in the standard classification setting. We adapt this backward-correction technique to handle noisy SD data. A key step necessary to correct the loss is to write the posterior over Q in terms of the posterior over Y.

We now present the posterior SD probabilities in terms of the ordinary class posterior probabilities. The detailed derivation of the following equations can be found in Appendix 2. For the pairing corruption noise model we obtain:

$$
\begin{aligned}
P(Q = 1|X) &= P(Y = 1|X)[(1 - \rho_S)\pi + \rho_D(1 - \pi)] \\
&\quad + P(Y = -1|X)[\rho_D\pi + (1 - \rho_S)(1 - \pi)], \\
P(Q = -1|X) &= P(Y = -1|X)[(1 - \rho_D)\pi + \rho_S(1 - \pi)] \\
&\quad + P(Y = 1|X)[\rho_S\pi + (1 - \rho_D)(1 - \pi)].
\end{aligned}
\tag{6}
$$

On the other hand, for the labeling corruption noise model we obtain:

$$
\begin{aligned}
P(Q = 1|X) &= P(Y = 1|X)[(1 - \rho_+)\tilde{\pi} + \rho_-(1 - \tilde{\pi})] \\
&\quad + P(Y = -1|X)[\rho_-\tilde{\pi} + (1 - \rho_+)(1 - \tilde{\pi})], \\
P(Q = -1|X) &= P(Y = -1|X)[(1 - \rho_-)\tilde{\pi} + \rho_+(1 - \tilde{\pi})] \\
&\quad + P(Y = 1|X)[\rho_+\tilde{\pi} + (1 - \rho_-)(1 - \tilde{\pi})].
\end{aligned}
\tag{7}
$$

Thus, in both noise models we can express:

$P(Q = 1|x) = \alpha_1 P(y = 1|x) + \alpha_2 P(y = -1|x)$ and

$P(Q = -1|x) = \beta_1 P(y = 1|x) + \beta_2 P(y = -1|x)$ for some coefficients $\alpha_1, \alpha_2, \beta_1, \beta_2$.

It is noteworthy that the structure of the posterior probabilities are remarkably similar between the two noise models once we introduce the modified class prior $\tilde{\pi}$, defined in (5). Hence, the labeling corruption noise model can be interpreted as the following two-step data generating process: pointwise labels following the modified class prior $\tilde{\pi}$ are observed first, then pairwise labels are observed via the pairing corruption noise model with noise rates ρ_+ and ρ_-.

Having expressed the posterior probabilities of the noisy SD data in terms of the original class posteriors, we can adopt the technique of backward correction to construct the modified loss function such that, the minimizer of the expected risk with this new loss function over the noisy SD data is the same as the minimizer of the expected risk with the original loss function over $\mathcal{Z}$ (test distribution).

Let $\ell : \mathbb{R} \times \mathcal{Y} \to \mathbb{R}_{\geq 0}$ be a loss function such that $\ell(t, y)$ measures discrepancy between the prediction t and the target label y and let $T = \begin{bmatrix} \alpha_1 & \alpha_2 \\ \beta_1 & \beta_2 \end{bmatrix}$. If $\tilde{\ell}(t) = T^{-1}\ell(t)$ denotes the backward corrected loss, then $\ell(t) = \mathbb{E}[T\tilde{\ell}(t)]$. Note that the expectation is taken w.r.t Q. $\tilde{\ell}(t, Q)$ can be obtained as the first or the second row of $T^{-1}\ell(t)$ corresponding to $Q = +1$ or $Q = -1$ respectively.

Remark 1. Here we have assumed T is invertible which almost always holds in practice. However, if the condition number is large, i.e., T is almost singular, we can mix T (with an appropriate value of λ) with the matrix T' which corresponds to the noise-free SD case (obtained by setting noise rates to 0 in (6), (7); refer to (13) below) before inverting it. This is essentially including a noise-free prior.

$$T \leftarrow T + \lambda T', \quad \text{where} \quad T' = \begin{bmatrix} \pi & 1 - \pi \\ 1 - \pi & \pi \end{bmatrix}$$

The following is the empirical $\tilde{\ell}$-risk on the observed pairwise training data $\mathcal{D}$ of n instances.

$$\hat{R}_{\tilde{\ell}}(f) = \frac{1}{n} \sum_{i=1}^{n} \tilde{\ell}(g(X_i, X_i'), Q_i), \tag{8}$$

where, $g \in \mathcal{G}$ is a real-valued decision function and $\mathcal{G} \subseteq \mathbb{R}^{\mathcal{X}} \times \mathbb{R}^{\mathcal{X}}$. Further, for ease of analysis we always deal with the pointwise view of the training dataset $\mathcal{D}'$ with $2n$ instances (refer to Sect. 2 above for more details). In this view, the empirical $\tilde{\ell}$-risk on the observed pointwise training data $\mathcal{D}'$ of $2n$ instances is:

$$\hat{R}_{\tilde{\ell}}(f) = \frac{1}{2n} \sum_{i=1}^{2n} \tilde{\ell}(f(X_i), Q_i), \tag{9}$$

where $f \in \mathcal{F}$ is a real-valued decision function ,$\mathcal{F} \subseteq \mathbb{R}^{\mathcal{X}}$ is a hypothesis class. We can use the corrected loss to train our classifier f on the noisy SD data directly by empirical risk minimization of (9).

Performance Bounds: Now we discuss the performance bounds for this approach. The following are some important notations used in the following results:

- $\hat{f} = \arg\min_{f \in \mathcal{F}} \hat{R}_{\tilde{\ell}}(f)$,
- $R_{\tilde{\ell}, \mathcal{Z}_Q}(f) = E_{(X,Q) \sim \mathcal{Z}_Q}[\tilde{\ell}(f(X), Q)]$,
- $R_{\ell, \mathcal{Z}}(f) = E_{(X,Y) \sim \mathcal{Z}}[\ell(f(X), Y)]$.

The empirical estimate of the risk is unbiased to $R_{\ell, \mathcal{Z}}(f)$ because $\tilde{\ell}$ is corrected appropriately. By performing ERM on the noisy SD data using the corrected loss, the empirical risk converges to the true risk on the standard class-labeled Positive(P)-Negative(N) data drawn from the underlying distribution $\mathcal{Z}$. Let L_Q be the Lipschitz constant of the loss $\tilde{\ell}$ in its first argument. Note that $L_Q \leq \max\{\alpha_1, \alpha_2, \beta_1, \beta_2\}L$. Let $\mathcal{R}(\mathcal{F}, 2n)$ be the Rademacher complexity [4] of the function class $\mathcal{F}$ for the $2n$ noisy SD instances.

Further, for the following theorem we also need the notion of classification calibrated surrogate losses. If a surrogate loss $\ell(\cdot, \cdot)$ is calibrated, then convergence of the surrogate excess risk $R_{\ell, \mathcal{Z}}(f) - \min_f R_{\ell, \mathcal{Z}}(f)$ to zero implies convergence of the target excess risk $R_{\mathcal{Z}}(f) - \min_f R_{\mathcal{Z}}(f)$ to zero. For more details refer to [3].

Lemma 1. *For any $\delta > 0$, we have the following: with probability at least $1 - \delta$,*

$$\max_{f \in \mathcal{F}} |\hat{R}_{\tilde{\ell}}(f) - R_{\tilde{\ell}, \mathcal{Z}_Q}(f)| \leq 2L_Q \mathcal{R}(\mathcal{F}, 2n) + \sqrt{\frac{\log(1/\delta)}{4n}}. \tag{10}$$

The proof of Lemma 1 can be found in Appendix 3. We see that the generalization error of any f w.r.t. $\mathcal{Z}_Q$ vanishes asymptotically if $\mathcal{R}(\mathcal{F}, 2n)$ is moderately controlled as is the case for linear-in-parameter models [18].

Theorem 1. *For any $\delta > 0$, we have the following: with probability at least $1 - \delta$,*

$$R_{\ell, \mathcal{Z}}(\hat{f}) \leq \min_{f \in \mathcal{F}} R_{\ell, \mathcal{Z}}(f) + 4L_Q \mathcal{R}(\mathcal{F}, 2n) + 2\sqrt{\frac{\log(1/\delta)}{4n}}. \tag{11}$$

If ℓ is classification calibrated [3], there exists a non-decreasing function ξ_ℓ with $\xi_\ell(0) = 0$ such that,

$$R_{\mathcal{Z}}(\hat{f}) - R^* \leq \xi_\ell^{-1} \left(\min_{f \in \mathcal{F}} R_{l, \mathcal{Z}}(f) - \min_f R_{l, \mathcal{Z}}(f) + 4L_Q \mathcal{R}(\mathcal{F}, 2n) + 2\sqrt{\frac{\log(1/\delta)}{4n}} \right). \tag{12}$$

Detailed proof is available in Appendix 4.

We see that the estimation error of $\hat{f}$ vanishes asymptotically if $\mathcal{R}(\mathcal{F}, 2n)$ is moderately controlled as is the case for linear-in-parameter models [18].

Special Case of Noise-Free SD Learning: When there is no noise $\rho_S = \rho_D = 0$ or $\rho_+ = \rho_- = 0$ and $\pi \neq 0.5$, for both the noise models,

$$T = \begin{bmatrix} \pi & 1 - \pi \\ 1 - \pi & \pi \end{bmatrix} \implies T^{-1} = \begin{bmatrix} \frac{\pi}{(2\pi-1)} & \frac{-(1-\pi)}{(2\pi-1)} \\ \frac{-(1-\pi)}{(2\pi-1)} & \frac{\pi}{(2\pi-1)} \end{bmatrix}. \tag{13}$$

We see this matches the loss function derived for noise-free SD learning in [23] (on setting X' as X and replacing $\pi_S E_{X_S}[1] = \pi_D E_{X_D}[1] = \frac{1}{2n}$):

$$\hat{R}_{\tilde{\ell}}(f) = \frac{1}{2n} \sum_{i=1}^{2n} [\mathcal{L}(f(X_i), Q_i)], \quad \text{where} \quad \mathcal{L}(z, t) = \frac{\pi}{2\pi - 1} \ell(z, t) - \frac{1 - \pi}{2\pi - 1} \ell(z, -t). \tag{14}$$

Our analysis through the lens of loss correction provides an estimation error bound for noise-free SD learning as a special case of noisy SD learning. The Lipschitz constant for the corrected loss in the noise-free SD case is $L_Q = \frac{L}{|2\pi - 1|}$, where L is the Lipschitz constant for ℓ. An estimation error bound for noise-free SD learning is:

Corollary 1. *For any $\delta > 0$, we have the following: with probability at least $1 - \delta$,*

$$R_{\ell,z}(\hat{f}) \leq \min_{f \in \mathcal{F}} R_{\ell,z}(f) + \frac{4L}{|2\pi - 1|}\mathcal{R}(\mathcal{F}, 2n) + 2\sqrt{\frac{\log(\frac{1}{\delta})}{4n}}. \tag{15}$$

$$R_z(\hat{f}) - R^* \leq \xi_\ell^{-1}\left(\min_{f \in \mathcal{F}} R_{l,z}(f) - \min_f R_{l,z}(f) + \frac{4L\mathcal{R}(\mathcal{F}, 2n)}{|2\pi - 1|} + 2\sqrt{\frac{\log(\frac{1}{\delta})}{4n}}\right). \tag{16}$$

This is directly obtained from Theorem 1 by plugging in the value of L_Q.

Optimization: While we have a performance guarantee, efficient optimization is a concern especially because the corrected loss $\tilde{\ell}(\cdot, \cdot)$ may not be convex. We present a condition which will guarantee the corrected loss to be convex.

Theorem 2. *If $\ell(t, y)$ is convex and twice differentiable almost everywhere in t (for each y) and also satisfies:*

- *$\forall t \in \mathbb{R}$, $\ell'(t, y) = \ell'(t, -y)$, where the differentiation is w.r.t. t.*
- *$\mathrm{sign}(\alpha_1 - \beta_1) = \mathrm{sign}(\beta_2 - \alpha_2)$, where $\alpha_1, \alpha_2, \beta_1, \beta_2$ are elements of T.*

then $\tilde{\ell}(t, y)$ is convex in t.

Proof of Theorem 2 is available in Appendix 5.

The first condition is satisfied by several common losses such as squared loss $\ell(t, y) = (1 - ty)^2$ and logistic loss $\ell(t, y) = \log(1 + \exp(-ty))$. The second condition depends on the noise rates and the class prior. We can simplify this for the pairing corruption noise model as:

$$\frac{1 - 2\rho_S}{1 - 2\rho_D} \in \left[\frac{1 - \pi}{\pi}, \frac{\pi}{1 - \pi}\right] \quad \text{if} \quad \pi \geq 0.5 \ ,$$
$$\frac{1 - 2\rho_S}{1 - 2\rho_D} \in \left[\frac{\pi}{1 - \pi}, \frac{1 - \pi}{\pi}\right] \quad \text{if} \quad \pi \leq 0.5 \ . \tag{17}$$

In the case of the labeling corruption noise model, this condition reduces to

$$\frac{1 - 2\rho_+}{1 - 2\rho_-} \in \left[\frac{1 - \tilde{\pi}}{\tilde{\pi}}, \frac{\tilde{\pi}}{1 - \tilde{\pi}}\right] \quad \text{if} \quad \tilde{\pi} \geq 0.5 \ ,$$
$$\frac{1 - 2\rho_+}{1 - 2\rho_-} \in \left[\frac{\tilde{\pi}}{1 - \tilde{\pi}}, \frac{1 - \tilde{\pi}}{\tilde{\pi}}\right] \quad \text{if} \quad \tilde{\pi} \leq 0.5 \ . \tag{18}$$

For all cases of noise-free or symmetric-noise ($\rho_S = \rho_D$ or $\rho_+ = \rho_-$) SD learning, any noise rates will satisfy this condition and thus, we can always perform efficient optimization. For cases where the above condition is not satisfied, i.e., $\tilde{l}$ is not guaranteed to be convex, this is often not a problem in practice since, neural networks optimized by stochastic gradient descent (our setup in Sect. 5) converges efficiently to a globally optimal solution, under certain conditions [6].

4   Weighted Classification Approach

Now we develop our second algorithm for dealing with noisy S and D data. One key issue that we investigate here is how the Bayes classifier learned from noisy SD data relates to the traditional Bayes classifier.

Lemma 2. *Denote the modified posterior under a SD noise model as* $P(Q = 1|x) = \eta_Q(x)$ *and* $P(Y = 1|x) = \eta(x)$. *Then the Bayes classifier under the noisy SD distribution* $\tilde{f}^* = \arg\min_{f \in \mathcal{F}} \mathbb{E}_{(X,Q) \sim \mathcal{Z}_Q} [1_{\{\text{sign}(f(X)) \neq Q\}}]$ *is given by*

$$\tilde{f}^*(x) = \text{sign}\left(\eta_Q(x) - \frac{1}{2}\right) = \text{sign}\left(\eta(x) - \tau\right), \tag{19}$$

where, τ *depends on the noise model, noise rates and* π *and is presented below. For the pairwise corruption case, assuming* $\pi \neq 0.5$,

$$\tau = \frac{\frac{1}{2} - [(1 - \rho_S)(1 - \pi) + \rho_D \pi]}{(1 - \rho_S - \rho_D)(2\pi - 1)}.$$

For the label corruption case, threshold τ *is:*

$$\tau = \frac{\frac{1}{2} - \pi(\rho_+ + \rho_- - \rho_+\rho_-) - (1 - \pi)(\rho_+^2 + (1 - \rho_-)^2)}{(1 - \rho_+ - \rho_-)[\pi(1 - 2\rho_+) - (1 - \pi)(1 - 2\rho_-)]},$$

$$= \frac{\frac{1}{2} - \pi(\rho_+ + \rho_- - \rho_+\rho_-) - (1 - \pi)(\rho_+^2 + (1 - \rho_-)^2)}{(1 - \rho_+ - \rho_-)(2\tilde{\pi} - 1)}$$

assuming $\tilde{\pi} \neq 0.5$ *where* $\tilde{\pi}$ *is defined in* (5).

These expressions can be derived by using (6) and (7) in (19) and the detailed proof of Lemma 2 is available in Appendix 6. They give us an important insight:

Remark 2. The Bayes classifier for noisy SD learning uses a different threshold from $\frac{1}{2}$ while the traditional Bayes classifier has $\eta(x)$ thresholded at $\frac{1}{2}$.

Towards designing an algorithm we note that we can also obtain this Bayes classifier by minimizing the weighted 0-1 risk defined as follows:

$$U_\alpha(t, y) = (1 - \alpha)1_{\{y=1\}}1_{\{t \leq 0\}} + \alpha 1_{\{y=-1\}}1_{\{t>0\}}.$$

The following lemma from [22] is crucial in connecting the Bayes classifier threshold with the weight α in weighted 0-1 classification.

Lemma 3. *[22]: Denote the* U_α *risk under distribution* $\mathcal{Z}$ *as*

$$R_{\alpha, \mathcal{Z}}(f) = E_{(x,y) \sim \mathcal{Z}}[U_\alpha(f(x), y)].$$

Then $f_\alpha^*(x) = \text{sign}(\eta(x) - \alpha)$ *minimizes* $R_{\alpha, \mathcal{Z}}(f)$.

We now show that there exists a choice of weight α such that the weighted risk under the noisy SD distribution is linearly related to the ordinary risk under distribution $\mathcal{Z}$.

Theorem 3. *There exist constants α and A and a function $B(X)$ that only depends on X but not on f, such that*

$$R_{\alpha, \mathcal{Z}_Q}(f) = AR_{\mathcal{Z}}(f) + E_X[B(X)].$$

For the pairing corruption case:

$$\alpha = \frac{1 - \rho_S + \rho_D}{2}, \quad A = \frac{1 - \rho_S - \rho_D}{2}(2\pi - 1). \tag{20}$$

For the label corruption case:

$$\alpha = \pi(1 - \rho_+ + \rho_+^2 - \rho_+\rho_-) - \frac{1}{2}(1 - \rho_+ - \rho_-) + (1 - \pi)(1 - \rho_- + \rho_-^2 - \rho_+\rho_-),$$

$$A = \frac{(1 - \rho_+ - \rho_-)}{2}[\pi(1 - 2\rho_+) - (1 - \pi)(1 - 2\rho_-)]. \tag{21}$$

Proof of Theorem 3 is available in Appendix 7.

Remark 3. The α-weighted Bayes optimal classifier under the noisy SD distribution coincides with the Bayes classifier of the 0-1 loss under the standard distribution $\mathcal{Z}$.

$$\arg\min_f R_{\alpha*, \mathcal{Z}_Q}(f) = \arg\min_f R_{\mathcal{Z}}(f) = \text{sign}\left(\eta(x) - \frac{1}{2}\right).$$

Performance Bounds and Optimization: For the ease of optimization, we will use a surrogate loss instead of the 0-1 loss to perform weighted ERM. Any surrogate loss can be used as long as it can be decomposed as $\ell(t, Q) = 1_{\{Q=1\}}\ell_1(t) + 1_{\{Q=-1\}}\ell_{-1}(t)$, for partial losses ℓ_1, ℓ_{-1} of ℓ [3,22]. The margin-based surrogate loss functions ℓ such that $\ell(t, Q) = 1_{\{Q=1\}}\phi(t) + 1_{\{Q=-1\}}\phi(-t)$ for some $\phi : \mathbb{R} \to \mathbb{R}_{\geq 0}$ [18] is expressible in this form. The commonly used surrogate losses such as the squared, hinge, and logistic losses are encompassed in the margin-based surrogate loss. We want to minimize the following empirical risk using the weighted surrogate loss l_α:

$$\min_{g \in \mathcal{G}} \frac{1}{n} \sum_{i=1}^{n} \ell_\alpha(g(X_i, X_i'), Q_i). \tag{22}$$

Similar to (9) we consider the pointwise version of the empirical risk using the weighted surrogate loss l_α:

$$\min_{f \in \mathcal{F}} \frac{1}{2n} \sum_{i=1}^{2n} \ell_\alpha(f(X_i), Q_i), \tag{23}$$

and let $\hat{f}_\alpha$ denote the minimizer of (23).

$$\hat{f}_\alpha = \arg\min_{f \in \mathcal{F}} \frac{1}{n} \sum_{i=1}^{n} \ell_\alpha(f(x_i), Q_i). \tag{24}$$

We already discussed classification calibration in Sect. 3. For the following theorem we need the notion of α-classification calibrated losses, developed in [22], that extends [3] to the asymmetric classification setting where the misclassification costs are unequal for the two classes.

Theorem 4. *If ℓ_α is an α-weighted margin loss [22] of the form: $l_\alpha(t, Q) = (1 - \alpha)1_{\{Q=1\}}\ell(t) + \alpha 1_{\{Q=-1\}}\ell(-t)$ and ℓ is convex, classification calibrated ($\ell'(0) < 0$ where the derivative is w.r.t. t) and L-Lipschitz, then for the choices of α and A in (20), (21) (assuming $\pi \neq 0.5$ or $\hat{\pi} \neq 0.5$ for the corresponding noise model), there exists a non-decreasing function ξ_{ℓ_α} with $\xi_{\ell_\alpha}(0) = 0$ such that the following bound holds with probability at least $1 - \delta$:*

$$R_{\mathcal{Z}}(\hat{f}_\alpha) - R^* \leq A^{-1}\xi_{\ell_\alpha}\left(\min_{f \in \mathcal{F}} R_{\alpha, z_Q}(f) - \min_f R_{\alpha, z_Q}(f) + 4LR(\mathcal{F}, n) + 2\sqrt{\frac{\log(\frac{1}{\delta})}{2n}}\right), \tag{25}$$

where R^* denotes the corresponding Bayes risk under $\mathcal{Z}$.

Note that using Corollary 4.1 from [22] we know l_α is α-classification calibrated. The right-side in (25) is finite because $A \neq 0$ whenever $\pi, \hat{\pi} \neq 0.5$. Proof of Theorem 4 is available in Appendix 8.

Remark 4. For a fixed Lipschitz constant L, as A decreases we get a weaker excess risk bound. For the pairing corruption noise model, its easy to see that as noise rates increase, A decreases. On the other hand, the relationship is more complicated for the labeling corruption noise model. When the noise is symmetric $(\rho_+ = \rho_- = \rho)$, $A = \frac{(1-2\rho)^2(2\pi-1)}{2}$. In this case, again we observe as ρ increases, A decreases and we get a weaker bound.

Remark 5. When $\rho_S = \rho_D$ or $\rho_+ = \rho_-$, we see that the optimal Bayes classifier for the (noisy) SD learning problem is the same as the Bayes classifier for the standard class-labeled binary classification task under distribution $\mathcal{Z}$. In these settings, this result allows us to learn a classifier for standard class-labeled binary classification from (noisy) SD data simply by treating the similar and dissimilar classes as the positive and negative class for any chosen classifier.

Estimation of Prior and Noise Parameters: We briefly discuss the parameters (the class prior π and the noise rates ρ_S and ρ_D in the pairing corruption noise model and $\rho_\pm$ in the labeling corruption noise model) that we need to know or estimate to apply each method for each noise model.

(I) Loss Correction Approach: The noise rate parameters can be tuned by cross-validation on the noisy SD data. We also need to estimate the class prior to construct the loss correction matrix T, under both noise models. Let n_S be the number of similar pairs and n_D be the number of dissimilar pairs in the training dataset. The class prior π can be estimated from the following equations:

- For the pairing corruption noise model:

$$\frac{n_S}{n_D} \approx \frac{(1-\rho_S)(\pi^2 + (1-\pi)^2) + 2\rho_D\pi(1-\pi)}{\rho_S(\pi^2 + (1-\pi)^2) + 2(1-\rho_D)\pi(1-\pi)} \qquad (26)$$

- For the labeling corruption noise model:

$$\frac{n_S}{n_D} \approx \frac{(1-\rho_+)(\pi\tilde{\pi} + (1-\pi)(1-\tilde{\pi})) + \rho_-(\pi(1-\tilde{\pi}) + \tilde{\pi}(1-\pi))}{\rho_+(\pi\tilde{\pi} + (1-\pi)(1-\tilde{\pi})) + (1-\rho_-)(\pi(1-\tilde{\pi}) + \tilde{\pi}(1-\pi))} \qquad (27)$$

From each of the above equations we can obtain an estimate $\hat{\pi}$ of the class prior π. The above equations can be derived from (6) and (7) by marginalizing out X and using $\frac{n_S}{n_D} \approx \frac{P(Q=1)}{P(Q=-1)}$ (equality holds for the population).

(II) Weighted Classification Approach: The class prior only appears in the weight α in the labeling corruption model. In the pairing corruption model, knowledge of the class prior is not needed to calculate α. However, since we just have one parameter α for the optimization problem, in practice we can obtain α directly by cross-validation under both noise models. Note that if we are given the noise rates, in the pairing corruption noise model we can calculate the optimum α exactly but in the labeling corruption noise model we still get only an estimate of the optimum α since, $\hat{\pi} \approx \pi$.

5 Experiments

We empirically verify that the proposed algorithms are able to learn a classifier for the underlying distribution $\mathcal{Z}$ from only noisy similar and dissimilar training data. All experiments are repeated 3 times on random train-test splits of 75:25 and the average accuracies are shown. We conduct experiments on two noise models independently. In the learning phase, the noise parameters and the weight α is tuned by cross-validation for the Loss Correction Approach and the Weighted Classification Approach respectively, for both noise models, by searching in $[0, 0.5]$ in increments of 0.1. Evaluation is done on the standard class-labeled test dataset using standard classification accuracy (Eq. 1) as evaluation metric which is averaged over the test datasets to reduce variance across the corruption in the training data. We use a multi-layer perceptron (MLP) with two hidden layers of 100 neurons, ReLU activation and a single logistic sigmoid output, as our model architecture for all experiments trained using the squared loss: $\ell(t, y) = (t - y)^2$. We use stochastic gradient descent with momentum of 0.9 with a mini-batch size of 32 and a learning rate of 0.001, for 500 epochs.

Synthetic Data: We use a non-separable benchmark "banana" dataset which has two dimensional attributes and two classes. We perform two kinds of experiments. In the first experiment, for a given noise model, for different settings

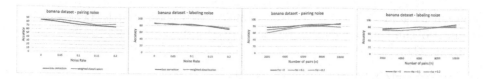

Fig. 2. The left two images depict the gradual decrease in classification accuracy of the learned classifier (from either algorithm) as the noise rate in the noisy SD training data increases, for each noise model. The right two images depict the increase in classification accuracy of the learned classifier (from the weighted classification method) as the number of noisy SD training samples increases, for different noise rates in each noise model. The accuracy achieved by training on standard P-N training data provided by the banana dataset is 90.8%.

of symmetric noise parameters ($\rho_S = \rho_D$ and $\rho_+ = \rho_-$) we plot the variation of standard test accuracy with the number of noisy SD pairs (n) sampled for training. For this experiment setting, we show the results for the weighted classification algorithm in Fig. 2. Since the Bayes classifier under the symmetric noise is identical to that of noise-free case under both the noise models (see Remark 4), we see that the accuracy improves as we get a better approximation of the Bayes classifier as we have more SD data-points in training. Note that the number of original training points in the dataset is fixed—what changes is only the number of SD points we sample from them. In the second experiment, for each noise model, for a fixed n we show the gradual degradation of performance of the proposed algorithms (loss correction approach as well as the weighted classification approach) with increasing symmetric noise rates. These experiments confirm that higher noise hurts accuracy and more pairwise samples helps it.

Real World Datasets. We further conduct experiments on several benchmark datasets from the UCI classification tasks.[3] All tasks are binary classification tasks of varying dimensions, class priors, and sample sizes. We compare the performance of our proposed approaches against two kinds of baselines.

(A) Supervised Baselines: The state-of-the-art algorithm [23] for learning from pairwise similar-dissimilar data is used and this provides a strong baseline for the loss-correction approach. We also compare the performance of the weighted classification approach thresholded at $\frac{1}{2}$, i.e., under the noise-free assumption. While these baselines have been proved to perform very well in the noise-free scenario (both theoretically and empirically), here we investigate if they are robust to noisy annotations, for varying noise rates.

(B) Unsupervised Baselines: We also compare against unsupervised clustering and semi-supervised clustering based methods. For unsupervised clustering, pairwise information is ignored KMeans [16] is applied with $K = 2$ clusters, directly on the noisy SD datapoints and the obtained clusters are used to classify the test data. We also use constrained KMeans clustering [25], where we treat the

---

[3] Available at https://archive.ics.uci.edu/ml/datasets.php.

Table 1. PAIRING NOISE: The best P-N column denotes the test accuracy after training on the standard class-labeled train dataset provided. d, π, N denote the feature dimension, class prior and the size of the entire class-labeled data respectively. Clean S-D denotes test accuracy after training on noise-free S-D data generated from the train dataset. T-Loss indicates the test accuracy after training on S-D data with the loss correction approach (by the matrix T) and SD-Loss denotes the non-corrected variant of [23]. Similarly, weighted and unweighted denotes the test accuracy after training on S-D data using weighted ERM and normal ERM respectively—note, they are identical for symmetric noise. KM denotes the KMeans baseline and KM-COP is the KMeans with constraints. Accuracies within 1% of the best in each row are bolded.

Dataset (d, π) N	best P-N	clean S-D	Noise Rates (ρ_S, ρ_D)	T-Loss	SD-Loss	Weighted	Unweighted	KM	KM-COP
DIABETES	77	77	$(0.2, 0.2)$	**76.57**	75	74.95	74.95	65.63	64.58
(8,0.35)			$(0.1, 0.2)$	74.95	75.52	**77.61**	76.04		65.10
768			$(0.3, 0.3)$	**75.52**	73.95	74.48	74.48		64.06
ADULT	83.09	83.03	$(0.2, 0.2)$	**82.49**	**82.22**	**82.35**	**82.35**	71.25	71.25
(106,0.24)			$(0.1, 0.2)$	77.8	76.26	**82.41**	75.92		71.25
48842			$(0.3, 0.3)$	**81.42**	80.26	**81.10**	**81.10**		53.14
CANCER	97.2	97.2	$(0.2, 0.2)$	**97.18**	**96.47**	95.78	95.78	88.7	**92.95**
(30,0.37)			$(0.1, 0.2)$	**97.18**	**97.18**	95.78	95.78		91.54
569			$(0.3, 0.3)$	**97.18**	95.07	95.78	95.78		92.25

Table 2. LABELING NOISE: The setup is same as Table 1 but now we use the labeling corruption noise model to generate the noisy S-D data.

Dataset (d, π) N	best P-N	clean S-D	Noise Rates (ρ_+, ρ_-)	T-Loss	SD-Loss	Weighted	Unweighted	KM	KM-COP
IONOSPHERE	90.91	90.91	$(0.2, 0.2)$	**88.67**	85.23	86.4	86.4	70.45	71.59
(34,0.64)			$(0.1, 0.2)$	85.24	80.68	**90.91**	85.23		71.59
351			$(0.3, 0.3)$	87.5	80.7	**88.64**	**88.64**		71.59
SPAMBASE	91.83	89.74	$(0.2, 0.2)$	**87.56**	83.22	82.78	82.78	78.08	78.96
(57,0.39)			$(0.1, 0.2)$	83.74	84.15	**86.78**	85.56		79.13
4601			$(0.3, 0.3)$	**85.304**	75.65	78.44	78.44		78.61
MAGIC	84.12	83.40	$(0.2, 0.2)$	80.06	81.13	**82.21**	**82.21**	59.09	63.28
(10,0.65)			$(0.2, 0.1)$	73.27	73.61	**81.67**	79.70		66.03
19020			$(0.3, 0.3)$	**79.39**	78.42	**79.50**	**79.50**		62.34

SD pairs as *must-link* and *cannot-link* constraints to supervise the clustering of the SD data pooled together. While constrained clustering is a strong baseline for pairwise learning [23], here we investigate if it is robust to noisy annotations.

In Tables 1 and 2, we show the performance of our proposed algorithms versus the baselines. We observe that for both noise models and for almost all noise rates, our proposed approaches *significantly* outperform the baselines. We also observe, that as the noise rates increase, performance degrades for all the methods. Further, we see that the noise-free SD performances match the best P-N performance which empirically verifies the optimal classifiers for learning from noise-free standard P-N and pairwise SD data coincide. The complete set of experiments along with additional details are provided in Appendix 9.

6 Conclusion and Future Work

In this paper we theoretically investigated a novel setting that is commonly encountered in several applications—learning from noisy pairwise labels and studied it under two distinct noise models. We showed the connections of this problem to standard class-labeled binary classification, proposed two algorithms and derived their performance bounds. We empirically showed that they outperform state-of-the-art supervised and unsupervised baselines and are able to handle severe noise corruption. For future work, it is worthwhile to investigate more complicated noise models such as instance-dependent noise [17] in this setting.

Acknowledgement. HB was supported by JSPS KAKENHI Grant Number 19J21094. MS was supported by JST AIP Acceleration Research Grant Number JPMJCR20U3, Japan. SD was an intern at RIKEN-AIP.

References

1. Bao, H., Niu, G., Sugiyama, M.: Classification from pairwise similarity and unlabeled data. In: International Conference on Machine Learning, pp. 461–470 (2018)
2. Bartlett, P.L., Bousquet, O., Mendelson, S., et al.: Local rademacher complexities. Ann. Stat. **33**(4), 1497–1537 (2005)
3. Bartlett, P.L., Jordan, M.I., McAuliffe, J.D.: Convexity, classification, and risk bounds. J. Am. Stat. Assoc. **101**(473), 138–156 (2006)
4. Bartlett, P.L., Mendelson, S.: Rademacher and gaussian complexities: risk bounds and structural results. J. Mach. Learn. Res. **3**(Nov), 463–482 (2002)
5. Basu, S., Davidson, I., Wagstaff, K.: Constrained Clustering: Advances in Algorithms, Theory, and Applications. CRC Press, Boca Raton (2008)
6. Du, S.S., Zhai, X., Poczos, B., Singh, A.: Gradient descent provably optimizes over-parameterized neural networks. arXiv preprint arXiv:1810.02054 (2018)
7. Elkan, C.: The foundations of cost-sensitive learning. In: International Joint Conference on Artificial Intelligence, vol. 17, pp. 973–978 (2001)
8. Eric, B., Freitas, N.D., Ghosh, A.: Active preference learning with discrete choice data. In: Advances in Neural Information Processing Systems, pp. 409–416 (2008)
9. Fisher, R.J.: Social desirability bias and the validity of indirect questioning. J. Consum. Res. **20**(2), 303–315 (1993)
10. Fürnkranz, J., Hüllermeier, E.: Preference Learning. Springer, Heidelberg (2010). https://doi.org/10.1007/978-3-642-14125-6
11. Gomes, R., Welinder, P., Krause, A., Perona, P.: Crowdclustering. In: NIPS (2011)
12. Han, B., et al.: Co-teaching: robust training of deep neural networks with extremely noisy labels. In: Advances in Neural Information Processing Systems, pp. 8527–8537 (2018)
13. Hsu, Y.C., Lv, Z., Schlosser, J., Odom, P., Kira, Z.: Multiclass classification without multiclass labels. In: International Conference on Learning Representations (2018)
14. Jamieson, K.G., Nowak, R.: Active ranking using pairwise comparisons. In: Advances in Neural Information Processing Systems, pp. 2240–2248 (2011)
15. Jiang, L., Zhou, Z., Leung, T., Li, L.J., Fei-Fei, L.: Mentornet: regularizing very deep neural networks on corrupted labels. arXiv preprint arXiv:1712.05055 (2017)

16. MacQueen, J., et al.: Some methods for classification and analysis of multivariate observations. In: Proceedings of the Fifth Berkeley Symposium on Mathematical Statistics and Probability, Oakland, CA, USA, vol. 1, pp. 281–297 (1967)
17. Menon, A.K., Van Rooyen, B., Natarajan, N.: Learning from binary labels with instance-dependent corruption. arXiv preprint arXiv:1605.00751 (2016)
18. Mohri, M., Rostamizadeh, A., Talwalkar, A.: Foundations of Machine Learning. MIT Press, Cambridge (2018)
19. Natarajan, N., Dhillon, I.S., Ravikumar, P.K., Tewari, A.: Learning with noisy labels. In: Advances in Neural Information Processing Systems, pp. 1196–1204 (2013)
20. Patrini, G., Rozza, A., Krishna Menon, A., Nock, R., Qu, L.: Making deep neural networks robust to label noise: a loss correction approach. In: Proceedings of the IEEE Conference on Computer Vision and Pattern Recognition (2017)
21. Saaty, T.L.: Decision Making for Leaders: The Analytic Hierarchy Process for Decisions in a Complex World. RWS Publications (1990)
22. Scott, C., et al.: Calibrated asymmetric surrogate losses. Electron. J. Stat. 6, 958–992 (2012)
23. Shimada, T., Bao, H., Sato, I., Sugiyama, M.: Classification from pairwise similarities/dissimilarities and unlabeled data via empirical risk minimization. arXiv preprint arXiv:1904.11717 (2019)
24. Thurstone, L.L.: A law of comparative judgment. Psychol. Rev. 34(4) (1927)
25. Wagstaff, K., Cardie, C., Rogers, S., Schrödl, S., et al.: Constrained k-means clustering with background knowledge. In: ICML, vol. 1, pp. 577–584 (2001)
26. Yi, J., Jin, R., Jain, A.K., Jain, S.: Crowdclustering with sparse pairwise labels: a matrix completion approach. In: HCOMP@ AAAI. Citeseer (2012)

Knowledge Distillation with Distribution Mismatch

Dang Nguyen[✉], Sunil Gupta, Trong Nguyen, Santu Rana, Phuoc Nguyen, Truyen Tran, Ky Le, Shannon Ryan, and Svetha Venkatesh

Applied Artificial Intelligence Institute (A^{2}I^2), Deakin University, Geelong, Australia
{d.nguyen,sunil.gupta,trong.nguyen,santu.rana,phuoc.nguyen,
truyen.tran,k.le,shannon.ryan,svetha.venkatesh}@deakin.edu.au

Abstract. Knowledge distillation (KD) is one of the most efficient methods to compress a large deep neural network (called *teacher*) to a smaller network (called *student*). Current state-of-the-art KD methods assume that the distributions of training data of teacher and student are identical to maintain the student's accuracy close to the teacher's accuracy. However, this strong assumption is not met in many real-world applications where the distribution mismatch happens between teacher's training data and student's training data. As a result, existing KD methods often fail in this case. To overcome this problem, we propose a novel method for KD process, which is still effective when the distribution mismatch happens. We first learn a distribution based on student's training data, from which we can sample images well-classified by the teacher. By doing this, we can discover the data space where the teacher has good knowledge to transfer to the student. We then propose a new loss function to train the student network, which achieves better accuracy than the standard KD loss function. We conduct extensive experiments to demonstrate that our method works well for KD tasks with or without distribution mismatch. To the best of our knowledge, our method is the first method addressing the challenge of distribution mismatch when performing KD process.

Keywords: Knowledge distillation · Model compression · Distribution mismatch · Distribution shift · Mismatched teacher

1 Introduction

Recently, deep learning has become one of the most successful machine learning techniques [16], and it has been applied widely to many real-world applications including face recognition [8], security systems [21], disease detection [18], recommended systems [25], etc. Deep neural networks (the main component of deep learning) often have millions of parameters (aka weights) to train, thus require heavy computation and storage, which can only be executed on powerful servers. This characteristic renders deep networks inapplicable to many real-time devices, especially for those edge devices with limited resources such as smart phones, autonomous cars, and micro robots.

© Springer Nature Switzerland AG 2021
N. Oliver et al. (Eds.): ECML PKDD 2021, LNAI 12976, pp. 250–265, 2021.
https://doi.org/10.1007/978-3-030-86520-7_16

A well-known solution in machine learning to compress a large deep network to a smaller network is *knowledge distillation* (KD) [7,9]. The main goal of KD is to transfer the knowledge learned by a large pre-trained deep network (called *teacher*) to a smaller network (called *student*) such that the student network can mimic the teacher network, resulting in a comparable classification performance [9]. Many methods have been proposed for KD, and most of them follow the method introduced in Hinton et al.'s paper [9], which attempts to map the predictions of student to both the true labels and the predictions of teacher on the student's training data (called *student-data*). The intuition behind this method is that the student will improve its classification performance when it not only learns from its training data but also is guided by a powerful teacher that was often trained on a larger data (called *teacher-data*) and achieved very good performance due to its generalization ability. The idea of standard KD method is illustrated in Fig. 1.

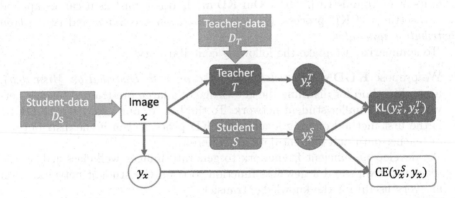

Fig. 1. Illustration of standard KD method. Given a student-data D_S and a teacher network T pre-trained on teacher-data D_T, the student network S is trained on each image $x \in D_S$ such that its output y_x^S matches both the true label y_x via a cross-entropy loss and the output of teacher y_x^T via a Kullback–Leibler (KL) divergence loss.

All current methods following the standard KD method shown in Fig. 1 have a significant constraint – they assume that both teacher-data D_T and student-data D_S come from the same distribution [2,9,10,22]. This strong assumption is not realistic in real-world applications, where the distribution shift often happens between D_T and D_S, and in some cases D_T is different from D_S e.g. teacher pre-trained on ImageNet while student trained on CIFAR-100. As a result, the teacher network performs very poorly on D_S. All existing KD methods will fail when using the knowledge transferred from such mismatched teachers, and the classification accuracy of student network often drops significantly. Thus, *the ability of applying the KD process when the distribution mismatch between teacher-data D_T and student-data D_S exists, is an open problem.*

Our Method. To solve the above problem, we propose a novel KD method that is still effective when the distribution mismatch happens. In particular, we first train a generative model to obtain the distribution of a latent variable representing the student-data. We then adjust this distribution using Bayesian optimization [14,19] to find an optimized distribution from which we can sample images that are well-classified by the teacher. By doing this, we can replace the *original student-data* where the teacher performs poorly by a *new student-data* where the teacher achieves a good performance. Finally, we propose a novel KD loss function to train the student network to match its predictions to the true labels of *original student-data* and to the predictions of teacher network on *new student-data*. The intuition behind our method is that the teacher network should be given the data points on which it has good knowledge and achieves accurate predictions. By that way, the teacher's good knowledge on such data points will be useful when transferred to the student. For other data points on which the teacher has little/wrong knowledge, the student should learn their information from its own ground-truth labels. Our KD method is robust as it can be applied to two settings of KD process: (1) *with distribution mismatch* and (2) *without distribution mismatch*.

To summarize, we make the following contributions:

1. We propose **KDDM** (*Knowledge Distillation with Distribution Mismatch*), a novel method for distilling the knowledge of a large pre-trained teacher network into a smaller student network. To the best of our knowledge, **KDDM** is the first method offering a successful KD process even if the distributions of teacher-data and student-data are different.
2. We develop an efficient framework to generate images well-classified by the teacher network and a new loss function to train the student network. Both are very useful for the knowledge transfer.
3. We only treat the teacher as a *black-box* model where we require no internal information (e.g. model architecture, weights, etc.) from the teacher except its probabilistic outputs.
4. We demonstrate the benefits of **KDDM** in two cases of KD process, namely *with distribution mismatch* and *without distribution mismatch*. Our method significantly outperforms the standard KD method and is comparable with recent state-of-the-art KD methods that train teacher and student networks on data with matching distributions.

2   Related Works

Knowledge distillation (KD) has become an attractive research topic since 2015 when Hinton et al. introduced the concept of KD in their teacher-student framework [9]. The main goal of KD is to transfer the knowledge learned from a teacher network to a student network such that the student can mimic the teacher, resulting in an improvement in its classification performance. In recent years, many methods have been proposed for KD, which can be categorized into three groups: relation-based, feature-based, and response-based KD methods.

Relation-Based KD. These methods not only use the output of teacher but also explore the relationships between different layers of teacher when training the student network. Examples include [11,15,24]. One key challenge of these methods is how to define the correlation function between the teacher and student layers.

Feature-Based KD. These methods leverage both the output of last layer and the output of intermediate layers (i.e. feature map) of teacher when training student [1,10,15]. The main benefit of these approaches is that deep neural networks are often good at representation learning, therefore not only the predictions but also representations learned by teacher network are useful knowledge to transfer to student network.

Response-Based KD. These methods directly mimic the final prediction of teacher network [4,9,12,23]. Compared to relation-based and feature-based methods, response-based methods have a significant advantage that they only treat the teacher as a *black-box* model without requiring access to its internal information (e.g. model parameters, feature maps, or derivatives), thus are applicable to *any* type of deep network.

To successfully train student with the knowledge distilled from teacher, most KD methods assume that both teacher-data and student-data come from the same distribution. For example, [3,9] pointed out that the student only achieved its best accuracy when it had access to the teacher's original training data. Similarly, [13] mentioned the typical setting in existing KD methods is the student network trained on the teacher-data. Recent state-of-the-art methods [2,10,22] also train both teacher and student networks on the same dataset.

Although these methods can distill a large deep network into a smaller one, their success relies on the strong assumption that both teacher-data and student-data are identical, a condition often not met in many real-world applications. As far as we know, no KD approach has been proposed to explicitly address the challenge of distribution mismatch between teacher-data and student-data.

3 Framework

3.1 Problem Definition

Given a student-data D_S and a teacher network T pre-trained on a teacher-data D_T, the goal of KD method is to train a student network S on D_S such that S can mimic the prediction of T.

The standard KD method minimizes the following loss function:

$$\mathcal{L}_{KD} = \sum_{x \in D_S} \alpha \text{CE}(y_x^S, y_x) + (1 - \alpha)\text{KL}(y_x^S, y_x^T), \tag{1}$$

where $\text{CE}(y_x^S, y_x)$ is the cross-entropy loss between the output (i.e. the probabilities for all classes) of student and true label, $\text{KL}(y_x^S, y_x^T)$ is the Kullback–Leibler (KL) divergence loss between the output of student and the output of teacher, and α is a trade-off factor to balance the two loss terms. Note that in Eq. (1) we do not use the *temperature* factor as in Hinton's KD method [9] because this requires access to the pre-softmax activations of teacher, which violates our assumption of "black-box" teacher.

From Eq. (1), since the true labels y_x in the first loss term are fixed, the improvement of a KD method mainly relies on the second loss term $\mathrm{KL}(y_x^S, y_x^T)$. The predictions of teacher y_x^T are assumed to be highly accurate on the student-data D_S so that the classification accuracy of student can be improved. In many real cases, this assumption is not true, where the performance of teacher on D_S is not good due to the distribution mismatch between D_T and D_S. For example, D_T and D_S can come from two different distributions or D_T and D_S can be two different datasets. As a result, the standard KD process makes harmful effects to student, where the classification accuracy of student drops significantly compared to the same model trained from scratch on D_S (we call this model *student-alone*).

Problem Statement. Given a student-data D_S and a *black-box* teacher network T pre-trained on a teacher-data D_T, we assume there is a distribution mismatch between D_T and D_S. Our goal is to train a student network S on D_S, which achieves two objectives: (1) our student's classification accuracy is better than that of student network trained with the standard KD loss in Eq. (1), and (2) our student's classification accuracy is better than that of student-alone trained from scratch on D_S.

3.2 Proposed Method KDDM

To improve the classification performance of student network, one direct solution is to adjust the trade-off factor $\alpha \in [0, 1]$ in Eq. (1), where a small value for α means the KD process will rely more on the predictions of teacher whereas a large value for α means the KD process will rely more on the true labels. Typically, existing KD methods choose $\alpha = 0.5$ to balance these two objectives. Since the distribution mismatch exists, a reasonable thinking is that we should trust the true labels more than the predictions of teacher, leading to choosing large values for α (e.g. 0.7 or 0.9). Although this simple approach can improve the accuracy of student, it cannot boost the student's performance better than the accuracy of student-alone network that uses $\alpha = 1.0$. We can see the accuracy of student-alone network serves as an upper bound on the accuracy obtained by the standard KD methods when the distribution mismatch occurs.

Our framework to solve the problem in Sect. 3.1 is novel, which has three main steps: (1) we train a generative model to obtain the distribution of a latent variable representing the student-data D_S, (2) we adjust this distribution using Bayesian optimization (BO) to find an optimized distribution to generate new images well-classified by the teacher, and (3) we perform the KD process to train the student with a new loss function. The overview of our proposed framework is shown in Fig. 2.

Learning the Distribution of Latent Variable z. We train Conditional Variational Autoencoder (CVAE) [20] to learn the distribution of latent variable

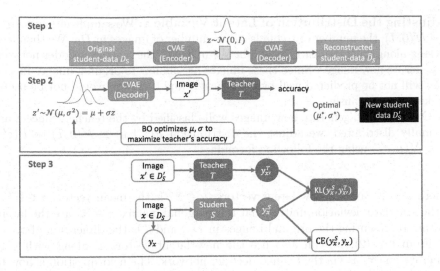

Fig. 2. Our framework **KDDM** consists of three steps. **Step 1:** it trains a CVAE model to learn the standard normal distribution of latent variable z representing the student-data D_S. **Step 2:** it adjusts this distribution using BO to generate the new student-data D'_S where the teacher achieves the best accuracy. **Step 3:** it trains the student to match its output to the true labels in original student-data D_S and to the predictions of teacher on new student-data D'_S.

z that is close to the distribution of student-data D_S. CVAE is a generative model consisting of an *encoder* and a *decoder*. We use the encoder network to map an image along with its label $(x, y) \in D_S$ to a latent vector z that follows a standard normal distribution $\mathcal{N}(0, I)$. From the latent vector z conditioned on the label y, we use the decoder network to reconstruct the input image x. Following [20], we train the CVAE by maximizing the variational lower bound objective:

$$\log P(x \mid y) \geq \mathbb{E}(\log P(x \mid z, y)) - \mathrm{KL}(Q(z \mid x, y), P(z \mid y)),$$

where $Q(z \mid x, y)$ is the encoder network mapping input image x and its label y to the latent vector z, $P(x \mid z, y)$ is the decoder network reconstructing input image x from the latent vector z and label y, $\mathbb{E}(\log P(x \mid z, y))$ is the expected likelihood, which is implemented by a cross-entropy loss between input image and reconstructed image, and $P(z \mid y) \equiv \mathcal{N}(0, I)$ is the prior distribution of z conditioned on y.

After training the CVAE model, we obtain the distribution of latent variable z, which is a standard normal distribution $\mathcal{N}(0, I)$. If we sample any $z \sim \mathcal{N}(0, I)$ and feed it along with a label y to the decoder network, then we can generate an image that follows the distribution of D_S. Since we use CVAE to generate images with labels, we can find a distribution to sample images that are well-classified by the teacher in the next step.

Adjusting the Distribution of Latent Variable z. We sample latent vectors $z \sim \mathcal{N}(0, I)$; the number of z equals to the number of images in D_S. We then feed these z along with the list of true labels y in D_S to the trained decoder network to generate new images. Since these images are similar to original images in D_S, they will not be predicted well by the teacher, and therefore would not be useful for knowledge transfer.

Our goal is to generate new images well-classified by the teacher. Since z are normally distributed, we adjust the distribution $P(z \mid y) \equiv \mathcal{N}(0, I)$ to $P(z' \mid y) \equiv \mathcal{N}(\mu, \sigma^2)$ using the following formula:

$$z' = \mu + \sigma z, \tag{2}$$

where $z' \in \mathbb{R}^d$ are the new latent vectors, $\mu \in \mathbb{R}^d$ is the mean vector, $\sigma \in \mathbb{R}^{d \times d}$ is the standard deviation matrix that is a diagonal matrix, $z \in \mathbb{R}^d$ are the latent vectors representing the original images in D_S, and d is the dimension of z.

From Eq. (2), we can also use the new latent vectors z' along with y to generate images x' via the trained decoder network. The next question is how to generate relevant images x' on which the teacher predicts well. Since the output of generated images x' depends on z' that can be controlled by μ and σ, we use BO to optimize μ and σ by maximizing the following objective function:

$$(\mu^*, \sigma^*) = \arg\max_{\mu, \sigma} f_{acc}(y, T(x')), \tag{3}$$

where $x' = G(z', y)$ are the generated images via the trained decoder network G, $T(x')$ are the predictions of teacher for generated images x', and f_{acc} is the function computing the accuracy between true labels y and predicted labels of teacher on x'. Note that since we set y during the image generation, y are the true labels of x'. We use BO to optimize Eq. (3) since it is a black-box and expensive-to-evaluate function.

After the optimization finishes, we obtain the optimal distribution indicated by μ^* and σ^*. We compute optimal latent vectors $z' = \mu^* + \sigma^* z$. Using z' along with the list of true labels y in D_S, we generate images x' whose labels are y, and x' have a higher chance to be correctly predicted by the teacher. On one hand, we can generate a new student-data D'_S containing images x' that are well-classified by the teacher. On the other hand, we can discover the data space where the teacher has relevant knowledge for the student and achieves accurate predictions. Note that for each image $x \in D_S$ we have its corresponding image $x' \in D'_S$, and both x and x' have the same true label y.

Training the Student S Using a New Knowledge Distillation Loss. After generating the new student-data D'_S, we then use the knowledge of teacher on D'_S to transfer to the student. Precisely, we train a student network whose predictions match both the true labels in the original student-data D_S and the predictions of teacher on the new student-data D'_S.

Our new loss function to train the student network is:

$$\mathcal{L}_S = \sum_{x \in D_S, x' \in D'_S} \beta \mathrm{CE}(y_x^S, y_x) + (1 - \beta)\mathrm{KL}(y_x^S, y_{x'}^T), \tag{4}$$

where $\mathrm{CE}(y_x^S, y_x)$ is the cross-entropy loss between the outputs of student and the true labels in the original student-data D_S, $\mathrm{KL}(y_x^S, y_{x'}^T)$ is the KL divergence loss between the outputs of student and the outputs of teacher on the new student-data D_S', and β is a trade-off factor to balance the two loss terms. Note that the number of images in both D_S and D_S' are the same, and each $x' \in D_S'$ is the corresponding image of each $x \in D_S$.

Compared to the standard KD loss function in Eq. (1), our loss function uses the knowledge of teacher on the new student-data D_S' to transfer to the student, which is much better than using the knowledge of teacher on the original student-data D_S. This is because D_S' contains images on which the teacher has good knowledge and predicts accurately, as ensured by the BO process. For the knowledge on D_S where the teacher does not have good knowledge, our student network only learns from the true labels of D_S. Consequently, our method is much better than the standard KD method.

Discussion. To create the new student-data D_S', one can simply select images in the original student-data D_S on which the teacher has accurate predictions (i.e. the actual classes having the highest probabilities). However, this naive approach has two disadvantages. First, since the teacher's accuracy on D_S is low e.g. its accuracy on CIFAR-100 is only 14.46% in our experiment, very few images in D_S provide useful knowledge to transfer to the student via the teacher's predictions. Second, since the number of images in D_S' is less than that in D_S, D_S' cannot be used in Eq. (4) to train the student network, which requires $|D_S'| = |D_S|$.

4   Experiments and Discussions

We conduct extensive experiments on three benchmark image datasets to evaluate the classification performance (accuracy and F1-score) of our method **KDDM**, compared with strong baselines.

4.1   Datasets

We run experiments on MNIST, CIFAR-10, and CIFAR-100. These datasets are commonly used to evaluate the performance of a KD method [2,5,9,22].

4.2   Baselines

Since there are no existing KD methods dealing with distribution mismatch, we compare **KDDM** with two baselines.

- *Student-Alone*: the student network is trained on the student-data D_S from scratch.
- *Standard-KD*: the student network is trained with the standard KD loss in Eq. (1) and with the trade-off factor $\alpha = 0.5$. We also use larger values $\alpha = 0.7$ and $\alpha = 0.9$ to improve the performance of student, as discussed in Sect. 3.2.

As a reference, we also report the accuracy of current state-of-the-art KD methods, namely KD [9], VID [2], and CRD [22]. Note that these three methods are not comparable with ours since they train both teacher and student networks on the same dataset whereas our method is dealing with the teacher and student networks trained on teacher- and student-datasets with different distributions.

To have a fair comparison, we use the same teacher-student network architecture and hyper-parameters (e.g. batch size and the number of epochs) across all the methods i.e. Student-Alone, Standard-KD, and our **KDDM**. For other baselines KD, VID, and CRD, we obtain their accuracy from related papers[1].

4.3   Results on MNIST

The first experiment shows the KD results on MNIST dataset, where the distribution shift happens between teacher-data D_T and student-data D_S.

Experiment Settings. Following [5], we use the HintonNet-1200 network for the teacher, which has two hidden layers of 1,200 units and the HintonNet-800 network for the student, which has two hidden layers of 800 units. These two network architectures are feed-forward neural networks (FNNs), where the student has much fewer parameters than the teacher. We train the teacher network with a batch size of 256 and 20 epochs. We train the student network with a batch size of 64 and 20 epochs. We train the CVAE with FNNs for both encoder and decoder, using a latent dimension of 2, a batch size of 256, and 100 epochs. To find the optimal values (μ^*, σ^*), we use Thomson sampling implemented in the Turbo package [6], with 2,000 iterations, 50 suggestions per batch, and 20 trusted regions. The searching ranges for $\mu = [\mu_1, \mu_2], \sigma = \begin{bmatrix} \sigma_{11} & 0 \\ 0 & \sigma_{22} \end{bmatrix}$ are $\mu_1, \mu_2 \in [-6.0, 6.0]$ and $\sigma_{11}, \sigma_{22} \in [0, 6.0]$. The trade-off β in our new loss function (see Eq. (4)) is set to 0.5, and $\beta = 0.5$ is used across all experiments.

The MNIST dataset has 28×28 images from 10 classes ($[0, 1, ..., 9]$), containing 60K training images (called D_{train}) and 10K testing images (called D_{test}). We use D_{train} as the student-data D_S while we create the teacher-data D_T by randomly removing 98% of images in even classes $0, 2, 4, 6, 8$ in D_{train}, leading to two different distributions for D_T and D_S. We train the teacher network on D_T and the student network on D_S, then evaluate both teacher and student networks on *hold-out* D_{test}. The distributions of three datasets D_T, D_S, and D_{test} are illustrated in Fig. 3.

Quantitative Results. From Table 1(a), we can see our **KDDM** outperforms both Student-Alone and Standard-KD ($\alpha = 0.5$). **KDDM** achieves 98.30%, which is much better than Standard-KD achieving only 96.61%. The teacher's accuracy achieves only 91.52% due to the distribution mismatch between D_T

---

[1] This is possible because we use benchmark datasets, and the training and test splits are fixed.

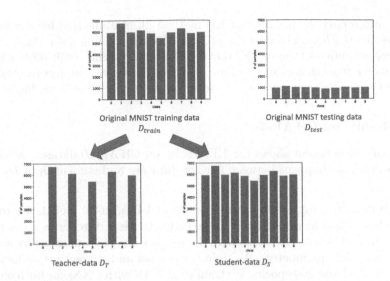

Fig. 3. Distributions of teacher-data D_T, student-data D_S, and test data D_{test}. The distribution of D_T is much different from that of D_S while the distributions of D_S and D_{test} are quite similar (i.e. the number of samples among classes is balanced). *We expect that the teacher trained on D_T will achieve low accuracy on D_S and D_{test} due to the distribution mismatch.*

Table 1. Classification results on MNIST: **(a)** teacher-data D_T and student-data D_S come from two different distributions and **(b)** D_T and D_S are identical. The results of **(b)** are obtained from [5]. "-" means the result is not available in the original paper.

(a) $P(D_T) \neq P(D_S)$			(b) $D_T \equiv D_S \equiv D_{train}$		
Model	Accuracy	F1-score	Model	Accuracy	F1-score
Teacher	91.52%	91.40%	Teacher	98.39%	-
Student-Alone	98.21%	98.19%	Student-Alone	98.11%	-
Standard-KD ($\alpha = 0.5$)	96.61%	96.60%	KD [9]	98.39%	-
Standard-KD ($\alpha = 0.7$)	97.93%	97.92%	VID [2]	-	-
Standard-KD ($\alpha = 0.9$)	98.17%	98.15%	CRD [22]	-	-
KDDM (Ours)	**98.30%**	**98.29%**			

versus D_S and D_{test}, leading to the poor performance of Standard-KD, as discussed in Sect. 1. When we increase the value of α to 0.7 and 0.9 (i.e. the student relies more on the true labels of D_S), the accuracy of Standard-KD is improved as expected (see Sect. 3.2), and approaches the accuracy of Student-Alone.

We also report the accuracy of KD method [9] in Table 1(b) for a reference. KD uses the teacher trained on the original MNIST training data D_{train}, which has more advantages than our **KDDM**. KD's teacher achieves 98.39%, which is much higher than that of our teacher (91.52%). However, our KD framework – **KDDM** still achieves a comparable accuracy with KD (98.30% vs. 98.39%).

4.4   Results on CIFAR-10

The second experiment shows the KD results on CIFAR-10 dataset, where the distribution shift happens between teacher-data D_T and student-data D_S.

Experiment Settings. We use ResNet-50 and ResNet-20 models for teacher and student networks respectively. These two network architectures are convolutional neural networks (CNNs), where the teacher has 763K parameters and the student has 274K parameters. We train the teacher and student networks with a batch size of 32 and 200 epochs. We train the CVAE with CNNs for both encoder and decoder, using a latent dimension of 2, a batch size of 64, and 600 epochs. We find the optimal values (μ^*, σ^*) in the same way as in our experiments on MNIST.

The CIFAR-10 dataset is set of 32×32 pixel RGB images with 10 classes, and contains 50K training images (called D_{train}) and 10K testing images (called D_{test}). We use D_{train} as the student-data D_S while we create the teacher-data D_T by randomly removing 98% of images in even classes in D_{train}, leading to $P(D_T) \neq P(D_S)$. We train the teacher network on D_T and the student network on D_S, then evaluate both teacher and student networks on *hold-out* D_{test}.

Quantitative Results. From Table 2(a), we can see the accuracy of teacher on D_{test} is only 70.73%, which is much lower than that of Student-Alone trained on D_S. This is because of the distribution mismatch between D_T versus D_S and D_{test}. Consequently, when this poor teacher is used for KD, it makes harmful effects to Standard-KD ($\alpha = 0.5$), where the accuracy significantly drops around 6% from Student-Alone. Even if we use larger values for α to shift the predictions of Standard-KD to the true labels of D_S, its accuracy is only improved to 90.90%, which is still worse than that of Student-Alone. Our **KDDM** achieves 91.54%, which is better than both Student-Alone and Standard-KD baselines, thanks to the new loss function to train the student in Eq. (4).

Compared to state-of-the-art KD methods in Table 2(b), **KDDM** is comparable with KD and VID. Note that both KD and VID train teacher and student networks on the same D_{train}, which obtains a much stronger teacher than our method. Their teacher achieves 94.26% whereas our teacher achieves only 70.73% on the test data D_{test}. However, the accuracy of our **KDDM** and those of KD and VID are comparable (91.54% vs. 91.27% and 91.85%).

Table 2. Classification results on CIFAR-10: **(a)** teacher-data D_T and student-data D_S come from two different distributions and **(b)** D_T and D_S are identical. The results of **(b)** are obtained from [2]. "-" means the result is not available in the original paper.

(a) $P(D_T) \neq P(D_S)$

Model	Accuracy	F1-score
Teacher	70.73%	69.36%
Student-Alone	91.22%	91.20%
Standard-KD ($\alpha = 0.5$)	85.49%	85.60%
Standard-KD ($\alpha = 0.7$)	89.34%	89.40%
Standard-KD ($\alpha = 0.9$)	90.90%	90.91%
KDDM (Ours)	**91.54%**	**91.56%**

(b) $D_T \equiv D_S \equiv D_{train}$

Model	Accuracy	F1-score
Teacher	94.26%	-
Student-Alone	90.72%	-
KD [9]	91.27%	-
VID [2]	91.85%	-
CRD [22]	-	-

Qualitative Results. In Fig. 4, we show the original images in student-data D_S versus the images generated by our method corresponding to four true labels "ship", "deer", "bird", and "automobile". We can see our generated images have comparable qualities with original images, where the objects are easily recognized and clearly visualized. These generated images are also correctly classified by the teacher. In contrast, the original images are wrongly predicted by the teacher since they may not be similar to any images in D_T – the training data of teacher. Remind that D_T has much fewer samples in classes 2 ("bird"), 4 ("deer"), and 8 ("ship"). Consequently, the teacher does not observe a diversity of samples and often predicts wrong labels for the images belonging to these three classes. To overcome this problem, our method tries to generate only images best fitting to the teacher (i.e. the images well-classified by the teacher), and use them for the KD process.

ship   deer

bird   automobile

Fig. 4. Original images in original student-data D_S (left) versus generated images in new student-data D'_S (right) in 4 classes "ship", "deer", "bird", "automobile".

4.5   Results on CIFAR-100

The third experiment shows the KD results on CIFAR-100 dataset, where teacher-data D_T and student-data D_S are two different datasets.

Experiment Settings. We use a *pre-trained* ResNet-50 model[2] on ImageNet (i.e. teacher-data D_T) for the teacher network and a ResNet-20 model for the student network. We train the student network on the original training set D_{train} of CIFAR-100 (i.e. student-data D_S) with a batch size of 16 and 200 epochs. D_T and D_S differ in their distribution as ImageNet has much higher resolution than CIFAR-100, and two datasets cover different sub-types of a class in general e.g. "green snake" and "grass snake" in ImageNet vs. "snake" in CIFAR-100. We train the CVAE and find the optimal values (μ^*, σ^*) in the same way as in our experiments on CIFAR-10. We evaluate both teacher and student networks on the original testing set D_{test} of CIFAR-100. Note that the classification performance is only measured on 56 overlapped classes between ImageNet and CIFAR-100.

Table 3. Classification results on CIFAR-100: **(a)** teacher-data D_T and student-data D_S are two different datasets and **(b)** D_T and D_S are identical. The results of **(b)** are obtained from [22]. "-" means the result is not available in the original paper.

(a) D_T is ImageNet while D_S is CIFAR-100

Model	Accuracy	F1-score
Teacher	14.46%	12.70%
Student-Alone	71.09%	71.03%
Standard-KD ($\alpha = 0.5$)	68.21%	68.27%
Standard-KD ($\alpha = 0.7$)	69.57%	69.53%
Standard-KD ($\alpha = 0.9$)	70.75%	70.78%
KDDM (Ours)	**71.14%**	**71.13%**

(b) Both D_T and D_S are CIFAR-100

Model	Accuracy	F1-score
Teacher	72.34%	-
Student-Alone	69.06%	-
KD [9]	70.66%	-
VID [2]	70.38%	-
CRD [22]	71.16%	-

Quantitative Results. From Table 3(a), we can observe similar results as those in the experiments on MNIST and CIFAR-10, where our **KDDM** outperforms both Student-Alone and Standard-KD methods. **KDDM** achieves 71.14% of accuracy, which is better than 71.09% of Student-Alone and 68.21% of Standard-KD ($\alpha = 0.5$). Again, using large values for α helps Standard-KD to improve its performance, but it is still worse than our method. In this experiment, the teacher performs much worse than student models, achieving only 14.46% of accuracy on the testing set of CIFAR-100. This behavior can be explained by the fact that the teacher is pre-trained on ImageNet containing high-resolution images (224×224 pixel images) whereas CIFAR-100 only contains low-resolution images (32×32 pixel images). This resolution mismatch makes the teacher difficult to

---

[2] https://keras.io/api/applications/.

recognize the true objects in CIFAR-100 images [17]. Even using only a very poor performing teacher, **KDDM** still achieves an impressive accuracy.

As a reference, we report the accuracy of three current state-of-the-art KD methods in Table 3(b). Different from our method, these methods train both teacher and student networks on the same training set of CIFAR-100, leading to a good performance for the teacher network (72.34%).

4.6 Distillation When Teacher-Data and Student-Data Are Identical

In this experiment, we show how our method **KDDM** performs when there is no distribution mismatch between teacher-data D_T and student-data D_S, a common setting used in state-of-the-art KD methods. Here, we use the original training set D_{train} of CIFAR-10 for both D_T and D_S.

Table 4 reports the accuracy of our **KDDM** and other baselines on CIFAR-10 when both D_T and D_S are identical. We can see that our teacher improves its accuracy significantly with the full training set of CIFAR-10. It achieves 92.29% of accuracy compared to 70.73% when it was trained on D_T with distribution mismatch (see Table 2(a)). This stronger teacher helps Standard-KD ($\alpha = 0.5$), which boosts its accuracy to 91.50%. **KDDM** can also benefit from this strong teacher, which achieves 91.88% of accuracy slightly better than 91.85% of VID.

Table 4. Classification results on CIFAR-10 when D_T and D_S are identical (i.e. there is no distribution mismatch between D_T and D_S). The accuracy of VID is obtained from [2]. "-" means the result is not available in the original paper.

$D_T \equiv D_S \equiv D_{train}$		
Model	Accuracy	F1-score
Teacher	92.29%	92.28%
Student-Alone	91.22%	91.20%
Standard-KD ($\alpha = 0.5$)	91.50%	91.48%
VID [2]	91.85%	-
KDDM (Ours)	**91.88%**	**91.86%**

To summarize, the results in Table 4 confirm the real benefit of our method not only for the KD process when the distribution mismatch happens but also for the KD process when both teacher-data and student-data are identical. In both cases, our method is always better than the standard KD method and the student network trained from scratch on the student-data.

5 Conclusion

We have presented **KDDM** – a novel KD method for distilling a large pre-trained *black-box* deep network into a smaller network while maintaining a high accuracy.

Different from existing KD methods, our proposed method is still effective when the teacher-data and the student-data differ in their distribution. **KDDM** uses an efficient framework to generate images well-classified by the teacher network and a new loss function to train the student network. We demonstrate the real benefits of **KDDM** on three standard benchmark image datasets. The empirical results show that **KDDM** outperforms both the standard KD method and the student model trained from scratch. Our accuracy is also comparable with those of state-of-the-art KD methods that assume no distribution mismatch between teacher-data and student-data.

Our future work will study how to extend our method to the setting of *white-box* teacher, which is expected to further improve our student's accuracy since we now can also transfer the representation learned by the teacher to the student.

Acknowledgment. This research was a collaboration between the Commonwealth Australia (represented by Department of Defence) and Deakin University, through a Defence Science Partnerships agreement.

References

1. Adriana, R., Nicolas, B., Ebrahimi, S., Antoine, C., Carlo, G., Yoshua, B.: FitNets: hints for thin deep nets. In: ICLR (2015)
2. Ahn, S., Hu, X., Damianou, A., Lawrence, N., Dai, Z.: Variational information distillation for knowledge transfer. In: CVPR, pp. 9163–9171 (2019)
3. Chawla, A., Yin, H., Molchanov, P., Alvarez, J.: Data-free knowledge distillation for object detection. In: CVPR, pp. 3289–3298 (2021)
4. Chen, G., Choi, W., Yu, X., Han, T., Chandraker, M.: Learning efficient object detection models with knowledge distillation. In: NIPS, pp. 742–751 (2017)
5. Chen, H., et al.: Data-free learning of student networks. In: ICCV, pp. 3514–3522 (2019)
6. Eriksson, D., Pearce, M., Gardner, J., Turner, R., Poloczek, M.: Scalable global optimization via local bayesian optimization. In: NIPS, pp. 5496–5507 (2019)
7. Gou, J., Yu, B., Maybank, S.J., Tao, D.: Knowledge distillation: a survey. arXiv preprint arXiv:2006.05525 (2020)
8. Guo, G., Zhang, N.: A survey on deep learning based face recognition. Comput. Vis. Image Underst. **189**, 102805 (2019)
9. Hinton, G., Vinyals, O., Dean, J.: Distilling the knowledge in a neural network. arXiv preprint arXiv:1503.02531 (2015)
10. Kim, J., Park, S., Kwak, N.: Paraphrasing complex network: network compression via factor transfer. In: NIPS, pp. 2760–2769 (2018)
11. Lee, S., Song, B.C.: Graph-based knowledge distillation by multi-head attention network. arXiv preprint arXiv:1907.02226 (2019)
12. Meng, Z., Li, J., Zhao, Y., Gong, Y.: Conditional teacher-student learning. In: ICASSP, pp. 6445–6449. IEEE (2019)
13. Nayak, G.K., Mopuri, K.R., Chakraborty, A.: Effectiveness of arbitrary transfer sets for data-free knowledge distillation. In: CVPR, pp. 1430–1438 (2021)
14. Nguyen, D., Gupta, S., Rana, S., Shilton, A., Venkatesh, S.: Bayesian optimization for categorical and category-specific continuous inputs. In: AAAI, pp. 5256–5263 (2020)

15. Passalis, N., Tzelepi, M., Tefas, A.: Heterogeneous knowledge distillation using information flow modeling. In: CVPR, pp. 2339–2348 (2020)
16. Pouyanfar, S., et al.: A survey on deep learning: algorithms, techniques, and applications. ACM Comput. Surv. **51**(5), 1–36 (2018)
17. Salman, H., Ilyas, A., Engstrom, L., Kapoor, A., Madry, A.: Do adversarially robust ImageNet models transfer better? In: NIPS, pp. 3533–3545 (2020)
18. Shen, L., Margolies, L., Rothstein, J., Fluder, E., McBride, R., Sieh, W.: Deep learning to improve breast cancer detection on screening mammography. Sci. Rep. **9**(1), 1–12 (2019)
19. Snoek, J., Larochelle, H., Adams, R.: Practical Bayesian optimization of machine learning algorithms. In: NIPS, pp. 2951–2959 (2012)
20. Sohn, K., Lee, H., Yan, X.: Learning structured output representation using deep conditional generative models. In: NIPS, pp. 3483–3491 (2015)
21. Sreenu, G., Durai, S.: Intelligent video surveillance: a review through deep learning techniques for crowd analysis. J. Big Data **6**(1), 1–27 (2019)
22. Tian, Y., Krishnan, D., Isola, P.: Contrastive representation distillation. In: ICLR (2020)
23. Wang, D., Li, Y., Wang, L., Gong, B.: Neural networks are more productive teachers than human raters: active mixup for data-efficient knowledge distillation from a blackbox model. In: CVPR, pp. 1498–1507 (2020)
24. Yim, J., Joo, D., Bae, J., Kim, J.: A gift from knowledge distillation: fast optimization, network minimization and transfer learning. In: CVPR, pp. 4133–4141 (2017)
25. Zhang, S., Yao, L., Sun, A., Tay, Y.: Deep learning based recommender system: a survey and new perspectives. ACM Comput. Surv. **52**(1), 1–38 (2019)

Certification of Model Robustness in Active Class Selection

Mirko Bunse[(✉)] and Katharina Morik

Artificial Intelligence Group, TU Dortmund University, 44221 Dortmund, Germany
{mirko.bunse,katharina.morik}@tu-dortmund.de

Abstract. Active class selection provides machine learning practitioners with the freedom to actively choose the class proportions of their training data. While this freedom can improve the model performance and decrease the data acquisition cost, it also puts the practical value of the trained model into question: is this model really appropriate for the class proportions that are handled during deployment? What if the deployment class proportions are uncertain or change over time? We address these questions by *certifying* supervised models that are trained through active class selection. Specifically, our certificate declares a set of class proportions for which the certified model induces a training-to-deployment gap that is small with a high probability. This declaration is theoretically justified by PAC bounds. We apply our proposed certification method in astro-particle physics, where a simulation generates telescope recordings from actively chosen particle classes.

Keywords: Active class selection · Label shift · Model certification · Learning theory · Classification · Validation · Imbalanced learning

1 Introduction

The increasing adoption of machine learning in practice motivates model performance reports [2,17,22] that are easily accessible by a diverse group of stakeholders. One particular concern of this trend is the robustness of trained models with regard to changing deployment conditions, like distribution shifts [26], input perturbations [12], or adversarial attacks [25,30]. Ideally, robustness criteria are *certified* in the sense of being formally proven or thoroughly tested [12].

The framework of active class selection (ACS; see Fig. 1) [13,16] presumes a *class-conditional* training data generator, e.g. an experiment or a simulation that produces feature vectors for arbitrarily chosen classes. As a consequence, the developer of a machine learning model must actively decide for the class proportions of the training data set. This freedom can benefit the learning process in terms of data acquisition cost and model performance. However, the active decision for class proportions is difficult if the class proportions that occur during deployment are not precisely known or are subject to changes. In astro-particle physics, for instance, the ratio between the signal and the background class is

© Springer Nature Switzerland AG 2021
N. Oliver et al. (Eds.): ECML PKDD 2021, LNAI 12976, pp. 266–281, 2021.
https://doi.org/10.1007/978-3-030-86520-7_17

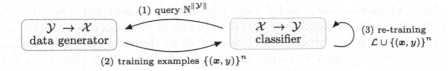

Fig. 1. Active class selection optimizes class-conditioned data acquisition [5].

only roughly estimated as $1 : 10^3$ or even $1 : 10^4$ [4]. Other use cases of ACS are brain computer interaction [10,21,29] and gas sensor arrays [16].

Recently, we have studied ACS from an information-theoretic viewpoint [5]. This viewpoint suggests that, in ACS, the training class proportions should be chosen identically to the deployment class proportions, at least when the sample size is sufficiently large. However, we also know from imbalanced classification [7] that highly imbalanced class proportions, as in the astro-particle use case, are often far from optimal. Moreover, the deployment class proportions may be uncertain at training time or may be subject to change during deployment. Therefore, we make the following contributions in the present paper:

- We study ACS through the lens of learning theory. This view-point provides us with PAC bounds that are more nuanced than previous [5] results. Namely, our bounds are applicable also to extremely imbalanced domains and they account for finite data volumes.
- Through these bounds, we quantify the *domain gap* which results from the label shift between the ACS-generated training data and the data that is predicted during deployment.
- We refine these results in a *certificate* for binary classifiers under label shift. This certificate verifies the range of class proportions for which an ACS-trained classifier is accurate (i.e. has a domain-induced error smaller than some $\varepsilon > 0$) with a high probability (i.e. with probability at least $1 - \delta$). This certificate is specific, understandable, theoretically justified, and applicable to any learning method. Users specify ε and δ according to their demands.

In the following, we briefly review the ACS problem statement (Sect. 1.1) and previous work on the topic (Sect. 1.2). Section 2 presents our theoretical contributions, followed by their experimental verification in Sect. 3. We present additional related work in Sect. 4 and conclude with Sect. 5.

1.1 Active Class Selection Constitutes a Domain Gap

Following the terminology of domain adaptation [20,27], we consider a *domain* as a probability density function over the labeled data space $\mathcal{X} \times \mathcal{Y}$. This density function stems from some particular data-generating process; a different process will induce a different domain. To this end, let $\mathcal{S}$ be the *source* domain where a machine learning model is trained and let $\mathcal{T}$ be the *target* domain where the trained model is required to be accurate. We are interested in the impact of deviations $\mathcal{S} \neq \mathcal{T}$ on the deployment performance.

Here, we assume that S and T *only* differ in their class proportions, to study the effect of ACS in isolation from other potential deviations between S and T. Put differently, we let all data be generated by the same causal mechanism $Y \to X$, according to the factorization $\mathbb{P}(x, y) = \mathbb{P}(x \mid y) \cdot \mathbb{P}(y)$. More formally:

Definition 1 (Identical mechanism assumption [5] a.k.a. label shift or target shift [31]). *Assume that all data in the domains S and T is generated independently by the same class-conditioned mechanism, i.e.*

$$\mathbb{P}_S(X = x \mid Y = y) \; = \; \mathbb{P}_T(X = x \mid Y = y) \qquad \forall x \in \mathcal{X}, \, \forall y \in \mathcal{Y}$$

1.2   A Qualitative Intuition from Information Theory

We have recently studied the domain gap $S \neq T$ in the limit of data acquisition, i.e. when the sample size $m \to \infty$ [5]. In this limit, the deployment proportions should also be reflected in the ACS-generated training data. However, we have also observed that certain deviations from this fixed-point are feasible without impairing the classifier; the range of these feasible deviations depends on the correlation between features and labels.

Proposition 2 (Information-theoretical bound [5]). *The domain S misrepresents the prediction function $\mathbb{P}_T(Y \mid X)$ by the KL divergence $d_{Y|X}$, which is bounded above by the KL divergence d_Y between $\mathbb{P}_S(Y)$ and $\mathbb{P}_T(Y)$:*

$$d_{Y|X} \; = \; d_Y - d_X \; \leq \; d_Y$$

Remarkably, the more data is being acquired by ACS, the less beneficial will any class proportion other than $\mathbb{P}_T(Y)$ be. Beyond these qualitative insights, however, the information-theoretic perspective has not allowed us to provide quantitative bounds which precisely assess the impact of the sample size. In the following, we therefore employ a different perspective on ACS from PAC learning theory. This perspective accounts for the sample size m, an error margin $\varepsilon > 0$ and a desired probability $1 - \delta < 1$.

2   A Quantitative Perspective from Learning Theory

We start by recalling a standard i.i.d. bound from learning theory, which we then extend to the domain gap induced by ACS. The standard i.i.d. bound quantifies the probability that the estimation error of the loss $L_S(h)$, induced by the finite amount m of data in a data set $\mathrm{D} = \{(x_i, y_i) \in \mathcal{X} \times \mathcal{Y} : 1 \leq i \leq m\} \sim S^m$ relative to the training domain S, is bounded above by some $\varepsilon > 0$:

Proposition 3 (I.i.d. bound [24]). *For any $\varepsilon > 0$ and any fixed $h \in \mathcal{H}$, it holds with probability at least $1 - \delta$, where $\delta = 2e^{-2m\varepsilon^2}$, that:*

$$|L_{\mathrm{D}}(h) - L_S(h)| \; \leq \; \varepsilon$$

Proof. We repeat the proof by Shalev-Shwartz and Ben-David [24, Sec. 4.2] here to extract Corollary 4 for later reference. Let $L_D(h) = \frac{1}{m}\sum_{i=1}^{m}\ell(y_i, h(x_i))$ be the empirical loss over a data set D and let $L_S(h) = \mathbb{E}_{(x,y)\sim\mathbb{P}_S}[\ell(y, h(x))]$ be the expected value of $L_D(h)$ and every $\ell(y_i, h(x_i))$. Then, by letting $\theta_i = \ell(y_i, h(x_i))$ and $\mu = L_S(h)$, we apply Hoeffding's inequality for $0 \le \theta_i \le 1$:

$$\mathbb{P}_{D\sim S^m}\left(\left|\frac{1}{m}\sum_{i=1}^{m}\theta_i - \mu\right| > \varepsilon\right) \le 2e^{-2m\varepsilon^2} = \delta \tag{1}$$

We see that the converse, i.e. $\left|\frac{1}{m}\sum_{i_1}^{m}\theta_i - \mu\right| \le \varepsilon$, holds with probability at least $1 - \delta$; taking Eq. 1 for granted would therefore already yield our claim ($\square$). Instead, however, we take another step back and prove Eq. 1 from Hoeffding's Lemma, which states that for every $\lambda > 0$ and any random variable $X \in [a, b]$ with $\mathbb{E}[X] = 0$ it holds that:

$$\mathbb{E}[e^{\lambda X}] \le e^{\frac{\lambda^2(b-a)^2}{8}} \tag{2}$$

Letting $X_i = \theta_i - \mu$ and $\bar{X} = \frac{1}{m}\sum_{i=1}^{m}X_i$, we use *i)* monotonicity, *ii)* Markov's inequality, *iii)* independence, and *iv)* Eq. 2 with $a = 0$, $b = 1$, and $\lambda = 4m\varepsilon$:

$$\mathbb{P}[\bar{X} \ge \varepsilon] \overset{i)}{=} \mathbb{P}[e^{\lambda\bar{X}} \ge e^{\lambda\varepsilon}] \overset{ii)}{\le} e^{-\lambda\varepsilon}\mathbb{E}[e^{\lambda\bar{X}}] \overset{iii)}{=} e^{-\lambda\varepsilon}\prod_{i=1}^{m}\mathbb{E}[e^{\lambda X_i/m}] \overset{iv)}{=} e^{-2m\varepsilon^2} \tag{3}$$

We apply Eq. 3 to $\bar{X}$ and $-\bar{X}$ to yield Eq. 1 via the union bound. $\square$

Corollary 4 (Asymmetric i.i.d. bound). *For any $\varepsilon^{(l)}, \varepsilon^{(u)} > 0$ and any fixed $h \in \mathcal{H}$, each of the following bounds holds with probability at least $1 - \delta^{(i)}$ respectively, where $\delta^{(i)} = e^{-2m(\varepsilon^{(i)})^2}$ and $i \in \{l, u\}$:*

i) $L_D(h) - L_S(h) \le \varepsilon^{(l)}$
ii) $L_S(h) - L_D(h) \le \varepsilon^{(u)}$

Proof. The claim follows from applying Eq. 3 to $\bar{X}$ and $-\bar{X}$, just like in the proof of Proposition 3. This time, however, we use two different $\varepsilon^{(l)}$, $\varepsilon^{(u)}$ for the two sides of the bound and we do not combine them via the union bound. $\square$

To study the ACS problem, we now replace the i.i.d. assumption above with the identical mechanism assumption from Definition 1. The result is Theorem 5, in which the factor 4 in δ, as compared to the factor 2 in the δ of Lemma 3, stems from the fact that either the upper bound or the lower bound might be violated, each time with at most the same probability. Figure 2 illustrates the idea.

Theorem 5 (Identical mechanism bound). *For any $\varepsilon > 0$ and any fixed $h \in \mathcal{H}$, it holds with probability at least $1 - \delta$, where $\delta = 4e^{-2m\varepsilon^2}$, that:*

$$|L_T(h) - L_S(h)| - \varepsilon \le |L_T(h) - L_D(h)| \le |L_T(h) - L_S(h)| + \varepsilon$$

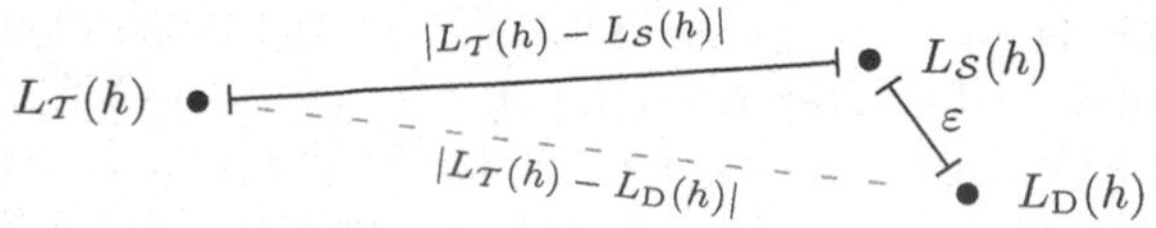

$$|L_\mathcal{T}(h) \bullet \vdash \frac{|L_\mathcal{T}(h) - L_\mathcal{S}(h)|}{} \dashv \bullet L_\mathcal{S}(h)$$

Fig. 2. Illustration of Theorems 5 and 6. Keeping $\delta > 0$ fixed, we can choose $\varepsilon \to 0$ as $m \to \infty$. What remains is the inter-domain gap $|L_\mathcal{T}(h) - L_\mathcal{S}(h)|$.

Proof. We employ Proposition 3 through the triangle inequality (see Fig. 2):

$$
\begin{aligned}
|L_\mathrm{D}(h) - L_\mathcal{T}(h)| &\leq |L_\mathrm{D}(h) - L_\mathcal{S}(h)| + |L_\mathcal{S}(h) - L_\mathcal{T}(h)| \\
&\leq \varepsilon + |L_\mathcal{S}(h) - L_\mathcal{T}(h)|,
\end{aligned}
$$

where the second inequality holds with probability at least $1 - 2e^{-2m\varepsilon^2}$.

Likewise, and with the same probability, we use the triangle inequality for the other side of the claim:

$$
\begin{aligned}
|L_\mathcal{T}(h) - L_\mathcal{S}(h)| &\leq |L_\mathcal{T}(h) - L_\mathrm{D}(h)| + |L_\mathrm{D}(h) - L_\mathcal{S}(h)| \\
\Leftrightarrow |L_\mathcal{T}(h) - L_\mathrm{D}(h)| &\geq |L_\mathcal{T}(h) - L_\mathcal{S}(h)| - |L_\mathrm{D}(h) - L_\mathcal{S}(h)| \\
&\geq |L_\mathcal{T}(h) - L_\mathcal{S}(h)| - \varepsilon \qquad \square
\end{aligned}
$$

The above bound addresses a single fixed hypothesis $h \in \mathcal{H}$, which suits our goal of certifying any given prediction model. For completeness, however, let us also mention that Theorem 5 can be extended to entire hypothesis classes $\mathcal{H}$. As an example, we obtain the following result for finite classes, i.e. for $|\mathcal{H}| < \infty$:

Theorem 6 (Identical mechanism bound; finite hypothesis class). *With probability at least $1-\delta$, where $\delta = 4|\mathcal{H}|e^{-2m\varepsilon^2}$, the upper and lower bounds from Theorem 5 hold for all $h \in \mathcal{H}$.*

Proof. The claim follows from the union bound, being detailed in Appendix 1.

The lower and upper bounds in Theorems 5 and 6 quantify how the total error $|L_\mathcal{T}(h) - L_\mathrm{D}(h)|$ approaches the inter-domain gap $|L_\mathcal{T}(h) - L_\mathcal{S}(h)|$ in dependence of the interplay between ε, δ, m, L, and $\mathcal{H}$. It is therefore a quantitative and thus more nuanced equivalent of Proposition 2. The inter-domain gap is constant w.r.t. the random draw of the training sample $\mathrm{D} \sim \mathcal{S}^m$ and is therefore independent of ε, of δ, and of m. Consequently, it remains even with an infinite amount of training data. Depending on the choice of $\mathcal{H}$ and L, and in dependence of the data distribution, it may be large or negligible. In the following, we will therefore study this error in more detail.

2.1   Quantification of the Domain Gap

We begin by factorizing the total error $L(h)$ into label-dependent losses $\ell_X(h, y)$ that are marginalized over the entire feature space $\mathcal{X}$. These losses only depend on the hypothesis h and on the label y and are, under the identical mechanism assumption from Definition 1, identical among $\mathcal{S}$ and $\mathcal{T}$.

$$
\begin{aligned}
L(h) &= \int_y \int_{\mathcal{X}} \mathbb{P}(x, y)\ell(y, h(x)) \, dx \, dy \\
&= \int_y \mathbb{P}(y) \underbrace{\int_{\mathcal{X}} \mathbb{P}(x \mid y)\ell(y, h(x)) \, dx}_{= \, \ell_X(h, y)} \, dy
\end{aligned}
$$

Plugging $\ell_X(h, y)$ into the domain gap $|L_{\mathcal{T}}(h) - L_{\mathcal{S}}(h)|$ from the Theorems 5 and 6 allows us to marginalize the label-dependent losses over the label space:

$$
|L_{\mathcal{T}}(h) - L_{\mathcal{S}}(h)| = \left| \int_y \mathbb{P}_{\mathcal{T}}(y)\,\ell_X(h, y) \, dy \; - \; \int_y \mathbb{P}_{\mathcal{S}}(y)\,\ell_X(h, y) \, dy \right|
$$

For classification tasks, i.e. $\mathcal{Y} = \{1, 2, \ldots, N\}$ with $N \geq 2$, we define the vectors $\mathbf{p}_{\mathcal{S}}, \mathbf{p}_{\mathcal{T}} \in [0, 1]^N$ through $[\mathbf{p}.]_i = \mathbb{P}.(Y = i)$, i.e. through the label probabilities in the domains $\mathcal{S}$ and $\mathcal{T}$. Furthermore, we define a vector $\boldsymbol{\ell}_h \in \mathbb{R}_+^N$ of class-wise losses through $[\boldsymbol{\ell}_h]_i = \ell_X(h, i)$. The computation of the ACS-induced domain gap then simplifies to an absolute difference between scalar products $L.(h) = \sum_{i \in \mathcal{Y}}[\mathbf{p}.]_i[\boldsymbol{\ell}_h]_i = \langle \mathbf{p}., \boldsymbol{\ell}_h \rangle$. Namely, for classification tasks:

$$
|L_{\mathcal{T}}^{\mathrm{clf}}(h) - L_{\mathcal{S}}^{\mathrm{clf}}(h)| = |\langle \mathbf{p}_{\mathcal{T}}, \boldsymbol{\ell}_h \rangle - \langle \mathbf{p}_{\mathcal{S}}, \boldsymbol{\ell}_h \rangle| \tag{4}
$$

Before we move on to a theorem about what Eq. 4 can mean in practice, let us build an intuition about the implications of this equation in a more simple setting: in binary classification.

Example 7 (Binary classification). In binary classification, the situation from Eq. 4 simplifies to $\mathcal{Y} = \{1, 2\}$ with $\mathbb{P}.(Y = 1) = p.$ and $\mathbb{P}.(Y = 2) = 1 - p.$. Let $\Delta p = |p_{\mathcal{T}} - p_{\mathcal{S}}|$ be be the absolute difference of the binary class proportions between the two domains and let $\Delta \ell_X = |\ell_X(h, 2) - \ell_X(h, 1)|$ be the absolute difference between the class-wise losses. The difference $\Delta \ell_X$ is independent of the class proportions and can be defined over any loss function ℓ. Rearranging Eq. 4 for binary classification, we obtain

$$
\begin{aligned}
&|L_{\mathcal{T}}^{\mathrm{bin}}(h) - L_{\mathcal{S}}^{\mathrm{bin}}(h)| \\
&= |(p_{\mathcal{T}}\ell_X(h, 2) + (1 - p_{\mathcal{T}})\ell_X(h, 1)) - (p_{\mathcal{S}}\ell_X(h, 2) + (1 - p_{\mathcal{S}})\ell_X(h, 1))| \\
&= |(p_{\mathcal{T}} - p_{\mathcal{S}}) \cdot (\ell_X(h, 2) - \ell_X(h, 1))| \\
&= \Delta p \cdot \Delta \ell_X,
\end{aligned} \tag{5}
$$

from which we see that in binary classification, for any loss function, the domain gap induced by ACS is simply the product of the class proportion difference Δp and the (true) class-wise loss difference $\Delta \ell_X$. If one of these terms is zero, so is the inter-domain gap. If one of these terms is non-zero but fixed, the domain gap will grow linearly with the other term.

Example 8 (Binary classification with zero-one loss). Let us illustrate Eq. 5 a little further. The zero-one loss is defined by $\ell(y, h(x)) = 0$ if the prediction is correct, i.e. if $y = h(x)$, and $\ell(y, h(x)) = 1$ otherwise. Consequently, $\ell_X(h, 2)$ is the true rate of false positives and $\ell_X(h, 1)$ is the true rate of false negatives. The more similar these rates are, the smaller will the inter-domain gap be for any distribution of classes in the target domain. Supposing that balanced training sets tend to balance $\ell_X(h, 2)$ and $\ell_X(h, 1)$, we can argue that balanced training sets (supposedly) maximize the range of feasible target domains with respect to the zero-one loss.

Example 9 (Cost-sensitive learning). The situation is quite different if the binary zero-one loss is weighted by the class, i.e. $\ell(y, h(x)) = w_y$ for $y \neq h(x)$. Such a weighting is common in cost-sensitive and imbalanced classification [7]. Here, Eq. 5 illustrates how counteracting class imbalance with weights can increase the robustness of the model: balancing $\ell_X(h, 2)$ and $\ell_X(h, 1)$ will increase the range of target domains that are feasible under the class-based weighting.

For completeness, we extend a part of this intuition from binary classification to classification tasks with an arbitrary number of classes:

Theorem 10. *In classification, the inter-domain gap $|L_T^{clf}(h) - L_S^{clf}(h)|$ from Theorem 5 is equal to zero if one of the following conditions holds:*

i) $\mathbf{p}_S = \mathbf{p}_T$
ii) $\ell_X(h, i) = \ell_X(h, j) \; \forall \, i, j \in \mathcal{Y}$

Proof. Condition i) trivially yields the claim through Eq. 4. Condition ii) means that $\Delta\ell = |\ell_X(h, i) - \ell_X(h, j)| = 0$ for every binary sub-task in a one-vs-one decomposition of the label set $\mathcal{Y}$. The domain gap of each binary sub-task, and therefore the total domain gap, is then zero according to Eq. 5.   □

Despite the fact that condition 10.ii) yields a domain gap of zero, one should not prematurely jump to the conclusion that a learning algorithm should enforce this condition necessarily. Recall that Theorem 10 addresses the domain gap, but not the deployment loss which we actually want to minimize; if enforcing condition 10.ii) results in a high source domain error, all domain robustness will not help to find an accurate target domain model. We therefore advise practitioners to carefully weigh out the source domain error with the domain robustness of the model, depending on the requirements of the use case at hand. Bayesian classifiers, which allow practitioners to mimic arbitrary $\mathbf{p}_S$ even after training, can prove useful in this regard.

2.2   Certification of Domain Robustness for Binary Predictors

We certify the set of class proportions to which a fixed hypothesis h, trained on $\mathcal{S}$, is safely applicable. By "safely", we mean that during the deployment on $\mathcal{T}$, h induces only a small domain-induced error with a high probability.

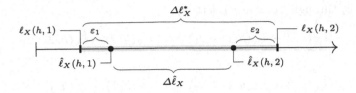

Fig. 3. Estimation of the minimum upper bound $\Delta\ell_X^*$ from data.

Definition 11 (Certified hypothesis). *A hypothesis $h \in \mathcal{H}$ is (ε, δ)-certified for all class proportions in the set $\mathcal{P} \subseteq [0, 1]^N$ if with probability at least $1 - \delta$ and $\varepsilon, \delta > 0$:*

$$|L_T(h) - L_S(h)| \leq \varepsilon \quad \forall \, \mathbf{p}_T \in \mathcal{P}$$

For simplicity, we limit our presentation to binary classification, i.e. $N = 2$ (see Example 7). In this case, $\mathcal{P}$ is simply a range $[p_T^{\min}, p_T^{\max}]$ of class proportions. According to Eq. 5, this range is defined by the largest Δp^* for which

$$\Delta p \cdot \Delta\ell_X \leq \varepsilon \quad \forall \, \Delta p \leq \Delta p^*. \tag{6}$$

Keep in mind that $\Delta\ell_X$ is defined over the *true* class-wise losses. If we knew them, we could simply rearrange Eq. 6 to find the largest Δp for a given ε; the equation would then hold with probability one. However, we do not know the true class-wise losses; instead, we estimate an upper bound that is only exceeded by the true $\Delta\ell_X$ with a small probability of at most $\delta > 0$. Particularly, to maximize Δp^*, we find the *smallest* upper bound $\Delta\ell_X^*$ among all such upper bounds.

An empirical estimate $\Delta\hat{\ell}_X$ of the true $\Delta\ell_X$ is given by the empirical class-wise losses $\hat{\ell}_X(h, y)$ observed in an ACS-generated validation sample D:

$$\Delta\hat{\ell}_X = \left| \hat{\ell}_X(h, 1) - \hat{\ell}_X(h, 2) \right|, \quad \text{where} \quad \hat{\ell}_X(h, y) = \frac{1}{m_y} \sum_{i \,:\, y_i = y} \ell(y, h(x_i))$$

Here, each $\hat{\ell}_X(h, y)$ can be associated with maximum lower and upper errors $\varepsilon_y^{(l)}, \varepsilon_y^{(u)} > 0$ that are not exceeded with probabilities at least $1 - \delta_y^{(l)}$ and $1 - \delta_y^{(u)}$. By choosing $\varepsilon_y^{(l)}, \varepsilon_y^{(u)}$ for both classes, we can thus find all upper bounds of the true $\Delta\ell_X$ that hold with at least the desired probability $1 - \delta$.

Figure 3 sketches our estimation of the *smallest* upper bound $\Delta\ell_X^*$. For simplicity, we assume that $\hat{\ell}_X(h, 2) \geq \hat{\ell}_X(h, 1)$. This assumption comes without loss of generality because we can otherwise simply switch the labels to make the assumption hold. Now, $\hat{\ell}_X(h, 1)$ shrinks at most by ε_1 and $\hat{\ell}_X(h, 2)$ grows at most by ε_2. Minimizing ε_1 and ε_2 simultaneously, within a user-specified probability budget δ, yields the desired minimum upper bound $\Delta\ell_X^*$ which the true $\Delta\ell_X$ only exceeds with probability at most $\delta = \delta_1 + \delta_2 - \delta_1\delta_2$. We find the values of δ_1 and δ_2 through Corollary 4, letting

$$\underbrace{-(\hat{\ell}_X(h,2) - \hat{\ell}_X(h,1) + \varepsilon_1)}_{= \varepsilon_2^{(l)}} \leq \ell_X(h,2) - \hat{\ell}_X(h,2) \leq \underbrace{\varepsilon_2}_{= \varepsilon_2^{(u)}}$$

$$\text{and} \quad \underbrace{-(\hat{\ell}_X(h,2) - \hat{\ell}_X(h,1) + \varepsilon_2)}_{= \varepsilon_1^{(u)}} \leq \hat{\ell}_X(h,1) - \ell_X(h,1) \leq \underbrace{\varepsilon_1}_{= \varepsilon_1^{(l)}},$$

so that $\delta_y = \delta_y^{(l)} + \delta_y^{(u)} - \delta_y^{(l)}\delta_y^{(u)}$ and $\delta_y^{(i)} = e^{-2m_y(\varepsilon_y^{(i)})^2}$.

During the optimization, strict inequalities are realized through non-strict inequalities with some sufficiently small $\tau > 0$:

$$\min_{\varepsilon_1, \varepsilon_2 \in \mathbb{R}} \varepsilon_2 + \varepsilon_1, \quad \text{s.t.} \quad \begin{cases} \varepsilon_1, \varepsilon_2 & \geq \tau \\ \delta - (\delta_1 + \delta_2 - \delta_1\delta_2) & \geq 0 \end{cases} \tag{7}$$

The minimizer $(\varepsilon_1^*, \varepsilon_2^*)$ of this optimization problem defines the smallest upper bound $\Delta\ell_X^* = (\hat{\ell}_X(h,2) + \varepsilon_2^*) - (\hat{\ell}_X(h,1) - \varepsilon_1^*)$ that is not exceeded by the true $\Delta\ell_X$ with probability at least $1 - \delta$. Choosing $\Delta p^* = \varepsilon/\Delta\ell_X^*$ will make Eq. 6 hold with the same probability, so that the range $[p_S - \Delta p^*,\ p_S + \Delta p^*]$ of binary deployment class proportions p_T is (ε, δ)-certified according to Definition 11.

If only small data volumes are available, it can happen that ϵ_1 must exceed $\hat{\ell}_X(h,1)$ to stay within the user-specified probability budget δ. This situation would mean that the lower bound $\ell_X(h,1) = \hat{\ell}_X(h,1) - \varepsilon_1$ is below zero, which does not reflect the basic loss property $\ell(h,y) \geq 0$. If the estimation of $\Delta\ell^*$ fails in this way, we fall back to a more simple, one-sided estimation. Namely, we only minimize the two upper bounds $\varepsilon_y^{(u)}$ that depend only on ε_2 and fix the two lower bounds to $\varepsilon_y^{(l)} = 0$. Doing so allows us to estimate a valid upper bound $\Delta\ell^*$ also for arbitrarily small data sets.

3   Experiments

In the following, we show that an (ε, δ) certified class proportion set $\mathcal{P}$ indeed characterizes an upper bound of the inter-domain gap. Our experiments even demonstrate that our certificate, being estimated only with source domain data, is very close to bounds that are obtained with labeled target domain data and are therefore not accessible in practice.

3.1   Binary (ϵ, δ) Certificates Are Tight

We randomly subsample the data to generate different deployment class proportions p_T while keeping $\mathbb{P}(x|y)$ fixed, in accordance to Definition 1. We compare two ways of estimating the target domain loss:

a) Our baseline is an empirical estimate $\hat{L}_T$ of the target domain loss that is computed with actual target domain data unavailable in practice.

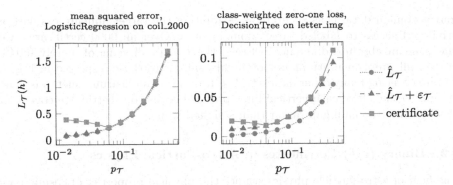

Fig. 4. The target domain loss $L_T(h)$ is upper-bounded by our (ε, δ) certificate and a baseline $\hat{L}_T + \varepsilon_T$ with privileged access to target domain data. Each of the above plots displays a different combination of loss function, learning method, and data set. The class proportions p_T of the target domain are varied over the x axis with a thin vertical line indicating the source domain proportions p_S.

b) We predict the target domain loss L_T from an (ε, δ) certificate by adding the domain gap parameter ε to the empirical source domain loss $\hat{L}_S$. We always choose the certificates such that they cover the class proportion difference $\Delta p = |p_T - p_S|$; in fact, we consider ε as a function of Δp in this experiment.

The certificate is *correct* if $\hat{L}_S + \varepsilon \geq \hat{L}_T$ holds, i.e. if ε indeed characterizes an upper bound of the inter-domain gap. If the two values are close to each other, i.e. if $\hat{L}_S + \varepsilon \approx \hat{L}_T$, we speak of a *tight* upper bound.

Correctness: Our experiments cover a repeated three-fold cross validation on eight imbalanced data sets, eight loss functions, and three learning algorithms, to represent a broad range of scenarios. Of all 9000 certificates, only 4.5% fail the test of ensuring $\hat{L}_S + \varepsilon \geq \hat{L}_T$. Since we have used $\delta = 0.05$ in these experiments, this amount of failures is actually foreseen by the statistical nature of our certificate: if it holds with probability at least $1 - \delta$, it is allowed to fail in 5% of all tests. This margin is almost completely used but not exceeded. Consequently, our certificate is correct in the sense of indeed characterizing an upper bound ε of the inter-domain gap with probability at least $1 - \delta$.

Tightness: A *fair* comparison between our certificate and our baseline $\hat{L}_T$ requires us to take the estimation error ε_T of the baseline into account. This necessity stems from the fact that $\hat{L}_T$ is also just an estimate from a finite amount of data. Having access to labeled target domain data will thus yield an upper bound $\hat{L}_T + \varepsilon_T$ of the true target domain error L_T, according to Proposition 3; this upper bound is then compared to our certificate, which has only seen the source domain data.

Figure 4 presents this comparison for two of our experiments. For most target domains p_T, the two predictions (■ and ▲) are almost indistinguishable from each other. This observation means that the certificate, which is based only on source domain data, is as accurate as estimating the target domain loss with a privileged access to labeled target domain data. Over all 9000 certificates, we find a mean absolute difference between the two predictions of merely 0.049; in fact, all supplementary plots look highly similar to those displayed in Fig. 4, despite covering many other data sets, loss functions, and learning methods. The margin to the left of each vertical line appears because our certificate covers an absolute inter-domain gap rather than a signed value.

3.2   Binary (ϵ, δ) Certificates in Astro-Particle Physics

The field of astro-particle physics studies the physical properties of cosmic particle accelerators such as active galactic nuclei and supernova remnants. Some of these accelerators produce gamma radiation, which physicists measure through imaging air Cherenkov telescopes (IACTs). Since IACTs also record non-gamma particles, it is necessary to separate the relevant gamma recordings from the non-gamma background. This task is commonly approached with classification models trained on simulated data [4]. The accurate physical simulations that are used for training produce telescope readings (feature vectors) from user-chosen particles (labels). The default approach to this ACS problem is to simulate a training set with balanced classes and to alter the decision threshold of the model after its training.

We apply our certification scheme to the FACT telescope [1], an IACT for which a big data set is publicly available. In particular, we reproduce the default analysis pipeline, fix δ to a small value (0.01 or 0.1), and select ε such that the resulting (ε, δ) certificate covers the anticipated class proportion difference $\Delta p = |p_T - p_S|$ between the simulated and the observed domain. For both δ values, we obtain similar ε values under the zero-one loss, namely $\varepsilon_{(\delta=0.01)} = 0.0315$ and $\varepsilon_{(\delta=0.1)} = 0.0313$. We conclude that the ACS-induced zero-one loss of the FACT pipeline is at most 3.15% with probability at least 99%, and at most 3.13% with probability at least 90%. The pipeline is trustworthy within these specific ranges and improvements to these certified values can be achieved by improving the performance of the pipeline. See Appendix 2 for additional details.

4   Related Work

Most of the previous work on ACS has focused on the empirical evaluation of heuristic data acquisition strategies [6,10,13,15,16,21,28,29]. A recent theoretical contribution by us [5] is only valid for infinite data and lacks applicability to imbalanced domains; both of these issues motivate our present paper.

Model certification, in the broad sense of performance reports [2,17,22] and formal proofs of robustness [11,25,26,30], has motivated us to study model robustness in ACS. Our certificate is only a single component in the more comprehensive reports that are conceived in the literature; yet, the certification of feasible class proportions is a trust-critical issue when the training class proportions are chosen arbitrarily. A related lane of research is concerned with the certification of learning algorithms [18,19] instead of trained models.

Domain adaptation [20,27] assumes data from $\mathcal{T}$ with which a source domain model can be transferred to the target domain. If the data from $\mathcal{T}$ are unlabeled, it becomes necessary to employ additional assumptions about the differences between $\mathcal{S}$ and $\mathcal{T}$. For instance, our identical mechanism assumption from Definition 1 has also been introduced as the *target shift* assumption [31]. In ACS, we are free to choose the shift between $\mathbb{P}_{\mathcal{S}}(Y)$ and $\mathbb{P}_{\mathcal{T}}(Y)$ as small as permitted by our knowledge about $\mathcal{T}$, instead of having to adapt to $\mathcal{T}$. We conceive combinations of ACS and domain adaptation for future work.

Imbalanced learning [7] handles majority and minority classes differently from each other, so that the resulting classifier is not impaired by the dispro-portion between these classes. For instance, over-sampling the minority class with synthetic instances [3,8] will achieve more balanced training sets in which the minority class is not "overlooked" by the learning algorithm. In ACS, we can generate *actual* instances instead of synthetic ones; still, the idea of over-sampling can guide us in selecting the class proportions for an imbalanced target domain $\mathcal{T}$. Conversely, our certificate can guide imbalanced learners in choos-ing the amount of over-sampling or under-sampling to apply: the certified class proportion range $[\, p_{\mathcal{T}}^{\min},\, p_{\mathcal{T}}^{\max}]$ should ideally cover the class proportions $p_{\mathcal{T}}$ that are expected during deployment; otherwise the sampling scheme can introduce a domain gap larger than ε, which impairs the target domain performance.

Cost-sensitive learning is often discussed as a means to tackle imbalanced learning (e.g. Chap. 4 in [7]) because many applications associate a dispropor-tionally high cost with mis-classifications in the minority class. Our certificate supports these settings, without loss of generality, via class-wise loss weights.

Active learning [23] assumes that an oracle $\mathcal{X} \to \mathcal{Y}$ (e.g. a human expert) can label feature vectors after their acquisition. This assumption is fundamen-tally different from ACS, where a data generator $\mathcal{Y} \to \mathcal{X}$ produces feature vec-tors from labels. Still, some acquisition heuristics for ACS borrow from active learning strategies by aggregating the scores of pseudo-instances [13,15].

Quantification Learning [9] estimates class prevalences in the target domain, which can help in assessing the amount of label shift that is to be expected.

5   Conclusion

Motivated by a limited trust in active class selection, we have developed an (ε, δ) certificate for classifiers, which declares a set of class proportions to which the certified model can be safely applied. "Safely" means that the inter-domain gap induced by active class selection (or any other reason for a shift in the class proportions) is at most ε with probability at least $1 - \delta$. Our experiments show that the certificate is correct and bounds the true domain gap tightly.

So far, we have assumed that the loss function is decomposable over $\mathcal{X} \times \mathcal{Y}$, like the (weighted) zero-one loss, the hinge loss, and the mean squared error are. Future work should extend these results to loss functions that do not have this property, like the F_{β} and AUROC scores. We are also looking forward to extensions of our certificate towards multi-class settings and regression.

Acknowledgments. This work has been supported by Deutsche Forschungsgemeinschaft (DFG) within the Collaborative Research Center SFB 876 "Providing Information by Resource-Constrained Data Analysis", project C3, and by the Federal Ministry of Education and Research of Germany as part of the competence center for machine learning ML2R (01IS18038 A/B).

Appendix 1: Proof of the Identical Mechanism Bound

We draw a training set D of size m, where each individual example is drawn from $\mathcal{X} \times \mathcal{Y}$, according to $\mathbb{P}_S$. Consequently, the full training set is drawn from $(\mathcal{X} \times \mathcal{Y})^m$, according to the probability density $\mathbb{P}_S^m$. We are now interested in the probability that $\mathbb{P}_S^m$ assigns to the event that all $h \in \mathcal{H}$ admit to the identical mechanism bound:

$$\mathbb{P}_S^m \Big(\{ D : \forall h \in \mathcal{H}, \ |L_T(h) - L_S(h)| \ - \ \varepsilon \ \leq |L_T(h) - L_D(h)|$$

$$\leq |L_T(h) - L_S(h)| \ + \ \varepsilon \} \Big)$$

We estimate the above probability from the probability of the converse event; if the above probability is p, then the following must be $1 - p$:

$$\mathbb{P}_S^m \Big(\{ D : \exists h \in \mathcal{H}, \ |L_T(h) - L_S(h)| \ - \ \varepsilon \ > \ |L_T(h) - L_D(h)|$$

$$\wedge \quad |L_T(h) - L_D(h)| \ > \ |L_T(h) - L_S(h)| \ + \ \varepsilon \} \Big)$$

We now apply the union bound twice. This bound states that $\mathbb{P}(A \wedge B) \leq \mathbb{P}(A) + \mathbb{P}(B)$ for any two events A and B:

$$\ldots \quad \leq \ \mathbb{P}_S^m \Big(\{ D : \exists h \in \mathcal{H}, \ |L_T(h) - L_S(h)| \ - \ \varepsilon \ > \ |L_T(h) - L_D(h)| \} \Big)$$

$$+ \ \mathbb{P}_S^m \Big(\{ D : \exists h \in \mathcal{H}, \ |L_T(h) - L_D(h)| \ > \ |L_T(h) - L_S(h)| \ + \ \varepsilon \} \Big)$$

$$\leq \ \sum_{h \in \mathcal{H}} \mathbb{P}_S^m \Big(\{ D : |L_T(h) - L_S(h)| \ - \ \varepsilon \ > \ |L_T(h) - L_D(h)| \} \Big)$$

$$+ \ \sum_{h \in \mathcal{H}} \mathbb{P}_S^m \Big(\{ D : |L_T(h) - L_D(h)| \ > \ |L_T(h) - L_S(h)| \ + \ \varepsilon \} \Big)$$

We have thus reduced the probability of the identical mechanism bound w.r.t. an entire hypothesis class $\mathcal{H}$ to a sum of probabilities for single hypotheses $h \in \mathcal{H}$. The single-hypothesis case has already been proven in Sect. 1. Let us restate this result here to clarify the connection: Each of the following statements describes a violation of the Sect. 1 bound, each having a probability of at most $2e^{-2m\varepsilon^2}$:

- $\quad |L_T(h) - L_S(h)| \ - \ \varepsilon \ > \ |L_T(h) - L_D(h)|$
- $\quad |L_T(h) - L_D(h)| \ > \ |L_T(h) - L_S(h)| \ + \ \varepsilon$

These two events, together with their probabilities, can be plugged into the above transformation, which proves the claim:

$$\ldots \quad \leq \ \sum_{h \in \mathcal{H}} 2e^{-2m\varepsilon^2} + \sum_{h \in \mathcal{H}} 2e^{-2m\varepsilon^2} \ = \ 4|\mathcal{H}|e^{-2m\varepsilon^2} \qquad \square$$

Appendix 2: Experimental Details and Reproducibility

We provide an implementation of our proposed (ε, δ) certificate with the supplementary material of this paper. This material also contains a 56-page supplement of plots that can be reproduced with this implementation. All supplements are hosted at https://github.com/mirkobunse/AcsCertificates.jl.

The experiment in Sect. 3.1 verifies that (ε, δ) certificates are indeed correct and tight. However, by choosing ε as a function of Δp, we have "turned the certificate around"; in a usual application, a user would rather fix the ε value and look for a certified range Δp of feasible class proportions. Therefore, we provide the certified Δp values for all experiments in the supplementary material. Table 1 provides an excerpt of these values in which the certified target domain ranges $[p_S - \Delta p^*, \, p_S + \Delta p^*]$ induce a domain gap of at most $\varepsilon = 0.01$ with a probability of at least $1 - \delta = 0.95$. Since the domain gap is at most 0.01, we can expect a target domain loss of at most $L_S(h) + 0.01$.

Table 1. Feasible class proportions Δp^*, according to (ε, δ) certificates that are computed for a class-weighted zero-one loss with $\varepsilon = 0.01$ and $\delta = 0.05$.

Data	Classifier	$L_S(h)$	p_S	Δp^*
coil_2000	LogisticRegression	0.0722	0.0597	0.0109
coil_2000	DecisionTree	0.0778	0.0597	0.0107
letter_img	LogisticRegression	0.0179	0.0367	0.0463
letter_img	DecisionTree	0.0139	0.0367	0.0504
optical_digits	LogisticRegression	0.0406	0.0986	0.0437
optical_digits	DecisionTree	0.0463	0.0986	0.0309
pen_digits	LogisticRegression	0.038	0.096	0.044
pen_digits	DecisionTree	0.0216	0.096	0.0695
protein_homo	LogisticRegression	0.0056	0.0089	0.036
protein_homo	DecisionTree	0.006	0.0089	0.0291
satimage	LogisticRegression	0.1205	0.0973	0.0118
satimage	DecisionTree	0.0763	0.0973	0.018

Table 2 presents the results of our astro-particle experiment. The significance of detection [14] is a domain-specific score which measures the effectiveness of the telescope. While higher values are better, 25σ are a usual value for accurate prediction models on the given data set. The fact that all ε values are close to each other stems from the large amount of source domain data (24000 examples) we use to certify the model.

Table 2. The parameters of an (ε, δ) certificate that covers the extreme class proportions $p_T = 10^{-4}$ in astro-particle physics.

Significance of detection [σ]	$L_S(h)$	δ	ϵ_δ
25.067 ± 0.268	0.058 ± 0.015	0.01	0.0315
		0.025	0.0314
		0.05	0.0314
		0.1	0.0313

References

1. Anderhub, H., et al.: Design and operation of FACT-the first G-APD Cherenkov telescope. J. Inst. **8**(06) (2013). https://doi.org/10.1088/1748-0221/8/06/p06008
2. Arnold, M., et al.: FactSheets: increasing trust in AI services through supplier's declarations of conformity. IBM J. Res. Dev. **63**(4/5) (2019). https://doi.org/10.1147/jrd.2019.2942288
3. Bellinger, C., Sharma, S., Japkowicz, N., Zaïane, O.R.: Framework for extreme imbalance classification: SWIM - sampling with the majority class. Knowl. Inf. Syst. **62**(3) (2020). https://doi.org/10.1007/s10115-019-01380-z
4. Bockermann, C., et al.: Online analysis of high-volume data streams in astroparticle physics. In: Bifet, A., et al. (eds.) ECML PKDD 2015. LNCS (LNAI), vol. 9286, pp. 100–115. Springer, Cham (2015). https://doi.org/10.1007/978-3-319-23461-8_7
5. Bunse, M., Weichert, D., Kister, A., Morik, K.: Optimal probabilistic classification in active class selection. In: International Conference on Data Mining. IEEE (2020). https://doi.org/10.1109/icdm50108.2020.00106
6. Cakmak, M., Thomaz, A.L.: Designing robot learners that ask good questions. In: International Conference on Human-Robot Interaction. ACM (2012). https://doi.org/10.1145/2157689.2157693
7. Fernández, A., García, S., Galar, M., Prati, R.C., Krawczyk, B., Herrera, F.: Learning from Imbalanced Data Sets. Springer, Cham (2018). https://doi.org/10.1007/978-3-319-98074-4_11
8. Fernández, A., García, S., Herrera, F., Chawla, N.V.: SMOTE for learning from imbalanced data: progress and challenges, marking the 15-year anniversary. J. Artif. Intell. Res. **61** (2018). https://doi.org/10.1613/jair.1.11192
9. González, P., Castaño, A., Chawla, N.V., del Coz, J.J.: A review on quantification learning. ACM Comput. Surv. **50**(5) (2017). https://doi.org/10.1145/3117807
10. Hossain, I., Khosravi, A., Nahavandi, S.: Weighted informative inverse active class selection for motor imagery brain computer interface. In: Canadian Conference on Electrical and Computer Engineering. IEEE (2017). https://doi.org/10.1109/ccece.2017.7946613
11. Huang, J., Smola, A.J., Gretton, A., Borgwardt, K.M., Schölkopf, B.: Correcting sample selection bias by unlabeled data. In: Advances in Neural Information Processing Systems. MIT Press (2006)
12. Huang, X., Kroening, D., Ruan, W., Sharp, J., Sun, Y., Thamo, E., et al.: A survey of safety and trustworthiness of deep neural networks: verification, testing, adversarial attack and defence, and interpretability. Comput. Sci. Rev. **37** (2020). https://doi.org/10.1016/j.cosrev.2020.100270

13. Kottke, D., et al.: Probabilistic active learning for active class selection. In: NeurIPS Workshop on the Future of Interactive Learning Machine (2016)
14. Li, T.P., Ma, Y.Q.: Analysis methods for results in gamma-ray astronomy. Astrophysical J. **272** (1983). https://doi.org/10.1086/161295
15. Liu, S., Ding, W., Gao, F., Stepinski, T.F.: Adaptive selective learning for automatic identification of sub-kilometer craters. Neurocomputing **92** (2012). https://doi.org/10.1016/j.neucom.2011.11.023
16. Lomasky, R., Brodley, C.E., Aernecke, M., Walt, D., Friedl, M.: Active class selection. In: Kok, J.N., Koronacki, J., Mantaras, R.L., Matwin, S., Mladenič, D., Skowron, A. (eds.) ECML 2007. LNCS (LNAI), vol. 4701, pp. 640–647. Springer, Heidelberg (2007). https://doi.org/10.1007/978-3-540-74958-5_63
17. Mitchell, M., et al.: Model cards for model reporting. In: Conference on Fairness, Accountability, and Transparency. ACM (2019). https://doi.org/10.1145/3287560.3287596
18. Chatila, R., et al.: Trustworthy AI. In: Braunschweig, B., Ghallab, M. (eds.) Reflections on Artificial Intelligence for Humanity. LNCS (LNAI), vol. 12600, pp. 13–39. Springer, Cham (2021). https://doi.org/10.1007/978-3-030-69128-8_2
19. Morik, K., et al.: Yes we care! - certification for machine learning methods through the care label framework (2021). https://arxiv.org/abs/2105.10197
20. Pan, S.J., Yang, Q.: A survey on transfer learning. IEEE Trans. Knowl. Data Eng. **22**(10) (2010). https://doi.org/10.1109/tkde.2009.191
21. Parsons, T.D., Reinebold, J.L.: Adaptive virtual environments for neuropsychological assessment in serious games. IEEE Trans. Consum. Electron. **58**(2) (2012). https://doi.org/10.1109/tce.2012.6227413
22. Raji, I.D., et al.: Closing the AI accountability gap: defining an end-to-end framework for internal algorithmic auditing. In: Conference on Fairness, Accountability, and Transparency. ACM (2020). https://doi.org/10.1145/3351095.3372873
23. Settles, B.: Active Learning. Morgan & Claypool (2012). https://doi.org/10.2200/s00429ed1v01y201207aim018
24. Shalev-Shwartz, S., Ben-David, S.: Understanding Machine Learning - From Theory to Algorithms. Cambridge University Press, Cambridge (2014)
25. Singh, G., Gehr, T., Mirman, M., Püschel, M., Vechev, M.T.: Fast and effective robustness certification. In: Advances in Neural Information Processing Systems (2018)
26. Taori, R., Dave, A., Shankar, V., Carlini, N., Recht, B., Schmidt, L.: Measuring robustness to natural distribution shifts in image classification. In: Advances in Neural Information Processing Systems (2020)
27. Wang, M., Deng, W.: Deep visual domain adaptation: a survey. Neurocomputing **312** (2018). https://doi.org/10.1016/j.neucom.2018.05.083
28. Weiss, G.M., Provost, F.J.: Learning when training data are costly: the effect of class distribution on tree induction. J. Artif. Intell. Res. **19** (2003). https://doi.org/10.1613/jair.1199
29. Wu, D., Lance, B.J., Parsons, T.D.: Collaborative filtering for brain-computer interaction using transfer learning and active class selection **8**(2) (2013). https://doi.org/10.1371/journal.pone.0056624
30. Zhang, D., Ye, M., Gong, C., Zhu, Z., Liu, Q.: Black-box certification with randomized smoothing: a functional optimization based framework. In: Advances in Neural Information Processing Systems (2020)
31. Zhang, K., Schölkopf, B., Muandet, K., Wang, Z.: Domain adaptation under target and conditional shift. In: International Conference on Machine Learning (2013)

Graphs and Networks

Inter-domain Multi-relational Link Prediction

Luu Huu Phuc[1]($\boxtimes$), Koh Takeuchi[1], Seiji Okajima[2], Arseny Tolmachev[2], Tomoyoshi Takebayashi[2], Koji Maruhashi[2], and Hisashi Kashima[1]

[1] Kyoto University, Kyoto, Japan
{phuc,koh,kashima}@ml.ist.i.kyoto-u.ac.jp
[2] Fujitsu Research, Fujitsu Ltd, Kanagawa, Japan
{okajima.seiji,t.arseny,takebayashi.tom,maruhashi.koji}@fujitsu.com

Abstract. Multi-relational graph is a ubiquitous and important data structure, allowing flexible representation of multiple types of interactions and relations between entities. Similar to other graph-structured data, link prediction is one of the most important tasks on multi-relational graphs and is often used for knowledge completion. When related graphs coexist, it is of great benefit to build a larger graph via integrating the smaller ones. The integration requires predicting hidden relational connections between entities belonged to different graphs (inter-domain link prediction). However, this poses a real challenge to existing methods that are exclusively designed for link prediction between entities of the same graph only (intra-domain link prediction). In this study, we propose a new approach to tackle the inter-domain link prediction problem by softly aligning the entity distributions between different domains with optimal transport and maximum mean discrepancy regularizers. Experiments on real-world datasets show that optimal transport regularizer is beneficial and considerably improves the performance of baseline methods.

Keywords: Inter-domain link prediction · Multi-relational data · Optimal transport

1 Introduction

Multi-relational data represents knowledge about the world and provides a graph-like structure of this knowledge. It is defined by a set of entities and a set of predicates between these entities. The entities can be objects, events, or abstract concepts while the predicates represent relationships involving two entities. A multi-relational data contains a set of facts represented as triplets (e_h, r, e_t) denoting the existence of a predicate r from subject entity e_h to object entity e_t. In a sense, multi-relational data can also be seen as a directed graph with multiple types of links (multi-relational graph).

A multi-relational graph is often very sparse with only a small subset of true facts being observed. Link prediction aims to complete a multi-relational graph by

© Springer Nature Switzerland AG 2021
N. Oliver et al. (Eds.): ECML PKDD 2021, LNAI 12976, pp. 285–301, 2021.
https://doi.org/10.1007/978-3-030-86520-7_18

predicting new hidden true facts based on the existing ones. Many existing methods follow an embedding-based approach which has been proved to be effective for multi-relational graph completion. These methods all aim to find reasonable embedding presentations for each entity (node) and each predicate (type of link). In order to predict if a fact (e_h, r, e_t) holds true, they use a scoring function whose inputs are embeddings of the entities e_h, e_t and the predicate r to compute a prediction score. Some of the most prominent methods in that direction are TransE [3], RESCAL [22], DisMult [35], and NTN [27], to name a few.

TransE [3] model is inspired by the intuition from Word2Vec [18,19] that many predicates represent linear translations between entities in the latent embedding space, e.g. $a_{Japan} - a_{Tokyo} \approx a_{Germany} - a_{Berlin} \approx a_{is\_capital\_of}$. Therefore, TransE tries to learn low-dimensional and dense embedding vectors so that $a_h + a_r \approx a_t$ for a true fact (e_h, r, e_t). Its scoring function is defined accordingly via $\|a_h + a_r - a_t\|_2$. RESCAL [22] is a tensor factorization-based method. It converts a multi-relational graph data into a 3-D tensor whose first two modes indicate the entities and the third mode indicates the predicates. A low-rank decomposition technique is employed by RESCAL to compute embedding vectors a of the entities and embedding matrices R of the predicates. Its scoring function is the bilinear product $a_h^\top R_r a_t$. DistMult [35] is also a bilinear model and is based on RESCAL where each predicate is only represented by a diagonal matrix rather than a full matrix. The neural tensor network (NTN) model [27] generalizes RESCAL's approach by combining traditional MLPs and bilinear operators to represent each relational fact.

Despite achieving state of the art for link prediction tasks, existing methods are exclusively designed and limited to intra-domain link prediction. They only consider the case in which both entities belong to the same relational graph (intra-domain). When the needs for predicting hidden facts between entities of different but related graphs (inter-domain) arise, unfortunately, the existing methods are inapplicable. One of such examples is when it is necessary to build a large relational graph by integrating several existing smaller graphs whose entity sets are related. This study proposes to tackle the inter-domain link prediction problem by learning suitable latent embeddings that minimize dissimilarity between the domains' entity distributions.

Two popular divergences, namely optimal transport's Wasserstein distance (WD) and the maximum mean discrepancy (MMD), are investigated. Given two probability distributions, optimal transport computes an optimal transport plan that gives the minimum total transport cost to relocate masses between the distributions. The minimum total transport cost is often known under the name of Wasserstein distance. In a sense, the computed optimal transport plan and the corresponding Wasserstein distance provide a reasonable alignment and quantity for measuring the dissimilarity between the supports/domains of the two distributions. Minimizing Wasserstein distance has been proved to be effective in enforcing the alignment of corresponding entities across different domains and is successfully applied in graph matching [34], cross-domain alignment [7], and multiple-graph link prediction problems [25]. As another popular statistical divergence between distributions, MMD computes the dissimilarity by comparing the kernel mean embeddings of two distributions in a reproducing kernel Hilbert space (RKHS). It has been widely applied in two-sample tests for differentiating distributions [12,13] and distribution matching in domain adaptation tasks [6], to name a few.

The proposed method considers a setting of two multi-relational graphs whose entities are assumed to follow the same underlying distribution. For example, the multi-relational graphs can be about relationships among users/items in different e-commerce flatforms of the same country. They could also be knowledge graphs of semantic relationships between general concepts that are built from different common-knowledge sources, e.g. Freebase and DBpedia. In both examples, it is safe to assume that the entity sets are distributionally identical. This assumption is fundamental for the regularizers to be effective in connecting the entity distributions of the two graphs.

2 Preliminary

This section briefly introduces the components that are employed in the proposed method.

2.1 RESCAL

RESCAL [22] formulates a multi-relational data as a three-way tensor $\mathbf{X} \in \mathbb{R}^{n \times n \times m}$, where n is the number of entities and m is the number of predicates. $\mathbf{X}_{i,j,k} = 1$ if the fact (e_i, r_k, e_j) exists and $\mathbf{X}_{i,j,k} = 0$ otherwise. In order to find proper latent embeddings for the entities and the predicates, RESCAL performs a rank-d factorization where each slice along the third mode $\mathcal{X}_k = \mathbf{X}_{\cdot,\cdot,k}$ is factorized as

$$\mathcal{X}_k \approx \mathbf{A}\mathbf{R}_k\mathbf{A}^\top, \quad \text{for } k = 1, ..., m.$$

Here, $\mathbf{A} = [\mathbf{a}_1, ..., \mathbf{a}_n]^\top \in \mathbb{R}^{n \times d}$ contains the latent embedding vectors of the entities and $\mathbf{R}_k \in \mathbb{R}^{d \times d}$ is an asymmetric matrix that represents the interactions between entities in the k-th predicate.

Originally, it is proposed to learn $\mathbf{A}$ and $\mathbf{R}_k$ with the regularized squared loss function

$$\min_{\mathbf{A}, \mathbf{R}_k} g(\mathbf{A}, \mathbf{R}_k) + \text{reg}(\mathbf{A}, \mathbf{R}_k),$$

where

$$g(\mathbf{A}, \mathbf{R}_k) = \frac{1}{2} \left(\sum_k \|\mathcal{X}_k - \mathbf{A}\mathbf{R}_k\mathbf{A}^\top\|_F^2 \right)$$

and reg is the following regularization term

$$\text{reg}(\mathbf{A}, \mathbf{R}_k) = \frac{1}{2}\mu \left(\|\mathbf{A}\|_F^2 + \sum_k \|\mathbf{R}_k\|_F^2 \right).$$

$\mu > 0$ is a hyperparameter.

It is later proposed by the authors of RESCAL to learn the embeddings with pairwise loss training [21], i.e. using the following margin-based ranking loss function

$$\min_{\mathbf{A}, \mathbf{R}_k} L(\mathbf{A}, \mathbf{R}_k) = \sum_{(e_i, r_k, e_j) \in \mathcal{D}^+} \sum_{(e_l, r_h, e_t) \in \mathcal{D}^-} \mathcal{L}(f_{ijk}, f_{lth}) + \mathrm{reg}(\mathbf{A}, \mathbf{R}_k), \tag{1}$$

where $\mathcal{D}^+$ and $\mathcal{D}^-$ are the sets of all positive triplets (true facts) and all negative triplets (false facts), respectively. f_{ijk} denotes the score of (e_i, r_k, e_j), $f_{ijk} = \mathbf{a}_i^\top \mathbf{R}_k \mathbf{a}_j$ and $\mathcal{L}$ is the ranking function

$$\mathcal{L}(f^+, f^-) = \max(1 + f^- - f^+, 0).$$

The negative triplet set $\mathcal{D}^-$ is often generated by corrupting positive triplets, i.e. replacing one of the two entities in a positive triplet (e_i, r_k, e_j) with a randomly sampled entity.

The pairwise loss training aims to learn $\mathbf{A}$ and $\mathbf{R}_k$ so that the score f^+ of a positive triplet is higher than the score f^- of a negative triplet. Moreover, the margin-based ranking function is more flexible and easier to optimize with *stochastic gradient descent* (SGD) than the original squared loss function. In the proposed method, the pairwise loss training is adopted.

2.2 Optimal Transport

Given two probability vectors $\boldsymbol{\pi}_1 \in \mathbb{R}_+^{n_1}$ and $\boldsymbol{\pi}_2 \in \mathbb{R}_+^{n_2}$ that satisfy $\boldsymbol{\pi}_1^\top \mathbb{1}_{n_1} = \boldsymbol{\pi}_2^\top \mathbb{1}_{n_2} = 1$, a matrix $\mathbf{P} \in \mathbb{R}_+^{n_1 \times n_2}$ is called a transport plan between $\boldsymbol{\pi}_1$ and $\boldsymbol{\pi}_2$ if $\mathbf{P}\mathbb{1}_{n_2} = \boldsymbol{\pi}_1$ and $\mathbf{P}^\top \mathbb{1}_{n_1} = \boldsymbol{\pi}_2$. Here, $\mathbb{1}_n$ indicates a n-dimensional vector of ones. Let's denote the supports of $\boldsymbol{\pi}_1$ and $\boldsymbol{\pi}_2$ as $\mathbf{A}^1 = [\mathbf{a}_1^1, ..., \mathbf{a}_{n_1}^1]^\top \in \mathbb{R}^{n_1 \times d}$ and $\mathbf{A}^2 = [\mathbf{a}_1^2, ..., \mathbf{a}_{n_2}^2]^\top \in \mathbb{R}^{n_2 \times d}$, respectively. A transport cost $\mathbf{C} \in \mathbb{R}_+^{n_1 \times n_2}$ can be defined as

$$C_{ij} = \|\mathbf{a}_i^1 - \mathbf{a}_j^2\|_2^2.$$

Given a transport matrix C, the transport cost of a transport plan $\mathbf{P}$ is computed by

$$\langle \mathbf{P}, \mathbf{C} \rangle = \sum_{i,j} P_{ij} C_{ij}.$$

A transport plan $\mathbf{P}^*$ that gives the minimum transport cost, $\mathbf{P}^* = \arg\min_{\mathbf{P}} \langle \mathbf{P}, \mathbf{C} \rangle$, is called an optimal transport plan and the corresponding minimum cost is called the Wasserstein distance. The optimal transport plan $\mathbf{P}^*$ gives a reasonable "soft" matching between the two distributions $(\boldsymbol{\pi}_1, \mathbf{A}_1)$ and $(\boldsymbol{\pi}_2, \mathbf{A}_2)$ while the Wasserstein distance provides a measurement of how far the two distributions are from each other.

In the scope of multi-relational graphs, $\boldsymbol{\pi}_1$ and $\boldsymbol{\pi}_2$ are predefined over the sets of entities, normally being set to be uniform and the supports $\mathbf{A}_1$ and $\mathbf{A}_2$ can be seen as embeddings of the entities.

The computational complexity of computing the optimal transport plan and Wasserstein distance is often prohibitive. An efficient approach to compute an approximation has been proposed by Cuturi et al. [9]. Instead of the exact optimal transport $\mathbf{P}^*$, they compute an entropic-regularized transport plan $\mathbf{P}^\lambda$ via minimizing a cost M as follows,

$$\mathbf{P}^\lambda = \arg\min_{\mathbf{P}} M(\mathbf{P}) = \langle \mathbf{P}, \mathbf{C} \rangle + \frac{1}{\lambda} \sum_{i,j} P_{ij} \log P_{ij}, \qquad (2)$$

where $\lambda > 0$ is a hyperparameter controlling the effect of the negative entropy of matrix $\mathbf{P}$. With large enough λ, emperically when $\lambda > 50$, $\mathbf{P}^*$ and the Wasserstein distance can be accurately approximated by $\mathbf{P}^\lambda$ and $M(\mathbf{P}^\lambda)$.

$\mathbf{P}^\lambda$ has a unique solution of the following form

$$\mathbf{P}^\lambda = \mathbf{diag}(\mathbf{u})\mathbf{K}\mathbf{diag}(\mathbf{v}),$$

where $\mathbf{diag}(\mathbf{u})$ indicates a diagonal matrix whose diagonal elements are elements of $\mathbf{u}$. The matrix $\mathbf{K} = e^{-\lambda \mathbf{C}}$ is the element-wise exponential of $-\lambda \mathbf{C}$. Vectors $\mathbf{u}$ and $\mathbf{v}$ can be initialized randomly and updated via Sinkhorn iteration

$$(\mathbf{u}, \mathbf{v}) \leftarrow \left(\frac{\boldsymbol{\pi}_1}{\mathbf{K}\mathbf{v}}, \frac{\boldsymbol{\pi}_2}{\mathbf{K}^\top \mathbf{u}} \right).$$

2.3 Maximum Mean Discrepancy

Maximum Mean Discrepancy (MMD) is originally introduced as a non-parametric statistic to test if two distributions are different [12,13]. It is defined as the difference between mean function values on samples generated from the distributions. If MMD is large, the two distributions are likely to be distinct. On the other hand, if MMD is small, the two distributions can be seen to be similar. Formally, let $\boldsymbol{\pi}_1$ and $\boldsymbol{\pi}_2$ be two distributions whose the supports are subsets of $\mathbb{R}^d$, and $\mathcal{F}$ be a class of functions $f : \mathbb{R}^d \to \mathbb{R}$. Usually, $\mathcal{F}$ is selected to be the unit ball in a universal RKHS $\mathcal{H}$. Then MMD is defined as

$$M(\mathcal{F}, \boldsymbol{\pi}_1, \boldsymbol{\pi}_2) = \sup_{f \in \mathcal{F}} \left(\mathbb{E}_{x \sim \boldsymbol{\pi}_1}[f(x)] - \mathbb{E}_{y \sim \boldsymbol{\pi}_2}[f(y)] \right).$$

From sample sets $\mathbf{A}^1 = \{\mathbf{a}_1^1, ..., \mathbf{a}_{n_1}^1\}$ and $\mathbf{A}^2 = \{\mathbf{a}_1^2, ..., \mathbf{a}_{n_2}^2\}$, $\mathbf{a}_i^t \in \mathbb{R}^d$, sampled from the two distributions, MMD can be unbiasedly approximated using Gaussian kernels $k(\cdot, \cdot)$ as follows [12,23].

$$
\begin{aligned}
M(\mathbf{A}^1, \mathbf{A}^2) = & \frac{1}{n_1(n_1 - 1)} \sum_{i,i'} k(\mathbf{a}_i^1, \mathbf{a}_{i'}^1) + \frac{1}{n_2(n_2 - 1)} \sum_{j,j'} k(\mathbf{a}_j^2, \mathbf{a}_{j'}^2) \\
& - \frac{2}{n_1 n_2} \sum_{i,j} k(\mathbf{a}_i^1, \mathbf{a}_j^2)
\end{aligned}
\qquad (3)
$$

When $\mathbf{A}^1$ and $\mathbf{A}^2$ are the embeddings of entities in two domains, MMD represents a dissimilarity between the domains' entity distributions.

3 Problem Setting and Proposed Method

3.1 Problem Setting

The formal problem setting considered in this study is stated as follows. Given two multi-relational graphs G^1 and G^2, each graph G^t is defined with a set of entities (nodes) $\mathcal{E}^t = \{e_1^t, ..., e_{n_t}^t\}$, a set of predicates (types of links) $\mathcal{R}^t = \{r_1^t, ..., r_{m_t}^t\}$, and a set of true facts (observed links) $\mathcal{T}^t = \{(e_i^t, r_k^t, e_j^t)\}$ for $t \in \{1, 2\}$. For simplicity, this study only considers the case where the two graphs share the same set of predicates, i.e. $\mathcal{R}^1 \equiv \mathcal{R}^2 \equiv \mathcal{R}$. The goal is to predict if an inter-domain fact (e_i^1, r_k, e_j^2) or (e_i^2, r_k, e_j^1) holds true or not.

The entity embeddings of the two graphs are assumed to follow the same distribution, i.e. there exists a distribution π such that $\mathbf{a}_i^t \sim \pi$ for embedding $\mathbf{a}_i^t$ of entity $e_i^t \in \mathcal{E}^t$. In the experiments, the entity sets $\mathcal{E}^1$ and $\mathcal{E}^2$ are controlled so that they are completely disjoint or partially overlapped with only a small amount of common entities. The common entities are known in overlapping settings.

3.2 Proposed Objective Function

The proposed method's objective function consists of two components. The first component is for learning embedding representations of the entities and the predicates of each multi-relational graph, which is based on an existing tensor-factorization method. RESCAL [22] is specifically chosen in the proposed method. The second component is a regularization term for enforcing the entity embedding distributions of the two graphs to become similar.

For each graph G^t, lets denote the entity embeddings as $\mathbf{A}^t = [\mathbf{a}_1^t, ..., \mathbf{a}_{n_t}^t]^\top \in \mathbb{R}^{n_t \times d}$, where d is the embedding dimension. If the entity sets $\mathcal{E}^1$ and $\mathcal{E}^2$ overlap, the embeddings of common entities are set to be identical in both domains, i.e. $\mathbf{A}^t = [\mathbf{A}'^t, \mathbf{A}_c]^\top$ where $\mathbf{A}_c \in \mathbb{R}^{d \times |\mathcal{E}^1 \cap \mathcal{E}^2|}$ is the embeddings of common entities. The embedding of predicate $r_k \in \mathcal{R}$ is denoted as $\mathbf{R}_k \in \mathbb{R}^{d \times d}$ for $k \in \{1, ..., m\}$. The objective function of the proposed method is given as

$$F(\mathbf{A}^1, \mathbf{A}^2, \mathbf{R}_k, [\mathbf{P}]) = L(\mathbf{A}^1, \mathbf{R}_k) + L(\mathbf{A}^2, \mathbf{R}_k) + \alpha M(\mathbf{A}^1, \mathbf{A}^2, [\mathbf{P}]). \quad (4)$$

In (4), the first two terms $L(\mathbf{A}^t, \mathbf{R}_k)$ are the loss functions of RESCAL and are defined as in (1). The third term $M(\mathbf{A}^1, \mathbf{A}^2, [\mathbf{P}])$ is the entropic-regularized Wasserstein distance (WD) or the MMD discrepancy between the entity distributions of the two graphs. In the case of WD regularizer, $M = M(\mathbf{A}^1, \mathbf{A}^2, \mathbf{P})$ is defined as in (2) with $\mathbf{P} \in \mathbb{R}_+^{n_1 \times n_2}$. In the case of MMD regularizer, $M = M(\mathbf{A}^1, \mathbf{A}^2)$ is defined as in (3).

Via $L(\mathbf{A}^t, \mathbf{R}_k)$, the underlying embedding distribution over each entity set $\mathcal{E}^t$ is learned and characterized into $\mathbf{A}^t$, while $M(\mathbf{A}^1, \mathbf{A}^2, [\mathbf{P}])$ helps to drive these two distributions to become similar. Through the objective function F, similar entities of G^1 and G^2 are expected to lie close to each other on the latent embedding space, which encourages similar entities to involve in similar relations/links. Specifically, if $e_i^1 \in \mathcal{E}^1$ and $e_i^2 \in \mathcal{E}^2$ have similar embeddings $\mathbf{a}_i^1$ and $\mathbf{a}_i^2$, the inter-domain fact (e_i^1, r_k, e_j^2) is likely to exist if the intra-domain fact (e_i^2, r_k, e_j^2) exists thanks to their similar scores $\mathbf{a}_i^{1\top} \mathbf{R}_k \mathbf{a}_j^2 \approx \mathbf{a}_i^{2\top} \mathbf{R}_k \mathbf{a}_j^2$.

The objective function $F(\mathbf{A}^1, \mathbf{A}^2, \mathbf{R}_k)$ (MMD regularizer) is directly optimized with SGD. On the other hand, $F(\mathbf{A}^1, \mathbf{A}^2, \mathbf{R}_k, \mathbf{P})$ (WD regularizer) is minimized iteratively. In each epoch, the transport plan $\mathbf{P}$ is fixed and the embedding vectors $\mathbf{A}^1$ and $\mathbf{A}^2$ are updated with SGD. At the end of each epoch, $\mathbf{A}^1$ and $\mathbf{A}^2$ are fixed and the plan $\mathbf{P}$ is sequentially updated via Sinkhorn algorithm [9].

4 Experiments

4.1 Datasets

The datasets used in the experiments are created from four popular knowledge graph datasets, namely FB15k-237 [30], WN18RR [10], DBbook2014, and ML1M [5]. The FB15k-237 dataset contains $272k$ facts about general knowledge. It has $14k$ entities and 237 predicates. The WN18RR dataset consists of $86k$ facts about 11 lexical relations between $40k$ word senses. The other two datasets represent interactions among users and items in e-commerce. The ML1M (MovieLens-1M) dataset composes of $434k$ facts with $14k$ users/items and 20 relations, while the DBbook2014 has $334k$ facts with $13k$ users/items and 13 relations. To create G^1 and G^2 for each dataset, two smaller sub-graphs of around $2k$ to $3k$ entities are randomly sampled from the original graph. The two graphs are controlled to share some amounts of common entities. Different levels of entity overlapping are investigated, from 0% (non-overlapping setting) to around 1.5%, 3%, and 5% (overlapping setting). Moreover, different predicates are removed so that G^1 and G^2 share the same predicate set, i.e. $\mathcal{R}^1 \equiv \mathcal{R}^2 \equiv \mathcal{R}$.

Intra-domain triplets (e_i, r_k, e_j) whose both entities e_i, e_j belong to the same graph are used for training. Inter-domain triplets (e_i, r_k, e_j) whose entities e_i, e_j belong to different graphs are used for validating and testing inter-domain performance. The validation and test ratio is 20 : 80. Even though the goal is to evaluate a model's ability to perform inter-domain link prediction, both inter-domain and intra-domain link prediction performances are evaluated. This is because the proposed method should improve inter-domain link prediction while does not harm intra-domain link prediction. Therefore, 5% of intra-domain triplets are further spared from the training data for monitoring intra-domain performance.

The details for the case of 3% overlapping are shown in Table 1. In other cases, the datasets share similar statistics.

4.2 Evaluation Methods and Baselines

In the experiments, Hit@10 score and ROC-AUC score are used for quantifying both inter-domain and intra-domain performances.

Table 1. Details of the datasets in the case of 3% overlapping. The other cases share similar statistics.

Datasets	#Ent G1	#Ent G2	#Rel	#Train	#Inter Valid	#Intra Test	#Inter Test
FB15k-237	2675	2677	179	24.3k	4.3k	1.3k	17.7k
WN18RR	2804	2720	10	5.1k	105	148	1.1k
DBbook2014	2932	2893	11	34.6k	6.5k	1.8k	26.8k
ML1M	2764	2726	18	39.3k	6.5k	2k	27k

Evaluation with Hit@10. The Hit@10 score is computed by ranking true entities based on their scores. For each true triplet (e_i, r_k, e_j) in the test sets, one entity e_i (or e_j) is hidden to create an unfinished triplet $(\cdot, r_k, e_j)$ (or $(e_i, r_k, \cdot)$). All entities e_{cand} are used as candidates for completing the unfinished triplet and the scores of $(e_{\text{cand}}, r_k, e_j)$ (or $(e_i, r_k, e_{\text{cand}})$) are computed. Note that the candidates e_{cand} are taken from the same entity set as e_i (or e_j), i.e. if e_i (or e_j) $\in \mathcal{E}^t$ then entities e_{cand} are taken from $\mathcal{E}^t$. The ranking of e_i (or e_j) is computed according to the scores. The higher "true" entities are ranked the better a model is at predicting hidden true triplets. Hit@10 score is used for quantifying the link prediction performance and is calculated as the percentage of "true" entities being ranked inside the top 10.

Evaluation with ROC-AUC. In order to compute the ROC-AUC score, triplets in the test set are treated as positive samples. An equal number of triplets are uniformly sampled from the entity sets and the predicate set to create negative samples. Due to the sparsity of each graph, it is safe to consider the sampled triplets as negative. During the sampling process, both sampled entities are controlled to belong to the same graph in the intra-domain case and belong to different graphs in the inter-domain case.

Evaluated Models. In the experiments, RESCAL is used as the baseline method. The proposed method with Wasserstein regularization is denoted as WD while the one with MMD regularization is denoted as MMD.

4.3 Implementation Details

Negative Sampling. Only intra-domain negative triplets are used in order to train the pairwise ranking loss (1) with SGD, i.e. negative triplet set $\mathcal{D}^-$ only contains negative triplets (e_l, r_h, e_t) whose both entities belong to the same graph.

Warmstarting. Completely learning from scratch might be difficult since the regularizer M can add noise at the early state. Instead, it is beneficial to warmstart the proposed method's embeddings with embeddings roughly learned by RESCAL. Specifically, we run RESCAL for 100 epochs to learn initial embeddings. After that, to maintain the fairness of equal training time, both the proposed method and RESCAL are warmstarted with the roughly learned embeddings.

Hyperparameters. In the implementation, the latent embedding dimension is set to equal 100. All experiments are run for 300 epochs. Early stopping is employed with a patience budget of 50 epochs. Other hyperparameters, namely α, learning rate, and batch size, are tuned on the inter-domain validation set using Optuna [1]. During the tuning process, α is sampled to be between 0.5 and 10.0, while the learning rate and batch size are chosen from $\{0.01, 0.005, 0.001, 0.0005\}$ and $\{100, 300, 500, 700\}$, respectively. The hyperparameters of RESCAL is tuned similarly with fixed $\alpha = 0.0$. The kernel used in MMD is set to be a mixture of Gaussian kernels with the bandwidth list of $[0.25, 0.5, 1., 2., 4.] * c$ where c is the mean Euclidean distance between the entities. All results are averaged over 10 random runs[1].

4.4   Experimental Results

The experimental results are shown in Tables 2, 3, 4, and 5. Note that a random predictor has a Hit@10 score of less than 0.004 and a ROC-AUC score of around 0.5.

Inter-domain Link Prediction. As being demonstrated in Tables 2 and 3, the proposed method with WD regularizer works well with the FB15k-237 dataset, which outperforms RESCAL in all settings. Especially in the overlapping cases where few entities are shared between the graphs, both Hit@10 and ROC-AUC scores are improved significantly. The WD regularizer also demonstrates its usefulness with the DBbook2014 and ML1M datasets. The Hit@10 scores are boosted up in most cases of overlapping settings, while the ROC-AUC scores are consistently enhanced over that of RESCAL. Most of the time, the improvements are considerable. However, for the case of the ML1M dataset with 3% overlapping entities, the WD regularizer causes the Hit@10 score to deteriorate, from 0.230 to 0.213. On the other hand, the MMD regularizer seems not to be beneficial for the task. Unexpectedly, the regularizer introduces noise and reduces the accuracy of inter-domain link prediction. In the case of the WN18RR dataset, both RESCAL and the proposed method fail to perform, in which all Hit@10 and ROC-AUC scores are close to random. This might be due to the extreme sparsity of the dataset, whose amount of observed triplets is only about one-fifth of that of the other datasets.

In all the four datasets, sharing some common entities, even with a small number, is helpful and important for predicting inter-domain links. These common entities act as anchors between the graphs, which guide the regularizer to learn similar embedding distributions. Without common entities, the learning process becomes more challenging and often results in uncertain predictors as being shown in the 0% overlapping cases. The overlapping setting is reasonable because, in practice, the two graphs often share some amounts of common entities, e.g. the same users and the same popular items reappear in different e-commerce platforms.

---

[1] The code is available at https://github.com/phucdoitoan/inter-domain_lp.

Table 2. Inter-domain Hit@10 scores. Italic numbers indicate better results while bold numbers and bold numbers with asterisk * indicate better results at significance level $p = 0.1$ and $p = 0.05$, respectively. The proposed method with WD regularizer achieves better scores in many settings.

Overlapping	Model	FB15k-237	WN18RR	DBbook2014	ML1M
0%	RESCAL	0.110 ± 0.038	0.027 ± 0.003	0.087 ± 0.058	0.062 ± 0.074
	MMD	0.111 ± 0.038	0.031 ± 0.004	0.085 ± 0.057	0.063 ± 0.072
	WD	**0.145 ± 0.063**	0.024 ± 0.004	0.084 ± 0.070	0.061 ± 0.067
1.5%	RESCAL	0.251 ± 0.031	0.025 ± 0.002	0.107 ± 0.035	0.210 ± 0.034
	MMD	0.237 ± 0.043	0.026 ± 0.003	0.109 ± 0.037	0.180 ± 0.067
	WD	**0.291 ± 0.031***	0.024 ± 0.002	*0.128 ± 0.059*	**0.240 ± 0.031***
3%	RESCAL	0.302 ± 0.020	0.028 ± 0.004	0.266 ± 0.056	**0.230 ± 0.003***
	MMD	0.292 ± 0.020	0.026 ± 0.004	0.227 ± 0.081	0.228 ± 0.002
	WD	**0.328 ± 0.011***	0.025 ± 0.004	**0.318 ± 0.066***	0.213 ± 0.006
5%	RESCAL	0.339 ± 0.007	0.027 ± 0.005	0.389 ± 0.032	0.237 ± 0.011
	MMD	0.334 ± 0.006	0.026 ± 0.004	0.388 ± 0.027	0.236 ± 0.010
	WD	**0.361 ± 0.010***	0.031 ± 0.004	0.389 ± 0.051	**0.256 ± 0.006***

Table 3. Inter-domain ROC-AUC scores. Italic numbers indicate better results while bold numbers and bold numbers with asterisk * indicate better results at significance level $p = 0.1$ and $p = 0.05$, respectively. The proposed method with WD regularizer achieves better scores in many settings.

Overlapping	Model	FB15k-237	WN18RR	DBbook2014	ML1M
0%	RESCAL	0.504 ± 0.092	0.504 ± 0.009	0.483 ± 0.097	0.464 ± 0.173
	MMD	0.507 ± 0.093	0.500 ± 0.010	0.485 ± 0.095	0.480 ± 0.172
	WD	*0.548 ± 0.118*	0.505 ± 0.009	0.488 ± 0.099	0.495 ± 0.179
1.5%	RESCAL	0.793 ± 0.044	0.512 ± 0.009	0.640 ± 0.066	0.805 ± 0.027
	MMD	0.770 ± 0.063	0.507 ± 0.009	0.632 ± 0.063	0.754 ± 0.087
	WD	**0.837 ± 0.033***	0.510 ± 0.007	*0.671 ± 0.087*	**0.842 ± 0.017***
3%	RESCAL	0.825 ± 0.022	0.503 ± 0.009	0.762 ± 0.032	0.832 ± 0.006
	MMD	0.813 ± 0.030	0.498 ± 0.011	0.714 ± 0.060	0.831 ± 0.007
	WD	**0.850 ± 0.013***	0.502 ± 0.013	**0.809 ± 0.030***	0.840 ± 0.008
5%	RESCAL	0.870 ± 0.008	0.498 ± 0.021	0.824 ± 0.012	0.845 ± 0.007
	MMD	0.875 ± 0.007	0.498 ± 0.012	0.823 ± 0.015	0.845 ± 0.006
	WD	**0.902 ± 0.010***	0.498 ± 0.013	**0.835 ± 0.020**	**0.867 ± 0.003***

Intra-domain Link Prediction. Even though the main goal is to predict inter-domain links, it is preferable that the regularizers do not harm performance on intra-domain link prediction when fusing the two domains' entity distributions. As being demonstrated in Table 5, the proposed method is able to maintain similar or better intra-domain ROC-AUC scores compared to RESCAL. However, it sometimes requires trade-offs in terms of the Hit@10 score, which is shown in

Table 4. Intra-domain Hit@10 scores. Bold numbers with asterisk * indicate better results at significance level $p = 0.05$. Generally, the proposed method with WD regularizer preserves the intra-domain Hit@10 scores despite requiring trade-offs in some cases.

Overlapping	Model	FB15k-237	WN18RR	DBbook2014	ML1M
0%	RESCAL	0.451 ± 0.031	0.418 ± 0.031	0.468 ± 0.011	0.302 ± 0.076
	MMD	0.461 ± 0.029	0.342 ± 0.086	0.449 ± 0.012	0.307 ± 0.070
	WD	0.469 ± 0.019	0.421 ± 0.032	0.472 ± 0.014	0.332 ± 0.027
1.5%	RESCAL	0.433 ± 0.008	0.390 ± 0.040	0.296 ± 0.039	**0.425 ± 0.006***
	MMD	0.438 ± 0.008	0.330 ± 0.067	0.328 ± 0.027	0.423 ± 0.036
	WD	0.427 ± 0.009	0.408 ± 0.035	0.291 ± 0.038	0.412 ± 0.008
3%	RESCAL	0.433 ± 0.009	0.476 ± 0.074	0.413 ± 0.008	**0.447 ± 0.006***
	MMD	0.447 ± 0.011	0.485 ± 0.074	0.411 ± 0.017	0.444 ± 0.008
	WD	0.439 ± 0.009	**0.620 ± 0.026***	0.412 ± 0.009	0.413 ± 0.021
5%	RESCAL	**0.433 ± 0.009***	0.455 ± 0.038	0.418 ± 0.010	0.408 ± 0.005
	MMD	0.421 ± 0.009	0.416 ± 0.058	0.420 ± 0.014	0407 ± 0.004
	WD	0.413 ± 0.007	0.479 ± 0.076	0.412 ± 0.022	0.401 ± 0.005

Table 5. Intra-domain ROC-AUC scores. Bold numbers with asterisk * indicate better results at significance level $p = 0.05$. The propose method maintains similar or better intra-domain ROC-AUC scores compared to RESCAL.

Overlapping	Model	FB15k-237	WN18RR	DBbook2014	ML1M
0%	RESCAL	0.925 ± 0.018	0.819 ± 0.018	0.915 ± 0.004	0.897 ± 0.022
	MMD	0.924 ± 0.018	0.818 ± 0.019	0.915 ± 0.005	0.897 ± 0.035
	WD	0.928 ± 0.006	0.811 ± 0.017	0.918 ± 0.005	**0.932 ± 0.004***
1.5%	RESCAL	0.929 ± 0.003	0.814 ± 0.018	0.871 ± 0.032	0.950 ± 0.003
	MMD	0.931 ± 0.003	0.807 ± 0.029	0.892 ± 0.009	0.954 ± 0.003
	WD	0.932 ± 0.006	0.818 ± 0.020	0.868 ± 0.040	0.954 ± 0.002
3%	RESCAL	0.922 ± 0.006	0.870 ± 0.018	0.885 ± 0.008	0.946 ± 0.005
	MMD	0.926 ± 0.005	0.861 ± 0.011	0.877 ± 0.026	0.948 ± 0.003
	WD	0.921 ± 0.007	0.860 ± 0.018	0.890 ± 0.005	0.949 ± 0.003
5%	RESCAL	0.927 ± 0.007	0.869 ± 0.007	0.878 ± 0.008	0.949 ± 0.003
	MMD	0.935 ± 0.005	0.835 ± 0.050	0.879 ± 0.008	0.952 ± 0.003
	WD	**0.937 ± 0.004***	0.860 ± 0.020	**0.885 ± 0.009***	0.953 ± 0.003

Table 4. Specifically, the WD regularizer worsens the intra-domain Hit@10 scores compared to RESCAL in FB15k-237 with 5% overlapping and ML1M with 1.5% overlapping settings despite helping improve the inter-domain counterparts. It also hurts the intra-domain Hit@10 score in ML1M with 3% overlapping setting.

Summary. The proposed method with WD regularizer significantly improves the performance of inter-domain link prediction over the baseline method while being able to preserve the intra-domain performance in the FB15k-237 and DBbook2014 datasets. In the ML1M dataset, it benefits the inter-domain performance at the risk of decreasing intra-domain Hit@10 scores. Unexpectedly, the MMD regularizer does not work well and empirically causes deterioration of the inter-domain performance. These negative results might be due to local optimal arising when minimizing MMD with a finite number of samples, as recently studied in [26]. Further detailed analysis would be necessary before one can firmly judge the performance of the MMD regularizer. We leave this matter for future works. It is also worth mentioning that, in the experiment setting, the sampling of G^1 and G^2 is repeated independently for each overlapping level. Therefore, it is not necessary for the link prediction scores to monotonically increase when the overlapping level increases.

Embedding Visualization. Figures 1 and 2 visualize the entity embeddings learned by RESCAL and the WD regularizer in the case of 3% overlapping. As being seen in Fig. 1, WD can learn more identical embedding distributions than RESCAL in the case of the FB15k-237 and DBbook2014 datasets. Especially, in the DBbook2014 dataset, RESCAL can only learn similar shape distributions, but the regularizer can learn distributions with both similar shape and close absolute position. However, as being shown in Fig. 2, in the WN18RR and ML1M datasets, the WD regularizer seems to only add noise when learning the embeddings, which results in no improvement or even degradation of both intra-domain and inter-domain Hit@10 scores.

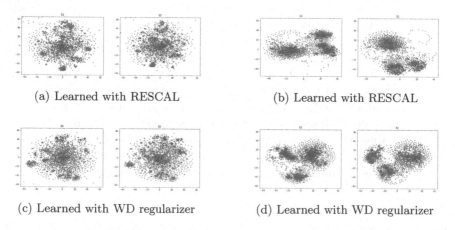

(a) Learned with RESCAL   (b) Learned with RESCAL

(c) Learned with WD regularizer   (d) Learned with WD regularizer

Fig. 1. Embedding visualization of FB15k-237 (subfigures a and c) and DBbook2014 (subfigures b and d) datasets with 3% overlapping. The proposed method learns more identical embedding distributions across both domains.

5   Related Work

In recent years, the embedding-based approach has become popular in dealing with the link prediction task on a multi-relational knowledge graph (intra-domain). One of the pioneering works in this direction is TransE [3]. It is a translation model whose each predicate type corresponds to a translation between the entities' embedding vectors. The model is suitable for 1-to-1 relationships only. Following models such as TransH, TransR, and TransD [14,16,33] are designed to deal with n-to-1, 1-to-n, and n-to-n relationships. Furthermore, tensor-based models such that RESCAL, DistMult, and SimplE [15,22,35] also gain huge interest. They interpret multi-relational knowledge graphs as 3-D tensors and employ tensor factorization to learn the entity and predicate embeddings. Besides, neural network and complex vector-based models [27,31] are also introduced in the literature. Further details can be found in [20].

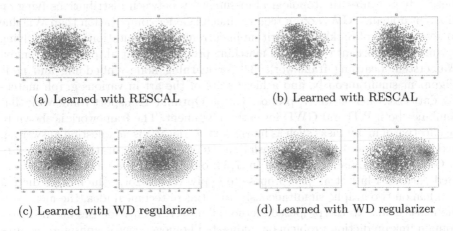

(a) Learned with RESCAL   (b) Learned with RESCAL

(c) Learned with WD regularizer   (d) Learned with WD regularizer

Fig. 2. Embedding visualization of WN18RR (subfigures a and c) and ML1M (subfigures b and d) datasets with 3% overlapping. RESCAL is able to learn similar embedding distributions between the two domains while the proposed method seems to add more noise.

To the best of our knowledge, the proposed method is the first to consider the inter-domain link prediction problem between multi-relational graphs. Existing methods in the literature do not directly deal with the problem. The closest line of research focuses on entity alignment in multilingual knowledge graphs, which often aims to match words of the same meanings between different languages. The first work in this line of research is MTransE [8]. It employs TransE to independently embed different knowledge graphs and perform matching on the embedding spaces. Other methods like JAPE [28] and BootEA [29] further improve MTransE by exploiting additional attributes or description information and bootstrapping strategy. MRAEA [17] directly learns multilingual entity embeddings by attending over the entities' neighbors and their meta semantic information. Other methods [4, 11] apply Graph Neural Networks for learning alignment-oriented embeddings and achieve state-of-the-art results in many datasets. All these entity-matching methods implicitly assume most entities in one graph to have corresponding counterparts in the other graph, e.g. words in one lingual graph to have the same meaning words in the other lingual graph. Meanwhile, the proposed method only assumes the similarity between entity distributions.

Minimizing a dissimilarity criterion between distributions is a popular strategy for distribution matching and entity alignment problems. Cao et al. propose Distribution Matching Machines [6] that optimizes maximum mean discrepancy (MMD) between source and target domains for unsupervised domain adaptation tasks. The criterion is successfully applied in distribution matching and domain confusion tasks as well [2, 32]. Besides Wasserstein distance (WD), Gromov-Wasserstein distance (GWD) [24] also is a popular optimal transport metric. It measures the topological dissimilarity between distributions lying on different domains. GWD often requires much heavier computation than WD due to nested loops of Sinkhorn algorithm in current implementations [24]. Applying optimal transport into the graph matching problem, Xu et al. propose Gromov-Wasserstein Learning framework [34] for learning node embedding and node alignment simultaneously, and achieve state of the art in various graph matching datasets. Chen et al. [7] propose Graph Optimal Transport framework that combines both WD and GWD for entity alignment. The framework is shown to be effective in many tasks such as image-text retrieval, visual question answering, text generation, and machine translation. Due to the computational complexity of GWD, each domain considered in [7, 34] only contains less than several hundred entities. Phuc et al. [25] propose to apply WD to solve the link prediction problem on two graphs simultaneously. In terms of technical idea, the method is the most similar to the proposed method; however, it only focuses on the intra-domain link prediction problem on undirected homogeneous graphs and requires most of the nodes in one graph to have corresponding counterparts in the other graph.

6   Conclusion and Future Work

Inter-domain link prediction is an important task for constructing large multi-relational graphs from smaller related ones. However, existing methods in the literature do not directly address this problem. In this paper, we propose a new approach for the problem via jointly minimizing a divergence between entity distributions during the embedding learning process. Two regularizers have been investigated, in which the WD-based regularizer shows promising results and improves inter-domain link prediction performance considerably. For future works, we would like to verify the proposed method's effectiveness using more baseline embedding methods besides RESCAL. Further analysis on the performance of the MMD-based regularizer will also be conducted. Moreover, the proposed method currently assumes that both domains share the same underlying entity distribution. This assumption is violated when the domains' entity distributions are not completely identical but partially different. One possible direction for further research is to adopt unbalanced optimal transport as the regularizer, which flexibly allows mass destruction and mass creation between distributions.

Acknowledgement. The authors would like to thank the anonymous reviewers for their insightful suggestions and constructive feedback.

References

1. Akiba, T., Sano, S., Yanase, T., Ohta, T., Koyama, M.: Optuna: a next-generation hyperparameter optimization framework. In: Proceedings of the 25th ACM SIGKDD International Conference on Knowledge Discovery & Data Mining (KDD), pp. 2623–2631 (2019)
2. Baktashmotlagh, M., Harandi, M.T., Salzmann, M.: Distribution-matching embedding for visual domain adaptation. J. Mach. Learn. Res. **17**, 108:1–108:30 (2016)
3. Bordes, A., Usunier, N., García-Durán, A., Weston, J., Yakhnenko, O.: Translating embeddings for modeling multi-relational data. Adv. Neural Inform. Process. Syst. **26**, 2787–2795 (2013)
4. Cao, Y., Liu, Z., Li, C., Liu, Z., Li, J., Chua, T.: Multi-channel graph neural network for entity alignment. In: Proceedings of the 57th Conference of the Association for Computational Linguistics (ACL), pp. 1452–1461 (2019)
5. Cao, Y., Wang, X., He, X., Hu, Z., Chua, T.: Unifying knowledge graph learning and recommendation: towards a better understanding of user preferences. In: Proceedings of the World Wide Web Conference (WWW), pp. 151–161 (2019)
6. Cao, Y., Long, M., Wang, J.: Unsupervised domain adaptation with distribution matching machines. In: Proceedings of the Thirty-Second AAAI Conference on Artificial Intelligence (AAAI), pp. 2795–2802 (2018)
7. Chen, L., Gan, Z., Cheng, Y., Li, L., Carin, L., Liu, J.: Graph optimal transport for cross-domain alignment. In: Proceedings of the 37th International Conference on Machine Learning (ICML), pp. 1542–1553 (2020)
8. Chen, M., Tian, Y., Yang, M., Zaniolo, C.: Multilingual knowledge graph embeddings for cross-lingual knowledge alignment. In: Proceedings of the 26th International Joint Conference on Artificial Intelligence (IJCAI), pp. 1511–1517. ijcai.org (2017)
9. Cuturi, M.: Sinkhorn distances: lightspeed computation of optimal transport. Adv. Neural Inform. Process. Syst. **26**, 2292–2300 (2013)
10. Dettmers, T., Minervini, P., Stenetorp, P., Riedel, S.: Convolutional 2D knowledge graph embeddings. In: Proceedings of the 32nd AAAI Conference on Artificial Intelligence (AAAI), pp. 1811–1818 (2018)
11. Fey, M., Lenssen, J.E., Morris, C., Masci, J., Kriege, N.M.: Deep graph matching consensus. In: Proceedings of th 8th International Conference on Learning Representations (ICLR) (2020)
12. Gretton, A., Borgwardt, K.M., Rasch, M.J., Schölkopf, B., Smola, A.J.: A kernel method for the two-sample-problem. Adv. Neural Inform. Process. Syst. **19**, 513–520 (2006)

13. Gretton, A., Borgwardt, K.M., Rasch, M.J., Schölkopf, B., Smola, A.J.: A kernel approach to comparing distributions. In: Proceedings of the 22nd AAAI Conference on Artificial Intelligence (AAAI), pp. 1637–1641 (2007)
14. Ji, G., He, S., Xu, L., Liu, K., Zhao, J.: Knowledge graph embedding via dynamic mapping matrix. In: Proceedings of the 53rd Annual Meeting of the Association for Computational Linguistics (ACL), pp. 687–696 (2015)
15. Kazemi, S.M., Poole, D.: Simple embedding for link prediction in knowledge graphs. Adv. Neural Inform. Process. Syst. **31**, 4289–4300 (2018)
16. Lin, Y., Liu, Z., Sun, M., Liu, Y., Zhu, X.: Learning entity and relation embeddings for knowledge graph completion. In: Proceedings of the 29th AAAI Conference on Artificial Intelligence (AAAI), pp. 2181–2187 (2015)
17. Mao, X., Wang, W., Xu, H., Lan, M., Wu, Y.: MRAEA: an efficient and robust entity alignment approach for cross-lingual knowledge graph. In: Proceedings of the 13th ACM International Conference on Web Search and Data Mining (WSDM), pp. 420–428 (2020)
18. Mikolov, T., Chen, K., Corrado, G., Dean, J.: Efficient estimation of word representations in vector space. In: Proceedings of the First International Conference on Learning Representations (ICLR) (2013)
19. Mikolov, T., Sutskever, I., Chen, K., Corrado, G.S., Dean, J.: Distributed representations of words and phrases and their compositionality. Adv. Neural Inform. Process. Syst. **26**, 3111–3119 (2013)
20. Nguyen, D.Q.: An overview of embedding models of entities and relationships for knowledge base completion. CoRR abs/1703.08098 (2017)
21. Nickel, M., Murphy, K., Tresp, V., Gabrilovich, E.: A review of relational machine learning for knowledge graphs. Proc. IEEE **104**(1), 11–33 (2016)
22. Nickel, M., Tresp, V., Kriegel, H.: A three-way model for collective learning on multi-relational data. In: Proceedings of the 28th International Conference on Machine Learning (ICML), pp. 809–816 (2011)
23. Peyré, G., Cuturi, M.: Computational optimal transport. Found. Trends Mach. Learn. **11**(5–6), 355–607 (2019)
24. Peyré, G., Cuturi, M., Solomon, J.: Gromov-Wasserstein averaging of kernel and distance matrices. In: Proceedings of the 33nd International Conference on Machine Learning (ICML), pp. 2664–2672 (2016)
25. Phuc, L.H., Takeuchi, K., Yamada, M., Kashima, H.: Simultaneous link prediction on unaligned networks using graph embedding and optimal transport. In: Proceedings of the Seventh IEEE International Conference on Data Science and Advanced Analytics (DSAA), pp. 245–254 (2020)
26. Sansone, E., Ali, H.T., Sun, J.: Coulomb autoencoders. In: ECAI 2020–24th European Conference on Artificial Intelligence, vol. 325, pp. 1443–1450 (2020)
27. Socher, R., Chen, D., Manning, C.D., Ng, A.Y.: Reasoning with neural tensor networks for knowledge base completion. Adv. Neural Inform. Process. Syst. **26**, 926–934 (2013)
28. Sun, Z., Hu, W., Li, C.: Cross-lingual entity alignment via joint attribute-preserving embedding. In: Proceedings of the 16th International Semantic Web Conference (ISWC), pp. 628–644 (2017)
29. Sun, Z., Hu, W., Zhang, Q., Qu, Y.: Bootstrapping entity alignment with knowledge graph embedding. In: Proceedings of the Twenty-Seventh International Joint Conference on Artificial Intelligence (IJCAI), pp. 4396–4402 (2018)

30. Toutanova, K., Chen, D., Pantel, P., Poon, H., Choudhury, P., Gamon, M.: Representing text for joint embedding of text and knowledge bases. In: Proceedings of the 2015 Conference on Empirical Methods in Natural Language Processing (EMNLP), pp. 1499–1509 (2015)
31. Trouillon, T., Welbl, J., Riedel, S., Gaussier, É., Bouchard, G.: Complex embeddings for simple link prediction. In: Proceedings of the 33rd International Conference on Machine Learning (ICML), pp. 2071–2080 (2016)
32. Tzeng, E., Hoffman, J., Zhang, N., Saenko, K., Darrell, T.: Deep domain confusion: Maximizing for domain invariance. CoRR abs/1412.3474 (2014)
33. Wang, Z., Zhang, J., Feng, J., Chen, Z.: Knowledge graph embedding by translating on hyperplanes. In: Proceedings of the 28th AAAI Conference on Artificial Intelligence (AAAI), pp. 1112–1119 (2014)
34. Xu, H., Luo, D., Zha, H., Carin, L.: Gromov-Wasserstein learning for graph matching and node embedding. In: Proceedings of the 36th International Conference on Machine Learning (ICML), pp. 6932–6941 (2019)
35. Yang, B., Yih, W., He, X., Gao, J., Deng, L.: Embedding entities and relations for learning and inference in knowledge bases. In: Proceedings of the Third International Conference on Learning Representations (ICLR) (2015)

GraphSVX: Shapley Value Explanations for Graph Neural Networks

Alexandre Duval$^{(\boxtimes)}$ and Fragkiskos D. Malliaros

Université Paris-Saclay, CentraleSupélec, Inria, Giff-sur-Yvette, France
{alexandre.duval,fragkiskos.malliaros}@centralesupelec.fr

Abstract. Graph Neural Networks (GNNs) achieve significant performance for various learning tasks on geometric data due to the incorpora-
tion of graph structure into the learning of node representations, which
renders their comprehension challenging. In this paper, we first propose
a unified framework satisfied by most existing GNN explainers. Then,
we introduce GraphSVX, a post hoc local model-agnostic explanation
method specifically designed for GNNs. GraphSVX is a decomposition
technique that captures the "fair" contribution of each feature and node
towards the explained prediction by constructing a surrogate model on a
perturbed dataset. It extends to graphs and ultimately provides as explanation the Shapley Values from game theory. Experiments on real-world
and synthetic datasets demonstrate that GraphSVX achieves state-of-
the-art performance compared to baseline models while presenting core
theoretical and human-centric properties.

1 Introduction

Many aspects of the everyday life involve data without regular spatial structure, known as non-euclidean or geometric data, such as social networks, molecular structures or citation networks [1,10]. These datasets, often represented
as graphs, are challenging to work with because they require modelling rich
relational information on top of node feature information [37]. Graph Neural
Networks (GNNs) are powerful tools for representation learning of such data.
They achieve state-of-the-art performance on a wide variety of tasks [8,36] due
to their recursive message passing scheme, where they encode information from
nodes and pass it along the edges of the graph. Similarly to traditional deep learning frameworks, GNNs showcase a complex functioning that is rather opaque to
humans. As the field grows, understanding them becomes essential for well known
reasons, such as ensuring privacy, fairness, efficiency, and safety [20].

While there exist a variety of explanation methods [25,27,29], they are not
well suited for geometric data as they fall short in their ability to incorporate
graph topology information. [2,21] have proposed extensions to GNNs, but in
addition to limited performance, they require model internal knowledge and show
gradient saturation issues due to the discrete nature of the adjacency matrix.

GNNExplainer [33] is the first explanation method designed specifically for
GNNs. It learns a continuous (and a discrete) mask over the edges (and features)

© Springer Nature Switzerland AG 2021
N. Oliver et al. (Eds.): ECML PKDD 2021, LNAI 12976, pp. 302–318, 2021.
https://doi.org/10.1007/978-3-030-86520-7_19

of the graph by formulating an optimisation process that maximizes mutual information between the distribution of possible subgraphs and GNN prediction. More recently, PGExplainer [16] and GraphMask [26] generalize GNNExplainer to an inductive setting; they use re-parametrisation tricks to alleviate the "introduced evidence" problem [5]—i.e. continuous masks deform the adjacency matrix and introduce new semantics to the generated graph. Regarding other approaches; GraphLIME [12] builds on LIME [22] to provide a non-linear explanation model; PGM-Explainer [32] learns a simple Bayesian network handling node dependencies; XGNN [34] produces model-level insights via graph generation trained using reinforcement learning.

Despite recent progress, existing explanation methods do not relate much and show clear limitations. Apart from GNNExplainer, none consider node features together with graph structure in explanations. Besides, they do not present core properties of a "good" explainer [19] (see Sect. 2). Since the field is very recent and largely unexplored, there is little certified knowledge about explainers' characteristics. It is, for instance, unclear whether optimising mutual information is pertinent or not. Overall, this often yields explanations with a poor signification, like a probability score stating how essential a variable is [16,26,33]. Existing techniques not only lack strong theoretical grounds, but also do not showcase an evaluation that is sophisticated enough to properly justify their effectiveness or other desirable aspects [24]. Lastly, little importance is granted to their human-centric characteristics [18], limiting the comprehensibility of explanations from a human perspective.

In light of these limitations, first, we propose a unified explanation framework encapsulating recently introduced explainers for GNNs. It not only serves as a connecting force between them but also provides a different and common view of their functioning, which should inspire future work. In this paper, we exploit it ourselves to define and endow our explainer, GraphSVX, with desirable properties. More precisely, GraphSVX carefully constructs and combines the key components of the unified pipeline so as to jointly capture the average marginal contribution of node features and graph nodes towards the explained prediction. We show that GraphSVX ultimately computes, via an efficient algorithm, the *Shapley values* from game theory [28], that we extend to graphs. The resulting unique explanation, thus, satisfy several theoretical properties by definition, while it is made more human-centric through several extensions. In the end, we evaluate GraphSVX on real-world and synthetic datasets for node and graph classification tasks. We show that it outperforms existing baselines in explanation accuracy, and verifies further desirable aspects such as robustness or certainty.

2   Related Work

Explanations methods specific to GNNs are classified into five categories of methods according to [35]: gradient-based, perturbation, decomposition, surrogate, and model-level. We utilise the same taxonomy in this paper to position GraphSVX.

Decomposition methods [2,21] distribute the prediction score among input features using the weights of the network architecture, through backpropagation. Despite offering a nice interpretation, they are not specific to GNNs and present several major limits such as requiring access to model parameters or being sensitive to small input changes, like **gradient-based methods** discussed in Sect. 1.

Perturbation methods [16,26,33] monitor variations in model prediction with respect to different input perturbations. Such methods provide as explanation a continuous mask over edges (features) holding importance probabilities learned via a simple optimisation procedure, affected by the introduced-evidence problem.

Surrogate methods [12,32] approximate the black box GNN model locally by learning an interpretable model on a dataset built around the instance of interest v (e.g., neighbours). Explanations for the surrogate model are used as explanations for the original model. For now, such approaches are rather intuition-based and consider exclusively node features or graph topology, not both.

Model level methods [34] provide general insights on the model functioning. It supports only graph classification, requires passing a candidate node set as input and is challenged by local methods also providing global explanations [16].

As we will show shortly, GraphSVX bridges the gap between these categories by learning a surrogate explanation model on a perturbed dataset that ultimately decomposes the explained prediction among the nodes and features of the graph, depending on their respective contribution. It also derives model-level insights by explaining subsets of nodes, while avoiding the respective limits of each category.

Desirable properties of explanations have received subsequent attention from the social sciences and the machine learning communities, but are often overlooked when designing an explainer. From a theoretical perspective, good explanations are accurate, fidel (*truthful*), and reflect the proportional importance of a feature on prediction (*meaningful*) [4,35]. They also are stable and consistent (*robust*), meaning with a low variance when changing to a similar model or a similar instance [19]. Besides, they reflect the certainty of the model (*decomposable*) and are as representative as possible of its (*global*) functioning [17]. Finally, since their ultimate goal is to help humans understand the model, explanations should be intuitive to comprehend (*human-centric*). Many sociological and psychological studies emphasise key aspects: only a few motives (*selective*) [31], comparable to other instances (*contrastive*) [14], and interactive with the explainee (*social*).

3   Preliminary Concepts and Background

Notation. We consider a graph $\mathcal{G}$ with N nodes and F features defined by $(\mathbf{X}, \mathbf{A})$ where $\mathbf{X} \in \mathbb{R}^{N \times F}$ is the feature matrix and $\mathbf{A} \in \mathbb{R}^{N \times N}$ the adjacency matrix. $f(\mathbf{X}, \mathbf{A})$ denotes the prediction of the GNN model f, and $f_v(\mathbf{X}, \mathbf{A})$ the score of the predicted class for node v. Let $\mathbf{X}_{*j} = (X_{1j}, \ldots, X_{Nj})$ with feature values $\mathbf{x}_{*j} = (x_{1j}, \ldots, x_{Nj})$ represent feature j's value vector across all nodes. Similarly, $\mathbf{X}_i = \mathbf{X}_{i*} = (X_{i1}, \ldots, X_{iF})$ stands for node i's feature vector, with $\mathbf{X}_{iS} = \{X_{ik}|k \in S\}$. $\mathbf{1}$ is the all-ones vector.

3.1   Graph Neural Networks

GNNs adopt a message passing mechanism [11] where the update at each GNN layer ℓ involves three key calculations [3]: (i) The propagation step. The model computes a message $m_{ij}^\ell = \mathrm{MSG}(\mathbf{h}_i^{\ell-1}, \mathbf{h}_j^{\ell-1}, a_{ij})$ between every pair of nodes (v_i, v_j), that is, a function MSG of v_i's and v_j's representations $\mathbf{h}_i^{\ell-1}$ and $\mathbf{h}_j^{\ell-1}$ in the previous layer and of the relation a_{ij} between the nodes. (ii) The aggregation step. For each node v_i, GNN calculates an aggregated message M_i from v_i's neighbourhood $\mathcal{N}_{v_i}$, whose definition vary across methods. $M_i^\ell = \mathrm{AGG}(m_{ij}^\ell|v_j \in \mathcal{N}_{v_i})$. (iii) The update step. GNN non-linearly transforms both the aggregated message M_i^ℓ and v_i's representation $\mathbf{h}_i^{\ell-1}$ from the previous layer, to obtain v_i's representation $\mathbf{h}_i^\ell$ at layer ℓ: $\mathbf{h}_i^\ell = \mathrm{UPD}(M_i^\ell, \mathbf{h}_i^{\ell-1})$. The representation $\mathbf{z}_i = \mathbf{h}_i^L$ of the final GNN layer L serves as final node embedding and is used for downstream machine learning tasks.

3.2   The Shapley Value

The Shapley value is a method from Game Theory. It describes how to fairly distribute the total gains of a game to the players depending on their respective contribution, assuming they all collaborate. It is obtained by computing the average marginal contribution of each player when added to any possible coalition of players [28]. This method has been extended to explain machine learning model predictions on tabular data [13,30], assuming that each feature of the explained instance ($\mathbf{x}$) is a player in a game where the prediction is the payout.

The characteristic function $val : S \rightarrow \mathbb{R}$ captures the marginal contribution of the coalition $S \subseteq \{1, \ldots, F\}$ of features towards the prediction $f(\mathbf{x})$ with respect to the average prediction: $val(S) = \mathbb{E}[f(\mathbf{X})|\mathbf{X}_S = \mathbf{x}_s] - \mathbb{E}[f(\mathbf{X})]$. We isolate the effect of a feature j via $val(S \cup \{j\}) - val(S)$ and average it over all possible ordered coalitions S to obtain its Shapley value as:

$$\phi_j(val) = \sum_{S \subseteq \{1, \ldots, F\} \setminus \{j\}} \frac{|S|! \, (F - |S| - 1)!}{F!} (val(S \cup \{j\}) - val(S)).$$

The notion of fairness is defined by four axioms (*efficiency, dummy, symmetry, additivity*), and the Shapley value is the unique solution satisfying them. In practice, the sum becomes impossible to compute because the number of possible coalitions (2^{F-1}) increases exponentially by adding more features. We thus approximate Shapley values using sampling [6,15].

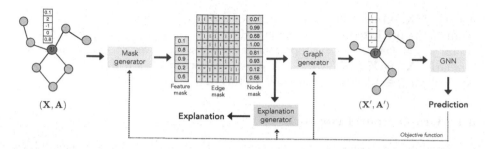

Fig. 1. Overview of unified framework. All methods take as input a given graph $\mathcal{G} = (\mathbf{X}, \mathbf{A})$, feed it to a mask generator (MASK) to create three masks over nodes, edges, and features. These masks are then passed to a graph generator (GEN) that converts them to the original input space $(\mathbf{X}', \mathbf{A}')$ before feeding them to the original GNN model f. The resulting prediction $f(\mathbf{X}', \mathbf{A}')$ is used to improve the mask generator, the graph generator or the downstream explanation generator (EXPL), which ultimately provides the desired explanation using masks and $f(\mathbf{X}', \mathbf{A}')$. This passage through the framework is repeated many times so as to create a proper dataset $\mathcal{D}$ from which each generator block learns. Usually, only one is optimised with a carefully defined optimisation process involving the new and original GNN predictions.

4   A Unified Framework for GNN Explainers

As detailed in the previous section, existing interpretation methods for GNNs are categorised and often treated separately. In this paper, we approach the explanation problem from a new angle, proposing a unified view that regroups existing explainers under a single framework: GNNExplainer, PGExplainer, GraphLIME, PGM-Explainer, XGNN, and the proposed GraphSVX. The key differences across models lie in the definition and optimisation of the three main blocks of the pipeline, as shown in Fig. 1:

- MASK generates discrete or continuous masks over features $\mathbf{M}_F \in \mathbb{R}^F$, nodes $\mathbf{M}_N \in \mathbb{R}^N$ and edges $\mathbf{M}_E \in \mathbb{R}^{N \times N}$, according to a specific strategy.
- GEN outputs a new graph $\mathcal{G}' = (\mathbf{X}', \mathbf{A}')$ from the masks $(\mathbf{M}_E, \mathbf{M}_N, \mathbf{M}_F)$ and the original graph $\mathcal{G} = (\mathbf{X}, \mathbf{A})$.
- EXPL generates explanations, often offered as a vector or a graph, using a function g whose definition vary across baselines.

In the following, we show how each baseline fits the pipeline. $\odot$ stands for the element wise multiplication operation, σ the softmax function, $\|$ the concatenation operation, and $\mathbf{M}^{\text{ext}}$ describes the extended vector $\mathbf{M}$ with repeated entries, whose size makes the operation feasible. All three masks are not considered for a single method; some are ignored as they have no effect on final explanations; one often studies node feature $\mathbf{M}_F$ or graph structure ($\mathbf{M}_E$ or $\mathbf{M}_N$).

GNNExplainer's key component is the mask generator. It generates both $\mathbf{M}_F$ and $\mathbf{M}_E$, where $\mathbf{M}_E$ has continuous values and $\mathbf{M}_F$ discrete ones. They are both randomly initialised and jointly optimised via a mutual information loss function $MI(Y, (\mathbf{M}_E, \mathbf{M}_F)) = H(Y) - H(Y|\mathbf{A}', \mathbf{X}')$, where GEN gives $\mathbf{A}' = \mathbf{A} \odot \sigma(\mathbf{M}_E)$ and $\mathbf{X}' = \mathbf{X} \odot \mathbf{M}_F^{\text{ext}}$. Y represents the class label and $H(\cdot)$ the entropy term. EXPL simply returns the learned masks as explanations, via the identity function $g(\mathbf{M}_E, \mathbf{M}_F) = (\mathbf{M}_E, \mathbf{M}_F)$.

PGExplainer is very similar to GNNExplainer. MASK generates only an edge mask $\mathbf{M}_E$ using a multi-layer neural network MLP_ψ and the learned matrix $\mathbf{Z}$ of node representations: $\mathbf{M}_E = \text{MLP}_\psi(\mathcal{G}, \mathbf{Z})$. The new graph is constructed with $\text{GEN}(\mathbf{X}, \mathbf{A}, \mathbf{M}_E) = (\mathbf{X}, \mathbf{A} \odot \rho(\mathbf{M}_E))$, where ρ denotes a reparametrisation trick. The obtained prediction $f_v(\mathbf{X}, \mathbf{A}')$ is also used to maximise mutual information with $f_v(\mathbf{X}, \mathbf{A})$ and backpropagates the result to optimise MASK. As for GNNExplainer, EXPL provides $\mathbf{M}_E$ as explanations.

GraphLIME is a surrogate method with a simple and not optimised mask generator. Although it measures feature importance, it creates a node mask $\mathbf{M}_N$ using the neighbourhood of v (i.e., $\mathcal{N}_v$). The kth mask (or sample) is defined as $\mathbf{M}_{N,i}^k = 1$ if $v_i = \mathcal{N}_v[k]$ and 0 otherwise. $\text{GEN}(\mathbf{X}, \mathbf{A}, \mathbf{M}_N) = (\mathbf{X}, \mathbf{A})$, so in fact, it computes and stores the original model prediction. $\mathbf{X}$ and $f(\mathbf{X}, \mathbf{A})$ are then combined with the mask $\mathbf{M}_N$ via simple dot products $\mathbf{M}_N^\top \cdot \mathbf{X}$ and $\mathbf{M}_N^\top \cdot f(\mathbf{X}, \mathbf{A})$ respectively, to isolate the original feature vector and prediction of the kth neighbour of v. These two elements are treated as input and target of an HSIC Lasso model g, trained with an adapted loss function. The learned coefficients constitute importance measures that are given as explanations by EXPL.

PGM-Explainer builds a probabilistic graphical model on a local dataset that consists of random node masks $\mathbf{M}_N \in \{0,1\}^N$. The associated prediction $f_v(\mathbf{X}', \mathbf{A}')$ is obtained by posing $\mathbf{A}' = \mathbf{A}$ and $\mathbf{X}' = \mathbf{M}_N^{\text{ext}} \odot \mathbf{X} + (1 - \mathbf{M}_N^{\text{ext}} \odot \boldsymbol{\mu}^{\text{ext}})$, with $\boldsymbol{\mu} = (E[\mathbf{X}_{*1}], \ldots, E[\mathbf{X}_{*F}])^\top$. This means that each excluded node feature ($\mathbf{M}_{N,j} = 0$) is set to its mean value across all nodes. This dataset is fed sequentially to the main component EXPL, which learns and outputs a Bayesian Network g with input $\mathbf{M}_N$ (made sparser by looking at the Markov-blanket of v), BIC score loss function, and target $I(f_v(\mathbf{X}', \mathbf{A}'))$, where $I(\cdot)$ is a specific function that quantifies the difference in prediction between original and new prediction.

XGNN is a model-level approach that trains an iterative graph generator (add one edge at a time) via reinforcement learning. This causes two key differences with previous approaches: (1) the input graph at iteration t ($\mathcal{G}_t$) is obtained from the previous iteration and is initialised as the empty graph; (2) we also pass a candidate node set $\mathcal{C}$, such that $\mathbf{X}_\mathcal{C}$ contains the feature vector of all distinct nodes across all graphs in dataset. MASK generates an edge mask $\mathbf{M}_E = \mathbf{A}_t$ and a node mask $\mathbf{M}_{N_t} \in \{0,1\}^{|\mathcal{C}|}$ specifying the latest node added to $\mathcal{G}_t$, if any. GEN produces a new graph $\mathcal{G}_{t+1}$ from $\mathcal{G}_t$ by predicting a new edge, possibly creating a new node from $\mathcal{C}$. This is achieved by applying a GCN and two MLP networks. Then, $\mathcal{G}_{t+1}$ is fed to the explained GNN. The resulting prediction is used to update model parameters via a policy gradient loss function. EXPL stores nonzero $\mathbf{M}_{N_t}$ at each time step and provides $g(\{\mathbf{M}_{N_t}\}_t, \mathbf{X}_C, \mathbf{M}_E)) = (\|\|_t \mathbf{M}_{N_t} \cdot \mathbf{X}_C, \mathbf{M}_E)$ as explanation—i.e. the graph generated at the final iteration, written $\mathcal{G}_T$.

GraphSVX. As we will see in the next section, the proposed GraphSVX model carefully exploits the potential of this framework through a better design and combination of complex mask, graph and explanation generators–in the perspective of improving performance and embedding desirable properties in explanations.

5 Proposed Method

GraphSVX is a post hoc model-agnostic explanation method specifically designed for GNNs, that jointly computes graph structure and node feature explanations for a single instance. More precisely, GraphSVX constructs a perturbed dataset made of binary masks for nodes and features $(\mathbf{M}_N, \mathbf{M}_F)$, and computes their marginal contribution $f(\mathbf{X}', \mathbf{A}')$ towards the prediction using a graph generator $\text{GEN}(\mathbf{X}, \mathbf{A}, \mathbf{M}_F, \mathbf{M}_N) = (\mathbf{X}', \mathbf{A}')$. It then learns a carefully defined explanation model on the dataset $(\mathbf{M}_N \| \mathbf{M}_F, f(\mathbf{X}', \mathbf{A}'))$ and provides it as explanation. Ultimately, it produces a unique deterministic explanation that decomposes the original prediction and has a real signification (Shapley values) as well as other desirable properties evoked in Sect. 2. Without loss of generality, we consider a node classification task for the presentation of the method.

5.1 Mask and Graph Generators

First of all, we create an efficient mask generator algorithm that constructs discrete feature and node masks, respectively denoted by $\mathbf{M}_F \in \{0,1\}^F$ and $\mathbf{M}_N \in \{0,1\}^N$. Intuitively, for the explained instance v, we aim at studying the joint influence of a subset of features and neighbours of v towards the associated prediction $f_v(\mathbf{X}, \mathbf{A})$. The mask generator helps us determine the subset being studied. Associating 1 with a variable (node or feature) means that it is considered, 0 that it is discarded. For now, we let MASK randomly sample from all possible (2^{F+N-1}) pairs of masks $\mathbf{M}_F$ and $\mathbf{M}_N$, meaning all possible coalitions S of features and nodes (v is not considered in explanations). Let $\mathbf{z}$ be the random variable accounting for selected variables, $\mathbf{z} = (\mathbf{M}_F \| \mathbf{M}_N)$. This is a simplified version of the true mask generator, which we will come back to later, in Sect. 5.4.

We now would like to estimate the joint effect of this group of variables towards the original prediction. We thus isolate the effect of selected variables marginalised over excluded ones, and observe the change in prediction. We define $\text{GEN} : (\mathbf{X}, \mathbf{A}, \mathbf{M}_F, \mathbf{M}_N) \rightarrow (\mathbf{X}', \mathbf{A}')$, which converts the obtained masks to the original input space, in this perspective. Due to the message passing scheme of GNNs, studying jointly node and features' influence is tricky. Unlike GNNExplainer, we avoid any overlapping effect by considering feature values of v (instead of the whole subgraph around v) and all nodes except v. Several options are possible to cancel out a node's influence on the prediction, such as replacing its feature vector by random or expected values. Here, we decide to isolate the node in the graph, which totally removes its effect on the prediction. Similarly, to neutralise the effect of a feature, as GNNs do not handle missing values, we set it to the dataset expected value. Formally, it translates into:

$$\mathbf{X}' = \mathbf{X} \text{ with } \mathbf{X}'_v = \mathbf{M}_F \odot \mathbf{X}_v + (\mathbf{1} - \mathbf{M}_F) \odot \boldsymbol{\mu} \tag{1}$$

$$\mathbf{A}' = (\mathbf{M}_N^{\text{ext}\top} \cdot \mathbf{A} \cdot \mathbf{M}_N^{\text{ext}}) \odot I(\mathbf{A}), \tag{2}$$

where $\boldsymbol{\mu} = (\mathbb{E}[\mathbf{X}_{*1}], \ldots, \mathbb{E}[\mathbf{X}_{*F}])^\top$ and $I(\cdot)$ captures the indirect effect of k-hop neighbours of v $(k > 1)$, which is often underestimated. Indeed, if a 3-hop neighbour w is considered alone in a coalition, it becomes disconnected from v in the new graph $\mathcal{G}'$. This prevents us from capturing its indirect impact on the prediction since it does not pass information to v anymore. To remedy this problem, we select one shortest path $\mathcal{P}$ connecting w to v via *Dijkstra's* algorithm, and include $\mathcal{P}$ back in the new graph. To keep the influence of the new nodes (in $\mathcal{P} \setminus \{w, v\}$) switched off, we set their feature vector to mean values obtained by Monte Carlo sampling.

To finalize the perturbation dataset, we pass $\mathbf{z}' = (\mathbf{X}', \mathbf{A}')$ to the GNN model f and store each sample $(\mathbf{z}, f(\mathbf{z}'))$ in a dataset $\mathcal{D}$. $\mathcal{D}$ associates with a subset of nodes and features of v their estimated influence on the original prediction.

5.2   Explanation Generator

In this section, we build a surrogate model g on the dataset $\mathcal{D} = \{(\mathbf{z}, f(\mathbf{z}'))\}$ and provide it as explanation. More rigorously, an explanation ϕ of f is normally drawn from a set of possible explanations, called interpretable domain Ω. It is the solution of the following optimisation process: $\phi = \arg\min_{g \in \Omega} \mathcal{L}_f(g)$, where the loss function attributes a score to each explanation. The choice of Ω has a large impact on the type and quality of the obtained explanation. In this paper, we choose broadly Ω to be the set of interpretable models, and more precisely the set of Weighted Linear Regression (WLR).

In short, we intend our model to learn to calculate the individual effect of each variable towards the original prediction from the joint effect $f(\mathbf{z}')$, using many different coalitions S of nodes and features. This is made possible by the definition of the input dataset $\mathcal{D}$ and is enforced by a cross entropy loss function, as follows:

$$\mathcal{L}_{f,\pi}(g) = \sum_{\mathbf{z}} \left[g(\mathbf{z}) - f(\mathbf{z}') \right]^2 \pi_{\mathbf{z}},$$

$$\text{where} \quad \pi_{\mathbf{z}} = \frac{F + N - 1}{(F + N) \cdot |\mathbf{z}|} \cdot \binom{F + N - 1}{|\mathbf{z}|}^{-1}. \tag{3}$$

π is a kernel weight that attributes a high weight to samples $\mathbf{z}$ with small or large dimension, or in different terms, groups of features and nodes with few or many elements—since it is easier to capture individual effects from the combined effect in these cases.

In the end, we provide the learned parameters of g as explanation. Each coefficient corresponds to a node of the graph or a feature of v and represents its estimated influence on the prediction $f_v(\mathbf{X}, \mathbf{A})$. In fact, it approximates the extension of the Shapley value to graphs, as shown in next paragraph.

5.3 Decomposition Model

We first justify why it is relevant to extend the Shapley value to graphs. Looking back at the original theory, each player contributing to the total gain is allocated a proportion of that gain depending on its fair contribution. Since a GNN model prediction is fully determined by node feature information ($\mathbf{X}$) and graph structural information ($\mathbf{A}$), both edges/nodes and node features are players that should be considered in explanations. In practice, we extend to graphs the four Axioms defining fairness (please see the extended version [9]), and redefine how is captured the influence of players (features and nodes) towards the prediction as $val(S) = \mathbb{E}_{\mathbf{X}_v}[f_v(\mathbf{X}, \mathbf{A}_S)|\mathbf{X}_{vS} = \mathbf{x}_{vS}] - \mathbb{E}[f_v(\mathbf{X}, \mathbf{A})]$. $\mathbf{A}_S$ is the adjacency matrix where all nodes in $\overline{S}$ (not in S) have been isolated.

Assuming model linearity and feature independence, we show that GraphSVX, in fact, captures via $f(\mathbf{z}')$ the marginal contribution of each coalition S towards the prediction:

$$\mathbb{E}_{\mathbf{X}_v}[f_v(\mathbf{X}, \mathbf{A}_S)|\mathbf{X}_{vS}] = \mathbb{E}_{\mathbf{X}_{v\overline{S}}|\mathbf{X}_{vS}}[f_v(\mathbf{X}, \mathbf{A}_S)]$$
$$\approx \mathbb{E}_{\mathbf{X}_{v\overline{S}}}[f_v(\mathbf{X}, \mathbf{A}_S)] \qquad \text{by independence}$$
$$\approx f_v(\mathbb{E}_{\mathbf{X}_{v\overline{S}}}[\mathbf{X}], \mathbf{A}_S) \qquad \text{by linearity}$$
$$= f_v(\mathbf{X}', \mathbf{A}'),$$

where $\mathbf{A}' = \mathbf{A}_S$ and $\mathbf{X}'_{ij} = \begin{cases} \mathbb{E}[\mathbf{X}_{*j}] \text{ if } i = v \text{ and } j \in \overline{S} \\ \mathbf{X}_{ij} \text{ otherwise.} \end{cases}$

Using the above, we prove that GraphSVX calculates the Shapley values on graph data. This builds on the fact that Shapley values can be expressed as an additive feature attribution model, as shown by [15] in the case of tabular data.

In this perspective, we set $\boldsymbol{\pi}_v$ such that $\boldsymbol{\pi}_v(\mathbf{z}) \to \infty$ when $|\mathbf{z}| \in \{0, F+N\}$ to enforce the *efficiency* axiom: $g(\mathbf{1}) = f_v(\mathbf{X}, \mathbf{A}) = \mathbb{E}[f_v(\mathbf{X}, \mathbf{A})] + \sum_{i=1}^{F+N} \phi_i$. This holds due to the specific definition of GEN and g (i.e., EXPL), where $g(\mathbf{1}) = f_v(\mathbf{X}, \mathbf{A})$ and the constant ϕ_0, also called base value, equals $\mathbb{E}_{\mathbf{X}_v}[f_v(\mathbf{X}, \mathbf{A}_v)] \approx \mathbb{E}[f_v(\mathbf{X}, \mathbf{A})]$, so the mean model prediction. $\mathbf{A}_v$ refers to $\mathbf{A}_\emptyset$, where v is isolated.

Theorem 1. *With the above specifications and assumptions, the solution to $\min_{g \in \Omega} \mathcal{L}_{f,\pi}(g)$ under Eq. (3) is a unique explanation model g whose parameters compute the extension of the Shapley values to graphs.*

Proof. Please see the extended version of this paper [9].

5.4   Efficient Approximation Specific to GNNs

Similarly to the euclidean case, the exact computation of the Shapley values becomes intractable due to the number of possible coalitions required. Especially that we consider jointly features and nodes, which augments exponentially the complexity of the problem. To remedy this, we derive an efficient approximation via a smart mask generator.

Firstly, we reduce the number of nodes and features initially considered to $D \leq N$ and $B \leq F$ respectively, without impacting performance. Indeed, for a GNN model with k layers, only k-hop neighbours of v can influence the prediction for v, and thus receive a non-zero Shapley value. All others are allocated a null importance according to the *dummy* axiom[1] and can therefore be discarded. Similarly, each feature j of v whose value is comprised in the confidence interval $I_j = [\mu_j - \lambda \cdot \sigma_j, \mu_j + \lambda \cdot \sigma_j]$ around the mean value μ_j can be discarded, where σ_j is the corresponding standard deviation and λ a constant.

The complexity is now $\mathcal{O}(2^{B+D})$ and we further drive it down to $\mathcal{O}(2^B + 2^D)$ by sampling separately masks of nodes and features, while still considering them jointly in g. In other words, instead of studying the influence of possible combinations of nodes and features, we consider all combinations of features with no nodes selected, and all combinations of nodes with all features included: $(2^B + 2^D)$. We observe empirically that it achieves identical explanations with fewer samples, while it seems to be more intuitive to capture the effect of nodes and features on prediction (expressed by Axiom 1).

Axiom 1. (Relative efficiency). *Node contribution to predictions can be separated from feature contribution, and their sum decomposes the prediction with respect to the average one, as* $\begin{cases} \sum_{j=1}^{B} \phi_j = f_v(\mathbf{X}, \mathbf{A}_v) - \mathbb{E}[f_v(\mathbf{X}, \mathbf{A})] \\ \sum_{i=1}^{D} \phi_{B+i} = f_v(\mathbf{X}, \mathbf{A}) - f_v(\mathbf{X}, \mathbf{A}_v). \end{cases}$

Lastly, we approximate explanations using $P \ll 2^B + 2^D$ samples, where P is sufficient to obtain a good approximation. We reduce P by greatly improving MASK, as evoked in Sect. 5.1. Assuming we have a budget of P samples, we develop a smart space allocation algorithm to draw in priority coalitions of order k, where k starts at 0 and is incremented when all coalitions of the order are sampled. This means that we sample in priority coalitions with high weight, so with nearly all or very few players. If they cannot all be chosen (for current k) due to space constraints, we proceed to a smart sampling that favours unseen players. The pseudocode and an efficiency evaluation lie in the extended version [9].

5.5   Desirable Properties of Explanations

In the end, GraphSVX generates fairly distributed explanations $\sum_j \phi_j = f_v(\mathbf{X}, \mathbf{A})$, where each ϕ_j approximates the average marginal contribution of a node or feature j towards the explained GNN prediction (with respect to the

---

[1] Axiom: If $\forall S \in \mathcal{P}(\{1, \ldots, p\})$ and $j \notin S$, $val(S \cup \{j\}) = val(S)$, then $\phi_j(val) = 0$.

average prediction ϕ_0). By definition, the resulting explanation is unique, consistent, and stable. It is also truthful and robust to noise, as shown in Sect. 6.2. The last focus of this paper is to make them more selective, global, contrastive and social; as we aim to design an explainer with desirable properties. A few aspects are detailed here.

Contrastive. Explanations are contrastive already as they yield the contribution of a variable with respect to the average prediction $\phi_0 = \mathbb{E}[f(\mathbf{X}, \mathbf{A})]$. To go futher and explain an instance with respect to another one, we could substitute $\mathbf{X}_v$ in Eq. (1) by $\mathbf{X}'_v = \mathbf{M}_F \odot \mathbf{X}_v + (1 - \mathbf{M}_F) \odot \boldsymbol{\xi}$, with $\boldsymbol{\xi}$ being the feature vector of a specific node w, or of a fictive representative instance from class C.

Global. We derive explanations for a subset U of nodes instead of a single node v, following the same pipeline. The neighbourhood changes to $\bigcup_i^U \mathcal{N}_i$, Eq. (1) now updates $\mathbf{X}_U$ instead of $\mathbf{X}_v$ and $f(\mathbf{z}')$ is calculated as the average prediction score for nodes in U. Also, towards a more global understanding, we can output the global importance of each feature j on v's prediction by enforcing in Eq. (1) $\mathbf{X}_{\mathcal{N}_v \cup \{v\}, j}$ to a mean value obtained by Monte Carlo sampling on the dataset, when $\mathbf{z}_j = 0$. This holds when we discard node importance, otherwise the overlapping effects between nodes and features render the process obsolete.

Graph Classification. Until now, we had focused on node classification but the exact same principle applies for graph classification. We simply look at $f(\mathbf{X}, \mathbf{A}) \in \mathbb{R}$ instead of $f_v(\mathbf{X}, \mathbf{A})$, derive explanations for all nodes or all features (not both) by considering features across the whole dataset instead of features of v, like our global extension.

6   Experimental Evaluation

In this section, we conduct several experiments designed to determine the quality of our explanation method, using synthetic and real world datasets, on both node and graph classification tasks. We first study the effectiveness of GraphSVX in presence of ground truth explanations. We then show how our explainer generalises to more complex real world datasets with no ground truth, by testing GraphSVX's ability to filter noisy features and noisy nodes from explanations. Detailed dataset statistics, hyper-parameter tuning, properties' check and further experimental results including ablation study, are given in the extended version [9]. The source code is available at https://github.com/AlexDuvalinho/GraphSVX.

6.1   Synthetic and Real Datasets with Ground Truth

Synthetic Node Classification Task. We follow the same setting as [16,33], where four kinds of datasets are constructed. Each input graph is a combination of a base graph together with a set of motifs, which both differ across datasets.

Table 1. Evaluation of GraphSVX and baseline GNN explainers on various datasets. The top part describes the construction of each dataset, with its base graph, the motif added, and the node features generated. Node labels are represented by colors. Then, we provide a visualisation of GraphSVX's explanations, where an important substructure is drawn in bold, as well as a quantitative evaluation based on the accuracy metric.

| | Node Classification | | | | Graph Classification | |
	BA-Shapes	BA-Community	Tree-Cycles	Tree-Grid	BA-2motifs	MUTAG
Base						
Motifs						
Features	None	$\mathcal{N}(\mu_l, \sigma_l)$	None	None	None	Atom types

Visualization

	Explanation Accuracy					
GNNExplainer	0.83	0.75	0.86	0.84	0.68	0.65
PGM-Explainer	0.96	0.92	0.95	0.87	0.91	0.72
PGExplainer	0.92	0.81	0.96	0.88	0.85	**0.79**
GraphSVX	**0.99**	**0.93**	**0.97**	**0.93**	**0.99**	0.77

The label of each node is determined based on its belonging and role in the motif. As a consequence, the explanation for a node in a motif should be the nodes in the same motif, which creates ground truth explanation. This ground truth can be used to measure the performance of an explainer via an accuracy metric.

Synthetic and Real-World Graph Classification Task. With a similar evaluation perspective, we measure the effectiveness of our explainer on graph classification, also using ground truth. We use a synthetic dataset *BA-2motifs* that resembles the previous ones, and a real life dataset called *MUTAG*. It consists of 4,337 molecule graphs, each assigned to one of 2 classes based on its mutagenic effect [23]. As discussed in [7], carbon rings with groups NH_2 or NO_2 are known to be mutagenic, and could therefore be used as ground truth.

Baselines. We compare the performance of GraphSVX to the main explanation baselines that incorporate graph structure in explanations, namely GNNExplainer, PGExplainer and PGM-Explainer. GraphLIME and XGNN are not applicable here, since they do not provide graph structure explanations for such tasks.

Experimental Setup and Metrics. We train the same GNN model – 3 graph convolution blocks with 20 hidden units, (maxpooling) and a fully connected classification layer – on every dataset during 1,000 epochs, with relu activation, Adam optimizer and initial learning rate 0.001. The performance is measured with an accuracy metric (node or edge accuracy depending on the nature of explanations) on top-k explanations, where k is equal to the ground truth dimension. More precisely, we formalise the evaluation as a binary classification of nodes (or edges) where nodes (or edges) inside motifs are positive, and the rest negative.

Results. The results on both synthetic and real-life datasets are summarized in Table 1. As shown both visually and quantitatively, GraphSVX correctly identifies essential graph structure, outperforming the leading baselines on all but one task, in addition to offering higher theoretical guarantees and human-friendly explanations. On *MUTAG*, the special nature of the dataset and ground truth favours edge explanation methods, which capture slightly more information than node explainers. Hence, we expect PGExplainer to perform better. For *BA-Community*, GraphSVX demonstrates its ability to identify relevant features and nodes together, as it also identifies important node features with 100% accuracy. In terms of efficiency, our explainer is slower than the scalable PGExplainer despite our efficient approximation, but is often comparable to GNNExplainer.

6.2   Real-World Datasets Without Ground Truth

Previous experiments involve mostly synthetic datasets, which are not totally representative of real-life scenarios. Hence, in this section, we evaluate GraphSVX on two real-world datasets without ground truth explanations: *Cora* and *PubMed*. Instead of looking if the explainer provides the correct explanation, we check that it does not provide a bad one. In particular, we introduce noisy features and nodes to the dataset, train a new GNN on the latter (which we verify do not leverage these noisy variables) and observe if our explainer includes them in explanations. In different terms, we investigate if the explainer filters useless features/nodes in complex datasets, selecting only relevant information in explanations.

Datasets. *Cora* is a citation graph where nodes represent articles and edges represent citations between pairs of papers. The task involved is document classification where the goal is to categorise each paper into one out of seven categories. Each feature indicates the absence/presence of the corresponding term in its abstract. *PubMed* is also a publication dataset with three classes and 500 features, each indicating the TF-IDF value of the corresponding word.

Noisy Features. Concretely, we artificially add 20% of new "noisy" features to the dataset. We define these new features using existing ones' distribution. We re-train a 2-layer GCN and a 2-layer GAT model on this noisy data, whose test accuracy is above 75%. We then produce explanations for 50 test samples using different explainer baselines, on *Cora* and *PubMed*, and we compare their performance by assessing how many noisy features are included in explanations among top-k features. Ultimately, we compare the resulting frequency distributions using a kernel density estimator (KDE). Intuitively, since features are noisy,

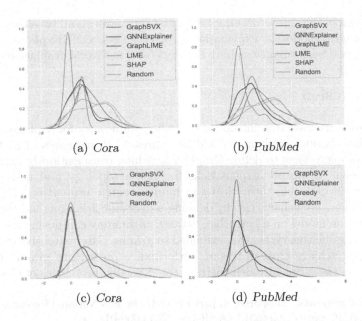

Fig. 2. Frequency distributions of noisy features (a), (b) and nodes (c), (d) using a GAT model on *Cora* and *PubMed*.

they are not used by the GNN model, and thus are unimportant. Therefore, the less noisy features are included in the explanation, the better the explainer.

Baselines include GNNExplainer, GraphLIME (described previously) as well as the well-known SHAP [15] and LIME [22] models. We also compare GraphSVX to a method based on a Greedy procedure, which greedily removes the most contributory features/nodes of the prediction until the prediction changes, and to the Random procedure, which randomly selects k features/nodes as the explanations for the prediction being explained.

The results are depicted in Fig. 2 (a)–(b). For all GNNs and on all datasets, the number of noisy features selected by GraphSVX is close to zero, and in general lower than existing baselines—demonstrating its robustness to noise.

Noisy Nodes. We follow a similar idea for noisy neighbours instead of noisy features. Each new node's connectivity and feature vector are determined using the dataset's distribution. Only a few baselines (GNNExplainer, Greedy, Random) among the ones selected previously can be included for this task since GraphLIME, SHAP, and LIME do not provide explanations for nodes.

As before, this evaluation builds on the assumption that a well-performing model will not consider as essential these noisy variables. We check the validity of this assumption for the GAT model by looking at its attention weights. We retrieve the average attention weight of each node across the different GAT layers and compare the one attributed to noisy nodes versus normal nodes. We expect it to be lower for noisy nodes, which proves to be true: 0.11 vs. 0.15.

As shown in Fig. 2 (c)–(d), GraphSVX also outperforms all baselines, showing nearly no noisy nodes in explanations. Nevertheless, GNNExplainer achieves almost as good performance on both datasets (and in several evaluation settings).

7 Conclusion

In this paper, we have first introduced a unified framework for explaining GNNs, showing how various explainers could be expressed as instances of it. We then use this complete view to define GraphSVX, which conscientiously exploits the above pipeline to output explanations for graph topology and node features endowed with desirable theoretical and human-centric properties, eligible of a good explainer. We achieve this by defining a decomposition method that builds an explanation model on a perturbed dataset, ultimately computing the Shapley values from game theory, that we extended to graphs. After extensive evaluation, we not only achieve state-of-the-art performance on various graph and node classification tasks but also demonstrate the desirable properties of GraphSVX.

Acknowledgements. Supported in part by ANR (French National Research Agency) under the JCJC project GraphIA (ANR-20-CE23-0009-01).

References

1. Backstrom, L., Leskovec, J.: Supervised random walks: predicting and recommending links in social networks. In: WSDM (2011)
2. Baldassarre, F., Azizpour, H.: Explainability techniques for graph convolutional networks. arXiv (2019)
3. Battaglia, P.W., Hamrick, J.B., Bapst, V., Sanchez-Gonzalez, A., et al.: Relational inductive biases, deep learning, and graph networks. arXiv (2018)
4. Burkart, N., Huber, M.F.: A survey on the explainability of supervised machine learning. JAIR **70**, 245–317 (2021)
5. Dabkowski, P., Gal, Y.: Real time image saliency for black box classifiers. In: NeurIPS (2017)
6. Datta, A., Sen, S., Zick, Y.: Algorithmic transparency via quantitative input influence: theory and experiments with learning systems. In: IEEE Symposium on Security and Privacy (SP) (2016)
7. Debnath, K., et al.: Structure-activity relationship of mutagenic aromatic and heteroaromatic nitro compounds. Correlation with molecular orbital energies and hydrophobicity. J. Med. Chem. **34**(2), 786–797 (1991)
8. Defferrard, M., Bresson, X., Vandergheynst, P.: Convolutional neural networks on graphs with fast localized spectral filtering. In: NeurIPS (2016)
9. Duval, A., Malliaros, F.D.: GraphSVX: shapley value explanations for graph neural networks. arXiv preprint arXiv:2104.10482 (2021)
10. Duvenaud, D., et al.: Convolutional networks on graphs for learning molecular fingerprints. In: NeurIPS (2015)

11. Hamilton, W., Ying, Z., Leskovec, J.: Inductive representation learning on large graphs. In: NeurIPS (2017)
12. Huang, Q., Yamada, M., Tian, Y., Singh, D., Yin, D., Chang, Y.: GraphLIME: local interpretable model explanations for graph neural networks. arXiv (2020)
13. Lipovetsky, S., Conklin, M.: Analysis of regression in game theory approach. ASMBI **17**(4), 319–330 (2001)
14. Lipton, P.: Contrastive explanation. Roy. Inst. Philos. Suppl. **27**, 247–266 (1990)
15. Lundberg, S.M., Lee, S.I.: A unified approach to interpreting model predictions. In: NeurIPS (2017)
16. Luo, D., et al.: Parameterized explainer for graph neural network. In: NeurIPS (2020)
17. Miller, T.: Explanation in artificial intelligence: insights from the social sciences. Artif. Intell. **267**, 1–38 (2019)
18. Miller, T., Howe, P., Sonenberg, L.: Explainable AI: beware of inmates running the asylum or: How I learnt to stop worrying and love the social and behavioural sciences. arXiv (2017)
19. Molnar, C.: Interpretable Machine Learning (2020). Lulu.com
20. O'neil, C.: Weapons of Math Destruction: How Big Data Increases Inequality and Threatens Democracy. Broadway Books (2016)
21. Pope, P.E., Kolouri, S., Rostami, M., Martin, C.E., Hoffmann, H.: Explainability methods for graph convolutional neural networks. In: CVPR (2019)
22. Ribeiro, M.T., Singh, S., Guestrin, C.: Why should I trust you? Explaining the predictions of any classifier. In: KDD (2016)
23. Riesen, K., Bunke, H.: IAM graph database repository for graph based pattern recognition and machine learning. In: da Vitoria Lobo, N., et al. (eds.) SSPR /SPR 2008. LNCS, vol. 5342, pp. 287–297. Springer, Heidelberg (2008). https:// doi.org/10.1007/978-3-540-89689-0_33
24. Robnik-Šikonja, M., Bohanec, M.: Perturbation-based explanations of prediction models. In: Zhou, J., Chen, F. (eds.) Human and Machine Learning. HIS, pp. 159–175. Springer, Cham (2018). https://doi.org/10.1007/978-3-319-90403-0_9
25. Saltelli, A.: Sensitivity analysis for importance assessment. Risk Anal. **22**(3), 579–590 (2002)
26. Schlichtkrull, M.S., Cao, N.D., Titov, I.: Interpreting graph neural networks for NLP with differentiable edge masking. In: ICLR (2021)
27. Selvaraju, R., et al.: Grad-CAM: visual explanations from deep networks via gradient-based localization. In: ICCV (2017)
28. Shapley, L.S.: A value for n-person games. Contrib. Theory Games **2**(28), 307–317 (1953)
29. Shrikumar, A., Greenside, P., Kundaje, A.: Learning important features through propagating activation differences. In: ICML (2017)
30. Strumbelj, E., Kononenko, I.: An efficient explanation of individual classifications using game theory. JMLR **11**, 1–18 (2010)
31. Ustun, B., Rudin, C.: Methods and models for interpretable linear classification. arXiv (2014)
32. Vu, M.N., Thai, M.T.: PGM-explainer: probabilistic graphical model explanations for graph neural networks. In: NeurIPS (2020)
33. Ying, Z., Bourgeois, D., You, J., Zitnik, M., Leskovec, J.: GNNExplainer: generating explanations for graph neural networks. In: NeurIPS (2019)
34. Yuan, H., Tang, J., Hu, X., Ji, S.: XGNN: towards model-level explanations of graph neural networks. In: KDD (2020)

35. Yuan, H., Yu, H., Gui, S., Ji, S.: Explainability in graph neural networks: a taxonomic survey. arXiv preprint arXiv:2012.15445 (2020)
36. Zhang, M., Chen, Y.: Link prediction based on graph neural networks (2018)
37. Zhou, J., et al.: Graph neural networks: a review of methods and applications. arXiv (2018)

Multi-view Self-supervised Heterogeneous Graph Embedding

Jianan Zhao[1] ⓘ, Qianlong Wen[1] ⓘ, Shiyu Sun[1] ⓘ, Yanfang Ye[1](✉) ⓘ, and Chuxu Zhang[2](✉) ⓘ

[1] Case Western Reserve University, Cleveland, OH, USA
{jxz1244,qxw294,sxs2293,yanfang.ye}@case.edu
[2] Brandeis University, Waltham, MA, USA
chuxuzhang@brandeis.edu

Abstract. Graph mining tasks often suffer from the lack of supervision from labeled information due to the intrinsic sparseness of graphs and the high cost of manual annotation. To alleviate this issue, inspired by recent advances of self-supervised learning (SSL) on computer vision and natural language processing, graph self-supervised learning methods have been proposed and achieved remarkable performance by utilizing unlabeled information. However, most existing graph SSL methods focus on homogeneous graphs, ignoring the ubiquitous heterogeneity of real-world graphs where nodes and edges are of multiple types. Therefore, directly applying existing graph SSL methods to heterogeneous graphs can not fully capture the rich semantics and their correlations in heterogeneous graphs. In light of this, we investigate self-supervised learning on heterogeneous graphs and propose a novel model named Multi-View Self-supervised heterogeneous graph Embedding (MVSE). By encoding information from different views defined by meta-paths and optimizing both intra-view and inter-view contrastive learning tasks, MVSE comprehensively utilizes unlabeled information and learns node embeddings. Extensive experiments are conducted on various tasks to show the effectiveness of the proposed framework.

Keywords: Self-supervised learning · Heterogeneous graph embedding · Graph neural network

1 Introduction

With the proliferation of real-world interaction systems, graph mining has been a popular topic with many real-world applications such as node classification, graph classification, and recommendation. Due to the ubiquitous sparseness of graphs and the deficiency of label supervision, it is vital to fully utilize the unlabeled information on graphs. However, the current state-of-the-art algorithms, which are mostly based on Graph Neural Networks (GNNs) [24,36,41], mainly utilize unlabeled information by simply aggregating their features and cannot thoroughly take advantage of the abundant unlabeled data [20]. Recently, aiming to fully exploit the unlabeled information for GNNs, self-supervised learning

© Springer Nature Switzerland AG 2021
N. Oliver et al. (Eds.): ECML PKDD 2021, LNAI 12976, pp. 319–334, 2021.
https://doi.org/10.1007/978-3-030-86520-7_20

(SSL) is naturally harnessed for providing additional supervision and achieves impressive improvements on various graph learning tasks [27].

The existing graph SSL methods fall into two categories, generative and contrastive [27]. However, they mainly focus on designing self-supervised tasks on homogeneous graphs, overlooking the ubiquitous heterogeneity and rich semantics in graphs. Unlike homogeneous graphs, a heterogeneous graph [34] is composed of multiple types of nodes and edges. To illustrate, consider a bibliography graph with its network schema shown in Fig. 1 (a), where four types of nodes: Author (A), Paper (P), Venue (V), and Term (T) along with three types of edges: an author writes a paper, a paper is published in a venue, and a paper contains a term.

To fully capture the rich heterogeneity and complex semantics inside heterogeneous graph data, we are motivated to study the problem of self-supervised learning on heterogeneous graphs. However, this is a non-trivial task as there are several challenges to be addressed. Above all, *how to deal with the intrinsic heterogeneity of heterogeneous graphs?* Different from homogeneous graphs, heterogeneous graph contains rich semantics for each node. For example, in the example bibliography graph mentioned above, we can introduce two meta-paths APA and APVPA to capture the co-author and co-venue semantics respectively. Therefore, how to design self-supervised tasks to fully capture the rich semantic information is a critical yet challenging problem. What's more, *how to effectively model the complex correlations between these different semantics?* Previous works mainly focus on discriminating the heterogeneous context instances [2,3,5,12], e.g. whether two authors have a co-author relationship, preserving the intra-context proximity [43]. However, the complex correlations between these contexts (inter-context), e.g. whether two authors with co-venue relationships have co-author relationships, remain less explored. Modeling these interactions not only encourages the embedding to preserve these interactions between semantics, pushing the model to extract useful information and encode them in node embeddings, but also alleviates the negative impact of the intrinsic sparseness of heterogeneous graphs [49].

To address the challenges mentioned above, we study self-supervised learning on heterogeneous graphs and focus on comprehensively encode the semantics and their correlations into node embeddings. In particular, we propose a novel model named **M**ulti-**V**iew **S**elf-supervised heterogeneous graph **E**mbedding (MVSE). MVSE firstly samples semantic subgraphs of different views defined by meta-paths. Then, each semantic subgraph is encoded to its own semantic latent space and further decoded to other semantic spaces to capture the semantic correlations. Finally, the embeddings are optimized by a contrastive loss preserving both intra-view, and inter-view interactions of semantic contexts. Our major contributions are highlighted as follows:

- We propose a novel self-supervised heterogeneous graph embedding model, in which some delicate designs, e.g., heterogeneous context encoding and multi-view contrastive learning are proposed to comprehensively learn good heterogeneous graph embeddings. Our work is among the earliest works that study self-supervised learning on heterogeneous graphs.

– While intra-semantic relationships are widely utilized, few works have attempted to model the correlations between the semantics in heterogeneous graphs. We design self-supervised learning tasks that preserve both intra- and inter-semantic information in node embeddings.

– We conduct extensive experiments on three real-world datasets to validate the effectiveness of MVSE compared with state-of-the-art methods. Through parameter analysis and ablation study, we further demonstrate that though often overlooked, preserving inter-view interactions is beneficial for heterogeneous graph embedding.

2   Related Work

2.1   Self-supervised Learning on Graphs

To fully exploit the ample unlabeled information, self-supervised learning (SSL) on graphs has become a promising research topic and achieved impressive improvements on various graph learning tasks [20]. Existing graph SSL methods design generative or contrastive tasks [27] to better harness the unlabeled graph data. On the one hand, generative graph SSL models learn graph embedding by recovering graph structure and attributes. For example, VGAE [23] applies GCN-based variational auto-encoder [22] to recover the adjacency matrix of the graph by measuring node proximity. GraphRNN [44] uses a graph-level RNN and reconstructs adjacency matrix iteratively. GPT-GNN adopts GCNs [24] to reconstruct both graph structure and attribute information. On the other hand, contrastive graph SSL models learn graph embedding by discriminating positive and negative samples generated from graphs. To illustrate, Context Prediction and Attribute Mask [15] are proposed to preserve the structural and attribute information. DGI [37] contrasts local (node) and global (graph) embedding via mutual information maximization. MVGRL [8] contrasts embeddings from first-order and high-order neighbors by maximizing mutual information. GCC [32] performs subgraph instance discrimination across different graphs. Though graph SSL works have achieved significant performance improvements, most of the existing Graph SSL works focus on homogeneous graphs and can not address the complex semantics of heterogeneous graphs.

2.2   Heterogeneous Graph Embedding

Our work is also related to heterogeneous graph embedding (HGE), which encodes nodes in a graph to low-dimensional representations while effectively preserving the heterogeneous graph structure. HGE methods can be roughly divided into three categories [43]: proximity-preserving methods, relation learning methods, and message passing methods. The proximity-preserving HGE methods [3,10,18,47,47] are mostly random walk [31] based and optimized by (heterogeneous) skip-gram. The relation-learning HGE methods [1,26,28,35,40,42] construct head, tail, and relation triplets and optimize embedding by a relation-specific scoring function that evaluates an arbitrary triplet and outputs a scalar to measure the acceptability of this triplet. Recently, with the proliferation of graph neural networks [7,24,36], message-passing HGE methods are brought forward and have achieved remarkable improvements on series of applications [4,13,14,25,38,38]. These message-passing HGEs learn graph embedding by aggregating and transforming the embeddings of the original neighbors [11,14,17,46,48] or metapath-based neighbors [6,39,45].

Nevertheless, most of the existing HGE methods follow a unified framework [43] which learns embedding by minimizing the distance between the node embeddings of target node and its context nodes, preserving the heterogeneous semantics. However, the underlying rich correlations [49] between these rich semantics are seldom discussed and explored.

3   The Proposed Model

3.1   Model Framework

Consider a heterogeneous graph $G = (\mathcal{V}, \mathcal{E}, \mathbf{X})$ composed of a node set $\mathcal{V}$, an edge set $\mathcal{E}$, and a feature matrix $\mathbf{X} \in \mathbb{R}^{|\mathcal{V}| \times d_F}$ (d_F: feature dimension) along with the node type mapping function $\phi : \mathcal{V} \to \mathcal{A}$, and the edge type mapping function $\psi : \mathcal{E} \to \mathcal{R}$, where $\mathcal{A}$ and $\mathcal{R}$ denotes the node and edge types, and $|\mathcal{A}| + |\mathcal{R}| > 2$. The task of heterogeneous graph embedding is to learn the representation of nodes $\mathbf{Z} \in \mathbb{R}^{|\mathcal{V}| \times d}$, where d is the dimension of representation.

The key idea of MVSE is to capture the rich heterogeneous semantics and their correlations by self-supervised contrastive learning. As shown in Fig. 1 (c), given a node in heterogeneous graph, MVSE firstly samples several metapath-based semantic subgraphs and encodes them to its semantic space by semantic-specific encoders. Then, the semantic embeddings are further decoded to other semantic spaces to model the correlations between different semantics. Finally, the semantic embeddings and the decoded embeddings are optimized by intra-view, and inter-view contrastive learning losses.

3.2   Heterogeneous Context Encoding

A node in heterogeneous graph is associated with rich semantic information defined by meta-paths, providing different views of node property. Therefore, it is vital to encode the metapath-based neighbor information into node embeddings. Inspired by the recent advances of contrastive learning [8,32], we propose heterogeneous subgraph instance discrimination as our self-supervised contrastive learning task. In this section, we elaborate how MVSE constructs multi-view heterogeneous subgraphs and encodes them as heterogeneous context embeddings.

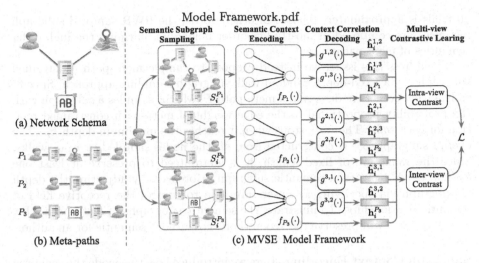

Fig. 1. (a) The network schema of an example bibliography heterogeneous graph with four types of nodes: Author (A), Paper (P), Venue (V), and Term/keyword (T) along with three types of edges: an author writes a paper, a paper is published in a venue, and a paper contains a term. (b) The template meta-paths (views of semantics) APVPA, APA, APTPA. (c) The model framework of the proposed model MVSE.

Semantic Subgraph Sampling. Given a node v_i in heterogeneous graph G and a meta-path set $\mathcal{P}$, MVSE samples a subgraph instance set $\mathcal{S}_i^{\mathcal{P}} = \{S_i^P, P \in \mathcal{P}\}$ and further encodes them to semantic embeddings. In the homogeneous graph, an effective way of generating subgraph instances for contrastive learning is to apply RWR (random walk with restart), by iteratively generating subgraph structure via random walk with a restart probability γ [32]. Therefore, a straightforward idea to construct heterogeneous subgraph instances would be applying meta-path constrained RWR.

However, this straightforward extension can not well preserve the metapath-based context for heterogeneous graphs due to the intrinsic lack of high-order neighbors preservation of RWR. Specifically, each random walk trace is a Bernoulli trial with probability $(1 - \gamma)^k$ sampling k-hop neighbors. Therefore, the number of time n_s that k-hop neighbors is sampled after n_{RW} number of restart time, is a binomial distribution:

$$P(n_s|k, n_{RW}) \sim B(n_{RW}, (1 - \gamma)^k), \tag{1}$$

Hence, we can obtain that the expectation of number of times that k-hop neighbors are sampled a subgraph sampled by RWR :

$$E(n_s|k, n_{RW}) = n_{RW}(1 - \gamma)^k, \tag{2}$$

which decreases exponentially when k increases, harming the high-order preservation. Specifically, with the recommended setting [32,33], i.e. $\gamma = 0.8$, the probability of at least one 4-hop neighbor (which is the maximum depth of commonly used meta-paths e.g. APVPA, APTPA) is sampled in subgraphs within

20 trials is approximately 0.0315. In other words, the RWR sampled subgraph instances are composed of mostly low-order neighbors, harming the high-order semantics of meta-paths.

To address this issue, we instead propose to sample meta-path constrained subgraphs by a fixed-depth random walk subgraph sampling approach. Specifically, given a center node v_i and a meta-path P, MVSE samples a subgraph with the probability proportional to the edge weight of meta-path constrained relation for each walk. The walk stops when it reaches the maximum depth k_P. The overall subgraph S_i^P is constructed from all the nodes sampled in n_{RW} walks. Since the walks are of fixed length, it is guaranteed to preserve at least one k_P-hop neighbor in each semantic subgraph. Moreover, by adjusting the depth of subgraph via specifying k_P, users are able to control the receptive field of semantic relationships. To illustrate, the semantic subgraphs of meta-path APA with $k_{APA} = 4$ will preserve the "co-authors' co-author" semantic for an author.

Subgraph Context Encoding. Here we introduce how to encode the semantic subgraphs to obtain a semantic embedding for each node. Specifically, given a node v_i and the sampled semantic subgraph set $\mathcal{S}_i^P = \{S_i^P, P \in \mathcal{P}\}$, the task is to encode subgraphs into multi-view embeddings $\mathcal{H}_i = \{\mathbf{h}_i^P \in \mathbb{R}^{1 \times d_s}, P \in \mathcal{P}\}$, where d_s stands for the hidden dimension of subgraph embeddings.

To fully capture the heterogeneity of different semantics [2,28], we propose to use a semantic-specific encoder for each meta-path. Therefore, the semantic embedding of node v_i in the view of meta-path P denoted as $\mathbf{h}_i^P$ is obtained by:

$$\mathbf{h}_i^P = f_P(S_i^P), \tag{3}$$

where $f_P(\cdot)$ stands for the semantic-specific encoder for meta-path P. The choice of encoder can be any graph neural networks [24]. We adopt the Graph Isomorphism Network (GIN) [41] as the graph encoder. Hence, the semantic embedding is calculated by:

$$\mathbf{h}_i^P = \text{CONCAT}\left(\text{SUM}(\{\mathbf{h}_v^{P,(l)} \mid v \in S_i^P\}) \mid l = 0, 1, \dots, L\right),$$

$$\mathbf{h}_v^{P,(l)} = \text{MLP}^{P,(l)}\left((1 + \epsilon) \cdot \mathbf{h}_v^{P,(l-1)} + \sum_{u \in \mathcal{N}_i^P(v)} \mathbf{h}_u^{P,(l-1)}\right), \tag{4}$$

where $\text{MLP}^{P,(l)}$ stands for the semantic-specific encoder for meta-path P at l-th layer, $\mathcal{N}_i^P(v)$ stands for the neighbors of node v in S_i^P, $\mathbf{h}_v^{P,(l)}$ is the l-th layer node representation of node v in semantic subgraph S_i^P, and the input is set as the node feature, i.e. $\mathbf{h}_v^{P,(0)} = \mathbf{x}_v$, ϵ is a fixed scalar.

3.3 Multi-view Contrastive Learning

At this point, we have obtained multi-view embeddings $\mathcal{H}_i$ of each node v_i. Here, we elaborate how to perform self-supervised contrastive learning on these embeddings to comprehensively learn the heterogeneous semantics and their correlations.

Preservation of Semantic Contexts. We utilize MoCo [9] as the contrastive learning framework where a query node and a set of key nodes are contrasted in each epoch. MoCo maintains a dynamic dictionary of keys (nodes) and encodes the new keys on-the-fly by a momentum-updated encoder. In each epoch of MVSE, a query node is contrasted with K nodes, where K is the size of the dynamic dictionary. Here, as each node is encoded as multi-view embeddings $\mathcal{H}_i = \{\mathbf{h}_i^P \in \mathbb{R}^{1 \times d_s}, P \in \mathcal{P}\}$, we perform multi-view contrastive learning on each view separately by the InfoNCE [30] loss, preserving the intra-semantic information:

$$\mathcal{L}_{intra} = \frac{1}{|\mathcal{P}|} \sum_{P \in \mathcal{P}} -\log \frac{\exp\left(\mathbf{h}_q^P \cdot \mathbf{h}_{k+}^P / \tau\right)}{\sum_{j=0}^{K} \exp\left(\mathbf{h}_q^P \cdot \mathbf{h}_{k_j}^P / \tau\right)}, \tag{5}$$

where $\mathbf{h}_q^P$ is the query node's semantic embedding of metapath P calculated by Eq. 3, $\mathbf{h}_k^P$ stands for the key node's semantic embedding encoded by momentum encoders [9], $k+$ stands for the positive key in the dictionary, τ is the temperature hyper-parameter. Thus, by minimizing $\mathcal{L}_{intra}$, MVSE is able to distinguish subgraph instances of different nodes using each meta-path in $\mathcal{P}$.

Preservation of Semantic Correlations. As discussed in the Introduction, most existing HGE methods focus on discriminating the heterogeneous context instances, e.g. whether two authors have a co-author relationship, preserving the intra-context relationships. However, few works have explored the complex interactions (inter-context) [49] between these contexts, e.g. whether two authors with co-venue relationships have co-author relationships.

In light of this, we propose to explicitly capture these correlations by inter-view contrastive learning. Specifically, for each semantic embedding $\mathbf{h}_i^P$ of node v_i of meta-path P, we model the correlations between semantics by decoding them to other semantic embeddings:

$$\hat{\mathbf{h}}_i^{s,t} = g^{s,t}(\mathbf{h}_i^{P_s}) \tag{6}$$

where $g^{s,t}(\cdot)$ stands for the decoder that decodes the semantic embedding from source view P_s to target view P_t. $\hat{\mathbf{h}}_i^{s,t}$ stands for the semantic embedding of target view P_s decoded from source view P_t. In this way, the correlation between source view and target view is preserved. For example, if we set source view as APVPA and target view as APA, the decoder attempts to predict the co-author relationships using the co-venue relationships, modeling the interactions between these two semantics. Hence, the complex correlations between semantics can be well preserved by the inter-view contrastive loss defined as follows:

$$\mathcal{L}_{inter} = \frac{1}{|\mathcal{P}| * (|\mathcal{P}| - 1)} \sum_{P_s, P_t \in \mathcal{P}, s \neq t} -\log \frac{\exp\left(\hat{\mathbf{h}}_i^{s,t} \cdot \mathbf{h}_{k+}^{P_t} / \tau\right)}{\sum_{j=0}^{K} \exp\left(\hat{\mathbf{h}}_i^{s,t} \cdot \mathbf{h}_{k_j}^{P_t} / \tau\right)}, \tag{7}$$

Finally, MVSE optimizes the overall loss $\mathcal{L}$ to comprehensively learn representations considering both the intra-view and inter-view semantics:

$$\mathcal{L} = \alpha\mathcal{L}_{intra} + (1 - \alpha)\mathcal{L}_{inter} \tag{8}$$

where α is the hyper-parameter for balancing different loss functions.

4   Experiment

To demonstrate the effectiveness of our proposed model, we conduct comprehensive experiments on three public benchmark heterogeneous graph datasets. We firstly evaluate our model on two downstream tasks (node classification and link prediction). Then, we perform ablation study to further demonstrate the effectiveness of the designs in MVSE. Visualization experiments are also conducted to show the effectiveness of our model intuitively.

4.1   Experimental Setup

Datasets. We employ the following real-world heterogeneous graph datasets to evaluate our proposed model. **DBLP** [28]: We extract a subset of DBLP which includes 4,057 authors (A), 20 conferences (C), 14,328 papers (P) and four types of edges (AP, PA, CP, and PC). The target nodes are authors and they are divided into four areas: database, data mining, machine learning, and information retrieval. The node features are the terms related to authors, conferences and papers respectively.

ACM [45]: We extract papers published in KDD, SIGMOD, SIGCOMM, Mobi-COMM, and VLDB and construct a heterogeneous graph which includes 5,912 authors (A), 3,025 papers (P), 57 conference subjects (S) and four types of edges (AP, PA, SP, and PS). The target nodes are papers and they are divided into three classes according to their conferences: database, data mining, and wireless communication. The node features are the terms related to authors, papers and subjects respectively.

IMDB [39]: We extract a subset of IMDB which includes 4,461 movies (M), 2,270 actors (A), 5,841 directors (D), and four types of edges (AM, MA, DM, and MD). The target nodes are movies labeled by genre (action, comedy, and drama). The movie features are bag-of-words representation of plot keywords.

Baselines. To comprehensively evaluate our model, we compare MVSE with ten graph embedding methods. Based on their working mechanisms, these baselines can be divided into three categories: The unsupervised representation learning methods, i.e.DeepWalk [31], MP2Vec [3], DGI [37], and HeGAN [12]; the semi-supervised representation learning methods, i.e. GCN [24], GIN [41], HAN [39], and GTN [45], and the self-supervised learning methods, i.e. GCC [32] and GPT-GNN [16]. For unsupervised baselines, the embeddings are learned without label supervision and then fed into a logistic classifier to perform the downstream tasks. The semi-supervised methods are optimized through an end-to-end supervised manner, e.g. cross entropy loss in node classification tasks. The self-supervised methods are firstly pre-trained to fully encode the unlabeled information and then fine-tuned by labeled information via cross entropy loss in node classification.

Implementation Details. Here, we briefly introduce the experimental settings. For MVSE, in each epoch, we construct semantic subgraphs by performing 3 times of the meta-path constrained fixed-depth random walk (in Sect. 3.2) with the maximum depth set as twice the depth of meta-path, i.e. $n_{RW} = 3, k_P = 2|P|$, where $|P|$ stands for the depth of meta-path P. The decoders for modeling the semantic correlations are 2-layer MLPs. We use Adam [21] optimizer with learning rate set as 0.005. The semantic embedding dimension d_s is set as 64, therefore the dimension d of final node embedding $\mathbf{Z}$ is $64|\mathcal{P}|$. For MoCo-related settings, the dynamic dictionary size K is set as 4096 with $\tau = 0.07$. For all GNN related models, we use 2-layer GCNs [24] with weight decay set as 1e-5. The code and data to reproduce our results is publicly available at Github[1].

4.2   Node Classification

As a common graph application, node classification is widely used to evaluate the performance of the graph embedding algorithms. Given a graph with some labeled nodes, the task of node classification is to predict the labels of unlabeled nodes. Here, we evaluate the performance of node classification on the three datasets mentioned above in this section. For each dataset, the percentage of training labeled nodes are set as 1%, 3%, and 5%, and the rest of the labeled nodes are used as test nodes. We adopt Macro-F1 and Micro-F1 as metrics and report the node classification performance on the test set. The results (in percentage) of the three datasets are shown in Table 1, Table 2 and Table 3, respectively, from which we have the following observations: (1) By comprehensively preserving the rich semantics and their correlations inside heterogeneous graphs, our proposed MVSE outperforms other baselines, demonstrating the effectiveness of our proposed model. (2) Most self-supervised learning models (MVSE and GPT-GNN) generally achieve better performance than other baselines, since the pre-training of self-supervised tasks extract robust embedding with rich semantics and structural information and provide a better initialization for the fine-tuning process. The performance improvement is more significant when the ratio of labeled information is low. (3) Since the node features are of vital importance in node classification tasks, GNN-based models generally outperform the random walk-based models due to their ability to utilize node features.

4.3   Link Prediction

The objective of link prediction is to predict unobserved edges using the observed graph. To evaluate the effectiveness of semantic preservation, we use metapath-based link prediction task [19] on three datasets and evaluate the metapath-based link prediction performance on 2-hop symmetric meta-paths. Specifically, in each task, the meta-path instances are firstly randomly splitted as training and test set with 1:1 ratio. Then, the self-supervised/unsupervised models are applied to learn the node representations. Finally, the embeddings are fed to

---

[1] https://github.com/Andy-Border/MVSE.

Table 1. Performance of node classification experiment on DBLP dataset in percentage (Micro-F1 and Macro-F1), MVSE outperforms the baselines in all the settings.

Models	DBLP (1%)		DBLP (3%)		DBLP (5%)	
	Micro-F1	Macro-F1	Micro-F1	Macro-F1	Micro-F1	Macro-F1
GCN [24]	72.63	70.86	79.36	78.22	82.08	81.17
GIN [41]	71.22	67.26	88.23	88.28	88.31	88.32
HAN [39]	85.65	85.24	89.42	88.83	89.72	89.36
GTN [45]	85.99	85.45	88.66	88.13	89.73	89.37
DeepWalk [31]	86.58	87.31	87.48	87.95	87.72	88.34
DGI [37]	87.73	86.44	90.01	89.33	91.22	89.57
MP2Vec [3]	86.33	85.87	88.16	87.82	88.91	88.63
HeGAN [12]	79.12	77.73	81.66	80.25	83.78	82.44
GPT_GNN [16]	86.61	86.33	90.62	89.26	90.91	89.43
GCC [32]	78.92	77.94	81.78	81.11	82.67	81.89
MVSE	**90.46**	**89.27**	**91.57**	**90.97**	**91.96**	**89.83**

Table 2. Performance of node classification experiment on ACM dataset in percentage (Micro-F1 and Macro-F1), MVSE outperforms the baselines in all the settings.

Models	ACM (1%)		ACM (3%)		ACM (5%)	
	Micro-F1	Macro-F1	Micro-F1	Macro-F1	Micro-F1	Macro-F1
GCN [24]	85.78	85.87	88.97	89.01	89.55	89.62
GIN [41]	78.40	77.99	84.62	84.69	87.09	87.17
HAN [39]	85.77	85.92	87.41	87.62	88.37	88.58
GTN [45]	80.08	79.54	84.56	84.16	88.71	88.24
DeepWalk [31]	79.02	79.28	80.75	81.03	80.15	80.57
DGI [37]	84.99	85.21	88.93	89.06	89.36	89.50
MP2Vec [3]	80.74	80.36	82.42	81.87	82.63	82.12
HeGAN [12]	78.23	78.67	80.84	81.35	81.95	82.52
GPT_GNN [16]	84.62	84.86	88.90	89.22	89.27	89.54
GCC [32]	80.14	78.84	83.91	82.35	84.72	83.17
MVSE	**86.14**	**86.17**	**89.43**	**89.44**	**89.74**	**89.64**

logistic regression classifiers and predict whether the test edges exist by training edges. We use F1 and AUC-ROC as our evaluation metrics, the results in percentage are shown in Table 4, from which we have the following observations: (1) MVSE consistently outperforms other baselines on all the metapath-based link prediction tasks. The reason is that MVSE is able to capture the correlations between meta-paths, thus alleviates the impact of intrinsic sparseness in graphs [49], e.g. APA link prediction can be enhanced by APCPA relationship, and further improve the link prediction performance. (2) Models that consider heterogeneity show better performance than their counterparts since they are able to extract the rich semantic contexts from the different meta-paths.

Table 3. Performance of node classification experiment on IMDB dataset in percentage (Micro-F1 and Macro-F1), MVSE outperforms the baselines in most of the settings.

Models	IMDB (1%)		IMDB (3%)		IMDB (5%)	
	Micro-F1	Macro-F1	Micro-F1	Macro-F1	Micro-F1	Macro-F1
GCN [24]	51.71	42.76	55.06	45.88	58.05	50.69
GIN [41]	48.87	43.24	53.70	48.38	58.03	53.42
HAN [39]	50.76	43.47	52.87	48.46	56.43	51.25
GTN [45]	51.66	45.87	57.83	49.31	59.49	53.58
DeepWalk [31]	53.92	49.34	54.44	49.85	54.48	49.88
DGI [37]	53.02	44.61	56.62	48.29	57.93	50.15
MP2Vec [3]	54.50	**49.82**	55.10	50.34	56.97	52.79
HeGAN [12]	47.70	41.47	49.98	44.29	51.04	46.46
GPT_GNN [16]	55.17	48.30	58.78	52.69	61.24	56.74
GCC [32]	52.33	47.29	53.68	48.82	53.85	49.08
MVSE	**55.61**	44.25	**60.15**	**53.95**	**63.32**	**58.42**

Table 4. Performance of link prediction experiment on different datasets and meta-paths in percentage (Micro-F1 and ROC-AUC), MVSE outperforms the baselines on all the datasets and meta-paths.

Models	DBLP (APA)		ACM (PAP)		ACM (PSP)		IMDB (MAM)		IMDB (MDM)	
	F1	AUC	F1	AUC	F1	AUC	F1	AUC	F1	AUC
DeepWalk [31]	79.08	77.72	72.99	72.73	78.65	69.85	72.62	75.37	59.16	60.37
DGI [37]	80.59	81.28	73.31	72.41	84.87	74.4	70.13	69.54	61.37	59.24
MP2Vec [3]	81.64	80.66	75.64	74.82	82.88	75.16	82.55	81.06	64.39	63.78
HeGAN [12]	80.75	80.26	78.67	78.51	81.25	71.02	80.27	80.11	67.75	68.13
GPT-GNN [16]	86.84	86.02	80.94	80.55	86.31	78.25	91.88	91.31	68.54	68.12
GCC [32]	79.15	78.63	73.62	72.98	74.22	68.17	81.63	82.37	64.71	63.85
MVSE	**88.09**	**87.92**	**81.18**	**80.73**	**87.72**	**79.22**	**98.25**	**98.23**	**69.51**	**69.54**

4.4   Ablation Study

In order to verify the effectiveness of the delicate designs in MVSE, we design five variants of MVSE and compare their node classification performance against MVSE on three datasets. The results in terms of Micro-F1 are shown in Fig. 2 (a), Fig. 2 (b) and Fig. 2 (c), respectively.

Effectiveness of Heterogeneous Context Encoders. As discussed in Sect. 3.2, to capture the intrinsic heterogeneity [2,28] in different metapath-based semantics, MVSE use semantic-specific encoders to embed the contexts of different meta-paths. To verify the effectiveness of semantic specific encoders, we propose a variant of MVSE which uses metapath-shared encoders, namely MVSE-MP-Shared. The results in Fig. 2 show MVSE outperforms the variant on the three datasets since MVSE-MP-Shared ignores the heterogeneity of semantics by modeling them using an unified (homogeneous) model. This phenomenon further demonstrates the importance of considering heterogeneity in heterogeneous graph contrastive learning.

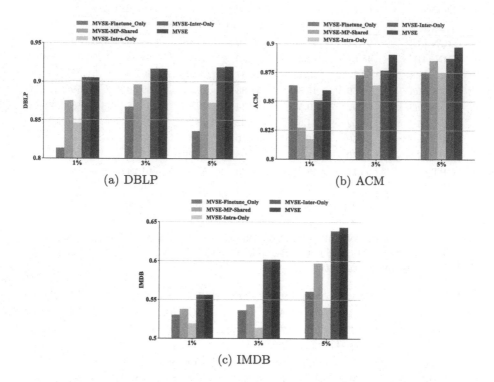

(a) DBLP

(b) ACM

(c) IMDB

Fig. 2. Performance of MVSE variants on different datasets (Micro-F1), MVSE outperforms the variants over all the datasets and settings. MVSE-Inter-Only has better performance than other variants, which demonstrates the importance of preserving semantic correlations.

Effectiveness of Multi-view Contrastive Learning. As discussed in Sect. 3.3, MVSE comprehensively learns the heterogeneous semantics and their correlations by intra-view and inter-view contrastive learning tasks. To investigate the effects of these contrastive learning tasks, we propose two variants of MVSE which only consider intra-view (MVSE-Intra-Only) and inter-view (MVSE-Inter-Only) respectively and evaluate their node classification performance. From the results shown in Fig. 2, we can find that MVSE beats all variants on every task, which indicates the effectiveness of performing multi-view contrastive learning by optimizing both intra- and inter-semantic SSL tasks. Besides, the phenomenon that MVSE-Inter-Only outperforms the other two variants further demonstrates the importance of preserving semantic correlations.

Effectiveness of Unlabeled Data Utilization. To investigate the ability of utilizing unlabeled information, we propose MVSE-Finetune-Only which skips the pre-training process and trains the model from scratch. As shown in Fig. 2, MVSE consistently outperforms this variant in all tasks since MVSE-Finetune-Only cannot fully utilize the unlabeled information by optimizing objective that considers labeled information only [20]. The self-supervised pre-training strategy provides a better start point than the random initialization and further improves the classification performance. In addition, the performance improvement of MVSE over MVSE-Finetune-Only is generally more significant when the percentage of labeled nodes is low, demonstrating the superiority of SSL in tasks with little label supervision.

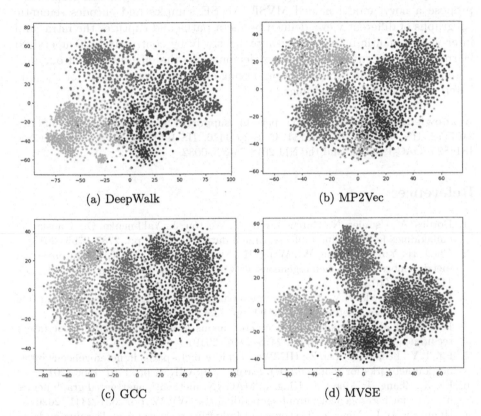

(a) DeepWalk   (b) MP2Vec

(c) GCC   (d) MVSE

Fig. 3. Node embedding visualization of different methods on DBLP dataset. Each point indicates one author and its color indicates the research area. MVSE has least overlapping area and largest cluster-wise distance. (Color figure online)

4.5   Visualization

To examine the graph representation intuitively, we visualize embeddings of author nodes in DBLP using the t-SNE [29] algorithm. Here, we choose Deep-Walk, MP2Vec, and GCC as the representatives of homogeneous embedding, heterogeneous embedding, and self-supervised based embedding methods, respectively. The visualization results are shown in Fig. 3, from which we can find that although all of the baselines can roughly embed the authors with same research fields into same clusters, the heterogeneous models generate more distinct boundaries and less overlapping area between clusters. What's more, among all of the unsupervised graph learning algorithms, MVSE generates embeddings with the largest cluster-wise distance, indicating better embeddings are learned.

5   Conclusion

In this paper, we study self-supervised learning on heterogeneous graphs and propose a novel model named MVSE. MVSE samples and encodes semantic subgraphs of different views defined by meta-paths and captures the intra- and inter-view semantic information comprehensively by contrastive self-supervised learning. Our extensive experiments demonstrate the effectiveness of our proposed model and the necessity of preserving cross-view interactions for learning heterogeneous graph embeddings.

Acknowledgements. This work is partially supported by the NSF under grants IIS-2107172, IIS-2027127, IIS-2040144, CNS-2034470, IIS-1951504, CNS-1940859, CNS-1814825, OAC-1940855, and the NIJ 2018-75-CX-0032.

References

1. Bordes, A., Usunier, N., Garca-Durán, A., Weston, J., Yakhnenko, O.: Translating embeddings for modeling multi-relational data. In: NIPS, pp. 2787–2795 (2013)
2. Chen, H., Yin, H., Wang, W., Wang, H., Nguyen, Q.V.H., Li, X.: PME: projected metric embedding on heterogeneous networks for link prediction. In: KDD, pp. 1177–1186 (2018)
3. Dong, Y., Chawla, N.V., Swami, A.: metapath2vec: scalable representation learning for heterogeneous networks. In: KDD, pp. 135–144 (2017)
4. Fan, S., et al.: Metapath-guided heterogeneous graph neural network for intent recommendation. In: KDD, pp. 2478–2486 (2019)
5. Fu, T.Y., Lee, W.C., Lei, Z.: HIN2Vec: explore meta-paths in heterogeneous information networks for representation learning. In: CIKM, pp. 1797–1806 (2017)
6. Fu, X., Zhang, J., Meng, Z., King, I.: MAGNN: metapath aggregated graph neural network for heterogeneous graph embedding. In: WWW, pp. 2331–2341 (2020)
7. Hamilton, W.L., Ying, Z., Leskovec, J.: Inductive representation learning on large graphs. In: NIPS, pp. 1024–1034 (2017)
8. Hassani, K., Ahmadi, A.H.K.: Contrastive multi-view representation learning on graphs. In: ICML, pp. 4116–4126 (2020)
9. He, K., Fan, H., Wu, Y., Xie, S., Girshick, R.B.: Momentum contrast for unsupervised visual representation learning. In: CVPR, pp. 9726–9735 (2020)
10. He, Y., Song, Y., Li, J., Ji, C., Peng, J., Peng, H.: HeteSpaceyWalk: a heterogeneous spacey random walk for heterogeneous information network embedding. In: CIKM, pp. 639–648 (2019)
11. Hong, H., Guo, H., Lin, Y., Yang, X., Li, Z., Ye, J.: An attention-based graph neural network for heterogeneous structural learning. In: AAAI, pp. 4132–4139 (2020)

12. Hu, B., Fang, Y., Shi, C.: Adversarial learning on heterogeneous information networks. In: KDD, pp. 120–129 (2019)
13. Hu, B., Zhang, Z., Shi, C., Zhou, J., Li, X., Qi, Y.: Cash-out user detection based on attributed heterogeneous information network with a hierarchical attention mechanism. In: AAAI, pp. 946–953 (2019)
14. Hu, L., Yang, T., Shi, C., Ji, H., Li, X.: Heterogeneous graph attention networks for semi-supervised short text classification. In: EMNLP-IJCNLP, pp. 4820–4829 (2019)
15. Hu, W., et al.: Strategies for pre-training graph neural networks. In: ICLR (2020)
16. Hu, Z., Dong, Y., Wang, K., Chang, K.W., Sun, Y.: GPT-GNN: generative pre-training of graph neural networks. In: KDD, pp. 1857–1867 (2020)
17. Hu, Z., Dong, Y., Wang, K., Sun, Y.: Heterogeneous graph transformer. In: WWW, pp. 2704–2710 (2020)
18. Hussein, R., Yang, D., Cudré-Mauroux, P.: Are meta-paths necessary?: Revisiting heterogeneous graph embeddings. In: CIKM, pp. 437–446 (2018)
19. Hwang, D., Park, J., Kwon, S., Kim, K.M., Ha, J.W., Kim, H.J.: Self-supervised auxiliary learning with meta-paths for heterogeneous graphs. arXiv preprint arXiv:2007.08294 (2020)
20. Jin, W., et al.: Self-supervised learning on graphs: deep insights and new direction. arXiv preprint arXiv:2006.10141 (2020)
21. Kingma, D.P., Ba, J.: Adam: a method for stochastic optimization. In: ICLR (2015)
22. Kingma, D.P., Welling, M.: Auto-encoding variational Bayes. In: ICLR (2014)
23. Kipf, T.N., Welling, M.: Variational graph auto-encoders. arXiv preprint arXiv:1611.07308 (2016)
24. Kipf, T.N., Welling, M.: Semi-supervised classification with graph convolutional networks. In: ICLR (2017)
25. Li, A., Qin, Z., Liu, R., Yang, Y., Li, D.: Spam review detection with graph convolutional networks. In: CIKM, pp. 2703–2711 (2019)
26. Lin, Y., Liu, Z., Sun, M., Liu, Y., Zhu, X.: Learning entity and relation embeddings for knowledge graph completion. In: AAAI, pp. 2181–2187 (2015)
27. Liu, X., et al.: Self-supervised learning: Generative or contrastive. arXiv preprint arXiv:2006.08218 (2020)
28. Lu, Y., Shi, C., Hu, L., Liu, Z.: Relation structure-aware heterogeneous information network embedding. In: AAAI, pp. 4456–4463 (2019)
29. Maaten, L.V.D., Hinton, G.: Visualizing data using t-SNE. JMLR 9(11), 2579–2605 (2008)
30. Oord, A.V.D., Li, Y., Vinyals, O.: Representation learning with contrastive predictive coding. arXiv preprint arXiv:1807.03748 (2018)
31. Perozzi, B., Al-Rfou, R., Skiena, S.: DeepWalk: online learning of social representations. In: KDD, pp. 701–710 (2014)
32. Qiu, J., et al.: GCC: graph contrastive coding for graph neural network pre-training. In: KDD, pp. 1150–1160 (2020)
33. Qiu, J., Tang, J., Ma, H., Dong, Y., Wang, K., Tang, J.: DeepInf: social influence prediction with deep learning. In: KDD, pp. 2110–2119 (2018)
34. Sun, Y., Han, J., Yan, X., Yu, P.S., Wu, T.: PathSim: meta path-based top-K similarity search in heterogeneous information networks. VLDB 4(11), 992–1003 (2011)
35. Trouillon, T., Welbl, J., Riedel, S., Gaussier, É., Bouchard, G.: Complex embeddings for simple link prediction. In: ICML. pp. 2071–2080 (2016)
36. Velickovic, P., Cucurull, G., Casanova, A., Romero, A., Liò, P., Bengio, Y.: Graph attention networks. In: ICLR (2018)

37. Velickovic, P., Fedus, W., Hamilton, W.L., Liò, P., Bengio, Y., Hjelm, R.D.: Deep graph infomax. In: ICLR (2019)
38. Wang, X., Bo, D., Shi, C., Fan, S., Ye, Y., Yu, P.S.: A survey on heterogeneous graph embedding: methods, techniques, applications and sources. arXiv preprint arXiv:2011.14867 (2020)
39. Wang, X., et al.: Heterogeneous graph attention network. In: WWW, pp. 2022–2032 (2019)
40. Wang, Z., Zhang, J., Feng, J., Chen, Z.: Knowledge graph embedding by translating on hyperplanes. In: AAAI, pp. 1112–1119 (2014)
41. Xu, K., Hu, W., Leskovec, J., Jegelka, S.: How powerful are graph neural networks? In: ICLR (2019)
42. Yang, B., Yih, W., He, X., Gao, J., Deng, L.: Embedding entities and relations for learning and inference in knowledge bases. In: ICLR (2015)
43. Yang, C., Xiao, Y., Zhang, Y., Sun, Y., Han, J.: Heterogeneous network representation learning: a unified framework with survey and benchmark. TKDE (2020)
44. You, J., Ying, R., Ren, X., Hamilton, W.L., Leskovec, J.: GraphRNN: generating realistic graphs with deep auto-regressive models. In: ICML, pp. 5694–5703 (2018)
45. Yun, S., Jeong, M., Kim, R., Kang, J., Kim, H.J.: Graph transformer networks. In: NIPS, pp. 11960–11970 (2019)
46. Zhang, C., Song, D., Huang, C., Swami, A., Chawla, N.V.: Heterogeneous graph neural network. In: KDD, pp. 793–803 (2019)
47. Zhang, C., Swami, A., Chawla, N.V.: SHNE: representation learning for semantic-associated heterogeneous networks. In: WSDM, pp. 690–698 (2019)
48. Zhao, J., Wang, X., Shi, C., Liu, Z., Ye, Y.: Network schema preserving heterogeneous information network embedding. In: IJCAI, pp. 1366–1372 (2020)
49. Zhao, K., et al.: Deep adversarial completion for sparse heterogeneous information network embedding. In: WWW, pp. 508–518 (2020)

Semantic-Specific Hierarchical Alignment Network for Heterogeneous Graph Adaptation

YuanXin Zhuang[1], Chuan Shi[1($\boxtimes$)], Cheng Yang[1] and Fuzhen Zhuang[2,3], and Yangqiu Song[4]

[1] Beijing University of Posts and Telecommunications, Beijing 100876, China
{zhuangyuanxin,shichuan,yangcheng}@bupt.edu.cn
[2] Institute of Artificial Intelligence, Beihang University, Beijing 100191, China
zhuangfuzhen@buaa.edu.cn
[3] SKLSDE, School of Computer Science, Beihang University, Beijing 100191, China
[4] Hong Kong University of Science and Technology, Hong Kong 999077, China
yqsong@cse.ust.hk

Abstract. Node classification has been substantially improved with the advent of Heterogeneous Graph Neural Networks (HGNNs). However, collecting numerous labeled data is expensive and time-consuming in many applications. Domain Adaptation (DA) tackles this problem by transferring knowledge from a label-rich domain to a label-scarce one. However the heterogeneity and rich semantic information bring great challenges for adapting HGNN for DA. In this paper, we propose a novel semantic-specific hierarchical alignment network for heterogeneous graph adaptation, called HGA. HGA designs a sharing-parameters HGNN aggregating path-based neighbors and hierarchical domain alignment strategies with the MMD and L_1 normalization term. Extensive experiments on four datasets demonstrate that the proposed model can achieve remarkable results on node classification.

Keywords: Heterogeneous graph · Domain adaptation · Graph neural network

1 Introduction

Graph Neural Networks (GNNs) have attracted much attention as it can be applied to many applications where the data can be represented as graphs [10,22]. Heterogeneous Graphs (HGs), where nodes and edges can be categorized into multiple types, has also been proven to be effective to model many real-world applications, such as social networks and recommender systems [1,17]. In order to learn representations of HG, there is a surge of Heterogeneous Graph Neural Networks (HGNNs) in the last few years, which employ graph neural network for heterogeneous graph to capture features from various types of nodes and relations. Different from traditional GNNs aggregating adjacent neighbors on homogeneous graph,

© Springer Nature Switzerland AG 2021
N. Oliver et al. (Eds.): ECML PKDD 2021, LNAI 12976, pp. 335–350, 2021.
https://doi.org/10.1007/978-3-030-86520-7_21

HGNNs usually aggregates heterogeneous neighbors along meta-paths with a two-level attention mechanism [2,25]. For example, HAN [25] designs node-level and semantic-level attention on meta-path based neighbors. MAGNN [2] employs the intra-metapath aggregation to incorporate intermediate semantic nodes and the inter-metapath aggregation to combine messages from multiple metapaths.

Recent advances in HGNNs have achieved remarkable results on node classification task which usually requires large amounts of labeled data to train a good network. However, In the real HGs, it is often expensive and laborsome to collect enough labeled data. A potential solution is to transfer knowledge from a related HG with rich labeled data (called source graph) to another HG with the shortage of labeled data (called target graph). Existing HGNNs are mostly developed for a single graph which has similar distribution in training and test data. However, different graphs generally have varied data distributions in real applications, which is usually called domain shift phenomenon [13]. Domain shift will undermine the generalization ability of learning models. Thus, those single graph based HGNNs which without addressing domain shift would fail to learn transferable representations.

Domain Adaptation (DA) [30] has shown promising advances for learning a discriminative model in the presence of the shift between the training and test data distributions. Given a target domain short of labels, DA aims to leverage the abundant labeled data from a source domain to help target domain learning, which has already attracted a lot of interests from the fields of Computer Vision [13,23] and Natural Language Processing [9,14]. The newest deep domain adaptation algorithms learn domain-invariant feature representations to mitigate domain shift with the Maximum Mean Discrepancy (MMD) metric [5,12] or Generative Adversarial Net (GAN) [4,21]. In recent years, there have been several attempts to apply domain adaptation to graph structure data. Some methods employ stacked autoencoders and MMD to learn network-invariant node representations [15,16], while some methods apply graph convolutional network and adversarial learning to learn transferable embeddings [26,29]. However, these techniques primarily focus on domain adaptation across homogeneous graphs, which cannot be directly applicable to heterogeneous graph. More recently, a heterogeneous graph domain adaptation method has been proposed to handle heterogeneity with multi-channel GCNs and two-level selection mechanisms [27]. But this method is not designed based on HGNN framework, which reduces its versatility. In addition, its performance improvement could be limited, because of lacking semantic-specific domain alignment mechanism to align the rich semantics of heterogeneous graphs separately.

Motivated by these observations, we make the first attempt to design a HGNN for DA, which is not a trivial task, due to the following two challenges: (1) How to adopt existing HGNNs to fully learn the knowledge of source graph and migrate to the target graph for the category-discriminative representations. We know that existing HGNNs are designed for single graph, we need to design an effective HGNN for knowledge transfer when adopting it for multiple graphs, (2) How to diminish the distribution discrepancy between source and target graphs to learn domain-invariant representations. Because of the domain shift among different semantics in source and target graphs, we need to design diverse domain alignment strategies to align distribution in source and target graphs intra- and inter-semantics.

In this paper, we propose a semantic-specific hierarchical alignment network for Heterogeneous Graph Adaptation (called HGA). The basic framework of HGA is a sharing-parameters HGNN which use hierarchical attentions to aggregate neighbor information via different meta-paths, to transfer knowledge from source graph to target graph. To be specific, HGA aggregates path-based neighbors with semantic-specific feature extractor and then classify and fuse these embeddings of different semantics with semantic-specific classifiers. In order to eliminate the distribution shift, a MMD normalization term is designed to align the feature distribution of nodes in source and target graph of every semantic path, and a L_1 normalization term is designed to align the class scores of nodes in target graph.

The contributions of this paper are summarized as follows:

- We study an important but seldom exploited problem of adopting DA to HGNN. The solution to this problem is crucial for label-absent HG representation.
- We design a novel heterogeneous graph adaptation method, called HGA, which employs a sharing-parameters HGNN with the MMD and L_1 normalization terms for domain-invariant and category-discriminative node representations.
- Experiments on eight transfer learning tasks show that the proposed HGA achieves significant performance improvements, compared to other state-of-the-art baselines.

2 Related Work

In this section, we briefly overview methods that are related to heterogeneous graph neural network and graph domain adaptation.

2.1 Heterogeneous Graph Neural Network

HGNN is designed to use GNN on heterogeneous graph, it can be divided into unsupervised and semi-supervised settings [24]. HetGNN [28] is the representative work of unsupervised HGNNs. It uses type specific RNNs to encode features for each type of neighbor vertices, followed by another RNN to aggregate the encoded neighbor representations of different types. Semi-supervised HGNNs prefer to use attention mechanism to capture the most relevant structural and attribute information. There are a series of attention-based HGNNs was proposed [2,7,25]. HAN [25] uses a hierarchical attention mechanism to capture both node and semantic importance. MAGNN [2] extends HAN by considering both the meta-path based neighborhood and the nodes along the meta-path. HGT [7] uses each edge's meta relation to parameterize the Transformer-like self-attention architecture.

These HGNNs are designed for a single graph, and thus they can not be directly applied for knowledge transfer among multiple graphs.

2.2 Graph Domain Adaptation

There have been several attempts in the literature to apply domain adaptation to graph structure data. CDNE [16] incorporate MMD-based domain adaptation technique into deep network embedding to learn label-discriminative and network-invariant representations. ACDNE [15] integrate deep network embedding with the emerging adversarial domain adaptation technique to address cross-network node classification. DANE [29] applies graph convolutional network with constraints of adversarial learning regularization to learn transferable embeddings. UDA-GCN [26] used a dual graph convolutional networks to exploit both local and global relations of the graphs. However, these methods only consider knowledge transfer among homogeneous graphs. Recently, a heterogeneous graph domain adaptation method is proposed [27], which utilizes multi-channel GCNs to project nodes into multiple spaces, and proposes two-level selection mechanisms to choose the combination of channels and fuse the selected channels. Unfortunately, this method has limited performance improvement, due to lack semantic-specific domain alignment strategies.

3 Preliminaries

Definition 1. Heterogeneous Graph *[17]. A heterogeneous graph, denoted as $\mathcal{G} = (\mathcal{V}, \mathcal{E})$, consists of an object set $\mathcal{V}$ and a link set $\mathcal{E}$. Each node $v \in \mathcal{V}$ and each link $e \in \mathcal{E}$ are associated with their node type mapping function $\phi : \mathcal{V} \to \mathcal{A}$ and their link type mapping function $\psi : \mathcal{E} \to \mathcal{R}$. $\mathcal{A}$ and $\mathcal{R}$ denote the sets of predefined object types and link types, where $|\mathcal{A}| + |\mathcal{R}| > 2$.*

In heterogeneous graph, two objects can be connected via different semantic paths, which are called meta-paths.

Definition 2. Meta-path *[19]. A meta-path Φ is defined as a path in the form of $A_1 \xrightarrow{R_1} A_2 \xrightarrow{R_2} \cdots \xrightarrow{R_l} A_{l+1}$ (simplified to $A_1 A_2 \cdots A_{l+1}$), which describes a composite relation $R = R_1 \circ R_2 \circ \cdots \circ R_l$ between objects A_1 and A_{l+1}, where $\circ$ denotes the composition operator on relations.*

Definition 3. Domain Adaptation (DA) *[30]. Given a labeled source domain $\mathcal{D}_S$ and a unlabeled target domain $\mathcal{D}_T$, assume that their feature spaces and their class spaces are the same, i.e. $\mathcal{X}_S = \mathcal{X}_T$, $\mathcal{Y}_S = \mathcal{Y}_T$. The goal of domain adaptation is to use labeled data $\mathcal{D}_S$ to learn a classifier $f : \mathbf{x}_T \mapsto \mathbf{y}_T$ to predict the label $\mathbf{y}_T \in \mathcal{Y}_T$ of the target domain $\mathcal{D}_T$.*

Definition 4. Heterogeneous Graph Domain Adaptation. *Given a source heterogeneous graph $\mathcal{G}_S = (\mathcal{V}_S, \mathcal{E}_S, \mathcal{X}_S, \mathcal{Y}_S)$, and a target heterogeneous graph $\mathcal{G}_T = (\mathcal{V}_T, \mathcal{E}_T, \mathcal{X}_T)$, where $\mathcal{A}_S \cap \mathcal{A}_T \neq \oslash$ and $\mathcal{R}_S \cap \mathcal{R}_T \neq \oslash$. $\mathcal{X}$ represents the* *features of $\mathcal{V}$, $\mathcal{Y}$ indicates the labels of $\mathcal{V}$. The goal of heterogeneous graph domain adaptation is to build a classifier f to predict the labels on $\mathcal{V}_T$ through reducing the domain shifts in different graphs and utilizing the structural information on both graphs, as well as $\mathcal{Y}_S$.*

Figure 1(a) demonstrates HGs on bibliographic data, where two authors can be connected via multiple meta-paths, e.g., Author-Paper-Author (APA) and Author-Paper-Conference-Paper-Author (APCPA). The meta-path APA depicts the co-author relation, whereas the APCPA depicts the co-conference relation. A task on heterogeneous graph domain adaptation is to predict the label of nodes in the target graph, with the help of the labeled source graph.

4   The Proposed Model

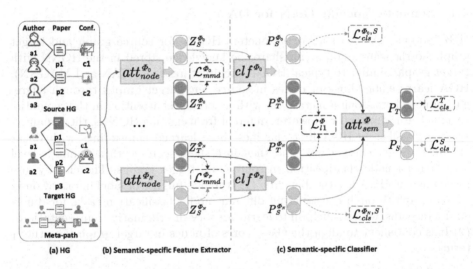

Fig. 1. An overview of the proposed hierarchical alignment network for Heterogeneous Graph domain Adaptation (HGA). HGA receives source graph instances with annotated ground truth and adapts to classifying the target samples. There are semantic-specific feature extractor and classifier for each meta-path.

In this paper, we propose a novel semantic-specific hierarchical alignment network for Heterogeneous Graph domain Adaptation (called HGA), whose basic idea is to adopt DA to HGNNs. As we know, existing HGNNs are designed for learning category-discriminative embeddings for node classification in single graph. That is, the learned embeddings can distinguish the category of nodes in a graph. Most HGNNs (e.g., HAN and MAGNN) employ node-level (also called intra-metapath) and semantic-level (also called inter-metapath) attention mechanism to aggregate node embeddings along different meta-paths. Unfortunately, these HGNNs cannot be directly applied to transfer knowledge among multiple graphs, because of domain shift.

In order to solve this obstacle, the proposed HGA adopts DA to HGNN with the goal of learning domain-invariant representations, as well as category-discriminative representations. HGA designs a shared parameters HGNN for source graph and target graph to aggregate path-based neighbors with semantic-specific feature extractor and then classify and fuse these embeddings of different meta-paths with semantic-specific classifiers. Furthermore, two normalized terms in HGA (i.e., mmd and l_1 terms) are proposed to hierarchically align the domain distribution of nodes intra- and inter-metapaths for domain-invariant representations. Concretely, the mmd term aligns the feature distribution of nodes in source and target graph of every semantic path, while the l_1 term aligns the class scores of nodes in target graph. The overall architecture of HGA is shown in Fig. 1.

4.1 Semantic-Specific GNN for DA

HGA adopts DA to a shared parameters HGNN for source graph and target graph, so the source graph can share the knowledge stored in the HGNN with target graph. Similar to typical HGNN architectures (e.g., HAN and MAGNN), HGA learns embeddings of nodes in source and target graphs through aggregating neighbors along a meta-path with a node-level attention in the semantic-specific feature extractor. However, different from existing HGNNs, the semantic-specific classifier in HGA first classifies these learned embeddings with linear classifiers to get class scores, and then fuse these scores with a semantic-level attention for node classification in source and target graphs. The classify-fuse mechanism in HGA has two benefits: (1) It makes full use of label information in source graph through constructing different node classification tasks for different meta-paths, which is helpful to learn category-discriminative representations. (2) It is convenient to align the class scores of nodes in target graph (i.e., the l_1 term).

Semantic-Specific Feature Extractor. Given a meta-path Φ, similar to typical HGNN architectures, the embedding of node i can aggregated from its meta-path based neighbors $\mathcal{N}_i^\Phi = \{i\} \cup \{j | j$ connects with i via the meta-path $\Phi\}$ like HAN [25]:

$$\mathbf{z}_i^\Phi = att_{node}^\Phi \left(\mathbf{h}_j, j \in \mathcal{N}_i^\Phi \right), \tag{1}$$

where $\mathbf{z}_i^\Phi$ denotes the learned embedding of node i based on meta-path Φ, while att_{node}^Φ is the feature extractor of meta-path Φ which is a general component to aggregate neighbors. For example, att_{node}^Φ can be the node-level attention in HAN which simply aggregates meta-path based neighbors, as well as the intra-metapath aggregation in MAGNN which also considers the nodes along the meta-path instances.

Semantic-Specific Classifier. Given an embedding $\mathbf{z}_i^\Phi$ of node i based on meta-path Φ, the class scores $\mathbf{p}_i^\Phi$ of node i in meta-path Φ can be obtained by a classifier clf^Φ, such as linear classifier or softmax classifier:

$$\mathbf{p}_i^\Phi = clf^\Phi \left(\mathbf{z}_i^\Phi \right). \tag{2}$$

As we know, semantic-specific embedding of nodes under a meta-path only reflect node characteristics from one aspect, while nodes contain multiple aspects of semantic information under different meta-paths. To learn a more comprehensive node embeddings, we need to fuse multiple semantics which can be revealed by meta-paths. To address the challenge of meta-path selection and semantic fusion in a heterogeneous graph, we adopt a semantic attention to automatically learn the importance of different meta-paths and fuse them for the specific task.

Given a set of meta-paths $\{\Phi_0, \Phi_1, \cdots, \Phi_N\}$, after feeding the feature of node i into semantic-specific feature extractors and semantic-specific classifiers, it has N semantic-specific node embeddings $\left\{ \mathbf{p}_i^{\Phi_0}, \mathbf{p}_i^{\Phi_1}, \cdots, \mathbf{p}_i^{\Phi_N} \right\}$. To effectively aggregate different semantic embeddings, we use a semantic fusion mechanism:

$$\mathbf{p}_i = att_{sem} \left(\mathbf{p}_i^{\Phi_j} \right) = \sum_{j=1}^{N} \beta_j \cdot \mathbf{p}_i^{\Phi_j}, \tag{3}$$

where

$$\beta_j = \frac{\exp \left(\frac{1}{|\mathcal{V}|} \sum_{i \in \mathcal{V}} \mathbf{q}^{\mathrm{T}} \cdot \tanh \left(\mathbf{M} \cdot \mathbf{p}_i^\Phi + \mathbf{b} \right) \right)}{\sum_{i=1}^{N} \exp \left(\frac{1}{|\mathcal{V}|} \sum_{i \in \mathcal{V}} \mathbf{q}^{\mathrm{T}} \cdot \tanh \left(\mathbf{M} \cdot \mathbf{p}_i^\Phi + \mathbf{b} \right) \right)} \tag{4}$$

can be interpreted as the contribution of meta-path Φ_j for the specific task. Respectively, $\mathbf{q}$ is the semantic attention vector; $\mathbf{M}$ and $\mathbf{b}$ denote the weight matrix and bias vector; $\mathbf{p}_i$ denotes the final embedding of node i, and att_{sem} denotes the semantic aggregator which aggregates embeddings of different meta-paths. Then we can apply the final embeddings to specific tasks and design different loss functions.

In order to obtain category-discriminative representations and facilitate knowledge transfer between graphs, we optimize three different loss functions as follows to reduce the domain discrepancy and enable efficient domain adaptation, and thus our model can differentiate class labels in the source graph and target graph, respectively.

- Semantic-specific source classifier minimizes the cross-entropy loss for the source graph in a mate-path Φ:

$$\mathcal{L}_{cls}^{\Phi,S} \left(\mathcal{P}_S^\Phi, \mathcal{Y}_S \right) = -\frac{1}{N_S} \sum_{i=1}^{N_S} y_i^S \log \left(\hat{y}_i^S \right), \tag{5}$$

- Source classifier minimizes the cross-entropy loss for the source graph after semantic fusion:

$$\mathcal{L}_{cls}^{S} \left(\mathcal{P}_S, \mathcal{Y}_S \right) = -\frac{1}{N_S} \sum_{i=1}^{N_S} y_i^S \log \left(\hat{y}_i^S \right), \tag{6}$$

- Target classifier minimizes the entropy loss for target graph information absorption. Here we employ the predicted labels of target nodes obtained by the shared classifiers:

$$\mathcal{L}_{cls}^T \left(\mathcal{P}_T\right) = -\frac{1}{N_T} \sum_{i=1}^{N_T} \hat{y}_i^T \log\left(\hat{y}_i^T\right), \tag{7}$$

where y_i^S denotes the label of the i-th node in the source graph, $\hat{y}_i^S$ is the classification prediction for the i-th node in source graph, $\hat{y}_i^T$ is the classification prediction for the i-th node in target graph, N_S is the node number of source graph and N_T is the node number of target graph.

The total classification loss of HGA can be represented by Eq. 8, which can learn category-discriminative embeddings for source and target graph.

$$\mathcal{L}_C \left(\mathcal{G}_S, \mathcal{G}_T\right) = \mathcal{L}_{cls}^{\Phi,S} \left(\mathcal{P}_S^\Phi, \mathcal{Y}_S\right) + \mathcal{L}_{cls}^S \left(\mathcal{P}_S, \mathcal{Y}_S\right) + \mathcal{L}_{cls}^T \left(\mathcal{P}_T\right). \tag{8}$$

4.2 Hierarchical Domain Alignment

Although the target graph can share knowledge from source graph with the shared parameters HGNN, the above model cannot solve the domain shift problem in domain adaptation. In order to learn domain-invariant representations, we furtherly propose semantic-specific hierarchical alignment mechanism, which includes intra-semantic feature alignment and inter-semantic label alignment. The intra-semantic feature alignment aims to map each pair of semantic between source and target graph into multiple different feature spaces and align semantic-specific distributions to learn multiple semantic-invariant representations. Since the target samples near semantic-specific decision boundary predicted by different classifiers might get different labels, the inter-semantic label alignment is designed to align the classifiers' output for the target nodes.

Intra-semantic Feature Alignment. To learn domain-invariant representations, we need to match the distributions of source graph and target graph. In domain adaptation, the MMD [6] is a widely adopted nonparametric metric. We use the following term as the estimate of the discrepancy between source graph and target graph:

$$\mathcal{L}_{mmd} \left(\mathcal{G}_S, \mathcal{G}_T\right) = \left\| \frac{1}{N_S} \sum \phi\left(\mathbf{z}_S^\Phi\right) - \frac{1}{N_T} \sum \phi\left(\mathbf{z}_T^\Phi\right) \right\|_{\mathcal{H}}^2, \tag{9}$$

where $\phi\left(\cdot\right)$ denotes some feature projection function to map the original samples to reproducing kernel hilbert space. Through minimizing the Eq. 9, the specific-semantic feature extractor could align the domain distributions between source domain and target domain under meta-path Φ.

Inter-semantic Label Alignment. The classifiers are trained based on different meta-paths, hence they might have a disagreement on the prediction for target samples. Intuitively, the same target node predicted by different classifiers should get the same prediction. Hence, we need to minimize the classification discrepancy of nodes in target graph among all classifiers. Here we define the discrepancy loss as the differences of classification probability of nodes under different meta-paths with l_1 normalization.

$$\mathcal{L}_{l_1}\left(\mathcal{G}_T\right) = \frac{2}{N \times (N-1)} \sum_{j=1}^{N-1} \sum_{i=j+1}^{N} \mathbf{E}\left[|\mathbf{p}_T^{\Phi_i} - \mathbf{p}_T^{\Phi_j}|\right], \tag{10}$$

where N is the number of meta-path. By minimizing the Eq. 10, the probabilistic outputs of all classifiers tend to be similar, which enforces the domain alignment under different semantic paths.

4.3 Optimization Objective

For HGA, a label prediction function f is trained by minimizing the overall objective as shown in Eq. 11:

$$\mathcal{L}\left(\mathcal{G}_S, \mathcal{G}_T\right) = \mathcal{L}_C\left(\mathcal{G}_S, \mathcal{G}_T\right) + \lambda\left(\mathcal{L}_{mmd}\left(\mathcal{G}_S, \mathcal{G}_T\right) + \mathcal{L}_{l_1}\left(\mathcal{G}_T\right)\right), \tag{11}$$

where λ is the balance parameters. $\mathcal{L}_{mmd}$ and $\mathcal{L}_{l1}$ represent the intra-semantic feature alignment loss and the inter-semantic label alignment loss, respectively.

4.4 Discussion of the Proposed Model

Here we give the discussion of the proposed HGA as follows:

- From the optimization objective function Eq. 11, we can find that HGA provides a general framework to adopt DA to HGNN. If we do not consider target graph $\mathcal{G}_T$, HGA degrades into tradition HGNNs for single graph. If we do not consider multiple meta-paths, HGA can be used for homogeneous graph domain adaptation. If ignoring the mmd and l_1 normalization terms, HGA becomes a simple DA-version of HGNN without considering domain shift. What's more, the $\mathcal{L}_{mmd}$ could be replaced by other adaptation methods, such as adversarial loss [3], coral loss [18]. And the $\mathcal{L}_{l1}$ could be replaced by other loss, such as l_2 regularization.
- Compared to traditional HGNNs, the additional complexity of HGA mainly lies on the mmd and l_1 normalization term. The complexity of mmd term is linear to the size of nodes in graphs, while the complexity of l_1 term is the square of the number of meta-paths, which is very small. And thus HGA has the same complexity with traditional HGNNs. Experiments also validate this point.

– The proposed HGA is highly efficient and can be easily parallelized. In the shared parameters HGNNs, the complexity is linear to the number of nodes and meta-path based node pairs. HGA can be easily parallelized, because att_{node}^{Φ} and att_{sem} can be parallelized across node pairs and meta-paths, respectively. The overall complexity is linear to the number of nodes and meta-path based node pairs.

5   Experiments

5.1   Datasets

We evaluate all the models on three academic attributed networks constructed from AMiner [20], DBLP [8] and ACM [11], and the detailed description is shown in Table 1. First, we adopt the constructed datasets from the work [27], i.e., ACM_A vs. ACM_B, DBLP_A vs. DBLP_B, AMiner_A vs. AMiner_B. For each pair of graphs, e.g., ACM_A vs. ACM_B, the density of meta-path edges is quite different between each other, which means they have domain discrepancy. (More statistics can be found in [27]).

Furthermore, we construct another pair of much larger graphs, i.e., ACM vs. DBLP. For ACM, we collected the papers published in SIGMOD, KDD, COLT and WWW, and divided them into four classes (*Database, data mining, machine learning, information retrieval*). The attributes of each paper in ACM are extracted from the paper title and abstract. For DBLP, we collected the papers published in ICDE, ICDM, PAKDD, PKDD, AAAI and SIGIR, and also divided them into the same classes. The attributes of each paper in DBLP are extracted from the paper title. Note that, DBLP has no overlapping nodes with ACM, and it is sparser than ACM. Finally, we have four pairs of datasets.

Table 1. Statistics of the experimental datasets.

Dataset	# Nodes	# Meta-path edges	Dataset	# Nodes	# Meta-path edges
ACM_A	1,500	4,960	ACM_B	1,500	759
		6,691			3,996
		26,748			75,180
DBLP_A	1,496	2,602	DBLP_B	1,496	3,460
		673,730			744,994
		977,348			1,068,250
AMiner_A	1,500	4,360	AMiner_B	1,500	462
		554			3,740
		89,274			67,116
ACM	4,177	34,638	DBLP	4,154	38,966
		15,115,590			1,496,938

5.2   Baselines and Implementation Details

Baselines. In order to make a fair comparison and demonstrate the effectiveness of our proposed model, we compare our approach with both state-of-the-art single-domain methods as well as some domain adaptation methods on graphs.

State-of-the-Art Single-Domain Methods:

- **GCN** [10]: a typical deep convolutional network designed for homogeneous graphs.
- **HAN** [25]: a heterogeneous graph embedding method uses meta-paths as edges to augment the graph, and maintains different weight matrices for each meta-path-defined edge. And uses semantic-level attention to differentiate and aggregate information from different meta-paths.
- **MAGNN** [2]: a heterogeneous graph embedding method uses intra-metapath aggregation to sample some meta-path instances surrounding the target node and use an attention layer to learn the importance of different instances. And uses inter-metapath aggregation to learn the importance of different meta-paths.

Domain Adaptation Methods on Graphs:

- **UDAGCN** [26]: a homogeneous graph domain adaptation method uses a dual graph convolutional networks to exploit both local and global relations of the graphs. And uses a domain adversarial loss for domain discrimination.
- **MuSDAC** [27]: a heterogeneous graph domain adaptation method uses multi-channel shared weight GCNs and a Two-level Selection strategy to aggregate embedding spaces to ensure both domain similarity and distinguishability.
- **HAN+MMD**: The feature generator is a shared parameters HAN architecture [25] for source and target graph. And a MMD [5,12] regularization term is added on the final embedding.
- **MAGNN+MMD**: The feature generator is a shared parameters MAGNN architecture [2] for source and target graph. And a MMD [5,12] regularization term is added on the final embedding.
- **HGA-HAN**: The att^{Φ}_{node} and att_{sem} in HGA framework is using the node-level attention and semantic-level attention in HAN [25].
- **HGA-MAGNN**: The att^{Φ}_{node} and att_{sem} in HGA framework is using the Intra-metapath aggregation and Inter-metapath aggregation in MAGNN [2].

To further validate the effectiveness of mmd loss and l_1 loss, we also evaluate several variants of HGA: (1) HGA$_{\neg l_1}$, only considers mmd loss; (2) HGA$_{\neg mmd}$, only considers l_1 loss; (3) HGA$_{\neg mmd \wedge \neg l_1}$, only has the shared weight architecture of HGNN.

Implementation Details. All deep learning algorithms are implemented in Pytorch and trained with Adam optimizer. In the experiment we employ linear classifier. The learning rate is using the following formula: $\eta_p = \frac{\eta_0}{(1+\alpha p)^\beta}$, where

p is the training progress linearly changing from 0 to 1, $\eta_0 = 0.01$, $\alpha = 10$ and $\beta = 0.75$, which is optimized to promote convergence and low error on the source domain. To suppress noisy activations at the early stages of training, instead of fixing the adaptation factor λ, we gradually change it from 0 to 1 by a progressive schedule: $\lambda_p = \frac{2}{\exp(-\theta p)} - 1$, and $\theta = 10$ is fixed throughout the experiments [3]. This progressive strategy significantly stabilizes parameter sensitivity and eases model selection for HGA. As for single-domain network methods, we take the data from source graph as training set and the one from target graph as test set. As for domain adaptation method which acts on homogeneous graph, we ignore multiple semantics in HGs.

5.3   Results

We compare HGA with the baselines on four pairs of datasets and the results are shown in Table 2. From these results, we have the following insightful observations:

Table 2. Performance comparison on classification accuracy.

| Source | ACM | DBLP | ACM_B | ACM_A | AMiner_B | AMiner_A | DBLP_B | DBLP_A | AVG |
Target	DBLP	ACM	ACM_A	ACM_B	AMiner_A	AMiner_B	DBLP_A	DBLP_B	
GCN	0.472	0.517	0.580	0.698	0.755	0.481	0.357	0.459	0.540
HAN	0.632	0.694	0.687	0.686	0.676	0.698	0.768	0.812	0.707
MAGNN	0.678	0.702	0.713	0.693	0.703	0.717	0.772	0.817	0.724
UDAGCN	0.673	0.696	0.654	0.687	0.792	0.712	0.693	0.723	0.704
MuSDAC	0.704	0.764	0.788	0.730	0.810	0.761	0.795	0.819	0.771
HAN+MMD	0.724	0.712	0.727	0.706	0.832	0.745	0.774	0.817	0.755
MAGNN+MMD	0.735	0.728	0.739	0.721	0.843	0.749	0.781	0.820	0.765
HGA-HAN	0.785	0.759	0.791	0.757	0.929	0.83	0.828	0.835	0.814
HGA-MAGNN	**0.793**	**0.771**	**0.798**	**0.765**	**0.937**	**0.838**	**0.833**	**0.840**	**0.822**

- Because MAGNN not only considers the meta-path based neighborhood, but also consider the nodes along the meta-path. So the effect of HGA-MAGNN is better than that of HGA-HAN.
- HGA-MAGNN (or HGA-HAN) outperforms all compared baseline methods over all tasks. These encouraging results indicate that the proposed intra-semantic feature alignment mechanism can learn semantic-invariant representations for each pair of source and target graphs effectively, and inter-semantic label alignment mechanism can control all the classifiers to learn a consensus label for each target node.
- HAN+MMD and MAGNN+MMD are the simplest way to apply DA to HGNN. By comparing HAN+MMD (or MAGNN+MMD) and HAN (or MAGNN), we can see that traditional HGNNs cannot deal with the problem of domain shift. By comparing HAN+MMD (or MAGNN+MMD) and HGA-MAGNN (or HGA-HAN), we can observe that HGA is more effectively

in transferring the knowledge of the source domain to the target domain by intra-semantic feature alignment mechanism and inter-semantic label alignment mechanism.

- Compared to HAN and MAGNN which do not consider the domain discrepancy between different graphs, HGA achieves better performance, especially on the pair of AMiner_A vs. AMiner_B where the density of meta-path edges is significantly different between them.

- For single-domain methods, HAN and MAGNN perform better GCN on most of tasks, which implies the superiority of considering heterogeneous graphs rather than homogeneous ones. The similar conclusion also can be concluded for domain adaptation methods.

Table 3. Performance comparison on classification accuracy between HGA variants.

| Source | ACM | DBLP | ACM_B | ACM_A | AMiner_B | AMiner_A | DBLP_B | DBLP_A | AVG |
Target	DBLP	ACM	ACM_A	ACM_B	AMiner_A	AMiner_B	DBLP_A	DBLP_B	
HGA-HAN$_{\neg mmd \wedge \neg l_1}$	0.667	0.676	0.752	0.746	0.835	0.814	0.813	0.824	0.766
HGA-HAN$_{\neg mmd}$	0.774	0.742	0.790	0.749	0.931	0.819	0.820	0.830	0.807
HGA-HAN$_{\neg l_1}$	0.765	0.739	0.784	0.751	0.920	0.822	0.826	0.833	0.805
HGA-HAN	**0.785**	**0.759**	**0.791**	**0.757**	**0.929**	**0.830**	**0.828**	**0.835**	**0.814**
HGA-MAGNN$_{\neg mmd \wedge \neg l_1}$	0.681	0.698	0.744	0.748	0.837	0.820	0.819	0.821	0.771
HGA-MAGNN$_{\neg mmd}$	0.782	0.764	0.788	0.752	0.935	0.827	0.825	0.832	0.813
HGA-MAGNN$_{\neg l_1}$	0.784	0.770	0.791	0.761	0.932	0.834	0.829	0.838	0.817
HGA-MAGNN	**0.793**	**0.771**	**0.798**	**0.765**	**0.937**	**0.838**	**0.833**	**0.840**	**0.822**

The ablation study results are shown in Table 3, From Table 3, we can easily observe that both HGA-MAGNN$_{\neg l_1}$ and HGA-MAGNN$_{\neg mmd}$ (or HGA-HAN$_{\neg l_1}$ and HGA-HAN$_{\neg mmd}$) outperform HGA-MAGNN$_{\neg mmd \wedge \neg l_1}$ (or HGA-HAN$_{\neg mmd \wedge \neg l_1}$), which verifies that on one hand the effectiveness of aligning the intra-semantic distributions of each pair of semantic in the source and target domains, and on the other hand the consideration of the inter-semantic label alignment to reduce the gap between all classifiers can help each classifier learn the knowledge from other classifiers.

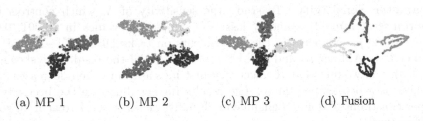

(a) MP 1   (b) MP 2   (c) MP 3   (d) Fusion

Fig. 2. The visualization of classifier's output and after fusion in target domain ('MP' means meta-path)

5.4   Analysis

Classification Output Visualization. We visualize the outputs of each classifier and the output after meta-path fusion on the target domain of the task Aminer_B → Aminer_A with the model of HGA-HAN. From Fig. 2, we can observe that the results in Fig. 2(d) are better than the ones in Fig. 2(a)(b)(c), which show that by fusing more information from meta-paths can lead to performance improvement. What's more, we can see that the target nodes near the class boundaries are more likely to be misclassified by the classifiers learned from single meta-path of source graph, while we can minimize the discrepancy among all classifiers by using inter-semantic label alignment.

Algorithm Convergence. To investigate the convergence of our algorithm, we record the performance of target domain over all meta-path classifiers and the fusion one during the iterating on the task Aminer_B → Aminer_A. The results are shown in Fig. 3(a). We can find that all algorithms can converge very fast, e.g., less than 20 iterations. Particularly, the fusion one is more stable with better accuracy, which illustrates the benefits of fusing multiple meta-paths again.

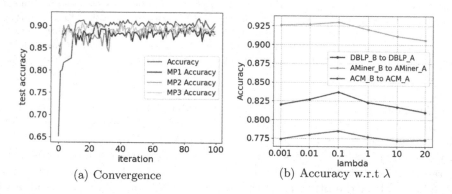

(a) Convergence   (b) Accuracy w.r.t λ

Fig. 3. Algorithm convergence and parameter sensitivity.

Parameter Sensitivity. To study the sensitivity of λ, which controls the importance of mmd loss and l_1 loss. We sample the values in {0.001, 0.01, 0.1, 1, 10, 20}, and perform the experiments on tasks DBLP_B → DBLP_A, AMiner_B → AMiner_A, and ACM_B → ACM_A. All the results are shown in Fig. 3(b), and we find that the accuracy first increases and then decreases, and displays as a bell-shaped curve. The results further illustrate the necessity of proper constraint of domain alignments. Finally, we set $\lambda = 0.1$ to achieve good performance.

6 Conclusion

Most previous heterogeneous graph neural networks focus on a single graph and fail to consider the knowledge transfer across graphs. In this paper, we study the problem of HGNN for domain adaptation, and propose a semantic-specific hierarchical alignment network for heterogeneous graph adaptation, called HGA. The HGA employs a shared parameters HGNN with the mmd and $l1$ normalization terms for domain-invariant and category-discriminative node representations. Experiments on eight transfer learning tasks validate the effectiveness of the proposed HGA.

Acknowledgments. This work is supported in part by the National Natural Science Foundation of China (No. U20B2045, 61772082, 61702296, 62002029). It is also supported by "The Fundamental Research Funds for the Central Universities 2021RC28".

References

1. Fan, S., et al.: MetaPath-guided heterogeneous graph neural network for intent recommendation. In: KDD, pp. 2478–2486 (2019)
2. Fu, X., Zhang, J., Meng, Z., King, I.: MAGNN: metapath aggregated graph neural network for heterogeneous graph embedding. In: WWW, pp. 2331–2341 (2020)
3. Ganin, Y., Lempitsky, V.: Unsupervised domain adaptation by backpropagation. In: ICML, pp. 1180–1189 (2015)
4. Goodfellow, I., et al.: Generative adversarial nets. Adv. Neural. Inf. Process. Syst. **27**, 2672–2680 (2014)
5. Gretton, A., Borgwardt, K., Rasch, M., Schölkopf, B., Smola, A.: A kernel method for the two-sample-problem. Adv. Neural. Inf. Process. Syst. **19**, 513–520 (2006)
6. Gretton, A., Borgwardt, K.M., Rasch, M.J., Schölkopf, B., Smola, A.: A kernel two-sample test. J. Mach. Learn. Res. **13**(1), 723–773 (2012)
7. Hu, Z., Dong, Y., Wang, K., Sun, Y.: Heterogeneous graph transformer. In: WWW, pp. 2704–2710 (2020)
8. Ji, M., Sun, Y., Danilevsky, M., Han, J., Gao, J.: Graph regularized transductive classification on heterogeneous information networks. In: Balcázar, J.L., Bonchi, F., Gionis, A., Sebag, M. (eds.) ECML PKDD 2010. LNCS (LNAI), vol. 6321, pp. 570–586. Springer, Heidelberg (2010). https://doi.org/10.1007/978-3-642-15880-3_42
9. Jiang, J., Zhai, C.: Instance weighting for domain adaptation in NLP. In: ACL (2007)
10. Kipf, T.N., Welling, M.: Semi-supervised classification with graph convolutional networks. arXiv preprint arXiv:1609.02907 (2016)
11. Kong, X., Yu, P.S., Ding, Y., Wild, D.J.: Meta path-based collective classification in heterogeneous information networks. In: Chen, X., Lebanon, G., Wang, H., Zaki, M.J. (eds.) CIKM, pp. 1567–1571. ACM (2012)
12. Long, M., Zhu, H., Wang, J., Jordan, M.I.: Deep transfer learning with joint adaptation networks. ICML **70**, 2208–2217 (2017)
13. Luo, Y., Zheng, L., Guan, T., Yu, J., Yang, Y.: Taking a closer look at domain shift: category-level adversaries for semantics consistent domain adaptation. In: CVPR, pp. 2507–2516 (2019)

14. Ramponi, A., Plank, B.: Neural unsupervised domain adaptation in NLP–a survey. arXiv preprint arXiv:2006.00632 (2020)
15. Shen, X., Dai, Q., Chung, F., Lu, W., Choi, K.S.: Adversarial deep network embedding for cross-network node classification. In: AAAI, pp. 2991–2999 (2020)
16. Shen, X., Dai, Q., Mao, S., Chung, F., Choi, K.S.: Network together: node classification via cross-network deep network embedding. TNNLS **32**, 1935–1948 (2020)
17. Shi, C., Li, Y., Zhang, J., Sun, Y., Philip, S.Y.: A survey of heterogeneous information network analysis. TKDE **29**(1), 17–37 (2016)
18. Sun, B., Saenko, K.: Deep CORAL: correlation alignment for deep domain adaptation. In: Hua, G., Jégou, H. (eds.) ECCV 2016. LNCS, vol. 9915, pp. 443–450. Springer, Cham (2016). https://doi.org/10.1007/978-3-319-49409-8_35
19. Sun, Y., Han, J., Yan, X., Yu, P.S., Wu, T.: PathSIM: meta path-based top-k similarity search in heterogeneous information networks. Proc. VLDB Endow. **4**(11), 992–1003 (2011)
20. Tang, J., Zhang, J., Yao, L., Li, J., Zhang, L., Su, Z.: ArnetMiner: extraction and mining of academic social networks. In: Li, Y., Liu, B., Sarawagi, S. (eds.) KDD, pp. 990–998. ACM (2008)
21. Tzeng, E., Hoffman, J., Saenko, K., Darrell, T.: Adversarial discriminative domain adaptation. In: CVPR, pp. 2962–2971 (2017)
22. Veličković, P., Cucurull, G., Casanova, A., Romero, A., Lio, P., Bengio, Y.: Graph attention networks. arXiv preprint arXiv:1710.10903 (2017)
23. Wang, M., Deng, W.: Deep visual domain adaptation: a survey. Neurocomputing **312**, 135–153 (2018)
24. Wang, X., Bo, D., Shi, C., Fan, S., Ye, Y., Yu, P.S.: A survey on heterogeneous graph embedding: methods, techniques, applications and sources. arXiv preprint arXiv:2011.14867 (2020)
25. Wang, X., et al.: Heterogeneous graph attention network. In: WWW, pp. 2022–2032 (2019)
26. Wu, M., Pan, S., Zhou, C., Chang, X., Zhu, X.: Unsupervised domain adaptive graph convolutional networks. In: WWW, pp. 1457–1467 (2020)
27. Yang, S., Song, G., Jin, Y., Du, L.: Domain adaptive classification on heterogeneous information networks. In: IJCAI, pp. 1410–1416 (2020)
28. Zhang, C., Song, D., Huang, C., Swami, A., Chawla, N.V.: Heterogeneous graph neural network. In: KDD, pp. 793–803 (2019)
29. Zhang, Y., Song, G., Du, L., Yang, S., Jin, Y.: Dane: domain adaptive network embedding. arXiv preprint arXiv:1906.00684 (2019)
30. Zhuang, F., et al.: A comprehensive survey on transfer learning. CoRR abs/1911.02685 (2019). http://arxiv.org/abs/1911.02685

The KL-Divergence Between a Graph Model and its Fair I-Projection as a Fairness Regularizer

Maarten Buyl$^{(\boxtimes)}$ ⓘ and Tijl De Bie ⓘ

IDLab, Ghent University, Ghent, Belgium
{maarten.buyl,tijl.debie}@ugent.be

Abstract. Learning and reasoning over graphs is increasingly done by means of probabilistic models, e.g. exponential random graph models, graph embedding models, and graph neural networks. When graphs are modeling relations between people, however, they will inevitably reflect biases, prejudices, and other forms of inequity and inequality. An important challenge is thus to design accurate graph modeling approaches while guaranteeing fairness according to the specific notion of fairness that the problem requires. Yet, past work on the topic remains scarce, is limited to debiasing specific graph modeling methods, and often aims to ensure fairness in an indirect manner.

We propose a generic approach applicable to most probabilistic graph modeling approaches. Specifically, we first define the class of fair graph models corresponding to a chosen set of fairness criteria. Given this, we propose a fairness regularizer defined as the KL-divergence between the graph model and its I-projection onto the set of fair models. We demonstrate that using this fairness regularizer in combination with existing graph modeling approaches efficiently trades-off fairness with accuracy, whereas the state-of-the-art models can only make this trade-off for the fairness criterion that they were specifically designed for.

Keywords: Fairness · I-projection · Link prediction · Graph · Regularizer

1 Introduction

Graphs are flexible data structures, naturally suited for representing relations between people (e.g. in social networks) or between people and objects (e.g. in recommender systems). Here, links between nodes may represent any kind of relation, such as interest or similarity. It is common in real-world relational data that the corresponding graphs are often imperfect or only partially observed. For example, it may contain spurious or missing edges, or some node pairs may be explicitly marked as having unknown status. In such cases, it is often useful to correct or predict the link status between any given pair of nodes. This task is known as *link prediction*: predicting the link status between any pair of nodes, given the known part of the graph and possibly any node or edge features [23].

© Springer Nature Switzerland AG 2021
N. Oliver et al. (Eds.): ECML PKDD 2021, LNAI 12976, pp. 351–366, 2021.
https://doi.org/10.1007/978-3-030-86520-7_22

Methods for link prediction are typically based on machine learning. A first class of methods constructs a set of features for each node-pair, such as their number of common neighbors, the Jaccard similarity between their neighborhoods, and more [24]. Other methods are based on probabilistic models, with exponential random graph models as a notable class originating mostly from the statistics and physics communities [30]. More recently, the machine learning community has proposed graph embedding methods [14], which represent each node as a point in a vector space, from which a probabilistic model for the graph's edges can be derived (among other possible uses). Related to this, graph neural network models [33] have been proposed which equally can be used to probabilistically model the presence or absence of edges in a graph [35].

The use of such models can have genuine impact on the lives of the individuals concerned. For example, a graph of data on job seekers and job vacancies can be used to determine which career opportunities an individual will be recommended. If it is a social network, it may determine which friendships are being recommended. The existence of particular undesirable biases in such networks (e.g. people with certain demographics being recommended only certain types of jobs, or people with a certain social position only being recommended friendships with people of similar status) may result in biased link predictions that perpetuate inequity in society. Yet, graph models used for link prediction typically exploit properties of graphs that are a direct or indirect result of those existing biases. For example, many will exploit the presence of *homophily*: the tendency of people to associate with similar individuals [27]. However, homophily leads to segregation, which often adversely affects minority groups [16,19].

The mitigation of bias in machine learning algorithms has been studied quite extensively for classification and regression models in the fairness literature, both in formalizing a range of fairness measures [13,15] and in developing methods that ensure fair classification and regression [28]. However, despite the existence of biases, such as homophily, that are specific to relational datasets, fairness has so far received limited attention in the graph modeling and link prediction literature. Current approaches focus on resolving bias issues for *specific algorithms* [4,6], or use adversarial learning to improve a *specific notion of fairness* [4,25].

Contributions. In this paper, we introduce a regularization approach to ensure fairness in link prediction that is *generically* applicable across *different link prediction fairness notions* and *different network models*.

To that end, in Sect. 3 we first express the set of all *fair probabilistic network models*. For any possibly biased network model, we can then compute the *I-projection* [11] onto this class: the distribution within the class of fair models that has the smallest KL-divergence with the biased model. In an information-theoretic sense, this I-projection can be seen as the *fair* distribution that is closest to the considered biased model. We also show that for common fairness metrics, the set of fair graph models is a linear set, for which the computation of the I-projection is well-studied and easy to compute in practice.

In Sect. 4, we then propose the KL-divergence between a (possibly biased) fitted probabilistic network model and its fair I-projection as a generic *fairness regularizer*, to be minimized in combination with the usual cost function for the network model. We also propose and analyze a generic algorithmic approach to efficiently solve the resulting fairness-regularized optimization problem.

Finally, our empirical results in Sect. 5 demonstrate that our proposed fairness regularizer *can be applied to a wide diversity of probabilistic network models* such that the desired fairness score is improved. In terms of that fairness criterion, our fairness modification outperforms *DeBayes* and *Compositional Fairness Constraints*, even on the models these baselines were specifically designed for.

2   Related Work

Fairness-aware machine learning is traditionally divided into three types [28]: *pre-processing* methods that involve transforming the dataset to remove bias [7], *in-processing* methods that try to modify the algorithm itself and *post-processing* methods that transform the predictions of the model [15]. Our method belongs to the in-processing category, because we directly modify the objective function with the aim of improving fairness. Here, one approach is to enforce constraints that keep the algorithm fair throughout the learning process [32].

The fairness-constrained optimization problem can also be solved using the method of Lagrange multipliers [2,8,17,31]. This is related to the problem of finding the fair I-projection [11]: the distribution from the set of fair distributions with the smallest KL-divergence to a reference distribution, e.g. an already trained (biased) model [3]. While we also compute the I-projection of the model onto the class of fair link predictors, we do not use it to transform the model directly. Instead, we consider the distance to that I-projection as a regularization term.

The work on applying fairness methods to the task of link prediction is limited. Methods *DeBayes* [6], *Fairwalk* [29] and *FairAdj* [22] all adapt specific graph embedding models to make them more fair. Other approaches, e.g. *FLIP* [25] and *Compositional Fairness Constraints* [4], rely on adversarial learning to remove sensitive information from node representations.

3   Fair Information Projection

After discussing some notation in Sect. 3.1, we characterize the set of fair graph models in Sect. 3.2. In Sect. 3.3, we will leverage this characterization to discuss the *I-projection* onto the set of fair graph models, i.e. the distribution belonging to the set with the smallest KL-divergence to a reference distribution.

3.1   Notation

We denote a random unweighted and undirected graph without self-loops as $G = (V, E)$, with $V = \{1, 2, \ldots, n\}$ the set of n vertices and $E \subseteq \binom{V}{2}$ the set of edges. It is often convenient to represent the set of edges also by a symmetric adjacency matrix with zero diagonal $\mathbf{A} \in \{0, 1\}^{n \times n}$ with element a_{ij} at row i and column j equal to 1 if $\{i, j\} \in E$ and 0 otherwise. An empirical graph over the same set of vertices will be denoted as $\hat{G} = (V, \hat{E})$ with adjacency matrix $\hat{\mathbf{A}}$ and adjacency indicator variables $\hat{a}_{ij}$. In some applications, $\hat{a}_{ij}$ may be unobserved and thus unknown for some $\{i, j\}$.

A probabilistic graph model p for a given vertex set V is a probability distribution over the set of possible edge sets E, or equivalently over the set of adjacency matrices $\mathbf{A}$, with $p(\mathbf{A})$ denoting the probability of the graph with adjacency matrix $\mathbf{A}$. Probabilistic graph models are used for various purposes, but one important purpose is link prediction: the prediction of the existence of an edge (or not) connecting any given pair of nodes i and j. This is particularly important when some elements from $\hat{\mathbf{A}}$ are unknown. But it is also useful when the empirical adjacency matrix is assumed to be noisy, in which case link prediction is used to reduce the noise. Link prediction can be trivially done by making use of the marginal probability distribution p_{ij}, defined as $p_{ij}(x) = \sum_{\mathbf{A}:a_{ij}=x} p(\mathbf{A})$.

Note that many practically useful probabilistic graph models are dyadic independence models: they can be written as the product of the marginal distributions: $p(\mathbf{A}) = \prod_{i<j} p_{ij}(a_{ij})$. This is true for the models evaluated in our empirical results section, but the approach proposed in this paper is conceptually applicable also where this is not the case (e.g. for more complex random graph models), albeit at the cost of greater mathematical and computational complexity.

Finally, we assume vertices belong to one of a set of *sensitive groups*, defined by categorical attributes with respect to which discrimination is undesirable or forbidden. These sensitive groups are denoted as V_s with $s \in S$ for some finite set S. The sets V_s with $s \in S$ form a partition of V. For notational convenience, we also introduce the notation $U_{st} \triangleq \{\{i, j\} | i \in V_s, j \in V_t, i \neq j\}$, the set of possible unordered pairs of distinct vertex pairs between V_s and V_t. Thus, $|U_{ss}| = \binom{|V_s|}{2}$ and $|U_{st}| = |V_s| \times |V_t|$ for $s \neq t$. Similarly, we write $U \triangleq \binom{V}{2}$ for the set of all (unordered) vertex pairs.

3.2   Fairness Constraints

Here we take inspiration from prior work [6,21,22] on translating two classification fairness criteria to the graph setting: *demographic parity* and *equalized opportunity*. We then formalize a general definition for such fairness criteria.

Demographic Parity (DP). A classifier could be thought of as non-discriminatory when its expected score of an individual is the same regardless of which sensitive group they belong to. This traditional criterion of fairness is referred to as *demographic* or *statistical parity* (DP) [13].

We generalize this to the graph setting by requiring that the expected proportion of vertex pairs belonging to any two sensitive groups V_s and V_t that are connected, is constant over all pairs of sensitive groups. More formally, the probabilistic graph model p satisfies the DP fairness criterion iff:

$$\exists d \in \mathbb{R} : \forall s, t \in S : \mathop{\mathbb{E}}_{\mathbf{A}\sim p} \left[\frac{1}{|U_{st}|} \sum_{\{i,j\}\in U_{st}} a_{ij} \right] = d,$$

where choices for d are discussed in Sect. 4.2. (Note that this criterion also ensures that the average expected vertex degree is the same for all sensitive groups.)

Thanks to linearity of the expectation operator, and with p_{ij} the marginal distribution for the edge indicator variable a_{ij}, this can be simplified as follows:

$$\exists d \in \mathbb{R} : \forall s,t \in S : \sum_{\{i,j\} \in U_{st}} \mathop{\mathbb{E}}_{a_{ij} \sim p_{ij}} [a_{ij}] = d|U_{st}|.$$

We thus define the set $\mathbb{P}_{\text{DP}}$ of distributions satisfying these constraints as fair with respect to DP. The DP fairness criterion is notable for diminishing the effect of homophily, since it encourages inter-group ($s \neq t$) interaction to have the same expected score as intra-group ($s = t$) interactions, thereby reducing segregation based on the nodes' sensitive traits. We note that some previous definitions [21,22] enforce a weaker form of demographic parity that only requires balance between the set of all intra-group connections and the set of all inter-group connections. Quite trivially, our approach could handle this weaker form as well. However, in our experiments we maintain the stronger definition of DP fairness (defined for all pairs $\forall s,t \in S$) in order to penalize situations where one type of inter-group connections U_{ss} is discriminated against in favor of a second type of inter-group connections $U_{tt} \neq U_{ss}$.

Equalized Opportunity (EO). A drawback of the DP fairness notion is that it disregards the possibility that there are justifiable reasons for some sensitive groups to be scored higher [15]. For example, in the social graph context one sensitive group s may generally have more social interactions with others, regardless of their sensitive group $t \neq s$ [6]. Depending on the application, it may then be deemed fair to predict inter-group edges (U_{st}) from this more social group as more probable than intra-group edges between nodes in other groups (U_{tt}).

A fairness criterion that takes this into account is *equalized opportunity* (EO) [15]. EO requires that the true positive rate, and consequently also the false negative rate, is equal across groups. In other words, and applied to the graph context: when averaging the probability under the model of edge-connected vertex-pairs $\hat{E}$ between two sensitive groups V_s and V_t, the result should always be the same irrespective of s and t. More formally:

$$\exists d \in \mathbb{R} : \forall s,t \in S : \mathop{\mathbb{E}}_{A \sim p} \left[\frac{1}{|\hat{E} \cap U_{st}|} \sum_{\{i,j\} \in \hat{E} \cap U_{st}} a_{ij} \right] = d,$$

where $\hat{E}$ is the fixed empirical set of edges.

Thanks to linearity of the expectation operator, and with p_{ij} the marginal distribution for the edge indicator variable a_{ij}, this can be simplified as follows:

$$\exists d \in \mathbb{R} : \forall s,t \in S : \sum_{\{i,j\} \in \hat{E} \cap U_{st}} \mathop{\mathbb{E}}_{a_{ij} \sim p_{ij}} [a_{ij}] = d|\hat{E} \cap U_{st}|.$$

We thus define the set $\mathbb{P}_{\text{EO}}$ of distributions satisfying these constraints as fair with respect to EO.

General Sets of Fair Graph Distributions. Both the DP and EO criteria are thus formalized as a constraint that is linear in the probability distribution p. Using $\mathbf{1}$ to denote the indicator function, the DP and EO constraints on p can both be formalized in the following form:

$$F_c(p) \triangleq \sum_{\{i,j\} \in U} \mathop{\mathbb{E}}_{a_{ij} \sim p_{ij}} [f_c(\{i,j\}, a_{ij})] = d_c, \tag{1}$$

where for DP the functions $f_c : U \times \{0,1\} \to \mathbb{R}$ and corresponding constants d_c are given by:

$$f_{st}(\{i,j\}, x) = x\mathbf{1}(\{i,j\} \in U_{st}),$$
$$d_{st} = d|U_{st}|,$$

for all $s, t \in S$ and for some $d \in \mathbb{R}$. Similarly, for EO:

$$f_{st}(\{i,j\}, x) = x\mathbf{1}(\{i,j\} \in \hat{E} \cap U_{st}),$$
$$d_{st} = d|\hat{E} \cap U_{st}|.$$

As a matter of fact, many other statistical fairness criteria, such as *equalized odds*, *accuracy equality* or *churn equality* can formalized in this manner, with different choices for f_c and d_c [2,3,8].

Thus, although our implementation and experiments are focused on DP and EO only, we develop the theory in this paper for the general formulation of a set of fair probabilistic graph models as:[1]

$$\mathbb{P}_{\mathcal{F}} := \{p \in \mathbb{P} \mid \forall c \in \mathcal{C}_{\mathcal{F}} : F_c(p) = d_c\}, \tag{2}$$

with $\mathbb{P}$ the set of all possible distributions over $\mathbf{A}$, and $\mathcal{C}_{\mathcal{F}}$ a countable (and typically finite) set indexing the constraints that enforce fairness criterion $\mathcal{F}$. Importantly, F_c as defined in Eq. (1) is a linear function of p, such that I-projecting any distribution onto $\mathbb{P}_{\mathcal{F}}$ is a mathematically elegant operation. This is the subject of the following.

3.3   Information Projection

We now show how to find, for any possibly unfair distribution h, the fair distribution $p \in \mathbb{P}_{\mathcal{F}}$ that is as close to h as possible. When that closeness is computed in terms of the KL-divergence, then the desired distribution, denoted by $h_{\mathcal{F}}$, is known as the *I-projection* [10,11]:

$$h_{\mathcal{F}} = \arg\min_{p \in \mathbb{P}_{\mathcal{F}}} D_{KL}(p \parallel h),$$

---

[1] In our proposed framework, we require these constraints to be satisfied exactly in order for p to be fair. However, prior work has also allowed for a percentage-wise deviation [34].

where it is assumed that $\mathbb{P}_{\mathcal{F}} \neq \emptyset$ and $D_{KL}(p \parallel h) < \infty$. Since $\mathbb{P}_{\mathcal{F}}$ is linear and thus convex, the I-projection $h_{\mathcal{F}}$ is unique [11].

Finding the I-projection of model h under linear constraints $\mathcal{C}_{\mathcal{F}}$ is a convex optimization problem[2]. Although it is straightforward to generalize this, let us assume that h is a dyadic independence model. This is justified as many contemporary probabilistic graph models (including graph embedding methods and graph neural networks) are dyadic independence models, and because it simplifies notation. Then, the I-projection of h is the product distribution of the marginal distributions for the vertex pairs $\{i, j\}$, given by [9]:

$$h_{\mathcal{F},ij}(x) = \frac{h_{ij}(x)}{Z_{\mathcal{F},ij}(\lambda)} \exp\left(\sum_{c \in \mathcal{C}_{\mathcal{F}}} \lambda_c f_c(\{i,j\}, x)\right),$$

with

$$Z_{\mathcal{F},ij}(\lambda) = \sum_{x \in \{0,1\}} h_{ij}(x) \exp\left(\sum_{c \in \mathcal{C}_{\mathcal{F}}} \lambda_c f_c(\{i,j\}, x)\right).$$

the *log-partition* function and with λ denoting the vector of λ_c values. Let $Z_{\mathcal{F}}(\lambda) = \prod_{\{i,j\} \in U} Z_{\mathcal{F},ij}(\lambda)$. The values of the λ_c are found by maximizing:

$$L_h(\lambda) = -\log Z_{\mathcal{F}}(\lambda) + \sum_{c \in \mathcal{C}_{\mathcal{F}}} \lambda_c d_c. \tag{3}$$

This function $L_h(\lambda)$ is the Lagrange dual of the KL-divergence minimization problem with reference model h, and λ is the set of Lagrange multipliers corresponding to the fairness constraints.

4 The KL-Divergence to the I-projection as a Fairness Regularizer

We argue that the KL-divergence $D_{KL}(h_{\mathcal{F}} \parallel h)$ between a probabilistic model h and its fair I-projection $h_{\mathcal{F}}$ is an adequate measure of the unfairness of h.

Indeed, suppose that $h_{\mathcal{F}}$ represents an idealized version of reality that is free from undue bias (i.e. fair). Specifically, it is the idealized version of reality that is closest to the model h, which, in turn, can be seen as the unfairly biased version of the reality $h_{\mathcal{F}}$. For example, it may be the result of discrimination and cultural social biases in historical data. Then the KL-divergence $D_{KL}(h_{\mathcal{F}} \parallel h)$ quantifies the amount of information lost when using the biased model h instead of the idealized model $h_{\mathcal{F}}$ [5]. In other words, it is the information lost due to any unfairness in the model h, and thus, informally speaking, the amount of 'unfair information' in h.

---

[2] The distribution that results from the reverse KL-divergence formulation $\arg\min_{p \in \mathbb{P}_{\mathcal{F}}} D_{KL}(h \parallel p)$ is much less practical to compute and was therefore not further considered for this work.

Algorithm 1: Optimizing $\mathcal{L}$ with respect to link predictor h, in the case where DP is the fairness criterion.

Data: possible distinct vertex pairs U, empirical adjacency matrix $\hat{\mathbf{A}}$, and fairness strength parameter γ

initialize model h and I-projection parameters λ;

for $t = 1$ **to** T **do**

$\quad\mathcal{L}_{\mathcal{A}} \leftarrow -\log h\left(\hat{\mathbf{A}}\right);$

$\quad d \leftarrow \frac{1}{|U|}\mathbb{E}_{\mathbf{A}\sim h}\left[\mathbf{A}\right];$

$\quad\mathcal{L}_{\mathcal{F}} \leftarrow \max_\lambda\left[-\log Z_{h_{\mathcal{F}}}(\lambda) + \sum_{s,t \in S}\lambda_{st}d|U_{st}|\right];$

$\quad\mathcal{L} \leftarrow \mathcal{L}_{\mathcal{A}} + \gamma\mathcal{L}_{\mathcal{F}};$

$\quad\text{UPDATE}(h, \nabla_h\mathcal{L});$

end

Moreover, the KL-divergence, in being a measure of information, is commensurate with commonly used loss terms in machine learning, in particular with the cross-entropy between the empirical distribution and the learned model, which is equivalent to the KL-divergence between those two up to a constant. This is the topic of the next subsection.

4.1 I-Projection Regularization

Let $\hat{p}$ represent the empirical distribution, i.e. $\hat{p}(\mathbf{A} = \hat{\mathbf{A}}) = 1$ and $\hat{p}(\mathbf{A} \neq \hat{\mathbf{A}}) = 0$. The common machine learning objective is then to minimize the KL-divergence $D_{KL}(\hat{p} \| h)$, denoted by $\mathcal{L}_{\mathcal{A}}$, which is equivalent to maximizing the log-likelihood of h under $\hat{p}$, or equivalently the cross-entropy. We propose to add the KL-divergence $D_{KL}(h_{\mathcal{F}} \| h)$ as an extra loss term $\mathcal{L}_{\mathcal{F}}$. The overall objective function $\mathcal{L}$ to find h is thus:

$$\mathcal{L} = \min_h\left[\mathcal{L}_{\mathcal{A}} + \gamma\mathcal{L}_{\mathcal{F}}\right]$$
$$= \min_h\left[D_{KL}(\hat{p} \| h) + \gamma D_{KL}(h_{\mathcal{F}} \| h)\right]$$

with γ a hyperparameter that controls the strength of the loss term. Recall that, for a parameter λ that satisfies the fairness constraints, $D_{KL}(h_{\mathcal{F}} \| h)$ is equivalent to the loss function in Eq. (3):

$$\mathcal{L} = \min_h\left[D_{KL}(\hat{p} \| h) + \gamma \min_{p \in \mathbb{P}_{\mathcal{F}}} D_{KL}(p \| h)\right]$$
$$= \min_h\left[D_{KL}(\hat{p} \| h) + \gamma \max_\lambda L_h(\lambda)\right].$$

4.2 Practical Considerations

So far, we did not yet specify the choice of d in the DP and EO constraints. To enforce $p \in \mathbb{P}_{\mathcal{DP}}$, a straightforward option is to set d equal to the mean of p.

However, d is then no longer constant with respect to p and instead depends on changes in the λ parameters. The gradient of the second term of the loss function $L_h(\lambda)$ in Eq. (3) is then more complicated. Alternatively, setting d equal to the mean of the empirical distribution $\hat{p}$ forces p to adopt the same mean as the empirical one, even though there is no specific reason that $h_{\mathcal{F}}$ or consequently h should match the empirical mean. We finally chose to set d equal to the mean of h, such that when optimizing λ, we can treat d as a fixed, constant value.

Furthermore, out of several ways to optimize $\mathcal{L}$, we opted to fully optimize λ for every parameter update of h. On the one hand, the λ parameters are typically very few in number (for DP and EO, there are only $\mathcal{C}_{\mathcal{F}} = |S|^2$), making it cheap to store them. On the other hand, optimizing λ exactly requires the repeated evaluation of the probability under h of all unordered vertex pairs U. With $|U| = \frac{n(n-1)}{2}$, this is infeasible for large n. However, for $|S| \ll n$, using a relatively small subsample of all unordered vertex pairs will suffice in practice to obtain a good estimate for the optimal λ, dramatically enhancing scalability. Moreover, using the optimal λ of the previous iteration's h as a starting guess for the next iteration also speeds up computations in practice.

For concreteness, the use of the proposed generic fairness regularizer to the DP fairness criterion is summarized in Algorithm 1.

5 Experiments

Our experiments were performed on three datasets, described in Sect. 5.1. We applied our proposed fairness regularizer on four simple, yet diverse methods explained in Sect. 5.2. Though the method variants without fairness regularizer are already baselines, we additionally compared our results with state-of-the-art approaches for link prediction based on fair graph embedding in Sect. 5.3. All methods went through the same evaluation pipeline described in Sect. 5.4. The results of which were discussed in Sect. 5.5.

5.1 Datasets

The methods were evaluated on three attributed graph datasets, summarized in Table 1. They were chosen for their diverse properties and manageable size.

Polblogs: The POLBLOGS [1] dataset was constructed from blogs discussing United States politics in 2005. In the undirected version, there is an edge between blogs if either of them had a hyperlink to the other. The sensitive attribute is the US political *party* (the *Republican* or *Democratic Party*) that the blog supported, either by their own admission or through manual labeling from the dataset creators. Intra-group links are heavily favored over inter-group links.

ML100k: Movielens datasets are often used as a benchmark for recommender systems. The data contains users' movie ratings on a five-star scale. An unweighted, bipartite graph is formed by considering the users and movies as nodes and an edge between them if the user rated the movie. While the data

Table 1. Properties of the datasets. The dataset names are URLs to hosts of the datasets.

DATASET	#NODES	#EDGES	S	$|S|$
POLBLOGS	1,222	16,714	PARTY	2
ML100K	2,625	100,000	AGE	7
FACEBOOK	3,955	85,482	GENDER	2

contains several types of sensitive attributes, we opted to group the *age* attribute into seven bins, delineated by the ages $[18, 25, 35, 45, 50, 56]$. There are only user-movie edges, so the domain of sensitive value of an edge is only affected by the user's sensitive value. Note that all methods were adapted such that they took the bipartiteness of the graph into account when sampling negative training edges.

Facebook [26]: The FACEBOOK graph consists of user nodes that are linked if they are 'friends'. Each user either has *gender* feature '0', '1' or neither. For the last group of users, of which there are 84, it is unclear whether their gender is unknown or non-binary. Their nodes and edges were removed from the dataset. Only 3 undirected attribute pairs thus remain in the data. In contrast to POLBLOGS, the bias effect is much weaker.

5.2   Algorithms

The proposed fairness regularizer was applied to four relatively simple graph models. A *PyTorch* implementation was sought or implemented for each of them, such that the fairness loss can easily be added.

MaxEnt: We will refer to the MAXENT model as the maximum entropy graph model under which the expected degree of each node matches its empirical degree [12]. The solution is a simple exponential random graph model [30].

Dot-Product: Given a set of embeddings, one for every node, taking the DOT-PRODUCT an embedding pair is a straightforward way to perform link prediction [14]. In this simple model, the 'decoder' for edge (i, j) is the dot product operator, while the 'encoder' for node i just looks up its representation in a learned table of embeddings.

CNE: A method that combines both the MAXENT model and the DOT-PRODUCT decoder is the *Conditional Network Embedding* (CNE) model [18]. Instead of the Dot-Product, it 'decodes' the distance between nodes (i, j). Moreover, it uses the MAXENT model as a prior distribution over the graph data.

GAE: The Graph Auto-Encoder (GAE) [20] is also a DOT-PRODUCT model, though it uses a Graph Convolutional Network (GCN) as its encoder. As such, it is an example of a graph neural network [33]. In our implementation we used two layers for the GCN and used the identity matrix as the node feature matrix.

5.3   Fair Graph Embedding Baselines

In part, the algorithms from Sect. 5.2 were chosen such that they allow for easy comparison with two recent methods in the field of fair graph embedding.

CFC: The Compositional Fairness Constraints (CFC) method [4] aims to generate fair embeddings by learning filters that mask the sensitive attribute information. This is done through adversarial learning. When applied to link prediction, it also uses the DOT-PRODUCT decoder. Note that our implementation of the basic DOT-PRODUCT differs from the source code of CFC, causing differences in performance between our DOT-PRODUCT experiments and CFC with a fairness regularization strength of zero.

DeBayes: Finally, DEBAYES [6] is an adaptation of CNE where the bias in the data is used as additional prior information when learning the embeddings, such that the embeddings are debiased. By using a prior without this biased information at testing time, the link prediction using these embeddings is expected to at least not be less fair than the standard CNE.

5.4   Evaluation

Every method was run for 10 different random seeds on each dataset. Those 10 seeds each had a different train/test split, where the latter consisted of around 20% of the edges in the data. The test set was extended with the same amount of non-edges. However, it was made sure that the test set did not contain nodes unknown in the train set, since the graph models in our evaluation are transductive methods. Only test set results are reported.

Hyperparameter tuning in order to improve the performance of the considered methods was minimal, as our aim is to show the effect of the fairness regularization and not the predictive quality of the methods themselves. As such, we did no hyperparameter sweep with the aim of improving AUC, and instead only deviated from default parameters when it could allow for an easier comparison between models, e.g. the dimensionality of DOT-PRODUCT and CFC embeddings. We only report results of our proposed method with a fairness regularization strength of $\gamma = 100$, because this parameter almost always caused a significant effect on the fairness measures while not diminishing predictive power too strongly. For DEBAYES the default values were used, while for CFC we report the results for the regularization strength $\lambda \in \{10, 100, 1000\}$. Smaller values did not cause a noticeable effect on fairness, while larger values caused a strong degradation in terms of AUC.

Along with the link prediction AUC score, all methods were tested for their deviation from Demographic Parity (DP) and Equalized Opportunity (EO). The calculation of those measures follows [6], where DP is the maximal difference between the mean predicted value of any subgroup. Similarly, the EO measure refers to the maximal difference between true positive rates of subgroups. Lower DP and EO scores therefore imply a fairer model. Note that the test set contains proportionally less negative edges than the overall dataset, possibly skewing the DP score. This effect was compensated for by proportionately increasing the contribution of negative samples when calculating DP. Furthermore, in the Appendix additional measures are reported on the diversity in the ranking of prediction scores, as well as diversity in the embeddings.

5.5 Results

The test set results[3] are reported in Fig. 1. We reiterate that our intention is not to find the specific link prediction method with the best trade-off in terms of AUC and fairness. Rather, we want to verify that our proposed regularizer can be applied to a variety of methods and fairness criteria, with an efficient AUC-fairness trade-off for the considered criterion.

Fairness Quality: In many cases and across all four methods, it can indeed be observed that the use of our proposed fairness regularizer significantly reduces the link prediction bias, according to the employed fairness criteria. This is in contrast to the baselines DEBAYES and CFC. The former did not improve fairness scores over CNE, while the latter could only become more fair at a significant cost to AUC.

There are a few exceptions where our method does not reduce unfairness according to the fairness criterion. First, there are some cases where an already low DP score for the base method can not be improved further by adding the DP regularizer. This happens for MAXENT in Fig. 1a, GAE and DOT-PRODUCT in Fig. 1b and for CNE in Fig. 1c. A second kind of exception is where the method with the DP regularizer is less EO-unfair than with the EO regularizer. It occurs for the DOT-PRODUCT (EO) variant in Fig. 1a and 1c, possibly because the former had a larger reduction in predictive power overall. In both these cases, DOT-PRODUCT (EO) still significantly reduces EO compared to the DOT-PRODUCT model without fairness regularizer.

Predictive Quality: Moreover, the decrease in AUC is fairly minimal with our fairness regularizer, especially compared to an adversarial approach like CFC. While the addition of the EO regularizer has no noticeable effect on the AUC, the DP variant does cause strong reduction on some models in Fig. 1a. This is to be expected, because enforcing DP can cause a significant loss in predictive power if the subgroups in the underlying data have different base rates [15]. For a network like POLBLOGS, which strongly favors intra-group connections, encouraging the inter-group connections therefore results in AUC loss.

Runtimes: Runtimes[4] of each method are listed in Table 2. In our experiments, the addition of our regularizer causes a large increase in runtime. However, several easy speed improvements are available to make the method scale to large graphs. For example, the optimal λ parameters of $h_{\mathcal{F}}$ can be approximated by only fitting them on a subsample of the vertex pairs that h is trained on. As

---

[3] A table with the results in text format is provided in the Appendix.
[4] All experiments were conducted using half the hyperthreads on a machine equipped with a 12 Core Intel(R) Xeon(R) Gold processor and 256 GB of RAM.

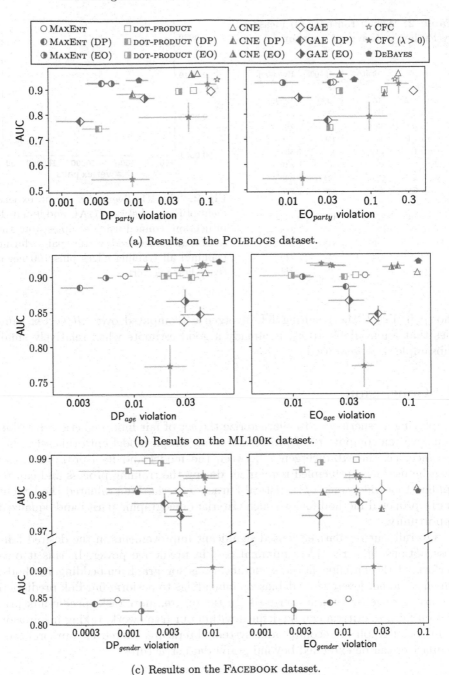

(a) Results on the POLBLOGS dataset.

(b) Results on the ML100K dataset.

(c) Results on the FACEBOOK dataset.

Fig. 1. Markers display the mean over ten identical experiment runs with different random seeds. Error bars horizontally and vertically show the standard deviation. Completely empty markers refer to methods without any fairness modification. Methods with a fairness regularizer that enforces the DP or EO fairness criterion are left-filled or right-filled respectively. On the x-axis, unfairness is measured, so **more left is better**. On the y-axis, AUC is measured, so **higher is better**.

Table 2. Median runtimes (s) measured by Python's `time.perf_counter`.

Dataset	PolBlogs	ML100K	Facebook
MaxEnt	14	68	158
with MA	707	3050	1924
with EO	170	773	1191
Dot-Product	60	62	200
with DP	349	456	1169
with EO	135	239	531
CNE	105	307	349
with DP	574	1417	2065
with EO	286	843	865
CNE	28	26	101
with DP	278	437	1072
with EO	92	255	388
CFC	280	843	1601
CFC ($\lambda > 0$)	242	2623	3494
DeBayes	98	305	343

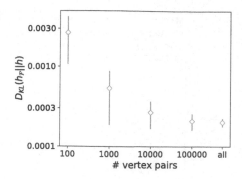

Fig. 2. The KL-divergence in the experiment of Fig. 1c between GAE and its fair I-projection, trained using samples from the set of all considered vertex pairs during training: all training edges plus 100 negative edges per vertex.

shown in Fig. 2, the resulting KL-divergence (computed over *all* vertex samples that are available to h), is already a good estimate when relatively small subsample sizes were used.

6   Conclusion

Employing a generic way to characterize the set of fair link prediction distributions, we can compute the I-projection of any graph model onto this set. That distance, i.e. the KL-divergence between the model and its I-projection, can then be used as a principled regularizer during the training process and can be applied to a wide range of statistical fairness criteria. We evaluated the benefit of our proposed method for two such criteria: demographic parity and equalized opportunity.

Overall, our regularizer caused significant improvements in the desired fairness notions, at a relatively minimal cost in predictive power. In this it outperformed the baseline fairness modifications for graph embedding methods, which could not leverage its debiased embeddings to perform fair link prediction according to generic fairness criteria. In the future, more task-specific link prediction fairness criteria can be defined within our framework, taking inspiration from social graph or recommender systems literature. Moreover, our proposed regularizer can be extended beyond graph data structures.

Acknowledgments. This research was funded by the ERC under the EU's 7th Framework and H2020 Programmes (ERC Grant Agreement no. 615517 and 963924), the Flemish Government (AI Research Program), the BOF of Ghent University (PhD scholarship BOF20/DOC/144), and the FWO (project no. G091017N, G0F9816N, 3G042220).

References

1. Adamic, L.A., Glance, N.: The political blogosphere and the 2004 US election: divided they blog. In: Proceedings of the 3rd International Workshop on Link Discovery, pp. 36–43 (2005)
2. Agarwal, A., Beygelzimer, A., Dudík, M., Langford, J., Wallach, H.: A reductions approach to fair classification. In: International Conference on Machine Learning, pp. 60–69. PMLR (2018)
3. Alghamdi, W., Asoodeh, S., Wang, H., Calmon, F.P., Wei, D., Ramamurthy, K.N.: Model projection: theory and applications to fair machine learning. In: 2020 IEEE International Symposium on Information Theory (ISIT), pp. 2711–2716. IEEE (2020)
4. Bose, A., Hamilton, W.: Compositional fairness constraints for graph embeddings. In: International Conference on Machine Learning, pp. 715–724 (2019)
5. Burnham, K.P., Anderson, D.R.: Practical use of the information-theoretic approach. In: Model selection and inference, pp. 75–117. Springer, New York (1998). https://doi.org/10.1007/978-1-4757-2917-7_3
6. Buyl, M., De Bie, T.: DeBayes: a Bayesian method for debiasing network embeddings. In: International Conference on Machine Learning, pp. 1220–1229. PMLR (2020)
7. Calmon, F.P., Wei, D., Vinzamuri, B., Ramamurthy, K.N., Varshney, K.R.: Optimized pre-processing for discrimination prevention. In: Proceedings of the 31st International Conference on Neural Information Processing Systems, pp. 3995–4004 (2017)
8. Cotter, A., et al.: Optimization with non-differentiable constraints with applications to fairness, recall, churn, and other goals. J. Mach. Learn. Res. **20**(172), 1–59 (2019)
9. Cover, T.M.: Elements of Information Theory. Wiley, Hoboken (1999)
10. Csiszár, I.: I-divergence geometry of probability distributions and minimization problems. Annals Prob. 146–158 (1975)
11. Csiszár, I., Matus, F.: Information projections revisited. IEEE Trans. Inf. Theory **49**(6), 1474–1490 (2003)
12. De Bie, T.: Maximum entropy models and subjective interestingness: an application to tiles in binary databases. Data Min. Knowl. Discov. **23**(3), 407–446 (2011)
13. Dwork, C., Hardt, M., Pitassi, T., Reingold, O., Zemel, R.: Fairness through awareness. In: Proceedings of the 3rd Innovations in Theoretical Computer Science Conference, pp. 214–226 (2012)
14. Hamilton, W.L., Ying, R., Leskovec, J.: Representation learning on graphs: methods and applications. arXiv preprint arXiv:1709.05584 (2017)
15. Hardt, M., Price, E., Srebro, N.: Equality of opportunity in supervised learning. In: Proceedings of the 30th International Conference on Neural Information Processing Systems, pp. 3323–3331 (2016)
16. Hofstra, B., Corten, R., Van Tubergen, F., Ellison, N.B.: Sources of segregation in social networks: a novel approach using Facebook. Am. Sociol. Rev. **82**(3), 625–656 (2017)
17. Jiang, H., Nachum, O.: Identifying and correcting label bias in machine learning. In: International Conference on Artificial Intelligence and Statistics, pp. 702–712. PMLR (2020)
18. Kang, B., Lijffijt, J., De Bie, T.: Conditional network embeddings. In: International Conference on Learning Representations (2018)

19. Karimi, F., Génois, M., Wagner, C., Singer, P., Strohmaier, M.: Homophily influences ranking of minorities in social networks. Sci. Reports **8**(1), 1–12 (2018)
20. Kipf, T.N., Welling, M.: Variational graph auto-encoders. arXiv preprint arXiv:1611.07308 (2016)
21. Laclau, C., Redko, I., Choudhary, M., Largeron, C.: All of the fairness for edge prediction with optimal transport. In: International Conference on Artificial Intelligence and Statistics, pp. 1774–1782. PMLR (2021)
22. Li, P., Wang, Y., Zhao, H., Hong, P., Liu, H.: On dyadic fairness: exploring and mitigating bias in graph connections. In: International Conference on Learning Representations (2021)
23. Liben-Nowell, D., Kleinberg, J.: The link-prediction problem for social networks. J. Am. Soc. Inf. Sci. Technol. **58**(7), 1019–1031 (2007)
24. Martínez, V., Berzal, F., Cubero, J.C.: A survey of link prediction in complex networks. ACM Comput. Surv. (CSUR) **49**(4), 1–33 (2016)
25. Masrour, F., Wilson, T., Yan, H., Tan, P.N., Esfahanian, A.: Bursting the filter bubble: fairness-aware network link prediction. In: Proceedings of the AAAI Conference on Artificial Intelligence, vol. 34, pp. 841–848 (2020)
26. McAuley, J., Leskovec, J.: Learning to discover social circles in ego networks. In: Proceedings of the 25th International Conference on Neural Information Processing Systems, pp. 539–547 (2012)
27. McPherson, M., Smith-Lovin, L., Cook, J.M.: Birds of a feather: homophily in social networks. Annual Rev. Sociol. **27**(1), 415–444 (2001)
28. Mehrabi, N., Morstatter, F., Saxena, N., Lerman, K., Galstyan, A.: A survey on bias and fairness in machine learning. arXiv preprint arXiv:1908.09635 (2019)
29. Rahman, T., Surma, B., Backes, M., Zhang, Y.: Fairwalk: towards fair graph embedding. In: Proceedings of the Twenty-Eighth International Joint Conference on Artificial Intelligence, IJCAI 2019, pp. 3289–3295. International Joint Conferences on Artificial Intelligence Organization (2019)
30. Robins, G., Pattison, P., Kalish, Y., Lusher, D.: An introduction to exponential random graph (p*) models for social networks. Soc. Netw. **29**(2), 173–191 (2007)
31. Wei, D., Ramamurthy, K.N., Calmon, F.: Optimized score transformation for fair classification. In: International Conference on Artificial Intelligence and Statistics, pp. 1673–1683. PMLR (2020)
32. Woodworth, B., Gunasekar, S., Ohannessian, M.I., Srebro, N.: Learning non-discriminatory predictors. In: Conference on Learning Theory, pp. 1920–1953. PMLR (2017)
33. Wu, Z., Pan, S., Chen, F., Long, G., Zhang, C., Philip, S.Y.: A comprehensive survey on graph neural networks. IEEE Trans. Neural Netw. Learn. Syst. **32**, 4–24 (2020)
34. Zafar, M.B., Valera, I., Rogriguez, M.G., Gummadi, K.P.: Fairness constraints: mechanisms for fair classification. In: Artificial Intelligence and Statistics, pp. 962–970. PMLR (2017)
35. Zhang, M., Chen, Y.: Link prediction based on graph neural networks. In: Proceedings of the 32nd International Conference on Neural Information Processing Systems, pp. 5171–5181 (2018)

On Generalization of Graph Autoencoders with Adversarial Training

Tianjin Huang$^{(\boxtimes)}$, Yulong Pei, Vlado Menkovski, and Mykola Pechenizkiy

Department of Mathematics and Computer Science,
Eindhoven University of Technology, 5600 Eindhoven, MB, The Netherlands
{t.huang,y.pei.1,v.menkovski,m.pechenizkiy}@tue.nl

Abstract. Adversarial training is an approach for increasing model's resilience against adversarial perturbations. Such approaches have been demonstrated to result in models with feature representations that generalize better. However, limited works have been done on adversarial training of models on graph data. In this paper, we raise such a question – does adversarial training improve the generalization of graph representations. We formulate L_2 and L_∞ versions of adversarial training in two powerful node embedding methods: graph autoencoder (GAE) and variational graph autoencoder (VGAE). We conduct extensive experiments on three main applications, i.e. link prediction, node clustering, graph anomaly detection of GAE and VGAE, and demonstrate that both L_2 and L_∞ adversarial training boost the generalization of GAE and VGAE.

Keywords: Graph autoencoders · Variational graph autoencoders · Adversarial training · Node embedding · Generalization

1 Introduction

Networks are ubiquitous in a plenty of real-world applications and they contain relationships between entities and attributes of entities. Modeling such data is challenging due to its non-Euclidean characteristic. Recently, graph embedding that converts graph data into low dimensional feature space has emerged as a popular method to model graph data, For example, DeepWalk [14], node2vec [7] and LINE [23] learn graph embedding by extracting patterns from the graph. Graph Convolutions Networks (GCNs) [9] learn graph embedding by repeated multiplication of normalized adjacency matrix and feature matrix. In particular, graph autoencoder (GAE) [10,24,27] and graph variational autoencoder (VGAE) [10] have been shown to be powerful node embedding methods as unsupervised learning. They have been applied to many machine learning tasks, e.g. node clustering [16,20,24], link prediction [10,19], graph anomaly detection [4,13] and etc.

Adversarial training is an approach for increasing model's resilience against adversarial perturbations by including adversarial examples in the training set [11]. Several recent studies demonstrate that adversarial training improves feature representations leading to better performance for downstream tasks [18,26].

© Springer Nature Switzerland AG 2021
N. Oliver et al. (Eds.): ECML PKDD 2021, LNAI 12976, pp. 367–382, 2021.
https://doi.org/10.1007/978-3-030-86520-7_23

However, little work in this direction has been done for GAE and VGAE. Besides, real-world graphs are usually highly noisy and incomplete, which may lead to a sub-optimal results for standard trained models [32]. Therefore, we are interested to seek answers to the following two questions:

- *Does adversarial training improve generalization, i.e. the performance in applications of node embeddings learned by GAE and VGAE?*
- *Which factors influence this improvement?*

In order to answer the first question above, we firstly formulate L_2 and L_∞ adversarial training for GAE and VGAE. Then, we select three main tasks of VGAE and GAE: link prediction, node clustering and graph anomaly detection for evaluating the generalization performance brought by adversarial training. Besides, we empirically explore which factors affect the generalization performance brought by adversarial training.

Contributions: To the best of our knowledge, we are the first to explore generalization for GAE and VGAE using adversarial training. We formulate L_2 and L_∞ adversarial training, and empirically demonstrate that both L_2 and L_∞ adversarial training boost the generalization with a large margin for the node embeddings learned by GAE and VGAE. An additional interesting finding is that the generalization performance of the proposed adversarial training is more sensitive to attributes perturbation than adjacency matrix perturbation and not sensitive to the degree of nodes.

2   Related Work

Adversarial training has been extensively studied in images. It has been important issues to explore whether adversarial training can help generalization. Tsipras et al. [25] illustrates that adversarial robustness could conflict with model's generalization by a designed simple task. However, Stutz et al. [21] demonstrates that adversarial training with on-manifold adversarial examples helps the generalization. Besides, Salman et al. [18] and Utrera et al. [26] show that the latent features learned by adversarial training are improved and boost the performance of their downstream tasks.

Recently, few works bring adversarial training in graph data. Deng, Dong and Zhu [3] and Sun et al. [22] propose virtual graph adversarial training to promote the smoothness of model. Feng et al. [5] propose graph adversarial training by inducing dynamical regularization. Dai et al. [2] formulate an interpretable adversarial training for DeepWalk. Jin and Zhang [8] introduce latent adversarial training for GCN, which train GCN based on the adversarial perturbed output of the first layer. Besides, several studies explored adversarial training based on adversarial perturbed edges for graph data [1,28,31]. Among these works,part of studies pay attention to achieving model's robustness while ignoring the effect of generalization [1,2,8,28,31] and the others simply utilize perturbations on nodal attributes while not explore the effect of perturbation on edges [3,5,22].

The difference between these works and ours is two-fold: (1) We extend both L_∞ and L_2 adversarial training for graph models while the previous studies only explore L_2 adversarial training. (2) We focus on the generalization effect brought by adversarial training for unsupervised deep learning graph models, i.e. GAE and VGAE while most of the previous studies focus on adversarial robustness for supervised/semi-supervised models.

3 Preliminaries

We first summarize some notations and definitions used in this paper. Following the commonly used notations, we use bold uppercase characters for matrices, e.g. X, bold lowercase characters for vectors, e.g. b, and normal lowercase characters for scalars, e.g. c. The ith row of a matrix A is denoted by $A_{i,:}$ and $(i,j)^{th}$ element of matrix A is denoted as $A_{i,j}$. The ith row of a matrix X is denoted by x_i. We use **KL** for Kullback-Leibler divergence.

We consider an attributed network $\mathcal{G} = \{V, E, X\}$ with $|V| = n$ nodes, $|E| = m$ edges and X node attributed matrix. A is the binary adjacency matrix of $\mathcal{G}$.

3.1 Graph Autoencoders

Graph autoencoders is a kind of unsupervised learning models on graph-structure data [10], which aim at learning low dimensional representations for each node by reconstructing inputs. It has been demonstrated to achieve competitive results in multiple tasks, e.g. link prediction [10,16,17], node clustering [16,20,24], graph anomaly detection [4,13]. Generally, graph autoencoder consists of a graph convolutional network for encoder and an inner product for decoder [10]. Formally, it can be expressed as follows:

$$Z = GCN(A, X) \tag{1}$$

$$\hat{A} = \sigma(ZZ^T), \tag{2}$$

where σ is the sigmoid function, GCN is a graph convolutional network, Z is the learned low dimensional representations and $\hat{A}$ is the reconstructed adjacency matrix.

During the training phase, the parameters will be updated by minimizing the reconstruction loss. Usually, the reconstruction loss is expressed as cross-entropy loss between A and $\hat{A}$ [10]:

$$\mathcal{L}^{ae} = -\frac{1}{n^2} \sum_{(i,j)\in V\times V} \left[A_{i,j} log\hat{A}_{i,j} + (1 - A_{i,j})log(1 - \hat{A}_{i,j}) \right]. \tag{3}$$

3.2 Graph Variational Autoencoders

Kipf and Welling [10] introduced variational graph autoencoder (VGAE) which is a probabilistic model. VGAE is consisted of inference model and generative model. In their approach, the inference model, i.e. corresponding to the encoder of VGAE, is expressed as follows:

$$q(\boldsymbol{Z}|\boldsymbol{X}, \boldsymbol{A}) = \prod_{i=1}^{N} q(\boldsymbol{z}_i|\boldsymbol{X}, \boldsymbol{A}), \ with \ q(\boldsymbol{z}_i|\boldsymbol{X}, \boldsymbol{A}) = \mathcal{N}(\boldsymbol{z}_i|\boldsymbol{\mu}_i, diag(\boldsymbol{\sigma}_i^2)), \quad (4)$$

where $\boldsymbol{\mu}_i$ and $\boldsymbol{\sigma}_i$ are learned by a graph neural network respectively. That is, $\boldsymbol{\mu} = GCN_{\mu}(\boldsymbol{X}, \boldsymbol{A})$ and $log\boldsymbol{\sigma} = GCN_{\delta}(\boldsymbol{X}, \boldsymbol{A})$, with $\boldsymbol{\mu}$ is the matrix of stacking vectors $\boldsymbol{\mu}_i$; likewise, $\boldsymbol{\sigma}$ is the matrix of stacking vectors δ_i.

The generative model, i.e. corresponding to the decoder of autoencoder, is designed as an inner product between latent variables $\boldsymbol{Z}$, which is formally expressed as follows:

$$p(\boldsymbol{A}|\boldsymbol{Z}) = \prod_{i=1}^{n} \prod_{j=1}^{n} p(\boldsymbol{A}_{i,j}|\boldsymbol{z}_i, \boldsymbol{z}_j), \ with \ p(\boldsymbol{A}_{i,j} = 1|\boldsymbol{z}_i, \boldsymbol{z}_j) = \sigma(\boldsymbol{z}_i^T \boldsymbol{z}_j). \quad (5)$$

During the training phase, the parameters will be updated by minimizing the variational lower bound $\mathcal{L}^{vae}$:

$$\mathcal{L}^{vae} = \mathbf{E}_{q(\boldsymbol{Z}|\boldsymbol{X}, \boldsymbol{A})}[logp(\boldsymbol{A}|\boldsymbol{Z})] - \mathbf{KL}[q(\boldsymbol{Z}|\boldsymbol{X}, \boldsymbol{A})||p(\boldsymbol{Z})], \quad (6)$$

where a Gaussian prior is adopted for $p(\boldsymbol{Z}) = \prod_i p(\boldsymbol{z}_i) = \prod_i \mathcal{N}(\mathbf{z}_i|0, \mathbf{I})$.

3.3   Adversarial Training

By now, multiple variants of adversarial training has been proposed and most of them are built on supervised learning and Euclidean data, e.g. FGSM-adversarial training [6], PGD-adversarial training [11], Trades [33], MART [29] and etc. Here we introduce Trades that will be extended to GAE and VGAE settings in Sect. 4. Trades [33] separates loss function into two terms:1) Cross-Entropy Loss for achieving natural accuracy; 2) Kullback-Leibler divergence for achieving adversarial robustness. Formally, given inputs (X, Y), it can be expressed as follows [33]:

$$\min_{\theta} \mathbf{E}_{(X,Y)}[L(f_{\theta}(X), Y) + \lambda \cdot \mathbf{KL}(P(Y|X')||P(Y|X))], \quad (7)$$

where f_{θ} is a supervised model, X' is the adversarial examples that maximize $\mathbf{KL}$ divergence and $P(Y|X)$ is the output probability after softmax. λ is a tunable hyperparameter and it controls the strength of the $\mathbf{KL}$ regularization term.

4   Graph Adversarial Training

In this section, we formulate L_2 and L_{∞} adversarial training for GAE and VGAE respectively.

4.1 Adversarial Training in Graph Autoencoder

Considering that: (1) the inputs of GAE contains adjacency matrix and attributes, (2) the latent representation Z is expected to be invariant to the input perturbation, we reformulate the loss function in Eq. 3 as follows:

$$\min_{\theta} \mathcal{L}^{ae} + \lambda \cdot \mathbf{KL}(P(Z|A', X')\|P(Z|A, X)) \tag{8}$$

$$X' = arg \max_{\|X'-X\|\leq \epsilon} \mathcal{L}^{ae}(A, X), \ A' = arg \max_{\|A'-A\|\leq \epsilon} \mathcal{L}^{ae}(A, X) \tag{9}$$

where A' is the adversarial perturbed adjacency matrix and X' is the adversarial perturbed attributes. Here the important question is how to generate the perturbed adjacency matrix A' and attributes X' in Eq. 9.

Attributes Perturbation X'. We generate the perturbed X' by projection gradient descent (PGD) [11]. We denote total steps as T.

For X' bounded by L_2 norm ball, the perturbed data in t-th step X^t is expressed as follows:

$$X^t = \prod_{\mathcal{B}(X, \epsilon\|X\|_2)} (X^{t-1} + \alpha \cdot g \cdot \|X\|_2 / \|g\|_2) \tag{10}$$

$$g = \nabla_{X^{t-1}} \mathcal{L}^{ae}(A, X^{t-1}) \tag{11}$$

where $\prod$ is the projection operator and $\mathcal{B}(X, \epsilon\|X\|_2)$ is the L_2 norm ball of nodal attributes $x_i : \{x_i' : \|x_i' - x_i\|_2 \leq \epsilon \|x_i\|_2\}$.

For X' bounded by L_∞ norm ball, the perturbed data in t-th step X^t is expressed as follows:

$$X^t = \prod_{\mathcal{B}(X, \epsilon)} (X^{t-1} + \alpha \cdot g) \tag{12}$$

$$g = sgn(\nabla_{X^{t-1}} \mathcal{L}^{ae}(A, X^{t-1})), \tag{13}$$

where $\mathcal{B}(X, \epsilon)$ is the L_∞ norm ball of nodal attributes $x_i : \{x_i' : \|x_i' - x_i\|_\infty \leq \epsilon\}$ and $sgn(\cdot)$ is the sign function.

Adjacency Matrix Perturbation A'. Adjacency matrix perturbation includes two-fold: (1) perturb node connections, i.e. Adding or dropping edges, (2) perturb the strength of information flow between nodes, i.e. the strength of correlation between nodes. Here we choose to perturb the strength of information flow between nodes and leave the perturb of node connections for future work. Specifically, we add weight for each edge and change these weights in order to perturb the strength of information flow. Formally, given the adjacency matrix A, the weighted adjacency matrix $\tilde{A}$ is expressed as $A \odot M$ where the elements of M are continuous and its values are initialized as same value as A. $\odot$ denotes the element-wise product. Formally, A' is expressed as follows:

$$M' = arg \max_{\|M'-M\|\leq \epsilon} \mathcal{L}^{ae}(\tilde{A}, X) \tag{14}$$

$$A' = A \odot M'. \tag{15}$$

For $\boldsymbol{A}'$ bounded by L_2 norm ball, the perturbed data in t-th step $\boldsymbol{A}^t$ is expressed as follows:

$$g = \nabla_{\boldsymbol{M}^{t-1}} \mathcal{L}^{ae}(\tilde{\boldsymbol{A}}^{t-1}, \boldsymbol{X}) \tag{16}$$

$$\boldsymbol{M}^t = \prod_{\mathcal{B}(M, \epsilon \|M\|_2)} (\boldsymbol{M}^{t-1} + \alpha \cdot g \cdot \|M\|_2 / \|g\|_2) \tag{17}$$

$$\boldsymbol{A}^t = \tilde{\boldsymbol{A}}^t = \boldsymbol{A} \odot \boldsymbol{M}^t. \tag{18}$$

For $\boldsymbol{A}'$ bounded by L_∞ norm ball, the perturbed data in t-th step $\boldsymbol{A}^t$ is expressed as follows:

$$g = sgn(\nabla_{\boldsymbol{M}^{t-1}} \mathcal{L}^{ae}(\tilde{\boldsymbol{A}}^{t-1}, \boldsymbol{X})) \tag{19}$$

$$\boldsymbol{M}^t = \prod_{\mathcal{B}(M, \epsilon)} (\boldsymbol{M}^{t-1} + \alpha \cdot g) \tag{20}$$

$$\boldsymbol{A}^t = \tilde{\boldsymbol{A}}^t = \boldsymbol{A} \odot \boldsymbol{M}^t. \tag{21}$$

4.2   Adversarial Training in Variational Graph Autoencoder

Similarly to GAE, we reformulate the loss function for training VGAE (Eq. 6) as follows:

$$\min_{\theta} \mathcal{L}^{vae} + \lambda \cdot \mathbf{KL}(P(\boldsymbol{Z}|\boldsymbol{A}', \boldsymbol{X}') \| P(\boldsymbol{Z}|\boldsymbol{A}, \boldsymbol{X})) \tag{22}$$

$$\boldsymbol{X}' = arg \max_{\|\boldsymbol{X}'-\boldsymbol{X}\| \le \epsilon} \mathcal{L}^{vae}(\boldsymbol{A}, \boldsymbol{X}), \ \boldsymbol{A}' = arg \max_{\|\boldsymbol{A}'-\boldsymbol{A}\| \le \epsilon} \mathcal{L}^{vae}(\boldsymbol{A}, \boldsymbol{X}) \tag{23}$$

We generate $\boldsymbol{A}'$ and $\boldsymbol{X}'$ exactly the same way as with GAE (replacing $\mathcal{L}^{ae}$ with $\mathcal{L}^{vae}$ in Eq. 10–21.)

For convenience, we abbreviate L_2 and L_∞ adversarial training as AT-2 and AT-Linf respectively in the following tables and figures where L_2/L_∞ denote both attributes and adjacency matrix perturbation are bounded by L_2/L_∞ norm ball.

In practice, we train models by alternatively adding adjacency matrix perturbation and attributes perturbation[1].

5   Experiments

In this section, we present the results of the performance evaluation of L_2 and L_∞ adversarial training under three main applications of GAE and VGAE: link prediction, node clustering, and graph anomaly detection. Then we conduct parameter analysis experiments to explore which factors influence the performance.

Datasets. We used six real-world datasets: Cora, Citeseer and PubMed for link prediction and node clustering tasks, and BlogCatalog, ACM and Flickr for the

---

[1] We find that optimizing models by alternatively adding these two perturbation is better than adding these two perturbation together (See Appendix).

graph anomaly detection task. The detailed descriptions of the six datasets are showed in Table 1.

Model Architecture. All our experiments are based on the GAE/VGAE model where the encoder/inference model is consisted with a two-layer GCN by default.

Table 1. Datasets descriptions.

DataSets	Cora	Citeseer	PubMed	BlogCatalog	ACM	Flickr
#Nodes	2708	3327	19717	5196	16484	7575
#Links	5429	4732	44338	171743	71980	239738
#Features	1433	3703	500	8189	8337	12074

5.1 Link Prediction

Metrics. Following [10], we use the area under a receiver operating characteristic curve (AUC) and average precision (AP) as the evaluation metric. We conduct 30 repeat experiments with random splitting datasets into 85%, 5% and 10% for training sets, validation sets and test sets respectively. We report the mean and standard deviation values on test sets.

Parameter Settings. We train models on Cora and Citeseer datasets with 600 epochs, and PubMed with 800 epochs. All models are optimized with Adam optimizer and 0.01 learning rate. The λ is set to 4. For attributes perturbation, the ϵ is set to 3e−1 and 1e−3 on Citeseer and Cora, 1 and 5e−3 on PubMed for L_2 and L_∞ adversarial training respectively. For adjacency matrix perturbation, the ϵ is set to 1e−3 and 1e−1 on Citeseer and Cora, and 1e−3 and 3e−1 on PubMed for L_2 and L_∞ adversarial training respectively. The steps T is set to 1. The α is set to $\frac{\epsilon}{T}$.

For standard training GAE and VGAE, we run the official Pytorch geometric code[2] with 600 epochs for Citeseer and Cora datasets, 1000 epochs[3] for PubMed dataset. Other parameters are set the same as in [10].

Experimental Results. The results are showed in Table 2. It can be seen that both L_2 and L_∞ Adversarial trained GAE and VGAE models consistently boost their performance for both AUC and AP metrics on Cora, Citeseer and PubMed datasets. Specifically, the improvements on Cora and Citeseer dataset reaches at least 2% for both GAE and VGAE (Table 2). The improvements on PubMed is relative small with around 0.3%.

---

[2] https://github.com/rusty1s/pytorch_geometric/blob/master/examples/autoencode r.py.

[3] Considering PubMed is big graph data, we use more epochs in order to avoiding underfitting.

Table 2. Results for link prediction.

Methods	Cora		Citeseer		PubMed	
	AUC (in%)	AP (in%)	AUC (in%)	AP (in%)	AUC (in%)	AP (in%)
GAE	90.6 ± 0.9	91.2 ± 1.0	88.0 ± 1.2	89.2 ± 1.0	96.8 ± 0.2	97.1 ± 0.2
AT-L2-GAE	$\mathbf{93.0 \pm 0.9}$	$\mathbf{93.5 \pm 0.6}$	$\mathbf{92.5 \pm 0.7}$	$\mathbf{93.2 \pm 0.6}$	$\mathbf{97.2 \pm 0.2}$	$\mathbf{97.4 \pm 0.2}$
AT-Linf-GAE	92.8 ± 1.1	93.4 ± 1.0	92.3 ± 0.9	92.6 ± 1.1	96.9 ± 0.2	97.3 ± 0.2
VGAE	89.8 ± 0.9	90.3 ± 1.0	86.6 ± 1.4	87.6 ± 1.3	96.2 ± 0.4	96.3 ± 0.4
AT-L2-VGAE	$\mathbf{92.8 \pm 0.6}$	$\mathbf{93.1 \pm 0.6}$	90.7 ± 1.1	91.1 ± 0.9	$\mathbf{96.6 \pm 0.2}$	$\mathbf{96.7 \pm 0.2}$
AT-Linf-VGAE	92.2 ± 1.2	92.3 ± 1.3	$\mathbf{91.9 \pm 0.8}$	$\mathbf{92.0 \pm 0.6}$	96.5 ± 0.2	96.6 ± 0.3

5.2   Node Clustering

Metrics. Following [12,30], we use accuracy (ACC), normalized mutual information (NMI), precision, F-score (F1) and average rand index (ARI) as our evaluation metrics. We conduct 10 repeat experiments. For each experiment, datasets are random split into training sets (85% edges), validation sets (5% edges) and test sets (10% edges). We report the mean and standard deviation values on test sets.

Parameter Settings. We train GAE models on Cora and Citeseer datasets with 400 epochs, and PubMed dataset with 800 epochs. We train VGAE models on Cora and Citeseer datasets with 600 epochs and PubMed dataset with 800 epochs. All models are optimized by Adam optimizer with 0.01 learning rate. The λ is set to 4. For attributes perturbation, the ϵ is set to 5e−1 and 1e−3 on both Cora and Citeseer dataset, and 1 and 5e−3 on PubMed dataset for L_2 and L_∞ adversarial training respectively. For adjacency matrix perturbation, the ϵ is set to 1e−3 and 1e−1 on Cora and CiteSeer, 1e−3 and 3e−1 on PubMed for L_2 and L_∞ adversarial training respectively. The steps T is set to 1. The α is set to $\frac{\epsilon}{T}$.

Likewise, for standard GAE and VGAE, we run the official Pytorch geometric code with 400 epochs for Citeseer and Cora datasets, 800 epochs for PubMed dataset.

Experimental Results. The results are showed in Table 3, Table 4 and Table 5. It can be seen that both L_2 and L_∞ adversarial trained models consistently outperform the standard trained models for all metrics. In particular, on Cora and Citeseer datasets, both L_2 and L_∞ adversarial training improve the performance with large margin for all metrics, i.e. at least +5.4% for GAE, +6.7% for VGAE on Cora dataset (Table 3), and at least +5.8% for GAE, +5.6% for VGAE on Citeseer dataset (Table 4).

5.3   Graph Anomaly Detection

We exactly follow [4] to conduct experiments for graph anomaly detection. In [4], the authors take reconstruction errors of attributes and links as the anomaly

Table 3. Results for node clustering on cora.

Methods	Acc (in%)	NMI (in%)	F1 (in%)	Precision (in%)	ARI (in%)
GAE	61.6 ± 3.4	44.9 ± 2.3	60.8 ± 3.4	62.5 ± 3.5	37.2 ± 3.2
AT-L2-GAE	67.0 ± 3.0	50.8 ± 1.7	66.6 ± 1.7	69.4 ± 1.7	$\mathbf{44.1 \pm 4.1}$
AT-Linf-GAE	$\mathbf{67.1 \pm 3.8}$	$\mathbf{51.4 \pm 1.9}$	$\mathbf{67.5 \pm 2.8}$	$\mathbf{70.7 \pm 2.2}$	43.4 ± 4.3
VGAE	58.7 ± 2.7	42.3 ± 2.2	57.3 ± 3.2	58.8 ± 3.5	34.6 ± 2.8
AT-L2-VGAE	$\mathbf{67.3 \pm 3.8}$	$\mathbf{50.5 \pm 2.1}$	$\mathbf{66.1 \pm 4.1}$	$\mathbf{67.5 \pm 3.8}$	$\mathbf{44.3 \pm 3.3}$
AT-Linf-VGAE	65.4 ± 2.3	49.5 ± 1.6	64.0 ± 2.3	65.8 ± 3.0	42.9 ± 2.8

Table 4. Results for node clustering on citeseer.

Methods	Acc (in%)	NMI (in%)	F1 (in%)	Precision (in%)	ARI (in%)
GAE	51.8 ± 2.6	28.0 ± 1.9	50.6 ± 3.1	55.1 ± 3.1	22.8 ± 2.3
AT-L2-GAE	$\mathbf{61.6 \pm 2.3}$	36.3 ± 1.4	$\mathbf{58.8 \pm 2.1}$	60.9 ± 1.4	$\mathbf{34.6 \pm 2.3}$
AT-Linf-GAE	60.2 ± 2.8	$\mathbf{38.0 \pm 2.3}$	57.0 ± 2.7	$\mathbf{61.1 \pm 1.6}$	34.1 ± 3.4
VGAE	53.6 ± 3.5	28.4 ± 3.3	51.1 ± 3.8	53.2 ± 4.1	26.1 ± 3.5
AT-L2-VGAE	59.2 ± 2.3	35.1 ± 2.3	57.3 ± 2.3	60.4 ± 3.1	33.0 ± 2.4
AT-Linf-VGAE	$\mathbf{60.4 \pm 1.5}$	$\mathbf{36.5 \pm 1.4}$	$\mathbf{58.2 \pm 1.4}$	$\mathbf{61.1 \pm 1.4}$	$\mathbf{34.7 \pm 2.0}$

Table 5. Results for node clustering on PubMed.

Methods	Acc (in%)	NMI (in%)	F1 (in%)	Precision (in%)	ARI (in%)
GAE	66.2 ± 2.0	27.9 ± 3.7	65.0 ± 2.3	68.8 ± 2.2	27.1 ± 3.3
AT-L2-GAE	67.5 ± 2.9	30.4 ± 5	66.7 ± 3.3	70.2 ± 3.1	28.9 ± 4.8
AT-Linf-GAE	$\mathbf{68.4 \pm 1.6}$	31.9 ± 3.2	$\mathbf{67.7 \pm 1.9}$	$\mathbf{70.9 \pm 1.8}$	$\mathbf{30.2 \pm 2.8}$
VGAE	67.5 ± 2.0	29.4 ± 3.2	66.5 ± 2.2	69.9 ± 2.2	28.4 ± 3.2
AT-L2-VGAE	$\mathbf{69.8 \pm 2.0}$	$\mathbf{33.2 \pm 3.4}$	$\mathbf{69.4 \pm 2.3}$	$\mathbf{71.7 \pm 2.5}$	$\mathbf{32.5 \pm 3.2}$
AT-Linf-VGAE	68.5 ± 1.2	30.7 ± 2.5	67.4 ± 1.5	70.1 ± 1.5	30.4 ± 2.0

scores. Specifically, the node with larger scores are more likely to be considered as anomalies.

Model Architecture. Different from link prediction and node clustering, the model architecture in graph anomaly detection not only contains structure reconstruction decoder, i.e. link reconstruction, but also contains attribute reconstruction decoder. We adopt the same model architecture as in the official code of [4] where the encoder is consisted of two GCN layers, and the decoder of structure reconstruction decoder is consisted of a GCN layer and a InnerProduction layer, and the decoder of attributes reconstruction decoder is consisted of two GCN layers.

Metrics. Following [4,13], we use the area under the receiver operating characteristic curve (ROC-AUC) as the evaluation metric.

Parameter Settings. We set the α in anomaly scores to 0.5 where it balances the structure reconstruction errors and attributes reconstruction errors. We train the GAE model on Flickr, BlogCatalog and ACM datasets with 300 epochs. We set λ to 5. For adjacency matrix perturbation, we set ϵ to 3e−1, 5e−5 on both BlogCatalog and ACM datasets, 1e−3 and 1e−6 on Flickr dataset for L_∞ and L_2 adversarial training respectively. For attributes perturbations, we set ϵ to 1e−3 on BlogCatalog for both L_∞ and L_2 adversarial training, 1e−3 and 1e−2 on ACM for L_∞ and L_2 adversarial training respectively, 5e−1 and 3e−1 on Flickr for L_∞ and L_2 adversarial training respectively. We set steps T to 1 and the α to $\frac{\epsilon}{T}$

Anomaly Generation. Following [4], we inject two kinds of anomaly by perturbing structure and nodal attributes respectively:

- Structure anomalies. We randomly select s nodes from the network and then make those nodes fully connected, and then all the s nodes forming the clique are labeled as anomalies. t cliques are generated repeatedly and totally there are $s \times t$ structural anomalies.
- Attribute anomalies. We first randomly select $s \times t$ nodes as the attribute perturbation candidates. For each selected node v_i, we randomly select another k nodes from the network and calculate the Euclidean distance between v_i and all the k nodes. Then the node with largest distance is selected as v_j and the attributes of node v_j is changed to the attributes of v_i.

In this experiments, we set $s = 15$ and $t = 10, 15, 20$ for BlogCatalog, Flickr and ACM respectively which are the same to [4,13].

Experimental Results. From Table 6, it can be seen that both L_2 and L_∞ adversarial training boost the performance in detecting anomalous nodes. Since adversarial training tend to learn feature representations that are less sensitive to perturbations in the inputs, we conjecture that the adversarial trained node embeddings are less influenced by the anomalous nodes, which helps the graph anomaly detection. A similar claim are also made in image domain [15] where they demonstrate adversarial training of autoencoders are beneficial to novelty detection.

Table 6. Results w.r.t. AUC (in%) for graph anomaly detection.

Methods	Flickr	BlogCatalog	ACM
GAE	80.2 ± 1.3	82.9 ± 0.3	72.5 ± 0.6
AT-L2-GAE	$\mathbf{84.9 \pm 0.2}$	$\mathbf{84.7 \pm 1.4}$	74.2 ± 1.7
AT-Linf-GAE	81.1 ± 1.1	82.8 ± 1.3	$\mathbf{75.3 \pm 0.9}$

6   Understanding Adversarial Training

In this section, we explore the impact of three hyper-parameters on the performance of GAE and VGAE with adversarial training, i.e. the ϵ, λ and T in generating A' and X'. These three hyper-parameters are commonly considered to control the strength of regularization for adversarial robustness [33]. Besides, we explore the relationship between the improvements achieved by adversarial training and node degree.

6.1   The Impact of ϵ

The experiments are conducted on link prediction and node clustering tasks based on Cora dataset. We fix ϵ to 5e−1 and 1e−3 on adjacency matrix perturbation for L_∞ and L_2 adversarial training respectively when vary ϵ on attributes perturbation. We fix ϵ to 1e−3 and 3e−1 on attributes perturbation for L_∞ and L_2 adversarial training respectively when vary ϵ on adjacency matrix perturbation.

The results are showed in Fig. 1. From Fig. 1, we can see that the performance are less sensitive to adjacency matrix perturbation and more sensitive to attributes perturbation. Besides, it can be seen that there is an increase and then a decrease trend when increasing ϵ for attributes perturbation. We conjecture that it is because too large perturbation on attributes may destroy useful information in attributes. Therefore, it is necessary to carefully adapt the perturbation magnitude ϵ when we apply adversarial training for improving the generalization of model.

6.2   The Impact of T

The experiments are conducted on link prediction and node clustering tasks based on Cora dataset. For L_2 adversarial training, we set ϵ to 1e−3 and 5e−1 for adjacency matrix perturbation and attributes perturbation respectively. For L_∞ adversarial training, we set ϵ to 1e−1 and 1e−3 for adjacency matrix perturbation and attributes perturbation respectively. We set λ to 4.

Results are showed in Fig. 2. From Fig. 2, we can see that there is a slightly drop on both link prediction and node clustering tasks when increasing T from 2 to 4, which implies that a big T is not helpful to improve the generalization of node embeddings learned by GAE and VGAE. We suggest that one step is good choice for generating adjacency matrix perturbation and attributes perturbation in both L_2 and L_∞ adversarial training.

6.3   The Impact of λ

The experiments are conducted on link prediction and node clustering task based on Cora dataset. Likewise, for L_2 adversarial training, ϵ is set to 1e−3 and 5e−1 for adjacency matrix perturbation and attributes perturbation respectively. For L_∞ adversarial training, ϵ is set to 1e−1 and 1e−3 for adjacency matrix perturbation and attributes perturbation respectively. T is set to 1.

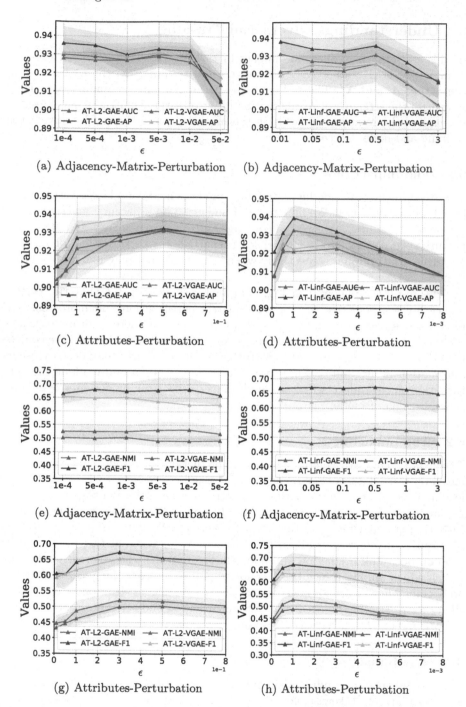

(a) Adjacency-Matrix-Perturbation   (b) Adjacency-Matrix-Perturbation

(c) Attributes-Perturbation   (d) Attributes-Perturbation

(e) Adjacency-Matrix-Perturbation   (f) Adjacency-Matrix-Perturbation

(g) Attributes-Perturbation   (h) Attributes-Perturbation

Fig. 1. The impact of ϵ in adjacency matrix perturbation and attributes perturbation. (a)–(d) show AUC/AP values for link prediction task and (e)–(h) show NMI/F1 values for node clustering task. Dots denote mean values with 30 repeated runs.

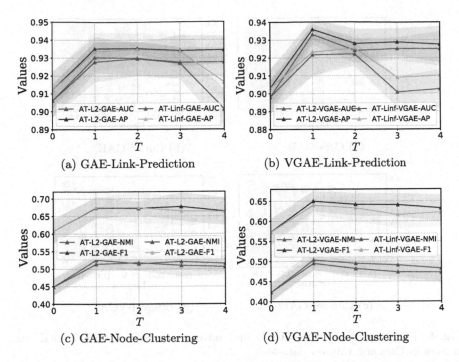

Fig. 2. The impact of steps T. Dots denote mean AUC/AP values for link prediction task and mean NMI/F1 values for node clustering task.

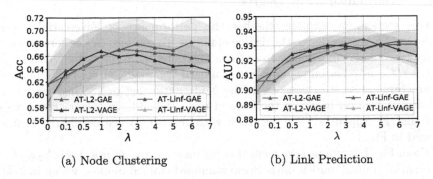

Fig. 3. The impact of λ. λ is varied from 0 to 7. Dots denote mean Acc for node clustering task and mean AUC for link prediction task. Experiments are conducted with 30 repeated runs.

Results are showed in Fig. 3. From Fig. 3, it can be seen that there is a significant increasing trend with the increase of λ, which indicates the effectiveness of both L_2 and L_∞ adversarial training in improving the generalization of GAE and VGAE. Besides, we also notice that a too large λ is not necessary and may lead to a negative effect in generalization of GAE and VGAE.

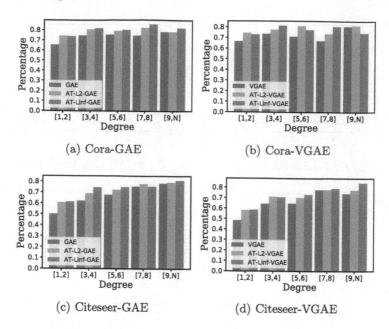

(a) Cora-GAE   (b) Cora-VGAE

(c) Citeseer-GAE   (d) Citeseer-VGAE

Fig. 4. Performance of GAE/VGAE and adversarial trained GAE/VGAE w.r.t. degrees in Cora and Citeseer datasets.

6.4   Performance w.r.t. Degree

In this section, we explore whether the performance of adversarial trained GAE/VGAE is sensitive to the degree of nodes. To conduct this experiments, we firstly learn node embeddings from Cora and Citeseer datasets by GAE/VGAE with L_2/L_∞ adversarial training and standard training respectively. The hyper-parameters are set the same as in the Node clustering task. Then we build a linear classification based on the learned node embeddings. Their accuracy can be found in Appendix. The accuracy with respect to degree distribution are showed in Fig. 4.

From Fig. 4, it can be seen seem that for most degree groups, both L_2 and L_∞ adversarial trained models outperform standard trained models, which indicates that both L_2 and L_∞ adversarial training improve the generalization of GAE and VGAE with different degrees. However, we also notice that adversarial training does not achieve a significant improvement on [9, N] group. We conjecture that it is because node embeddings with very large degrees already achieve a high generalization.

7   Conclusion

In this paper, we formulated L_2 and L_∞ adversarial training for GAE and VGAE, and studied their impact on the generalization performance. We conducted experiments on link prediction, node clustering and graph anomaly detection tasks.

The results show that both L_2 and L_∞ adversarial trained GAE and VGAE outperform GAE and VGAE with standard training. This indicates that L_2 and L_∞ adversarial training improve the generalization of GAE and VGAE. Besides, we showed that the generalization performance achieved by the L_2 and L_∞ adversarial training is more sensitive to attributes perturbation than adjacency matrix perturbation, and not sensitive to node degree. In addition, the parameter analysis suggest that a too large λ, ϵ and T would lead to a negative effect on the performance w.r.t. generalization.

References

1. Chen, J., Wu, Y., Lin, X., Xuan, Q.: Can adversarial network attack be defended? arXiv preprint arXiv:1903.05994 (2019)
2. Dai, Q., Shen, X., Zhang, L., Li, Q., Wang, D.: Adversarial training methods for network embedding. In: The World Wide Web Conference, pp. 329–339 (2019)
3. Deng, Z., Dong, Y., Zhu, J.: Batch virtual adversarial training for graph convolutional networks. arXiv preprint arXiv:1902.09192 (2019)
4. Ding, K., Li, J., Bhanushali, R., Liu, H.: Deep anomaly detection on attributed networks. In: Proceedings of the 2019 SIAM International Conference on Data Mining, pp. 594–602. SIAM (2019)
5. Feng, F., He, X., Tang, J., Chua, T.S.: Graph adversarial training: dynamically regularizing based on graph structure. IEEE Trans. Knowl. Data Eng. **PP**, 1 (2019)
6. Goodfellow, I.J., Shlens, J., Szegedy, C.: Explaining and harnessing adversarial examples. arXiv preprint arXiv:1412.6572 (2014)
7. Grover, A., Leskovec, J.: node2vec: scalable feature learning for networks. In: Proceedings of the 22nd ACM SIGKDD International Conference on Knowledge Discovery and Data Mining, pp. 855–864 (2016)
8. Jin, H., Zhang, X.: Latent adversarial training of graph convolution networks. In: ICML Workshop on Learning and Reasoning with Graph-Structured Representations (2019)
9. Kipf, T.N., Welling, M.: Semi-supervised classification with graph convolutional networks. arXiv preprint arXiv:1609.02907 (2016)
10. Kipf, T.N., Welling, M.: Variational graph auto-encoders. arXiv preprint arXiv:1611.07308 (2016)
11. Madry, A., Makelov, A., Schmidt, L., Tsipras, D., Vladu, A.: Towards deep learning models resistant to adversarial attacks. arXiv preprint arXiv:1706.06083 (2017)
12. Pan, S., Hu, R., Long, G., Jiang, J., Yao, L., Zhang, C.: Adversarially regularized graph autoencoder for graph embedding. arXiv preprint arXiv:1802.04407 (2018)
13. Pei, Y., Huang, T., van Ipenburg, W., Pechenizkiy, M.: ResGCN: attention-based deep residual modeling for anomaly detection on attributed networks. arXiv preprint arXiv:2009.14738 (2020)
14. Perozzi, B., Al-Rfou, R., Skiena, S.: DeepWalk: online learning of social representations. In: Proceedings of the 20th ACM SIGKDD International Conference on Knowledge Discovery and Data Mining, pp. 701–710 (2014)
15. Salehi, M., ARAE: adversarially robust training of autoencoders improves novelty detection. arXiv preprint arXiv:2003.05669 (2020)
16. Salha, G., Hennequin, R., Vazirgiannis, M.: Simple and effective graph autoencoders with one-hop linear models. arXiv preprint arXiv:2001.07614 (2020)

17. Salha, G., Limnios, S., Hennequin, R., Tran, V.A., Vazirgiannis, M.: Gravity-inspired graph autoencoders for directed link prediction. In: Proceedings of the 28th ACM International Conference on Information and Knowledge Management, pp. 589–598 (2019)
18. Salman, H., Ilyas, A., Engstrom, L., Kapoor, A., Madry, A.: Do adversarially robust ImageNet models transfer better? arXiv preprint arXiv:2007.08489 (2020)
19. Schlichtkrull, M., Kipf, T.N., Bloem, P., van den Berg, R., Titov, I., Welling, M.: Modeling relational data with graph convolutional networks. In: Gangemi, A., Navigli, R., Vidal, M.-E., Hitzler, P., Troncy, R., Hollink, L., Tordai, A., Alam, M. (eds.) ESWC 2018. LNCS, vol. 10843, pp. 593–607. Springer, Cham (2018). https://doi.org/10.1007/978-3-319-93417-4_38
20. Shi, H., Fan, H., Kwok, J.T.: Effective decoding in graph auto-encoder using triadic closure. In: Proceedings of the AAAI Conference on Artificial Intelligence, vol. 34, pp. 906–913 (2020)
21. Stutz, D., Hein, M., Schiele, B.: Disentangling adversarial robustness and generalization. In: Proceedings of the IEEE/CVF Conference on Computer Vision and Pattern Recognition, pp. 6976–6987 (2019)
22. Sun, K., Lin, Z., Guo, H., Zhu, Z.: Virtual adversarial training on graph convolutional networks in node classification. In: Lin, Z., et al. (eds.) PRCV 2019. LNCS, vol. 11857, pp. 431–443. Springer, Cham (2019). https://doi.org/10.1007/978-3-030-31654-9_37
23. Tang, J., Qu, M., Wang, M., Zhang, M., Yan, J., Mei, Q.: Line: large-scale information network embedding. In: Proceedings of the 24th International Conference on World Wide Web, pp. 1067–1077 (2015)
24. Tian, F., Gao, B., Cui, Q., Chen, E., Liu, T.Y.: Learning deep representations for graph clustering. In: Proceedings of the AAAI Conference on Artificial Intelligence, vol. 28 (2014)
25. Tsipras, D., Santurkar, S., Engstrom, L., Turner, A., Madry, A.: Robustness may be at odds with accuracy. arXiv preprint arXiv:1805.12152 (2018)
26. Utrera, F., Kravitz, E., Erichson, N.B., Khanna, R., Mahoney, M.W.: Adversarially-trained deep nets transfer better. arXiv preprint arXiv:2007.05869 (2020)
27. Wang, D., Cui, P., Zhu, W.: Structural deep network embedding. In: Proceedings of the 22nd ACM SIGKDD International Conference on Knowledge Discovery and Data Mining, pp. 1225–1234 (2016)
28. Wang, X., Liu, X., Hsieh, C.J.: GraphDefense: towards robust graph convolutional networks. arXiv preprint arXiv:1911.04429 (2019)
29. Wang, Y., Zou, D., Yi, J., Bailey, J., Ma, X., Gu, Q.: Improving adversarial robustness requires revisiting misclassified examples. In: International Conference on Learning Representations (2019)
30. Xia, R., Pan, Y., Du, L., Yin, J.: Robust multi-view spectral clustering via low-rank and sparse decomposition. In: Proceedings of the AAAI Conference on Artificial Intelligence, vol. 28 (2014)
31. Xu, K., et al.: Topology attack and defense for graph neural networks: an optimization perspective. arXiv preprint arXiv:1906.04214 (2019)
32. Yu, D., Zhang, R., Jiang, Z., Wu, Y., Yang, Y.: Graph-revised convolutional network. arXiv preprint arXiv:1911.07123 (2019)
33. Zhang, H., Yu, Y., Jiao, J., Xing, E., El Ghaoui, L., Jordan, M.: Theoretically principled trade-off between robustness and accuracy. In: International Conference on Machine Learning, pp. 7472–7482. PMLR (2019)

Inductive Link Prediction
with Interactive Structure Learning
on Attributed Graph

Shuo Yang, Binbin Hu, Zhiqiang Zhang, Wang Sun, Yang Wang, Jun Zhou[✉],
Hongyu Shan, Yuetian Cao, Borui Ye, Yanming Fang, and Quan Yu

Ant Group, Hangzhou, China
{kexi.ys,bin.hbb,lingyao.zzq,sunwang.sw,zuoxu.wy,jun.zhoujun,
xinzong.shy,yuetian.cyt,borui.ybr,yanming.fym,jingmin.yq}@antgroup.com

Abstract. Link prediction is one of the most important tasks in graph machine learning, which aims at predicting whether two nodes in a network have an edge. Real-world graphs typically contain abundant node and edge attributes, thus how to perform link prediction by simultaneously learning structure and attribute information from both interactions/paths between two associated nodes and local neighborhood among node's ego subgraph is intractable.

To address this issue, we develop a novel **P**ath-aware **G**raph **N**eural **N**etwork (PaGNN) method for link prediction, which incorporates interaction and neighborhood information into graph neural networks via broadcasting and aggregating operations. And a cache strategy is developed to accelerate the inference process. Extensive experiments show a superior performance of our proposal over state-of-the-art methods on real-world link prediction tasks.

1 Introduction

Graph-structured data are ubiquitous in a variety of real-world scenarios, e.g., social networks, protein-protein interactions, supply chains, and so on. As one of the most common and important tasks of graph mining, *link prediction*, which aims at predicting the existence of edges connecting a pair of nodes in a graph, has become an impressive way to solve various crucial problems such as friend recommendation [24,26], supply chain mining [29], entity interactions prediction [23,28], and knowledge graph completion [7,17,30].

In general, current researches towards link prediction can be categorized into three lines: heuristic methods, network embedding based methods and graph neural network based methods. *Heuristic methods* [2,13] focus on estimate the likelihood of the edge through different heuristic similarities between nodes under certain assumptions, which, unfortunately, may fail when their assumptions do not hold true in the targeted scenario [11]. *Network Embedding* (NE) based methods [3] learn node representation with context (e.g., random walk [5,14] or neighborhood [22]) in the graph, followed by a well-trained classifier for link

© Springer Nature Switzerland AG 2021
N. Oliver et al. (Eds.): ECML PKDD 2021, LNAI 12976, pp. 383–398, 2021.
https://doi.org/10.1007/978-3-030-86520-7_24

prediction. However, most of them fail to take into account the rich attributes of nodes and edges when learning node representation, so they cannot obtain better performance and are not suitable for inductive link prediction. By making full use of the structure and attribute information in an inductive manner, the emerging *Graph Neural Network* (GNN) based methods [25] achieve the state-of-the-art performance in link prediction [12].

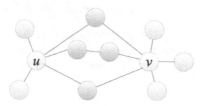

Fig. 1. Node roles for link prediction. (Color figure online)

Nevertheless, we believe that the effectiveness and efficiency of current GNN-based methods for link prediction are still unsatisfactory. Current GNN-based methods can be categorised into two lines: node-centric and edge-centric. *Node-centric* GNN-based methods [32] learn representations of two targeted nodes via certain GNN architecture independently, followed by a pairwise prediction function (e.g., MLP, dot product, etc.). Such a two-tower architecture is good at modeling the surrounding context centered at each targeted node, but fails to perceive **interactions** (or **paths**, red nodes in Fig. 1) between two targeted nodes (yellow nodes in Fig. 1), which are essential for the *effectiveness* of some real-world link prediction scenarios. Recently, several *edge-centric* GNN-based methods propose to adopt a different technique (e.g., node labeling function [34], between-node path reachability [27], enclosing-subgraph level pooling [20], meta-graph [35], etc.) to model such interactions or paths to some extent, and achieve better performance. However, on the one hand, they still do not explicitly integrate the structure and attribute information of interactions between targeted nodes. For example, the node labeling function [20,34] only models the structure information to some extent, while the between-node path reachability [27] only estimates reachable probability from one node to another via random walk. Both of them neglect the abundant *attributes* of interactions/paths. On the other hand, all of the above approaches are time-consuming in the training or inference phase, thus the *efficiency* becomes a great challenge when scaling up to industrial graphs with billions of nodes and tens of billions edges. For example, Graph learning framework [27] computes the reachability via random walks, which need to be repeated many times until convergence, and this process also has to be redone for any newly emerging edge. Therefore, it is time-consuming when performing inference in huge graphs.

Here, we summarize the challenges facing by current link prediction methods in three aspects:

- *Integrating the structure and attribute information of interaction.* Since most real-world graphs contain abundant node and edge attributes, and most scenarios can benefit from such information, the link prediction model should be able to subtly model the structure information and attribute information simultaneously.
- *Scalability and Efficiency.* Existing works that model interactions by calculating high-order information are usually time-consuming. How to scale up and become more efficient is another challenge when performing over real-world huge graphs.
- *Inductive ability.* In real-world graphs, nodes and edges may emerge at any time. To handle such newly emerging nodes and edges requires the inductive ability of the link prediction model.

Addressing the above challenges, we propose a **P**ath-aware **G**raph **N**eural **N**etwork (PaGNN) towards link prediction. Considering the motivation of both node-centric and edge-centric methods, PaGNN jointly learns the structure and attribute information from both interactions/paths between two targeted nodes (edge-centric) and the local neighborhood of each targeted node (node-centric), through the novel *broadcasting* and *aggregation* operation. The broadcasting operation responds to "send" information from one targeted node to all other nodes in its local neighborhood and generate the *broadcasted embeddings*, while the aggregation operation aims to aggregate information (including to "receive" the broadcasted embeddings) for another targeted node from its local neighborhood and generate its final embedding. Note that the destination node can perceive all paths connecting two targeted nodes via aggregating broadcasted embeddings. Thus, such a broadcasting and aggregation operation can explicitly integrate structure and attribute information of interactions and local neighborhood of targeted nodes. In addition, addressing the poor scalability and efficiency of edge-centric methods in the inference phase, we propose a *cache embedding strategy* that nearly doubles the speed of the inference phase. Note that by leveraging the native inductive power of GNNs, PaGNN can handle newly emerging nodes and edges naturally. At last, We conduct extensive experiments on several public datasets to demonstrate the effectiveness and efficiency of PaGNN compared with state-of-the-art baselines.

2   Model Formulation

In this section, we introduce the proposed path-aware graph neural network (PaGNN) model towards link prediction. First, we briefly exhibit the overall architecture of PaGNN. Then, we elaborate the detail implementation of PaGNN, including the broadcasting operator, the aggregation operation, edge representation learning, and at last the loss function.

2.1   Notations and Definitions

Before diving into PaGNN, we first give the definition of link prediction.

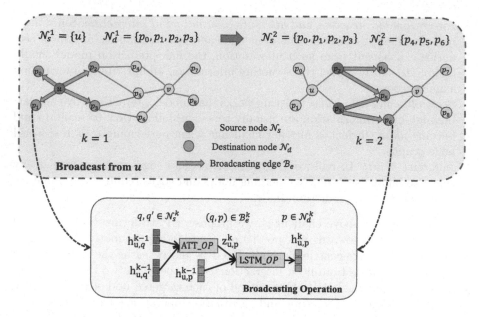

Fig. 2. Broadcasting operation

Definition 1. *Link Prediction.* *Given an attributed graph* $\mathcal{G} = (\mathcal{V}, \mathcal{E}, \mathbf{X})$, *where* $\mathcal{V}$ *is the set of nodes,* $\mathcal{E} \subseteq \mathcal{V} \times \mathcal{V}$ *is the set of observed edges, and* $\mathbf{X} \in R^{|\mathcal{V}| \times d}$ *consists of d-dimensional feature vectors of all nodes, as well as a set of labeled edges* $\mathcal{L} = \{(\langle u, v \rangle, y) | u, v \in \mathcal{V}, y \in \{0, 1\}\}$, $y = 1$ *denotes that there exists an edge between u and v (i.e.,* $(u, v) \in \mathcal{E}$), *otherwise* $y = 0$, *for an unlabeled edge set* $\mathcal{U} = \{(\langle u, v \rangle | u, v \in \mathcal{V}\}$, *the goal of link prediction is to predict the existence probability of edges in* $\mathcal{U}$.

2.2   Overview of PaGNN

PaGNN aims to learn the representations on the centralized subgraph of two associated nodes, Fig. 2 and 3 show the overall workflow of PaGNN. By leveraging the broadcasting and aggregation operations, PaGNN can model all interaction (i.e., paths) and neighborhood information between two associated nodes of the targeted edge, and generate their embeddings for link prediction. PaGNN first performs broadcasting and then aggregation operation on both two targeted nodes. More specifically, one node broadcasts information (called broadcasted embedding) to nodes in its local neighborhood, then the other node aggregates information from its neighborhood If there is any overlap between the broadcasted neighborhood of two nodes, the information from one node can be aggregated by the other node via the paths between them, which means two nodes can perceive each other. Therefore, the structure and attribute information of the interactions, as well as the local neighborhood, of the two nodes, can be subtly modeled through such broadcasting and aggregation operations, and encoded

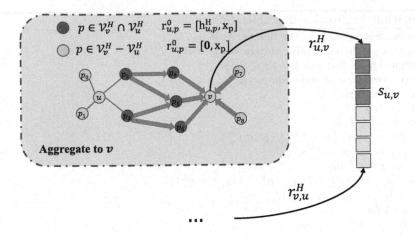

Fig. 3. Aggregation operation

into their final embeddings. At last, PaGNN simply employs the concatenation of final embeddings as the edge representation, and performs a MLP on it to do binary classification. Moreover, our framework is operated on the centralized subgraph of associated nodes within fixed steps, thus it naturally provides inductive ability.

2.3 Broadcasting Operation

The broadcasting operation aims to "send" a message from a node to other nodes in its local neighborhood. Formally, given a pair of nodes $\langle u, v \rangle$, the goal is to predict whether an edge exists between them. Without loss of generality, we set node u to broadcast information to other nodes within the H-hops ego-subgraph centered on u, denoted as $\mathcal{G}_u^H$, details of broadcasting operation are shown in Fig. 2.

Broadcasting operation is performed in a breadth first search (BFS) style that attempts to "spread out" from the source nodes. At each step, nodes (called *source nodes*) broadcast information to their directed neighbors (called *destination nodes*). Specifically, in the k-th step, we maintain a source node set $\mathcal{N}_s^k$, a destination node set $\mathcal{N}_d^k$ and a broadcasting edge set $\mathcal{B}_e^k$. $\mathcal{N}_s^k$ contains the nodes that will broadcast information (the initial $\mathcal{N}_s^1$ only contains node u), $\mathcal{N}_d^k$ consists of directed neighbors of nodes in $\mathcal{N}_s^k$, $\mathcal{B}_e^k$ are edges connecting nodes in $\mathcal{N}_s^k$ and $\mathcal{N}_d^k$, in other words, $\mathcal{B}_e^k = \{(q, p) | (q, p) \in \mathcal{E}_u^H, q \in \mathcal{N}_s^k, p \in \mathcal{N}_d^k\}$.

Each node in $\mathcal{N}_s^k$ first *broadcasts* information to their directed neighborhoods in destination nodes set $\mathcal{N}_d^k$ via the edges in $\mathcal{B}_e^k$. Next, each node in $\mathcal{N}_d^k$ *integrates* the broadcasted information together with its own embedding. Then, all destination nodes form the new $\mathcal{N}_s^{k+1}$ of the $k + 1$-th step. This process will be repeated H times until all nodes in $\mathcal{G}_u^H$ receive broadcasted information from u.

As shown in Algorithm 1, $\mathcal{V}_u^H$ and $\mathcal{E}_u^H$ are the node set and edge set in $\mathcal{G}_u^H$ respectively. $\mathbf{h}_{u,p}^k \in \mathbb{R}^d$ denotes the broadcasted information (called *broadcasted*

Algorithm 1. Broadcasting Operation

Require: Source node u, the number of hops H,
  node u's H-hop enclosing subgraph $\mathcal{G}_u^H = (\mathcal{V}_u^H, \mathcal{E}_u^H)$,
  original node features $\{\mathbf{x}_p \mid p \in \mathcal{V}_u^H\}$.
Ensure: broadcasted embeddings $\{\mathbf{h}_{u,p}^H \mid p \in \mathcal{V}_u^H\}$.
1: $\mathbf{h}_{u,p}^0 \leftarrow \mathbf{x}_p, \forall p \in \mathcal{V}_u^H$
2: $\mathcal{N}_s^1 \leftarrow \{u\}$
3: **for** $k = 1...H$ **do**
4:   $\mathcal{B}_e^k \leftarrow \{(q,p) \mid q \in \mathcal{N}_s^k, (q,p) \in \mathcal{E}_u^H\}$
5:   $\mathcal{N}_d^k \leftarrow \{p \mid (q,p) \in \mathcal{B}_e^H\}$
6:   **for** $p \in \mathcal{N}_d^k$ **do**
7:     $\mathbf{z}_{u,p}^k \leftarrow ATT\_OP(\mathbf{x}_p, \{\mathbf{h}_{u,q}^{k-1} \mid (q,p) \in \mathcal{B}_e^k\})$
8:     $\mathbf{h}_{u,p}^k \leftarrow LSTM\_OP(\mathbf{h}_{u,p}^{k-1}, \mathbf{z}_{u,p}^k)$
9:   **end for**
10:   **for** $p \in \mathcal{V}_u^H - \mathcal{N}_d^k$ **do**
11:     $\mathbf{h}_{u,p}^k \leftarrow \mathbf{h}_{u,p}^{k-1}$
12:   **end for**
13:   $\mathcal{N}_s^{k+1} \leftarrow \mathcal{N}_d^k$
14: **end for**

---

embeddings) that starts from u and ends with p at the k-th step. The information broadcasted from u to node p is initialized as p's original feature $\mathbf{x}_p$ ($\mathbf{h}_{u,p}^0$, line 1). At the k-th step, $\mathcal{B}_e^k$ is updated as the directed edges of $\mathcal{N}_s^k$ and $\mathcal{N}_d^k$ is extracted from $\mathcal{G}_u^H$, which are the directed neighbors of $\mathcal{N}_s^k$ (line 5). In order to handle the situation that different source nodes in $\mathcal{N}_s^k$ broadcast information to the same destination node, we first employ the attention mechanism (ATT_OP in line 7) to aggregate the broadcasted embedding, and then we employs a LSTM-like operator [6] (LSTM_OP in line 8) to combine embeddings of the $k-1$-th step and the k-th step. Note that the integrated embedding will become the propagated embedding of the next step. And for nodes in $\mathcal{V}_u^H$ but not in $\mathcal{N}_d^k$, the broadcasted embeddings stay the same as previous step (line 10 to 12). The attention operator ATT_OP of the k-th step is defined as:

$$\alpha_{q,p}^k = \frac{\exp(\mathbf{v}_\phi^{k\mathrm{T}} \sigma(\mathbf{W}_{\phi 1}^k [\mathbf{x}_p, \mathbf{h}_{u,q}^{k-1}]))}{\sum_{q' \in \mathcal{N}_s^k, (q',p) \in \mathcal{E}_u^H} \exp(\mathbf{v}_\phi^{k\mathrm{T}} \sigma(\mathbf{W}_{\phi 1}^k [\mathbf{x}_p, \mathbf{h}_{u,q'}^{k-1}]))}, \tag{1}$$

$$\mathbf{z}_{u,p}^k = \sigma(\mathbf{W}_{\phi 2}^k [\sum_{q \in \mathcal{N}_s^k, (q,p) \in \mathcal{E}_u^H} \alpha_{q,p}^k \mathbf{h}_{u,q}^{k-1}, \mathbf{x}_p]) \tag{2}$$

where $\alpha_{q,p}^k$ is the attention value, $\mathbf{z}_{q,p}^k$ is the intermedia embedding after attention calculation and the input of the next LSTM-like operator, $[\cdot, \cdot]$ denotes the concatenation of embeddings, $\mathbf{v}_\phi^k, \mathbf{W}_{\phi 1}^k, \mathbf{W}_{\phi 2}^k$ are learnable parameters. Next, the LSTM_OP integrates the information of the $k-1$-th step and the k-th step:

$$i_{u,p}^k = \sigma(\mathbf{W}_{\varphi i}[\mathbf{h}_{u,p}^{k-1}, \mathbf{z}_{u,p}^k]),$$

$$\mathbf{f}_{u,p}^k = \sigma(\mathbf{W}_{\varphi f}[\mathbf{h}_{u,p}^{k-1}, \mathbf{z}_{u,p}^k]),$$

$$\mathbf{c}_{u,p}^k = \mathbf{f}_{u,p}^{k-1} \odot \mathbf{c}_{u,p}^{k-1} + \mathbf{i}_{u,p}^k \odot \tanh(\mathbf{W}_{\varphi c}[\mathbf{h}_{u,p}^{k-1}, \mathbf{z}_{u,p}^k]),$$

$$\mathbf{o}_{u,p}^k = \sigma(\mathbf{W}_{\varphi o}[\mathbf{h}_{u,p}^{k-1}, \mathbf{z}_{u,p}^k]),$$

$$\mathbf{h}_{u,p}^k = \mathbf{o}_{u,p}^k \odot \tanh(\mathbf{c}_{u,p}^k),$$

where $\mathbf{i}_{u,p}, \mathbf{f}_{u,p}, \mathbf{o}_{u,p}$ are input gate, forget gate and output gate in LSTM respectively. $\mathbf{W}_{\varphi i}, \mathbf{W}_{\varphi f}, \mathbf{W}_{\varphi o}$ are learnable parameters and $\mathbf{c}_{u,p}$ is the cell state with $\mathbf{c}_{u,p}^0 = \mathbf{0}$. After H step, the final broadcasted embedding $\mathbf{h}_{u,p}^H$ is taken as the output which may contains the structure and attribute information from node u to node p.

2.4 Aggregation Operation

Different from conventional GCN-based model, the aggregation operation in PaGNN not only recursively aggregates neighbor attributes but also aims to "receive" the broadcasted embeddings from the other node. Suppose information is broadcasted from u and aggregated to v, we will introduce details of aggregation operation, as illustrated in Fig. 3.

First, node u broadcasts information to nodes among its centralized subgraph $\mathcal{G}_u^H$, then $\mathbf{h}_{u,p}^H, \forall p \in \mathcal{V}_u^H$ is obtained. Afterwards, nodes in v's centralized subgraph $\mathcal{G}_v^H$ aggregate broadcasted embeddings and initial node attribute to v. Before aggregation, initial embedding $\mathbf{r}_{u,p}^0$ is set to combination of its broadcasted embedding and initial node attributes. In particular, if there is no path between u and p, the broadcasted information is set to $\mathbf{0}$, which is defined as:

$$\mathbf{r}_{u,p}^0 = \begin{cases} [\mathbf{h}_{u,p}^H, \mathbf{x}_p], & p \in \mathcal{V}_v^H \cap \mathcal{V}_u^H \\ [\mathbf{0}, \mathbf{x}_p], & p \in \mathcal{V}_v^H - \mathcal{V}_u^H \end{cases}, \forall p \in \mathcal{V}_v^H \tag{3}$$

At last, for each node $p \in \mathcal{V}_v^H$, p aggregates information from its neighbors in a GCN style:

$$\mathbf{r}_{u,p}^k \leftarrow AGG(\mathbf{r}_{u,p}^{k-1}, \{\mathbf{r}_{u,i}^{k-1} \mid (i,p) \in \mathcal{E}_v^H\}), \forall p \in \mathcal{V}_v^H \tag{4}$$

where AGG is the aggregation function, $\mathbf{r}_{u,v}^H$ represents information that broadcasted from u and aggregated to v at H-th step is taken as output.

2.5 Edge Representation Learning

With the broadcasting and the aggregation operation mentioned above, edge $\langle u, v \rangle$ is represented from two ways. PaGNN first broadcasts information from u among $\mathcal{G}_u^H$ and aggregates information to v among $\mathcal{G}_v^H$, $\mathbf{r}_{u,v}^H$ is obtained. Meanwhile, we broadcast information from v among $\mathcal{G}_v^H$ and aggregate to u among

$\mathcal{G}_u^H$, $\mathbf{r}_{v,u}^H$ is obtained. Concatenation of two embeddings is taken as the edge representation:

$$\mathbf{s}_{u,v} = [\mathbf{r}_{u,v}^H, \mathbf{r}_{v,u}^H] \tag{5}$$

Loss Function. On the basis of edge representation, we employ a cross entropy loss based on the edge representation:

$$loss = -\frac{1}{|\mathcal{L}|} \sum_{(\langle u,v \rangle, y) \in \mathcal{L}} y \log(\hat{y}) + (1 - y) \log(1 - \hat{y}), \tag{6}$$

Where $\hat{y} = MLP(\mathbf{s}_{u,v})$ and $MLP(\cdot)$ is a multi-layer perception with two fully-connected layers.

2.6   Cache Strategy in Inference

As PaGNN is operated on subgraph of each candidate edge, it's more time-consuming than node-centric model. To address this, we design a cache mechanism to accelerate the inference.

For two associated nodes $\langle u, v \rangle$, since u only broadcasts information to neighbors among its ego subgraph $\mathcal{G}_u^H$, it's obviously that no matter where v is, the broadcasted embeddings of u are the same. In other words, different v doesn't affect any broadcasted embeddings of u. Base on this property, we can pre-calculate broadcasted embeddings $\mathbf{h}_{u,p}^H$ for u and cache it in storage. When u appears as a targeted node in inference, the cached broadcasted embeddings can be reused. As a result, only aggregation operation is necessary by leveraging cache strategy.

Time Complexity. For PaGNN, suppose information is broadcasted from u and aggregated to v, the time complexity is $O(H(|\mathcal{E}_u^H| + |\mathcal{V}_u^H|))$ for broadcasting and $O(H(|\mathcal{E}_v^H| + |\mathcal{V}_v^H|))$ for aggregating. Based on above analysis, the overall time complexity for training and inference stage is $O(H(|\mathcal{E}_u^H| + |\mathcal{V}_u^H| + |\mathcal{V}_v^H| + |\mathcal{V}_v^H|))$.

With the cache strategy, the inference time complexity decreases to $O(H(|\mathcal{E}_v^H| + |\mathcal{V}_v^H|))$, since only aggregation operation is required. As a result, our cache strategy has a nearly two times speed-up for PaGNN.

2.7   Summary

Comparing with other GNN models, PaGNN integrates the broadcasted embeddings into the aggregation process (Eq. 3 and 4). The broadcasting mechanism guarantees that information is broadcasted from a subset of nodes $\mathcal{N}_s$, and other nodes in $\mathcal{N}_d$ can only absorb information from nodes in the subset. This mechanism guarantees that we only aggregate the information from the path between two target nodes.

On one hand, nodes on the overlap of two ego-subgraphs ($p \in \mathcal{V}_v^H \cap \mathcal{V}_u^H$) not only receive broadcasted information from u, but also aggregate this information to v, these nodes are called "bridge nodes". And if $p \in \mathcal{V}_v^H - \mathcal{V}_u^H$, the "interaction information" is set to $\mathbf{0}$, indicating that node p isn't on any path between u and

v, which is also useful to label the node role. Therefore node v *integrates all the structure and attribute information from interaction* between u and v through the "bridge nodes". On the other hand, all nodes in $\mathcal{G}_v^H$ aggregate its attributes to v, thus *representation of node v also embeds the structure and attribute information from its local neighborhood*.

In summary, PaGNN jointly learns structure and attribute information from both interaction and local neighborhood. And since PaGNN is operated on subgraphs of associated nodes, it provides inductive ability.

3 Experiments

3.1 Experiment Setup

Data Protection Statement. (1) The data used in this research does not involve any **P**ersonal **I**dentifiable **I**nformation (PII). (2) The data used in this research were all processed by data abstraction and data encryption, and the researchers were unable to restore the original data. (3) Sufficient data protection was carried out during the process of experiments to prevent the data leakage and the data was destroyed after the experiments were finished. (4) The data is only used for academic research and sampled from the original data, therefore it does not represent any real business situation in Ant Financial Services Group.

Datasets. We adopt four real-world dataset from different domains to evaluate the effectiveness of our model, consisting of **Collab**[1] and **PubMeb** [31] from bibliographic domain, **Facebook**[2] from social domain and **SupChain** [29] from E-commerce domain. The statistics of these datasets are illustrated in Table 1.

Table 1. The statistics of the datasets.

Datasets	PubMed	Facebook	Collab	SupChain
Nodes	19.7K	4.0K	235.8K	23.4M
Edges	44.3K	66.2K	1.2M	103.2M
Node features	500	161	53	95

Baseline. We compare our proposal with following three categories of link prediction methods:

Heuristic Methods. Two heuristic methods are implemented: the **C**ommon **N**eighbors (CN) [13] and the Jaccard [19].

Network Embedding Methods. We include two network embedding methods, DeepWalk [14] and Node2vec [5]. Network embedding based methods are

---

[1] https://snap.stanford.edu/ogb/data/linkproppred.
[2] https://snap.stanford.edu/data/egonets-Facebook.html.

implemented based on open source code[3], which can not be applied to large scale graph, as a consequence, experiment on dataset *SupChain* is unable to be conducted.

Node-Centric GNN Methods. We take **PinSage** [32] as baseline, which encodes node information via a GNN model, and then predicts the edges based on a pairwise decoder (a MLP layer). In our experiment, **GCN** and **GAT** [21] models are chosen as encoders, which are represented as $\mathbf{PinSage}_{GCN}$, $\mathbf{PinSage}_{GAT}$.

Edge-Centric GNN Methods. We also compare our model with link prediction models that learn high order interactive structure on targeted edge's subgraph, such as SEAL [34], and GraIL [20]. Note that, SEAL and GraIL not only label node roles but also *integrate it with node attributes.*

Ablation Studies. To study the effectiveness of broadcasted embeddings, we remove it before information aggregation, i.e. Eq. 3 is set as $\mathbf{r}_{u,p}^0 = [\mathbf{0}, \mathbf{x}_p]$. This method is represented as $\mathbf{PaGNN}_{broadcast}$. We further quantitatively analyze whether only broadcasting information from one node is enough for predicting the edges, i.e., Eq. 5 is changed to $\mathbf{s}_{u,v} = \mathbf{r}_{u,v}^H$, it's represented as $\mathbf{PaGNN}_{two\_way}$. And in order to demonstrate effectiveness of LSTM_OP in broadcasting, we change LSTM_OP to concatenation, i.e. $\mathbf{h}_{u,p}^k = [\mathbf{h}_{u,p}^{k-1}, \mathbf{z}_{u,p}^k]$ in Algorithm 1 (line 8), and it is represented as $\mathbf{PaGNN}_{lstm}$.

Specifically, *GAT* is chosen as the aggregation function for all edge-centric GNN methods.

Parameter Setting. For all GNN models, we set the scale of the enclosing subgraph H to 2, embedding size to 32, and other hyper-parameters are set to be the same. For all labeled candidate edges, we randomly sampled 75% of the them as the training set, 5% as the validation set, and 20% as the test set. The *negative injection trick* mentioned in previous work [1] is also employed, i.e. the negative samples are also inserted into original graph $\mathcal{G}$. We adopt Adam optimizer for parameter optimization with an initial learning rate 0.001. The other hyper-parameters are set to be the same. All GNN based models are trained on a cluster of 10 Dual-CPU servers with AGL [33] framework.

Metric. In the link prediction task, we adopt the Area Under Curve (AUC) and F1-score as metrics to evaluate all these models, as done in many other link prediction work.

3.2 Performance Comparison

Table 2 summarizes the performance of the proposed method and other methods on four datasets. Based on the results of the experiment, we summarize the following points:

---
[3] https://github.com/shenweichen/GraphEmbedding.

Table 2. Performance comparison on four Datasets

Types	Model	AUC				F1			
		Facebook	PubMed	Collab	SupChain	Facebook	PubMed	Collab	SupChain
Heuristic	CN	0.927	0.662	0.771	0.601	0.869	0.492	0.703	0.593
Heuristic	Jaccard	0.938	0.630	0.801	0.622	0.874	0.527	0.719	0.601
NE	DeepWalk	0.884	0.842	0.864	–	0.826	0.826	0.811	–
NE	Node2vec	0.902	0.897	0.857	–	0.855	0.867	0.821	–
Node	PinSage$_{GCN}$	0.924	0.823	0.851	0.941	0.900	0.766	0.910	0.763
Node	PinSage$_{GAT}$	0.917	0.832	0.833	0.968	0.902	0.774	0.909	0.822
Edge	SEAL (attributed)	0.963	0.898	0.909	0.977	0.915	0.841	0.923	0.862
Edge	GraIL (attributed)	0.971	0.904	0.947	0.979	0.928	0.848	0.964	0.864
Edge	PaGNN$_{two\_way}$	0.909	0.762	0.902	0.976	0.820	0.720	0.921	0.861
Edge	PaGNN$_{broadcast}$	0.940	0.866	0.853	0.970	0.917	0.790	0.910	0.848
Edge	PaGNN$_{lstm}$	0.969	0.934	0.958	0.978	0.932	0.852	0.976	0.868
Edge	PaGNN	**0.972**	**0.944**	**0.967**	**0.987**	**0.933**	**0.878**	**0.979**	**0.897**

Heuristic-Based Methods achieve considerable performance on the social network dataset (i.e., Facebook), but poor performance on the other three datasets. It indicates that the assumptions of heuristic methods are not applicable in many types of network data. Compared with the heuristic method, the *network embedding methods* achieve better results, but the effectiveness are worse than *node-centric GNN models*. One reason is that the GNN method straightforwardly leverages the node attributes and learns representations in a supervised manner.

All *edge-centric GNN models* achieve better performance since they take the interaction structure into account. And **PaGNN** performs better than the other edge-centric GNN models: **SEAL** and **GraIL**, although they have combined high-order interactions with node attributes. This is because **PaGNN** explicitly incorporates interaction and node attribute information into GNN. Labeling nodes with the structure may not achieve satisfied performance when the attribute on path is crucial.

From the experimental results of *ablation studies*, **PaGNN**$_{two\_way}$ is worse than **PaGNN** and it's unstable on some datasets. It is due to when there is no path between two nodes, the final representation only contains local neighborhood information of one node, performance of **PaGNN**$_{two\_way}$ drops significantly. For example, PubMed data is relatively sparse and only 18.2% samples are connected, the performance of **PaGNN**$_{two\_way}$ is poor. Comparing the performance of **PaGNN** and **PaGNN**lstm, it can be observed that LSTM_OP is effective to integrate the interactions and node attributes, since LSTM forgets useless information. In particular, by observing results of **PaGNN**$_{broadcast}$ and **PaGNN**, the broadcasted embeddings significantly improve the performance, which verify the effectiveness of the broadcasting operation.

On average, **PaGNN** improved upon the best baseline of every dataset by 1.7% in AUC and by 1.4% in F1. In summary, **PaGNN** outperforms all baselines, and every studied component has a positive impact on final results.

Parameter Sensitivity. We also evaluate model performance on the Collab dataset as H ranges from 1 to 5, shown in Fig. 4(a). All models achieved the best performance when H = 2. The AUC value decreases when $H > 2$, which indicates a large H may bring in noises from distant nodes. $\mathbf{PaGNN}_{lstm}$ shows better performance than **PaGNN** when H = 1, since the LSTM_OP has no advantage when path is short. When H becomes larger, **PaGNN** shows better performance, since LSTM forgets useless path information of distant nodes.

The convergence of different models is also evaluated. Figure 4(b) records the validation AUC of varying training steps. $\mathbf{PinSage}_{GAT}$ first achieves the best performance at 6K training steps, as it learns minimum parameters. **GraIL** and **PaGNN** need more time to be converged, which takes about 8K to 10K training steps. Figure 4(c) compares the performance of varying node embedding size, three models are over-fitting when embedding size is larger than 32.

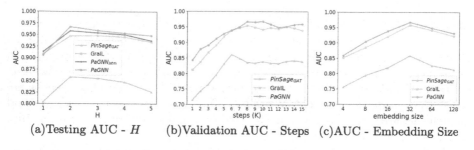

(a)Testing AUC - H (b)Validation AUC - Steps   (c)AUC - Embedding Size

Fig. 4. Parameter sensitivity.

Case Study. To examine whether **PaGNN** learns interaction information, we illustrate an example of *Facebook* dataset in Fig. 5, the goal is to predict whether two yellow nodes exist an edge, nodes on the path between them are colored with green and other nodes are colored gray, and the value is the attention of corresponding edge in last aggregation step. We can learn that the neighbor attention output by $\mathbf{PinSage}_{GAT}$ between associated node and its neighbors is almost the same, without regard to where the neighbor is. Attention value of nodes on the path of edge-centric methods is larger, which indicates they have ability to find the pattern that nodes on path are more important. And comparing with **GraIL**, attention value of green nodes for **PaGNN** is larger, shown in Fig. 5(b) and (c).

To further prove the impact of paths, in *Facebook* and *SupChain* dataset, we categorize candidate relations into different groups according to the number of paths between two target persons or enterprises, proportion of each group is illustrated in Fig. 6. It can be learned that observed friends in social network and enterprises with supply-chain relationship tend to have more paths, which drives to effectively capture fine-grained interactions between target node pairs.

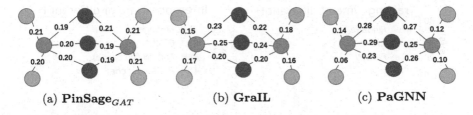

| (a) PinSage$_{GAT}$ | (b) GraIL | (c) PaGNN |

Fig. 5. Attention of different models. (Color figure online)

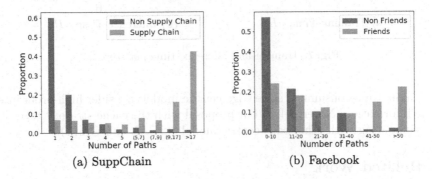

| (a) SuppChain | (b) Facebook |

Fig. 6. Relation distribution of different path number.

3.3 Efficiency Analysis

Training Phase. We calculate the training time of three models with varying H on Collab dataset (Fig. 7(a)). The training time increases rapidly as H increases, and all edge-centric models are more time-consuming than node-centric models. **PaGNN** takes about 1.5 longer the training time comparing with **PinSage**$_{GAT}$ when $H = 1$, but about 3.5 times the training time when $H = 5$. It is due to that as H increases, the subgraphs become larger, for node-centric methods, different edges' subgraphs in same batch share more common nodes, which avoids duplicate computation comparing with edge-centric methods.

Inference Phase. Efficiency evaluation of inference phase is illustrated in Fig. 7(b). It can be inferred that edge-centric methods need more time. For example, **PaGNN** takes 2.4 times longer than **PinSage**$_{GAT}$ when $H = 5$, and **GraIL** takes 2.0 times longer. Fortunately, the cache strategy significantly improves the efficiency of **PaGNN**, which takes 1.7 times longer than **PinSage**$_{GAT}$ when $H = 5$ and has a 30% speed-up compared to **PaGNN** without cache strategy. In particular, although the time complexity of cache strategy analysed in Sect. 2.6 has a two times speed-up theoretically, the statistics reported here are the time cost of the whole inference phase (also including subgraph extraction, input data preparing).

To summarize, comparing with node-centric models (e.g. PinSage), edge-centric models (e.g. PaGNN, SEAL, GraIL) achieve better performance despite

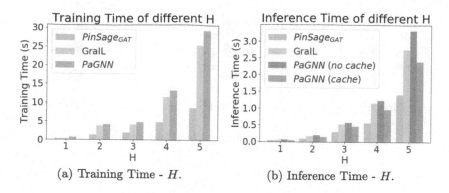

(a) Training Time - H.   (b) Inference Time - H.

Fig. 7. Training and inference time per step.

being more time-consuming, since edge-centric models consider high-order inter-active information. Nevertheless, our proposal with the cache strategy takes less time comparing with other edge-centric models (SEAL and GraIL).

4   Related Work

Previous work for link prediction can be divided into the heuristic methods, network embedding, and supervised methods. Heuristic methods usually assume that nodes with links satisfy some specific properties. For instance, common neighbor [13] defines the similarity as the number of shared neighbors of two nodes. Katz index [9] calculates similarity by counting the number of reachable paths. The rooted PageRank [2] calculates the stationary distribution from one node to other nodes through random walk. These heuristic methods rely on hand-crafted rules and have strong assumptions, which can not be applied to all kinds of networks.

Recently, a number of researchers propose to perform link prediction through constructing node latent features, which are learned via classical matrix factorization [10,18] and shallow graph representation learning (e.g., Deepwalk [14] and Node2Vec [5]). And there is also research [20,34] inductively predicts node relations by leveraging graph neural networks, which combine high-order topological and initial features in the form of graph patterns. However, it learns high-order information and node attributes separately. Some literature [4,8,15,16,35] also considers integrating target behavior information in a heterogeneous graph, which requires nodes and edges belong to a specific type. Research work [27] also employs characteristics of path reachability to represent high-order information, but fails to integrate the path structure and attributes.

5   Conclusion

In this paper, we aim at simultaneously leveraging the structure and attribute information from both interactions/paths and local neighborhood, to predict the edge between two nodes. A novel PaGNN model is proposed which first broadcasts information from one node, afterwards aggregates broadcasted embeddings and node attributes to the other node from its ego subgraph. PaGNN inductively learns representation from node attributes and structures, which incorporates high-order interaction and neighborhood information into GNN. And we also employ a cache strategy to accelerate inference stage. Comprehensive experiments show the effectiveness and efficiency of our proposal on real-world datasets.

References

1. AbuOda, G., Morales, G.D.F., Aboulnaga, A.: Link prediction via higher-order motif features. In: ECML PKDD, pp. 412–429 (2019)
2. Brin, S., Page, L.: Reprint of: the anatomy of a large-scale hypertextual web search engine. Comput. Netw. **56**(18), 3825–3833 (2012)
3. Cui, P., Wang, X., Pei, J., Zhu, W.: A survey on network embedding. IEEE Trans. Knowl. Data Eng. **31**(5), 833–852 (2018)
4. Feng, Y., Hu, B., Lv, F., Liu, Q., Zhang, Z., Ou, W.: ATBRG: adaptive target-behavior relational graph network for effective recommendation (2020)
5. Grover, A., Leskovec, J.: node2vec: scalable feature learning for networks. In: SIGKDD, pp. 855–864 (2016)
6. Hochreiter, S., Schmidhuber, J.: Long short-term memory. Neural Comput. **9**(8), 1735–1780 (1997)
7. Hu, B., Hu, Z., Zhang, Z., Zhou, J., Shi, C.: KGNN: distributed framework for graph neural knowledge representation. In: ICML Workshop (2020)
8. Hu, B., Shi, C., Zhao, W.X., Yu, P.S.: Leveraging meta-path based context for top-n recommendation with a neural co-attention model. In: SIGKDD, pp. 1531–1540 (2018)
9. Katz, L.: A new status index derived from sociometric analysis. Psychometrika **18**(1), 39–43 (1953)
10. Koren, Y., Bell, R., Volinsky, C.: Matrix factorization techniques for recommender systems. Computer **42**(8), 30–37 (2009)
11. Kovács, I.A., et al.: Network-based prediction of protein interactions. Nat. Commun. **10**(1), 1–8 (2019)
12. Kumar, A., Singh, S.S., Singh, K., Biswas, B.: Link prediction techniques, applications, and performance: A survey. Phys. A **553**, 124289 (2020)
13. Liben-Nowell, D., Kleinberg, J.: The link-prediction problem for social networks. J. Am. Soc. Inform. Sci. Technol. **58**(7), 1019–1031 (2007)
14. Perozzi, B., Al-Rfou, R., Skiena, S.: DeepWalk: online learning of social representations. In: SIGKDD, pp. 701–710 (2014)
15. Sankar, A., Zhang, X., Chang, K.C.C.: Meta-GNN: metagraph neural network for semi-supervised learning in attributed heterogeneous information networks. In: ASONAM, pp. 137–144 (2019)
16. Sha, X., Sun, Z., Zhang, J.: Attentive knowledge graph embedding for personalized recommendation. arXiv preprint arXiv:1910.08288 (2019)
17. Shi, B., Weninger, T.: ProjE: embedding projection for knowledge graph completion. In: AAAI, vol. 31 (2017)

18. Shi, C., Hu, B., Zhao, W.X., Philip, S.Y.: Heterogeneous information network embedding for recommendation. IEEE Trans. Knowl. Data Eng. **31**(2), 357–370 (2018)
19. Shibata, N., Kajikawa, Y., Sakata, I.: Link prediction in citation networks. J. Am. Soc. Inform. Sci. Technol. **63**(1), 78–85 (2012)
20. Teru, K., Denis, E., Hamilton, W.: Inductive relation prediction by subgraph reasoning. In: ICML, pp. 9448–9457 (2020)
21. Veličković, P., Cucurull, G., Casanova, A., Romero, A., Lio, P., Bengio, Y.: Graph attention networks. arXiv preprint arXiv:1710.10903 (2017)
22. Wang, D., Cui, P., Zhu, W.: Structural deep network embedding. In: SIGKDD, pp. 1225–1234 (2016)
23. Wang, H., Lian, D., Zhang, Y., Qin, L., Lin, X.: GOGNN: graph of graphs neural network for predicting structured entity interactions, pp. 1317–1323 (2020)
24. Wang, Z., Liao, J., Cao, Q., Qi, H., Wang, Z.: FriendBook: a semantic-based friend recommendation system for social networks. IEEE Trans. Mob. Comput. **14**(3), 538–551 (2014)
25. Wu, Z., Pan, S., Chen, F., Long, G., Zhang, C., Philip, S.Y.: A comprehensive survey on graph neural networks. IEEE Trans. Neural Netw. Learn. Syst. **32**, 4–24 (2020)
26. Wu, Z., Pan, S., Long, G., Jiang, J., Zhang, C.: Graph WaveNet for deep spatial-temporal graph modeling. arXiv preprint arXiv:1906.00121 (2019)
27. Xu, C., Cui, Z., Hong, X., Zhang, T., Yang, J., Liu, W.: Graph inference learning for semi-supervised classification. arXiv preprint arXiv:2001.06137 (2020)
28. Xu, N., Wang, P., Chen, L., Tao, J., Zhao, J.: MR-GNN: multi-resolution and dual graph neural network for predicting structured entity interactions, pp. 3968–3974 (2019)
29. Yang, S., et al.: Financial risk analysis for SMEs with graph-based supply chain mining. In: IJCAI, pp. 4661–4667 (2020)
30. Yang, S., Zou, L., Wang, Z., Yan, J., Wen, J.R.: Efficiently answering technical questions–a knowledge graph approach. In: AAAI, vol. 31 (2017)
31. Yang, Z., Cohen, W., Salakhudinov, R.: Revisiting semi-supervised learning with graph embeddings. In: ICML, pp. 40–48 (2016)
32. Ying, R., He, R., Chen, K., Eksombatchai, P., Hamilton, W.L., Leskovec, J.: Graph convolutional neural networks for web-scale recommender systems. In: SIGKDD, pp. 974–983 (2018)
33. Zhang, D., et al.: AGL: a scalable system for industrial-purpose graph machine learning. arXiv preprint arXiv:2003.02454 (2020)
34. Zhang, M., Chen, Y.: Link prediction based on graph neural networks. In: NeurIPS, pp. 5165–5175 (2018)
35. Zhang, W., Fang, Y., Liu, Z., Wu, M., Zhang, X.: mg2vec: learning relationship-preserving heterogeneous graph representations via metagraph embedding. IEEE Trans. Knowl. Data Eng. (2020)

Representation Learning on Multi-layered Heterogeneous Network

Delvin Ce Zhang$^{(\boxtimes)}$ ⓘ and Hady W. Lauw ⓘ

School of Computing and Information Systems, Singapore Management University,
Singapore, Singapore
{cezhang.2018,hadywlauw}@smu.edu.sg

Abstract. Network data can often be represented in a multi-layered structure with rich semantics. One example is e-commerce data, containing user-user social network layer and item-item context layer, with cross-layer user-item interactions. Given the dual characters of homogeneity within each layer and heterogeneity across layers, we seek to learn node representations from such a multi-layered heterogeneous network while jointly preserving structural information and network semantics. In contrast, previous works on network embedding mainly focus on single-layered or homogeneous networks with one type of nodes and links. In this paper we propose intra- and cross-layer proximity concepts. Intra-layer proximity simulates propagation along homogeneous nodes to explore latent structural similarities. Cross-layer proximity captures network semantics by extending heterogeneous neighborhood across layers. Through extensive experiments on four datasets, we demonstrate that our model achieves substantial gains in different real-world domains over state-of-the-art baselines.

Keywords: Representation learning · Heterogeneous network · Dimensionality reduction

1 Introduction

Much of the data on the Web can be represented in a network structure, ranging from social and biological to academic networks, etc. Network analysis recently attracts escalating research attention due to its importance and wide applicability. Diverse problems could be formulated as network tasks, e.g., recommending items to users on e-commerce [12]. As the primary information is the inherent structure of the network itself, one promising direction known as the *network embedding* problem is to learn the representation of each node, which could in turn fuel tasks such as node classification, node clustering, and link prediction.

Figure 1 illustrates an example network with various object types (users, movies, movie actors). These objects are connected via various links, e.g., a user may friend other users, favor some movies, and follow some actors, while a movie may share similar contexts as another (being added to the same preference

© Springer Nature Switzerland AG 2021
N. Oliver et al. (Eds.): ECML PKDD 2021, LNAI 12976, pp. 399–416, 2021.
https://doi.org/10.1007/978-3-030-86520-7_25

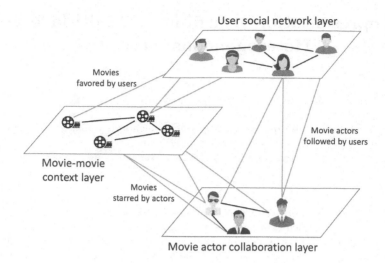

Fig. 1. Illustration of multi-layered heterogeneous network. Three homogeneous layers (user social network layer, *movie-movie* context layer, movie actor collaboration layer) are connected by heterogeneous interactions.

folder, or recommended in the same movie list) or feature some actors. Network embedding learns a low-dimensional representation for each node (user, movie, or actor), which preserves the network information. In turn, the node representations may be used in applications such as predicting whether a user is likely to favor a movie, or whether a user is likely to friend another user.

Present Work and Challenges. Previous works on network embedding focus on *homogeneous* networks [19,23]. They treat all nodes and edges as the same type, regardless of varying relations. Such homogeneous treatment may miss out on the nuances that arise from the diversity of associations (e.g., user favoring a movie has different semantics from movie featuring an actor).

More recent works recognize the value of absorbing the varying semantics into the node representation, modeling a *heterogeneous* network. However, to encode semantics, models such as Metapath2vec [6] rely on the notion of metapath scheme (sequence of node types that make up a path). These are to be prespecified in advance, requiring domain-specific knowledge (incurring manual costs) or exhaustive enumeration of schemes (incurring computational costs). Other models [11] only consider each edge relation as one type of connections, but ignore the two end-point nodes are sometimes mutually homogeneous, thereby losing structural information in node embeddings.

Proposed Approach and Contributions. We observe that complex networks simultaneously exhibit homogeneous and heterogeneous tendencies. The interplay between the two gives rise to a *multi-layered* structure, whereby each layer encodes the structural connectivity of objects of the same type, and connections across layers bear rich semantics between objects of different types. Figure 1 can be seen as a *multi-layered heterogeneous network* with three layers. We offer a formal definition in Sect. 3.

Given the dual characters of multi-layered network, we seek to learn node embeddings that preserve both structure and semantics to facilitate downstream tasks, e.g., item recommendation in e-commerce, user alignment across multiple social networks [15]. In contrast to heterogeneous models that rely on prespecified schemes [6], the cross-layer proximity of our model naturally 'induces' various schemes by how it models layers, and its maximum order controls the semantics learning. In contrast to heterogeneous models [11] that do not consider that the two end-point nodes are sometimes mutually homogeneous, we use nodes of the same type to jointly preserve semantics, so as to embody structural proximity.

In this paper, we propose **Multi-Layered** Heterogeneous Network Embedding, or (MULTILAYEREDHNE), describing how it models both intra-layer proximities to explore structural similarities in a breadth-wise propagation manner and cross-layer proximities for depth-wise semantics capture in Sect. 4. In a nutshell, like ripples expanding across the water, intra-layer proximities broadcast one node's homogeneous neighborhood hop by hop to investigate latent structural relationships. Cross-layer proximities iteratively extend heterogeneous relations layer by layer and leverage intra-layer proximities to *jointly* preserve network semantics.

Our contributions in this paper are as follows:

- Though "multi-layered" notion may have appeared in prior, here we articulate a concrete definition in the context of heterogeneous network. Importantly, we define the novel notions of intra- and cross-layer proximities underlining our approach.
- To capture network homogeneity and heterogeneity jointly, we propose a novel framework that encodes both structural proximity and network semantics into unified node embeddings by higher-order intra- and cross-layer proximities.
- We conduct extensive experiments on four real datasets, and the results validate the effectiveness of our model over state-of-the-art baselines.

2   Related Work

Here we review related research works for homogeneous, heterogeneous, and multi-layered network embedding.

Homogeneous Network Embedding. Homogeneous networks are those with one single type of nodes and links. DeepWalk [19] generates random walk on the network as corpus and applies skip-gram model to train the nodes. Node2vec [9] extends DeepWalk by simulating biased random walk to explore diverse neighborhoods. LINE [23] learns node representations by preserving first- and second-order proximities. GraRep [2] generalizes LINE to incorporate higher-order proximities, but may not scale efficiently to very large networks. These methods mainly focus on embedding network topology to preserve structural information.

Meanwhile, there are also models dealing with attributed homogeneous networks with task-specific supervision (e.g., GCN [26], GAT [24]). They are different from our model that embeds network in an unsupervised manner to support arbitrary downstream tasks. Others that operate on attributed graph for

Table 1. Summary of main notations.

Notation	Explanation
$\mathcal{G}$	The input network
$\mathcal{V}, \mathcal{E}$	The node set and edge set, resp.
$\mathcal{O}, \mathcal{R}$	The node type set and edge type set, resp.
$\mathcal{L}$	The layer set
$\mathcal{I}_v^m$	The mth-order intra-layer proximity of node v
$\mathcal{C}_v^n$	The nth-order cross-layer proximity of node v
M, N	Maximum order of intra- and cross-layer proximity, resp.
$\mathcal{K}$	Number of negative samples

multi-modal learning (EP [8]) and are designed specifically for document network (Adjacent-Encoder [27]) are also not directly comparable.

Heterogeneous Network Embedding. Some heterogeneous network models leverage meta-path-based random walks to capture network semantics, such as Metapath2vec [6] and HIN2vec [7]. The applications of meta-path-based models (e.g. recommender systems) are also widely studied [20]. Some of them simulate meta-paths of specified schemes on each network to preserve complex semantics. To this end, the cross-layer proximity of our model does not restrict to specific schemes, and its maximum order controls the semantics learning. There also exist some methods that do not require specific meta-paths, such as HeGAN [11], which utilizes GAN to generate fake nodes to train discriminator. More recently, Graph Neural Networks have been successfully applied to attributed heterogeneous networks with satisfactory results [25].

Multi-layered Network Embedding. Multi-layered networks, as a set of inter-dependent network layers, appear in real-world scenarios including recommender and academic systems, cross-platform social networks, etc. Previous works focus on cross-layer links inference [4,5] and network ranking [18]. MANE [14] studies representation learning on multi-layered networks by seeking low-dimensional node embeddings by modeling each intra-layer and cross-layer links. Our model has a couple of distinctions. For one, we incorporate higher-order proximities. For another, the manner in which we model proximities integrates nodes with similar structures, instead of predicting links individually. These differences do make a difference to the effectiveness of our node embeddings (see Sect. 5).

There exists another definition of "multi-layered" network [16], which really is *multiplex* or *multi-view* network [3], where there are multiple relationships between the same set of nodes. In contrast, multi-layered network in this paper refers to a set of interdependent network layers, each with a different set of nodes.

3   Definitions and Problem Formulation

We introduce intra- and cross-layer proximities, and formalize our problem. Table 1 lists the notations.

Definition 1. *A **Heterogeneous Information Network** (HIN)* $\mathcal{G} = \{\mathcal{V}, \mathcal{E}, \mathcal{O}, \mathcal{R}\}$ *consists of a node set* $\mathcal{V}$ *and an edge set* $\mathcal{E}$. *This network is associated with a node type mapping function* $\phi\colon \mathcal{V} \to \mathcal{O}$ *and an edge type mapping function* $\varphi\colon \mathcal{E} \to \mathcal{R}$. $\mathcal{O}$ *and* $\mathcal{R}$ *represent the sets of predefined node types and edge types respectively, where* $|\mathcal{O}| + |\mathcal{R}| > 2$.

We use the terms *edge* and *link* interchangeably, ditto for *network* and *graph*. Multi-layered network is defined over HIN, with an additional requirement of layers.

Definition 2. *A **Multi-Layered Heterogeneous Network** $\mathcal{G} = \{\mathcal{V}, \mathcal{E}, \mathcal{O}, \mathcal{R}, \mathcal{L}\}$ is a connected HIN that contains a layer set $\mathcal{L}$ of $|\mathcal{L}| > 1$ homogeneous network layers. In addition to ϕ and φ, we have two more mapping functions. The node mapping function $\theta\colon \mathcal{V} \to \mathcal{L}$ projects each node $v \in \mathcal{V}$ to a certain layer $l_v \in \mathcal{L}$. The edge mapping function $\vartheta\colon \mathcal{E} \to \mathcal{L} \times \mathcal{L}$ places each edge $e \in \mathcal{E}$ between two layers $(l_{e,1}, l_{e,2}) \in \mathcal{L} \times \mathcal{L}$.*

$\mathcal{L} \times \mathcal{L} = \{(l_{e,1}, l_{e,2})| l_{e,1}, l_{e,2} \in \mathcal{L}\}$ represents the Cartesian product of two sets. Thus $l_{e,1}$ and $l_{e,2}$ could be the same, and edge e is intra-layer, otherwise cross-layer.

Figure 1 illustrates a multi-layered network with three homogeneous layers (*user–user, movie–movie, actor–actor*) and three heterogeneous interactions (*user–movie, movie–actor, user–actor*). Intra-layer edges (black) connect nodes of the same type. Cross-layer edges (green) of different relations connect arbitrary type of nodes. Multi-layered networks are a subset of HIN, as each layer contains intra-layer edges.

Problem 1. Given a multi-layered heterogeneous network $\mathcal{G} = \{\mathcal{V}, \mathcal{E}, \mathcal{O}, \mathcal{R}, \mathcal{L}\}$, the goal of **Representation Learning on Multi-Layered Heterogeneous Network** is to learn a mapping function to project each node $v \in \mathcal{V}$ to a low-dimensional space $\mathbb{R}^d$ where $d \ll |\mathcal{V}|$. The node representation in the new space should preserve both structural proximities and network semantics within $\mathcal{G}$.

To harness contributions from intra-layer edges containing structural information within layers and from cross-layer edges capturing semantics, we propose the MULTILAYEREDHNE framework built on *intra-layer proximity* and *cross-layer proximity*.

Definition 3. *The mth-order **Intra-Layer Proximity** of node v is defined as the set of nodes that can be reached by m intra-layer edges from v:*

$$\mathcal{I}_v^m = \{v^m | v^{m-1} \in \mathcal{I}_v^{m-1}, (v^{m-1}, v^m) \in \mathcal{E}, l_{v^m} = l_{v^{m-1}}\}, \tag{1}$$

where $m = 1, 2, ..., M$, *and* $\mathcal{I}_v^0 = \{v\}$.

This concept is illustrated by Fig. 2(a) (best seen in color). Here we suppose this network is homogeneous within a network layer. Node v's first-order intra-layer proximity consists of four nodes inside the inner white circle with black links connecting them. Similarly nodes lying in the gray annulus represent v's second-order intra-layer proximity. We can extend this concept up to M^{th} order and obtain $\mathcal{I}_v^M$.

Definition 4. *The nth-order **Cross-Layer Proximity** of node v is defined as the set of nodes that can be reached by n cross-layer edges from v:*

$$C_v^n = \{v^n | v^{n-1} \in C_v^{n-1}, (v^{n-1}, v^n) \in \mathcal{E}, l_{v^n} \neq l_{v^{n-1}}\}, \qquad (2)$$

where $n = 1, 2, ..., N$, and $C_v^0 = \{v\}$.

To illustrate this concept, we use Fig. 2(d) (best seen in color). v_C^1 represents one node in v's first-order cross-layer proximity with a cross-layer green link connecting them. Extending this example up to N^{th} order, we obtain C_v^N.

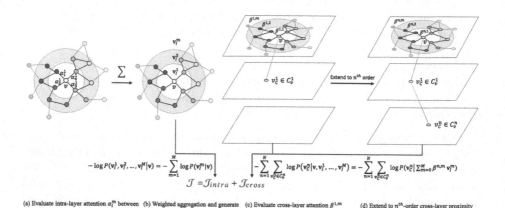

(a) Evaluate intra-layer attention α_i^m between v and nodes in each intra-layer proximity.   (b) Weighted aggregation and generate $\mathbf{v}_l^1, \mathbf{v}_l^2, ..., \mathbf{v}_l^M$ for each proximity.   (c) Evaluate cross-layer attention $\beta^{1,m}$ between v_C^1 and $\mathbf{v}, \mathbf{v}_l^1, ..., \mathbf{v}_l^M$.   (d) Extend to n^{th}-order cross-layer proximity and use $\mathbf{v}, \mathbf{v}_l^1, ..., \mathbf{v}_l^M$ to *jointly* evaluate loss.

Fig. 2. Illustration of intra- and cross-layer proximity modeling. (Color figure online)

4   Model Architecture

We now describe our proposed model MULTILAYEREDHNE. It consists of two modeling components. First, intra-layer proximity modeling (Figs. 2(a) and (b)) simulates the breadth-wise propagation across homogeneous neighbors within layers to explore structural similarities. Second, cross-layer proximity modeling (Fig. 2(c) and (d)) captures semantics by extending heterogeneous neighborhood across layers.

4.1   Intra-layer Proximity Modeling

Suppose that $\mathbf{v} \in \mathbb{R}^d$ is the embedding of a node v. This is the quantity that we seek to derive. Intra-layer proximity concerns the relationships between v and its homogeneous neighbors from the same layer. We first consider the first-order proximity ($m = 1$). This effectively concerns the direct neighbors of v, collectively denoted $\mathcal{I}_v^1$. The embedding of each first-order neighbor $v_i^1 \in \mathcal{I}_v^1$ is denoted $\mathbf{v}_i^1 \in \mathbb{R}^d$. Since intra-layer proximities contain nodes of the same type, it is reasonable to treat these nodes homogeneously. Thus we derive a representation of v's first-order intra-layer proximity $\mathbf{v}_I^1$ as a weighted aggregation of its neighbors' embeddings.

$$\mathbf{v}_I^1 = \sum_{i=1}^{|\mathcal{I}_v^1|} \alpha_i^1 \mathbf{v}_i^1. \tag{3}$$

Not all neighbors are equally important to v. Some may be of greater importance. Therefore, the aggregation in Eq. 3 factors in an attention coefficient w.r.t. v.

$$a_i^1 = \sigma(\mathbf{v}^T \mathbf{v}_i^1), \qquad \alpha_i^1 = \frac{\exp(a_i^1)}{\sum_{i=1}^{|\mathcal{I}_v^1|} \exp(a_i^1)}, \tag{4}$$

where $i = 1, 2, ..., |\mathcal{I}_v^1|$, and σ is sigmoid function. The attention values α_i^1 can be regarded as the similarity between v and v_i^1, as illustrated by Fig. 2(a). This aggregation is illustrated in Fig. 2(b). Here $\mathbf{v}_I^1$ can be seen as the first-order propagation of v, or the first "ripple" of intra-layer proximity.

Extending beyond the first-order proximity, we repeat the process above (Eqs. 3–4) for each mth-order intra-layer proximity up to the a specified maximum order M, propagating to the subsequent ripples. This generates a set of representations $\{\mathbf{v}_I^m\}_{m=1}^M = \{\mathbf{v}_I^1, \mathbf{v}_I^2, ..., \mathbf{v}_I^M\}$. For simplicity, we let $\mathbf{v}_I^0 = \mathbf{v}$.

Proximities indicate a shared relationship. Nodes within maximum proximity from v would likely have similar representation with v. Thus, given node v and $\{\mathcal{I}_v^m\}_{m=1}^M$, our objective is to minimize the following negative log-likelihood.

$$-\log P(\mathbf{v}_I^1, \mathbf{v}_I^2, ..., \mathbf{v}_I^M | \mathbf{v}) = -\sum_{m=1}^M \log P(\mathbf{v}_I^m | \mathbf{v}). \tag{5}$$

4.2   Cross-Layer Proximity Modeling

Beyond a single layer, connections across layers encode network semantics. Node v and its intra-layer proximities $\{\mathcal{I}_v^m\}_{m=1}^M$ are mutually homogeneous, they are expected to reflect the same identity. They would carry information to *jointly* preserve semantics w.r.t. v's cross-layer proximities. Formally, for each node $v_C^n \in \mathcal{C}_v^n$ in n^{th}-order cross-layer proximity, we have the following negative log-likelihood.

$$-\log P(\mathbf{v}_C^n | \mathbf{v}_I^0, ..., \mathbf{v}_I^M) = -\log P(\mathbf{v}_C^n | \sum_{m=0}^M \beta^{n,m} \mathbf{v}_I^m). \tag{6}$$

Given $\mathbf{v}$ and its intra-layer proximities $\{\mathbf{v}_I^m\}_{m=1}^M$, we force them to predict the observation of cross-layer node $\mathbf{v}_C^n$ together. As we increase the order of intra-layer proximity M and reach more nodes, more noisy nodes may inadvertently be included. Thus we expect that different $\mathbf{v}_I^m$ may affect this prediction to different degrees. This is the intuition behind $\beta^{n,m}$, which measures the relative importance of $\mathbf{v}_I^m$.

$$b^{n,m} = \sigma(\mathbf{v}_C^{nT} \mathbf{v}_I^m), \qquad \beta^{n,m} = \frac{\exp(b^{n,m})}{\sum_{m=0}^M \exp(b^{n,m})}. \tag{7}$$

This is illustrated by Fig. 2(c), where we evaluate attention between $\mathbf{v}_C^1$ and $\{\mathbf{v}_I^m\}_{m=0}^M$. $\beta^{n,m}$ is specific to each cross-layer node, since different nodes capture semantics from different aspects. For example, a user and his friends may like superhero movies, but suppose he is the only one who likes it because of the actors. In this case, $\beta^{n,m}$ should be assigned equally between the user and his friends in terms of superhero genre, but biased to only the user in terms of actors.

Similarly as for intra-layer, we extend cross-layer proximity to specified maximum N^{th} order, and obtain the following objective, which is also illustrated by Fig. 2(d).

$$-\sum_{n=1}^N \sum_{v_C^n \in \mathcal{C}_v^n} \log P(\mathbf{v}_C^n | \sum_{m=0}^M \beta^{n,m} \mathbf{v}_I^m). \tag{8}$$

Table 2. Dataset statistics.

Dataset	#nodes	#intra-layer links	#cross-layer links	#layers	#labels
ACM	30,126	77,484	38,772	3	7
Aminer	17,504	72,237	35,229	3	8
TF	3,218	25,811	1,609	2	N.A.
LastFM	19,524	301,015	92,834	2	N.A.

4.3 Learning Strategy

As in [6], the conditional probabilities in Eq. 5 and 8 are defined as the heterogeneous softmax function.

$$P(\mathbf{v}_j | \mathbf{v}_i) = \frac{\exp(\mathbf{v}_j^T \mathbf{v}_i)}{\sum_{l_{v_k} = l_{v_j}} \exp(\mathbf{v}_k^T \mathbf{v}_i)}, \tag{9}$$

where $\mathbf{v}_k$ comes from the same network layer as $\mathbf{v}_j$. Here we use $P(\mathbf{v}_j | \mathbf{v}_i)$ to denote both conditional probabilities for simplicity. Finally, we leverage heterogeneous negative sampling to approximate both objective functions, and obtain Eq. 10, where $\mathcal{K}$ is the number of negative samples. v_k is a negative sample, randomly drawn from a noise distribution $P_l(v_k)$ defined on the node set of each proximity's corresponding layer. v_C^n is one node from v's n^{th}-order cross-layer proximity, sampled at each iteration.

$$\begin{aligned}
\mathcal{J} =& \mathcal{J}_{intra} + \mathcal{J}_{cross} \\
=& -\sum_{m=1}^M \left(\log \sigma(\mathbf{v}_I^{mT} \mathbf{v}) + \sum_{k=1}^{\mathcal{K}} \mathbb{E}_{v_k \sim P_l(v_k)} \log \sigma(-\mathbf{v}_k^{mT} \mathbf{v}) \right) \\
& -\sum_{n=1}^N \mathbb{E}_{v_C^n \sim \mathcal{C}_v^n} \left(\log \sigma(\mathbf{v}_C^{nT} \sum_{m=0}^M \beta^{n,m} \mathbf{v}_I^m) + \sum_{k=1}^{\mathcal{K}} \mathbb{E}_{v_k \sim P_l(v_k)} \log \sigma(-\mathbf{v}_k^T \sum_{m=0}^M \beta^{n,m} \mathbf{v}_I^m) \right).
\end{aligned} \tag{10}$$

Complexity. We use $\mathcal{I}_{\max}$ to denote the maximum size of intra-layer proximity, and $|\mathcal{E}_{cross}|$ to denote the number of cross-layer links in the network, thus we have $O(|\mathcal{E}_{cross}|Md(\mathcal{I}_{\max} + \mathcal{K}))$ per iteration for intra-layer proximity modeling, where d represents the dimensionality of node embeddings. The complexity of cross-layer proximity modeling is $O(|\mathcal{E}_{cross}|Nd(M + \mathcal{K}))$ on a training iteration. Putting two components together, we have $O(|\mathcal{E}_{cross}|d(M\mathcal{I}_{\max} + MN + \mathcal{K}M + \mathcal{K}N))$ per iteration.

5 Experiments

Our experimental objective is to validate the node embeddings learned by MUL-TILAYEREDHNE as compared to baselines.

5.1 Setup

We conduct experiments on four publicly available datasets from different domains. Table 2 summarizes their statistics. ACM [21] and Aminer [14] are two academic datasets with three network layers: co-authorship, paper citation, and venue citation layer. Two types of cross-layer links are *author-paper* and *paper-venue* links. Twitter-Foursquare (TF) [29] is a cross-platform social network dataset, containing two social networks: Twitter and Foursquare. Each node only has one cross-layer link, representing his identity across two platforms. LastFM [12] is a recommendation dataset with two layers: *user-user* social network and *artist-artist* context network. TF and LastFM are reserved for link prediction task only, since they do not have labels for nodes.

Baselines. To investigate the efficacy of modeling heterogeneity, we compare to two *homogeneous baselines* that treat all nodes and links as the same type: **DeepWalk** [19] and **LINE** [23]. For LINE, we consider the advanced version with first- and second-order proximities with $d/2$ dimensions each. To study the effects of homogeneity in addition to heterogeneity, we compare to three

Table 3. Micro-F1 and Macro-F1 scores of node classification on ACM.

Model	Micro-F1				Macro-F1			
	20%	40%	60%	80%	20%	40%	60%	80%
DeepWalk	0.916	0.920	0.919	0.919	0.872	0.877	0.876	0.878
LINE (1st+2nd)	0.924	0.926	0.927	0.927	0.875	0.879	0.880	0.879
Metapath2vec	0.921	0.921	0.922	0.926	0.887	0.887	0.888	0.887
HIN2vec	0.936	0.938	0.938	0.937	0.902	0.908	0.907	0.906
HeGAN	0.938	0.940	0.941	0.941	0.903	0.908	0.910	0.918
MANE	0.842	0.850	0.854	0.859	0.711	0.742	0.756	0.759
GAT	0.867	0.908	0.927	0.921	0.786	0.853	0.883	0.875
HAN	0.828	0.869	0.900	0.905	0.728	0.773	0.824	0.834
MULTILAYEREDHNE	0.951*	0.954*	0.956*	0.953*	0.919*	0.928*	0.933*	0.929*

Table 4. Micro-F1 and Macro-F1 scores of node classification on Aminer.

Model	Micro-F1				Macro-F1			
	20%	40%	60%	80%	20%	40%	60%	80%
DeepWalk	0.959	0.962	0.963	0.964	0.922	0.930	0.934	0.931
LINE (1st+2nd)	0.964	0.967	0.968	0.969	0.925	0.930	0.933	0.935
Metapath2vec	0.962	0.963	0.964	0.964	0.870	0.876	0.886	0.893
HIN2vec	0.960	0.962	0.963	0.963	0.922	0.925	0.926	0.927
HeGAN	0.955	0.960	0.963	0.966	0.875	0.892	0.895	0.905
MANE	0.949	0.953	0.956	0.955	0.876	0.893	0.900	0.903
GAT	0.946	0.958	0.965	0.969	0.867	0.919	0.927	0.925
HAN	0.908	0.942	0.956	0.959	0.888	0.911	0.918	0.931
MultiLayeredHNE	**0.972***	**0.974***	**0.975***	**0.975***	**0.926***	**0.935***	**0.943***	**0.944***

heterogeneous baselines: **Metapath2vec** [6], **HIN2vec** [7], and **HeGAN** [11]. To see if higher-order proximities are useful, we compare to a *multi-layer baseline*: **MANE** [14]. Although GCN-based models are designed with task-specific supervision, and different from our unsupervised model, for completeness, we still compare to **GAT** [24] and **HAN** [25].

Implementation Details. Hyperparameters are chosen based on validation set. For MultiLayeredHNE, intra-layer proximity order M is 1 on all datasets. The cross-layer proximity order N is 4 for ACM and Aminer, 1 for TF and LastFM. The number of negative samples K is 16. For random walk models, as in [25], the number of walks per node is 40, the walk length is 100, the window size is 5. For Metapath2vec, the combination of meta-path schemes APVPA and APPVPPA has the best performance on ACM and Aminer. TTTF and TFFF produce the best results on TF, while for LastFM we combine UAUA, UUUA, and UAAA. For other baselines, we follow the hyperparameter settings in the original paper. For fair comparison, as in [11], the embedding dimension is set to 64 for all methods.

Table 5. NMI on node clustering.

Model	ACM	Aminer
DeepWalk	0.519	0.787
LINE (1st+2nd)	0.458	0.800
Metapath2vec	0.358	0.570
HIN2vec	0.201	0.589
HeGAN	0.322	0.586
MANE	0.473	0.789
GAT	0.497	0.832
HAN	0.499	0.781
MultiLayeredHNE	**0.534***	**0.862***

5.2 Node Classification

We expect a good model to embed nodes from the same category closely, while separating different categories. We train a logistic regression based on the embeddings, varying the ratio of the training set from 20% to 80% (of these, 10% is further reserved for validation). We report *Micro-F1* and *Macro-F1* scores on the testing sets in Tables 3 and 4 for the two respective datasets that are applicable. In this paper we use "*" to denote that the performance of our model is significantly different from the best baseline model's based on the paired t-test at the significance level of 0.01.

MULTILAYEREDHNE consistently outperforms the baselines across all training splits. As the training ratio increases, all models tend to perform better, as expected. It is worth noting that HIN-based models, including MULTILAYEREDHNE, generally classify nodes more accurately than those models working solely on homogeneous networks, highlighting the effectiveness of modeling network semantics. Among HIN embedding models, Metapath2vec based on only specific cross-layer links performs the worst, emphasizing the necessity of modeling both intra-layer and cross-layer links.

5.3 Node Clustering

Intuitively, good node embeddings would put "similar" nodes together. We apply K-means algorithm [1] to perform clustering on the node embeddings. Since for ACM and Aminer, nodes are labeled, we can assess whether nodes in a cluster tend to share the same labels. We evaluate the clustering quality using *Normalized Mutual Information (NMI)* w.r.t. the true labels (not used in training).

Table 6. Intra-layer link prediction results.

Model	ACM			Aminer			TF			LastFM		
	AUC	AP	F1	AUC	AP	F1	AUC	AP	F1	AUC	AP	F1
DeepWalk	0.900	0.908	0.688	0.880	0.860	0.685	0.678	0.651	0.683	0.587	0.600	0.693
LINE (1st+2nd)	0.962	0.972	0.668	0.811	0.731	0.739	0.701	0.725	0.661	0.665	0.729	0.666
Metapath2vec	0.786	0.830	0.672	0.851	0.840	0.667	0.620	0.621	0.666	0.789	0.778	0.502
HIN2vec	0.871	0.872	0.671	0.579	0.544	0.667	0.770	0.749	0.667	**0.884**	0.876	0.682
HeGAN	0.509	0.517	0.667	0.641	0.626	0.667	0.512	0.510	0.667	0.507	0.504	0.665
MANE	0.973	0.978	0.675	0.871	0.858	0.688	0.750	0.693	0.679	0.864	0.873	0.667
GAT	0.674	0.675	0.589	0.854	0.812	0.583	-	-	-	-	-	-
HAN	0.592	0.607	0.585	0.647	0.638	0.608	-	-	-	-	-	-
MULTILAYEREDHNE	**0.979***	**0.983***	**0.799***	**0.897***	**0.890***	**0.795***	**0.798***	**0.823***	**0.722***	0.880	**0.892***	**0.768***

Table 5 presents the results. Overall, MULTILAYEREDHNE outperforms baseline models significantly. In comparison to DeepWalk and LINE that model all nodes and links homogeneously, we observe that our distinctive treatment of intra-layer and cross-layer proximities is helpful. MULTILAYEREDHNE also clusters nodes more effectively than MANE, demonstrating that higher-order proximities could help better explore network structure. Overall, MULTILAY-EREDHNE achieves performance gains over the closest baseline by 2.8% and 7.7%, respectively.

5.4 Link Prediction

Here we predict intra- and cross-layer links, respectively. For *intra-layer link prediction*, we predict the *author-author* link on ACM and Aminer [22], *user-user* link on Twitter of TF [28], and *artist-artist* link on LastFM. As in leave-one-out evaluation [10], for nodes with more than one intra-layer links, we hide one as the ground truth positives, and randomly sample the same number of disconnected node pairs as negative instances. The remaining network is our training set. Since this is a binary classification for the held-out links, we adopt inner product [13] to make predictions, and report *AUC, Average Precision (AP)*, and *F1 score* in Table 6. For *cross-layer link prediction*, we predict *author-paper* links on ACM and Aminer [11], *user-user* links on TF, and *user-artist* links on LastFM. We hide cross-layer links similarly with intra-layer. Table 7 presents the results. Since GAT and HAN are designed with label supervision to learn embeddings, they do not have link prediction results on TF and LastFM.

MULTILAYEREDHNE generally outperforms baselines significantly on all evaluation metrics, except for the sole case of the LastFM dataset. For intra-layer link prediction, compared with DeepWalk and LINE, this task verifies the effectiveness of MULTILAYEREDHNE on predicting links between homogeneous nodes. We attribute this to the network heterogeneity captured by our model. For cross-layer link prediction, MULTILAYEREDHNE benefits from the structure-preserving embeddings learned via intra-layer proximity as compared with heterogeneous baselines.

5.5 Network Visualization

Visualization provides an intuitive sense of how nodes are embedded. We visualize node embeddings using t-SNE [17], and color nodes using their corresponding labels. Figure 3 presents four models on ACM dataset. By encoding network

Table 7. Cross-layer link prediction results.

Model	ACM			Aminer			TF			LastFM		
	AUC	AP	F1	AUC	AP	F1	AUC	AP	F1	AUC	AP	F1
DeepWalk	0.891	0.887	0.690	0.923	0.919	0.687	0.707	0.720	0.669	0.606	0.617	0.647
LINE (1st+2nd)	0.912	0.920	0.685	0.894	0.842	0.781	0.725	0.750	0.663	0.714	0.746	0.646
Metapath2vec	0.780	0.798	0.704	0.819	0.773	0.691	0.916	0.911	0.752	**0.923**	0.906	0.476
HIN2vec	0.929	0.939	0.223	0.921	0.924	0.173	0.524	0.561	0.173	0.265	0.397	0.276
HeGAN	0.530	0.534	0.268	0.683	0.683	0.545	0.705	0.685	0.168	0.535	0.527	0.027
MANE	0.923	0.913	0.670	0.906	0.857	0.673	0.724	0.727	0.646	0.736	0.765	0.645
GAT	0.653	0.629	0.549	0.880	0.832	0.574	-	-	-	-	-	-
HAN	0.606	0.590	0.554	0.690	0.669	0.667	-	-	-	-	-	-
MULTILAYEREDHNE	**0.950***	**0.948***	**0.812***	**0.936***	**0.925***	**0.827***	**0.954***	**0.952***	**0.799***	0.919	**0.911***	**0.837***

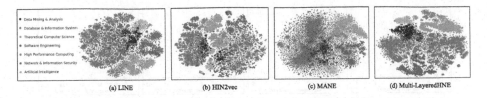

Fig. 3. t-SNE visualization on ACM dataset.

structural proximity and semantics, MULTILAYEREDHNE provides denser clusters with clearer category boundaries than others.

5.6 Model Analysis

Here we conduct several analysis on MULTILAYEREDHNE to better understand the underlying mechanism of it.

Homogeneity and Heterogeneity. To investigate if MULTILAYEREDHNE effectively leverages network homogeneity and heterogeneity, we conduct ablation analysis here. MULTILAYEREDHNE-homo removes the intra-layer proximity modeling, and only maintains cross-layer proximity. Conversely, MULTILAYEREDHNE-hetero assumes all nodes and links are of the same type, and discards network layer concept to investigate network semantics.

Results in Fig. 4 reveal three insights. First, MULTILAYEREDHNE-homo performs worse than MULTILAYEREDHNE, showcasing the advantage of modeling structural information. Second, MULTILAYEREDHNE can indeed encode semantics, since MULTILAYEREDHNE-hetero, which ignores heterogeneity, leads to worse performance compared to MULTILAYEREDHNE. Third, by comparing MULTILAYEREDHNE-homo and MULTILAYEREDHNE-hetero, we conclude that network structural proximity is more informative than semantics, as MULTILAYEREDHNE-homo drops more than MULTILAYEREDHNE-hetero from MULTILAYEREDHNE.

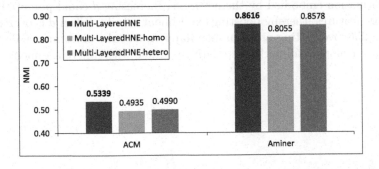

Fig. 4. Impact of homogeneity and heterogeneity

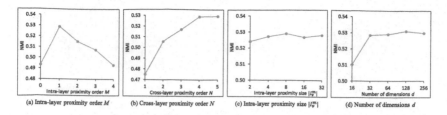

(a) Intra-layer proximity order M (b) Cross-layer proximity order N (c) Intra-layer proximity size $|\mathcal{I}_v^m|$ (d) Number of dimensions d

Fig. 5. MULTILAYEREDHNE parameter sensitivity

Parameter Sensitivity. We vary maximum order of intra- and cross-layer proximity to investigate performance sensitivity. We report the results of clustering (NMI) on ACM dataset in Fig. 5. We first test intra-layer proximity order M. Compared with $M = 0$ where no intra-layer proximity is modeled, MULTILAY-EREDHNE achieves notable performance gain at $M = 1$. However, our model deteriorates its clustering when M is greater than 1, since more noisy neighbors are involved.

We then vary the order of cross-layer proximity N. Too small N apparently could not effectively explore network semantics. The clustering quality is boosted as the order increases, emphasizing the efficacy of modeling cross-layer proximity to capture network heterogeneity and semantics.

Intra-layer Proximity Size $|\mathcal{I}_v^m|$. We limit the size of each intra-layer proximity to further investigate the robustness of MULTILAYEREDHNE on sparse scenarios. Figure 5(c) shows the results. With the increase of the size of intra-layer proximity, the performance of MULTILAYEREDHNE is improved at first, because a larger set of neighbors can encode more structural information on the network. But the clustering results decrease slightly and then stay flat when the size is too large. Overall, the performance is stable w.r.t. different sizes.

Number of Dimensions d. To check the impact of different embedding dimensions d on model performance, we vary the value of d and report the results (Fig. 5 (d)). With the growth of d from 16 to 32, NMI rises at first, and fluctuates slightly when $d > 32$. Since small dimensions cannot fully encode the rich information embodied by the networks, increasing d could potentially capture more features, thereby boosting experiment results. When d is overly large, e.g., $d = 256$, over-fitting problem may happen, and the performance decreases. Overall, our model still performs relatively stable with different dimensions.

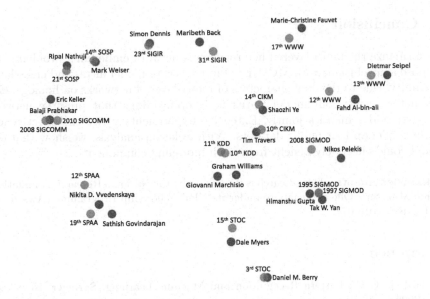

Fig. 6. Case study on ACM dataset (best seen in color). t-SNE visualization of various conferences (orange) from different years and most active authors (green) in those years. (Color figure online)

5.7   Case Study

As an illustration of how MULTILAYEREDHNE encodes homogeneity and heterogeneity, we conduct a case study on ACM dataset. We randomly select two or three years of each conference, and draw the most active authors in those years. Figure 6 shows the t-SNE visualization. Interestingly, the distance between 12th and 13th WWW conferences (top right corner) is shorter than their distances to 17th WWW conference. SIGMOD (bottom right corner) also has similar observations, where 1995 and 1997 are almost overlapping, but far from 2008. That closer years are more related is quite intuitive. Researchers tend to cite more recent papers, authors also collaborate with recently active researchers. Due to intra-layer modeling, our model is able to capture these homogeneous connections.

Figure 6 also depicts close relationships between conferences and their highly-profiled authors. Moreover, different areas tend to display some separation. Data points from Data Mining, Databases, and Artificial Intelligence dominate the right-hand side, while left-hand side has more from Information Security, Operating Systems, and Computer Architecture. This layout among conferences from diverse domains, and among authors actively involved in conferences of different years, demonstrates the embedding ability of MULTILAYEREDHNE to preserve network heterogeneity.

6   Conclusion

We formalize the multi-layered heterogeneous network embedding problem, and propose a novel framework MULTILAYEREDHNE to model intra- and cross-layer proximity. Due to the dual characters of multi-layered networks on homogeneity and heterogeneity, our model learns node embeddings that preserve network topology and semantics jointly. Extensive experiments verify the effectiveness of our model on four public datasets. With ablation analysis, we show that our model could effectively benefit from both modeling components.

Acknowledgments. This research is supported by the National Research Foundation, Prime Minister's Office, Singapore under its NRF Fellowship Programme (Award No. NRF-NRFF2016-07).

References

1. Bishop, C.M.: Pattern Recognition and Machine Learning. Springer, New York (2006)
2. Cao, S., Lu, W., Xu, Q.: GraRep: learning graph representations with global structural information. In: Proceedings of the 24th ACM International on Conference on Information and Knowledge Management, pp. 891–900 (2015)
3. Cen, Y., Zou, X., Zhang, J., Yang, H., Zhou, J., Tang, J.: Representation learning for attributed multiplex heterogeneous network. In: Proceedings of the 25th ACM SIGKDD International Conference on Knowledge Discovery and Data Mining, pp. 1358–1368 (2019)
4. Chen, C., Tong, H., Xie, L., Ying, L., He, Q.: FASCINATE: fast cross-layer dependency inference on multi-layered networks. In: Proceedings of the 22nd ACM SIGKDD International Conference on Knowledge Discovery and Data Mining, pp. 765–774 (2016)
5. Chen, C., Tong, H., Xie, L., Ying, L., He, Q.: Cross-dependency inference in multi-layered networks: a collaborative filtering perspective. ACM Trans. Knowl. Discov. Data (TKDD) **11**(4), 1–26 (2017)
6. Dong, Y., Chawla, N.V., Swami, A.: metapath2vec: scalable representation learning for heterogeneous networks. In: Proceedings of the 23rd ACM SIGKDD International Conference on Knowledge Discovery and Data Mining, pp. 135–144 (2017)
7. Fu, T.Y., Lee, W.C., Lei, Z.: Hin2vec: explore meta-paths in heterogeneous information networks for representation learning. In: Proceedings of the 2017 ACM on Conference on Information and Knowledge Management, pp. 1797–1806 (2017)
8. Garcia Duran, A., Niepert, M.: Learning graph representations with embedding propagation. In: Advances in Neural Information Processing Systems, pp. 5119–5130 (2017)
9. Grover, A., Leskovec, J.: node2vec: scalable feature learning for networks. In: Proceedings of the 22nd ACM SIGKDD International Conference on Knowledge Discovery and Data Mining, pp. 855–864 (2016)
10. He, X., Liao, L., Zhang, H., Nie, L., Hu, X., Chua, T.S.: Neural collaborative filtering. In: Proceedings of the 26th International Conference on World Wide Web, pp. 173–182 (2017)

11. Hu, B., Fang, Y., Shi, C.: Adversarial learning on heterogeneous information networks. In: Proceedings of the 25th ACM SIGKDD International Conference on Knowledge Discovery and Data Mining, pp. 120–129 (2019)
12. Hu, B., Shi, C., Zhao, W.X., Yu, P.S.: Leveraging meta-path based context for top-n recommendation with a neural co-attention model. In: Proceedings of the 24th ACM SIGKDD International Conference on Knowledge Discovery and Data Mining, pp. 1531–1540 (2018)
13. Kipf, T.N., Welling, M.: Variational graph auto-encoders. In: NIPS workshop on Bayesian Deep Learning (2016)
14. Li, J., Chen, C., Tong, H., Liu, H.: Multi-layered network embedding. In: Proceedings of the 2018 SIAM International Conference on Data Mining, pp. 684–692 (2018)
15. Liu, L., Cheung, W.K., Li, X., Liao, L.: Aligning users across social networks using network embedding. In: Proceedings of the Twenty-Fifth International Joint Conference on Artificial Intelligence, pp. 1774–1780 (2016)
16. Liu, W., Chen, P.Y., Yeung, S., Suzumura, T., Chen, L.: Principled multilayer network embedding. In: 2017 IEEE International Conference on Data Mining Workshops (ICDMW), pp. 134–141 (2017)
17. Van der Maaten, L., Hinton, G.: Visualizing data using t-SNE. J. Mach. Learn. Res. **9**, 2579–2605 (2008)
18. Ni, J., Tong, H., Fan, W., Zhang, X.: Inside the atoms: ranking on a network of networks. In: Proceedings of the 20th ACM SIGKDD International Conference on Knowledge Discovery and Data Mining, pp. 1356–1365 (2014)
19. Perozzi, B., Al-Rfou, R., Skiena, S.: DeepWalk: online learning of social representations. In: Proceedings of the 20th ACM SIGKDD International Conference on Knowledge Discovery and Data Mining, pp. 701–710 (2014)
20. Shi, C., Hu, B., Zhao, W.X., Philip, S.Y.: Heterogeneous information network embedding for recommendation. IEEE Trans. Knowl. Data Eng. **31**(2), 357–370 (2018)
21. Shi, C., Kong, X., Huang, Y., Philip, S.Y., Wu, B.: HeteSim: a general framework for relevance measure in heterogeneous networks. IEEE Trans. Knowl. Data Eng. **26**(10), 2479–2492 (2014)
22. Sun, Y., Barber, R., Gupta, M., Aggarwal, C.C., Han, J.: Co-author relationship prediction in heterogeneous bibliographic networks. In: 2011 International Conference on Advances in Social Networks Analysis and Mining, pp. 121–128 (2011)
23. Tang, J., Qu, M., Wang, M., Zhang, M., Yan, J., Mei, Q.: Line: large-scale information network embedding. In: Proceedings of the 24th International Conference on World Wide Web, pp. 1067–1077 (2016)
24. Veličković, P., Cucurull, G., Casanova, A., Romero, A., Lio, P., Bengio, Y.: Graph attention networks. In: Proceedings of International Conference on Learning Representations (2018)
25. Wang, X., et al.: Heterogeneous graph attention network. In: The World Wide Web Conference, pp. 2022–2032 (2019)
26. Ying, R., He, R., Chen, K., Eksombatchai, P., Hamilton, W.L., Leskovec, J.: Graph convolutional neural networks for web-scale recommender systems. In: Proceedings of the 24th ACM SIGKDD International Conference on Knowledge Discovery and Data Mining, pp. 974–983 (2018)
27. Zhang, C., Lauw, H.W.: Topic modeling on document networks with adjacent-encoder. Proc. AAAI Conf. Artif. Intell. **34**(04), 6737–6745 (2020)

28. Zhang, J., Kong, X., Philip, S.Y.: Predicting social links for new users across aligned heterogeneous social networks. In: 2013 IEEE 13th International Conference on Data Mining, pp. 1289–1294 (2013)
29. Zhang, J., Philip, S.Y.: Integrated anchor and social link predictions across social networks. In: Twenty-Fourth International Joint Conference on Artificial Intelligence, pp. 2125–2131 (2015)

Adaptive Node Embedding Propagation for Semi-supervised Classification

Yuya Ogawa[1]([✉]), Seiji Maekawa[1], Yuya Sasaki[1], Yasuhiro Fujiwara[2], and Makoto Onizuka[1]

[1] Graduate School of Information Science of Technology, Osaka University,
1-5 Yamadaoka, Suita, Osaka, Japan
{maekawa,sasaki,onizuka}@ist.osaka-u.ac.jp
[2] NTT Communication Science Laboratories, 3-1, Morinosato Wakamiya,
Atsugi-shi, Kanagawa, Japan
yasuhiro.fujiwara.kh@hco.ntt.co.jp

Abstract. Graph Convolutional Networks (GCNs) are state-of-the-art approaches for semi-supervised node classification task. By increasing the number of layers, GCNs utilize high-order relations between nodes that are more than two hops away from each other. However, GCNs with many layers face three drawbacks: (1) over-fitting due to the increasing number of parameters, (2) over-smoothing in which embeddings converge to similar values, and (3) the difficulty in selecting the appropriate number of propagation hops. In this paper, we propose ANEPN that effectively utilizes high-order relations between nodes by overcoming the above drawbacks of GCNs. First, we introduce Embedding Propagation Loss which increases the number of propagation hops while keeping the number of parameters constant for mitigating over-fitting. Second, we propose Anti-Smoothness Loss (ASL) that prevents embeddings from converging to similar values for avoiding over-smoothing. Third, we introduce a metric for predicted class labels for adaptively controlling the number of propagation hops. We show that ANEPN outperforms ten state-of-the-art approaches on three standard datasets.

1 Introduction

Semi-supervised node classification aims to predict the class labels of nodes in a graph, a popular task in graph analysis. Graph Convolutional Networks (GCNs) [12], which are a family of neural networks, have achieved the state-of-the-art performance on the task. GCNs combine graph structure and node features by propagating node embeddings to neighbors on a graph. Recently, many variants of GCNs have been developed, including Graph attention networks [20,26], GraphSage [11], GraphUnet [8], and LNet [14].

In the previous approaches, GCNs typically consist of two layers in performing semi-supervised node classification; they do not utilize high-order relations between nodes more than two hops away from each other. We should utilize high-order relations by stacking many layers of GCNs in semi-supervised node

N. Oliver et al. (Eds.): ECML PKDD 2021, LNAI 12976, pp. 417–433, 2021.
https://doi.org/10.1007/978-3-030-86520-7_26

classification, since high-order relations are known to be beneficial when labeled data size is small [24]. However, GCNs with many layers have the following three drawbacks. First, GCNs with many layers are prone to over-fitting to labeled dataset [28] since the number of parameters in GCNs increases with the number of layers. Second, stacking many layers of GCNs causes over-smoothing of node embeddings, i.e., a phenomenon in which the embeddings of all connected nodes converge to similar values [13]. Third, existing GCNs need to predefine the number of layers, i.e., the number of propagation hops. This indicates that they require careful effort to select the number of propagation hops since the appropriate number of hops is not known in advance depending on input graphs.

To address the above three drawbacks, this paper presents Adaptive Node Embedding Propagation Network (ANEPN) that utilizes high-order relations between nodes by adaptively increasing the number of propagation hops. ANEPN is effective for semi-supervised node classification, since it can leverages high-order relations that is more beneficial as labeled data size gets smaller. First, to avoid over-fitting, we introduce Embedding Propagation Loss (EPL), which increases the number of propagation hops while keeping the number of parameters constant. The novel idea of EPL is that node embeddings are propagated by an update rule of the gradient descent. Thus, ANEPN can propagate node embeddings to neighbors within many hops by iteratively applying the update rule without increasing the number of layers. Second, to avoid over-smoothing, we additionally introduce Anti-Smoothness Loss (ASL). ASL imposes a penalty on structurally distant nodes (not directly linked nodes) in a graph, so that those nodes have dissimilar embeddings. Third, ANEPN automatically selects the appropriate number of propagation hops by employing the idea of metric learning [6]. That is, ANEPN controls the number of propagation hops by using a metric that evaluates predicted class labels; it minimizes within-class variance of node embeddings and maximizes between-classes variance of node embeddings. Our experimental results on three datasets show that ANEPN outperforms ten state-of-the-art approaches under various label rates.

The rest of the paper is organized as follows. Section 2 introduces preliminaries. In Sect. 3, we propose ANEPN to address the drawbacks of GCNs. In Sect. 4, we conduct experiments to validate the effectiveness of ANEPN on semi-supervised node classification. Section 5 introduces related work. Finally, Sect. 6 concludes this paper.

2 Preliminaries

2.1 Problem Definition

An attribute graph $\mathcal{G}$ with N nodes is represented by an adjacency matrix $\boldsymbol{A}$, a feature matrix $\boldsymbol{X}$ and a class matrix $\boldsymbol{T}$. In the adjacency matrix $\boldsymbol{A} \in \mathbb{R}^{N \times N}$, the (i,j)-th element $A_{ij} = 1$ if there is an edge between node i and node j, $A_{ij} = 0$ otherwise. $\boldsymbol{D} \in \mathbb{R}^{N \times N} = diag(d_1, \cdots, d_N)$ denotes the degree matrix of $\boldsymbol{A}$, where $d_i = \sum_j A_{ij}$ is the degree of node i. The normalized graph laplacian

Table 1. Notations and definitions

N	Number of nodes
F	Dimension of features
C	Number of classes
E	Number of edges
H	Dimension of hidden layer
$\boldsymbol{A} \in \mathbb{R}^{N \times N}$	Adjacency matrix
$\boldsymbol{X} \in \mathbb{R}^{N \times F}$	Feature matrix
$\boldsymbol{T} \in \mathbb{R}^{N \times C}$	Class matrix
$\boldsymbol{D} \in \mathbb{R}^{N \times N}$	Degree matrix
$\tilde{\boldsymbol{L}} \in \mathbb{R}^{N \times N}$	Normalized graph laplacian matrix
$\boldsymbol{G} \in \mathbb{R}^{N \times N}$	Graph filter matrix
$\boldsymbol{W}^{(0)} \in \mathbb{R}^{F \times H}, \boldsymbol{W}^{(1)} \in \mathbb{R}^{H \times C}$	Weight matrices
$\boldsymbol{B}^{(0)} \in \mathbb{R}^{N \times H}, \boldsymbol{B}^{(1)} \in \mathbb{R}^{N \times C}$	Bias matrices
$\boldsymbol{Z} \in \mathbb{R}^{N \times H}$	Node embeddings
$\boldsymbol{Y} \in \mathbb{R}^{N \times C}$	Predicted label matrix
α, β	Coefficients of losses
μ	Margin
$\Delta \alpha$	Increasing step
T_i	Interval of increasing
T_p	Iteration to pre-train
T_{max}	Maximum iteration
$patience$	Patience for training stop

matrix $\tilde{\boldsymbol{L}}$ are defined as $\tilde{\boldsymbol{L}} = \boldsymbol{I} - \boldsymbol{D}^{-\frac{1}{2}} \boldsymbol{A} \boldsymbol{D}^{-\frac{1}{2}}$, where $\boldsymbol{I} \in \mathbb{R}^{N \times N}$ is an identity matrix. The feature matrix $\boldsymbol{X} \in \mathbb{R}^{N \times F}$ represents node features information, where the i-th row vector $\boldsymbol{X}_{i,:}$ is the F-dimensional feature vector of node i. The class matrix $\boldsymbol{T} \in \mathbb{R}^{N \times C}$ contains class information of each node, where C is the number of classes. Nodes in a given attributed graph are divided into two sets, the labeled node set $\mathcal{V}^L$ and the unlabeled node set $\mathcal{V}^U$. $\boldsymbol{T}_{i,:} \in \{0,1\}^C$ is a one-hot vector if node i is included in the labeled node set $\mathcal{V}^L$, and $\boldsymbol{T}_{i,:}$ is a zero vector otherwise. We summarize notations and their definitions in Table 1. For a given attributed graph $\mathcal{G}$, semi-supervised node classification aims to predict the class labels of unlabeled nodes in the unlabeled node set $\mathcal{V}^U$.

2.2 Graph Convolutional Networks

A two-layer GCN [12] is a standard GCN model. It has a single hidden layer and an output layer, and applies graph convolution in each layer. The GCN propagates the embedding of a node to its neighbors within two hops since the number of propagation hops corresponds to the number of layers. Graph convolution is defined as the multiplication of a graph filter matrix $G \in \mathbb{R}^{N \times N}$ and the feature matrix X (or node embeddings). The graph filter is known as a low-pass filter that filters out noises in node features [23]. Kipf and Welling (2017) defines the graph filter as:

$$G = I + \tilde{L}. \tag{1}$$

They further applied a normalization trick to the graph filter G as:

$$I + \tilde{L} \rightarrow \tilde{D}^{-\frac{1}{2}} \tilde{A} \tilde{D}^{-\frac{1}{2}}, \tag{2}$$

where $\tilde{A} = A + I$ and $\tilde{D}_{ii} = \sum_j \tilde{A}_{ij}$. Let H be the dimension of the hidden layer. Let $W^{(0)} \in \mathbb{R}^{F \times H}$ and $B^{(0)} \in \mathbb{R}^{N \times H}$ denote the weight matrix and the bias matrix[1] in the hidden layer, respectively. In the hidden layer, the GCN outputs node embeddings $Z \in \mathbb{R}^{N \times H}$ as:

$$Z = \max(GXW^{(0)} + B^{(0)}, 0). \tag{3}$$

In the output layer, the GCN outputs a predicted label matrix $Y \in \mathbb{R}^{N \times C}$ as:

$$Y = \text{softmax}(GZW^{(1)} + B^{(1)}), \tag{4}$$

where $W^{(1)} \in \mathbb{R}^{H \times C}$ and $B^{(1)} \in \mathbb{R}^{N \times C}$ denote the weight matrix and the bias matrix in the output layer, respectively. Besides, $\text{softmax}(P)_{ic} = \frac{P_{ic}}{\sum_c P_{ic}}$ for a matrix P. In Eq. (4), the GCN propagates the embedding of a node to its neighbors by the multiplication of the graph filter G and node embeddings Z as:

$$(GZ)_{ij} = \frac{Z_{ij}}{d_i + 1} + \sum_{k \in \Gamma(i)} \frac{Z_{kj}}{\sqrt{d_i + 1}\sqrt{d_k + 1}}, \tag{5}$$

where $\Gamma(i)$ denotes the set of neighbors of node i. In Eq. (5), the first and second terms on the right side work as propagating node embeddings from a node and its neighbors to the node itself. So, the coefficients $\frac{1}{d_i + 1}$ and $\frac{1}{\sqrt{d_i + 1}\sqrt{d_k + 1}}$ indicate propagation weights for self loops and neighbors, respectively. The loss function of the GCN is defined as the cross entropy loss to minimize the difference between the class labels of labeled nodes in predicting class labels:

$$L_{ce} = - \sum_{i,c=1}^{N,C} T_{ic} \log(Y_{ic}). \tag{6}$$

The parameters ($W^{(0)}$, $B^{(0)}$, $W^{(1)}$ and $B^{(1)}$) are trained to decrease the loss.

3   Our Approach

This section presents ANEPN, a new node embedding propagation approach that effectively utilizes high-order relations between nodes by overcoming three

---

[1] We express bias in a matrix form by expanding a bias vector.

issues: over-fitting, over-smoothing, and the difficulty in selecting the appropriate number of propagation hops. The loss L for ANEPN consists of cross entropy loss L_{ce} and two additional losses: Embedding Propagation Loss (EPL) L_{ep} and Anti-Smoothness Loss (ASL) L_{asm} as follows;

$$L = L_{ce} + \alpha L_{ep} + \beta L_{asm}, \tag{7}$$

where α and β are the coefficients of EPL and ASL, respectively. EPL enables ANEPN to increase the number of propagation hops while keeping the number of parameters constant, so ANEPN avoids over-fitting. ANEPN also avoids over-smoothing of node embeddings by using ASL since it prevents embeddings of all connected nodes from converging to similar values. Furthermore, ANEPN automatically selects the appropriate number of propagation hops by introducing a metric that evaluates the quality of predicted class labels.

3.1 Embedding Propagation Loss

The idea of EPL is that node embeddings are propagated by an update rule of the gradient descent. To this end, we design EPL as a function of node embeddings Z and then formulate the update rule of Z that works as graph convolution to Z. By iteratively applying the update rule, ANEPN propagates node embeddings without increasing the number of layers, i.e., increasing the number of parameters. Hence, ANEPN can propagate node embeddings to neighbors within many hops without over-fitting. First, we define a general form of update rule for node embeddings Z. Then, we derive EPL from the update rule.

Update Rule for Node Embeddings. Let $\frac{\partial L_{ep}}{\partial W^{(0)}_{:,h}}$ and $\frac{\partial L_{ep}}{\partial B^{(0)}_{:,h}}$ be the gradient of L_{ep} with respect to $W^{(0)}_{:,h}$ and $B^{(0)}_{:,h}$, respectively, since $Z = GXW^{(0)} + B^{(0)}$.[2] Let η denote a learning rate. The h-th column vector $Z_{:,h} \in \mathbb{R}^N$ is updated according to the gradient descent algorithm as follows:

$$Z_{:,h} \leftarrow GX(W^{(0)}_{:,h} - \eta \frac{\partial L_{ep}}{\partial W^{(0)}_{:,h}}) + B^{(0)}_{:,h} - \eta \frac{\partial L_{ep}}{\partial B^{(0)}_{:,h}}. \tag{8}$$

Let $\frac{\partial L_{ep}}{\partial Z_{:,h}}$ be the gradient of L_{ep} with respect to $Z_{:,h}$. By the chain rule (i.e., $\frac{\partial L_{ep}}{\partial W^{(0)}_{:,h}} = \frac{\partial L_{ep}}{\partial Z_{:,h}} \frac{\partial Z_{:,h}}{\partial W^{(0)}_{:,h}}$ and $\frac{\partial L_{ep}}{\partial B^{(0)}_{:,h}} = \frac{\partial L_{ep}}{\partial Z_{:,h}} \frac{\partial Z_{:,h}}{\partial B^{(0)}_{:,h}}$), Eq. (8) is rewritten as:

$$Z_{:,h} \leftarrow Z_{:,h} - \eta GX(GX)^T \frac{\partial L_{ep}}{\partial Z_{:,h}} - \eta \frac{\partial L_{ep}}{\partial Z_{:,h}}, \tag{9}$$

In order to propagate node embeddings, we design $\frac{\partial L_{ep}}{\partial Z_{:,h}}$ to express graph convolution to node embeddings, i.e., the multiplication of a graph filter ($\gamma \tilde{L} - \nu I$)

---

[2] We ignore $\max(\cdot, 0)$ since we remove it later (see Sect. 3.4).

and node embeddings where γ and ν are scalar parameters:

$$\frac{\partial L_{ep}}{\partial \mathbf{Z}_{:,h}} = (\gamma \tilde{\mathbf{L}} - \nu \mathbf{I}) \mathbf{Z}_{:,h}. \tag{10}$$

By using Eq. (10), the third term in Eq. (9) updates $\mathbf{Z}$ as follows:

$$\mathbf{Z}_{ih} \leftarrow (\nu - \gamma) \mathbf{Z}_{ih} + \gamma \sum_{j \in \Gamma(i)} \frac{\mathbf{Z}_{jh}}{\sqrt{d_i} \sqrt{d_j}}. \tag{11}$$

The update rule of Eq. (11) corresponds to a single hop propagation of node embeddings $\mathbf{Z}$. The first and second terms in the update rule work as propagating node embeddings from a node and its neighbors to the node itself.[3] Therefore, ANEPN propagates node embeddings in multiple hops by iteratively applying the update rule instead of stacking many layers including parameters, e.g., weight matrices. As a result, ANEPN can utilize high-order relations by increasing the number of iterations t while keeping the number of the parameters constant.

Derivation of EPL. We derive EPL L_{ep} by integrating both sides in Eq. (10):

$$L_{ep}(\gamma, \nu) = \int (\gamma \tilde{\mathbf{L}} - \nu \mathbf{I}) \mathbf{Z}_{:,h} d\mathbf{Z}_{:,h}. \tag{12}$$

Equation (12) shows a general form of embedding propagation loss, which is parameterized by the propagation weights (γ and ν) in graph convolution. This equation suggests that we can have several variations of EPL by assigning the scalar parameters γ and ν. As an example, we consider a smoothness loss, which measures the smoothness of embeddings between adjacent nodes; it corresponds to the smoothness used in [7,27]. Specifically, by assigning ν to $\frac{2\mathbf{Z}_{:,h}^T \mathbf{Z}_{:,h}}{H(\mathbf{Z}_{:,h}^T \mathbf{Z}_{:,h})^2}$ and γ to $\frac{2\mathbf{Z}_{:,h}^T \tilde{\mathbf{L}} \mathbf{Z}_{:,h}}{H(\mathbf{Z}_{:,h}^T \mathbf{Z}_{:,h})^2}$, we have:

$$L_{ep} = \frac{1}{2H} \sum_{h,i,j=1}^{H,N,N} \mathbf{A}_{ij} \left(\frac{\hat{\mathbf{Z}}_{ih}}{\sqrt{d_i}} - \frac{\hat{\mathbf{Z}}_{jh}}{\sqrt{d_j}} \right)^2, \tag{13}$$

where $\hat{\mathbf{Z}}_{:,h} = \frac{\mathbf{Z}_{:,h}}{||\mathbf{Z}_{:,h}||}$. By minimizing EPL (smoothness), the node embeddings become smoother, i.e., structurally near nodes on a graph have more similar embeddings. We use this variation of EPL in our experiments. It is our future work to investigate other variations.

3.2 Anti-smoothness Loss

To avoid over-smoothing of node embeddings, we introduce Anti-Smoothness Loss (ASL) which increases the smoothness of embeddings between structurally

---

[3] From the aspect of node embedding propagation, the update rule in Eq. (11) has a similar effect to the GCN propagation rule in Eq. (5).

distant nodes so that the nodes have dissimilar node embeddings. We refer to the smoothness of embeddings between structurally distant nodes as anti-smoothness. It is defined as the sum of distances between embeddings of structurally distant nodes as follows:

$$\frac{1}{2H} \sum_{h,i,j=1}^{H,N,N} \bar{A}_{ij} (\frac{\hat{Z}_{ih}}{\sqrt{\bar{d}_i}} - \frac{\hat{Z}_{jh}}{\sqrt{\bar{d}_j}})^2, \tag{14}$$

where $\bar{d}_i = \sum_j \bar{A}_{ij}$. In Eq. (14), $\bar{A} = J - A$ where $J_{ij} = 0$ when $i = j$, and $J_{ij} = 1$ otherwise. $\bar{A}$ is a negative edge matrix expressed with virtual edges between structurally distant nodes. In order to reduce computational cost, we under-sample $2E$ negative edges from $\bar{A}$ where E is the number of edges in A by following [16], since the number of negative edges in $\bar{A}$ is very large. For simplification, we adopt random uniform sampling[4]. Replacing $\bar{A}$ with the sampled negative edge matrix $\bar{A}^{sam}$, we formulate ASL function as follows:

$$L_{asm} = \max(\mu - \frac{1}{2H} \sum_{h,i,j=1}^{H,N,N} \bar{A}_{ij}^{sam} (\frac{\hat{Z}_{ih}}{\sqrt{\bar{d}_i}} - \frac{\hat{Z}_{jh}}{\sqrt{\bar{d}_j}})^2, 0). \tag{15}$$

In Eq. (15), to ensure that ASL is positive value, we introduce a margin parameter μ indicating the degree of the anti-smoothness and use $\max(\cdot, 0)$ by utilizing the idea used in [10]. We use the same coefficient both for EPL and ASL by assigning $\beta = \alpha$ (see Eq. (7)), since these losses adversarially work to each other [9].

3.3   Adaptive Propagation Control

To adaptively control the number of iterations t, i.e., the number of propagation hops in semi-supervised node classification, we introduce a metric to evaluate predicted class labels. This metric is beneficial in particular when labeled data size is small, since it does not use given labels. We can adaptively control t based on the metric, depending on input graphs.

To introduce the metric, we employ the idea of Siamese Network [6], which proposes a metric learning. Our idea is that we select the appropriate number of propagation hops by leveraging class variance used in Siamese Network. It minimizes within-class variance of node embeddings (i.e., the sum of squared distances between embeddings of nodes with the same class labels) and maximizes between-classes variance of node embeddings (i.e., the sum of squared distances between embeddings of nodes with different class labels). Let $m = \frac{1}{N} \sum_i Z_{i,:}$ denote the average vector of embeddings of all nodes. Besides, $m_q = \frac{1}{N_q} \sum_{i \in C_q} Z_{i,:}$ denote the average vector of embeddings of nodes in class

---

[4] We remove negative edges between nodes with one or more common neighbors from sampling because those nodes are considered to be structurally close to each other [19,21].

Algorithm 1 ANEPN

Input: attributed graph $\mathcal{G} = \{A, X, T\}$, number of sampled negative edge $\bar{E} = 2E$,
increasing step $\Delta\alpha$, interval of increasing T_{int}, iteration to pre-train T_p, maximum
iteration T_{max}, number of patience for training stop *patience*
Output: predicted class labels Y
1: ### **Parameter initialization and Pre-process** ###
2:   $\alpha \leftarrow 0$ ▷ coefficient for EPL and ASL
3:   $pat\_count \leftarrow 0$ ▷ Patience count for training stop
4:   $\bar{X} \leftarrow G^2 X$ ▷ pre-smoothing
5:   $\bar{A}^{sam} \leftarrow$ negative edge sampling$(A, \bar{E})$
6: ### **Pre-training** ###
7: **for** t from 1 to T_p **do**
8:     $Z \leftarrow \bar{X} W^{(0)} + B^{(0)}$
9:     $Y \leftarrow$ softmax$(Z W^{(1)} + B^{(1)})$
10:    $loss \leftarrow L_{ce}$
11:    Update parameters $(W^{(0)}, B^{(0)}, W^{(1)}, B^{(1)})$
12: ### **Training** ###
13: **for** t from T_p to T_{max} **do**
14:    $Z \leftarrow \bar{X} W^{(0)} + B^{(0)}$
15:    $Y \leftarrow$ softmax$(Z W^{(1)} + B^{(1)})$
16:    **if** $t \% T_{int} == 0$ **then**
17:        $\alpha+ = \Delta\alpha$
18:    **if** $\text{VR}^{(t)}(Z, Y) \leq \text{VR}^{(t-1)}(Z, Y)$ **then**
19:        $pat\_count+ = 1$
20:    **if** $patience == pat\_count$ **then**
21:        break
22:    $loss \leftarrow L_{ce} + \alpha L_{ep} + \alpha L_{asm}$
23:    Update parameters $(W^{(0)}, B^{(0)}, W^{(1)}, B^{(1)})$
24: **return** Y

C_q, where N_q is the number of nodes in class C_q. By using node embedding Z
and predicted labels Y, we define the within-class variance as trace $\text{Tr}(W_c)$ of
the within-class variance matrix $W_c \in \mathbb{R}^{N \times N}$ as follows:

$$\text{Tr}(W_c) = \text{Tr}(\sum_{q=1}^{C} \sum_{i \in C_q} (Z_{i,:} - m_q)^T (Z_{i,:} - m_q)). \tag{16}$$

Similarly, we define the between-classes variance as trace $\text{Tr}(B_c)$ of the between-
classes variance matrix $B_c \in \mathbb{R}^{N \times N}$ as follows:

$$\text{Tr}(B_c) = \text{Tr}(\sum_{q=1}^{C} N_q (m_q - m)^T (m_q - m)). \tag{17}$$

Finally, by extending the Calinski-Harabasz index [4], which is a clustering performance metric, we define the metric to evaluate all node embeddings and

predicted labels as ratio of the above two variances:

$$\mathrm{VR}(\boldsymbol{Z}, \boldsymbol{Y}) = \frac{\mathrm{Tr}(\boldsymbol{B}_c)}{\mathrm{Tr}(\boldsymbol{W}_c)} \times \frac{N - C}{C - 1} \tag{18}$$

We adaptively control the number of iterations t so that $\mathrm{VR}(\boldsymbol{Z}, \boldsymbol{Y})$ is higher. To improve classification accuracy, we also control the coefficient α of two additional losses, EPL and ASL. In the proposed approach, we iteratively increment t and the coefficient α, and compute $\mathrm{VR}^{(t)}(\boldsymbol{Z}, \boldsymbol{Y})$, which is $\mathrm{VR}(\boldsymbol{Z}, \boldsymbol{Y})$ with respect to t, in each iteration. If the variance ratio is smaller than that in previous iteration, i.e., if $\mathrm{VR}^{(t)}(\boldsymbol{Z}, \boldsymbol{Y}) \leq \mathrm{VR}^{(t-1)}(\boldsymbol{Z}, \boldsymbol{Y})$, we increment patience count p by 1. Then, we stop the iteration if p equals a given patience q. By using patience as in early stopping, we can avoid stopping the iteration when the variance ratio unexpectedly decreases.

3.4 Architecture of ANEPN

We change the architecture of the two-layer GCN [12] to improve classification performance. First, we remove $\max(\cdot, 0)$ in the hidden layer (see Eq. (3)) since the function could lead to the over-smoothing [17]. Second, because node embeddings are propagated by using EPL, we move the graph filter $\boldsymbol{G}$ in the output layer (see Eq. (4)) into the hidden layer as follows:

$$\boldsymbol{Z} = \boldsymbol{G}^2 \boldsymbol{X} \boldsymbol{W}^{(0)} + \boldsymbol{B}^{(0)}, \tag{19}$$

$$\boldsymbol{Y} = \mathrm{softmax}(\boldsymbol{Z} \boldsymbol{W}^{(1)} + \boldsymbol{B}^{(1)}). \tag{20}$$

Third, since the graph filter $\boldsymbol{G} = \tilde{\boldsymbol{D}}^{-\frac{1}{2}} \tilde{\boldsymbol{A}} \tilde{\boldsymbol{D}}^{-\frac{1}{2}}$ used in [12] may not appropriately filter out noises, we adopt the optimal low-pass filter $\boldsymbol{G} = \boldsymbol{I} - \frac{1}{2}\tilde{\boldsymbol{L}}$ used in [27] to filter out noises in the feature matrix (Table 2).

Table 2. Dataset statistics

Dataset	#Nodes	#Edges	#Features	#Classes
Cora	2708	5429	1433	7
Citeseer	3312	4732	3703	6
Pubmed	19717	44338	500	3

We show the algorithm of ANEPN in Algorithm 1. First, we initialize the parameters, filter the feature matrix, and obtain the sampled negative edge matrix (lines 1–5). In order to prevent ANEPN from using randomly initialized embeddings, we pre-train the parameters $\{\boldsymbol{W}^{(0)}, \boldsymbol{B}^{(0)}, \boldsymbol{W}^{(1)}, \boldsymbol{B}^{(1)}\}$ in ANEPN by using only cross entropy loss (lines 6–11). After pre-training, we increase the coefficient α with an interval of T_{int} (lines 16–17) and evaluate the variance ratio by using embedding $\boldsymbol{Z}$ and predicted labels $\boldsymbol{Y}$ (lines 18–19), in addition to training of parameters (lines 14–15 and 22–23). Finally, ANEPN outputs the predicted class labels (line 24).

In Algorithm 1, the complexity of the filtering of the feature matrix is in $\mathcal{O}(EF)$. The complexity of negative edge sampling is in $\mathcal{O}(Ed)$, where d is the average degree over all nodes. In the training of ANEPN, the complexity of the feed-forward is in $\mathcal{O}(FHC)$. In addition, ANEPN requires the calculations of EPL, ASL, and the variance ratio. Those complexities are in $\mathcal{O}(EH)$, $\mathcal{O}(EH)$, and $\mathcal{O}(NH)$, respectively. Thus, the overall time complexity of ANEPN is in $\mathcal{O}(t(FHC + EH) + EF)$, where we assume that E is larger than N since real-world graphs often follow such assumption. Therefore, ANEPN is more efficient than GCN: its complexity $\mathcal{O}(tEFHC)$ [12].

4   Experiments

We design this section to answer the following five questions; **Q1**: Does ANEPN outperform existing approaches in the term of accuracy?, **Q2**: Does ANEPN outperform existing approaches in the term of efficiency?, **Q3**: Is ASL effective to avoid over-smoothing?, **Q4**: Does ANEPN appropriately control the number of propagation hops?, and **Q5**: How does the margin parameter μ affect accuracy and variance ratio? **Q6**: How do EPL and ASL affect node embeddings?

4.1   Settings

By following the settings in [18], we use three datasets of attributed graphs: Cora, Citeseer, and Pubmed. Cora, Citeseer, and Pubmed are citation networks, in which the nodes and edges represent publications and citations, respectively. The nodes in Cora and Citeseer are associated with binary word vectors, and they in Pubmed are associated with tf-idf weighted word vectors. Publications in Cora, Citeseer and Pubmed are categorized by the research sub-fields. See Ref. [13] for more details.

We compare the proposed approach against ten state-of-the-art competitors. Label propagation (LP) [29] propagates labels on a graph. Graph Convolutional Network (GCN) [12] and Graph Attention Network (GAT) [20] are popular two-layer GCNs. Self-training, Co-training, Union, and Intersection [13] incorporate self-training and co-training into the training of GCN. Multi-Stage Self-Supervised Training Algorithm (M3S) [18] integrates self-supervised learning with GCN. ALaGCN and ALaGAT [24] utilize high-order relations by increasing the number of propagation hops. We implement ANEPN with Pytorch[5] 1.7.0. Our code can be found at https://github.com/suzu97t/ANEPN.

For ANEPN, we use Adam optimizer with a learning rate of 0.01 and set hidden layer size to 64 and weight decay to 5e-4 by following the same setting to [20]. For other our original parameters, we set margin μ, increasing step $\Delta\alpha$, interval of increasing T_i, iteration to pre-train T_p, maximum iteration T_{max}, and

---

[5] https://pytorch.org/.

patience for training stop *patience* to 1, 0.05, 10, 50, 500, and 10, respectively. We tune the above our original parameters so that we stop training the model when the variance ratio is the highest. For competitors, we follow the settings used in each original paper.

For each run, we randomly split nodes into a small set for training, and a set with 1000 samples for testing. By following the settings used in [18], we test our approach and competitors under 0.5%, 1%, 2%, 3%, 4% label rates on Cora and Citeseer, and 0.03%, 0.05%, 0.1% label rates on Pubmed.

Table 3. Test accuracy (%) on Cora, Citeseer and Pubmed, where bold numbers indicate the best results. Gain-GCN and Gain-SOTA represent the difference between the results of our approach and GCN, and that between the results of our approach and the best results in competitors.

Label rate	Cora					Citeseer					Pubmed		
	0.5%	1%	2%	3%	4%	0.5%	1%	2%	3%	4%	0.03%	0.05%	0.1%
LP	54.3	60.1	64.0	65.3	66.5	37.7	42.0	44.2	45.7	46.3	58.6	61.9	66.9
GCN	44.5	59.8	68.7	74.4	77.0	43.6	47.4	61.7	66.8	68.6	45.6	55.0	64.9
GAT	41.1	50.2	54.2	60.3	77.0	40.1	46.2	62.8	67.0	68.7	50.2	53.0	60.5
Self-training	55.4	62.5	73.0	76.4	79.1	48.4	59.5	65.4	66.0	70.2	58.7	59.2	66.6
Co-training	50.1	60.3	69.5	76.2	77.8	39.5	53.2	63.5	66.6	69.8	53.3	59.2	63.4
Union	45.7	57.3	72.5	76.3	77.2	41.2	52.9	62.7	65.6	68.1	47.2	59.1	66.3
Intersection	48.7	60.9	73.0	77.3	79.8	49.1	60.1	63.7	68.3	69.4	49.2	54.1	69.7
M3S	59.9	66.7	75.8	77.4	79.2	54.2	62.7	66.2	69.8	70.4	57.0	62.9	68.4
ALaGCN	57.9	66.7	73.7	74.6	78.5	41.0	49.7	59.3	63.5	67.2	57.1	63.0	**71.4**
ALaGAT	48.2	62.4	73.5	75.0	77.3	38.4	52.3	58.6	66.7	68.4	56.8	62.4	69.3
ANEPN (ours)	**66.1**	**73.2**	**77.6**	**78.3**	**79.9**	**60.5**	**64.8**	**68.8**	**70.5**	**71.0**	**60.8**	**69.5**	**71.4**
Gain-GCN	+21.6	+13.4	+8.9	+3.9	+2.9	+16.9	+17.4	+7.1	+3.7	+2.4	+15.2	+14.5	+6.5
Gain-SOTA	+6.2	+6.5	+1.8	+0.9	+0.1	+6.3	+2.1	+2.6	+0.7	+0.6	+2.1	+6.5	+0.0

4.2 Results

Node Classification (Q1). Table 3 shows the classification results, where the results are averaged over 10 runs. ANEPN consistently outperforms all competitors in all the cases. Although ALaGCN and ALaGAT utilize high-order relations, ANEPN outperforms them since it can utilize much more higher-order relations than the approaches. Especially under lower label rates (e.g., 0.5% in Cora and Citeseer, and 0.03% in Pubmed), ANEPN achieves larger performance gains (see Gain-GCN and Gain-SOTA in Table 3). This result demonstrates that high-order relations are effective in semi-supervised node classification since the model can utilize high-order relations in order to predict the labels of nodes.

Table 4. Training time (second)

Dataset	Cora	Citeseer	Pubmed
LP	0.0063	0.0063	0.028
GCN	0.78	0.80	0.86
GAT	2.11	2.22	38.35
Self-training	1.51	1.54	1.66
Co-training	1.71	2.46	290.27
Union	2.37	3.25	291.11
Intersection	2.42	3.23	291.18
M3S	3.62	3.73	4.13
ALaGCN	126.34	89.80	241.82
ALaGAT	50.67	150.39	309.17
ANEPN (ours)	0.74	0.69	0.70

Table 5. Ablation study on Cora. "w/o ASL", "w/o EPL", and "w/o EPL or ASL" indicate the variant removing ASL, EPL, and both losses, respectively.

	Cora				
Label rate	0.5%	1%	2%	3%	4%
ANEPN	**66.1**	**73.2**	**77.6**	**78.3**	**79.9**
w/o ASL	39.1	60.8	68.2	75.3	78.6
w/o EPL	52.6	63.5	71.5	73.0	73.3
w/o EPL or ASL	49.4	62.6	70.0	74.4	76.2
GCN	44.5	59.8	68.7	74.4	77.0

Training Time (Q2). In Table 4, we report the training time of each approach on a single Tesla V100 GPU with 16GB RAM, where the results are averaged over 10 runs. ANEPN finishes the model training faster than other GCN-based approaches while achieving higher classification accuracy. This result shows that ANEPN efficiently propagates node embeddings than GCNs. Although LP finishes the model training fastest of all approaches, the classification performance is significantly worse than our approach, as shown in Table 3.

Ablation Study (Q3). We evaluate the effect on classification performance of EPL and ASL. Table 5 shows the classification results of the variants of ANEPN on Cora. In this table, "w/o ASL", "w/o EPL", and "w/o EPL or ASL" indicate the variant removing ASL, EPL, and both losses, respectively. As shown in Table 5, ANEPN outperforms other variants in all the cases, which shows that EPL and ASL are effective for classification performance. Especially, compared to "w/o ASL", this result demonstrates that ANEPN improves the classification accuracy by avoiding over-smoothing with ASL. The classification accuracy of

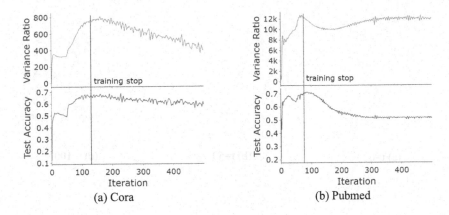

(a) Cora   (b) Pubmed

Fig. 1. Test accuracy and variance ratio w.r.t. training iteration

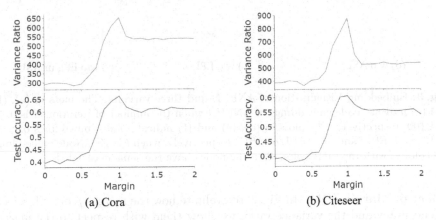

(a) Cora   (b) Citeseer

Fig. 2. Test accuracy and variance ratio w.r.t. margin

"w/o EPL or ASL" is similar to that of GCN since the loss of "w/o EPL or ASL" and GCN consists only of the cross entropy loss.

Control of Propagation (Q4). We verify that ANEPN appropriately controls the propagation of node embeddings. Figure 1 shows the test accuracy and the variance ratio with respect to training iteration on Cora and Pubmed, where training stop indicates the number of iterations on which ANEPN stops the model training. The result on Citeseer is removed since the training stop iteration is close to that on Cora. Note that, in this experiment, we continue the model training after the training stop. Figure 1 demonstrates that ANEPN appropriately stops the model training, that is, it controls the propagation of node embeddings since the test accuracy is close to the best on each graph (see training stop in Fig. 1).

430   Y. Ogawa et al.

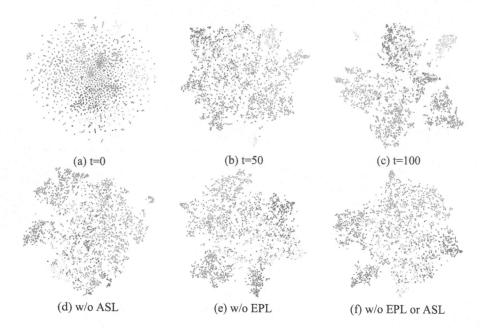

(a) t=0   (b) t=50   (c) t=100

(d) w/o ASL   (e) w/o EPL   (f) w/o EPL or ASL

Fig. 3. Embedding Visualization of ANEPN and three variants. The plots in (a), (b) and (c) indicate node embeddings of ANEPN when the number of iterations t is 0, 50 and 100, respectively. The plots in (d), (e) and (f) indicate node embeddings of 'w/o ASL', 'w/o EPL' and 'w/o EPL or ASL', respectively, when t is 100. Node embeddings are colored with the same color when the nodes have the same class labels.

Effect of Margin (Q5). In Fig. 2, to evaluate how the margin μ of ASL affects the accuracy and the variance ratio, we show them with respect to the margin on Cora and Citeseer. The result on Pubmed is removed since we obtain a similar result on Pubmed. As shown in Fig. 2, when the margin is around 0, the accuracy and the variance ratio are very low since ASL does not work (due to being always 0) so it causes over-smoothing. The accuracy and the variance ratio are improved as the margin increases to around 1 since ASL effectively works to avoid over-smoothing. However, when the margin is relatively large (larger than 1.5 in Fig. 2), the accuracy and the variance ratio decrease since ASL is expected to prevent the propagation of node embeddings by EPL.

Embedding Visualization (Q6). In order to verify how EPL and ASL affect node embeddings, we visualize the node embeddings on Cora. In Fig. 3, we plot the node embeddings in a two-dimensional space by applying a dimensional reduction technique, t-SNE [15], to the embeddings. The plots in Fig. 3(a), (b) and (c) depict the node embeddings of ANEPN when the number of iterations t is before the training stop, 0, 50 and 100, respectively. The plots in Fig. 3(d), (e) and (f) depict the node embeddings of 'w/o ASL', 'w/o EPL' and 'w/o EPL or ASL' when t is 100. The node embeddings are colored with the same color when the nodes have the same class label.

Figure 3(a), (b) and (c) reveal that, as t approaches to the training stop, the node embeddings are learned more appropriately: the nodes with the same class label have more similar embeddings and those with different class labels have more dissimilar embeddings. Also, observe that the node embeddings in Fig. 3(c) are better than those in Fig. 3(d), (e) and (f). In detail, the node embeddings with different class labels are more clearly separated in (c) than for 'w/o ASL' in (d). This clearly indicates that ASL is effective for avoiding over-smoothing. Next, (e) and (f) reveal that the node embeddings with the same class label are not similar due to not using EPL. To summarize, these results indicate that ANEPN improves the classification accuracy (see Table 5) by utilizing both of EPL and ASL.

5 Related Work

5.1 Graph-Based Semi-supervised Learning

Graph-based semi-supervised learning has been a popular research topic. Many classical approaches [1–3,29] assume that adjacent nodes tend to have the same label, which is called cluster assumption [5]. These approaches focus on graph structure, so they ignore node features. However, real-world graphs (or networks) often contain node features. Thus, some researchers seek to utilize both graph structure and node features in order to improve classification accuracy. For example, SemiEmb [22] and Planetoid [25] encode node features to embeddings with neural networks and incorporate graph structure into the embeddings with a regularizer. On the other hand, graph convolutional networks (GCNs) encode graph structure directly using a neural network model.

5.2 Graph Convolutional Networks

A two-layer GCN [12] is a standard GCN model. Recently, many variants of GCN have been developed, including Graph Attention Networks [20], Multi-Stage Self-Supervised Training Algorithm [18]. Most GCN models, including the above GCNs models, do not utilize high-order relations due to the two-layer architecture. However, if we simply stack many layers in order to utilize high-order relations, GCNs with many layers suffer from three drawbacks: overfitting, over-smoothing, and the difficulty in selecting the appropriate number of propagation hops.

ALaGCN proposed by Xie et al. can utilize high-order relations by increasing the number of propagation hops [24]. In addition, ALaGCN adaptively controls the number of propagation hops. However, it propagates node embeddings to neighbors only within ten hops since too many graph convolutions cause over-smoothing. On the other hand, ANEPN can propagate node embeddings to neighbors within more than ten hops, e.g., around 100 hops on Cora (see Sect. 4), since our approach avoids over-smoothing.

Adaptive Graph Convolution (AGC) [27] also utilizes high-order relations and controls the number of propagation hops. However, we cannot apply this approach to semi-supervised classification task since AGC is an approach only for graph clustering task.

6 Conclusion

In this paper, we argued that most GCNs do not utilize high-order relations between nodes. Although GCNs with many layers can utilize the high-order relations, the GCNs suffer from three drawbacks: over-fitting, over-smoothing, and the difficulty in selecting the appropriate number of propagation hops. To address the above drawbacks, we proposed ANEPN which effectively utilizes the high-order relations. By using EPL and ASL, ANEPN can propagate node embeddings to neighbors within many hops without suffering from over-fitting and over-smoothing. Furthermore, ANEPN controls the number of propagation hops based on the variance ratio that evaluates predicted class labels. The experimental results demonstrate that ANEPN is effective for semi-supervised node classification.

Acknowledgements. This work was supported by JSPS KAKENHI Grant Numbers JP20H00583.

References

1. Bengio, Y., Delalleau, O., Le Roux, N.: Label propagation and quadratic criterion. In: Semi-supervised Learning, pp. 193–216 (2006)
2. Blum, A., Chawla, S.: Learning from labeled and unlabeled data using graph min-cuts. In: ICML (2001)
3. Blum, A., Lafferty, J., Rwebangira, M.R., Reddy, R.: Semi-supervised learning using randomized mincuts. In: ICML, p. 13 (2004)
4. Caliński, T., Harabasz, J.: A dendrite method for cluster analysis. Commun. Stat. Theory Methods **3**(1), 1–27 (1974)
5. Chapelle, O., Zien, A.: Semi-supervised classification by low density separation. In: AISTATS (2005)
6. Liu, X., Tang, X., Chen, S.: Learning a similarity metric discriminatively with application to ancient character recognition. In: Qiu, H., Zhang, C., Fei, Z., Qiu, M., Kung, S.-Y. (eds.) KSEM 2021. LNCS (LNAI), vol. 12815, pp. 614–626. Springer, Cham (2021). https://doi.org/10.1007/978-3-030-82136-4_50
7. Cui, G., Zhou, J., Yang, C., Liu, Z.: Adaptive graph encoder for attributed graph embedding. In: KDD (2020)
8. Gao, H., Ji, S.: Graph u-nets. In: ICML (2019)
9. Goodfellow, I.J., et al.: Generative adversarial nets. In: NIPS (2014)
10. Hadsell, R., Chopra, S., LeCun, Y.: Dimensionality reduction by learning an invariant mapping. CVPR **2**, 1735–1742 (2006)
11. Hamilton, W.L., Ying, R., Leskovec, J.: Inductive representation learning on large graphs. In: NIPS (2017)

12. Kipf, T.N., Welling, M.: Semi-supervised classification with graph convolutional networks. In: ICLR (2017)
13. Li, Q., Han, Z., Wu, X.M.: Deeper insights into graph convolutional networks for semi-supervised learning. In: AAAI, pp. 3538–3545 (2018)
14. Liao, R., Zhao, Z., Urtasun, R., Zemel, R.S.: LanczosNet: multi-scale deep graph convolutional networks. In: ICLR (2019)
15. van der Maaten, L., Hinton, G.: Visualizing data using t-SNE. J. Mach. Learn. Res. **9**(Nov), 2579–2605 (2008)
16. Mikolov, T., Sutskever, I., Chen, K., Corrado, G.S., Dean, J.: Distributed representations of words and phrases and their compositionality. In: NIPS (2013)
17. Sun, K., Lin, Z., Zhu, Z.: ADAGCN: adaboosting graph convolutional networks into deep models. ArXiv abs/1908.05081 (2019)
18. Sun, K., Lin, Z., Zhu, Z.: Multi-stage self-supervised learning for graph convolutional networks on graphs with few labels. In: AAAI (2020)
19. Tang, J., Qu, M., Wang, M., Zhang, M., Yan, J., Mei, Q.: Line: large-scale information network embedding. In: WWW, pp. 1067–1077 (2015)
20. Veličković, P., Cucurull, G., Casanova, A., Romero, A., Liò, P., Bengio, Y.: Graph attention networks. In: ICLR (2018). https://openreview.net/forum?id=rJXMpikCZ
21. Wang, D., Cui, P., Zhu, W.: Structural deep network embedding. In: SIGKDD, pp. 1225–1234 (2016)
22. Weston, J., Ratle, F., Mobahi, H., Collobert, R.: Deep learning via semi-supervised embedding. In: Montavon, G., Orr, G.B., Müller, K.-R. (eds.) Neural Networks: Tricks of the Trade. LNCS, vol. 7700, pp. 639–655. Springer, Heidelberg (2012). https://doi.org/10.1007/978-3-642-35289-8_34
23. Wu, F., Zhang, T., de Souza, Jr., A.H., Fifty, C., Yu, T., Weinberger, K.Q.: Simplifying graph convolutional networks. In: ICML, pp. 6861–6871 (2019)
24. Xie, Y., Li, S., Yang, C., Wong, R.C.W., Han, J.: When do GNNs work: understanding and improving neighborhood aggregation. In: IJCAI (2020)
25. Yang, Z., Cohen, W.W., Salakhutdinov, R.: Revisiting semi-supervised learning with graph embeddings. In: ICML (2016)
26. Zhang, J., Shi, X., Xie, J., Ma, H., King, I., Yeung, D.: GaAN: gated attention networks for learning on large and spatiotemporal graphs. In: UAI, pp. 339–349 (2018)
27. Zhang, X., Liu, H., Li, Q., Wu, X.M.: Attributed graph clustering via adaptive graph convolution. In: IJCAI, pp. 4327–4333. AAAI Press (2019). http://dl.acm.org/citation.cfm?id=3367471.3367643
28. Zhao, L., Akoglu, L.: PairNorm: tackling oversmoothing in GNNs. In: ICLR (2020)
29. Zhou, D., Bousquet, O., Lal, T.N., Weston, J., Schölkopf, B.: Learning with local and global consistency. In: NIPS, pp. 321–328 (2004)

Probing Negative Sampling for Contrastive Learning to Learn Graph Representations

Shiyi Chen, Ziao Wang, Xinni Zhang, Xiaofeng Zhang$^{(\boxtimes)}$, and Dan Peng

School of Computer Science, Harbin Institute of Technology (Shenzhen), Shenzhen, China
{19s151083,20s051054,pengdan}@stu.hit.edu.cn, zhangxiaofeng@hit.edu.cn

Abstract. Graph representation learning has long been an important yet challenging task for various real-world applications. However, its downstream tasks are mainly performed in the settings of supervised or semi-supervised learning. Inspired by recent advances in unsupervised contrastive learning, this paper is thus motivated to investigate how the node-wise contrastive learning could be performed. Particularly, we respectively resolve the class collision issue and the imbalanced negative data distribution issue. Extensive experiments are performed on three real-world datasets and the proposed approach achieves the SOTA model performance.

Keywords: Graph neural network · Contrastive learning · Negative sampling

1 Introduction

In the literature, various graph neural network (GNN) models have been proposed for graph analysis tasks, such as node classification [14], link prediction [33] and graph classification [31]. Generally, most existing GNN-based approaches are proposed to train, in a semi-supervised manner, graph encoder to embed localized neighboring nodes and node attributes for a graph node into the low-dimensional feature space. By convoluting K-hops neighboring nodes, adjacent nodes naturally have similar feature representations. Notably, the consequent downstream tasks inevitably rely on the quality of the learnt node embeddings.

For many real graph applications, e.g., protein analysis [34], they intuitively requires an unavoidable cost or even the specialized domain knowledge to manually annotate sufficient data to well train the graph encoders with the specified supervised learning loss. Alternatively, a number of milestone unsupervised random walk based GNNs, including but not limited to node2vec [8] and graph2vec [19], are consequently proposed towards training the universal node embeddings and then various supervised downstream tasks are directly applied on these node embeddings. Similarly, another line of unsupervised graph representation

© Springer Nature Switzerland AG 2021
N. Oliver et al. (Eds.): ECML PKDD 2021, LNAI 12976, pp. 434–449, 2021.
https://doi.org/10.1007/978-3-030-86520-7_27

learning approaches, i.e., graph kernel-based methods, also utilizes the graph structural information to embed graph nodes with similar structures into similar representations. Most recently, the contrastive learning [11,12] is originally proposed to learn feature embeddings for each image in a self-supervised manner. To this end, these proposed approaches first generate two random augmentations for the same image and define these two as a pair of positive samples, and simply treat samples augmented from other images as negative samples. Then, the contrastive loss is designed to maximize the mutual information between each pair of positive samples. The learnt embeddings are believed to well preserve its inherent discriminative features. Research attempts are then made to adapt the successful contrastive learning approaches to unsupervised graph representation learning problem [10,29]. In [29], the graph-level representation is generated to contrast with each node representation to acquire node representations fitting for diverse downstream tasks. [10] adopts diffusion graph as another view of the given graph, and the contrast is performed between graph representation of one view and node representation of another view. As all the node embeddings and the graph embedding are forced to be close to each other, intuitively a coarser level graph analysis task, e.g., graph classification, would benefit a lot from such kind of contrastive learning, whereas a fine-grained level task, e.g., node classification, might not benefit that large.

To address aforementioned research gap, this work is thus motivated to investigate whether the unsupervised contrastive learning could be effectively carried on in a node-wised manner. That is, for each graph node x to be embedded, our desired contrastive learning is to maximize the mutual information between x and its positive examples x^+ instead of a graph representation, and simultaneously to minimize the mutual information between x and its negative examples x^-. Meanwhile, there exist two research challenges to be addressed. First, the sampled negative examples x^- might contain some true positive examples x^+, which belong to the same category as x in some downstream tasks, which is known as class collision issue. Second, how the density of negative samples will affect the contrastive learning has not been studied. We assume that the underlying true positive examples could be statistically similar to x, i.e., unseen positive examples should obey the same prior probability distribution as x. Similarly, the multiple typed negative examples x^- are assumed to obey different probability distributions. With this assumption, the class collision issue could be intuitively resolved by removing those examples from the set of x^- they are more likely generated by the assumed positive data distribution. For the second point, it can be known from the contrastive loss that x will be farther away from feature area with dense x^- than area with sparse x^-. The density distribution of x^- is used as factor to determine the distance of x and x^- is questionable. Therefore, after removing negative examples in doubt, a subset of negative examples should be diversely sampled for the contrastive learning. Thus, this paper proposed an adaptive negative sampling strategy for the learning of the node embedding in a node-wised contrastive learning manner. The major contributions of this paper are summarized as follows.

- To the best of our knowledge, this paper is among the first attempts to propose a node-wise contrastive learning approach to learn node embedding in an unsupervised manner. In the proposed approach, positive samples and negative samples are assumed to obey different data distributions, and the class collision issue could be addressed by eliminating "in doubt" negative samples if they are more likely generated by a positive data distribution.
- We propose a determinantal point process based negative instances sampling strategy which is believed to be able to sample diverse negative examples.
- We perform extensive experiments on several benchmark datasets and the promising results have demonstrated that the proposed approach is superior to both baseline and the state-of-the-art approaches.

2   Related Work

2.1   Graph Representation Learning

Supervised Methods. The earlier graph representation learning attempts have been made in the supervised settings. ChebyNet [5] leverages graph Fourier transformations to convert graph signals into spectral domain. Kipf and Welling [14] propose graph convolutional network (GCN) via a localized first-order approximation to ChebyNet [5], and extend graph convolution operations to spatial domain. To further the success of GCN, GAT [28] and GeniePath [16] are proposed to sample more informative neighboring nodes for aggregation. There also exist some approaches targeting at resolving efficiency issues [3,9].

Unsupervised Methods. The unsupervised graph representation learning methods could be classified into random walk-based methods [21,27] and graph kernel-based methods [2,24]. The random walk-based methods are applied for each graph node to generate the corresponding node sequences. By doing so, those nodes that have similar "context" nodes are trained to have similar embeddings regardless of the graph structural information as well as the node attributes. Such kinds of methods are usually transductively performed and thus need to be re-trained to represent the unseen nodes which inevitably limits their wide applicability. Graph Kernel [24,25] methods decompose graphs into well-designed substructures and use kernel function to measure graph similarity between them. Nevertheless, the design of these substructure requires a full understanding and professional knowledge of graph.

2.2   Contrastive Learning and Negative Sampling

Contrastive learning is recently proposed to learn feature embeddings in a self-supervised manner. The quality of the learned embeddings largely replies on the generated positive instance set and the negative instance set. Accordingly, various approaches have been proposed with a focus on constructing positive samples. In the domain of NLP, [18] treats the contextual words as positive pairs. In the domain of image recognition, [12] generates more difficult negative samples by mixing approach to improve model representation ability. [4] eliminates the sampling bias by increasing the number of positive samples to reduce false negative samples. For graph data, node2vec [8] apply two search strategies, BFS and DFS, to redefine neighbors, i.e., positive samples. In struc2vec [23], two nodes that have similar local structure are considered as a pair of positive samples despite the node position and their attributes. Then, several SOTA approaches have been proposed to adapt contrastive learning on graph data [10,22,26,29]. DGI [29] maximizes the mutual information between the local node embeddings and global graph embeddings. InfoGraph [26] treats nodes that are virtually generated by shuffling feature matrix or corrupting adjacency matrix as negative samples. Mvgrl [10] further defines two views on graph data and the encoder is trained to maximize mutual information between node representations of one view and graph representations of another view and vice versa. GCC [22] consider two subgraphs augmented from the same r-ego network as a positive instance pair and these subgraphs from different r-ego network as negative sample pairs, where r-ego represents the induced subgraph containing the set of neighbor nodes of a given node within r hops.

3   Preliminaries and Problem Formulation

In this section, we first briefly review the Determinantal Point Process (DPP) [17] adopted to diversely sample negative instances, then we describe the notations as well as the problem setup.

3.1   Determinantal Point Process

The original DPP is proposed to model negatively correlated random variables, and then it is widely adopted to sample a subset of data where each datum in this set is required to be correlated with the specified task, and simultaneously be far away from each other. Formally, let $\mathcal{P}$ denotes a probability distribution defined on a power set 2^Y of a discrete finite point set $Y = \{1, 2, ..., M\}$. $Y \sim \mathcal{P}$ is a subset composed of data items randomly generated from $\mathcal{P}$. Let A be a subset of Y and $B \in \mathbb{R}^{M \times M}$ be a real positive semi-definite similarity matrix, then we have $\mathcal{P}(A \subseteq Y) = |B_A|$, where B_A is a sub-matrix of B indexed by the elements of subset A. $|\cdot|$ denotes the determinant operator. If $A = \{i\}$, $\mathcal{P}(A \subseteq Y) = B_{i,i}$; and if $A = \{i, j\}$, $\mathcal{P}(A \in Y)$ can be written as

$$\mathcal{P}(A \subseteq Y) = \begin{vmatrix} B_{i,i} & B_{i,j} \\ B_{j,i} & B_{j,j} \end{vmatrix} = \mathcal{P}(i \in Y)\mathcal{P}(j \in Y) - B_{i,j}^2, \tag{1}$$

Thus, the non-diagonal matrix entries represent the correlation between a pair of data items. The larger the value of $B_{i,j}$, the less likely that i and j appear at the same time. Accordingly, the diversity of entries in the subset A could be calculated. As for our approach, DPP is adapted to sample a evenly distributed subset from a negative instance set.

3.2   Notations and Definitions

Let $G = (V, E)$ denotes a graph, V denotes the node set containing N nodes, $E \subseteq V \times V$ denotes the edge set where $e = (v_i, v_j) \in E$ denotes an edge between two graph nodes, $X = \{x_1, ..., x_d\}$ denotes the node feature set where $x_i \in \mathbb{R}^{d_{in}}$ represents the features of node v_i. The adjacency matrix is denoted as $A \in \mathbb{R}^{N \times N}$, where $A_{ij} = 1$ represents that there is an edge between v_i and v_j in the graph and 0 otherwise. For a given node v_i, its K-hops neighbor set is denoted as $\mathcal{N}_K(v_i)$, which contains all neighboring nodes of v_i within K hops, defined as $\mathcal{N}_K(v_i) = \{v_j : d(v_i, v_j) \leq K\}$ where $d(v_i, v_j)$ is the shortest path distance between v_i and v_j in the graph G. Then, the induced subgraph is defined as follows.

Definition 1. *Induced subgraph* s. *Given $G = (V, E)$, a subgraph $s = (V', E')$ of G is said to be an induced subgraph of G if all the edges between the vertices in V' belong to E'.*

3.3   Problem Setup

Given a G, our goal is to train a graph encoder $\mathcal{G} : \mathbb{R}^{N \times d_{in}} \times \mathbb{R}^{N \times N} \rightarrow \mathbb{R}^{N \times d}$, such that $H = \mathcal{G}(X, A) = \{h_1, ..., h_N\}$ represents the low-dimensional feature representations, where h_i denotes the embedding of v_i. Then, the learned $\mathcal{G}$ is used to generate node embeddings for downstream tasks, e.g., node classification.

The purpose of our approach is to maximize the mutual information between a pair of positive instances. Similar to the infoNCE [20], the general form of our unsupervised contrastive learning is to minimize the contrastive loss, given as

$$L = -\sum_{i=1}^{N} \log \frac{e^{f(h_i^q, h_i^k)/\tau}}{e^{f(h_i^q, h_i^k)/\tau} + \sum_{j \neq i}^{N} e^{f(h_i^q, h_j^k)/\tau}}, \qquad (2)$$

where $f(\cdot, \cdot)$ is a function to score the agreement of two embeddings, and in our approach the score function is simply the dot product calculated as $f(h_i^q, h_j^k) = h_i^{q T} \cdot h_j^k$, and τ is the temperature hyper-parameter. h_i^k is a positive instance of h_i^q, and two of them are usually defined as two different random augmentations of the same data. In our model, we define h_i^q and h_i^k are two random augmentation embeddings of node v_i. By optimizing Eq. 2, the employed model is believed to be able to learn the features that are invariant in positive instances.

Nevertheless, there are some problems that must be addressed. Intuitively, we hope that two nodes that have the same label in the classification task should have similar embeddings. However, we treat $h_{j \neq i}^k$ as negative sample of h_i^q and try to be away from them in feature space while v_j may have the same label as v_i, which is called class collision. At the same time, we noticed that the current negative sampling strategy ignores the influence of the density of embedding distribution of negative samples. A node embedding will be updated to be farther away from feature subspace where its negative nodes are more densely distributed. Therefore, we designed a approach to adaptively sample negative instances to avoid those problems described above.

4 The Proposed Approach

4.1 Graph Embeddings

The proposed node-wised contrastive learning scheme allows various choices of graph neural network architectures. We opt for simplicity reason to adopt graph convolution network (GCN) [14] as our graph encoder $\mathcal{G}$.

Augmentation. By following [22], we first employ a k-steps random walk on G starting from a specific node v_i, and a sequence of walking nodes $seq_i = \{t_1, ..., t_k\}$ is used to form the set of vertices V'. The subgraph s_i induced by V' is regarded as a random augmentation of node v_i. Then, we repeat aforementioned procedure and eventually we generate two induced subgraphs s_i^q, s_i^k, those embeddings are respectively denoted as h_i^q and h_i^k and regarded as a positive pair.

Encoder. The employed GCN layers are defined as $\sigma(\widetilde{A}_i X_i W)$ which is used to embed node v_i, where $\widetilde{A}_i = \widehat{D}_i^{-\frac{1}{2}} \widehat{A}_i \widehat{D}_i^{-\frac{1}{2}} \in \mathbb{R}^{n_i \times n_i}$ is symmetrically normalized adjacency matrix of a subgraph s_i. $\widehat{D}_i$ is the degree matrix of $\widehat{A}_i = A_i + I_{n_i}$, where A_i is the original adjacency matrix of s_i, $X_i \in \mathbb{R}^{n_i \times d_{in}}$ is the initial features of nodes in s_i, $W \in \mathbb{R}^{d_{in} \times d}$ is network parameters, σ is a ReLU [7] non-linearity and n_i is the number of nodes in s_i. Putting $\widetilde{A}_i, X_i$ into graph layer and then we could acquire node embeddings $H_i \in \mathbb{R}^{n_i \times d}$ of subgraph s_i.

Readout. After aggregation operation of GCN layers, we feed the embedding set H_i into the readout function $\mathcal{R}(\cdot)$ to compute an embedding of v_i. The readout function adopted in the experiments is given as follows

$$\mathcal{R}(H_i) = \sigma(\frac{1}{n_i} \sum_{j=1}^{n_i} h_{i,j} + \max(H_i)), \tag{3}$$

where $h_{i,j}$ represents the j-th node embedding in H_i, $\max(\cdot)$ simply takes the largest element along the column-wise and σ is the non-linear sigmoid function. Eventually, the node embedding is acquired as $h_i = \mathcal{R}(H_i)$.

4.2 Resolving Class Collision

Given a node v_i, its positive and negative sample set are respectively denoted as $S_i^+ = \{v_i\}$ and $S_i^- = \{v_1, ..., v_{i-1}, v_{i+1}, v_N\}$. To alleviate the class collision issue, it is desired to remove those "in doubt" negative samples that are more likely to belong to the same class of v_i. The overall procedure is depicted as below.

We assume that S_i^+ and S_i^- respectively obey different prior probability distributions. The "in-doubt" negative examples are removed if they are more likely to be generated by the data distribution of positive instances. To fit the

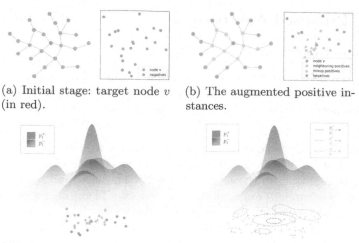

(a) Initial stage: target node v (in red).

(b) The augmented positive instances.

(c) Fit the embedding distribution for positive and negative samples.

(d) Resample the positive and negative samples.

Fig. 1. Process of generating positive and negative samples. In subfigure **(a)** and **(b)**, the left depicts the graph topological structure and the right plots the feature embedding space where each colored dot represent the embeddings of a positive or negative node. In the initial stage, plotted in **(a)**, there is only one target node treated as positive node (in red) and the rest are negative nodes (in blue). In **(b)**, the positive nodes are augmented by adding "in-doubt" nodes (in orange dot) and mixup positive nodes (in orange triangle). In **(c)**, the underlying positive instance distribution and negative instance distribution could be well fit using these data. In **(d)**, the dashed loop is the contour of $\frac{p_i^+(v_j)}{p_i^-(v_j)}$. Note that the smaller the orange dashed loop, the more confident that datum falling a positive instance. (Color figure online)

embedding distribution p_i^+ and p_i^-, we employ two independent neural networks, i.e., $\mathcal{F}_i^+$ and $\mathcal{F}_i^-$ to fit distributions. If the probability that v_j belongs to S_i^+ is higher than the probability of being a negative instance, i.e., $\frac{p_i^+(v_j)}{p_i^-(v_j)} > \alpha$, we remove v_j from S_i^-, where α is the soft-margin to discriminate an instance. Detailed steps are illustrated in the following paragraphs.

Forming the Sample Sets S_i^+ and S_i^-. Initially, the positive instance of a given node v_i is also augmented by v_i plotted in orange and the rest nodes plotted in blue are considered as negative instances, which is shown in Fig. 1(a). Apparently, not all the blue data are the true negative instances. To consider the unsupervised settings, it is acceptable to assume that the "closest" node to v_i should have the same underlying class label. Therefore, a few nearest neighbor nodes $v_j \in \mathcal{N}_K$ are transited to S_i^+ from S_i^-. Using the mixup algorithm [32], more positive samples are generated to further augment the positive instance set S_i^+ and the results are illustrated in Fig. 1(b).

Fitting the Positive and Negative Instance Distribution. With the augmented S_i^+ and S_i^-, we employ two independent two neural networks $\mathcal{F}_i^+$, $\mathcal{F}_i^-$ to respectively fit the embeddings distribution for $v \in S_i^+$ and $v \in S_i^-$ as plotted in Fig. 1(c). To train $\mathcal{F}_i^+$, we treat it as a classifier, and data belong to S_i^+ is assigned with a virtual label *class* 1, and data belong to S_i^- is virtually assigned with *class* 0. $\mathcal{F}_i^-$ uses the opposite settings of $\mathcal{F}_i^+$

For a node $v_j \in S_i^-$, the output of $\mathcal{F}_i^+(h_j^k)$ and $\mathcal{F}_i^+(h_j^k)$ are respectively the probability that v_j is a positive or negative instance of v_i. The ratio of these two probabilities with a soft-margin, calculated as $\frac{p_i^+(v_j)}{p_i^-(v_j)} > \alpha$, is adopted to determine whether v_j should be removed from S_i^- or not, and this soft-margin α is plotted in the small orange dashed circle as shown in Fig. 1(d).

(a) Unevenly sampled instances.   (b) Diversely sampled instances.

Fig. 2. The illustration of different sampling strategies. The dashed circle denotes that the corresponding node will be sampled. The orange colored dot is the embedding of target node v. **(a)** shows that for the current strategy, random sampling, the updated embedding of v will close to the node at the lower left corner whereas we desire that the embedding of v should, simultaneously, stay away from all negative instances embeddings as much as possible, as plotted in **(b)** where a diverse sampling strategy is applied on the embeddings space and reasonably ignores embedding density distribution. (Color figure online)

4.3 Sampling Diverse Negative Examples

As illustrated in Fig. 2, we can regard the process of contrastive learning as the interaction of forces between positive and negative samples. For the worst case of randomly sampling negative instances in S_i^- for comparison, where the embeddings distribution is seriously imbalanced as Fig. 2(a), the updated h_i will be farther away from the feature subspace where sampled negative instances densely distributed. Intuitively, the comparison result between positive and negative samples should not be related to the density distribution of negative samples. To cope with this distorted result, we adapt the Determinant point process (DPP) [17] to our problem. In Fig. 2(b), the DPP algorithm is applied to S^- to sample a negative instances subset, where sampled negative instances spread across the entire feature space. In this case, the node embedding can avoid the influence of the density of feature space and be evenly away from each negative sample. We set the correlation between h_i^q and each negative instance in S_i^- equally to a constant. To calculate the similarity between negative instances, the Euclidean distance is adopted to measure the pair-wise distance, computed as $d(h_i^q, h_j^k) = \sqrt{\sum_{l=1}^{d}(h_{i,l}^q - h_{j,l}^k)^2}$.

4.4   Node-Wise Contrastive Learning Loss

As pointed in [12], different nodes contribute differently to the unsupervised contrastive learning. We are therefore inspired to further differentiate the importance of the diversely sampled negative instances. For those negative instances that are far away from the query instance h_i^q, the contributions of these nodes are rather limited as they could be easily distinguished w.r.t. h_i^q. However for those close negative instances, it is hard for the model to discriminate them and thus their contributions should be assigned with higher weights.

Accordingly, the weight of the j-th negative instance's embedding h_j^k w.r.t. the query embedding h_i^q is calculated as $w_{i,j} = h_i^q \cdot h_j^k / \tau_w$, where τ_w is a temperature hyper-parameter. Thus, the overall node-wise contrastive loss could be written as

$$L = -\sum_{i=1}^{N} \log \frac{e^{f(h_i^q, h_i^k)}}{e^{f(h_i^q, h_i^k)} + \sum_{j \in S_i^-} w_{i,j} e^{f(h_i^q, h_j^k)}}. \tag{4}$$

5   Experimental Results

In this section, we first briefly introduce experimental datasets, evaluation metrics as well as the experimental settings. Then, to evaluate the model performance, we not only compare our method with unsupervised models, but also some semi-supervised models to fully demonstrate the effectiveness of our approach. Extensive experiments are evaluated on several real-world datasets to answer following research questions:

- **RQ1:** Whether the proposed approach outperforms the state-of-the-art semi-supervised and unsupervised methods or not?
- **RQ2:** Whether the proposed components could affect the model performance or not (ablation study)?
- **RQ3:** Whether the proposed approach is sensitive to model hyper-parameters or not?
- **RQ4:** The visualization results of the learned node embeddings.

5.1   Experimental Setup

Datasets. In the experiments, three real-world datasets are adopted to evaluate the model performance including *Cora*, *Citeseer* and *Pubmed*. We follow the work [28] to partition each dataset into training set, validation set and test set. The statistics of these datasets are reported in Table 1.

Algorithm 1: Generating positive and negative instances set

Input: Adjacency matrix A, node embeddings H^q, H^k,
　　　hyper-parameter K, hyper-parameter α,
　　　K-hops neighboring nodes set $\{\mathcal{N}_K(v_1), ..., \mathcal{N}_K(v_N)\}$,
　　　DPP sampler Γ_{dpp}, mixup operator Mix,
　　　node embedding set $S = \{h_1^k, ..., h_N^k\}$,
　　　neural network set $\mathcal{F}^+ = \{\mathcal{F}_1^+, ..., \mathcal{F}_N^+\}$, $\mathcal{F}^- = \{\mathcal{F}_1^-, ..., \mathcal{F}_N^-\}$.
Output: Positive samples sets $\{S_1^+, ..., S_N^+\}$,
　　　　negative samples sets $\{S_1^-, ..., S_N^-\}$.

1　initialization;
2　**for** $i = 1$ to N **do**
3　　$S_i^- = S \setminus \{h_i^k, i \in \mathcal{N}_K(v_i)\}$
4　　$S_i^+ = \{h_i^k, i \in \mathcal{N}_K(v_i)\}$
5　　$S_i^+ = Mix(S_i^+)$
6　**end**
7　**for** $i = 1$ to N **do**
8　　$p_i^- = \mathcal{F}_i^-(S_i^-)$, $p_i^+ = \mathcal{F}_i^+(S_i^+)$
9　　**for** $j = 1$ to N **do**
10　　　**if** $j \notin \mathcal{N}_K(v_i)$ and $\frac{p_i^+(v_j)}{p_i^-(v_j)} > \alpha$ **then**
11　　　　$S_i^- = S_i^- \setminus \{h_j^k\}$, $S_i^+ = S_i^+ \cup \{h_j^k\}$
12　　　**end**
13　　**end**
14　　$S_i^- = \Gamma_{dpp}(S_i^-)$
15　**end**
16　**return** $\{S_1^+, ..., S_N^+\}$, $\{S_1^-, ..., S_N^-\}$

Baseline Models. To evaluate the model performance of the proposed approach on node classification task, both the unsupervised and semi-supervised methods are compared in the experiments.

The **unsupervised** models we used in the experiment are as follows

- *Deepwalk* [21] first deploys random walk on each node to generate node walks sequences, and then input these sequences to skip-gram model to acquire node embeddings.
- *GAE* [15] is considered as the SOTA approach which applies variational auto-encoder to graphs.

Table 1. The statistics of experimental datasets.

Dataset	# of nodes	# of edges	# of features	# of classes
Cora	2708	5429	1433	7
Citeseer	3327	4732	3703	6
Pubmed	19717	44338	500	3

- **GraphSAGE** [9] learns a function for generating low-dimensional embeddings by aggregating the embeddings of neighboring nodes. We use the unsupervised loss function mentioned in [9] to train the model.
- **DGI** [29] is considered as the SOTA unsupervised learning approach which maximizes the mutual information between the node-level and the graph-level feature embeddings.
- **Mvgrl** [10] is the SOTA self-supervised method proposed to learn node and graph embeddings by optimizing the contrast between node and graph representations from different graph views.

The **semi-supervised** models we used in the experiment are as follows

- **GCN** [14] is one of the milestone GNN models originally proposed for node classification problem.
- **Chebyshev** [5] designs the convolution kernel using the Chebyshev inequality to speed up the Fourier transformation for the graph convolution process.
- **GAT** [28] is essentially an attention based approach. GAT designs a multi-head self-attention layer to assign different weights to neighboring nodes.
- **GeniePath** [16] samples neighboring nodes which contribute a lot to the target node via a hybrid of BFS and DFS search strategies.
- **JK-Net** [30] adaptively uses different neighborhood ranges for each node to perform aggregation operations.
- **MixHop** [1] proposes to perform multi-order convolution to aggregate the mixing of neighborhood information.

Setting of Model Parameters. We set the same experimental settings as the SOTA [10,29] and report the mean classification results on the testing set after 50 runs of training followed by a linear model. We initialize the parameters using Xavier initialization [6] and train the model using Adam optimizer [13] with an initial learning rate of 0.001. We set the number of epochs to 2000. We vary the batch size from 50 to 2000, The early stopping with a patience of 20 is adopted. The embedding dimension is set to 512. Unlike DGI, we use two layers of GCN. We set the step of random walk as 25, soft-margin α as 0.9, dropout rate as 0.7.

5.2 RQ1: Performance Comparison

The results of node classification task are reported in Table 2. Obviously, the proposed approach achieves the best results both in comparison with SOTA models, except for Cora dataset where the Mvgrl achieves the best result, and ours is the second best one. Particularly, the accuracy on Pubmed dataset, which has the most nodes, is improved by 81.5%. As the Mvgrl method could make full use of global diffusion information **S**, the model performance is expected to be superior to ours. But due to the use of diffusion information, Mvgrl cannot be used in inductive manner, which limits its applicability. However, it is well noticed that our approach is better than the SOTA DGI trained with **S**, which verifies the effectiveness of our approach. It is also noteworthy that our model has already performed well on the Cora and Pubmed datasets without adding any modules, that is, only applying node-level comparison between nodes.

Table 2. The average node classification results for both supervised and unsupervised models. The available data column highlights the data available to each model during the model training process (**X**: features, **A**: adjacency matrix, **S**: diffusion matrix, **Y**: labels).

Available data	Method	Cora	Citeseer	Pubmed
X, Y	Raw features	55.1%	46.5%	71.4%
X, A, Y	GCN	81.5%	70.3%	79.0%
X, A, Y	Chebyshev	81.2%	69.8%	74.4%
X, A, Y	GAT	83.0%	72.5%	79.0%
X, A, Y	GeniePath	75.5%	64.3%	78.5%
X, A, Y	JK-Net	82.7%	73.0%	77.9%
X, A, Y	MixHop	81.9%	71.4%	80.8%
A	Deepwalk	67.2%	43.2%	65.3%
X, A	GAE	71.5%	65.8%	72.1%
X, A	GraphSAGE	68.0%	68.0%	68.0%
X, A	DGI	82.3%	71.8%	76.8%
X, S	DGI	83.8%	72.0%	77.9%
X, A,S	Mvgrl	**86.8%**	73.3%	80.1%
X, A	Ours w/o all	83.5%	69.3%	80.6%
X, A	Ours	84.3%	**73.5%**	**81.5%**

5.3 RQ2: Ablation Study

In this experiment, we investigate the effectiveness of the proposed components. We respectively remove the component of soft-margin sampling, DPP sampling and node weights, and report the results in Table 3.

Table 3. The ablation study results. In this table, ours w/o all denotes that we remove all proposed components. And ours with α, ours with DPP, ours with w denote the model with soft-margin sampling, DPP sampling and node weights, respectively.

Variants	Cora	Citeseer	Pubmed
Ours w/o all	83.5%	69.3%	80.6%
Ours with α	83.8%	70.9%	80.8%
Ours with DPP	83.9%	71.8%	81.2%
Ours with w	83.8%	70.1%	80.9%
Ours	**84.3%**	**73.5%**	**81.5%**

Effect of Soft-Margin Sampling. It is noticed that the performance of "ours with α", i.e., repartitioning positive and negative samples by ratio α, improves. This suggests the proposed module indeed contributes to choose positive samples from "in-doubt" negative samples, and thus partially alleviates class collision.

Effect of DPP Sampling. It could be observed that the "ours with DPP" achieves the second best results w.r.t. all evaluation criteria. This verifies our proposed assumption that the data distribution of negative examples is a key factor in affecting the model performance.

Effect of Node Weights. Compared to the other two modules, especially on Citeseer dataset, this module has the least significant effect. The reason may be the weight function we designed is too simple. How to design a more effective weight function is our next research direction.

5.4   RQ3: Parameter Analysis

In the section, we evaluate how the model parameters, e.g., batch size, the step of random walk and soft-margin α, affect the model performance, and the corresponding results are plotted in Fig. 3 and Fig. 4.

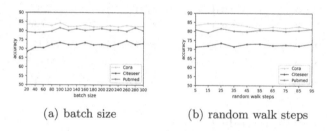

(a) batch size   (b) random walk steps

Fig. 3. Parameter analysis of batch size and random walk length.

From these figures, we have following observations. First, we highlight that our model is insensitive to parameter "batch size" and "random walk steps", as shown in Fig. 3(a) and 3(b). Second, our model is obviously sensitive to the soft-margin α which controls the ratio of selecting potential positive samples from the negative samples set. It is also well noticed that when α is close to 1, the model performance dramatically drops. This verifies that the number of positive examples is also crucial to the model performance of the contrastive learning.

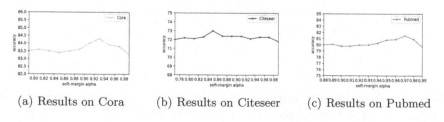

(a) Results on Cora   (b) Results on Citeseer   (c) Results on Pubmed

Fig. 4. Parameter analysis of soft-margin α.

5.5 RQ4: Visualization Results

Due to page limitations, we choose to visualize the node embeddings learned on Pubmed dataset to provide a vivid illustration of the proposed model performance. We also visualized the results of the Mvgrl method for comparison as this approach achieves the superior performance. The visualization results are plotted in Fig. 5. Obviously, our model achieves the best visualization result. It is noticed that in Fig. 5(a), most of the "red" and "yellow" class are mixed up together, which makes the classification task difficult. It is also noticed that the three classes could be well separated and spread over the whole data space by our model, whilst the "red" class and "yellow" class are still mixed up in Fig. 5(b).

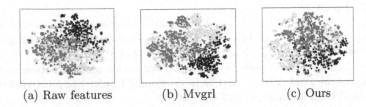

(a) Raw features   (b) Mvgrl   (c) Ours

Fig. 5. Visualizing the learned embeddings of nodes on Pubmed dataset. In this figure, each color represents one class, and each colored point represents the node embedding. (Color figure online)

6 Conclusion

In this paper, we propose a novel node-wise unsupervised contrastive learning approach to learn node embeddings for a supervised task. Particularly, we propose to resolve class collision issue by transiting the detected "in doubt" negative instances from the negative instance set to the positive instance set. Furthermore, a DPP-based sampling strategy is proposed to evenly sample negative instances for the contrastive learning. Extensive experiments are evaluated on three real-world datasets and the promising results demonstrate that the proposed approach is superior to both the baseline and the SOTA approaches.

Acknowledgements. This work was supported in part by the National Key Research and Development Program of China under Grant No. 2018YFB1003800, 2018YFB1003804, the National Natural Science Foundation of China under Grant No. 61872108, and the Shenzhen Science and Technology Program under Grant No. JCYJ20200109113201726, JCYJ20170811153507788.

References

1. Abu-El-Haija, S., et al.: MixHop: higher-order graph convolutional architectures via sparsified neighborhood mixing. In: Proceedings of the 36th International Conference on Machine Learning, vol. 97, pp. 21–29 (2019)

2. Borgwardt, K.M., Kriegel, H.P.: Shortest-path kernels on graphs. In: Fifth IEEE International Conference on Data Mining (ICDM 2005), pp. 8-pp (2005)
3. Chen, J., Ma, T., Xiao, C.: FastGCN: fast learning with graph convolutional networks via importance sampling. In: International Conference on Learning Representations (2018)
4. Chuang, C.Y., Robinson, J., Yen-Chen, L., Torralba, A., Jegelka, S.: Debiased contrastive learning. In: Advances in Neural Information Processing Systems, vol. 33, pp. 8765–8775 (2020)
5. Defferrard, M., Bresson, X., Vandergheynst, P.: Convolutional neural networks on graphs with fast localized spectral filtering. In: Advances in Neural Information Processing Systems, vol. 29 (2016)
6. Glorot, X., Bengio, Y.: Understanding the difficulty of training deep feed forward neural networks. In: Proceedings of the Thirteenth International Conference on Artificial Intelligence and Statistics, Chia Laguna Resort, Sardinia, Italy, vol. 9, pp. 249–256 (2010)
7. Glorot, X., Bordes, A., Bengio, Y.: Deep sparse rectifier neural networks. In: Proceedings of the Fourteenth International Conference on Artificial Intelligence and Statistics, Fort Lauderdale, FL, USA, pp. 315–323 (2011)
8. Grover, A., Leskovec, J.: node2vec: scalable feature learning for networks. In: Proceedings of the 22nd ACM SIGKDD International Conference on Knowledge Discovery and Data Mining, pp. 855–864 (2016)
9. Hamilton, W.L., Ying, R., Leskovec, J.: Inductive representation learning on large graphs. In: Proceedings of the 31st International Conference on Neural Information Processing Systems, pp. 1025–1035 (2017)
10. Hassani, K., Khasahmadi, A.H.: Contrastive multi-view representation learning on graphs. In: International Conference on Machine Learning, pp. 4116–4126 (2020)
11. He, K., Fan, H., Wu, Y., Xie, S., Girshick, R.: Momentum contrast for unsupervised visual representation learning. In: Proceedings of the IEEE/CVF Conference on Computer Vision and Pattern Recognition, pp. 9729–9738 (2020)
12. Kalantidis, Y., Sariyildiz, M.B., Pion, N., Weinzaepfel, P., Larlus, D.: Hard negative mixing for contrastive learning. In: Advances in Neural Information Processing Systems, vol. 33, pp. 21798–21809 (2020)
13. Kingma, D.P., Ba, J.: Adam: a method for stochastic optimization. In: International Conference on Learning Representations, ICLR 2015, San Diego, CA, USA, May 7–9, 2015, Conference Track Proceedings (2015)
14. Kipf, T., Welling, M.: Semi-supervised classification with graph convolutional networks. In: International Conference of Learning Representations (2017)
15. Kipf, T.N., Welling, M.: Variational graph auto-encoders. arXiv preprint arXiv:1611.07308 (2016)
16. Liu, Z., et al.: GeniePath: graph neural networks with adaptive receptive paths. In: Proceedings of the AAAI Conference on Artificial Intelligence, vol. 33, pp. 4424–4431 (2019)
17. Macchi, O.: The coincidence approach to stochastic point processes. Adv. Appl. Probab. **7**, 83–122 (1975)
18. Mikolov, T., Sutskever, I., Chen, K., Corrado, G.S., Dean, J.: Distributed representations of words and phrases and their compositionality. In: Advances in Neural Information Processing Systems, vol. 26 (2013)
19. Narayanan, A., Chandramohan, M., Chen, L., Liu, Y., Saminathan, S.: subgraph2vec: learning distributed representations of rooted sub-graphs from large graphs (2016)

20. van den Oord, A., Li, Y., Vinyals, O.: Representation learning with contrastive predictive coding (2019)
21. Perozzi, B., Al-Rfou, R., Skiena, S.: Deepwalk: online learning of social representations. In: Proceedings of the 20th ACM SIGKDD International Conference on Knowledge Discovery and Data Mining, pp. 701–710 (2014)
22. Qiu, J., et al.: GCC. In: Proceedings of the 26th ACM SIGKDD International Conference on Knowledge Discovery & Data Mining (2020)
23. Ribeiro, L.F., Saverese, P.H., Figueiredo, D.R.: struc2vec: learning node representations from structural identity. In: Proceedings of the 23rd ACM SIGKDD International Conference on Knowledge Discovery and Data Mining, pp. 385–394 (2017)
24. Shervashidze, N., Vishwanathan, S., Petri, T., Mehlhorn, K., Borgwardt, K.: Efficient graphlet kernels for large graph comparison. In: Proceedings of the Twelfth International Conference on Artificial Intelligence and Statistics, pp. 488–495 (2009)
25. Shervashidze, N., Vishwanathan, S., Petri, T., Mehlhorn, K., Borgwardt, K.: Efficient graphlet kernels for large graph comparison. In: Proceedings of the Twelfth International Conference on Artificial Intelligence and Statistics, vol. 5, pp. 488–495 (2009)
26. Sun, F.Y., Hoffmann, J., Verma, V., Tang, J.: Infograph: unsupervised and semi-supervised graph-level representation learning via mutual information maximization. In: International Conference on Learning Representations (2019)
27. Tang, J., Qu, M., Wang, M., Zhang, M., Yan, J., Mei, Q.: Line: large-scale information network embedding. In: Proceedings of the 24th International Conference on World Wide Web, pp. 1067–1077. International World Wide Web Conferences Steering Committee (2015)
28. Veličković, P., Cucurull, G., Casanova, A., Romero, A., Liò, P., Bengio, Y.: Graph attention networks. In: International Conference of Learning Representations (2018)
29. Veličković, P., Fedus, W., Hamilton, W.L., Liò, P., Bengio, Y., Hjelm, R.D.: Deep graph infomax. In: International Conference on Learning Representations (2019)
30. Xu, K., Li, C., Tian, Y., Sonobe, T., Kawarabayashi, K.I., Jegelka, S.: Representation learning on graphs with jumping knowledge networks. In: International Conference on Machine Learning, pp. 5453–5462 (2018)
31. Yuan, H., Ji, S.: Structpool: structured graph pooling via conditional random fields. In: International Conference on Learning Representations (2020)
32. Zhang, H., Cisse, M., Dauphin, Y.N., Lopez-Paz, D.: mixup: beyond empirical risk minimization. In: International Conference on Learning Representations (2018)
33. Zhang, M., Chen, Y.: Link prediction based on graph neural networks. In: Proceedings of the 32nd International Conference on Neural Information Processing Systems, pp. 5171–5181 (2018)
34. Zitnik, M., Leskovec, J.: Predicting multicellular function through multi-layer tissue networks. Bioinformatics **33**, i190–i198 (2017)

Beyond Low-Pass Filters: Adaptive Feature Propagation on Graphs

Shouheng Li[1,3]([✉]), Dongwoo Kim[2], and Qing Wang[1]

[1] The Australian National University, Canberra, Australia
shouheng.li@anu.edu.au
[2] GSAI, POSTECH, Gyeongbuk, South Korea
[3] CSIRO Data61, Eveleigh, Australia

Abstract. Graph neural networks (GNNs) have been extensively studied for prediction tasks on graphs. As pointed out by recent studies, most GNNs assume local homophily, i.e., strong similarities in local neighborhoods. This assumption however limits the generalizability power of GNNs. To address this limitation, we propose a flexible GNN model, which is capable of handling any graphs without being restricted by their underlying homophily. At its core, this model adopts a node attention mechanism based on multiple learnable spectral filters; therefore, the aggregation scheme is learned adaptively for each graph in the spectral domain. We evaluated the proposed model on node classification tasks over eight benchmark datasets. The proposed model is shown to generalize well to both homophilic and heterophilic graphs. Further, it outperforms all state-of-the-art baselines on heterophilic graphs and performs comparably with them on homophilic graphs.

Keywords: Graph neural network · Representation learning · Spectral methods

1 Introduction

Graph neural networks (GNNs) have recently demonstrated great power in graph-related learning tasks, such as node classification [12], link prediction [43] and graph classification [15]. Most GNNs follow a message-passing architecture where, in each GNN layer, a node aggregates information from its direct neighbors indifferently. In this architecture, information from long-distance nodes is propagated and aggregated by stacking multiple GNN layers together [4,12,38]. However, this architecture underlies the assumption of local homophily, i.e. proximity of similar nodes. While this assumption seems reasonable and helpful to achieve good prediction results on homophilic graphs such as citation networks [26], it limits GNNs' generalizability to heterophilic graphs. Heterophilic graphs commonly exist in the real-world, for instance, people tend to connect to opposite gender in dating networks, and different amino acid types are more likely to form connections in protein structures [45]. Moreover, determining whether a graph is homophilic or not is a

© Springer Nature Switzerland AG 2021
N. Oliver et al. (Eds.): ECML PKDD 2021, LNAI 12976, pp. 450–465, 2021.
https://doi.org/10.1007/978-3-030-86520-7_28

challenge by itself. In fact, strong and weak homophily can both exhibit in different parts of a graph, which makes a learning task more challenging.

Pei et al. [26] proposed a metric to measure local node homophily based on how many neighbors of a node are from the same class. Using this metric, they categorized graphs as homophilic (strong homophily) or heterophilic (weak homophily), and showed that classical GNNs such as GCN [12] and GAT [38] perform poorly on heterophilic graphs. Liu et al. [20] further showed that GCN and GAT are outperformed by a simple multi-layer perceptron (MLP) in node classification tasks on heterophilic graphs. This is because the naive local aggregation of homophilic models brings in more noise than useful information for such graphs. These findings indicate that these GNN models perform sub-optimally when the fundamental assumption of homophily does not hold.

Based on the above observation, we argue that a well-generalized GNN should perform well on graphs regardless of homophily. Furthermore, since a real-world graph can exhibit both strong and weak homophily in different node neighborhoods, a powerful GNN model should be able to aggregate node features using different strategies accordingly. For instance, in heterophilic graphs where a node shares no similarity with any of its direct neighbors, such a GNN model should be able to ignore direct neighbors and reach farther to find similar nodes, or at least, resort to the node's attributes to make a prediction. Since the validity of the assumption about homophily is often unknown, such aggregation strategies should be learned from data rather than decided upfront.

To circumvent this issue, in this paper, we propose a novel GNN model with attention-based adaptive aggregation, called ASGAT. Most existing attention-based aggregation architectures perform self-attention to the local neighborhood of a node [38]. Unlike these approaches, we aim to design an aggregation method that can gather informative features from both close and far-distant nodes. To achieve this, we employ graph wavelets under a relaxed condition of localization, which enables us to learn attention weights for nodes in the spectral domain. In doing so, the model can effectively capture information from frequency components and thus aggregate both local information and global structure into node representations.

To further improve the generalizability of our model, instead of using predefined spectral kernels, we propose to use multi-layer perceptrons (MLP) to learn desired spectral filters without limiting their shapes. Existing works on graph wavelet transform choose wavelet filters heuristically, such as heat kernel, wave kernel and personalized page rank kernel [13,14,41]. They are mostly low-pass filters, which means that these models implicitly treat high-frequency components as "noises" and have them discarded [2,9,25,34]. However, this may hinder the generalizability of models since high-frequency components can carry meaningful information about local discontinuities, as analyzed in [34]. Our model overcomes these limitations using node attentions derived from fully learnable spectral filters.

To summarize, the main contributions of this work are as follows:

1. We empirically show that high-frequency components carry important information on heterophilic graphs which can be used to improve prediction performance.
2. We propose a generalized GNN model which performs well on both homophilic and heterophilic graphs, regardless of graph homophily.
3. We exhibit that multi-headed attention produced by multiple spectral filters work better than attention obtained from a single filter, as it enables flexibility to aggregate features from different frequency components.

We conduct extensive experiments to compare ASGAT with well-known baselines on node classification tasks. The experimental results show that ASGAT significantly outperforms the state-of-the-art methods on heterophilic graphs where local node homophily is weak, and performs comparably with the state-of-the-art methods on homophilic graphs where local node homophily is strong. This empirically verifies that ASGAT is a general model for learning on different types of graphs.[1]

2   Preliminaries

Let $\mathcal{G} = (V, E, A, x)$ be an undirected graph with N nodes, where V, E, and A are the node set, edge set, and adjacency matrix of $\mathcal{G}$, respectively, and $x : V \mapsto \mathbb{R}^m$ is a graph signal function that associates each node with a feature vector. The normalized Laplacian matrix of $\mathcal{G}$ is defined as $L = I - D^{-1/2}AD^{-1/2}$, where $D \in \mathbb{R}^{N \times N}$ is the diagonal degree matrix of $\mathcal{G}$. In spectral graph theory, the eigenvalues $\Lambda = \mathrm{diag}(\lambda_1, ..., \lambda_N)$ and eigenvectors U of $L = U\Lambda U^H$ are known as the graph's spectrum and spectral basis, respectively, where U^H is the Hermitian transpose of U. The graph Fourier transform of x is $\hat{x} = U^H x$ and its inverse is $x = U\hat{x}$.

The spectrum and spectral basis carry important information on the connectivity of a graph [34]. Intuitively, lower frequencies correspond to global and smooth information on the graph, while higher frequencies correspond to local information, discontinuities and possible noise [34]. One can apply a spectral filter and use graph Fourier transform to manipulate signals on a graph in various ways, such as smoothing and denoising [32], abnormally detection [22] and clustering [39]. Spectral convolution on graphs is defined as the multiplication of a signal x with a filter $g(\Lambda)$ in the Fourier domain, i.e.

$$g(L)x = g(U\Lambda U^H)x = Ug(\Lambda)U^H x = Ug(\Lambda)\hat{x}. \tag{1}$$

When a spectral filter is parameterized by a scale factor, which controls the radius of neighbourhood aggregation, Eq. 1 is also known as the Spectral Graph Wavelet Transform (SGWT) [9,34]. For example, Xu et al. [41] uses a small scale parameter $s < 2$ for a heat kernel, $g(s\lambda) = e^{-\lambda s}$, to localize the wavelet at a node.

---

[1] The extended version of this work is available on arXiv [18]. Our open-sourced code is available at https://github.com/seanli3/asgat.

3   Proposed Approach

Original input graph   Adaptive spectral filters via MLPs   Aggregation via adaptive attention   Multi-head concatenation

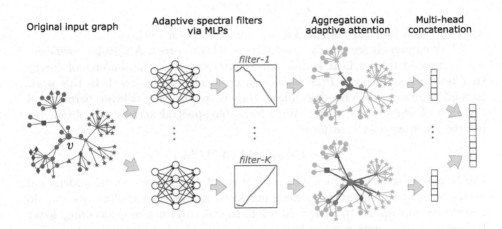

Fig. 1. Illustration of a spectral node attention layer on a three-hop ego network of the central node v from the CITESEER dataset. Shape and color indicate node classes. Passing the graph through two learned spectral filters place attention scores on nodes, including node v itself. Nodes with positive attention scores are presented in color. Node features are aggregated for node v according to attention scores. The low-pass filter attends to local neighbors (filter 1), while the high-pass filter skips the first hop and attends the nodes in the second hop (filter K). The resulting embeddings from multiple heads are then concatenated before being sent to the next layer (multi-head concatenation). Note that we have visualized learned filters from experiments.

Graph neural networks (GNNs) learn lower-dimensional embeddings of nodes from graph structured data. In general, given a node, GNNs iteratively aggregate information from its neighbor nodes, and then combine the aggregated information with its own information. An embedding of node v at the lth layer of GNN is typically formulated as

$$m_v = \text{aggregate}(\{h_u^{(l-1)} | u \in \mathcal{N}_v\})$$
$$h_v^{(l)} = \text{combine}(h_v^{(l-1)}, m_v),$$

where $\mathcal{N}_v$ is the set of neighbor nodes of node v, m_v is the aggregated information from the neighbors, and $h_v^{(l)}$ is the embedding of node v at the lth layer ($h_v^{(0)} = x_v$). The embedding $h_v^{(L)}$ of node v at the final layer is then used for some prediction tasks. In most GNNs, $\mathcal{N}_v$ is restricted to a set of one-hop neighbors of node v. Therefore, one needs to stack multiple aggregation layers in order to collect the information from more than one-hop neighborhood within this architecture.

Adaptive Spectral Filters. Instead of stacking multiple aggregation layers, we introduce a spectral attention layer that rewires a graph based on spectral graph wavelets. A spectral graph wavelet $\boldsymbol{\psi}_v$ at node v is a modulation in the spectral domain of signals centered around the node v, given by an N-dimensional vector

$$\boldsymbol{\psi}_v = \boldsymbol{U}g(\Lambda)\boldsymbol{U}^H\delta_v, \tag{2}$$

where $g(\cdot)$ is a spectral filter and δ_v is a one-hot vector for node v.

The common choice of a spectral filter is a heat kernel. A wavelet coefficient ψ_{vu} computed from a heat kernel can be interpreted as the amount of energy that node v has received from node u in its local neighborhood. In this work, instead of using pre-defined localized kernels, we use multi-layer perceptrons (MLP) to learn spectral filters. With learnable spectral kernels, we obtain the inverse graph wavelet transform

$$\boldsymbol{\psi}_v = \boldsymbol{U}\mathrm{diag}(\mathrm{MLP}(\Lambda))\boldsymbol{U}^H\delta_v. \tag{3}$$

Unlike a low-pass heat kernel, where the wavelet coefficients can be understood as the amount of energy after heat diffusion, the learned coefficients ψ_{vu} do not always correspond to energy diffusion. In spectral imaging processing, lower frequency components preserve an image's background, while higher frequency components are useful to detect object edges or outlines. Similarly, in spectral graph theory, lower-frequency components carry smoothly changing signals. Therefore a low-pass filter is a reasonable choice to extract features and denoise a homophilic graph. On the contrary, higher-frequency components carry abruptly changing signals, corresponding to the discontinuities and "opposite attraction" characteristics of heterophilic graphs. In our experiments, the trained MLP resembles a low-pass filter, working as a diffusion operator, with homophilic graphs. In contrast, with heterophilic graphs, the trained MLP reaches a high-pass filter at most times (Sect. 4).

Note that we use the terminology wavelet and spectral filter interchangeably as we have relaxed the wavelet definition from [9] so that learnable spectral filters in our work are not necessarily localized in the spectral and spatial domains.

Remark 1. Equation 3 requires the eigen-decomposition of a Laplacian matrix, which is expensive and infeasible for large graphs. To address this computational issue, one may use well-studied methods such as Chebyshev [9,14,41] and Auto-Regressive Moving-Average (ARMA) [11,19] to efficiently compute an approximate the graph filtering of MLP in Eq. 3.

Attention Mechanism. Unlike the previous work [41] where the output of inverse graph wavelet transform are directly used to compute node embeddings, we normalize the output through a softmax layer

$$\boldsymbol{a}_v = \mathrm{softmax}(\boldsymbol{\psi}_v), \tag{4}$$

where $\boldsymbol{a}_v \in \mathbb{R}^N$ is an attention weight vector. With attention weights, an update layer is then formalized as

$$\boldsymbol{h}_v^{(l)} = \sigma\left(\sum_{u=1}^{N} a_{vu}\boldsymbol{h}_u^{(l-1)}\boldsymbol{W}^{(l)}\right), \tag{5}$$

where $\boldsymbol{W}^{(l)}$ is a weight matrix shared across all nodes at the lth layer and σ is ELU nonlinear activation.

Note that the update layer is not divided into aggregation and combine steps in our work. Instead, we compute the attention a_{vv} directly from a spectral filter. Unlike heat kernel and other spectral filters, the output of inverse graph wavelet transform with a learnable spectral kernel are not always localized. Hence, the model can adaptively aggregate information from both close and far-distant nodes, depending on their attention weights.

Sparsified Node Attentions. With predefined localized spectral filters such as a heat kernel, most wavelet coefficients are zero due to their locality. In our work, spectral filters are fully determined by data. Consequently, attention weights obtained from learnable spectral filters do not impose any sparsity. This means we need to retrieve all possible nodes in a graph, which is inefficient, to perform an aggregation operation. From our experiments, we observe that most attention weights are negligible after softmax. Thus, we consider a sparsification technique to keep only the largest k entries of Eq. 3 for each node, i.e.

$$\bar{\psi}_{vu} = \begin{cases} \psi_{vu} & \text{if } \psi_{vu} \in \text{topK}(\{\psi_{v0}, ..., \psi_{vN}\}, k) \\ -\infty & \text{otherwise}, \end{cases} \tag{6}$$

where topK is a partial sorting function that returns the largest k entries from a set of wavelet bases $\{\psi_{v0}, ..., \psi_{vN}\}$. This technique guarantees attention sparsity such that the embedding of each node can be aggregated from at most k other nodes with a time complexity trade-off of $O(N + k \log N)$. The resulting $\bar{\psi}$ is then fed into the softmax layer to compute attention weights.

We adopt multi-head attention to model multiple spectral filters. Each attention head aggregates node information with a different spectral filter, and the aggregated embedding is concatenated before sent to the next layer. To reduce redundancy, we adopt a single MLP: $\mathbb{R}^N \to \mathbb{R}^{N \times M}$, where M is the number of attention heads, and each column of the output corresponds to one adaptive spectral filter.

We name the multi-head spectral attention architecture as a *adaptive spectral graph attention network* (ASGAT). The design of ASGAT is easily generalizable, and many existing GNNs can be expressed as special cases of ASGAT (see Appendix [18]). Figure 1 illustrates how ASGAT works with two attention heads learned from the CITESEER dataset. As shown in the illustration, the MLP learns adaptive filters such as low-pass and high-pass filters. A low-pass filter assigns high attention weights in local neighborhoods, while a high-pass filter assigns high attention weights on far-distant but similar nodes, which a traditional hop-by-hop aggregation scheme cannot capture.

4 Experiments

To evaluate the performance of our proposed model, we conduct experiments on node classification tasks with homophilic graph datasets, and heterophilic graph

Table 1. Micro-F1 results for node classification. The proposed model consistently outperforms the GNN methods on heterophilic graphs and performs comparably on homophilic graphs. Results marked with † are obtained from Pei et al. [26]. Results marked with ‡ are obtained from Zhu et al. [45].

| | Homophily ⟸══════════════════════⟹ Heterophily | | | | | | | |
	CORA	PUBMED	CITESEER	CHAMELEON	SQUIRREL	WISCONSIN	CORNELL	TEXAS
β	0.83	0.79	0.71	0.25	0.22	0.16	0.11	0.06
#Nodes	2,708	19,717	3,327	2,277	5,201	251	183	183
#Edges	5,429	44,338	4,732	36,101	217,073	515	298	325
#Features	1,433	500	3,703	2,325	2,089	1,703	1,703	1,703
#Classes	7	3	6	5	5	5	5	5
GCN	87.4 ± 0.2	87.8 ± 0.2	78.5 ± 0.5	$59.8 \pm 2.6^{\ddagger}$	$36.9 \pm 1.3^{\ddagger}$	64.1 ± 6.3	59.2 ± 3.2	64.1 ± 4.9
ChevNet	88.2 ± 0.2	89.3 ± 0.3	79.4 ± 0.4	66.0 ± 2.3	39.6 ± 3.0	82.5 ± 2.8	76.5 ± 9.4	79.7 ± 5.0
ARMANet	85.2 ± 2.5	86.3 ± 5.7	76.7 ± 0.5	62.1 ± 3.6	47.8 ± 3.5	78.4 ± 4.6	74.9 ± 2.9	82.2 ± 5.1
GAT	87.6 ± 0.3	83.0 ± 0.1	77.7 ± 0.3	$54.7 \pm 2.0^{\ddagger}$	$30.6 \pm 2.1^{\ddagger}$	62.0 ± 5.2	58.9 ± 3.3	60.0 ± 5.7
SGC	87.2 ± 0.3	81.1 ± 0.3	78.8 ± 0.4	33.7 ± 3.5	46.9 ± 1.7	51.8 ± 5.9	58.1 ± 4.6	58.9 ± 6.1
GraphSAGE	86.3 ± 0.6	89.2 ± 0.5	77.4 ± 0.5	51.1 ± 0.5	$41.6 \pm 0.7^{\ddagger}$	77.6 ± 4.6	67.3 ± 6.9	82.7 ± 4.8
APPNP	$\mathbf{88.4} \pm 0.3$	86.0 ± 0.2	77.6 ± 0.6	45.3 ± 1.6	31.0 ± 1.6	81.2 ± 2.5	70.3 ± 9.3	79.5 ± 4.6
Geom-GCN	86.3 ± 0.3	89.1 ± 0.1	$\mathbf{81.4} \pm 0.3$	$60.9^{\dagger}$	$38.1^{\dagger}$	$64.1^{\dagger}$	$60.8^{\dagger}$	$67.6^{\dagger}$
H$_2$GCN	88.3 ± 0.3	89.1 ± 0.4	78.4 ± 0.5	59.4 ± 2.0	37.9 ± 2.0	86.5 ± 4.4	82.2 ± 6.0	82.7 ± 5.7
MLP	72.1 ± 1.3	88.6 ± 0.2	74.9 ± 1.8	45.7 ± 2.7	28.1 ± 2.0	82.7 ± 4.5	81.4 ± 6.3	79.2 ± 6.1
Vanilla ASGAT	–	–	–	–	–	86.9 ± 4.2	84.6 ± 5.8	82.2 ± 3.2
ASGAT-Cheb	87.5 ± 0.5	$\mathbf{89.9} \pm 0.9$	79.3 ± 0.6	$\mathbf{66.5} \pm 2.8$	$\mathbf{55.8} \pm 3.2$	86.3 ± 3.7	82.7 ± 8.3	$\mathbf{85.1} \pm 5.7$
ASGAT-ARMA	87.4 ± 1.1	88.3 ± 1.0	79.2 ± 1.4	65.8 ± 2.2	51.4 ± 3.2	84.7 ± 4.4	83.2 ± 5.5	79.5 ± 7.7

datasets. Further ablation study highlights the importance of considering the entire spectral frequency.

4.1 Experimental Setup

Baseline Methods. An exact computation of Eq. 3 requires to compute the eigenvectors of the Laplacian matrix, which is often infeasible due to a large graph size. To overcome this issue, we approximate graph wavelet transform response of MLP with Chebyshev polynomial, dubbed as ASGAT-Cheb, and ARMA rational function, dubbed as ASGAT-ARMA. We also report the results from the exact computation of eigenvectors whenever possible, which is dubbed as vanilla ASGAT.

We compare all variants against 10 benchmark methods, they are vanilla GCN [12] and its simplified version SGC [40]; two spectral methods: ChevNet [4] and ARMANet [1]; the graph attention model GAT [38][2]; APPNP, which also adopts adaptive aggregation [13]; the neighbourhood-sampling method Graph-Sage [8]; Geom-GCN [26] and H$_2$GCN [45], both also target prediction on heterophilic graphs. We also include MLP in the baselines since it performs better than many GNN methods on some heterophilic graphs [20].

---

[2] It was reported in Velickovic et al. [38] that GAT does not always outperform GCN when using different data splittings, and similar results have been reported by Zhu et al. [45].

Datasets. We evaluate our model and the baseline methods on node classification tasks over three citation networks: CORA, CITESEER and PUBMED [33], three webgraphs from the WebKB dataset[3]: WISCONSIN, TEXAS and CORNELL, and webgraphs from Wikipedia called CHAMELEON and SQUIRREL [30].

To quantify the homophily of graphs, we use the metric β introduced by Pei et al. [26],

$$\beta = \frac{1}{N} \sum_{v \in V} \beta_v \quad \text{and} \quad \beta_v = \frac{|\{u \in \mathcal{N}_v | \ell(u) = \ell(v)\}|}{|\mathcal{N}_v|}, \tag{7}$$

where $\ell(v)$ refers to the label of node v. β measures the degree of homophily of a graph, and β_v measures the homophily of node v in the graph. A graph has strong local homophily if β is large and vice versa. Details of these datasets are summarized in Table 1.

Experimental Settings. For citation networks, we follow the experimental setup for node classification from [3,8,10] and report the results averaged on 10 runs. For webgraphs, we run each model on the 10 splits provided by [26] and take the average, where each split uses 60%, 20%, and 20% nodes of each class for training, validation and testing, respectively. The results we report on GCN and GAT are better than Pei et al. [26] as a result of converting the graphs to undirected before training[4]. Geom-GCN uses node embeddings pretrained from different embedding methods such as Isomap [37], Poincare [24] and struc2vec [29]. We report the best micro-F1 results among all three variants for Geom-GCN.

We use the best-performing hyperparameters specified in the original papers of baseline methods. For hyperparameters not specified in the original papers, we use the parameters from Fey and Lenssen [6]. We report the test accuracy results from epochs with both the smallest validation loss and highest validation accuracy. Early termination is adopted for both validation loss and accuracy, thus training is stopped when neither validation loss or accuracy improve for 100 consecutive epochs. For ASGAT, we use a two-layer architecture where multi-headed filters are learned using a MLP of 2 hidden layers. Each layer of the MLP consists of a linear function and a ReLU activation. To avoid overfitting, dropout is applied in each ASGAT layer on both attention weights and inputs equally. Results for vanilla ASGAT are only reported for small datasets where eigendecomposition is feasible. Other hyperparameters are obtained by grid search, where details are given in Appendix [18].

4.2 Results and Discussion

We use two evaluation metrics to evaluate the performance of node classification tasks: micro-F1 and macro-F1. The results with micro-F1 are summarized in Table 1. Overall, on homophilic citation networks, ASGAT performs comparably with the state-of-the-art methods, ranking first on PUBMED and second

---

[3] http://www.cs.cmu.edu/afs/cs.cmu.edu/project/theo-11/www/wwkb/.
[4] https://openreview.net/forum?id=S1e2agrFvS.

on CORA and CITESEER in terms of micro-F1 scores. On heterophilic graphs, ASGAT outperforms all other methods by a margin of at least 2.4% on 3 out of 4 datasets. These results indicate that ASGAT generalizes well on different types of graphs. The results with macro-F1 are summarized in Appendix [18]. Macro-F1 scores have not been reported widely in the literature yet. Here, we report the macro-F1 since the heterophilic graphs have imbalanced class distributions than the homophilic graphs. As the results show, ASGAT outperforms all other methods across all heterophilic graphs in macro-F1. The difference between the two approximation methods is not significant. Except for a few cases, the difference comes from hyperparameters selection. The vanilla ASGAT gives more consistent results than the approximations, although the difference seems marginal.

Although ASGAT performs well on both homophilic and heterophilic graphs, it is unclear how ASGAT performs on heterophilic neighbourhoods of an homophilic graph where nodes are mostly of different classes. Thus, we report an average classification accuracy on nodes at varying levels of β_v in Fig. 2 on the homophilic graphs CITESEER and PUBMED. The nodes are binned into five groups based on β_v. For example, all nodes with $0.3 < \beta_v \leq 0.4$ belong to the bin at 0.4. We have excluded CORA from the report since it has very few heterophilic neighbourhoods.

The results in Fig. 2 show that all models except ASGAT perform poorly when β_v is low. One may argue that the performance on heterophilic graphs might improve by stacking multiple GNN layers together to obtain information from far-distant nodes. However, it turns out that this approach introduces an oversmoothing problem [17] that degrades performance. On the other hand, the better performance of ASGAT on heterophilic nodes suggests the adaptive spectral filters reduce noise aggregated locally while allowing far-distant nodes to participate.

Attention Sparsification. The restriction on top k entries in Eq. 6 guarantees a certain level of sparsification. Nonetheless, ASGAT requires a partial sorting which adds an overhead of $O(n + k \log N)$. To further analyze the impact of attention sparsity on run-time, we plot the density of an attention matrix with varying k in Fig. 3 along with its running time. The results are drawn from two datasets: the heterophilic dataset CHAMELEON and the homophilic dataset CORA. As expected, ASGAT shows a stable growth in the attention density as the value of k increases. It also shows that ASGAT runs much faster when attention weights are well-sparsified. In our experiments, we find the best results are achieved on $k < 20$. The impact of k on classification performance is further analyzed in Appendix [18].

Frequency Range Ablation. To understand how adaptive spectral filters contribute to ASGAT's performance on heterophilic graphs, we conduct an ablation study on spectral frequency ranges. We first divide the entire frequency range (0–2) into a set of predefined sub-ranges exclusively. Then we manually set the filter frequency responses to zero for each sub-range to check its impact on the classification. The frequencies within a selected sub-range contribute to neither node attention nor feature aggregation, therefore helping to reveal the

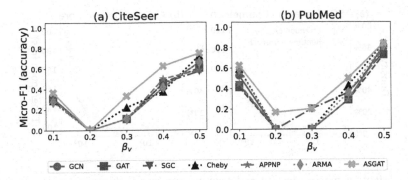

Fig. 2. Micro-F1 results for classification accuracy on heterophilic nodes ($\beta_v \leq 0.5$). ASGAT shows better accuracy on classifying heterophilic nodes than the other methods.

importance of the sub-range. We consider three different lengths of sub-ranges, i.e., step = 1.0, step = 0.5, and step = 0.25. The results of frequency ablation on the three homophilic graphs are summarized in Fig. 4.

The results for step = 1.0 reveal the importance of high-frequency range (1–2) on node classification of heterophilic graphs. The performance is significantly dropped by ablating high-frequency range on all datasets. Further investigation at the finer-level sub-ranges (step = 0.5) shows that ablating sub-range 0.5–1.5 has the most negative impact on performance, whereas the most important sub-range varies across different datasets at the finest level (step = 0.25). This finding matches our intuition that low-pass filters used in GNNs underlie a homophily assumption similar to naive local aggregation. We suspect the choice of low-pass filters also relates to oversmoothing issues in spectral methods [17], but we leave it for future work.

Attention Head Ablation. In ASGAT, each head uses a spectral filter to produce attention weights. To delve into the importance of a spectral filter, we further follow the ablation method used by Michel et al. [21]. Specifically, we ablate one or more filters by manually setting their attention weights to zeros. We then measure the impact on performance using micro-F1. If the ablation results in a large decrease in performance, the ablated filters are considered important. We observe that all attention heads (spectral filters) in ASGAT are of similar importance, and only all attention heads combined produce the best performance. Please check Appendix [18] for the detailed results.

Time Complexity. In vanilla ASGAT, eigen-decomposition is required for Eq. 3 which has a time complexity of $O(N^3)$. ASGAT-Cheb and ASGAT-ARMA avoid eigen-decomposition and are able to scale to large graphs as their time complexities are $O(R \times |E|)$ and $O((P \times T + Q) \times |E|)$ respectively, where R, P and Q are polynomial orders that are normally less than 30, T is the number of iterations that is normally less than 50. Therefore, both ASGAT-Cheb and ASGAT-ARMA scale linearly with the number of edges $|E|$. Readers can refer to Appendix [18]

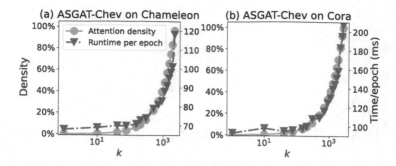

Fig. 3. Attention matrix density and training runtime with respect to k. Attention matrix sparsified by keeping the top k elements at each row, which effectively improves runtime efficiency.

for a more detailed introduction of these two methods. Secondly, partial sorting used in the attention sparsification of Eq. 6 requires $O(N + k \log N)$. Lastly, Eq. 4 is performed on a length-k vector for N rows; therefore, a time complexity of $O(k \times N)$ is needed. In practice, we have $R \sim P \sim T \sim Q \sim k \ll N \ll |E|$ for most graphs, therefore, for a model with M heads, the overall time complexity is $O(M \times R \times |E|)$ for ASGAT-Cheb and $O(M \times (P \times T + Q) \times |E|)$ for ASGAT-ARMA.

5   Related Work

Graph neural networks have been extensively studied recently. We categorize work relevant to ours into three perspectives and summarize the key ideas.

Attention on Graphs. Graph attention networks (GAT) [38] was the first to introduce attention mechanisms on graphs. GAT assigns different importance scores to local neighbors via an attention mechanism. Similar to other GNN variants, long-distance information propagation in GAT is realized by stacking multiple layers together. Therefore, GAT suffers from the oversmoothing issue [44]. Zhang et al. [42] improve GAT by incorporating both structural and feature similarities while computing attention scores.

Spectral Graph Filters and Wavelets. Some GNNs also use graph wavelets to extract information from graphs. Xu et al. [41] applied graph wavelet transform defined by Shuman et al. [34] to GNNs. Klicpera et al. [14] proposed a general GNN argumentation using graph diffusion kernels to rewire the nodes. Donnat et al. [5] used heat wavelet to learn node embeddings in unsupervised ways and showed that the learned embeddings closely capture structural similarities between nodes. Other spectral filters used in GNNs can also be viewed as special forms of graph wavelets [1,4,12]. Coincidentally, Chang et al. [2] also noticed useful information carried by high-frequency components from a graph Laplacian. Similarly, they attempted to utilize such components using

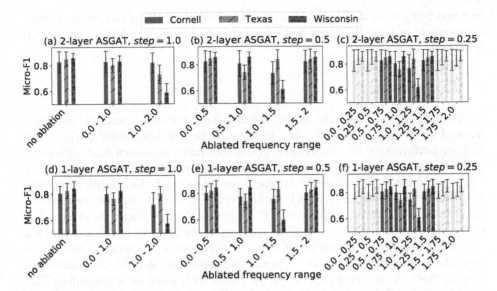

Fig. 4. Micro-F1 with respect to ablated frequency sub-ranges on heterophilic graphs. We divide the frequency range into a set of sub-ranges with different lengths. The results (a) and (d) reveal the importance of high-frequency range (1–2). Further experiments show that there is a subtle difference in the most important range across datasets, but it ranges between (0.75–1.25).

node attentions. However, they resorted to the traditional choice of heat kernels and applied such kernels separately to low-frequency and high-frequency components divided by a hyperparameter. In addition to this, their work did not link high-frequency components to heterophilic graphs.

Prediction on Heterophilic Graphs. Pei et al. [26] have drawn attention to GCN and GAT's poor performance on heterophilic graphs very recently. They try to address the issue by essentially pivoting feature aggregation to structural neighborhoods from a continuous latent space learned by unsupervised methods. Liu et al. [20] proposed another attempt to address the issue. They proposed to sort locally aggregated node embeddings along with a one-dimensional space and used a one-dimensional convolution layer to aggregate embeddings a second time. By doing so, non-local but similar nodes can attend to the aggregation. Very recently, Zhu et al. [45] showed that a heuristic combination of ego-, neighbor and higher-order embedding improves GNN performance on heterophilic graphs. Coincidentally, they also briefly mentioned the importance of higher-frequency components on heterophilic graphs, but they did not provide an empirical analysis.

Although our method shares some similarities in motivation with the work above, it is fundamentally different in several aspects. To the best of our knowledge, our method is the first architecture we know that computes multi-headed node attention weights purely from learned spectral filters. As a result, in contrast to the commonly used heat kernel, our method utilizes higher-frequency components of a graph, which helps predict heterophilic graphs and neighbourhoods.

6 Conclusion

In this paper, we study the node classification tasks on graphs where local homophily is weak. We argue that the assumption of homophily is the cause of poor performance on heterophilic graphs. In order to design more generalizable GNNs, we suggest that a more flexible and adaptive feature aggregation scheme is needed. To demonstrate, we have introduced the adaptive spectral graph attention network (ASGAT), which achieves flexible feature aggregation using learnable spectral graph filters. By utilizing the full graph spectrum adaptively via the learned filters, ASGAT can aggregate features from close and far nodes. For node classification tasks, ASGAT outperforms all benchmarks on heterophilic graphs and performs comparably on homophilic graphs. On homophilic graphs, ASGAT also performs better for nodes with weak local homophily. We find that the performance gain is closely linked to the higher end of the frequency spectrum through our analysis.

Acknowledgement. This work was partly supported by the National Research Foundation of Korea (NRF) grant funded by the Korea government (MSIT) (No. 2020R1F1A1061667).

References

1. Bianchi, F.M., Grattarola, D., Livi, L., Alippi, C.: Graph neural networks with convolutional ARMA filters. CoRR arxiv:1901.01343 (2019)
2. Chang, H., et al.: Spectral graph attention network. CoRR arxiv:2003.07450 (2020)
3. Chen, J., Ma, T., Xiao, C.: FastGCN: fast learning with graph convolutional networks via importance sampling. In: 6th International Conference on Learning Representations, ICLR. OpenReview.net (2018)
4. Defferrard, M., Bresson, X., Vandergheynst, P.: Convolutional neural networks on graphs with fast localized spectral filtering. In: Advances in Neural Information Processing Systems (NeurIPS), pp. 3837–3845 (2016)
5. Donnat, C., Zitnik, M., Hallac, D., Leskovec, J.: Learning structural node embeddings via diffusion wavelets. In: Proceedings of the 24th ACM International Conference of Knowledge Discovery & Data Mining (KDD), pp. 1320–1329 (2018)
6. Fey, M., Lenssen, J.E.: Fast graph representation learning with pytorch geometric. CoRR arxiv:1903.02428 (2019)
7. Grover, A., Leskovec, J.: node2vec: scalable feature learning for networks. In: Krishnapuram, B., Shah, M., Smola, A.J., Aggarwal, C.C., Shen, D., Rastogi, R. (eds.) Proceedings of the 22nd ACM SIGKDD International Conference on Knowledge Discovery and Data Mining, pp. 855–864. ACM (2016)
8. Hamilton, W.L., Ying, Z., Leskovec, J.: Inductive representation learning on large graphs. In: Guyon, I., et al. (eds.) Annual Conference on Neural Information Processing Systems, pp. 1024–1034 (2017)

9. Hammond, D.K., Vandergheynst, P., Gribonval, R.: Wavelets on graphs via spectral graph theory. Appl. Comput. Harmonic Anal. **30**(2), 129–150 (2011). ISSN 10635203

10. Huang, W., Zhang, T., Rong, Y., Huang, J.: Adaptive sampling towards fast graph representation learning. In: Bengio, S., Wallach, H.M., Larochelle, H., Grauman, K., Cesa-Bianchi, N., Garnett, R. (eds.) Annual Conference on Neural Information Processing Systems, pp. 4563–4572 (2018)

11. Isufi, E., Loukas, A., Simonetto, A., Leus, G.: Autoregressive moving average graph filtering. IEEE Trans. Sig. Process. **65**(2), 274–288 (2017). ISSN 1053587X

12. Kipf, T.N., Welling, M.: Semi-supervised classification with graph convolutional networks. In: Proceedings of the 5th International Conference on Learning Representations (ICLR) (2017)

13. Klicpera, J., Bojchevski, A., Günnemann, S.: Predict then propagate: graph neural networks meet personalized PageRank. In: Proceedings of the 7th International Conference on Learning Representations (ICLR) (2019)

14. Klicpera, J., Weißenberger, S., Günnemann, S.: Diffusion improves graph learning. In: Advances in Neural Information Processing Systems (NeurIPS), pp. 13333–13345 (2019)

15. Lee, J.B., Rossi, R.A., Kong, X.: Graph classification using structural attention. In: Proceedings of the 24th ACM International Conference on Knowledge Discovery & Data Mining (KDD), pp. 1666–1674 (2018)

16. Levy, O., Goldberg, Y.: Neural word embedding as implicit matrix factorization. In: Ghahramani, Z., Welling, M., Cortes, C., Lawrence, N.D., Weinberger, K.Q. (eds.) Annual Conference on Neural Information Processing Systems, pp. 2177–2185 (2014)

17. Li, Q., Han, Z., Wu, X.: Deeper insights into graph convolutional networks for semi-supervised learning. In: Proceedings of the 32nd Conference on Artificial Intelligence (AAAI), pp. 3538–3545 (2018)

18. Li, S., Kim, D., Wang, Q.: Beyond low-pass filters: adaptive feature propagation on graphs. Preprint arxiv:2103.14187 (2021)

19. Liu, J., Isufi, E., Leus, G.: Filter design for autoregressive moving average graph filters. IEEE Trans. Sig. Inf. Process. Netw. **5**(1), 47–60 (2019). ISSN 2373776X

20. Liu, M., Wang, Z., Ji, S.: Non-local graph neural networks. CoRR arxiv:2005.14612 (2020)

21. Michel, P., Levy, O., Neubig, G.: Are sixteen heads really better than one? In: Wallach, H.M., Larochelle, H., Beygelzimer, A., d'Alché-Buc, F., Fox, E.B., Garnett, R. (eds.) Annual Conference on Neural Information Processing Systems 2019, pp. 14014–14024 (2019)

22. Miller, B.A., Beard, M.S., Bliss, N.T.: Matched filtering for subgraph detection in dynamic networks. In: 2011 IEEE Statistical Signal Processing Workshop (SSP), pp. 509–512 (2011)

23. Napoli, E.D., Polizzi, E., Saad, Y.: Efficient estimation of eigenvalue counts in an interval. Numer. Linear Algebra Appl. **23**(4), 674–692 (2016)

24. Nickel, M., Kiela, D.: Poincaré embeddings for learning hierarchical representations. Adv. Neural Inf. Process. Syst. 30, 6338–6347 (2017). ISSN 10495258

25. NT, H., Maehara, T.: Revisiting graph neural networks: all we have is low-pass filters. CoRR arxiv:1905.09550 (2019)

26. Pei, H., Wei, B., Chang, K.C., Lei, Y., Yang, B.: Geom-GCN: geometric graph convolutional networks. In: 8th International Conference on Learning Representations, ICLR 2020, Addis Ababa, Ethiopia, 26–30 April 2020. OpenReview.net (2020)

27. Perozzi, B., Al-Rfou, R., Skiena, S.: DeepWalk: online learning of social represen-tations. In: Macskassy, S.A., Perlich, C., Leskovec, J., Wang, W., Ghani, R. (eds.) The 20th ACM SIGKDD International Conference on Knowledge Discovery and Data Mining, KDD 2014, New York, NY, USA, 24–27 August 2014, pp. 701–710. ACM (2014)

28. Qiu, J., Dong, Y., Ma, H., Li, J., Wang, K., Tang, J.: Network embedding as matrix factorization: unifying deepWalk, LINE, PTE, and node2vec. In: Chang, Y., Zhai, C., Liu, Y., Maarek, Y. (eds.) Proceedings of the Eleventh ACM International Conference on Web Search and Data Mining, WSDM 2018, Marina Del Rey, CA, USA, 5–9 February 2018, pp. 459–467. ACM (2018)

29. Ribeiro, L.F., Saverese, P.H., Figueiredo, D.R.: Struc2vec: learning node representa-tions from structural identity. In: Proceedings of the ACM SIGKDD International Conference on Knowledge Discovery and Data Mining (2017). ISBN 9781450348874

30. Rozemberczki, B., Allen, C., Sarkar, R.: Multi-scale attributed node embedding. CoRR arxiv:1909.13021 (2019)

31. Sakiyama, A., Watanabe, K., Tanaka, Y.: Spectral graph wavelets and filter banks with low approximation error. IEEE Trans. Sig. Inf. Process. Netw. $2(3)$, 230–245 (2016). ISSN 2373776X

32. Schaub, M.T., Segarra, S.: Flow smoothing and denoising: graph signal processing in the edge-space. In: 2018 IEEE Global Conference on Signal and Information Processing (GlobalSIP), pp. 735–739 (2018)

33. Sen, P., Namata, G., Bilgic, M., Getoor, L., Gallagher, B., Eliassi-Rad, T.: Collec-tive classification in network data. AI Mag. $29(3)$, 93–106 (2008)

34. Shuman, D.I., Narang, S.K., Frossard, P., Ortega, A., Vandergheynst, P.: The emerging field of signal processing on graphs: extending high-dimensional data analysis to networks and other irregular domains. IEEE Sig. Process. Mag. $30(3)$, 83–98 (2013)

35. Tang, J., Qu, M., Mei, Q.: PTE: predictive text embedding through large-scale heterogeneous text networks. In: Cao, L., Zhang, C., Joachims, T., Webb, G.I., Margineantu, D.D., Williams, G. (eds.) Proceedings of the 21th ACM SIGKDD International Conference on Knowledge Discovery and Data Mining, pp. 1165–1174. ACM (2015)

36. Tang, J., Qu, M., Wang, M., Zhang, M., Yan, J., Mei, Q.: LINE: large-scale infor-mation network embedding. In: Gangemi, A., Leonardi, S., Panconesi, A. (eds.) Proceedings of the 24th International Conference on World Wide Web, WWW, pp. 1067–1077. ACM (2015)

37. Tenenbaum, J.B., De Silva, V., Langford, J.C.: A global geometric framework for nonlinear dimensionality reduction. Science $290(5500)$, 2319–2323 (2000). ISSN 00368075

38. Velickovic, P., Cucurull, G., Casanova, A., Romero, A., Liò, P., Bengio, Y.: Graph attention networks. In: Proceedings of the 6th International Conference on Learn-ing Representations (ICLR) (2018)

39. Wai, H., Segarra, S., Ozdaglar, A.E., Scaglione, A., Jadbabaie, A.: Community detection from low-rank excitations of a graph filter. In: 2018 IEEE International Conference on Acoustics, Speech and Signal Processing (ICASSP), pp. 4044–4048 (2018)

40. Wu, F., Jr., A.H.S., Zhang, T., Fifty, C., Yu, T., Weinberger, K.Q.: Simplifying graph convolutional networks. In: Proceedings of the 36th International Conference on Machine Learning (ICML), vol. 97, pp. 6861–6871 (2019)

41. Xu, B., Shen, H., Cao, Q., Qiu, Y., Cheng, X.: Graph wavelet neural network. In: Proceedings of the 7th International Conference on Learning Representations (ICLR) (2019)
42. Zhang, K., Zhu, Y., Wang, J., Zhang, J.: Adaptive structural fingerprints for graph attention networks. In: Proceedings of the 8th International Conference on Learning Representations (ICLR) (2020)
43. Zhang, M., Chen, Y.: Link prediction based on graph neural networks. In: Bengio, S., Wallach, H.M., Larochelle, H., Grauman, K., Cesa-Bianchi, N., Garnett, R. (eds.) Advances in Neural Information Processing Systems 31: Annual Conference on Neural Information Processing Systems 2018, NeurIPS 2018, 3–8 December 2018, Canada, Montréal, pp. 5171–5181 (2018)
44. Zhao, L., Akoglu, L.: PairNorm: tackling oversmoothing in GNNs. In: Proceedings of the 8th International Conference on Learning Representations (ICLR) (2020)
45. Zhu, J., Yan, Y., Zhao, L., Heimann, M., Akoglu, L., Koutra, D.: Beyond homophily in graph neural networks: current limitations and effective designs. In: Larochelle, H., Ranzato, M., Hadsell, R., Balcan, M., Lin, H. (eds.) Advances in Neural Information Processing Systems 33: Annual Conference on Neural Information Processing Systems 2020, NeurIPS 2020, 6–12 December 2020 (2020). Virtual

Zero-Shot Scene Graph Relation Prediction Through Commonsense Knowledge Integration

Xuan Kan, Hejie Cui, and Carl Yang[(✉)]

Department of Computer Science, Emory University, Atlanta, USA
{xuan.kan,hejie.cui,j.carlyang}@emory.edu

Abstract. Relation prediction among entities in images is an important step in scene graph generation (SGG), which further impacts various visual understanding and reasoning tasks. Existing SGG frameworks, however, require heavy training yet are incapable of modeling unseen (*i.e.*, zero-shot) triplets. In this work, we stress that such incapability is due to the lack of commonsense reasoning, *i.e.*, the ability to associate similar entities and infer similar relations based on general understanding of the world. To fill this gap, we propose Comm**O**nsense-integr**A**ted s**C**ene grap**H** r**E**lation p**R**ediction (**COACHER**), a framework to integrate commonsense knowledge for SGG, especially for zero-shot relation prediction. Specifically, we develop novel graph mining pipelines to model the neighborhoods and paths around entities in an external commonsense knowledge graph, and integrate them on top of state-of-the-art SGG frameworks. Extensive quantitative evaluations and qualitative case studies on both original and manipulated datasets from Visual Genome demonstrate the effectiveness of our proposed approach. The code is available at https://github.com/Wayfear/Coacher.

Keywords: Scene graph generation · Relation prediction · Zero-shot · Commonsense · Knowledge graph · Reasoning · Graph mining

1 Introduction

With the unprecedented advances of computer vision, visual understanding and reasoning tasks such as Image Captioning and Visual Question Answer (VQA) have attracted increasing interest recently. Scene graph generation (SGG), which predicts all relations between detected entities from images, distills visual information in structural understanding. With clear semantics of entities and relations, scene graphs are widely used for various downstream tasks like VQA [7,22,36], image captioning [4,31,32], and image generation [9–11].

A critical and challenging step in SGG is *relation prediction*. A relation instance in scene graph is defined as a triplet ⟨*subject, relation, object*⟩. Given two detected entities, which relation exists between them is predicted based on the probability score from the learned relation prediction model. However, most

© Springer Nature Switzerland AG 2021
N. Oliver et al. (Eds.): ECML PKDD 2021, LNAI 12976, pp. 466–482, 2021.
https://doi.org/10.1007/978-3-030-86520-7_29

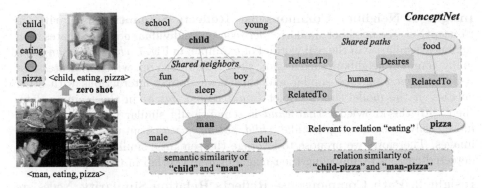

Fig. 1. Toy example of zero-shot relation prediction in scene graph generation and insights about commonsense knowledge integration.

of the existing scene graph generation models rely on heavy training to memorize seen triplets, which limits their utility in reality, because many real triples are never seen during training.

Challenge: Performance Deterioration on Zero-Shot Triplets. When evaluating the performance of relation prediction model in practice, there are two types of triplets, the ones seen in the training data and the ones unseen. Those unseen ones are called *zero-shot* triplets. As shown in Fig. 1, triplet ⟨*man, eating, pizza*⟩ is observed in the training data. If this triplet appears again in the testing phase, then it is called a non-zero-shot triplet. In contrast, a triplet ⟨*child, eating, pizza*⟩ with a new entity-relation combination not observed in training data is called a zero-shot triplet.

Although several existing scene graph generation methods have achieved decent performance on the whole testing data, little analysis has been done on the performance on zero-shot triplets. Unfortunately, based on our preliminary analysis (Sect. 2.1), the performance of existing methods degrades remarkably when solely tested on zero-shot triplets, while many of such triplets are rather common in real life. Such reliance on seeing all triplets in training data is problematic, as possible triplets in the wild are simply inexhaustible, which requires smarter models with better generalizability.

Motivation: Commonsense as a "COACHER" for Zero-Shot Relation Prediction. Commonsense knowledge refers to general facts about the world that empower human beings to reason over unfamiliar scenarios. Motivated by this process from humans' perspective, in this work, we propose to integrate commonsense knowledge to alleviate the inexhaustible-triplet problem and improve the performance of zero-shot relation prediction in SGG.

The illustration of our problem and insights are illustrated in Fig. 1. Specifically, the commonsense knowledge utilized in this paper comes from Concept-Net [13], a crowd-sourced semantic knowledge graph containing rich structured knowledge regarding real-world concepts.

Insight 1: Neighbor Commonsense Reflects Semantic Similarity. In ConceptNet, the neighbor similarity between two individual nodes indicates their semantic similarity in the real world. For example, in Fig. 1, *child* and *man* share many common neighbors such as *fun*, *sleep*, *boy* and so on, which indicates that *child* and *man* may be similar and thus have similar interactions with other entities. If the model sees a triplet ⟨*man, eating, pizza*⟩ in the training data, then with the knowledge that *child* is semantically similar to *man*, it should more easily recognize triplets like ⟨*child, eating, pizza*⟩ from unseen but similar images. Therefore, we propose to leverage the semantic similarity between two detected entities by modeling their neighborhood overlap in ConceptNet.

Insight 2: Path Commonsense Reflects Relation Similarity. Nodes are connected by paths composed of multiple consecutive edges in ConceptNet. As is shown on the right in Fig. 1, the entity pairs of (*child, pizza*) and (*man, pizza*) share common intermediate paths like ⟨*RelatedTo, human, Desires, food, RelatedTo*⟩. This similarity of intermediate paths indicates that the relations between *man* and *pizza* may be similar to those between *child* and *pizza*. If there is a triplet ⟨*man, eating, pizza*⟩ in the training data, then the model should tend to predict the relation *eating* given (*child, pizza*) in an unseen but similar image. Following the idea above, we propose to infer the relation between two entities by modeling their path coincidence with other entity pairs in ConceptNet.

Approach: Scene Graph Relation Prediction through Commonsense Knowledge Integration. In this work, we propose a novel framework that integrates external commonsense knowledge into SGG for relation prediction on zero-shot triplets, which we term as CommOnsense-integrAted sCene grapH rElation pRediction, COACHER for brevity (Fig. 3). To be specific, we investigate the utility of external commonsense knowledge through real data analysis. With the validated effectiveness of both neighbor- and path-based commonsense knowledge from ConceptNet, we design three modules for different levels of knowledge integration, which generate auxiliary knowledge embedding for generalizable relation prediction.

In summary, our main contributions are three-fold.

- We analyze the ignorance of zero-shot triplets by existing SGG models and validate the potential utility of commonsense knowledge from ConceptNet through real data analysis (Sect. 2).
- Based on the state-of-the-art SGG framework, we integrate external commonsense knowledge regarding ConceptNet neighbors and paths to improve relation prediction on zero-shot triplets (Sect. 3).
- Extensive quantitative experiments and qualitative analyses on the widely-used SGG benchmark dataset of Visual Genome demonstrate the effectiveness of our proposed COACHER framework. Particularly, COACHER achieves consistently better performance over state-of-the-art baselines on the original dataset, and outperforms them significantly on amplified datasets towards more severe zero-shot learning settings (Sect. 4).

2   Motivating Analysis

2.1   Ignorance yet Importance of Zero-Shot Triplets

In SGG, triplets are used to model entities with their relations. Among the great number of possible rational triplets in the wild, some of them exist in the training data while more others do not. The ability of correctly inferring zero-shot triplets can be extremely important to reflect the generalization capability of the model and its real utility in practice.

Although the performance of zero-shot scene graph generation was once studied in the early days on a small dataset [15], later researchers do not pay much attention to this setting. Until 2020, Tang et al. [21] first reported zero-shot performance on Visual Genome. However, they have not proposed any particular solutions to improve it.

Table 1. Performance (%) of three state-of-the-art models on non-zero-shot and zero-shot triplets on Visual Genome (Please refer to Sect. 4 for more details about the presented models).

Methods	NM			NM+			TDE		
MeanRecall@K	$K=20$	$K=50$	$K=100$	$K=20$	$K=50$	$K=100$	$K=20$	$K=50$	$K=100$
None-zero-shot	25.12	33.32	37.06	25.08	33.69	37.54	26.26	35.93	40.27
Zero-shot	12.85	18.93	21.84	12.28	18.28	21.30	5.84	11.68	15.10

The ignorance of zero-shot settings causes existing methods a dramatic descent on the relation prediction on zero-shot triplets. Table 1 shows the performance of three state-of-the-art models on Visual Genome, the most widely used benchmark dataset for SGG. Note that *mean recall* is used here for performance evaluation, which is the average result of *triplet-wise* recall. Consistently, the mean recall of non-zero-shot triplets under different values of K can achieve almost twice of that on zero-shot ones, which demonstrates a concerning performance deterioration on zero-shot relation prediction.

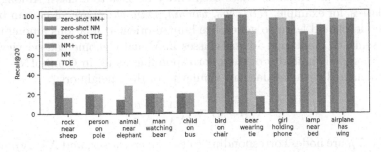

Fig. 2. Recall@20 performance on two different types of triplets. The five triplets on the left are zero-shot ones while the other five on the right are non-zero-shot.

Furthermore, although zero-shot triplets are not labeled in the training data, some of them are actually no less common in reality compared to the labeled ones. To give a more concrete illustration, we visualize the *recall@20* on five actual zero-shot triplets and another five non-zero-shot triplets in Fig. 2. Consistent with the results in Table 1, performance of existing models on the five zero-shot triplets turns out to be much lower than that on the five non-zero-shot triplets. However, these zero-shot triplets represent very common relations such as ⟨*child, on, bus*⟩ which are in fact more common than some non-zero-shot ones such as ⟨*bear, wearing, tie*⟩. The performance on certain triplets like ⟨*bear, wearing, tie*⟩ is much better simply because they appear in the training data and got memorized by the model, but the utility of such memorization is rather limited in reality without the ability of generalization.

Lastly, due to the inexhaustibility of triplets in the real world, labeled datasets can only cover a limited portion of the extensive knowledge. Many common triplets simply do not appear in the dataset. For example, ⟨*men, holding, umbrella*⟩, ⟨*vehicle, parked on, beach*⟩, ⟨*women, in, boat*⟩ and ⟨*fence, across, sidewalk*⟩, to name a few, are common triplets that never appear in Visual Genome, even though all involved objects and relations appear in other labeled triplets. Therefore, to propose a practical SGG model that can be widely used and assist various downstream tasks, we need to pay more attention to zero-shot triplets.

Motivated by the ignorance and importance of zero-shot triplets, in this work, we focus on integrating commonsense knowledge from external resources to improve the relation prediction performance on zero-shot triplets. Specifically, we leverage ConceptNet as the external knowledge resource from several other alternatives due to its wide coverage of concepts and accompanying semantic embeddings of concepts as useful features [13]. In ConceptNet, each concept (word or phrase) is modeled as a node and each edge represents the relation between two concepts. Thanks to its wide coverage, we are able to link each entity class in Visual Genome to one node in ConceptNet.

2.2   Commonsense Knowledge from ConceptNet Neighbors

Many entities in real life share similar semantic meanings, which can potentially support zero-shot relation prediction. Given an image from Visual Genome, we can detect multiple entities, where each entity belongs to a class. Among these entity classes, for example, *(girl, boy, woman, man, child)* are all human beings, so the model should learn to generalize human-oriented relations among them.

The semantic similarity among classes in Visual Genome can be viewed as the neighborhood similarity of their corresponding nodes in ConceptNet, which can be calculated with the Jaccard similarity of their neighbors:

$$J(\mathcal{V}_A, \mathcal{V}_B) = \frac{|\mathcal{N}_A \cap \mathcal{N}_B|}{|\mathcal{N}_A \cup \mathcal{N}_B|}, \tag{1}$$

where $\mathcal{V}_A$, $\mathcal{V}_B$ are nodes corresponding to two given classes, and $\mathcal{N}_A$, $\mathcal{N}_B$ are the neighbors of them in ConceptNet, respectively.

In order to validate the effectiveness of utilizing neighborhood similarity in ConceptNet as a measurement of semantic similarity in Visual Genome, we calculate the similarity between each pair of the top 150 mostly observed classes

Table 2. Top 20 pairs of similar entity classes.

1	chair-seat	2	shoe-sock	3	hill-mountain	4	coat-jacket	5	house-building
6	airplane-plane	7	woman-girl	8	desk-table	9	window-door	10	men-man
11	shirt-jacket	12	house-room	13	room-building	14	girl-boy	15	plate-table
16	arm-leg	17	chair-table	18	tree-branch	19	cow-sheep	20	arm-hand

in Visual Genome, and rank their similarity in descending order. Results of the top 20 most similar pairs are shown in Table 2. As can be seen, the top similar pairs such as *chair − seat* indeed capture the semantic similarity between two classes, which validates the virtue of using ConceptNet neighbors to bring in commonsense knowledge for modeling the semantic similarity among entities.

2.3   Commonsense Knowledge from ConceptNet Paths

In ConceptNet, besides the one-hop information from neighbors, paths composed by multiple edges can further encode rich multi-hop information. Specifically, if two pairs of entities are connected by many same paths in ConceptNet, they are more likely to share similar relations. In order to investigate such path-relation correlation between node pairs on ConceptNet, we define MidPath as follows:

Definition 1 (MidPath). *Given two nodes $\mathcal{V}_A$ and $\mathcal{V}_B$ in the graph, a MidPath between $\mathcal{V}_A$ and $\mathcal{V}_B$ is defined as the sequence of all intermediate edges and nodes on a path from $\mathcal{V}_A$ to $\mathcal{V}_B$, excluding both the head and tail nodes.*

For example, given a path ⟨*people, RelatedTo, automobile, AtLocation, street*⟩ between nodes *people* and *street*, the corresponding MidPath is ⟨*RelatedTo, automobile, AtLocation*⟩.

Table 3. Top 3 related MidPaths for 10 relations.

Relation	Top1 MidPath	Top2 MidPath	Top3 MidPath
Parked on	RelatedTo-cars-RelatedTo	RelatedTo-driven-ReceivesAction	AtLocation-automobile-RelatedTo
Says	RelatedTo	RelatedTo-communication_device-RelatedTo	RelatedTo-command-RelatedTo
Laying on	RelatedTo-legs-RelatedTo	AtLocation	RelatedTo-human-RelatedTo
Wearing	RelatedTo-body-RelatedTo	RelatedTo-dress-RelatedTo	RelatedTo-clothing-Desires
Against	AtLocation-garage-AtLocation	RelatedTo-wall-RelatedTo	RelatedTo
Sitting on	AtLocation	AtLocation-human-RelatedTo	RelatedTo-legs-RelatedTo
Walking in	RelatedTo-home-RelatedTo	UsedFor-children-RelatedTo	RelatedTo-crowd-RelatedTo
Growing on	RelatedTo-growth-RelatedTo	RelatedTo-leaves-RelatedTo	RelatedTo-stem-RelatedTo
Watching	RelatedTo-human-RelatedTo	RelatedTo-date-RelatedTo	RelatedTo-female-RelatedTo
Playing	RelatedTo-human-RelatedTo	RelatedTo	RelatedTo-clown-RelatedTo

For each relation in Visual Genome, probability analysis is done to investigate the related MidPaths. The smallest unit in the dataset is a triplet ⟨*subject, relation, object*⟩. With nodes *subject* and *object*, we can extract a set of Mid-Paths $\mathcal{P}$ connecting them from ConceptNet. For each path $p_i \in \mathcal{P}$ and relation

$r_j \in \mathcal{R}$, the number of co-occurrence $I(\mathcal{MP} = p_i, \mathcal{R} = r_j)$ can be counted. Following the formula below, we calculate the conditional probability of observing MidPath p_i given a specific relation r_j:

$$P(\mathcal{MP} = p_i | \mathcal{R} = r_j) = \frac{I(\mathcal{MP} = p_i, \mathcal{R} = r_j)}{\sum_{p_k \in \mathcal{P}} I(\mathcal{MP} = p_k, \mathcal{R} = r_j)}. \tag{2}$$

To eliminate random effects, the probability of observing a random MidPath p_i is also calculated, as follows

$$P(\mathcal{MP} = p_i) = \frac{I(\mathcal{MP} = p_i)}{\sum_{p_k \in \mathcal{P}} I(\mathcal{MP} = p_k)}, \tag{3}$$

where $I(\mathcal{MP} = p_i)$ is the occurrence number of MidPath p_i. Now we can measure how significant MidPath p_i is given a specific relation r_j by

$$Score(p_i, r_j) = P(\mathcal{MP} = p_i | \mathcal{R} = r_j) - P(\mathcal{MP} = p_i). \tag{4}$$

The higher the score is, the more significantly MidPath p_i can imply relation r_j. Table 3 shows top three MidPaths with highest scores for 10 relations. Clearly, these top MidPaths are semantically meaningful and potentially beneficial for generalizing relation predictions among entity pairs.

3   COACHER

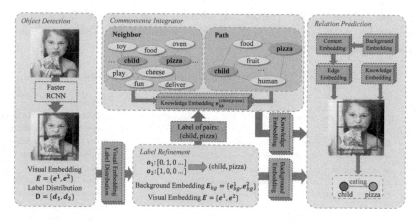

Fig. 3. The overall framework of our proposed COACHER.

In this section, we present the proposed SGG with commonsense knowledge integration model COACHER in detail. Figure 3 shows the major components in COACHER: (a) object detection and refinement, (b) commonsense knowledge embedding generation, and (c) commonsense enhanced relation prediction.

3.1   Backbone Scene Graph Generation Pipeline

Scene graph generation is a task aiming to understand visual scenes by extracting entities from images and predicting the semantic relations between them. Various technique has been explored for scene graph generation (Sect. 5). Here we adopt one of the state-of-the-art pipelines from Neural Motif [34] as the backbone framework. We mainly focus on commonsense-integrated relation prediction for zero-shot triplets, which is crucial for practice as discussed in Sect. 2.

The backbone scene graph generation pipeline includes three components:

Step1: Object Detection. With the development of deep learning, some popular frameworks such as R-CNN [3,19], YOLO [18] and SSD [14] achieve impressive performance on this task. Following previous literature [5,21], we adopt a pre-trained Faster R-CNN model as the detector in our framework. In this step, with the input of a single image $\mathcal{I}$, the output includes: a set of region proposals $B = \{b_1, \cdots, b_n\}$, a set of distribution vectors $D = \{d_1, \cdots, d_n\}$, where $d_i \in \mathbb{R}^{|\mathcal{C}|}$ is a label probabilities distribution and $|\mathcal{C}|$ is the number of classes, as well as a visual embedding $E = \{e^1, \cdots, e^n\}$ for each detected object.

Step2: Label Refinement. Based on the label distribution D generated from Step1, we conduct label refinement to generate a one-hot vector of entity classes for each region proposal, which will be used for relation prediction.

First, background embeddings E_{bg} containing information from both region proposal level and global level of the image are generated using a bi-LSTM:

$$E_{bg} = \text{biLSTM}([e^i; \text{MLP}(d_i)]_{i=1,\cdots,n}), \tag{5}$$

where $E_{bg} = [e_{bg}^1, e_{bg}^2, \cdots, e_{bg}^n]$ is the hidden state of the last layer in LSTM and n is the number of region proposals. Then, an LSTM is used to decode each region proposal embedding e_{bg}^i:

$$h_i = \text{LSTM}([e_{bg}^i; o_{i-1}]), \tag{6}$$

$$o_i = \text{argmax}(\text{MLP}(h_i)). \tag{7}$$

o_i is the one-hot vector representing the refined class label of a region proposal.

Step3: Relation Prediction. After obtaining refined object labels for all region proposals, we use them to further generate context embeddings E_{ct}:

$$E_{ct} = \text{biLSTM}([e_{bg}^i; \text{MLP}(o_i)]_{i=1,\cdots,n}), \tag{8}$$

where $E_{ct} = [e_{ct}^1, e_{ct}^2, \cdots, e_{ct}^n]$, which are then used to extract edge embeddings e_{eg} and predict the relation between each pair of bounding boxes:

$$e_{eg}^{(i,j)} = \text{MLP}(e_{ct}^i) \circ \text{MLP}(e_{ct}^j) \circ (e^i \cup e^j), \tag{9}$$

$$r_{(i,j)} = \text{argmax}(\text{MLP}([e_{eg}^{(i,j)}; e_{kb}^{(i,j)}])), \tag{10}$$

where $\circ$ represents element-wise product, $e_{kb}^{(i,j)}$ is the commonsense knowledge embedding obtained from ConceptNet, which we will introduce next.

3.2 Commonsense Integrator

Commonsense knowledge integration is achieved by computing e_{kb} from external resources. Specifically, we use ConceptNet [13] here as the source of external commonsense knowledge. ConceptNet is a knowledge graph that connects words and phrases of natural language with labeled edges. It is constructed from rich resources such as Wiktionary and WordNet. With the combination of these resources, ConceptNet contains over 21 million edges and over 8 million nodes, covering all of the entity classes in Visual Genome. Besides, it also provides a semantic embedding for each node, which can serve as a semantic feature. Here we develop three types of integrators to generate commonsense knowledge embeddings from ConceptNet.

Neighbor Integrator. ConceptNet is a massive graph $\mathcal{G} = (\mathcal{V}, \mathcal{E})$, where $\mathcal{V}$ and $\mathcal{E}$ are the node set and edge set, respectively. Each detected entity can be seen as a node in ConceptNet. Given a class $c \in \mathcal{C}$ of a detected entity, its neighborhood information can be collected as follows:

$$c \xrightarrow{\text{link}} \mathcal{V}_c \xrightarrow{\text{retrieve neighbors}} \mathcal{N}_c = \{\mathcal{V}_n | (\mathcal{V}_c, \mathcal{V}_n) \in \mathcal{E}\}. \tag{11}$$

Denote $\boldsymbol{F} \subset \mathbb{R}^{|\mathcal{V}| \times k}$ as the feature matrix of all nodes in ConceptNet from [20], where k is the dimension of the feature vector. Neighbor embedding e_{nb}^c of node $\mathcal{V}_c$ is calculated as the average over all of its neighbors' embeddings:

$$e_{nb}^c = \frac{1}{|\mathcal{N}_c|} \sum_{\mathcal{V}_n \in \mathcal{N}_c} \boldsymbol{F}_n, \tag{12}$$

where $\boldsymbol{F}_n$ is the n_{th} row of $\boldsymbol{F}$. For relation prediction, given a pair of detected entities with classes (a, b), the neighbor-based commonsense knowledge embedding $e_{kb}^{(a,b)}$ of this pair is calculated as:

$$e_{kb}^{(a,b)} = \text{ReLU}(\text{MLP}([e_{nb}^a; e_{nb}^b])), \tag{13}$$

where MLP denotes the multi-layer perceptron.

Path Integrator. Given a pair of entities with classes (a, b) recognized from the object detection model, a set of paths connecting them can be obtained following the procedure below:

$$(a, b) \xrightarrow{\text{link}} (\mathcal{V}_a, \mathcal{V}_b) \xrightarrow{\text{retrieve paths}} \mathcal{P}_{(a,b)} = \{p | p = \{\mathcal{V}_a, \mathcal{V}_1, \cdots, \mathcal{V}_b\}\}. \tag{14}$$

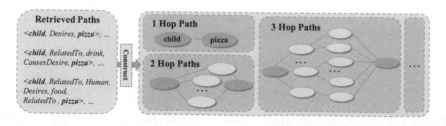

Fig. 4. Graphic representation construction from retrieved paths.

We classify these extracted paths based on their number of hops. The graphical representation of paths is shown in Fig. 4. Each set of *l-Hop Paths* naturally constitutes a small graph $G_l^{(a,b)}$, where l is the number of hops on the paths. The goal is to learn a representation of all graphs $\{G_l^{(a,b)}\}_{l=1}^L$, where L is often a small number like 2 or 3, since longer paths are too noisy and hard to retrieve.

Classical sequence models such as LSTM cannot handle very short paths effectively. Inspired by message passing network for graph representation learning [2,16], we design a neural message passing mechanism to learn a representation for each set of *l-Hop Paths*, and then combine them as the final path-based commonsense knowledge embedding for pair (a, b).

Specifically, we design the message passing mechanism as follows:

$$\mathrm{MSG}_v^t(\mathcal{V}_a) = \mathrm{MSG}_v^{t-1}(\mathcal{V}_a) + \sum_{v \in \mathcal{N}_a} \mathrm{MSG}_e^t(v), \tag{15}$$

$$\mathrm{MSG}_e^t(v) = \mathrm{MLP}(\mathrm{MSG}_v^{t-1}(v)), \tag{16}$$

where $\mathrm{MSG}_v^0(v)$ is initialized as $\boldsymbol{F}_v$, *i.e.*, the original node features from [20].

In order to update the embedding of each node with all hops on the corresponding paths, given $G_l^{(a,b)}$, we iterate the above process $t = l$ times to obtain the final node embeddings $\boldsymbol{T}_l^{(a,b)}$ on each set of *l-Hop Paths* in $G_l^{(a,b)}$.

To join the information of multiple paths on each graph, we select and aggregate the most important paths. As a simplification, we adopt the GlobalSortPool operator [35] on the embeddings of all nodes in $G_l^{(a,b)}$ as follows:

$$e_g^{(a,b),l} = \mathrm{GlobalSortPool}(\boldsymbol{T}_l^{(a,b)}), \tag{17}$$

which learns to sort all nodes by a specific embedding channel and pool the top K (*e.g.*, K = 5) nodes on all embedding channels.

Finally, we aggregate the path embeddings $\{e_g^{(a,b),l}\}_{l=1}^L$ of different length l through vector concatenation.

Fused Integrator. To fuse neighbor- and path-based commonsense knowledge, we inject the neighbor-based knowledge into the path-based knowledge by initializing $\mathrm{MSG}_v^0(v)$ in Eq. 15 as the element-wise mean between e_{nb} from Eq. 12 and the original node features $\boldsymbol{F}$ as follows:

$$\mathrm{MSG}_v^0(v) = \mathrm{MEAN}(\boldsymbol{F}_v, e_{nb}^v), \tag{18}$$

where v can be replaced as a and b for the entity class pair (a, b).

4   Experiments

4.1   Experimental Settings

Original Whole Dataset. For scene graph generation, we use the Visual
Genome dataset [8], a commonly used benchmark for SGG, to train and test our
framework. This dataset contains 108,077 images, where the number of classes
and relations are 75,729 and 40,480, respectively. However, 92% of relationships
have no more than 10 instances. Therefore, we follow the widely used split strategy on Visual Genome [5, 21, 24] that selects the most frequent 150 object classes
and 50 relations as a representative. Besides, we use 70% images as well as their
corresponding entities and relations as the training set, and the other 30% of
images are left out for testing. A 5k validation set is split from the training set
for parameter tuning.

Zero-Shot Amplified Dataset. To further investigate the model's generalization ability in more severe zero-shot settings, we reduce the information that
the model can leverage during training by constructing another zero-shot amplified dataset. This is achieved by simply removing images containing less common
relations from the training data. As a result, the triplet numbers of the last thirty
common relations are halved, while the triplet numbers of the first twenty common relations mostly remain the same. In this way, we exacerbate the difficulty
for the model, especially on predicting relations for zero-shot triplets.

Compared Algorithms. We compare COACHER with four baseline methods.

- **NeuralMotifs (NM)** [34] is a strong baseline which is widely-compared on
  Visual Genome for SGG.
- **NeuralMotifs with Knowledge Refinement Network (NM+)** [5] is
  the only existing method that leverages external knowledge for SGG, which
  is the closest to ours. This method mainly contains two new parts, knowledge
  refinement and image reconstruction. We add its knowledge refinement part
  on top of Neural Motifs, which we call NM+.
- **TDE** [21] is the current state-of-the-art method for scene graph generation.
  This work is also the first one reporting zero-shot performance on Visual
  Genome but it does not take effort to improve it.
- **CSK-N** is a baseline based on our framework which makes predictions without visual information. Given a pair of entities, we predict their relation using
  only the neighbor-based commonsense knowledge embedding.

Evaluation Metrics. Since the purpose of this work is to improve performance
on zero-shot triplets, we follow the relation classification setting for evaluation
[21]. Specifically, we use the following two metrics and focus on their evaluations
on the zero-shot subset of the whole testing data:

- **zR@K**(zero-shot Recall@K): Recall@K is the earliest and the most widely
  accepted metric in SGG, first used by Lu et al. [15]. Since the relation between
  a pair of entities is not complete, treating it as a retrieval problem is more

proper than a classification problem. Here we take the Recall@K on zero-shot subset and shorten it as zR@K.

- **ng-zR@K**(zero-shot no-graph-constraint Recall@K): No-graph-constraint Recall@K is first used by Newell et al. for SGG [17]. It allows a pair of entities to have multiple relations, which significantly improves the recall value. Here we take the ng-R@K on zero-shot subset and shorten it as ng-zR@K.

4.2 Hyper-parameter Setting

A pre-trained and frozen Faster-RCNN [19] equipped with the ResNeXt-101-FPN [6] backbone is used as the object detector for all models. Batch size and the max iteration number are set as 12 and 50,000 respectively. The learning rate begins with 1.2×10^{-1} with a decay rate of 10 and a stepped scheduler based on the performance stability on the validation set. A Quadro RTX 8000 GPU with 48 GB of memory is used for our model training. For path length, here we only use 1 and 2 hop paths ($L = 2$) due to the GPU memory limitation. Based on observation from humans' perspective, these short paths are also the more informative ones compared with longer paths.

4.3 Performance Evaluations

Table 4 and Table 5 show performance of three variants of COACHER with different knowledge integrators (COACHER-N, COACHER-P, COACHER-N+P) as well as four baselines on the original and amplified datasets, respectively.

Table 4. Zero-shot performance (%) on the original whole dataset.

Method	zR@20	zR@50	zR@100	ng-zR@20	ng-zR@50	ng-zR@100
NM	13.05 ± 0.06	19.03 ± 0.22	21.98 ± 0.22	15.16 ± 0.49	28.78 ± 0.57	41.52 ± 0.79
NM+	12.35 ± 0.28	18.10 ± 0.13	21.13 ± 0.24	14.47 ± 0.11	27.93 ± 0.15	40.84 ± 0.28
TDE	8.36 ± 0.25	14.35 ± 0.27	18.04 ± 0.46	9.84 ± 0.33	19.28 ± 0.56	28.99 ± 0.44
CSK-N	5.95 ± 0.62	10.12 ± 0.79	13.05 ± 0.64	8.15 ± 0.57	16.79 ± 0.62	26.19 ± 1.19
COACHER-N	12.73 ± 0.22	18.88 ± 0.12	21.88 ± 0.11	15.10 ± 0.47	28.73 ± 0.21	41.06 ± 0.26
COACHER-P	12.24 ± 0.17	18.12 ± 0.16	21.55 ± 0.39	14.39 ± 0.46	28.90 ± 0.43	40.98 ± 0.45
COACHER-N+P	13.42 ± 0.28	19.31 ± 0.27	22.22 ± 0.29	15.54 ± 0.27	29.31 ± 0.27	41.39 ± 0.22

Table 5. Zero-shot performance (%) on the zero-shot amplified dataset.

Method	zR@20	zR@50	zR@100	ng-zR@20	ng-zR@50	ng-zR@100
NM	11.98 ± 0.09	17.86 ± 0.13	20.48 ± 0.02	13.98 ± 0.33	27.43 ± 0.72	39.33 ± 0.77
NM+	11.82 ± 0.06	17.27 ± 0.34	20.10 ± 0.31	13.83 ± 0.18	26.92 ± 0.25	38.71 ± 0.32
TDE	5.67 ± 0.03	11.08 ± 0.40	14.20 ± 0.22	6.51 ± 0.05	18.61 ± 0.05	32.20 ± 0.47
CSK-N	5.65 ± 0.34	9.55 ± 0.40	11.74 ± 0.98	7.35 ± 0.45	15.04 ± 0.85	23.91 ± 1.05
COACHER-N	11.79 ± 0.70	17.42 ± 0.88	20.08 ± 1.11	14.14 ± 0.78	26.74 ± 1.38	38.44 ± 1.52
COACHER-P	12.17 ± 0.84	18.02 ± 1.23	20.58 ± 1.52	14.53 ± 0.92	27.66 ± 1.21	38.86 ± 1.48
COACHER-N+P	12.29 ± 0.17	17.85 ± 0.33	20.26 ± 0.46	14.71 ± 0.58	27.67 ± 0.82	39.57 ± 0.54

The results from baseline methods are reported under their best settings. Using a Bayesian correlated t-Test [1], there is 98% probability that COACHER-N+P is better than baseline methods. It is shown that on both datasets, our methods COACHER-N+P and COACHER-P surpass all baselines and achieve by far the highest results on both zR@K and ng-zR@K. The performance gains in Table 5 are more significant than Table 4. Such observations directly support the effectiveness of COACHER in zero-shot relation prediction, indicating its superior generalizability as we advocate in this work. Note that, consistent with their own report and our analysis in Table 1, TDE reaches state-of-the-art on non-zero-shot triplets, but performs rather poor on zero-shot ones.

Further Amplifications on Testing Data. In order to observe how powerful our model is on harder zero-shot triplets, we further manipulate the testing data by removing zero-shot triplets whose relations are more commonly observed in other triplets in the training data. As is shown in Fig. 5, our proposed method COACHER-N+P shows the least drop compared to other methods as triplets of the top 5 common relations are removed one-by-one from the testing data, which again indicates the advantageous generalization ability of our model.

4.4   Case Studies

In this subsection, we illustrate the contributions of both neighbor- and path-based commonsense knowledge integrators for SGG with real zero-shot relation prediction examples, shown in Fig. 6.

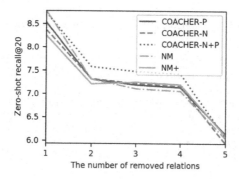

Fig. 5. Performance on manipulated testing data (we removed TDE and CSK-N here due to their rather poor performance on zR@20).

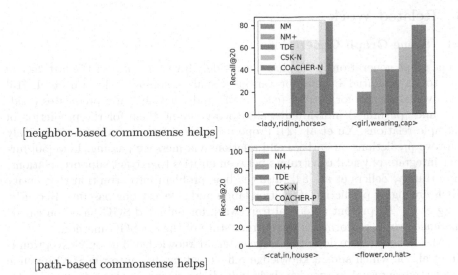

[neighbor-based commonsense helps]

[path-based commonsense helps]

Fig. 6. Case studies on integration of neighbor- and path-based commonsense.

Neighbor. The contribution of neighbor-based commonsense mainly comes from the semantic similarity implied by the neighborhood overlap between entities. As is shown in Fig. 6a, although triplet ⟨*lady, riding, horse*⟩ has never appeared in the training data, several similar triplets such as ⟨*man, riding, horse*⟩ and ⟨*person, riding, horse*⟩ are observed in the training data, where their occurrences reach 194 times and 83 times respectively. Estimated from the nodes neighborhood similarity in ConceptNet, the semantic meaning between *lady*, *man* and *person* are similar to each other. COACHER can leverage this external knowledge to improve prediction performance on zero-shot triplets. Similarly, in another case of ⟨*girl, wearing, cap*⟩, where *cap* is semantically similar with *hat*, since ⟨*girl, wearing, hat*⟩ is commonly observed in the training data, COACHER can use the neighbor-based commonsense knowledge to make better relation predictions on zero-shot triplets. In contrast, other baseline methods fail in both cases because the generalizable semantic knowledge is hard to directly learn from limited visual information.

Path. The contribution of path-based commonsense mainly comes from the relations implied by *MidPaths* between entities. As shown in Fig. 6b, for the triplet ⟨*flower, on, hat*⟩, we can find an inductive path ⟨*flower, RelatedTo, decoration, UsedFor, hat*⟩, which promotes the prediction of relation *on*. Similarly, given the pair (*cat, house*), multiple paths like ⟨*cat, AtLocation, home, RelatedTo, house*⟩, ⟨*cat, RelatedTo, house*⟩ and ⟨*cat, AtLocation, apartment, Antonym, house*⟩ can be found. All of them are inductive towards the correct relation of *in*.

5   Related Work

5.1   Scene Graph Generation

Scene graph generation (SGG) has been widely investigated over the last decade due to its potential benefit for various visual reasoning tasks. Lu et al. [15] train visual models for entities (e.g. "man" and "bicycle") and predicates ("riding" and "pushing") individually, and then combine them for the prediction of multiple relations. Xu et al. [24] propose a joint inference model to iteratively improve predictions on entities and relations via message passing. Liao [30] further integrate physical constraints between entities to extract support relations. Importantly, Zellers et al. [34] raise the bias problem into attention that entity labels are highly predictive of relation labels and give a strong baseline. Recently, Tang et al. [21] present a general framework for unbiased SGG based on causal inference, which performs as the current state-of-the-art SGG method.

All methods above do not leverage external knowledge. The only exception is Gu et al. [5], which adds a knowledge refinement module into the SGG pipeline to leverage external knowledge. It is indeed the closest work to ours and compared as a major baseline. However, they do not consider the zero-shot relation prediction problem and their integration of external knowledge is rather coarse and limited compared with ours.

5.2   External Knowledge Enhanced Deep Neural Networks

Knowledge Bases (KBs) can be used as an external knowledge to improve various down-stream tasks. The natural language processing and computer vision communities have proposed several ways to benefit deep neural networks from KBs. The most straightforward and efficient way is to represent external knowledge as an embedding, and then combine it with other features to improve model's performance. For example, in order to incorporate external knowledge to answer open-domain visual questions with dynamic memory networks, Li et al. [12,23] extract the most informative knowledge and feed them into the neural network after embedding the candidate knowledge into a continuous feature space to enhance QA task. The graph community also utilizes external knowledge to enhance graph learning [25–29]. Besides, external knowledge can be added to loss function or perform as a regularization in the training process. For example, Yu et al. [33] obtain linguistic knowledge by mining from internal training annotations as well as external knowledge from publicly available text.

However, external knowledge has hardly been utilized in SGG, mainly because the direct entity-level embeddings can hardly help in object detection, whereas what kind of embeddings are helpful in what kind of relation prediction has remained unknown before our exploration.

6   Conclusion

Scene graph generation has been intensively studied recently due to its potential benefit in various downstream visual tasks. In this work, we focus on the key challenge of SGG, *i.e.*, relation prediction on zero-shot triplets. Inspired by the natural ability of human beings to predict zero-shot relations from learned commonsense knowledge, we design integrators to leverage neighbor- and path-based commonsense from ConceptNet. We demonstrate the effectiveness of our proposed COACHER through extensive quantitative and qualitative experiments on the most widely used benchmark dataset of Visual Genome.

For future works, more in-depth experiments can be done to study the external knowledge graphs for SGG. Although the current ConceptNet is comprehensive enough to cover all entities detected from Visual Genome images, the relations it models are more from the factual perspectives, such as *has property*, *synonym*, *part of*, whereas the relations in scene graphs are more from the actional perspectives, such as the spatial or dynamical interactions among entities. One promising direction based on this study is to construct a scene-oriented commonsense knowledge graph specifically for visual tasks, while the downstream training process can further refine the graph construction. In this way, the gap between these two isolated communities, *i.e.*, visual reasoning and knowledge extraction, can be bridged and potentially enhance each other.

References

1. Benavoli, A., Corani, G., Demšar, J., Zaffalon, M.: Time for a change: a tutorial for comparing multiple classifiers through Bayesian analysis. JMLR **18**, 2653–2688 (2017)
2. Gilmer, J., Schoenholz, S.S., Riley, P.F., Vinyals, O., Dahl, G.E.: Neural message passing for quantum chemistry. In: ICML (2017)
3. Girshick, R.: Fast R-CNN. In: ICCV (2015)
4. Gu, J., Cai, J., Wang, G., Chen, T.: Stack-captioning: coarse-to-fine learning for image captioning. In: AAAI (2018)
5. Gu, J., Zhao, H., Lin, Z., Li, S., Cai, J., Ling, M.: Scene graph generation with external knowledge and image reconstruction. In: CVPR (2019)
6. He, K., Zhang, X., Ren, S., Sun, J.: Deep residual learning for image recognition. In: CVPR (2016)
7. Hudson, D.A., Manning, C.D.: GQA: a new dataset for real-world visual reasoning and compositional question answering. In: CVPR (2019)
8. Krishna, R., et al.: Visual genome: connecting language and vision using crowd-sourced dense image annotations. IJCV **123**, 32–73 (2017)
9. Li, B., Qi, X., Lukasiewicz, T., Torr, P.: Controllable text-to-image generation. In: NeurIPS (2019)
10. Li, B., Qi, X., Torr, P., Lukasiewicz, T.: Lightweight generative adversarial networks for text-guided image manipulation. In: NeurIPS (2020)
11. Li, B., Qi, X., Torr, P., Lukasiewicz, T.: ManiGAN: text-guided image manipulation. In: CVPR (2020)
12. Li, G., Su, H., Zhu, W.: Incorporating external knowledge to answer open-domain visual questions with dynamic memory networks (2017)
13. Liu, H., Singh, P.: ConceptNet – a practical commonsense reasoning tool-kit. BT Technol. J. **22**, 211–226 (2004)
14. Liu, W., et al.: SSD: single shot multibox detector. In: Leibe, B., Matas, J., Sebe, N., Welling, M. (eds.) ECCV 2016, Part I. LNCS, vol. 9905, pp. 21–37. Springer, Cham (2016). https://doi.org/10.1007/978-3-319-46448-0_2

15. Lu, C., Krishna, R., Bernstein, M., Fei-Fei, L.: Visual relationship detection with language priors. In: Leibe, B., Matas, J., Sebe, N., Welling, M. (eds.) ECCV 2016, Part I. LNCS, vol. 9905, pp. 852–869. Springer, Cham (2016). https://doi.org/10.1007/978-3-319-46448-0_51
16. Murphy, K., Weiss, Y., Jordan, M.I.: Loopy belief propagation for approximate inference: an empirical study. In: UAI (1999)
17. Newell, A., Deng, J.: Pixels to graphs by associative embedding. In: NeurIPS (2017)
18. Redmon, J., Farhadi, A.: Yolov3: an incremental improvement. arXiv (2018)
19. Ren, S., He, K., Girshick, R., Sun, J.: Faster R-CNN: towards real-time object detection with region proposal networks. In: NeurIPS (2015)
20. Speer, R., Chin, J., Havasi, C.: ConceptNet 5.5: an open multilingual graph of general knowledge. In: AAAI (2017)
21. Tang, K., Niu, Y., Huang, J., Shi, J., Zhang, H.: Unbiased scene graph generation from biased training. In: CVPR (2020)
22. Teney, D., Liu, L., van Den Hengel, A.: Graph-structured representations for visual question answering. In: CVPR (2017)
23. Wu, Q., Shen, C., Wang, P., Dick, A., van den Hengel, A.: Image captioning and visual question answering based on attributes and external knowledge. TPAMI 40(6), 1367–1381 (2018)
24. Xu, D., Zhu, Y., Choy, C.B., Fei-Fei, L.: Scene graph generation by iterative message passing. In: CVPR (2017)
25. Yang, C.: Multi-facet graph mining with contextualized projections. Ph.D. thesis (2020)
26. Yang, C., Xiao, Y., Zhang, Y., Sun, Y., Han, J.: Heterogeneous network representation learning: a unified framework with survey and benchmark. TKDE (2020)
27. Yang, C., et al.: Multisage: empowering graphsage with contextualized multi-embedding on web-scale multipartite networks. In: KDD (2020)
28. Yang, C., Zhang, J., Han, J.: Co-embedding network nodes and hierarchical labels with taxonomy based generative adversarial networks. In: ICDM (2020)
29. Yang, C., Zhuang, P., Shi, W., Luu, A., Pan, L.: Conditional structure generation through graph variational generative adversarial nets. In: NeurIPS (2019)
30. Yang, M.Y., Liao, W., Ackermann, H., Rosenhahn, B.: On support relations and semantic scene graphs. ISPRS J. Photogramm. Remote Sens. 131, 15–25 (2017)
31. Yang, X., Tang, K., Zhang, H., Cai, J.: Auto-encoding scene graphs for image captioning. In: CVPR (2019)
32. Yao, T., Pan, Y., Li, Y., Mei, T.: Exploring visual relationship for image captioning. In: Ferrari, V., Hebert, M., Sminchisescu, C., Weiss, Y. (eds.) Computer Vision – ECCV 2018. LNCS, vol. 11218, pp. 711–727. Springer, Cham (2018). https://doi.org/10.1007/978-3-030-01264-9_42
33. Yu, R., Li, A., Morariu, V.I., Davis, L.S.: Visual relationship detection with internal and external linguistic knowledge distillation. In: ICCV (2017)
34. Zellers, R., Yatskar, M., Thomson, S., Choi, Y.: Neural Motifs: scene graph parsing with global context. In: CVPR (2018)
35. Zhang, M., Cui, Z., Neumann, M., Chen, Y.: An end-to-end deep learning architecture for graph classification. In: AAAI (2018)
36. Zhu, Y., Zhang, C., Ré, C., Fei-Fei, L.: Building a large-scale multimodal knowledge base system for answering visual queries. arXiv preprint arXiv:1507.05670 (2015)

Graph Fraud Detection Based on Accessibility Score Distributions

Minji Yoon[(✉)]

Carnegie Mellon University, Pittsburgh, USA
minjiy@cs.cmu.edu

Abstract. Graph fraud detection approaches traditionally present frauds as subgraphs and focus on characteristics of the fraudulent subgraphs: unexpectedly high densities or sparse connections with the rest of the graph. However, frauds can easily circumvent such approaches by manipulating their subgraph density or making connections to honest user groups. We focus on a trait that is hard for fraudsters to manipulate: the unidirectionality of communication between honest users and fraudsters. We define an accessibility score to quantify the unidirectionality, then prove the unidirectionality induces skewed accessibility score distributions for fraudsters. We propose SKEWA, a novel fraud detection method that measures the skewness in accessibility score distributions and uses it as an honesty metric. SKEWA is (a) robust to frauds with low density and various types of camouflages, (b) theoretically sound: we analyze how the unidirectionality brings skewed accessibility score distributions, and (c) effective: showing up to 95.6% accuracy in real-world data where all competitors fail to detect any fraud.

1 Introduction

Various online platforms allow people to share their thoughts and recommend products and services to each other. Users rely on reviews with the belief they are written by disinterested people, thus more objective and unbiased. Fraudsters exploit people's trust on these platforms and derive benefits from fake followers and reviews. These frauds hinder and mislead people's decision making, thus detecting these actions is crucial for companies and customers alike.

Various graph-based approaches have been proposed to detect frauds. Most of them [2,5,7] focus on dense interconnections among fraudsters (dense sub-block/subtensor/subgraph). Another popular approach focuses on the isolation of fraud communities [1,11]. However, those methods have vulnerabilities. To evade the density-based methods, frauds generate a number of bot accounts, make their subgraph sparse, and their density low. To circumvent the isolation-based algorithms, frauds camouflage themselves as honest users by writing reviews on normal products or hijacking honest accounts.

In this paper, we focus on a characteristic that is hard for frauds to manipulate: the unidirectionality of communication between honest users and fraudsters. Honest users rarely communicate with fraudsters while fraudsters write

© Springer Nature Switzerland AG 2021
N. Oliver et al. (Eds.): ECML PKDD 2021, LNAI 12976, pp. 483–498, 2021.
https://doi.org/10.1007/978-3-030-86520-7_30

reviews or follow honest users for camouflage. This unidirectionality is generated by honest users, thus hard for fraudsters to manipulate like densities or connections. To quantify the unidirectionality, we first define accessibility scores that estimate how easily other nodes can access a given node (Sect. 4.1). Fraudsters show skewed accessibility score distributions—high accessibility scores from each other but low accessibility scores from honest users (Sect. 4.2). We prove this skewness in the accessibility score distributions theoretically and empirically (Sect. 4.3). Finally, we propose SKEWA, a novel approach to detect frauds. SKEWA defines a novel metric for honesty that measures the skewness then spots frauds with lowest honesty scores (Sect. 4.4). Through extensive experiments, we demonstrate the superior performance of SKEWA over existing methods.

The main contributions of this paper are as follows:

- **Insight:** The unidirectionality of communication results in skewness in accessibility score distributions for fraudsters: high scores on fraud groups and low scores on honest groups.
- **Robustness:** SKEWA is based on the unidirectionality generated by honest users, thus hard for fraudsters to manipulate.
- **Theoretical guarantees:** SKEWA proves how the skewed accessibility score distributions are generated and preserved under camouflages.
- **Effectiveness:** SKEWA presents up to 95.6% accuracy in public benchmarks, where all competitors fail to detect any fraud.

Reproducibility: our code is publicly available[1].

2   Related Work

Graph fraud detection algorithms could be classified into supervised and unsupervised methods based on whether a method requires labels of fraudulent or benign users/products. See [1] for an extensive survey.

Supervised methods model a fraud detection task as a binary classification problem for nodes on graphs. [3,16] leverage either labeled normal nodes or labeled fraudulent nodes. They exploit random walks to propagate the initial normalness/badness scores to the remaining nodes. [6,14,15] leverage both fraudulent and normal users. [6] is based on random walks, while [15] exploits pairwise Markov Random Field (pMRF). GANG [14] leverages pMRF and Loopy Belief Propagation to detect fraudsters.

Unsupervised methods measure suspicious scores based on graph topology. [13] factorizes the adjacency matrix and flags edges, which introduce high reconstruction error as outliers. SpokEN [11] and [12] focuses on singular vectors of a graph, which are clearly separated when plotted against each other. Fraudar [5] adapts the theoretical perspective to fraud detection and camouflage resistance and achieves meaningful bounds for applications. DeFraudar [4] presents six

---

[1] https://github.com/minjiyoon/PKDD21-SkewA.

Table 1. Table of symbols.

Symbol	Definition
G	Bipartite graph $G = (V, E)$
n_1, n_2	Numbers of products and users in G
m	Number of edges in G
$\tilde{A}_C$	$(n_1 \times n_2)$ column-normalized adjacency matrix
$\tilde{A}_R$	$(n_2 \times n_1)$ column-normalized adjacency matrix
c	Restart probability of RWR
b	$(n_1 \times 1)$ starting vector of RWR

Table 2. Comparison between methods.

Property	Method					
	GANG [14]	HoloScope [9]	SpokEN [11]	DeFraudar [4]	Fraudar [5]	**SkewA**
Unsupervised		✓	✓	✓	✓	✓
Robust to density						✓
Camouflage-resistant	✓	?		?	✓	✓
Theoretical guarantees					✓	✓

fraud indicators that measure the spamicity of a group. HoloScope [9] penalizes nodes with many connections from other nodes based on the unidirectional communication between fraudulent and honest users (Table 1).

Several methods have used PageRank or Random Walk to detect frauds. However, most of them [6,15] are supervised learning requiring labels to assign initial scores to propagates. One of our design goals is to avoid the requirement for sources other than the graph topology to measure anomalousness. [9] and [14] exploit the unidirectional communication between honest users and frauds, but both lack theoretical guarantees on how their metric preserves the unidirectionality under fraud's camouflage. In this paper, we propose an unsupervised fraud detection method SKEWA with theoretical analysis on robustness to fraud's camouflage. Table 2 compares SKEWA to existing methods.

3 Preliminaries

We review Random Walk with Restart (RWR) [10] which is used in accessibility score computation then describe how to compute RWR in a bipartite graph.

3.1 Random Walk with Restart

RWR measures each node's relevance w.r.t. a seed node s in a graph. It assumes a random walker starting from s, who traverses edges in the graph with probability $1 - c$ and occasionally restarts at the seed node s with probability c. Then the frequency of visitation of the walker on each node becomes its relevance

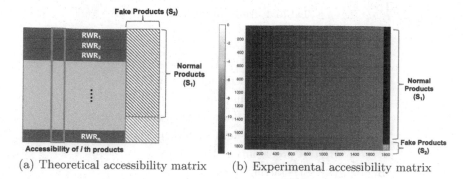

(a) Theoretical accessibility matrix   (b) Experimental accessibility matrix

Fig. 1. In an RWR matrix stacking n RWR row vectors, each column corresponds to an accessibility column vector.

score w.r.t. the seed node. From [17], the RWR score vector $\mathbf{r}_{\mathrm{RWR}}$ is presented as $\mathbf{r}_{\mathrm{RWR}} = c \sum_{i=0}^{\infty} \left((1-c)\tilde{\mathbf{A}} \right)^i \mathbf{b}$ where $\tilde{\mathbf{A}}$ is the column-normalized adjacency matrix, c is the restart probability and $\mathbf{b}$ is the seed vector with the seed node's index s set to 1 and others to 0. If $0 < c < 1$ and $\tilde{\mathbf{A}}$ is irreducible and aperiodic, $\mathbf{r}_{\mathrm{RWR}}$ is guaranteed to converge to a unique solution [8].

3.2 RWR for Bipartite Graphs

In a bipartite graph, we have two adjacency matrices, $\mathbf{A_C}$ and $\mathbf{A_R}$, which are transpose to each other. $\mathbf{A_C}$ puts products in its rows and users in its columns, while $\mathbf{A_R}$ puts users in its rows and products in its columns. $\mathbf{A_C}(i,j)$ and $\mathbf{A_R}(j,i)$ are set to 1 when j-th user writes a review on i-th product and 0 otherwise. Then $\tilde{\mathbf{A}}_{\mathbf{C}}$ and $\tilde{\mathbf{A}}_{\mathbf{R}}$ become column-normalized $(n_1 \times n_2)$ and $(n_2 \times n_1)$ matrices where n_1 and n_2 denote the total numbers of products and users, respectively. One iteration in RWR computation in a unipartite graph is divided into two sub-steps in a bipartite graph. From the original equation, we replace $\tilde{\mathbf{A}}$ with $\tilde{\mathbf{A}}_{\mathbf{C}}\tilde{\mathbf{A}}_{\mathbf{R}}$. By multiplying with $\tilde{\mathbf{A}}_{\mathbf{R}}$, scores are propagated from products to users. Then, by multiplying with $\tilde{\mathbf{A}}_{\mathbf{C}}$, the scores are propagated from the user nodes back to the product nodes. Other components are identical to the regular RWR computation.

4   Proposed Method

On a review website, fraudsters write a number of reviews on normal products to disguise themselves as honest users. In contrast, normal users purchase and review fake products only accidentally. When abstracting this phenomenon to a user-product bipartite graph, fraudulent user nodes are connected to normal product nodes, while honest user nodes rarely make connections to fake product nodes. This **unidirectionality of communication** is decided by honest users; thus, frauds cannot manipulate or dissimulate it. Based on this unidirectionality, we propose a robust fraud detection method SKEWA.

To quantify the unidirectionality, we first define accessibility scores for each node as how easily other nodes could reach to the node (Sect. 4.1). Then we show how the unidirectionality of communication leads to the skewed accessibility score distributions for fraudsters (Sects. 4.2 and 4.3). Finally, we propose our novel algorithm SKEWA to detect frauds (Sect. 4.4).

4.1 Accessibility

RWR scores with seed node i measure how easily the seed node i could reach other nodes. The scores are measured in the perspective of the seed node; thus easily manipulated by the seed node by adding edges to target nodes to increase their RWR scores. Here we define accessibility scores that measure how easily other nodes could reach the seed node i. The accessibility scores appear to be identical to the RWR scores at first glance. However, the probability of crossing an edge (i,j) from node i is different from the probability of crossing the same edge from the node j. When source node i has a larger number of out-edges than target node j, the probability of crossing the edge (i,j) is smaller since a random walker has more options to choose. This results in the different RWR and accessibility scores for each target node given the same seed node. Contrary to RWR scores, accessibility scores are estimated by target nodes and hard for the seed node to control. This explains why we choose accessibility scores as a measurement for detecting frauds.

Definition 1 (Accessibility score vector). *In an n-dimensional accessibility score vector of node i, the j-th component contains the probability that a random walker starts from node j and reaches node i.*

Accessibility score computation is based on RWR computation. We vertically stack n RWR row vectors with n different seed nodes (Fig. 1(a)). Then the i-th column in this $(n \times n)$ matrix becomes an accessibility score vector for node i, presenting how easily other nodes could reach to node i. We exploit that the accessibility score matrix is the transpose of the RWR score matrix.

4.2 Skewness in Accessibility Score Distributions

In Fig. 2(a), a graph is partitioned into two disjoint groups, the honest group A and the fraud group B. The honest group A has the most nodes and edges of the graph. Then, fraudsters in B add a few edges towards the normal group A to camouflage themselves as honest users (green dashed line). With the camouflage edges, crossing these two communities becomes possible for a random walker. However, the possibilities for the walker to pass from A to B is still small: a larger number of edges in A implies more options for the random walker to choose; thus,

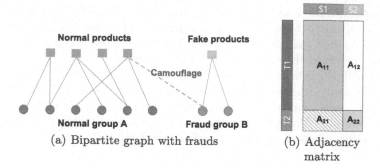

(a) Bipartite graph with frauds   (b) Adjacency matrix

Fig. 2. User-product bipartite graph: an edge is generated when a user writes a review on a product. (Color figure online)

the random walker starting from A is more likely to select the honest edges (blue line) than the camouflage edges (green dashed line). In short, honest users in the group A are less likely to reach out to a fraudster in the group B, resulting in low accessibility scores. In contrast, fraud colleagues in group B access to the target fraudster easily with help of dense interconnection (green line). Then the fraud colleagues have high accessibility scores. This pattern results in the skewness in the accessibility score distributions for the fraudster: low scores from the honest group while high scores from the fraudulent group. On the other hand, honest users have weak skewness in accessibility score distributions. A random walker starting from fraud group B is more likely to choose the camouflage edges (green dashed line) than a walker starting from group A because group B has fewer inter-connected edges than group A. This pattern brings the moderate accessibility scores from B to A, thus less skewed distributions for honest users.

4.3   Theoretical Analysis

In this Section, we prove how skewness is generated in accessibility score distributions of frauds and preserved under the camouflage of the frauds. In Fig. 2(b), S_1 (orange part in X-axis) indicates the normal products while S_2 (yellow part in X-axis) denotes the fake products for which fraudsters write fake reviews. T_1 (blue part in Y-axis) denotes the honest users while T_2 (green part in Y-axis) denotes the fraudsters. In an $(n_2 \times n_1)$ adjacency matrix $\mathbf{A}$, $\mathbf{A}_{11}$ (blue part in the matrix) corresponds to edges (reviews) between honest users and normal products. $\mathbf{A}_{22}$ (plain green part in the matrix) contains edges from fraudsters to their target products; we call these edges fake edges. $\mathbf{A}_{22}$ is dense due to a large number of fake reviews. $\mathbf{A}_{21}$ (hatched green part) corresponds to camouflage edges from fraudsters to normal products. Finally, $\mathbf{A}_{12}$ contains reviews written by honest users on fake products. $\mathbf{A}_{12}$ has almost no edge since honest users purchase fake products only accidentally. m_{ij} denotes the total number of edges in the sub-block $\mathbf{A}_{ij}$ where $i, j \in 0, 1$.

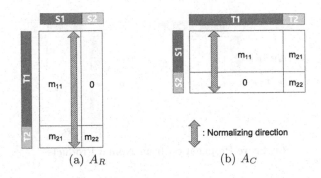

(a) A_R (b) A_C

Fig. 3. A_R and A_C are column-normalized adjacency matrices of a bipartite graph.

We analyze accessibility score distributions based on the RWR computation—accessibility score vectors are computed from columns of the corresponding RWR matrix. When a vector is multiplied with a column-normalized adjacency matrix, the amount of scores in the input vector is preserved in the output vector. Based on this characteristic, we model the ratio of propagated scores in the output vector as follows:

Assumption 1 (Ratio of Propagated Scores). *I denotes a group of nodes in an input vector, while O_1 and O_2 denote two disjoint groups in an output vector. The numbers of edges from I to O_1 and O_2 are m_1 and m_2, respectively. When I with total scores s is multiplied with a column-normalized matrix, O_1 receives $\frac{m_1}{m_1+m_2}s$ while O_2 receives the remaining $\frac{m_2}{m_1+m_2}s$.*

Based on Assumption 1, when S_1 starts with total scores s, T_1 receives $\frac{m_{11}}{m_{11}+m_{21}}s$ while T_2 receives the remaining $\frac{m_{21}}{m_{11}+m_{21}}s$ (Fig. 3(a)). Similarly, when S_2 starts with total scores s, T_1 receives $\frac{m_{12}}{m_{12}+m_{22}}s$ while T_2 receives the remaining $\frac{m_{22}}{m_{12}+m_{22}}s$. However, since $m_{12} \approx 0$ (honest users rarely purchase fake products), T_2 receives the whole score s from S_2. Under the same assumption, when T_1 starts with total scores s, S_1 receives $\frac{m_{11}}{m_{11}+m_{12}}s$ while S_2 receives the remaining $\frac{m_{12}}{m_{11}+m_{12}}s$ (Fig. 3(b)). However, since $m_{12} \approx 0$, S_1 receives the whole score s from T_1. Similarly, when T_2 starts with total scores s, S_1 receives $\frac{m_{21}}{m_{21}+m_{22}}s$ while S_2 receives the remaining $\frac{m_{22}}{m_{21}+m_{22}}s$.

We show the effectiveness of Assumption 1 empirically on real-world data in Sect. 5. In the following Section, we analyze the ratio of propagated scores after two sub-steps of RWR computation varying the location of a seed node. We define two ratio parameters: the ratio of camouflage edges to honest edges $\rho_a = \frac{m_{21}}{m_{11}+m_{21}}$, and the ratio of camouflage edges to fake edges $\rho_c = \frac{m_{21}}{m_{21}+m_{22}}$.

Seed Node from Normal Products (S_1): In Fig. 4(a), by multiplying with A_R, score s from S_1 is propagated into T_1 and T_2 with scores $(1-\rho_a)s$ and $\rho_a s$, respectively. Then these scores are propagated back to group S_1 and S_2 by multiplying with A_C. All scores $(1-\rho_a)s$ in group T_1 are propagated into only group S_1, while score $\rho_a s$ in group T_2 is divided into $\rho_a\rho_c s$ and $\rho_a(1-\rho_c)s$ and propagated into group S_1 and S_2, respectively. In short, score s starting from

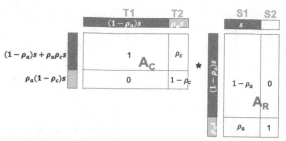

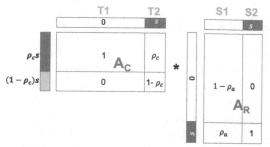

(a) Score Propagation from Nomal Products

(b) Score Progatation from Fake Products

Fig. 4. Two sub-steps in score propagation.

normal product group $S1$ will be propagated into S_1 with $(1 - \rho_a)s + \rho_a \rho_c s$ and S_2 with $\rho_a(1 - \rho_c)s$ after two sub-steps in one iteration of RWR computation.

Seed Node from Fake Products (S_2)**:** In Fig. 4(b), by multiplying with A_R, score s from S_2 is propagated into only T_2. Then, by multiplying with A_C, the score s in T_2 is propagated back to S_1 and S_2 with $\rho_c s$ and $(1 - \rho_c)s$, respectively. In summary, score s starting from the fake products $S2$ is propagated into S_1 with the score $\rho_c s$ and S_2 with the score $(1 - \rho_c)s$ after one iteration of RWR computation.

Ratio of Propagated Scores after One RWR Iteration: Score $s_1(k)$ and $s_2(k)$ denote scores propagated into group S_1 and S_2 at the k-th iteration of RWR computation. When the seed node is located at S_1, $s_1(0) = 1$ and $s_2(0) = 0$. Otherwise, $s_1(0) = 0$ and $s_2(0) = 1$. We present $s_1(k)$ and $s_2(k)$ in the iterative equation forms as follows:

Theorem 2 (Ratio of Propagated Scores). *Given ratio of camouflage edges to honest edges ρ_a and ratio of camouflage edges to fake edges ρ_c, scores propagated into group S_1 and S_2 at the k-th iteration of RWR computation are:*

$$s_1(k) = (1 - \rho_a)s_1(k - 1) + \rho_a \rho_c s_1(k - 1) + \rho_c s_2(k - 1)$$
$$s_2(k) = \rho_a(1 - \rho_c)s_1(k - 1) + (1 - \rho_c)s_2(k - 1)$$

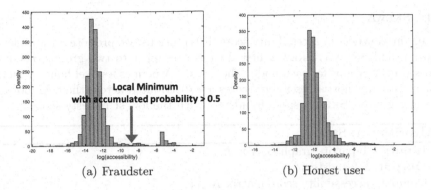

(a) Fraudster   (b) Honest user

Fig. 5. Probability density function of accessibility scores.

Proof. $s_1(k)$ and $s_2(k)$ are the sum of scores propagated from $s_1(k-1)$ and $s_2(k-1)$ to each group, respectively. We simply apply the same rule as above.

Camouflage edges generated by frauds are much fewer than the total number of edges in real-world graphs, thus $\rho_a = \frac{m_{21}}{m_{11}+m_{21}}$ has small values ($\rho_a \ll 1$). Then Theorem 2 is approximated as follows:

$$s_1(k) \approx s_1(k-1) + \rho_c s_2(k-1)$$
$$s_2(k) \approx (1-\rho_c)s_2(k-1)$$

When a seed node is located in S_1 ($s_1(0) = 1, s_2(0) = 0$), S_2 rarely receives scores ($s_2(k) \approx 0$). In other words, the accessibility scores from S_1 to S_2 are small. On the other hand, when a seed score is located in S_2 ($s_1(0) = 0, s_2(0) = 1$), S_2 receives large scores, resulting in high accessibility scores from S_2 to S_2. Then the fraud group (S_2) has skewed accessibility score distributions: small scores from the honest group (S_1) while large scores from the fraud group (S_2).

Real-World Graphs: We reproduce our theoretical analysis on the Tripadvisor dataset. We inject a fraudulent block with size of 5% of total users and products. We inject fake edges randomly to the block with 5% density, then add camouflage edges amounting to 10% of the fake edges. Figure 1(b) shows the resulting accessibility score matrix. The last 90 columns correspond to the accessibility score vectors of the injected fraud group and show clear skewness: low scores (dark-colored) for normal products and high scores (blight-colored) for fake products as we analyzed. Figure 5 shows two sampled distributions from the same dataset. In a fraudster's distribution (Fig. 5(a)), the neighbor group has high scores around e^{-5}, while the stranger group has low scores around e^{-13}. On the other hand, the distribution of an honest user (Fig. 5(b)) is less skewed with majority gathered around e^{-10}. This shows the effectiveness of our theoretical analysis on the real-world graph—for fraudulent nodes, skewness in the accessibility distribution between two groups is apparent; for honest nodes, there is no clear disparity in accessibility scores between the neighbor and stranger groups.

4.4 SKEWA

Based on skewness in accessibility score distributions, we propose a fraud detection method SKEWA. SKEWA first divides a graph into two groups, neighbor and stranger groups for each node. Then SKEWA defines a novel honesty metric which measures how accessibility scores are distributed across the neighbor and stranger groups. SKEWA spots fraudsters with the lowest honesty scores.

Algorithm 1: SKEWA

Input: A bipartite graph G, Top k
Output: k fraudsters
Compute accessibility score matrix $\mathbf{A}_{acc}$;
Compute $\alpha = log(\frac{m}{n_1})$;
foreach *column vector* $\mathbf{a}$ *in* $\mathbf{A}_{acc}$ **do**
  $\lfloor$ *ComputeHonesty*($\mathbf{a}$, α)
return k *nodes with lowest honesty scores*

Algorithm 2: ComputeHonesty

Input: Accessibility score vector $\mathbf{a}$, parameter α
Output: Honesty score s_{honest}
Find local minimum in pdf;
Divide into S_1 and S_2 by the local minimum;
Compute sum and variance of S_1 and S_2;
$s_{honest} = (var_1 var_2)^{\frac{\alpha}{2}} (sum_2)^{-\frac{2}{\alpha}}$;
return s_{honest}

Clustering. We divide nodes into the neighbor and stranger groups based on the probability density function (pdf) of the accessibility score distribution (Fig. 5). We first find local minimums in pdf whose accumulated probabilities from zero are larger than 0.5 then choose the one who has the smallest accessibility score. Based on the local minimum, we partition nodes into two groups, those accessibility scores are less or greater than the score of the minimum, then classify them as stranger and neighbor groups, respectively. We exploit that the neighbor group has high accessibility scores, while the stranger group has low scores. We consider the local minimums whose accumulated probabilities are larger than 0.5 because the neighbor group is smaller than half of the graph.

Metric for Honesty. Given the stranger and neighbor groups, we measure sum_1, var_1 and sum_2, var_2 denoting sum and variance of stranger and neighbor groups, respectively. We define a metric for honesty as follows:

$$honesty = (var_1 var_2)^{\frac{\alpha}{2}} (sum_2)^{-\frac{2}{\alpha}} \tag{1}$$

where α is defined as $log(\frac{m}{n_1})$, the ratio of the number of edges to the number of product nodes. The lower the honesty score, the more likely a node is to be a fraud. We describe each component in the honesty metric.

$var_1 var_2$ has small values for frauds. Accessibility scores from honest users toward a fraudster are all small, resulting in small values of var_1. Accessibility scores among fraud colleagues are similar with each other due to dense interconnections, resulting in small var_2. In contrast, honest users has variable accessibility scores across the graph, resulting in large values of var_1 and var_2.

Isolated honest users who have few connections with the rest part of the graph have small accessibility scores for all nodes, resulting in small values of $var_1 var_2$. To deal with isolated users, we introduce the second term.

sum_2 has large values with frauds. Dense interconnections in the fraud group result in high accessibility scores among them. In contrast, the isolated honest users have small-sized neighbor groups, resulting in small sum_2. sum_1 is not a good metric for honesty—both fraudsters and isolated honest users have small sum_1 with low accessibility scores for the stranger group.

Parameter $\alpha = log(\frac{m}{n_1})$ regulates the effects of sum_2 and $var_1 var_2$ on the honesty estimation. The density of a graph $(\frac{m}{n_1 n_2})$ is a good indicator of the number of isolated users in the graph—when a graph has low density, it implies that there are many isolated users. With more isolated honest users, we need to put more priority on sum_2 than $var_1 var_2$.

Algorithm. Algorithm 1 describes how we spot frauds based on the skewness in accessibility score distributions. We first compute an accessibility score matrix $\mathbf{A}_{acc}$ and the parameter α. Then we measure the honesty score based on Eq. 1 in Algorithm 2. Finally, SKEWA chooses top-k nodes with the lowest honesty scores as fraudsters.

5 Experiments

In this Section, we evaluate the performance of SKEWA compared to state-of-the-art fraud detection methods. We aim to answer the following questions:

- **Q1. Robustness to sparse frauds:** Does SKEWA outperform state-of-the-art competitors under various densities of frauds? (Sect. 5.2)
- **Q2. Camouflage-resistance:** How accurately does SKEWA detect frauds under various types of camouflages? (Sect. 5.3)
- **Q3. Effects of camouflage ratio:** How does the camouflage ratio affect on the performance of SKEWA? (Sect. 5.4)
- **Q4. Effectiveness of theoretical analysis:** Does our analysis on the accessibility score distributions coincide with the real-world datasets? (Sect. 5.5)

5.1 Setup

We implement SKEWA in C++; all experiments are carried out on a 2.2 GHz Intel Core i7 Macbook Pro, 16 GB RAM.

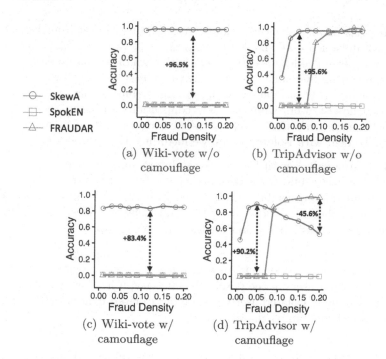

(a) Wiki-vote w/o camouflage

(b) TripAdvisor w/o camouflage

(c) Wiki-vote w/ camouflage

(d) TripAdvisor w/ camouflage

Fig. 6. Robustness to sparse and camouflaged frauds.

Dataset: we use two real-world datasets, Wiki-vote and TripAdvisor[2]. Wiki-vote is a who-trust-whom voting bipartite graph with $16K$ nodes ($8K$ for source and $8K$ for target) and $103K$ edges. TripAdvisor is a bipartite review graph with $147K$ nodes ($145K$ for users and $2K$ for products) and $176K$ edges. Parameter $\alpha = log(\frac{num.edges}{num.products})$ is set approximately with 1 and 2 on the Wiki-vote and TripAdvisor datasets, respectively.

Fraud Injection: we inject a fraudulent block into each dataset. The numbers of fraudsters and fake products are 5% of total users and total items, respectively. The density of the block is set to 5%, and the corresponding number of edges are randomly generated among them. We inject four types of camouflage scenarios: 1) fraud with no camouflage, 2) random camouflage, 3) biased camouflage, and 4) hijacked accounts. In scenario 2), frauds write reviews on randomly chosen normal products. In scenario 3), frauds write reviews on normal products chosen with probability proportional to each product's degree. Finally, in scenario 4), frauds hijack honest accounts randomly and add reviews on fake products. The number of camouflage edges is decided by the camouflage ratio ρ_c (ratio of camouflage edges to fake edges). In our experiment, ρ_c is set to 0.1.

Baseline: we compare SKEWA to state-of-the-art fraud detection methods, FRAUDAR [5] and SpokEN [11] described in Sect. 3.

---

[2] http://snap.stanford.edu/data/.

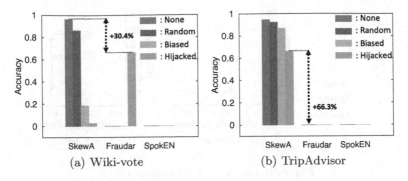

Fig. 7. Robustness to various camouflages of frauds.

5.2 Robustness to Sparse Frauds

We examine the robustness of SKEWA varying density of frauds from 1% to 20%. We inject $1st$ and $2nd$ camouflage scenarios ('No Camo' and 'Random Camo') on the datasets. We compute honesty scores by each method, then choose bottom-k honest nodes where k is the number of injected frauds.

In Fig. 6, SKEWA shows consistently high accuracy under various densities of frauds on both datasets, while FRAUDAR and SpokEN barely detect frauds. FRAUDAR shows high accuracy only with high-density frauds on the TripAdvisor dataset. Since FRAUDAR focuses on dense subgraphs to detect fraud groups, sparse graphs (e.g., TripAdvisor) which make dense fraudulent subgraphs more noticeable are helpful for FRAUDAR. SpokEN relies on SVD to detect frauds, thus it is vulnerable to low-density and camouflages of frauds.

SKEWA's accuracy decreases at a high density of frauds on the TripAdvisor dataset. TripAdvisor dataset has more isolated honest users with its low density. Then, high-density frauds result in higher var_2 (variance among colleagues) than var_2 of the isolated honest users. With lower var_2 than frauds, the isolated honest users has lower honest scores then become false positives. Overall, SKEWA shows consistently high accuracy across all settings.

5.3 Camouflage-Resistance

In this Section, we demonstrate the camouflage-resistance of SKEWA. We change the camouflage scenarios: 1) 'No Camo', 2) 'Random Camo', 3) 'Baised Camo', and 4) 'Hijacked'. Other settings are same as described in Sect. 5.1.

In Fig. 7, SKEWA is resistant to various types of camouflage attacks, while FRAUDAR and SpokEN miss most of the frauds. One exception is on the Wiki-vote dataset with the 'Hijacked' scenario where FRAUDAR shows high accuracy. On the Wiki-vote dataset, which has high density, frauds are likely to hijack honest users that are part of dense subgraphs. Then FRAUDAR, which focuses on dense subgraphs, is more likely to detect the frauds.

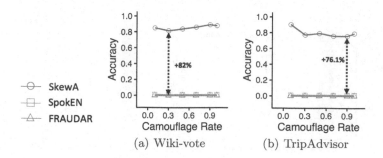

Fig. 8. Robustness to camouflage ratios.

SKEWA shows low accuracy in the 'Biased' and 'Hijacked' scenarios. In the 'Biased' scenario, a fraud makes connections to popular nodes which are connected with most honest nodes. Then any honest node connected to the popular nodes can reach the fraud groups easily through the popular nodes. Then accessibility scores from honest users to fraudsters increase, resulting in less skewed distributions. In the 'Hijacked' scenario, hijacked accounts are originally honest ones, thus already connected to other honest users. This brings high accessibility scores from honest users to fraudsters, resulting in less skewed accessibility score distributions. However, SKEWA still spots some skewness in accessibility score distributions, showing higher accuracy than its competitors.

5.4 Effects of Camouflage Ratio

We discuss the effects of the camouflage ratio on the performance of SKEWA. The camouflage ratio denotes the ratio between the number of camouflage edges and the number of fake edges. We vary the camouflage ratio from 0.1 to 1.0 under the same experimental setting described in Sect. 5.1. The camouflage type is set with the 'Random Camo' scenario. In Fig. 8, as the camouflage ratio increases, SKEWA shows consistently high accuracy, while FRAUDAR and SpokEN fail to detect frauds. SKEWA exploits the unidirectionality of communication between frauds and honest users, thus not affected by the camouflage ratio.

5.5 Effectiveness of Theoretical Analysis

Theorem 2 describes the ratio of propagated scores into an honest group and a fraud group. Based on this theorem, we find out the skewness in accessibility score distributions of fraudsters. Here, we verify the effectiveness of Theorem 2 empirically on the TripAdvisor dataset. We compute the sum of scores propagated into each group based on Theorem 2 and compare with the experimental values. Under the same experimental setting described in Sect. 5.1, we notate the ratio of camouflage edges to fake edges as ρ_c and vary ρ_c from 0.1 to 1.0. Then the ratio of camouflage edges to honest edges ρ_a is decided by ρ_c and other parameters. The camouflage type is set with the 'Random Camo' scenario.

Table 3. $\frac{\sum_i s_1(i)}{\sum_i s_2(i)}$ on the TripAdvisor dataset: we compute ratio of sums of propagated scores into honest group S_1 and fraud group S_2 varying the seed node location.

Seed location		Theoretical ratio		Experimental ratio	
		S_1	S_2	S_1	S_2
$\rho_c = 0$	$\rho_a = 0$	∞	0	∞	0
$\rho_c = 0.1$	$\rho_a = 2.2e{-}4$	2288.4	0.25	177.1	0.36
$\rho_c = 0.3$	$\rho_a = 6.6e{-}4$	1028	0.79	153.1	0.91
$\rho_c = 0.5$	$\rho_a = 1.1e{-}3$	789	1.36	148.9	1.38
$\rho_c = 0.7$	$\rho_a = 1.6e{-}3$	688.8	1.94	142.5	1.73
$\rho_c = 1$	$\rho_a = 2.2e{-}3$	614.9	2.83	135.7	1.96

Score $s_1(k)$ and $s_2(k)$ denote scores propagated into the honest group S1 and the fraud group S2 at the k-th propagation step, respectively. We measure the sum of scores $\sum_i s_1(i)$ and $\sum_i s_2(i)$ propagated into each group and compute the ratio $(\frac{\sum_i s_1(i)}{\sum_i s_2(i)})$. In Table 3, the theoretical ratio and the experimental ratio show similar tendencies. When a seed node is chosen from normal product group S_1, fake product group S_2 receives only small amounts of scores, resulting in high ratios. This coincides with the skewed accessibility score distributions of fraudsters—low accessibility scores from normal users to fraudsters. On the other hand, when the seed node is chosen from S_2, S_1 receives moderate amounts of scores, leading to low ratios. This shows the weak skewness in the accessibility score distributions of normal users.

The differences between theoretical and experimental ratios come from dead-ends in real-world graphs. Scores could not be propagated further on dead-end nodes, and this leads to the score leak. Differences between theoretical ratios and experimental ratios are much smaller when a seed is located in S_2. The fraud group has fewer dead-ends than the honest group since they intentionally create accounts to make as many connections as possible for frauds. When scores are started from S_1, scores are more likely to meet dead-ends (then diminished) and it leads to a larger gap between theoretical and experimental values.

Similar tendencies in theoretical and empirical ratios prove our analysis on the accessibility score distributions is effective on the real-world datasets.

6 Conclusion

In this paper, we propose a novel algorithm SKEWA for graph fraud detection. Due to the unidirectionality of communication between frauds and honest users, fraudsters show skewness in the accessibility score distributions. SKEWA measures honesty based on this skewness. SKEWA presents up to 95.6% accuracy in the public benchmarks where all competitors fail to detect any fraud. Future works include ensembling SKEWA with density-focused fraud detection methods. The ensemble will make SKEWA more robust to adversarial attacks with a high density of frauds.

References

1. Akoglu, L., Tong, H., Koutra, D.: Graph based anomaly detection and description: a survey. Data Min. Knowl. Disc. **29**(3), 626–688 (2014). https://doi.org/10.1007/s10618-014-0365-y
2. Beutel, A., Xu, W., Guruswami, V., Palow, C., Faloutsos, C.: Copycatch: stopping group attacks by spotting lockstep behavior in social networks. In: Proceedings of the 22nd International Conference on World Wide Web (2013)
3. Cao, Q., Sirivianos, M., Yang, X., Pregueiro, T.: Aiding the detection of fake accounts in large scale social online services. In: Presented as part of the 9th {USENIX} Symposium on Networked Systems Design and Implementation ({NSDI} 2012) (2012)
4. Dhawan, S., Gangireddy, S.C.R., Kumar, S., Chakraborty, T.: Spotting collective behaviour of online frauds in customer reviews. arXiv preprint arXiv:1905.13649 (2019)
5. Hooi, B., Song, H.A., Beutel, A., Shah, N., Shin, K., Faloutsos, C.: Fraudar: bounding graph fraud in the face of camouflage. In: KDD (2016)
6. Jia, J., Wang, B., Gong, N.Z.: Random walk based fake account detection in online social networks. In: 2017 47th Annual IEEE/IFIP International Conference on Dependable Systems and Networks (DSN). IEEE (2017)
7. Jiang, M., Cui, P., Beutel, A., Faloutsos, C., Yang, S.: Catchsync: catching synchronized behavior in large directed graphs. In: KDD (2014)
8. Langville, A.N., Meyer, C.D.: Google's PageRank and Beyond: The Science of Search Engine Rankings. Princeton University Press, Princeton (2011)
9. Liu, S., Hooi, B., Faloutsos, C.: Holoscope: topology-and-spike aware fraud detection. In: Proceedings of the 2017 ACM on Conference on Information and Knowledge Management (2017)
10. Pan, J.Y., Yang, H.J., Faloutsos, C., Duygulu, P.: Automatic multimedia cross-modal correlation discovery. In: KDD (2004)
11. Prakash, B.A., Seshadri, M., Sridharan, A., Machiraju, S., Faloutsos, C.: Eigenspokes: surprising patterns and community structure in large graphs (2010)
12. Shah, N., Beutel, A., Gallagher, B., Faloutsos, C.: Spotting suspicious link behavior with fbox: an adversarial perspective. In: ICDM (2014)
13. Tong, H., Lin, C.Y.: Non-negative residual matrix factorization with application to graph anomaly detection. In: SDM (2011)
14. Wang, B., Gong, N.Z., Fu, H.: Gang: detecting fraudulent users in online social networks via guilt-by-association on directed graphs. In: 2017 IEEE International Conference on Data Mining (ICDM). IEEE (2017)
15. Wang, B., Zhang, L., Gong, N.Z.: Sybilscar: Sybil detection in online social networks via local rule based propagation. In: IEEE INFOCOM 2017-IEEE Conference on Computer Communications. IEEE (2017)
16. Yang, C., Harkreader, R., Zhang, J., Shin, S., Gu, G.: Analyzing spammers' social networks for fun and profit: a case study of cyber criminal ecosystem on twitter. In: Proceedings of the 21st International Conference on World Wide Web (2012)
17. Yoon, M., Jung, J., Kang, U.: TPA: fast, scalable, and accurate method for approximate random walk with restart on billion scale graphs. In: ICDE (2018)

Correlation Clustering with Global Weight Bounds

Domenico Mandaglio[1], Andrea Tagarelli[1]([✉]), and Francesco Gullo[2]

[1] DIMES Department, University of Calabria, Rende, CS, Italy
{d.mandaglio,tagarelli}@dimes.unical.it
[2] UniCredit, Rome, Italy
gullof@acm.org

Abstract. Given a set of objects and nonnegative real weights expressing "positive" and "negative" feeling of clustering any two objects together, *min-disagreement correlation clustering* partitions the input object set so as to minimize the sum of the intra-cluster negative-type weights plus the sum of the inter-cluster positive-type weights. Min-disagreement correlation clustering is **APX**-hard, but efficient constant-factor approximation algorithms exist if the weights are bounded in some way. The weight bounds so far studied in the related literature are mostly *local*, as they are required to hold for every object-pair. In this paper, we introduce the problem of min-disagreement correlation clustering with *global weight bounds*, i.e., constraints to be satisfied by the input weights altogether. Our main result is a sufficient condition that establishes when any algorithm achieving a certain approximation under the probability constraint keeps the same guarantee on an input that violates the constraint. This extends the range of applicability of the most prominent existing correlation-clustering algorithms, including the popular Pivot, thus providing benefits, both theoretical and practical. Experiments demonstrate the usefulness of our approach, in terms of both worthiness of employing existing efficient algorithms, and guidance on the definition of weights from feature vectors in a task of *fair clustering*.

1 Introduction

Correlation clustering [8] is a popular clustering formulation that has received considerable attention from both theoreticians and practitioners, and has found application in several contexts, including document clustering, duplicate detection, computational biology, image segmentation [10,22].

The input of correlation clustering is a set V of objects, and two nonnegative, real-valued weights w_{uv}^+, w_{uv}^- for every (unordered) object pair $u, v \in V$. Any "positive" w_{uv}^+ (resp. "negative" w_{uv}^-) weight expresses the benefit of clustering u and v together (resp. separately). This input can equivalently be represented as a graph G with vertex set V and edge weights w_{uv}^+, w_{uv}^-, for all $u, v \in V$, and with edge (u, v) being drawn only if at least one among w_{uv}^+ and w_{uv}^- is nonzero.

The objective of correlation clustering is to partition V so as to either minimize the sum of intra-cluster negative-type weights plus the sum of inter-cluster

© Springer Nature Switzerland AG 2021
N. Oliver et al. (Eds.): ECML PKDD 2021, LNAI 12976, pp. 499–515, 2021.
https://doi.org/10.1007/978-3-030-86520-7_31

positive-type weights (*min-disagreement*), or maximize the sum of intra-cluster positive-type weights plus the sum of inter-cluster negative-type weights (*max-agreement*). The two formulations are equivalent in terms of exact optimization and complexity class (both **NP**-hard [8,25]), but they have different approximation properties, with the maximization variant being easier in this respect.

Apart from being more theoretically appealing, min-disagreement correlation clustering tends to be more relevant than the maximization counterpart in practice too. The reason is twofold. First, the best known approximation algorithms for max-agreement correlation clustering either yield trivial solutions (single-cluster and all-singletons solutions are $\frac{1}{2}$-approximate solutions for complete graphs with binary weights [26]), or are inefficient and provide unpractical clusterings with a fixed number of clusters (like semidefinite-programming Swamy's algorithm for general graphs [26], which is very expensive and always yields a 6-cluster solution). Second, more importantly, among the algorithms for the minimization version is the popular Pivot [5], which provides the best tradeoff between theoretical guarantees (it achieves constant-factor expected approximation guarantee), efficiency (it takes linear time), and ease of implementation.

Correlation-Clustering with Local Weight Bounds. The seminal work by Bansal *et al.* [8] limits the input graph to be complete, with binary weights, and with exactly one nonzero weight for each weight pair (i.e., $(w_{uv}^+, w_{uv}^-) \in \{(0,1),(1,0)\}$, for all $u,v \in V$). Even for this particular input, min-disagreement correlation clustering is **APX**-hard [11], although it admits constant-factor approximation algorithms [5,8,11,12,27]. Since then, less restrictive inputs have been considered. With no constraints on the input weights, the best known approximation factor is $\mathcal{O}(\log |V|)$ [11,16], and is unlikely to be meliorable [11,16].

Motivated by this and the above arguments in favor of the minimization version, the research community has focused on weight bounds that go beyond Bansal *et al.*'s ones, but are still restrictive enough to allow constant-factor guarantee. In this regard, the *probability constraint* (i.e., $w_{uv}^+ + w_{uv}^- = 1, \forall u, v \in V$) has received significant attention. Under this constraint, Pivot is recognized as a (randomized expected) 5-approximation algorithm [5]. Coupling the probability constraint with *triangle inequality* (i.e., $w_{uz}^- \leq w_{uv}^- + w_{vz}^-, \forall u, v, z \in V$) makes Pivot's approximation factor become 2. Further algorithms achieve a factor-4 guarantee under the probability constraint [11], and $(5 - \frac{1}{h})$-approximation for a generalization of the probability constraint (i.e., $\forall u, v \in V$, $w_{uv}^+ \leq 1$, $w_{uv}^- \leq h$ for some $h \in [1, +\infty)$, and $w_{uv}^+ + w_{uv}^- \geq 1$) [23]. Those two algorithms however are based on rounding the solution to a (large) linear program, thus they do not possess Pivot's nice peculiarities of efficiency and ease of implementation.

This Work: Correlation Clustering with *Global* Weight Bounds. Regardless of the type, the weight bounds that have been so far studied are *local bounds*, i.e., constraints that are required to hold *for every object pair in isolation*.

In this work, we are the first to consider *global weight bounds* in min-disagreement correlation clustering. We derive bounds on edge weights' aggregate functions that are sufficient to lead to proved quality guarantees. Specifically, let avg^+ and avg^- be the average of the positive-type weights and negative-type weights over all the input vertex pairs, respectively. Let also Δ_{max} be the maximum absolute difference between the positive-type weight and the negative-type weight of a vertex pair. Our result is: if the condition $avg^+ + avg^- \geq \Delta_{max}$ holds for a graph G, then it is possible to construct a graph G' (in linear time and space) such that (i) the probability constraint holds on G', and (ii) an α-approximate clustering on G' (i.e., a clustering whose objective-function value is no more than α times G''s optimum) is an α-approximate clustering on G too.

A noteworthy consequence of this result is that, if a graph G satisfies our condition, then the Pivot algorithm can be used to get (in linear time and space) a clustering achieving a 5-approximation guarantee on G.[1] This corresponds to extending the range of validity of Pivot's guarantee beyond the probability constraint: our global-weight-bounds condition now suffices for the 5-approximation to hold. A key advantage of this finding is that our condition is milder than the probability constraint, thus more likely to be satisfied. For instance, it may happen that a bunch of edges are missing from the input graph (meaning violation of the probability constraint for at least those unlinked vertex pairs), but, if our condition holds, still one can get a 5-approximate clustering with Pivot.

We point out that our result is general and holds for *any* min-disagreement correlation-clustering algorithm achieving approximation guarantees under the probability constraint. However, the contextualization to the Pivot algorithm is relevant and worth to be emphasized, because, as said above, Pivot achieves the best tradeoff between quality guarantees, efficiency, and ease of implementation.

Benefits of Our Result. We believe that the findings of this work can be tremendously useful, from several perspectives.

Practical Benefits. Our result can be exploited to quickly yet easily recognize whether employing probability-constraint-aware approximation algorithms is a worth choice even if the probability constraint is not met. As an example, consider a graph that violates the probability constraint. So far, that graph would have likely been handled with linear-programming (LP) algorithms [11,16], as they achieve (factor-$\mathcal{O}(\log|V|)$) approximation guarantees on general graphs/weights (whereas algorithms like Pivot are just heuristics if the probability constraint does not hold). Instead, our condition can be used as an indicator of whether Pivot can still achieve guarantees even if the probability constraint is violated, thus being preferred over the LP algorithms. This has important practical implications, as Pivot is much faster and easier-to-implement than the LP counterparts. In our evaluation we experimentally confirm this theoretical finding, by showing that a better fulfilment of our condition corresponds to better performance of Pivot with respect to the LP algorithms, and vice versa.

A second practical exploitability of our result concerns the task of feature selection for clustering. In the context of correlation clustering this corresponds to selecting features that lead edge weights to express the best tradeoff between

---

[1] In fact, a probability-constraint-compliant graph G' can be derived from G in linear time and space (statement (i) of our result). Pivot on G' yields a 5-approximate clustering [5]. A 5-approximate clustering on G' is a 5-approximate clustering on G (statement (ii) of our result).

an accurate representation of the objects' vectors (i.e., discarding not too many features), and the way how the weights facilitate the downstream correlation-clustering algorithm performing well (e.g., by making it achieve approximation guarantees). Our global-weight-bounds condition can be an effective yet easy-to-use guiding principle to the achievement of this tradeoff. Being less restrictive than local weight bounds, our condition can be fulfilled more easily (e.g., in case of probability constraint, it is hard to find a subset of features leading to positive-type and negative-type weights summing *exactly* to one *for all the object pairs*). In our experiments we showcase this capability in a task of *fair clustering*.

Theoretical Benefits. This work extends the validity range of the approximation guarantees of algorithms for min-disagreement correlation clustering. This extension can pave the way for more advanced theoretical results. As an example, it is not uncommon that correlation clustering is a building block of a more complex problem [17,20,21]. Thus, more general guarantees in correlation clustering may enable better theoretical results on those complex problems too.

Benefits for the Research Community. To the best of our knowledge, global weight bounds for correlation clustering have never been studied so far. We believe this work can pioneer a brand new line of research, and stimulate the community to go beyond our initial results.

Summary of Contributions and Outline. The contributions we achieve in this work can be summarized as follows. We focus for the first time on global weight bounds in (the minimization formulation of) correlation clustering (Sect. 3). We derive a sufficient condition on input weights' aggregate functions to extend the validity range of the approximation guarantees of existing correlation-clustering algorithms beyond the probability constraint (Sect. 4). We experimentally assess that our condition is an effective indicator of the empirical performance of existing probability-constraint-aware correlation-clustering algorithms (Sect. 5.1). We showcase our results in a real-world scenario of fair clustering (Sect. 5.2).

2   Related Work

Correlation Clustering. The literature on min-disagreement correlation clustering that is functional to our work has been presented in the Introduction. As a complement, we (briefly) overview the main results on the maximization formulation (not a focus of this work), and extensions to the basic formulations.

For original Bansal *et al.*'s input of unweighted and complete graphs [8], max-agreement correlation clustering admits a PTAS [8]. On general graphs/weights, it becomes **APX**-hard [11], but admits constant-factor approximation algorithms, achieving factor-0.7664 [11] and factor-0.7666 [26] guarantees. Extensions to the basic correlation-clustering formulations include constrained/relaxed formulations, and adaptations to nonconventional types of graph or computational settings. We point the interested reader to [10,22] for more details.

In this work we shift the attention from local to *global* weight bounds in min-disagreement correlation clustering. To the best of our knowledge, this is a completely novel perspective that has never been considered so far.

Fair Clustering. Roughly speaking, the problem of fair clustering consists in partitioning a set of objects based on both clustering quality and *fairness*, i.e., limiting as much as possible the bias against/towards particular objects' subsets.

Chierichetti *et al.*'s seminal work [14] formulates fair versions of the traditional k-center and k-median problems. Since then, research has focused on generalizing those formulations [9,24], incorporating fairness constraints into k-center [18], scalability of fair k-median [7], different fairness measures [3,13], and fair versions of other traditional problems, i.e., k-means [1], spectral clustering [19], hierarchical clustering [2]. As for correlation clustering, Ahmadian *et al.* [4] study the problem where vertices of a *complete and unweighted* graph are assigned a *single label* representing a protected class attribute (e.g., gender, ethnicity), and every cluster is constrained to fairly represent each label.

In this work we showcase our theoretical results in a task of fair clustering. We pick a scenario where positive-type and negative-type edge weights express similarities on non-sensitive and sensitive features assigned to the input vertices, respectively. The goal is to define such weights so as to account for both an effective representation of the semantics underlying objects' features, and the peculiarities that make the downstream correlation-clustering algorithm effective. Thus, our setting differs from Ahmadian *et al.*'s one, where the graph is complete and unweighted, and vertices are not assigned feature vectors, but just a class label. In any case, the focus on fair clustering in this work is just on the application side: advancing the fair-clustering literature is beyond our scope.

3   Problem Definition

In this work we tackle the problem of *min-disagreement correlation clustering*:

*Problem 1 (*MIN-CC [5]*).* Given an undirected graph $G = (V, E)$, with vertex set V and edge set $E \subseteq V \times V$, and nonnegative weights $w_e^+, w_e^- \in \mathbb{R}_0^+$ for all edges $e \in E$, find a clustering (i.e., an injective function expressing cluster-membership) $\mathcal{C} : V \to \mathbb{N}^+$ that minimizes

$$\sum_{(u,v) \in E, \mathcal{C}(u)=\mathcal{C}(v)} w_{uv}^- + \sum_{(u,v) \in E, \mathcal{C}(u) \neq \mathcal{C}(v)} w_{uv}^+. \tag{1}$$

For the sake of presentation, we assume $w_e^+ = w_e^- = 0$, for all $e \notin E$, and non-trivial MIN-CC instances, i.e., $w_e^+ \neq w_e^-$, for some $e \in E$.

MIN-CC is **NP**-hard [8,25] yet difficult to approximate, being it **APX**-hard even for complete graphs and edge weights $(w_e^+, w_e^-) \in \{(0,1),(1,0)\}$, $\forall e \in E$ [11]. For general (i.e., not necessarily complete) graphs and general (i.e., unconstrained) weights, the best known approximation factor is $\mathcal{O}(\log |V|)$ [11,16]. This factor improves if restrictions on edge weights are imposed. A constraint that has received considerable attention is the *probability constraint* (PC):

Definition 1 (Probability constraint). *A* MIN-CC *instance is said to satisfy the* probability constraint (PC) *if* $w_{uv}^+ + w_{uv}^- = 1$, *for all vertex pairs* $u, v \in V$.

Algorithm 1. Pivot [5]
___
Input: Graph $G = (V, E)$; nonnegative weights $w_e^+, w_e^-, \forall e \in E$
Output: Clustering $\mathcal{C}$ of V
1: $\mathcal{C} \leftarrow \emptyset, \quad V' \leftarrow V$
2: **while** $V' \neq \emptyset$ **do**
3:      pick a pivot vertex $u \in V'$ uniformly at random
4:      add $\mathcal{C}_u = \{u\} \cup \{v \in V' \mid (u, v) \in E, w_{uv}^+ > w_{uv}^-\}$ to $\mathcal{C}$ and remove $\mathcal{C}_u$ from V'
___

A MIN-CC instance obeying the PC necessarily corresponds to a complete graph (otherwise, any missing edge would violate the PC). Under the PC, MIN-CC admits constant-factor guarantees. The best known approximation factor is 4, achievable – as shown in [23] – by Charikar *et al.*'s algorithm [11]. That algorithm is based on rounding the solution to a large linear program (with a number $\Omega(|V|^3)$ of constraints), thus being feasible only on small graphs.

Here, we are particularly interested in the Pivot algorithm [5], due to its theoretical properties – it achieves a factor-5 expected guarantee for MIN-CC under the PC – and practical benefits – it takes $\mathcal{O}(|E|)$ time, and is easy-to-implement. Pivot simply picks a random vertex u, builds a cluster as composed of u and all the vertices v such that an edge with $w_{uv}^+ > w_{uv}^-$ exists, and removes that cluster. The process is repeated until the graph has become empty (Algorithm 1).

4 Theoretical Results and Algorithms

Let MIN-PC-CC denote the version of MIN-CC operating on instances that satisfy the PC. The main theoretical result of this work is a *sufficient condition* – to be met *globally* by the input edge weights – on the existence of a *strict approximation-preserving* (SAP) reduction from MIN-CC to MIN-PC-CC. In the remainder of this section we detail our findings, presenting partial results (Sects. 4.1–4.2), our overall result (Sect. 4.3), and algorithms (Section 4.4).

4.1 PC-Reduction

As a first partial result, in this subsection we define the proposed reduction from MIN-CC instances to MIN-PC-CC ones, and the condition that makes it yield valid (i.e., nonnegative) edge weights. We start by recalling some basic notions, including the one of strict approximation-preserving (SAP) reduction.

Definition 2 (Minimization problem, optimum, performance ratio [6]).
A minimization problem Π is a triple $(\mathcal{I}, sol, obj)$, where $\mathcal{I}$ is the set of problem instances; for every $I \in \mathcal{I}$, $sol(I)$ is the set of feasible solutions of I; obj is the objective function, i.e., given $I \in \mathcal{I}$, $S \in sol(I)$, $obj(I, S)$ measures the quality of solution S to instance I.

$OPT_\Pi(I)$ denotes the objective-function value of an optimal solution to I.
Given $I \in \mathcal{I}$, $S \in sol(I)$, $R_\Pi(I, S) = obj(I, S)/OPT_\Pi(I)$ denotes the performance ratio of S with respect to I.

Definition 3 (Reduction and SAP-reduction [15]**).** *Let* $\Pi_1 = (\mathcal{I}_1, sol_1,$ $obj_1)$ *and* $\Pi_2 = (\mathcal{I}_2, sol_2, obj_2)$ *be two minimization problems.*

A reduction *from* Π_1 *to* Π_2 *is a pair* (f, g) *of polynomial-time-computable functions, where* $f : \mathcal{I}_1 \rightarrow \mathcal{I}_2$ *maps* Π_1*'s instances to* Π_2*'s instances, and, given* $I_1 \in \mathcal{I}_1$, $g : sol_2(f(I_1)) \rightarrow sol_1(I_1)$ *maps back* Π_2*'s solutions to* Π_1*'s solutions.*

A reduction *is said* strict approximation-preserving *(*SAP*) if, for any* $I_1 \in \mathcal{I}_1$, $S_2 \in sol_2(f(I_1))$, *it holds that* $R_{\Pi_1}(I_1, g(I_1, S_2)) \leq R_{\Pi_2}(f(I_1), S_2)$.

The proposed reduction is as follows. To map MIN-CC instances to MIN-PC-CC ones, we adopt a function f that simply redefines edge weights, while leaving the underlying graph unchanged. Function g is set to the identity function. That is, a MIN-PC-CC solution is interpreted as a solution to the original MIN-CC instance as is. Function f makes use of two constants $M, \gamma > 0$ (which will be better discussed later), and σ_e quantities, $\forall e \in E$, which are defined as:

$$\sigma_e = \gamma \, (w_e^+ + w_e^-) - M. \tag{2}$$

We term our reduction PC-*reduction* and define it formally as follows.

Definition 4 (PC-reduction). *The* PC-reduction *is a reduction* (f, g) *from* MIN-CC *to* MIN-PC-CC*, where* g *is the identity function, while* f *maps a* MIN-CC *instance* $\langle G = (V, E), \{w_e^+, w_e^-\}_{e \in E} \rangle$ *to a* MIN-PC-CC *instance* $\langle G' = (V', E'), \{\tau_e^+, \tau_e^-\}_{e \in E'} \rangle$*, such that* $V' = V$, $E' = V \times V$, *and*

$$\tau_e^+ = \tfrac{1}{M} \left(\gamma \, w_e^+ - \tfrac{\sigma_e}{2} \right), \quad \tau_e^- = \tfrac{1}{M} \left(\gamma \, w_e^- - \tfrac{\sigma_e}{2} \right), \quad \forall e \in E'. \tag{3}$$

Note that the proposed PC-reduction is always guaranteed to yield τ_e^+, τ_e^- weights satisfying the PC (i.e., $\tau_e^+ + \tau_e^- = 1$), for any M and γ. As a particular case, recalling the assumption $w_e^+ = w_e^- = 0$ for $e \notin E$, the PC-reduction yields weights $\tau_e^+ = \tau_e^- = 0.5$ for any $e \notin E$. However, not every choice of M and γ leads to nonnegative τ_e^+, τ_e^- weights, as stated next.

Lemma 1. *The* PC-*reduction yields nonnegative* τ_e^+, τ_e^- *weights if and only if* $M - \gamma \, \Delta_{max} \geq 0$, *where* $\Delta_{max} = \max_{e \in E} |w_e^+ - w_e^-|$.

Proof. By simple math on the formula of τ_e^+, τ_e^- in Definition 4, it follows that $\tau_e^+, \tau_e^- \geq 0$ holds if and only if the conditions $w_e^+ - w_e^- \geq -M/\gamma$ and $w_e^+ - w_e^- \leq M/\gamma$ are simultaneously satisfied. This in turn corresponds to have $|w_e^+ - w_e^-| \leq M/\gamma$ satisfied. As the latter must hold for all $e \in E$, then the lemma. $\square$

Constraining M and γ as in Lemma 1 is a key ingredient of our ultimate global-weight-bounds condition. We will come back to it in Sect. 4.3.

4.2   Preserving the Approximation Factor Across PC-Reduction

Let I be a MIN-CC instance and I' be the MIN-PC-CC instance derived from I via PC-reduction. Here, we present a further partial result, i.e., a sufficient condition according to which an approximation factor holding on I' is preserved on I. We state this result in Lemma 3. Before that, we provide the following auxiliary lemma, which shows the relationship between the objective-function values of a clustering $\mathcal{C}$ on I and on I'.

Lemma 2. *Let $I = \langle G = (V, E), \{w_e^+, w_e^-\}_{e \in E}\rangle$ be a* MIN-CC *instance, and $I' = \langle G' = (V, E' = V \times V), \{\tau_e^+, \tau_e^-\}_{e \in E'}\rangle$ be the* MIN-PC-CC *instance derived from I via* PC-reduction. *Let also $\mathcal{C}$ be a clustering of V. The following relationship holds between the objective-function value $obj(I, \mathcal{C})$ of $\mathcal{C}$ on I and the objective-function value $obj(I', \mathcal{C})$ of $\mathcal{C}$ on I':*

$$obj(I, \mathcal{C}) = \tfrac{M}{\gamma} obj(I', \mathcal{C}) + \tfrac{1}{2\gamma} \sum\nolimits_{u,v \in V} \sigma_{uv}. \tag{4}$$

Proof.

$$obj(I', \mathcal{C}) = \sum_{\substack{(u,v) \in E', \\ \mathcal{C}(u) = \mathcal{C}(v)}} \tfrac{1}{M}\left(\gamma w_{uv}^+ - \tfrac{\sigma_{uv}}{2}\right) + \sum_{\substack{(u,v) \in E', \\ \mathcal{C}(u) = \mathcal{C}(v)}} \tfrac{1}{M}\left(\gamma w_{uv}^- - \tfrac{\sigma_{uv}}{2}\right)$$

$$= \tfrac{\gamma}{M}\left(\sum_{\substack{(u,v) \in E, \\ \mathcal{C}(u) = \mathcal{C}(v)}} w_{uv}^+ + \sum_{\substack{(u,v) \in E, \\ \mathcal{C}(u) \neq \mathcal{C}(v)}} w_{uv}^-\right) - \tfrac{1}{2M}\left(\sum_{\substack{(u,v) \in E', \\ \mathcal{C}(u) = \mathcal{C}(v)}} \sigma_{uv} + \sum_{\substack{(u,v) \in E', \\ \mathcal{C}(u) \neq \mathcal{C}(v)}} \sigma_{uv}\right)$$

$$= \tfrac{\gamma}{M} obj(I, \mathcal{C}) - \tfrac{1}{2M} \sum\nolimits_{u,v \in V} \sigma_{uv}.$$

$\square$

Lemma 3. *Let I and I' be the two instances of Lemma 2. Let also $\mathcal{C}$ be an α-approximate solution to I', i.e., a clustering achieving objective-function value no more than α times I''s optimum, for any $\alpha > 1$. It holds that: if $\tfrac{1}{2\gamma} \sum_{e \in E} \sigma_e \geq 0$, then $\mathcal{C}$ is an α-approximate solution to I too.*

Proof. Let OPT and OPT' be the optima of I and I', respectively. It holds that:

$$obj(I', \mathcal{C}) \leq \alpha\, OPT'$$

$$\Rightarrow \tfrac{M}{\gamma} obj(I', \mathcal{C}) + \tfrac{1}{2\gamma} \sum\nolimits_{u,v \in V} \sigma_{uv} \leq \alpha\left(\tfrac{M}{\gamma} OPT' + \tfrac{1}{2\gamma} \sum\nolimits_{u,v \in V} \sigma_{uv}\right)$$

$$\Leftrightarrow obj(I, \mathcal{C}) \leq \alpha\, OPT,$$

where the second step holds because $\tfrac{1}{2\gamma} \sum_{e \in E} \sigma_e \geq 0$ and $\alpha > 1$ by hypothesis, while the last step holds because of Lemma 2. $\square$

4.3   Ultimate Global Weight Bounds

With the above partial results in place, we can now present our ultimate result, i.e., a sufficient condition to guarantee that the PC-reduction is a SAP-reduction. To show our result, for a MIN-CC instance $\langle G = (V, E), \{w_e^+, w_e^-\}_{e \in E}\rangle$ we define:

$$avg^+ = \binom{|V|}{2}^{-1} \sum\nolimits_{e \in E} w_e^+, \quad avg^- = \binom{|V|}{2}^{-1} \sum\nolimits_{e \in E} w_e^-. \tag{5}$$

Theorem 1. *If $avg^+ + avg^- \geq \Delta_{max}$, then the* PC-reduction *is a* SAP-reduction.

Proof. Lemma 3 provides a (sufficient) condition to have an approximation factor on a MIN-PC-CC instance carried over to the original MIN-CC instance. Thus, that condition suffices to make the PC-reduction a SAP-reduction according to Definition 3. The condition in Lemma 3 has to be coupled with the one in

---

Algorithm 2. GlobalCC

Input: Graph $G = (V, E)$; nonnegative weights w_e^+, w_e^-, $\forall e \in E$, satisfying Theorem 1;
    algorithm A achieving α-approximation guarantee for MIN-PC-CC
Output: Clustering $\mathcal{C}$ of V
1: choose M, γ s.t. $\frac{M}{\gamma} \in [\Delta_{max}, avg^+ + avg^-]$ {Theorem 1}
2: compute τ_{uv}^+, τ_{uv}^-, $\forall u, v \in V$, as in Equation (3) (using M, γ defined in Step 1)
3: $\mathcal{C} \leftarrow$ run A on MIN-PC-CC instance $\langle G' = (V, V \times V), \{\tau_e^+, \tau_e^-\}_{e \in V \times V}\rangle$

Lemma 1, which guarantees nonnegativity of the edge weights of the yielded MIN-PC-CC instance. To summarize, we thus require the following:

$$\begin{cases} \frac{M}{\gamma} \geq \Delta_{max}, & \text{\{Lemma 1\}} \\ \frac{1}{2\gamma}\sum_{u,v \in V} \sigma_{uv} \geq 0 \iff \frac{M}{\gamma} \leq avg^+ + avg^-, & \text{\{Lemma 3\}} \end{cases}$$

which corresponds to $\frac{M}{\gamma} \in [\Delta_{max}, avg^+ + avg^-]$, i.e., to $avg^+ + avg^- \geq \Delta_{max}$. $\quad\square$

4.4 Algorithms

According to Theorem 1, if $avg^+ + avg^- \geq \Delta_{max}$ for a MIN-CC instance, then any α-approximation algorithm for MIN-PC-CC can be employed – *as a black box* – to get an α-approximate solution to that MIN-CC instance. The algorithm for doing so is simple: get a MIN-PC-CC instance via PC-reduction, and run the black-box algorithm on it (Algorithm 2). Being the PC-reduction SAP, the guarantee of this algorithm straightforwardly follows as a corollary of Theorem 1.

Corollary 1. *Let I be a MIN-CC instance, and A be an α-approximation algorithm for MIN-PC-CC. Algorithm 2 on input $\langle I, A\rangle$ achieves factor-α guarantee on I.*

Let $T(A)$ be the running time of the black-box algorithm A. The time complexity of Algorithm 2 is $\mathcal{O}(\max\{|E|, T(A)\})$, assuming that there is no need to materialize edge weights τ_e^+, τ_e^- for missing edges $e \notin E$. This is an assumption valid in most cases: we recall that $e \notin E \Rightarrow \tau_e^+ = \tau_e^- = 0.5$, thus it is likely that their definition can safely be kept implicit. For instance, this assumption holds if Pivot [5] is used as a black-box algorithm (although with Pivot the picture is much simpler, see below). Instead, the assumption is not true for the LP algorithms in [11,16]. In that case, however, the time complexity of Algorithm 2 would correspond to the running time of those LP algorithms nevertheless, as they take (at least) $\Omega(|V|^3)$ time to build their linear programs.

Using Pivot in Algorithm 2. It is easy to see that $w_e^+ > w_e^- \Leftrightarrow \tau_e^+ > \tau_e^-$, $\forall e \in E$. As Pivot makes its choices based on the condition $w_e^+ > w_e^-$ solely, the output of Algorithm 2 equipped with Pivot corresponds to the output of Pivot run directly on the input MIN-CC instance. Thus, to get the 5-approximation guaranteed by Pivot, it suffices to run Pivot on the original input, without explicitly performing the PC-reduction. This finding holds in general for any algorithm whose output is determined by the condition $w_e^+ > w_e^-$ only. It does not hold for the LP algorithms: in that case, the general Algorithm 2 is still needed.

Table 1. Main characteristics of real-world graph datasets (left) and relational datasets (right) used in our evaluation stages.

	$|V|$	$|E|$	den.	a_deg	a_pl	diam	cc
Karate	34	78	0.14	4.59	2.41	5	0.26
Dolphins	62	159	0.08	5.13	3.36	8	0.31
Adjnoun	112	425	0.07	7.59	2.54	5	0.16
Football	115	613	0.09	10.66	2.51	4	0.41

	#objs.	#attrs.	fairness-aware (sensitive) attributes
Adult	32 561	7/8	race, sex, country, education, occupation, marital-status, workclass, relationship
Bank	41 188	18/3	job, marital-status, education
Credit	10 127	17/3	gender, marital-status, education-level
Student	649	28/5	sex, male_edu, female_edu, male_job, female_job

Role of M and γ. According to Theorem 1, τ_e^+, τ_e^- weights can be defined by picking any values of M and γ such that $\frac{M}{\gamma} \in [\Delta_{max}, avg^+ + avg^-]$. The condition $avg^+ + avg^- \geq \Delta_{max}$ ensures that the $[\Delta_{max}, avg^+ + avg^-]$ range is nonempty, while the assumption made in Sect. 3 that our input MIN-CC's instances are nontrivial (thus, $\Delta_{max} > 0$) guarantees $M, \gamma > 0$.

From a theoretical point of view, all valid values of M and γ are the same. The choice of M and γ may instead have practical implications. Specifically, M and γ determine the difference between the resulting positive-type and negative-type edge weights. This may influence the empirical performance of those algorithms (e.g., the LP algorithms) for which the weight values matter. However, we remark that, in the case of Pivot –which just depends on whether the positive-type weight is more than the negative-type one— M and γ do not play any role, not even empirically. Being Pivot the main object of our practical focus, we defer a deeper investigation on M and γ to future work (see Sect. 6).

5   Experiments

5.1   Analysis of the Global-Weight-Bounds Condition

Settings. We selected four real-world graphs,[2] whose summary is reported in Table 1-(left). Note that the small size of such graphs is not an issue because this evaluation stage involves, among others, linear-programming correlation-clustering algorithms, whose time complexity $(\Omega|V|^3)$ makes them unafford-able for graphs larger than that. We augmented these graphs with artificially-generated edge weights, to test different levels of fulfilment of our global-weight-bounds condition stated in Theorem 1. We controlled the degree of compliance of the condition by a *target ratio* parameter, defined as $t = \Delta_{max}/(avg^+ + avg^-)$. The condition is satisfied if and only if $t \in [0, 1]$, and smaller target-ratio values correspond to better fulfilment of the condition, and vice versa.

Given a desired target ratio, edge weights are generated as follows. First, all weights are drawn uniformly at random from a desired $[lb, ub]$ range. Then, the weights are adjusted in a two-step iterative fashion, until the desired target ratio is achieved: (i) keeping the maximum gap Δ_{max} fixed, the weights are changed

---

[2] Publicly available at http://konect.cc/networks/.

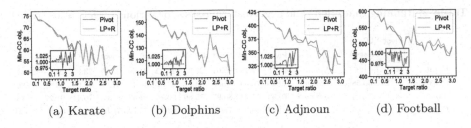

(a) Karate   (b) Dolphins   (c) Adjnoun   (d) Football

Fig. 1. MIN-CC objective by varying the target ratio.

for pairs that do not contribute to Δ_{max} so as to reflect a change in avg^+, avg^-; (*ii*) keeping avg^+, avg^- fixed, Δ_{max} is updated by randomly modifying pairs that contribute to Δ_{max}. Once properly adjusted to meet the desired target ratio, weight pairs are randomly assigned to the edges of the input graph.

We compared the performance of Pivot (Algorithm 1 [5]) to one of the state-of-the-art algorithms achieving factor-$\mathcal{O}(\log|V|)$ guarantee on general graphs/ weights [11]. We dub the latter LP+R, alluding to the fact that it rounds the solution of a linear program. We evaluated correlation-clustering objective, number of output clusters, and runtimes of these algorithms.

Results. Figure 1 shows the quality (i.e., MIN-CC objective) of the clusterings produced by the selected algorithms, with the bottom-left insets reporting the ratio between the performance of Pivot and LP+R. Results refer to target ratios t varied from $[0, 3]$, with stepsize 0.1, and weights generated with $lb = 0, ub = 1$. For each target ratio, all reported measurements correspond to averages over 10 weight-generation runs, and each of such runs in turn corresponds to averages over 50 runs of the tested algorithms (being them both randomized).

The main goal here is to have experimental evidence that a better fulfilment of our global condition leads to Pivot's performance closer to LP+R's one, and vice versa. This would attest that our condition is a reliable proxy to the worthiness of employing Pivot. Figure 1 confirms this claim: in all datasets, Pivot performs more closely to LP+R as the target ratio gets smaller. In general, Pivot performs similarly to LP+R for $t \in [0, 1]$, while being outperformed for $t > 1$. This conforms with the theory: on these small graphs, factor-5 Pivot's approximation is close to factor-$\mathcal{O}(\log|V|)$ LP+R's approximation. Pivot achieves the best performance on *Football*, where it outperforms LP+R even if the condition is not met. This is motivated by *Football*'s higher clustering coefficient and average degree, which help Pivot sample vertices (and, thus, build clusters) in dense regions of the graph. This is confirmed by the number of clusters (Table 2-(right)): Pivot yields more clusters than LP+R on all datasets but *Football*.

As far as runtimes (Table 2-(left)),[3] Pivot is extremely faster than LP+R, as expected. The inefficiency of LP+R further emphasizes the importance of our result in extending the applicability of faster algorithms like Pivot.

We complement this stage of evaluation by testing different graph densities. We synthetically added edges with uniform probability, ranging from 0

---

[3] Experiments were carried out on the Cresco6 cluster https://www.eneagrid.enea.it.

Table 2. Running times (left) and avg. clustering-sizes for various target ratios (right).

	Pivot (secs.)	LP+R (secs.)		0.1		0.5		1		2		3	
				Pivot	LP+R	Pivot	LP+R	Pivot	LP+R	Pivot	LP+R	Pivot	LP+R
Karate	< 1	1.9	*Karate*	21.75	17.18	29.61	27.93	27.22	24.66	25.55	23.82	28.17	26.81
Dolphins	< 1	36.58	*Dolphins*	49.25	50.59	45.3	38.67	49.57	44.45	47.91	48.05	48.89	43.66
Adjnoun	< 1	775.4	*Adjnoun*	70.35	65.93	80.97	75.86	90.76	84.93	85.83	70.41	91.27	79.78
Football	< 1	819.8	*Football*	64.43	84.91	77.14	96.43	68.35	78.72	78.65	85.31	90.87	100.31

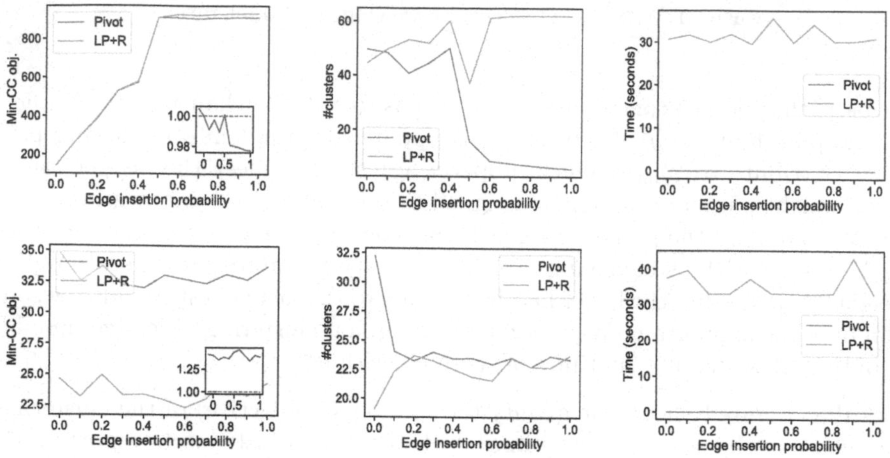

(a) Min-CC objective   (b) Number of clusters   (c) Running time

Fig. 2. Varying graph density: target ratio 1 (top) and 20 (bottom), on *Dolphins*.

(no insertions) to 1 (complete graph). Figure 2 shows the results on *Dolphins* (similar results are found in all the other datasets, here omitted for the sake of brevity), and for target ratios $t = 1$ (borderline satisfaction of our condition) and $t = 20$ (far fulfilment of the condition). Again, the results meet the expectations: in terms of clustering quality, Pivot performs closely to or better than LP+R for $t = 1$, while the opposite happens for $t = 20$. Denser graphs correspond to better Pivot performance. This is again motivated by the above argument that higher densities favor better Pivot's random choices. Runtimes are not affected by the differences in graph density. This is expected as well, as LP+R runtimes are dominated by the time spent in building and solving the linear program, which depends on the number of vertices only, whereas variations in the runtimes of Pivot cannot be observed due to the small size of the datasets at hand.

5.2   Application to Fair Clustering

Let $\mathcal{X}$ be a set of objects defined over a set of attributes $\mathcal{A}$. The latter is assumed to be divided into two sets, $\mathcal{A}^F$ and $\mathcal{A}^{\neg F}$, where $\mathcal{A}^F$ contains *fairness-aware*, or *sensitive* attributes (e.g., gender, race, religion), and $\mathcal{A}^{\neg F}$ denotes the remaining, *non-sensitive* attributes. In both cases, we assume that part of the attributes might be numerical, and the others as categorical; we will use superscripts N and C to distinguish the two types, therefore $\mathcal{A}^F = \mathcal{A}^F_N \cup \mathcal{A}^F_C$ and $\mathcal{A}^{\neg F} = \mathcal{A}^{\neg F}_N \cup \mathcal{A}^{\neg F}_C$.

We consider a twofold fair-clustering objective: cluster the objects such that (i) the intra-cluster similarity and the inter-cluster similarity are maximized and minimized, respectively, according to the non-sensitive attributes; (ii) the intra-cluster similarity and the inter-cluster similarity are minimized and maximized, respectively, according to the sensitive attributes. Pursuing this second objective would help distribute similar objects (in terms of sensitive attributes) across different clusters, thus helping the formation of diverse clusters. This is beneficial to ensure that the distribution of groups defined on sensitive attributes within each cluster approximates the distribution across the dataset.

The task of fair clustering can be mapped to a MIN-CC instance where the positive-type and negative-type weights, respectively, can be defined as follows:

$$w^+_{uv} := \psi^+ \left(\alpha^{\neg F}_N \cdot sim_{\mathcal{A}^{\neg F}_N}(u,v) + (1 - \alpha^{\neg F}_N) \cdot sim_{\mathcal{A}^{\neg F}_C}(u,v) \right) \tag{6}$$

$$w^-_{uv} := \psi^- \left(\alpha^F_N \cdot sim_{\mathcal{A}^F_N}(u,v) + (1 - \alpha^F_N) \cdot sim_{\mathcal{A}^F_C}(u,v) \right) \tag{7}$$

where $\alpha^F_N = |\mathcal{A}^F_N|/(|\mathcal{A}^F_N| + |\mathcal{A}^F_C|)$ and $\alpha^{\neg F}_N = |\mathcal{A}^{\neg F}_N|/(|\mathcal{A}^{\neg F}_N| + |\mathcal{A}^{\neg F}_C|)$ are coefficients to weight similarities proportionally to the size of the involved set of attributes, $\psi^+ = exp(|\mathcal{A}^F|/(|\mathcal{A}^F| + |\mathcal{A}^{\neg F}|) - 1)$ and $\psi^- = exp(|\mathcal{A}^{\neg F}|/(|\mathcal{A}^F| + |\mathcal{A}^{\neg F}|) - 1)$ are smoothing factors to penalize correlation-clustering weights that are computed on a small number of attributes (which is usually the case for sensitive attributes, and hence negative-type weights), and $sim_S(\cdot)$ denotes any object similarity function defined over the subspace S of the attribute set.

Problem 2 (Attribute Selection for Fair Clustering). Given a set of objects $\mathcal{X}$ defined over the attribute sets $\mathcal{A}^F$, $\mathcal{A}^{\neg F}$, find maximal subsets $S^F \subseteq \mathcal{A}^F$ and $S^{\neg F} \subseteq \mathcal{A}^{\neg F}$, with $|S^F| \geq 1, |S^{\neg F}| \geq 1$, s.t. the correlation-clustering weights in Eqs. (6)–(7) satisfy the global-weight-bounds condition in Theorem 1.

Heuristics. Our first proposal to solve Problem 2 is a greedy heuristic, dubbed Greedy, which iteratively removes the attribute that leads to the correlation-clustering weights with the lowest target ratio until our global condition is satisfied. This algorithm runs in $\mathcal{O}(|\mathcal{X}|^2|\mathcal{A}|^2)$ time since, at each iteration, for each candidate attribute to be removed $\mathcal{O}(|\mathcal{X}|^2)$ similarities are computed to quantify the decrease of the target ratio. We also devised other heuristics which, like Greedy, remove one attribute at time, but exploit some easy-to-compute proxy measures to select the attribute that avoid the pairwise similarity computation for each candidate attribute. The Hlv (resp. Hmv) heuristic removes the least (resp. most) variable attribute where the variability is measured through normalized entropy for categorical attributes and with variation coefficient (capped to 1 if above 1) for numerical features. Hlv_B and Hmv_B, like the previous two heuristics, remove the least and most variable attribute, respectively, but the selection is constrained to the biggest set of features among $\mathcal{A}^F$ and $\mathcal{A}^{\neg F}$, in order to try to balance their size. Finally, Hlv_BW removes the least variable attribute from the set ($\mathcal{A}^F$ or $\mathcal{A}^{\neg F}$) which induces the highest average similarity value using the current weights, whereas Hmv_SW removes the most variable attribute from the set which induces the lowest average similarity value using the current weights. Note that all these heuristics (but Greedy) run in $\mathcal{O}(|\mathcal{X}|^2|\mathcal{A}|)$ time.

Table 3. Fair clustering results.

	#it	Target ratio	%(w^+ > w^-)	Orig.-weights Min-CC obj.	Avg. Eucl. fairness	Avg. #clusts.	Intra-clust $\mathcal{A}^{\neg F}$	Intra-clust $\mathcal{A}^F$	Inter-clust $\mathcal{A}^{\neg F}$	Inter-clust $\mathcal{A}^F$	Time (seconds)
					Adult						
initial	–	1.086	90.34	1.1915E+08	0.082	77	0.699	0.672	0.378	0.181	–
Hlv	12	0.986	93.19	1.122659E+08	**0.031**	9	0.465	0.326	0.347	0.194	545.249
Hlv_B	12	0.765	78.09	1.119757E+08	0.039	69	0.608	0.547	0.375	0.184	529.674
Hmv	5	0.974	90.83	1.21187E+08	0.094	79	0.689	0.687	0.373	**0.203**	220.056
Hmv_B	4	0.936	87.39	1.25516E+08	0.109	905	0.963	0.96	0.377	0.199	**178.813**
Hlv_BW	5	0.963	83.17	1.343503E+08	0.152	1479	**0.969**	0.964	0.384	0.199	217.333
Hmv_SW	9	0.926	91.41	1.159874E+08	0.037	5	0.451	**0.308**	**0.329**	0.195	380.875
Greedy	2	0.967	92.36	**1.094787E+08**	0.036	32	0.668	0.654	0.361	0.195	595.610
					Bank						
initial	–	1.612	98.84	7.738171E+07	0.019	9	0.593	0.466	0.413	0.083	–
Hlv	19	0.95	99.88	7.063441E+07	0.001	3	0.52	0.209	0.368	**0.082**	1289.785
Hlv_B	16	0.906	97.19	8.489668E+07	0.038	752	0.859	0.818	0.456	0.077	1223.205
Hmv	17	0.972	100.0	**7.032421E+07**	**0.0**	2	0.497	**0.136**	**0.151**	0.03	1254.341
Hmv_B	16	0.981	97.19	8.250374E+07	0.032	35	0.775	0.665	0.451	0.079	**1143.517**
Hlv_BW	17	0.984	92.87	1.163447E+08	0.095	1048	**0.997**	0.996	0.444	0.076	1212.091
Hmv_SW	17	0.972	100.0	7.032421E+07	0.0	2	0.497	**0.136**	**0.151**	0.03	1336.888
Greedy	13	0.981	99.57	7.240143E+07	0.006	3	0.508	0.371	0.381	0.076	11978.472
					CreditCardCustomers						
initial	–	1.415	96.97	7.556837E+06	0.050	13	0.586	0.53	0.397	0.133	–
Hlv	18	0.935	75.51	1.234939E+07	0.121	4	0.452	**0.176**	0.402	0.114	75.252
Hlv _B	17	0.981	85.64	1.013557E+07	0.153	1210	**0.996**	0.994	0.414	0.113	78.471
Hmv	15	0.985	99.41	**6.674586E+06**	**0.002**	3	0.461	0.225	**0.343**	0.132	72.112
Hmv _B	13	0.977	97.37	7.498595E+06	0.045	12	0.601	0.559	0.402	**0.134**	**58.486**
Hlv_BW	16	0.926	85.81	9.636214E+06	0.125	571	0.986	0.982	0.409	0.123	75.484
Hmv_SW	15	0.985	99.41	**6.674586E+06**	**0.002**	3	0.461	0.225	**0.343**	0.132	72.109
Greedy	14	0.941	95.5	7.584107E+06	0.049	20	0.612	0.57	0.406	0.115	714.02
					Student						
initial	–	1.042	96.79	4.307303E+04	0.034	4	0.568	0.479	0.315	0.17	–
Hlv	30	0.968	84.18	5.236701E+04	0.064	2	0.407	**0.213**	0.392	0.189	14.838
Hlv_B	22	0.967	70.09	5.828042E+04	0.143	11	0.581	0.459	0.392	0.190	10.551
Hmv	8	0.994	96.28	4.303145E+04	0.031	5	0.577	0.484	0.379	0.189	3.91
Hmv_B	8	0.974	96.94	**4.260863E+04**	**0.030**	5	**0.588**	0.507	0.364	0.184	3.809
Hlv_BW	22	0.967	70.09	5.828042E+04	0.143	11	0.581	0.459	0.392	0.190	10.923
Hmv_SW	3	0.938	94.97	4.382731E+04	0.035	4	0.561	0.446	**0.350**	0.188	**1.543**
Greedy	2	0.975	94.8	4.595454E+04	0.059	5	0.535	0.434	0.376	**0.193**	9.980

Data and Results. We considered 4 real-world relational datasets: *Adult*,[4] *Bank*, (see footnote 4) *CreditCardCustomers*,[5] and *Student* (see footnote 4). For each of them, we report in Table 1-(right) the number of objects, a pair of values corresponding to the count of non-sensitive and sensitive attributes, and a description of the latter.

---

[4] https://archive.ics.uci.edu/ml/index.php.
[5] https://www.kaggle.com/sakshigoyal7/credit-card-customers.

Table 3 summarizes results achieved by each of the above heuristics, on the various datasets, according to the following criteria (columns from left to right): number of iterations at convergence, target ratio, percentage of pairs u, v having $w_{uv}^+ > w_{uv}^-$; also, computed w.r.t. the full attribute space are: value of the objective function, average Euclidean fairness[6] (the lower, the better), average number of clusters, intra-cluster and inter-cluster similarities according to either the subset of sensitive attributes or the subset of non-sensitive attributes, and running time. (see footnote 3) Euclidean and Jaccard similarity functions are used for numerical and categorical attributes, resp., and the overall similarity is obtained by linear combination analogously to Eqs. (6)–(7). Note that higher values correspond to better performance for $\mathcal{A}^F$-based intra-cluster and $\mathcal{A}^{\neg F}$-based inter-cluster similarities, while the opposite holds for the other two measures. The first row in each table refers to the initial, full-attribute-space status of the relational network, as a *baseline*, whereby the global-weight-bounds condition is not satisfied.

Hlv_BW and Hlv_B tend to produce solutions that correspond to the highest (i.e., worst) value of the objective function and by far the highest clustering size; this should be ascribed to the fact that both heuristics favor the removal of the least variable attributes. By contrast, Hmv_SW and Hmv are the best performing in terms of objective function and, on average, also in terms of Euclidean fairness; moreover, they tend to produce very few clusters. Remarkably, while a higher number of clusters is found to be coupled with a worsening of the objective function, the opposite does not hold in general. Also, contrarily to the intuition that a higher percentage of pairs having $w^+ > w^-$ should favor the grouping into fewer clusters, we observed that an ordering of the clustering size is not aligned with the percentage ordering. As far as efficiency, Greedy tends to converge in less iterations, i.e., it removes fewer attributes than the other methods. In some cases (e.g., *Student, Adult*), this allows Greedy for compensating its expected higher cost per iteration. Hmv_B mostly provides the best time performance.

Notably, each method lowers the initial target ratio below 1 so as to satisfy the global condition, and the per-dataset best-performing method improves all intra-/inter-cluster similarities and Euclidean fairness w.r.t. the baseline.

6   Conclusions

We have studied for the first time global weight bounds in correlation clustering. We have derived a sufficient condition to extend the range of validity of approximation guarantees beyond local weight bounds, such as the probability constraint. Extensive experiments have attested the usefulness of our condition.

We believe this work offers a new perspective on correlation clustering which opens stimulating yet challenging opportunities for further research, such as

---

[6] The average weighted by cluster-size of the per-attribute averages of the Euclidean distances between the frequency attribute vector computed over the set of objects of a cluster and the frequency attribute vector over the whole set of objects [1].

investigating the role of M and γ constants, extending our results to other constraints (e.g., triangle inequality), and studying the by-product problem of feature selection guided by our condition.

For reproducibility purposes, we make source code and data available at: https://github.com/Ralyhu/globalCC and http://people.dimes.unical.it/andreatagarelli/globalCC/.

References

1. Abraham, S.S., Sundaram, S.S.: Fairness in clustering with multiple sensitive attributes. In: Proceedings of EDBT Conference, pp. 287–298 (2020)
2. Ahmadian, S., et al.: Fair hierarchical clustering. In: Proceedings of NIPS Conference (2020)
3. Ahmadian, S., Epasto, A., Kumar, R., Mahdian, M.: Clustering without over-representation. In: Proceedings of ACM KDD Conference, pp. 267–275 (2019)
4. Ahmadian, S., Epasto, A., Kumar, R., Mahdian, M.: Fair correlation clustering. In: Proceedings of AISTATS Conference, pp. 4195–4205 (2020)
5. Ailon, N., Charikar, M., Newman, A.: Aggregating inconsistent information: ranking and clustering. JACM **55**(5), 23:1–23:27 (2008)
6. Ausiello, G., Marchetti-Spaccamela, A., Crescenzi, P., Gambosi, G., Protasi, M., Kann, V.: Complexity and Approximation: Combinatorial Optimization Problems and Their Approximability Properties. Springer, Heidelberg (1999). https://doi.org/10.1007/978-3-642-58412-1
7. Backurs, A., Indyk, P., Onak, K., Schieber, B., Vakilian, A., Wagner, T.: Scalable fair clustering. In: Proceedings of ICML Conference, pp. 405–413 (2019)
8. Bansal, N., Blum, A., Chawla, S.: Correlation clustering. Mach. Learn. **56**(1), 89–113 (2004)
9. Bera, S.K., Chakrabarty, D., Flores, N., Negahbani, M.: Fair algorithms for clustering. In: Proceedings of NIPS Conference, pp. 4955–4966 (2019)
10. Bonchi, F., García-Soriano, D., Liberty, E.: Correlation clustering: from theory to practice. In: Proceedings of ACM KDD Conference, p. 1972 (2014)
11. Charikar, M., Guruswami, V., Wirth, A.: Clustering with qualitative information. JCSS **71**(3), 360–383 (2005)
12. Chawla, S., Makarychev, K., Schramm, T., Yaroslavtsev, G.: Near optimal LP rounding algorithm for correlation clustering on complete and complete k-partite graphs. In: Proceedings of ACM STOC Symposium, pp. 219–228 (2015)
13. Chen, X., Fain, B., Lyu, L., Munagala, K.: Proportionally fair clustering. In: Proceedings of ICML Conference, pp. 1032–1041 (2019)
14. Chierichetti, F., Kumar, R., Lattanzi, S., Vassilvitskii, S.: Fair clustering through fairlets. In: Proceedings of NIPS Conference, pp. 5029–5037 (2017)
15. Crescenzi, P.: A short guide to approximation preserving reductions. In: Proceedings of IEEE CCC Conference, pp. 262–273 (1997)
16. Demaine, E.D., Emanuel, D., Fiat, A., Immorlica, N.: Correlation clustering in general weighted graphs. TCS **361**(2–3), 172–187 (2006)
17. Gionis, A., Mannila, H., Tsaparas, P.: Clustering aggregation. ACM TKDD **1**(1), 4 (2007)
18. Kleindessner, M., Awasthi, P., Morgenstern, J.: Fair k-center clustering for data summarization. In: Proceedings of ICML Conference, pp. 3448–3457 (2019)

19. Kleindessner, M., Samadi, S., Awasthi, P., Morgenstern, J.: Guarantees for spectral clustering with fairness constraints. In: Proceedings of ICML Conference, pp. 3458–3467 (2019)
20. Kollios, G., Potamias, M., Terzi, E.: Clustering large probabilistic graphs. IEEE TKDE **25**(2), 325–336 (2013)
21. Mandaglio, D., Tagarelli, A., Gullo, F.: In and out: optimizing overall interaction in probabilistic graphs under clustering constraints. In: Proceedings of ACM KDD Conference, pp. 1371–1381 (2020)
22. Pandove, D., Goel, S., Rani, R.: Correlation clustering methodologies and their fundamental results. Expert. Syst. **35**(1), e12229 (2018)
23. Puleo, G.J., Milenkovic, O.: Correlation clustering with constrained cluster sizes and extended weights bounds. SIAM J. Optim. **25**(3), 1857–1872 (2015)
24. Rösner, C., Schmidt, M.: Privacy preserving clustering with constraints. In: Proceedings of ICALP Colloquium, vol. 107, pp. 96:1–96:14 (2018)
25. Shamir, R., Sharan, R., Tsur, D.: Cluster graph modification problems. Discret. Appl. Math. **144**(1–2), 173–182 (2004)
26. Swamy, C.: Correlation clustering: maximizing agreements via semidefinite programming. In: Proceedings of ACM-SIAM SODA Conference, pp. 526–527 (2004)
27. van Zuylen, A., Williamson, D.P.: Deterministic algorithms for rank aggregation and other ranking and clustering problems. In: Proceedings of WAOA, pp. 260–273 (2007)

Modeling Multi-factor and Multi-faceted Preferences over Sequential Networks for Next Item Recommendation

Yingpeng Du[1], Hongzhi Liu[1(✉)], and Zhonghai Wu[2,3(✉)]

[1] School of Software and Microelectronics, Peking University,
Beijing, People's Republic of China
{dyp1993,liuhz}@pku.edu.cn

[2] National Engineering Research Center of Software Engineering, Peking University,
Beijing, People's Republic of China
wuzh@pku.edu.cn

[3] Key Lab of High Confidence Software Technologies (MOE), Peking University,
Beijing, People's Republic of China

Abstract. Attributes of items carry useful information for accurate recommendations. Existing methods which tried to use items' attributes relied on either 1) feature-level compression which may introduce much noise information of irrelevant attributes, or 2) item- and attribute-level transition modeling which ignored the mutual effects of multi-factor for users' behaviors. In addition, these methods failed to capture multi-faceted preferences of users, therefore, the prediction for the next behavior may be affected or misled by the irrelevant facets of preferences. To address these problems, we propose a S̲equential N̲etwork based R̲ecommendation model, named SNR, to extract and utilize users' multi-factor and multi-faceted preferences for next item recommendation. To model users' multi-factor preferences, we organize the item- and attribute- level sequences of users' behaviors as unified sequential networks, and propose an attentional gated Graph Convolutional Network model to explore the mutual effects of the preference factors contained in sequential networks. To capture users' multi-faceted preferences, we propose a multi-faceted preference learning model to simulate the decision-making process of users with the Gumbel sotfmax trick. Finally, we fuse the multi-factor and multi-faceted preferences in a unified latent space for next item recommendation. Extensive experiments on four real-world data sets show that the proposed model SNR consistently outperforms several state-of-the-art methods.

Keywords: Sequential recommendation · Sequential networks · Preference learning · Multi-factor preference · Multi-faceted preference

1 Introduction

Recommender systems help users to discover the items that they may prefer from numerous choices, which can enhance both users' satisfaction and platforms' profits. In real-world scenarios, users' behavior data, e.g. click or purchase,

© Springer Nature Switzerland AG 2021
N. Oliver et al. (Eds.): ECML PKDD 2021, LNAI 12976, pp. 516–531, 2021.
https://doi.org/10.1007/978-3-030-86520-7_32

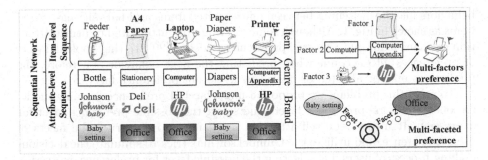

Fig. 1. The left part is an illustrating example that a user purchases items sequentially. The right part concludes the multi-factor and multi-faceted preferences of users.

always appears as a sequence and every behavior is linked with a timestamp. Predicting and recommending the next item user may act on, known as sequential recommendation, has been becoming a hot research topic in recent years.

Most of existing research about sequential recommendation focused on modeling item-level sequences of users' behaviors, which ignores the attribute information of items which is beneficial for accurate recommendations [1]. Therefore, how to utilize attribute information effectively for sequential recommendations remains a challenge. Some methods tried to combine attribute information and item information with compression strategies, e.g. feature-level aggregation or concatenation, and adopted traditional sequential models like LSTM for recommendation [2,3]. However, not all attributes are relevant for recommending the target, which may introduce noise information and degrade the performance of recommender systems. Some methods tried to model the item- and attribute- level transition patterns in the sequential behaviors of users [1,4,5], which ignored the mutual effects of multi-factor for users' next behavior. Figure 1 shows the user's purchase records in an online shopping website. The user may prefer the HP printer because he/she requires a printer (Factor1: Computer → Computer Appendix → Printer and Factor2: A4 Paper→ Printer) and prefers Brand HP that produces printer (Factor3: HP laptop → HP → HP printer). These factors consist of useful correlations and high-order dependencies between attributes and items in users' sequential behaviors, whose mutual effects can help with users' fine-grained preference modeling.

To deal with this issue, we first combine the item- and attribute- level sequences into a unified sequential network for each user. Then, we design an attentional gated graph convolutional network (agGCN) model to explore the sequential networks, which can extract useful correlations and high-order dependencies for multi-factor preference modeling. Compared to constant compression strategies such as traditional GCN, the agGCN can alleviate the noise problem with attentional and gated mechanisms.

In real-world scenarios, each user may play multiple different roles in his/her life, which makes his/her preference shows multi-faceted correlations. For example, Bob is a teacher and meanwhile a baby's father, therefore, his purchasing behaviors may be switching between office-related items and babysitting-related items as shown in Fig. 1. Although some existing work attempted to utilize attention mechanism for multi-faceted preference modeling [20,21], they took all facets of preferences into account when predicting the next behavior. Therefore, the predictions may be affected or misled by irrelevant preference facets.

To solve this problem, we propose a multi-faceted preference learning model to construct users' preference facets and select the possible preference facet for expression. It utilizes the self-attention mechanism to construct the facets of users' preference, and utilizes the Gumbel softmax trick to simulate the decision-making process of users, i.e. selecting the possible facet for preference expression w.r.t different items.

2   Related Work

In this section, we review the related work from four aspects: item-aware sequential recommendation, attribute-aware sequential recommendation, GCN based recommendation, and attention based recommendation.

Item-Aware Sequential Recommendation. Most of the existing work for sequential recommendation focused on modeling item-level transition patterns of users' behaviors. For example, FPMC [7] directly combined matrix factorization (MF) with Markov chains (MC) to predict the next item based on users' recent engaged items (e.g. purchased items). Hierarchical Representation Model (HRM) [8] extended FPMC by applying aggregation operations to explore more complex interactions (e.g. non-linear interactions) for sequential recommendation. These MC based methods focus on the sequential patterns between two adjacent behaviors or baskets while ignoring the long-term dependencies of the whole sequences [9]. Recurrent neural networks (RNN), which are good at modeling long-term dependencies in sequence data, have been adopted for sequential recommendation. Various RNN variants have been extended for different scenarios [10,11]. For example, [12] proposed a gated recurrent unit (GRU) based model GRURec with a ranking loss for session-based recommendation. In addition, graph neural networks were adopted to learn the representations of the session interaction graph for session-based recommendation [13]. These methods rely on the user-item binary relation sequences, which ignores item attributes that are useful for users' preference modeling.

Attribute-Aware Sequential Recommendation. Recently, attribute information is used by several sequential models to improve the performance of sequential recommendation [4,14]. Several methods tried to combine attributes and items information with compression strategies, and adopted the sequential models for recommendation. For example, HA-RNN [3] combined the embeddings of items and their attributes into a unified sequential representations, then feed them into an LSTM model for next item recommendation. ANAM [2] utilized a hierarchical architecture to incorporate the attribute information of items, and adopted an attention mechanism to explicitly model users' evolving preferences on items. However, the feature-level compression may aggregate irrelevant attributes and introduce noise, which degrades the performance of sequential recommendation.

Several methods tried to explore the sequence data with attribute information by mining their transition patterns of adjacent behaviors or baskets. For example, FDSA [1] tried to model item transition patterns and attribute transition patterns from adjacent behaviors for next item recommendation. KA-MemNN [4] utilized the memory networks to store the categories of items in the last basket, and inferred the categories that user may need for next basket prediction. However, these methods fail to explore the sequential networks of users, which contain useful correlations and high-order dependencies between items and attributes for users' multi-factor preference modeling.

GCN Based Recommendation. In recent years, GCN based techniques gain their popularity in graph-based RSs [15], which organize the input data as graphs and try to extract the structure information for recommendation [16]. Ying et al. [17] proposed to apply GCN to the graph with neighborhood sampling technique, which can be adopted for web-scale recommendation tasks. Wang et al. [29] adopted GCN to discover high-order structure and semantic information in the knowledge graph, and utilized these information to enrich users' preference modeling for recommendation. These methods achieved good performance. However, these methods are designed for static graphs and not available for sequential recommendation, and most of them assume equal or constant importance during convolution which may bring more noise for recommendation.

Attention Based Recommendation. Attention mechanism has been widely used in natural language processing such as text matching [18], and reading comprehension [19], due to its excellent theory and good performance. For sequential recommendation, several methods utilized attention mechanism to aggregate users' preferences or behaviors for unified representations. For example, DMFP [20] assumed users' long-term preferences are multi-faceted, adopted the multi-hops attention mechanism to model users' multi-faceted preferences. MANN [21] stored users' historical records explicitly by the memory networks, and adopted the attention mechanism to learn the different importance of behavioral records for recommendation. These attention based methods can be seen as the aggregation of the multi-faceted preferences or behaviors. However, these methods are essentially weighted aggregating all facets of preferences or behaviors, which make the prediction for the next behavior may be affected by irrelevant facets.

3   Problem Formulation

Let $\mathcal{U} = \{u_1, \cdots, u_N\}$, $\mathcal{I} = \{i_1, \cdots, i_M\}$ and $\mathcal{F} = \{f_1, \cdots, f_L\}$ denote the sets of N users, M items and L attributes, respectively. Each user $u \in U$ engaged a series of items $S_u(T) = [s_t | t = 1, \cdots, T]$ in the chronological order, where $s_t \in \mathcal{I}$ denotes the t-th item he/she engaged. For each item $i \in \mathcal{I}$, we collect its K types of attributes, i.e. $F_i = \{f_i^1, \cdots, f_i^K\}$, where $f_i^j \in \mathcal{F}$ denotes the j-typed attribute of item i. For example, we can collect the items' attributes with types *Genre, Brand, Color* and etc.

The goal of sequential recommendation is to learn a prediction function $f(\cdot)$ based on all users' sequential behaviors and other related information. Formally, the function can predict the next item that the target user may act on. In this paper, we take the attribute information of items into consideration and define the prediction function as $s(u, T) = \max_{i \in \mathcal{I}} f(u, i, S_u(T), F_{\mathcal{I}})$.

4   The Proposed Method

Figure 2 shows the architecture of the Sequential Networks based Recommendation model, named SNR, which is built on the sequential networks (the left part of Fig. 2) that organize users' engaged items and attributes in a unified form. SNR consists of two main parts, i.e. the multi-factor preference modeling (the top part of Fig. 2) and multi-faceted preference modeling (the bottom part of Fig. 2). First, we design an agGCN model to learn the mutual effects of factors with the help of sequential networks. Then, we propose a multi-faceted preference learning model to simulate the decision-making process of users. Finally, we fuse them in a unified latent space to learn users' hybrid preferences for next item recommendations.

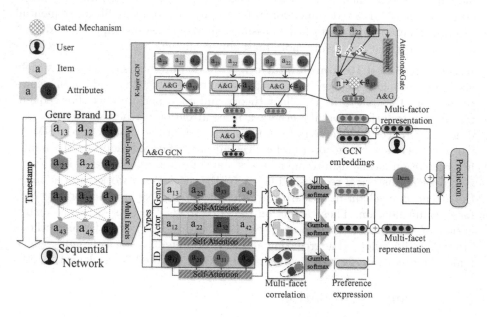

Fig. 2. The architecture of the proposed model SNR.

4.1   Construction and Embedding of Sequential Networks

To make use of the sequential information of users' behaviors, we combine the **item-level sequence** $S_u(T) = [s_t | t = 1, \cdots, T]$ with the corresponding

attribute-level sequence $A_u(T) = [(g_{s_t}^1, \cdots, g_{s_t}^K)|t = 1, \cdots, T]$ into a unified sequence $B_u(T) = [B_u^t|t = 1, \cdots, T]$, where $g_{s_t}^j$ denotes the j-typed attribute of items s_t and $B_u^t = (s_t, g_{s_t}^1, \cdots, g_{s_t}^K)$ denotes the engaged items and attributes in his/her t-th behavior. We organize the unified sequence as the **Sequential Network** for user u, which is a directed graph $\mathcal{G}_u = (\mathcal{V}_u, \mathcal{E}_u)$ as shown in Fig. 2. The node (or entity[1]) set $\mathcal{V}_u$ consists of the items and attributes engaged by user u, i.e., $\mathcal{V}_u = \bigcup_{t=1,\cdots,T} B_u^t$. The edges set $\mathcal{E}_u$ consists of the chronological relations between adjacent behaviors, i.e., $\mathcal{E}_u = \{(a,b)|a \in B_u^{t-1}, b \in B_u^t, t = 2, \cdots, T\}$. We define the neighbors of $b \in B_u^t$ as its adjacent nodes before time t, i.e. $J(b) = \{a|a \in B_u^{t-1}\}$.

To make use of information shared in different users' sequential networks, we embed users, items and attributes into the same latent space, which is denoted as $\boldsymbol{W} = \mathbb{R}^{N \times d}$, $\boldsymbol{Q} = \mathbb{R}^{M \times d}$ and $\boldsymbol{P} = \mathbb{R}^{L \times d}$ respectively, where d is the dimensionality of latent space. With the embedding of items and attributes, we denote the embedding of sequence network $\mathcal{G}_u$ as E_u for the user u:

$$E_u = \begin{bmatrix} q_{s_1} & p_{s_1}^1 & \cdots & p_{s_1}^K \\ \vdots & \vdots & \vdots & \vdots \\ q_{s_T} & p_{s_T}^1 & \cdots & p_{s_T}^K \end{bmatrix} = \begin{bmatrix} e_{11} & \cdots & e_{1J} \\ \vdots & \cdots & \vdots \\ e_{T1} & \cdots & e_{TJ} \end{bmatrix} \tag{1}$$

where $q_{s_t} = \mathrm{emb}(s_t) \in \mathbb{R}^d$ and $p_{s_t}^1 = \mathrm{emb}(f_{s_t}^j) \in \mathbb{R}^d$. As items can be treated as a special type (e.g. ID) of attribute, we simplify E_u with the unified form $E_u = [e_{tj}]_{T \times J}$, where $J = K + 1$ and $e_{tj} \in E_u$ denotes the embedding of node $a_{tj} \in B_u^t$. With the matrix representations of sequential networks, we can utilized GCN based method to explore and make use of information contained in the sequential networks.

4.2 Multi-factor Preference Modeling

To model users' multi-factor preferences, we propose an agGCN model to explore the sequential networks. An attentional mechanism is utilized to model the correlations between adjacent entities in the sequential network, and a gated mechanism is utilized to explore the high-order dependencies between entities which may be nonadjacent in the sequential networks. Combining them can explore the mutual effects of multi-factor for fine-grained preference modeling.

Attention Mechanism. To extract useful correlations between adjacent items and attributes in the sequential network, we design an attentional mechanism to learn the importance of neighbors when aggregating them in the GCN framework. Therefore, we can reduce the noise caused by constantly or equally aggregating all neighbor nodes in traditional GCNs.

Specifically at the k-th layer of GCN, we aggregate the neighbors of node a_{tj} in $\mathcal{G}_u$ for user u in the $(k-1)$-th layer, which is denoted as the $n_{tj}^{(k)}$:

$$n_{tj}^{(k)} = \sum_{b \in J(a_{tj})} \alpha(a_{tj}, b) h_b^{(k-1)} \tag{2}$$

---

[1] In this paper, nodes, entities and attributes&items are used interchangeably.

where $J(a_{tj})$ denotes the neighbor set of $a_{tj} \in \mathcal{V}_u$ in $\mathcal{G}_u$ and $h_b^{(k-1)} \in \mathbb{R}^d$ denotes the $(k-1)$-th layer GCN representation of node b. The $\alpha(a_{tj}, b)$ denotes the attentional weight that models the similarity between a_{tj} and its neighbor b for user u, i.e.

$$\alpha(a_{tj}, b) = \frac{\exp <T_1 \cdot e_{tj}, T_1 \cdot h_b^{(k-1)}>}{\sum_{b \in J(a_{tj})} \exp <T_1 \cdot e_{tj}, T_1 \cdot h_b^{(k-1)}>} \tag{3}$$

where $< \cdot, \cdot >$ denotes the inner product of two vectors, and $T_1 \in \mathbb{R}^{d \times d}$ denotes the projection matrix that transforms the two kinds of embedding into the same latent space for attention modeling. Intuitively, the high attentional weights indicate useful correlations information between entities contained in sequential networks for recommendation.

Gated Mechanism. Although traditional GCNs can capture the dependencies of nodes among multi-layers, they usually assume the constant combination between neighborhood information $n_{tj}^{(k)}$ and feature node e_{tj} (i.e. the 0-th layer representation $h_{tj}^{(0)} = e_{tj}$) [22]. Therefore, the high-order dependent nodes will be affected or misled by irrelevant attributes or items between these nodes in the sequential networks. Inspired by the LSTM model [23], we propose a gated mechanism that learns to incorporate useful entities information for GCN:

$$h_{tj}^{(k)} = \psi(\beta_{tj}^{(k)} \odot e_{tj} + (1 - \beta_{tj}^{(k)}) \odot n_{tj}^{(k)}) \tag{4}$$

where $\psi(\cdot)$ denotes an activation function, for which we adopt the identity map as in [24]. $\beta_{tj}^{(k)} \in (0, 1)$ denotes the "gate" of information fusion which can be formulated as follows:

$$\beta_{tj}^{(k)} = \sigma(< T_2 \cdot e_{tj}, T_2 \cdot n_{tj}^{(k)} > +b^{(k)}) \tag{5}$$

where $\sigma(\cdot)$ denotes the sigmoid function, i.e., $\sigma(x) = 1/(1 + \exp(-x))$ and $b^{(k)} \in \mathbb{R}^p$ denotes the bias for the k-th layer, and $T_2 \in \mathbb{R}^{d \times d}$ denotes the projection matrix that transforms the two kinds of embedding into the same space for gate modeling.

Once we obtain the k-th layer GCN embeddings $H_t^{(k)} = [h_{t,1}^{(k)}; \cdots ; h_{t,J}^{(k)}] \in \mathbb{R}^{J \times d}$ of entities B_u^t in the sequential network, we aggregate them as the multi-factor preferences of user u for next item prediction at time $t + 1$, i.e.

$$p_{u,t+1}^{\text{factor}} = w_u + \sum_{j=1}^{J} h_{t,j}^{(k)} \tag{6}$$

where $w_u \in \mathbb{R}^d$ denotes the personalized preference representation of user u.

4.3   Multi-faceted Preference Modeling

Generally, users' sequential behaviors are the mixture of behaviors that switch between different preference facets. We assume the behaviors from similar preference facets have high correlations. To construct users' preference facets w.r.t each typed attributes, we utilize a self-attention mechanism to learn the correlations of the user engaged entities:

$$\bar{M}_{uj}^t = \text{Self-Attention}(M_{uj}^t) \tag{7}$$

where $M_{uj}^t = [e_{1j}, \cdots, e_{tj}]$ denotes the j-typed attributes user u engaged before time t. Self-Attention(E) is the special case of the scaled dot-product attention SDP-Attention(Q, K, V) when $Q = K = V = M$:

$$\text{SDP-Attention}(Q, K, V) = softmax(\frac{QK^T}{\sqrt{d_k}})V \tag{8}$$

where d denotes the dimension of the queries Q. The self-attention mechanism can learn the embeddings of similar entities with the high correlation weights, i.e. $softmax(\frac{QK^T}{\sqrt{d_k}})$, and thus make them be close in the latent space, which makes similar entities belong to the same or similar preference facets.

To simulate the process of users' decision-making, we adopt the Gumbel softmax trick to select the possible facet of preferences that user may express for next behavior. Gumbel softmax trick [6] provides a differentiable method for categorical distribution sampling by using inverse transform sampling and reparameterization trick. For user's multi-faceted preferences, we first model the expression probability $\pi_{uj}^t(i) \in \mathbb{R}^t$ of each facet for item i as follows:

$$\pi_{uj}^t(i) = softmax(\bar{M}_{uj}^t \cdot emb_{ij}) \tag{9}$$

where $emb_i = [q_i, p_i^1, \cdots, p_i^K] \in \mathbb{R}^{J \times d}$ denotes the embeddings of item $i \in \mathcal{I}$ and its attributes. Then, we select the preference facet $m_{uj}^t(i) \in \mathbb{R}^d$ that user may express for next behavior among multiple facets $\bar{M}_{uj}^t$:

$$m_{uj}^t(i) = G_{uj}^t(i) \cdot \bar{M}_{uj}^t \tag{10}$$

where $G_{uj}^t(i) \in \mathbb{R}^t$ approximates to one-hot encoding with selected facet as 1 otherwise 0, which derives from the categorical distribution based on Gumbel softmax trick:

$$G_{uj}^t(i) = Gumbel(\pi_{uj}^t(i)) = softmax(-\log(-\log(\epsilon)) + \pi_{uj}^t(i)) \tag{11}$$

where $\epsilon = [\epsilon_1, \cdots, \epsilon_t]$ derives from uniform distribution $U(0, 1)$.

Once we select user u's preference facets for all attribute types $m_u^t(i) = [m_{u1}^t(i), \cdots, m_{uJ}^t(i)]$, we fuse them as the multi-faceted preferences w.r.t item i for next item prediction at time $t + 1$ as follows:

$$p_{u,t+1}^{facet}(i) = \sum_{j=1}^J m_{uj}^t(i) \tag{12}$$

4.4 Recommendation Model

To make use of both users' multi-factor and multi-faceted preferences for next item recommendation, we fuse them in the same latent space. Specifically, we aggregate them as the hybrid preferences of users w.r.t item i at time $t + 1$:

$$p_{u,t+1}^{hybrid}(i) = (1 - \mu) \cdot p_{u,t+1}^{factor} + \mu \cdot p_{u,t+1}^{facet}(i) \tag{13}$$

where $0 \leq \mu \leq 1$ is a fusion coefficient which is used to control the relative importance of the two kinds of representations.

With hybrid preference representation of users, we calculate user u's relative preference score for item i at time t as follows:

$$s(u, i, t+1) = <p_{u,t+1}^{\text{hybrid}}(i), v_i> \tag{14}$$

where $v_i \in \mathbb{R}^d$ denotes the embedding of item i for score prediction.

For model learning, we adopt the pairwise loss to define the objective function as follows:

$$\text{SNR-opt} = \max_{\theta} \sum_{(u,i,j,t) \in D} \ln \sigma(\hat{y}(u,i,j,t)) - \lambda_{\theta}||\theta||^2 \tag{15}$$

where $D = \{(u,t,i,j)|u \in \mathcal{U}, t = 1, \cdots, |B_u|\}$ means that user u gave positive feedback to item i instead of item j at time t. We denote the relative score of user u's on item i and item j at time t, i.e. $\hat{y}(u,i,j,t) = s(u,i,t) - s(u,j,t)$. The θ denotes all parameters need to be learned in the proposed model and λ_{θ} is the regularization coefficient of L2 norm $|| \cdot ||^2$. The objective function shows that the item i with positive feedback should have a higher score than the item j without feedback for user u at time t.

As the objective function is differentiable, we optimize it by stochastic gradient descent (SGD) and adaptively adjust the learning rate by AdamGrad [25], which can be automatically implemented by TensorFlow[2]. The implementation of our method will be publicly available after the paper is accepted.

5   Experiment

In this section, we aim to evaluate the performance and effectiveness of the proposed method. Specifically, we conduct several experiments to study the following research questions:

- **RQ1**: Whether the proposed method outperforms state-of-the-art methods in sequential recommendation?
- **RQ2**: Whether the proposed sequential networks help with modeling the mutual effects of multi-factor, and outperform methods that explore individual transition patterns of attributes and items?
- **RQ3**: Whether the agGCN model can reduce the noise and outperform traditional GCNs?
- **RQ4**: Whether the proposed method benefits from the multi-facet preference learning model which simulates the decision-making process of users?

---

[2] https://www.tensorflow.org/.

5.1   Experimental Setup

Datasets

We adopt four public data sets with the attributes of items as the experimental data, including ML[3] (Hetrec-MovieLens), Dianping[4], Amazon_Kindle (Kindle Store), and Amazon_App (Apps for Android)[5]. The ML data set contains users' ratings on movies with timestamps and the attributes of movies (e.g. Actors, Director, Genres, and Countries). The Dianping dataset consists of the users' ratings on restaurants in China with timestamps and attributes of restaurants (e.g. City, Business_district, Average_cost, and Style). The Amazon_App and Amazon_Kindle data sets contain users' ratings on APPs and books with timestamps and metadata of items. For these data sets, we filter users and items which have less than 10 records except 20 records for the Dianping and Amazon_Kindle as in [26]. We consider ratings higher than 3.5 points as positive interactions as in [27]. The characteristics of the four data sets are summarized in Table 1.

Table 1. Statistics of the experimental data sets

Dataset	#User	#Item	#Interaction	Sparsity
ML	2,059	4,220	287,033	3.30%
Dianping	9,329	18,036	190,503	0.11%
Amazon_App	10,724	5,995	120,856	0.19%
Amazon_Kindle	11,677	24,097	206,575	0.07%

Evaluation Methodology and Metrics

We create the real-world scenario by stimulating the dynamic data stream. We sort all interactions chronologically, then reserve the first 80% interactions as the train set and hold the last 20% for testing. We test the interaction from the hold-out data one by one correspondingly. Experimental results are recorded as the average of the five runs with different random initializations of model parameters.

To evaluate the performance of our method and the baseline methods, we adopt two widely used evaluation metrics for top-N sequential recommendation [32], i.e. hit ratio (hr) and normalized discounted cumulative gain (ndcg), which can be formulated as follows:

$$hr@K = \frac{1}{|test|} \sum_{s \in test} \sum_{i=1}^{K} r_i(s) \qquad (16)$$

$$ndcg@K = \frac{1}{|test|} \sum_{s \in test} \sum_{i=1}^{K} \frac{2^{r_i(s)} - 1}{\log_2(i+1)} \qquad (17)$$

---

[3] https://grouplens.org/datasets/movielens/.
[4] https://www.dianping.com/.
[5] http://jmcauley.ucsd.edu/data/amazon/links.html.

For each test sample $s \in test$, $r_i(s) = 1$ means the ranking list hits the ground truth item at i-th position while 0 otherwise. HR measures the ratio of ground-truth (GT) item set is hit, while NDCG focuses on the position of hit.

Baselines
We take the following state-of-the-art methods as the baselines.

- **BPR** [30]: It is a recommendation method for implicit feedback, which proposes a pair-wise loss function to model the relative preferences of users.
- **FPMC** [7]: It combines MF and MC to capture users' dynamic preferences.
- **Caser** [31]: It adopts the convolutional filters to learn users' union and skip patterns for sequential recommendation, which incorporates the convolutional neural network with a latent factor model.
- **SASRec** [26] It is a self-attention based sequential model, and it can consider engaged item-level sequences for next item recommendation.
- **SASRec+**: It is an extension to the SASRec method, which concatenates item vector representations and category vector representations together as the input of the item-level self-attention network.
- **FM_BPR** [28,29]: It models the interactions between each pair of features to estimate the target. We treat users' recent behaviors and items' attributes as the features of FM and adopt the pair-wise loss for objective function as in [29].
- **ANAM** [2]. It utilizes a hierarchical architecture to incorporate the attribute information by an attention mechanism for sequential recommendation.
- **HA-RNN** [3]. It combines the representation of items and their attributes, then fits them into the LSTM model for sequential recommendation.
- **FDSA** [1]. It models item and attribute transition patterns for next item recommendation based on the Transformer model [33].
- **SNR**: It is the proposed method in this paper.

Experimental Design
We set the learning rating $\alpha = 0.001$ and the regularization coefficient $\lambda = 0.0001$ on all methods for a fair comparison. We set the window size $k = 5$ to construct the sequential networks, i.e. keep only the latest k behaviors for next recommendation, based on which we involve all nodes by the k-layer GCN. We set $\mu = 0.5$ for equal importance of multi-factor and multi-faceted preferences modeling. For baseline models, we set their parameters as authors' implementation if they exist, otherwise we tune them to their best. To be fair, we keep users' recent k behaviors as the last basket or sequential behaviors for baseline models.

5.2   Model Comparison

Table 2 shows the performance of different methods for sequential recommendation. To make the table more notable, we bold the best results and underline the best baseline results for each data set with a specific evaluation metric. From the experimental results, we can get the following conclusions. First, the proposed method SNR performs better than all baselines in all cases, which proves the effectiveness of our proposed model (RQ1). Second, some methods with consideration

Table 2. Performance of different methods. * indicates statistically significant improvement on an independent-samples t-test ($p < 0.01$).

Method	ML		Dianping		Aamazon_APP		Aamazon_Kindle	
	hr@10	ndcg@10	hr@10	ndcg@10	hr@10	ndcg@10	hr@10	ndcg@10
BPR	0.035	0.017	0.024	0.011	0.067	0.034	0.040	0.020
PFMC	0.051	0.028	0.026	0.013	0.093	0.048	0.067	0.035
Caser	0.044	0.021	0.020	0.010	0.093	0.050	0.055	0.030
SASRec	0.059	0.024	0.024	0.010	0.087	0.043	0.066	0.0297
SASRec+	0.039	0.016	0.023	0.010	0.078	0.036	0.058	0.027
FM_BPR	0.058	0.029	0.028	0.013	0.094	0.050	0.100	0.054
ANAM	0.034	0.016	0.023	0.011	0.066	0.033	0.041	0.021
HA-RNN	0.053	0.025	0.028	0.013	0.097	0.053	0.049	0.025
CFSA	0.047	0.022	0.018	0.009	0.072	0.038	0.066	0.036
SNR (Ours)	**0.077***	**0.038***	**0.032***	**0.015***	**0.120***	**0.064***	**0.110***	**0.061***
Improve.	29.5%	33.8%	14.8%	15.9%	24.8%	20.7%	10.2%	14.0%

of attribute information, e.g. HA-RNN and FM_BPR, achieve the best performance in most cases, e.g. Amazon_App and Dianping, among baseline models, which confirms the necessity of extracting the rich information from attributes. Third, we also notice some attribute based methods, e.g. ANAM, SASRec+ and FDSA, show low accuracy in some cases, which indicates that the attribute data may contain lots of irrelevant information for recommendation. For example, SASRec+ with consideration of attribute information shows worse performance than SASRec without the attribute information, which indicates simple compression strategy may introduce more noise rather than useful information.

5.3   Ablation Studies

Table 3. Performance of variant SNR for ablation studies

Dataset	Method	hr@10	ndcg@10	Dataset	Method	hr@10	ndcg@10
ML	SNR	**0.0768**	**0.0384**	Dianping	SNR	**0.0319**	0.0153
	SNR-noSN	0.0738	0.0364		SNR-noSN	0.0314	0.0153
	SNR-noAG	0.0681	0.0332		SNR-noAG	0.0267	0.0129
	SNR-noGS	0.0750	0.0376		SNR-noGS	0.0318	**0.0155**
Amazon_App	SNR	**0.1204**	**0.0636**	Amazon_Kindle	SNR	**0.1101**	**0.0611**
	SNR-noSN	0.1189	0.0630		SNR-noSN	0.1041	0.0584
	SNR-noAG	0.1110	0.0593		SNR-noAG	0.0963	0.0519
	SNR-noGS	0.1174	0.0624		SNR-noGS	0.1060	0.0597

To evaluate the effectiveness of module design of the proposed method, we take some special cases of the proposed method as the comparisons.

- **SNR-noSN**: It removes the unified Sequential Networks, and replaces it with several independent item-level sequences and attribute-level sequences to explore the individual transition patterns.
- **SNR-noAG**: It removes the Attentional Gated mechanism in SNR and utilizes the average and accumulative convolution as in the traditional GCNs.
- **SNR-noGS**: It removes the Gumbel Sotfmax trick, and adopts the attentional mechanism for users' preference expression.

Table 3 shows the performance of ablation models, i.e. SNR, SNR-noAG, SNR-noSN and SNR-noGS, for sequential recommendation. First, SNR consistently outperforms SNR-noSN on all data sets, which indicates that exploring the mutual effects of multi-factor in sequential networks shows the priority to exploring individual transition patterns of attributes and items (RQ2). Second, SNR outperforms SNR-noAG that adopts the traditional GCN, which confirms the effectiveness of the agGCN method can reduce the noise and explore the useful correlations and high-order dependencies of items and attributes in sequential networks (RQ3). Third, SNR outperforms SNR-noGS in most cases except for the NDCG@10 on the Dianping data set. It indicates that modeling users' multifacet preferences by sampling strategy can alleviate the affection or misleading by the irrelevant facets of preferences. (RQ4).

5.4   Hyper-Parameter Study

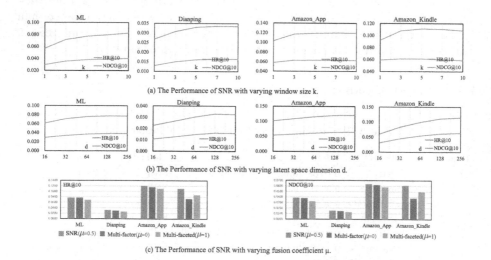

(a) The Performance of SNR with varying window size k.

(b) The Performance of SNR with varying latent space dimension d.

(c) The Performance of SNR with varying fusion coefficient μ.

Fig. 3. The Performance of SNR with varying window size k, latent space dimension d and fusion coefficient μ on all data sets.

There are several key parameters for SRN, including the window size k, the dimension of latent space d, and the trade-off coefficient μ which is used to control the importance of the multi-factor and multi-faceted preference modeling parts.

First, we explore how the window size k influences our model. Figure 3 (a) shows the rapid growth of HR and NDCG when $k \leq 5$ and the moderate or inapparent growth of HR and NDCG when $k > 5$. It indicates that incorporating too little historical information of users' behaviors, i.e. $k < 5$, suffers from insufficient information utilization for SNR. Meanwhile, incorporating too much historical information of users' behaviors, i.e. $k > 7$, may lead no improvement but high complexity for SNR. We suggest to set $k = 5$ with consideration of both accuracy and efficiency. Then, we explore how dimension latent space d influences our model. Figure 3 (b) shows that SNR perform well when $d = 128$, which indicates too large or too small dimension d may be less predictive or calculative complexity for SNR. Finally, to explore how fusion coefficient μ influences our model, we set $\mu = [0, 0.5, 1]$ to evaluate individual and hybrid performance of multi-factor and multi-faceted preference modeling parts respectively. Figure 3 (c) shows that combining both them can achieve better performance than any individual one, which indicates the necessity of modeling both multi-factor and multi-faceted preferences of users.

6   Conclusion

In this paper, we propose to extract and utilize multi-factor and multi-faceted preference based on the sequential networks. To model users' multi-factor preference, we design an agGCN that can explore the useful correlations and high-order dependencies between entities in the sequential networks. To capture users' multi-faceted preferences, we propose a multi-faceted preference learning model to simulate the decision-making process of users. Extensive experiments show our model consistently outperforms state-of-the-art methods. In addition, the ablation experiments prove the effectiveness of module design of the proposed method and our motivations. In this paper, we only make use of the behavior sequences of users and the attribute information of items, while ignoring other useful external information, such as knowledge graph about the items and context information of user behaviors. In the future, we will study how to extend the proposed model SNR to make use of these external information to further improve the performance of sequential recommendation.

Acknowledgments. This work was supported by Peking University Education Big Data Project (Grant No. 2020YBC10).

References

1. Zhang, T., Zhao, P., Liu, Y., et al.: Feature-level deeper self-attention network for sequential recommendation. In: IJCAI International Joint Conference on Artificial Intelligence, pp. 4320–4326. AAAI Press (2019)

2. Bai, T., Nie, J.Y., Zhao, W.X., et al.: An attribute-aware neural attentive model for next basket recommendation. In: The 41st International ACM SIGIR Conference on Research & Development in Information Retrieval, pp. 1201–1204 (2018)
3. Liu, K., Shi, X., Natarajan, P.: Sequential heterogeneous attribute embedding for item recommendation. In: 2017 IEEE International Conference on Data Mining Workshops (ICDMW), pp. 773–780. IEEE (2017)
4. Zhu, N., Cao, J., Liu, Y., et al.: Sequential modeling of hierarchical user intention and preference for next-item recommendation. In: Proceedings of the 13th International Conference on Web Search and Data Mining, pp. 807–815 (2020)
5. Zhou, K., Wang, H., Zhao, W.X., et al.: S3-rec: self-supervised learning for sequential recommendation with mutual information maximization. In: Proceedings of the 29th ACM International Conference on Information & Knowledge Management, pp. 1893–1902 (2020)
6. Jang, E., Gu, S., Poole, B.: Categorical reparameterization with gumbel-softmax. arXiv preprint arXiv:1611.01144 (2016)
7. Rendle, S., Freudenthaler, C., Schmidt-Thieme, L.: Factorizing personalized Markov chains for next-basket recommendation. In: Proceedings of the 19th International Conference on World Wide Web, pp. 811–820 (2010)
8. Wang, P., Guo, J., Lan, Y., et al.: Learning hierarchical representation model for nextbasket recommendation. In: Proceedings of the 38th International ACM SIGIR Conference on Research and Development in Information Retrieval, pp. 403–412 (2015)
9. Xu, C., Zhao, P., Liu, Y., et al.: Recurrent convolutional neural network for sequential recommendation. In: The World Wide Web Conference, pp. 3398–3404 (2019)
10. Jannach, D., Ludewig, M.: When recurrent neural networks meet the neighborhood for session-based recommendation. In: Proceedings of the Eleventh ACM Conference on Recommender Systems, pp. 306–310 (2017)
11. Quadrana, M., Karatzoglou, A., Hidasi, B., et al.: Personalizing session-based recommendations with hierarchical recurrent neural networks. In: Proceedings of the Eleventh ACM Conference on Recommender Systems, pp. 130–137 (2017)
12. Hidasi, B., Karatzoglou, A., Baltrunas, L., et al.: Session-based recommendations with recurrent neural networks. arXiv preprint arXiv:1511.06939 (2015)
13. Wu, S., Tang, Y., Zhu, Y., et al.: Session-based recommendation with graph neural networks. In: Proceedings of the AAAI Conference on Artificial Intelligence, vol. 33, no. 01, pp. 346–353 (2019)
14. Huang, X., Qian, S., Fang, Q., et al.: CSAN: contextual self-attention network for user sequential recommendation. In: Proceedings of the 26th ACM International Conference on Multimedia, pp. 447–455 (2018)
15. Kipf, T.N., Welling, M.: Semi-supervised classification with graph convolutional networks. arXiv preprint arXiv:1609.02907 (2016)
16. Jin, B., Gao, C., He, X., et al.: Multi-behavior recommendation with graph convolutional networks. In: Proceedings of the 43rd International ACM SIGIR Conference on Research and Development in Information Retrieval, pp. 659–668 (2020)
17. Ying, R., He, R., Chen, K., et al.: Graph convolutional neural networks for web-scale recommender systems. In: Proceedings of the 24th ACM SIGKDD International Conference on Knowledge Discovery & Data Mining, pp. 974–983 (2018)
18. Bahdanau, D., Cho, K.H., Bengio, Y.: Neural machine translation by jointly learning to align and translate. In: Proceedings of the 3rd International Conference on Learning Representations, ICLR 2015 (2015)

19. Cui, Y., Chen, Z., Wei, S., et al.: Attention-over-attention neural networks for reading comprehension. In: Proceedings of the 55th Annual Meeting of the Association for Computational Linguistics (vol. 1: Long Papers), pp. 593–602 (2017)
20. Wang, H., Liu, G., Zhao, Y., et al.: DMFP: a dynamic multi-faceted fine-grained preference model for recommendation. In: 2019 IEEE International Conference on Data Mining (ICDM), pp. 608–617. IEEE (2019)
21. Chen, X., Xu, H., Zhang, Y., et al.: Sequential recommendation with user memory networks. In: Proceedings of the Eleventh ACM International Conference on Web Search and Data Mining, pp. 108–116 (2018)
22. Li, Q., Han, Z., Wu, X.M.: Deeper insights into graph convolutional networks for semi-supervised learning. In: Proceedings of the AAAI Conference on Artificial Intelligence, vol. 32, no. 1 (2018)
23. Hochreiter, S., Schmidhuber, J.: Long short-term memory. Neural Comput. $9(8)$, 1735–1780 (1997)
24. He, X., Deng, K., Wang, X., et al.: LightGCN: simplifying and powering graph convolution network for recommendation. In: Proceedings of the 43rd International ACM SIGIR Conference on Research and Development in Information Retrieval, pp. 639–648 (2020)
25. Kingma, D.P., Adam, J.B.: A Method for Stochastic Optimization. arXiv e-prints, arXiv preprint arXiv:1412.6980 (2014)
26. Kang, W.C., McAuley, J.: Self-attentive sequential recommendation. In: 2018 IEEE International Conference on Data Mining (ICDM), pp. 197–206. IEEE (2018)
27. Zhou, K., Zha, H.: Learning binary codes for collaborative filtering. In: Proceedings of the 18th ACM SIGKDD International Conference on Knowledge Discovery and Data Mining, pp. 498–506 (2012)
28. Rendle, S.: Factorization machines. In: 2010 IEEE International Conference on Data Mining (ICDM), pp. 995–1000. IEEE (2010)
29. Wang, Z., Liu, H., Du, Y., et al.: Unified embedding model over heterogeneous information network for personalized recommendation. In: IJCAI International Joint Conference on Artificial Intelligence. AAAI Press (2019)
30. Rendle, S., Freudenthaler, C., Gantner, Z., et al.: BPR: Bayesian personalized ranking from implicit feedback. In: Proceedings of the Twenty-Fifth Conference on Uncertainty in Artificial Intelligence, pp. 452–461 (2009)
31. Tang, J., Wang, K.: Personalized top-n sequential recommendation via convolutional sequence embedding. In: Proceedings of the Eleventh ACM International Conference on Web Search and Data Mining, pp. 565–573 (2018)
32. Du, Y., Liu, H., Wu, Z., et al.: Hierarchical hybrid feature model for top-N context-aware recommendation. In: 2018 IEEE International Conference on Data Mining (ICDM), pp. 109–116. IEEE (2018)
33. Vaswani, A., Shazeer, N., Parmar, N., et al.: Attention is all you need. In: Proceedings of the 31st International Conference on Neural Information Processing Systems, pp. 6000–6010 (2017)

PATHATTACK: Attacking Shortest Paths in Complex Networks

Benjamin A. Miller[1]([✉]) [iD], Zohair Shafi[1] [iD], Wheeler Ruml[2] [iD],
Yevgeniy Vorobeychik[3] [iD], Tina Eliassi-Rad[1] [iD], and Scott Alfeld[4] [iD]

[1] Northeastern University, Boston, MA 02115, USA
{miller.be,shafi.z,t.eliassirad}@northeastern.edu
[2] University of New Hampshire, Durham, NH 03824, USA
ruml@cs.unh.edu
[3] Washington University in St. Louis, St. Louis, MO 63130, USA
yvorobeychik@wustl.edu
[4] Amherst College, Amherst, MA 01002, USA
salfeld@amherst.edu

Abstract. Shortest paths in complex networks play key roles in many
applications. Examples include routing packets in a computer network,
routing traffic on a transportation network, and inferring semantic distances between concepts on the World Wide Web. An adversary with the
capability to perturb the graph might make the shortest path between
two nodes route traffic through advantageous portions of the graph (e.g.,
a toll road he owns). In this paper, we introduce the Force Path Cut problem, in which there is a specific route the adversary wants to promote by
removing a low-cost set of edges in the graph. We show that Force Path
Cut is NP-complete. It can be recast as an instance of the Weighted Set
Cover problem, enabling the use of approximation algorithms. The size
of the universe for the set cover problem is potentially factorial in the
number of nodes. To overcome this hurdle, we propose the PATHATTACK
algorithm, which via constraint generation considers only a small subset
of paths—at most 5% of the number of edges in 99% of our experiments.
Across a diverse set of synthetic and real networks, the linear programming formulation of Weighted Set Cover yields the optimal solution in
over 98% of cases. We also demonstrate running time vs. cost tradeoff
using two approximation algorithms and greedy baseline methods. This
work expands the area of adversarial graph mining beyond recent work
on node classification and embedding.

Keywords: Adversarial graph perturbation · Shortest path ·
Constraint generation

Electronic supplementary material The online version of this chapter (https://
doi.org/10.1007/978-3-030-86520-7_33) contains supplementary material, which is
available to authorized users.

N. Oliver et al. (Eds.): ECML PKDD 2021, LNAI 12976, pp. 532–547, 2021.
https://doi.org/10.1007/978-3-030-86520-7_33

1   Introduction

In a variety of applications, finding shortest paths among interconnected entities is an important task. Whether routing traffic on a road network, packets in a computer network, ships in a maritime network, or identifying the "degrees of separation" between two actors, locating the shortest path is often key to making efficient use of the interconnected entities. By manipulating the shortest path between two popular entities—e.g., people or locations—those along the altered path could have much to gain from the increased exposure. Countering such behavior is important, and understanding vulnerability to such manipulation is a step toward more robust graph mining.

In this paper, we present the *Force Path Cut* problem in which an adversary wants the shortest path between a source node and a target node in an edge-weighted network to go through a preferred path. The adversary has a fixed budget and achieves this goal by cutting edges, each of which has a cost for removal. We show that this problem is NP-complete via a reduction from the 3-Terminal Cut problem [5]. To solve Force Path Cut, we recast it as a Weighed Set Cover problem, which allows us to use well-established approximation algorithms to minimize the total edge removal cost. We propose the PATHATTACK algorithm, which combines these algorithms with a constraint generation method to efficiently identify paths to target for removal. While these algorithms only guarantee an approximately optimal solution in general, PATHATTACK yields the lowest-cost solution in a large majority of our experiments.

The main contributions of the paper are as follows: (1) We formally define Force Path Cut and show that it is NP complete. (2) We demonstrate that approximation algorithms for Weighted Set Cover can be leveraged to solve the Force Path Cut problem. (3) We identify an oracle to judiciously select paths to consider for removal, avoiding the combinatorial explosion inherent in naïvely enumerating all paths. (4) We propose the PATHATTACK algorithm, which integrates these elements into an attack strategy. (5) We summarize the results of over 20,000 experiments on synthetic and real networks, in which PATHATTACK identifies the optimal attack in over 98% of the time.

2   Problem Statement

We are given a graph $G = (V, E)$, where the vertex set V is a set of N entities and E is a set of M undirected edges representing the ability to move between the entities. In addition, we have nonnegative edge weights $w : E \to \mathbb{R}_{\geq 0}$ denoting the expense of traversing edges (e.g., distance or time).

We are also given two nodes $s, t \in V$. An adversary has the goal of routing traffic from s to t along a given path p^*. This adversary removes edges with full knowledge of G and w, and each edge has a cost $c : E \to \mathbb{R}_{\geq 0}$ of being removed. Given a budget b, the adversary's objective is to remove a set of edges $E' \subset E$ such that $\sum_{e \in E'} c(e) \leq b$ and p^* is the exclusive shortest path from s to t in the resulting graph $G' = (V, E \setminus E')$. We refer to this problem as *Force Path Cut*.

We show that this problem is computationally intractable in general by reducing from the 3-Terminal Cut problem, which is known to be NP-complete [5]. In 3-Terminal Cut, we are given a graph $G = (V, E)$ with weights w, a budget $b \geq 0$, and three terminal nodes $s_1, s_2, s_3 \in V$, and are asked whether a set of edges can be removed such that (1) the sum of the weights of the removed edges is at most b and (2) s_1, s_2, and s_3 are disconnected in the resulting graph (i.e., there is no path connecting any two terminals). Given that 3-Terminal Cut is NP-complete, we prove the following theorem.

Theorem 1. *Force Path Cut is NP-complete for undirected graphs.*

Here we provide an intuitive sketch of the proof; the formal proof is included in the supplementary material.

Proof Sketch. Suppose we want to solve 3-Terminal Cut for a graph $G = (V, E)$ with weights w, where the goal is to find $E' \subset E$ such that the terminals are disconnected in $G' = (V, E \setminus E')$ and $\sum_{e \in E'} w(e) \leq b$. We first consider the terminal nodes: If any pair of terminals shares an edge, that edge must be included in E' regardless of its weight; the terminals would not be disconnected if this edge remains. Note also that for 3-Terminal Cut, edge weights are edge removal costs; there is no consideration of weights as distances. If we add new edges between the terminals that are costly to both traverse and remove, then forcing one of these new edges to be the shortest path requires removing any other paths between the terminal nodes. This causes the nodes to be disconnected in the original graph. We will use a large weight for this purpose: $w_{\text{all}} = \sum_{e \in E} w(e)$, the sum of all weights in the original graph.

We reduce 3-Terminal Cut to Force Path Cut as follows. Create a new graph $\hat{G} = (V, \hat{E})$, where $\hat{E} = E \cup \{\{s_1, s_2\}, \{s_1, s_3\}, \{s_2, s_3\}\}$—i.e., $\hat{G}$ is the input graph with edges between the terminals added if they did not already exist. In addition, create new weights $\hat{w}$ where, for some $\epsilon > 0$, $\hat{w}(\{s_1, s_2\}) = \hat{w}(\{s_2, s_3\}) = w_{\text{all}} + 2\epsilon$ and $\hat{w}(\{s_1, s_3\}) = 2w_{\text{all}} + 3\epsilon$, and $\hat{w}(e) = w(e)$ for all other edges. Let the edge removal costs in the new graph be equal to the weights, i.e., $\hat{c}(e) = \hat{w}(e)$ for all $e \in \hat{E}$. Finally, let the target path consist only of the edge from s_1 to s_3, i.e., $s = s_1$, $t = s_3$, and $p^* = (s, t)$.

If we could solve Force Path Cut on $\hat{G}$ with weights $\hat{w}$ and costs $\hat{c}$, it would yield a solution to 3-Terminal Cut. We can assume the budget b is at most w_{all}, since this would allow the trivial solution of removing all edges and any additional budget would be unnecessary. If any edges exist between terminals in the original graph G, they must be included in the set of edges to remove, and their weights must be removed from the budget, yielding a new budget $\hat{b}$. Using this new budget for Force Path Cut, we will find a solution $\hat{E}' \subset \hat{E}$ if and only if there is a solution $E' \subset E$ for 3-Terminal Cut. A brief explanation of the reasoning is as follows:

– When we solve Force Path Cut, we are forcing an edge with a very large weight to be on the shortest path. If any path from s_1 to s_3 from the original graph remained, it would be shorter than (s_1, s_3). In addition, if any path from G

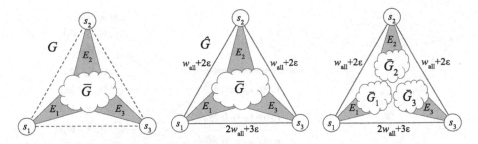

Fig. 1. Conversion from input to 3-Terminal Cut to Force Path Cut. The initial graph (left) includes 3 terminal nodes s_1, s_2, and s_3, which are connected to the rest of the graph by edges E_1, E_2, and E_3, respectively. The dashed lines indicate the possibility of edges between terminals. The input to Force Path Cut, $\hat{G}$ (center), includes the original graph plus high-weight, high-cost edges between terminals. A single edge comprising p^* is indicated in red. The result of Force Path Cut (right) is that any existing paths between the terminals have been removed, thus disconnecting them in the original graph and solving 3-Terminal Cut. (Color figure online)

between s_1 and s_2 remained, its length would be at most w_{all}, and thus a path from s_1 to s_3 that included s_2 would have length at most $2w_{\text{all}} + 2\epsilon$. This would mean (s_1, s_3) is not the shortest path between s_1 and s_3. A similar argument holds for paths between s_2 and s_3. Thus, no paths can remain between the terminals if we find a solution for Force Path Cut.

- If a solution exists for 3-Terminal Cut in G, it will yield the solution for Force Path Cut in $\hat{G}$. Any edge added to the graph to create $\hat{G}$ would be more costly to remove than removing all edges from the original G, so none will be removed. With all original paths between terminals removed, the only ones remaining from s_1 to s_3 are (s_1, s_3) and (s_1, s_2, s_3), the former of which is shortest, thus yielding a solution to Force Path Cut.

Figure 1 illustrates the aforementioned procedure. Note that the procedure would yield a solution to 3-Terminal Cut even if Force Path Cut allows for ties with p^*, so Force Path Cut is NP-complete in this case as well.   □

3   Proposed Method: PATHATTACK

While solving Force Path Cut is computationally intractable, we formulate the problem in a way that enables the use of established approximation algorithms.

3.1   Path Cutting as Set Cover

The success condition of Force Path Cut is that all paths from s to t aside from p^* must be strictly longer than p^*. This is an example of the (Weighted) Set Cover problem. In Weighted Set Cover, we are given a discrete universe $\mathcal{U}$ and a set of

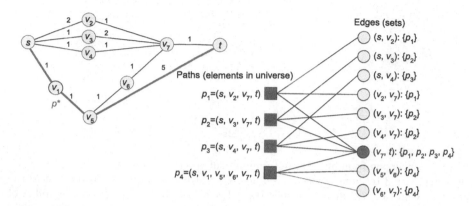

Fig. 2. The Force Path Cut problem is an example of the Weighted Set Cover problem. In the bipartite graph on the right, the square nodes represent paths and the circle nodes represent edges. Note that edges along p^* are not included. When the red-colored circle (i.e., edge (v_7, t)) is removed, then the red-colored squares (i.e., paths p_1, p_2, p_3, and p_4) are removed. (Color figure online)

subsets of the universe $\mathcal{S}$, $S \subset \mathcal{U}$ for all $S \in \mathcal{S}$, where each set has a cost $c(S)$. The goal is to choose those subsets whose aggregate cost is within a budget yet whose union equals the universe. In Force Path Cut, the elements of the universe to cover are the paths and the sets represent edges: each edge corresponds to a set containing all paths from s to t on which it lies. Including this set in the cover implies removing the edge, thus covering the elements (i.e., cutting the paths). Figure 2 shows how Force Path Cut is an example of Weighted Set Cover.

While Set Cover is NP-complete, there are known approximation algorithms to get a solution within a factor of $O(\log |\mathcal{U}|)$ of the optimal cost. The challenge in our case is that the universe may be extremely large. We address this challenge over the remainder of this section.

3.2   Linear Programming Formulation

In this section, we focus on minimizing cost without explicitly considering a budget. In practice, the adversary would run one of the optimization algorithms, compare budget and cost, and decide whether the attack is possible given resource constraints. Let $\mathbf{c} \in \mathbb{R}_{\geq 0}^M$ be a vector of edge costs, where each entry in the vector corresponds to an edge in the graph. We want to minimize the sum of the costs of edges that are cut, which is the dot product of $\mathbf{c}$ with a binary vector indicating which edges are cut, denoted by $\Delta \in \{0, 1\}^M$. This means that we optimize over values of Δ under constraints that (1) p^* is not cut and (2) all other paths from s to t not longer than p^* are cut. We represent paths in this formulation by binary indicator vectors—i.e., the vector $\mathbf{x}_p \in \{0, 1\}^M$ that represents path p is 1 at entries corresponding to edges in p and 0 elsewhere. Since any edge can only occur once, we only consider simple paths—those without cycles—which is sufficient for our purposes. If there is one index that is one in both Δ and $\mathbf{x}_p$, the path p is cut. Let P_p be the set of all paths in G from p's source to its destination that are no longer than p. The integer linear program formulation of Force Path Cut is as follows:

$$\hat{\Delta} = \arg\min_{\Delta} \mathbf{c}^{\mathsf{T}} \Delta \tag{1}$$

$$\text{s.t. } \Delta \in \{0,1\}^M \tag{2}$$

$$\mathbf{x}_p^{\mathsf{T}} \Delta \geq 1 \; \forall p \in P_{p^*} \setminus \{p^*\} \tag{3}$$

$$\mathbf{x}_{p^*}^{\mathsf{T}} \Delta = 0. \tag{4}$$

Constraint (3) ensures that any path not longer than (thus competing with) p^* will be cut, and constraint (4) forbids cutting p^*. As mentioned previously, P_{p^*} may be extremely large, which we address in Sect. 3.3.

The formulation (1)–(4) is analogous to the formulation of Set Cover as an integer program [19]. The goal is to minimize the cost of covering the universe— i.e., for each element $x \in \mathcal{U}$, at least one set $S \in \mathcal{S}$ where $x \in S$ is included. Letting δ_S be a binary indicator of the inclusion of subset S, the integer program formulation of Set Cover is

$$\hat{\hat{\delta}} = \arg\min_{\delta} \sum_{S \in \mathcal{S}} c(S)\delta_S \tag{5}$$

$$\text{s.t. } \delta_S \in \{0,1\} \; \forall S \in \mathcal{S} \tag{6}$$

$$\sum_{S \in \{S' \in \mathcal{S} | x \in S\}} \delta_S \geq 1 \; \forall x \in \mathcal{U}. \tag{7}$$

Equations (1), (2), and (3) are analogous to (5), (6), and (7), respectively. The constraint (4) can be incorporated by not allowing some edges to be cut, which manifests itself as removing some subsets from $\mathcal{S}$.

With Force Path Cut formulated as Set Cover, we consider two approximation algorithms. The first method, `GreedyPathCover`, iteratively adds the most cost-effective subset: that with the largest number of uncovered elements per cost. In Force Path Cut, this is equivalent to iteratively cutting the edge that removes the most paths per cost. The pseudocode is shown in Algorithm 1. We have a fixed set of paths $P \subset P_{p^*} \setminus \{p^*\}$. Note that this algorithm only uses costs, not weights: the paths of interest have already been determined and we only need to determine the cost of breaking them. `GreedyPathCover` performs a constant amount of work at each edge in each path in the initialization loop and the edge and path removal. We use lazy initialization to avoid initializing entries in the tables associated with edges that do not appear in any paths. Thus, populating the tables and removing paths takes time that is linear in the sum of the number of edges over all paths, which in the worst case is $O(|P|N)$. Finding the most cost-effective edge takes $O(M)$ time with a naïve implementation, and this portion is run at most once per path, leading to an overall running time of $O(|P|(N + M))$. Using a more sophisticated data structure, like a Fibonacci heap, to hold the number of paths for each edge would enable finding the most cost effective edge in constant time, but updating the counts when edges are removed would take $O(\log M)$ time, for an overall running time of $O(|P|N \log M)$. The worst-case approximation factor is the harmonic function of the size of the universe [19], i.e., $H_{|\mathcal{U}|} = \sum_{n=1}^{|\mathcal{U}|} 1/n$, which implies that the GreedyPathCover algorithm has a worst-case approximation factor of $H_{|P|}$. As we discuss in Sect. 3.4, this approximation factor extends to the overall Force Path Cut problem.

Input: Graph $G = (V, E)$, costs c, target path p^*, path set P
Output: Set E' of edges to cut
$T_P \leftarrow$ empty hash table; // set of paths for each edge
$T_E \leftarrow$ empty hash table; // set of edges for each path
$N_P \leftarrow$ empty hash table; // path count for each edge
foreach $e \in E$ **do**
   | $T_P[e] \leftarrow \emptyset$;
   | $N_P[e] \leftarrow 0$;
end
foreach $p \in P$ **do**
   | $T_E[p] \leftarrow \emptyset$;
   | **foreach** *edges e in p and not p^** **do**
      | $T_P[e] \leftarrow T_P[e] \cup \{p\}$;
      | $T_E[p] \leftarrow T_E[p] \cup \{e\}$;
      | $N_P[e] \leftarrow N_P[e] + 1$;
   | **end**
end
$E' \leftarrow \emptyset$;
while $\max_{e \in E} N_P[e] > 0$ **do**
   | $e' \leftarrow \arg\max_{e \in E} N_P[e]/c(e)$; // find most cost-effective edge
   | $E' \leftarrow E' \cup \{e'\}$;
   | **foreach** $p \in T_P[e']$ **do**
      | **foreach** $e_1 \in T_E[p]$ **do**
         | $N_P[e_1] \leftarrow N_P[e_1] - 1$; // decrement path count
         | $T_P[e_1] \leftarrow T_P[e_1] \setminus \{p\}$; // remove path
      | **end**
      | $T_E[p] \leftarrow \emptyset$; // clear edges
   | **end**
end
return E'

Algorithm 1: GreedyPathCover

The second approximation algorithm we consider involves relaxing the integer constraint into the reals and rounding the resulting solution. We refer to this algorithm as LP-PathCover. In this case, we replace (2) with the condition $\Delta \in [0,1]^M$ and get a $\hat{\Delta}$ that may contain non-integer entries. Following the procedure in [19], we apply randomized rounding as follows for each edge e:

1. Treat the corresponding entry $\hat{\Delta}_e$ as a probability.
2. Draw $\lceil \ln(4|P|) \rceil$ independent Bernoulli random variables w/ probability $\hat{\Delta}_e$.
3. Cut e if and only if at least one random variable from step 2 is 1.

If the result either does not cut all paths or is too large—i.e., greater than $4 \ln(4|P|)$ times the fractional (relaxed) cost—the procedure is repeated. These conditions are both satisfied with probability greater than $1/2$, so the expected number of attempts to get a valid solution is less than 2. By construction, the approximation factor is $4 \ln(4|P|)$ in the worst case. The running time is dominated by running the linear program; the remainder of the algorithm is (with high probability) linear in the number of edges and logarithmic in the number of constraints $|P|$. Algorithm 2 provides the pseudocode for LP-PathCover.

Input: Graph $G = (V, E)$, costs $\mathbf{c}$, path p^*, path set P
Output: Binary vector Δ denoting edges to cut
$\hat{\Delta} \leftarrow$ relaxed cut solution to (1)–(3) with paths P;
$\Delta \leftarrow \mathbf{0}$;
$E' \leftarrow \emptyset$;
not_cut$\leftarrow$ **True**;
while $\mathbf{c}^{\mathsf{T}} \Delta > \mathbf{c}^{\mathsf{T}} \hat{\Delta}(4 \ln(4|P|))$ **or** not_cut **do**
  $\quad E' \leftarrow \emptyset$;
  $\quad$**for** $i \leftarrow 1$ *to* $\lceil \ln(4|P|) \rceil$ **do**
    $\quad\quad$// randomly select edges based on $\hat{\Delta}$
    $\quad\quad E_1 \leftarrow \{e \in E$ with probability $\hat{\Delta}_e\}$;
    $\quad\quad E' \leftarrow E' \cup E_1$;
  $\quad$**end**
  $\quad \Delta \leftarrow$ indicator vector for E';
  $\quad$not_cut$\leftarrow (\exists p \in P$ where p has no edge in $E')$;
end
return Δ

Algorithm 2: LP-PathCover

3.3 Constraint Generation

In general, it is intractable to include every path from s to t. Take the example of an N-vertex clique (a.k.a. complete graph) in which all edges have weight 1 except the edge from s to t, which has weight N, and let $p^* = (s, t)$. Since all simple paths other than p^* are shorter than N, all of those paths will be included as constraints in (3), including $(N - 2)!$ paths of length $N - 1$. If we only explicitly include constraints corresponding to the two- and three-hop paths (a total of $(N - 2)^2 + (N - 2)$ paths), then the optimal solution will be the same as if we had included all constraints: cut the $N - 2$ edges around either s or t that do not directly link s and t. Optimizing using only necessary constraints is the other technique we use to make an approximation of Force Path Cut tractable.

Constraint generation is a technique for automatically building a relatively small set of constraints when the total number is extremely large or infinite [2,12]. The method requires an oracle that, given a proposed solution, returns a constraint that is being violated. This constraint is then explicitly incorporated into the optimization, which is run again and a new solution is proposed. This procedure is repeated until the optimization returns a feasible point or determines there is no feasible region.

Given a proposed solution to Force Path Cut—obtained by either approximation algorithm from Sect. 3.2—we have an oracle to identify unsatisfied constraints in polynomial time. We find the shortest path p in $G' = (V, E \setminus E')$ aside from p^*. If p is not longer than p^*, then cutting p is added as a constraint. We combine this constraint generation oracle with the approximation algorithms to create our proposed method PATHATTACK.

3.4   PATHATTACK

Combining the above techniques, we propose the PATHATTACK algorithm, which enables flexible computation of attacks to manipulate shortest paths. Starting with an empty set of path constraints, PATHATTACK alternates between finding edges to cut and determining whether removal of these edges results in p^* being the shortest path from s to t. Algorithm 3 provides PATHATTACK's pseudocode. Depending on time or budget considerations, an adversary can vary the underlying approximation algorithm.

Input: Graph $G = (V, E)$, cost function c, weights w, target path p^*, flag l
Output: Set E' of edges to cut
$E' \leftarrow \emptyset$;
$P \leftarrow \emptyset$;
$\mathbf{c} \leftarrow$ vector from costs $c(e)$ for $e \in E$;
$G' \leftarrow (V, E \setminus E')$;
$s, t \leftarrow$ source and destination nodes of p^*;
$p \leftarrow$ shortest path from s to t in G' (not including p^*);
while *p is not longer than p^** **do**
  $\quad P \leftarrow P \cup \{p\}$;
  $\quad$**if** *l* **then**
    $\quad\quad \Delta \leftarrow$ LP-PathCover$(G, \mathbf{c}, p^*, P)$;
    $\quad\quad E' \leftarrow$ edges from Δ;
  $\quad$**end**
  $\quad$**else**
    $\quad\quad E' \leftarrow$ GreedyPathCover(G, c, p^*, P);
  $\quad$**end**
  $\quad G' \leftarrow (V, E \setminus E')$;
  $\quad p \leftarrow$ shortest path from s to t in G' (not including p^*) using weights w;
end
return E'

Algorithm 3: PATHATTACK

While the approximation factor for Set Cover is a function of the size of the universe (all paths that need to be cut), this is not the fundamental factor in the approximation in our case. The approximation factor for PATHATTACK-Greedy is based only on the paths we consider explicitly. Using only a subset of constraints, the optimal solution could potentially be lower-cost than when using all constraints. By the final iteration of PATHATTACK, however, we have a solution to Force Path Cut that is within $H_{|P|}$ of the optimum of the less constrained problem, using $|P|$ from the final iteration. This yields the following proposition:

Proposition 2. *The approximation factor of PATHATTACK-Greedy is at most $H_{|P|}$ times the optimal solution to Force Path Cut.*

A similar argument holds for PATHATTACK-LP, applying the results of [19]:

Proposition 3. *PATHATTACK-LP yields a worst-case $O(\log |P|)$ approximation to Force Path Cut with high probability.*

4   Experiments

This section presents baselines, datasets, experimental setup, and results.

4.1   Baseline Methods

We consider two simple greedy methods as baselines for assessing performance. Each of these algorithms iteratively computes the shortest path p between s and t; if p is not longer than p^*, it uses some criterion to cut an edge from p. When we cut the edge with minimum cost, we refer to the algorithm as GreedyCost. We also consider a version where we cut the edge in p with the largest ratio of eigenscore[1] to cost, since edges with high eigenscores are known to be important in network flow [18]. This version of the algorithm is called GreedyEigenscore. In both cases, edges from p^* are not allowed to be cut.

4.2   Synthetic and Real Networks

Our experiments are on synthetic and real networks. All networks are undirected.

For the synthetic networks, we run five different random graph models to generate 100 synthetic networks of each model. We pick parameters to yield networks with similar numbers of edges ($\approx$ 160K). We use 16,000-node Erdős–Rényi (ER) and Barabási–Albert (BA) graphs, 2^{14}-node stochastic Kronecker graphs, 285 × 285 lattices, and 565-node complete graphs.

We use seven weighted and unweighted networks. The unweighted networks are Wikispeedia graph (WIKI) [21], Oregon autonomous system network

---

[1] The eigenscore of an edge is the product of the entries in the principal eigenvector of the adjacency matrix corresponding to the edge's vertices.

(AS) [10], and Pennsylvania road network (PA-ROAD) [11]. The weighted networks are Central Chilean Power Grid (GRID) [9], Lawrence Berkeley National Laboratory network data (LBL), the Northeast US Road Network (NEUS), and the DBLP coauthorship graph (DBLP) [3]. The networks range from 444 edges on 347 nodes to over 8.3M edges on over 1.8M nodes, with average degree ranging from over 2.5 to over 46.5 nodes and number of triangles ranging from 40 to close to 27M. Further details on the real and synthetic networks—including URLs to the real data—are provided in the supplementary material.

For synthetic networks and unweighted real networks, we try three different edge-weight initialization schemes: Poisson, uniform random, or equal weights. For Poisson weights, each edge e has an independently random weight $w_e = 1 + w'_e$, where w'_e is drawn from a Poisson distribution with rate parameter 20. For uniform weights, each weight is drawn from a discrete uniform distribution of integers from 1 to 41. This yields the same average weight as Poisson weights.

4.3   Experimental Setup

For each graph—considering graphs with different edge-weighting schemes as distinct—we run 100 experiments unless otherwise noted. For each graph, we select s and t uniformly at random among all nodes, with the exception of LAT, PA-ROAD, and NEUS, where we select s uniformly at random and select t at random among nodes 50 hops away from s^2. Given s and t, we identify the shortest simple paths and use the 100th, 200th, 400th, and 800th shortest as p^* in four experiments. For the large grid-like networks (LAT, PA-ROAD, and NEUS), this procedure is run using only the 60-hop neighborhood of s. We focus on the case where the edge removal cost is equal to the weight (distance).

The experiments were run on Linux machines with 32 cores and 192 GB of memory. The LP in PATHATTACK-LP was implemented using Gurobi 9.1.1, and shortest paths were computed using shortest_simple_paths in NetworkX.^{3}

4.4   Results

Across over 20,000 experiments, PATHATTACK-LP finds the optimal solution (where the relaxed LP yields only integers) in over 98% of cases. In addition, the number of constraints used by PATHATTACK is typically a small fraction of the number of edges (M): at most 5% of M in 99% of our experiments. For brevity, we highlight a few results in this section. See the supplementary material for more results on each network and weighting scheme.

We treat the result of GreedyCost as our baseline cost and report the cost of other algorithms' solutions as a reduction from the baseline. With one exception^4, GreedyCost outperforms GreedyEigenscore in both running time and

---

^{2} This alternative method of selecting the destination was used due to the computational expense of identifying successive shortest paths in large grid-like networks.

^{3} Gurobi is at https://www.gurobi.com. NetworkX is at https://networkx.org. Code from the experiments is at https://github.com/bamille1/PATHATTACK.

^{4} GreedyEigenscore only outperforms GreedyCost in COMP with uniform weights.

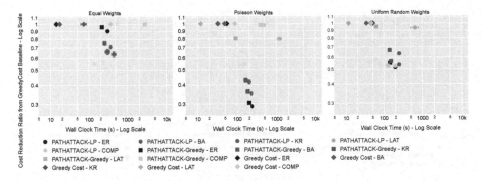

Fig. 3. Results on synthetic networks. Shapes represent different algorithms and colors represent different networks. The horizontal axis represents wall clock time in seconds and the vertical axis represents edge removal cost as a proportion of the cost required by the GreedyCost baseline. Lower cost reduction ratio and lower wall clock time is better. PATHATTACK yields a substantial cost reduction for weighted ER, BA, and KR graphs, while the baseline achieves nearly optimal performance for LAT.

edge removal cost, so we omit the GreedyEigenscore results for clarity of presentation. Figure 3 shows the results on synthetic networks, Fig. 4 shows the results on real networks with synthetic edge weights, and Fig. 5 shows the results on real weighted networks. In these figures, the 800th shortest path is used as p^*; other results were similar and omitted for brevity.

Comparing the cost achieved by PATHATTACK to those obtained by the greedy baseline, we observe some interesting phenomena. Across the synthetic networks in Fig. 3, the real graphs with synthetic weights in Fig. 4, and the graphs with real weights in Fig. 5, lattices and road networks have a similar tradeoff: PATHATTACK provides a mild improvement in cost at the expense of an order of magnitude additional processing time. Considering that PATHATTACK-LP typically results in the optimal solution, this means that the baselines are achieving near-optimal cost with a naïve algorithm. On the other hand, ER, BA, and KR graphs follow a trend more similar to the AS and WIKI networks, particularly in the randomly weighted cases: The cost is cut by a substantial fraction—enabling the attack with a smaller budget—for a similar or smaller time increase. This suggests that the time/cost tradeoff is much less favorable for less clustered, grid-like networks.

Cliques (COMP, yellow in Fig. 3) are particularly interesting in this case, showing a phase transition as the entropy of the weights increases. When edge weights are equal, cliques behave like an extreme version of the road networks: an order of magnitude increase in run time with no decrease in cost. With Poisson weights, PATHATTACK yields a slight improvement in cost, whereas when uniform random weights are used, the clique behaves much more like an ER or BA graph. In the unweighted case, p^* is a three-hop path, so all other two- and three-hop paths from s to t must be cut, which the baseline does efficiently. Adding Poisson weights creates some randomness, but most edges have a weight that is about average, so it is still similar to the unweighted scenario. With uniform random

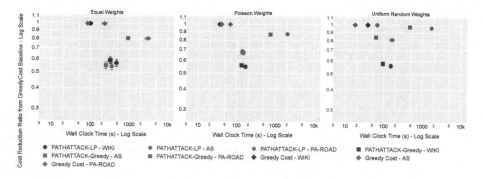

Fig. 4. Results on unweighted real networks. Shapes represent different algorithms and colors represent different networks. The horizontal axis represents wall clock time in seconds and the vertical axis represents edge removal cost as a proportion of the cost required by the `GreedyCost` baseline. Lower cost reduction ratio and lower wall clock time is better. As with synthetic networks, `PATHATTACK` significantly reduces cost in networks other than those that are grid-like, where the baseline is nearly optimal.

weights, we get the potential for much different behavior (e.g., short paths with many edges) for which the greedy baseline's performance suffers.

There is an opposite, but milder, phenomenon with PA-ROAD and LAT: using higher-entropy weights *narrows* the cost difference between the baseline and `PATHATTACK`. This may be due to the source and destination being many hops away. With the terminal nodes many hops apart, many shortest paths between them could go through a few low-weight (thus low-cost) edges. A very low weight edge between two nodes would be very likely to occur on many of the shortest paths, and would be found in an early iteration of the greedy algorithm and removed, while considering more shortest paths at once would yield a similar result. We also note that, in the weighted graph data, LBL and GRID behave similarly to road networks. Among our real datasets, these have a low clustering coefficient (see supplementary material). This lack of overlap in nodes' neighborhoods may lead to better relative performance with the baseline, since there may not be a great deal of overlap between candidate paths.

5   Related Work

Early work on attacking networks focused on disconnecting them [1]. This work demonstrated that targeted removal of high-degree nodes was highly effective against networks with powerlaw degree distributions (e.g., BA networks), but far less so against random networks. This is due to the prevalence of hubs in networks with such degree distributions. Other work has focused on disrupting shortest paths via edge removal, but in a narrower context than ours. Work on the most vital edge problem (e.g., [13]) attempts to efficiently find the single edge whose removal most increases the distance between two nodes. In contrast, we consider a devious adversary that wishes a certain path to be shortest.

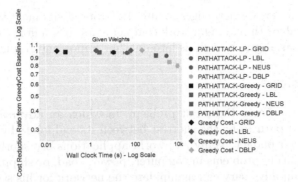

Fig. 5. Results on weighted real networks. Shapes represent different algorithms and colors represent different networks. The horizontal axis represents wall clock time in seconds and the vertical axis represents edge removal cost as a proportion of the cost required by the GreedyCost baseline. Lower cost reduction ratio and lower wall clock time is better. PATHATTACK reduces the cost of attacking the DBLP social network, while the other networks (those with low clustering) achieve high performance with the baselines. Note: the range of the time axis is lower than that of the previous plots.

There are several other adversarial contexts in which path-finding is highly relevant. Some work is focused on traversing hostile territory, such as surreptitiously planning the path of an unmanned aerial vehicle [7]. The complement of this is work on network interdiction, where the goal is to intercept an adversary who is attempting to traverse the graph while remaining hidden. This problem has been studied in a game theoretic context for many years [20], and has expanded into work on disrupting attacks, with the graph representing an attack plan [12]. In this work, as in ours, oracles can be used to avoid enumerating an exponentially large number of possible strategies [6].

Work on Stackelberg planning [17] is also relevant, though somewhat distinct from our problem. This work adopts a leader-follower paradigm, where rather than forcing the follower to make a specific set of actions, the leader's goal is to make whatever action the follower takes as costly as possible. This could be placed in our context by having the leader (adversary) attempt to make the follower take the longest path possible between the source and the destination, though finding this path would be NP-hard in general.

Another related area is the common use of heuristics, such as using Euclidean distances to approximate graph distances [16]. Exploiting deviations in the heuristic enables an adversary to manipulate automated plans. Fuzzy matching has been used to quickly solve large-scale problems [15]. Attacks and defenses in this context is an interesting area for inquiry. A problem similar to Stackelberg planning is the adversarial stochastic shortest path problem, where the goal is to maximize reward while traversing over a highly uncertain state space [14].

There has recently been a great deal of work on attacking machine learning methods where graphs are part of the input. Attacks against vertex classification [22,23] and node embeddings [4] consider attackers that can manipulate edges, node attributes, or both in order to affect the outcome of the learning method. In addition, attacks against community detection have been proposed where a node can create new edges to alter its group assignment from a community detection algorithm [8]. Our work complements these efforts, expanding the space of adversarial graph analysis into another important graph mining task.

6   Conclusions

We introduce the Force Path Cut problem, in which an adversary's aim is to force a specified path to be the shortest between its endpoints by cutting edges within a required budget. Many real-world applications use shortest-path algorithms (e.g., routing problems in computer, power, road, or shipping networks). We show that an adversary can manipulate the network for his strategic advantage. While Force Path Cut is NP-complete, we show how it can be translated into Weighted Set Cover, thus enabling the use of established approximation algorithms to optimize cost within a logarithmic factor of the true optimum. With this insight, we propose the PATHATTACK algorithm, which uses a natural oracle to generate only those constraints needed to execute the approximation algorithms. Across various synthetic and real networks, we find that the PATHATTACK-LP variant identifies the optimal solution in over 98% of more than 20,000 randomized experiments. Another variant, PATHATTACK-Greedy, has very similar performance and typically runs faster than PATHATTACK-LP, while a greedy baseline method is faster still but with much higher cost.

Ethical Implications: This work demonstrates how an adversary can attack shortest paths in complex networks. Appropriate defenses include building resilient network structures (e.g., adding redundancy to form cliques around key communication channels) and developing methods that not only detect attacks, but also identify the most likely source of the attack (e.g., whether an edge failed due to a random outage or a malicious destruction).

Acknowledgments. BAM was supported by the United States Air Force under Contract No. FA8702-15-D-0001. TER was supported in part by the Combat Capabilities Development Command Army Research Laboratory (under Cooperative Agreement No. W911NF-13-2-0045) and by the Under Secretary of Defense for Research and Engineering under Air Force Contract No. FA8702-15-D-0001. YV was supported by grants from the Army Research Office (W911NF1810208, W911NF1910241) and National Science Foundation (CAREER Award IIS-1905558). Any opinions, findings, conclusions or recommendations expressed in this material are those of the authors and should not be interpreted as representing the official policies, either expressed or implied, of the funding agencies or the U.S. Government. The U.S. Government is authorized to reproduce and distribute reprints for Government purposes not withstanding any copyright notation here on.

References

1. Albert, R., Jeong, H., Barabási, A.L.: Error and attack tolerance of complex networks. Nature **406**(6794), 378–382 (2000)
2. Ben-Ameur, W., Neto, J.: A constraint generation algorithm for large scale linear programs using multiple-points separation. Math. Program. **107**(3), 517–537 (2006)

3. Benson, A.R., Abebe, R., Schaub, M.T., Jadbabaie, A., Kleinberg, J.: Simplicial closure and higher-order link prediction. Proc. Nat. Acad. Sci. **115**(48), E11221–E11230 (2018)
4. Bojchevski, A., Günnemann, S.: Adversarial attacks on node embeddings via graph poisoning. In: ICML, pp. 695–704 (2019)
5. Dahlhaus, E., Johnson, D.S., Papadimitriou, C.H., Seymour, P.D., Yannakakis, M.: The complexity of multiterminal cuts. SIAM J. Comput. **23**(4), 864–894 (1994)
6. Jain, M., et al.: A double oracle algorithm for zero-sum security games on graphs. In: AAMAS, pp. 327–334 (2011)
7. Jun, M., D'Andrea, R.: Path planning for unmanned aerial vehicles in uncertain and adversarial environments. In: Butenko, S., Murphey, R., Pardalos, P.M. (eds.) Cooperative Control: Models, Applications and Algorithms, pp. 95–110. Springer, Boston (2003). https://doi.org/10.1007/978-1-4757-3758-5_6
8. Kegelmeyer, W.P., Wendt, J.D., Pinar, A.: An example of counter-adversarial community detection analysis. Technical report, SAND2018-12068, Sandia National Laboratories (2018). https://doi.org/10.2172/1481570
9. Kim, H., Olave-Rojas, D., Álvarez-Miranda, E., Son, S.W.: In-depth data on the network structure and hourly activity of the central Chilean power grid. Sci. Data **5**(1), 1–10 (2018)
10. Leskovec, J., Kleinberg, J., Faloutsos, C.: Graphs over time: densification laws, shrinking diameters and possible explanations. In: KDD, pp. 177–187 (2005)
11. Leskovec, J., Lang, K.J., Dasgupta, A., Mahoney, M.W.: Community structure in large networks: natural cluster sizes and the absence of large well-defined clusters. Internet Math. **6**(1), 29–123 (2009)
12. Letchford, J., Vorobeychik, Y.: Optimal interdiction of attack plans. In: AAMAS, pp. 199–206 (2013)
13. Nardelli, E., Proietti, G., Widmayer, P.: Finding the most vital node of a shortest path. Theoret. Comput. Sci. **296**(1), 167–177 (2003)
14. Neu, G., Gyorgy, A., Szepesvari, C.: The adversarial stochastic shortest path problem with unknown transition probabilities. In: AISTATS, pp. 805–813 (2012)
15. Qiao, M., Cheng, H., Yu, J.X.: Querying shortest path distance with bounded errors in large graphs. In: Bayard Cushing, J., French, J., Bowers, S. (eds.) SSDBM 2011. LNCS, vol. 6809, pp. 255–273. Springer, Heidelberg (2011). https://doi.org/10.1007/978-3-642-22351-8_16
16. Rayner, D.C., Bowling, M., Sturtevant, N.: Euclidean heuristic optimization. In: AAAI (2011)
17. Speicher, P., Steinmetz, M., Backes, M., Hoffmann, J., Künnemann, R.: Stackelberg planning: towards effective leader-follower state space search. In: AAAI, pp. 6286–6293 (2018)
18. Tong, H., Prakash, B.A., Eliassi-Rad, T., Faloutsos, M., Faloutsos, C.: Gelling, and melting, large graphs by edge manipulation. In: CIKM, pp. 245–254 (2012)
19. Vazirani, V.V.: Approximation Algorithms. Springer, Heidelberg (2003). https://doi.org/10.1007/978-3-662-04565-7
20. Washburn, A., Wood, K.: Two-person zero-sum games for network interdiction. Oper. Res. **43**(2), 243–251 (1995)
21. West, R., Pineau, J., Precup, D.: Wikispeedia: an online game for inferring semantic distances between concepts. In: IJCAI (2009)
22. Zügner, D., Akbarnejad, A., Günnemann, S.: Adversarial attacks on neural networks for graph data. In: KDD, pp. 2847–2856 (2018)
23. Zügner, D., Günnemann, S.: Certifiable robustness and robust training for graph convolutional networks. In: KDD, pp. 246–256 (2019)

Embedding Knowledge Graphs Attentive to Positional and Centrality Qualities

Afshin Sadeghi[1,2]([✉])(iD), Diego Collarana[2,4](iD), Damien Graux[3](iD),
and Jens Lehmann[1,2](iD)

[1] Smart Data Analytics Group, University of Bonn, Bonn, Germany
[2] Fraunhofer IAIS, Sankt Augustin, Germany
{afshin.sadeghi,diego.collarana.vargas,jens.lehmann}@iais.fraunhofer.de
[3] Inria, Université Côte d'Azur, CNRS, I3S, Nice, France
damien.graux@inria.fr
[4] Universidad Privada Boliviana, Cochabamba, Bolivia

Abstract. Knowledge graphs embeddings (KGE) are lately at the center of many artificial intelligence studies due to their applicability for solving downstream tasks, including link prediction and node classification. However, most Knowledge Graph embedding models encode, into the vector space, only the local graph structure of an entity, i.e., information of the 1-hop neighborhood. Capturing not only local graph structure but global features of entities are crucial for prediction tasks on Knowledge Graphs. This work proposes a novel KGE method named Graph Feature Attentive Neural Network (GFA-NN) that computes graphical features of entities. As a consequence, the resulting embeddings are attentive to two types of global network features. First, nodes' relative centrality is based on the observation that some of the entities are more "prominent" than the others. Second, the relative position of entities in the graph. GFA-NN computes several centrality values per entity, generates a random set of reference nodes' entities, and computes a given entity's shortest path to each entity in the reference set. It then learns this information through optimization of objectives specified on each of these features. We investigate GFA-NN on several link prediction benchmarks in the inductive and transductive setting and show that GFA-NN achieves on-par or better results than state-of-the-art KGE solutions.

1 Introduction

Knowledge graphs (KGs) are capable of integrating heterogeneous data sources under the same graph data model. Thus KGs are at the center of many artificial intelligence studies. KG nodes represent concepts (entities), and labeled edges represent the relation between these entities[1]. KGs such as Wikidata, WordNet, Freebase, and Nell include millions of entities and relations representing the current knowledge about the world. KGs in combination with Machine Learning models are used for refining the Knowledge Graph itself and for downstream tasks, like

---

[1] E.g. (Berlin, CapitalOf, Germany) is a fact stating Berlin is the capital of Germany.

© Springer Nature Switzerland AG 2021
N. Oliver et al. (Eds.): ECML PKDD 2021, LNAI 12976, pp. 548–564, 2021.
https://doi.org/10.1007/978-3-030-86520-7_34

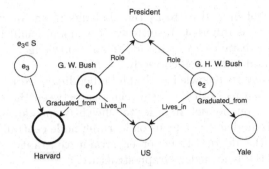

Fig. 1. Example Knowledge Graph in which nodes e_1 and e_2 are difficult to distinguish by a KGE model only using their neighborhood information.

link prediction and node classification. However, to use KGs in Machine Learning methods, we need to transform graph representation into a vector space presentation, named Knowledge Graph embeddings (KGE).

KGE have many applications including analysis of social networks and biological pathways. Thus, many approaches have been proposed ranging from translation methods, e.g., Trans* family [3,13,29]; Rotation-based methods, e.g., RotatE [20]; Graph Convolutional methods, e.g., R-GCN [19], COMPGCN [25], and TransGCN [4]; and Walk-based methods, e.g., RDF2Vec [16]. Traditional graph embedding methods, however, rely exclusively on facts (triples) that are explicitly present in a Knowledge Graph. Therefore, their prediction ability is limited to a set of incomplete facts. A means of improvement is to incorporate complementary information in the embeddings. A class of methods applies external knowledge such as entity text descriptions [30] and text associations related to entities [26] into the KG modeling. In contrast, intrinsic methods extract complementary knowledge from the same KG. For example, the algorithms that derive logical rules from a KG and combine them with embeddings of the KG [6,28]. Analogously recent studies [35] consider graph structural features as an intrinsic aspect of KGs in the embedding.

We motivate our model by addressing a challenge of most KGE models; These methods independently learn the existence of relation from an entity to its hop-1 neighborhood. This learning strategy neglects the fact that entities located at a distance can still affect an entity's role in the graph. Besides that, the location of the entities in the network can be useful to distinguish nodes. Figure 1 illustrates such an example where the goal is to learn embeddings for e_1 and e_2 entities in the KG. Distinguishing between the two candidates, i.e., George W. Bush and George H. W. Bush, is challenging for previous methods since e_1 and e_2 have almost the same neighbors, except George W. Bush graduated from Harvard University while George H. W. Bush did not.

However, If we compare e_1 to e_2 using their eigenvector centrality, we can easily distinguish them. e_1 has a greater centrality than e_2 since e_1 is connected to Harvard that has a high eigenvector centrality. Analogously, if we consider the shortest path of e_1 and e_2 to e_3 that belongs to set of reference node S, their distance to e_3 is different. Intuitively, if a model could beforehand know the centrality and distance to e_3 as additional knowledge, it can more easily model e_1 and e_2 and rank them correctly.

With a new view to Knowledge Graph embeddings, we propose GFA-NN[2], an approach that learns both the local relations between the entities and their global properties in one model. In order to efficiently encode entity indicators in Knowledge Graph modeling, we focus on learning node centrality and positional indicators, (e.g., the degree, Katz, or eigenvalue centrality of entities in the graph) as well as the Knowledge Graph structure.

For this purpose, we fuse the modeling of each entity indicator in the style of Multiple Distance Embedding (MDE) [17] where distinct views to Knowledge Graphs are modeled through independent embedding weights.

GFA-NN extracts positional information and four centrality indicators of nodes from the KG and defines a learning function for each one. Then GFA-NN scores their aggregation with MDE.

Previously, different leanings were applied to embedding models using constraints in the loss function. Now that MDE has broken the limitation of using more than one objective function on independent embeddings, we directly add new extracted information about the entities as aggregated objective functions.

Centrality values and position of nodes in graphs are global measurements for nodes across the whole graph. If we use a local assignment, for example the number of paths between specific nodes, this measurement may have different wights based on what portion of the network is considered in the calculation.

Despite the exciting recent advancements, most of the previous works fail to learn the relation between entities regarding the whole graph. Therefore, we define relative position attentive and relative centrality attentive functions for embedding the relative importance of nodes and their position relative to the whole network. In the following section, we discuss the relation between our work and the current state-of-the-art. Later in Sect. 3, we introduce the preliminaries and notations required to explain our chosen method. We outline in Sect. 4 the idea of centrality and positional qualities learning and explain our approach. In Sect. 5, we mention the model's theoretical analysis; and we continue with experiments that evaluate our model in Sect. 6.

2   Related Work

A large and growing body of literature has investigated KGE models. A typical KGE model consists of three main elements: (1) entities and relations representation in a continuous vector space, (2) a scoring function to measure KG's facts plausibility, and (3) a loss function that allows learning KGE in a supervised manner. Based on this formulation, we classify KGE models in: latent distance approaches, tensor factorization and multiplicative models, and neural networks.

---

[2] Source code is available at: https://github.com/afshinsadeghi/GFA-NN.

Latent Distance Models, e.g., Trans* [3,13,29] family, measure a fact's plausibility by scoring the distance between the two entities, usually after a translation carried out by the relation. RotatE [20] combines translation and rotation. RotatE models relations as rotations from head to tail entities in the complex space and uses the Hadamard product in the score function to do these rotations.

Tensor factorization and multiplicative approaches define the score of triples via pairwise multiplication of embeddings. DistMult [33], for example, multiplies the embedding vectors of a triple element by element (h, r, t) as the objective function. However, DistMult fails to distinguish displacement of head relation and tail entities, and therefore, it cannot model anti-symmetric relations. ComplEx [23] solves DistMult's issue.

Unlike previous methods, the *neural network-based methods* learn KGE by connecting artificial neurons in different layers. Graph Neural Network (GNN) aggregate node formation using a message-passing architecture. Recently, hybrid neural networks such as CompGCN [24] and MDE_{nn} [17] have raised. These methods benefit from neural network architectures to model relations with (anti)symmetry, inversion, and composition patterns.

Several studies have investigated the benefits of using graph features to bridge the graph structure gap and the numeric vector space. Muzzamil et al. [14] defined a Fuzzy Multilevel Graph Embedding (FMGE), an embedding of attributed graphs with many numeric values. P-GNN [35] incorporates positional information by sampling anchor nodes and calculating their distance to a given node (see Sect. 5.1 for an in-depth comparison with GFA-NN). Finally, it learns a non-linear distance weighted aggregation scheme over the anchor nodes.

This effort's main difference with previous approaches is in the message passing mechanism. Traditionally in GNNs, approaches learn just nodes' local features (similar to the modeling schema of KGEs) while focusing on neighbor nodes; here, our approach also learns nodes' features regarding the whole graph, known as global graph properties.

3 Background and Notation

A Knowledge graph KG, is comprised of a set of entities $e \in \mathcal{E}$ and a set of relations $r \in \mathcal{R}$. A fact in a Knowledge Graph is a triple of the form (h, r, t) in which h (head) and t (tail) are entities and r is a relation. A KG is a subset of all true facts KG $\subset \xi$. A KG can be conceived as a multi-relational graph. An entity in such formulation is equivalent to a node in graph theory, and an edge represents a relation. In this study, we use Node and Entity interchangeably. We use the term "Node" to emphasize its graphical properties. We use the term "Entity" to highlight the entity's concept.

Link prediction on Knowledge Graphs is made by a Siamese classifier that embeds KG's entities and relations into a low-dimensional space. Thus, a Knowledge Graph embedding model is a function $f : \mathcal{E}, \mathcal{R} \rightarrow \mathcal{Z}$, that maps entities $\mathcal{E}$ and relations $\mathcal{R}$ to d-dimensional vectors $\mathcal{Z} = \{z_1, \ldots, z_n\}$, $z_i \in \mathbb{R}$.

Centrality value of a node designates the importance of the node with regard to the whole graph. For instance, *degree* is a centrality attribute of a node that indicates the number of links incident upon it. When we consider degree as centrality value, the higher the degree of a node is, the greater is its importance in a graph. We provide a generalization of the position-aware embedding definition [35] that distinguishes our method from the previous works.

Structure-Based Embedding: A KG embedding $z_i = f : \mathcal{E}, \mathcal{R} \to \mathcal{Z}$ is attentive to network structure if it is a function of entities and relations such that it models the existence of a neighborhood of an entity e_i using relations r_i and other entities $e_j \in \mathcal{E}$. Most Knowledge Graph embedding methods like QuatE and RotatE compute embeddings using the information describing connections between entities and, therefore, structure-based.

Property-Attentive Embedding: A KG embedding $z_i = f : \mathcal{E}, \mathcal{R} \to \mathcal{Z}$ is attentive to network properties of an entity if there exists a function $g_p(., ., ...)$ such that $d_p(v_i, v_j, ...) = g_p(z_i, z_j)$, where $d_p(,)$ is a graphical property in G. This definition includes both the property of a sole node such as its centrality and the properties that describe the inter-relation of two nodes such as their shortest path. Examples of Property-Attentive Embedding are P-GNNs and RDF2Vec, which their objective function incorporates the shortest path between nodes into embedding computation.

We show that current KGE methods cannot recover global graph properties, such as path distances between entities and centrality of nodes, limiting the performance in tasks where such information is beneficial. Principally, structure-aware embeddings cannot be mapped to property-aware embeddings. Therefore, only using structure-aware embeddings as input is not sufficient when the learning task requires node property information. This work focuses on learning KGEs capturing both entities' local network structures conjointly with the global network properties. We validate our hypothesis that a trait between local and global network features is crucial for link prediction and node classification tasks. A KGE is attentive to node network properties if the embedding of two entities and their relation can be used to approximately estimate their network feature, e.g., their degree relative to other entities in the network.

You *et al.* [35] show for position attentive networks, there exists a mapping g that maps structure-based embeddings $f_{st}(v_i)$, $\forall v_i \in V$ to position attentive embeddings $f_p(v_i)$, $\forall v_i \in V$, if and only if no pair of nodes have isomorphic local q-hop neighborhood graphs. This proposition justifies the good performance of KGE models in tasks requiring graphical properties and their under-performance in real-world graphs such as biological and omniscience KGs (e.g., Freebase, DBpedia), in which the structure of local neighborhoods are quite common. This proposition, however, does not hold for centrality attentive embeddings. The reason is that if no pair of nodes have isomorphic local q-hop neighborhood graphs, it is still possible for them to have the same centrally attentive embeddings. For example, two nodes with the same number of neighbors consisting of different nodes have the same degree; however, their neighborhoods are non-isometric. We show in Sect. 4 how we address this challenge for centrality learning.

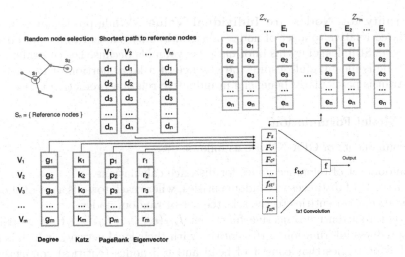

Fig. 2. Architecture of GFA-NN. GFA-NN first pre-computes the centrality property of nodes and their distance to a set of to randomly selected reference nodes (**Left**). Then, node centrality and position embeddings attentive to position z_{v_m} are computed via scores $F_1, ..., F_k$ from the distance between a given node v_i and the reference-sets S_i which are shared across all the entities (**Top-middle**). To compute the embedding z_{v_1} for node v_1, a score of GFA-NN first computes via function F_i and then aggregates the F_i scores via 1×1 convolution and an activation function over obtains a vector of final scores. Inside 1×1 a vector w learned, which is used to reduce scores into one centrality and position-aware score and produces embeddings z_{v_1} which is the output of the GFA-NN (**Right**).

4   Method

This Section details our proposed method for generating entity network properties attentive embeddings from Knowledge Graphs. We generalize the concept of Knowledge Graph embeddings with a primary insight that incorporating centrality and distance values enables KGE models to compute embeddings with respect to the graphical proprieties of entities relative to the whole network instead of only considering the direct local neighbors (Fig. 2, left side).

When modeling the positional information, instead of letting each entity model the information independently and selecting a new reference set per iteration, we keep a set of reference entities through training iterations and across all the networks in order to create comparable embeddings. This design choice enables the model to learn the position of nodes with respect to the spectrum of different reference node positions and makes each embedding attentive to position (Fig. 2, top left). GFA-NN models each graphical feature with a dedicated objective function, meaning that the information encrypted in centrality attentive embeddings does not interfere with the embedding vectors that keep the positional information (Fig. 2, top right).

Centrality for Nodes are Individual Values. While positional values are calculated relative to a set of nodes in a graph, only one centrality per entity is extracted. Still, learning this information is valuable because the centrality value of a node is meaningful despite the absence of a large portion of the network. This trait is particularly beneficial in inductive relation prediction tasks.

4.1   Model Formulation

The components of GFA-NN are as follows:

- Random set of reference nodes for distance calculations.
- Matrix M of distances to random entities, where each row i is a set of shortest distance of an entities to the selected set of random nodes.
- Structure-attentive objective functions $f_{st^1}(v_i), \ldots, f_{st^k}(v_i)$ that model the relatedness information of two entities with their local network, which is indicated by triples that consist of head and tail nodes (entities) connected by an edge (relation).
- Position-attentive objective function F_s that models the position of a node (entity) in the graph with respect to its distance to other nodes. This objective considers these distances as a factor of relatedness of entities.
- Centrality attentive objective functions F_c that model the relatedness information of two entities according to centrality properties of nodes (entities). In this setting, the global importances of nodes are learned relatively to the centrality of other nodes.
- Trainable aggregation function $f_{1\times 1}$ is a 1×1 convolution [12] that fuses the modeling of the structure-based connectivity information of the entities and relations with their position aware and centrality attentive scoring.
- Trainable vectors r_d, h_d, t_d that project distance matrix M to a lower dimensional embedding space $z \in \mathcal{R}_k$.

Our approach consists of several centrality and position-attentive phases that each of which learns an indicator in a different metric of the status for entities relative to the network.

In the first phase, GFA-NN performs two types of computation to determine the position status and the centrality status of entities. The unit for centrality status computes the relative significance of entities as a vector of length one c_i^j, where j represents each of the centrality metrics. The unit for position status embedding samples n random reference-entities S_n, and computes an embedding for entities. Each dimension i of the embedding is obtained by a function F that computes the shortest path to the i-th reference entity relative to the maximum shortest path in the network.

Then objective functions $F_s, F_c^1, \ldots, F_c^4$ apply an entity interaction model to enforce the property features e_i^s into entity embeddings e_i, which in the next phase makes a 1×1 convolution [12] over the scores via weights $w \in \mathbb{R}^r$ and non-linear transformation $Tanhshrink$.

Specifically, each entity earns an embedding per attribute that includes values that reveal the relative status information from input entity network properties information. Calculation of the centrality for all nodes in the network leads to a vector representation of the graph for each measure, while the distances to the reference nodes S generate a dense matrix representation.

The network property attentive modeling functions are the same class of functions as used by existing translational KGEs plus a modeling function of embeddings that we extended to be performed in 3D using rotation matrix. In the following, we further elaborate on the design choices.

4.2 Centrality-Attentive Embedding

As shown in Sect. 3, the centrality values are not canonical. Therefore, the model learns their difference in a normal form, in which the equality of their norm does not mean they are equal. Degree centrality is defined as: $C_d(n) = \deg(n)$.

Katz centrality [8] extends degree centrality from counting neighbor nodes to nodes that can be connected through a path, where the contribution of distant nodes are reduced:

$$C_k(n) = \sum_{k=1}^{\infty} \sum_{j=1}^{N} \alpha^k A_{j,i}^k$$

where A is the adjacency matrix and α is attenuation factor in the range $(0, 1)$. Another included centrality measure is PageRank with the following formulation:

$$C_p(n) = \alpha \sum_j a_{j,i} \frac{C_p(j)}{L(j)} + \frac{1-\alpha}{N}$$

where N is $|V|$, the number of nodes in the graph, and $L(j)$ is the degree of node j. Relative eingenvector centrality score of a node n is defined as:

$$C_e i(n) = \frac{1}{\lambda} \sum_{m \in KG} a_{m,n} x_m$$

where $A = (a_{v,t})$ is the adjacency matrix such that $a_{v,t} = 1$ if node n is linked to node m, and $a_{v,t} = 0$ otherwise. λ is a constant which fulfils the eingenvector formulation $Ax = \lambda x$. Note that the method in first phase normalizes each of the centrality values. The normalization occurs with respect to minimum and the maximum value for nodes in the network and makes attributes relative to the whole network. For example, degree centrality is normalized as follows:

$$C_i^d = \frac{degree(i) - degree_{min}}{degree_{max} - degree_{min}}$$

The centrality-attentive modeling embeddings functions are the same class of dissimilarity functions used by existing KGEs plus a penalty we define on the difference of the entity embeddings as:

$$F_{c^d} = \|h_i - t_i\|_2 - \| \cos(log(C_h^d)) - \cos(log(C_t^d))\|_2 \tag{1}$$

where the function is normalized with the l_2 norm, h_i and t_i represent the vector representation of head and tail in a triple and lastly, C_h^d and C_t^d respectively denote the centrality values of the head and tail entities in that triple.

4.3 Position-Attentive Embedding:

GFA-NN models the neighborhood structure using rotations in 3D space and a penalty that forces the method to encode the difference of distances of entities to the reference nodes. The formulation for the structure-attentive part is:

$$F_{rot} = \parallel v_h - v_r \otimes v_t \parallel_2 \tag{2}$$

where $\otimes$ represents a rotation using a rotation matrix of Euler angles with the formulation of direction cosine matrix (DCM):

$$\begin{bmatrix} \cos\theta\cos\psi & -\cos\phi\sin\psi + \sin\phi\sin\theta\cos\psi & \sin\phi\sin\psi + \cos\phi\sin\theta\cos\psi \\ \cos\theta\sin\psi & \cos\phi\cos\psi + \sin\phi\sin\theta\sin\psi & -\sin\phi\cos\psi + \cos\phi\sin\theta\sin\phi \\ -\sin\theta & \sin\phi\cos\theta & \cos\phi\cos\theta \end{bmatrix} \tag{3}$$

where ϕ, θ and ψ are Euler angles. The modeling of positional information is performed by a score function made from rotation matrices and a penalty:

$$F_p = F_{rot} - \parallel \cos(S_i^h) - \cos(S_i^t)) \parallel_2 \tag{4}$$

where S_C^i is the calculated distance from the head and tail nodes to the reference nodes. Hence, the score enforces to learn structure-attentive embeddings with a penalty that is the normalized scalar difference of distance to reference nodes. Here we use the l_2 norm to regularize the F_i score functions and apply negative adversarial sampling [20]. We utilise Adam [9] for optimization.

Reference-set selection relies on a Gaussian random number generator to select normally distributed random reference nodes from the network. GFA-NN keeps a fixed set of reference nodes during the training of different entities through different iterations to generate embeddings attentive to the position that are in the same space and, hence, comparable to each other.

Multiple Property aware scores can be naturally fused to achieve higher expressive power. This happens in $f_{1\times1}$.

Since canonical position-attentive embeddings do not exist, GFA-NN also computes structure-attentive embeddings h_v via the common distance-based modelings of MDE. These scores are aggregated with attribute attentive scores, and then the model using a linear combination of these scores forms a 1×1 convolution to produce only one value that contains both properties. The output of this layer is then fed into the nonlinear activation function.

It is notable that independent weights in MDE formulation allow restricting solution space without limiting the learnability power of the model. Note also that the method is still Semi-supervised learning, where the train and test data are disjoint, and the centrality and path information computation do not consider the portion of the unknown network to the model and only exist in the test data.

5 Theoretical Analysis

5.1 Connection to Preceding KGE Methods

GFA-NN generalizes the existing Knowledge Graph embedding models. Taking the definition for the structure-aware and node properties attentive models into perspective, existing knowledge embedding models use the same information of connecting entities through different relations techniques, but use different neighborhood selection scoring function and sampling strategies, and they only output the structure-aware embeddings.

GFA-NN shares the score function aggregate training with MDE [17]. There, a linear combination of scores $f_{1\times1} = \sum w_i F_i$ is trained, where w_i weights are learnt together with the embeddings in the score functions F_i. GFA-NN also shares the concept of training independent embeddings with MDE. The direction cosine matrix used in modeling positional information is convertible into a four-element unit quaternion vector (q_0, q_1, q_2, q_3). The quaternions are the center of the structure-based model QuatE [36], where the relations are models as rotations in the quaternion space. Here, besides modeling rotation, we formulated the score to include a translation as well. RotatE [20] similarly, formulates the relations with a rotation and reduction in $\| v_h \circ v_r - v_t \|$, however RotatE models rotation in the complex space. In the branch of Graph neural networks, the aggregate information of a node's neighborhood in one-hop [10,25,27] or nodes in the higher hops [32] is used in message passing mechanism.

P-GNN [35] explicitly learns the shortest path of random nodes for simple graphs. However, it takes a new set of reference nodes in each iteration, which makes the learning of shortest paths local and incremental. In addition, it makes it difficult to retain the structural information from positional embedding. GFA-NN generalizes positional learning by learning the distances to a fixed set of random nodes through the whole network, which makes the positional embedding vectors globally comparable. From the point of view of graph type, GFA-NN generalizes the positional learning to multi-relational graphs to support KGs.

GFA-NN not only learns a weight for each of the network features, but it also associates it with the existing relation types between the two entities that their features are being learned. By including the relation type into position-attentive embeddings, the position also is encoded into relation vectors that connect the entities. Note that relation type learning is sub-optimal for learning centrality values because the dimension of relation types is much more higher than dimension of the node property values (one integer value), which makes the centrality value differentiation diminish when learnt together with the association information belonging to relations. Another aspect that GFA-NN generalize the existing graph learning algorithms is that this method learns several centrality aspect and positional information at the same time.

5.2 Expressive Power

In this Section we explain how GFA-NN generalizes the expressive power of Knowledge Graph embedding methods in the perspective of a broader Inductive bias. Generally, inductive bias in a learning algorithm allows it to better prioritize one solution over another, independent of the observed data [2].

Assuming that a labeling function y labels a triple (h, r, t) as $d_y^r(h, t)$, we predict y^r, similar to [35] from the prospective of representation learning, which is by learning an embedding function f, where $v_h = f(v, G)$ and f computes the entity embeddings for v_h, v_r and v_t. Thus, the objective becomes the task of maximizing the probability of the conditional distribution $p(y|v_h, v_r, v_t)$. This probability can be designated by a distance function $d_v(v_h, v_r, v_t)$ in the embedding space, which usually is an l_p norm of the objective function of the model.

A KGE model, with a goal to predict the existence of an unseen triple (h, r, t) learns embeddings weights v_h and v_t for the entities h and t and v_r for a relation r that lies between them. In this formulation, the embedding for an entity e is computed based on its connection through its one-hop neighborhood, which we express that by structural information S_e, and optimization over the objective function $f_\theta(e, S_e)$. Hereby, the neighborhood information of two entities S_{e_1} and S_{e_2} is computed independently. However, the network feature attentive objective function f_ϕ in GFA-NN poses a more general inductive bias that takes in the distance from a random shared set of reference-nodes, which are common across all entities, and the centrality values, which are relative to all nodes. In this setting, any pair of entity embeddings are correlated through the reference-set and the spectrum of relative centrality and therefore are not independent anymore. We call this feature attentive information I.

Accordingly, we define a joint distribution $p(w_{e_1}, w_{e_2})$ over node embeddings, where $w_{e_i} = f_\phi(e_i, I)$. We formalize the problem of KG representation learning by minimizing the expected value of the likelihood of the objective function in margin-based ranking setting, in the following for a structure base KGE:

$$\min_\theta \mathbb{E}_{e_1, e_2, e_3, S_{e_1}, S_{e_2}, S_{e_3}}$$
$$\mathcal{L}(d_v^+(f_\theta(e_1, S_{e_1}), f_\theta(e_2, S_{e_2})) - d_v^-(f_\theta(e_1, S_{e_1}), f_\theta(e_3, S_{e_3})) - m) \quad (5)$$

and in GFA-NN:

$$\min_\theta \mathbb{E}_{e_1, e_2, e_3, I} \quad \mathcal{L}(d_v^+(f_\phi(e_1, I), f_\phi(e_2, I)) - d_v^-(f_\phi(e_1, I), f_\phi(e_3, I)) - m) \quad (6)$$

where d_v^+ is the similarity metric determined by the objective function for a positive triple, indicating existing a predicate between entities and by optimizing converges to the target label function $d_y(e_1, e_2) = 0$ for positive samples(existing triples) and $d_y(e_1, e_3) = m$ on negative samples. Here, m is the margin value in the margin ranking loss optimization setting. Note that the representations of entities are calculated using joint and marginal distributions, respectively.

Similar to the proof of expressive power in [35], considering the selection of entities $e_1, ..., e_i \in G$ as random variables to form any triples, the mutual information between the joint distribution of entity embeddings and any $Y = d_y(e_1, e_2)$ is greater than that between the marginal distributions. $Y : I(Y; X_{joint}) \geq I(Y; X_{marginal})$. Where,
$X_{joint} = (f_\phi(e_1, S_{e_1}), f_\phi(e_2, S_{e_2})) \sim p(f_\phi(e_1, S_{e_1}), f_\phi(e_2, S_{e_2}))$
$X_{marginal} = (f_\theta(e_1, I), f_\theta(e_2, I))$

Because the gap of mutual information is large when the targeted task is related to positional and centrality information of the network, we deduce that KGE embedding based on the joint distribution of distances to reference nodes and relative centrality values have more expressive power than the current structure-based KGE models.

Table 1. Statistics of the data sets used in the experiments.

Dataset	#entities	#relations	#train	#validation	#test
WN18RR	40943	11	86835	3034	3134
FB15k-237	14541	237	272115	17535	20466
ogbl-biokg	45085	51	4762678	162886	162870
WN18RR-v_3-ind	5084	11	6327	538	605
WN18RR-v_4-ind	7084	9	12334	1394	1429
NELL-995-v_1-ind	225	14	833	101	100
NELL-995-v_4-ind	2795	61	7073	716	731

5.3 Complexity Analysis

Next, we explain the complexity of the method and show its complexity compared to the structure-based models. When the shortest paths are calculated on the fly, the learning complexity is added up by $O(b\ log(b))$ for finding the shortest paths on b entities in each batch, and similarly, the centrality computation aggregates to the complexity. We, therefore, pre-calculate this information to separate them from the learning complexity. The complexity of each of the objective functions on a batch with size b is $O(b)$, and suppose n property attentive features and m structure-aware scores be involved, the overall complexity becomes $O((n+m)\,b)$. Note that the larger number here is b and the complexity increases by b times when a graphical feature is involved in the learning.

6 Experiments

We evaluate the performance of our model with two link prediction experiments; First, the traditional transductive ranking evaluation, which is originally introduced in [3], and second, inductive relation prediction experiment. In the inductive setting, the experiment evaluates a models' ability to generalize the link prediction task to unseen entities. Table 1 shows the statistics of the datasets used in the experiments.

Metrics and Implementation: We evaluate the link prediction performance by ranking the score of each test triple against all possible derivable negative samples by once replacing its head with all entities and once by replacing its tail. We then calculate the hit at N (Hit@N), mean rank (MR), and mean reciprocal rank (MRR) of these rankings. We report the evaluations in the filtered setting. We determine the hyper-parameters by using grid search. We select the testing models which give the best results on the validation set. In general, we fix the learning rate on 0.0005 and search the embedding size amongst {200, 300, 400, 500}. We search the batch size from {250, 300, 500, 800, 1000}, and the number of negative samples amongst {10, 100, 200, 400, 600, 800, 1000}.

Table 2. Results on WN18RR and FB15k-237. Best results are in bold.

Model	WN18RR			FB15k-237		
	MR	MRR	Hit@10	MR	MRR	Hit@10
ComplEx-N3	–	0.48	0.57	–	0.37	0.56
QuatE^2	–	0.482	0.572	–	**0.366**	**0.556**
TuckER	–	0.470	0.526	–	0.358	0.544
CompGCN	3533	0.479	0.546	197	0.355	0.535
RotatE	3340	0.476	0.571	177	0.338	0.533
MDE	3219	0.458	0.536	203	0.344	0.531
GFA-NN	3390	**0.486**	**0.575**	186	0.338	0.522

Table 3. MRR Results for ogbl-biokg. (Results of previous models are from [7].)

Method	Validation	Test
TRANSE	0.7456	0.7452
DISTMULT	0.8055	0.8043
COMPLEX	0.8105	0.8095
ROTATE	0.7997	0.7989
GFA-NN	**0.9011**	**0.9011**

6.1 Transductive Link Prediction Experiment

Datasets: We perform experiments on three benchmark datasets: WN18RR [5], FB15k-237 [22], and ogbl-biokg [7], which is comparably a sizeable Knowledge Graph assembled from a large number of biomedical repositories.

Baselines: We compare our model with several state-of-the-art structure-based embedding approaches. Our baselines include RotatE [20], TuckER [1], ComplEx-N3 [11], QuatE [36], MDE [17] and the recent graph neural network CompGCN [25]. We report results of each method on WN18RR and FB15k-237 from their respective papers, while the results of the other models in ogbl-biokg are from [7]. For RotatE, we report its best results with self-adversarial negative sampling, and for QuatE, we report the results with N3 regularization. For our model, we use the same self-adversarial negative sampling introduced in RotatE. This negative sampling schema is also applied to all the other models in the ogbl-biokg benchmark.

Results and Discussion: Table 2 and Table 3 summarize the performance of GFA-NN and other KGE models in the transductive link prediction task. We observe that GFA-NN outperforms other state-of-the-art KGEs on WN18RR and is producing competitive results on FB15k-237.

Our analysis shows that the standard deviation of different positional and centrality measures through the network in WN18RR is ≈0.009, while in FB15k-237, it is ≈0.002, which is 4.5 times smaller. This comparison indicates that in

Table 4. Hit@10 results for inductive datasets. (Other models' results are from [21].)

Model	WN18RR-v_3-ind	WN18RR-v_4-ind	NELL-995-v_1-ind	NELL-995-v_4-ind
NeuralLP	0.4618	0.6713	0.4078	**0.8058**
DRUM	0.4618	0.6713	0.5950	**0.8058**
RuleN	0.5339	0.7159	0.5950	0.6135
GraiL	0.5843	0.7341	0.5950	0.7319
GFA-NN	**0.5893**	**0.7355**	**0.9500**	0.7722

WN18RR, these features are more diversified, but in FB15k237, they are close to each other. This analysis suggests the crucial impact of learning centrality and positional-attentive embeddings on the superiority of the GFA-NN on the WN18RR benchmark. While the result on the FB15k-237 is still very competitive to the state-of-the-art, as a lesson learned, we can declare it as a fixed procedure to perform the standard deviation analysis on a dataset before determining how much the network property attentive embedding learning method would be beneficial.

Table 3 shows the **MRR** evaluation results on the comparably large biological dataset named as `ogbl-biokg`. In this benchmark, the number of entity and training samples is much larger than the WN18rr and FB15k-237 datasets. The capability of learning feature attentive embeddings is crucial in this transductive link prediction task. While the best KGEs can only achieve the MRR of 0.8105 on the validation and 0.8095 on the test dataset, GFA-NN reaches 0.901 on both datasets, improving state-of-the-art by 9%. This wide gap between the results supports the assumption that property-attentive embeddings surpass prior methods in larger-scale real-world networks. This improvement in such a small-world structured network is because of its significant entity-to-relation ratio, which causes a large standard deviation of positional and centrality qualities. As indicated earlier, this feature is beneficial to the efficiency of the model.

6.2   Inductive Link Prediction Experiment

Datasets: For evaluations in the inductive setting, we select four variant datasets which Komal et al. [21] extracted from WN18RR and NELL-995 [31].

Baselines: Inductive baselines include GraIL [21], which uses sub-graph reasoning for inductive link prediction. RuleN [15] that applies a statistical rule mining method, and two differentiable methods of rule learning NeuralLP [34] and DRUM [18]. We report the results of these state-of-the-art models from Komal et al. [21].

Results: Table 4 summarizes the GFA-NN's Hit@10 ranking performance against methods specified on the inductive link prediction task. Although we did not explicitly design GFA-NN for this task, we observe GFA-NN performs very competitively in this setting and outperforms the best inductive learning models in most cases. This result supports our hypothesis that the Knowledge Graph embeddings attentive to positional and centrality qualities are beneficial for prediction tasks in challenging settings, i.e., inductive link prediction task.

7 Conclusion

In this article, with a new view to the relational learning algorithms, we propose to learn the structural information of the network conjointly with the learning of the centrality and positional properties of the Knowledge Graph entities in one model. We provide theoretical analyses and empirical evaluations to identify the improvements and constraints in the expressive power for this class of KGEs. In particular, we demonstrate that with proper formulation, the learning of these global features is beneficial to the link prediction task, given that GFA-NN performs highly efficiently in a variety of benchmarks and often outperforms current state-of-the-art solutions in both inductive and transductive settings. Since GFA-NN is efficient on networks with a higher entity-to-relation ratio, applications of the approach can be considered on biological, chemical, and social networks in future works.

Acknowledgments. First author thanks Firas Kassawat for related discussions. This study was supported by MLwin project grant 01IS18050F of the Federal Ministry of Education and Research of Germany, the EU H2020 Projects Opertus Mundi (GA 870228), and the Federal Ministry for Economic Affairs and Energy (BMWi) project SPEAKER (FKZ 01MK20011A).

References

1. Balazevic, I., Allen, C., Hospedales, T.: Tucker: tensor factorization for knowledge graph. In: EMNLP-IJCNLP, pp. 5185–5194 (2019)
2. Battaglia, P.W., et al.: Relational inductive biases, deep learning, and graph networks. Preprint arXiv:1806.01261 (2018)
3. Bordes, A., Usunier, N., Garcia-Duran, A., Weston, J., Yakhnenko, O.: Translating embeddings for modeling multi-relational data. In: NIPS, pp. 2787–2795 (2013)
4. Cai, L., Yan, B., Mai, G., Janowicz, K., Zhu, R.: TransGCN: coupling transformation assumptions with graph convolutional networks for link prediction. In: K-CAP, pp. 131–138. ACM (2019)
5. Dettmers, T., Minervini, P., Stenetorp, P., Riedel, S.: Convolutional 2D knowledge graph embeddings. In: AAAI, pp. 1811–1818 (2018)
6. Guo, S., Wang, Q., Wang, L., Wang, B., Guo, L.: Knowledge graph embedding with iterative guidance from soft rules. In: AAAI, pp. 4816–4823 (2018)
7. Hu, W., et al.: Open graph benchmark: datasets for machine learning on graphs. In: NeurIPS (2020)
8. Katz, L.: A new status index derived from sociometric analysis. Psychometrika **18**(1), 39–43 (1953)
9. Kingma, D.P., Ba, J.: Adam: a method for stochastic optimization. In: Bengio, Y., LeCun, Y. (eds.) ICLR (2015)

10. Kipf, T.N., Welling, M.: Semi-supervised classification with graph convolutional networks. In: ICLR (2017)
11. Lacroix, T., Usunier, N., Obozinski, G.: Canonical tensor decomposition for knowledge base completion. In: ICML, pp. 2863–2872 (2018)
12. Lin, M., Chen, Q., Yan, S.: Network in network. Preprint arXiv:1312.4400 (2013)
13. Lin, Y., Liu, Z., Sun, M., Liu, Y., Zhu, X.: Learning entity and relation embeddings for knowledge graph completion. In: AAAI, pp. 2181–2187 (2015)
14. Luqman, M.M., Ramel, J.Y., Lladós, J., Brouard, T.: Fuzzy multilevel graph embedding. Pattern Recognit. **46**(2), 551–565 (2013)
15. Meilicke, C., Fink, M., Wang, Y., Ruffinelli, D., Gemulla, R., Stuckenschmidt, H.: Fine-grained evaluation of rule- and embedding-based systems for knowledge graph completion. In: ISWC, pp. 3–20 (2018)
16. Ristoski, P., Paulheim, H.: RDF2Vec: RDF graph embeddings for data mining. In: ISWC, pp. 498–514 (2016)
17. Sadeghi, A., Graux, D., Shariat Yazdi, H., Lehmann, J.: MDE: multiple distance embeddings for link prediction in knowledge graphs. In: ECAI (2020)
18. Sadeghian, A., Armandpour, M., Ding, P., Wang, D.Z.: DRUM: end-to-end differentiable rule mining on knowledge graphs. In: NeurIPS, pp. 15321–15331 (2019)
19. Schlichtkrull, M.S., Kipf, T.N., Bloem, P., van den Berg, R., Titov, I., Welling, M.: Modeling relational data with graph convolutional networks. In: ESWC (2018)
20. Sun, Z., Deng, Z.H., Nie, J.Y., Tang, J.: RotatE: knowledge graph embedding by relational rotation in complex space. In: ICLR (2019)
21. Teru, K., Denis, E., Hamilton, W.: Inductive relation prediction by subgraph reasoning. In: ICML, pp. 9448–9457 (2020)
22. Toutanova, K., Chen, D.: Observed versus latent features for knowledge base and text inference. In: CVSC, pp. 57–66 (2015)
23. Trouillon, T., Welbl, J., Riedel, S., Gaussier, É., Bouchard, G.: Complex embeddings for simple link prediction. In: ICML, pp. 2071–2080 (2016)
24. Vashishth, S., Sanyal, S., Nitin, V., Talukdar, P.: Composition-based multi-relational graph convolutional networks. In: ICLR (2020)
25. Vashishth, S., Sanyal, S., Nitin, V., Talukdar, P.P.: Composition-based multi-relational graph convolutional networks. In: ICLR (2020)
26. Veira, N., Keng, B., Padmanabhan, K., Veneris, A.G.: Unsupervised embedding enhancements of knowledge graphs using textual associations. In: IJCAI (2019)
27. Velickovic, P., Cucurull, G., Casanova, A., Romero, A., Liò, P., Bengio, Y.: Graph attention networks. In: ICLR (2018)
28. Wang, W.Y., Cohen, W.W.: Learning first-order logic embeddings via matrix factorization. In: IJCAI, pp. 2132–2138 (2016)
29. Wang, Z., Zhang, J., Feng, J., Chen, Z.: Knowledge graph embedding by translating on hyperplanes. In: AAAI (2014)
30. Xie, R., Liu, Z., Jia, J., Luan, H., Sun, M.: Representation learning of knowledge graphs with entity descriptions. In: AAAI, pp. 2659–2665 (2016)
31. Xiong, W., Hoang, T., Wang, W.Y.: DeepPath: a reinforcement learning method for knowledge graph reasoning. In: EMNLP, pp. 564–573 (2017)
32. Xu, K., Li, C., Tian, Y., Sonobe, T., Kawarabayashi, K., Jegelka, S.: Representation learning on graphs with jumping knowledge networks. In: Dy, J.G., Krause, A. (eds.) ICML, pp. 5449–5458 (2018)
33. Yang, B., Yih, W., He, X., Gao, J., Deng, L.: Embedding entities and relations for learning and inference in knowledge bases. In: ICLR (2015)
34. Yang, F., Yang, Z., Cohen, W.W.: Differentiable learning of logical rules for knowledge base reasoning. In: NeurIPS, pp. 2319–2328 (2017)

35. You, J., Ying, R., Leskovec, J.: Position-aware graph neural networks. In: ICML, pp. 7134–7143 (2019)
36. Zhang, S., Tay, Y., Yao, L., Liu, Q.: Quaternion knowledge graph embeddings. In: NeurIPS, pp. 2731–2741 (2019)

Interpretation, Explainability, Transparency, Safety

Reconnaissance for Reinforcement Learning with Safety Constraints

Shin-ichi Maeda[1]([✉]), Hayato Watahiki[2], Yi Ouyang[3], Shintarou Okada[1], Masanori Koyama[1], and Prabhat Nagarajan[1]

[1] Preferred Networks, Inc., Tokyo, Japan
{ichi,okada,masomatics,prabhat}@preferred.jp
[2] The University of Tokyo, Tokyo, Japan
watahiki@logos.t.u-tokyo.ac.jp
[3] Preferred Networks America, Inc., Burlingame, CA, USA
ouyangyi@preferred-america.com

Abstract. As RL algorithms have grown more powerful and sophisticated, they show promise for several practical applications in the real world. However, safety is a necessary prerequisite to deploying RL systems in real world domains such as autonomous vehicles or cooperative robotics. Safe RL problems are often formulated as constrained Markov decision processes (CMDPs). In particular, solving CMDPs becomes challenging when safety must be ensured in rare, dangerous situations in stochastic environments. In this paper, we propose an approach for CMDPs where we have access to a generative model (e.g. a simulator) that can preferentially sample rare, dangerous events. In particular, our approach, termed the *RP algorithm* decomposes the CMDP into a pair of MDPs which we term a *reconnaissance* MDP (R-MDP) and a *planning* MDP (P-MDP). In the R-MDP, we leverage the generative model to preferentially sample rare, dangerous events and train a *threat function*, the Q-function analog of danger that can determine the safety level of a given state-action pair. In the P-MDP, we train a reward-seeking policy while using the trained threat function to ensure that the agent considers only safe actions. We show that our approach, termed the *RP algorithm* enjoys several useful theoretical properties. Moreover, we present an approximate version of the *RP algorithm* that can significantly reduce the difficulty of solving the R-MDP. We demonstrate the efficacy of our method over classical approaches in multiple tasks, including a collision-free navigation task with dynamic obstacles.

Keywords: Safe reinforcement learning · Constrained MDPs · Safety

1 Introduction

With recent advances in reinforcement learning (RL), we can train complex, reward-maximizing policies in increasingly complex environments. However, in general, it is difficult to assess whether the policies found by RL algorithms are physically safe when applied to real world scenarios such as autonomous driving

© Springer Nature Switzerland AG 2021
N. Oliver et al. (Eds.): ECML PKDD 2021, LNAI 12976, pp. 567–582, 2021.
https://doi.org/10.1007/978-3-030-86520-7_35

These actions are no more safe after updating the policy $\pi \to \pi'$

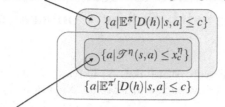

These actions are always safe irrespective of the policy π

Fig. 1. illustration of the idea: Once the baseline policy η and its threat function $\mathscr{T}^\eta(s,a) \equiv \mathbb{E}^\eta[D(h)|s,a]$ are found, we can find the safe actions without re-evaluating the expectation w.r.t the policy π after every update of π.

or cooperative robotics. Safety has long been one of the greatest challenges in the application of RL to mission-critical systems.

The safety problem in RL in often formulated as a constrained Markov decision process (CMDP). This setup assumes a Markovian system together with a predefined measure of danger. That is, using classical RL notation in which π represents an agent's decision-making policy, we optimize:

$$\max_\pi \mathbb{E}^\pi[R(h)] \quad \text{s.t.} \quad \mathbb{E}^\pi[D(h)] \le c, \tag{1}$$

where h is a trajectory of state-action pairs, $R(h)$ is the total return obtained by h, and $D(h)$ is the cumulative danger of trajectory h. To solve this problem, one must monitor the value of $\mathbb{E}^\pi[D(h)]$ throughout training. Methods like [1,3,13,14,20] use sampling to approximate $\mathbb{E}^\pi[D(h)]$ or its Lyapunov function at every update. However, the sample-based evaluation of $\mathbb{E}^\pi[D(h)]$ is particularly difficult when the system involves "rare" catastrophic events, because an immense number of samples will be required to collect information about the cause of such an accident.

This problem can be partially resolved if we can use a generative model to predict the outcome of any given initial state and sequence of actions. Model Predictive Control (MPC) [6,15,33,34] is a method that at each timestep uses a generative model to predict the outcome of some horizon of future actions (finite horizon control) in order to determine the agent's next action. However, MPC suffers from some practical drawbacks. In order to find a good, feasible solution, a long horizon optimization problem must be solved repeatedly at each timestep. Particularly, when the task complexity increases to nonlinear, stochastic dynamics, this quickly becomes intractable.

We propose a generative model-based approach to overcome the difficulty of sample-based evaluation methods without the drawbacks of MPC. In particular we search for a solution to the CMDP problem by decomposing the CMDP into a pair of MDPs: a reconnaissance MDP (R-MDP) and a planning MDP (P-MDP). The purpose of the R-MDP is to (1) *recon* the state space using the generative model and (2) train a baseline policy for the *threat function*, which is a Q-function analogue of D. We show that once we obtain the threat function for an appropriate baseline policy, we can construct the upper bound of the threat function for any policy and can construct the set of actions that are guaranteed to satisfy the safety constraints. To efficiently learn the threat function, our method requires access to a generative model that can preferentially sample dangerous states. This assumption is satisfied in many problem settings, including safety-critical problems of practical interest. For example, the CARLA simulator [16] is an autonomous car simulator that can artificially generate dangerous events such as pedestrians close to cars or adverse weather conditions. In the R-MDP, we use the generative model to preferentially sample trajectories containing rare dangerous events, and learn the threat function for the baseline policy through supervised learning. Once we obtain a good approximation of the threat function for the baseline policy, we can determine whether a given action is safe at each state or not by simply evaluating the threat function. This process does not involve prediction, which can be computationally demanding. The P-MDP is essentially the original MDP except that the agent can only select actions from the set of safe policies induced by the threat function. The P-MDP can be solved with standard RL methods. With our framework, the user need not monitor $\mathbb{E}^\pi[D]$ throughout the training process. We show that our approach enjoys several useful theoretical properties:

1. The learning process is guaranteed to be safe when training in the P-MDP stage if the baseline policy from the R-MDP stage is safe.
2. We can increase the size of the set of safe actions by improving the safety level of the baseline policy.
3. After solving one CMDP, we can solve other CMDPs with different reward functions and constraint thresholds by reusing the learned threat function, so long as the CMDPs share the same D.
4. In problem settings with multiple sources of danger, the threat function can be upper-bounded by the sum of sub-threat functions, corresponding to the threat for each source of danger.

Our experiments demonstrate the efficacy of our approach on multiple tasks, including a challenging dynamic-obstacle 2D navigation task. We show that the RP algorithm can successfully learn high-reward and safe policies in these tasks. Additionally, we show in these experiments how our RP algorithm exhibits properties 2 and 3 above.

The remainder of this paper is structured as follows. Section 2 presents the requisite background on RL and constrained MDPs. Section 3 presents basic definitions and theoretical results. Section 4 includes additional theory and introduces the RP algorithms. Section 5 shows our experimental results. Section 6 covers related work. Section 7 concludes our work.

2   Background

We assume that the system in consideration is a discrete-time CMDP with finite horizon, defined by a tuple (S, A, r, d, P, P_0), where S is the set of states, A is the set of actions, $P(s'|s,a)$ is the density of the state transition probability from s to s' when the action is a, $r(s,a)$ is the reward obtained by action a at state s, $d(s,a)$ is the non-negative stepwise danger of taking action a at state s, and P_0 is the distribution of the initial state. We use $\pi(a|s)$ to denote the policy π's probability of taking an action a at a state s. Finally, let 1_B represent the indicator function of an event B. Formally, the optimization problem in Eq. 1 is:

$$\arg\max_{\pi}\mathbb{E}^{\pi}\left[\sum_{t=0}^{T-1}\gamma^t r(s_t,a_t)\right]\ \text{s.t.}\ \mathbb{E}^{\pi}\left[\sum_{t=0}^{T-1}\beta^t d(s_t,a_t)\right]\leq c, \qquad (2)$$

where $c \geq 0$ specifies the safety level, $\gamma, \beta \in [0,1)$ are the discount factors, and $\mathbb{E}^{\pi}[\cdot]$ denotes the expectation with respect to π, P and P_0. $\mathbb{E}$ alone denotes the expectation with respect to P_0.

3   Theory

Just as traditional RL has action-value functions and state-value functions to represent cumulative discounted rewards under a policy η, we can define analogous functions for danger. In our formulation, we term these the *action-threat function* and the *state-threat function*, respectively:

$$\mathscr{T}_t^{\eta}(s_t,a_t) = \mathbb{E}^{\eta}\left[\sum_{k=t}^{T-1}\beta^{k-t}d(s_k,a_k)\mid s_t,a_t\right], \qquad (3)$$

$$\mathscr{D}_t^{\eta}(s_t) = \mathbb{E}^{\eta}\left[\mathscr{T}_t^{\eta}(s_t,a_t)\right]. \qquad (4)$$

We say that a policy η is *safe* if $\mathbb{E}[\mathscr{D}_0^{\eta}(s_0)] \leq c$. Indeed, the set of safe policies is the set of feasible policies for the CMDP (1). Before we proceed further, we describe several definitions and theorems that stem from the threat function. For now, let us consider a time-dependent safety threshold x_t defined at each time t, and let η be any policy. Let us also use $\boldsymbol{x}$ to denote $(x_0,\ldots,x_{T-1})$. Then the set of $(\eta,\boldsymbol{x})$-secure actions is the set of actions that are deemed *safe* by η for the safety threshold $\boldsymbol{x}$ in the sense of the following definition;

Definition 1 ($(\eta,\boldsymbol{x})$-secure actions and $(\eta,\boldsymbol{x})$-secure states). *Let $A^{\eta,\boldsymbol{x}}(s,t) = \left\{a;\ \mathscr{T}_t^{\eta}(s,a) \leq x_t\right\}$.*

$$A^{\eta,\boldsymbol{x}}(s) = \bigcap_{t\in\{0,\cdots,T-1\}} A^{\eta,\boldsymbol{x}}(s,t),$$

$$S^{\eta,\boldsymbol{x}} = \left\{s \in S;\ A^{\eta,\boldsymbol{x}}(s) \neq \emptyset\right\}. \qquad (5)$$

This $(\eta,\boldsymbol{x})$-secure set of actions $A^{\eta,\boldsymbol{x}}(s)$ represents the agent's freedom in seeking the reward under the safety protocol created by the policy η. The set of secure actions for an arbitrary η could be empty for some states. $(\eta,\boldsymbol{x})$-secure states, $S^{\eta,\boldsymbol{x}}$ is defined as a set of states for which there is non-empty $(\eta,\boldsymbol{x})$-secure actions. If we use supp(p) to denote the support of a distribution p, we can use this definition to define a set of policies that is at least as safe as η. First, let us define the set of distributions,

$$\mathcal{F}^\eta(s) = \left\{ p(\cdot); \int_a p(a)\mathcal{T}_t^\eta(s,a)da \leq \mathbb{E}^\eta[\mathcal{T}_t^\eta(s,a)] \; \forall t \right\}.$$

Then the following set of policies are at least as safe as η.

Definition 2 (General (η, x)-secure policies).

$$\Pi_G^{\eta,x} = \{\pi; \; for \; s \in S^{\eta,x}, \; supp(\pi(\cdot|s)) \subseteq A^{\eta,x}(s), \; otherwise \; \pi(\cdot|s) \in \mathcal{F}^\eta(s)\}.$$

Now, we are ready to develop our theory for determining when a given policy is safe. The following theorem enables us to bound $\mathcal{D}_t^\pi(s_t)$ without evaluating the expectation with respect to π.

Theorem 1. *For a given policy η and a sequence of safety thresholds $x = (x_0, \ldots, x_{T-1})$, let π be a policy in $\Pi_G^{\eta,x}$. Let us use $d_{TV}(p,q)$ to denote the total variation distance[1] between two distributions p and q. Then for all $t \in \{0, \ldots, T-1\}$*

$$\mathcal{D}_t^\pi(s_t) \leq \mathcal{D}_t^\eta(s_t) + 2\sum_{k=t}^{T-1} \beta^{k-t} x_k \mathbb{E}^\pi[z_k \mid s_t]. \tag{6}$$

where $z_t = 1_{s_t \in S^{\eta,x}} d_{TV}(\pi(\cdot|s_t), \eta(\cdot|s_t))$ is a distance measure of the two policies.

The proof of this result uses practically same logic as the one used for Theorem 1 in [1]. Please see the Appendix [30] in for more details. In practical applications, it is more convenient to set $x_t = x$ for all t. If we also bound z_t from above by 1, we obtain the following useful result.

Corollary 1. *If $\mathbb{E}[\mathcal{D}_0^\eta(s_0)] \leq c$, let $x = (x_c^\eta, \ldots, x_c^\eta)$ with $x_c^\eta = \frac{1}{2}(c - \mathbb{E}[\mathcal{D}_0^\eta(s_0)])\frac{1-\beta}{1-\beta^T}$. Then a policy π is safe if $\pi \in \Pi_G^{\eta,x_c^\eta}$, (i.e., $\mathbb{E}[\mathcal{D}_0^\pi(s_0)] \leq c$).*

Corollary 1 provides a safety guarantee for π when η itself is safe. But in fact, if we restrict our view to a smaller subset of $\Pi_G^{\eta,x}$, we can guarantee safety even when η itself may not be safe.

Definition 3 $((\eta, x)$-secure policies). *Let greedy-$\eta(a|s) = 1_{a=\arg\min_{a'} \mathcal{T}_t^\eta(s,a')}$.*

$$\Pi^{\eta,x} = \{\pi; for \; s \in S^{\eta,x}, \; supp(\pi(\cdot|s)) \subseteq A^{\eta,x}(s) \; otherwise \; \pi(\cdot|s) = greedy\text{-}\eta(\cdot|s)\}.$$

The set of (η, x)-secure policies $\Pi^{\eta,x}$ is indeed a subset of $\Pi_G^{\eta,x}$ because greedy-η is just the one-step policy improvement from η. It turns out that we can construct a pool of absolutely safe policies *explicitly* from $\mathcal{T}_t^\eta(s,a)$ and c alone even when η itself is not necessarily safe.

---

[1] Total variation distance is defined as $d_{TV}(p(a), q(a)) = \frac{1}{2}\sum_a |p(a) - q(a)|$.

Corollary 2. *If* $\mathbb{E}[\mathscr{D}_0^{greedy-\eta}(s_0)] \leq c$, *by setting* $\boldsymbol{x} = (x_{c,g}^\eta, \dots, x_{c,g}^\eta)$ *with* $x_{c,g}^\eta = \frac{1}{2}(c - \mathbb{E}[\mathscr{D}_0^{greedy-\eta}(s_0)])\frac{1-\beta}{1-\beta^T}$, *any policy* $\pi \in \Pi^{\eta,\boldsymbol{x}_{c,g}^\eta}$ *is safe, i.e.,* $\mathbb{E}[\mathscr{D}_0^\pi(s_0)] \leq c$.

This result follows from Theorem 2, which is a similar result to Theorem 1 (see Appendix B in [30] for the statement and proof of Theorem 2 and Corollary 2. In the next section, we use $\Pi^{\eta,\boldsymbol{x}_{c,g}^\eta}$ to construct a pool of safe policies from which we search for a good and safe reward-seeking policy. Now, several remarks are in order. First, if we set $\beta = 1$, then $x_c^\eta \to 0$ as $T \to \infty$. This is in agreement with the law of large numbers; that is, any accident with positive probability is bound to happen at some point. Also, note that we have $A^{\eta,\boldsymbol{x}}(s) \subseteq A^{\eta',\boldsymbol{x}}(s)$ whenever $\mathscr{T}_t^{\eta'}(s,a) \leq \mathscr{T}_t^\eta(s,a)$ for any t. Thus, by finding the risk-minimizing η, we can maximize the pool of safe policies. Therefore, when possible we seek not just a safe η, but also for the threat-minimizing policy. Finally and most importantly, note that the threshold expression in Corollary 2 is not dependent on π. We can use this result to tackle the CMDP problem by solving two separate MDP problems. In particular, in seeking a solution to the CMDP problem we can (i) first look for an η satisfying $\mathbb{E}[\mathscr{D}_0^{greedy-\eta}(s_0)] \leq c$, and then (ii) look for a safe reward maximizing policy π in Π^η. We will further articulate this procedure in the next section. Hereafter unless otherwise noted, we will use Π^η to denote $\Pi^{\eta,\boldsymbol{x}_{c,g}^\eta}$, and use $S^\eta, A^\eta(s)$ to denote $S^{\eta,\boldsymbol{x}_{c,g}^\eta}, A^{\eta,\boldsymbol{x}_{c,g}^\eta}(s)$.

4   Reconnaissance-MDP and Planning-MDP

4.1   The RP Algorithm

In the previous section, we showed that a pool of safe policies can be created using a safe baseline policy. Also, by training a policy to minimize the threat, we can find a safe policy that corresponds to a larger pool of secure policies, and we search within this pool for a possibly better reward-seeking policy. This insight motivates the decomposition of the constrained MDP problem into two consecutive MDP sub-problems: one which minimizes threat, and one which maximizes cumulative reward.

The purpose of the Reconnaissance MDP (R-MDP) is thus to *reconnoiter* the system prior to the reward maximization process and to find the threat-minimizing safe policy η^* that solves

$$\eta^* = \arg \min_\eta \mathbb{E}^\eta \left[\sum_{t=0}^{T-1} \beta^t d(s_t, a_t) \right]. \tag{7}$$

Indeed, solution η^* is not unique, up to freedom of the actions on the states unreachable by any optimal policy η^*. For our ensuing discussions, we will choose η^* to be a version whose policy on each unreachable state s^* is computed by Eq. 7 with initial state being s^*. If the problem is of infinite horizon, then under standard assumptions of irreducibility and aperiodicity, η^* computed from any initial state will be the same due to the fact that such finite MDPs have unique steady state distributions.

An R-MDP is the same as the original MDP except that we have a danger function in lieu of a reward function, and the agent's goal to minimize the cumulative danger instead of maximize the cumulative reward. The following is true about any CMDP.

Corollary 3. *If the set of feasible policies of the original CMDP is nonempty, the optimal R-MDP policy η^* is safe. Thus, every policy in Π^{η^*} is safe.*

After the R-MDP is solved, the Planning MDP (P-MDP) searches within Π^{η^*} for a good reward-seeking policy π^*. The P-MDP is similar to the original MDP except that the agent is only allowed to take actions from A^{η^*} when $s \in S^{\eta^*}$ and that it follows greedy-η^* at non-secure states $s \notin S^{\eta^*}$.

$$\pi^* = \arg\max_{\pi \in \Pi^{\eta^*}} \mathbb{E}^{\pi} \left[\sum_{t=0}^{T-1} \gamma^t r(s_t, a_t) \right]. \tag{8}$$

In implementation, we do not explicitly construct Π^{η^*}. Instead, we evaluate $\mathcal{T}_t^{\eta^*}(s, a)$ for every considered state-action pair in the P-MDP and ensure that all suggested reward-seeking actions are in A^{η^*}. Note that, if policy η^* is safe, every policy in Π^{η^*} is guaranteed to be safe. In particular, in such a case, any policy in the entire learning process of the P-MDP is safe. We refer to the whole procedure of solving the R-MDP and P-MDP as *Reconnaissance and Planning (RP) algorithm*. Algorithm 1 summarizes the RP algorithm. Naturally, whether

---

Algorithm 1. RP algorithm

1: Obtain the baseline policy η^* by either solving R-MDP or selecting a heuristic policy
2: Estimate $\mathcal{T}_t^{\eta^*}(\cdot, \cdot)$
3: Solve the P-MDP (8) while referring the evaluation of $\mathcal{T}_t^{\eta^*}$ at every considered state-action pair so that all actions will be chosen from $A^{\eta^*, x}$

---

this algorithm works in an application depends on how well we can evaluate the threat in the R-MDP. If we wish to avoid any dangerous events, even those with extremely low probability, guaranteeing safety becomes difficult. A good policy that works well in most cases may be dangerous in some rare scenarios, but such scenarios are difficult to find because they require an intractable number of runs to detect such rare, dangerous events. This makes the empirical approximation $\mathbb{E}^{\pi}[\cdot]$ difficult. We resolve this problem by evaluating $\mathcal{T}_t^{\eta}(s, a)$ with the generative model. With the generative model, we can freely explore the system from any arbitrary initial state, and evaluate $\mathcal{T}_t^{\eta}(s, a)$ for any (s, a) and η. To facilitate the learning of $\mathcal{T}_t^{\eta^*}(s, a)$, we use the generative model to preferentially sample the states with relatively high estimated $\mathcal{T}_t^{\eta^*}(s, a)$ more frequently. We will next introduce a technique to approximate the threat function.

4.2   Threat Decomposition and the Approximate RP Algorithm

In this subsection, we present a useful bound on the threat function that can be used when the danger is described in terms of multiple risky events. Suppose that there are N risky events $\{E_1, ..., E_N\}$ to consider, and that the goal is optimize the reward while avoiding *any* risky event. Formally, we represent the indicator function of E_n by $d^n(s_t^{(o)}, s_t^{(n)}, a_t)$ where $s_t^{(n)}$ is the state of the system relevant to E_n and $s^{(o)}$ is the state of the system not relevant to any risky events.

Assumption 1. The transition probabilities can be decomposed in the following way:

$$p(s_{t+1}|s_t, a_t) = p(s_{t+1}^{(o)}|s_t^{(o)}, a_t) \prod_{n=1}^{N} p(s_{t+1}^{(n)}|s_t^{(n)}, s_t^{(o)}, a_t).$$

If d is the indicator function of $\cup_{n=1}^{m} E_n$, we can formalize this result as follows:

Theorem 3. *Let η be a policy that takes actions based solely on $s_t^{(o)}$, and let T^η be the threat function defined for the indicator function of $\cup_{n=1}^{N} E_n$. Then* $\mathscr{T}_t^\eta(s_t, a_t) \leq \sum_{n=1}^{N} \mathscr{T}_t^{\eta,n}(s_t^{(o)}, s_t^{(n)}, a_t)$.

This result is especially useful in navigation-like tasks. For example, when E_n is the collision event with the nth obstacle, s_n will be the state of nth obstacle, and s_o will be the aggregate state of the system that is not related to any obstacles (state of the agent, etc.). In such cases, each $\mathscr{T}^{n,n}$ can be estimated using the simulator containing just the nth obstacle and the agent. Algorithm 2 is the algorithm that uses Theorem 3. In the context of collision-free planning,

---

Algorithm 2. Approximate RP algorithm

1: Pick a heuristic policy $\eta^{(o)}$ which depends on $s^{(o)}$ only.
2: Estimate the threat functions $\mathscr{T}_t^{\eta^{(o)},n}(\cdot, \cdot)$ for all sub R-MDPs.
3: Solve the P-MDP (8) while referring to the evaluation of $\sum_n \mathscr{T}_t^{\eta^{(o)},n}$ at every
   considered state-action pair so that all chosen actions will satisfy $\sum_n \mathscr{T}_t^{\eta^{(o)},n} \leq x_t$.

Theorem 3 suggests that the threat function is similar to a risk potential [21,29,35]. Risk-potential based methods for collision-free planning also evaluate the overall risk of colliding with any obstacle by summing the evaluated risk potential for each object. However, most risk-potential methods are designed for specific task settings (e.g. collision-free planning) and thus assume the risk potential has a specific heuristically defined structure. By contrast, our method can be interpreted as *learning* the risk potential.

5 Experiments

We conducted a series of experiments to analyze (i) the nature of the threat function, (ii) the effect of the choice of the baseline policy η on P-MDP performance, and (iii) the efficacy of the approximate RP algorithm. We also show that the threat function computed for one CMDP can be reused to solve another CMDP problem with similar safety constraints[2].

5.1 Experimental Setup

Environments and Baselines. We summarize our environments here, but more in-depth descriptions of the environments are available in Appendix F [30].

Frozen Lake Frozen lake environment from OpenAI Gym [7]. In this environment, the agent navigates an 8×8 gridworld (depicted in Fig. 2) consisting of frozen surfaces and holes. The agent starts from 'S' and must navigate through the frozen surfaces in the gridworld to reach the goal 'G' while avoiding holes. Due to the slippery surface, the agent's actual movement may differ from its intended one.

Point Gather The Point Gather environment is a common benchmark task used for safe RL [1]. In this task, the agent, a point mass, must navigate a bounded region and collect as many green apples as possible while avoiding red bombs.

Circuit In this 364-dimensional task, the agent's goal is to cover as much distance as possible in a circuit within 200 timesteps without crashing into the wall.

Jam In the 396-dimensional Jam task, the agent must navigate its way out of a square room of size 3.0×3.0 through an exit located at the top right corner in the presence of dynamic obstacles. There are safety zones in the corners of the environment (one of which is the exit). This task is more challenging due to the dynamism of the obstacles. The Jam task is illustrated in Fig. 4.

We use three baselines. We use deep Q-networks (DQN) [26], a standard off-policy RL algorithm. We use constrained policy optimization (CPO) [1], a popular safe RL algorithm. Lastly, we use model predictive control (MPC) with α-β pruning. Appendix G in [30] contains more information on the baselines.

5.2 Frozen Lake Results

For this task, both the R-MDP and P-MDP are tabular and can be solved through value iteration. We selected threshold x in accordance with the safety guarantee given by Corollary 2. The threat function of the optimal R-MDP policy is shown in Fig. 2. Note that the threat values are indeed higher at states closer to holes. Moreover, we can see that there is a safe path from 'S' to 'G'

---
[2] The code and videos are available at https://github.com/pfnet-research/rp-safe-rl.

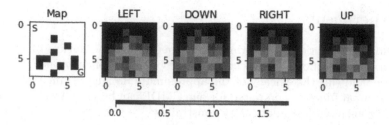

Fig. 2. Map and action-threat function. In the map, Black squares represent holes and white squares represent the frozen surface. Warmer colors represent higher values of the threshold function at the corresponding position and action. (Color figure online)

following the rightmost column. The challenging positions are (6, 8) and (7, 8), where the only safe action is 'RIGHT', which is not the best reward-seeking action ('DOWN').

Figures 3(a) and 3(b) compare the performance of a policy learned in the P-MDP from the optimal R-MDP policy η^* as a baseline policy against those learned from sub-optimal R-MDP baseline policies. The policy η^0 represents a uniformly random policy and η^i represents the policy after i steps of policy iteration on η^0. In Fig. 3(a) we also show the CMDP solution found using the Lagrange multiplier method, which represents a near optimal solution to the CMDP task. As Fig. 3(b) shows, the RP algorithm satisfies the danger constraints for every baseline policy η, by a large margin. Because η^0 is a naive, dangerous policy, our RP algorithm was unable to reach the goal and receives no reward with baseline policy η^0. Although the RP algorithm produces conservative results, as evidenced by its large safety margin, it can achieve the highest reward if we set the constraint c to be larger. We can see that, even though η^0 is completely random, the P-MDP constructed from η^0 generates safe policies when the threat threshold x is small. When we optimize the baseline policy η through policy iteration, the set of safe actions (secure actions defined in Definition 1) increases in size, enabling the agent to attain more rewards, as seen in Fig. 3(a). However, these differences become negligible after two policy iteration steps. We see that two steps of policy iteration are enough to obtain a near-optimal policy for this problem. This observation suggests that using a heuristic baseline without optimizing in the R-MDP may suffice, and in fact we use heuristic policies in the higher-dimensional environments.

5.3 High-Dimensional Tasks: Point Gather, Circuit, and Jam

Before delving into the experimental results, note that the total reward attained changes depending on how severely the agent is penalized. For reference, we show results for heavier danger penalties in Appendix H [30]. Since the constraint in the Point Gather, Circuit, and Jam environments is to avoid collision with *any* obstacles, we can use the approximate RP algorithm introduced in Sect. 4.2 which uses a set of sub-R-MDPs each containing one obstacle and one

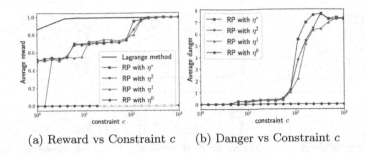

(a) Reward vs Constraint c (b) Danger vs Constraint c

Fig. 3. FrozenLake results. (a) Average reward obtained by RP method with various baseline policies vs safety constraint c. (b) Average danger suffered by RP method with various baseline policies vs safety constraint c.

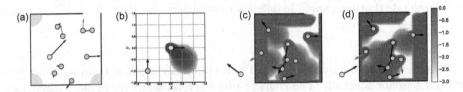

Fig. 4. (a) Illustration of Jam task. The light blue circles are obstacles and the yellow circle is the agent. Three shaded corners are safe-zones. The arrow attached to each object shows its direction of movement. (b) Heat map of the trained threat function whose value at point (x, y) represent the threat when the obstacle (light blue) is placed at (x, y) with the same velocity. (c) and (d) are the heat maps of the sum of the threat functions of all moving obstacles. (Color figure online)

agent. Thus, to guarantee safety in P-MDP, we only learn a threat function for a heuristic baseline policy $\eta^{(o)}$ in a sub-R-MDP with a single obstacle. For Circuit and Jam, we treat the *wall* as a set of immobile obstacles so that we can construct the threat function of any shape. After 10 min of data collection with the generative model, this threat function can be learned in less than half an hour on a CPU. Please see the Appendix in [30] for more detailed experimental conditions and for videos of the agent trained by each method. Figure 4(c), (d) are the heat maps for the upper bound of the threat function computed by way of Theorem 3. The color of each heat map at pixel z represents $\sum_n \mathcal{T}_0^{\eta, n}(s(z), a)$, where $s(z)$ represents the state at which the agent's current location is z and its velocity and direction is given by the picture located at the left corner of the heat map. We see that our threat function is playing a role similar to an artificial potential field [8,19,23]. Because our threat function is computed using all aspects of the agent's state (acceleration, velocity, location), we can provide a more comprehensive measure of risk compared to other risk metrics such as TTC (Time To Collision) [24] used in smart automobiles that consider only 1D movement.

578   S. Maeda et al.

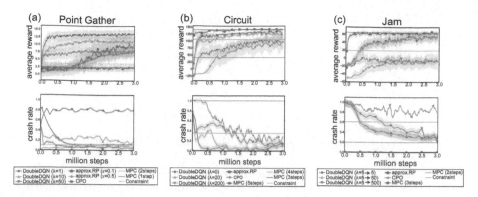

Fig. 5. Comparison of multiple CMDP methods in terms of rewards (upper panels) and crash rate (bottom panels) in different environments (a) Point Gather, (b) Circuit and (c) Jam.

Figure 5 plots the average reward and crash rate against training timesteps for our baseline methods. The DQN results are representative of model-free deep RL approaches applied to the CMDP task. To evaluate the threat in MPC, we take the same approximation strategy of threat decomposition as our method (see Appendix G.4 in [30]) in order to make the prediction efficient.

The mean values and the 90% confidence intervals in the plot are computed over 10 seeds. The curve plotted for our approximate RP algorithm corresponds to the result obtained by a DQN-based P-MDP solution. The plot does not include the R-MDP phase. As we can see, our method achieves the highest reward in almost all phases of training for both Circuit and Jam while maintaining the lowest crash rate. Our method is safer than the 3-step MPC, a method with significantly higher runtime computational costs. For Point Gather, the RP algorithm with $x = 0.1$ performs worse in terms of reward than the penalized DQN and CPO. Fortunately, however, we can fine-tune the threshold to increase the pool of allowed actions in the P-MDP without re-learning the threat function. Our RP algorithm with $x = 0.5$ performs competitively while satisfying the prescribed constraint. That we can fine-tune the threshold without retraining is a significant advantage of our method over Lagrange multiplier-type methods. Also, by virtue of Corollary 2, that our policy is experimentally safe suggests that greedy-$\eta^{(0)}$ is also safe for the choice of the baseline policy $\eta^{(0)}$.

5.4 Recycling the Threat Function for a Different Environment

As mentioned in Sect. 1, learned threat functions can be reused on different CMDP tasks if their safety constraints are defined similarly. To verify this feature of our algorithm, we train a safe policy on an instance of the Jam task, and test its performance on other Jam instances with different numbers of randomly moving obstacles. The results are shown in Table 1. For the modified Jam instances, we have no results for MPC with more than 3-step prediction since the exhaustive search cannot be completed within a reasonable time frame. We also conducted experiments for Circuit environment. In particular, we evaluate on two modified instances: (1) a narrowed circuit with the original shape, and (2) a circular-shaped circuit with same width. The results are shown in Appendix H [30].

In both the Circuit and Jam experiments, we find that, even in different task instances, the policy obtained by the RP algorithm can achieve safety with high probability while attaining high reward. On the other hand, reusing a DQN policy in different task instances fails both in terms of reward and safety in both the Jam and Circuit tasks.

Table 1. Performance of trained policies on known and unknown Jam environments. The values in the table are the obtained rewards, and inside the parentheses is the percentage of episodes in which a collision occurs. Computation Time is the time used for 100 actions.

Environment	Approx. RP	MPC (3 steps)	DQN ($\lambda = 5$)	DQN ($\lambda = 500$)
$N = 3$	78.2 (**0**)	77.5 (0.05)	77.2 (0.04)	4.4 (0.17)
$N = 8$ (training env.)	69.1 (**0**)	65.3 (0.2)	47.1 (0.38)	−1.0 (0.24)
$N = 15$	33.0 (**0.02**)	36.6 (0.45)	16.5 (0.66)	−16.8 (0.51)
Computation time (s)	1.2	285	0.4	0.4

6 Related Work

6.1 Optimization-Based Approaches

Many optimization-based approaches such as MPC [6,15,25,33,34] are available to tackle CMDPs by solving certain optimization problems. However, *nonlinear* and *stochastic* state transitions, as in the Jam experiments, make these optimization problems extremely difficult to solve. To apply these approaches to nonlinear and/or stochastic environments, special assumptions are typically made to make the optimization tractable.

Exact Bound-Based MPC. MPC approaches solve CMDPs by solving finite horizon control problems constructed by a predictive model. Wabersich and Zeilinger [32] assume that the danger constraint is given by a linear inequality with respect to the state. Summers et al. [31] consider a discrete state space and use dynamic programming to obtain a reach-avoid control policy under a stochastic time-varying environment. Ames et al. [4] surveys control barrier function methods in deterministic nonlinear affine control systems.

Reachability Analysis. Collision avoidance is also considered in reachability analysis where a *backward reachable set* is computed to represent the states from which the agent can reach the desired state despite worst-case disturbances [10]. By considering the worst case disturbance for stochastic or unknown dynamics, a CMDP can be formulated as a minimax problem [2,5,28]. Unfortunately, considering worst-case disturbances can lead to overly conservative solutions. To balance the safety guarantee and conservatism in the environment where the model includes uncertainty, Fisac et al. [18] introduces online learning based on Gaussian processes. Although reachability analysis requires exponentially more computation as the system dimensionality increases, some studies overcome this issue by considering the conditionally independent subsystem similar to ones considered in Sect. 4.2 [10,11].

6.2   Sampling and Learning-Based Approaches

Sampling-Based MPC. Monte Carlo sampling can be used as a predictive model which allows MPC to scale to high-dimensional state spaces and can handle nonlinear dynamics and non-Gaussian transition probabilities [9,17]. However, Monte Carlo sampling also requires per-timestep sampling, and the methods become computationally intractable when the time horizon is large.

Learning-Based Methods. Learning-based methods [12,20,22,27] optimize the parameters of the policy, which is used for all states. In contrast, MPC methods perform optimization for only the current state. Learning-based methods also make use of sampling to mitigate the computational difficulty of computing cumulative danger or total reward. We can apply learning-based methods on environments with high-dimensional state-action spaces. Typically a large set of samples is required in order to accurately estimate the expected reward and danger, especially when rare events must be taken into account. This estimation is necessary for every policy update. Approximating expected reward and danger is difficult for both learning-based and sampling-based MPC approaches.

7   Conclusion

In this study we proposed a method that isolates the safety-seeking procedure from the reward-seeking procedure in solving a CMDP. Although our method is not guaranteed to find the optimal reward-seeking safe policy, it can perform significantly better than classical methods both in terms of safety and reward in high-dimensional dynamic environments like Jam. Our method is designed so that *more training* in the R-MDP will always expand the search space in the P-MDP. Our treatment of the threat function not only allows us to solve similar problems without the need to retrain, it also helps us obtain a more comprehensive measure of danger at each state than conventional methods. Overall, we find that utilizing threat functions is a promising approach to safe RL. We expect that further research on our framework may lead to new CMDP methods applicable to complex, real-world environments.

References

1. Achiam, J., Held, D., Tamar, A., Abbeel, P.: Constrained policy optimization. In: ICML, pp. 22–31 (2017)

2. Akametalu, A.K., Fisac, J.F., Gillula, J.H., Kaynama, S., Zeilinger, M.N., Tomlin, C.J.: Reachability-based safe learning with Gaussian processes. In: CDC, pp. 1424–1431 (2014)
3. Altman, E.: Constrained Markov Decision Processes, vol. 7. CRC Press, Boca Raton (1999)
4. Ames, A.D., Coogan, S., Egerstedt, M., Notomista, G., Sreenath, K., Tabuada, P.: Control barrier functions: theory and applications. In: ECC, pp. 3420–3431 (2019)
5. Bansal, S., Chen, M., Herbert, S.L., Tomlin, C.J.: Hamilton-Jacobi reachability: a brief overview and recent advances. In: CDC, pp. 2242–2253 (2017)
6. Blake, R.J., Mayne David, Q.: Model Predictive Control: Theory and Design. Nob Hill Pub., Madison (2009)
7. Brockman, G., et al.: OpenAI gym (2016)
8. Cetin, O., Kurnaz, S., Kaynak, O., Temeltas, H.: Potential field-based navigation task for autonomous flight control of unmanned aerial vehicles. Int. J. Autom. Control 5(1), 1–21 (2011)
9. Chang, P., Mertz, C.: Monte Carlo sampling based imminent collision detection algorithm. In: ICTIS, pp. 368–376 (2017)
10. Chen, M., Herbert, S., Tomlin, C.J.: Fast reachable set approximations via state decoupling disturbances. In: CDC, pp. 191–196 (2016)
11. Chen, M., Herbert, S., Tomlin, C.J.: Exact and efficient Hamilton-Jacobi-based guaranteed safety analysis via system decomposition. In: ICRA (2017)
12. Chow, Y., Ghavamzadeh, M., Janson, L., Pavone, M.: Risk-constrained reinforcement learning with percentile risk criteria. JMLR (2018)
13. Chow, Y., Nachum, O., Duenez-Guzman, E., Ghavamzadeh, M.: A Lyapunov-based approach to safe reinforcement learning. In: NeurIPS (2018)
14. Chow, Y., Nachum, O., Faust, A., Ghavamzadeh, M., Duenez-Guzman, E.: Lyapunov-based safe policy optimization for continuous control. In: ICML (2019)
15. Di Cairano, S., Bernardini, D., Bemporad, A., Kolmanovsky, I.V.: Stochastic MPC with learning for driver-predictive vehicle control and its application to HEV energy management. IEEE Trans. Control Syst. Technol. 22(3), 1018–1031 (2013)
16. Dosovitskiy, A., Ros, G., Codevilla, F., Lopez, A., Koltun, V.: CARLA: an open urban driving simulator. In: CoRL, pp. 1–16 (2017)
17. Eidehall, A., Petersson, L.: Statistical threat assessment for general road scenes using Monte Carlo sampling. IEEE Trans. Intell. Transp. Syst. 9(1), 137–147 (2008)
18. Fisac, J.F., Akametalu, A.K., Zeilinger, M.N., Kaynama, S., Gillula, J., Tomlin, C.J.: A general safety framework for learning-based control in uncertain robotic systems. IEEE Trans. Autom. Control 64(7), 2737–2752 (2019)
19. Ge, S.S., Cui, Y.J.: New potential functions for mobile robot path planning. IEEE Trans. Robot. Autom. 16(5), 615–620 (2000)
20. Geibel, P.: Reinforcement learning for MDPs with constraints. In: Fürnkranz, J., Scheffer, T., Spiliopoulou, M. (eds.) ECML 2006. LNCS (LNAI), vol. 4212, pp. 646–653. Springer, Heidelberg (2006). https://doi.org/10.1007/11871842_63
21. Ji, J., Khajepour, A., Melek, W.W., Huang, Y.: Path planning and tracking for vehicle collision avoidance based on model predictive control with multiconstraints. IEEE Trans. Veh. Technol. 66(2), 952–964 (2016)
22. Koller, T., Berkenkamp, F., Turchetta, M., Krause, A.: Learning-based model predictive control for safe exploration. In: CDC, pp. 6059–6066 (2018)
23. Lam, C.P., Chou, C.T., Chiang, K.H., Fu, L.C.: Human-centered robot navigation-towards a harmoniously human-robot coexisting environment. T-RO 27(1), 99–112 (2010)

24. Lee, D.N.: A theory of visual control of braking based on information about time-to-collision. Perception **5**(4), 437–459 (1976)
25. Maciejowski, J.M.: Predictive Control: With Constraints. Pearson Education, London (2002)
26. Mnih, V., et al.: Human-level control through deep reinforcement learning. Nature **518**(7540), 529–533 (2015)
27. Moldovan, T.M., Abbeel, P.: Safe exploration in Markov decision processes. In: ICML (2012)
28. Prajna, S., Jadbabaie, A., Pappas, G.J.: A framework for worst-case and stochastic safety verification using barrier certificates. IEEE Trans. Autom. Control **52**(8), 1415–1428 (2007)
29. Rasekhipour, Y., Khajepour, A., Chen, S.K., Litkouhi, B.: A potential field-based model predictive path-planning controller for autonomous road vehicles. IEEE Trans. Intell. Transp. Syst. **18**(5), 1255–1267 (2016)
30. Maeda, S., Watahiki, H., Ouyang, Y., Okada, S., Koyama, M., Nagarajan, P.: Supplementary of reconnaissance for reinforcement learning with safety constraints (2021). https://github.com/pfnet-research/rp-safe-rl
31. Summers, S., Kamgarpour, M., Lygeros, J., Tomlin, C.: A stochastic reach-avoid problem with random obstacles. In: 14th International Conference on Hybrid Systems: Computation and Control, pp. 251–260 (2011)
32. Wabersich, K.P., Zeilinger, M.N.: Linear model predictive safety certification for learning-based control. In: CDC, pp. 7130–7135 (2018)
33. Wang, Y., Boyd, S.: Fast model predictive control using online optimization. IEEE Trans. Control Syst. Technol. **18**(2), 267–278 (2010)
34. Weiskircher, T., Wang, Q., Ayalew, B.: Predictive guidance and control framework for (semi-) autonomous vehicles in public traffic. IEEE Trans. Control Syst. Technol. **25**(6), 2034–2046 (2017)
35. Wolf, M.T., Burdick, J.W.: Artificial potential functions for highway driving with collision avoidance. In: ICRA, pp. 3731–3736 (2008)

VeriDL: Integrity Verification of Outsourced Deep Learning Services

Boxiang Dong[1], Bo Zhang[2], and Hui (Wendy) Wang[3(✉)]

[1] Montclair State University, Montclair, NJ, USA
dongb@montclair.edu
[2] Amazon Inc., Seattle, WA, USA
bzhanga@amazon.com
[3] Stevens Institute of Technology, Hoboken, NJ, USA
Hui.Wang@stevens.edu

Abstract. Deep neural networks (DNNs) are prominent due to their superior performance in many fields. The deep-learning-as-a-service (DLaaS) paradigm enables individuals and organizations (clients) to outsource their DNN learning tasks to the cloud-based platforms. However, the DLaaS server may return incorrect DNN models due to various reasons (e.g., Byzantine failures). This raises the serious concern of how to verify if the DNN models trained by potentially untrusted DLaaS servers are indeed correct. To address this concern, in this paper, we design VERIDL, a framework that supports efficient correctness verification of DNN models in the DLaaS paradigm. The key idea of VERIDL is the design of a small-size cryptographic proof of the training process of the DNN model, which is associated with the model and returned to the client. Through the proof, VERIDL can verify the correctness of the DNN model returned by the DLaaS server with a deterministic guarantee and cheap overhead. Our experiments on four real-world datasets demonstrate the efficiency and effectiveness of VERIDL.

Keywords: Deep learning · Integrity verification · Deep-learning-as-a-service

1 Introduction

The recent abrupt advances in deep learning (DL) [1,10] have led to breakthroughs in many fields such as speech recognition, image classification, text translation, etc. However, this success crucially relies on the availability of both hardware and software resources, as well as human expertise for many learning tasks. As the complexity of these tasks is often beyond non-DL-experts, the rapid growth of DL applications has created a demand for cost-efficient, off-shelf solutions. This motivated the emerge of the *deep-learning-as-a-service* (DLaaS) paradigm which enables individuals and organizations (clients) to outsource their data and deep learning tasks to the cloud-based service providers for their needs of flexibility, ease-of-use, and cost efficiency.

© Springer Nature Switzerland AG 2021
N. Oliver et al. (Eds.): ECML PKDD 2021, LNAI 12976, pp. 583–598, 2021.
https://doi.org/10.1007/978-3-030-86520-7_36

Despite providing cost-efficient DL solutions, outsourcing training to DLaaS service providers raises serious security and privacy concerns. One of the major issues is the *integrity* of the deep neural network (DNN) models trained by the server. For example, due to Byzantine failures such as software bugs and network connection interruptions, the server may return a DNN model that does not reach its convergence. However, it is difficult for the client to verify the correctness of the returned DNN model easily due to the lack of hardware resources and/or DL expertise.

In this paper, we consider the *Infrastructure-as-a-Service* (IaaS) setting where the DLaaS service provider delivers the computing infrastructure including servers, network, operating systems, and storage as the services through virtualization technology. Typical examples of IaaS settings are Amazon Web Services[1] and Microsoft Azure[2]. In this setting, a client outsources his/her training data T to the DLaaS service provider (server). The client does not need to store T locally after it is outsourced (i.e., the client may not have access to T after outsourcing). The client also has the complete control of the infrastructure. He can customize the configuration of the DNN model M, including the network topology and hyperparameters of M. Then the server trains M on the outsourced T and returns the trained model M to the client. As the client lacks hardware resources and/or DL expertise, a third-party verifier will authenticate on behalf of the client if M returned by the server is *correct*, i.e., M is the same as being trained locally with T under the same configuration. Since the verifier may not be able to access the private training data owned by the client, our goal is to design a lightweight verification mechanism that enables the verifier to authenticate the correctness of M without full access to T.

A possible solution is that the *verifier* executes the training process independently. Since the verifier does not have the access to the client's private data, he has to execute training on the private data encrypted by homomorphic encryption (HE) [6,9]. Though correct, this solution can incur expensive overhead due to the high complexity of HE. Furthermore, since HE only supports polynomial functions, some activation functions (e.g., ReLU, Sigmoid, and Tanh) have to be approximated by low-degree polynomials when HE is used, and thus the verifier cannot compute the exact model updates. On the other hand, the existing works on verifying the integrity of DNNs (e.g., SafetyNets [5] and VeriDeep [8] hold a few restrictions on the activation function (e.g., it must be polynomials with integer coefficients) and data type of weights/inputs (e.g., they must be integers). We do not have any assumption on activation functions and input data types. Furthermore, these existing works have to access the original data, which is prohibited in our setting due to privacy protection.

Our Contributions. We design VERIDL, a framework that supports efficient verification of outsourced DNN model training by a potentially untrusted DLaaS server which may return wrong DNN model as the result. VERIDL provides the *deterministic* correctness guarantee of remotely trained DNN models without

---

[1] Amazon Web Services: https://aws.amazon.com/.

[2] Microsoft Azure: https://azure.microsoft.com/en-us/.

any constraint on the activation function and the types of input data. The key idea of VERIDL is that the server constructs a cryptographic proof of the model updates, and sends the proof along with the model updates to the verifier. Since the proof aggregates the intermediate model updates (in compressed format) during training, the verifier can authenticate the correctness of the trained model by using the proof only. In particular, we make the following contributions. First, we design an efficient procedure to construct the cryptographic proof whose size is significantly smaller than the training data. The proof is constructed by using *bilinear pairing*, which is a cryptographic protocol commonly used for aggregate signatures. Second, we design a lightweight verification method named VERIDL that can authenticate the correctness of model updates through the cryptographic proof. By using the proof, VERIDL does not need access to the training data for correctness verification. Third, as the existing bilinear mapping methods cannot deal with the weights in DNNs that are decimal or negative values, we significantly extend the bilinear mapping protocol to handle decimal and negative values. We formally prove that VERIDL is secure against the attacker who may have full knowledge of the verification methods and thus try to escape from verification. Last but not least, we implement the prototype of VERIDL, deploy it on a DL system, and evaluate its performance on four real-world datasets that are of different data types (including non-structured images and structured tabular data). Our experimental results demonstrate the efficiency and effectiveness of VERIDL. The verification by VERIDL is faster than the existing DNN verification methods [6,9] by more than three orders of magnitude.

2   Preliminaries

Bilinear Mapping. Let G and G_T be two multiplicative cyclic groups of finite order p. Let g be a generator of G. A bilinear group mapping e is defined as $e : G \times G \rightarrow G_T$, which has the following property: $\forall a, b \in \mathbb{Z}_p$, $e(g^a, g^b) = e(g, g)^{ab}$. In the following discussions, we use the terms bilinear group mapping and bilinear mapping interchangeably. The main advantage of bilinear mapping is that determining whether $c \equiv ab \bmod p$ *without the access to a, b and c* can be achieved by checking whether $e(g^a, g^b) = e(g, g^c)$, by given g, g^a, g^b, g^c.

Outsourcing Framework. We consider the outsourcing paradigm that involves three parties: (1) a *data owner* (client) $\mathcal{O}$ who holds a private training dataset T; (2) a third-party service provider (server) $\mathcal{S}$ who provides infrastructure services to $\mathcal{O}$; and (3) a third-party verifier $\mathcal{V}$ who authenticates the integrity of $\mathcal{S}'$ services. In this paradigm, $\mathcal{O}$ outsources T to $\mathcal{S}$ for training of a DNN model M. Meanwhile $\mathcal{O}$ specifies the configuration of M on $\mathcal{S}'$ infrastructure for training of M. After $\mathcal{S}$ finishes training of M, it sends M to $\mathcal{V}$ for verification. Due to privacy concerns, $\mathcal{V}$ cannot access the private training data T for verification.

Basic DNN Operations. In this paper, we only focus on deep feedforward networks (DNNs), and leave more complicated structures like convolutional and recurrent networks for the future work. In this section, we present the basic operations of training a DNN model. We will explain in Sect. 4 how to verify the output of these operations. In this paper, we only concentrate on fully-connected neural networks, and refrain from convolutional networks or recurrent networks. However, our design can be adapted to more advanced network structures.

A DNN consists of several layers, including the input layer (data samples), the output layer (the predicted labels), and a number of hidden layers. During the feedforward computation, for the neuron n_k^ℓ, its weighted sum z_k^ℓ is defined as:

$$z_k^\ell = \begin{cases} \sum_{i=1}^{m} x_i w_{ik}^\ell & \text{if } \ell = 1 \\ \sum_{j=1}^{d_{\ell-1}} a_j^{\ell-1} w_{jk}^\ell & \text{otherwise,} \end{cases} \tag{1}$$

where x_i is the i-th feature of the input $\vec{x}$, and d_i is the number of neurons on the i-th hidden layer. The activation a_k^ℓ is calculated as follows:

$$a_k^\ell = \sigma(z_k^\ell), \tag{2}$$

where σ is the activation function. We allow a broad class of activation functions such as sigmoid, ReLU (rectified linear unit), and hyperbolic tangent.

On the output layer, the output o is generated by following:

$$o = \sigma(z^o) = \sigma(\sum_{j=1}^{d_L} a_j^L w_j^o), \tag{3}$$

where w_j^o is the weight that connects n_j^ℓ to the output neuron.

In this paper, we mainly consider the mean square error (MSE) as the cost function. For any sample $(\vec{x}, y) \in T$, the cost $C(\vec{x}, y; W)$ is measured as the difference between the label y and the output o:

$$C(\vec{x}, y; W) = C(o, y) = \frac{1}{2}(y - o)^2. \tag{4}$$

Then the error E is calculated as the average error for all samples:

$$E = \frac{1}{N} \sum_{(\vec{x}, y) \in T} C(\vec{x}, y; W). \tag{5}$$

In the backpropagation process, gradients are calculated to update the weights in the neural network. According to the chain rule of backpropagation [10], for any sample $(\vec{x}, y)$, the error signal δ^o on the output neuron is

$$\delta^o = \nabla_o C(o, y) \odot \sigma'(z^o) = (o - y)\sigma'(z^o). \tag{6}$$

While the error signal δ_k^ℓ at the ℓ-th hidden layer is

$$\delta_k^\ell = \begin{cases} \sigma'(z_k^\ell) w_k^o \delta^o & \text{if } \ell = L, \\ \sigma'(z_k^\ell) \sum_{j=1}^{d_{\ell+1}} w_{kj}^{\ell+1} \delta_j^{\ell+1} & \text{otherwise.} \end{cases} \tag{7}$$

The derivative for each weight w_{jk}^{ℓ} is computed as:

$$\frac{\partial C}{\partial w_{jk}^{\ell}} = \begin{cases} x_j \delta_k^{\ell} & \text{if } \ell = 1 \\ a_j^{\ell-1} \delta_k^{\ell} & \text{otherwise.} \end{cases} \tag{8}$$

Then the weight increment Δw_{jk}^{ℓ} is

$$\Delta w_{jk}^{\ell} = -\frac{\eta}{N} \sum_{(\vec{x},y) \in T} \frac{\partial C}{\partial w_{jk}^{\ell}}, \tag{9}$$

where η is the learning rate. Finally, the weight is updated as

$$w_{jk}^{\ell} = w_{jk}^{\ell} + \Delta w_{jk}^{\ell}. \tag{10}$$

The DNN is iteratively optimized by following the above feedforward and backpropagation process until it reaches convergence, $|E_1 - E_2| \leq \theta$, where E_1 and E_2 are the error/loss of two consecutive epochs in the optimization process, and θ is a small constant.

Verification Protocol. We adapt the definition of the integrity verification protocol [11] to our setting:

Definition 21 (Deep Learning Verification Protocol). *Let W be the set of weight parameters in a DNN, and T be a collection of data samples. Let ΔW be the parameter update after training the DNN on T. The authentication protocol is a collection of the following four polynomial-time algorithms:* **genkey** *for key generation,* **setup** *for initial setup,* **certify** *for verification preparation, and* **verify** *for verification.*

- *$\{s_k, p_k\} \leftarrow$ **genkey**(): It outputs a pair of secret and public key;*
- *$\{\gamma\} \leftarrow$ **setup**$(\mathbf{T}, \mathbf{s_k}, \mathbf{p_k})$: Given the dataset T, the secret key s_k and the public key p_k, it returns a single signature γ of T;*
- *$\{\pi\} \leftarrow$ **certify**$(\mathbf{T}, \mathbf{W_0}, \mathbf{\Delta W}, \mathbf{p_k})$: Given the data collection T, the initial DNN model parameters W_0, the model update ΔW, and a public key p_k, it returns the proof π;*
- *$\{$accept, reject$\} \leftarrow$ **verify**$(\mathbf{W_0}, \mathbf{\Delta W}, \pi, \gamma, \mathbf{p_k})$: Given the initial DNN model parameters W_0, the model update ΔW, the proof π, the signature γ, and the public key p_k, it outputs either accept or reject.*

In this paper, we consider the adversary who has full knowledge of the authentication protocol. Next, we define the security of the authentication protocol against such adversary.

Definition 22 *Let* **Auth** *be an authentication scheme $\{$**genkey**, **setup**, **certify**, **verify**$\}$. Let* **Adv** *be a probabilistic polynomial-time adversary that is only given p_k and has unlimited access to all algorithms of* **Auth**. *Then, given a DNN with initial parameters W_0 and a dataset T,* **Adv** *returns a wrong model update $\Delta W'$ and a proof π': $\{\Delta W', \pi'\} \leftarrow$ **Adv**(D, W_0, p_k). We say* **Auth** *is* *secure if for any p_k generated by the* **genkey** *routine, for any γ generated by the* **setup** *routine, and for any probabilistic polynomial-time adversary* **Adv**, *it holds that*

$$Pr(accept \leftarrow \textbf{verify}(W_0, \Delta W', \pi', \gamma, p_k)) \leq negli(\lambda),$$

where $negli(\lambda)$ is a negligible function in the security parameter λ. Intuitively, **Auth** is secure if with negligible probability the incorrect model update can escape from verification.

3   Problem Statement

Threat Model. In this paper, we consider the server $\mathcal{S}$ that may return incorrect trained model due to various reasons. For example, the learning process might be terminated before it reaches convergence due to the system's Byzantine failures (e.g., software bugs and network issues). $\mathcal{S}$ may also be incentivized to halt the training program early in order to save the computational cost and seek for a higher profit. Given the untrusted nature of the remote server, it is thus crucial for the client to verify the correctness of the returned DNN model before using the model for any decision-making task.

Problem Statement. We consider the problem setting in which the data owner $\mathcal{O}$ outsources the training set T on the server. $\mathcal{O}$ also can specify the configuration of the DNN model M whose initial parameters are specified by W_0. The server $\mathcal{S}$ trains M until it reaches convergence (a local optima), and outputs the model update $\Delta W = f(T; W_0)$. However, with the presence of security threats, the model update ΔW returned by the server may not be a local optima. Therefore, our goal is to design an integrity verification protocol (Definition 21) that enables a third-party verifier $\mathcal{V}$ to verify if ΔW helps the model reach convergence without the access to the private training data.

4   Authentication Method

In this section, we explain the details of our authentication protocol. The **genkey** protocol is straightforward: the data owner $\mathcal{O}$ picks a pairing function e on two sufficiently large cyclic groups G and G_T of order p, a generator $g \in G$, and a secret key $s \in \mathbb{Z}_p$. Then it outputs a pair of secrete and public key (s_k, p_k), where $s_k = s$, and $p_k = \{g, G, G_T, e, v, H(\cdot)\}$, where $v = g^s \in G$, and $H(\cdot)$ is a hash function whose output domain is $\mathbb{Z}_p$. $\mathcal{O}$ keeps s_k private and distributes p_k to the other involved parties. In the following discussions, we only focus on the **setup**, **certify** and **verify** protocols.

Overview of Our Approach. We design a verification method that only uses a short proof of the results for verification. Consider a data owner $\mathcal{O}$ that has a private dataset T. Before transferring T to the server, $\mathcal{O}$ executes the *setup* protocol to generate a short signature γ of T, and disseminate γ to the verifier

V. $\mathcal{O}$ also sets up a DNN model M with initial weights W_0. Then $\mathcal{O}$ outsources M (with W_0) and the training dataset T to $\mathcal{S}$. After receiving T and M with its initial setup, the server $\mathcal{S}$ optimizes M and obtains the model updates ΔW. Besides returning ΔW to the verifier $\mathcal{V}$, $\mathcal{S}$ sends two errors E_1 and E_2, where E_1 is the error when the model reaches convergence as claimed (computed by Eq. 5) and E_2 is the error by running an additional round of backpropagation and feedforward process after convergence. Furthermore, $\mathcal{S}$ follows the *certify* protocol and constructs a short cryptographic proof π of E_1 and E_2. The proof π includes: (1) the cryptographic digest π_T of the samples, and (2) the intermediate results of feedforward and backpropagation processes in computing E_1 and E_2. The verifier $\mathcal{V}$ then runs the *verify* protocol and checks the correctness of ΔW by the following three steps:

- *Authenticity verification of π_T*: $\mathcal{V}$ checks the integrity of π_T against the dataset signature γ that is signed by $\mathcal{O}$;
- *Authenticity verification of E_1 and E_2*: Without access to the private data T, $\mathcal{V}$ verifies if both errors E_1 and E_2 are computed honestly from T, by using π_T and the other components in the proof π;
- *Convergence verification*: $\mathcal{V}$ verifies if E_1 and E_2 satisfy the convergence condition (i.e., whether ΔW helps the model to reach convergence).

Next, we discuss the **Setup**, **Certify** and **Verify** protocols respectively. Then we discuss how to deal with decimal and negative weights.

4.1 Setup Protocol

Based on the public key, we define the following function for the data owner $\mathcal{O}$ to calculate a synopsis for each sample $(\vec{x}, y)$ in T. In particular,

$$d(\vec{x}, y)) = H(g^{x_1}||g^{x_2}||\cdots||g^{x_m}||g^y), \tag{11}$$

where $x_1, x_2, \ldots, x_m$ are the features, y is the label, and g is the group generator.

With the help the secret key s, $\mathcal{O}$ generates the signature γ for $(\vec{x}, y)$ with $\tau = d(\vec{x}, y))^s$. Then instead of sharing the large amount of signatures with the verifier, $\mathcal{O}$ creates an aggregated signature $\gamma = \pi_{i=1}^n \tau_i$, where τ_i is the signature for the i-th sample in the training data T. Then γ serves as a short signature of the whole dataset T.

4.2 Certify Protocol

To enable the verifier to verify E_1 and E_2 without access to the private samples $T = \{(\vec{x}, y)\}$, our Certify protocol construct a *proof* π as following: $\pi = \{\pi_E, \pi_W, \pi_T\}$, where

- $\pi_E = \{E_1, E_2\}$, i.e., π_E stores the errors of the model.

- $\pi_T = \{\{g^{x_i}\}, g^y | \forall (\vec{x}, y) \in T\}$, i.e., π_T stores the digest of original data $\{\vec{x}\}$ and $\{y\}$. Storing the digest but not the original data is to due to the privacy concern in the outsourcing setting (Sect. 2).
- $\pi_W = \{\{\Delta w_{jk}^1\}, \{z_k^1\}, \{\hat{z}_k^1\}, g^{\delta^o}, \{\delta_k^L\}\}$, where Δw_{jk}^1 is the weight updated between the input and *first* hidden layer by one round of backpropagation after the model reaches convergence, z_k^1 and $\hat{z}_k^1$ are the weighted sum of the neuron n_k^1 (Eq. 1) at convergence and one round after convergence respectively, δ^o and $\{\delta_k^L\}$ are the error signals at output and the last hidden layer at convergence respectively. Intuitively, π_W stores a subset of model outputs at the final two rounds (i.e., the round reaching convergence and one additional round afterwards).

4.3   Verify Protocol

The verification process consists of four steps: (1) authenticity verification of π_T; (2) one feedforward to verify the authenticity of E_1; (3) one backpropagation to update weights and another feedforward to verify the authenticity of E_2; and (4) verification of convergence, i.e. if $|E_1 - E_2| \leq \theta$, where θ is a pre-defined threshold for termination condition. Next, we discuss these steps in details.

Step 1. Verification of π_T: The verifier firstly verifies the authenticity of π_T, i.e., the digest of training samples. In particular, the verifier checks whether the following is true: $\Pi_{d(\vec{x},y) \in \pi_T} e((\vec{x}, y), v) \overset{?}{=} e(\gamma, g)$, where $d(\cdot)$ is the synopsis function (Eq. (11)), $v = g^s$ is a part of the public key, γ is the aggregated signature provided by the data owner. If π_T passes the verification, $\mathcal{V}$ is assured that the digests in π_T are calculated from the intact dataset T.

Step 2. Verification of E_1: First, the verifier $\mathcal{V}$ verifies if the weighted sum $\{z_k^1\}$ at the final round is correctly computed. Note that $\mathcal{V}$ is aware of w_{ik}^1. $\mathcal{V}$ also obtains $\{g^{x_i}\}$ and $\{z_k^1\}$ from π_W in the proof. Then to verify the correctness of $\{z_k^1\}$, for each z_k^1, $\mathcal{V}$ checks if the following is true:

$$\Pi e(g^{x_i}, g^{w_{ik}^1}) \overset{?}{=} e(g, g)^{z_k^1}. \tag{12}$$

Once $\mathcal{V}$ verifies the correctness of $\{z_k^1\}$, it calculates the activation of the hidden layers and thus the output o (Eqs. (2) and (3)). Next, $\mathcal{V}$ checks if the following is true:

$$\Pi_{(\vec{x},y) \in D} e(g^{y-o}, g^{y-o}) \overset{?}{=} e(g, g)^{2NE_1}, \tag{13}$$

where $g^{y-o} = g^y * g^{-o}$. Note that g^y is included in the proof. $\mathcal{V}$ can compute g^{-o} by using o computed previously.

Step 3. Verification of E_2: This step consists of five-substeps. The first four substeps verify the correctness of weight increment in the backpropagation process, including the verification of error signal at the output layer, the verification of error signal at the last hidden layer, the verification of weight increments between all hidden layers, and verification of weight increments between the input and the first hidden layer. The last substep is to verify the authenticity of E_2 based on the updated weights. Next, we discuss the details of these five substeps.

First, $\mathcal{V}$ verifies the correctness of g^{δ^o}. Following Eq. (6), $\mathcal{V}$ can easily predict label y with δ^o. Therefore, π_W only includes g^{δ^o}. $\mathcal{V}$ verifies the following:

$$e(g^{-o}g^y, g^{-\sigma'(z^o)}) \overset{?}{=} e(g, g^{\delta^o}), \tag{14}$$

where g^{-o} and $g^{-\sigma'(z^o)}$ are computed by $\mathcal{V}$, and g^y and g^{δ^o} are from the proof.

Second, $\mathcal{V}$ verifies the correctness of δ_k^L (Eq. (7)), i.e., the error signal on the k-th neuron on the last hidden layer, by checking if $e(g^{w_k^o \sigma'(z_k^L)}, g^{\delta^o}) \overset{?}{=} e(g, g)^{\delta_k^L}$, where $g^{w_{kj}^o \sigma'(z_k^L)}$ is computed by $\mathcal{V}$, and δ_k^L and g^{δ^o} are obtained from the proof.

Third, $\mathcal{V}$ calculates the error signal of other hidden layers by following Eq. (7). Then with the knowledge of the activation on every hidden layer (by Step 2), $\mathcal{V}$ computes the derivatives of the weights (Eq. 8) on the hidden layers to update the weights between consecutive hidden layers (Eqs. 9–10).

Fourth, $\mathcal{V}$ verifies the weight increment between input and the first hidden layer. We must note that $\mathcal{V}$ cannot compute $\frac{\partial C}{\partial w_{jk}^1}$ (Eq. (8)) and Δw_{jk}^1 (Eq. (9)) as it has no access to the input feature x_j. Thus $\mathcal{V}$ obtains Δw_{jk}^1 from the proof π and verifies its correctness by checking if the following is true:

$$\Pi_{(\vec{x}, y) \in D} e(g^{x_j}, g^{\eta \delta_k^1}) \overset{?}{=} e(g^{\Delta w_{jk}^1}, g^{-N}). \tag{15}$$

Note that g^{x_j} and Δw_{jk}^1 are included in the proof, and $g^{\eta \delta_k^1}$ and g^{-N} are calculated by $\mathcal{V}$. After Δw_{jk}^1 is verified, $\mathcal{V}$ updates the weight by Eq. (10). Finally, $\mathcal{V}$ verifies E_2 by following the same procedure of Step 2 on the updated weights.

Step 4. Verification of Convergence: If E_1 and E_2 pass the authenticity verification, the verifier verifies the convergence of training by checking if $|E_1 - E_2| \le \theta$, i.e., it reaches the termination condition.

We have the following theorem to show the security of VERIDL.

Theorem 1. *The authentication protocols of* VERIDL *is secure (Definition 22).*

We refer the readers to the extended version [3] of the paper for the detailed proof.

4.4   Dealing with Decimal and Negative Values

One weaknesses of bilinear pairing is that it cannot use decimal and negative values as the exponent in g^e. Therefore, the verification in Eqs. 12–15 cannot be performed easily. To address this problem, we extend the bilinear pairing protocol to handle decimal and negative values.

Decimal Values. We design a new method that conducts decimal arithmetic in an integer field without accuracy loss. Consider the problem of checking if $b * c \overset{?}{=} e$, where b, c and e are three variables that may hold decimal values. Let L_T be the maximum number of bits after the decimal point allowed for any value. We define a new operator $f(\cdot)$ where $f(x) = x * 2^{L_T}$. Obviously, $f(x)$ must be an integer. We pick two cyclic groups G and G_T of sufficiently large order p such that $f(x)f(y) < Z_p$. Thus, we have $g^{f(x)} \in G$, and $e(g^{f(x)}, g^{f(y)}) \in G_T$. To make the verification in Eq. (14) applicable with decimal values, we check if $e(g^{f(b)}, g^{f(c)}) \overset{?}{=} e(g, g)^{f(e)}$. Obviously, if $e(g^{f(b)}, g^{f(c)}) = e(g, g)^{f(e)}$, it is natural that $b * c = e$. The verification in Eq. (12), (13) and (15) is accomplished in the same way, except that the involved values should be raised by 2^{L_T} times.

Negative Values. Equations (12–15) check for a given pair of vectors $\vec{u}, \vec{v}$ of the same size, whether $\sum u_i v_i = z$. Note that the verification in Eq. (14) can be viewed as a special form in which both $\vec{u}$ and $\vec{v}$ only include a single scalar value. Also note that u_i, v_i or z may hold negative values. Before we present our methods to deal with negative values, we first define an operator $[\cdot]$ such that $[x] = x \bmod p$. Without loss of generality, we assume that for any $\sum u_i v_i = z$, $-p < u_i, v_i, z < p$. We have the following lemma.

Lemma 2. *For any pair of vectors $\vec{u}, \vec{v}$ of the same size, and $z = \sum u_i v_i$, we have*

$$\left[\sum [u_i][v_i] \right] = \begin{cases} z & \text{if } z \geq 0 \\ z + p & \text{otherwise.} \end{cases}$$

We omit the proof of Lemma 2 here due to the limited space; the proof can be found in the extended version [3] of the paper. Following Lemma 2, we have Theorem 3 to verify vector dot product operation in case of negative values based on bilinear pairing.

Theorem 3. *To verify $\sum u_i v_i \overset{?}{=} z$, it is equivalent to checking if*

$$\Pi e(g^{[u_i]}, g^{[v_i]}) \overset{?}{=} \begin{cases} e(g, g)^z & \text{if } z \geq 0 \\ e(g, g)^{(z+p)} & \text{otherwise.} \end{cases} \tag{16}$$

We omit the proof due to the limited space, and include it in the extended version [3]. Next, we focus on Eq. (12) and discuss our method to handle negative values. First, based on Lemma 2, we can see that for any x_i and w_{ik}^1, if $x_i w_{ik}^1 \geq 0$, then $[x_i][w_{ik}^1] = x_i w_{ik}^1$; otherwise, $[x_i][w_{ik}^1] = x_i w_{ik}^1 + p$. Therefore, to prove $z_k^1 = \sum x_i w_{ik}^1$, the server includes a flag $sign_i$ for each x_i in the proof, where

$$sign_i = \begin{cases} + & \text{if } x_i \geq 0 \\ - & \text{otherwise.} \end{cases}$$

Meanwhile, for each z_k^1, the server prepares two values $p_k^1 = \sum_{i:x_iw_{ik}^1 \geq 0} x_iw_{ik}^1$ and $n_k^1 = \sum_{i:x_iw_{ik}^1 < 0} x_iw_{ik}^1$, and includes them in the proof.

In the verification phase, since the client is aware of w_{ik}^1, with the knowledge of $sign_i$ in the proof, it can tell if $x_iw_{ik}^1 \geq 0$ or not. So the client first verifies if

$$\Pi_{i:x_iw_{ik}^1 \geq 0} e(g^{[x_i]}, g^{[w_{ik}^1]}) \overset{?}{=} e(g,g)^{p_k^1}, \Pi_{i:x_iw_{ik}^1 < 0} e(g^{[x_i]}, g^{[w_{ik}^1]}) \overset{?}{=} e(g,g)^{n_k^1+p},$$

where $g^{[x_i]}$ is included in the proof, and $g^{[w_{ik}^1]}$ is computed by the client. Next, the client checks if $p_k^1 + n_k^1 \overset{?}{=} z_k^1$.

5   Experiments

5.1   Setup

Hardware and Platform. We implement VERIDL in C++. We use the implementation of bilinear mapping from PBC library[3]. The DNN model is implemented in Python on TensorFlow. We simulate the server on a computer of 2.10 GHz CPU, 48 cores and 128 GB RAM, and the data owner and the verifier on 2 computers of 2.7 GHz Intel CPU and 8 GB RAM respectively.

Datasets. We use the following four datasets that are of different data types: (1) **MNIST** dataset that contains 60,000 image samples and 784 features; (2) **TIMIT** dataset that contains 4,620 samples of broadband recordings and 100 features; (3) **ADULT** dataset that includes 45,222 records and 14 features; and (4) **HOSPITAL** dataset that contains 230,000 records and 33 features.

Neural Network Architecture. We train a DNN with four fully connected hidden layers for the MNIST, ADULT and HOSPITAL datasets. We vary the number of neurons on each hidden layer from 10 to 50, and the number of parameters from 20,000 to 100,000. We apply sigmoid function on each layer, except for the output layer, where we apply softmax function instead. We optimize the network by using gradient descent with the learning rate $\eta = 0.1$. By default, the minibatch size is 100. We use the same DNN structure for the TIMIT dataset with ReLU as the activation function.

Basic and Optimized Versions of VERIDL. We implement two versions of VERIDL: (1) Basic approach (**B-VERIDL**): the proof of model updates is generated for every single input example $(\vec{x}, y)$; and (2) Optimized approach (**O-VERIDL**): the proof is generated for every unique value in the input $\{(\vec{x}, y)\}$.

Existing Verification Approaches for Comparison. We compare the performance of VERIDL with two alternative approaches: (1) C_1. **Homomorphic encryption (LHE) vs. bilinear mapping:** When generating the proof, we use LHE to encrypt the plaintext values in the proof instead of bilinear mapping; (2) C_2. **Result verification vs. re-computation of model updates**

---

[3] https://crypto.stanford.edu/pbc/.

by privacy-preserving DL: The server encrypts the private input samples with homomorphic encryption. The verifier executes the learning process on the encrypted training data, and compares the computed results with the server's returned updates. For both comparisons, we use three different implementations of HE. The first implementation is the Brakerski-Gentry-Vaikuntanathan (BGV) scheme provided by HElib library[4]. The second implementation is built upon the PALISADE library[5] that uses primitives of lattice-based cryptography for implementation of HE. The last one is built upon the Microsoft SEAL project [12], which provides a programming interface to lightweight homomorphic encryption.

5.2   Efficiency of VERIDL

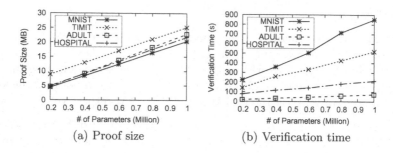

(a) Proof size   (b) Verification time

Fig. 1. Performance of VERIDL (minibatch size 100)

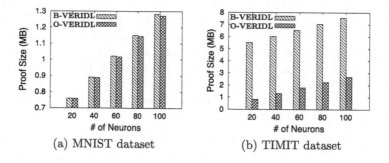

(a) MNIST dataset   (b) TIMIT dataset

Fig. 2. Proof size

Proof Size. The results of proof size of VERIDL on four datasets, with various number of neurons at each hidden layer, are shown in Fig. 1 (a). In all settings, the proof size is small (never exceeds 25 MB even with one million parameters).

---

[4] https://github.com/shaih/HElib.
[5] https://git.njit.edu/palisade/PALISADE/wikis/home.

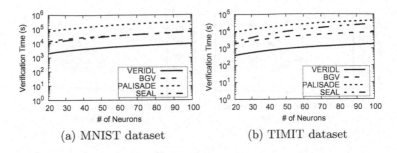

(a) MNIST dataset   (b) TIMIT dataset

Fig. 3. Verification time (minibatch size 100)

This demonstrates the benefit of using bilinear pairing for proof construction. Second, we observe a linear increase in the proof size with the growth of the number of parameters. This is because the dominant components of proof size is the size of $\{\Delta w_{jk}^1\}$, $\{z_k^1\}$ and $\{\hat{z}_k^1\}$, which grows with the number of parameters.

Verification Time. The results of verification time on all four datasets are shown in Fig. 1 (b). First, the verification time is affordable even on the datasets of large sizes. Second, the verification time grows linearly with the number of hyperparameters. The reason is that the number of neurons on the first hidden layer increases linearly with the growth of parameters in the neuron network, while the verification time linearly depends on the input dimension and the number of neurons in the first hidden layer.

B-VERIDL VS. O-VERIDL. We compare the performance of the basic and optimized versions of VERIDL. Figure 2 demonstrates the proof size of B-VERIDL and O-VERIDL with various number of neurons at each hidden layer in the DNN model. In general, the proof size is small (less than 1.3 MB and 8 MB for MNIST and TIMIT datasets respectively). Furthermore, the proof size of O-VERIDL can be smaller than B-VERIDL; it is 20%–26% of the size by B-VERIDL on TIMIT dataset. This demonstrates the advantage of O-VERIDL. The results also show that the proof size of both O-VERIDL and B-VERIDL gradually rises when the number of neurons increases. However, the growth is moderate. This shows that VERIDL can be scaled to large DNNs.

Comparison with Existing Approaches. We evaluate the verification time of different proof generation methods (defined by the comparison C_1 in Sect. 5.1) for various numbers of neurons on all four datasets, and report the results of MNIST and TIMIT datasets in Figs. 3. The results on ADULT and HOSPITAL datasets are similar; we omit them due to the limited space. we observe that for all four datasets, VERIDL (using bilinear mapping) is more efficient than using HE (i.e., BGV, PALISADE and SEAL) in the proof. Thus bilinear mapping is a good choice as it enables the same function over ciphertext with cheaper cost. Besides, the time performance of both VERIDL and HE increases when the number of neurons in the network grows. This is expected as it takes more time to verify a more complex neural network. We also notice that all approaches take

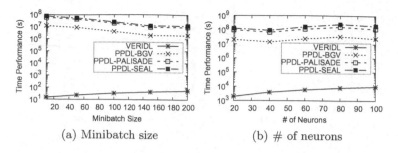

Fig. 4. Verification vs. re-computation of model updates

longer time on the MNIST dataset than the other datasets. This is because the MNIST dataset includes more features than the other datasets; it takes more time to complete the verification in Eqs. 12–15.

5.3 Verification vs. Re-computation of Model Updates

We perform the comparison C_2 (defined in Sect. 5.1) by implementing the three HE-based privacy-preserving deep learning (PPDL) approaches [6,9,12] and comparing the performance of VERIDL with them. To be consistent with [6,9], we use the approximated ReLU as the activation function due to the fact that HE only supports low degree polynomials. Figure 4 shows the comparison results. In Fig. 4 (a), we observe that VERIDL is faster than the three PPDL methods by more than three orders of magnitude. An interesting observation is that VERIDL and PPDL take opposite pattern of time performance when the minibatch size grows. The main reason is that when the minibatch size grows, VERIDL has to verify E_1 and E_2 from more input samples (thus takes longer time), while PPDL needs fewer epochs to reach convergence (thus takes less time). Figure 4 (b) shows the impact of the number of neurons on the time performance of both VERIDL and PPDL. Again, VERIDL wins the three PPDL methods by at least three orders of magnitude.

5.4 Robustness of Verification

To measure the robustness of VERIDL, we implement two types of server's misbehavior, namely *Byzantine failures* and *model compression attack*, and evaluate if VERIDL can catch the incorrect model updates by these misbehavior.

Byzantine Failure. We simulate the Byzantine failure by randomly choosing 1% neurons and replacing the output of these neurons with random values. We generate three types of wrong model updates: (1) the server sends the wrong error E_1 with the proof constructed from correct E_1; (2) the server sends wrong E_1 with the proof constructed from wrong E_1; (3) the server sends correct E_1 and wrong E_2. Our empirical results demonstrate that VERIDL caught all wrong model updates by these Byzantine failures with 100% guarantee.

Model Compression Attack. The attack compresses a trained DNN network with small accuracy degradation [2,7]. To simulate the attack, we setup a fully-connected network with two hidden layers and sigmoid activation function. The model parameters are set by randomly generating 32-bits weights. We use ADULT dataset as the input. We simulate two types of model compression attacks: (1) the *low-precision floating points attack* that truncates the initial weights to 8-bits and 16-bits respectively and train the truncated weights; and (2) the *network pruning attack* that randomly selects 10%–25% weights to drop out during training. For both attacks, we run 50 times and calculate the absolute difference between the error E_1' computed from the compressed model and the error E_1 of the correct model. From the results, we observe that the error difference produced by the low-precision attack is relatively high (with a 35% chance of larger than or equal to 0.02), and can be as large as 0.2. While the error differences of the network pruning attack are all between 0.002 and 0.01. In all cases, we have $|E_1' - E_1| \geq 10^{-9}$. We omit the results due to the limited space. We must note that given the DNN model is a 32-bit system, VERiDL can determine that $E_1' \neq E_1$ as long as $|E_1' - E_1| \geq 10^{-9}$. Therefore, VERiDL can detect the incorrect model updates by both network compression attacks, even though the attacker may forge the proof of E_1 to make E_1' pass the verification.

6 Related Work

Verified artificial intelligence (AI) [13] aims to design AI-based systems that are provably correct with respect to mathematically-specified requirements. SafetyNet [5] provides a protocol that verifies the execution of DL on an untrusted cloud. The verification protocol is built on top of the *interactive proof* (IP) protocol and arithmetic circuits. It places a few restrictions on DNNs, e.g., the activation functions must be polynomials with integer coefficients, which disables the activation functions that are commonly used in DNNs such as ReLU, sigmoid and softmax. Recent advances in zero-knowledge (ZK) proofs significantly reduce the verification and communication costs, and make the approach more practical to verify delegated computations in public [14]. ZEN [4] is the first ZK-based protocol that enables privacy-preserving and verifiable inferences for DNNs. However, ZEN only allows ReLU activation functions. We remove such strict assumption. *VeriDeep* [8] generates a few minimally transformed inputs named *sensitive samples* as fingerprints of DNN models. If the adversary makes changes to a small portion of the model parameters, the outputs of the sensitive samples from the model also change. However, VeriDeep only can provide a probabilistic correctness guarantee.

7 Conclusion and Future Work

In this paper, we design VERiDL, an authentication framework that supports efficient integrity verification of DNN models in the DLaaS paradigm. VERiDL extends the existing bilinear grouping technique significantly to handle the verification over DNN models. The experiments demonstrate that VERIDL can verify the correctness of the model updates with cheap overhead.

While VERIDL provides a deterministic guarantee by verifying the output of *all* neurons in DNN, generating the proof for such verification is time costly. Thus an interesting direction to explore in the future is to design an alternative *probabilistic* verification method that provides high guarantee (e.g., with 95% certainty) but with much cheaper verification overhead.

References

1. Bengio, Y.: Learning deep architectures for AI. Found. Trends Mach. Learn. **2**(1), 1–127 (2009)
2. Courbariaux, M., Bengio, Y., David, J.-P.: Training deep neural networks with low precision multiplications. arXiv preprint arXiv:1412.7024 (2014)
3. Dong, B., Zhang, B., (Wendy) Wang, H.: VeriDL: integrity verification of outsourced deep learning services (extended version version). arXiv preprint arXiv:2107.00495 (2021)
4. Feng, B., Qin, L., Zhang, Z., Ding, Y., Chu, S.: Zen: efficient zero-knowledge proofs for neural networks. IACR Cryptology ePrint Archive 2021, 87 (2021)
5. Ghodsi, Z., Gu, T., Garg, S.: SafetyNets: verifiable execution of deep neural networks on an untrusted cloud. In: Advances in Neural Information Processing Systems, pp. 4675–4684 (2017)
6. Gilad-Bachrach, R., Dowlin, N., Laine, K., Lauter, K., Naehrig, M., Wernsing, J.: CryptoNets: applying neural networks to encrypted data with high throughput and accuracy. In: International Conference on Machine Learning, pp. 201–210 (2016)
7. Gong, Y., Liu, L., Yang, M., Bourdev, L.: Compressing deep convolutional networks using vector quantization. arXiv preprint arXiv:1412.6115 (2014)
8. He, Z., Zhang, T., Lee, R.B.: VeriDeep: verifying integrity of deep neural networks through sensitive-sample fingerprinting. arXiv preprint arXiv:1808.03277 (2018)
9. Hesamifard, E., Takabi, H., Ghasemi, M.: CryptoDL: deep neural networks over encrypted data. arXiv preprint arXiv:1711.05189 (2017)
10. LeCun, Y., Bengio, Y., Hinton, G.: Deep learning. Nature **521**(7553), 436 (2015)
11. Papamanthou, C., Tamassia, R., Triandopoulos, N.: Optimal verification of operations on dynamic sets. In: Annual Cryptology Conference, pp. 91–110 (2011)
12. Microsoft SEAL (release 3.5). https://github.com/Microsoft/SEAL, April 2020. Microsoft Research, Redmond
13. Seshia, S.A., Sadigh, D., Shankar Sastry, S.: Towards verified artificial intelligence. *arXiv preprint*arXiv:1606.08514 (2016)
14. Yang, K., Sarkar, P., Weng, C., Wang, X.: QuickSilver: efficient and affordable zero-knowledge proofs for circuits and polynomials over any field. IACR Cryptology ePrint Archive, 2021:76 (2021)

A Unified Batch Selection Policy
for Active Metric Learning

K. Priyadarshini[1]([✉]), Siddhartha Chaudhuri[2], Vivek Borkar[1],
and Subhasis Chaudhuri[1]

[1] IIT Bombay, Bombay, India
priyadarshini.k@iitb.ac.in, {borkar,sc}@ee.iitb.ac.in
[2] Adobe Research, Bangalore, India
sidch@adobe.com

Abstract. Active metric learning is the problem of incrementally selecting high-utility batches of training data (typically, ordered triplets) to
annotate, in order to progressively improve a learned model of a metric over some input domain as rapidly as possible. Standard approaches,
which independently assess the informativeness of each triplet in a batch,
are susceptible to highly *correlated* batches with many redundant triplets
and hence low overall utility. While a recent work [20] proposes *batch-
decorrelation* strategies for metric learning, they rely on ad hoc heuristics
to estimate the correlation between two triplets at a time. We present a
novel batch active metric learning method that leverages the Maximum
Entropy Principle to learn the least biased estimate of triplet distribution
for a given set of prior constraints. To avoid redundancy between triplets,
our method collectively selects batches with maximum *joint entropy*,
which simultaneously captures both informativeness *and* diversity. We
take advantage of the submodularity of the joint entropy function to construct a tractable solution using an efficient greedy algorithm based on
Gram-Schmidt orthogonalization that is provably $(1 - \frac{1}{e})$-optimal. Our
approach is the first batch active metric learning method to define a unified score that balances informativeness and diversity for an entire batch
of triplets. Experiments with several real-world datasets demonstrate
that our algorithm is robust, generalizes well to different applications
and input modalities, and consistently outperforms the state-of-the-art.

Keywords: Batch active learning · Perceptual metric · Submodular
optimization · Maximum Entropy Principle

1 Introduction

Understanding similarity between two objects is fundamental to many vision
and machine learning tasks, e.g. object retrieval [33], clustering [35] and classification [30]. Most existing methods model a *discrete* measure of similarity

Electronic supplementary material The online version of this chapter (https://
doi.org/10.1007/978-3-030-86520-7_37) contains supplementary material, which is
available to authorized users.

N. Oliver et al. (Eds.): ECML PKDD 2021, LNAI 12976, pp. 599–616, 2021.
https://doi.org/10.1007/978-3-030-86520-7_37

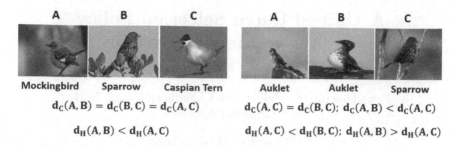

A   B   C   A   B   C

Mockingbird   Sparrow   Caspian Tern   Auklet   Auklet   Sparrow

$d_C(A, B) = d_C(B, C) = d_C(A, C)$ $d_C(A, C) = d_C(B, C); d_C(A, B) < d_C(A, C)$

$d_H(A, B) < d_H(A, C)$ $d_H(A, C) < d_H(B, C); d_H(A, B) > d_H(A, C)$

Fig. 1. Difference between class-based and perceptual distances on two different types of triplets. In each case, the class-based metric d_C fails to capture intra-class variations and inter-class similarities and is not compatible with the perceptual metric d_H.

based on class labels: all inter-class samples are considered equally dissimilar, even though their features differ by different degrees. But human estimation of perceptual (dis)similarity is often more fine-grained. We may choose, for example, *continuous* measures such as the *degree* of perceived similarity in taste or visual appearance for comparing two food dishes, rather than discrete categorical labels (Fig. 1). Thus, it is important to build a *continuous* perceptual space to model human-perceived similarity between objects. Recent studies demonstrate the importance of perceptual metrics in several tasks in computer vision and cognitive science [15, 19, 36].

Early work on perceptual metric learning focuses on non-parametric methods (e.g. Multidimensional scaling (MDS) [18]) which use numerical measurements of pairwise similarity for training. These are hard to gather and suffer from inconsistency. Instead, similarity *comparisons* of the form "Is object x_i more similar to object x_j than object x_k?" are easier to gather and more stable [14]. They form a useful foundation for several tasks, including perceptual metric learning. However, the number of possible triplets of n objects is $O(n^3)$, making it infeasible to label even a significant fraction of them. Fortunately, many triplets are redundant and we can effectively model the metric using only a few high-utility triplets (Fig. 2). Thus it is imperative to identify and annotate a subset of high-quality triplets that are jointly informative for the model, *without knowing the annotations of any triplets in advance*. We stress this last point since it renders common triplet sampling strategies such as (semi-)hard negative mining, which rely on access to a fully annotated dataset, inadmissible.

Active learning is a standard technique that addresses this issue by iteratively identifying small batches of informative samples and soliciting labels for them. While extensively studied for class label-based learning tasks, there exists very little literature [9, 20, 31] on active learning which focuses on perceptual/general metric learning. Further, these works merely assess the informativeness of *individual* triplets with uncertainty measures, which assume a triplet with high prediction uncertainty is more crucial to label. Although effective in many scenarios, such an uncertainty measure makes a myopic decision based solely on the current model's prediction and fails to capture the triplets' collective distribution as a

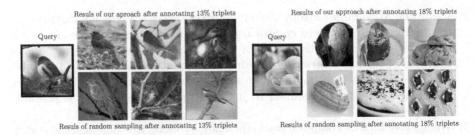

Fig. 2. Top-3 retrieved images ranked from most to least similar by a perceptual metric (visual appearance for birds, and taste for food) trained on randomly selected (but correctly annotated) triplets vs. high-quality triplets identified for annotation by our method. For a fair comparison, both methods run for equal training rounds and solicit annotations for equal amounts of training data – 13% of the CUB-200 bird dataset on the left, and 18% of the Yummly-Food dataset on the right.

whole. Independently assessed triplets may themselves have much redundancy even if they are individually informative. Hence the triplets should be not merely informative but also *diverse* or *decorrelated*.

Kumari *et al.* [20] proposed a method for selecting informative *and* decorrelated batches of triplets for active metric learning. However, their approach suffers from three major limitations: (1) The active learning strategy is based on a two stage optimization for informativeness (choice of an overcomplete batch-pool of individually informative triplets) and diversity (subsequent trimming of the batchpool), applied sequentially. It does not always ensure an optimal trade-off between the two criteria. (2) The proposed diversity measures are all ad-hoc with no principled connection to informativeness. Being heuristic, no single measure works consistently well in all cases, making it harder for a user to select which measure to use in practice. (3) The informativeness of a triplet is determined using a point estimate of the perceptual metric. Bias in the latter, e.g., because of suboptimal batch selection in prior iterations, directly translates to bias in informativeness, which can misguide the strategy.

To mitigate these issues, we propose a new batch active learning algorithm developed specifically for triplet-based metric learning. Our **key insight** is to express a set of (unannotated) triplets as a vector of random variables, and select batches of triplets that maximize the *joint entropy* measure. Thus, instead of separately expressing and optimizing informativeness for individual triplets and diversity for pairs of triplets, we develop a single probabilistic informativeness measure *for a batch of triplets*. We also provide computationally efficient approximate solutions with provable guarantees. Specifically, our main technical contributions are:

1. We propose to use the joint entropy of the distribution of triplet margins to rank a batch of unannotated triplets. We estimate the second-order statistics (mean and covariance) of *triplet margins* by randomly perturbing the current model trained on prior batches as in [7], to characterize the distribution.

2. Using the Maximum Entropy Principle, we arrive at a Gaussian distribution compatible with the given empirical mean and covariance, whose entropy is characterized by the determinant of the covariance matrix. As exact maximization of the joint entropy is prohibitively expensive (there are $\binom{m}{b}$ possible batches of size b from m triplets), we use the fact that entropy is monotone increasing and submodular to justify a greedy policy which is provably $(1 - \frac{1}{e})$-optimal [22].
3. We achieve further computational efficiency by using the fact that the covariance matrix is a Gram matrix, and its determinant can be computed recursively using a greedy policy. Our method recursively maximizes successive projection errors of a set of vectors, picked one at a time, when projected onto the span of previous choices. This amounts to successive maximization of the conditional entropy, and is easily implemented using Gram-Schmidt orthogonalization.

We demonstrate the effectiveness of our approach through extensive experiments on different applications and data in different modalities (image, taste and haptic). In addition to having a sound theoretical justification, our method provides a significant performance gain over the current state-of-the-art.

2   Related Work

The prior work can be roughly divided into three categories. We review representative techniques in each and discuss how our work differs from the existing methods.

2.1   Perceptual Metric Learning

While there is extensive recent research on distance metric learning, most of the algorithms are specific to class-based learning tasks such as classification [30] and clustering [35], which consider two objects similar if they belong to the same class. See Bellet *et al.* [3] for a comprehensive review. In contrast, our goal is to define a perceptual distance that captures the degree of similarity between any two objects irrespective of their classes. Recently, a whole new literature has emerged that emphasizes the importance of learning such continuous measures of similarity for various applications, e.g. for measuring image similarity [36], face recognition [5], concept learning [32,33] and perceptual embedding of objects [1,11,19]. The closest application to ours is perceptual embedding of objects, where the embedding function is learned so as to model the human-perceived inter-object similarity. While multidimensional scaling (MDS) techniques have been extensively applied for this [1,11,18], they are non-parametric and require numerical similarity measurement as inputs, which are hard to gather [14]. Recent works [21,36] address these limitations by developing parametric models using non-numeric relative comparisons. A relevant method is the triplet-based deep metric learning method of Kumari *et al.* [19].

Although our method borrows base metric learning architectures from [19], [19] doesn't aim to make the metric learning algorithm data-efficient by developing an active data sampling technique.

2.2   Active Learning for Classification

Active learning (AL) methods have been well explored for vision and learning tasks, see Settles [27] for a detailed review of active learning methods for class-based learning. Typically, the AL methods select a single instance with the maximum individual utility for annotation in each iteration. The utility of an instance is decided by different heuristics, e.g. uncertainty sampling [20], query-by-committee (QBC) [8], expected gradient length (EGL) [2], and model-output-change (MOC) [6]. The simplest and most widely applicable uncertainty sampling approach has been extended to modern deep learning frameworks and variational inference [29]. However, in all these methods, each sample's utility is evaluated independently without considering dependence between them.

In batch-mode active learning, data items are assessed not one at a time but in batches, to reduce the number of times the model is retrained. To avoid selecting correlated batches, some recent attempts evaluate the whole batch's utility by taking mutual information between samples into account. In contrast to our work, most of them are developed for classification tasks [2,17,24,26]. For example, Kirch et al. [17] define the utility score as the mutual information between data points in a batch and model parameters and then pick a subset with the maximum score. Pinsler et al. [24] formulate the active learning problem as a sparse subset selection problem approximating the complete data posterior of the model parameters. Both methods have a similar motivation to our work, but they are developed for the classification task, and their informativeness measures are not easy to extend to the metric learning task. Ash et al. [2] use the norm of the gradient at a sample to implicitly capture both informativeness and diversity, and select a subset of the farthest samples in the gradient space. This ensures both informativeness and diversity by a single gradient-based measure, which does not work well in the metric learning task, as shown by Kumari et al. [20]. Sener and Savarese [26] follow a similar strategy in a different feature space. Shui et al. [28] introduce a unified approach for training and batch selection process and explicitly define uncertainty-diversity trade-off by adopting Wasserstein distance.

2.3   Active Learning of Perceptual Metrics

There are only a few works on active learning of a perceptual metric. Most of these, e.g. [9,31], are based on a single instance evaluation criterion. They define the utility of a single triplet and select a batch of the individually highest-utility triplets to annotate. In contrast, we define a utility score for a batch taking joint information between triplets into account. The closest work to ours is a very recent paper by Kumari et al. [20]. The algorithm involves a two-stage process. First, it selects an overcomplete set of individually highly informative samples, and then subsamples a less correlated subset, using different triplet-based decorrelation heuristics, as the current batch. This method, in essence, is still based on a single triplet selection strategy. In contrast, we present a new, rigorous approach to define utility for a batch as a whole based on *joint entropy*, providing a unified utility function to balance both informativeness and diversity.

3   Proposed Method

In this section, we first briefly describe the perceptual metric learning setup and the underlying neural network-based learner called PerceptNet [19]. Next, we introduce our novel batch selection policy explicitly designed for triplet-based active metric learning.

3.1   Triplet-Based Active Metric Learning

Let $X = \{x_i\}_1^n$ represent a set of n objects, each described by a d-D feature vector x_i. Also, let T_L be a set of ordered triplets, where each triplet (x_i, x_j, x_k) indicates that the object x_i is more similar to object x_j than to x_k. For brevity we denote (x_i, x_j, x_k) by ijk. We frame the perceptual metric learning problem as learning an embedding $\phi : R^d \rightarrow R^{\hat{d}}$, s.t. the L_2 distance between any two objects in the embedding space $d_\phi(x, y) = \|\phi(x) - \phi(y)\|$ reflects the perceptual distance between them. In recent work, ϕ is typically modeled with a neural network: in our experiments, we choose the existing PerceptNet model [19], where three copies of the same network, with shared weights, process three objects x_i, x_j and x_k during training. The output is optimized with an exponential triplet loss $\mathcal{L} = \sum_{T_L} e^{-\left(d_\phi^2(x_i, x_k) - d_\phi^2(x_i, x_j)\right)}$ to maximize the distance margins (a.k.a "triplet margins"), as defined by the exponent, for training triplets.

The number of possible triplets is cubic in the number of objects, so annotating a significant fraction of them is often intractable, e.g. in domains such as haptics and food tasting where annotation is especially slow. However, an effective embedding can be modeled with far fewer comparisons if triplets are sampled selectively based on *how much information* they would provide if annotated. This calls for active learning. The model is trained iteratively: batches of triplets informative to the current model are selected for annotation in each iteration, after which the model is retrained. However, the efficiency gain of selecting larger batches may be undone by *correlation* among triplets in a batch implying low overall information, a common issue in independent optimization of individual informativeness of each triplet. To mitigate this, prior works have studied *batch decorrelation* strategies for classification [2,17,24]. Recently, Kumari *et al.* [20] developed a decorrelation strategy for metric learning with separate steps for optimizing individual triplet informativeness and then batch diversity. However, as already noted, this work suffers from limitations related to their design choices. In contrast, we develop a method that jointly defines and optimizes the informativeness of an entire batch while implicitly ensuring diversity. The method is grounded in the Maximum Entropy Principle and leads to an attractive computational scheme.

3.2   Joint Entropy Measure for Batch Selection

The key to a good batch mode active learning is an effective informativeness measure for a batch of triplets. For tractability, earlier work typically defines a measure adding up the individual informativeness scores of triplets. A popular score is the Shannon entropy of the prediction probability p_y of the current model trained on prior batches, for a triplet t taking one of two possible orderings $y \in \{ijk, ikj\}$, $H(t) = -\sum_{y \in \{ijk, ikj\}} p_y \log p_y$ [31]. While often termed "uncertainty", this is not a good predictor of actual model uncertainty due to possible bias in the current model [7]. Further, individually high-entropy triplets may also have high mutual information, hence simply adding up the scores may overestimate the actual utility of the batch.

We propose a novel batch selection algorithm based on the *joint entropy* of an entire batch of triplets B, capturing their mutual dependence. We define the joint probability distribution of a set of unannotated triplets on some feature space such as their distance margins. This probability is defined using the *distribution of likely models* given prior batches, reducing any bias due to model training. Note that this is **quite different** from the *prediction probability* of a single fixed model, described above. The joint distribution over a set of triplets naturally captures the notion of interdependence among them.

3.3   Maximum-Entropy Model of the Joint Distribution

Our goal is to postulate the joint probability distribution of unannotated triplets in a batch, preferably in a form that allows efficient computation of its entropy. We represent each triplet t by its distance margin $\xi_t = d_\phi^2(x_i, x_k) - d_\phi^2(x_i, x_j)$. Then a batch, denoted $B = \{t_1, t_2, \ldots, t_b\}$ is represented by the vector of distance margins given by $\boldsymbol{\xi}_B = [\xi_{t_1}, \xi_{t_2}, \ldots, \xi_{t_b}]$.[1] We assume there is uncertainty about these margin predictions arising from the fact that there is a *distribution of plausible models* given the previously annotated data. Hence, each distance margin ξ_{t_i} is a 1D random variable taking different values for different choices of model parameters ϕ. As discussed above, simply looking at the predicted ordering probabilities of individual triplets is both error-prone and fails to consider correlation between triplets. Fortunately, if the model is a neural network, it has been shown that random dropout yields a good Bayesian approximation of model uncertainty [7]. We stochastically apply the dropout K times to the model, evaluating $\boldsymbol{\xi}_B$ each time, to sample the joint margin vector distribution of the batch and to compute the corresponding b-dimensional mean and covariance matrix. We invoke the Maximum Entropy Principle [12] which maximizes the Shannon entropy subject to constraints on prescribed averages. The maximum entropy distribution, consistent with all prior constraints, ensures the largest amount of uncertainty with respect to unknown, and hence introduces no additional biases in the estimation. Empirical estimates of the entropy of the batch from samples are susceptible to noise, and lead to a hard combinatorial optimization over

---

[1] Other triplet based representations are possible: we found the above to be a consistent and more useful feature in practice.

Algorithm 1. Greedy algorithm to maximize $H(B)$

Input: Unlabeled triplets T_U, batch size b, entropy function $H : 2^{T_U} \to \mathbb{R}$ as in Eq. 1.
Output: Batch B that is an $(1 - \frac{1}{e})$-approximation to $\arg\max_{B \subset T_U, |B|=b} H(B)$.

1: $B_0 \leftarrow \emptyset$, $H(B_0) = 0$
2: **for** k=1,..., b **do**
3:      $t_k \leftarrow \arg\max_{t \in T_U \setminus B_{k-1}} \log\left(\det\left(\Sigma_{B_{k-1} \cup \{t\}}\right)/\det\left(\Sigma_{B_{k-1}}\right)\right)$
4:      $B_k \leftarrow B_{k-1} \cup \{t_k\}$
5: **end for**
6: **return** Final subset B_k

---

batches. So we constrain the mean and covariance matrix of the triplet margins to match their empirical values μ_B and Σ_B and maximize the differential entropy $H(B) = -\int p(\boldsymbol{\xi}_B) \log p(\boldsymbol{\xi}_B) d\boldsymbol{\xi}_B$ subject to these constraints. This leads to a multivariate gaussian distribution $N(\mu_B, \Sigma_B)$ with entropy

$$H(B) = \frac{1}{2} \log\left((2\pi e)^b \det(\Sigma_B)\right) \tag{1}$$

Note that this score takes into account inter-triplet correlation, unlike measures depending only on individual marginals. The next task is to efficiently select an optimum batch of size b with maximum informativeness: $B^* = \arg\max_{B \subset T_U, |B|=b} H(B)$, where T_U is the set of currently unannotated triplets.

3.4   Greedy Algorithm for Batch Selection

Since the maximization of the joint entropy function $H(B)$ over subsets is computationally prohibitive, we use the fact that entropy is monotone increasing and submodular to justify a greedy policy which is provably $(1 - \frac{1}{e})$-optimal by the results of Nemhauser *et al.* [22]. The greedy algorithm builds up the set B^* incrementally. In step k, we pick the triplet t_k which has maximum conditional entropy given triplets B_{k-1} selected in previous steps. Specifically,

$$
\begin{aligned}
t_k &= \arg\max_{t \in T_U \setminus B_{k-1}} H(\{t\} \mid B_{k-1}) \\
&= \arg\max_{t \in T_U \setminus B_{k-1}} H(B_{k-1} \cup \{t\}) - H(B_{k-1}) \\
&= \arg\max_{t \in T_U \setminus B_{k-1}} \log\left(\frac{\det\left(\Sigma_{B_{k-1} \cup \{t\}}\right)}{\det\left(\Sigma_{B_{k-1}}\right)}\right).
\end{aligned} \tag{2}
$$

This step is repeated $|T_U|$ times. The greedy policy has low complexity (quantified later) and scales well to large datasets. The overall batch selection algorithm is listed in Algorithm 1. The remaining challenge is to efficiently compute the increment in the determinant of the covariance matrix in each step. We present a recursive algorithm for this, which also clarifies why the method selects a decorrelated batch.

3.5 Recursive Computation of Determinant of Covariance Matrix

The covariance matrix is a Gram matrix, i.e. its $(i, j)^{\text{th}}$ element can be written as the dot product of the ith and j^{th} vectors from a given family of vectors. This allows us to recursively compute its determinant and choose the recursion order according to the greedy policy for approximate optimization. Let u_t denote the zero-mean vector of all sampled distance margins, $[\xi_t(\phi_1), \cdots, \xi_t(\phi_K)] - [\mu_t, \cdots, \mu_t]$, for a single triplet t. The covariance matrix $\Sigma_{B_{k-1}}$ has the form UU^T, where each column of U is $u_s, s \in B_{k-1}$. In the kth step, a new row and column vector for a new triplet t are appended to U. Using the Gram matrix property, we have $\det(\Sigma_{B_{k-1} \cup \{t\}}) - \det(\Sigma_{B_{k-1}}) = \|\tilde{u}_k\|^2$, where $\tilde{u}_k$ is the normal from u_t onto $\text{span}\{u_s \mid s \in B_{k-1}\}$. Thus the greedy scheme successively maximizes the squared projection error $\|\tilde{u}_k\|^2$, over the remaining vectors $\{u_t \mid t \in T_U \backslash B_{k-1}\}$. Thus we select at each step the triplet that is least correlated with the already chosen triplets. The orthogonal projections are computed using the modified Gram-Schmidt orthogonalization scheme from [10], with complexity dn^2, where d is the dimension of the ambient vector space and n the number of vectors. Since we compute the projection error for all $|T_U| - n \approx |T_U|$ remaining triplets at each step (because $|T_U| \gg n$), the overall complexity of the scheme is $dn^2 |T_U|$.

In summary, the submodularity of the joint entropy function naturally combines informativeness, diversity, and representativeness, which are precisely the desired properties for batch mode active learning.

4 Experiments

We perform several experiments to answer the following questions: (1) Is our method competitive with standard baselines, including the state-of-the-art method(s), for different choices of hyperparameters, feature dimension, applications, and datasets? (2) How good is our assumption that the second-order statistics (mean and covariance) are sufficient statistics for estimating the reasonable distribution? (3) How robust is our method to labeling error? We address these questions by conducting several experiments on real-world datasets with different modalities: image, food and haptic. For each of these datasets, we select an appropriate neural network architecture – for the haptic and food datasets we ensure that these architectures exactly match those of Kumari et al. [20] so that the comparison is fair ([20] did not present any result on images, requiring us to implement their method on image databases). We test with different initial pools and varying batch sizes. We also simulated random errors in the triplet orderings to test robustness to labeling error.

Datasets. We evaluate the performance of our method on five real-world datasets for which triplets defining perceptual metrics are available: Yummly food dataset [34]; TUM haptic texture dataset [30]; Abstract500 image dataset [25]; CUB-200 image dataset [32], and Scoot facial sketch dataset [5]. The **Yummly-Food** dataset has 72148 triplets defined over 73 food items based on taste similarity. Each food item is represented by a 6D feature vector (this is an experiment with a low feature dimension) with each component indicating different taste properties. We use 20K training and 20K test triplets sampled from the entire set of triplets. The **TUM-Haptic** dataset contains signals from 108 different types of surface materials. Each type of material has 32-D spectral feature vectors for 10 representative acceleration traces. The triplets are generated from a given ground-truth perceptual matrix, which has user-recorded perceived similarity responses. Like the Yummly-Food dataset, we have training and test sets of 20K triplets each. We also evaluate our method on a comparatively larger dataset (but relatively small for image data), the **Abstract500** image dataset [25], which contains 500 images of 128×128 pixels, with pairwise perceptual similarities between them. Each image is represented by a 512-D GIST feature (an example of a relatively high-dimensional feature vector) extracted using 32 Gabor filters at four scales and eight orientations [23]. We use perceptual matrix to generate 20K training and 20K test triplets. Next, we use the popular and much larger **CUB-200** bird database that contains 200 bird species with roughly 30 images in each class. We choose five representative images for each class and generate its features using a pretrained ResNeXt-101-32x8d model. The network takes segmented images as input and outputs 2048-D feature vectors. The training and test sets each have 10K triplets sampled from the entire set of 93530 triplets. Finally, the results on the **Scoot** dataset, which is relatively small, consisting of just 1282 triplets, are presented in the supplementary material because of space constraints.

Baselines. We compare our method with five baselines, including the state-of-the-art method: (1) **US-⟨Dist⟩**: A batch of individually high-entropy triplets is pruned subjected to different (denoted by ⟨Dist⟩) decorrelation measures to select a diverse batch of informative triplets [20]. It is the current state-of-the-art for batch mode active metric learning, and outperforms other alternatives like BADGE [2] (adapted to metric learning). We pick ⟨Dist⟩ to be the highest-performing variant in each individual experiment. (2) **Variance:** Triplets with the highest individual distance-margin variance across a collection of models generated using dropout [13]. This method simulates the effect of replacing the joint entropy of a batch with the sum of individual entropies of triplets in the batch. (3) **Random:** A passive learning strategy that uniformly samples each batch of triplets at random. Though naïve, this choice often results in reasonably good accuracy. (4) **US:** Uncertainty method, which picks the top b triplets with highest uncertainty in predicted triplet ordering (i.e. the model's (lack of) ordering confidence), without taking correlation among them into account [31]. (5) **BADGE:** A diverse set of triplets with maximum loss gradients for the most probable label, selected using k-means ([2] adapted to the triplet scenario).

Active Learning Setup. For the CUB-200 dataset, we begin each experiment with an initial pool of 1000 annotated triplets, and for the other three datasets with 600 annotated triplets, and pretrain the model ϕ_0, which is used as a common starting point for all compared methods. In each active learning iteration, we select the best fixed-size batch of unannotated triplets, using the chosen batch

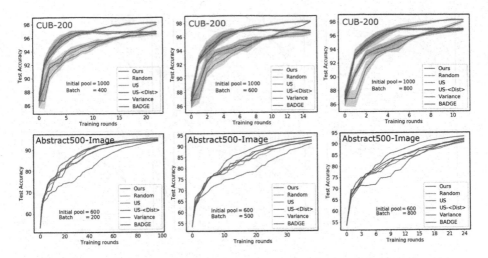

Fig. 3. Performance of different active learning methods on two image datasets CUB-200 and Abstract500 for increasing batch sizes (400/600/800 or 200/500/800, from left to right). Here accuracy means what fraction of test triplets have been detected with the correct ordering. To avoid clutter, standard deviations are shown only for the CUB-200 dataset and the rest are shown in the supplementary material.

selection method, and acquire their orderings. To make convergence faster, we update the current model ϕ_i to obtain ϕ_{i+1} using the available additional annotated triplets instead of training *ab initio*. The performance of the learned model is evaluated by its *triplet generalization accuracy*, which denotes the fraction of triplets whose ordering is correctly predicted [19]. Each experiment is repeated with five random train/test splits, and the average performance along with the standard deviation is reported. (For most plots, the standard deviation is shown in supplementary material, for clarity.)

Implementation Details. The architecture and training hyperparameters used for different datasets are as follows: Yummly-Food: 3 fully-connected (FC) layers with 6, 12 and 12 neurons; TUM-Haptic: 4 FC layers with 32, 32, 64 and 32 neurons; Abstract500-Image: 6 FC layers with 512, 256, 128, 64, 32 and 16 neurons; CUB-200: 3 FC layers with 2048, 512 and 32 neurons. Each layer is followed by a dropout layer with a dropout probability of 0.02. The Adam optimizer [16] is used for training all models with a learning rate of 10^{-4} for Yummly-Food, TUM-Haptic, and Abstract500-Image dataset, and 10^{-5} for CUB-200 dataset. The model is trained with an SGD batch size of 500 for 1000 epochs for all four datasets.

Active Learning Performance. The performance of our method against the baselines described earlier is plotted in Fig. 3 for the image datasets CUB-200 and Abstract500-Image, shown as a function of the number of active learning

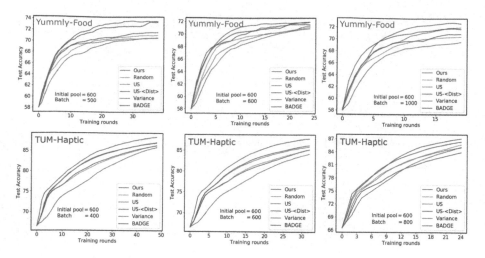

Fig. 4. Performance of different active learning methods on Yummly-Food and TUM-Haptic texture datasets for varying batch sizes (500/800/1000 or 400/600/800).

iterations. In Fig. 4, we compare the performances of all methods for the data from other modalities, i.e., haptic and food. We observe that our method is consistently better than the state-of-the-art **US-⟨Dist⟩** method (for clarity, we only show the specific variant offering the best performance in each experiment). Our method reaches higher accuracies quicker and also tends to converge to a higher final accuracy on both Yummly-Food and TUM-Haptic datasets. For the large CUB-200 image dataset, our method is neck-and-neck with the state-of-the-art for the first few iterations and then rapidly overtakes it, widening the gap with additional iterations. For the smaller Abstract500 image dataset, the improvements are more prominent with larger batch sizes, reflecting the focus of our work on batch-mode learning. Additionally, for the CUB-200 dataset, we plot the standard deviation in the same plot as the shaded region (of the same color) around the performance curves for different methods (standard deviations on other datasets are shown in supplementary). Even though the figure looks a little cluttered, one can see that the standard deviation for the proposed method is better than that of the next-best method, signifying a more consistent performance. This substantiates our claim that joint entropy is a better batch score than an ad-hoc combination of independent informativeness and diversity heuristics. Further, our method does not require the user to select a suitable decorrelation heuristic to manually fine-tune the performance.

We also outperform the other two baselines: **Random** and **Variance**. It is particularly informative to see the generally poor performance of **Variance** (lower than the Random). Because of high correlations among informative triplets, individually selecting the most informative triplets does not learn the entire metric space as well as just picking triplets at random. In contrast, our method as well as that of Kumari et al. [20] both incorporate batch decorrelation

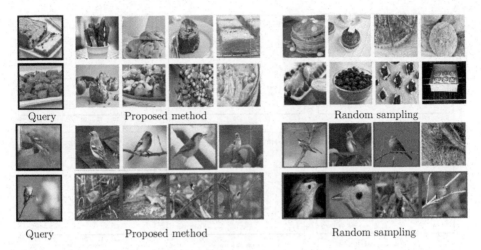

Fig. 5. Top-4 retrieved images in the order of increasing perceptual distance, left to right, using our method and random sampling (randomly-selected batches are annotated for training) on different modalities datasets. On both datasets, each model is trained for twelve training rounds, constituting 18% of training triplets. Images different from the query class are bounded by a red box, substantiating that two images from different classes can be perceptually more similar than two from the same class. The *triplet order accuracy*, defined here as the number of test triplets whose order is preserved by the ranked list of retrieved images, for our method vs random sampling after the 12th round of training is $M_{\text{Ours}}^{12} = 96.7\%$, $M_{\text{Random}}^{12} = 92\%$ for image dataset and $M_{\text{Ours}}^{12} = 72.9\%$, $M_{\text{Random}}^{12} = 69.3\%$ for food dataset. More results with different queries and learned metrics are shown in the supplementary material. (Color figure online)

and outperform random sampling. This shows the critical importance of batch diversity in an active learning strategy.

Next, we evaluate the effectiveness of our method for an **object retrieval** task. Specifically, we compare our method with the random sampling baseline at different training rounds. We show the retrieval results on two different modalities, food and image. We split the Yummly-Food dataset into 40000 training and 32148 test triplets, and the CUB-200 dataset into 40000 training and 33000 test triplets. On both datasets, we perform active learning with a batch size of 600 and an initial pool of 500 triplets. For a given query image, the top four instances from the retrieval set are shown (ranked from most similar to least similar) in Fig. 5. As we can see, retrieval results of our method resemble the query in taste or visual appearance better than the random sampling. Please see the supplementary material for further results from this experiment.

Ablation Study. In order to get an estimate of the covariance matrix, we perform random dropouts in the neural network K times. Naturally, as K increases, one gets a better estimate of the covariance matrix. However, this may increase the computation time. We perform an ablation study to see how this hyper-

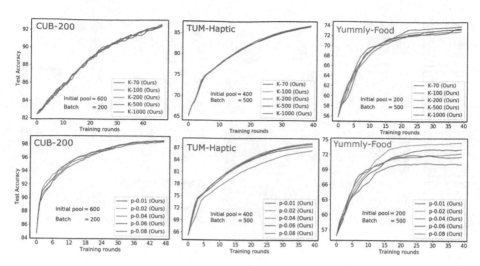

Fig. 6. Ablation study to analyze the robustness of our method to different values of K (# of prior models sampled by dropout) and p (dropout probability) on different datasets.

parameters (i.e., variation in K and dropout probability p) affect the triplet order accuracy, and the results are shown in Fig. 6 for three different modalities: image, haptic and food. It can be seen from the plots that a moderate value of about $K = 70$ or 100 is good enough as the performance is not significantly dependent on the choice of K. We also observe that the performance is robust to variation in dropout probability; however, there is a significant variation for the Yummly-Food dataset, with optimal $p = 0.02$.

Runtime Analysis. We also compare the computational requirement of the proposed method with that of Kumari *et al.* [20]. The key computational step in [20] involves searching for the subset of maximally apart (in the feature space) triplet at each training round, apart from the computation of the gradients. They also use a greedy search technique for subset selection. For the proposed method, the subset selection process is efficient, but the computation of determinant of covariance matrix at each iteration does consume a good amount of time. Overall, both the methods were found to consume a nearly equal amount of computation time when the Gram-Schmidt orthogonalization is used. For instance runtimes (in secs) of different batch selection policies to select a 500-triplet batch from Yummly-food are: US: 0.109, Variance: 0.083, BADGE: 60.104, US-⟨Dist⟩: 8.110, Ours: 7.803. While the computation complexity varies with the feature dimension and model size, the relative performance remains similar.

Robustness to Labeling Error.
To evaluate the robustness of our method against labeling error, we corrupt 10% and 30% of the ground-truth training triplets in the food and image datasets by flipping their orders. Figure 7 shows how the noisy training set affects the performance of our method vs the random sampling

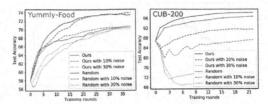

Fig. 7. Performance of our method vs random sampling in the presence of labeling error.

baseline. For the food data (top), with a relatively low 10% labeling error, our method takes slightly more iterations to gain accuracy, but eventually converges to a comparably high accuracy as the noise-free case, while random batch selection fails to achieve the same performance even with clean data. As the percentage of noisy triplets increases, the performance of both methods degrade, showing vulnerability to large scale labeling error. In the absence of abnormally high levels of outliers, our method shows robust performance. For the more complex image dataset (bottom), labeling error has a stronger negative effect (the first selected batch actually decreases overall accuracy), but at each noise level our method still outperforms the baseline.

Comparison of Data Distribution to Theoretical Distribution. We study the validity of the Gaussian embedding, though it already has justification as "worst-case analysis" due to the Maximum Entropy Principle. A standard test is the quantile-quantile (QQ) plot [4], which indicates how close the empirical distribution is to the theoretical distribution. For ease of visualization, we show the QQ plot and histogram for a single randomly-selected unlabeled triplet, for a particular model trained on the initial triplet pool in each dataset. (We cannot visualize a full multivariate QQ plot over all possible batches.) In the QQ plot, the x-axis denotes the theoretical quantiles, which in our case is a Gaussian distribution with the empirical mean and variance, and the observed ordered distance margins are on the y-axis. The *goodness of fit* is indicated by the alignment of points with the straight line having a unit slope. As shown in Fig. 8, in all four datasets, the plotted curve closely approximates the corresponding straight lines shown in red. Our approximation is further validated in the histogram, where our data distribution shows a reasonable fit with the theoretical distribution (shown in green) for the most part, except that the actual distribution is a little more peaked.

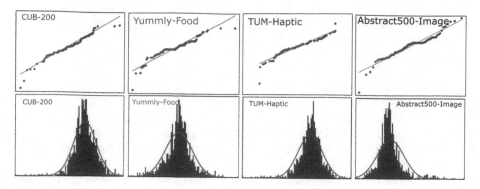

Fig. 8. QQ plot and histogram for all four datasets to demonstrate how closely the actual distribution follows the theoretical distribution.

5   Conclusion, Limitations, and Future Work

We have introduced a novel approach for batch-mode active metric learning based on maximizing the joint entropy of a batch. We found that a batch of individually informative triplets does not form an optimal subset, even if decorrelation heuristics are applied to reduce their correlation. Instead of defining separate measures for informativeness and diversity, our method defines the joint entropy of a batch of triplets as a unified measure that jointly optimizes both. The overall method involves no heuristic parameter selection and has no control parameter to tweak, other than the number of dropout samples and dropout probability, once the network architecture is chosen.

While our method shows promising results, it does have a few limitations. First, approximating the joint distribution of data using the Maximum Entropy Principle gives the most general distribution for a given prior, which in the case of second-order statistics as constraints is a Gaussian. However, in some cases, where the actual distribution may be quite non-Gaussian, the joint entropy measure defined with the second-order statistics may misguide the batch selection policy. One important direction for future work is extending our framework beyond second-order statistics to learn the joint distribution of data closer to empirical distribution. Another important extension would be to modify our framework to dynamically learn the optimal batch size and batch selection policy, which we believe would further improve the performance and generalize well to diverse inputs and applications.

References

1. Agarwal, S., Wills, J., Cayton, L., Lanckriet, G., Kriegman, D., Belongie, S.: Generalized non-metric multidimensional scaling. In: AIS (2007)
2. Ash, J.T., Zhang, C., Krishnamurthy, A., Langford, J., Agarwal, A.: Deep batch active learning by diverse, uncertain gradient lower bounds. In: ICLR (2020)

3. Bellet, A., Habrard, A., Sebban, M.: A survey on metric learning for feature vectors and structured data. arXiv preprint arXiv:1306.6709 (2013)
4. Ben, M.G., Yohai, V.J.: Quantile-quantile plot for deviance residuals in the generalized linear model. J. Comput. Graph. Stat. **13**, 36–47 (2004)
5. Fan, D.P., et al.: Scoot: a perceptual metric for facial sketches. In: ICCV (2019)
6. Freytag, A., Rodner, E., Denzler, J.: Selecting influential examples: active learning with expected model output changes. In: Fleet, D., Pajdla, T., Schiele, B., Tuytelaars, T. (eds.) ECCV 2014, Part IV. LNCS, vol. 8692, pp. 562–577. Springer, Cham (2014). https://doi.org/10.1007/978-3-319-10593-2_37
7. Gal, Y., Ghahramani, Z.: Dropout as a Bayesian approximation: representing model uncertainty in deep learning. In: ICML (2016)
8. Gilad-Bachrach, R., Navot, A., Tishby, N.: Query by committee made real. In: NeurIPS (2006)
9. Heim, E., Berger, M., Seversky, L., Hauskrecht, M.: Active perceptual similarity modeling with auxiliary information. In: AAAI (2015)
10. Hoffmann, W.: Iterative algorithms for Gram-Schmidt orthogonalization. Computing **41**, 335–348 (1989)
11. Jamieson, K.G., Nowak, R.D.: Low-dimensional embedding using adaptively selected ordinal data. In: Allerton (2011)
12. Jaynes, E.T.: Information theory and statistical mechanics. Phys. Rev. **106**, 620 (1957)
13. Kendall, A., Badrinarayanan, V., Cipolla, R.: Bayesian segnet: model uncertainty in deep convolutional encoder-decoder architectures for scene understanding. arXiv preprint arXiv:1511.02680 (2015)
14. Kendall, M.G.: Rank correlation methods. Griffin (1948)
15. Kim, S., Seo, M., Laptev, I., Cho, M., Kwak, S.: Deep metric learning beyond binary supervision. In: CVPR (2019)
16. Kingma, D.P., Ba, J.: Adam: a method for stochastic optimization. In: ICLR (2015)
17. Kirsch, A., van Amersfoort, J., Gal, Y.: BatchBALD: efficient and diverse batch acquisition for deep Bayesian active learning. In: NeurIPS (2019)
18. Kruskal, J.B., Wish, M.: Multidimensional Scaling. Elsevier, Amsterdam (1978)
19. Kumari, P., Chaudhuri, S., Chaudhuri, S.: PerceptNet: learning perceptual similarity of haptic textures in presence of unorderable triplets. In: WHC (2019)
20. Kumari, P., Goru, R., Chaudhuri, S., Chaudhuri, S.: Batch decorrelation for active metric learning. In: IJCAI-PRICAI (2020)
21. McFee, B., Lanckriet, G.: Learning multi-modal similarity. JMLR **12**, 491–523 (2011)
22. Nemhauser, G.L., Wolsey, L.A., Fisher, M.L.: An analysis of approximations for maximizing submodular set functions - I. Math. Program. **14**, 265–294 (1978)
23. Oliva, A., Torralba, A.: Modeling the shape of the scene: a holistic representation of the spatial envelope. IJCV **42**, 145–175 (2001)
24. Pinsler, R., Gordon, J., Nalisnick, E., Hernández-Lobato, J.M.: Bayesian batch active learning as sparse subset approximation. In: NeurIPS (2019)
25. Robb, D.A., Padilla, S., Kalkreuter, B., J. Chantler, M.: Crowdsourced feedback with imagery rather than text: would designers use it? In: SIGCHI (2015)
26. Sener, O., Savarese, S.: Active learning for convolutional neural networks: a core-set approach. In: ICLR (2018)
27. Settles, B.: Active learning. SLAIML (2012)
28. Shui, C., Zhou, F., Gagné, C., Wang, B.: Deep active learning: unified and principled method for query and training. In: AISTATS (2020)

29. Sinha, S., Ebrahimi, S., Darrell, T.: Variational adversarial active learning. In: ICCV (2019)
30. Strese, M., Boeck, Y., Steinbach, E.: Content-based surface material retrieval. In: WHC (2017)
31. Tamuz, O., Liu, C., Belongie, S., Shamir, O., Kalai, A.T.: Adaptively learning the crowd kernel. In: ICML (2011)
32. Wah, C., Maji, S., Belongie, S.: Learning localized perceptual similarity metrics for interactive categorization. In: WACV (2015)
33. Wang, J., et al.: Learning fine-grained image similarity with deep ranking. In: CVPR (2014)
34. Wilber, M.J., Kwak, I.S., Belongie, S.J.: Cost-effective hits for relative similarity comparisons. In: AAAI (2014)
35. Xing, E.P., Jordan, M.I., Russell, S.J., Ng, A.Y.: Distance metric learning with application to clustering with side-information. In: NeurIPS (2003)
36. Zhang, R., Isola, P., Efros, A.A., Shechtman, E., Wang, O.: The unreasonable effectiveness of deep features as a perceptual metric. In: CVPR (2018)

Off-Policy Differentiable Logic Reinforcement Learning

Li Zhang[1], Xin Li[1(✉)], Mingzhong Wang[2], and Andong Tian[3]

[1] School of Computer Science, Beijing Institute of Technology, Beijing, China
{3120181069,xinli}@bit.edu.cn
[2] USC Business School, University of the Sunshine Coast, Sippy Downs, Australia
mwang@usc.edu.au
[3] Ubisoft China AI & Data Lab, Chengdu, China
an-dong.tian@ubisoft.com

Abstract. In this paper, we proposed an Off-Policy Differentiable Logic Reinforcement Learning (OPDLRL) framework to inherit the benefits of interpretability and generalization ability in Differentiable Inductive Logic Programming (DILP) and also resolves its weakness of execution efficiency, stability, and scalability. The key contributions include the use of approximate inference to significantly reduce the number of logic rules in the deduction process, an off-policy training method to enable approximate inference, and a distributed and hierarchical training framework. Extensive experiments, specifically playing real-time video games in Rabbids against human players, show that OPDLRL has better or similar performance as other DILP-based methods but far more practical in terms of sample efficiency and execution efficiency, making it applicable to complex and (near) real-time domains.

Keywords: Deep reinforcement learning · Interpretable reinforcement learning · Neural-Symbolic AI

1 Introduction

Despite the advantages and benefits of Deep Reinforcement Learning (DRL), its successful application and deployment need to address the challenges including: (1) Interpretability. The use of deep neural networks makes the learned policies difficult to interpret and verify, restricting the application of DRL in many real-world domains which require clear scientific interpretation, e.g., healthcare and medical systems. (2) Generalization. The learned policies tend to "over-fit" the training environment, leading the performance of learned polices to drastically decrease even when the test environment slightly changes from the training environment. (3) Sample efficiency. DRL methods generally require massive numbers of samples to explore the environment.

Differentiable Inductive Logic Programming (DILP) [6–8,22,29] has been integrated into DRL frameworks to achieve better interpretability and generalization. Trained with standard back-propagation, DILP provides a special formulation of a function approximator, which generates interpretable and verifiable

© Springer Nature Switzerland AG 2021
N. Oliver et al. (Eds.): ECML PKDD 2021, LNAI 12976, pp. 617–632, 2021.
https://doi.org/10.1007/978-3-030-86520-7_38

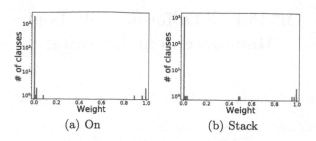

(a) On   (b) Stack

Fig. 1. Weight distributions on two benchmark tasks.

logic rules via training samples. The logic rules also behave as a form of regularization, helping to mitigate over-fitting and improve the generalization ability. Integrating DILP into DRL helps to interpret policies, making the agent's behavior more verifiable and robust.

[7,8] proposed ∂ILP to learn logic rules from noisy data, demonstrating the strength of DILP in interpretability and generalization. [17] introduced Neural Logic Reinforcement Learning (NLRL) which applies DILP in sequential decision making tasks and trains it via vanilla policy gradient [38]. [6] proposed Neural Logic Machines which trades off some interpretability for better scalability in comparison with NLRL.

However, the integration of DILP and DRL suffers from its execution efficiency, stability, and scalability, making it infeasible to many applications requiring real-time or near real-time responses, such as autonomous driving and game playing.

- **Execution efficiency.** DILP takes a top-down and generate-and-test approach, which first generates all potential logic rules and then finds the optimal subset. In general, the number of potential rules is relatively large and the computational cost of DILP is much higher than Multi-layer Perceptron (MLP) networks for policy learning. For an MLP with n linear layers and h hidden neurons at each layer, e.g. in a medium-size problem where $n = 3$ and $h = 256$, its forward computational complexity is $O(nh^2) \approx 2 \times 10^5$. In comparison, the complexity of the DILP, which deduces n steps with l logic rules and each rule matches k cases, m ground atoms, is $O(nmlk) \approx 2 \times 10^7$ for a medium-size problem where $n = 5$, $l = 2000$, $k = 10$, and $m = 200$.
  The number of logic rules in DILP is generally relatively large, resulting in long periods of policy response. This prohibits its application in real-time or near real-time domains. To address this issue, we proposed to reduce the number of rules by at least an order of magnitude. In fact, our study of all potential rules after training revealed that only a small number of rules are non-trivial to induction. Figure 1 depicts the weight distribution, in which the experiment task "On" showed that 99.60%(1981/1989) rules have weight less than 0.01, thus being negligible. Therefore, we proposed the solution of *approximate inference* which extends the technique of network pruning [9,20,21,40] to DILP. It measures the importance of logic rules and maintains a dynamic

set of rules at run time. Due to the boost of execution efficiency, our model can provide (near) real-time policy responses, which is difficult for previous Neuro-Symbolic methods.

In addition, we developed a distributed reinforcement learning framework which decouples the learning (experience utilization) and acting (experience collection) processes, enabling their parallel execution on different machines to reduce the training time.

- **Stability**. In comparison with MLP, DILP, specifically, after the use of approximate inference, is much harder to be optimized and its learning curve oscillates intensely (see Sect. 4). Therefore, we proposed to adopt Maximum Entropy Reinforcement Learning (MERL) approach [10–12,39,42] which augments an entropy term to the objective of standard reinforcement learning (cumulative reward), encouraging policies to consider both optimal and suboptimal actions. MERL can decrease overall estimation errors to stabilize the training, as demonstrated by [41].
- **Scalability**. It is difficult to apply DILP in large-scale/continuous domains due to its high computational cost. Thus, we further extended DILP with hierarchical reinforcement learning [1,5,28,35], to decompose the entire task to simpler sub-tasks, making it possible to employ DILP in complex domains, such as video games.

Although approximate inference helps to significantly reduce the time required for the policy response, it also causes existing on-policy RL algorithms to fail as agents cannot sample actions from online-policy due to non-trivial feature of approximate inference. Our empirical results showed that the errors can be ignored if the model is trained sufficiently. However, errors are inevitable in the early learning stages. Thus, a well-designed off-policy training method becomes the key to the success of approximate inference. Besides, off-policy training also helps to greatly improve sample efficiency by reusing the samples in the experience replay buffer [15,16,25,37].

The natural solutions of policy gradient algorithms [24,31–33] are not directly feasible to train the policy expressed by DILP, or differentiable logic policy (DLP), due to the requirement of off-policy training. Therefore, Q-learning [15, 25,37], a classic off-policy algorithm, is adopted. To enable Q-learning to work seamlessly with DLP in the MERL framework, we used the Soft Q-Learning and Soft Policy Iteration theorem [10–12] as the bridge connecting the Q-value and policy.

In summary, we proposed Off-Policy Differentiable Logic Reinforcement Learning (OPDLRL), which inherits the benefits of interpretability and generalization ability from DILP but also resolves its weakness of execution efficiency, stability, and scalability, making OPDLRL applicable to complex and (near) real-time domains. The key contributions of the paper include:

- We proposed the use of approximate inference to significantly improve the execution efficiency, making our model feasible in (near) real-time applications. To achieve approximate inference and improve sample efficiency, we proposed an off-policy training method in MERL framework, which uses the

soft Q-learning and soft policy iteration theorem to connect the policy and Q-value.
- We developed a distributed and hierarchical training framework for DILP, which significantly improves the training efficiency and makes our model feasible in complex application domains.
- We tested OPDLRL extensively with both benchmark tasks and complex domain tasks. The results showed that OPDLRL significantly outperformed other DILP-based DRL algorithms regarding both performance and sample efficiency in Block Manipulation and Car Avoiding tasks, and OPDLRL could learn to play Rabbids[1] video game and competed with human players successfully while MLP-based and other DILP-based solutions failed.

2   Preliminary

2.1   First-Order Logic Programming

An **atom** $\alpha = p(t_1, \ldots, t_n)$ consists of the **predicate** (relation) p and **terms** (entities) $t_1, \ldots, t_n$, where t_i is a **variable** or a **constant**. A **ground** atom has all terms as constants. An **extensional** predicate is defined by a set of ground atoms while an **intensional** predicate is defined by some clauses. A **clause** $\alpha \leftarrow \alpha_1, \ldots, \alpha_n$ consists of the **head** α and **body** $\alpha_1, \ldots, \alpha_n$, meaning that the head is true if all atoms of body are true. A **deduction** starts from a set of ground atoms, and applies a set of clauses to generate more ground atoms.

2.2   Differentiable Inductive Logic Programming

An Inductive Logic Programming (ILP) problem [26,27] is a tuple $\{\mathcal{B}, \mathcal{P}, \mathcal{N}\}$, in which $\mathcal{B}, \mathcal{P}, \mathcal{N}$ are sets of **background**, **positive** and **negative** atoms, respectively. DILP denotes the truth of an atom as $p \in [0, 1]$, representing the probability that the atom is true. Let G be the set of all ground atoms, and a **valuation** is a vector $\mathbf{v} \in [0, 1]^{||G||}$ representing the probability of all ground atoms. A **language frame** defines target predicates (objective of ILP), extensional predicates, their arity and constants. A **program template** defines available auxiliary predicates, their arity and rule templates. A **rule template** describes whether intensional predicates can be used, and claims the number of existentially quantified variables. With language frames and program templates specified, the set of all possible clauses can be generated, and DILP assigns a probability/confidence for each clause or combination of clauses. The deduction (forward computation) of DILP evaluates the results of all clauses and weights them by corresponding confidence, which is trained to maximize the probability for positive and negative atoms to be satisfied via standard back-propagation. Please refer to [7] for more details.

---

[1] https://en.wikipedia.org/wiki/Raving_Rabbids.

2.3 Maximum Entropy Reinforcement Learning

A Markov Decision Process (MDP) is defined as a tuple $\{S, A, T, R\}$, where S is the state space, A is the action space, T is the transition function, and R is the reward function. In standard RL, the objective is to find a policy that can maximize the expectation of cumulative discount reward $E[\sum_{t=0}^{T} \gamma^t r_t^\pi]$, where γ is the discount factor and r_t^π is the reward at time step t. In maximum entropy RL, the objective is augmented with an entropy term: $E[\sum_{t=0}^{T} \gamma^t [r_t^\pi + \alpha \mathcal{H}(\pi(\cdot|s_t))]]$, where α is the temperature/weighting parameter and $\mathcal{H}(\pi(\cdot|s_t))$ is the entropy of action distribution of a policy π given state s_t.

The entropy augmented objective can be optimized via soft policy iteration [10–12]. In the soft policy evaluation step, soft Q-value is computed based on policy π to evaluate its performance:

$$
\begin{aligned}
Q^\pi(s,a) &= r(s,a) + \gamma E_{s' \sim T}[V^\pi(s')] \\
V^\pi(s) &= E_{a \sim \pi}[Q^\pi(s,a) - \alpha \log \pi(a|s)]
\end{aligned}
\tag{1}
$$

In the soft policy improvement step, a new policy is generated by minimizing the Kullback–Leibler (KL) divergence between the action distribution and the exponential of soft Q-value under the old policy π for each state:

$$
\pi_{new} = \arg\min_{\pi' \in \Pi} D_{KL}(\pi'(\cdot|s) \| \frac{\exp(\frac{1}{\alpha} Q^\pi(s,\cdot))}{Z^\pi(s)})
\tag{2}
$$

where Π is the set of all potential policies and $Z^\pi(s)$ is the partition function normalizing the distribution.

3 Off-Policy Differentiable Logic Reinforcement Learning

Figure 2 shows an overview of the OPDLRL framework. The following sections explain its components.

3.1 Differentiable Logic Policy

To implement logic programming in DRL, symbolic compilers are first required to align logic expressions with reinforcement learning environments. In this paper, a symbolic state is represented by a set of ground atoms $\bar{s} \subseteq G_s$, where G_s is the set of all ground state atoms. A state compiler $S \to \{0,1\}^{\|G_s\|}$ translates environment states into symbolic states. A symbolic action is represented by a ground atom $\bar{a} \in G_a$ where G_a is the set of all ground action atoms. Once a symbolic action is decided by policies, it needs to be translated to an environment action by an action compiler $G_a \to A$. These compilers are usually hand-crafted or initiated by a pre-trained neural network. A set of background atoms is provided to describe the relations of constants regarding the task.

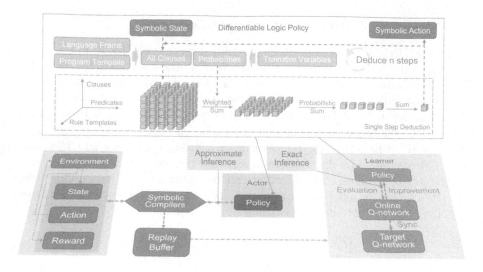

Fig. 2. Framework of Off-Policy differentiable logic reinforcement learning.

With the symbolic compilers, a MDP becomes a First-order MDP (FOMDP) problem [2,18,30] in a stricter form. We define the target and extensional predicates of the language frame as the action and state predicates of FOMDP respectively, and define the auxiliary predicates that help to represent policies in the program template. Thereafter, all possible clauses can be generated with the language frame and program template.

We denote the policy of the FOMDP as π_θ, the i-th rule template of predicate e as τ_i^e, and the corresponding set of clauses as Γ_i^e. For each Γ_i^e, we initialize a vector θ_i^e of length $\|\Gamma_i^e\|$ by Gaussian distribution with mean 0.00 and standard deviation 0.05. The probability (confidence) of the j-th clause in Γ_i^e is parameterized as:[2]

$$p_i^e(j) = \frac{\theta_i^e(j)^2}{\|\theta_i^e\|^2}. \tag{3}$$

In a single deduction step, the output valuation vector is computed by:

$$\mathbf{v} = \min(1, \mathbf{v}_0 + \sum_e \bigoplus_i \sum_j p_i^e(j)\mathbf{v}_{e,i,j}) \tag{4}$$

where $\mathbf{v}_{e,i,j}$ denotes the valuation inferred by the j-th clause in the clause set Γ_i^e according to input valuation of current deduction step. The weighted sum amalgamates the valuations for different clauses but with the same rule template, the probabilistic sum $\oplus$ $(x \oplus y = x + y - xy)$ amalgamates the valuations for different rule templates, and finally, the valuations for different predicates are summed up. $\mathbf{v}_0$ is the initial valuation which is a multi-hot vector defined

---

[2] The comparison of different parameterization methods can be found in the Supplementary Material.

jointly by the current symbolic state and background atoms, and it is added in each deduction step to steer off the local optima as suggested in [17]. The final valuation is clipped to $[0, 1]$ to prevent the overflow and treated as input valuation to the next deduction step.

3.2 Approximate Inference

The inference process in DLP involves all clauses, thus significantly more expensive than an MLP policy regarding the computational cost. We propose approximate inference to reduce the response time of the inference process, given as:

$$\mathbf{v} = \min(1, \mathbf{v}_0 + \sum_e \bigoplus_i \sum_{k \in \{j | p_i^e(j) > \eta\}} p_i^e(k) \mathbf{v}_{e,i,k}) \tag{5}$$

where η is the threshold to filter out the negligible clauses to the deduction.

3.3 Off-Policy Training

Section 1 has shown approximate inference can significantly reduce the computation time of the deduction, but cannot leverage existing on-policy RL algorithms due to its non-trivial feature. Thus, we propose an off-policy approach to train the approximate inference-facilitated DLP.

Our off-policy training process stems from soft policy iteration theorem, which connects Q-learning and policy gradient method in MERL framework. Therefore, the solution benefits from both Q-learning, which implements off-policy to enable approximate inference and improve sample efficiency, and policy gradient, which applies a separate policy network to achieve better interpretability.

For the soft policy evaluation, we use MLP networks to approximate soft Q-value and train it by minimizing TD-error (with an augmented entropy term), given as:

$$J_Q(\vartheta) = E_{(s,a,r,s') \sim D}[\frac{1}{2}(r + \gamma V(s') - Q_\vartheta(s, a))^2]$$
$$V(s) = \sum_a \pi_\theta(a|s)[Q_{\bar{\vartheta}}(s, a) - \alpha \log \pi_\theta(a|s)] \tag{6}$$

where D is the replay buffer, ϑ and $\bar{\vartheta}$ are the parameters of online and target Q-network respectively, and $\pi_\theta(a|s)$ is the action distribution computed by the exact inference of DLP. Target Q-network [25] periodically syncs with online Q-network to stabilize the learning of soft Q-value. Note that our method has two expectation terms in Eq. (1): $E_{s' \sim T}$ is computed by Monte-Carlo estimation (sampling from replay buffer), and $E_{a \sim \pi}$ is computed as the weighted sum of networks' output values for all actions (discrete space).

For the soft policy improvement, we ignore the constant partition function $Z^\pi(s)$ and take the logarithm of Eq. (2) to jointly optimize π_θ and Q_ϑ, given as:

$$J_\pi(\theta) = E_{s \sim D}[\sum_a \pi_\theta(a|s)(\alpha \log \pi_\theta(a|s) - Q_\vartheta(a|s))] \tag{7}$$

The expectation term in KL-divergence in Eq. (2) is computed as weighted sum.

3.4 Hierarchical Policy Implementation

To enable OPDLRL to work effectively for complex tasks with large/continuous state/action space and (near) real-time response requirement, we further develop a hierarchical policy implementation in which the agent's policy is formulated as being hierarchical. OPDLRL works at the top level to determine which sub-policy should be taken according to the symbolic state (high-level state) and tries to minimize external reward. A sub-policy (skill) is pre-trained to achieve a sub-goal and it takes the primary states and outputs primary actions. Symbolic compilers perform the translation between different hierarchies.

4 Experiments

OPDLRL, its variants, and other relevant methods have been extensively tested in two relational RL benchmark tasks, namely Block Manipulation and Car Avoiding, and one real-time task Rabbids, which is a popular video game consisting of many mini-games. In this paper, we used OPDLRL to play the bumper-car game in Rabbids against behavior tree [4] agent and human players.

Table 1. Performance comparison on different tasks. Items in table are the mean and standard deviation for 300 different runs.

	On	Stack	Unstack	Car Avoiding
OPDLRL⁻	0.936(±0.01)	0.941(±0.03)	0.950(±0.01)	0.960(±0.00)
OPDLRL	0.934(±0.01)	0.942(±0.03)	0.950(±0.01)	0.960(±0.00)
NLRL	0.881(±0.05)	0.908(±0.05)	0.909(±0.05)	0.947(±0.16)
NLRL(AI)	0.816(±0.08)	0.817(±0.12)	0.902(±0.05)	0.488(±0.82)
SAC	0.933(±0.01)	0.947(±0.02)	0.953(±0.01)	0.960(±0.00)
PPO	0.940(±0.00)	0.957(±0.05)	0.960(±0.00)	0.960(±0.00)
PPO(DLP)	0.940(±0.00)	0.870(±0.09)	0.897(±0.06)	0.952(±0.11)
PPO(DLPAI)	0.888(±0.05)	0.862(±0.09)	0.898(±0.05)	0.953(±0.11)
	On+	Stack+	Unstack+	Car Avoiding+
OPDLRL⁻	0.936(±0.01)	0.953(±0.03)	0.950(±0.02)	0.960(±0.00)
OPDLRL	0.935(±0.01)	0.948(±0.04)	0.949(±0.02)	0.960(±0.00)
NLRL	0.880(±0.05)	0.908(±0.05)	0.896(±0.06)	0.947(±0.16)
NLRL(AI)	0.805(±0.09)	0.800(±0.13)	0.894(±0.06)	0.465(±0.84)
SAC	−0.985(±0.23)	−0.390(±0.79)	0.295(±0.66)	−0.896(±0.44)
PPO	−0.821(±0.49)	−0.437(±0.77)	−0.989(±0.22)	−0.974(±0.22)
PPO(DLP)	0.940(±0.00)	0.906(±0.08)	0.894(±0.06)	0.946(±0.16)
PPO(DLPAI)	0.877(±0.06)	−0.866(±0.46)	0.905(±0.06)	0.946(±0.16)

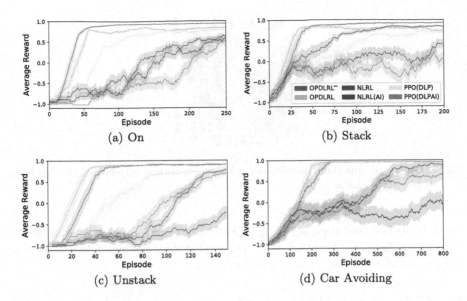

(a) On

(b) Stack

(c) Unstack

(d) Car Avoiding

Fig. 3. Learning curve in different tasks.

4.1 Methods for Evaluation

8 methods have been experimented, including: i) **OPDLRL**: the method we proposed. ii) **OPDLRL⁻**: OPDLRL without approximate inference. iii) **NLRL**: refer to Sect. 1. iv) **NLRL(AI)**: NLRL with approximate inference. v) **SAC**: Soft Actor Critic [3,11,12], a state-of-the-arts off-policy RL algorithm based on MLP networks. vi) **PPO**: Proximal Policy Optimization [32], a state-of-the-arts on-policy RL algorithm based on MLP networks. vii) **PPO(DLP)**: PPO with differentiable logic policy. viii) **PPO(DLPAI)**: PPO(DLP) with approximate inference. Note that all the compared methods share the same observation space, task hierarchies and pre-trained subpolicies for fairness.

4.2 Experiment Setting

Block Manipulation. In Block Manipulation, the agent keeps on moving the top block of a pile until the goal block state is achieved. The constants include $\{a, b, c, d, floor\}$ where a, b, c, d are blocks. The predicates include a state predicate $on(X, Y)$ representing block X is on top of Y and Y can be a block or $floor$, an action/target predicate $move(X, Y)$ representing moving block X to the top of Y, and $floor(X)$ representing whether X is $floor$ and the background is $\{floor(floor)\}$. The agent will get -0.02 reward for each step if it cannot achieve the goal within 50 steps, and $+1$ reward when it achieves.

There are three kinds of goals in the experiments: (1) **On** task: The goal state is having block X right on top of block Y which is represented as an additional background predicate $goal(X, Y)$. The initial state has all blocks in a pile. (2)

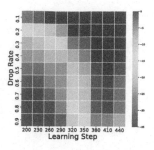

Fig. 4. Impact of approximate inference. Colors are computed via Eq. (10) with log scale. (Color figure online)

Stack task: The goal state is having all blocks in a pile and the initial state has all blocks on floor. (3) **Unstack** task: The goal state is having all blocks on floor and the initial state has all blocks in a pile.

Car Avoiding. A Car Avoiding task has two cars in a circular platform. The goal of the agent is to control one car to occupy the center without colliding with the opponent car. The platform is divided into 12 regions which are denoted as (X, Y) where $X \in \{a, b, c\}$ describes the distance levels from center and $Y \in \{0, 1, 2, 3\}$ describes quadrants. The predicates include: me(X, Y) and enemy(X, Y) for the position of agent and opponent respectively, forward(), backward(), left(), and right() for actions of going forward, backward, turning left and right respectively, and outer(X, Y) for the relation between a, b, c with the background as $\{\text{outer}(a, b), \text{outer}(b, c)\}$. In the initial state, the agent is at $(c, 0)$ and opponent is at $(a, 0)$. The opponent car will move along $(a, 0) \rightarrow (b, 0) \rightarrow (c, 0)$ and remain at $(c, 0)$. The agent gets -1 reward if it is hit by its opponent and $+1$ reward if it occupies the center $((a, *))$. The game terminates if the agent is hit or fails to reach the center within 50 steps.

4.3   Results and Analysis

Performance. Table 1 shows the performance results of OPDLRL[3] and other methods. The top section shows the experiments with the same training and test environments. The performance of OPDLRL and OPDLRL$^-$ is very close to the best in all tasks. MLP-based methods (SAC and PPO) perform slightly better than OPDLRL in this setting as MLP has a more flexible structure and is easy to be optimized. Note that the number of trained parameters in MLP is much larger than that in DILP. In the experiments, MLP-based approaches have about 5×10^4 parameters while DILP requires only 2×10^3, indicating a better fitting ability in MLP. In comparison with the DILP-based methods (NLRL, NLRL(AI), PPO(DLP) and PPO(DLPAI)), OPDLRL and OPDLRL$^-$

---

[3] We used the same setting of hyper-parameters for all tasks.

show evident advantages as they adopt the entropy-augmented reward and optimize with soft policy iteration to reduce the overall estimation error and stabilize the training, as discussed by [41].

Interpretability. The logic rules learned by our method in tasks **On** and **Car Avoiding** are as follows (for concise, we only show the rules with probability greater than 0.1):

On:

$$1.00 : \mathrm{aux}2(X) \leftarrow \mathrm{on}(X,Y), \mathrm{on}(Y,Z)$$
$$1.00 : \mathrm{aux}1(X) \leftarrow \mathrm{aux}2(X), \mathrm{top}(X)$$
$$0.97 : \mathrm{move}(X,Y) \leftarrow \mathrm{floor}(Y), \mathrm{aux}1(X) \qquad (8)$$
$$0.90 : \mathrm{move}(X,Y) \leftarrow \mathrm{goal}(X,Y), \mathrm{top}(X)$$

where aux1 and aux2 are the auxiliary predicates induced by our method. $\mathrm{aux}2(X)$ is true if there is a Y such that X is on the top of Y and Y is on the top of Z. Y on Z means Y is not *floor* (Y is a block), thus $\mathrm{aux}2(X)$ means X is on the top of a block (X is not on *floor*). $\mathrm{aux}1(X)$ is true if X is not on *floor* ($\mathrm{aux}2(X)$) and X is the top of a pile ($\mathrm{top}(X)$). The behavior of an agent can be interpreted as: 1) For a pile with 2 or more blocks, move the top block to *floor*. 2) If the goal is to have X on Y and X is movable (X is the top block), then move X to Y.

Car Avoiding:

$$1.00 : \mathrm{forward}() \leftarrow \mathrm{enemy}(X,Y), \mathrm{outer}(Z,X)$$
$$0.58 : \mathrm{left}() \leftarrow \mathrm{me}(X,Y), \mathrm{enemy}(Z,Y) \qquad (9)$$
$$0.59 : \mathrm{right}() \leftarrow \mathrm{me}(X,Y), \mathrm{enemy}(Z,Y)$$

The goal of an agent is to control one car to occupy the center without colliding with the opponent car. The behavior of an agent can be interpreted as: 1) If the agent and opponent are in the same quadrant, then move left or right. 2) If the opponent is not in $(a, *)$ then move forward.

Generalization. The bottom section of Table 1 shows the experiments with different training and test initial state, thus evaluating the generalization ability of models. Specifically, (1) **On+**: changing the order of blocks in the test. (2) **Stack+**: using 2 piles in the test while 4 piles for the training. (3) **Unstack+**: using 2 piles in the test while 1 pile for the training. (4) **Car Avoiding+**: changing the initial position of two cars in the test. OPDLRL and OPDLRL⁻ achieve the best performance in this new task settings as our methods utilize logic expressions to capture the essence of task operations, resulting in better generalization. The performance of MLP-based methods (SAC and PPO) decreases drastically due to their over-fitting to the training environment. PPO(DLP) has a slightly better performance than ours in task **On** and **On+** but lags in all other tasks. Note that the loss function of PPO also includes an entropy term but it is used as a regularization term to help exploration.

Off-Policy Training. Figure 3 depicts the learning curves of all DILP-based methods on four tasks. OPDLRL and OPDLRL⁻ substantially outperform others in all tasks regarding sample efficiency (convergence), indicating the effectiveness of the off-policy training and the approximate inference strategy in our framework. For each task, NLRL(AI)/PPO(DLPAI) perform remarkably worse than their corresponding variant without approximate inference because NLRL and PPO(DLP) apply on-policy training but the use of approximate inference requires off-policy, resulting in the outstanding performance loss. The learning curves of OPDLRL and OPDLRL⁻ are similar as they both apply off-policy training. In fact, OPDLRL converges faster than OPDLRL⁻ in **Stack**, **Unstack**, and **Car Avoiding** tasks because of efficient *Exploration-Exploitation*: the noisy produced by approximate inference helps policy to diverge from the local-optima in early training stages. Moreover, the use of approximate inference in OPDLRL can significantly reduce the response/deduction time of policies due to dropping a large number of negligible rules (90% potential rules were discarded in the experiments without compromising the performance). SAC and PPO have similar convergence patterns as our methods but with worse generalization. Therefore, they are not included in the comparison.

Impact of Approximate Inference. Figure 4 illustrates the impact of approximate inference by measuring the difference between the policy π with full inference and $\hat{\pi}$ with approximate inference on task **On** for given training steps and drop rates[4]. 1000 states $\{s_1, \cdots, s_{1000}\}$ were sampled from the environment based on π. For each s_i, the KL-divergence between $\pi(\cdot|s_i)$ and $\hat{\pi}(\cdot|s_i)$ was used. The difference between π and $\hat{\pi}$ is computed as:

$$E_{s_i \sim \pi}[D_{KL}(\pi(\cdot|s_i)||\hat{\pi}(\cdot|s_i)) + D_{KL}(\hat{\pi}(\cdot|s_i)||\pi(\cdot|s_i))] \tag{10}$$

The heatmap depicts the log scale value of Eq. (10). The difference caused by approximate inference is only notable in the early training stages. The difference becomes trivial after 440 learning steps even with a large drop rate ($e^{-20} \approx 2 \times 10^{-9}$ when $x = 440, y = 0.9$). The observation supports the discussion in Sect. 1 and the key motivation of the paper: approximate inference with off-policy training can significantly boost the execution efficiency without sacrificing the benefits of DILP.

4.4   Rabbids Game

Experiments Setting. To test OPDLRL's capability of (near) real-time inferences in complex domains, we developed a real-time game-play agent with OPDLRL to compete with other models and human players in the bumper-car mini-game of the Rabbids, which requires a policy response within the interval between frames. The state (vector) implies the position coordinates, velocity,

---

[4] Drop rate represents the percentage of rules ignored in the approximate inference, see Eq. (5).

and orientation of four cars, and the actions are forward, backward, left, right, as well as their combination. The state compiler translates the continuous state vector to high-level symbolic states, and the action compiler translates the high-level symbolic actions to a sub-goal, which a sub-policy tries to achieve. The high-level symbolic environment of Rabbids is a super-set of Car Avoiding with the following extra predicates: danger(X, Y) for whether a region (X, Y) can be reached by an opponent, reach(X, Y) for whether a region (X, Y) can be reached by agent, to(X) for a sub-goal of moving to the opponent X, avoid() for a sub-goal of avoiding the opponent, and tocenter() for a sub-goal of moving to the center. The action predicates in Car Avoiding are also considered as sub-goals to make high-level policy more flexible. danger(X, Y) and reach(X, Y) contains the velocity and orientation information, computed by a pre-trained network. Our agent was trained against behavior tree agents. Agent gets $+1$ reward when an opponent falls from the platform and get -1 reward when agent itself falls. Within one episode, each car has three chances to re-spawn. Agent wins the episode if and only if it stays alive until the end.

Results. After the training, both OPDLRL agent and SAC agent achieved higher win rates against the behavior tree agent (OPDLRL:100.0%(124/124), SAC:97.1%(99/102)). To evaluate the generalization ability, we organized a series of competitions against human players, and OPDLRL agent kept 100.0%(21/21) win rate while SAC agent decreased to only 20.0%(4/20). We observed that most loss of SAC agent was caused by states which were unseen during training. For example, SAC agent may act a self-killing behavior when an opponent stays still, as staying still has never occurred during the training against a behavior tree agent. The higher win rates of OPDLRL agent show that our method successfully captures/induces *the key point* via DILP. Unfortunately, other DILP-based approaches failed this test as they cannot compute a practical response within the time limit.

5   Discussion

Mirror Descent. Reinforcement learning with interpretable policy representation is challenging because of its highly structured nature of the policy space. Thus, the training of interpretable policies cannot be seen as an unconstrained policy optimization. Mirror descent is an efficient method to solve a constrained optimization problem and [34] has demonstrated how to learn a programmatic policy via mirror descent. SAC is known as a special form of mirror descent [23,36]. The use of SAC to optimize DILP in our paper can thus be interpreted from a perspective of mirror descent: i) learning an unconstrained policy via MLP-based Q-function (Eq. (6)); ii) projecting the unconstrained policy into constrained DILP-based policy space (Eq. (7)). From a theoretical perspective, this view may help understand the advantages of our method and provide insight to further improving Neuro-Symbolic methods with structured nature.

Multi-Agent. Rabbids can also be seen as a multi-agent game. Generally, single-agent RL algorithms cannot converge to a robust policy which can rationally respond to various opponent's policy due to the non-stationary feature caused by the change of opponent's policy. Currently, the most popular and successful solution to multi-agent game is self-play [13,14,19]. Instead of using self-play, in this paper, we explored to learn a robust/well-performed single-agent policy for a multi-agent video game via a specified regularization, which restricts the policy within the interpretability structure. The results proved the effectiveness of our solution when dealing with the multi-agent environment.

6   Conclusion

In this paper, we proposed Off-Policy Differentiable Logic Reinforcement Learning (OPDLRL) framework to inherit the benefits of interpretability and generalization ability in DILP and also resolve its weakness of execution efficiency, stability, and scalability. OPDLRL has similar or better performance than other DILP-based methods but far more practical in terms of sample efficiency and execution efficiency, making it applicable to complex and (near) real-time domains. The key contributions include the use of approximate inference to significantly reduce the number of logic rules required for inferences and the well-designed off-policy training process to enable approximate inference. Various experiments, specifically playing real-time video games in Rabbids against human players, demonstrated its strength and practicability.

Acknowledgement. Their work is partially supported by NSFC under Grant (U19B2020 and 61772074) and National Key R&D Program of China under Grant (2017YFB0803300).

References

1. Barto, A.G., Mahadevan, S.: Recent advances in hierarchical reinforcement learning. Discrete Event Dyn. Syst. **13**(1–2), 41–77 (2003)
2. Boutilier, C., Reiter, R., Price, B.: Symbolic dynamic programming for first-order MDPs. IJCAI **1**, 690–700 (2001)
3. Christodoulou, P.: Soft actor-critic for discrete action settings. arXiv preprint arXiv:1910.07207 (2019)
4. Colledanchise, M., Ögren, P.: Behavior trees in robotics and AI: an introduction. CoRR abs/1709.00084 (2017)
5. Dietterich, T.G.: Hierarchical reinforcement learning with the MAXQ value function decomposition. J. Artif. Intell. Res. **13**, 227–303 (2000)
6. Dong, H., Mao, J., Lin, T., Wang, C., Li, L., Zhou, D.: Neural logic machines. arXiv preprint arXiv:1904.11694 (2019)
7. Evans, R., Grefenstette, E.: Learning explanatory rules from noisy data. J. Artif. Intell. Res. **61**, 1–64 (2018)

8. Evans, R., Grefenstette, E.: Learning explanatory rules from noisy data (extended abstract). In: Proceedings of the Twenty-Seventh International Joint Conference on Artificial Intelligence, IJCAI-18, pp. 5598–5602. International Joint Conferences on Artificial Intelligence Organization (2018). https://doi.org/10.24963/ijcai.2018/792

9. Frankle, J., Carbin, M.: The lottery ticket hypothesis: finding sparse, trainable neural networks. In: International Conference on Learning Representations (2019)

10. Haarnoja, T., Tang, H., Abbeel, P., Levine, S.: Reinforcement learning with deep energy-based policies. In: Proceedings of the 34th International Conference on Machine Learning-Volume 70, pp. 1352–1361 (2017). JMLR.org

11. Haarnoja, T., Zhou, A., Abbeel, P., Levine, S.: Soft actor-critic: Off-policy maximum entropy deep reinforcement learning with a stochastic actor. arXiv preprint arXiv:1801.01290 (2018)

12. Haarnoja, T., et al.: Soft actor-critic algorithms and applications. arXiv preprint arXiv:1812.05905 (2018)

13. Heinrich, J., Lanctot, M., Silver, D.: Fictitious self-play in extensive-form games. In: Proceedings of the 32nd International Conference on International Conference on Machine Learning - Volume 37, ICML 2015, pp. 805–813 (2015). JMLR.org

14. Heinrich, J., Silver, D.: Deep reinforcement learning from self-play in imperfect-information games. CoRR abs/1603.01121 (2016)

15. Hessel, M., et al.: Rainbow: combining improvements in deep reinforcement learning. In: Thirty-Second AAAI Conference on Artificial Intelligence (2018)

16. Horgan, D., et al.: Distributed prioritized experience replay. In: International Conference on Learning Representations (2018)

17. Jiang, Z., Luo, S.: Neural logic reinforcement learning. In: International Conference on Machine Learning. pp. 3110–3119 (2019)

18. Kersting, K., Otterlo, M.V., De Raedt, L.: Bellman goes relational. In: Proceedings of the Twenty-First International Conference on Machine Learning, p. 59 (2004)

19. Lanctot, M., et al.: A unified game-theoretic approach to multiagent reinforcement learning. In: Guyon, I., et al. (eds.) Advances in Neural Information Processing Systems, vol. 30, pp. 4190–4203. Curran Associates, Inc. (2017)

20. Lin, J., Rao, Y., Lu, J., Zhou, J.: Runtime neural pruning. In: Advances in Neural Information Processing Systems, pp. 2181–2191 (2017)

21. Liu, Z., Sun, M., Zhou, T., Huang, G., Darrell, T.: Rethinking the value of network pruning. arXiv preprint arXiv:1810.05270 (2018)

22. Manhaeve, R., Dumancic, S., Kimmig, A., Demeester, T., De Raedt, L.: DeepProbLog: neural probabilistic logic programming. In: Advances in Neural Information Processing Systems, pp. 3749–3759 (2018)

23. Mei, J., Xiao, C., Huang, R., Schuurmans, D., Müller, M.: On principled entropy exploration in policy optimization. In: Proceedings of the Twenty-Eighth International Joint Conference on Artificial Intelligence, IJCAI-19, pp. 3130–3136. International Joint Conferences on Artificial Intelligence Organization (2019). https://doi.org/10.24963/ijcai.2019/434

24. Mnih, V., et al.: Asynchronous methods for deep reinforcement learning. In: International Conference on Machine Learning, pp. 1928–1937 (2016)

25. Mnih, V., et al.: Human-level control through deep reinforcement learning. Nature 518(7540), 529–533 (2015)

26. Muggleton, S.: Inductive logic programming. New Generation Comput. 8(4), 295–318 (1991)

27. Muggleton, S., et al.: Stochastic logic programs. Adv. Inductive Logic Programm. 32, 254–264 (1996)

632 L. Zhang et al.

28. Nachum, O., Gu, S.S., Lee, H., Levine, S.: Data-efficient hierarchical reinforcement learning. In: Advances in Neural Information Processing Systems, pp. 3303–3313 (2018)
29. Payani, A., Fekri, F.: Inductive logic programming via differentiable deep neural logic networks. arXiv preprint arXiv:1906.03523 (2019)
30. Sanner, S., Boutilier, C.: Practical solution techniques for first-order MDPs. Artif. Intell. **173**(5–6), 748–788 (2009)
31. Schulman, J., Levine, S., Abbeel, P., Jordan, M., Moritz, P.: Trust region policy optimization. In: International Conference on Machine Learning, pp. 1889–1897 (2015)
32. Schulman, J., Wolski, F., Dhariwal, P., Radford, A., Klimov, O.: Proximal policy optimization algorithms. arXiv preprint arXiv:1707.06347 (2017)
33. Sutton, R.S., McAllester, D.A., Singh, S.P., Mansour, Y.: Policy gradient methods for reinforcement learning with function approximation. In: Advances in Neural Information Processing Systems, pp. 1057–1063 (2000)
34. Verma, A., Le, H., Yue, Y., Chaudhuri, S.: Imitation-projected programmatic reinforcement learning. In: Wallach, H., Larochelle, H., Beygelzimer, A., d'Alché-Buc, F., Fox, E., Garnett, R. (eds.) Advances in Neural Information Processing Systems, vol. 32, pp. 15752–15763. Curran Associates, Inc. (2019)
35. Vezhnevets, A.S., et al.: Feudal networks for hierarchical reinforcement learning. In: Proceedings of the 34th International Conference on Machine Learning-Volume 70, pp. 3540–3549 (2017). JMLR.org
36. Wang, Q., Li, Y., Xiong, J., Zhang, T.: Divergence-augmented policy optimization. In: Wallach, H., Larochelle, H., Beygelzimer, A., d'Alché-Buc, F., Fox, E., Garnett, R. (eds.) Advances in Neural Information Processing Systems, vol. 32, pp. 6099–6110. Curran Associates, Inc. (2019)
37. Wang, Z., Schaul, T., Hessel, M., Van Hasselt, H., Lanctot, M., De Freitas, N.: Dueling network architectures for deep reinforcement learning. arXiv preprint arXiv:1511.06581 (2015)
38. Williams, R.J.: Simple statistical gradient-following algorithms for connectionist reinforcement learning. Mach. Learn. **8**(3–4), 229–256 (1992)
39. Wulfmeier, M., Ondruska, P., Posner, I.: Maximum entropy deep inverse reinforcement learning. arXiv preprint arXiv:1507.04888 (2015)
40. Zhou, H., Lan, J., Liu, R., Yosinski, J.: Deconstructing lottery tickets: zeros, signs, and the supermask. In: Advances in Neural Information Processing Systems, pp. 3592–3602 (2019)
41. Ziebart, B.D.: Modeling purposeful adaptive behavior with the principle of maximum causal entropy (2018). https://doi.org/10.1184/R1/6720692.v1
42. Ziebart, B.D., Maas, A.L., Bagnell, J.A., Dey, A.K.: Maximum entropy inverse reinforcement learning. In: AAAI, vol. 8, pp. 1433–1438. Chicago (2008)

Causal Explanation of Convolutional Neural Networks

Hichem Debbi[(⊠)] [ID]

Department of Computer Science, University of M'sila, M'sila, Algeria
hichem.debbi@univ-msila.dz

Abstract. In this paper we introduce an explanation technique for Convolutional Neural Networks (CNNs) based on the theory of causality by Halpern and Pearl [12]. The causal explanation technique (CexCNN) is based on measuring the filter importance to a CNN decision, which is measured through counterfactual reasoning. In addition, we employ extended definitions of causality, which are responsibility and blame to weight the importance of such filters and project their contribution on input images. Since CNNs form a hierarchical structure, and since causal models can be hierarchically abstracted, we employ this similarity to perform the most important contribution of this paper, which is localizing the important features in the input image that contributed the most to a CNN's decision. In addition to its ability in localization, we will show that CexCNN can be useful as well for model compression through pruning the less important filters. We tested CexCNN on several CNNs architectures and datasets. (The code is available on https://github.com/HichemDebbi/CexCNN)

Keywords: Explainable Artificial Intelligence (XAI) · Convolutional Neural Networks (CNNs) · Saliency maps · Causality · Object localization · Pruning

1 Introduction

Convolution Neural networks (CNNs) [16,17] represent a class of Deep Neural Networks (DNNs) that focus mainly on image data. Employing CNN made breakthroughs in computer vision tasks, such as image classification [8], object detection [10] and semantic segmentation [21]. To get a deep insight on CNN's behavior and explain their decisions, recently many works have been proposed.

To get a deep insight on CNNs' behavior and understand their decisions, recently many works have investigated the visualization of their internal structure, through visualizing CNNs filters in the aim of exploring hidden visual patterns. The Gradient-based methods [29,33] made a breakthrough in CNN visualization. These methods aim to compute the gradient of the class score with respect to the input image, where the method of [33] gives top-down projections from a layer to another enabling hierarchical visualization of the features in the network. While

© Springer Nature Switzerland AG 2021
N. Oliver et al. (Eds.): ECML PKDD 2021, LNAI 12976, pp. 633–649, 2021.
https://doi.org/10.1007/978-3-030-86520-7_39

gradient-based methods have shown a great success, they lack actually expressing the causes for such features classification. To this end, recently a class of causal explanation methods [15,27] has emerged in order to give causal interpretation for CNNs decisions, or abstract the CNN models into causal models.

Pearl suggests that truly machine learning models should provide counterfactual interpretations of the form: "I have done $X = x$, and the outcome was $Y = y$, but if I had acted differently, say $X = x'$, then the outcome would have been better, perhaps $Y = y'$." The Causal hierarchy as described by Pearl consists of three main classes. At the first level comes association, which is used mainly to express statistical relationships, and then comes interventions, which is based on changing what we observe, not just seeing it as it is, so it answers the question, what if I do X. Then at the top level of this hierarchy we find counterfactuals, which combine both association and intervention. So it would help to answer the questions, was it X that caused Y? What if I had acted differently?

Due to the importance of counterfactuals, Halpern and Pearl have extended the definition of counterfactuals by Lewis [19] to build a rigorous mathematical model of causation, which they refer to as structural equations [12,13]. Based on this definition, Halpern and Chockler [6] introduced the definition of responsibility. Responsibility extends the concept of all-or-nothing of the actual cause $X = x$ for the truth value of Boolean formula φ. It measures the number of changes that have to be made in a context u in order to make φ counterfactually depends on X. When we have an uncertainty around the context, we face in addition to the question of responsibility the question of blame [6].

Recently Beckers and Halpern [5] addressed the problem of abstracting causal models, through arising the question regarding the human behavior: does a high-level "macro" causal model that describes for instance beliefs, is a faithful abstraction of a low-level "micro" model that describes the neuronal level. They concluded that abstracting causal models is very relevant to the increasing demand for explainable AI, building on the fact that the only available causal model for such ML model is too complicated for humans to understand.

In this paper, we propose a causal explanation technique for CNNs (CexCNN) that adopts all the definitions of causality, responsibility and blame in a complementary way. Through this paper, we will show how these definitions are very appropriate for explaining CNNs predictions. The key concept of this adoption is about identifying actual causes. In CNNs, it is evident that the main building blocks that derive the output are the filters. So, we consider filters as actual causes for deriving such a decision. Once the causal learning process is complete, for each outcome we obtain some filters that have more importance than others. For each filter we assign a degree of responsibility as a measure for its importance to the related class. Then, the responsibilities of these filters are projected back to compute the blame for each region in the input image. The regions with the most blame are returned then as the most important explanations.

Since filters are identified at the level of each convolutional layer, and since convolutional layers in CNNs have a hierarchical form, we can say that the filters at low levels have a causal effect on the output of the last layer. So CNNs actually express causality abstraction. This fact drives us to consider this abstraction as well for our explanation technique. To our knowledge, this is the first application in which all the definitions of causality, responsibility and blame, in addition to abstracting causality are brought together.

To prove the effectiveness of CexCNN, we conducted several experiments on different CNNs architectures and datasets. The results obtained showed the good quality of the explanations generated, and how they highlight only the most important regions. We describe in the following the main contributions of this paper:

- We propose CexCNN, a causal explanation framework for CNNs, which combines different notions on causality: responsibility, blame, and causal abstraction.
- CexCNN identifies salient regions in input images based on the most responsible filters of the last convolution layer
- CexCNN does need to neither modify the input image, nor the network. Moreover, no retraining is needed.
- CexCNN allows the identification of all salient regions, from the most important to the least ones, thus it can be used for object localization.
- Comparing to existing Weakly Supervised Object Localization(WSOL) methods, CexCNN shows better results.
- Through the causal information learned of each filter, CexCNN can be used as well for compressing CNNs architectures through pruning the least responsible filters.

2   Related Work

Visualizing and Explaining CNNs

Gradient-based methods: DeepLIFT [4], CAM [34], GradCAM [28], Integrated gradients [32] and SmoothGrad [31] have been proposed recently in the aim of localizing neurons having more effect, and then assign scores to the inputs for a given output. The latter provides saliency maps that can be obtained by testing the network repeatedly, and trying to find the smallest regions on input images whose removal causes the classification score to drop significantly. Some of these methods have been implemented and regrouped together in different visualization toolboxes [1]. While these models are mainly applicable to CNNs, and most of them are very fast, which enable them to be employed in real-time applications, there exist models such as LIME [25] and SHAP [22] that can be used for interpreting decisions of any ML model, but unfortunately they are slow to compute since they require multiple evaluations.

Our technique has many similarities to CAM [34] and Grad-CAM [28], since we employ a score-based technique, however, it is represented here in term of filters responsibilities. As main differences, the scores of filters based on counterfactual information are computed for only one instance of the class category, which can be then used for localizing the most discriminative regions of any instance of this class. However, for Grad-CAM, the scores might change depending on the instance in use. This important feature would lead to consistent explanations for different instances. In addition, CexCNN does not require any architectural changes or retraining. Although Grad-CAM does not require architectural changes as well, it could be affected by some modifications, such as: Grad-CAM has been found to be more effective with global average pooling than global max pooling. CexCNN is not affected by such a modification. Actually, based on an experiment on MNIST dataset, we will also show that the results of CexCNN are consistent despite the visualization method in use, in contrast to Grad-CAM, where we will have different attention results.

Finally, both Grad-CAM and CexCNN can be used for Weakly Supervised Object Localization (WSOL), since they provide attention maps representing the most important regions. However, our method could give better results in this regard, since it can be easily extended to localize less important features, thus, resulting in identifying the entire object, not only its important features.

Causal Explanation of CNNs: The causal-based explanation methods suggest to rely on the cause-effect principle. By employing either statistical, intervention or counterfactual approaches, many causal explanation methods have been proposed in order to give causal interpretation for CNNs decisions, or abstract the CNN models into causal models.

Harradon et al. [15] have considered building a causal model for CNNs, which is a casual Bayesian model built based on extracting salient concepts. These concepts represent then the variables in the Bayesian model. Based on the definition of causal interventions, Narendra et al. [24] proposed structural causal models (SEM) as an abstraction of CNNs. This method is based on considering filters as causes, and then ranking them by their counterfactual influence with respect to each convolution layer. Given the SEM, the user would be able to get an answer for the following query or question: what is the impact of the n-th filter on the m-th layer on the model's predictions?". The main drawback of this approach is the size of the causal model generated, which could consist of thousands of nodes and edges.

Our approach to causality differs to previous methods in different ways. While these methods attempt to build an equivalent causal model that acts as an explanatory model for CNNs, our approach learns only the most important features, thus enabling class discrimination through estimating the responsibility of each causal filter in the last convolution layer, where the rest of non-causal filters are omitted. From another side, our work employs counterfactual reasoning, which comes at the top of the causal hierarchy, in contrast to [15] and [27], which employ interventions, and association respectively. Only [24] adopted counterfactuals similarly to our work, however, with major differences. First, with respect to counterfactual adoption, they measure the importance of filters in term of variance, not as we do here, in term of responsibility, which represents the quantitative extension of counterfactual causality. Besides, we adopt here in addition to responsibility, blame and causal abstraction, which make CexCNN a robust causal framework. With respect to the use of causality for explaining CNNS, [24] does not aim to identify salient regions, but rather, it helps to understand the inner working of CNNs.

3 Causal Explanation of CNNs

Gradient-based methods of explanation such as CAM [34] and Grad–CAM [28] are very useful explanation methods for CNNs. However, they only explain the output activations of a specific class in terms of the input activations, thus they are dependent on gradients with the absence of any causal interpretation. In this section we will show that CNNs have actually a causal structure, where the final decision at the final layer has causal dependencies on all the previous layers. So, what we are about to do, is to show how causality can be interpreted in CNNs, and how to define it in terms of the definition of causality by Halpern and Pearl. Moreover, we will show how we can benefit from its quantitative measures responsibility and blame to give robust explanations for CNNs decisions.

It is well known that the most important layers of CNNs models are the convolution layers, which include the filters. A filter actually represents the basic element of the network that gives activations for different regions in the image. In this section, we will show that addressing causality in CNNs should be built upon filters. Although some works that investigated causality on CNNs through couterfactuals addressed the perturbation of the input image [15], we will show that in our technique, we let both the input image as well as the network intact, without any modification and retraining.

3.1 Filters as Actual Causes

In this section we will show how we can define a probabilistic causal model [13] for CNNs. Each input image consists of a specific number of pixels. In CNNs, we move filters of specific sizes on sets of pixels, and the values obtained refer to match probabilities of these filters on every region in the input image. Then, these values serve as inputs for the flowing convolution layers. Filters with higher matches on the image's regions (or convoluted images in next layers) give higher activations, which will help to decide on the final class in the last layer. So, if we think of the nature of filters in terms of causality with respect to a CNN architecture, we find that filters represent actual causes, since their match on input image regions give rise to the activations that lead to the final decision.

With respect to the definition of causality by Halpern and Pearl, the actual cause X is defined on a context. So, we need to reason on contexts in CNNs. Actually, given an input/convolved image, we have different contexts, which represent the different regions of the image on which we apply these filters. After defining a context u, we can also reason on the probability function of the context $Pr(u)$, which refers to the probability of the filter's match given a context u. So we can say that a filter at a specific layer f_l is a cause in a region u with a probability $Pr_{f_l}(u)$ for such a decision, where the filters probabilities can be returned based on the activations of the feature maps of the last convolutional layer. Now the remaining task is to define φ in CNNs, what would φ represent? It is evident that φ would represent the final class identified by a CNN, for instance a hummingbird (see Fig. 1). However, with CNNs, the class predicted is returned with a probability, thus φ should not be just a Boolean formula that refers to the class predicted being a hummingbird, but rather is returned with a probability P.

We should recall that an actual cause is defined based on counterfactual theory. Formally, let us denote by F the set of all filters. Now we can introduce the definition of an actual cause in CNNs.

Definition 1. *Actual cause let us consider a filter at a specific layer l denoted $f_l \in F$. We say that f_l is an actual cause for a decision φ, if its own removal, where all the filters are kept the same, decreases the prediction probability P of φ by p.*

The set of filters F can be partitioned into two sets F_W^φ and F_Z^φ, where the set F_W^φ refers to filters causing the prediction φ, i.e. their removal will decrease the prediction probability P, whereas F_Z^φ are not causal, in way that removal of a filter in F_Z^φ does not affect P, or increases it. When the removal of a filter leads to increasing P, this means that this filter has a negative effect on the decision, which means that it is responsible for increasing the prediction probability of another class, not the current class (see Fig. 1). With this definition in hand, we can now introduce the definition of a filter's responsibility.

Definition 2. *Responsibility The degree of responsibility of a filter f_l for a prediction φ denoted $dr(f_l, \varphi)$ is 0 if $f_l \in F_Z^\varphi$, i.e. f_l is not a cause, and otherwise is $1/(|f_W| + 1)$, where $f_W \subseteq F_W^\varphi$ refers to the minimal subset such that their removal $F_W^\varphi - f_W$ makes decreasing the probability P below P_t counterfactually depends on f_l, where P_t refers to a probability threshold.*

So the filter's importance for such a prediction is measured with regard to its responsibility, where the most responsible filter would be the one whose removal results in decreasing P significantly. A most responsible filter would have 1 as a degree of responsibility, i.e. $|f_W| = 0$, if its own removal decreases P below the probability threshold P_t.

The definitions of actual cause and responsibility are very useful for getting an insight on how CNN decisions were derived, thus they could be mainly useful for model diagnosis. However, for the main purpose of this paper, which is providing interpretable explanations of CNNs decisions, these definitions are not sufficient. In order for the explanation to be interpretable, it should highlight the features or regions of interest in the input image. Here comes the definition of blame, which can be used to localize the sets of pixels in the input image that should be blamed the most for the decision φ. Once we answer this question, the explanations can be given to the user by highlighting discriminative regions. So, the remaining task is to relate the definition of filters *responsibilities* to regions *blame*.

Definition 3. *Blame Let us denote by K the set of regions in an input image. The degree of blame for a region $\sigma \in K$ for a prediction φ denoted $db(\sigma, \varphi)$ is*

$$db(\sigma, \varphi) = \sum_{f_l \in F_W^\varphi, \sigma in K} dr(f_l, \varphi) Pr_{f_l}(\sigma) \tag{1}$$

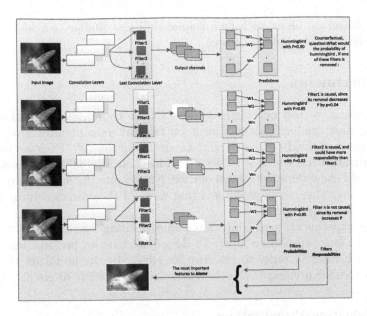

Fig. 1. CexCNN architecture: Localizing the most discriminative regions to blame, which is performed by projecting back the product of filters responsibilities and their activations. Based on causality abstraction, only the last convolution layer is considered for this computation.

So the degree of blame db for a region σ is computed as the sum of the products of every causal filter match's probability in this region with its responsibility. In other words, the regions that should be blamed the most for a decision φ, would be the ones having the highest match probabilities with the most responsible filters. Those regions are identified critical for the decision φ, and when returned together they show interpretable features on the input image.

Abstraction of Causality

Beckers and Halpern [5] investigated the problem of abstracting causal models, when micro variables have causal effect on macro variables at higher levels. This intuition behind abstraction of causal models leads us to consider for explaining such a decision, only the filters of the last convolution layer, since they represent the top macro-variables. We will show that they will be sufficient to estimate the degree of blame for every region in the input image.

What we aim to do through explanation is to identify at the final layer which important features led to such a decision. For instance, forehead and tail in the input image have led to the decision of a hummingbird (see Fig. 1). In other words, which features should be blamed for such a decision.

The importance of an image's regions are identified by projecting back the products of the activations of the last convolutional feature maps that refer to filters match probabilities and the responsibilities associated with these filters. In other words, among all outputs of the last convolution layer, we project back only the output of filters with high responsibilities. Similar to CAM, CexCNN provides heatmaps as grids of scores.

4 Experiments

In this section we test our method CexCNN on many datasets and architectures. We first conduct an experiment on two different architectures: VGG16 and Inception trained on the same dataset: ImageNet [9]. Then we test it on a third architecture, which is LeNet[17] trained on MNIST [18]. These experiments would show the effectiveness of CexCNN despite the architecture and the dataset in use. One common thing is that none of these architectures needs to be modified or retrained.

All the experiments are mainly based on measuring the responsibility of each filter of the last convolution layer. To do so, we have to perform removing or zeroing each filter in order to measure its effect. In the literature there exist some libraries for pruning CNNs through removing specific filters [2, 20].

4.1 Evaluating Visualizations

Since we introduce CexCNN as an explanation technique, we choose to evaluate it against some desirable properties that every explanation method for ML models should meet. Some properties are summarized in [23].

In this experiment we first test our method on VGG16 trained on the ImageNet dataset [9]. We choose the last convolution layer of the model to be the top of the causal hierarchy. This layer includes 512 filters, among them we consider only the most responsible ones. Filters whose removing has no effect on the predicted class probability, or has a positive effect, are not causes, and thus they are not considered for computing responsibility and blame. For the rest of causal filters, we compute their responsibilities, and then project their values on the original image through all the previous layers, in order to obtain the blame of each region. The most blamed regions in the image would have the highest heatmaps.

Examples of some classes and the heatmaps generated by CexCNN are depicted in Fig. 2. The explanations provided by our method are compared to those provided by Grad-CAM in the same figure.

Comparing to Grad-CAM, we find that our method sometimes localizes nearly the same important features that Grad-CAM localizes, like umbrella, and sometimes it localizes different features. For instance, with boxer, our method detects the face as the most important feature(k), however with Grad-CAM, the chest is considered more important (l).

(a) Image   (b) CexCNN   (c) Grad-CAM

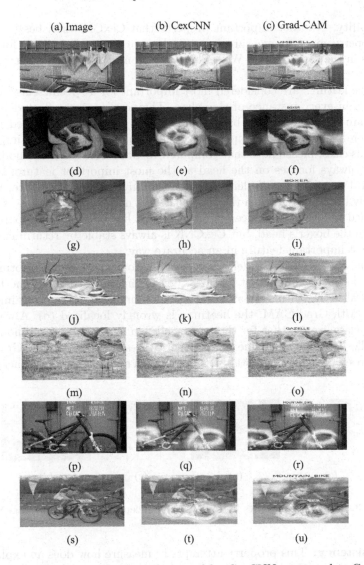

Fig. 2. Discriminative image regions returned by CexCNN compared to Grad-CAM. The images are from ImageNet ILSVRC [26], on which VGG16 is trained. The heatmaps generated by CexCNN refer to the most important features to blame.

A) Discriminative Ability. We see that CexCNN localizes important features that should be blamed the most. Besides, it is highly sensitive, where critical regions are clearly visualized, without highlighting regions on the other objects. For instance, the boxer with a hat (e), the hat is not highlighted. In the second image of boxer (h), the belt is excluded as well. Other examples include umbrella and mountain bike, whose only discriminative features are highlighted despite the presence of other objects.

B) Stability. Another important property that CexCNN satisfies is stability. This property aims to measure how similar are the explanations for similar instances in the same model. We see that our framework satisfies very well this property.

If some features are considered most important for a specific class, they are considered in every instance of this class. They can be just affected by which features are in foreground or background pixels. This can be noticed in boxer class, where the face is most important, in addition to chest with low importance (h). In this class, please note that compared to Grad-CAM, our method is more stable, it always focuses on the head as the most important feature, however, Grad-CAM sometimes considers chest as the only important feature, ignoring completely the head (i). With the same class, we notice that in figure (f), Grad-CAM failed to localize the face feature, where it has been fooled apparently by the hat on the boxer's head, but CexCNN is always stable by returning the face as the most important feature in an accurate way.

In other classes such as gazelle, it is clear that our method outperforms Grad-CAM for localizing important features. See for instance (m), our method was able to localize the features of all visible gazelles similarly in the input image (n). However, with Grad-CAM, the heatmap is wrongly localized (o). Always with the same class, we see that for the two gazelles in figure (j), our method for the two gazelles localizes the same important features (k), whereas for Grad-CAM the features returned are not stable (l), with much noise on the grass background.

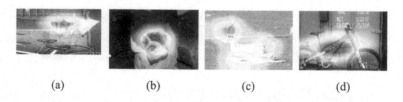

(a)   (b)   (c)   (d)

Fig. 3. Explanation results by CexCNN for Inception trained on ImageNet

C) Consistency: This property attempts to measure how does an explanation differ between models that have been trained on the same dataset. To this end, in addition to VGG16, we tested CexCNN on Inception trained on Imagenet as well.

Comparing to VGG16, providing explanations for Inception is more challenging, since Inception consists of many modules, which consist of multiple convolution layers. The convolution layers considered by CexCNN are the last concatenated ones. While the last convolution layer of VGG16 consists of 512 filters, the last convolution layers of Inception when concatenated consist of 2048 filters. Similar to VGG16, we have to measure the responsibility of each filter and use this information to find the most important features to blame.

Before presenting the results of this experiment, we should note here that providing similar explanations for different models on the same instances is only possible if the two models see the world similarly, i.e. if the models focus on the same features to make a decision. Actually, providing explanations could be itself useful for identifying if different models see the world in a similar way. However, in order to evaluate consistency, it is necessary for the explanations to be stable with respect to each model. The results of stable explanations provided by CexCNN on Inception for some classes compared to those returned by VGG16 are presented in Fig. 3. We see that for some instances such as umbrella and boxer, the explanations are very similar, which means that both VGG16 and Inception focus on the same features for predicting these classes. However, with other classes such as mountain bike, Inception focuses on a different important feature, which absolutely discriminates mountain bike as well from other classes.

D) Robustness against Adversarial Attacks. Goodfellow et al. [11] have discovered a critical weakness for DNNs models, which is adversarial examples or adversarial attacks. Goodfellow et al. showed that perturbating input images through introducing noise to the original images, in way that they look identical to the human eye, such perturbation could fool the network to completely misclassify the input image. In this section, we will show how CexCNN performs against perturbed images that have been misclassified by VGG16. For generating perturbations we use the FGSM method [11]. Some qualitative examples are presented in Fig. 4. As we see, although CexCNN is affected slightly by the perturbations, it always try to focus on the same discriminative regions.

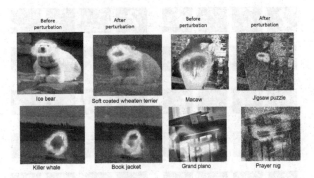

Fig. 4. Robustness against adversarial attacks:heatmaps returned by CexCNN

4.2   MNIST

We tested our method on LeNet architecture trained on MNIST dataset [18]. We consider the last layer of the LeNet architecture with 64 filters. In this experiment we tried another strategy for computing counterfactual information and measuring responsibility and blame. Since this dataset is small (10 classes), with simple classes (digits), not complex ones comparing ImageNet, we measured the causal effect on the accuracy of the entire model, not just the prediction probability of an individual class. The heatmaps as generated by our method on every digit compared to heatmaps generated by Grad-CAM are presented in Fig. 5. These results are obtained using the Keras-vis tool [3].

We see that our method clearly outperforms the important regions localized by Grad-CAM. The quality of the explanations generated is clearly better than Grad-CAM, because for every digit it is sufficient to decide on the input image just by considering the highlighted features related to this digit. For some digits, 9 for instance, Grad-CAM fails to generate explanations at all. Another important thing about the effectiveness of our method compared to Grad-CAM, is that our method returns stable explanations despite the modifier in use.

4.3   Weakly Supervised Object Localization (WSOL)

WSOL has been recently addressed by many researchers for different visual tasks [7,30,34]. WSOL aims to localize objects with only image-level labels, under a

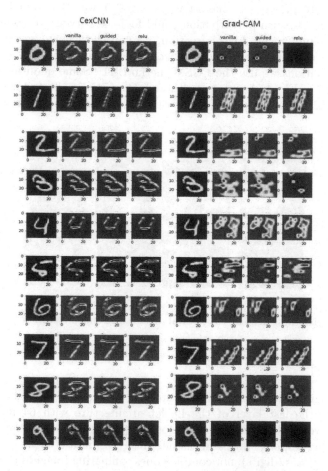

Fig. 5. CexCNN results on MNIST compared to Grad-CAM

general assumption that only one object of the specific category is present in the input image. One main drawback of existing WSOL methods such as CAM, which gave a rise to other WSOL methods, is that it localizes only the most discriminative features of an object rather than all the features, which leads logically to a less performance in object localization task. Kumar and Lee [30] for instance tried to tackle this issue by modifying the input image by randomly removing grid patches, thus forcing the network to be meticulous for WSOL, by identifying all the object's features not only the most discriminative ones. In this section, we will show that CexCNN provides good results in this regard. To make CexCNN good in WSOL, we should consider all the causal filters F_W^{φ}, not just the most responsible ones. This results in identifying all the object's parts from the most discriminative (having the highest blame dB) to parts with low discrimination (having the lowest blame dB). To show the effectiveness of CexCNN for WSOL, we provide qualitative (See Fig. 6) as well as quantitative results (See Table 1) on the ILSVRC validation dataset, which consists of 50.000 images of 1000 categories. The results are compared to CAM. We see in Fig. 6 that CexCNN localizes in a better way the concerned object, since it could look on all its features, from the most important to the less important ones, whereas CAM is designed to focus only on the most discriminative regions. Both CexCNN and CAM are evaluated in the same setting, on VGG16 without any architectural modifications.

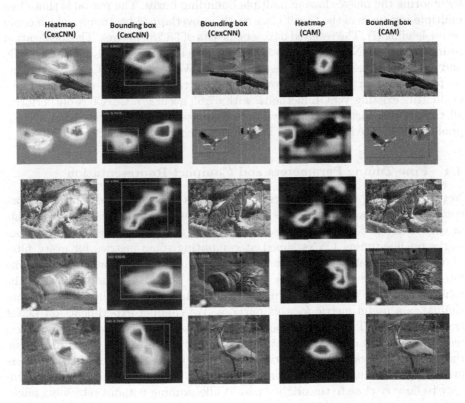

Fig. 6. Qualitative object localization results

Table 1. Quantitative object localization results for CexCNN compared to CAM on VGG16

Method	GT-known Loc
CAM	31.71
CexCNN	67.65

For quantitative evaluation, a set of evaluation metrics are commonly used to access the performance of WSOL methods. These metrics are *Top-1/Top-5* localization accuracy and localization accuracy with known ground truth class (GTKnown Loc). While Top-1/Top-5 represent both the classification and localization, GT-Known Loc is true for an input image if given the ground truth class, the intersection over union (IoU) between the ground truth bounding box and the predicted box is at least 50%. Since we are interested here in evaluating CexCNN only for object localization, not for classification, we decided to consider only GTKnown Loc. Besides, Choe et al. [7] have shown in their evaluative study of existing WSOL methods that *Top-1/Top-5* localization might be misleading, and thus they suggest to consider the *GT-known* metric. The experiments are conducted by performing a small modification on the dataset by ignoring the images having multiple bounding boxes. The reason is that close multiple instances of the target class result in overlapped heatmaps, which could be misleading [7]. The resulted dataset consists of 38.285 images. The localization results of CexCNN compared to CAM are provided in Table 1. Both CexCNN and CAM were executed on the same dataset. We see that CexCNN clearly outperforms CAM in WSOL. Moreover, based on the reproduced results in [7] for evaluating existing WSOL methods, with 67.65 accuracy, CexCNN outperforms all the existing methods that have been evaluated on VGG with global average pooling (VGG-GAP), where the best WSOL accuracy reported was 62.2.

4.4 Fine-Tuning Parameters and Compact Representation

Network pruning aims to compress CNN models in order to reduce the network complexity and over-fitting. This requires pruning and compressing the weights of various layers without affecting the accuracy of these models. Some previous works on fine-tuning CNNs aimed at computing some metrics for every filter in order to generate a compact network [20]. Li et al. [20] showed that pruning filters of multiple layers at once can be useful and gives a good view on the robustness of the network. In this regard, we may consider the filter's responsibility as a useful metric for this purpose. We conduct experiments on LeNet trained on MNIST. The experiments are based on removing the filters with low responsibilities at each layer, and then calculate the impact of their removal on the model's accuracy. The results presented focus on the number of the filters pruned and the number of network's parameters reduced, by allowing the accuracy to be very close to the original one. While pruning without retraining might

Table 2. Results after pruning less responsible filters for LeNet trained on MNIST

Parameters	Filters pruned conv1	Filters pruned conv2	total pruned	Pruned parameters %	Error%
639.760	2/20	12/50	14/70(20%)	152.876(23.90%)	0.05
639.760	13/20	28/50	41/70(58%)	355.068(55.92%)	5.1
639.760	0/20	36/50	36/70(51.42%)	456.516(71.35%)	3.03
639.760	0/20	39/50	39/70(55.71%)	494.559(77.30%)	4.5

be harmful for the model accuracy [14, 20], we will show that our results are good enough without retraining.

LeNet consists of two convolution layers: the first consists of 20 filters, and the last one consists of 50 filters. For LeNet on MNIST, the best results in terms of accuracy are obtained by removing (2/20) filters (10%) of the first convolution layer, and (12/50) filters (24%) of the last one. This operation results in reducing the network's parameters from 639.760 (original model) to 486, 884 (23.90%), and resulting in a good accuracy, which has been reduced just from 0.9933 to 0.9928 (See Table 2).

We notice that conv1 is more sensible for filters pruning, thus we want to challenge our technique on accuracy against the number of reduced parameters by considering only conv2. The results are presented in the same table. While most of pruning techniques consider retraining the model, Han et al. [14] have analyzed in addition the trade-off between accuracy and number of parameters without retraining. The results as described here are very close to the ones reported in [14]. We see that we are able to reduce the number of parameters by 77.30% , which results in dropping the accuracy from 0.9933 to 0.9483.

5   Conclusion and Future Work

In this paper we provided CexCNN, a causal explanation technique for CNNs. CexCNN employs the theory of causality by Halpern and Pearl, in addition to their quantitative measures: responsibility and blame, as well as causality abstraction. We showed that weighting filters by their responsibilities and then projecting this information back in the input image, allows the localization of the most important features to blame for such a decision.

We evaluated CexCNN on many datasets and architectures and it has shown good results given a set of evaluation properties for explanation methods. Although the main concern is to localize the most discriminative regions, we showed that CexCNN outperforms Grad-CAM and known existing methods for WSOL. In addition, we showed that CexCNN could stand as a good pruning technique.

As future work, we aim to address the usefulness of CexCNN for defending against adversarial attacks, as well as its application in transfer learning.

References

1. Investigate. https://github.com/albermax/innvestigate
2. keras-surgeon. https://github.com/BenWhetton/keras-surgeon
3. Keras visualization toolkit. https://github.com/raghakot/keras-vis
4. Avanti, S., Peyton, G., Anshul, K.: Learning important features through propagating activation differences, pp. 3145–3153. ICML'17 (2017)
5. Beckers, S., Halpern, J.Y.: Abstracting causal models. In: AAAI (2017)
6. Chockler, H., Halpern, J.Y.: Responsibility and blame: a structural-model approach. J. Artif. Int. Res. **22**(1), 93–115 (2004)
7. Choe, J., Oh, S.J., Lee, S., Chun, S., Akata, Z., Shim, H.: Evaluating weakly supervised object localization methods right. In: CVPR, pp. 3130–3139 (2020)
8. Gordon, D., Kembhavi, A., Rastegari, M., Redmon, J., Fox, D., Farhadi, A.: Iqa: visual question answering in interactive environments. In: In arXiv:1712.03316 (2017)
9. Deng, J., Dong, W., Socher, R., Li, L., Li, K., Fei-Fei, L.: Imagenet: a large-scale hierarchical image database. In: CVPR, pp. 248–255 (2009)
10. Girshick, R., Donahue, J., Darrell, T., Malik, J.: Rich feature hierarchies for accurate object detection and semantic segmentation. In: CVPR, pp. 580–587 (2014)
11. Goodfellow, I.J., Shlens, J., Szegedy, C.: Explaining and harnessing adversarial examples. In: ICLR (2015)
12. Halpern, J., Pearl, J.: Causes and explanations: a structural-model approach part i: Causes. In: Proceedings of the 17th UAI, pp. 194–202 (2001)
13. Halpern, J.Y., Pearl, J.: Causes and explanations: a structural-model approach. part ii: Explanations. Br. J. Philos. Sci. **56**(4), 889–911 (2008)
14. Han, S., Pool, J., Tran, J., Dally, W.J.: Learning both weights and connections for efficient neural networks. In: NIPS (2015)
15. Harradon, M., Druce, J., Ruttenberg, B.E.: Causal learning and explanation of deep neural networks via autoencoded activations. In: CoRR abs/1802.00541 (2018)
16. Krizhevsky, A., Sutskever, I., Hinton, G.E.: Imagenet classification with deep convolutional neural networks. Commun. ACM **60**(6), 84–90 (2017)
17. Lecun, Y., Bottou, L., Bengio, Y., Haffner, P.: Gradient-based learning applied to document recognition. Proc. IEEE **86**(11), 2278–2324 (1998)
18. LeCun, Y., Cortes, C., Burges, C.: Mnist handwritten digit database. ATT Labs [Online]. http://yann.lecun.com/exdb/mnist 2 (2010)
19. Lewis, D.: Causation. J. Philos. **70**, 556–567 (1972)
20. Li, H., Kadav, A., Durdanovic, I., Samety, H.: Pruning filters for efficient convnets. In: ICLR 2017, pp. 1–13 (2017)
21. Long, J., Shelhamer, E., Darrell, T.: Fully convolutional networks for semantic segmentation. In: CVPR, pp. 3431–3440 (2015)
22. Lundberg, S.M., Lee, S.I.: A unified approach to interpreting model predictions. In: NIPS, pp. 4768–4777 (2017)
23. Molnar, C.: Interpretable Machine Learning A Guide for Making Black Box Models Explainable (2018). https://christophm.github.io/interpretable-ml-book/
24. Narendra, T., Sankaran, A., Vijaykeerthy, D., Mani, S.: Explaining deep learning models using causal inference. In: arXiv:1811.04376 (2018)
25. Ribeiro, M.T., Singh, S., Guestrin, C.: "why should i trust you?": Explaining the predictions of any classifier, pp. 1135–1144. KDD '16 (2016)
26. Russakovsky, O., et al.: Imagenet large scale visual recognition challenge. Int. J. Comput. Vis. **115**(3), 211–252 (2015)

27. Schwab, P., Karlen, W.: Cxplain: causal explanations for model interpretation under uncertainty. NeurIPS, pp. 10220–10230 (2019)
28. Selvaraju, R.R., Cogswell, M., Das, A., Vedantam, R., Parikh, D., Batra, D.: Grad-cam: visual explanations from deep networks via gradient-based localization. In: ICCV, pp. 618–626 (2017)
29. Simonyan, K., Vedaldi, A., Zisserman, A.: Deep inside convolutional networks: visualising image classification models and saliency maps. In: In arXiv:1312.6034 (2013)
30. Singh, K.K., Lee, Y.J.: Forcing a network to be meticulous for weakly-supervised object and action localization. In: CVPR (2017)
31. Smilkov, D., Thorat, N., Kim, B., Viegas, F.B., Wattenberg, M.: Smoothgrad: removing noise by adding noise. In: CoRR, vol. abs/1706.03825 (2017)
32. Sundararajan, M., Taly, A., Yan, Q.: Axiomatic attribution for deep networks. In: ICML, pp. 3319–3328 (2017)
33. Zeiler, M.D., Fergus, R.: Visualizing and understanding convolutional networks. In: Fleet, D., Pajdla, T., Schiele, B., Tuytelaars, T. (eds.) ECCV 2014. LNCS, vol. 8689, pp. 818–833. Springer, Cham (2014). https://doi.org/10.1007/978-3-319-10590-1_53
34. Zhou, B., Khosla, A., Lapedriza, A., Oliva, A., Torralba, A.: Learning deep features for discriminative localization. In: CVPR, pp. 2921–2929 (2016)

Interpretable Counterfactual Explanations Guided by Prototypes

Arnaud Van Looveren[✉][iD] and Janis Klaise[iD]

Seldon Technologies, 41 Luke Street, London EC2A 4AR, UK
{avl,jk}@seldon.io

Abstract. We propose a fast, model agnostic method for finding interpretable counterfactual explanations of classifier predictions by using class prototypes. We show that class prototypes, obtained using either an encoder or through class specific k-d trees, significantly speed up the search for counterfactual instances and result in more interpretable explanations. We quantitatively evaluate interpretability of the generated counterfactuals to illustrate the effectiveness of our method on an image and tabular dataset, respectively MNIST and Breast Cancer Wisconsin (Diagnostic). Additionally, we propose a principled approach to handle categorical variables and illustrate our method on the Adult (Census) dataset. Our method also eliminates the computational bottleneck that arises because of numerical gradient evaluation for *black box* models.

Keywords: Interpretation · Transparency/Explainability · Counterfactual explanations

1 Introduction

Humans often think about how they can alter the outcome of a situation. *What do I need to change for the bank to approve my loan?* or *Which symptoms would lead to a different medical diagnosis?* are common examples. This form of counterfactual reasoning comes natural to us and explains how to arrive at a desired outcome in an interpretable manner. Moreover, examples of counterfactual instances resulting in a different outcome can give powerful insights of what is important to the underlying decision process, making it a compelling method to explain predictions of machine learning models (Fig. 1).

In the context of predictive models, given a test instance and the model's prediction, a counterfactual instance describes the necessary change in input features that alter the prediction to a predefined output [21]. For classification models the predefined output can be any target class or prediction probability distribution. Counterfactual instances can then be found by iteratively perturbing the input features of the test instance until the desired prediction is reached.

Electronic supplementary material The online version of this chapter (https:// doi.org/10.1007/978-3-030-86520-7_40) contains supplementary material, which is available to authorized users.

© Springer Nature Switzerland AG 2021
N. Oliver et al. (Eds.): ECML PKDD 2021, LNAI 12976, pp. 650–665, 2021.
https://doi.org/10.1007/978-3-030-86520-7_40

In practice, the counterfactual search is posed as an optimization problem—we want to minimize an objective function which encodes desirable properties of the counterfactual instance with respect to the perturbations. The key insight of this formulation is the need to design an objective function that allows us to generate high quality counterfactual instances. A counterfactual instance x_{cf} should have the following desirable properties:

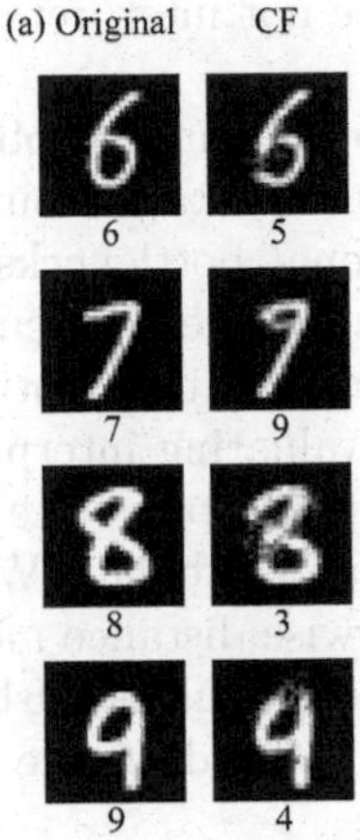

(a) Original	CF	(b)	Original	CF
6	6	Workclass	Private	State-gov
		Education	High school	Bachelors
6	5	Marital Status	Married	Married
		Occupation	Blue-Collar	Blue-Collar
7	7	Relationship	Husband	Husband
		Race	White	White
7	9	Sex	Male	Male
		Country	United-States	United-States
8	8	Age	46	46
		Capital Gain	0	0
8	3	Capital Loss	0	0
		Hours p/w	40	40
9	4	Prediction	$\leq \$50k/y$	$> \$50k/y$
9	4			

Fig. 1. (a) Examples of original and counterfactual instances on the MNIST dataset along with predictions of a CNN model. (b) A counterfactual instance on the Adult (Census) dataset highlighting the feature changes required to alter the prediction of an NN model.

1. The model prediction on x_{cf} needs to be close to the predefined output.
2. The perturbation δ changing the original instance x_0 into $x_{\text{cf}} = x_0 + \delta$ should be sparse.
3. The counterfactual x_{cf} needs to be interpretable. We consider an instance x_{cf} interpretable if it lies close to the model's training data distribution. This definition does not only apply to the overall data set, but importantly also to the *training instances that belong to the counterfactual class*. Let us illustrate this with an intuitive example. Assume we are predicting house prices with features including the square footage and the number of bedrooms. Our house is valued below £500,000 and we would like to know what needs to change about the house in order to increase the valuation above £500,000. By simply increasing the number of bedrooms and leaving the other features unchanged, the model predicts that our *counterfactual house* is now worth more than £500,000. This sparse counterfactual instance lies fairly close to the overall training distribution since only one feature value was changed. The counterfactual is however out-of-distribution with regards to the subset of houses in the training data valued above £500,000 because other relevant

features like the square footage still resemble a typical house valued below £500,000. As a result, we do not consider this counterfactual to be very interpretable. We show in the experiments that there is often a trade-off between sparsity and interpretability.

4. The counterfactual instance x_{cf} needs to be found fast enough to ensure it can be used in a real life setting.

An overly simplistic objective function may return instances which satisfy properties 1. and 2., but where the perturbations are not interpretable with respect to the counterfactual class.

In this paper we propose using class prototypes in the objective function to guide the perturbations quickly towards an interpretable counterfactual. The prototypes also allow us to remove computational bottlenecks from the optimization process which occur due to numerical gradient calculation for black box models. In addition, we propose two novel metrics to quantify interpretability which provide a principled benchmark for evaluating interpretability at the instance level. We show empirically that prototypes improve the quality of counterfactual instances on both image (MNIST) and tabular (Wisconsin Breast Cancer) datasets. Finally, we propose using pairwise distance measures between categories of categorical variables to define meaningful perturbations for such variables and illustrate the effectiveness of the method on the Adult (Census) dataset.

2   Related Work

Counterfactual instances—synthetic instances of data engineered from real instances to change the prediction of a machine learning model—have been suggested as a way of explaining individual predictions of a model as an alternative to feature attribution methods such as LIME [23] or SHAP [19].

Wacther et al. [27] generate counterfactuals by minimizing an objective function which sums the squared difference between the predictions on the perturbed instance and the desired outcome, and a scaled L_1 norm of the perturbations. Laugel et al. [15] find counterfactuals through a heuristic search procedure by growing spheres around the instance to be explained. The above methods do not take local, class specific interpretability into account. Furthermore, for black box models the number of prediction calls during the search process grows proportionally to either the dimensionality of the feature space [27] or the number of sampled observations [9,15], which can result in a computational bottleneck. Dhurandhar et al. [7,9] propose the framework of *Contrastive Explanations* which find the minimal number of features that need to be changed/unchanged to keep/change a prediction.

A key contribution of this paper is the use of prototypes to guide the counterfactual search process. Kim et al. [14], Gurumoorthy et al. [11] use prototypes as example-based explanations to improve the interpretability of complex datasets. Besides improving interpretability, prototypes have a broad range of applications like clustering [13], classification [4,26], and few-shot learning [25]. If we have access to an encoder [24], we follow the approach of [25] who define a class prototype as the mean encoding of the instances which belong to that class. In the absence of an encoder, we find prototypes through class specific k-d trees [3].

To judge the quality of the counterfactuals we introduce two novel metrics which focus on local interpretability with respect to the training data distribution. This is different from [8] who define an interpretability metric relative to a target model. Kim et al. [14] on the other hand quantify interpretability through a human pilot study measuring the accuracy and efficiency of the humans on a predictive task. Luss et al. [20] also highlight the importance of good local data representations in order to generate high quality explanations.

Another contribution of this paper is a principled approach to handling categorical variables during the counterfactual generation process. Some previously proposed solutions are either computationally expensive [27] or do not take relationships between categories into account [9,22]. We propose using pairwise distance measures to define embeddings of categorical variables into numerical space which allows us to define meaningful perturbations when generating counterfactuals.

3 Methodology

3.1 Background

The following section outlines how the prototype loss term is constructed and why it improves the convergence speed and interpretability. Finding a counterfactual instance $x_{cf} = x_0 + \delta$, with both x_{cf} and $x_0 \in \mathcal{X} \subseteq \mathbb{R}^D$ where $\mathcal{X}$ represents the D-dimensional feature space, implies optimizing an objective function of the following form:

$$\min_{\delta} c \cdot f_\kappa(x_0, \delta) + f_{\text{dist}}(\delta). \tag{1}$$

$f_\kappa(x_0, \delta)$ encourages the predicted class i of the perturbed instance x_{cf} to be different than the predicted class t_0 of the original instance x_0. Similar to [7], we define this loss term as:

$$\begin{aligned} L_{\text{pred}} &:= f_\kappa(x_0, \delta) \\ &= \max([f_{\text{pred}}(x_0 + \delta)]_{t_0} - \max_{i \neq t_0}[f_{\text{pred}}(x_0 + \delta)]_i, -\kappa), \end{aligned} \tag{2}$$

where $[f_{\text{pred}}(x_0 + \delta)]_i$ is the i-th class prediction probability, and $\kappa \geq 0$ caps the divergence between $[f_{\text{pred}}(x_0 + \delta)]_{t_0}$ and $[f_{\text{pred}}(x_0 + \delta)]_i$. The term $f_{\text{dist}}(\delta)$ minimizes the distance between x_0 and x_{cf} with the aim to generate sparse counterfactuals. We use an elastic net regularizer [28]:

$$f_{\text{dist}}(\delta) = \beta \cdot \|\delta\|_1 + \|\delta\|_2^2 = \beta \cdot L_1 + L_2. \tag{3}$$

While the objective function (1) is able to generate counterfactual instances, it does not address a number of issues:

654   A. Van Looveren and J. Klaise

1. x_{cf} does not necessarily respect the training data manifold, resulting in out-of-distribution counterfactual instances. Often a trade off needs to be made between sparsity and interpretability of x_{cf}.
2. The scaling parameter c of $f_\kappa(x_0, \delta)$ needs to be set within the appropriate range before a potential counterfactual instance is found. Finding a good range can be time consuming.

[7] aim to address the first issue by adding in an additional loss term L_{AE} which represents the L_2 reconstruction error of x_{cf} evaluated by an autoencoder AE which is fit on the training set:

$$L_{AE} = \gamma \cdot \|x_0 + \delta - AE(x_0 + \delta)\|_2^2. \tag{4}$$

The loss L to be minimized now becomes:

$$L = c \cdot L_{pred} + \beta \cdot L_1 + L_2 + L_{AE}. \tag{5}$$

The autoencoder loss term L_{AE} penalizes out-of-distribution counterfactual instances, but does not take the data distribution for each prediction class i into account. This can lead to sparse but uninterpretable counterfactuals, as illustrated by Fig. 2. The first row of Fig. 2(b) shows a sparse counterfactual 3 generated from the original 5 using loss function (5). Both visual inspection and reconstruction of the counterfactual instance using AE in Fig. 2(e) make clear however that the counterfactual lies closer to the distribution of a 5 and is not interpretable as a 3. The second row adds a prototype loss term to the objective function, leading to a less sparse but more interpretable counterfactual 6.

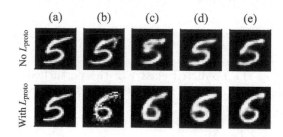

Fig. 2. First row: (a) original instance and (b) uninterpretable counterfactual 3. (c), (d) and (e) are reconstructions of (b) with respectively AE3, AE5 and AE. Second row: (a) original instance and (b) interpretable counterfactual 6. (c), (d) and (e) are reconstructions of (b) with respectively AE6, AE5 and AE.

The L_{AE} loss term also does not consistently speed up the counterfactual search process since it imposes a penalty on the distance between the proposed x_{cf} and its reconstruction by the autoencoder without explicitly guiding x_{cf} towards an interpretable solution. We address these issues by introducing an additional loss term, L_{proto}.

3.2   Prototype Loss Term

By adding in a prototype loss term L_{proto}, we obtain the following objective function:

$$L = c \cdot L_{\text{pred}} + \beta \cdot L_1 + L_2 + L_{\text{AE}} + L_{\text{proto}}, \tag{6}$$

where L_{AE} becomes optional. The aim of L_{proto} is twofold:

1. Guide the perturbations δ towards an interpretable counterfactual x_{cf} which falls in the distribution of counterfactual class i.
2. Speed up the counterfactual search process without too much hyperparameter tuning.

To define the prototype for each class, we can reuse the encoder part of the autoencoder from L_{AE}. The encoder $\text{ENC}(x)$ projects $x \in \mathcal{X}$ onto an E-dimensional latent space $\mathbb{R}^E$. We also need a representative, unlabeled sample of the training dataset. First the predictive model is called to label the dataset with the classes predicted by the model. Then for each class i we encode the instances belonging to that class and order them by increasing L_2 distance to $\text{ENC}(x_0)$. Similar to [25], the class prototype is defined as the average encoding over the K nearest instances in the latent space with the same class label:

$$\text{proto}_i := \frac{1}{K} \sum_{k=1}^{K} \text{ENC}(x_k^i) \tag{7}$$

for the ordered $\{x_k^i\}_{k=1}^K$ in class i. It is important to note that the prototype is defined in the latent space, not the original feature space.

The Euclidean distance is part of a class of distance functions called *Bregman divergences*. If we consider that the encoded instances belonging to class i define a cluster for i, then proto_i equals the cluster mean. For Bregman divergences the cluster mean yields the minimal distance to the points in the cluster [1]. Since we use the Euclidean distance to find the closest class to x_0, proto_i is a suitable class representation in the latent space. When generating a counterfactual instance for x_0, we first find the nearest prototype proto_j of class $j \neq t_0$ to the encoding of x_0:

$$j = \arg \min_{i \neq t_0} \|\text{ENC}(x_0) - \text{proto}_i\|_2. \tag{8}$$

The prototype loss L_{proto} can now be defined as:

$$L_{\text{proto}} = \theta \cdot \|\text{ENC}(x_0 + \delta) - \text{proto}_j\|_2^2, \tag{9}$$

where $\text{ENC}(x_0 + \delta)$ is the encoding of the perturbed instance. As a result, L_{proto} explicitly guides the perturbations towards the nearest prototype $\text{proto}_{j \neq t_0}$, speeding up the counterfactual search process towards the average encoding of class j. This leads to more interpretable counterfactuals as illustrated by the experiments. Algorithm 1 summarizes this approach.

Algorithm 1. Counterfactual search with encoded prototypes

1: **Parameters:** β, θ (required) and c, κ and γ (optional)
2: **Inputs:** AE (optional) and ENC models. A sample $X = \{x_1, \ldots, x_n\}$ from training set. Instance to explain x_0.
3: Label X and x_0 using the prediction function f_{pred}:
   $X^i \leftarrow \{x \in X \mid \text{argmax} \, f_{\text{pred}}(x) = i\}$ for each class i $t_0 \leftarrow \text{argmax} \, f_{\text{pred}}(x_0)$
4: Define prototypes for each class i:
   $\text{proto}_i \leftarrow \frac{1}{K} \sum_{k=1}^{K} \text{ENC}(x_k^i)$ for $x_k^i \in X^i$ where x_k^i is ordered by increasing $\|\text{ENC}(x_0) - \text{ENC}(x_k^i)\|_2$ and $K \leq |X^i|$
5: Find nearest prototype j to instance x_0 but different from original class t_0:
   $j \leftarrow \text{argmin}_{i \neq t_0} \|\text{ENC}(x_0) - \text{proto}_i\|_2$.
6: Optimize the objective function:
   $\delta^* \leftarrow \text{argmin}_{\delta \in \mathcal{X}} \, c \cdot L_{\text{pred}} + \beta \cdot L_1 + L_2 + L_{\text{AE}} + L_{\text{proto}}$ where $L_{\text{proto}} = \theta \cdot \|\text{ENC}(x_0 + \delta) - \text{proto}_j\|_2^2$.
7: **Return** $x_{\text{cf}} = x_0 + \delta^*$

Algorithm 2. Counterfactual search with k-d trees

1: **Parameters:** β, θ, k (required) and c, κ (optional)
2: **Input:** A sample $X = \{x_1, \ldots, x_n\}$ from training set. Instance to explain x_0.
3: Label X and x_0 using the prediction function f_{pred}:
   $X^i \leftarrow \{x \in X \mid \text{argmax} \, f_{\text{pred}}(x) = i\}$ for each class i $t_0 \leftarrow \text{argmax} \, f_{\text{pred}}(x_0)$
4: Build separate k-d trees for each class i using X_i
5: Find nearest prototype j to instance x_0 but different from original class t_0:
   $j \leftarrow \text{argmin}_{i \neq t_0} \|x_0 - x_{i,k}\|_2$ where $x_{i,k}$ is the k-th nearest item to x_0 in the k-d tree of class i.
   $\text{proto}_j \leftarrow x_{j,k}$
6: Optimize the objective function:
   $\delta^* \leftarrow \text{argmin}_{\delta \in \mathcal{X}} \, c \cdot L_{\text{pred}} + \beta \cdot L_1 + L_2 + L_{\text{proto}}$ where $L_{\text{proto}} = \theta \cdot \|x_0 + \delta - \text{proto}_j\|_2^2$.
7: **Return** $x_{\text{cf}} = x_0 + \delta^*$

3.3   Using K-D Trees as Class Representations

If we do not have a trained encoder available, we can build class representations using k-d trees [3]. After labeling the representative training set by calling the predictive model, we can represent each class i by a separate k-d tree built using the instances with class label i. This approach is similar to [12] who use class specific k-d trees to measure the agreement between a classifier and a modified nearest neighbour classifier on test instances. For each k-d tree $j \neq t_0$, we compute the Euclidean distance between x_0 and the k-nearest item in the tree $x_{j,k}$. The closest $x_{j,k}$ across all classes $j \neq t_0$ becomes the class prototype proto_j. Note that we are now working in the original feature space. The loss term L_{proto} is equal to:

$$L_{\text{proto}} = \theta \cdot \|x_0 + \delta - \text{proto}_j\|_2^2. \tag{10}$$

Algorithm 2 outlines the k-d trees approach.

3.4   Categorical Variables

Creating meaningful perturbations for categorical data is not straightforward as
the very concept of perturbing an input feature implies some notion of rank and
distance between the values a variable can take. We approach this by inferring
pairwise distances between categories of a categorical variable based on either
model predictions (Modified Value Distance Metric) [6] or the context provided
by the other variables in the dataset (Association-Based Distance Metric) [16].
We then apply multidimensional scaling [5] to project the inferred distances into
one-dimensional Euclidean space, which allows us to perform perturbations in
this space. After applying a perturbation in this space, we map the resulting
number back to the closest category before evaluating the classifier's prediction.

3.5   Removing L_{pred}

In the absence of L_{proto}, only L_{pred} encourages the perturbed instance to predict
class $i \neq t_0$. In the case of black box models where we only have access to the
model's prediction function, L_{pred} can become a computational bottleneck. This
means that for neural networks, we can no longer take advantage of automatic
differentiation and need to evaluate the gradients numerically. Let us express the
gradient of L_{pred} with respect to the input features x as follows:

$$\frac{\partial L_{\text{pred}}}{\partial x} = \frac{\partial f_\kappa(x)}{\partial x} = \frac{\partial f_\kappa(x)}{\partial f_{\text{pred}}} \frac{\partial f_{\text{pred}}}{\partial x}, \tag{11}$$

where f_{pred} represents the model's prediction function. The numerical gradient
approximation for f_{pred} with respect to input feature k can be written as:

$$\frac{\partial f_{\text{pred}}}{\partial x_k} \approx \frac{f_{\text{pred}}(x + \epsilon_k) - f_{\text{pred}}(x - \epsilon_k)}{2\epsilon}, \tag{12}$$

where ϵ_k is a perturbation with the same dimension as x and taking value ϵ
for feature k and 0 otherwise. As a result, the prediction function needs to be
evaluated twice for each feature per gradient step just to compute $\frac{\partial f_{\text{pred}}}{\partial x_k}$. For a
28×28 MNIST image, this translates into a batch of $28 \cdot 28 \cdot 2 = 1568$ prediction
function calls. Eliminating L_{pred} would therefore speed up the counterfactual
search process significantly. By using the prototypes to guide the counterfactuals,
we can remove L_{pred} and only call the prediction function once per gradient
update on the perturbed instance to check whether the predicted class i of $x_0 + \delta$
is different from t_0. This eliminates the computational bottleneck while ensuring
that the perturbed instance moves towards an interpretable counterfactual x_{cf}
of class $i \neq t_0$.

3.6   FISTA Optimization

Like [7], we optimize our objective function by applying a fast iterative shrinkage-
thresholding algorithm (FISTA) [2] where the solution space for the output $x_{\text{cf}} =$

$x_0 + \delta$ is restricted to $\mathcal{X}$. The optimization algorithm iteratively updates δ with momentum for N optimization steps. It also strips out the $\beta \cdot L_1$ regularization term from the objective function and instead shrinks perturbations $|\delta_k| < \beta$ for feature k to 0. The optimal counterfactual is defined as $x_{\mathrm{cf}} = x_0 + \delta^{n^*}$ where $n^* = \arg\min_{n \in 1,\dots,N} \beta \cdot \|\delta^n\|_1 + \|\delta^n\|_2^2$ and the predicted class on x_{cf} is $i \neq t_0$.

4   Experiments

The experiments are conducted on an image and tabular dataset. The first experiment on the MNIST handwritten digit dataset [17] makes use of an autoencoder to define and construct prototypes. The second experiment uses the Breast Cancer Wisconsin (Diagnostic) dataset [10]. The latter dataset has lower dimensionality so we find the prototypes using k-d trees. Finally, we illustrate our approach for handling categorical data on the Adult (Census) dataset [10].

4.1   Evaluation

The counterfactuals are evaluated on their interpretability, sparsity and speed of the search process. The sparsity is evaluated using the elastic net loss term $\mathrm{EN}(\delta) = \beta \cdot \|\delta\|_1 + \|\delta\|_2^2$ while the speed is measured by the time and the number of gradient updates required until a satisfactory counterfactual x_{cf} is found. We define a satisfactory counterfactual as the optimal counterfactual found using FISTA for a fixed value of c for which counterfactual instances exist.

In order to evaluate interpretability, we introduce two interpretability metrics IM1 and IM2. Let AE_i and AE_{t_0} be autoencoders trained specifically on instances of classes i and t_0, respectively. Then IM1 measures the ratio between the reconstruction errors of x_{cf} using AE_i and AE_{t_0}:

$$\mathrm{IM1}(\mathrm{AE}_i, \mathrm{AE}_{t_0}, x_{\mathrm{cf}}) := \frac{\|x_0 + \delta - \mathrm{AE}_i(x_0 + \delta)\|_2^2}{\|x_0 + \delta - \mathrm{AE}_{t_0}(x_0 + \delta)\|_2^2 + \epsilon}. \tag{13}$$

A lower value for IM1 means that x_{cf} can be better reconstructed by the autoencoder which has only seen instances of the counterfactual class i than by the autoencoder trained on the original class t_0. This implies that x_{cf} lies closer to the data manifold of counterfactual class i compared to t_0, which is considered to be more interpretable.

The second metric IM2 compares how similar the reconstructed counterfactual instances are when using AE_i and an autoencoder trained on all classes, AE. We scale IM2 by the L_1 norm of x_{cf} to make the metric comparable across classes:

$$\mathrm{IM2}(\mathrm{AE}_i, \mathrm{AE}, x_{\mathrm{cf}}) := \frac{\|\mathrm{AE}_i(x_0 + \delta) - \mathrm{AE}(x_0 + \delta)\|_2^2}{\|x_0 + \delta\|_1 + \epsilon}. \tag{14}$$

A low value of IM2 means that the reconstructed instances of x_{cf} are very similar when using either AE_i or AE. As a result, the data distribution of the counterfactual class i describes x_{cf} as good as the distribution over all classes. This implies that the counterfactual is interpretable. Figure 2 illustrates the intuition behind IM1 and IM2.

The uninterpretable counterfactual 3 ($x_{cf,1}$) in the first row of Fig. 2(b) has an IM1 value of 1.81 compared to 1.04 for $x_{cf,2}$ in the second row because the reconstruction of $x_{cf,1}$ by AE_5 in Fig. 2(d) is better than by AE_3 in Fig. 2(c). The IM2 value of $x_{cf,1}$ is higher as well—0.15 compared to 0.12 for $x_{cf,2}$—since the reconstruction by AE in Fig. 2(e) yields a clear instance of the original class 5.

Finally, for MNIST we apply a multiple model comparison test based on the maximum mean discrepancy [18] to evaluate the relative interpretability of counterfactuals generated by each method.

4.2   Handwritten Digits

The first experiment is conducted on the MNIST dataset. The experiment analyzes the impact of L_{proto} on the counterfactual search process with an encoder defining the prototypes for K equal to 5. We further investigate the importance of the L_{AE} and L_{pred} loss terms in the presence of L_{proto}. We evaluate and compare counterfactuals obtained by using the following loss functions:

$$
\begin{aligned}
A &= c \cdot L_{pred} + \beta \cdot L_1 + L_2 \\
B &= c \cdot L_{pred} + \beta \cdot L_1 + L_2 + L_{AE} \\
C &= c \cdot L_{pred} + \beta \cdot L_1 + L_2 + L_{proto} \\
D &= c \cdot L_{pred} + \beta \cdot L_1 + L_2 + L_{AE} + L_{proto} \\
E &= \beta \cdot L_1 + L_2 + L_{proto} \\
F &= \beta \cdot L_1 + L_2 + L_{AE} + L_{proto}
\end{aligned}
\tag{15}
$$

For each of the ten classes, we randomly sample 50 numbers from the test set and find counterfactual instances for 3 different random seeds per sample. This brings the total number of counterfactuals to 1,500 per loss function.

The model used to classify the digits is a convolutional neural network with 2 convolution layers, each followed by a max-pooling layer. The output of the second pooling layer is flattened and fed into a fully connected layer followed by a softmax output layer over the 10 possible classes. For objective functions B to F, the experiment also uses a trained autoencoder for the L_{AE} and L_{proto} loss terms. The autoencoder has 3 convolution layers in the encoder and 3 deconvolution layers in the decoder. Full details of the classifier and autoencoder, as well as the hyperparameter values used can be found in the supplementary material.

Results. Table 1 summarizes the findings for the speed and interpretability measures.

Speed. Figure 3(a) shows the mean time and number of gradient steps required to find a satisfactory counterfactual for each objective function. We also show

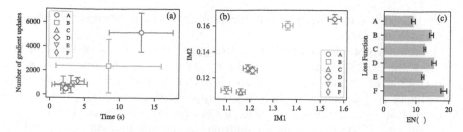

Fig. 3. (a) Mean time in seconds and number of gradient updates needed to find a satisfactory counterfactual for objective functions A to F across all MNIST classes. The error bars represent the standard deviation to illustrate variability between approaches. (b) Mean IM1 and IM2 for objective functions A to F across all MNIST classes (lower is better). The error bars represent the 95% confidence bounds. (c) Sparsity measure EN(δ) for loss functions A to F. The error bars represent the 95% confidence bounds.

the standard deviations to illustrate the variability between the different loss functions. For loss function A, the majority of the time is spent finding a good range for c to find a balance between steering the perturbed instance away from the original class t_0 and the elastic net regularization. If c is too small, the L_1 regularization term cancels out the perturbations, but if c is too large, x_{cf} is not sparse anymore.

The aim of L_{AE} in loss function B is not to speed up convergence towards a counterfactual instance, but to have x_{cf} respect the training data distribution. This is backed up by the experiments. The average speed improvement and reduction in the number of gradient updates compared to A of respectively 36% and 54% is significant but very inconsistent given the high standard deviation. The addition of L_{proto} in C however drastically reduces the time and iterations needed by respectively 77% and 84% compared to A. The combination of L_{AE} and L_{proto} in D improves the time to find a counterfactual instance further: x_{cf} is found 82% faster compared to A, with the number of iterations down by 90%.

Table 1. Summary statistics with 95% confidence bounds for each loss function for the MNIST experiment.

Method	Time (s)	Gradient steps	IM1	IM2 ($\times 10$)
A	13.06 ± 0.23	5158 ± 82	1.56 ± 0.03	1.65 ± 0.04
B	8.40 ± 0.38	2380 ± 113	1.36 ± 0.02	1.60 ± 0.03
C	3.06 ± 0.11	835 ± 36	1.16 ± 0.02	1.09 ± 0.02
D	2.31 ± 0.04	497 ± 10	1.21 ± 0.02	1.26 ± 0.03
E	1.93 ± 0.10	777 ± 44	1.10 ± 0.02	1.10 ± 0.03
F	4.01 ± 0.05	1116 ± 14	1.19 ± 0.02	1.27 ± 0.03

So far we have assumed access to the model architecture to take advantage of automatic differentiation during the counterfactual search process. L_{pred} can however form a computational bottleneck for black box models because numerical gradient calculation results in a number of prediction function calls proportionate to the dimensionality of the input features. Consider A' the equivalent of loss function A where we can only query the model's prediction function. E and F remove L_{pred} which results in approximately a 100x speed up of the counterfactual search process compared to A'. The results can be found in the supplementary material.

Quantitative Interpretability. IM1 peaks for loss function A and improves by respectively 13% and 26% as L_{AE} and L_{proto} are added (Fig. 3(b)). This implies that including L_{proto} leads to more interpretable counterfactual instances than L_{AE} which explicitly minimizes the reconstruction error using AE. Removing L_{pred} in E yields an improvement over A of 29%. While L_{pred} encourages the perturbed instance to predict a different class than t_0, it does not impose any restrictions on the data distribution of x_{cf}. L_{proto} on the other hand implicitly encourages the perturbed instance to predict $i \neq t_0$ while minimizing the distance in latent space to a representative distribution of class i.

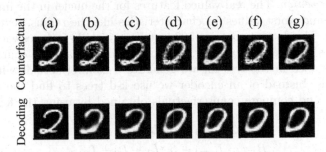

Fig. 4. (a) Shows the original instance, (b) to (g) on the first row illustrate counterfactuals generated by using loss functions A to F. (b) to (g) on the second row show the reconstructed counterfactuals using AE.

The picture for IM2 is similar. Adding in L_{proto} brings IM2 down by 34% while the combination of L_{AE} and L_{proto} only reduces the metric by 24%. For large values of K the prototypes are further from $\text{ENC}(x_0)$ resulting in larger initial perturbations towards the counterfactual class. In this case, L_{AE} ensures the overall distribution is respected which makes the reconstructed images of AE_i and AE more similar and improves IM2. The impact of K on IM1 and IM2 is illustrated in the supplementary material. The removal of L_{pred} in E and F has little impact on IM2. This emphasizes that L_{proto}—optionally in combination with L_{AE}—is the dominant term with regards to interpretability.

Finally, performing kernel multiple model comparison tests [18] indicates that counterfactuals generated by methods not including the prototype term (A and B) result in high rejection rates for faithfully modelling the predicted class distribution (see supplementary material).

Visual Interpretability. Figure 4 shows counterfactual examples on the first row and their reconstructions using AE on the second row for different loss functions. The counterfactuals generated with A or B are sparse but uninterpretable and are still close to the manifold of a 2. Including L_{proto} in Fig. 4(d) to (g) leads to a clear, interpretable 0 which is supported by the reconstructed counterfactuals on the second row. More examples can be found in the supplementary material.

Sparsity. The elastic net evaluation metric $\mathrm{EN}(\delta)$ is also the only loss term present in A besides L_{pred}. It is therefore not surprising that A results in the most sparse counterfactuals (Fig. 3(c)). The relative importance of sparsity in the objective function goes down as L_{AE} and L_{proto} are added. L_{proto} leads to more sparse counterfactuals than L_{AE} (C and E), but this effect diminishes for large K.

4.3 Breast Cancer Wisconsin (Diagnostic) Dataset

The second experiment uses the Breast Cancer Wisconsin (Diagnostic) dataset which describes characteristics of cell nuclei in an image and labels them as *malignant* or *benign*. The real-valued features for the nuclei in the image are the mean, error and worst values for characteristics like the radius, texture or area of the nuclei. The dataset contains 569 instances with 30 features each. The first 550 instances are used for training, the last 19 to generate the counterfactuals. For each instance in the test set we generate 5 counterfactuals with different random seeds. Instead of an encoder we use k-d trees to find the prototypes. We evaluate and compare counterfactuals obtained by using the following loss functions:

$$A = c \cdot L_{\mathrm{pred}} + \beta \cdot L_1 + L_2$$
$$B = c \cdot L_{\mathrm{pred}} + \beta \cdot L_1 + L_2 + L_{\mathrm{proto}} \tag{16}$$
$$C = \beta \cdot L_1 + L_2 + L_{\mathrm{proto}}$$

The model used to classify the instances is a 2 layer feedforward neural network with 40 neurons in each layer. More details can be found in the supplementary material.

Results. Table 2 summarizes the findings for the speed and interpretability measures.

Speed. L_{proto} drastically reduces the time and iterations needed to find a satisfactory counterfactual. Loss function B finds x_{cf} in 13% of the time needed compared to A while bringing the number of gradient updates down by 91%. Removing L_{pred} and solely relying on the prototype to guide x_{cf} reduces the search time by 92% and the number of iterations by 93%.

Quantitative Interpretability. Including L_{proto} in the loss function reduces IM1 and IM2 by respectively 55% and 81%. Removing L_{pred} in C results in similar improvements over A.

Sparsity. Loss function A yields the most sparse counterfactuals. Sparsity and interpretability should however not be considered in isolation. The dataset has 10 attributes (e.g. radius or texture) with 3 values per attribute (mean, error and worst). B and C which include L_{proto} perturb relatively more values of the same attribute than A which makes intuitive sense. If for instance the worst radius increases, the mean should typically follow as well. The supplementary material supports this statement.

Table 2. Summary statistics with 95% confidence bounds for each loss function for the Breast Cancer Wisconsin (Diagnostic) experiment.

Method	Time (s)	Gradient steps	IM1	IM2 ($\times 10$)
A	2.68 ± 0.20	2752 ± 203	2.07 ± 0.16	7.65 ± 0.79
B	0.35 ± 0.03	253 ± 33	0.94 ± 0.10	1.47 ± 0.15
C	0.22 ± 0.02	182 ± 30	0.88 ± 0.10	1.41 ± 0.15

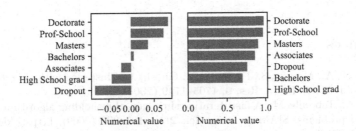

Fig. 5. Left: Embedding of the categorical variable "Education" in numerical space using association based distance metric (ABDM). Right: Frequency based embedding.

4.4 Adult (Census) Dataset

The Adult (Census) dataset consists of individuals described by a mixture of numerical and categorical features. The predictive task is to determine whether a person earns more than \$50k/year. As the dataset contains categorical features, it is important to use a principled approach to define perturbations over these features. Figure 5 illustrates our approach using the association based distance metric [16] (ABDM) to embed the feature "Education" into one dimensional numerical space over which perturbations can be defined. The resulting embedding defines a natural ordering of categories in agreement with common sense for this interpretable variable. By contrast, the frequency embedding method as proposed by [9] does not capture the underlying relation between categorical values.

Since ABDM infers distances from other variables by computing dissimilarity based on the K-L divergence, it can break down if there is independence between categories. In such cases one can use MVDM [6] which uses the difference between the conditional model prediction probabilities of each category. A counterfactual example changing categorical features is shown in Fig. 1.

5   Discussion

In this paper we introduce a model agnostic counterfactual search process guided by class prototypes. We show that including a prototype loss term in the objective results in more interpretable counterfactual instances as measured by two novel interpretability metrics. We demonstrate that prototypes speed up the search process and remove the numerical gradient evaluation bottleneck for black box models thus making our method more appealing for practical applications. By fixing selected features to the original values during the search process we can also obtain *actionable counterfactuals* which describe concrete steps to take to change a model's prediction. To facilitate the practical use of counterfactual explanations we provide an open source library with our implementation of the method.

References

1. Banerjee, A., Merugu, S., Dhillon, I.S., Ghosh, J.: Clustering with Bregman divergences. J. Mach. Learn. Res. **6**, 1705–1749 (2005)
2. Beck, A., Teboulle, M.: A fast iterative shrinkage-thresholding algorithm for linear inverse problems. SIAM J. Imag. Sci. **2**(1), 183–202 (2009). https://doi.org/10.1137/080716542
3. Bentley, J.L.: Multidimensional binary search trees used for associative searching. Commun. ACM **18**(9), 509–517 (1975). https://doi.org/10.1145/361002.361007
4. Bien, J., Tibshirani, R.: Prototype selection for interpretable classification. Ann. Appl. Stat. **5**(4), 2403–2424 (2011). https://doi.org/10.1214/11-AOAS495
5. Borg, I., Groenen, P.: Modern Multidimensional Scaling: Theory and Applications. Springer, New York (2005). https://doi.org/10.1007/0-387-28981-X
6. Cost, S., Salzberg, S.: A weighted nearest neighbor algorithm for learning with symbolic features. Mach. Learn. **10**(1), 57–78 (1993). https://doi.org/10.1023/A:1022664626993
7. Dhurandhar, A., et al.: Explanations based on the missing: towards contrastive explanations with pertinent negatives. Adv. Neural. Inf. Process. Syst. **31**, 592–603 (2018)
8. Dhurandhar, A., Iyengar, V., Luss, R., Shanmugam, K.: Tip: typifying the interpretability of procedures. arXiv preprint arXiv:1706.02952 (2017)

9. Dhurandhar, A., Pedapati, T., Balakrishnan, A., Chen, P.Y., Shanmugam, K., Puri, R.: Model agnostic contrastive explanations for structured data. arXiv preprint arXiv:1906.00117 (2019)
10. Dua, D., Graff, C.: UCI machine learning repository (2017)
11. Gurumoorthy, K.S., Dhurandhar, A., Cecchi, G.: Protodash: fast interpretable prototype selection. arXiv preprint arXiv:1707.01212 (2017)
12. Jiang, H., Kim, B., Guan, M., Gupta, M.: To trust or not to trust a classifier. Adv. Neural. Inf. Process. Syst. **31**, 5541–5552 (2018)
13. Kaufmann, L., Rousseeuw, P.: Clustering by means of medoids. Data Analysis based on the L1-Norm and Related Methods, pp. 405–416 (1987)
14. Kim, B., Khanna, R., Koyejo, O.O.: Examples are not enough, learn to criticize! criticism for interpretability. Adv. Neural. Inf. Process. Syst. **29**, 2280–2288 (2016)
15. Laugel, T., Lesot, M.-J., Marsala, C., Renard, X., Detyniecki, M.: Comparison-based inverse classification for interpretability in machine learning. In: Medina, J., et al. (eds.) IPMU 2018. CCIS, vol. 853, pp. 100–111. Springer, Cham (2018). https://doi.org/10.1007/978-3-319-91473-2_9
16. Le, S.Q., Ho, T.B.: An association-based dissimilarity measure for categorical data. Pattern Recogn. Lett. **26**(16), 2549–2557 (2005). https://doi.org/10.1016/j.patrec.2005.06.002
17. LeCun, Y., Cortes, C.: MNIST handwritten digit database (2010)
18. Lim, J.N., Yamada, M., Schölkopf, B., Jitkrittum, W.: Kernel stein tests for multiple model comparison. In: Advances in Neural Information Processing Systems 32, pp. 2240–2250. Curran Associates, Inc. (2019)
19. Lundberg, S.M., Lee, S.I.: A unified approach to interpreting model predictions. Adv. Neural. Inf. Process. Syst. **30**, 4765–4774 (2017)
20. Luss, R., Chen, P.Y., Dhurandhar, A., Sattigeri, P., Shanmugam, K., Tu, C.C.: Generating contrastive explanations with monotonic attribute functions. arXiv preprint arXiv:1905.12698 (2019)
21. Molnar, C.: Interpretable Machine Learning (2019). https://christophm.github.io/interpretable-ml-book/. Accessed 22 Jan 2020
22. Mothilal, R.K., Sharma, A., Tan, C.: Explaining machine learning classifiers through diverse counterfactual explanations. In: Proceedings of the 2020 Conference on Fairness, Accountability, and Transparency (2020). https://doi.org/10.1145/3351095.3372850
23. Ribeiro, M.T., Singh, S., Guestrin, C.: "Why should I trust you": explaining the predictions of any classifier. In: Proceedings of the 22Nd ACM SIGKDD International Conference on Knowledge Discovery and Data Mining, pp. 1135–1144 (2016). https://doi.org/10.1145/2939672.2939778
24. Rumelhart, D.E., Hinton, G.E., Williams, R.J.: Learning internal representations by error propagation. In: Parallel Distributed Processing: Explorations in the Microstructure of Cognition, vol. 1, pp. 318–362. MIT Press, Cambridge (1986)
25. Snell, J., Swersky, K., Zemel, R.: Prototypical networks for few-shot learning. Adv. Neural. Inf. Process. Syst. **30**, 4077–4087 (2017)
26. Takigawa, I., Kudo, M., Nakamura, A.: Convex sets as prototypes for classifying patterns. Eng. Appl. Artif. Intell. **22**(1), 101–108 (2009). https://doi.org/10.1016/j.engappai.2008.05.012
27. Wachter, S., Mittelstadt, B., Russell, C.: Counterfactual explanations without opening the black box: automated decisions and the GDPR. Harvard J. Law Technol. **31**, 841–887 (2018)
28. Zou, H., Hastie, T.: Regularization and variable selection via the elastic net. J. Roy. Stat. Soc. Ser. B (Stat. Methodol.) **67**(2), 301–320 (2005)

Finding High-Value Training Data Subset Through Differentiable Convex Programming

Soumi Das[1](✉), Arshdeep Singh[1], Saptarshi Chatterjee[1],
Suparna Bhattacharya[2], and Sourangshu Bhattacharya[1]

[1] Indian Institute of Technology, Kharagpur, Kharagpur, West Bengal, India
soumi_das@iitkgp.ac.in
[2] Hewlett Packard Labs, Hewlett Packard Enterprise, Bangalore, India

Abstract. Finding valuable training data points for deep neural networks has been a core research challenge with many applications. In recent years, various techniques for calculating the "value" of individual training datapoints have been proposed for explaining trained models. However, the value of a training datapoint also depends on other selected training datapoints - a notion which is not explicitly captured by existing methods. In this paper, we study the problem of selecting high-value subsets of training data. The key idea is to design a learnable framework for online subset selection, which can be learned using minibatches of training data, thus making our method scalable. This results in a parameterised convex subset selection problem that is amenable to a differentiable convex programming paradigm, thus allowing us to learn the parameters of the selection model in an end-to-end training. Using this framework, we design an online alternating minimization based algorithm for jointly learning the parameters of the selection model and ML model. Extensive evaluation on a synthetic dataset, and three standard datasets, show that our algorithm finds consistently higher value subsets of training data, compared to the recent state of the art methods, sometimes $\sim 20\%$ higher value than existing methods. The subsets are also useful in finding mislabelled training data. Our algorithm takes running time comparable to the existing valuation functions.

Keywords: Data valuation · Subset selection · Convex optimisation · Explainability

1 Introduction

Estimation of "value" of a training datapoint from a Machine Learning model point of view, broadly called *data valuation* [10,16,19], has become an important problem with many applications. One of the early applications included explaining and debugging training of deep neural models, where the influence of the training data points on the test loss is estimated [13,16]. Another application involves estimating the expected marginal value of a training datapoint

© Springer Nature Switzerland AG 2021
N. Oliver et al. (Eds.): ECML PKDD 2021, LNAI 12976, pp. 666–681, 2021.
https://doi.org/10.1007/978-3-030-86520-7_41

w.r.t. general value functions e.g. accuracy on a test set [10], which can be used in buying and selling of data in markets. [19] lists a number of other applications, including detecting domain shift between training and test data, suggesting adaptive data collection strategy, etc.

Most existing formulations for data valuation assume a supervised machine learning model, $y = f(x, \theta)$ with parameters θ. For all the above mentioned applications, the key technical question is to determine the improvement in the value function (usually test set loss) given that a new set of datapoints are added to the training set. The influence function based methods [13,16] aim to estimate the influence of each individual datapoint, on the test loss of a trained model. While these techniques are computationally efficient, they ignore an important fact: *the value of a training datapoint depends on the other datapoints in the training dataset.* For a given training data point x, another datapoint y may add to the value of x, possibly because it is very different from x. Alternately, y may reduce the value of x if it is very similar to x.

Shapley value based methods [10] also estimate the value of individual datapoints using the expected marginal gain in the value function while adding the particular datapoint. Hence, this method considers the effect of other datapoints, but only in an aggregate sense. Also, this method is computationally expensive. DVRL [19] is the closest to our approach. It learns a parameterised selection function which maximizes the reward of selecting training datapoints using Reinforcement Learning. The reward function is the marginal improvement in value function after selection of a training datapoint over the previous running average of value function. Unfortunately, this reward leads to selection of many training datapoints which are similar to each other, if their inclusion contributes to making the loss better than past average. To the best of our knowledge, none of the existing data valuation techniques explicitly consider similarities between training points, when selecting high value subsets.

In this paper, we consider a set value function for assigning value to a subset of training datapoints. A proxy for the value function is learned though an embedding of datapoints, such that distance between the embedded datapoints represent the inverse value of replacing one of the datapoints with another. In this framework, we propose the problem of *high-value data subset selection*, which computes the optimal subset of a given size maximizing the value of the subset, or minimizing the total distance between selected points and other training datapoints in the embedding space.

In order to efficiently compute the high value data subsets, we address two challenges. Firstly, the embedding function of datapoints needs to be learned by optimizing the value of the selected subset, which itself is the output of an optimization problem. Hence, our formulation should be able to propagate the gradient of value-function back through the optimization problem. Secondly, for practicality, our formulation should be able to select the datapoints in an online manner from minibatches of data. We address both these challenges by designing a novel learned online subset selection formulation for selecting subsets of training datapoints from minibatches. Our formulation is inspired by the online subset selection technique based on facility location objective [6], but uses a learned distance function using a parameterised embedding of the training datapoints. In order to learn the embedding parameters, our formulation is also compatible with the *differentiable convex programming* paradigm [1], which allows us to propagate the gradients of the value function (loss on test set) back to the embedding layer. We propose an online alternating minimization based algorithm to jointly learn the embedding function for selection of optimal training data subset, and learn the optimal model parameters for optimizing the loss on the selected subset. Our algorithm scales linearly in training data size for a fixed number of epochs.

We benchmark our algorithm against state of the art techniques. e.g. DVRL [19], Data Shapley [10], etc. using extensive experiments on multiple standard real world datasets, as well as a synthetic dataset. We show that selection of high value subset of training data points using the proposed method consistently leads to better test set accuracy, for similar sized selected training dataset, with upto 20% difference in test accuracy. Also, removal of high value subset of training datapoints leads to a much more significant drop in test accuracy, compared to baselines. We also show that the proposed method can detect higher fraction of incorrectly labelled data, for all sizes of selected subsets. Moreover, correcting the detected mislabelled points leads to higher increase in test set accuracy compared to baseline. We observe that DVRL [19], which is the closest competitor, selects datapoints in a biased manner, possibly because there, it selects many similar high-performing datapoints at each stage.

To summarise, our key contributions are:

1. We highlight the problem of high-value data subset selection, which selects a high-value subset, rather than assigning valuation to individual datapoints.
2. We formulate a novel learned online data subset selection technique which can be trained jointly with the ML model.
3. We propose an online alternating minimization based algorithm which trains the joint model using mini-batches of training data, and scales linearly with training data.
4. Results with extensive experimentation show that our method consistently outperforms recent baselines, sometimes with an accuracy improvement of $\sim 20\%$ for the same size subsets.

2 Related Work

The exponential rise in quantity of data and depth of deep learning models have given rise to the question of scalability. Several subset selection methods have been designed using submodular [2] and convex approaches [7], to combat the problem. However, all these techniques only rely on some measure of similarity (pairwise, pointwise) [5,6] and do not really account for the explanation behind its choice in view of the end-task. Hence the other question which arises is explainability. A set of methods for explainability is prototype based approaches which finds the important examples aimed towards the optimisation of a given value function, or in other words, improvement of a given performance metric (e.g. test set accuracy). Pioneering studies on influence functions [4] have been used to estimate the change in parameter θ, given an instance x_i is absent and is used to obtain an estimate of $\theta_{-i} - \theta$. This aids in approximating the importance of the instance x_i. Following the works of [13] on influence functions, there have been several other works which try to measure the influence of training points on overall loss by tracking the training procedure [12] while some try to measure the influence of training points on a single test point [16]. A recent work by [18] performs rapid retraining by reusing information cached during training phase.

Shapley value, an idea originated from game theory has been a driving factor in the field of economics. It was used as a feature ranking score for explaining black-box models [14,15]. However, [10] were the first to use shapley value to quantify data points where they introduce several properties of data valuation and use Monte Carlo based methods for approximation. There has also been a recent study in using shapley for quantifying the importance of neurons of a black box model [11]. A follow-up work by [9] aims to speed up the computation of data shapley values. However, all the above methods do not have a distinctive selection mechanism for providing an explainable representative set of points.

Our work is in close proximity to the prototype based approaches. While [9] aims to work with training set distribution, our work is essentially inclined towards selection from training datasets, hence leading to choose [10] as one of our baselines. Among the methods around influence functions, we use the work of [13,16] as two of our baselines. A recent work by [19] attempts to adaptively learn data values using reinforcement signals, along with the predictor model. This work is the closest in terms of the problem objective we are intending to solve and hence we use it as one of our baselines. However, they intend to optimize the selection using reinforcement trick which requires a greater number of runs, while we use a learnable convex framework for the purpose of selection. To the best of our knowledge, this is the first of its kind in using a differentiable convex programming based learning framework to optimize the selection mechanism in order to improve on a given performance metric.

3   High-Value Data Subset Selection

In this section, we formally define the problem of high-value data subset selection and propose a learned approximate algorithm for the same.

3.1   Motivation and Problem Formulation

Data valuation for deep learning is an important tool for generating explanations of deep learning models in terms of training data. Let $\mathcal{D} = \{(x_i, y_i) | i = 1, \dots, n\}$ be the training dataset with features $x_i \in \mathcal{X}$, and labels $y_i \in \mathcal{Y}$, and $\mathcal{D}^t = \{(x_i^t, y_i^t) | i = 1, \dots, m\}$ be a test dataset, which is used for valuation of the training data in $\mathcal{D}$. Also, let $s \in \mathcal{S}$ denote a subset of of the training data $\mathcal{D}$ ($\mathcal{S}$ is the power set of $\mathcal{D}$). Let $f(\theta)$ denote a parameterized family of learnable functions (typically implemented with neural networks), parameterized by a set of parameters θ. Given an average loss function $\mathcal{L}(f(\theta), s)$, defined as $\mathcal{L}(f(\theta), s) = \frac{1}{|s|} \sum_{(x_i, y_i) \in s} L(y_i, f(x_i; \theta))$, we define the optimal parameters $\theta^*(s)$ for a training data subset s as:

$$\theta^*(s) = \arg\min_{\theta} \mathcal{L}(f(\theta), s) \tag{1}$$

Given an optimal parameter $\theta^*(s)$ and the test dataset $\mathcal{D}^t$, we define the value function $v(s)$, which provides a valuation of the subset s as:

$$v(s) = v(\theta^*(s), \mathcal{D}^t) = -\mathcal{L}(f(\theta^*(s)), \mathcal{D}^t) \tag{2}$$

The *high-value subset selection problem* can be defined as:

$$\max_{s \in \mathcal{S}} v(s) \text{ sub. to } |s| \leq \gamma n \tag{3}$$

where γ is the fraction of retained datapoints in the selected subset.

This problem is NP-Hard for a general value function $v(s)$. Next, we describe our broad framework to approximate the above problem with a two step approach:

1. Learn an embedding function $h(x, \phi)$ of the data point x, such that the distance between datapoints x_i and x_j, $d_{ij} = d(x_i, x_j) = D(h(x_i, \phi), h(x_j, \phi))$ represents their inverse-ability to replace each other for a given learning task. Here, D is a fixed distance function.
2. Select the set of most important datapoints s^* by minimizing the sum of distance between each datapoint and its nearest selected neighbor, and update the parameters ϕ of the embedding function such that $v(s*)$ is maximized.

The objective function for selection problem in second step can be written as:

$$s^* = \min_{s \in \mathcal{S}} \sum_{(x,y) \in \mathcal{D}} \min_{(x',y') \in s} d(x, x') \tag{4}$$

This objective function corresponds to the facility location problem [6], which can be relaxed to the following convex linear programming problem:

$$\min_{z_{ij} \in [0,1]} \sum_{i,j=1}^{n} z_{ij} d(x_i, x_j) \tag{5}$$

$$\text{sub. to} \sum_{j=1}^{n} z_{ij} = 1, \ \forall i = \{1, \ldots, n\}$$

Here $z_{ij} = 1$ indicate that the datapoint j is the nearest selected point or "representative" for data point i. Hence, the objective function sums the total distance of each point from its representative point, and the constraint enforces that every datapoint i should have exactly one representative point. A point j is selected if it is representative to at least one point, i.e. $\sum_{i=1}^{n} z_{ij} \geq 1$. Let u_j be the indicator variable for selection of datapoint j. For robustness, we calculate u_j as:

$$u_j = \frac{1}{\xi} \max\{\sum_{i=1}^{n} z_{ij}, \xi\} \tag{6}$$

where $0 \leq \xi \leq 1$ is a constant.

There are two main challenges in utilizing this formulation for the problem of optimizing the value function. Firstly, the optimal parameters $\theta^*(s)$ depend on the variables z_{ij} through variables u_j, which are output of an optimization problem. Hence, optimizing $v(s^*)$ w.r.t parameters (ϕ) of the embedding function requires differentiating through the optimization problem. Secondly, for most practical applications calculating the subset of training data, s, by optimizing the above optimization problem is too expensive for an iterative optimization algorithm which optimizes ϕ. Moreover, optimizing the parameters (θ) of the neural network model $f(\theta)$, typically involves an online stochastic optimization algorithm (e.g. SGD) which operates with mini-batches of training data points. In the next section, we describe the proposed technique for joint online training of both the learning model parameters θ and embedding model parameters ϕ.

3.2   Joint Online Training of Subset Selection and Learning Model

Our problem of selection of high-value subsets from training data can be described as joint optimization of value function for selection of best subset, and optimization of loss function on the selected subset computing the best model. Given training and test datasets $\mathcal{D}$ and $\mathcal{D}^t$, overall value function parameterised by the embedding model parameters ϕ for selection was defined as (Eq. 2):

$$v(s(\phi)) = v(\theta^*(s(\phi)), \mathcal{D}^t) = \frac{1}{|\mathcal{D}^t|} \sum_{(x,y) \in \mathcal{D}^t} L(y, f(\theta^*(s(\phi)), x)) \tag{7}$$

where $\theta^*(s(\phi)) = \arg\min_\theta \mathcal{L}(f(\theta), s(\phi))$ (from Eq. 1). Representing $s(\phi)$ in terms of the selection variables u_j (Eq. 6), we can write the model loss on selected training set using embedding model parameter ϕ as:

$$\mathcal{L}(\theta; \phi, \mathcal{D}) = \frac{1}{\gamma |\mathcal{D}|} \sum_{(x_j, y_j) \in \mathcal{D}} u_j(\phi) L(y_j, f(x_j; \theta)) \tag{8}$$

We note that the above objective functions $\mathcal{L}(\theta)$ and $v(s(\phi))$ are coupled in the variables ϕ and θ. On one hand, the loss on the selected training set $\mathcal{L}(\theta; \phi, \mathcal{D})$ depends on the current embedding model parameters ϕ. On the other hand, intuitively, the value $v(s(\phi))$ of a selected set of datapoints $s(\phi)$ should also depend on the current model parameter value θ', which maybe be updated using loss from the selected set of datapoints $s(\phi)$. Hence, we define the cumulative value function $\mathcal{V}(\phi; \theta', \mathcal{D}, \mathcal{D}^t)$ as:

$$\mathcal{V}(\phi; \theta', \mathcal{D}, \mathcal{D}^t) = v(\hat{\theta}, \mathcal{D}^t), \text{ where } \hat{\theta} = \theta' - \alpha \nabla_\theta \mathcal{L}(\theta; \phi, \mathcal{D}) \tag{9}$$

Here, $\hat{\theta}$ is the one step updated model parameter using the selected examples from dataset $\mathcal{D}$ using parameter ϕ, and α is the stepsize for the update. We combine the two objectives into a joint objective function, $J(\theta, \phi; \mathcal{D}, \mathcal{D}^t)$. The combined optimization problem becomes:

$$\theta^*, \phi^* = \arg\min_{\theta,\phi} J(\theta, \phi; \mathcal{D}, \mathcal{D}^t) = \arg\min_{\theta,\phi}(\mathcal{V}(\phi; \theta', \mathcal{D}, \mathcal{D}^t) + \mathcal{L}(\theta; \phi, \mathcal{D})) \quad (10)$$

As discussed above, since the training dataset $\mathcal{D}$ is normally much bigger in size, we design our algorithm to optimize the above objective function in an online manner. Hence the training dataset $\mathcal{D}$ is split into k equal-sized minibatches $\mathcal{D} = \{D_1, \ldots, D_k\}$. The above joint objective function can be decomposed as:

$$J(\theta, \phi; \mathcal{D}, \mathcal{D}^t) = \frac{1}{k}\sum_{i=1}^{k}\{\mathcal{L}(\theta; \phi, D_i) + \mathcal{V}(\phi; \theta', D_i, \mathcal{D}^t)\} \quad (11)$$

We solve the above problem using an online alternating minimization method similar to the one described in [3]. Our algorithm updates θ and ϕ as:

> Initialize θ^1 and ϕ^1 randomly
> for $t = 1, \ldots, T$:
> 1 : $\qquad\qquad \theta^{t+1} = \theta^t - \alpha\frac{1}{k}\nabla_\theta \mathcal{L}(\theta; \phi^t, D_{i(t)})$
> 2 : $\qquad\qquad \phi^{t+1} = \phi^t - \beta\frac{1}{k}\nabla_\phi \mathcal{V}(\phi; \theta^t, D_{i(t)}, \mathcal{D}^t)$

Here $i(t)$ is the index of the minibatch chosen at update number t. Step 1 is the standard SGD update minimizing the loss over a subset of the minibatch $D_{i(t)}$ chosen using embedding parameters ϕ^t. Hence, it can be implemented using standard deep learning platforms. Implementation of step 2 poses two challenges: (1) computation of $\nabla_\phi \mathcal{V}(\phi; \theta^t, D_{i(t)}, \mathcal{D}^t)$ requires differentiation through an optimization problem, since $u_j(\phi)$ is the result of an optimization. (2) the datapoints in $D_{i(t)}$ must also be compared with datapoints in a previous minibatch for selecting the best representative points. Hence we need an online version of the optimization problem defined in Eq. 5. We describe solutions to these challenges in the next section.

3.3   Trainable Convex Online Subset Selection Layer

In order to implement the algorithm outlined in the previous section, we have to compute $\nabla_\phi \mathcal{V}(\phi; \theta^t, D_i, \mathcal{D}^t)$, which can be done by combining Eq. 7, 8 and 9 as:

$$\nabla_\phi \mathcal{V}(\phi; \theta^t, D_i, \mathcal{D}^t) = -\frac{1}{|\mathcal{D}^t|}\sum_{(x_j, y_j) \in \mathcal{D}^t} \nabla_{\hat{\theta}} l(y_j, f(\hat{\theta}, x_j)) \cdot \quad (12)$$

$$\frac{1}{\gamma|D_i|}\sum_{(x_j, y_j) \in D_i}(\nabla_\phi u_j(\phi)\nabla_\theta L(y_j, f(x_j, \theta)))$$

The key challenge here is computation of $\nabla_\phi u_j(\phi)$, since $u_j(\phi)$ is the output of an optimization problem. Also, the optimization problem in Eq. 5 selects from all datapoints in $\mathcal{D}$, whereas efficient loss minimization demands that we only select from the current minibatch $D_{i(t)}$. We use the online subset selection formulation, proposed in [6]. Let $\mathcal{O}(t)$ denote the set of old datapoints at time t which have already been selected, and let $D_{i(t)}$ denote the new set from which points will be selected. We use two sets of indicator variables: $z_{ij}^o = 1$ indicating that old datapoint j is a representative of new datapoint i, and $z_{ij}^n = 1$ indicating that new datapoint j is a representative of new datapoint i. Under this definition the optimization problem can written as:

$$\min_{z_{ij}^o, z_{ij}^n \in [0,1]} \sum_{x_i \in D_{i(t)}, x_j \in \mathcal{O}(t)} z_{ij}^o d(x_i, x_j) + \sum_{x_i \in D_{i(t)}, x_j \in D_{i(t)}} z_{ij}^n d(x_i, x_j) \quad (13)$$

$$\text{sub. to} \sum_{x_j \in \mathcal{O}(t)} z_{ij}^o + \sum_{x_j \in D_{i(t)}} z_{ij}^n = 1, \ \forall i = \{1, \ldots, n\}$$

$$\sum_{x_j \in D_{i(t)}} u_j \le \gamma |D_{i(t)}|$$

where $u_j = \frac{1}{\xi} \max\{\sum_{x_i \in D_{i(t)}} z_{ij}^n, \xi\}$. The first constraint enforces that all new points must have a representative, and the second constraint enforces that the total number of representatives from the new set is less than $\gamma |D_{i(t)}|$. The objective function minimizes the total distance of between all points and their representatives. As previously, $d(x_i, x_j) = D(h(x_i, \phi), h(x_j, \phi))$, where $h(x, \phi)$ is the embedding of the point x using parameters ϕ.

In order to compute $\nabla_\phi u_j(\phi)$ we use the differentiable convex programming paradigm [1]. The DCP framework allows us to embed solutions to optimization problems as layers in neural networks, and also calculate gradients of output of the layers w.r.t. parameters in the optimization problem, so long as the problem formulation satisfies all the disciplined parameterized programming (DPP) properties. The above convex optimization problem generally satisfies all the properties of disciplined parameterized programming (DPP) [1], with parameter ϕ, except for the objective function. Specifically, the constraints are all convex and are not functions of parameters. The objective function is also linear in variables z_{ij}^o, but should be an affine function of the parameters ϕ in order to satisfy the DPP constraints. Hence, we use the affine distance function $D(a, b) = |a - b|$. Hence, the DCP framework allows us to calculate $\nabla_\phi u_j(\phi) = \nabla_{z_{ij}^n} u_j \nabla_\phi z_{ij}^n(\phi)$. Algorithm 1 summarizes the proposed framework for jointly learning a deep learning model (parameterized by θ) and a selection function for selecting a fraction γ of the training examples, parameterized by ϕ.

Computational Complexity: Running time of the proposed algorithm scales linearly with the training dataset size. The algorithm also works at a time with a minibatch of training data, and hence is memory efficient. These two properties make the current algorithm work on very large training datasets, and complex deep learning models, provided there is enough computational power for inference of the model. A significant overhead of the current algorithm comes from solving an optimization problem for each parameter update. However, this cost can be controlled by keeping the minibatch sizes $|D_i|$ small, typically $|D_i| \leq 100$, which results in a linear optimization in 10000 variables. Another practical trick is to use a fixed number of datapoints for the old set $\mathcal{O}(t)$, where the datapoints can be chosen to be most similar to points in the minibatch D_i using a locality sensitive hashing based nearest neighbor search algorithm. Experimental results show that our algorithm takes a comparable amount of time as our closest competitor DVRL [19], while being much more accurate.

Algorithm 1. *HOST-CP*: High-value Online Subset selection of Training samples through differentiable Convex Programming

---

1: **Input:**
2:     Training dataset $\mathcal{D}$ with minibatches $D_1, \ldots, D_k$, Test dataset $\mathcal{D}^t$, fraction of selected instances γ
3: **Output:**
4:     Model Parameter θ, Embedding model for selection parameter ϕ
5: **Algorithm:**
6: Initialize old set $\mathcal{O}(0) \rightarrow \Phi$ (Null set)
7: Initialize parameters θ^0 and ϕ^0 randomly.
8: **for** each timestep $t = 1, \ldots, T$ **do**
9:     Let $D_{i(t)}$ be the current minibatch.
10:     for each datapoint $x_i \in D_{i(t)}$ and $x_j \in \mathcal{O}(t)$, calculate embeddings $h(x_i, \phi)$ and $h(x_j, \phi)$
11:     for each pair (i, j) , $x_i \in \mathcal{O}$ and $x_j \in D_{i(t)}$ calculate $d^o(x_i, x_j)$
12:     for each pair (i, j) , $x_i \in D_{i(t)}$ and $x_j \in D_{i(t)}$ calculate $d^n(x_i, x_j)$
13:     Calculate z_{ij}^n and z_{ij}^o by solving optimization problem in Equation 13
14:     Calculate $u_j = \frac{1}{\xi} \max\{\sum_{x_i \in D_{i(t)}} z_{ij}^n, \xi\}$ for all $x_j \in D_{i(t)}$
15:     Include the selected points in the old set $\mathcal{O}(t+1)$ for which $u_j = 1$
16:     Calculate $\mathcal{L}(\theta^t; \phi^t, D_{i(t)})$ on the selected set by forward propagation.
17:     Calculate $\theta^{t+1} = \theta^t - \alpha \frac{1}{k} \nabla_\theta \mathcal{L}(\theta; \phi^t, D_{i(t)})$ by backpropagation.
18:     Calculate $\hat{\theta} = \theta^t - \alpha \nabla_\theta \mathcal{L}(\theta; \phi^t, D_{i(t)})$ followed by $\mathcal{V}(\phi; \theta', \mathcal{D}, \mathcal{D}^t)$ using Equation 9, where $\nabla_\phi u_j(\phi)$ is calculated using DCP as described in Section 3.3
19:     Update the embedding model parameter as: $\phi^{t+1} = \phi^t - \beta \frac{1}{k} \nabla_\phi \mathcal{V}(\phi; \theta^t, D_{i(t)}, \mathcal{D}^t)$
20: **end for**

---

4   Experiment

In this section, we provide our experimental setup by describing the datasets and the different data valuation baselines. In Sect. 4.2, we show the effectiveness of our proposed method - High-value Online Subset selection of Training samples through differentiable Convex Programming (*HOST-CP*) over the baselines in terms of performance metric with addition and removal of high value data points. Next in Sect. 4.3, we show the use case of diagnosing mislabelled examples which, when fixed can improve the performance. Following that in Sect. 4.4, we provide an analysis of the quality of subsets obtained by one of the baselines, DVRL [19] and the proposed method. We also show a ranking evaluation of selected points and their correspondence with ground truth relevance in Sect. 4.5.

Lastly, Sect. 4.6 describes the performance of the proposed method in terms of its computational complexities.

4.1 Experimental Setup

Dataset: We have considered datasets of different modalities for experiments. We use four types of datasets: CIFAR10 - a multi-class dataset for image classification, protein dataset[1] for multi-class protein classification, 20newsgroups[2] for text classification and a synthetic dataset for binary classification. Following [10], we generate 10 synthetic datasets, using third-order polynomial function.

For CIFAR10, we obtain the image features from a pre-trained ResNet-18 model and train a 2 layer neural network on the learned features. In case of synthetic data and protein data, we pass the datapoints through a 3 layer neural network, whereas in case of 20newsgroups, we pass the TF-IDF features of the documents through a 3 layer network for classification.

Baselines: We consider four state-of-the-art techniques as our baselines. Two of the baselines - Influence Function - IF [13] and TracIn-CP [16] use the influence function method for data valuation, one uses data shapley (DS) value [10] for selecting high value data points while the fourth one, DVRL [19] uses reinforcement technique while sampling, to rank the data points.

4.2 Valuable Training Data

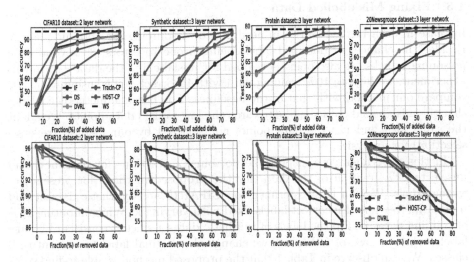

Fig. 1. Addition (top) and removal (bottom) of valuable training datapoints: We compare HOST-CP with the aforementioned baselines (IF [13], DS [10], TracIN-CP [16], DVRL [19]) in terms of accuracy on test set. For each method, we select (for addition) or discard (for removal) top k% important training points and assess their importance. HOST-CP performs better (*highest for addition and lowest for removal*) across all fractions and all datasets.

---

[1] https://www.csie.ntu.edu.tw/~cjlin/libsvmtools/datasets/.
[2] https://scikit-learn.org/0.19/datasets/twenty_newsgroups.html.

We compare our method[3]with state-of-the-art methods in terms of finding the most valuable training data points. We show the effectiveness of the proposed method by addition and removal of an incremental fraction of training instances to and from the dataset. We examine the test set performance for each fraction of selected or removed training points. The baseline methods are used to assign values to each data point, following which the top $k\%$ fraction of data are selected/removed from the whole dataset to train the classification network models. Our proposed method has the flexibility to optimally select the significant $k\%$ of data from the entire given dataset. We use the subsets of varying k or $(100 - k)$ fractions for training the networks and report the test set accuracies.

In Fig. 1, we report the test set accuracies after addition and removal of high-value training data points, that get selected using the respective methods. We can observe in Fig. 1 *(top)* that our method surpasses the baselines for all fractions and it approximately performs equivalent to the whole set (WS) within a fraction of 50–65%. We also note a similar trend in removal of valuable data points. We can observe in Fig. 1 *(bottom)* that our method shows a significant drop in accuracy with removal of high-value data points. All the baselines show varying performances on different datasets. However, the proposed method consistently outperforms across all of them. This clearly suggests that HOST-CP is efficient enough to provide the optimal subset of training data points necessary for explaining the predictions on a given test set.

4.3   Fixing Mislabelled Data

Increase in data has led to rise in need of crowd sourcing for obtaining labels. Besides the chances of obtained labels being prone to errors [8], several data poisoning attacks are also being developed for mislabelling the data [17]. In this experiment, we impute 10% training data with incorrect labels, using which we compare our method with that of the baselines. The baseline methods consider that the mislabelled data shall lie towards the low ranked data points. We adopt the same idea and use the data points not selected by our method (reverse-selection) for analysis.

In this experiment, we inspect the bottom $k\%$ fraction (for baselines) and unselected $k\%$ fraction of the data (for the proposed method) for presence of wrong labels. We run this experiment on CIFAR10 and synthetic dataset and report the fraction of incorrect labels fixed with varying k. While imputing 10% of training data, we flip the labels of synthetic dataset since it consists of binary labels, while in case of CIFAR10, we change the original label to one out of 9 classes. We can observe in Table 1 that the proposed method is able to find comparable mislabelled fraction of incorrect labels in comparison to the baselines. We also run another experiment where, we fix the incorrect labels in the considered $k\%$ fraction of data and train the entire data. Figure 2 shows a rise in test set accuracy with a significant margin using the proposed method. This denotes the effectiveness of HOST-CP, in diagnosing mislabelled examples which turns out to be an important use-case in data valuation techniques.

---

[3] https://github.com/SoumiDas/HOST-CP.

Table 1. Diagnosing mislabelled data: We inspect the training points starting from the least significant data and report the fraction of labels fixed from the inspected data. Overall, HOST-CP detects a higher rate of mislabelled data.

Fraction of data checked (%)	Fraction of incorrect labels fixed									
	CIFAR10					Synthetic data				
	IF	DS	DVRL	TracIn-CP	HOST-CP	IF	DS	DVRL	TracIn-CP	HOST-CP
20	0.21	0.19	0.195	0.20	**0.23**	0.22	0.22	0.19	0.23	0.23
35	0.35	0.34	0.336	0.35	**0.36**	0.34	0.37	0.33	0.35	**0.39**
50	0.50	0.50	0.49	0.50	**0.51**	0.43	0.54	0.46	0.51	**0.55**
65	0.61	0.61	0.65	0.63	**0.67**	0.56	0.67	0.58	0.68	**0.68**
80	0.74	0.78	0.80	0.79	**0.81**	0.66	0.81	0.79	0.81	**0.83**

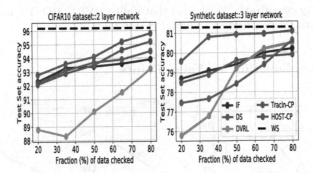

Fig. 2. Accuracy after fixing mislabelled data: For every k% fraction of inspected data, we fix the mislabelled instances and re-train. HOST-CP shows improved performance with each fixing of labels.

4.4 Qualitative Analogy of Data Valuation

As we had mentioned earlier, methodologically, DVRL [19] is the closest method to our approach. Hence, in order to qualitatively explain the difference in performance between that of DVRL and the proposed method, we try to analyse the quality of subsets obtained by both of them on CIFAR10 dataset. Figure 3 shows the fraction of selected images across classes by both the methods. We compare the top 5% selected subset by DVRL and the proposed method in terms of class distribution. Unlike HOST-CP which compares the past selected data points with the incoming points to compute the new subset, DVRL keeps track of past average test losses and values the instances based on the marginal difference with the current loss. This leads to a certain bias in case of DVRL, towards some particular class of images or similar high-performing datapoints which are more contributory towards reduction in test set losses. Following this, we observe that HOST-CP selects a diverse set of samples, thus justifying its better performance.

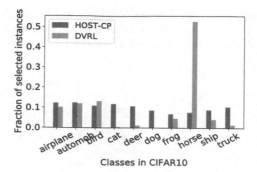

Classes in CIFAR10

Fig. 3. Quality of subset: We observe that unlike DVRL, HOST-CP selects diverse images across the classes leading to better performance.

4.5   Ranking Function for Value Points

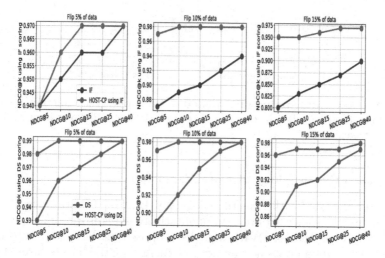

Fig. 4. Ranking of data points: We select k% fraction of data for each method and provide scores for each datum. While IF and DS subsets inherently come along with scores, the proposed method, HOST-CP returns a subset which is scored using IF and DS. Following the acquisition of scores for each data point, we use NDCG@k to calculate the ranking values. The proposed method is found to have a higher NDCG score.

Unlike the baseline methods which provide a score to each data point, HOST-CP has no such analogous scoring function. It is designed to return the explainable subset necessary for optimum performance on a given test set. We design this experiment to assess the significance of the selected data points in terms of ranking function values. We use NDCG score for reporting the ranking value.

We use the synthetic dataset and create a series of ground-truth values by flipping n% {=5, 10, 15} of the data. We assign ground-truth to the flipped and unflipped points with 0 and 1 respectively. We examine the bottom(or low-value) k% of the data for computing the ranking values since the flipped points are more inclined to lie towards the bottom. As we have observed in Sect. 4.3, we use the reverse-selection procedure to obtain the bottom k% of data for the proposed method, while for the baselines, the low-valued k% points occupy the bottom rung. In order to compute NDCG scores, we adopt the two baselines [10,13] to provide scores to the subset of points obtained using the proposed method. In case of baselines, we use the k-lowest scoring points for computing rank values.

We flip 5%, 10% and 15% of the data and compute NDCG@k for k ∈ {5, 10, 15, 25, 40}. Here k refers to the bottom k% subset from the methods. In Fig. 4, we compare the respective ranking values(NDCG) of subsets obtained using influence function (IF) or data shapley (DS) with the subset obtained using the proposed method followed by usage of the corresponding scoring method (IF or DS) on this acquired subset. We can observe that NDCG values obtained by the proposed method's subset followed by scoring using IF or DS always stays ahead than using the subsets obtained by IF or DS, starting from an early fraction. This shows that the subset obtained using HOST-CP is efficient enough in keeping the proportions of relevant(unflipped) and non-relevant(flipped) points in their respective positions leading to a higher-NDCG score.

4.6 Computational Complexity

We analysed the computational complexity of the proposed method on a 64-bit machine with one Quadro-P5000 GPU. For this experiment, we generated synthetic datasets with varying number of training datapoints (500, 2000, 4000, 6000, 10000). Keeping the network architecture fixed at a 3 layer network, we varied the size of the training data, on which the subset is meant to be computed. We also varied the size of the test datapoints for which explanations are sought to be found from the training datapoints. We report the time taken to perform the joint subset-training over an epoch, using the proposed method, which aims at finding the best set of explanations from the training set for a given set of test data. Figure 5 shows a linear pattern in time (in minutes) consumed by the proposed method across the varying size of datapoints.

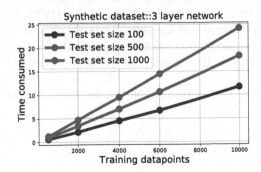

Fig. 5. Scaling with data: We show that the change in time consumed by HOST-CP with variation in training data points and test data points using a fixed network is linear with increasing data.

Table 2. Time comparison: We compare the times consumed by DVRL and HOST-CP. We observe that HOST-CP has a lower running time than DVRL.

Accuracy (5%)		Accuracy (20%)		Epochs (HOST-CP)/Iterations (DVRL)		Time (mins)	
HOST-CP	DVRL	HOST-CP	DVRL	HOST-CP	DVRL	HOST-CP	DVRL
62.46	57.8	71.32	66.4	1	25	1.1	3.40
63.34	57.2	73.82	66.8	2	50	2.01	5.31
65.14	57.4	74.89	66.5	3	100	3.05	6.50
65.71	57.6	75.0	66.6	4	150	4.08	10.57
65.72	57.4	75.0	66.4	5	200	5.12	14.26

We have earlier observed that the proposed method is consistently performing better than all the baselines. Since DVRL is the closest to our approach, we also record the time taken by DVRL and the proposed method on the synthetic dataset (Table 2). We increase the number of iterations for DVRL to trace if the accuracy obtained by the proposed method is achievable by DVRL under a certain subset cardinality (5%, 20%). We can observe that DVRL saturates at a certain accuracy value even after increasing the number of iterations to a much higher value. Thus, our method, HOST-CP besides attaining a higher accuracy, also takes considerably lower running time than that of DVRL.

5   Conclusion

In this paper, we propose a technique for finding high-value subsets of training datapoints essential for explaining the test set instances. We design a learning convex framework for subset selection, which is then used to optimise the subset with respect to a differentiable value function. The value function is essentially the performance metric which helps to evaluate the trained models. We compare our method against the state-of-the-art baselines, influence functions [13], data shapley [10], TracIn-CP [16] and DVRL [19] across different datasets in a range of applications like addition or removal of data, detecting mislabelled examples. We also analyse the quality of obtained subsets from DVRL and the proposed method. Lastly, we show that the proposed method scales linearly with the dataset size and takes a lower running time than DVRL which is the closest method to our approach methodologically. Thus, we are able to show that HOST-CP outperforms the baselines consistently in all the applications used for the experiments, thus proving its efficiency in terms of providing high-value subsets.

Acknowledgements. This project is funded by Hewlett Packard Labs, Hewlett Packard Enterprise.

References

1. Agrawal, A., Amos, B., Barratt, S., Boyd, S., Diamond, S., Kolter, J.Z.: Differentiable convex optimization layers. In: NeurIPS (2019)

2. Buchbinder, N., Feldman, M., Naor, J., Schwartz, R.: Submodular maximization with cardinality constraints. In: ACM-SIAM SODA (2014)
3. Choromanska, A., et al.: Beyond backprop: online alternating minimization with auxiliary variables. In: ICML. PMLR (2019)
4. Cook, R.D., Weisberg, S.: Residuals and Influence in Regression. Chapman and Hall, New York (1982)
5. Das, S., et al.: Multi-criteria online frame-subset selection for autonomous vehicle videos. Pattern Recognit. Lett. **133**, 349–355 (2020)
6. Elhamifar, E., Kaluza, M.C.D.P.: Online summarization via submodular and convex optimization. In: CVPR (2017)
7. Elhamifar, E., Sapiro, G., Sastry, S.S.: Dissimilarity-based sparse subset selection. IEEE TPAMI **38**(11), 2182–2197 (2015)
8. Frénay, B., Verleysen, M.: Classification in the presence of label noise: a survey. IEEE Trans. Neural Netw. Learn. Syst. **25**(5), 845–869 (2013)
9. Ghorbani, A., Kim, M., Zou, J.: A distributional framework for data valuation. In: ICML. PMLR (2020)
10. Ghorbani, A., Zou, J.: Data shapley: equitable valuation of data for machine learning. In: ICML. PMLR (2019)
11. Ghorbani, A., Zou, J.Y.: Neuron shapley: discovering the responsible neurons. In: NeurIPS (2020)
12. Hara, S., Nitanda, A., Maehara, T.: Data cleansing for models trained with SGD. In: NeurIPS (2019)
13. Koh, P.W., Liang, P.: Understanding black-box predictions via influence functions. In: ICML. PMLR (2017)
14. Lundberg, S., Lee, S.I.: A unified approach to interpreting model predictions. arXiv preprint arXiv:1705.07874 (2017)
15. Lundberg, S.M., Erion, G.G., Lee, S.I.: Consistent individualized feature attribution for tree ensembles. arXiv preprint arXiv:1802.03888 (2018)
16. Pruthi, G., Liu, F., Kale, S., Sundararajan, M.: Estimating training data influence by tracing gradient descent. NeurIPS (2020)
17. Steinhardt, J., Koh, P.W., Liang, P.: Certified defenses for data poisoning attacks. In: NIPS 2017 (2017)
18. Wu, Y., Dobriban, E., Davidson, S.: Deltagrad: rapid retraining of machine learning models. In: ICML. PMLR (2020)
19. Yoon, J., Arik, S., Pfister, T.: Data valuation using reinforcement learning. In: ICML. PMLR (2020)

Consequence-Aware Sequential Counterfactual Generation

Philip Naumann[1,2(✉)] and Eirini Ntoutsi[1,2]

[1] Freie Universität Berlin, Berlin, Germany
[2] L3S Research Center, Leibniz Universität Hannover, Hanover, Germany
{philip.naumann,eirini.ntoutsi}@fu-berlin.de

Abstract. Counterfactuals have become a popular technique nowadays for interacting with black-box machine learning models and understanding how to change a particular instance to obtain a desired outcome from the model. However, most existing approaches assume instant materialization of these changes, ignoring that they may require effort and a specific order of application. Recently, methods have been proposed that also consider the order in which actions are applied, leading to the so-called sequential counterfactual generation problem.

In this work, we propose a model-agnostic method for sequential counterfactual generation. We formulate the task as a multi-objective optimization problem and present a genetic algorithm approach to find optimal sequences of actions leading to the counterfactuals. Our cost model considers not only the direct effect of an action, but also its consequences. Experimental results show that compared to state-of-the-art, our approach generates less costly solutions, is more efficient and provides the user with a diverse set of solutions to choose from.

Keywords: Sequential counterfactuals · Multi-objective optimization · Genetic algorithms · Model-agnostic

1 Introduction

Due to the increasing use of machine learning algorithms in sensitive areas such as law, finance or labor, there is an equally increased need for transparency and so-called *recourse* options [12,24]. It is no longer sufficient to simply deliver a decision, but moreover to be able to explain it and, ideally, offer assistance if one feels unfairly treated by the algorithm. Hiding the decision-making algorithm behind a (virtual) wall like an API makes these issues especially intransparent and problematic in case of *black-box* models, as they are not able to communicate with the end user beyond the provided decision (e.g. `reject` or `accept`).

For this reason, algorithms emerged that aim to explain a (black-box) decision or even provide essential recourse information in order to change an undesired outcome in one's favor. The latter of these methods is of particular interest, since it has the capability to improve a bad decision for someone into a good one by giving explicit directions on what to change with respect to the provided

© Springer Nature Switzerland AG 2021
N. Oliver et al. (Eds.): ECML PKDD 2021, LNAI 12976, pp. 682–698, 2021.
https://doi.org/10.1007/978-3-030-86520-7_42

information. These methods are commonly referred to as *counterfactual explanations* [24]. The goal here is to change (tweak) characteristics (features) of a provided input so that the black-box decision turns out in favor afterwards (e.g. increase your level of education to get a higher salary). The result of applying these changes on the input is commonly referred to as a *counterfactual* [24].

Usually, those changes are atomic operations, meaning each feature is tweaked independently [4,15,17,24]. Thus, feature interrelationships, e.g. of causal nature, are not considered. Recent approaches propose to define changes in (multiple) features through so-called *actions*, which can be thought of as instructions on how to apply the modifications and their consequences (e.g. increasing the level of education has an impact on age because it takes time to obtain a degree). Actions further help to describe what features are actionable, mutable or immutable [12]. However, these approaches, like the traditional counterfactual methods, still assume that all these changes happen instantly and do not consider implications and consequences of the *order* of their application.

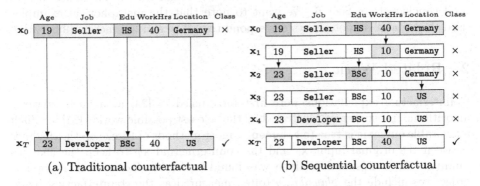

(a) Traditional counterfactual   (b) Sequential counterfactual

Fig. 1. The difference between traditional counterfactual generation (a) and the sequential approach (b). Although the generated counterfactual x_T is the same, the process and implied knowledge/information is different.

For this reason, recent works regard the application of feature altering changes as a *sequential* process [19,20]. The problem is then shifted from simply computing the counterfactual, to finding an ordered *sequence* of actions that accomplishes it. In Fig. 1 we visualized this difference for a person x_0 wishing to attain a higher salary (Class: $\times \Rightarrow \checkmark$) for which getting a higher education level, switching the job and changing the location is necessary. In this case, e.g., it was assumed that decreasing one's working hours is *beneficial* in order to obtain a higher degree, which is a relationship that traditional approaches do not model. Each state x_t in Fig. 1b describes the result of applying an action on the previous state x_{t-1} and x_T denotes the counterfactual. As we can see, the actions have different effects in the feature space and may alter more than one feature at a time. Moreover, the order makes a difference as increasing the level of education before changing the job is usually more beneficial, as well as decreasing the working hours in order to attain the degree (representing a focus on the education).

Later on, this change is reset to its original value through another action since an increased value is now more *plausible* again with respect to the job change. The whole process of Fig. 1b can be seen as a *sequential counterfactual* as each state is an important part of it, whereas in the traditional setting in Fig. 1a only the final result is important. Thus, a sequential counterfactual allows us to look beyond just flipping the class label and provides further information about the underlying process. It is not the absolute main goal anymore to only switch the class label, but to take consequential effects into account in order to improve the overall benefit of the actions. For the end user, this also means getting concrete information on the actions and their order. Our work is in the direction of sequential counterfactual generation and inspired by [19], which we will further elaborate in Sect. 2.

The rest of this paper is organized as follows: we will first give an overview on related work in Sect. 2. Then, we introduce the *consequence-aware counterfactual generation problem* in Sect. 3, for which our proposed method follows in Sect. 4. Lastly, we evaluate it in Sect. 5 to a state-of-the-art method and give final conclusions in Sect. 6. We want to note that the metaphorical examples used throughout do not always correspond to reality and are merely illustrative.

2   Related Work

Counterfactual explanations were first introduced in [24] as a mean towards algorithmic transparency. Motivated by the "closest possible world" [24] in which a favorable outcome is true (e.g. accept), most methods optimize on the *distance* between the original input x_0 and the counterfactual x_T to keep the changes minimal. Since this notion alone was found to be insufficient, other popular objectives include the *plausibility* (often measured as the counterfactual being within the class distribution [10] or being close to instances from the dataset [4, 18, 23]) and *sparsity* [4, 17, 23] (measuring how many features had to change) of solutions. Desirable criteria regarding the algorithms are, e.g., being model-agnostic [4, 11, 14, 15] (e.g. by using genetic algorithms) or providing a diverse set of solutions [4, 5, 16, 17] (e.g. by using multi-objective optimization).

More recently, works [13, 19, 22] began to replace the distance function with a *cost* in order to express aspects such as the *effort* of a change. In alignment to this, the term *recourse* [5, 10, 12, 13, 18, 22] has attracted attention to describe the counterfactual generation. It can be defined as "the ability of a person to change the decision of a model by altering actionable input variables" [22] and thus emphasizes on the *actionability* of the features to provide comprehensible recommendations that can be acted upon [12]. In addition, more attention has been paid on the causal nature of feature interrelationships, e.g. by including causal models in the generation process to assess mutual effects [5, 13, 16].

Lastly, motivated by the fact that in reality most changes do not happen instantly, but are rather part of a *process*, there have been works that extend the formulation of actions and their consequences by incorporating them in a sequential setting [19, 20]. In contrast to simply finding the counterfactual that switches the class label, the focus is on finding a subset of actions that, applied in a specific order, accomplishes the counterfactual while accounting for potential consequences of prior actions on subsequent ones (cf. Fig. 1).

In this work, we also focus on sequential counterfactual generation. The advantages of our method in comparison to [19] can be summarized as follows: our method is model-agnostic and not bound to differentiable (cost) functions. It finds diverse sequences instead of a single solution, thus giving more freedom of choice to the end user. Moreover, it is efficient in pruning and exploring the search space due to using already found knowledge (exploitation) and the ability to optimize all sub-problems (cf. Sect. 3.3) at once, while [19] breaks these down into separate problems. This efficiency allows us to find sequences of any length, whereas [19] requires multiple runs and more time for it (cf. Sect. 5). Another difference is our action-cost model (cf. Sect. 3.2). We regard consequences not only in the feature space (e.g. age increases as a consequence of obtaining a higher degree), but also explicitly model their effects in the objective or cost space (e.g. changing the job becomes *easier* with a higher degree). This way we extend the cost formulation of [19], which only proposes to model (feature) relationships through (boolean) pre- and post-requirement constraints (e.g. you *must* at least be 18 to get a driver's license). These constraints are also possible in our model. Finally, we note that the work of [20] also discusses consequential effects. However, since no cost function is optimized, but instead the target class likelihood of the counterfactual, we do not compare with [20] in this work.

3   Problem Statement

We assume a static *black-box classifier* $f : \mathcal{X} \to \mathcal{Y}$ where $\mathcal{X} = \mathcal{X}_1 \times \cdots \times \mathcal{X}_d$ is the feature space and $\mathcal{Y}$ is the class space. The notation $\ddot{\mathcal{X}}_h$ is used to refer to the feature itself (e.g. $\ddot{\mathcal{X}}_{\text{Edu}}$ denotes Education, whereas $\mathcal{X}_{\text{Edu}} = \{\ldots, \text{HS}, \text{BSc}, \ldots\}$ is the domain). For simplicity and without loss of generality, we assume f is a binary classifier with $\mathcal{Y} = \{\text{reject}, \text{accept}\}$. Let $\mathbf{x}_0 \in \mathcal{X}$ be an instance of the problem, e.g. a person seeking to receive an annual salary of more than \$50k, and the current decision of f based on $\mathbf{x}_0$ is $f(\mathbf{x}_0) = \text{reject}$. The goal is to find a counterfactual example $\mathbf{x}_T$ for $\mathbf{x}_0$ such that $f(\mathbf{x}_T) = \text{accept}$. In other words, we want to change the original instance so that it will receive a positive decision from the black-box. The sort of changes we refer to are in the feature description of the instance, like increasing Age by 5 years or decreasing Work Hours to 20 per week. Our problem formulation builds upon [19]. We introduce actions in Sect. 3.1 along with their associated cost to implement the suggested changes in Sect. 3.2. Finally, we formulate the generation of sequential counterfactuals as a multi-objective optimization problem in Sect. 3.3.

3.1   Actions, Sequences of Actions and States

Let $\mathcal{A} = \{a_1, \ldots, a_n\}$ be a problem-specific, manually-defined set of *actions*. Each action is a function $a_i : \mathcal{X} \times \mathcal{V} \to \mathcal{X}$ that modifies a subset of features

$\mathcal{I}_{a_i} = \{\ddot{\mathcal{X}}_h, \ddot{\mathcal{X}}_k, \dots\} \subseteq \ddot{\mathcal{X}}$ in a given input instance $\mathbf{x}_{t-1} \in \mathcal{X}$ in order to realize a new instance (which we refer to as a *state*) $a_i(\mathbf{x}_{t-1}, v_i) = \mathbf{x}_t \in \mathcal{X}$. An action *directly* affects one feature $\ddot{\mathcal{X}}_h \in \mathcal{I}_{a_i}$ (e.g. Education) based on a tweaking value $v_i \in \mathcal{V}_h \subseteq \mathcal{X}_h$ and may have *indirect* effects on other features $\ddot{\mathcal{X}}_{k \neq h} \in \mathcal{I}_{a_i}$ as a consequence (e.g. Age). Here, $\mathcal{V} \subseteq \mathcal{X}$ describes the *feasible value space* that restricts the tweaking values based on the given $\mathbf{x}$ (e.g. Age may only increase).

For example, $\mathbf{x}_2$ in Fig. 1b is the result of applying $a_{\text{Edu}}(\mathbf{x}_1, \text{BSc}) = \mathbf{x}_2$ which changes the Education (HS $\Rightarrow$ BSc) and affects the Age (19 $\Rightarrow$ 23) as a consequence. In this case, $\mathcal{V}_{\text{Edu}}$ and $\mathcal{V}_{\text{Age}}$ restrict the tweaking values so that they can only increase. It is possible to use a causal model as in [13] for evaluating how the indirectly affected features have to be changed. An action-value pair (a_i, v_i) thus represents a specification how $\mathcal{I}_{a_i}$ of $\mathbf{x}$ is affected with respect to the tweaking value v_i of feature $\ddot{\mathcal{X}}_h$.

Additionally, each action a_i may be subject to boolean constraints $\mathbb{C}_i : \mathcal{X} \times \mathcal{V} \to \mathbb{B}$ such as pre- and post-requirements as proposed in [19] (cf. Sect. 2), which can also be used to validate the feasibility of indirectly affected features. Each action-value pair is considered *valid*, if it satisfies the associated constraints and $\mathcal{V}$. An ordered sequence $\mathcal{S}$ of valid action-value pairs (a_i^t, v_i) leading to the counterfactual $\mathbf{x}_T$ is called a *sequential counterfactual*. Here, $t \in \{1, \dots, T\}$ is the order of applying the actions and $T \in \{1, \dots, |\mathcal{A}|\}$ is the number of used actions in $\mathcal{S}$, i.e. the sequence length.

3.2   Consequence-Aware Cost Model

Our goal is to assess the direct effort (which can, e.g., be abstract, as based on personal preferences, or concrete, like money or time etc.) of an action while considering possible consequences. We define the cost of an action a_i as a function of two components: $c_i(\cdot) = b_i(\cdot) \cdot g_i(\cdot)$. Here, b_i represents the *direct effort*, whereas g_i acts as a discount of it in order to express (beneficial) *consequences* of prior actions. Please note that each action has its *own* cost function. Summing up all action costs of a sequence $\mathcal{S}$ yields the *sequence cost*: $C_\mathcal{S} = \sum_{a_i \in \mathcal{S}} c_i(\cdot)$. In the following we will explain the components in more detail.

Action Effort b_i: First, we introduce $b_i : \mathcal{X} \times \mathcal{X} \to \mathbb{R}_+$, which is assumed to be an action-specific measure of the direct effort caused by an action a_i. Therefore, this is specified as a function between two consecutive states $\mathbf{x}_{t-1}, \mathbf{x}_t \in \mathcal{X}$, representing the direct effect of applying that action on $\mathbf{x}_{t-1}$. This function can be thought of as a typical cost function as in e.g. [19]. As an example, b_i could be specified linear and time based, whereby an effort caused by an action addEdu would be represented by the years required (e.g. four to progress from HS to BSc). Alternatively, monetary costs could be used (i.e. tuition costs), or a combination of both. Besides, there are no particular conditions on this function, so it can be defined arbitrarily (e.g. return a constant value).

Consequential Discount g_i: To assess a possible (beneficial) consequential effect of previous actions on applying the current one a_i^t, we introduce a so-called *consequential discount* $g_i : \mathcal{X} \to [0, 1]$ that affects the action effort b_i based on the current state $\mathbf{x}_{t-1}$ (i.e. before applying a_i^t). Such effects can be, e.g., "the higher the Education, the easier it is to increase Capital" or "increasing Education in Germany is cheaper than in the US (due to lower tuition fees)". This discount therefore describes a value in $[0, 1]$, where 0 would mean that the current state is so beneficial that the effort of the action to be applied is completely cancelled out, and 1 that there is no advantageous effect. We derive the aforementioned consequential effect on an action from consequential relationships between feature pairs. This is provided as a graph $\mathcal{G} = (X, E)$ where the nodes $X \subseteq \ddot{\mathcal{X}}$ are a subset of the features and edges $e_{kh} \in E$ between each two nodes $\ddot{\mathcal{X}}_k, \ddot{\mathcal{X}}_h \in X$ describe a function $\tau_{kh} : \mathcal{X} \to [0, 1]$ that models a consequential effect between one feature $\ddot{\mathcal{X}}_k$ to another $\ddot{\mathcal{X}}_h$. For the given features $\ddot{\mathcal{X}}_1 :=$ Job, $\ddot{\mathcal{X}}_2 :=$ Education and $\ddot{\mathcal{X}}_3 :=$ Location we have exemplified $\mathcal{G}$ in Fig. 2a by the following relations:

1. The Education cost depends on the Location $(\ddot{\mathcal{X}}_3 \xrightarrow{\tau_{32}} \ddot{\mathcal{X}}_2)$. E.g., it is *cheaper* to get a degree in Germany than the US because of lower tuition fees.
2. The easiness of getting a Job depends on the Location $(\ddot{\mathcal{X}}_3 \xrightarrow{\tau_{31}} \ddot{\mathcal{X}}_1)$. E.g., it is *easier* to get a Developer job in the US than in other locations.
3. The higher the Education, the *easier* it is to change the Job $(\ddot{\mathcal{X}}_2 \xrightarrow{\tau_{21}} \ddot{\mathcal{X}}_1)$.

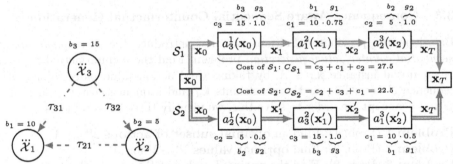

(a) Feature relationship graph $\mathcal{G}$ (b) Different sequences $\mathcal{S}_1$ (red) and $\mathcal{S}_2$ (blue)

Fig. 2. For simplicity, the $\tau(\cdot)$ functions in (a) are based on binary conditions: $\tau_{32} = 1.0$ if $\mathcal{X}_3 :=$ US, else 0.5. $\tau_{31} = 0.5$ if $\mathcal{X}_3 :=$ US, else 1.0. $\tau_{21} = 0.5$ if $\mathcal{X}_2 \geq$ BSc, else 1.0. As a reference, the action efforts b_i are provided above each feature in (a).

Based on this modeling in $\mathcal{G}$, we can then derive the consequential discount g_i of an action a_i (Eq. 2) by averaging the consequential effect $\hat{g}_h$ of *each affected feature* $\ddot{\mathcal{X}}_h \in \mathcal{I}_{a_i}$ (Eq. 1). It is assumed that g_i evaluates to 1.0 if no feature in $\mathcal{I}_{a_i}$ is influenced by another one in $\mathcal{G}$ (e.g. $\ddot{\mathcal{X}}_3$ in Fig. 2a).

$$\hat{g}_h(\mathbf{x}_{t-1}) = \text{avg}(\{\tau_{kh}(\mathbf{x}_{t-1}) \, \forall \, \ddot{\mathcal{X}}_k \in X \mid \exists \, e_{kh} \in E\}) \tag{1}$$

$$g_i(\mathbf{x}_{t-1}) = \text{avg}(\{\hat{g}_h(\mathbf{x}_{t-1}) \, \forall \, \ddot{\mathcal{X}}_h \in \mathcal{I}_{a_i} \mid \exists \, \ddot{\mathcal{X}}_h \in X\}) \tag{2}$$

In order to understand the benefit of the consequential discount on the sequence order, we exemplify in Fig. 2b the situation from Fig. 1b (for simplicity the working hours altering actions are omitted). The available actions are thus: "change `Job` to `Developer`" (a_1), "get a BSc degree" (a_2) and "change `Location` to `US`" (a_3) (notice all $\mathcal{V}_i$ are fixed to a single value). Each a_i alters their respective feature counterpart $\ddot{\mathcal{X}}_i$ (e.g. $\mathcal{I}_{a_1} = \{\ddot{\mathcal{X}}_1\}$). We can see the cost computations in Fig. 2b for two differently ordered sequences $S_1 = \langle a_3^1, a_1^2, a_2^3 \rangle$ and $S_2 = \langle a_2^1, a_3^2, a_1^3 \rangle$ that achieve the same final outcome $\mathbf{x}_T$ (i.e. only the application order is different). To compute the consequential discount for action a_1^2 in S_1, e.g., we consider the relations $\ddot{\mathcal{X}}_3 \xrightarrow{\tau_{31}} \ddot{\mathcal{X}}_1$ and $\ddot{\mathcal{X}}_2 \xrightarrow{\tau_{21}} \ddot{\mathcal{X}}_1$ with respect to $\mathbf{x}_1$ to derive the *feature* discount (Eq. 1): $\hat{g}_1(\mathbf{x}_1) = \frac{0.5+1.0}{2} = 0.75$. Since no other feature is affected by a_1 according to $\mathcal{I}_{a_1}$, the *action* discount (Eq. 2) evaluates to $g_1(\mathbf{x}_1) = \hat{g}_1(\mathbf{x}_1)$. After computing all action costs c_i, we can derive the sequence costs $C_{S_1} = 27.5$ and $C_{S_2} = 22.5$, which shows that S_2 would be preferred here as it benefits more from the consequential discount effects of $\mathcal{G}$. Note, that if we leave out the consequential discounts completely, i.e. $c_i = b_i$, then there would be no notion of order here as each sequence would receive the same costs (assuming the same tweaking values). Furthermore, our consequence-aware formulation means, that additional actions are only used if their induced effort is lower than the consequential benefit they provide (as this would otherwise make C_S worse than if the action was not used).

3.3   Consequence-Aware Sequential Counterfactual Generation

Based on the previous definitions, we now introduce the *consequence-aware sequential counterfactual generation* problem. Find the counterfactual $\mathbf{x}_T \in \mathcal{X}$ of an initial instance $\mathbf{x}_0 \in \mathcal{X}$ by taking valid action-value pairs (a_i^t, v_i) of a sequence S according to the constraints $\mathbb{C}_i$ and sequence cost C_S such that $f(\mathbf{x}_T) = $ accept. In order to solve this, we identify three sub-problems:

Problem 1 (Prob. 1): Find an **optimal subset of actions** $\mathcal{A}^* \subseteq \mathcal{A}$.
Problem 2 (Prob. 2): Find **optimal values** $\mathcal{V}^* \in \mathcal{V}$ for $\mathcal{A}^*$.
Problem 3 (Prob. 3): Find the **optimal order** of S to apply the actions.

For an arbitrary set of actions $\mathcal{A}$ and feasible value space $\mathcal{V}$ many sequences can be generated, therefore it is important to assess their quality. For this purpose, we will use the sequence cost $o_1 := C_S$ as a *subjective* measure, as well as the Gower's distance [9] $o_2 := \text{dist}(\mathbf{x}_0, \mathbf{x}_T)$ to act as an *objective* assessment how much $\mathbf{x}_T$ differs from $\mathbf{x}_0$. The Gower's distance is able to combine numerical and categorical features and is thus an appropriate measure here [4]. The reason for using both is, that o_1 measures the *effort* of the *whole process*, whereas o_2 only considers the *difference* to the final counterfactual and is agnostic of the process.

In order to propose diverse solutions, we will formulate the problem as a multi-objective one and add, next to o_1 and o_2, the individual tweaking frequencies of each of the $1 \le h \le d$ features after unrolling the complete sequence, i.e. $o_{2+h} = \#(\ddot{\mathcal{X}}_h \in \mathcal{I}_{a_i} \; \forall \; a_i \in S)$. In other words, o_{2+h} measures how often a feature $\ddot{\mathcal{X}}_h$ was affected by all actions of S combined. E.g., the frequencies for $o_3, \ldots, o_7$ would be $\{1, 1, 1, 2, 1\}$ in case of Fig. 1b. In a way this can be thought of as the *sparsity* objective mentioned in Sect. 2, but aggregated individually per feature instead of a sum. The idea behind the diversity objectives is to keep the number of feature changes minimal and additionally force the optimization to seek alternative changes instead. This means, solutions mainly compete with those that change the same features with respect to o_1 and o_2, resulting in a diverse set of optimal options. Combining all the above yields the following multi-objective minimization problem:

$$\min_{\mathcal{S}}(\underbrace{o_1}_{\text{Sequence cost}} , \underbrace{o_2}_{\text{Gower's distance}} , \underbrace{o_{2+1}, \ldots, o_{2+h}, \ldots, o_{2+d}}_{\text{Feature tweaking frequencies}})$$

$$\text{s.t. } f(\mathbf{x}_T) = \texttt{accept} \text{ and } \bigwedge_{(a_i,v_i)\in\mathcal{S}} \mathbb{C}_i \tag{3}$$

4   Consequence-Aware Sequential Counterfactuals (CSCF)

In order to address the combinatorial (Prob. 1, Prob. 3) and continuous (Prob. 2) sub-problems with respect to Eq. 3, we used a *Biased Random-Key Genetic Algorithm* (BRKGA) [8] and adapted it for multi-objective optimization by using non-dominated sorting (NDS) [21]. NDS is preferred over a scalarization approach to avoid manual prioritization of the objectives and to address them equally. Moreover, by using BRKGA we avoid the manual definition of problem-specific operators, which is not trivial here. This choice allows to solve all sub-problems at once, is model-agnostic, derivative-free and provides multiple solutions.

BRKGA: The main idea behind BRKGA is that it optimizes on the genotype of the solutions and evaluates on their phenotype, making the optimization itself problem-independent [8]. A *genotype* is the internal representation of a solution, whereas the *phenotype* is the actual solution we wish to generate. The phenotype must always be deterministically derivable from the genotype through a *decoder* function [8]. Because of this, each solution genotype in BRKGA is encoded as a vector of real (random) values in the range of $[0,1]$ (the so-called *random-keys*) [8].

In each generation (iteration), the decoded solution phenotypes are evaluated based on their fitness (given by the vector of evaluating each objective of Eq. 3 individually) and the population is divided into two subsets by applying NDS: the so-called *elites*, which are the feasible (valid), non-dominated (i.e. best) solutions in the Pareto-front [7], and the remaining ones, called *non-elites*. Then, genetic mating is performed by selecting two parents, one from the elite sub-population and one from the non-elites. A new solution is created by applying biased crossover [8] on the two parents, which favors selecting the value of the elite solution with a certain biased probability. This step is repeated until sufficient new solutions have been created. Additionally, a number of completely random solutions are generated in order to preserve the exploration of the search space and the diversity in the population. Finally, the different sub-populations (*elites, crossovered* and *random* solutions) are merged and evaluated and form the new population for the next generation. This loop continues until some termination criterion is reached. The Pareto-front of the last generation then represents our final solution set. Note that the Pareto-front usually holds more than one solution (i.e. a diverse set of optimal sequences according to the objectives of Eq. 3).

Genotype: We wish to solve the three sub-problems from Sect. 3.3 at once. Inspired by similar, successful, encodings of problems (cf. [8]), we thus compose the genotype as $G = [\mathbb{A}_1, \ldots, \mathbb{A}_N, \mathbb{V}_{N+1}, \ldots, \mathbb{V}_{2N}] = [\mathbb{A}, \mathbb{V}]$, with $\mathbb{A}_i, \mathbb{V}_i \in [0, 1]$.

The first N values, $\mathbb{A}$, in G encode the action subset $\mathcal{A}$ (cf. Prob. 1) and their ordering $t \in \{1, \ldots, T\}$ in the sequence $\mathcal{S}$ (cf. Prob. 3). Each index position $i \in \{1, \ldots, |\mathbb{A}|\}$ corresponds to one of the actions in the action set $a_i \in \mathcal{A}$ (i.e. $\mathbb{A}_i \in \mathbb{A}$ encodes $a_i \in \mathcal{A}$). The other half, $\mathbb{V}$, encodes the tweaking values $\mathcal{V}$ (cf. Prob. 2) of each action, which is also referred to by the index position $i \in \{1, \ldots, |\mathbb{V}|\}$ (i.e. $\mathbb{V}_{N+i} \in \mathbb{V}$ encodes $v_i \in \mathcal{V}_h$). Figure 3 visualizes this composition of the genotype.

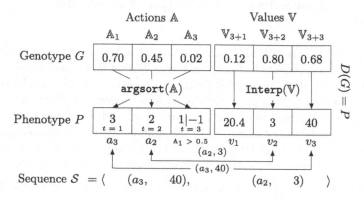

Fig. 3. Anatomy and representation of the solution decoding.

Decoder: Since BRKGA itself is problem-independent, we design a problem-specific *decoder* $D(G) = P$ to infer the phenotype P from G. Below we discuss its design, which is inspired by established concepts (cf. [8]).

The subset of actions (cf. Prob. 1) is decoded by identifying *inactive* actions in the actions part $\mathbb{A}$. As a simple heuristic, we define an action in G as inactive (denoted by "-1"), if its genotype value is greater than 0.5. This follows from the idea that an action has an equal chance to be active or inactive when chosen randomly. To get the *active* actions and their *order* (cf. Prob. 3), we apply the commonly used `argsort : A → A` decoding [1,8] on $\mathbb{A}$ and identify the inactive actions afterwards, which will always produce a non-repeating order. We find the actual actions by looking at the sorted index (cf. P in Fig. 3). Note that an action will be at an earlier position t in $\mathcal{S}$ the lower its genotype value is. Lastly, we decode the values part by applying a (linear) interpolation `Interp : V → V` on each of the genotype values in $\mathbb{V}$. Therefore, only the original value ranges need to be provided (via $\mathcal{V}$), or in case of categorical values a mapping between the interpolated value and the respective categorical value counterpart. The decoded value at position $i \in \{1, \ldots, |\mathbb{V}|\}$ then belongs to $a_i \in \mathcal{S}$.

An example of the full decoding process (from a genotype solution G to the actual solution sequence $\mathcal{S}$) is visualized in Fig. 3. As we can see, applying `argsort` on $\mathbb{A} \in G$ realizes an order (3,2,1) for $\mathcal{A}$ via P. Since $\mathbb{A}_1 > 0.5$, action a_1 is rendered inactive and the remaining, ordered, action set is $\langle a_3^1, a_2^2 \rangle$. The associated tweaking values are then decoded by interpolating $\mathbb{V} \in G$ and assigned to their action counterparts, thus creating the sequence $\mathcal{S}$. Note that decoding is necessary for all solutions in each iteration to evaluate the fitness and is repeated until the termination criterion (e.g. maximum number of iterations) is reached.

5 Experiments

The first goal of our experiments is to evaluate the costs of the generated sequences[1] in comparison to the state-of-the-art approach [19] (Sect. 5.1). Next, we analyze the diversity of the generated solutions in terms of the action space and the sequence orders (Sect. 5.2). Finally, we examine the effect of the action positions in a sequence for switching the class label (Sect. 5.3).

Datasets: We report on the datasets *Adult Census* [6] (for predicting whether income exceeds \$50k/year based on census data) and *German Credit* [6] (for predicting whether a loan application will be accepted).

Baselines: The following three methods were used for the evaluation. The first one acts as a direct competitor and the others are variations of our method.

SYNTH [19]: The competitor[2] only optimizes for finding a *single* minimal cost sequence and solves two separate sub-problems independently. First, they generate candidate action subsets according to Prob. 1 with respect to one of their proposed heuristics. Then, they perform an adversarial attack based on [3] in order to find the tweaking values for each candidate sequence to solve Prob. 2. There is no explicit sequence order notion as per Prob. 3 apart from pre- and post-requirements, thus it only optimizes on their provided cost function which is equivalent to the action effort that we introduced in Sect. 3.2, i.e. $c_i = b_i$. To make the comparison to [19] fair, we use their exact same action-cost model and provided pre-trained black-box classifiers (which are neural networks here) for all methods. That means, all action behavior is identical in this comparison (i.e. tweaking effects, constraints, conditions and costs). Hence, we use the costs of their model for the action effort b_i.

CSCF: Our method optimizes *all* sub-problems at once and provides *multiple* solutions. Regarding the cost, it considers the consequential discount and thus optimizes $c_i = b_i \cdot g_i$. For g_i, we provided a simple feature relationship graph $\mathcal{G}$ that models beneficial effects in *Adult Census* such as:

- The higher the `Education` level, the easier it gets to increase `Capital Gain`, change `Work-Class` and `Occupation`.

---

[1] Apart from the cost objective o_1, we will not report on the remaining objective space from Eq. 3 here, since the results were similar and thus not particularly informative.

[2] https://github.com/goutham7r/synth-action-seq.

- The lower the Work Hours, the easier it is to increase the Education.
- The higher the Work Hours, the easier it gets to increase Capital Gain.

We only use CSCF for *Adult Census* as it was not practical to create a meaningful graph based on the predefined actions from [19] for *German Credit*. Since the g_i part primarily affects the order of actions, we would generally expect CSCF to behave similarly to SCF in terms of b_i, though.

SCF: This is a variation of our proposed CSCF, leaving out the consequential discount and thus only optimizes on the action efforts from [19], i.e. $c_i = b_i$. When referring to findings that apply to both CSCF and SCF, we use "(C)SCF".

Implementations: We implemented our method in Python[3] using the *pymoo* [2] implementation of BRKGA. The parameters for BRKGA are mostly based on recommendations from [8]: the population size was set to 500, the mutant and offspring fractions to 0.2 and 0.8, respectively, and the crossover bias to 0.7. As the termination criterion for (C)SCF, we fixed the number of iterations to 150. From each dataset we chose 100 random instances that are currently classified by the black-box as the undesired class (i.e. Salary < $50k and Credit denied) with the intention of generating a sequential counterfactual to flip their class label. Each instance represents an experiment. We ran all methods on the same 100 experiments and fixed the maximum sequence length of SYNTH to $T = 2$ because of long runtimes for larger values which made the experiments of those unfeasible on our hardware[4]. The long runtimes of SYNTH were already mentioned in their paper: "Time/iteration is ~15s across instances" [19], which confirms our observations, since the algorithm may take up to a few 100 iterations according to [19] (running SYNTH for $T \leq 2$ took the same time as (C)SCF needed for all sequence lengths simultaneously). Because of this, we used the "*Vanilla*" heuristic for SYNTH as it was found to perform the best for shorter sequences based on [19]. Lastly, we had to filter out some experiments in the post-processing since SYNTH produced constraint violating solutions or did not find a feasible one. Consequently, the number of experiments for *German Credit* was post-hoc decreased to 85, but for *Adult Census* it did not change.

5.1   Sequence Costs of Sequential Counterfactuals

We show the *undiscounted* (i.e. only using the action effort share b_i) sequence costs of the solutions (o_1 objective) in Fig. 4 for *Adult Census* and *German Credit*. In the x-axis we see the individual, pair-wise relative differences between the computed minimal cost sequences for two methods and each of the valid initial inputs/experiments (which are represented by the y-axis). The green color indicates that the method mentioned first in the title (A) performed better, whereas red indicates that the other one (B) did. The blue line shows the point from which one method consistently outperforms the other.

---

[3] https://github.com/ppnaumann/CSCF.

[4] All experiments (competitor and our method) were executed on the free tier of *Google Colab* (https://colab.research.google.com/).

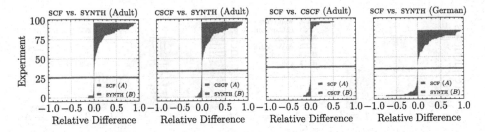

Fig. 4. Relative minimal sequence cost (o_1) differences between the three methods for both datasets and solutions with $T \leq 2$. It is computed as: $(B - A) / \max\{A, B\}$. (Color figure online)

Since our method finds *multiple* optimal sequences (on median 4 for *German Credit* and 7 for *Adult Census*) of different lengths per experiment, and SYNTH only finds a single one, we chose the least cost sequence in (C)SCF per set that satisfies $T \leq 2$ in order to make the comparison fair. Note, that there could be a longer sequence with less overall cost that was found by (C)SCF due to using more, but cheaper actions. Although CSCF has optimized on the discounted cost, we only use the effort share, b_i, of it in this analysis to guarantee comparability.

As we can see in Fig. 4, (C)SCF usually performed better in *Adult Census*. In *German Credit* it seems to be fairly even, but with a slight tendency towards SCF. The overall larger relative differences in favor of (C)SCF (green), with respect to SYNTH, appear to be the result of SYNTH selecting a different, but more expensive set of actions. By looking through the history trace, we identified that the same set of actions that (C)SCF found to be optimal was evaluated by SYNTH, although with different tweaking values. These values, however, seemed to produce a constraint-breaking solution that was either rendered invalid by SYNTH, or had high costs, since constraints are enforced as a penalty in SYNTH. The cases where SYNTH outperforms (C)SCF (red) show small cost differences only. Notably, the differences between CSCF and SCF are also minor, even though CSCF optimizes on the discounted costs. Thus, this suggests that the augmentation by g_i does not significantly interfere with the goal of keeping C_S minimal. In general, we conclude that (C)SCF is capable to find equivalent or better solutions in comparison to SYNTH in terms of costs.

5.2 Diversity of Sequential Counterfactuals

In Fig. 5 we illustrated the prevalence of actions at different positions t in a sequence (indicated by the height of each action in the pillars) along with the frequency of how often one action followed on from another (indicated by the widths of the flows). The whitespace of an action in the pillar shows that a sequence stopped there (i.e. had no subsequent actions). For this purpose, we aggregated over *all* optimal solution sequences (i.e. the whole final Pareto-fronts) from each experiment per method and dataset, respectively. Furthermore, we additionally show (C)SCF after filtering out all solutions with $T > 2$ (c, d, g in Fig. 5) for better comparability with SYNTH. The plots (a, b, c, d, e) in Fig. 5 belong to the *Adult Census* (A) and (f, g, h) to the *German Credit* (B) dataset.

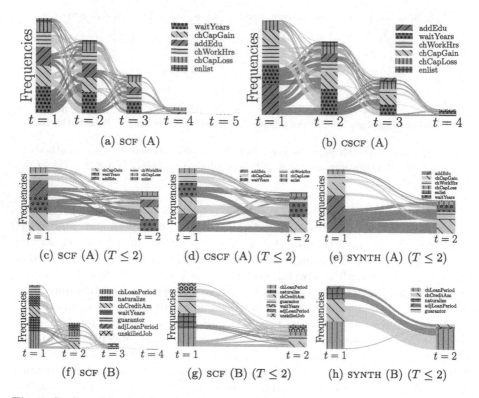

(a) SCF (A) (b) CSCF (A)

(c) SCF (A) ($T \leq 2$) (d) CSCF (A) ($T \leq 2$) (e) SYNTH (A) ($T \leq 2$)

(f) SCF (B) (g) SCF (B) ($T \leq 2$) (h) SYNTH (B) ($T \leq 2$)

Fig. 5. Sankey plots showing the sequence orders and flow between subsequent actions for differently aggregated solution (sub-)sets. The first two rows correspond to *Adult Census* (A) and the third to *German Credit* (B).

As we can see in Fig. 5 (a, b, c, d, f, g), (C)SCF makes use of the whole available action space $\mathcal{A}$, and more evenly utilizes the actions in each step than SYNTH (e, h). For *German Credit*, we notice that SYNTH only used five different actions, whereas SCF used all available (f, g). This observation is in alignment with our diversity goal and can be attributed to the tweaking frequency objectives (o_{2+h} from Eq. 3) which force the algorithm to seek alternative sequences that propose different changes while still providing minimal costs. Since SYNTH only finds the single least cost solution, the same actions were chosen as they appear to be less expensive than their alternatives. Regarding the lengths of the found sequences, we see that the minimal cost ones were usually found up to length three according to (a, b, f). After that, only few sequences still provide some sort of minimal cost. Because of this, we can say that (C)SCF does implicitly favor shorter sequences if they are in alignment with the costs. This behavior also follows from the tweaking frequency objectives, which minimize the number of times a feature was changed and thus the number of actions used (as these are directly related to another).

Looking at the particular differences between CSCF (b, d) and SCF (a, c), we can identify the distinct characteristics of discounting each action effort by g_i through $\mathcal{G}$. The relationship "the higher the Education, the easier it gets to attain Capital Gain" is reflected in (b, d) as addEdu is the most frequent action at $t = 1$. Moreover, there is no single sequence where addEdu would appear *after* chCapGain, indicating that the beneficial consequence was always used by CSCF. In comparison, SCF in (a, c) has a more equal spread as it has no knowledge of the relationships. Lastly, the same peculiarity can be observed for chWorkHrs, which was more often favored to be placed before chCapGain in CSCF than SCF because of the beneficial relation in $\mathcal{G}$. Even though it appears that the addEdu effect is also visible in SYNTH according to (e), this is an artifact since there is no explicit mechanism that would enforce it. The most likely reason for this is the order in which the actions were processed in the *Vanilla* heuristic.

Finally, looking at the $T \leq 2$ plots (c, d, e, g, h) specifically, we can see that some actions show a preferred co-occurrence. E.g., chCapGain and waitYears seem to appear more often subsequently than others (c, d). This is not visible in SYNTH (e), which on the other hand shows a distinct co-occurrence of chCreditAm and chLoanPeriod. The reason for this can be traced back to the cost model, which values these combinations as least expensive for sequence lengths of $T \leq 2$ (i.e. if we only use two actions). In case of (C)SCF this effect is weaker though, as it seeks for alternatives by design (cf. diversity principle from Sect. 3.3).

5.3   Effect of the Action Positions on Achieving the Counterfactual

Lastly, looking at Fig. 6 we can see how each action affects the target class probability of accept in the *Adult Census* with respect to their positional occurrence in a sequence. We again aggregated over *all* computed solutions here. The x-axis denotes the position t in the sequence and the y-axis shows the median probability of the target class based on the black-box and the bootstrapped 95% confidence interval (i.e. 2.5 & 97.5 percentiles).

As we can see, there are some actions that are almost able to switch the class on their own when used (chCapGain, addEdu), whereas the remaining ones only do it later at position two (waitYears, enlist) or three (chWorkHrs, chCapLoss). Based on this, it suggests that some actions are only of supportive nature, whereas others can be seen as the main driver behind class label changes. Additionally, the actions that affect a feature, which was attributed a consequential effect through $\mathcal{G}$ (chCapGain, addEdu, enlist), are the only ones that show a significant difference here. Furthermore, the effect of $\mathcal{G}$ in CSCF is visible. When $\mathcal{G}$ was used, chCapGain changed the class often on position one already, whereas in SCF it was usually not quite possible. Based on this, we can infer that chCapGain at position one in CSCF was typically used when it was able to change the class on its own. Moreover, it suggests that the Education level was already sufficiently high in the input instance, so that chCapGain was able to increase more while benefiting from the discount enough that the costs were kept low. The same might be the reason for enlist (i.e. joining the army),

P. Naumann and E. Ntoutsi

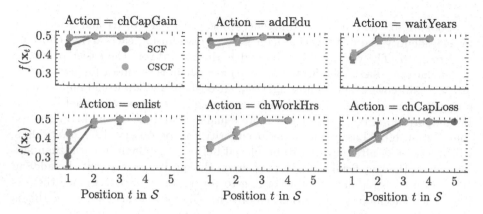

Fig. 6. Median effect with 95% confidence interval of each action at position t in a sequence on the target class (`accept`) probability for *Adult Census*. $f(\mathbf{x}_t) > 0.5$ indicates the class label switched at position t.

which was more able to change the class in CSCF. Since this action affects the `Occupation` feature, the beneficial edge of `Education` might have discounted the cost here again so much that it became a minimal cost sequence, whereas in SCF it might have been more expensive. The slightly lower probability of `addEdu` in CSCF may further suggest that `Education` was more commonly used as a supporting action in order to discount future actions (hence its disappearance after $t = 3$).

In summary, we saw that (C)SCF is able to find more diverse sequences while asserting the least cost objective with comparable or better performance to SYNTH and being more efficient. Furthermore, we demonstrated that the usage of $\mathcal{G}$ in CSCF provides the desired advantage of more meaningful sequence orders, according to the feature relationships, while maintaining minimal costs.

6 Conclusion and Future Work

We proposed a new method CSCF for sequential counterfactual generation that is model-agnostic and capable of finding multiple optimal solution sequences of varying sequence lengths. Our variants, CSCF and SCF, yield better or equivalent results in comparison to SYNTH [19] while being more efficient. Moreover, our extended consequence-aware cost model in CSCF, that considers feature relationships, provides more meaningful sequence orders compared to SCF and SYNTH [19]. In future work, we aim to incorporate causal models to estimate consequential effects in the feature and cost space. Additionally, we want to investigate alternative measures and objectives for evaluating sequence orders and develop a final selection guide for the end user for choosing a sequence from the solution set.

References

1. Bean, J.C.: Genetic algorithms and random keys for sequencing and optimization. ORSA J. Comput. **6**(2), 154–160 (1994)
2. Blank, J., Deb, K.: Pymoo: multi-objective optimization in python. IEEE Access **8**, 89497–89509 (2020)
3. Carlini, N., Wagner, D.: Towards evaluating the robustness of neural networks. In: 2017 IEEE Symposium on Security and Privacy (SP), pp. 39–57, May 2017
4. Dandl, S., Molnar, C., Binder, M., Bischl, B.: Multi-objective counterfactual explanations. In: Bäck, T., et al. (eds.) PPSN 2020. LNCS, vol. 12269, pp. 448–469. Springer, Cham (2020). https://doi.org/10.1007/978-3-030-58112-1_31
5. Downs, M., Chu, J.L., Yacoby, Y., Doshi-Velez, F., Pan, W.: CRUDS: counterfactual recourse using disentangled subspaces. In: ICML WHI 2020, pp. 1–23 (2020)
6. Dua, D., Graff, C.: UCI Machine Learning Repository. University of California, Irvine (2017). http://archive.ics.uci.edu/ml
7. Eiben, A., Smith, J.: Introduction to Evolutionary Computing. Natural Computing Series, Springer, Heidelberg (2015). https://doi.org/10.1007/978-3-662-44874-8
8. Gonçalves, J.F., Resende, M.G.C.: Biased random-key genetic algorithms for combinatorial optimization. J. Heuristics **17**(5), 487–525 (2011)
9. Gower, J.C.: A general coefficient of similarity and some of its properties. Biometrics **27**(4), 857–871 (1971)
10. Joshi, S., Koyejo, O., Vijitbenjaronk, W., Kim, B., Ghosh, J.: Towards Realistic Individual Recourse and Actionable Explanations in Black-Box Decision Making Systems. arXiv:1907.09615, July 2019
11. Karimi, A.H., Barthe, G., Balle, B., Valera, I.: Model-agnostic counterfactual explanations for consequential decisions. In: AISTATS, pp. 895–905. PMLR (2020)
12. Karimi, A.H., Barthe, G., Schölkopf, B., Valera, I.: A survey of algorithmic recourse: definitions, formulations, solutions, and prospects. arXiv:2010.04050 (2020)
13. Karimi, A.H., von Kügelgen, B.J., Schölkopf, B., Valera, I.: Algorithmic recourse under imperfect causal knowledge: a probabilistic approach. In: NeurIPS 33 (2020)
14. Lash, M.T., Lin, Q., Street, N., Robinson, J.G., Ohlmann, J.: Generalized inverse classification. In: SDM 2017, pp. 162–170. SIAM (2017)
15. Laugel, T., Lesot, M.-J., Marsala, C., Renard, X., Detyniecki, M.: Comparison-based inverse classification for interpretability in machine learning. In: Medina, J., et al. (eds.) IPMU 2018. CCIS, vol. 853, pp. 100–111. Springer, Cham (2018). https://doi.org/10.1007/978-3-319-91473-2_9
16. Mahajan, D., Tan, C., Sharma, A.: Preserving Causal Constraints in Counterfactual Explanations for Machine Learning Classifiers. arXiv:1912.03277, June 2020
17. Mothilal, R.K., Sharma, A., Tan, C.: Explaining machine learning classifiers through diverse counterfactual explanations. In: FAT* 2020, pp. 607–617. ACM, January 2020
18. Poyiadzi, R., Sokol, K., Santos-Rodriguez, R., De Bie, T., Flach, P.: FACE: feasible and actionable counterfactual explanations. In: Proceedings of the AAAI/ACM Conference on AI, Ethics, and Society, AIES 2020, pp. 344–350. ACM, February 2020
19. Ramakrishnan, G., Lee, Y.C., Albarghouthi, A.: Synthesizing action sequences for modifying model decisions. In: AAAI, vol. 34, no. 04, pp. 5462–5469 (2020)
20. Shavit, Y., Moses, W.S.: Extracting Incentives from Black-Box Decisions. arXiv:1910.05664, October 2019

21. Srinivas, N., Deb, K.: Muiltiobjective optimization using nondominated sorting in genetic algorithms. Evol. Comput. **2**(3), 221–248 (1994)
22. Ustun, B., Spangher, A., Liu, Y.: Actionable recourse in linear classification. In: Proceedings of the Conference on Fairness, Accountability, and Transparency, pp. 10–19. ACM, January 2019
23. Van Looveren, A., Klaise, J.: Interpretable Counterfactual Explanations Guided by Prototypes. arXiv:1907.02584, July 2019
24. Wachter, S., Mittelstadt, B., Russell, C.: Counterfactual Explanations Without Opening the Black Box: Automated Decisions and the GDPR. SSRN (2017)

Studying and Exploiting the Relationship Between Model Accuracy and Explanation Quality

Yunzhe Jia[1]([✉]), Eibe Frank[1,2], Bernhard Pfahringer[1,2], Albert Bifet[1,3], and Nick Lim[1]

[1] AI Institute, University of Waikato, Hamilton, New Zealand
{ajia,eibe,bernhard,abifet,nlim}@waikato.ac.nz
[2] Department of Computer Science, University of Waikato, Hamilton, New Zealand
[3] LTCI, Télécom Paris, IP Paris, Palaiseau, France

Abstract. Many explanation methods have been proposed to reveal insights about the internal procedures of black-box models like deep neural networks. Although these methods are able to generate explanations for individual predictions, little research has been conducted to investigate the relationship of model accuracy and explanation quality, or how to use explanations to improve model performance. In this paper, we evaluate explanations using a metric based on area under the ROC curve (AUC), treating expert-provided image annotations as ground-truth explanations, and quantify the correlation between model accuracy and explanation quality when performing image classifications with deep neural networks. The experiments are conducted using two image datasets: the CUB-200-2011 dataset and a Kahikatea dataset that we publish with this paper. For each dataset, we compare and evaluate seven different neural networks with four different explainers in terms of both accuracy and explanation quality. We also investigate how explanation quality evolves as loss metrics change through the training iterations of each model. The experiments suggest a strong correlation between model accuracy and explanation quality. Based on this observation, we demonstrate how explanations can be exploited to benefit the model selection process—even if simply maximising accuracy on test data is the primary goal.

Keywords: Interpretability · Explainability · Explanation quality

1 Introduction

Interpretability is considered an important characteristic of machine learning models, and it can be as crucial as accuracy in domains like medicine, finance, and criminal analysis. Recently, many methods [19,21,22,25,26] have been proposed to generate visual explanations for deep neural networks. Since both model accuracy and explanations are relevant for many practical applications of deep neural networks, it is important to study the relationship between them. This is challenging because (1) the lack of ground truth for explanations makes it difficult to quantify their quality—the evaluation of explanations is generally considered subject to users' visual judgement—and (2) no

---

Electronic supplementary material The online version of this chapter (https://doi.org/10.1007/978-3-030-86520-7_43) contains supplementary material, which is available to authorized users.

N. Oliver et al. (Eds.): ECML PKDD 2021, LNAI 12976, pp. 699–714, 2021.
https://doi.org/10.1007/978-3-030-86520-7_43

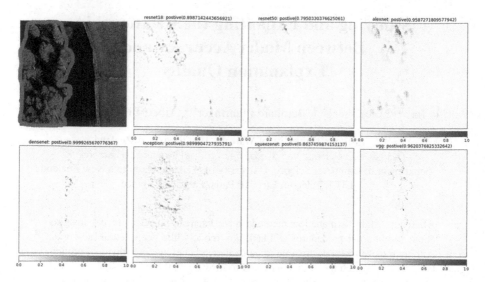

Fig. 1. Demonstration of different explanations by different models making the same prediction. From top-left to bottom-right: the original image with Kahikatea highlighted in the red region, the explanations from ResNet18, ResNet50, AlexNet, DenseNet, InceptionV3, SqueezeNet, and VGG11 generated by the Guided GradCAM explainer.

universal measurement has been agreed upon to evaluate explanations. Our work aims to address this problem and provide empirical results studying the correlation between model accuracy and explanation quality. Based on the observation that model accuracy and explanation quality are correlated, we examine a new model selection criterion combining both model accuracy and explanation quality on validation data.

1.1 Why It Is Important to Evaluate the Quality of Explanations?

Intuitively, a model that achieves competitive predictive performance and makes decisions based on reasonable evidence is better than one that achieves the same level of accuracy but makes decisions based on circumstantial evidence. Given a mechanism for extracting an explanation from a model, we can investigate what evidence the model uses for generating a particular prediction. If we consider the explanation to be of high quality if it is based on reasonable evidence and of low quality otherwise, we can attempt to use explanation quality to inform selection of an appropriate model.

An example of comparing models from the perspective of explanations is shown in Fig. 1. Given an image containing Kahikatea trees—a species of coniferous tree that is endemic to New Zealand—seven deep neural networks (ResNet-18, ResNet-50, AlexNet, DenseNet, Inception-V3, SqueezeNet and VGG11) correctly flag the presence of this type of tree in the image, but the explanations are different. For this particular image, and the visual explanations generated by Guided GradCAM [21] that are shown in the figure, one can argue that the explanations obtained from AlexNet, Inception-V3, ResNet-18, and VGG-11 are more reasonable than those from other models because they more closely align with the part of the image containing the species of tree (the red area marked in the photo).

If we are able to quantify the quality of an explanation, we can define a score for a model f with respect to both accuracy and explanation quality as

$$score(f) = \alpha \cdot score_{acc}(f) + (1 - \alpha) \cdot score_{explanation}(f) \qquad (1)$$

and use it for model selection instead of plain accuracy. Here, $score_{acc}$ represents the model performance in terms of accuracy, $score_{explanation}$ represents the model performance in terms of explanation quality, and $\alpha \in [0, 1]$ is a user-specified parameter.

In this paper, we propose a mechanism to measure the quality of explanations $score_{explanation}$ based on area under the ROC curve and perform a large number of experiments to test the hypothesis that model accuracy is positively correlated with explanation quality because a model tends to be accurate when it makes decisions based on reasonable evidence. Although some recent work [1,4] makes use of this intuition, there is no theoretical or empirical proof to support the claim. Our work makes a complementary contribution aimed to close this gap by providing empirical evidence for the relationship between model accuracy and explanation quality. We hope this will boost future research on how to use explanations to improve accuracy. As a first step in this direction, we use Eq. (1) as the selection criterion to choose deep image classification models from the intermediate candidates that are available at different epochs during the training process. The results show that the models chosen by considering the quality of explanations are consistently better than those chosen based on predictive accuracy alone—in terms of both accuracy and explanation quality on test data.

The main contributions of our work are:

- We show how to use a parameter-free AUC-based metric to evaluate explanation quality based on expert-provided annotations.
- We investigate the relationship between model accuracy and explanation quality by empirically evaluating seven deep neural networks and four explanation methods.
- We demonstrate that explanations can be useful for model selection especially when the validation data is limited.
- We publish a new Kahikatea image dataset together with expert explanations for individual images.

2 Background and Related Work

We first review work on explainability in neural networks and existing publications that consider the evaluation of explanation quality.

2.1 Explainability in Neural Networks

Early research on explainability of neural networks constructed a single tree to mimic the behaviour of a trained network [5] and uses the interpretable tree to explain the network. In contrast, recent research focuses on extracting explanations for individual predictions and can be categorized into two types of approaches: perturbation-based methods and gradient-based ones.

Perturbation-based methods generate synthetic samples of a given input and then extract explanations from the synthetic vicinity of the input. LIME and its variations

[19,20] train a local interpretable model (a linear model or anchors) from the perturbations. The approaches in [3,16] compute Shapley values based on perturbations to represent explanations, and KernelSHAP [16] estimates Shapley values with the LIME framework.

Gradient-based methods aim to estimate the gradient of a given input with respect to the target output or a specific layer, and visualize the gradient as an explanation. Saliency [23] generate gradients by taking a first-order Taylor expansion at the input layer. Backpropagation [23] and Guided Backpropagation [25] are proposed to compute the gradients of the input layer with respect to the prediction results. Class Activation Mapping (CAM) [31] and its variants Gradient-weighted Class Activation Mapping (GradCAM) and Guided Gradient-weighted Class Activation Mapping (GuidedGradCAM) [21] produce localization maps in the last intermediate layer before the output layer using gradients with respect to a specific class label. While perturbation-based methods are usually model-agnostic and can be applied for any model, gradient-based methods are often used in neural networks. A detailed discussion and comparison can be found in [7,17].

2.2 Evaluating Explanation Quality

Evaluating explanation quality is a challenging problem and, to the best of our knowledge, there is no universally recognized metric for this, mainly due to the variety of representations used for explanations. Manual evaluation [12,19,21] is commonly used for image explanations. However, evaluation by simple visual inspection is subject to potential bias [2]. In contrast, [31] computes top-1 error and top-5 error of the image segments generated by explanations provided by class activation mapping (CAM) technique. The publications on LIME [19] and LEAP [11] calculate precision, recall and F_1 score to measure explanation quality. The work in [6,15] converts the problem of generating image explanations to the problem of weakly-supervised object detection and adopts the Intersection over Union (IOU) metric that is used in object detection. All of these methods suffer from the problem that a user-specified threshold or trade-off parameter is implicitly assumed in the metrics they employ: top-N error, F-measure, and IOU. In this paper, we adopt a metric based on Area Under the ROC Curve (AUC) to evaluate explanation quality, which takes into account false positive and true positive rate for all possible thresholds, and perform an extensive empirical evaluation based on a large set of explanation methods and models.

3 Definitions

We first give definitions of key concepts that are used in this paper, focusing on the context of image classification.

Definition 1. *Given an image input x of size (M, N) and a model f, the explanation e for the prediction $f(x)$ is represented as a two dimensional array of the same size (M, N), where each entry in e is a real number and provides the attribution of the corresponding pixel in x.*

Definition 2. *Given an image input x of size (M, N) and a model f, an explainer is a procedure that takes x and f as inputs and returns an explanation e for the prediction $f(x)$.*

Fig. 2. Example of an explanation generated by GuidedGradCAM

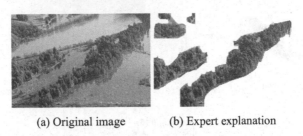

(a) Original image   (b) Expert explanation

Fig. 3. Example of expert explanation in Kahikatea dataset

An example of an explanation is given in Fig. 2. Given the image input shown on the left of Fig. 2, and a trained Resnet18 model [8], which makes the prediction that the input contains Kahikatea trees, the explanation for this prediction is given on the right of Fig. 2. In this example, the explanation is extracted using the explainer GuidedGrad-CAM and is visualized as a heat map.

Definition 3. *Given an image input x of size (M, N), the expert explanation e_{true} for x is an image of the same size and contains a subset of pixels of x. The pixels in x are present in e_{true} if and only if these pixels are selected by an expert based on their domain knowledge.*

The expert explanations for the Kahikatea dataset introduced in this paper are obtained by domain experts selecting the pixels that are part of Kahikatea trees; an example is given in Fig. 3. We also use a second dataset for our experiments in this paper: CUB-200-2011 [29]. The expert explanations for this dataset are extracted from the bounding box information that covers the locations of objects; an example is given in Fig. 4. Note that the expert explanations in this latter dataset may not consist exclusively of relevant information; however, crucially, all relevant object-specific information is included in the bounding box.

4   Investigating the Relationship Between Model Accuracy and Explanation Quality

We now discuss the experimental procedure used in our experiments to test the hypothesis that model accuracy and explanation quality are strongly related.

(a) Original image   (b) Expert explanation

Fig. 4. Example of expert explanation in CUB-200-2011 dataset

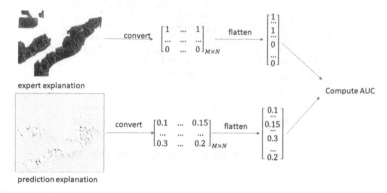

Fig. 5. Process for evaluating the explanation quality

4.1 Evaluating Explanation Quality

The key step in investigating the relationship between model accuracy and explanation quality is to quantitatively evaluate the quality of explanations. In this paper, we compute the Area under the ROC Curve (AUC) to quantify explanation quality because AUC is scale invariant and threshold invariant.

Given an image annotated with an expert explanation, and an explanation heat map generated by an explainer for the prediction of an image by a model, we compute AUC as follows (the procedure is illustrated in Fig. 5):

- Step 1: Convert the expert explanation e_{true} to a binary two-dimensional matrix e_{true}^{binary}, where each entry corresponds to a pixel in the image. The binary value is set to 1 if the corresponding pixel is selected in the expert explanation provided for the image, and to 0 otherwise.
- Step 2: Convert the prediction explanation to a two-dimensional matrix e, where each entry is the attribution of the corresponding pixel. The attributions are generated by an explainer and are normalized into the range [0, 1].
- Step 3: Flatten both e_{true}^{binary} and e into one-dimensional vectors.
- Step 4: Compute AUC using the e_{true}^{binary} and e vectors.

To show the benefit of using AUC, Fig. 6 shows the comparison of three metrics: precision, recall and AUC. These metrics are computed for the same generated explanation shown in Fig. 6c. To compute precision and recall, a threshold is required to

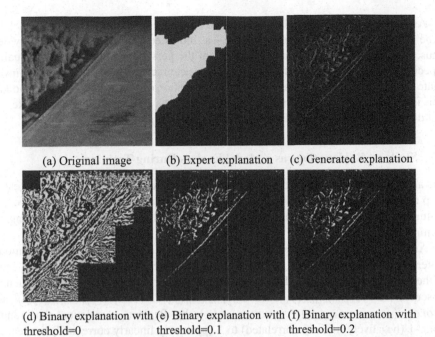

(a) Original image   (b) Expert explanation   (c) Generated explanation

(d) Binary explanation with (e) Binary explanation with (f) Binary explanation with
threshold=0   threshold=0.1   threshold=0.2

Fig. 6. Comparison of different evaluation metrics. The AUC for (c) is 0.64. (d)–(f) are the binary explanations that are converted from (c) with different thresholds. Precision and recall for the binary explanations are as follows: (d) precision = 0.29, recall = 0.48, (e) precision = 0.66, recall = 0.11, and (f) precision = 0.64, recall = 0.04

convert the generated explanation to a binary representation, thus these metrics change as different thresholds are applied, while AUC is threshold invariant.

In our experiments, we evaluate the quality of explanations generated by four different explainers from the literature, namely, Saliency [23], BackPropagation [25], GuidedGradCAM [21], and GradientShap [16].

4.2   Comparing Different Models

To study the interaction between explainer algorithm and deep neural network that it is applied to, we evaluate each of the four explainers for seven deep neural networks, namely, AlexNet [14], DenseNet [9], Inception-V3 [28], ResNet-18 [8], ResNet-50 [30], SqueezeNet [10], and VGG-11 [24].

We consider accuracy $score_{acc}$, loss, and explanation quality $score_{explanation}$. The primary metric of predictive performance is $score_{acc}$, which is calculated as the portion of correctly classified test instances, but we also report the results of test losses in the experiments (entropy losses are used for the neural networks in this paper). $score_{explanation}$ is calculated as the average AUC obtained across all test images.

As illustrated in the example in Fig. 1, it is clear that different models may reach the same decision based on different evidence—as indicated by the explanations provided. Thus, it is important to compare models from the perspective of explanation quality, especially when the models achieve comparable accuracy. However, we can also investigate the correlation between accuracy and explanation quality across different models. This is important to test the generality of our hypothesis that a model tends to make correct decisions when its prediction explanations are of high quality.

4.3 Studying Explanations as a Model Evolves During Training

It is also of interest to consider how explanations evolve during the training process of deep neural networks. To this end, instead of just comparing different models, we also evaluate the explanation quality obtained with a model at different epochs during the training process.

Assume a model f is trained with T iterations, and let f_t be the intermediate model at iteration t, $score_{explanation}(f_t)$ be the explanation metric for f_t, and $score_{acc}(f_t)$ be the accuracy metric for f_t. Then, we compute the correlation between the sequences of scores $[score_{explanation}(f_1), score_{explanation}(f_2), \ldots, score_{explanation}(f_T)]$ and $[score_{acc}(f_1), score_{acc}(f_2), \ldots, score_{acc}(f_T)]$ by Pearson correlation that ranges from -1 (negatively linearly correlated) to 1 (positively linearly correlated) to measure the strength of the statistical relationship.

If these two sequences are correlated, it means that the two tasks—learning accurate classifications and learning accurate explanations—are related. This would provide some empirical justification for multi-task learning frameworks [1,4] that jointly learn classifications and explanations.

4.4 Selecting Models Based on Explanation Quality

Assuming model accuracy is positively correlated with explanation quality, it is natural to consider whether we can choose models based on explanation quality. In the traditional model selection process, we choose the model that achieves best performance on validation data and hope it also performs well on test data or unseen data. This framework usually works if we have a sufficient amount of validation data. However, if the validation data is limited, a model that performs well on this data will not necessarily generalise well. In this case, it is worth considering whether the explanation quality on the validation data (or part of validation data) can be taken into consideration to inform model selection and thus choose a potentially better model.

A toy example, considering decision trees rather than neural networks, is given in Fig. 7. Assume there are two trained models (Fig. 7b and Fig. 7c) that perform equally well on the validation data (Fig. 7a). It is unclear which model will achieve better predictive accuracy on new data. However, if based on input from domain experts, we know that features F_1 and F_2 reflect the actual causes for the class label, then we can say that model 1 is better than model 2 because its explanation quality is better.

We explore this idea of using explanation quality for model selection in a case study with deep neural networks applied to image classification in Sect. 5.5.

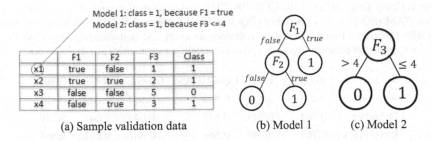

(a) Sample validation data   (b) Model 1   (c) Model 2

Fig. 7. Demonstration of how explanations help to choose a better model. Both models achieve the same accuracy on the validation data. Assuming expert knowledge that $F1$ and $F2$ are the actual causes for the class label, we can say that model 1 is better than model 2 as the explanation quality of model 1 is better.

5   Experimental Evaluation

We now discuss the empirical results obtained in our experiments, providing more detail on the two datasets used and the hardware and software set-up employed.

5.1   Data

The experiments are conducted on two datasets: the CUB-200-2011 dataset and the Kahikatea dataset. CUB-200-2011 [29] contains 11,788 images (5,994 for training and 5994 for testing) in 200 categories. The images are annotated with bounding boxes revealing the locations of objects, from where the expert explanations are extracted (see Fig. 4). The Kahikatea data contains 534 images (426 for training and 108 for testing) in two categories, and the classification problem is to predict whether an image contains Kahikatea or not. The expert explanations are generated by domain experts manually selecting the pixels that belong to Kahikatea trees (see Fig. 3). We publish the Kahikatea dataset with this paper, and the data can be found at https://doi.org/10.5281/zenodo.5059768.

5.2   Implementation

Our experiments use PyTorch [18] for training the neural networks and Captum [13] for implementations of the explainers. An NVIDIA GeForce RTX 2070 GPU and an Intel(R) Core(TM) i7-10750H CPU with 16 GB of memory are used as hardware platform. All neural networks are trained using the cross-entropy loss function with the SGD [27] optimizer using batch size = 16, learning rate = 0.001 and momentum = 0.9, while all explainers are applied with default parameters. The models are trained with 50 epochs as their performance becomes stable afterwards on the datasets we consider.

5.3   Results - Comparing Different Models

We first compare the models in terms of both classification performance and explanation quality. The accuracy and loss on the test data for the CUB-200-2011 dataset

Table 1. Comparing models on the CUB-200-2011 data. Explanation quality is shown for GuidedGradCAM (EQ-GGCAM), Saliency (EQ-SA), GradientShap (EQ-GS), and BackPropagation (EQ-BP). Corr(Acc): Pearson Correlation between accuracy and explanation quality; Corr(Loss): Pearson Correlation between loss and explanation quality. Best metrics are shown in bold.

Model	Accuracy	Loss	EQ-GGCAM	EQ-SA	EQ-GS	EQ-BP
AlexNet	0.495	3.326	0.508	0.507	0.622	0.521
DenseNet	0.764	0.943	0.649	0.641	0.641	**0.710**
Inception-V3	**0.765**	0.949	0.662	0.661	0.632	0.661
ResNet-18	0.705	1.141	0.622	0.681	0.644	0.657
ResNet-50	0.758	**0.934**	**0.681**	**0.687**	0.637	0.687
SqueezeNet	0.614	2.090	0.678	0.643	0.644	0.676
VGG-11	0.717	1.371	0.526	0.522	**0.650**	0.591
Corr(Acc)	-	-	0.563	0.620	0.516	**0.747**
Corr(Loss)	-	-	−0.606	−0.691	−0.566	**−0.784**

are reported in the second and third column in Table 1. The remaining columns in the table detail the quality of the explanations generated by the four explainers, measured using AUC. The correlation between accuracy and explanation quality and the correlation between loss and explanation quality across these models for each of the four explainers are reported in the last two rows. Similar results are shown in Table 2 for the Kahikatea dataset.

For both datasets, it can be seen that model accuracy is positively correlated with explanation quality, while the loss is negatively correlated with explanation quality. However, it is also worth noting that the model achieving the highest accuracy is not necessarily the model achieving the best explanation quality. For example, for the CUB-200-2011 dataset, the Inception-V3 model achieves the highest accuracy, but its explanation quality is not the best one using any of the explainers—in fact, the ResNet-50 explanations always achieve a better score. This observation highlights the fact that it may not be advisable to solely rely on accuracy when selecting models in some cases.

5.4 Results - Studying a Model at Different Iterations During Training

We now investigate the relationship between accuracy and explanation quality for the intermediate models obtained during the training process. Each model is trained with 50 iterations, which generates 50 intermediate models (including the last iteration). We compute the accuracy, loss, and explanation quality from four explainers for every intermediate model. For all intermediate models, we get an accuracy vector of size 50, a loss vector of size 50, and four explanation quality vectors of size 50. Then, we calculate the correlations between the accuracy vector and each explanation quality vector, and the correlations between each loss vector and each explanation quality vector.

The results for the CUB-200-2011 and Kahikatea datasets are reported in Table 3 and Table 4 respectively. It can be seen that during the training process of all seven models, the accuracy is positively correlated with the explanation quality, and the loss

Table 2. Comparing models on the Kahikatea data. Explanation quality is shown for Guided-GradCAM (EQ-GGCAM), Saliency (EQ-SA), GradientShap (EQ-GS), and BackPropagation (EQ-BP). Corr(Acc): Pearson Correlation between accuracy and explanation quality; Corr(Loss): Pearson Correlation between loss and explanation quality. Best metrics are shown in bold.

Model	Accuracy	Loss	EQ-GGCAM	EQ-SA	EQ-GS	EQ-BP
AlexNet	0.926	0.199	0.517	0.542	0.554	0.470
DenseNet	**0.981**	**0.072**	**0.615**	**0.587**	**0.619**	0.611
Inception-V3	0.954	0.166	0.526	0.519	0.532	0.472
ResNet-18	0.954	0.137	0.518	0.554	0.563	0.563
ResNet-50	0.972	0.137	0.545	0.566	0.570	0.617
SqueezeNet	0.935	0.227	0.536	0.538	0.558	0.525
VGG-11	0.963	0.118	0.587	0.580	0.600	**0.631**
Corr(Acc)	-	-	0.738	0.698	0.669	**0.787**
Corr(Loss)	-	-	−0.764	**−0.791**	−0.770	−0.731

Table 3. Results - studying models during training with the CUB-200-2011 dataset. Corr(A): Pearson Correlation between accuracy and explanation quality; Corr(L): Pearson Correlation between loss and explanation quality. Best metrics are shown in bold.

Model	GuidedGradCAM		Saliency		GradientShap		BackPropagation	
	Corr(A)	Corr(L)	Corr(A)	Corr(L)	Corr(A)	Corr(L)	Corr(A)	Corr(L)
AlexNet	0.707	−0.857	0.842	−0.790	0.827	−0.856	0.786	−0.652
DenseNet	0.840	−0.816	0.903	−0.908	0.759	−0.832	0.738	−0.705
Inception-V3	0.507	−0.603	0.758	−0.802	0.585	−0.661	**0.954**	**−0.934**
ResNet-18	0.673	−0.860	0.211	**−0.952**	0.782	−0.949	0.920	−0.923
ResNet-50	**0.921**	**−0.891**	0.891	−0.867	**0.974**	**−0.962**	0.905	−0.880
SqueezeNet	0.917	−0.708	**0.970**	−0.875	0.933	−0.743	0.872	−0.900
VGG-11	0.875	−0.476	0.701	−0.451	0.934	−0.773	0.637	−0.671

is negatively correlated with the explanation quality. This validates our intuition that the explanation quality improves as the accuracy increases.

5.5 Using Explanations for Model Selection

We now proceed to a case study[1] where we investigate whether explanations can be used to improve the model selection performance in the Kahikatea problem under the assumption that the validation data is limited.

Given training and validation data, in the traditional model selection setting, candidate models (i.e., different models structures, identical model structures trained with

---

[1] The code and supplementary material are available at https://bit.ly/3xdcrwS.

Table 4. Results - studying models during training with the Kahikatea dataset. Corr(A): Pearson Correlation between accuracy and explanation quality; Corr(L): Pearson Correlation between loss and explanation quality. Best metrics are shown in bold.

Model	GuidedGradCAM		Saliency		GradientShap		BackPropagation	
	Corr(A)	Corr(L)	Corr(A)	Corr(L)	Corr(A)	Corr(L)	Corr(A)	Corr(L)
AlexNet	0.507	**−0.602**	0.585	**−0.689**	0.530	−0.520	0.646	**−0.585**
DenseNet	**0.510**	−0.548	0.493	−0.427	0.550	−0.612	0.461	−0.423
Inception-V3	0.358	−0.421	0.475	−0.526	**0.780**	**−0.710**	0.576	−0.551
ResNet-18	0.423	−0.350	0.659	−0.460	0.706	−0.548	**0.801**	−0.562
ResNet-50	**0.510**	−0.454	0.499	−0.571	0.391	−0.311	0.394	−0.493
SqueezeNet	0.478	−0.281	0.498	−0.387	0.415	−0.535	0.498	−0.421
VGG-11	0.417	−0.511	**0.663**	−0.469	0.655	−0.384	0.722	−0.521

different hyper-parameters, or intermediate models from different training stages) are obtained on the training data, and the model that achieves the best performance in terms of accuracy or loss on the validation data is selected to later be applied on test data or unseen data.

Instead of using the accuracy metric as the selection criterion, we use $score(f) = \alpha \cdot score_{acc}(f) + (1 - \alpha) \cdot score_{explanation}(f)$ (see Eq. (1)), such that the model with the best $score(f)$ on the validation data is selected. This selection criterion is based on our previous observation that explanation quality and model accuracy are strongly correlated. $score_{explanation}$ can be viewed as a regularization term regarding explainability, and it helps to reduce variance and avoid overfitting by choosing models that make decisions based on reasonable evidence.

It is worth noting that in the case when $\alpha = 1$, the selection criterion only relies on accuracy, which is the way traditional model selection makes its choice, whilst in the case when $\alpha = 0$, the selection criterion only relies on explainability.

Given the Kahikatea dataset and a deep neural network model structure, we perform the following steps:

- Step 1: Randomly divide the Kahikatea dataset into three subsets such that 20% of the samples are for training, 10% are for validation, and the remaining 70% are for testing.
- Step 2: Train the model on the training data for $N = 50$ iterations to generate 50 model candidates $f_1, f_2, \ldots, f_N$.
- Step 3: Compute $score(f)$ in Eq. (1) on the validation data for $f_i, i \in [1, 2, \ldots, N]$, where $score_{acc}(f_i)$ is calculated as the percentage of correct predictions of f_i on the validation data, $score_{explanation}(f_i)$ is calculated using the AUC-based metric (see Sect. 4.1) with expert explanations for the validation data and model explanations generated with GuidedGradCAM.
- Step 4: Compute test accuracy (percentage of correct predictions on the test data) $Acc_{test}(f)$ for $f_i, i \in [1, 2, \ldots, N]$.

- Step 5: Calculate the Pearson correlation between $(score(f_1), \ldots, score(f_N))$ and $(Acc_{test}(f_1), \ldots, Acc_{test}(f_N))$. The correlation is 1 if the ranking of the candidate models based on $score(f)$ is the same as their ranking based on test accuracy.
- Step 6: Repeat step 1–5 for 10 times and compute the average correlation.

The procedure is applied on seven deep neural networks (AlexNet, DenseNet, Inception-V3, ResNet-18, ResNet-50, SqueezeNet and VGG-11) and α is varied from the list $(0, 0.1, 0.2, 0.3, 0.4, 0.5, 0.6, 0.7, 0.8, 0.9, 1.0)$ to cover both extreme cases.

The correlation between the scores of the selection criterion and test accuracy is reported in Table 5. It can be seen that, for all models, the highest correlations are achieved when α is neither 0 nor 1, which suggests that a combination of validation set accuracy and explanation quality is best for model selection.

Table 5. Results - Correlations of selection scores and test accuracy for different α. Best metrics for each model are shown in bold.

α	AlexNet	DenseNet	Inception-V3	ResNet-18	ResNet-50	SqueezeNet	VGG-11
0	0.5335	0.5076	0.4602	0.5914	0.4281	0.4448	0.6011
0.1	0.6006	0.5996	0.6393	0.6734	0.5345	0.5426	0.678
0.2	0.6629	0.6807	0.7627	0.7372	0.6289	0.6348	0.7438
0.3	0.717	0.7466	0.8332	0.7838	0.7057	0.7073	0.7947
0.4	0.7587	0.7964	0.8666	0.8156	0.7633	0.7578	0.8305
0.5	0.7844	0.8311	0.88	0.8354	0.8038	0.7899	0.8531
0.6	**0.7929**	0.8533	**0.8834**	0.8458	0.8305	0.8077	0.8653
0.7	0.7873	0.8655	0.882	**0.8494**	0.8465	0.8152	**0.8694**
0.8	0.7734	**0.8703**	0.8783	0.8479	0.8545	**0.8153**	0.8678
0.9	0.7561	0.8698	0.8735	0.8433	**0.8568**	0.8106	0.8619
1.0	0.7378	0.8657	0.8684	0.8366	0.8553	0.8028	0.8532

The comparison of the test accuracy of models selected with explanations and that of the models selected without explanations is shown in Table 6. When $\alpha = 1$, it is the case that we choose the models without considering explanation quality. Besides the test accuracy, the explanation quality on the test data of models selected with explanations is consistently better than that of models selected without explanations (see the supplementary material). It can be seen that the models selected by consulting explanation quality consistently outperform the models (except for SqueezeNet) selected using accuracy alone. It also shows that we cannot simply optimize the explanation quality (when $\alpha = 0$), and one possible reason is that the expert explanations can be biased and noisy.

Do We Need Expert Explanations for All Validation Data? It is interesting to consider how many instance-level expert explanations are sufficient to improve the performance of model selection if these are not available for the whole validation set.

Table 6. Results - Test accuracy of models selected using selection criterion with different α. Best metrics for each model are shown in bold.

α	AlexNet	DenseNet	Inception-V3	ResNet-18	ResNet-50	SqueezeNet	VGG-11
0	79.95%	81.37%	76.87%	**83.08%**	82.64%	77.28%	83.71%
0.1	79.97%	81.07%	77.28%	82.99%	82.61%	80.0%	83.71%
0.2	79.97%	82.39%	82.01%	82.8%	82.99%	80.96%	84.59%
0.3	82.01%	82.42%	82.34%	82.69%	83.1%	81.59%	**84.78%**
0.4	**82.06%**	82.17%	82.34%	82.99%	83.38%	81.18%	84.64%
0.5	**82.06%**	82.09%	82.23%	82.99%	**83.68%**	81.92%	84.64%
0.6	**82.06%**	82.99%	82.34%	82.94%	83.63%	81.81%	84.09%
0.7	**82.06%**	**83.24%**	**82.5%**	82.99%	83.49%	81.43%	84.42%
0.8	**82.06%**	**83.24%**	**82.5%**	82.99%	83.49%	81.54%	84.45%
0.9	81.54%	**83.24%**	**82.5%**	82.99%	83.49%	81.54%	84.45%
1.0	81.18%	82.42%	82.2%	82.72%	83.43%	**82.01%**	84.42%

We follow the setting described above and vary the availability of expert explanations from 10% to 100% of the validation set. The model selection result (test accuracy) for AlexNet with the GuidedGradCAM explainer is shown in Table 7. It can be seen that even with 10% expert explanations it is possible to improve the model selection performance. The results of other neural network structures (see the supplementary material) follow a similar trend: availability of expert explanations for 10% of the validation data (or more) can benefit the model selection process for the selected dataset.

Table 7. Results - Test accuracy of models selected with different percentages of expert explanations. Best results for each model are shown in bold.

α	Level of expert explanation availability									
	10%	20%	30%	40%	50%	60%	70%	80%	90%	100%
0	82.3%	81.8%	81.8%	82.0%	81.9%	81.3%	82.0%	80.7%	**82.6%**	79.9%
0.1	82.3%	81.8%	81.8%	82.0%	81.7%	81.4%	82.0%	80.7%	**82.6%**	80.0%
0.2	82.3%	82.0%	81.8%	82.0%	81.7%	81.4%	82.0%	80.7%	82.3%	80.0%
0.3	82.3%	82.0%	82.1%	82.0%	81.7%	81.4%	82.0%	81.3%	82.3%	82.0%
0.4	82.1%	82.1%	82.2%	82.0%	82.0%	82.0%	82.0%	82.0%	82.3%	**82.1%**
0.5	82.1%	82.1%	82.4%	82.0%	82.0%	82.1%	82.0%	82.2%	82.3%	**82.1%**
0.6	82.1%	82.1%	82.4%	**82.2%**	81.9%	82.4%	82.1%	82.2%	82.4%	**82.1%**
0.7	82.2%	**82.3%**	82.5%	**82.2%**	82.1%	**82.6%**	82.1%	82.2%	82.4%	**82.1%**
0.8	**82.4%**	81.8%	**82.6%**	81.8%	81.8%	**82.6%**	82.2%	**82.3%**	82.4%	**82.1%**
0.9	82.0%	81.9%	82.4%	81.8%	81.3%	**82.6%**	82.2%	81.9%	81.9%	81.5%
1.0	81.2%	81.2%	81.2%	81.2%	81.2%	81.2%	81.2%	81.2%	81.2%	81.2%

6   Conclusion

We empirically evaluate the relationship between model accuracy and explanation quality using seven deep neural networks and four explainers. To evaluate explanation quality, we adopt the Area under the ROC Curve (AUC), which is threshold invariant. The experimental results indicate that models tend to make correct predictions when these predictions are accompanied by explanations of high quality. Moreover, during a model's training process, predictive accuracy increases together with explanation quality. Our results provide strong empirical support for the claim that model accuracy and explanation quality are correlated. Exploiting this observation, we demonstrate how measuring the quality of explanations can help to improve the performance of model selection and also consider how this is affected by the number of available expert-provided explanations. To boost research in this area, we publish the Kahikatea dataset, which provides instance-level expert explanations for positive instances.

Acknowledgments. This work is partially supported by the TAIAO project (Time-Evolving Data Science / Artificial Intelligence for Advanced Open Environmental Science) funded by the New Zealand Ministry of Business, Innovation, and Employment (MBIE). URL https://taiao.ai/. This work is partially supported by Microsoft AI for Earth Grants for providing cloud computing resource. We thank Waikato Regional Council (WRC) staff and students that have worked on the Kahikatea project along with further assistance from Rebecca Finnerty and Paul Dutton for the Kahikatea dataset.

References

1. Avinesh, P., Ren, Y., Meyer, C.M., Chan, J., Bao, Z., Sanderson, M.: J3R: joint multi-task learning of ratings and review summaries for explainable recommendation. In: ECML-PKDD, pp. 339–355 (2019)
2. Buçinca, Z., Lin, P., Gajos, K.Z., Glassman, E.L.: Proxy tasks and subjective measures can be misleading in evaluating explainable AI systems. In: IUI, pp. 454–464 (2020)
3. Castro, J., Gómez, D., Tejada, J.: Polynomial calculation of the Shapley value based on sampling. Comput. Oper. Res. 36(5), 1726–1730 (2009)
4. Chen, Z., Wang, X., Xie, X., Wu, T., Bu, G., Wang, Y., Chen, E.: Co-attentive multi-task learning for explainable recommendation. In: IJCAI, pp. 2137–2143 (2019)
5. Craven, M., Shavlik, J.: Extracting tree-structured representations of trained networks. In: NIPS, pp. 24–30 (1995)
6. Guidotti, R.: Evaluating local explanation methods on ground truth. Artif. Intell. **291**, 103428 (2021)
7. Guidotti, R., Monreale, A., Ruggieri, S., Turini, F., Giannotti, F., Pedreschi, D.: A survey of methods for explaining black box models. ACM Comput. Surv. **51**(5), 1–42 (2018)
8. He, K., Zhang, X., Ren, S., Sun, J.: Deep residual learning for image recognition. In: CVPR, pp. 770–778 (2016)
9. Huang, G., Liu, Z., Van Der Maaten, L., Weinberger, K.Q.: Densely connected convolutional networks. In: CVPR, pp. 4700–4708 (2017)
10. Iandola, F.N., Han, S., Moskewicz, M.W., Ashraf, K., Dally, W.J., Keutzer, K.: SqueezeNet: AlexNet-level accuracy with 50x fewer parameters and < 0.5 MB model size. arXiv:1602.07360 (2016)

11. Jia, Y., Bailey, J., Ramamohanarao, K., Leckie, C., Houle, M.E.: Improving the quality of explanations with local embedding perturbations. In: KDD, pp. 875–884 (2019)
12. Kim, B., et al.: Interpretability beyond feature attribution: quantitative testing with concept activation vectors (TCAV). In: ICML, pp. 2668–2677 (2018)
13. Kokhlikyan, N., et al.: Captum: a unified and generic model interpretability library for PyTorch. arXiv:2009.07896 (2020)
14. Krizhevsky, A.: One weird trick for parallelizing convolutional neural networks. arXiv:1404.5997 (2014)
15. Lin, Y.S., Lee, W.C., Celik, Z.B.: What do you see? Evaluation of explainable artificial intelligence (XAI) interpretability through neural backdoors. arXiv:2009.10639 (2020)
16. Lundberg, S.M., Lee, S.I.: A unified approach to interpreting model predictions. In: NIPS, pp. 4765–4774 (2017)
17. Molnar, C.: Interpretable Machine Learning (2020). Lulu.com
18. Paszke, A., et al.: Pytorch: an imperative style, high-performance deep learning library. In: NeuRIPS, pp. 8026–8037 (2019)
19. Ribeiro, M.T., Singh, S., Guestrin, C.: "Why should I trust you?" Explaining the predictions of any classifier. In: KDD, pp. 1135–1144 (2016)
20. Ribeiro, M.T., Singh, S., Guestrin, C.: Anchors: high-precision model-agnostic explanations. In: AAAI, vol. 18, pp. 1527–1535 (2018)
21. Selvaraju, R.R., Cogswell, M., Das, A., Vedantam, R., Parikh, D., Batra, D.: Grad-CAM: visual explanations from deep networks via gradient-based localization. In: ICCV, pp. 618–626 (2017)
22. Shrikumar, A., Greenside, P., Kundaje, A.: Learning important features through propagating activation differences. In: ICML, pp. 3145–3153 (2017)
23. Simonyan, K., Vedaldi, A., Zisserman, A.: Deep inside convolutional networks: visualising image classification models and saliency maps. arXiv:1312.6034 (2013)
24. Simonyan, K., Zisserman, A.: Very deep convolutional networks for large-scale image recognition. arXiv:1409.1556 (2014)
25. Springenberg, J.T., Dosovitskiy, A., Brox, T., Riedmiller, M.: Striving for simplicity: the all convolutional net. arXiv:1412.6806 (2014)
26. Sundararajan, M., Taly, A., Yan, Q.: Axiomatic attribution for deep networks. In: ICML, pp. 3319–3328 (2017)
27. Sutskever, I., Martens, J., Dahl, G., Hinton, G.: On the importance of initialization and momentum in deep learning. In: ICML, pp. 1139–1147 (2013)
28. Szegedy, C., Vanhoucke, V., Ioffe, S., Shlens, J., Wojna, Z.: Rethinking the inception architecture for computer vision. In: CVPR, pp. 2818–2826 (2016)
29. Wah, C., Branson, S., Welinder, P., Perona, P., Belongie, S.: The Caltech-UCSD Birds-200-2011 Dataset. Technical report, CNS-TR-2011-001, California Institute of Technology (2011)
30. Xie, S., Girshick, R., Dollár, P., Tu, Z., He, K.: Aggregated residual transformations for deep neural networks. In: CVPR, pp. 1492–1500 (2017)
31. Zhou, B., Khosla, A., Lapedriza, A., Oliva, A., Torralba, A.: Learning deep features for discriminative localization. In: CVPR, pp. 2921–2929 (2016)

Explainable Multiple Instance Learning with Instance Selection Randomized Trees

Tomáš Komárek[1,2]($\boxtimes$), Jan Brabec[1,2], and Petr Somol[3]

[1] Czech Technical University in Prague, FEE, Prague, Czech Republic
[2] Cisco Systems, Cognitive Intelligence, Prague, Czech Republic
{tomkomar,janbrabe}@cisco.com
[3] Avast Software, Prague, Czech Republic
petr.somol@avast.com

Abstract. Multiple Instance Learning (MIL) aims at extracting patterns from a collection of samples, where individual samples (called bags) are represented by a group of multiple feature vectors (called instances) instead of a single feature vector. Grouping instances into bags not only helps to formulate some learning problems more naturally, it also significantly reduces label acquisition costs as only the labels for bags are needed, not for the inner instances. However, in application domains where inference transparency is demanded, such as in network security, the sample attribution requirements are often asymmetric with respect to the training/application phase. While in the training phase it is very convenient to supply labels only for bags, in the application phase it is generally not enough to just provide decisions on the bag-level because the inferred verdicts need to be explained on the level of individual instances. Unfortunately, the majority of recent MIL classifiers does not focus on this real-world need. In this paper, we address this problem and propose a new tree-based MIL classifier able to identify instances responsible for positive bag predictions. Results from an empirical evaluation on a large-scale network security dataset also show that the classifier achieves superior performance when compared with prior art methods.

Keywords: Explainable AI · Network security · Randomized trees

1 Introduction

Multiple Instance Learning (MIL) generalizes the traditional data representation as it allows individual data samples $\mathcal{B}_1, \mathcal{B}_2, \ldots$ (called bags) to be represented by a group of multiple d-dimensional feature vectors $\mathcal{B} = \{\mathbf{x}_1, \mathbf{x}_2, \ldots\}$, $\mathbf{x} \in \mathbb{R}^d$ (called instances), which are order independent and their counts may vary across bags. In the supervised classification, it is further assumed that each bag is associated with label y (e.g. $y \in \{-1, +1\}$ in the binary case), and the goal is to infer function $\mathcal{F}$ from dataset $\mathcal{D} = \{(\mathcal{B}, y)_1, (\mathcal{B}, y)_2, \ldots\}$, using algorithm $\mathcal{A}$, such that the function $\mathcal{F}$ can predict labels for new bags $\mathcal{F}(\mathcal{B}) = y$.

This formalism (originally introduced in [9]) has recently gained significant traction in domains dealing with complex data structures and/or labeling limitations [6]. A prime example of such a domain is network security and its problem

© Springer Nature Switzerland AG 2021
N. Oliver et al. (Eds.): ECML PKDD 2021, LNAI 12976, pp. 715–730, 2021.
https://doi.org/10.1007/978-3-030-86520-7_44

of detecting infected users in computer networks. While network traffic logs can be converted into feature vectors relatively easily, their labeling is often very labor-intensive and time-consuming. This is caused not only by the volume of logs that needs to be processed by experienced threat analysts (and can not be outsourced due to privacy issues), but also by the fact that the individual log records may not carry enough discriminatory information for making verdicts about them and a broader context of the communication must be considered by the analysts[1]. Previous works [10,13,18,22] have shown that MIL can greatly facilitate this problem as it enables to: *(i) formulate the problem more naturally:* users can be represented by bags of instances, where the instances correspond to users' communications with web servers; here the flexibility in bag sizes reflects the reality that each user can establish a different number of communications within a given time window, *(ii) reduce the label acquisition costs:* threat analysts do not have to pinpoint individual log records responsible for the infection; it is enough to provide labels for the whole users, *(iii) open new ways of acquiring labels:* since it is sufficient to know whether the user was infected in a particular time period or not, a completely separate source of data can be used for annotating (e.g. anti-virus reports) and thus benefit from cheaper and less ambiguous labels, *(iv) increase classification performance:* MIL classifiers can detect more types of infections as they make decisions by analysing the entire context of user's communications rather than individual log records in isolation.

The goal of this paper is to contribute to the above list of MIL advantages by further enabling to: *(v) explain the raised alerts:* although in the training phase the labels are supplied only for bags, in the application phase the model should be able to explain the raised alerts (positive bag predictions) by promoting instances responsible for the decisions. We argue that this capability is of great need, especially in applications where subsequent acting upon the raised alerts is associated with high costs (e.g. reimaging a user's computer) and therefore the verdicts need to be well justified. Time spent on the justification is usually strongly affected by the order in which the instances are investigated. Most prior approaches can not effectively prioritize instances of positive bags because they perform some sort of bag aggregation inside models to make learning on bags (i.e. sets of vectors) possible. The algorithm proposed in this paper works on an instance selection rather than the aggregation principle.

2   Instance Selection Randomized Trees

The algorithm for learning Instance Selection Randomized Trees (ISRT) follows the standard top-down greedy procedure for building an ensemble of unpruned decision trees. Every tree learner recursively divides the training sample set into two subsets until class homogeneity is reached or the samples can not be divided any further.

---

[1] For example, a seemingly legitimate request to google.com might be in reality related to malicious activity when it is issued by malware checking Internet connection. Similarly, requesting ad servers in low volumes is considered as a legitimate behavior, but higher numbers might indicate Click-fraud infection.

The main difference to the standard (single instance) tree-based learners applies in the way the conditions are evaluated inside the splitting nodes. In the MIL setting, the decision whether to send a sample (i.e. bag) to the left or right branch can no longer be based on a condition of type—if feature f is greater than value v—as the bag might contain multiple feature vectors (i.e. instances) that may or may not fulfill that condition. To cope with this problem, every node of ISRT (denoted as $\mathcal{N}_{\mathrm{ISRT}}$) is further parametrized with vector $\mathbf{w}$, called instance selector, in addition to the feature index f and the threshold value v (Eq. 1). The purpose of the instance selector is to select a single instance $\mathbf{x}^*$ from a bag $\mathcal{B}$ upon which the feature value comparison $x_f^* > v$ is made. The selection mechanism is implemented via calculating the inner product (denoted as $\langle \cdot, \cdot \rangle$) between the vector $\mathbf{w}$ and individual bag instances $\mathbf{x} \in \mathcal{B}$, followed by selecting the instance $\mathbf{x}^*$ associated with the maximum response. Note that if bags are of size one, then ISRT nodes behave like the traditional ones regardless of the extra parameter $\mathbf{w}$.

$$\underbrace{\mathcal{N}_{\mathrm{ISRT}}(\mathcal{B}; f, v, \mathbf{w})}_{\Phi} = \begin{cases} \text{left}, & \text{if } x_f^* > v, \ \mathbf{x}^* = \operatorname*{argmax}_{\mathbf{x} \in \mathcal{B}} \langle \mathbf{w}, \mathbf{x} \rangle, \\ \text{right}, & \text{otherwise}. \end{cases} \qquad (1)$$

Assuming the positive class is the class of interest, we would like to train an instance selector (on a local training subset available to the considered node) to give maximum values to the instances of positive bags (and thus cause their selection) that are most responsible for these bags being positive. More specifically, the selector should assign low (i.e. negative) values to all instances of negative bags and high (i.e. positive) values to at least one instance from each positive bag. We do not force the selector to assign high values to all instances of positive bags, since not all of them are necessarily relevant. For example, not all websites visited by an infected user within the last 24 h are automatically malicious. In fact, the vast majority of them will typically still be legitimate. These requirements lead to the following zero-one loss function for a single training data point $(\mathcal{B}, y)$:

$$\ell_{01}(\mathbf{w}; (\mathcal{B}, y)) = \mathbb{1}\left[y \max_{\mathbf{x} \in \mathcal{B}} \langle \mathbf{w}, \mathbf{x} \rangle < 0 \right], \qquad (2)$$

where $\mathbb{1}[\cdot]$ stands for an indicator function, which equals one if the argument is true and zero otherwise. If we approximate the indicator function $\mathbb{1}[z]$ with hinge loss surrogate $\max\{0, 1 - z\}$, take average over the local training subset $\mathcal{S} \subseteq \mathcal{D}$ and add regularization term λ, we obtain Multiple Instance Support Vector Machines [2] optimization problem:

$$\operatorname*{argmin}_{\mathbf{w}} \ \frac{\lambda}{2} \|\mathbf{w}\|^2 + \frac{1}{|\mathcal{S}|} \sum_{(\mathcal{B}, y) \in \mathcal{S}} \max\{0, 1 - y \max_{\mathbf{x} \in \mathcal{B}} \langle \mathbf{w}, \mathbf{x} \rangle\}. \qquad (3)$$

To approximately solve this non-convex optimization problem in linear time, we adapted Pegasos solver [21] (originally designed for conventional SVMs) to the

MIL setting. The resulting pseudo-code is given in Algorithm 1. It is a stochastic sub-gradient descent-based solver, which at each iteration t updates the current solution $\mathbf{w}^{t+1} \leftarrow \mathbf{w}^t - \eta^t \nabla^t$ (row 9 in Algorithm 1) using step size $\eta^t = 1/(t\lambda)$ and sub-gradient ∇^t of the objective function (Eq. 3) estimated on a single randomly chosen training sample. To avoid building strong classifiers inside nodes, which would go against the randomization principle for constructing diverse independent trees [4], we restrict the selectors to operate on random low-dimensional subspaces. Input zero-one vector $\mathbf{s} \in \{0, 1\}^d$ then serves as a mask defining the feature subspace. By taking element-wise product with that vector (i.e. $\mathbf{s} \odot \mathbf{w}$ or $\mathbf{s} \odot \mathbf{x}$), only feature positions occupied by ones remain effective. In Sect. 4.2, we empirically demonstrate that using sparse selectors, where the number of effective dimensions equals to the square root of the total dimensions d rounded to the closest integer (i.e. $\sum_f s_f = [\sqrt{d}]$), plays a crucial role in the overall ensemble performance. This subspace size ensures that in high dimensions the selectors will be approximately orthogonal and thus independent [12].

Algorithm 1: ISRT's routine for training selectors.

Function TrainSelector($\mathcal{S}; \lambda, E, \mathbf{s}$)

    Input : Training set of bags along with labels $\mathcal{S} = \{(\mathcal{B}, y)_1, \ldots\}$,
              the regularization $\lambda > 0$,
              the number of epochs $E > 0$,
              the zero-one vector defining feature subspace $\mathbf{s}$.
    Output: Instance-level selector $\mathbf{w}^t$ approximately solving Problem 3.

1   *extend all instances by a bias term* $[\mathbf{x}, 1]$ *including the subspace vector* $[\mathbf{s}, 1]$
2     $t \leftarrow 1$
3     $\mathbf{w}^0 \leftarrow$ *random vector* $w_f^0 \sim N(0, 1)$ // length($\mathbf{w}$) = length($\mathbf{x}$) = length($\mathbf{s}$)
4     $\mathbf{w}^1 \leftarrow \mathbf{s} \odot \mathbf{w}^0$ // $\odot$ element-wise product
5   **for** 1 in E **do**
6   **for** 1 in $|\mathcal{S}|$ **do**
7             $(\mathcal{B}, y) \leftarrow$ *(class-balanced) random draw (with replacement) from* $\mathcal{S}$
8             $\mathbf{x}^* \leftarrow \text{argmax}_{\mathbf{x} \in \mathcal{B}} \langle \mathbf{w}^t, \mathbf{x} \rangle$
9             $\mathbf{w}^{t+1} \leftarrow \mathbf{w}^t - \frac{1}{t\lambda}\left(\lambda\mathbf{w}^t - \mathbb{1}\left[y\langle\mathbf{w}^t, \mathbf{x}^*\rangle < 1\right] y(\mathbf{s} \odot \mathbf{x}^*)\right)$
10            $t \leftarrow t + 1$

11   **return** $\mathbf{w}^t[\text{start} : \text{end} - 1]$ // removing the bias term

---

    Being equipped with the routine for training selectors, we can represent bags in a local training subset $\{(\mathcal{B}, y)_1, \ldots\}$ with selected instances $\{(\mathbf{x}^*, y)_1, \ldots\}$. Now, on top of this representation, a standard search for the best splitting parameters, based on measuring purity of produced subsets (e.g. Information gain [20] or Gini impurity [5]), can be executed. This allows us to build an ensemble of ISRT in the same way as (Extremely) Randomized Trees [11] are. In particular, unlike e.g. Breiman's Random Forests [4], where each tree is grown on a bootstrap replica of the training data, the Randomized Trees (as well as our ISRT) are grown on the complete training set, which yields to a lower bias. Variance is then reduced by output aggregation of more diversified trees. The higher diversification is achieved through stronger randomization in splitting nodes, as for each feature f out of $[\sqrt{d}]$ randomly selected ones, only a limited number[2] T of uniformly drawn threshold values v from $[x_f^{\min}, x_f^{\max})$ is considered for splitting rather than every sample value as realized in Breiman's Random Forests. In the case of ISRT, the randomization is even stronger due to the fact that selected instances may vary from node to node. The whole training procedure of ISRT is summarized in Algorithm 2. Time complexity of the algorithm, assuming balanced trees, is $\Theta(M\sqrt{d}\,T\,E\,N_I \log N_B)$, where M is the number of constructed trees, E the number of epochs for training selectors, N_I the average number of instances in bags and N_B the number of training bags.

In the testing phase, a bag to be classified is propagated through individual trees and the final score $\hat{y} \in [-1, 1]$ is calculated as an average of leaves scores the bag falls into (Algorithm 3). However, besides the prediction score, the ISRT can also output a histogram of selection counts over the bag instances $\mathbf{i}$. This information might help to identify relevant instances upon which the decision was made and thus serve as an explanation for positive bag predictions. For example, an explanation—an user was found to be infected because it has communicated with these three hostnames (out of hundreds) within the last 24 h—can greatly speed up the work of threat analysts and reinforce their trust in the model if the hostnames will be shown to be indeed malicious. On the other side, it should be noted that this approach can not explain a positive prediction that would be based on an absence of some type of instance(s) in the bag. For example, there might be malware, hypothetically, its only visible behaviour would be preventing an operating system (or other applications like an anti-virus engine) from regular updates. For the same reason, negative predictions in general can not be explained with this approach.

3   Related Work

Surveys on MIL [1,6] categorize classification methods into two major groups according to the level at which they operate. Methods from the first *instance-level* group (proposed mostly by earlier works) construct instance-level classifiers $f(\mathbf{x}) \rightarrow \{-1, +1\}$ as they assume that the discriminative information lies on the level of instances. Meaning that a bag is positive if it contains at least one instance carrying a characteristics positive signal, and negative if does not contain any such instance. Bag-level predictions are then obtained by aggregating instance-level verdicts $\mathcal{F}(\mathcal{B}) = \max_{\mathbf{x} \in \mathcal{B}} f(\mathbf{x})$. MI-SVM [2], which is defined in its primary form in Eq. 3, is a representative example of this group.

The second *bag-level* group involves methods (proposed mainly by later works) assuming that the discriminative information lies at the level of bags. These methods build directly bag-level classifiers $\mathcal{F}(\mathcal{B})$ extracting information from whole bags to make decisions about their class rather than aggregating

---

[2] Term *extremely* in Extremely Randomized Trees [11] corresponds to setting $T = 1$.

Algorithm 2: ISRT's routine for building an ensemble of trees.

Function BuildTreeEnsemble($\mathcal{D}; M, T, E, \Lambda$)

    Input : Training set of bags along with labels $\mathcal{D} = \{(\mathcal{B}, y)_1, \ldots\}$,
            the number of trees to grow $M > 0$,
            the number of considered threshold values $T > 0$,
            the number of epochs $E > 0$ (for training selectors).
    Output: Ensemble of Instance Selection Randomized Trees $\mathcal{E}$.

1    $\mathcal{E} \leftarrow \emptyset$
2   **for** 1 in M **do**
3        $\mathcal{E} \leftarrow \mathcal{E} \cup \{\text{BuildTree}(\mathcal{D}; T, E, \Lambda)\}$
4   **return** $\mathcal{E}$

Function BuildTree($\mathcal{S}; T, E$)

    Input : Local training subset $\mathcal{S} \subseteq \mathcal{D}$.
    Output: Node with followers or Leaf with a prediction score.

5   **if** *all class labels y in $\mathcal{S}$ are equal* **then**
6   **return** Leaf (y)
7    $\Phi^* \leftarrow$ FindBestSplittingParameters($\mathcal{S}; T, E$)
8   **if** $\Phi^* = \emptyset$ **then**
9   **return** Leaf $(\frac{1}{|\mathcal{S}|} \sum_{y \in \mathcal{S}} y)$
10   $\mathcal{S}_{\text{left}} \leftarrow \{(\mathcal{B}, y) \in \mathcal{S} \mid \mathcal{N}_{\text{ISRT}}(\mathcal{B}; \Phi^*) = \text{left}\}$
11   $\mathcal{S}_{\text{right}} \leftarrow \mathcal{S} \setminus \mathcal{S}_{\text{left}}$
12   **if** $\mathcal{S}_{\text{left}} = \emptyset$ *or* $\mathcal{S}_{\text{right}} = \emptyset$ **then**
13   **return** Leaf $(\frac{1}{|\mathcal{S}|} \sum_{y \in \mathcal{S}} y)$
14   **return** Node $(\Phi^*, \text{BuildTree}(\mathcal{S}_{\text{left}}), \text{BuildTree}(\mathcal{S}_{\text{right}}))$

Function FindBestSplittingParameters($\mathcal{S}; T, E$)

    Output: Triplet of splitting parameters Φ as defined in Eq. 1

15   $\Phi^* \leftarrow \emptyset$
16   $\mathbf{s} \leftarrow$ *zero-one vector of length d with* $\lceil\sqrt{d}\rceil$ *ones at random positions*
17   $\mathbf{w} \leftarrow$ TrainSelector($\mathcal{S}; \lambda = 1, E, \mathbf{s}$)
18   $(\mathbf{X}^*, \mathbf{y}) \leftarrow$ *represent each pair $(\mathcal{B}, y) \in \mathcal{S}$ with* $(\text{argmax}_{\mathbf{x} \in \mathcal{B}} \langle \mathbf{w}, \mathbf{x} \rangle, y)$
19   **foreach** *feature f in* $\lceil\sqrt{d}\rceil$ *randomly selected ones (without replacement)*
       having non-constant values in $\mathbf{X}$ (i.e. $x_f^{min} \neq x_f^{max}$) **do**
20   **foreach** *value v in T uniformly drawn values from $[x_f^{min}, x_f^{max})$* **do**
21          $\Phi \leftarrow (f, v, \mathbf{w})$
22   *update* $\Phi^* \leftarrow \Phi$ *if* Score($\Phi, \mathbf{X}^*, \mathbf{y}$) *is the best so far found score*
23   **return** Φ^*

Algorithm 3: ISRT's prediction routine.

Function Predict($\mathcal{B}; \mathcal{E}$)

  Input : Bag to be classified $\mathcal{B}$ with ensemble of trees $\mathcal{E}$.
  Output: Bag score $\hat{y} \in [-1, 1]$ and histogram of selected instances i.

1 $\hat{y} \leftarrow 0$
2 i $\leftarrow$ *zero vector of length* $|\mathcal{B}|$
3 **foreach** Tree in $\mathcal{E}$ **do**
4   P $\leftarrow$ Tree   // pointer to Node and Leaf structures
5   i' $\leftarrow$ *zero vector of length* $|\mathcal{B}|$
6   **while** P *is of type* Node **do**
7      $(b, i_{\mathbf{x}^*}) \leftarrow \mathcal{N}_{\text{ISRT}}(\mathcal{B}; \text{P}.\Phi^*)$ // $i_{\mathbf{x}^*}$ index of selected instance
8      $i'[i_{\mathbf{x}^*}] \leftarrow i'[i_{\mathbf{x}^*}] + 1$
9   P $\leftarrow$ P.b // continue in left or right branch
10   $\hat{y} \leftarrow \hat{y} + \text{P}.y$
11   $i \leftarrow i \oplus \left(i' / (\sum_i i'[i])\right)$ // $\oplus$ element-wise addition
12 $\hat{y} \leftarrow \hat{y}/|\mathcal{E}|$
13 $i \leftarrow i/|\mathcal{E}|$
14 **return** $(\hat{y}, i)$

---

individual instance-level verdicts. Considering the global bag-level information is necessary, e.g., in a case when the positive label is caused by a co-occurrence of two specific types of instances. Since bags are non-vectorial objects, these methods typically first transform bags into single fixed-size vectors and then train an off-the-shelf classifier on top of them. Works of [13,22] use Bag-of-Words approach, where a vocabulary of prototype instances (i.e. words) is first found by a clustering algorithm and then each bag is represented by a histogram counting how many instances fall into each cluster. Work of [19] rather proposes to use the Neural Network (NN) formalism with pooling layers (e.g. based on max/mean aggregation) to achieve simultaneous optimization of the bag representation (first layer(s) followed by a pooling layer) and the classifier (one or more of subsequent layers) by means of back-propagation. The closest work to this paper is on Bag-level Randomized Trees (BLRT) [14]. The main difference to the ISRT is in the conditions evaluated inside the splitting nodes:

$$\mathcal{N}_{\text{BLRT}}(\mathcal{B}; \underbrace{f, v, r}_{\Phi}) = \begin{cases} \text{left}, & \text{if } \left[\frac{1}{|\mathcal{B}|} \sum_{\mathbf{x} \in \mathcal{B}} \mathbb{1}\left[x_f > v\right]\right] > r, \\ \text{right}, & \text{otherwise}, \end{cases} \tag{4}$$

where the additional parameter r determines a percentage of instances that must satisfy the inner condition $x_f > v$ to be the whole bag passed to the left branch. Surprisingly, this method has been shown to be significantly better than any of the prior 28 MIL classifiers on 29 datasets [14], although it can not extract a multivariate pattern from a single instance unless the bag contains only that one instance. This can be illustrated, e.g., on an inability to separate these two bags: $\{(0,0),(1,1)\}$ and $\{(0,1),(1,0)\}$. From none of the above bag-level methods can be trivially inferred which instances are responsible for positive bag predictions.

4   Experiments

In this section, we evaluate the proposed ISRT algorithm first on a private dataset from the network security field (Sect. 4.1) and then on 12 publicly available datasets from six other domains (Sect. 4.2). On both types of datasets, the algorithm is compared with state-of-the-art approaches: BLRT[3], NN[4] and MI-SVM[5] briefly reviewed in the previous section.

4.1   Private Dataset

The network security dataset represents a real-world problem of classifying users of computer networks as either infected or clean. The dataset contains metadata about network communications from more than 100 international corporate networks of various types and sizes.

Users in the dataset are represented as bags of instances, where the individual instances correspond to the established communications between users and hostnames within 24 h. High-level statistics about each such communication are computed from URL strings of made HTTP requests within that time window. The procedure is as follows. First, URL strings of all HTTP requests originating from a given user and targeting to a particular hostname are collected. Then, each URL string is converted into a feature vector by extracting a well-known set of URL features. The used feature set has been already described e.g. in works [10,15,16] that were primarily focused on detection of URLs generated by malicious applications. As such, the feature set incorporates a lot of domain knowledge. Examples of the features are: the number of occurrences of reserved URL characters (i.e. '_', '-', '?', '!', '@', '#', '&', '%'), the digit ratio, the lower/upper case ratio, the vowel change ratio, the number of non-base64 characters, the maximum length of lower/upper case stream, the maximum length of consonant/vowel/digit stream, etc. In sum, there are 359 features. Finally, to represent the communication with a single instance, the extracted URL feature vectors are aggregated using a maximum as the aggregation function.

The dataset consists of three sets: training, validation and testing. Training set is the largest one and was collected during the period of five working days in January 2021. Validation and testing sets then cover the first and the last Wednesday in February 2021, respectively. In total, there are 118,108 unique users. On average, each has 24 instances. Detailed statistics about the dataset along with the number of infected/clean users are given in Table 1.

---

[3] We used implementation from https://github.com/komartom/BLRT.jl.
[4] We used implementation from https://github.com/CTUAvastLab/Mill.jl.
[5] MI-SVM is trained with Algorithm 1 for complete feature space (s is vector of ones).

Table 1. Specification of the network security dataset. The aim is to train a user-level model. Counts of URL strings and communications correspond to the sum over all users and illustrate the potentially higher labeling requirements on these lower levels.

Dataset	Training set	Validation set	Testing set
Date range	18–22 Jan 2021	3 Feb 2021	24 Feb 2021
URL strings	1,186,465,181	239,079,385	264,342,073
Communications	2,829,316	581,826	554,425
Infected users	1,830	430	380
Clean users	115,700	24,987	23,695

We trained the proposed ISRT algorithm with the following hyper-parameter values: the number of trees to grow $M = 100$, the number of considered threshold values $T = 8$ and the number of epochs for training selectors $E = 10$. The computational time for training took 25 min on a single c4.8xlarge AWS instance[6]. We also used the same hyper-parameter settings (i.e. $M = 100$ and $T = 8$) to train the prior art tree-based algorithm BLRT[7]. In the case of MIL Neural Network (NN), we performed a grid search over the following configurations: the instance layer size $\{10, 30, 60\}$, the aggregation layer type $\{mean, max, mean\text{-}max\}$ and the bag layer size $\{5, 10, 20\}$. We used rectified linear units (ReLU), ADAM optimizer, mini-batch of size 32, and the maximum number of epochs 1000. MI-SVM classifier was trained for regularization values $\lambda \in \{10^{-5}, 10^{-4}, \ldots, 1\}$ and the number of epochs $E = 100$. The final configuration, in both cases, was selected based on the highest achieved performance on the validation set in terms of the area under the Precision-Recall curve.

Efficacy results of the above trained models on the validation and testing set are shown in Fig. 1 using the Precision-Recall and ROC curves. Different points on the Precision-Recall curve correspond to different decision threshold values of a particular model and indicate the percentage of alarms that are actually correct (precision) subject to the percentage of detected infected users in the whole dataset (recall/true positive rate). Perfect recall (score 1) means that all infected users are detected, while perfect precision means that there are no false alarms. ROC curve then provides information about the volume of false alarms as the percentage of monitored healthy users (false positive rate)[8]. Points on the ROC curve might also serve for calculating precision under different imbalance ratios of infected to clean users [3].

---

[6] 36 virtual Intel Xeon CPUs @ 2.9 GHz and 60 Gb of memory.

[7] It was shown in the work of BLRT [14], and we confirm that for ISRT in Sect. 4.2, that tuning of these parameters usually does not bring any additional performance.

[8] While precision answers to the question: "With how big percentage of false alarms the network administrators will have to deal with?", false positive rate gives answer to: "How big percentage of clean users will be bothered?".

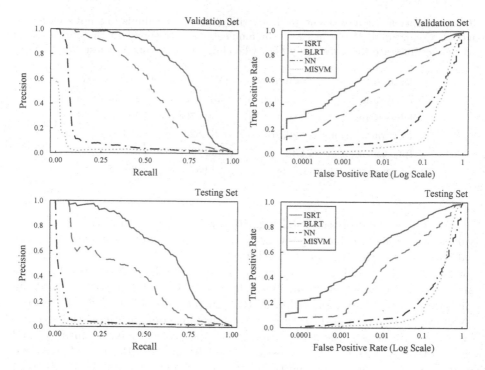

Fig. 1. Precision-recall and ROC curves of individual models on the validation (Feb 3, 2021) and testing (Feb 24, 2021) set of the private network security dataset.

As can be seen from Fig. 1, the proposed ISRT algorithm has a clear superior performance as it produces equal or less false alarms than any other involved method at any arbitrary recall on both sets. The second best method is the tree-based BLRT algorithm. It has a decent performance on the validation test (Feb 3, 2021), but on the testing set (Feb 24, 2021) the performance drops notably. This decrease in model performance over time is known as an aging effect—a model becomes obsolete as the distribution of incoming data shifts. Resistance to that is an important model characteristic, from an application point of view, albeit not always evaluated by researchers [17]. We attribute this decay to the fact that the BLRT model can extract only global bag-level uni-variate statistics computed across all instances within a bag. Because of this, it might be difficult to effectively separate a multivariate malicious signal hidden in a single instance from an abundant user background, which can evolve over time. On the other hand, the discriminative signal apparently does not lie on the local instance-level completely as the instance-based classifier MI-SVM per-forms poorly. Probably the ability to combine these two approaches, by selecting and judging individual instances according to the need while collecting global evidence, might be the reason why the ISTR model excels in this task. Inter-estingly, the NN model is able to reliably detect only a very limited number of infections, although in the recent work [18] it has been shown to perform well on a similar task. One reason might be that the NN model is sensitive to some hyper-parameters, which we did not tune. Another one might be that there is a substantial difference in the time window (5 min vs. 24 h) per which the users are classified. Shrinking the time window to five minutes on our data leads to one instance (i.e. communication with a unique hostname) per bag on average, which would eliminate the need of MIL and its benefits (i.e. lower labeling costs and richer contextual information).

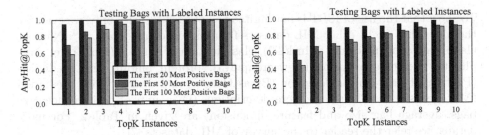

Fig. 2. Assessment of ISRT explanations as an information retrieval task. AnyHit@TopK (on the left) shows the percentage of bags for which at least one relevant (i.e. positively labeled) instance appeared among the top K instances. Since bags can contain multiple relevant instances, Recall@TopK (on the right) shows how many of them are among the top K instances. The perfect Recall@1 can not be achieved unless all bags have only one relevant instance.

As announced in the introduction to this paper, the main benefit of the newly proposed ISRT over the prior-art BLRT, besides higher performance on some datasets, is the ability to explain the positive bag predictions. The provided explanations are in the form of assigned scores to individual instances (according to the number of times they have been selected during the bag class prediction), upon which they can be sorted and presented to the end-user for judgment. Analogically to the information retrieval task, the goal is to place the most relevant instances at top positions. To assess this ability on the network security dataset, we used our internal deny list of known malicious hostnames to label instances[9] (i.e. communications with hostnames) of bags that have been classified as positive by the ISRT model on the testing set. On the first 20, 50 and 100 most positive bags (for which we had at least one positive instance-level label), we calculated AnyHit@TopK and Recall@TopK metrics. Figure 2 presents the results for the first top ten (TopK) instance positions.

It can be observed from the left subplot of Fig. 2 (AnyHit@TopK) that threat analysts investigating the first 20 most positive bags would encounter the first piece of evidence for the infection (i.e. any malicious hostname) just by ana- lyzing the top two recommended instances from each bag. This ability slightly

---

[9] This way of identifying malicious communications is not so effective in production, since new threats are not on the deny list yet and need to be first discovered.

decreases for the higher number of bags (i.e. 50 and 100), but it is still very useful considering the fact that the largest bags have over 100 instances, and on average only two are labeled as malicious—a needle in a haystack problem. This can be also seen from the right subplot (Recall@TopK) showing that about 50% of all positive instances in bags can be discovered just by verifying the first recommended instance ($K = 1$).

4.2 Public Datasets

To show that the use of ISRT is not limited to network security only, we evaluate the algorithm on 12 other MIL datasets from six different domains[10]. Namely, classification of molecules (Musk1-2), classification of images (Fox, Tiger, Elephant), text categorization (Newsgroups1-3), protein binding site prediction (Protein), breast cancer detection (BreastCancer) and drug activity prediction (Mutagenesis1-2). Their basic meta-descriptions (i.e. counts of positive/negative bags, average bag size and feature dimension) are given in Table 2. For more details, we refer the reader to the survey of MIL datasets [7].

Table 2. Metadata about 12 public datasets. Including the number of positive/negative bags, the average number of instances inside bags and the number of features. Plus evaluation results, measured in AUC × 100, for individual models. Best results are shown in bold face. Multiple models are highlighted if the difference is not statistically significant (at $\alpha = 0.05$) according to a paired t-test with Holm-Bonferroni correction (for multiple comparisons) [8] computed on the five runs of 10-fold cross-validation.

Dataset	Bags $+/-$	Inst.	Feat.	MI-SVM	NN	BLRT	ISRT [ours]
Musk1	47/45	5	166	85.9 (1.9)	91.9 (1.5)	**96.8** (1.6)	**97.2** (1.3)
Musk2	39/63	65	166	86.9 (1.5)	**90.3** (2.3)	**91.2** (1.8)	**92.3** (2.6)
Fox	100/100	7	230	55.2 (1.6)	65.9 (1.2)	**73.3** (1.4)	**74.0** (1.8)
Tiger	100/100	6	230	81.7 (2.7)	**90.7** (1.7)	**92.6** (1.0)	**92.5** (0.8)
Elephant	100/100	7	230	84.5 (0.3)	**93.9** (0.8)	**95.8** (0.9)	**95.0** (0.7)
Newsgroups1	50/50	54	200	**82.4** (4.9)	77.0 (3.6)	**78.8** (2.6)	55.4 (2.6)
Newsgroups2	50/50	31	200	**70.2** (3.7)	**63.3** (5.2)	**63.0** (4.0)	**63.8** (2.9)
Newsgroups3	50/50	52	200	54.5 (5.5)	63.9 (4.1)	**76.3** (4.1)	65.0 (2.6)
Protein	25/168	138	9	81.2 (1.7)	75.2 (4.2)	74.9 (2.3)	**85.8** (2.0)
BreastCancer	26/32	35	708	73.1 (2.6)	**76.7** (8.1)	**84.5** (2.5)	79.3 (1.9)
Mutagenesis1	125/63	56	7	53.4 (1.0)	**90.2** (1.0)	**92.1** (1.3)	88.6 (0.8)
Mutagenesis2	13/29	51	7	70.0 (8.2)	66.2 (2.6)	**86.0** (3.5)	70.0 (6.6)

Each dataset also contains a predefined list of splitting indices for 5-times repeated 10-fold cross-validation. Therefore, we followed this evaluation protocol precisely, similarly, as did the prior works of BLRT [14] and NN [19]. Since the

---

[10] Datasets are accessible at https://doi.org/10.6084/m9.figshare.6633983.v1.

protocol does not specify any approach for hyper-parameter optimization, we used the default values that are known to work well. In particular, to train ISRT, we set the ensemble size to $M = 500$, the number of considered threshold values to $T = 8$ and the number of epochs for training selectors to $E = 1$. The same setting of parameters (i.e. $M = 500$ and $T = 8$) is used for the BLRT model, which corresponds to the setting applied in the original work of BLRT during the evaluation. The NN architecture consists of a single instance layer of size 10 with rectified linear units (ReLU), followed by a mean-max aggregation layer, a single bag layer of size 10 and two output units. The weights are regularized with L1 regularization $\lambda = 10^{-3}$ to decrease overfitting as suggested in [19]. The training minimizes a cross-entropy loss function using ADAM optimizer, mini-batch of size 8, and the maximum number of epochs 1000. Finally, the MI-SVM model is trained with the regularization $\lambda = 10^{-3}$ and 100 epochs.

Table 2 shows the performance of each model on each dataset in terms of the average Area Under the ROC Curve (AUC)[11] $\pm$ one standard deviation. It can be seen that the proposed ISRT model significantly outperforms the other three models only on Protein dataset, whereas the prior-art BLRT model significantly wins on two datasets (Newsgroup3 and Mutagenesis2). The average ranks[12] of the models are: BLRT = 1.8, ISRT = 2.0, NN = 3.0 and MISVM = 3.2. According-ing to the non-parametric Friedman-Nemenyi test [8] (comparing all classifiers to each other based on the average ranks), there is no statistically significant difference[13] (at $\alpha = 0.05$) among the models, except for the pair BLRT and MI-SVM, where MI-SVM looses.

As the last experiment, we investigate the influence of individual model components/parameters on the final performance. Figure 3 shows results from this ablation study as a series of eight pair-wise comparisons. Each subplot compares two different variants (horizontal and vertical axis) of the proposed ISRT algo-rithm on the 12 datasets (dots on the scatter plot). X and Y coordinates of each dot are determined by the achieved AUCs of the corresponding variants on that particular dataset. Therefore, if a dot lies above the main diagonal, the variant associated with the vertical axis outperforms the other one associated with the horizontal axis and vice versa.

The first two **Subplots (A–B)** illustrate the effect of the ensemble size. While it is almost always better to build 100 trees than 5, building 500 trees usually does not bring any additional performance compared to 100. In **Sub-plot (C)**, we examine the model stability with respect to different random seeds (1234 vs. 42) and as can be seen, there is almost no difference. **Subplot (D)** shows a slight improvement that can be achieved by considering more thresholds for splitting ($T = 8$) than one as it is characteristic for Extremely Randomized Trees [11]. It is also worth experimenting with the sparse vs. dense selectors

---

[11] AUC is agnostic to class imbalance and classifier's decision threshold value.

[12] The best model is assigned the lowest rank (i.e. one).

[13] The performance of any two classifiers is significantly different if the corresponding average ranks differ by at least the critical difference, which is (for 12 datasets, four methods and $\alpha = 0.05$) approximately 1.35.

because, as can be observed from **Subplot (E)**, this option has different effects on different datasets. **Subplot (F)** then supports the idea that strongly randomized trees, unlike Breiman's Random Forests, do not have to be trained on the bootstrapped datasets. In **Subplot (G)** we analyze whether the search of splitting parameters $\Phi = (f, v, \mathbf{w})$ over multiple selectors with different regularization values $\lambda \in \{10^{-4}, 10^{-3} \ldots, 1\}$ instead of one $\lambda = 1$ can help. Finally, the last **Subplot (H)** indicates that there is no need to train selectors with a large number of epochs.

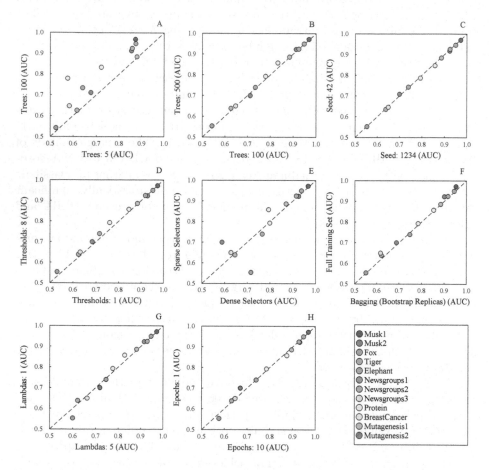

Fig. 3. Ablation study assessing influence of individual model components/parameters. It illustrates the effect of the ensemble size (A–B), the stability wrt. random seed (C), the slight improvement caused by considering more thresholds for splitting (D), the impact of sparse vs. dense selectors (E), the no need for using bagging (F), multiple regularization values (G) nor a large number of epochs for training selectors (H).

5   Conclusion

In this paper, we have proposed a new tree-based algorithm called Instance Selection Randomized Trees (ISRT)[14] for solving binary classification MIL problems. The algorithm naturally extends the traditional randomized trees, since bags of size one are processed in the standard way by evaluating single feature value conditions at each node. When bags contain multiple instances, every node selects one instance from the bag, upon which the decision whether to send the bag to the left or right branch is made. Making decisions upon deliberately selected instances at each step is essential for the algorithm as it enables to extract (even multivariate) discriminative information from both ends of the spectrum, the local instance-level and the global bag-level.

We demonstrated on the task of detecting infected users in computer networks that this capability may not only lead to superior performance when compared with three state-of-the-art methods, but it also may greatly help threat analysts with the post-alert analysis. This is because the positive bag predictions can be explained on the level of instances by their ranking according to how many times have been selected. The model also achieved competitive results on 12 publicly available datasets from six other domains. Finally, we conducted ablation experiments to understand contributions of individual algorithm parts.

References

1. Amores, J.: Multiple instance classification: review, taxonomy and comparative study. Artif. Intell. **201**, 81–105 (2013). http://dx.doi.org/10.1016/j.artint.2013. 06.003
2. Andrews, S., Tsochantaridis, I., Hofmann, T.: Support vector machines for multiple-instance learning. In: Proceedings of the 15th International Conference on Neural Information Processing Systems, pp. 577–584. NIPS 2002. MIT Press, Cambridge, MA, USA (2002). http://dl.acm.org/citation.cfm?id=2968618.2968690
3. Brabec, J., Komárek, T., Franc, V., Machlica, L.: On model evaluation under non-constant class imbalance. In: Krzhizhanovskaya, V.V., et al. (eds.) ICCS 2020. LNCS, vol. 12140, pp. 74–87. Springer, Cham (2020). https://doi.org/10.1007/978-3-030-50423-6_6
4. Breiman, L.: Random forests. Mach. Learn. **45**(1), 5–32 (2001). http://dx.doi.org/10.1023/A:1010933404324
5. Breiman, L., Friedman, J., Stone, C.J., Olshen, R.A.: Classification and Regression Trees. CRC Press, Boca Raton (1984)
6. Carbonneau, M.A., Cheplygina, V., Granger, E., Gagnon, G.: Multiple instance learning: a survey of problem characteristics and applications. Patt. Recogn. **77**, 329–353 (2018). https://www.sciencedirect.com/science/article/pii/S0031320317304065
7. Cheplygina, V., Tax, D.M.J.: Characterizing multiple instance datasets. In: Feragen, A., Pelillo, M., Loog, M. (eds.) SIMBAD 2015. LNCS, vol. 9370, pp. 15–27. Springer, Cham (2015). https://doi.org/10.1007/978-3-319-24261-3_2

---

[14] Source codes are available at https://github.com/komartom/ISRT.jl.

8. Demšar, J.: Statistical comparisons of classifiers over multiple data sets. J. Mach. Learn. Res. **7**, 1–30 (2006). http://dl.acm.org/citation.cfm?id=1248547.1248548
9. Dietterich, T.G., Lathrop, R.H., Lozano-Pérez, T.: Solving the multiple instance problem with axis-parallel rectangles. Artif. Intell. **89**(1), 31–71 (1997). http://www.sciencedirect.com/science/article/pii/S0004370296000343
10. Franc, V., Sofka, M., Bartos, K.: Learning detector of malicious network traffic from weak labels. In: Bifet, A., et al. (eds.) ECML PKDD 2015. LNCS (LNAI), vol. 9286, pp. 85–99. Springer, Cham (2015). https://doi.org/10.1007/978-3-319-23461-8_6
11. Geurts, P., Ernst, D., Wehenkel, L.: Extremely randomized trees. Mach. Learn. **63**(1), 3–42 (2006). https://doi.org/10.1007/s10994-006-6226-1
12. Ho, T.: The random subspace method for constructing decision forests. IEEE Trans. Pattern Anal. Mach. Intell. **20**, 832–844 (1998)
13. Kohout, J., Komárek, T., Čech, P., Bodnár, J., Lokoč, J.: Learning communication patterns for malware discovery in https data. Expert Syst. Appl. **101**, 129–142 (2018). http://www.sciencedirect.com/science/article/pii/S0957417418300794
14. Komárek, T., Somol, P.: Multiple instance learning with bag-level randomized trees. In: Berlingerio, M., Bonchi, F., Gärtner, T., Hurley, N., Ifrim, G. (eds.) ECML PKDD 2018. LNCS (LNAI), vol. 11051, pp. 259–272. Springer, Cham (2019). https://doi.org/10.1007/978-3-030-10925-7_16
15. Li, K., Chen, R., Gu, L., Liu, C., Yin, J.: A method based on statistical characteristics for detection malware requests in network traffic. In: 2018 IEEE Third International Conference on Data Science in Cyberspace (DSC), pp. 527–532 (2018). https://doi.org/10.1109/DSC.2018.00084
16. Machlica, L., Bartos, K., Sofka, M.: Learning detectors of malicious web requests for intrusion detection in network traffic (2017)
17. Pendlebury, F., Pierazzi, F., Jordaney, R., Kinder, J., Cavallaro, L.: TESSERACT: eliminating experimental bias in malware classification across space and time. In: 28th USENIX Security Symposium (USENIX Security 19), pp. 729–746. USENIX Association, Santa Clara, CA, August 2019. https://www.usenix.org/conference/usenixsecurity19/presentation/pendlebury
18. Pevny, T., Somol, P.: Discriminative models for multi-instance problems with tree structure. In: Proceedings of the 2016 ACM Workshop on Artificial Intelligence and Security, pp. 83–91. AISec 2016. Association for Computing Machinery, New York, NY, USA (2016). https://doi.org/10.1145/2996758.2996761
19. Pevný, T., Somol, P.: Using neural network formalism to solve multiple-instance problems. In: Cong, F., Leung, A., Wei, Q. (eds.) ISNN 2017. LNCS, vol. 10261, pp. 135–142. Springer, Cham (2017). https://doi.org/10.1007/978-3-319-59072-1_17
20. Quinlan, J.R.: Induction of decision trees. Mach. Learn. **1**(1), 81–106 (1986). http://dx.doi.org/10.1023/A:1022643204877
21. Shalev-Shwartz, S., Singer, Y., Srebro, N.: Pegasos: primal estimated sub-gradient solver for SVM. In: Proceedings of the 24th International Conference on Machine Learning, pp. 807–814. ICML 2007. Association for Computing Machinery, New York, NY, USA (2007). https://doi.org/10.1145/1273496.1273598
22. Stiborek, J., Pevný, T., Rehák, M.: Multiple instance learning for malware classification. Expert Syst. Appl. **93**, 346–357 (2018). http://www.sciencedirect.com/science/article/pii/S0957417417307170

Adversarial Representation Learning with Closed-Form Solvers

Bashir Sadeghi, Lan Wang, and Vishnu Naresh Boddeti[✉]

Michigan State University, East Lansing, MI 48823, USA
{sadeghib,wanglan3,vishnu}@msu.edu,
http://hal.cse.msu.edu

Abstract. Adversarial representation learning aims to learn data representations for a target task while removing unwanted sensitive information at the same time. Existing methods learn model parameters iteratively through stochastic gradient descent-ascent, which is often unstable and unreliable in practice. To overcome this challenge, we adopt closed-form solvers for the adversary and target task. We model them as kernel ridge regressors and analytically determine an upper-bound on the optimal dimensionality of representation. Our solution, dubbed OptNet-ARL, reduces to a stable one one-shot optimization problem that can be solved reliably and efficiently. OptNet-ARL can be easily generalized to the case of multiple target tasks and sensitive attributes. Numerical experiments, on both small and large scale datasets, show that, from an optimization perspective, OptNet-ARL is stable and exhibits three to five times faster convergence. Performance wise, when the target and sensitive attributes are dependent, OptNet-ARL learns representations that offer a better trade-off front between (a) utility and bias for fair classification and (b) utility and privacy by mitigating leakage of private information than existing solutions.

Code is available at https://github.com/human-analysis.

Keywords: Fair machine learning · Adversarial representation learning · Closed-form solver · Kernel ridge regression

1 Introduction

Adversarial Representation Learning (ARL) is a promising framework that affords explicit control over unwanted information in learned data representations. This concept has practically been employed in various applications, such as, learning unbiased and fair representations [7,28,29,37], learning controllable representations that are invariant to sensitive attributes [31,40], mitigating leakage of sensitive information [10,33–35], unsupervised domain adaption[13], learning flexibly fair representations [7,37], and many more.

The goal of ARL is to learn a data encoder $E : x \mapsto z$ that retains sufficient information about a desired *target* attribute, while removing information about a known *sensitive* attribute. The basic idea of ARL is to learn such a mapping under

© Springer Nature Switzerland AG 2021
N. Oliver et al. (Eds.): ECML PKDD 2021, LNAI 12976, pp. 731–748, 2021.
https://doi.org/10.1007/978-3-030-86520-7_45

an adversarial setting. The learning problem is setup as a three-player minimax game between three entities (see Fig. 1a, an encoder E, a predictor T, and a proxy adversary A. Target predictor T seeks to extract target information and make correct predictions on the target task. The proxy adversary A mimics an unknown real adversary and seeks to extract sensitive information from learned representation. As such, the proxy adversary serves only to aid the learning process and is not an end goal by itself. Encoder E seeks to simultaneously aid the target predictor and limit the ability of the proxy adversary to extract sensitive information from the representation z. By doing so, the encoder learns to remove sensitive information from the representation. In most ARL settings, while the encoder is a deep neural network, the target predictor and adversary are typically shallow neural networks.

The vanilla algorithm for learning the parameters of the encoder, target and adversary networks is gradient descent-ascent (GDA) [33,40], where the players take a gradient step simultaneously. However, applying GDA, including its stochastic version, is not an optimal strategy for ARL and is known to suffer from many drawbacks. Firstly, GDA has undesirable convergence properties; it fails to converge to a local minimax and can converge to fixed points that are not local minimax, while being very unstable and slow in practice [8,19]. Secondly, GDA exhibits strong rotation around fixed points, which requires using very small learning rates [3,30] to converge. Numerous solutions [14,30,32] have been proposed recently to address the aforementioned computational challenges. These approaches, however, seek to obtain solutions to the minimax optimization problem in the general case, where each player is modeled as a complex neural network.

In this paper, we take a different perspective and propose an alternative solution for adversarial representation learning. Our key insight is to replace the shallow neural networks with other analytically tractable models with similar capacity. We propose to adopt simple learning algorithms that admit closed-form solutions, such as linear or kernel ridge regressors for the target and adversary, while modeling the encoder as a deep neural network. Crucially, such models are particularly suitable for ARL and afford numerous advantages, including (1) closed-form solution allows learning problems to be optimized globally and efficiently, (2) analytically obtain upper bound on optimal dimensionality of the embedding z, (3) the simplicity and differentiability allows us to backpropagate through the closed-form solution, (4) practically it resolves the notorious rotational behaviour of iterative minimax gradient dynamics, resulting in a simple optimization that is empirically stable, reliable, converges faster to a local optima, and ultimately results in a more effective encoder E.

We demonstrate the practical effectiveness of our approach, dubbed OptNet-ARL, through numerical experiments on an illustrative toy example, fair classification on UCI Adult and German datasets and mitigating information leakage on the CelebA dataset. We consider two scenarios where the target and sensitive attributes are (a) dependent, and (b) independent. Our results indicate that, in comparison to existing ARL solutions, OptNet-ARL is more stable and converges faster while also outperforming them in terms of accuracy, especially in the latter scenario.

Notation: Scalars are denoted by regular lower case or Greek letters, e.g., n, λ. Vectors are boldface lowercase letters, e.g., $\boldsymbol{x}$, $\boldsymbol{y}$; Matrices are uppercase boldface letters, e.g., $\boldsymbol{X}$. A $n \times n$ identity matrix is denoted by $\boldsymbol{I}$, sometimes with a subscript indicating its size, e.g., $\boldsymbol{I}_n$. Centered (mean subtracted w.r.t columns) data matrix is indicated by "~", e.g., $\tilde{\boldsymbol{X}}$. Assume that $\boldsymbol{X}$ contains n columns, then $\tilde{\boldsymbol{X}} = \boldsymbol{X}\boldsymbol{D}$ where $\boldsymbol{D} = \boldsymbol{I}_n - \frac{1}{n}\boldsymbol{1}\boldsymbol{1}^T$ and $\boldsymbol{1}$ denotes a vector of ones with length of n. Given matrix $\boldsymbol{M} \in \mathbb{R}^{m \times m}$, we use $\mathrm{Tr}[\boldsymbol{M}]$ to denote its trace (i.e., the sum of its diagonal elements); its Frobenius norm is denoted by $\|\boldsymbol{M}\|_F$, which is related to the trace as $\|\boldsymbol{M}\|_F^2 = \mathrm{Tr}[\boldsymbol{M}\boldsymbol{M}^T]$. The pseudo-inverse of $\boldsymbol{M}$ is denoted by $\boldsymbol{M}^\dagger$. The subspace spanned by the columns of $\boldsymbol{M}$ is denoted by $\mathcal{R}(\boldsymbol{M})$ or simply $\mathcal{M}$ (in calligraphy); the orthogonal complement of $\mathcal{M}$ is denoted by $\mathcal{M}^\perp$. The orthogonal projector onto $\mathcal{M}$ is denoted by $P_\mathcal{M}$.

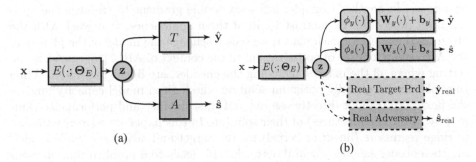

(a)   (b)

Fig. 1. Adversarial Representation Learning: (a) Consists of three players, an encoder E that obtains a compact representation $\boldsymbol{z}$ of input data $\boldsymbol{x}$, predictors T and S that seek to extract a desired target $\boldsymbol{y}$ and sensitive $\boldsymbol{s}$ attribute, respectively from the embedding. (b) OptNet-ARL adopts kernel regressors as proxy target predictor and adversary for learning the encoder. The learned encoder is evaluated against a real target predictor and adversary, which potentially can be neural networks.

2 Prior Work

Adversarial Representation Learning: The basic idea of learning data representations with controllable semantic information has been effective across multiple topics. Domain adaptation [12,13,38], where the goal is to learn representations that are invariant to the domain, is one of the earliest applications of ARL. More recently, adversarial learning has been extensively used [5–7,10,11,33,37,40,42] and advocated [29] for the task of learning fair, invariant or privacy preserving representations of data. All of the aforementioned approaches represent each entity in ARL by neural networks and optimize their parameters through stochastic gradient descent-ascent (SGDA). As we show in this paper, SGDA is unstable and sub-optimal for learning. Therefore, we trade-off model expressively for ease of learning through a hybrid approach of modeling the encoder by a deep neural network and target and adversary with closed-form regressors. Such a solution reduces alternating optimization into a simple optimization problem which is much more stable, reliable and effective. Table 1 shows a comparative summary of ARL approaches.

Optimization in Minmax Games: A growing class of learning algorithms, including ARL, GANs etc., involve more than one objective and are trained via games played by cooperating or dueling neural networks. An overview of the challenges presented by such algorithms and a plausible solution in general n-player games can be found in [26]. In the context of two-player minimax games such as GANs, a number of solutions [3,8,14,19,30,32] have been proposed to improve the optimization dynamics, many of them relying on the idea of taking an extrapolation step [23]. For example, [30] deploys some regularizations to encourage agreement between different players and improve the convergence properties. In another example, [32] uses double gradient to stabilize the optimization procedure. In contrast to all of these approaches, that work with the given fixed models for each player, we seek to change the model of the players in the ARL setup for ease of optimization. In the context of ARL, [35] considers the setting where all the players, including the encoder, are linear regressors. While they obtained a globally optimum solution, the limited model capacity hinders the flexibility (cannot directly use raw data), scalability and performance (limited by pre-trained features) of their solution. In this paper we advocate the use of ridge regressors (linear or kernel) for the target and adversary, while modeling the encoder as a deep neural network. This leads to a problem that obviates the need for gradient descent-ascent and can instead be easily optimized with standard SGD. Not only does this approach lead to stable optimization, it also scales to larger datasets and exhibits better empirical performance.

Table 1. Comparison between different ARL methods (n: sample size, b: batch size).

Method	Encoder/Target & Adversary	Optimization	Scalability	Enc soln	Input data
SGDA-ARL [29,40]	deep NN/shallow NN	Alternating SGD	$\geq \mathcal{O}(b^3)$	Unknown	Raw data
Kernel-SARL [35]	Kernel regressor/linear	Closed-form	$\mathcal{O}(n^3)$	Global optima	Features
OptNet-ARL (ours)	Deep NN/kernel regressor	SGD	$\mathcal{O}(b^3)$	Local optima	Raw data

Differentiable Solvers: A number of recent approaches have integrated differentiable solvers, both iterative as well as closed-form, within end-to-end learning systems. Structured layers for segmentation and higher order pooling were introduced by [17]. Similarly [39] proposed an asymmetric architecture which incorporates a Correlation Filter as a differentiable layer. Differential optimization as a layer in neural networks was introduced by [1,2]. More recently, differentiable solvers have also been adopted for meta-learning [4,25] as well. The primary motivation for all the aforementioned approaches is to endow deep neural networks with differential optimization and ultimately achieve faster convergence of the end-to-end system. In contrast, our inspiration for using differential closed-form solvers is to control the non-convexity of the optimization in ARL, in terms of stability, reliability and effectiveness.

3   Problem Setting

Let the data matrix $X = [x_1, \ldots, x_n] \in \mathbb{R}^{d \times n}$ be n realizations of d-dimensional data, $x \in \mathbb{R}^d$. Similarly, we denote n realizations of sensitive attribute vector $s \in \mathbb{R}^q$ and target attribute vector $y \in \mathbb{R}^p$ by matrices $S = [s_1, \cdots, s_n]$ and $Y = [y_1, \cdots, y_n]$, respectively. Treating the attributes as vectors enables us to consider both multi-class classification and regression under the same formulation. Each data sample x_k is associated with the sensitive attribute s_k and the target attribute y_k, respectively.

The ARL problem is formulated with the goal of learning parameters of an embedding function $E(\cdot; \Theta_E)$ that maps a data sample x to $z \in \mathbb{R}^r$ with two objectives: (i) aiding a target predictor $T(\cdot; \Theta_y)$ to accurately infer the target attribute y from z, and (ii) preventing an adversary $A(\cdot; \Theta_s)$ from inferring the sensitive attribute s from z. The ARL problem can be formulated as a bi-level optimization,

$$\min_{\Theta_E} \min_{\Theta_y} \mathcal{L}_y \left(T(E(x; \Theta_E); \Theta_y), y \right) \quad \text{s.t.} \quad \min_{\Theta_s} \mathcal{L}_s \left(A(E(x; \Theta_E); \Theta_s), s \right) \geq \alpha \tag{1}$$

where $\mathcal{L}_y$ and $\mathcal{L}_s$ are the loss functions (averaged over the training dataset) for the target predictor and the adversary, respectively; $\alpha \in (0, \infty)$ is a user defined value that determines the minimum tolerable loss α for the adversary on the sensitive attribute; and the minimization in the constraint is equivalent to the encoder operating against an optimal adversary. Denote the global minimums of the adversary and target estimators as

$$\begin{aligned} J_y(\Theta_E) &:= \min_{\Theta_y} \mathcal{L}_y \left(T(E(x; \Theta_E); \Theta_y), y \right) \\ J_s(\Theta_E) &:= \min_{\Theta_s} \mathcal{L}_s \left(A(E(x; \Theta_E); \Theta_s), s \right). \end{aligned} \tag{2}$$

The constrained optimization problem in (1) can be alternately solved through its Lagrangian version:

$$\min_{\Theta_E} \left\{ (1 - \lambda) J_y(\Theta_E) - \lambda J_s(\Theta_E) \right\}, \; 0 \leq \lambda \leq 1. \tag{3}$$

3.1   Motivating Exact Solvers

Most state-of-the-art ARL algorithms cannot solve the optimization problems in (2) optimally (e.g., SGDA). For any given Θ_E, denote the non-optimal adversary and target predictors loss functions as $J_y^{\text{approx}}(\Theta_E)$ and $J_s^{\text{approx}}(\Theta_E)$, respectively. It is obvious that for any given Θ_E, it holds

$$J_y^{\text{approx}}(\Theta_E) \geq J_y(\Theta_E) \quad \text{and} \quad J_s^{\text{approx}}(\Theta_E) \geq J_s(\Theta_E).$$

Note that the optimization problem raised from non-optimal adversary and target predictors is

$$\min_{\Theta_E} \left\{ (1 - \lambda) J_y^{\text{approx}}(\Theta_E) - \lambda J_s^{\text{approx}}(\Theta_E) \right\}, 0 \leq \lambda \leq 1. \tag{4}$$

Intuitively, solution(s) of (4) do not outperform that of (3). We now formulate this intuition more concretely.

Definition 1. *Let (a_1, a_2) and (b_1, b_2) be two arbitrary points in $\mathbb{R}^2$. We say (b_1, b_2) dominates (a_1, a_2) if and only if $b_1 > a_1$ and $b_2 < a_2$ hold simultaneously.*

Theorem 2. *For any $\lambda_1, \lambda_2 \in [0, 1]$, consider the following optimization problems*

$$\Theta_E^{exact} = \arg\min_{\Theta_E} \left\{ (1 - \lambda_1) J_y(\Theta_E) - \lambda_1 J_s(\Theta_E) \right\} \tag{5}$$

and

$$\Theta_E^{approx} = \arg\min_{\Theta_E} \left\{ (1 - \lambda_2) J_y^{approx}(\Theta_E) - \lambda_2 J_s^{approx}(\Theta_E) \right\}$$

Then, any adversary-target objective trade-off generated by $\left(J_s(\Theta_E^{exact}), J_y(\Theta_E^{exact}) \right)$ cannot be dominated by the trade-off generated by $\left(J_s(\Theta_E^{approx}), J_y(\Theta_E^{approx}) \right)$.

See supplementary material for the proof of all Lemmas and Theorems.

4 Approach

Existing instances of ARL adopt deep neural networks to represent E, T and A and learn their respective parameters $\{\Theta_E, \Theta_y, \Theta_s\}$ through stochastic gradient descent-ascent (SGDA). Consequently, the target and adversary in Eq. 2 are not solved to optimality, thereby resulting in a sub-optimal encoder.

4.1 Closed-Form Adversary and Target Predictor

The machine learning literature offers a wealth of methods with exact solutions that are appropriate for modeling both the adversary and target predictors. In this paper, we argue for and adopt simple, fast and differentiable methods such as kernel ridge regressors as shown in Fig. 1b. On one hand, such modeling allows us to obtain the optimal estimators globally for any given encoder $E(\cdot; \Theta_E)$.

On the other hand, kernelized ridge regressors can be stronger than the shallow neural networks that are used in many ARL-based solutions (e.g., [11,29,33,40]). Although it is not the focus of this paper, it is worth noting that even deep neural networks in the infinite-width limit reduce to linear models with a kernel called the neural tangent kernel [18], and as such can be adopted to increase the capacity of our regressors.

Consider two reproducing kernel Hilbert spaces (RKHS) of functions $\mathcal{H}_s$ and $\mathcal{H}_y$ for adversary and target regressors, respectively. Let a possible corresponding pair of feature maps be $\phi_s(\cdot) \in \mathbb{R}^{r_s}$ and $\phi_y(\cdot) \in \mathbb{R}^{r_y}$ where r_s and r_y are the dimensionality of the resulting features and can potentially approach infinity. The respective kernels for $\mathcal{H}_s$ and $\mathcal{H}_y$ can be represented as $k_s(z_1, z_2) = \langle \phi_s(z_1), \phi_s(z_2) \rangle_{\mathcal{H}_s}$ and $k_y(z_1, z_2) = \langle \phi_y(z_1), \phi_y(z_2) \rangle_{\mathcal{H}_y}$. Under this setting, we can relate the target and sensitive attributes to any given embedding z as,

$$\hat{y} = W_y \phi_y(z) + b_y, \qquad \hat{s} = W_s \phi_s(z) + b_s \qquad (6)$$

where $\Theta_y = \{W_y, b_y\}$ and $\Theta_s = \{W_s, b_s\}$ are the regression parameters, $W_y \in \mathbb{R}^{p \times r_y}$ and $W_s \in \mathbb{R}^{q \times r_s}$, $b_y \in \mathbb{R}^p$ and $b_s \in \mathbb{R}^q$ respectively.

Let the entire embedding of input data be denoted as $Z := [z_1, \cdots, z_n]$ and the corresponding features maps as $\Phi_y := [\phi_y(z_1), \cdots, \phi_y(z_n)]$ and $\Phi_s := [\phi_s(z_1), \cdots, \phi_s(z_n)]$, respectively. Furthermore, we denote the associated Gram matrices by $K_y = \Phi_y{}^T \Phi_y$ and $K_s = \Phi_s{}^T \Phi_s$. A centered Gram matrix $\tilde{K}$ corresponding to the Gram matrix K can be obtained [16] as,

$$\tilde{K} = \tilde{\Phi}^T \tilde{\Phi} = (\Phi D)^T (\Phi D) = D^T K D. \qquad (7)$$

Invoking the representer theorem [36], the regression parameters can be represented as $W_y = \Lambda_y \tilde{\Phi}_y^T$ and $W_s = \Lambda_s \tilde{\Phi}_s^T$ for target and adversary respectively, where $\Lambda_y \in \mathbb{R}^{p \times n}$ and $\Lambda_s \in \mathbb{R}^{n \times q}$ are new parameter matrices. As a result, the kernelized regressors in (6) can be equivalently expressed as

$$\hat{y} = \Lambda_y \tilde{\Phi}_y^T \phi_y(z) + b_y, \quad \hat{s} = \Lambda_s \tilde{\Phi}_s^T \phi_s(z) + b_s. \qquad (8)$$

In a typical ARL setting, once an encoder is learned (i.e., for a given fixed embedding z), we evaluate against the best possible adversary and target predictors. In the following Lemma, we obtain the minimum MSE for kernelized adversary and target predictors for any given embedding Z.

Lemma 3. *Let $J_y(Z)$ and $J_s(Z)$ be regularized minimum MSEs for adversary and target:*

$$J_y(Z) = \min_{\Lambda_y, b_y} \left\{ \mathbb{E}\{\|\hat{y} - y\|^2\} + \gamma_y \|\Lambda_y\|_F^2 \right\},$$

$$J_s(Z) = \min_{\Lambda_s, b_s} \left\{ \mathbb{E}\{\|\hat{s} - s\|^2\} + \gamma_s \|\Lambda_s\|_F^2 \right\}$$

where γ_y and γ_s are regularization parameters for target and adversary regressors, respectively. Then, for any given embedding matrix Z, the minimum MSE for kernelized adversary and target can be obtained as

$$J_y(Z) = \frac{1}{n} \|\tilde{Y}\|_F^2 - \frac{1}{n} \left\| P_{\mathcal{M}_y} \begin{bmatrix} \tilde{Y}^T \\ \mathbf{0}_n \end{bmatrix} \right\|_F^2,$$

$$J_s(Z) = \frac{1}{n} \|\tilde{S}\|_F^2 - \frac{1}{n} \left\| P_{\mathcal{M}_s} \begin{bmatrix} \tilde{S}^T \\ \mathbf{0}_n \end{bmatrix} \right\|_F^2 \qquad (9)$$

where

$$M_y = \begin{bmatrix} \tilde{K}_y \\ \sqrt{n\gamma_y} I_n \end{bmatrix}, \qquad M_s = \begin{bmatrix} \tilde{K}_s \\ \sqrt{n\gamma_s} I_n \end{bmatrix}$$

are both full column rank matrices and a projection matrix for any full column rank matrix M is

$$P_{\mathcal{M}} = M(M^T M)^{-1} M^T$$

It is straightforward to generalize this method to the case of multiple target and adversary predictors through Eq. (3). In this case we will have multiple λ's to trade-off between fairness and utility.

4.2 Optimal Embedding Dimensionality

The ability to effectively optimize the parameters of the encoder is critically dependent on the dimensionality of the embedding as well. Higher dimensional embeddings can inherently absorb unnecessary extraneous information in the data. Existing ARL applications, where the target and adversary are non-linear neural networks, select the dimensionality of the embedding on an ad-hoc basis.

Adopting closed-form solvers for the target and adversary enables us to analytically determine an upper bound on the optimal dimensionality of the embedding for OptNet-ARL. To obtain the upper bound we rely on the observation that a non-linear target predictor and adversary, by virtue of greater capacity, can learn non-linear decision boundaries. As such, in the context of ARL, the optimal dimensionality required by non-linear models is lower than the optimal dimensionality of linear target predictor and adversary. Therefore, we analytically determine the optimal dimensionality of the embedding in the following theorem.

Theorem 4. *Let z in Fig. 1b be disconnected from the encoder and be a free vector in $\mathbb{R}^r$. Further, assume that both adversary and target predictors are linear regressors. Then, for any $0 \leq \lambda \leq 1$ the optimal dimensionality of embedding vector, r is the number of negative eigenvalues of*

$$B = \lambda \tilde{S}^T \tilde{S} - (1 - \lambda) \tilde{Y}^T \tilde{Y}. \tag{10}$$

Given a dataset with the target and sensitive labels, Y and S respectively, the matrix B and its eigenvalues can be computed offline to determine the upper bound on the optimal dimensionality. By virtue of the greater capacity, the optimal dimensionality required by non-linear models is lower than the optimal dimensionality of linear predictors and therefore, Theorem 2 is a tight upper bound for the optimal embedding dimensionality. One large datasets where $B \in \mathbb{R}^{n \times n}$, the Nyström method with data sampling [24] can be adopted.

4.3 Gradient of Closed-Form Solution

In order to find the gradient of the encoder loss function in (3) with J_y and J_s given in (9), we can ignore the constant terms, $\|\tilde{Y}\|_F$ and $\|\tilde{S}\|_F$. Then, the optimization problem in (3) would be equivalent to

$$\min_{\Theta_E} \left\{ (1 - \lambda) \left\| P_{\mathcal{M}_s} \begin{bmatrix} \tilde{S}^T \\ 0_n \end{bmatrix} \right\|_F^2 - \lambda \left\| P_{\mathcal{M}_y} \begin{bmatrix} \tilde{Y}^T \\ 0_n \end{bmatrix} \right\|_F^2 \right\}$$

$$= \min_{\Theta_E} \left\{ (1 - \lambda) \sum_{k=1}^{p} \|P_{\mathcal{M}_s} u_s^k\|^2 - \lambda \sum_{m=1}^{q} \|P_{\mathcal{M}_y} u_y^m\|^2 \right\} \tag{11}$$

where the vectors $\boldsymbol{u}_s^k$ and $\boldsymbol{u}_y^m$ are the k-th and m-th columns of $\begin{bmatrix} \tilde{\boldsymbol{S}}^T \\ \boldsymbol{0}_n \end{bmatrix}$ and $\begin{bmatrix} \tilde{\boldsymbol{Y}}^T \\ \boldsymbol{0}_n \end{bmatrix}$, respectively. Let M be an arbitrary matrix function of $\boldsymbol{\Theta}_E$, and θ be arbitrary scalar element of $\boldsymbol{\Theta}_E$. Then, from [15] we have

$$\frac{1}{2}\frac{\partial \|P_{\mathcal{M}}\boldsymbol{u}\|^2}{\partial \theta} = \boldsymbol{u}^T P_{\mathcal{M}^\perp} \frac{\partial M}{\partial \theta} M^\dagger \boldsymbol{u} \qquad (12)$$

where

$$\left[\frac{\partial M}{\partial \theta}\right]_{ij} = \begin{cases} \nabla_{z_i}^T([M]_{ij})\nabla_\theta(z_i) + \nabla_{z_j}^T([M]_{ij})\nabla_\theta(z_j), & i \leq n \\ 0, & \text{else.} \end{cases}$$

Equation (12) can be directly used to obtain the gradient of objective function in (11).

Directly computing the gradient in Eq. (12) requires a pseudoinverse of the matrix $M \in \mathbb{R}^{2n \times n}$, which has a complexity of $\mathcal{O}(n^3)$. For large datasets this computation can get prohibitively expensive. Therefore, we approximate the gradient using a single batch of data as we optimize the encoder end-to-end. Similar approximations [24] are in fact commonly employed to scale up kernel methods. Thus, the computational complexity of computing the loss for OptNet-ARL reduces to $\mathcal{O}(b^3)$, where b is the batch size. Since maximum batch sizes in training neural networks are of the order of 10 s to 1000 s, computing the gradient is practically feasible. We note that, the procedure presented in this section is a simple SGD in which its stability can be guaranteed under Lipschitz and smoothness assumptions on encoder network [45].

5 Experiments

In this section we will evaluate the efficacy of our proposed approach, OptNet-ARL, on three different tasks; Fair Classification on UCI [9] datatset, mitigating leakage of private information on the CelebA dataset, and ablation study on a Gaussian mixture example. We also compare OptNet-ARL with other ARL baselines in terms of stability of optimization, the achievable trade-off front between the target and adversary objectives, convergence speed and the effect of embedding dimensionality. We consider three baselines, (1) **SGDA-ARL:** vanilla stochastic gradient descent-ascent that is employed by multiple ARL approaches including [11,20,29,33,40] etc., (2) **ExtraSGDA-ARL:** a state-of-the-art variant of stochastic gradient descent-ascent that uses an extra gradient step [23] for optimizing minimax games. Specifically, we use the ExtraAdam algorithm from [14], and (3) **SARL:** a global optimum solution for a kernelized regressor encoder and linear target and adversary [35]. Specifically, **hypervolume** (HV) [43], a metric for stability and goodness of trade-off (comparing algorithms under multiple objectives) is also utilized. A larger HV indicates a better Pareto front achieved and the standard deviation of the HV represents the stability.

In the training stage, the encoder, a deep neural network, is optimized **end-to-end** against kernel ridge regressors (RBF Gaussian kernel[1]) in the case of OptNet-ARL and multi-layer perceptrons (MLPs) for the baselines. Table 2 summarizes the network architecture of all experiments. We note that the optimal embedding dimensionality, r for binary target is equal to one which is consistent with Fisher's linear discriminant analysis [44]. The embedding is instance normalized (unit norm). So we adopted a fixed value of $\sigma = 1$ for Gaussian Kernel in all the experiments. We let the regression regularization parameter be 10^{-4} for all experiments. The learning rate is 3×10^{-4} with weight decay of 2×10^{-4} and we use Adam as optimizer for all experiments.

At the inference stage, the encoder is frozen, features are extracted and a new target predictor and adversary are trained. At this stage, for both OptNet-ARL and the baselines, the target and adversary have the same model capacity. Furthermore, each experiment on each dataset is repeated five times with different random seeds (except for SARL which has a closed-form solution for encoder) and for different trade-off parameters $\lambda \in [0, 1]$. We report the median and standard deviation across the five repetitions.

Table 2. Network architectures in experiments.

Method (ARL)	Encoder	Embd Dim	Target (Train)	Adversary (Train)	Target (Test)	Adversary (Test)
	Adult					
SGDA [29,40]	MLP-4-2	1	MLP-4	MLP-4	MLP-4-2	MLP-4-2
ExtraSGDA [23]	MLP-4-2	1	MLP-4	MLP-4	MLP-4-2	MLP-4-2
SARL [35]	RBF krnl	1	Linear	Linear	MLP-4-2	MLP-4-2
OptNet-ARL (ours)	MLP-4-2	1	RBF krnl	RBF krnl	MLP-4-2	MLP-4-2
	German					
SGDA [29,40]	MLP-4	1	MLP-2	MLP-2	Logistic	Logistic
ExtraSGDA [23]	MLP-4	1	MLP-2	MLP-2	Logistic	Logistic
SARL [35]	RBF krnl	1	Linear	Linear	Logistic	Logistic
OptNet-ARL (ours)	MLP-4	1	RBF krnl	RBF krnl	Logistic	Logistic
	CelebA					
SGDA [29,40]	ResNet-18	128	MLP-64	MLP-64	MLP-32-16	MLP-32-16
ExtraSGDA [23]	ResNet-18	128	MLP-64-32	MLP-64-32	MLP-32-16	MLP-32-16
OptNet-ARL (ours)	ResNet-18	[1, 128]	RBF krnl	RBF krnl	MLP-32-16	MLP-32-16
	Gaussian mixture					
SGDA [29,40]	MLP-8-4	2	MLP-8-4	MLP-8-4	MLP-4-4	MLP-4-4
ExtraSGDA [23]	MLP-8-4	2	MLP-8-4	MLP-8-4	MLP-4-4	MLP-4-4
SARL [35]	RBF krnl	2	Linear	Linear	MLP-4-4	MLP-4-4
RBF-OptNet-ARL (ours)	MLP-8-4	2	RBF krnl	RBF krnl	MLP-4-4	MLP-4-4
IMQ-OptNet-ARL (ours)	MLP-8-4	$[1, \cdots, 512]$	IMQ krnl	IMQ krnl	MLP-4-4	MLP-4-4

---

[1] $k(z, z') = \exp\left(-\frac{\|z - z'\|^2)}{2\sigma^2}\right).$

5.1 Fair Classification

We consider fair classification on two different tasks. **UCI Adult Dataset:** It includes 14 features from 45,222 instances. The task is to classify the annual income of each person as high (50K or above) or low (below 50K). The sensitive feature we wish to be fair with respect to is the gender of each person. **UCI German Dataset:** It contains 1000 instances of individuals with 20 different attributes. The target task is to predict their creditworthiness while being unbiased with respect to age. The correlation between target and sensitive attributes are 0.03 and 0.02 for the Adult and German dataset, respectively. This indicates that the target attributes are almost orthogonal to the sensitive attributes. Therefore, the sensitive information can be totally removed with only a negligible loss in accuracy for the target task.

Stability: Since there is no trade-off between the two attributes, we compare stability by reporting the median and standard deviation of the target and adversary performance in Table 3. Our results indicate that OptNet-ARL achieves a higher accuracy for target task and lower leakage of sensitive attribute and with less variance. For instance, in Adult dataset, OptNet-ARL method achieves 83.86% and 83.81% target accuracy with almost zero sensitive leakage. For OptNet-ARL the standard deviation of sensitive attribute is exactly zero, which demonstrates its effectiveness and stability in comparison to the baselines. Similarly for the German dataset, OptNet-ARL achieves 80.13% for sensitive accuracy, which is close to random chance (around 81%).

Fair Classification Performance: We compare our proposed approach with many baseline results on these datasets. The optimal dimensionality for OptNet-ARL is $r = 1$ as determined by Theorem 4 and $r = 50$ for the baselines (common choice in previous work). Diff value in Table 3 shows the difference between adversary accuracy and random guessing. On both datasets, both Linear-SARL and

Table 3. Fair classification on UCI dataset (in %)

Method	Adult dataset			German dataset		
	Target (income)	Sensitive (gender)	Diff 67.83	Target (credit)	Sensitive (age)	Diff 81
Raw Data	85.0	85.0	17.6	80.0	87.0	6.0
LFR [41]	82.3	67.0	0.4	72.3	80.5	0.5
AEVB [21]	81.9	66.0	1.4	72.5	79.5	1.5
VFAE [28]	81.3	67.0	0.4	72.7	79.7	1.3
SARL [35]	84.1	67.4	0.0	76.3	80.9	0.1
SGDA-ARL [40]	83.61 ± 0.38	67.08 ± 0.48	0.40	76.53 ± 1.07	87.13 ± 5.70	6.13
ExtraSGDA-ARL [14]	83.66 ± 0.26	66.98 ± 0.49	0.4	75.60 ± 1.68	86.80 ± 4.05	5.80
OptNet-ARL	83.81 ± 0.23	67.38 ± 0.00	0.00	76.67 ± 2.21	80.13 ± 1.48	0.87

OptNet-ARL can achieve high performance on target task with a tiny sensitive attribute leakage for the German dataset.

5.2   Mitigating Sensitive Information Leakage

The CelebA dataset [27] contains $202,599$ face images of $10,177$ celebrities. Each image contains 40 different binary attributes (e.g., gender, emotion, age, etc.). Images are pre-processed and aligned to a fixed size of 112×96 and we use the official train-test splits. The target task is defined as predicting the presence or absence of high cheekbones (binary) with the sensitive attribute being smiling/not smiling (binary). The choice of this attribute pair is motivated by the presence of a trade-off between them. We observe that the correlation between this attribute pair is equal to 0.45, indicating that there is no encoder that can maintain target performance without leaking the sensitive attribute.

For this experiment, we note that SARL [35] cannot be employed, since, (1) it does not scale to large datasets ($\mathcal{O}(n^3)$) like CelebA, and (2) it cannot be applied directly on raw images but needs features extracted from a pre-trained network. Most other attribute pairs in this dataset either suffer from severe class imbalance or small correlation, indicating the lack of a trade-off. Network architecture details are shown in Table 2.

Stability and Trade-off: Figure 2(a) shows the attainment surface [22] and hypervolume [43] (median and standard deviation) for all methods. SGDA spans only a small part of the trade-off and at the same time exhibits large variance around the median curves. Overall both baselines are unstable and unreliable when the two attributes are dependent on each other. On the other hand, OptNet-ARL solutions are very stable and while also achieving a better trade-off between target and adversary accuracy.

Optimal Embedding Dimensionality: Figure 2(b) compares the utility-bias trade-off the sub-optimal embedding dimensionality ($r = 128$) with that of the optimal dimensionality ($r = 1$). We can observe that optimal embedding dimensionality ($r = 1$) is producing a more stable trade-off between adversary and target accuracies.

Training Time: It takes five runs for SGDA-ARL and ExtraSGDA and two runs for OptNet-ARL to train a reliable encoder for overall 11 different values of $\lambda \in [0, 1]$. The summary of training time is given in Fig. 2(c). ExtraSGDA-ARL takes an extra step to update the weights and therefore, it is slightly slower than SGDA-ARL. OptNet-ARL on the other hand is significantly faster to obtain reliable results. Even for a single run, OptNet-ARL is faster than the baselines. This is because, OptNet-ARL uses closed-form solvers for adversary and target and therefore does not need to train any additional networks downstream to the encoder.

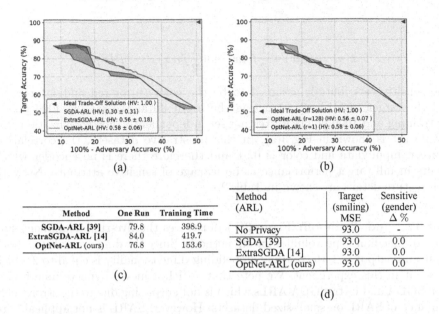

(a)

(b)

Method	One Run	Training Time
SGDA-ARL [39]	79.8	398.9
ExtraSGDA-ARL [14]	84.0	419.7
OptNet-ARL (ours)	76.8	153.6

(c)

Method (ARL)	Target (smiling) MSE	Sensitive (gender) Δ %
No Privacy	93.0	-
SGDA [39]	93.0	0.0
ExtraSGDA [14]	93.0	0.0
OptNet-ARL (ours)	93.0	0.0

(d)

Fig. 2. CelebA: (a) Trade-off between adversary and target accuracy for dependent pair (smiling/not-smiling, high cheekbones). (b) Comparison between the trade-offs of optimal embedding dimensionality $r = 1$ and that of $r = 128$. (c) Overall and single run training time for different ARL methods. (d) Trade-off between adversary and target for independent pair (smiling/not-smiling, gender).

Independent Features: We consider the target task to be binary classification of smiling/not smiling with the sensitive attribute being gender. In this case, the correlation between gender and target feature is 0.02, indicating that the two attributes are almost independent and hence it should be feasible for an encoder to remove the sensitive information without affecting target task. The results are presented in Fig. 2(d). In contrast to the scenario where the two attributes are dependent, we observe that all ARL methods can perfectly hide the sensitive information (gender) from representation without loss of target task. Therefore, OptNet-ARL is especially effective in a more practical setting where the target and sensitive attributes are correlated and hence can only attain a trade-off.

5.3 Ablation Study on Mixture of Four Gaussians

In this experiment we consider a simple example where the data is generated by a mixture of four different Gaussian distributions. Let $\{f_i\}_{i=1}^4$ be all Gaussian distributions with means at $(0,0)$, $(0,1)$, $(1,0)$, and $(1,1)$, respectively and covariance matrices all equal to $\boldsymbol{\Sigma} = 0.2^2 \boldsymbol{I}_2$. Denote by $f(\boldsymbol{x})$ the distribution of input data. Then,

$$f(\boldsymbol{x}|\bullet) = f_1(\boldsymbol{x}) + \frac{1}{2}f_2(\boldsymbol{x}) + +\frac{1}{2}f_3(\boldsymbol{x}), \qquad P\{\bullet\} = \frac{1}{2}$$

$$f(x|\bullet) = f_4(x) + \frac{1}{2}f_2(x) + +\frac{1}{2}f_3(x), \qquad P\{\bullet\} = \frac{1}{2}$$

The sensitive attribute is assumed to be the color (0 for red and 1 for blue) and the target task is reconstructing the input data. We sample 4000 points for training and 1000 points for testing set independently. For visualization, the testing set is shown in Fig. 3(a). In this illustrative dataset , the correlation between input data and color is 0.61 and therefore there is no encoder which results in full target performance at no leakage of sensitive attribute. Network architecture details are shown in Table 2.

Stability and Trade-off: Figure 3(b) illustrates the five-run attainment surfaces and median hypervolumes for all methods. Since the dimensionality of both input and output is 2, the optimal embedding dimensionality is equal to 2 which we set it in this experiment. We note that SARL achieves hypervolume better than SGDA and ExtraSGDA ARLs which is not surprising due to the strong performance of SARL on small-sized datasets. However, SARL is not applicable to large datasets. Among other baselines, ExtraSGDA-ARL appears to be slightly better. In contrast, the solutions obtained by RBF-OptNet-ARL (Gaussian kernel) outperform all baselines and are highly stable across different runs, which can be observed from both attainment surfaces and hypervolumes. Addition to Gaussian kernel, we also used inverse multi quadratic (IMQ) kernel [46]² for OptNet to examine the effect kernel of function. As we observe from Fig. 3(b), IMQ-OptNet-ARL performs almost similar to OptNet-ARR with Gaussian kernel in terms of both trade-off and stability.

Batch Size: In order to examine the effect of batch size on OptNet-ARL (with Gaussian kernel), we train the encoder with different values of batch size between 2 and 4000 (entire training data). The results are illustrated in Fig. 3(c). We observer that the trade-off HV is quite insensitive to batch sizes greater than 25 which implies that the gradient of min-batch is an accurately enough estimator of the gradient of entire data.

Embedding Dimensionality: We also study the effect of embedding dimensionality (r) by examining different values for r in $[1, 512]$ using RBF-OptNet-ARL. The results are illustrated in Fig. 3(d). It is evident that the optimal embedding dimensionality $(r = 2)$ outperforms other values of r. Additionally, HV of $r = 1$ suffers severely due to the information loss in embedding, while for $2 < r \leq 512$ the trade-off performance is comparable to that of optimal embedding dimensionality, $r = 2$.

---

² $k(z, z') = \frac{1}{\sqrt{\|z - z'\|^2 + c^2}}$.

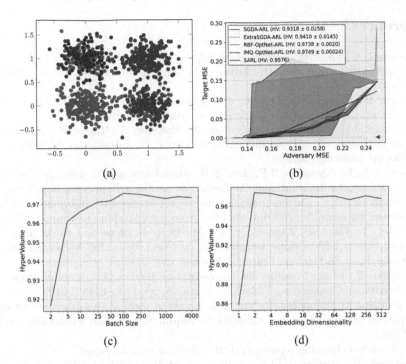

Fig. 3. Mixture of Gaussians: (a) Input data. The target task is to learn a representation which is informative enough to reconstruct the input data and at the same time hide the color information (● vs ●). (b) Trade-off between the MSEs of adversary and target task for different ARL methods. (c) The HVs of OptNet-ARL (Gaussian kernel) vs different batch size values in $[2, 4000]$. (d) The HV values of OptNet-ARL (Gaussian kernel) vs different values of r in $[1, 512]$. (Color figure online)

6   Concluding Remarks

Adversarial representation learning is a minimax theoretic game formulation that affords explicit control over unwanted information in learned data representations. Optimization algorithms for ARL such as stochastic gradient descent-ascent (SGDA) and their variants are sub-optimal, unstable and unreliable in practice. In this paper, we introduced OptNet-ARL to address this challenge by employing differentiable closed-form solvers, such as kernelized ridge regressors, to model the ARL players that are downstream from the representation. OptNet-ARL reduces iterative SGDA to a simple optimization, leading to a fast, stable and reliable algorithm that out-performs existing ARL approaches on both small and large scale datasets.

Acknowledgements. This work was performed under the following financial assistance award 60NANB18D210 from U.S. Department of Commerce, National Institute of Standards and Technology.

References

1. Agrawal, A., Amos, B., Barratt, S., Boyd, S., Diamond, S., Kolter, Z.: Differentiable convex optimization layers. In: Advances in Neural Information Processing Systems (2019)
2. Amos, B., Kolter, J.Z.: Optnet: differentiable optimization as a layer in neural networks. In: International Conference on Machine Learning (2017)
3. Balduzzi, D., Racaniere, S., Martens, J., Foerster, J., Tuyls, K., Graepel, T.: The mechanics of n-player differentiable games. In: International Conference on Machine Learning (2018)
4. Bertinetto, L., Henriques, J.F., Torr, P.H.: Meta-learning with differentiable closed-form solvers. In: International Conference on Learning Representations (2018)
5. Bertran, M., et al.: Adversarially learned representations for information obfuscation and inference. In: International Conference on Machine Learning (2019)
6. Beutel, A., Chen, J., Zhao, Z., Chi, E.H.: Data decisions and theoretical implications when adversarially learning fair representations. In: Accountability, and Transparency in Machine Learning, Fairness (2017)
7. Creager, E., et al.: Flexibly fair representation learning by disentanglement. In: International Conference on Machine Learning, pp. 1436–1445 (2019)
8. Daskalakis, C., Panageas, I.: The limit points of (optimistic) gradient descent in min-max optimization. In: Advances in Neural Information Processing Systems (2018)
9. UCI machine learning repository. http://archive.ics.uci.edu/ml
10. Edwards, H., Storkey, A.: Censoring representations with an adversary. In: International Conference on Learning Representations (2015)
11. Elazar, Y., Goldberg, Y.: Adversarial removal of demographic attributes from text data. In: Empirical Methods in Natural Language Processing (2018)
12. Ganin, Y., Lempitsky, V.: Unsupervised domain adaptation by backpropagation. In: International Conference on Machine Learning, pp. 1180–1189 (2015)
13. Ganin, Y., et al.: Domain-adversarial training of neural networks. J. Mach. Learn. Res. $17(1)$, 2030–2096 (2016)
14. Gidel, G., Berard, H., Vignoud, G., Vincent, P., Lacoste-Julien, S.: A variational inequality perspective on generative adversarial networks. In: International Conference on Learning Representations (2019)
15. Golub, G.H., Pereyra, V.: The differentiation of pseudo-inverses and nonlinear least squares problems whose variables separate. SIAM J. Numer. Anal. $10(2)$, 413–432 (1973)
16. Gretton, A., Herbrich, R., Smola, A., Bousquet, O., Schölkopf, B.: Kernel methods for measuring. J. Mach. Learn. Res. Ind. 6, 2075–2129 (2005)
17. Ionescu, C., Vantzos, O., Sminchisescu, C.: Training deep networks with structured layers by matrix backpropagation. In: IEEE International Conference on Computer Vision (2015)
18. Jacot, A., Gabriel, F., Hongler, C.: Neural tangent kernel: convergence and generalization in neural networks. In: Advances in Neural Information Processing Systems (2018)
19. Jin, C., Netrapalli, P., Jordan, M.: What is Local Optimality in Nonconvex-Nonconcave Minimax Optimization? arXiv preprint arXiv:1902.00618 (2019)
20. Kim, B., Kim, H., Kim, K., Kim, S., Kim, J.: Learning not to learn: training deep neural networks with biased data. In: IEEE Conference on Computer Vision and Pattern Recognition (2019)

21. Kingma, D.P., Welling, M.: Auto-encoding variational bayes. arXiv preprint arXiv:1312.6114 (2013)
22. Knowles, J.: A summary-attainment-surface plotting method for visualizing the performance of stochastic multiobjective optimizers. In: International Conference on Intelligent Systems Design and Applications (2015)
23. Korpelevich, G.M.: The extragradient method for finding saddle points and other problems. Matecon **12**, 747–756 (1976)
24. Kumar, S., Mohri, M., Talwalkar, A.: Sampling methods for the Nyström method. J. Mach. Learn. Res. **13**(4), 981–1006 (2012)
25. Lee, K., Maji, S., Ravichandran, A., Soatto, S.: Meta-learning with differentiable convex optimization. In: IEEE Conference on Computer Vision and Pattern Recognition (2019)
26. Letcher, A., Balduzzi, D., Racaniere, S., Martens, J., Foerster, J., Tuyls, K., Graepel, T.: Differentiable game mechanics. J. Mach. Learn. Res. **20**(84), 1–40 (2019)
27. Liu, Z., Luo, P., Wang, X., Tang, X.: Deep learning face attributes in the wild. In: IEEE International Conference on Computer Vision (2015)
28. Louizos, C., Swersky, K., Li, Y., Welling, M., Zemel, R.: The variational fair autoencoder. arXiv preprint arXiv:1511.00830 (2015)
29. Madras, D., Creager, E., Pitassi, T., Zemel, R.: Learning adversarially fair and transferable representations. In: International Conference on Machine Learning (2018)
30. Mescheder, L., Nowozin, S., Geiger, A.: The numerics of gans. In: Advances in Neural Information Processing Systems (2017)
31. Moyer, D., Gao, S., Brekelmans, R., Steeg, G.V., Galstyan, A.: Invariant Representations without Adversarial Training, In: Advances in Neural Information Processing Systems (2018)
32. Nagarajan, V., Kolter, J.Z.: Gradient descent GAN optimization is locally stable. In: Advances in Neural Information Processing Systems (2017)
33. Roy, P.C., Boddeti, V.N.: Mitigating information leakage in image representations: a maximum entropy approach. In: IEEE Conference on Computer Vision and Pattern Recognition (2019)
34. Sadeghi, B., Boddeti, V.N.: Imparting fairness to pre-trained biased representations. In: IEEE Conference on Computer Vision and Pattern Recognition Workshops (2020)
35. Sadeghi, B., Yu, R., Boddeti, V.: On the global optima of kernelized adversarial representation learning. In: IEEE International Conference on Computer Vision (2019)
36. Shawe-Taylor, J., Cristianini, N.: Kernel methods for pattern analysis. Cambridge University Press, Cambridge (2014)
37. Song, J., Kalluri, P., Grover, A., Zhao, S., Ermon, S.: Learning controllable fair representations. In: International Conference on Artificial Intelligence and Statistics (2019)
38. Tzeng, E., Hoffman, J., Saenko, K., Darrell, T.: Adversarial discriminative domain adaptation. In: IEEE Conference on Computer Vision and Pattern Recognition (2017)
39. Valmadre, J., Bertinetto, L., Henriques, J., Vedaldi, A., Torr, P.H.: End-to-end representation learning for correlation filter based tracking. In: IEEE Conference on Computer Vision and Pattern Recognition (2017)
40. Xie, Q., Dai, Z., Du, Y., Hovy, E., Neubig, G.: Controllable invariance through adversarial feature learning. In: Advances in Neural Information Processing Systems (2017)

41. Zemel, R., Wu, Y., Swersky, K., Pitassi, T., Dwork, C.: Learning fair representations. In: International Conference on Machine Learning (2013)
42. Zhang, B.H., Lemoine, B., Mitchell, M.: Mitigating unwanted biases with adversarial learning. In: AAAI/ACM Conference on AI, Ethics, and Society (2018)
43. Zitzler, E., Thiele, L.: Multiobjective optimization using evolutionary algorithms—a comparative case study. In: International Conference on Parallel Problem Solving from Nature (1998)
44. Fisher, R.A.: The use of multiple measurements in taxonomic problems. In: Annals of human eugenics. Wiley Online Library (1926)
45. Hardt, M., Recht, B., Singer, Y.: Train faster, generalize better: stability of stochastic gradient descent. In: International Conference on Machine Learning (2016)
46. Souza, C.R.: Kernel functions for machine learning applications. In: Creative commons attribution-noncommercial-share alike (2016)

Learning Unbiased Representations
via Rényi Minimization

Vincent Grari[1,3]([✉]), Oualid El Hajouji[2], Sylvain Lamprier[1],
and Marcin Detyniecki[3]

[1] Sorbonne Université LIP6/CNRS, Paris, France
{vincent.grari,sylvain.lamprier}@lip6.fr
[2] Ecole Polytechnique, Palaiseau, France
oualid.el-hajouji@polytechnique.edu
[3] AXA REV Research, Paris, France
marcin.detyniecki@axa.com

Abstract. In recent years, significant work has been done to include fairness constraints in the training objective of machine learning algorithms. Differently from classical prediction retreatment algorithms, we focus on learning fair representations of the inputs. The challenge is to learn representations that capture most relevant information to predict the targeted output Y, while not containing any information about a sensitive attribute S. We leverage recent work which has been done to estimate the Hirschfeld-Gebelein-Renyi (HGR) maximal correlation coefficient by learning deep neural network transformations and use it as a min-max game to penalize the intrinsic bias in a multi dimensional latent representation. Compared to other dependence measures, the HGR coefficient captures more information about the non-linear dependencies, making the algorithm more efficient in mitigating bias. After providing a theoretical analysis of the consistency of the estimator and its desirable properties for bias mitigation, we empirically study its impact at various levels of neural architectures. We show that acting at intermediate levels of neural architectures provides best expressiveness/generalization abilities for bias mitigation, and that using an HGR based loss is more efficient than more classical adversarial approaches from the literature.

1 Introduction

This recent decade, deep learning models have shown very competitive results by learning representations that capture relevant information for the learning task[1]. However, the representation learnt by the deep model may contain some bias from the training data. This bias can be intrinsic to the training data, and may therefore induce a generalisation problem due to a distribution shift between

---

[1] https://github.com/axa-rev-research/unbiased_representations_renyi.

Electronic supplementary material The online version of this chapter (https://doi.org/10.1007/978-3-030-86520-7_46) contains supplementary material, which is available to authorized users.

© Springer Nature Switzerland AG 2021
N. Oliver et al. (Eds.): ECML PKDD 2021, LNAI 12976, pp. 749–764, 2021.
https://doi.org/10.1007/978-3-030-86520-7_46

training and testing data. For instance, the color bias in the colored MNIST data set [25] can make models focus on the color of a digit rather than its shape for the classification task. The bias can also go beyond training data, so that inadequate representations can perpetuate or even reinforce some society biases [10] (e.g. gender or age). Since the machine learning models have far-reaching consequences in our daily lives (credit rating, insurance pricing, recidivism score, etc.), we need to make sure that the representation data contains as little bias as possible. A naive method to mitigate bias could be to simply remove sensitive attributes from the training data set [36]. However, this concept, known as "fairness through unawareness", is highly insufficient because any other non-sensitive attribute might indirectly contain significant sensitive information reflected in the deep learning representation. For example, the height of an adult could provide a strong indication about the gender. A new research field has emerged to find solutions to this problem: fair machine learning. Its overall objective is to ensure that the prediction model is not dependent on a sensitive attribute [46]. Many recent papers tackle this challenge using an adversarial neural architecture, which can successfully mitigate the bias. We distinguish two adversarial mitigation families. While prediction retreatment methods apply mitigation on the output prediction [48], fair representation methods consider sensitive bias in intermediary latent representations [1]. Our claim is that mitigating at intermediate levels of neural architectures allows a greater stability at test time, which we observe in our experiments (Sect. 6.3).

In this paper, we propose a new fair representation architecture by leveraging the recent Renyi neural estimator, previously used in a prediction retreatment algorithm [19] and we propose to study why such an architecture outperforms the state of the art. The contributions of this paper are:

- We provide a theoretical analysis of the consistency of the HGR estimator, along its nice properties compared to state-of-the-art for fair representation;
- We propose a neural network architecture which creates a fair representation by minimizing the HGR coefficient. The HGR network is trained to discover non-linear transformations between the multidimensional latent representation and the sensitive feature. Note that this is also the first use of a neural HGR estimator for multidimensional variables;
- We empirically demonstrate that our neural HGR-based approach is able to identify the optimal transformations with multidimensional features and present very competitive results for fairness learning;
- To the best of our knowledge, this is the first work to compare mitigation at different levels of neural architectures. We show that acting at intermediary levels of neural representations allows the best trade-off between expressiveness and generalisation for bias mitigation.

2   Related Work

Significant work has been done in the field of fair machine learning recently, in particular when it comes to quantifying and mitigating undesired bias. For the mitigation approaches, three distinct strategy groups exist. While pre-processing [7,11,23] and post-processing [13,21] approaches respectively act on the input or the output of a classically trained predictor, in-processing approaches mitigate the undesired bias directly during the training phase [12,30,46,48]. In this paper we focus on in-processing fairness, which proves to be the most powerful framework for settings where acting on the training process is an option.

Among the in-processing approaches, some of them, referred to as prediction retreatment, aim at directly modifying the prediction output by adversarial training. To ensure independence between the output and the sensitive attribute, Zhang et al. [48] feed the prediction output as input to an adversary network (upper right in Fig. 1 in appendix), whose goal is to predict the sensitive attribute, and update the predictor weights to fool the adversary. Grari et al. [19] minimize the HGR correlation between the prediction output and the sensitive attribute in an adversarial learning setting (middle right in Fig. 1 in appendix).

On the other hand, several research sub-fields in the in-processing family tackle the problem of learning unbiased representations. Domain adaptation [9,14] and domain generalization [28,35] consist in learning representations that are unbiased with respect to a source distribution, and can therefore generalize to other domains. Some of the works in these fields involve the use of adversarial methods [16,17], close to our work. Several strategies mitigate bias towards a sensitive attribute through representation. One approach [47] relies on a discriminative clustering model to learn a multinomial representation that removes information regarding a binary sensitive attribute. A different approach [2] consists in learning an unbiased representation by minimizing a confusion loss. Invariant representations can also be learnt using Variational Auto-Encoders [26], by adding a mutual information penalty term [34]. One of the first proposition by adversarial neural network for fair representation has been proposed by [32] by mitigating the bias on the latent space with an adversarial and decoding of X from Z and A. Adel et al. [1] learn also a fair representation by inputting it to an adversary network, which is prevented from predicting the sensitive attribute (upper left in Fig. 1 in appendix). Other papers minimize the mutual information between the representation and the sensitive attribute: Kim et al. [25] rely on adversarial training with a discriminator detecting the bias, while Ragonesi et al. [38] rely on an estimation by neural network of mutual information [6] (lower left in Fig. 1 in appendix). A kernelized version of such adversarial debiasing approach for fair representation is provided in [40].

3   Problem Statement

Throughout this document, we consider a supervised algorithm for regression or classification problems. The training data consists of n examples $(x_i, s_i, y_i)_{i=1}^n$, where $x_i \in \mathbb{R}^p$ is the feature vector with p predictors of the i-th example, s_i is its continuous sensitive attribute and y_i its continuous or discrete outcome. We address a common objective in fair machine learning, *Demographic Parity*, which ensures that the sensitive attribute S is independent of the prediction $\hat{Y}$.

3.1   Metrics for Continuous Statistical Dependence

In order to assess this fairness definition in the continuous case, it is essential to look at the concepts and measures of statistical dependence. Simple ways of measuring dependence are Pearson's rho, Kendall's tau or Spearman's rank. Those types of measure have already been used in fairness, with the example of mitigating the conditional covariance for categorical variables [46]. However, the major problem with these measures is that they only capture a limited class of association patterns, like linear or monotonically increasing functions. For example, a random variable with standard normal distribution and its cosine (non-linear) transformation are not correlated in the sense of Pearson.

Over the last few years, many non-linear dependence measures have been introduced like the Kernel Canonical Correlation Analysis (KCCA) [20], the Distance or Brownian Correlation (dCor) [41], the Hilbert-Schmidt Independence Criterion (HSIC and CHSIC) [37] or the Hirschfeld-Gebelein-Rényi (HGR) [39]. Comparing those non-linear dependence measures [29], the HGR coefficient seems to be an interesting choice: it is a normalized measure which is capable of correctly measuring linear and non-linear relationships, it can handle multidimensional random variables and it is invariant with respect to changes in marginal distributions.

Definition 1. *For two jointly distributed random variables $U \in \mathcal{U}$ and $V \in \mathcal{V}$, the Hirschfeld-Gebelein-Rényi maximal correlation is defined as:*

$$HGR(U,V) = \sup_{f:\mathcal{U}\to\mathbb{R}, g:\mathcal{V}\to\mathbb{R}} \rho(f(U), g(V)) = \sup_{\substack{f:\mathcal{U}\to\mathbb{R}, g:\mathcal{V}\to\mathbb{R} \\ E(f(U))=E(g(V))=0 \\ E(f^2(U))=E(g^2(V))=1}} E(f(U)g(V))$$

(1)

where ρ is the Pearson linear correlation coefficient with some measurable functions f and g with positive and finite variance.

The HGR coefficient is equal to 0 if the two random variables are independent. If they are strictly dependent the value is 1. The spaces for the functions f and g are infinite-dimensional. This property is the reason why the HGR coefficient proved difficult to compute.

Several approaches rely on Witsenhausen's linear algebra characterization [44] to compute the HGR coefficient. For discrete features, this characterization can be combined with Monte-Carlo estimation of probabilities [5], or with kernel density estimation (KDE) [33] to compute the HGR coefficient. We will refer to this second metric, in our experiments, as HGR_KDE. Note that this metric can be extended to the continuous case by discretizing the density computation. Another way to approximate this coefficient, Randomized Dependence Coefficient (RDC) [29], is to require that f and g belong to reproducing kernel Hilbert spaces (RKHS) and take the largest canonical correlation between two sets of copula random projections. We will make use of this approximated metric as HGR_RDC. Recently a new approach [19] proposes to estimate the HGR by deep neural network. The main idea is to use two inter-connected neural networks to approximate the optimal transformation functions f and g from 1. The $HGR_\Theta(U, V)$ estimator is computed by considering the expectation of the products of standardized outputs of both networks ($\hat{f}_{w_f}$ and $\hat{g}_{w_g}$). The respective parameters w_f and w_g are updated by gradient ascent on the objective function to maximize: $J(w_f, w_g) = E[\hat{f}_{w_f}(U)\hat{g}_{w_g}(V)]$. This estimation has the advantage of being estimated by backpropagation, the same authors therefore present a bias mitigation via a min-max game with an adversarial neural network architecture. However, this attenuation is performed on the predictor output only. Several recent papers [1,38] have shown that performing the attenuation on a representation tends to give better results in terms of prediction accuracy while remaining fair in complex real-world scenarios. In this work, we are interested in learning fair representations via this Renyi estimator.

4   Theoretical Properties

In this section we study the consistency of the HGR_NN estimator (referred to as $\widehat{HGR(U, V)}_n$), and provide a theoretical comparison with simple adversarial algorithms that rely on an adversary which predicts the sensitive attribute [1,48]. All the proofs can be found in the Supplementary Material.

4.1   Consistency of the HGR_NN

Definition 2 *(Strong consistency). The estimator $\widehat{HGR(U, V)}_n$ is strongly consistent if for all $\epsilon > 0$, there exists a positive integer N and a choice of statistics network such that:*

$$\forall n \geq N, |HGR(U, V) - \widehat{HGR(U, V)}_n| \leq \epsilon, a.s. \tag{2}$$

As explained in MINE [6], the question of consistency is divided into two problems: a deterministic approximation problem related to the choice of the statistics network, and an estimation problem related to the use of empirical measures.

The first lemma addresses the approximation problem using universal approximation theorems for neural networks [22]:

Lemma 1 *(Approximation). Let $\eta > 0$. There exists a family of continuous neural networks F_Θ parametrized by a compact domain $\Theta \subset \mathbb{R}^k$, such that*

$$|HGR(U, V) - HGR_\Theta(U, V)| \leq \eta. \tag{3}$$

The second lemma addresses the estimation problem, making use of classical consistency theorems for extremum estimators [18]. It states the almost sure convergence of HGR_NN to the associated theoretical neural HGR measure as the number of samples goes to infinity:

Lemma 2 *(Estimation). Let $\eta > 0$, and F_Θ a family of continuous neural networks parametrized by a compact domain $\Theta \subset \mathbb{R}^k$. There exists an $N \in \mathbb{N}$ such that:*

$$\forall n \geq N, |\widehat{HGR(U,V)}_n - HGR_\Theta(U,V)| \leq \eta, a.s. \tag{4}$$

It is implied here that, from rank N, all sample variances are positive in the definition of $\widehat{HGR(U,V)}_n$, which makes the latter well-defined.

We deduce from these two lemmas the following result:

Theorem 1. $\widehat{HGR(U,V)}_n$ *is strongly consistent.*

4.2 Theoretical Comparison Against Simple Adversarial Algorithms

Given X and Y two one-dimensional random variables, we consider the regression problem:

$$\inf_{f:\mathbb{R}\to\mathbb{R}} E((Y - f(X))^2) \tag{5}$$

The variable that minimizes the quadratic risk is $E(Y|X)$. Thus, prediction retreatment algorithms with predictive adversaries [48], which consider such optimization problems for mitigating biases, achieve the global fairness optimum when $E(S|\widehat{Y}) = E(S)$. This does **not generally imply demographic parity** when S is continuous. On the other hand, adversarial approaches based on the HGR_NN [19] achieve the optimum when $HGR(\widehat{Y}, S) = 0$, which is equivalent to demographic parity: $P(\widehat{Y}|S) = P(\widehat{Y})$.

To illustrate this, we consider the maximization problem $\sup_{f:\mathbb{R}\to\mathbb{R}} \rho(f(X), Y)$, which corresponds to the situation where the neural network g is linear in the HGR neural estimator. We have the following result:

Theorem 2. *If $E(Y|X)$ is constant, then $\sup_f \rho(f(X), Y) = 0$. Else, $f^* \in \arg\max_f \rho(f(X), Y)$ iff there exists $a, b \in \mathbb{R}$, with $a > 0$, such that:*

$$f^*(X) = aE(Y|X) + b \tag{6}$$

In other words, the simpler version of the HGR_NN, with g linear, finds the optimal function in terms of regression risk, up to a linear transformation that can be found by simple linear regression. The simplified HGR estimation module therefore captures the exact same non-linear dependencies as the predictive adversary in related work [1,48]. Thanks to the function g, in cases where Y cannot be expressed as a function of X only, the HGR neural network can capture more dependencies than a predictive NN (or equivalently a simplified HGR neural network).

Specific Example to Understand the Difference: Let us consider the following example below where:

$$Y \sim \mathcal{N}(\mu, \sigma^2) \qquad X = \arctan(Y^2) + U\pi \qquad (7)$$

where $U \perp Y$ and U follows a Bernoulli distribution with $p = \frac{1}{2}$. In this setting, we have $Y^2 = \tan(X)$, $HGR(X, Y) = 1$ and due to the hidden variable U, neither X nor Y can be expressed as a function of the other. In that case, the simplified maximal correlation, $\rho(E(Y|X), Y)$, has the following bounds, with $\alpha = \frac{\mu}{\sigma}$: $\sqrt{1 - e^{-\frac{\alpha^2}{2}}} \leq \rho(E(Y|X), Y) \leq \sqrt{1 - e^{-\frac{\alpha^2}{2}}(1 + \alpha^2)^{-\frac{3}{2}}}$. In the degenerate case $\alpha = 0$, we have $E(Y|X) = 0$: the predictive neural network cannot find any dependence. For non-zero values of α, the distribution of Y is no longer centered around the axis of symmetry of the square function, so that the prediction becomes possible. However, as shown in the inequality above, the simplified maximal correlation is less than 1, and close to 0 when $\mu \ll \sigma$.

In Fig. 1, we illustrate the bounds (proof in appendix), $\rho(E(Y|X), Y)$ being estimated by Monte-Carlo. First, we note that the upper bound is close to $\rho(E(Y|X), Y)$, whereas the lower bound $\sqrt{1 - e^{-\frac{\alpha^2}{2}}}$ is not as precise. For non-zero values of α, $\rho(E(Y|X), Y)$ is positive, so that a predictive neural network can capture some non-linear dependencies between Y and X.

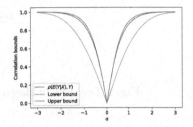

Fig. 1. Simplified HGR w.r.t α

This is due to the fact that, for $\alpha \neq 0$, the square function is bijective when restricted to some open interval containing the mean of Y, whereas when $\alpha = 0$, such an interval cannot be found. When this interval is large and the standard deviation of Y is not too large (which corresponds to high values of $|\alpha|$), $\rho(E(Y|X), Y)$ approaches 1 and the Y prediction error approaches 0. In the opposite case, $\rho(E(Y|X), Y)$ is close to 0 and a predictive neural network cannot capture dependencies.

Therefore, as shown by the example, the bilateral approach of the HGR, as opposed to the unilateral approach of predictive models, can capture more dependencies in complex regression scenarios. In adversarial bias mitigation settings, predictive adversaries might not be able to properly detect bias. Adversarial approaches based on the HGR_NN are better fitted for bias mitigation in such continuous complex settings.

5   Method

The objective is to find a latent representation Z which both minimizes the deviation between the target Y and the output prediction $\widehat{Y}$, provided by a function $\phi(Z)$, and does not imply too much dependence with the sensitive S. As explained above in Sect. 3, the HGR estimation by deep neural network [19] is a good candidate for standing as the adversary $HGR(Z, S)$ to plug in the global objective (8). Notice, we can consider the latent representation Z or even the sensitive attribute S as multi-dimensional. This can therefore provide a rich representation of the latent space or even take into account several sensitive features at the same time (for e.g. gender and age or the 3 channels of an image see Sect. 6.1). The HGR estimation paper [19] considers only the one-dimensional cases for both U and V but we can generalize to the multidimensional cases.

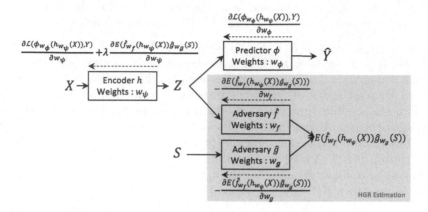

Fig. 2. Learning unbiased representations via Rényi Minimization

The mitigation procedure follows the optimization problem:

$$\underset{w_\phi, w_\psi}{\arg\min} \; \underset{w_f, w_g}{\max} \; \mathcal{L}(\phi_{\omega_\phi}(h_{\omega_\psi}(X)), Y) + \lambda E(\widehat{f}_{w_f}(h_{\omega_\psi}(X)) * \widehat{g}_{w_g}(S)) \qquad (8)$$

where $\mathcal{L}$ is the predictor loss function between the output prediction $\phi_{\omega_\phi}(h_{\omega_\psi}(X)) \in \mathbb{R}$ and the corresponding target Y, with ϕ_{ω_ϕ} the predictor neural network with parameters ω_ϕ and $Z = h_{\omega_\psi}(X)$ the latent fair representation with h_{ω_ψ} the encoder neural network, with parameters ω_ψ. The second term, which corresponds to the expectation of the products of standardized outputs of both networks ($\widehat{f}_{w_f}$ and $\widehat{g}_{w_g}$), represents the HGR estimation between the latent variable Z and the sensitive attribute S. The hyperparameter λ controls the impact of the correlation loss in the optimization.

Figure 2 gives the full architecture of our adversarial learning algorithm using the neural HGR estimator between the latent variable and the sensitive attribute. It depicts the encoder function h_{w_ψ}, which outputs a latent variable Z from X, the two neural networks f_{w_f} and g_{w_g}, which seek at defining the most strongly correlated transformations of Z and S and the neural network ϕ_{ω_ϕ} which outputs the prediction $\widehat{Y}$ from the latent variable Z. Left arrows represent gradients backpropagation. The learning is done via stochastic gradient, alternating steps of adversarial maximization and global loss minimization. The algorithm (more details in the supplementary) takes as input a training set from which it samples batches of size b at each iteration. At each iteration it first standardizes the output scores of networks f_{w_f} and g_{w_g} to ensure 0 mean and a variance of 1 on the batch. Then it computes the HGR neural estimate and the prediction loss for the batch. At the end of each iteration, the algorithm updates the parameters of the prediction parameters ω_ϕ as well as the encoder parameters ω_ψ by one step of gradient descent. Concerning the HGR adversary, the backpropagation of the parameters w_f and w_g is carried by multiple steps of gradient ascent. This allows us to optimize a more accurate estimation of the HGR at each step, leading to a greatly more stable learning process.

6 Experiments

6.1 Synthetic Scenario

We consider the following toy scenario in a binary target Y and continuous standard gaussian sensitive attribute S setting:

$$X|S = s \sim \mathcal{N}\left[\begin{pmatrix} 0 \\ 0 \end{pmatrix}, \begin{pmatrix} 1 & -\frac{1}{2} \\ -\frac{1}{2} & 1 \end{pmatrix}\right] \quad \text{when } Y = 0, \tag{9a}$$

$$X|S = s \sim \mathcal{N}\left[\begin{pmatrix} 1 \\ 1 + 3\sin s \end{pmatrix}, \begin{pmatrix} 1 & 0 \\ 0 & 1 \end{pmatrix}\right] \quad \text{when } Y = 1 \tag{9b}$$

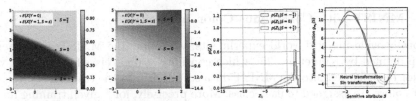

(a) Biased model: $\lambda = 0$; $HGR(Z, S) = 52\%$; $HGR(\widehat{Y}, S) = 30\%$; $Acc = 79\%$

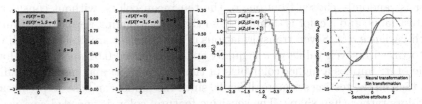

(b) Unbiased model: $\lambda = 13$; $HGR(Z, S) = 5\%$; $HGR(\widehat{Y}, S) = 4\%$; $Acc = 68\%$

Fig. 3. Toy example. (Left) Decision surface in the (X_1, X_2) plane. The figure (a) shows the decision surface for a biased model focused on a prediction loss. $\widehat{Y}$ values are highly correlated with S, samples with S around $\frac{\pi}{2}$ and $Y = 1$ being easier to classify than those with S between $-\frac{\pi}{2}$ and 0. The figure (b) shows decision surfaces for our fair model. These are vertical, meaning that only X_1 influences the classification, and therefore $\widehat{Y}$ is no longer biased w.r.t S. (Middle left) Z_1-slices in the (X_1, X_2) plane. The comparison between the figure below and above highlights the fact that adversarial training allows to create an unbiased representation Z. (Middle right) Conditional probability densities of Z_1 at $S = -\frac{\pi}{2}, 0, \frac{\pi}{2}$. With $\lambda = 0$, the densities are dependent on S, whereas they are not anymore with adversarial training. (Right) In blue, the function modeled by the neural network g in the HGR Neural Network. In red, the closest linear transformation of $\sin(S)$ to $g(S)$. (Color figure online)

Our goal is to learn a representation Z of the input data that is no longer biased w.r.t S, while still accurately predicting the target value Y. Figure 3 compares the results of both a biased model (a) with a hyperparameter $\lambda = 0$ and an unbiased model (b) with $\lambda = 13$ applied on the toy scenario data. In the context of the Rényi Minimization method, it is interesting to observe the maximal correlation functions learnt by the adversary. When $\lambda = 0$, the adversary with sensitive attribute input models the sin function up to a linear transformation, which also maximizes the correlation with the input data as shown in (9b). In that case, the representation Z still carries the bias of X w.r.t S, in the same sin shape. When $\lambda = 13$, the neural network g is unable to find the sin function, which seems to indicate that the representation Z does not carry the bias w.r.t S anymore. This is confirmed by the low HGR coefficient between Z and S, the Z_1-slices as well as the conditional densities of Z_1 at different values of S. Not only does the adversarial induces an unbiased representation, it also leads to an almost completely unbiased target $\widehat{Y}$, as shown by the vertical decision surfaces and the 4% HGR between $\widehat{Y}$ and S. This at the cost of a slight loss of accuracy, with an 11% decrease.

6.2 MNIST with Continuous Color Intensity

Before considering real-world experiments, we follow the MNIST experimental setup defined by Kim et al. [25], which considers a digit classification task with a color bias planted into the MNIST data set [24, 27]. In the training set, ten distinct colors are assigned to each class. More precisely, for a given training image, a color is sampled from the isotropic normal distribution with the corresponding class mean color, and a variance parameter σ^2. For a given test image, a mean color is randomly chosen from one of the ten mean colors, without considering the test label, and a color is sampled from the corresponding normal distribution (with variance σ^2). Seven transformations of the data set are designed with this protocol, with seven values of σ^2 equally spaced between 0.02 and 0.05. A lower value of σ^2 implies a higher color bias in the training set, making the classification task on the testing set more difficult, since the model can base its predictions on colors rather than shape. The sensitive feature, color, is encoded as a vector with 3 continuous coordinates. For each algorithm and for each data set, we obtain the best hyperparameters by grid search in five-fold cross validation.

Table 1. MNIST with continuous color intensity

Training	Color variance						
	$\sigma = 0.020$	$\sigma = 0.025$	$\sigma = 0.030$	$\sigma = 0.035$	$\sigma = 0.040$	$\sigma = 0.045$	$\sigma = 0.050$
ERM ($\lambda = 0.0$)	0.476 ± 0.005	0.542 ± 0.004	0.664 ± 0.001	0.720 ± 0.010	0.785 ± 0.003	0.838 ± 0.002	0.870 ± 0.001
Ragonesi et al. [38]	0.592 ± 0.018	0.678 ± 0.015	0.737 ± 0.028	0.795 ± 0.012	0.814 ± 0.019	0.837 ± 0.004	0.877 ± 0.010
Zhang et al. [48]	0.584 ± 0.034	0.625 ± 0.033	0.709 ± 0.027	0.733 ± 0.020	0.807 ± 0.013	0.803 ± 0.027	0.831 ± 0.027
Kim et al. [25]	0.645 ± 0.015	0.720 ± 0.014	0.787 ± 0.018	0.827 ± 0.012	0.869 ± 0.023	0.882 ± 0.019	0.900 ± 0.012
Grari et al. [19]	0.571 ± 0.014	0.655 ± 0.022	0.721 ± 0.030	0.779 ± 0.011	0.823 ± 0.013	0.833 ± 0.026	0.879 ± 0.010
Ours	$\mathbf{0.730 \pm 0.008}$	$\mathbf{0.762 \pm 0.021}$	$\mathbf{0.808 \pm 0.011}$	$\mathbf{0.838 \pm 0.010}$	$\mathbf{0.878 \pm 0.011}$	$\mathbf{0.883 \pm 0.012}$	$\mathbf{0.910 \pm 0.007}$

Results, in terms of accuracy, can be found in Table 1. Notice, the state-of-the-art obtains different results than reported ones because we consider a continuous sensitive feature and not a 24-bit binary encoding. Our adversarial algorithm achieves the best accuracy on the test set for the seven scenarios. The most important gap is for the smallest sigma where the generalisation is the most difficult. The larger number of degrees of freedom carried by the two functions f and g made it possible to capture more unbiased information than the other algorithms on the multidimensional variables Z and S.

6.3   Real-World Experiments

Our experiments on real-world data are performed on five data sets. In three data sets, the sensitive and the outcome true value are both continuous: the US Census data set [43], the Motor data set [42] and the Crime data set [15]. On two other data sets, the target is binary and the sensitive features are continuous: The COMPAS data set [3] and the Default data set [45]. For all data sets, we repeat five experiments by randomly sampling two subsets, 80% for the training set and 20% for the test set. Finally, we report the average of the mean squared error (MSE), the accuracy (ACC) and the mean of the fairness metrics HGR_NN [19], HGR_KDE [33], HGR_RDC [29] and MINE [6] on the test set. Since none of these fairness measures are fully reliable (they are only estimations which are used by the compared models), we also use the *FairQuant* metric [19], based on the quantization of the test samples in 50 quantiles w.r.t. to the sensitive attribute. The metric corresponds to the mean absolute difference between the global average prediction and the mean prediction of each quantile.

As a baseline, we use a classic, "unfair" deep neural network, Standard NN. We compare our approach with state-of-the-art algorithms. We also compare the Fair MINE NN [19] algorithm where fairness is achieved with the MINE estimation of the mutual information as a penalization in prediction retreatment (lower right in Fig. 1 in appendix).

For all the different fair representation algorithms, we assign the latent space with only one hidden layer with 64 units. Mean normalization was applied to all the outcome true values. Results of our experiments can be found in Table 2. For all of them, we attempted to obtain comparable results by giving similar accuracy to all models, via the hyperparameter λ (different for each model). For each algorithm and for each data set, we obtain the best hyperparameters by grid search in five-fold cross validation (specific to each of them). Notice that, as explained in Sect. 5, several optimization iterations are performed for the adversarial HGR neural estimation at each global backpropagation iteration (e.g., 50 iterations of HGR estimation at each step for the Compas dataset). For comparable results, we also optimize multiple iterations on the different adversarial state-of-the-art algorithms, we find the best number of adversarial backpropagation iterations by grid search between 1 to 300 by step of 25.

As expected, the baseline, Standard NN, is the best predictor but also the most biased one. It achieves the lowest prediction errors and ranks amongst the highest and thus worst values for all fairness measures for all data sets and

Table 2. Experimental results - Best performance among fair algorithms in bold.

		MSE	HGR_NN	HGR_KDE	HGR_RDC	MINE	FairQuant
US Census	Standard NN	0.274 ± 0.003	0.212 ± 0.094	0.181± 0.00	0.217 ± 0.004	0.023 ± 0.018	0.059 ± 0.00
	Grari et al. [19]	0.526 ± 0.042	0.057 ± 0.011	0.046 ± 0.030	0.042 ± 0.038	**0.001** ± 0.001	0.008 ± 0.015
	Mary et al. [33]	0.541 ± 0.015	0.075 ± 0.013	0.061 ± 0.006	0.078 ± 0.013	0.002 ± 0.001	0.019 ± 0.004
	Fair MINE NN	0.537 ± 0.046	0.058 ± 0.042	0.048 ± 0.029	0.045 ± 0.037	**0.001** ± 0.001	0.012 ± 0.016
	Adel et al. [1]	0.552 ± 0.032	0.100 ± 0.028	0.138 ± 0.042	0.146 ± 0.031	0.003 ± 0.003	0.035 ± 0.011
	Zhang et al. [48]	0.727 ± 0.264	0.097 ± 0.038	0.135 ± 0.036	0.165 ± 0.028	0.009 ± 0.005	0.022 ± 0.019
	Madras et al. [31]	0.525 ± 0.033	0.129 ± 0.010	0.158 ± 0.009	0.173 ± 0.012	0.007 ± 0.007	0.041 ± 0.003
	Sadeghi et al. [40]	0.526 ± 0.006	0.077 ± 0.031	0.136 ± 0.001	0.146 ± 0.001	0.008 ± 0.003	0.035 ± 0.000
	Ours	**0.523** ± 0.035	**0.054** ± 0.015	**0.044** ± 0.032	**0.041** ± 0.031	**0.001** ± 0.001	**0.007** ± 0.002
Motor	Standard NN	0.945 ± 0.011	0.201 ± 0.094	0.175 ± 0.0	0.200 ± 0.034	0.188 ± 0.005	0.008 ± 0.011
	Grari et al. [19]	0.971 ± 0.004	0.072 ± 0.029	0.058 ± 0.052	**0.066** ± 0.009	**0.000** ± 0.000	0.006 ± 0.02
	Mary et al. [33]	0.979 ± 0.119	0.077 ± 0.023	0.059 ± 0.014	0.067 ± 0.028	0.001 ± 0.001	0.006 ± 0.002
	Fair MINE NN	0.982 ± 0.003	0.078 ± 0.013	0.068 ± 0.004	0.069 ± 0.009	**0.000** ± 0.000	**0.004** ± 0.001
	Adel et al. [1]	0.979 ± 0.003	0.101 ± 0.04	0.09 ± 0.03	0.101 ± 0.04	0.002 ± 0.002	0.009 ± 0.004
	Zhang et al. [48]	0.998 ± 0.004	0.076 ± 0.034	0.091 ± 0.024	0.129 ± 0.08	0.001 ± 0.001	**0.004** ± 0.001
	Madras et al. [31]	0.978 ± 0.004	0.096 ± 0.035	0.083 ± 0.020	0.099 ± 0.030	0.004 ± 0.002	0.008 ± 0.001
	Sadeghi et al. [40]	0.975 ± 0.017	0.102 ± 0.020	0.115 ± 0.027	0.129 ± 0.039	0.001 ± 0.001	0.001 ± 0.001
	Ours	**0.962** ± 0.002	**0.070** ± 0.011	**0.055** ± 0.005	0.067 ± 0.006	**0.000** ± 0.000	**0.004** ± 0.001
Crime	Standard NN	0.384 ± 0.012	0.732 ± 0.013	0.525 ± 0.013	0.731 ± 0.009	0.315 ± 0.021	0.353 ± 0.006
	Grari et al. [19]	0.781 ± 0.016	0.356 ± 0.063	0.097 ± 0.022	**0.171** ± 0.03	**0.009** ± 0.008	**0.039**± 0.008
	Mary et al. [33]	0.778 ± 0.103	0.371 ± 0.116	0.115 ± 0.046	0.177 ± 0.054	0.024 ± 0.015	0.064 ± 0.023
	Fair MINE NN	0.782 ± 0.034	0.395 ± 0.097	0.110 ± 0.022	0.201 ± 0.021	0.032 ± 0.029	0.136 ± 0.012
	Adel et al. [1]	0.836 ± 0.005	0.384 ± 0.037	0.170 ± 0.027	0.371 ± 0.035	0.058 ± 0.027	0.057 ± 0.007
	Zhang et al. [48]	0.787 ± 0.134	0.377 ± 0.085	0.153 ± 0.056	0.313 ± 0.087	0.037 ± 0.022	0.063 ± 0.046
	Madras et al. [31]	**0.725** ± 0.023	**0.312** ± 0.022	0.290 ± 0.027	0.175 ± 0.016	0.036 ± 0.013	0.103 ± 0.015
	Sadeghi et al. [40]	0.782 ± 0.002	0.474 ± 0.006	0.123 ± 0.000	0.315 ± 0,009	0.098 ± 0.035	0.062 ± 0.001
	Ours	0.783 ± 0.031	0.369 ± 0.074	**0.087** ± 0.031	0.173 ± 0.044	0.011 ± 0.006	0.043 ± 0.012
		ACC	HGR_NN	HGR_KDE	HGR_RDC	MINE	FairQuant
COMPAS	Standard NN	68.7% ± 0.243	0.363 ± 0.005	0.326 ± 0.003	0.325 ± 0.008	0.046 ± 0.028	0.140 ± 0.001
	Grari et al. [19]	59.7% ± 2.943	0.147 ± 0.000	0.121 ± 0.002	0.101 ± 0.007	0.004 ± 0.001	0.018 ± 0.018
	Fair MINE NN	54.4% ± 7.921	0.134 ± 0.145	0.123 ± 0.111	0.141 ± 0.098	0.014 ± 0.023	0.038 ± 0.050
	Adel et al. [1]	55.4% ± 0.603	0.118 ± 0.022	0.091 ± 0.012	0.097 ± 0.034	0.006 ± 0.007	0.013 ± 0.016
	Zhang et al. [48]	51.0% ± 3.550	0.116 ± 0.000	0.081 ± 0.003	0.086 ± 0.010	0.002 ± 0.003	0.010 ± 0.005
	Madras et al. [31]	54.9% ± 2.221	0.175 ± 0.000	0.116 ± 0.015	0.107 ± 0.026	0.005± 0.003	0.011 ± 0.020
	Sadeghi et al. [40]	54.3% ± 0.024	0.194 ± 0.052	0.237 ± 0.040	0.264 ± 0.054	0.003 ± 0.003	**0.003** ± 0.003
	Ours	**60.2%** ± 3.076	**0.063** ± 0.024	**0.068** ± 0.018	**0.067** ± 0.014	**0.001** ± 0.002	0.011 ± 0.018
Default	Standard NN	82.1% ± 0.172	0.112 ± 0.013	0.067 ± 0.010	0.089 ± 0.014	0.002 ± 0.001	0.015 ± 0.002
	Grari et al. [19]	79.9% ± 2.100	0.082 ± 0.015	0.075 ± 0.019	0.072 ± 0.010	0.001 ± 0.001	0.007 ± 0.007
	Adel et al. [1]	79.2% ± 1.207	0.054 ± 0.025	0.048 ± 0.015	0.064 ± 0.009	0.001 ± 0.001	0.005 ± 0.002
	Fair MINE NN	80.1% ± 2.184	0.093 ± 0.020	0.057 ± 0.002	0.066 ± 0.012	0.001 ± 0.001	0.008 ± 0.001
	Zhang et al. [48]	77.9% ± 9.822	0.052 ± 0.017	**0.044** ± 0.013	0.056 ± 0.004	**0.000** ± 0.000	0.004 ± 0.000
	Madras et al. [31]	78.3% ± 0.605	0.064 ± 0.025	0.052 ± 0.018	0.061 ± 0.012	0.001 ± 0.001	**0.003** ± 0.005
	Sadeghi et al. [40]	79.7% ± 0.236	0.074 ± 0.019	0.062 ± 0.013	0.098 ± 0.041	0.002 ± 0.002	**0.003** ± 0.002
	Ours	**80.8%** ± 0.286	**0.041** ± 0.008	**0.044** ± 0.006	**0.047** ± 0.002	0.001 ± 0.002	0.005 ± 0.001

tasks. While being better in terms of accuracy, our fair representation algorithm achieves on four data sets (except on the Crime data set) the best level of fairness assessed by HGR estimation, MINE and FairQuant. On the Crime data set, the approach by Madras2018 [32] gets slightly better results on MSE and HGR estimation but not on the others metrics. Note, Adel [1] with the fair adversarial representation obtains (except on the Crime data set) better results than Zhang [48] which corresponds to the simple adversarial architecture.

What is the Impact of Mitigation Weight? In Fig. 4, we plot the performance of different scenarios by displaying the HGR against the Accuracy with different values of the hyperparameter λ.

This plot was obtained on the COMPAS data set with 4 algorithms: ours, Adel et al. [1], Grari et al. [19] and Zhang et al. [48]. The different curves is obtained by Nadaraya-Watson kernel regression [8] between the Accuracy of the model and the HGR. Varying the hyperparameter λ allows to control the fairness/accuracy trade-off. Here, we clearly observe for all algorithms that the Accuracy, or predictive performance, decreases when fairness increases.

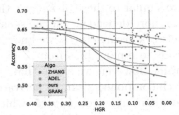

Fig. 4. Impact of hyperparameter λ (COMPAS data set)

Higher values of λ produce fairer predictions w.r.t the HGR, while near 0 values of the hyperparameter λ result in the optimization of the predictor loss with no fairness consideration (dots in the upper left corner of the graph). We note that, for all levels of predictive performance, our method outperforms the state of the art algorithms in terms of HGR.

Where Should We Apply Mitigation in Neural Architectures? In order to answer this question, and also further analyze the benefits of mitigation in neural representations compared to prediction retreatments as done in [19], we propose to consider various architectures of encoders h and predictors ϕ, with adversarial HGR mitigation being applied on the output of the encoder as depicted in Fig. 1. in appendix. To get comparable results between settings, we consider a constant full architecture (encoder + predictor), composed of 5 layers with 4 hidden layers with 32 units each.

In Fig. 5, we compare on the COMPAS dataset 5 different settings where mitigation is applied on a different layer of this full architecture: *LayerX* corresponds to a setting where mitigation is applied on the output of layer X (encoder of X layers, predictor of 5-X layers). *Layer5* thus corresponds to the prediction retreatment approach proposed in [19] (no predictor function, the encoder function h directly outputs the prediction). *Layer3* is the standard setting used for our approach in the remaining of this paper.

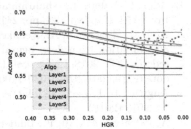

Fig. 5. Impact of hyperparameter λ (COMPAS data set) for various encoders h and predictors ϕ.

As in Fig. 4, plotted results correspond to fairness-accuracy trade-offs obtained with different values of λ. We notice that applying mitigation too early in the architecture (Layer1) leads to very poor results. This can be explained by the fact that for this simple encoding setting, the encoder expressiveness is to weak to effectively remove non-linear dependencies w.r.t. the sensitive attribute, without removing too much useful information for prediction. At the contrary, when mitigation is applied late in the architecture (*Layer4* and *Layer5*) we observe generalization limits of the approach. While results on the training set are similar to those of *Layer3*, these settings lead to predictions at test time that are more dependent on the sensitive attribute. Due to L-Lipschitzness of neural network architectures, we know that $HGR(Z, S) \geq HGR(\phi(Z), S)$. Acting on Z leads to remove bias from Z even for components ignored by the predictor ϕ in train. However, we argue that this allows to gain in stability at test time, when such components can be activated for new inputs, compared to late approaches, such as *Layer4* or *Layer5*, which induce a greater variance of sensitive dependence of the output $\hat{Y}$. Mitigation at intermediate levels, such as *Layer3*, appears to correspond to the best trade-off expressiveness/generalization.

7   Conclusion

We present a new adversarial learning approach to produce fair representations with a continuous sensitive attribute. We leverage the HGR measure, which is efficient in capturing non-linear dependencies, and propose to minimize a neural estimation of the HGR between the latent representation and the sensitive attributes. This method proved to be very efficient for different fairness metrics on various artificial and real-world data sets. For further investigation, we will apply this architecture for information bottleneck purposes (e.g. for data privacy), that might be improved with an HGR_NN penalization as suggested in [4].

References

1. Adel, T., Valera, I., Ghahramani, Z., Weller, A.: One-network adversarial fairness. In: AAAI 2019, vol. 33, pp. 2412–2420 (2019)
2. Alvi, M., Zisserman, A., Nellåker, C.: Turning a blind eye: explicit removal of biases and variation from deep neural network embeddings. In: Leal-Taixé, L., Roth, S. (eds.) ECCV 2018. LNCS, vol. 11129, pp. 556–572. Springer, Cham (2019). https://doi.org/10.1007/978-3-030-11009-3_34
3. Angwin, J., Larson, J., Mattu, S., Kirchner, L.: Machine bias. ProPublica, May 23, 2016 (2016)
4. Asoodeh, S., Alajaji, F., Linder, T.: On maximal correlation, mutual information and data privacy. In: 2015 IEEE 14th Canadian Workshop on Information Theory (CWIT), pp. 27–31. IEEE (2015)
5. Baharlouei, S., Nouiehed, M., Beirami, A., Razaviyayn, M.: Rényi fair inference. In: 8th International Conference on Learning Representations, ICLR 2020, Addis Ababa, Ethiopia, 26–30 April 2020. OpenReview.net (2020)
6. Belghazi, M.I., et al.: Mine: Mutual information neural estimation (2018)
7. Bellamy, R.K., et al.: AI fairness 360: an extensible toolkit for detecting, understanding, and mitigating unwanted algorithmic bias. arXiv preprint arXiv:1810.01943 (2018)
8. Bierens, H.J.: The Nadaraya-Watson kernel regression function estimator (1988)

9. Blitzer, J., McDonald, R., Pereira, F.: Domain adaptation with structural correspondence learning. In: Proceedings of the 2006 Conference on Empirical Methods in Natural Language Processing, pp. 120–128 (2006)
10. Bolukbasi, T., Chang, K.W., Zou, J.Y., Saligrama, V., Kalai, A.: Man is to computer programmer as woman is to homemaker? Debiasing word embeddings. In: NIPS (2016)
11. Calmon, F.P., Wei, D., Vinzamuri, B., Ramamurthy, K.N., Varshney, K.R.: Optimized pre-processing for discrimination prevention. In: Proceedings of the 31st International Conference on Neural Information Processing Systems, pp. 3995–4004 (2017)
12. Celis, L.E., Huang, L., Keswani, V., Vishnoi, N.K.: Classification with fairness constraints: a meta-algorithm with provable guarantees. In: Proceedings of the Conference on Fairness, Accountability, and Transparency, pp. 319–328 (2019)
13. Chen, J., Kallus, N., Mao, X., Svacha, G., Udell, M.: Fairness under unawareness: assessing disparity when protected class is unobserved. In: Proceedings of the Conference on Fairness, Accountability, and Transparency, pp. 339–348 (2019)
14. Daume, H., III., Marcu, D.: Domain adaptation for statistical classifiers. J. Artif. Intell. Res. **26**, 101–126 (2006)
15. Dua, D., Graff, C.: UCI ml repository (2017). http://archive.ics.uci.edu/ml
16. Ganin, Y., Lempitsky, V.: Unsupervised domain adaptation by backpropagation. In: International conference on machine learning. pp. 1180–1189. PMLR (2015)
17. Ganin, Y., et al.: Domain-adversarial training of neural networks. J. Mach. Learn. Res. **17**(1), 2030–2096 (2016)
18. Geer, S.A., van de Geer, S.: Empirical Processes in M-estimation, vol. 6. Cambridge University Press, Cambridge (2000)
19. Grari, V., Lamprier, S., Detyniecki, M.: Fairness-aware neural rényi minimization for continuous features. In: Proceedings of the Twenty-Ninth International Joint Conference on Artificial Intelligence, IJCAI 2020, pp. 2262–2268. ijcai.org (2020). https://doi.org/10.24963/ijcai.2020/313
20. Hardoon, D.R., Shawe-Taylor, J.: Convergence analysis of kernel canonical correlation analysis: theory and practice. Mach. Learn. **74**(1), 23–38 (2009)
21. Hardt, M., Price, E., Srebro, N.: Equality of opportunity in supervised learning. In: Advances in Neural Information Processing Systems, pp. 3315–3323 (2016)
22. Hornik, K., Stinchcombe, M., White, H., et al.: Multilayer feedforward networks are universal approximators. Neural Netw. **2**(5), 359–366 (1989)
23. Kamiran, F., Calders, T.: Data preprocessing techniques for classification without discrimination. Knowl. Inf. Syst. **33**(1), 1–33 (2012)
24. Kim, B., Kim, H., Kim, K., Kim, S., Kim, J.: Colored MNIST dataset (2019). https://github.com/feidfoe/learning-not-to-learn/tree/master/dataset/colored_mnist
25. Kim, B., Kim, H., Kim, K., Kim, S., Kim, J.: Learning not to learn: training deep neural networks with biased data. In: Proceedings of the IEEE Conference on Computer Vision and Pattern Recognition, pp. 9012–9020 (2019)
26. Kingma, D.P., Welling, M.: Auto-encoding variational Bayes. In: Bengio, Y., LeCun, Y. (eds.) 2nd International Conference on Learning Representations, ICLR 2014, Banff, AB, Canada, 14–16 April 2014, Conference Track Proceedings (2014). http://arxiv.org/abs/1312.6114
27. LeCun, Y., Cortes, C., Burges, C.: MNIST handwritten digit database (2010)
28. Li, D., Yang, Y., Song, Y.Z., Hospedales, T.M.: Deeper, broader and artier domain generalization. In: Proceedings of the IEEE International Conference on Computer Vision, pp. 5542–5550 (2017)

29. Lopez-Paz, D., Hennig, P., Schölkopf, B.: The randomized dependence coefficient. In: Advances in Neural Information Processing Systems, pp. 1–9 (2013)

30. Louppe, G., Kagan, M., Cranmer, K.: Learning to pivot with adversarial networks. In: Advances in Neural Information Processing Systems, pp. 981–990 (2017)

31. Madras, D., Creager, E., Pitassi, T., Zemel, R.: Learning adversarially fair and transferable representations. In: Dy, J., Krause, A. (eds.) Proceedings of the 35th ICML 2018, pp. 3384–3393. (2018)

32. Madras, D., Creager, E., Pitassi, T., Zemel, R.: Fairness through causal awareness: learning causal latent-variable models for biased data. In: Proceedings of the Conference on Fairness, Accountability, and Transparency, pp. 349–358 (2019)

33. Mary, J., Calauzènes, C., Karoui, N.E.: Fairness-aware learning for continuous attributes and treatments. In: Chaudhuri, K., Salakhutdinov, R. (eds.) Proceedings of the 36th ICML 2019, vol. 97, pp. 4382–4391. (2019). http://proceedings.mlr.press/v97/mary19a.html

34. Moyer, D., Gao, S., Brekelmans, R., Galstyan, A., Ver Steeg, G.: Invariant representations without adversarial training. In: Advances in Neural Information Processing Systems, pp. 9084–9093 (2018)

35. Muandet, K., Balduzzi, D., Schölkopf, B.: Domain generalization via invariant feature representation. In: ICML 2013, pp. 10–18 (2013)

36. Pedreshi, D., Ruggieri, S., Turini, F.: Discrimination-aware data mining. In: KDD 2008, p. 560 (2008). https://doi.org/10.1145/1401890.1401959. http://dl.acm.org/citation.cfm?doid=1401890.1401959

37. Póczos, B., Ghahramani, Z., Schneider, J.: Copula-based kernel dependency measures. In: Proceedings of the 29th ICML 2012, pp. 1635–1642 (2012)

38. Ragonesi, R., Volpi, R., Cavazza, J., Murino, V.: Learning unbiased representations via mutual information backpropagation. arXiv preprint arXiv:2003.06430 (2020)

39. Rényi, A.: On measures of dependence. Acta Math. Hungar. **10**(3–4), 441–451 (1959)

40. Sadeghi, B., Yu, R., Boddeti, V.: On the global optima of kernelized adversarial representation learning. In: Proceedings of the IEEE International Conference on Computer Vision, pp. 7971–7979 (2019)

41. Székely, G.J., Rizzo, M.L., et al.: Brownian distance covariance. Annals Appl. Stat. **3**(4), 1236–1265 (2009)

42. The Institute of Actuaries of France: Pricing game 2015. https://freakonometrics.hypotheses.org/20191. Accessed 14 Aug 2019

43. US Census Bureau: Us census demographic data. https://data.census.gov/cedsci/. Accessed 03 Apr 2019

44. Witsenhausen, H.S.: On sequences of pairs of dependent random variables. SIAM J. Appl. Math. **28**(1), 100–113 (1975)

45. Yeh, I.C., Lien, C.H.: The comparisons of data mining techniques for the predictive accuracy of probability of default of credit card clients. Expert Syst. Appl. **36**(2), 2473–2480 (2009). https://doi.org/10.1016/j.eswa.2007.12.020

46. Zafar, M.B., Valera, I., Rogriguez, M.G., Gummadi, K.P.: Fairness constraints: mechanisms for fair classification. In: AISTATS 2017, Fort Lauderdale, FL, USA, 20–22 April 2017, pp. 962–970 (2017)

47. Zemel, R., Wu, Y., Swersky, K., Pitassi, T., Dwork, C.: Learning fair representations. In: ICML 2013, pp. 325–333 (2013)

48. Zhang, B.H., Lemoine, B., Mitchell, M.: Mitigating unwanted biases with adversarial learning. In: AAAI 2018, pp. 335–340 (2018)

Diversity-Aware k-median: Clustering with Fair Center Representation

Suhas Thejaswi[1]([✉]), Bruno Ordozgoiti[1], and Aristides Gionis[1,2]

[1] Aalto University, Espoo, Finland
`bruno.ordozgoiti@aalto.fi`
[2] KTH Royal Institute of Technology, Stockholm, Sweden
`argioni@kth.se`

Abstract. We introduce a novel problem for diversity-aware clustering. We assume that the potential cluster centers belong to a set of groups defined by protected attributes, such as ethnicity, gender, etc. We then ask to find a minimum-cost clustering of the data into k clusters so that a specified minimum number of cluster centers are chosen from each group. We thus require that all groups are represented in the clustering solution as cluster centers, according to specified requirements. More precisely, we are given a set of clients C, a set of facilities $\mathcal{F}$, a collection $\mathscr{F} = \{F_1, \ldots, F_t\}$ of facility groups $F_i \subseteq \mathcal{F}$, a budget k, and a set of lower-bound thresholds $R = \{r_1, \ldots, r_t\}$, one for each group in $\mathscr{F}$. The *diversity-aware k-median problem* asks to find a set S of k facilities in $\mathcal{F}$ such that $|S \cap F_i| \geq r_i$, that is, at least r_i centers in S are from group F_i, and the k-median cost $\sum_{c \in C} \min_{s \in S} d(c, s)$ is minimized. We show that in the general case where the facility groups may overlap, the diversity-aware k-median problem is **NP**-hard, fixed-parameter intractable with respect to parameter k, and inapproximable to any multiplicative factor. On the other hand, when the facility groups are disjoint, approximation algorithms can be obtained by reduction to the *matroid median* and *red-blue median* problems. Experimentally, we evaluate our approximation methods for the tractable cases, and present a relaxation-based heuristic for the theoretically intractable case, which can provide high-quality and efficient solutions for real-world datasets.

Keywords: Algorithmic bias · Algorithmic fairness · Diversity-aware clustering · Fair clustering

1 Introduction

As many important decisions are being automated, algorithmic fairness is becoming increasingly important. Examples of critical decision-making systems include

This research is supported by the Academy of Finland projects AIDA (317085) and MLDB (325117), the ERC Advanced Grant REBOUND (834862), the EC H2020 RIA project SoBigData (871042), and the Wallenberg AI, Autonomous Systems and Software Program (WASP) funded by the Knut and Alice Wallenberg Foundation.

© Springer Nature Switzerland AG 2021
N. Oliver et al. (Eds.): ECML PKDD 2021, LNAI 12976, pp. 765–780, 2021.
https://doi.org/10.1007/978-3-030-86520-7_47

determining credit score for a consumer, computing risk factors for an insurance, pre-screening applicants for a job opening, dispatching patrols for predictive policing, and more. When using algorithms to make decisions for such critical tasks, it is essential to design and employ methods that minimize bias and avoid discrimination against people based on gender, race, or ethnicity.

Algorithmic fairness has gained wide-spread attention in the recent years [20]. The topic has been predominantly studied for *supervised machine learning*, while fairness-aware formulations have also been proposed for *unsupervised machine learning*, for example, *fair clustering* [3,5,9,23], *fair principal component analysis* [22], or *fair densest-subgraph mining* [1]. For the clustering problem the most common approach is to incorporate fairness by the means of *representation-based constraints*, i.e., requiring that all clusters contain certain proportions of the different groups in the data, where data groups are defined via a set of protected attributes, such as demographics. In this paper we introduce a novel notion for fair clustering based on *diversity constraints on the set of selected cluster centers*.

Research has revealed that bias can be introduced in machine-learning algorithms when bias is present in the input data used for training, and methods are designed without considerations for diversity or constraints to enforce fairness [11]. A natural solution is to introduce diversity constraints. We can look at diversification from two different perspectives: (*i*) avoiding over-representation; and (*ii*) avoiding under-representation. In this paper we focus on the latter requirement. Even though these two approaches look similar, a key contribution in our work is to observe that they are mathematically distinct and lead to computational problems having different complexity; for details see Sects. 3 and 4.

To motivate our work we present two application scenarios.

Committee Selection: We often select committees to represent an underlying population and work towards a task, e.g., a program committee to select papers for a conference, or a parliamentary committee to handle an issue. As we may require that each member of the population is represented by at least one committee member, it is natural to formalize the committee-selection task as a clustering problem, where the committee members will be determined by the centers of the clustering solution. In addition, one may require that the committee is composed by a diverse mix of the population with no under-represented groups, e.g., a minimum fraction of the conference PC members work in industry, or a minimum fraction of the parliamentary committee are women.

News-Articles Summarization: Consider the problem of summarizing a collection of news articles obtained, for example, as a result to a user query. Clustering these articles using a bag-of-words representation will allow us to select a subset of news articles that cover the different topics present in the collection. In addition, one may like to ensure that the representative articles comes from a diverse set of media sources, e.g., a minimum fraction of the articles comes from left-leaning media or from opinion columns.

To address the scenarios discussed in the previous two examples, we introduce a novel formulation of diversity-aware clustering with representation constraints

Table 1. An overview of our results. All problem cases we consider are **NP**-hard. FPT(k) indicates whether the problem is fixed-parameter tractable with respect to parameter k. *Approx. factor* shows the factor of approximation obtained, and *Approx. method* shows the method used.

Problem	**NP**-hard	FPT(k)	Approx. factor	Approx. method
Intractable case: intersecting facility groups				
General variant	✓	✗	inapproximable	
Tractable cases: disjoint facility groups				
$t > 2,\ \sum_{i \in [t]} r_i = k$	✓	Open	8	LP
$t > 2,\ \sum_{i \in [t]} r_i < k$	✓	Open	8	$\mathcal{O}(k^{t-1})$ calls to LP
$t = 2,\ r_1 + r_2 = k$	✓	Open	$5 + \epsilon$	Local search
$t = 2,\ r_1 + r_2 < k$	✓	Open	$5 + \epsilon$	$\mathcal{O}(k)$ calls to local search

on cluster centers. In particular, we assume that a set of groups is associated with the facilities to be used as cluster centers. Facility groups may correspond to demographic groups, in the first example, or to types of media sources, in the second. We then ask to cluster the data by selecting a subset of facilities as cluster centers, such that the clustering cost is minimized, and requiring that each facility group is *not under-represented* in the solution.

We show that in the general case, where the facility groups overlap, the diversity-aware k-median problem is not only **NP**-hard, but also fixed-parameter intractable with respect to the number of cluster centers, and inapproximable to any multiplicative factor. In fact, we prove it is **NP**-hard to even find a feasible solution, that is, a set of centers which satisfies the representation constraints, regardless of clustering cost. These hardness results set our clustering problem in stark contrast with other clustering formulations where approximation algorithms exist, and in particular, with the *matroid-median problem* [8,16,19], where one asks that facility groups are *not over-represented*. Unfortunately, however, the matroid-median problem does not ensure fairness for all facility groups.

On the positive side, we identify important cases for which the diversity-aware k-median problem is approximable, and we devise efficient algorithms with constant-factor approximation guarantees. These more tractable cases involve settings when the facility groups are disjoint. Even though the general variant of the problem in inapproximable, we demonstrate using experiments that we can obtain a desired clustering solution with representation constraints with almost the same cost as the unconstrained version using simple heuristics based on local-search. The hardness and approximability results for the diversity-aware k-median problem are summarized in Table 1.

In addition to our theoretical analysis and results we empirically evaluate our methods on several real-world datasets. Our experiments show that in many problem instances, both theoretically tractable and intractable, the *price of diversity* is low in practice (See Sect. 6.1 for a precise definition). In particular, our methods can be used to find solutions over a wide range of diversity requirements where the clustering cost is comparable to the cost of unconstrained clustering.

The rest of this paper is structured as follows. Section 2 discusses related work, Sect. 3 presents the problem statement and computational complexity results. Section 4 discusses special cases of the problem that admit polynomial-time approximations. In Sect. 5 we offer efficient heuristics and related tractable objectives, and in Sect. 6 we describe experimental results. Finally, Sect. 7 is a short conclusion.

2 Related Work

Algorithmic fairness has attracted a considerable amount of attention in recent years, as many decisions that affect us in everyday life are being made by algorithms. Many machine-learning and data-mining problems have been adapted to incorporate notions of fairness. Examples include problems in classification [4,13,17], ranking [24,28], recommendation systems [27], and more.

In this paper we focus on the problem of clustering and we consider a novel notion of fairness based on diverse representation of cluster centers. Our approach is significantly different (and orthogonal) from the standard notion of *fair clustering*, introduced by the pioneering work of Chierichetti et al. [9]. In that setting, data points are partitioned into groups (Chierichetti et al. considered only two groups) and the fairness constraint is that each cluster should contain a fraction of points from each group. Several recent papers have extended the work of Chierichetti et al. by proposing more scalable algorithms [3,18], extending the methods to accommodate more than two groups [6,23], or introducing privacy-preserving properties [21]. In this line of work, the fairness notion applies to the representation of data groups within each cluster. In contrast, in our paper the fairness notion applies to the representation of groups in the cluster centers.

The closest-related work to our setting are the problems of *matroid median* [8, 19] and *red-blue median* [15,16], which can be used to ensure that no data groups are over-represented in the cluster centers of the solution—in contrast we require that no data groups are under-represented. Although the two notions are related, in a way that we make precise in Sect. 4, in the general case they differ significantly, and they yield problems of greatly different complexity. Furthermore, in the general case, the matroid-median problem cannot be used to ensure fair representation, as upper-bound constraints cannot ensure that all groups will be represented in the solution. Although it is for those cases that our diversity-aware clustering problem is intractable, one can develop practical heuristics that achieve fair results, with respect to diverse representation, as shown in our experimental evaluation.

3 Problem Statement and Complexity

We consider a set of clients C and a set of facilities $\mathcal{F}$. In some cases, the set of facilities may coincide with the set of clients ($\mathcal{F} = C$), or it is a subset ($\mathcal{F} \subseteq C$). We assume a distance function $d : C \times \mathcal{F} \to \mathbb{R}_+$, which maps client–facility pairs into nonnegative real values. We also consider a collection $\mathscr{F} = \{F_1, \ldots, F_t\}$ of facility groups $F_i \subseteq \mathcal{F}$. During our discussion we distinguish different cases for the structure of $\mathscr{F}$. In the most general case the facility groups F_i may overlap. Two special cases of interest, discussed in Sect. 4, are when the facility groups F_i are disjoint and when there are only two groups. Finally, we are given a total budget k of facilities to be selected, and a set of lower-bound thresholds $R = \{r_1, \ldots, r_t\}$, i.e., one threshold for each group F_i in $\mathscr{F}$.

The diversity-aware k-median problem (DIV-k-MEDIAN) asks for a set S of k facilities in $\mathcal{F}$ subject to the constraint $|S \cap F_i| \geq r_i$, such that the k-median cost $\sum_{c \in C} \min_{s \in S} d(c, s)$ is minimized. Thus, we search for a minimum-cost clustering solution S where each group F_i is represented by at least r_i centers.

In the following sections, we study the computational complexity of the DIV-k-MEDIAN problem. In particular, we show that the general variant of the problem is (i) **NP**-hard; (ii) not fixed-parameter tractable with respect to parameter k, i.e., the size of the solution sought; and (iii) inapproximable to any multiplicative factor. In fact, we show that hardness results (i) and (ii) apply for the problem of simply finding a feasible solution. That is, in the general case, and assuming $\mathbf{P} \neq \mathbf{NP}$ there is no polynomial-time algorithm to find a solution $S \subseteq \mathcal{F}$ that satisfies the constraints $|S \cap F_i| \geq r_i$, for all $i \in [t]$. The inapproximability statement (iii) is a consequence of the **NP**-hardness for finding a feasible solution. These hardness results motivate the heuristics we propose later on.

3.1 NP-Hardness

We prove **NP**-hardness by reducing the dominating set problem to the problem of finding a *feasible solution* to DIV-k-MEDIAN.

Dominating Set Problem (DOMSET). Given a graph $G = (V, E)$ with $|V| = n$ vertices, and an integer $k \leq n$, decide if there exists a subset $S \subseteq V$ of size $|S| = k$ such that for each $v \in V$ it is either $\{v\} \cap S \neq \emptyset$ or $\{v\} \cap N(S) \neq \emptyset$, where $N(S)$ denotes the set of vertices adjacent to at least one vertex in S. In other words, each vertex in V is either in S or adjacent to at least one vertex in S.

Lemma 1. *Finding a feasible solution for* DIV-k-MEDIAN *is* **NP**-*hard.*

Proof. Given an instance of DOMSET ($G = (V, E), k$), we construct an instance of the DIV-k-MEDIAN problem $(C, \mathcal{F}, \mathscr{F}, d, k, R)$, such that $C = V$, $\mathcal{F} = V$, $d(u, v) = 1$ for all $(u, v) \in C \times F$, $\mathscr{F} = \{F_1, \ldots, F_n\}$ with $F_u = \{u\} \cup N(u)$, and $R = \{1, \ldots, 1\}$, i.e., the lower-bound thresholds are set to $r_u = 1$, for all $u \in V$.

Let $S \subseteq C$ be a feasible solution for DIV-k-MEDIAN. From the construction it is clear that S is a dominating set, as $|F_u \cap S| \geq 1$, and thus S intersects $\{u\} \cup N(u)$ for all $u \in V$. The proof that a dominating set is a feasible solution to DIV-k-MEDIAN is analogous. $\qquad\square$

The hardness of diversity-aware k-median follows immediately.

Corollary 1. *The* DIV-k-MEDIAN *problem is* **NP**-*hard.*

3.2   Fixed-Parameter Intractability

A problem P specified by input x and a parameter k is *fixed-parameter tractable* (FPT) if there exists an algorithm A to solve every instance $(x, k) \in P$ with running time of the form $f(k)|x|^{\mathcal{O}(1)}$, where $f(k)$ is function depending solely on the parameter k and $|x|^{\mathcal{O}(1)} = \mathrm{poly}(|x|)$ is a polynomial independent of the parameter k. A problem P is *fixed-parameter intractable* with respect to parameter k otherwise.

To show that the DIV-k-MEDIAN is fixed-parameter intractable we present a *parameterized reduction* from the DOMSET problem to DIV-k-MEDIAN.[1] The DOMSET problem is known to be fixed-parameter intractable [10, Theorem 13.9]. This means that there exists no algorithm with running time $f(k)|V|^{\mathcal{O}(1)}$ to solve DOMSET, where $f(k)$ is a function depending solely on the parameter k.

Theorem 1. *The* DIV-k-MEDIAN *problem is fixed-parameter intractable with respect to the parameter k, that is, the size of the solution sought.*

Proof. We apply the reduction from Lemma 1. It follows that (i) an instance (G, k) of the DOMSET problem has a feasible solution if and only if there exists a feasible solution for the DIV-k-MEDIAN problem instance $(C, \mathcal{F}, \mathscr{F}, d, k', R)$, with $k' = k$, and (ii) the reduction takes polynomial time in the size of the input. So there exists a parameterized reduction from the DOMSET problem to the DIV-k-MEDIAN problem. This implies that if there exists an algorithm with running time $f(k')|C|^{\mathcal{O}(1)}$ for the DIV-k-MEDIAN problem then there exits an algorithm with running time $f(k)|V|^{\mathcal{O}(1)}$ for solving the DOMSET problem. □

It would still be interesting to check whether there exists a parameter of the problem that can be used to design a solution where the exponential complexity can be restricted. We leave this as an open problem.

3.3   Hardness of Approximation

We now present hardness-of-approximation results for DIV-k-MEDIAN.

Theorem 2. *Assuming* **P** $\neq$ **NP**, *the* DIV-k-MEDIAN *problem cannot be approximated to any multiplicative factor.*

Proof. We apply the reduction of the DOMSET problem from the proof of Lemma 1. For the sake of contradiction let A be a polynomial-time approximation algorithm which gives a factor-c approximate solution for DIV-k-MEDIAN. Then we can employ algorithm A to obtain an exact solution to the DOMSET instance in polynomial time, by way of the aforementioned reduction. The reason is that an approximate solution for DIV-k-MEDIAN is also a feasible solution, which in turn implies a feasible solution for DOMSET. Thus, unless **P** $\neq$ **NP**, DIV-k-MEDIAN cannot be approximated to any multiplicative factor. □

---

[1] For a precise definition of parameterized reduction see Cygan et al. [10, Chapter 13].

4   Approximable Instances

In Sect. 3 we presented strong intractability results for the DIV-k-MEDIAN problem. Recall that inapproximability stems from the fact that satisfying the non under-representation constraints $|S \cap F_i| \geq r_i$ for $i \in [t]$ is **NP**-hard. *So the inapproximability holds even if we change the clustering cost, for instance, to k-center or a soft assignment variant of k-median.*[2] Fortunately, however, there are instances where finding a feasible solution is polynomial-time solvable, even if finding an minimum-cost clustering solution remains **NP**-hard. In this section we discuss such instances and give approximation algorithms.

4.1   Non-intersecting Facility Groups

We consider instances of DIV-k-MEDIAN where $F_i \cap F_j = \emptyset$ for all $F_i, F_j \in \mathscr{F}$, that is, the facility groups are disjoint. We refer to variants satisfying disjointness conditions as the DIV-k-MEDIAN$_\emptyset$ problem.

A feasible solution exists for DIV-k-MEDIAN$_\emptyset$ if and only if $|F_i| \geq r_i$ for all $i \in [t]$ and $\sum_{i \in [t]} r_i \leq k$. Furthermore, assuming that the two latter conditions hold true, finding a feasible solution is trivial: it can be done simply by picking r_i facilities from each facility group F_i.

It can be shown that the DIV-k-MEDIAN$_\emptyset$ problem can be reduced to the *matroid-median problem* [19], and use existing techniques for the latter problem to obtain an 8-approximation algorithm for DIV-k-MEDIAN$_\emptyset$ [25]. Before discussing the reduction we first introduce the matroid-median problem.

The Matroid-Median Problem (MATROIDMEDIAN) [19]. We are given a finite set of clients C and facilities $\mathcal{F}$, a metric distance function $d : C \times \mathcal{F} \rightarrow \mathbb{R}_+$, and a matroid $\mathcal{M} = (\mathcal{F}, \mathcal{I})$ with ground set $\mathcal{F}$ and a collection of independent sets $\mathcal{I} \subseteq 2^{\mathcal{F}}$. The problem asks us to find a subset $S \in \mathcal{I}(\mathcal{M})$ such that the cost function $\text{cost}(S) = \sum_{c \in C} \min_{s \in S} d(c, s)$ is minimized.

The MATROIDMEDIAN problem is a generalization of k-median, and has an 8-approximation algorithm based on LP relaxation [25]. Here we present a reduction of DIV-k-MEDIAN$_\emptyset$ to MATROIDMEDIAN. In this section we handle the case where $\sum_{i \in [t]} r_i = k$. In Sect. 4.3 we show that the case $\sum_{i \in [t]} r_i < k$ can be reduced to the former one with at most $\mathcal{O}(k^{t-1})$ calls. Approximating DIV-k-MEDIAN$_\emptyset$ in polynomial-time when $\sum_{i \in [t]} r_i < k$ is left open.

The Reduction. Given an instance $(C, \mathcal{F}, \mathscr{F}, d, k, R)$, of the DIV-$k$-MEDIAN$_\emptyset$ problem we generate an instance $(C', \mathcal{F}', \mathcal{I}', d')$ of the MATROIDMEDIAN problem as follows: $C' = C$, $\mathcal{F}' = \mathcal{F}$, $d' = d$, and $\mathcal{M} = (\mathcal{F}', \mathcal{I}')$ where $\mathcal{I}' \subseteq 2^{\mathcal{F}'}$ and $A \in \mathcal{I}'$ if $|A \cap F_i| \leq r_i$ for all $r_i \in R$. More precisely, the set of independent sets is comprised of all subsets of $\mathcal{F}'$ that satisfy *non over-representation* constraints. It is easy to verify that $\mathcal{M}$ is a matroid—it is a *partition matroid*. In the event that the algorithm for MATROIDMEDIAN outputs a solution where $|A \cap F_i| < r_i$,

---

[2] In soft clustering, each client is assigned to all cluster centers with a probability.

for some i, since $\sum_{i \in [t]} r_i = k$, it satisfies all the constraints with equality by completing the solution with facilities of the missing group(s) at no additional connection cost. Since we can ensure that $|A \cap F_i| = r_i$, for all i, it also holds $|A \cap F_i| \geq r_i$, for all i, that is, the DIV-k-MEDIAN$_\emptyset$ constraints.

Since the MATROIDMEDIAN problem has a polynomial-time approximation algorithm, it follows from our inapproximability results (Sect. 3) that a reduction of the general formulation of DIV-k-MEDIAN is impossible. We can thus conclude that allowing intersections between facility groups fundamentally changes the combinatorial structure of feasible solutions, interfering with the design of approximation algorithms.

4.2   Two Facility Groups

The approximation guarantee of the DIV-k-MEDIAN$_\emptyset$ problem can be further improved if we restrict the number of groups to two.

In particular, we consider instances of the DIV-k-MEDIAN problem where $F_i \cap F_j = \emptyset$, for all $F_i, F_j \in \mathscr{F}$, and $\mathscr{F} = \{F_1, F_2\}$. For simplicity, the facilities F_1 and F_2 are referred to as *red* and *blue* facilities, respectively.

As before, we can assume that $\sum_{i \in [t]} r_i = r_1 + r_2 \leq k$, otherwise the problem has no feasible solution. We first present a local-search algorithm for the case $r_1 + r_2 = k$. In Sect. 4.3 we show that the case with $r_1 + r_2 < k$ can be reduced to the former one with a linear number of calls for different values of r_1 and r_2. Before continuing with the algorithm we first define the rb-MEDIAN problem.

The Red-Blue Median Problem (rb-MEDIAN). We are given a set of clients C, two disjoint facility sets F_1 and F_2 (referred to as red and blue facilities, respectively), two integers r_1, r_2 and a metric distance function d : $C \times \{F_1 \cup F_2\} \to \mathbb{R}_+$. The problem asks to find a subset $S \subseteq F_1 \cup F_2$ such that $|F_1 \cap S| \leq r_1, |F_2 \cap S| \leq r_2$ and the cost function $\text{cost}(S) = \sum_{c \in C} \min_{s \in S} d(c, s)$ is minimized.

The rb-MEDIAN problem accepts a $5 + \epsilon$ approximation algorithm based on local-search [15]. The algorithm works by swapping a red-blue pair (r, b) with a red-blue pair (r', b') as long as the cost improves. Note that $(r' = r, b' \neq b)$, $(r' \neq r, b' = b)$ and $(r' \neq r, b' \neq b)$ are valid swap pairs. The reduction of DIV-k-MEDIAN to rb-MEDIAN is similar to the one given above for MATROIDMEDIAN. Thus, when the input consists of two non-intersecting facility groups we can obtain a $5 + \epsilon$ approximation of the optimum in polynomial time which follows from the local-search approximation of the rb-MEDIAN problem [15].

4.3   The Case $\sum_i r_i < k$

The reduction of DIV-k-MEDIAN$_\emptyset$ to MATROIDMEDIAN relies on picking exactly $\sum_i r_i$ facilities. This is because it is not possible to define a matroid that simultaneously enforces the desired lower-bound facility group constraints and the cardinality constraint for the solution. Nevertheless, we can overcome this obstacle at a bounded cost in running time.

1. Initialize S to be an arbitrary feasible solution.
2. While there exists a pair (s, s'), with $s \in S$ and $s' \in \mathcal{F}$ such that
   (a) $\mathrm{cost}(S \setminus \{s\} \cup \{s'\}) < \mathrm{cost}(S)$ and
   (b) $S \setminus \{s\} \cup \{s'\}$ is feasible i.e, $|S \setminus \{s\} \cup \{s'\} \cap F_i| \geq r_i$ for all $i \in [t]$,
   Set $S = S \setminus \{s\} \cup \{s'\}$.
3. Return S.

Fig. 1. Local search heuristic (LS-1) for DIV-k-MEDIAN$_\emptyset$.

So, in the case that $\sum_i r_i < k$, in order to satisfy the constraint $|S| = k$, we can simply increase the lower-bound group constraints $r_i \mapsto r_i' > r_i$, $i = 1, \ldots, t$ so that $\sum_i r_i' = k$. However, if we do this in an arbitrary fashion we might make a suboptimal choice. To circumvent this, we can exhaustively inspect all possible choices. For this, it suffices to construct $\binom{k - \sum_i r_i + t - 1}{t - 1} = \mathcal{O}(k^{t-1})$ instances of MATROIDMEDIAN. In the case of rb-MEDIAN discussed in Sect. 4.2, i.e., when $r_1 + r_2 < k$, the required number of instances is linear in k.

5 Proposed Methods

In this section we present practical methods to solve the diversity-aware clustering problem. In particular, we present local-search algorithms for DIV-k-MEDIAN$_\emptyset$ and a method based on relaxing the representation constraints for DIV-k-MEDIAN.

5.1 Local Search

Algorithms based on the local-search heuristic have been used to design approximation algorithms for many optimization problems, including facility location [2,7,14] and k-median [2,7,15] problems. In light of the inapproximability results presented in the previous section it comes as no surprise that any polynomial-time algorithm, including local-search methods, cannot be expected to find a feasible solution for the DIV-k-MEDIAN problem. Nevertheless, local-search methods are viable for the tractable instances discussed in Sect. 4, and can be shown to provide provable quality guarantees.

For solving the DIV-k-MEDIAN$_\emptyset$ problem we propose two algorithms based on local search.

Local Search Variant #1 (LS-1). We propose a single-swap local-search algorithm described in Fig. 1. The key difference with respect to vanilla local search is that we must ensure that a swap does not violate the representation constraints.

We stress that the proposed algorithm LS-1 is not viable for general instances of DIV-k-MEDIAN with intersecting facility groups. To illustrate, we present an example in Fig. 2. Let F_r, F_g, F_b, F_y be facility groups, corresponding to the colors red, green, blue and yellow, respectively. The intersection cardinality constraints $r_r = r_g = r_b = r_y = 1$ and the number of medians $k = 2$.

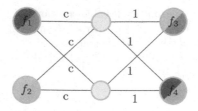

Fig. 2. An example illustrating the infeasibility of local search. (Color figure online)

1. Initialize — arbitrarily pick:
   (a) $S_i \subseteq F_i$ such that $|S_i| = r_i$ for all $i \in [t]$,
   (b) $S_{t+1} \subseteq \mathcal{F} \setminus \bigcup_{i \in [t]} S_i$ such that $|S_{t+1}| = k - \sum_{i \in [t]} r_i$, and
   (c) initial solution is $S = \bigcup_{i \in [t]} S_i \cup S_{t+1}$.
2. Iterate — while there exists tuples $(s_1, \dots, s_{t+1})$ and $(s'_1, \dots, s'_{t+1})$ such that:
   (a) $s_i \in S_i$, $s'_i \in F_i$ for all $i \in [t]$, $s_{t+1} \in S_{t+1}$, $s'_{t+1} \in \mathcal{F} \setminus \bigcup_{i \in [t]} S_i$
   (b) $S \setminus \{s_1, \dots, s_{t+1}\} \cup \{s'_1, \dots, s'_{t+1}\}$ is feasible, and
   (c) $\mathrm{cost}(S \setminus \{s_1, \dots, s_{t+1}\} \cup \{s'_1, \dots, s'_{t+1}\}) < \mathrm{cost}(S)$
   set $S = S \setminus \{s_1, \dots, s_{t+1}\} \cup \{s'_1, \dots, s'_{t+1}\}$.
3. Return S.

Fig. 3. Local-search heuristic (LS-2) for DIV-k-MEDIAN$_\emptyset$

Let $S = \{f_1, f_2\}$ be a feasible solution. It is trivial to see that we cannot swap f_1 with either f_3 or f_4, since both swaps violate the constraints $|S \cap F_b| \geq 1$ and $|S \cap F_r| \geq 1$, respectively. Likewise we cannot swap f_2 with either f_3 or f_4. So our local-search algorithm is stuck at a local optima and the approximation ratio is c, which can be made arbitrarily large. We can construct a family of infinitely many such problem instances where the local-search algorithm returns arbitrarily bad results. Similarly we can construct a family of infinitely many instances where the DIV-k-MEDIAN problem with t facility groups and $k < t$ would require at least $t - 1$ parallel swaps to ensure that local search is not stuck in a local optima. This example illustrates the limited applicability of the local-search heuristic for the most general variant of the DIV-k-MEDIAN problem, where the facility groups overlap in an arbitrary way.

Local Search Variant #2 (LS-2). Our second approach is the multi-swap local-search heuristic described in Fig. 3. The algorithm works by picking r_i facilities from F_i and $k - \sum_{i \in [t]} r_i$ from $\mathcal{F}$ as an initial feasible solution. We swap a tuple of facilities $(s_1, \dots, s_{t+1})$ with $(s'_1, \dots, s'_{t+1})$ as long as the cost improves. The algorithm has running time of $\mathcal{O}(n^t)$, and thus it is not practical for large values of t. The algorithm LS-2 is a $5 + \epsilon$ approximation for the DIV-k-MEDIAN$_\emptyset$ problem with two facility groups i.e., $t = 2$ (see Sect. 4.2). Bounding the approximation ratio of algorithm LS-2 for $t > 2$ is an open problem.

Note that the cost of the solution obtained by LS-1 and LS-2 is the k-median cost, that is, $\mathrm{cost}(S) = \sum_{v \in C} \min_{s \in S} d(v, s)$.

5.2   Relaxing the Representation Constraints

In view of the difficulty of solving the problem as formulated in Sect. 3, we explore alternative, more easily optimized formulations to encode the desired representation constraints. We first observe that a straightforward approach, akin to a Lagrangian relaxation, might result in undesirable outcomes. Consider the following objective function:

$$\text{cost}(S) = \sum_{v \in C} \min_{s \in S} d(v, s) + \lambda \sum_{i \in [t]} \max\{r_i - |F_i \cap S|, 0\}, \tag{1}$$

that is, instead of enforcing the constraints, we penalize their violations. A problem with this formulation is that every constraint satisfaction—up to r_i—counts the same, and thus the composition of the solution might be imbalanced.

We illustrate this shortcoming with an example. Consider $\mathscr{F} = \{F_1, F_2, F_3\}$, $k = 6$, $r_1 = 2$, $r_2 = 2$, $r_3 = 0$. Now consider two solutions: (i) 2 facilities from F_1, 0 from F_2, and 4 from F_3; and (ii) 1 facility from F_1, 1 from F_2, and 4 from F_3. Both solutions score the same in terms of number of violations. Nevertheless, the second one is more balanced in terms of group representation. To overcome this issue, we propose the following alternative formulation.

$$\text{cost}_f(S) = \sum_{v \in C} \min_{s \in S} d(v, s) + \lambda \sum_{i \in [t]} \frac{r_i}{|S \cap F_i| + 1}. \tag{2}$$

The second term that encodes the violations enjoys group-level diminishing returns. Thus, when a facility of a protected group is added, facilities from other groups will be favored. The cardinality requirements r_i act here as weights on the different groups.

We optimize the objective in Eq. 2 using vanilla local-search by picking an arbitrary initial solution with no restrictions.

6   Experimental Evaluation

In order to gain insight on the proposed problem and to evaluate our algorithms, we carried out experiments on a variety of publicly available datasets. Our objective is to evaluate the following key aspects:

Price of Diversity: What is the price of enforcing representation constraints? We measure how the clustering cost increases as more stringent requirements on group representation are imposed.

Relaxed Objective: We evaluate the relaxation-based method, described in Sect. 5, for the intractable case with facility group intersections. We evaluate its performance in terms of constraint satisfaction and clustering cost.

Running Time: Our problem formulation requires modified versions of standard local-search heuristics, as described in Sect. 5. We evaluate the impact of these variants on running time.

Table 2. Dataset statistics. n is the number of data points, D is dataset dimension, t is number of facility types. Columns $4, 5$ and $6, 7$ is the maximum and minimum size of facility groups when divided into two disjoint groups and four intersecting groups, respectively.

Dataset	n	D	$t = 2$		$t = 4$	
			Min	Max	Min	Max
heart-switzerland	123	14	10	113	–	–
heart-va	200	14	6	194	–	–
heart-hungarian	294	14	81	213	–	–
heart-cleveland	303	14	97	206	–	–
student-mat	395	33	208	187	–	–
house-votes-84	435	17	267	168	–	–
student-por	649	33	383	266	–	–
student-per2	666	12	311	355	–	–
autism	704	21	337	367	20	337
hcv-egy-data	1 385	29	678	707	40	698
cmc	1 473	10	220	1 253	96	511
abalone	4 177	9	1 307	1 342	–	–
mushroom	8 123	23	3 375	4 748	1 072	3 655
nursery	12 959	9	6 479	6 480	3 239	4 320

Datasets. We use datasets from the UCI machine learning repository [12]. We normalize columns to unit norm and use the L_1 metric as distance function. The dataset statistics are reported in Table 2.

Baseline. As a baseline we use a local-search algorithm with no cardinality constraints. We call this baseline LS-0. For each dataset we perform 10 executions of LS-0 with random initial assignments to obtain the solution with minimum cost ℓ_0 among the independent executions. LS-0 is known to provide a 5-approximation for the k-median problem [2]. We also experimented with *exhaustive enumeration* and *linear program solvers*, however these approaches failed to solve instances with modest size, which is expected given the inherent complexity of DIV-k-MEDIAN.

Experimental Setup. The experiments are executed on a desktop with 4-core *Haswell* CPU and 16 GB main memory. Our source code is written in Python and we make use of numpy to enable parallelization of computations. Our source code is available as open source [26].

6.1   Results

Price of Diversity. For each dataset we identify a protected attribute and classify data points into two disjoint groups. In most datasets we choose gender, except in house-votes dataset where we use party affiliation. We identify the smallest group in the dataset (*minority group*) and measure the fraction of the

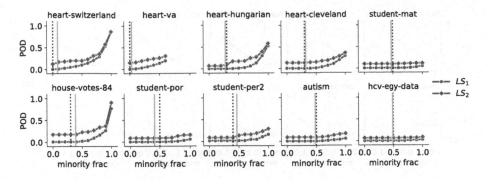

Fig. 4. Price of diversity ($k = 10$).

chosen facilities that belong to that group (*minority fraction*). When running LS-1 and LS-2 we enforce a specific minority fraction and repeat the experiments for ten iterations by choosing random initial assignments. We refer to the cost of the solutions obtained from LS-0, LS-1 and LS-2 as ℓ_0, ℓ_1 and ℓ_2, respectively.

The *price of diversity* (POD) is the ratio of increase in the cost of the solution to the cost of LS_0 i.e., $\text{POD(LS-1)} = \frac{\ell_1 - \ell_0}{\ell_0}$ and $\text{POD(LS-2)} = \frac{\ell_2 - \ell_0}{\ell_0}$. Recall that in theory the POD is unbounded. However, this need not be the case in practical scenarios. Additionally, we compute the differences in group representation between algorithms as follows. Let $R^i = \{r_1^i, \ldots, r_t^i\}$ be the set representing the number of facilities chosen from each group in $\mathscr{F}$ by algorithm $\text{LS}-j$. For $j = 1, 2$ we define $L_1(\text{LS}-j) = \sum_{i \in [t]} |r_i^j - r_i^0| / (kt)$.

In Fig. 4, we report the price of diversity (POD) as a function of the imposed minority fraction for LS-1 and LS-2. The blue and yellow vertical lines denote the minority fraction achieved by the baseline LS-0 and the fraction of minority facilities in the dataset, respectively. Notice that the minority fraction of the baseline is very close to the minority fraction of the dataset. With respect to our methods LS-1 and LS-2, we observe little variance among the independent executions. Most importantly, we observe that the price of diversity is relatively low, that is, for most datasets we can vary the representation requirements over a wide range and the clustering cost increases very little compared to the non-constrained version. An increase is observed only for a few datasets and only for extreme values of representation constraints. We also observe that LS-1 outperforms consistently LS-2. This is good news as LS-1 is also more efficient.

In Fig. 5, we report the L_1 measure as a function of the increase in the minority fraction. Note that we enforce a restriction that the ratio of minority nodes should be at least the minority fraction, however, the ratio of facilities chosen from the minority group can be more than the minority fraction enforced. In this experiment we measure the change in the type of facilities chosen. We observe more variance in L_1 score among the independent runs when the minority fraction of the solution is less than the minority fraction of the dataset. This shows that the algorithm has more flexibility to choose the different type of facilities. In Fig. 7 we report POD and L_1 measure for moderate size datasets.

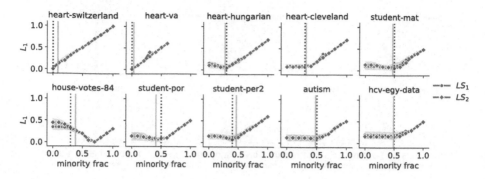

Fig. 5. L_1 distance of the chosen facility types ($k = 10$).

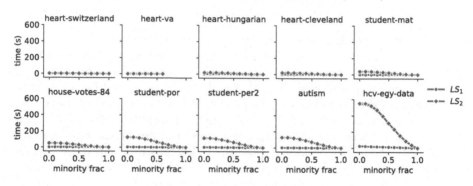

Fig. 6. Running Time ($k = 10$).

Running Time. In Fig. 6 we report the running time of LS-1 and LS-2 as a function of the minority fraction. For small datasets we observe no significant change in the running time of LS-1 and LS-2. However, the dataset size has a significant impact on running time of LS-2. For instance in the hcv-egy-data dataset, for $k = 10$ and minority fraction 0.1, LS-2 is 300 times slower than LS-1. This is expected, as the algorithm considers a quadratic number of replacements per iteration. Despite this increase in time, there is no significant improvement in the cost of the solution obtained, as observed in Fig. 4. This makes LS-1 our method of choice in problem instances where the facility groups are disjoint.

Relaxed Objective. In our final set of experiments we study the behavior of the LS-0 local-search heuristic with the relaxed objective function of Eq. (2). In Fig. 8 we report price of diversity (POD) and the fraction of violations of representation constraints L^* for each value of $\lambda = \{2^1, \ldots, 2^7\}$. For each dataset we choose four protected attributes to obtain intersecting facility types, and perform experiments with $k = 10$ and the representation constraints set $R = \{3, 3, 3, 3\}$. The value of L^* measures the fraction of violations of the representation constraints i.e., $L^* = \frac{\sum_{i \in [t]} \min(0, |S \cap F_i| - r_i)}{\sum_{i \in [t]} r_i}$. With the increase in the value of λ the value of L^* decreases and the value of POD increases, as expected. However,

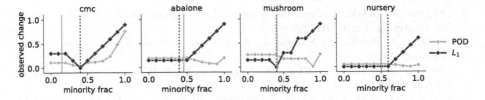

Fig. 7. Price of diversity for moderate size datasets ($k = 10$).

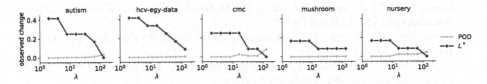

Fig. 8. Price of diversity for intersecting facility types ($k = 10$).

the increase in POD is very small and in all cases it is possible to find solutions where both POD and L^* are very close to zero, that is, solutions that have very few constraint violations and their clustering cost is almost as low as in the unconstrained version.

7   Conclusion

We introduce a novel formulation of diversity-aware clustering, which ensures fairness by avoiding under-representation of the cluster centers, where the cluster centers belong in different facility groups. We show that the general variant of the problem where facility groups overlap is **NP**-hard, fixed-parameter intractable with respect to the number of clusters, and inapproximable to any multiplicative factor. Despite such negative results we show that the variant of the problem with disjoint facility types can be approximated efficiently. We also present heuristic algorithms that practically solve real-world problem instances and empirically evaluated the proposed solutions using an extensive set of experiments. The main open problem left is to improve the run-time complexity of the approximation algorithm, in the setting of disjoint groups and $t > 2$, so that it does not use repeated calls to a linear program. Additionally, it would be interesting to devise FPT algorithms for obtaining exact solutions, again in the case of disjoint groups.

References

1. Anagnostopoulos, A., Becchetti, L., Fazzone, A., Menghini, C., Schwiegelshohn, C.: Spectral relaxations and fair densest subgraphs. In: CIKM (2020)
2. Arya, V., Garg, N., Khandekar, R., Meyerson, A., Munagala, K., Pandit, V.: Local-search heuristic for k-median and facility-location problems. In: STOC (2001)

3. Backurs, A., Indyk, P., Onak, K., Schieber, B., Vakilian, A., Wagner, T.: Scalable fair clustering. In: ICML (2019)
4. Barocas, S., Hardt, M., Narayanan, A.: Fairness and Machine Learning. fairmlbook.org (2019)
5. Bera, S., Chakrabarty, D., Flores, N., Negahbani, M.: Fair algorithms for clustering. In: NeurIPS (2019)
6. Bercea, I.O., et al.: On the cost of essentially fair clusterings. In: APPROX (2019)
7. Charikar, M., Guha, S.: Improved combinatorial algorithms for the facility location and k-median problems. In: FOCS (1999)
8. Chen, D.Z., Li, J., Liang, H., Wang, H.: Matroid and knapsack center problems. Algorithmica **75**(1), 27–52 (2016)
9. Chierichetti, F., Kumar, R., Lattanzi, S., Vassilvitskii, S.: Fair clustering through fairlets. In: NeurIPS (2017)
10. Cygan, M., et al.: Parameterized Algorithms, vol. 4. Springer, Cham (2015). https://doi.org/10.1007/978-3-319-21275-3
11. Danks, D., London, A.J.: Algorithmic bias in autonomous systems. In: IJCAI (2017)
12. Dua, D., Graff, C.: UCI machine learning repository (2017). http://archive.ics.uci.edu/ml
13. Dwork, C., Hardt, M., Pitassi, T., Reingold, O., Zemel, R.: Fairness through awareness. In: ITCS (2012)
14. Gupta, A., Tangwongsan, K.: Simpler analyses of local-search algorithms for facility location. arXiv:0809.2554 (2008)
15. Hajiaghayi, M., Khandekar, R., Kortsarz, G.: Local-search algorithms for the red-blue median problem. Algorithmica **63**(4), 795–814 (2012)
16. Hajiaghayi, M., Khandekar, R., Kortsarz, G.: Budgeted red-blue median and its generalizations. In: ESA (2010)
17. Hardt, M., Price, E., Srebro, N.: Equality of opportunity in supervised learning. In: NeurIPS (2016)
18. Huang, L., Jiang, S., Vishnoi, N.: Coresets for clustering with fairness constraints. In: NeurIPS (2019)
19. Krishnaswamy, R., Kumar, A., Nagarajan, V., Sabharwal, Y., Saha, B.: The matroid median problem. In: SODA (2011)
20. Pedreshi, D., Ruggieri, S., Turini, F.: Discrimination-aware data mining. In: KDD (2008)
21. Rösner, C., Schmidt, M.: Privacy preserving clustering with constraints. In: ICALP (2018)
22. Samadi, S., Tantipongpipat, U., Morgenstern, J., Singh, M., Vempala, S.: The price of fair PCA: one extra dimension. In: NeurIPS (2018)
23. Schmidt, M., Schwiegelshohn, C., Sohler, C.: Fair coresets and streaming algorithms for fair k-means. In: WAOA (2019)
24. Singh, A., Joachims, T.: Policy learning for fairness in ranking. In: NeurIPS (2019)
25. Swamy, C.: Improved approximation algorithms for matroid and knapsack median problems and applications. ACM Trans. Algorithms **12**(4), 1–22 (2016)
26. Thejaswi, S., Ordozgoiti, B., Gionis, A.: Diversity-aware k-median: experimental v1.0 (2021). https://github.com/suhastheju/diversity-aware-k-median
27. Yao, S., Huang, B.: Beyond parity: fairness objectives for collaborative filtering. In: NeurIPS (2017)
28. Zehlike, M., Bonchi, F., Castillo, C., Hajian, S., Megahed, M., Baeza-Yates, R.: Fa*ir: a fair top-k ranking algorithm. In: CIKM (2017)

Sibling Regression for Generalized Linear Models

Shiv Shankar[1(✉)] and Daniel Sheldon[1,2]

[1] University of Massachusetts, Amherst, MA 01003, USA
{sshankar,sheldon}@cs.umass.edu
[2] Mount Holyoke College, South Hadley, MA 01075, USA

Abstract. Field observations form the basis of many scientific studies, especially in ecological and social sciences. Despite efforts to conduct such surveys in a standardized way, observations can be prone to systematic measurement errors. The removal of systematic variability introduced by the observation process, if possible, can greatly increase the value of this data. Existing non-parametric techniques for correcting such errors assume linear additive noise models. This leads to biased estimates when applied to generalized linear models (GLM). We present an approach based on residual functions to address this limitation. We then demonstrate its effectiveness on synthetic data and show it reduces systematic detection variability in moth surveys.

Keywords: Sibling regression · GLM · Noise confounding

1 Introduction

Observational data is increasingly important across a range of domains and may be affected by measurement error. Failure to account for systemic measurement error can lead to incorrect inferences. Consider a field study of moth counts for estimating the abundance of different moth species over time. Figure 1(a) shows the counts of *Semiothisa burneyata* together with 5 other species. We see that *all* species had abnormally low counts on the same day. This suggests that the low count is due to a confounder and not an actual drop in the population. The same phenomenon is also prevalent in butterfly counts (Fig. 1(b)), where poor weather can limit detectability.

An abstract version of the aforementioned situation can be represented by Fig. 3. X here represents ecological factors such as temperature, season etc.; and Z_1, Z_2 represent the true abundance of species (such as moths). N is a corrupting noise (such as lunar phase) which affects the observable abundance θ of the organisms. Y represents an observation of the population and is modeled as a

Electronic supplementary material The online version of this chapter (https://doi.org/10.1007/978-3-030-86520-7_48) contains supplementary material, which is available to authorized users.

© Springer Nature Switzerland AG 2021
N. Oliver et al. (Eds.): ECML PKDD 2021, LNAI 12976, pp. 781–795, 2021.
https://doi.org/10.1007/978-3-030-86520-7_48

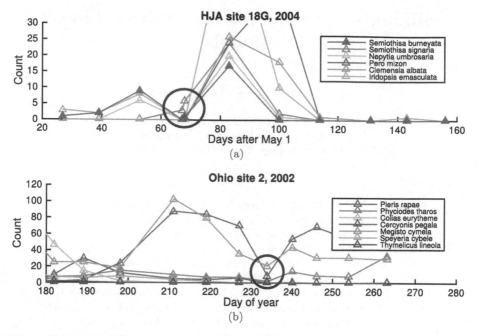

Fig. 1. Systematic detection error in moth and butterfly counts: (a) Correlated counts of other moths strongly suggest this is a detection problem. (b) Correlated detection errors in butterfly counts on day 234.

sample drawn distribution of observed abundance (for e.g. a Poisson distribution). Directly trying to fit a model ignoring the noise N can lead to erroneous conclusions about key factors such as effect of temperature on the population.

Distinguishing observational noise and measurement variability from true variability, often requires repeated measurements which in many cases can be expensive, if not impossible. However, sometimes even *in the absence of repeated measurements*, the effect of confounding noise can be estimated. Sibling regressions (Schölkopf et al. 2015) refer to one such technique that use auxiliary variables influenced by a shared noise factor to estimate the true value of the variable of interest. These techniques work without any parametric assumption about the noise distribution (Schölkopf et al. 2015; Shankar et al. 2019). However, these works assume an additive linear noise model, which limits their applications. For example in the aforementioned insect population case the effect of noise is more naturally modeled as multiplicative rather than additive.

We introduce a method that extends these ideas to generalized linear models (GLMs) and general exponential family models. First, we model non-linear effect of noise by considering it as an additive variable in the natural parameters of the underlying exponential family. Secondly, instead of joint inference of the latent variables, a stagewise approach is used. This stagewise approach is justified from a generalized interpretation of sibling regression. We provide justification behind our approach and then test it on synthetic data to quantitatively demonstrate the effectiveness of our proposed technique. Finally, we apply it to the moth survey data set used in Shankar et al. (2019) and show that it reduces measurement error more effectively than prior techniques.

2 Related Work

Estimation of observer variation can be framed as a causal effect estimation problem (Bang and Robins 2005; Athey and Imbens 2016). Such conditional estimates often require that all potential causes of confounding errors have been measured (Sharma 2018). However, real-life observational studies are often incomplete, and hence these assumptions are unlikely to hold.

Natarajan et al. (2013); Menon et al. (2015) develop techniques to handle measurement error as a latent variable. Similar approaches have been used to model observer noise as class-conditional noise (Hutchinson et al. 2017; Yu et al. 2014). One concern with such approaches is that they are generally unidentifiable.

A related set of literature is on estimation with missing covariates (Jones 1996; Little 1992). These are generally estimated via Monte-Carlo methods (Ibrahim and Weisberg 1992), Expectation-maximization like methods (Ibrahim et al. 1999) or by latent class analysis (Formann and Kohlmann 1996). Another set of approaches require strong parametric assumptions about the joint distribution (Little 1992). Like other latent variable models, these can be unidentifiable and often have multiple solutions (Horton and Laird 1999).

MacKenzie et al. (2002) and Royle (2004) learn an explicit model of the detection process to isolate observational error in ecological surveys using repeated measurements. Various identifiability criteria have also been proposed for such models (Sólymos and Lele 2016; Knape and Korner-Nievergelt 2016). Lele et al. (2012) extend these techniques to the case with only single observation. However, these models are only as reliable as the assumptions made about the noise variable. Our approach on the other hand makes weaker assumptions about the form of noise.

Schölkopf et al. (2015) introduced 'Half-sibling regression'; an approach for denoising of independent variables. This approach is both identifiable and does not make assumptions about the prior distribution of noise. Shankar et al. (2019) further extended the technique to the case when the variables of interest are only conditionally independent given an observed common cause.

3 Preliminaries

Exponential Family Distributions. A random variable Y is said to be from an exponential family (Kupperman 1958) if its density can be written as

$$p(y|\theta) = h(y)\exp(\theta^T T(y) - A(\theta))$$

where θ are the (natural) parameters of the distribution, $T(y)$ is a function of the value y called the *sufficient statistic*, and $A(\theta)$ is the log normalizing constant, and $h(y)$ is a base measure.

Given an exponential family density $p(y|\theta)$, let $L(y, \theta) = -\log p(y|\theta)$ be the negative log-likelihood loss function at data point y, and let $L(\theta) = \frac{1}{m} \sum_{i=1}^{m} L(y^{(j)}, \theta)$ be the overall loss of a sample $y^{(1)}, \ldots, y^{(m)}$ from the model. Also let $I(\theta) = \nabla^2 A(\theta)$ be the Fisher information.

We summarize few standard properties (e.g., see Koller and Friedman 2009) of exponential families that we will be of use later:

1. $\nabla A(\theta) = \mathbb{E}[T(Y)]$.
2. $\nabla_\theta L(y, \theta) = A(\theta) - T(y)$
3. $\forall y : \nabla_\theta^2 L(y, \theta) = \nabla^2 A(\theta) = I(\theta)$. This also implies that $\nabla^2 L(\theta) = \nabla^2 A(\theta) = I(\theta)$.

Generalized Linear Models. Generalized Linear Models (GLMs) (Nelder and Wedderburn 1972) are a generalization of linear regression models where the output variable is not necessarily Gaussian. In a GLM, the conditional distribution of Y given covariates X is an exponential family with mean $\mathbb{E}[Y|X]$ is related to a linear combination of the covariates by the link function g, and with the identity function as the sufficient statistic, i.e., $T(Y) = Y$. We focus on the special case of GLMs with *canonical link functions*, for which $\theta = X^T \beta$, i.e., the natural parameter itself is a linear function of covariates. For the canonical GLM, the following properties hold (Koller and Friedman 2009)[1]:

1. The link function is determined by $A : \mathbb{E}[Y|X] = g^{-1}(X^T \beta) = \nabla A(X^T \beta)$
2. The conditional Fisher information matrix $I_{Y|X} = I_{\cdot|X}(\theta) = \nabla^2 A(X^T \beta)$

GLMs are commonly used in many applications. For example, logistic regression for binary classification is identical to a Bernoulli GLM. Similarly, regressions where the response variable is of a count nature, such as the number of individuals of a species, are based on Poisson GLM.

Sibling Regression. Sibling regression (Schölkopf et al. 2015; Shankar et al. 2019) is a technique which detects and corrects for confounding by latent noise variable using observations of another variable influenced by the same noise variable.

The causal models depicted in Fig. 2 capture the essential use case of sibling regression techniques. Here, Z_1, Z_2 represent the unobserved variables of interest, which we would like to estimate using the observed variables Y_1, Y_2; however the observations are confounded by the common unobserved noise N. Schölkopf et al. (2015) use the model illustrated in Fig. 2(a) as the basis of their *half-sibling regression* approach to denoise measurements of stellar brightness. The *a priori* independence of Z_1, Z_2 implies that any correlation between Y_1, Y_2 is an artifact of the noise process.

The half-sibling estimator of Z_1 from (Schölkopf et al. 2015) is

$$\hat{Z}_1 = Y_1 - \mathbb{E}[Y_1|Y_2] + \mathbb{E}[Y_1] \tag{1}$$

---

[1] These properties can be obtained by applying the aforementioned exponential family properties.

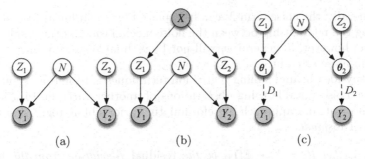

Fig. 2. (a) Half-sibling regression. (b) Three-quarter sibling regression. (c) Generalized linear sibling model without covariates.

This equation can be interpreted as removing the part of Y_1 that can be explained by Y_2, and then adding back the mean $\mathbb{E}[Y_1]$. Schölkopf et al. (2015) did not include the final $\mathbb{E}[Y_1]$ term, so our estimator differs from the original by a constant; we include this term so that $\mathbb{E}[\hat{Z}_1] = \mathbb{E}[Y_1]$. In practice, the expectations required are estimated using a regression model to predict Y_1 from Y_2, which gives rise to the name "sibling regression".

Three-quarter sibling regression (Shankar et al. 2019) generalizes the idea of half-sibling regression to the case when Z_1 and Z_2 are not independent but are conditionally independent given an observed covariate X. In the application considered by Shankar et al. (2019), these variables are population counts of different species in a given survey; the assumed dependence structure is shown in Fig. 2(b). This estimator has a similar form to the half sibling version, except the expectations now condition on X:

$$\hat{Z}_{1|X} = Y_1 - \mathbb{E}[Y_1|X,Y_2] + \mathbb{E}[Y_1|X] \qquad (2)$$

4 Sibling Regression for Generalized Linear Models

Model. We now formally specify the model of interest. We will focus first on the half-sibling style model with no common cause X, and revisit the more general case later. Figure 2(c) shows the assumed independence structure. As in the previous models, the true variables of interest are Z_1 and Z_2, and the observed variables are Y_1 and Y_2. The variable N is confounding noise that influences observations of both variables. Unlike half-sibling regression, exponential-family sampling distributions D_1 and D_2 mediate the relationship between the hidden and observed variables; the variables θ_1 and θ_2 are the parameters of these distributions. The dotted arrow indicate that the relation between Y_i and θ_i is via sampling from an exponential family distribution, and not a direct functional dependency. On the other hand θ_1, θ_2 are deterministic functions of (Z_1, N) and (Z_2, N), respectively. We assume the noise acts additively on the natural parameter of the exponential family. Mathematically the model is

$$Y_i \sim D_i(\theta_i), \quad \theta_i = Z_i + N \quad i \in \{1,2\}. \qquad (3)$$

More generally, the noise term N can be replaced by a non-linear function $\psi_i(N)$ mediating the relationship between the noise mechanism and the resultant additive noise; however, in general ψ_i will not be estimable so we prefer to directly model the noise as additive.

The key idea behind sibling regression techniques is to find a "signature" of the latent noise variable using observations of another variable. In this section we will motivate our approach by reformulating prior sibling regression methods in terms of *residuals*.

Lemma 1. *Let $R_i = Y_i - \mathbb{E}[Y_i]$ be the* residual *(deviation from the mean) of Y_i in the half-sibling model, and let $R_{i|X} = Y_i - \mathbb{E}[Y_i|X]$ be the residual relative to the conditional mean in the three-quarter sibling model. The estimators of Eqs. (1) and (2) can be rewritten as*

$$\hat{Z}_1 = Y_1 - \mathbb{E}[R_1|R_2], \quad \hat{Z}_{1|X} = Y_1 - \mathbb{E}\big[R_{1|X} \mid R_{2|X}\big].$$

Proof. For the half-sibling case, write

$$\begin{aligned}
\hat{Z}_1 &= Y_1 - \mathbb{E}[Y_1 \mid Y_2] + \mathbb{E}[Y_1] \\
&= Y_1 - \mathbb{E}\big[Y_1 - \mathbb{E}[Y_1] \mid Y_2\big] \\
&= Y_1 - \mathbb{E}\big[Y_1 - \mathbb{E}[Y_1] \mid Y_2 - \mathbb{E}[Y_2]\big] \\
&= Y_1 - \mathbb{E}[R_1 \mid R_2]
\end{aligned}$$

The three-quarter case is similar.

This formulation provides a concise interpretation of sibling regressions as regressing the residuals of Y_1 on those of Y_2.

4.1 Extension to GLM

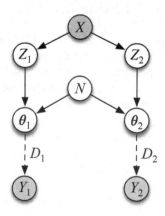

Problem Statement. The problem is to obtain estimates of Z_1 and Z_2 given m independent observations $(y_1^{(1)}, y_2^{(1)}), \dots (y_1^{(m)}, y_2^{(m)})$ of Y_1 and Y_2. The general idea is, as in prior sibling regressions, to model and remove the signature of the noise variable. Schölkopf et al. (2015); Shankar et al. (2019) solve this problem in the special case where D_i is Dirac. We wish to extend this to the case of other sampling distributions.

Fig. 3. Generalized sibling model with covariates

By symmetry, it suffices to focus on estimating only one variable, so we will henceforth consider Z_1 to be the estimation target. We allow the possibility that Z_2 is multivariate to model the case when there are many siblings that each contain a trace of the noise variable.

Inspiring from the residual form of sibling-regression 1, we derive a residual which can be used to approximate the confounding noise in the GLM case.

Computing a residual requires defining a reference, in analogy to the conditional global mean $\mathbb{E}[Y_1|X]$ used in Eq. 2. We will use the maximum-likelihood estimate under the "global" model $Y_1 \sim D_1(X^T \beta)$ as the reference. Let $\hat{\beta}$ be the estimated regression coefficients for the relationship between the covariates X and variable Y_1. The corresponding model predictions $\hat{Y}_1$ are then given by $g^{-1}(X^T \hat{\beta})$ where g is the link function. We define $\mathcal{R}$ as:

$$\mathcal{R}(Y) = I_{Y|X}^{-1}(Y - \hat{Y}) \tag{4}$$

where $I_{Y|X}$ is the conditional Fisher information.

Proposition 1

$$\mathbb{E}[\mathcal{R}(Y_1)|X, N] = Z_1 + N - X^T \hat{\beta} + O(|Z_1 + N - X^T \hat{\beta}|^2)$$
$$\approx Z_1 + N - X^T \hat{\beta}$$

Proof. We temporarily drop the subscript so that $(Y|Z, N) \sim D(\theta)$ where $\theta = Z + N$ is the natural parameter of the exponential family distribution D corresponding to the specified GLM.

$$\mathbb{E}[\mathcal{R}(Y)|X, N] = \mathbb{E}[I_{Y|X}^{-1}(Y - \hat{Y})|X, N]$$
$$= \mathbb{E}[[I_{Y|X}(\hat{\beta})]^{-1}(Y - \nabla A(X^T \hat{\beta})|X, N]$$
$$= [I_{Y|X}(\hat{\beta})]^{-1}(\mathbb{E}[Y|X, N] - \nabla A(X^T \hat{\beta}))$$
$$= [I_{Y|X}(\hat{\beta})]^{-1}(\nabla A(Z + N) - \nabla A(X^T \hat{\beta}))$$
$$= [I_{Y|X}(\hat{\beta})]^{-1}[\nabla^2 A(X^T \hat{\beta})(Z + N - X^T \hat{\beta})$$
$$+ O(|Z + N - X^T \hat{\beta}|^2)]$$
$$= [(Z + N - X^T \hat{\beta}) + O(|Z + N - X^T \hat{\beta}|^2)]$$

The second line used the definition of canonical link function. The third uses linearity of expectation. The fourth line uses the first property of exponential families from Sect. 3. The fifth line applies the Taylor theorem to ∇A about $X^T \hat{\beta}$. The last two lines use our previous definitions that $I_{Y|X} = \nabla^2 A(\theta)$ and $\theta = X^T \beta + N$. Restoring the subscripts on all variables except N, which is shared, gives the result.

In simpler terms, $\mathcal{R}(Y_i)$ can provide a useful approximation for $Z_i + N - X^T \hat{\beta}_i$. Moreover, if we assume that X largely explains Z_i, then the above expression is dominated by N. We would then like to use $\mathcal{R}$ to estimate and correct for the noise.

Proposition 2. *Let the approximation in Proposition 1 be exact, i.e.,* $\mathbb{E}[\mathcal{R}(Y_i)|X, N] = Z_i + N - X^T \hat{\beta}_i$, *then* $\mathbb{E}[\mathcal{R}(Y_1)|\mathcal{R}(Y_2), X] - \mathbb{E}[\mathcal{R}(Y_1)|X] = \mathbb{E}[N|X, \mathcal{R}(Y_2)]$ *upto a constant.*

The derivation is analogous to the one presented by Shankar et al. (2019) and is given in Appendix ??. While the higher order residual terms in Proposition 1 generally cannot be ignored, Proposition 2 informally justifies how we can estimate N by regressing $\mathcal{R}(Y_1)$ on $\mathcal{R}(Y_2)$.

Note that $\mathcal{R}(Y_i)$ is a *random variable* and that the above expression is for the expectation of $\mathcal{R}$ which may not be exactly known (due to limited number of samples drawn from each conditional distribution). However we observe that a) Z_1, Z_2 are independent conditional on the observed X and b) the noise N is independent of X and Z and is common between $\mathcal{R}(Y_1), \mathcal{R}(Y_2)$. This suggests that in presence of multiple auxiliary variables (let Y_{-i} denote all Y variables except Y_i), we can improve our estimate of N by combining information from all of them. Since Z_i's are conditionally independent (given X) of each other, we can expect their variation to cancel each other on averaging. The noise variable on the other hand being common will not cancel. Appendix ?? provides a justification for this statement.

Once we have an estimate $\hat{N}$ of the noise N, we can now estimate Z_1. However unlike standard sibling regression (Fig. 2(a), 2(b)), where the observed Y_1 was direct measurement of Z_1, in our case the relationship gets mediated via θ_1. If we had the true values of θ_1 one can directly use Eq. 2 to obtain Z_1. One can try to obtain θ_1 from Y_1; but since Y_1 includes sampling error affecting our estimates of Z_1. Instead we rely once again on the fact that the model is a GLM, which are efficient and unbiased when all covariates are available. We re-estimate our original GLM fit but now with additional covariate $\hat{N}$ which is a proxy for N.

Implementation. Based on the above insights, we propose Algorithm 1, labelled Sibling GLM (or SGLM) for obtaining denoised estimates from GLM models. In practice, the conditional expectations required are obtained by fitting regressors for the target quantity from the conditioning variables . As such we denote them by $\hat{E}[\cdot|\cdot]$): to distinguish them from true expectation values. Since, per Proposition 1, the true expected value is approximately linear in the conditioning variables, in our experiments we used ordinary least squares regression to estimate the conditional expectations in Step 3.

5   Experiments

In this section, we experimentally demonstrate the ability of our approach to reduce estimation errors in our proposed setting, first using simulations and semi-synthetic examples and then with a moth survey data set. Furthermore in our experiments we found the correlation between $\mathcal{R}(Y_i)$ and X to be small, and therefore simplified Step 3 and 4 in Algorithm 1 simplify to $N = \hat{\mathbb{E}}[\mathcal{R}(Y_1)|\mathcal{R}(Y_2)]$

Algorithm 1 SGLM Algorithm

Input: $(X^{(k)}, Y_1^{(k)}, Y_2^{(k)})_{k=1,\dots,n}$
Output: Estimates of latent variable $\hat{Z}_1$
Training: denote fitted regression models by $\hat{E}[\cdot|\cdot]$):

1. Compute $\hat{\mathbb{E}}[Y_1|X], \hat{\mathbb{E}}[Y_2|X]$ by training suitable GLM
2. Compute $\mathcal{R}(Y_1), \mathcal{R}(Y_2)$ as given by Eq. 4
3. Fit regression models for $\mathcal{R}(Y_1)$ using $\mathcal{R}(Y_2), X$ as predictors to obtain estimators of $\hat{\mathbb{E}}[\mathcal{R}(Y_1)|\mathcal{R}(Y_2), X], \hat{\mathbb{E}}[\mathcal{R}(Y_1)|X]$
4. Create $\hat{N}$ such that its kth value $\hat{N}^{(k)} = \hat{\mathbb{E}}[\mathcal{R}(Y_1)|\mathcal{R}(Y_2)^{(k)}, X^{(k)}] - \hat{\mathbb{E}}[\mathcal{R}(Y_1)|X^{(k)}]$ $\forall k \in [1, n]$
5. Estimate $\hat{Z}_1$ by fitting GLM models with n as an additional covariate i.e $\hat{E}[Y_1|X, n]$

5.1 Synthetic Experiment

We first conduct simulation experiments where, by design, the true value of Z_1 is known. We can then quantitatively measure the performance of our method across different ranges of available auxiliary information contained within Y_{-1}.

Description. We borrow the approach of Schölkopf et al. (2015), generalized to standard GLM distributions. We run these experiments for Poisson and Gamma distributions. This simulation was conducted by generating 120 observations of 20 different Poisson variables. Each individual observation was obtained via a noise-corrupted Poisson or Gamma distribution where, for each observation, the noise affected all variables simultaneously as described below. Specifically, each variable Y_i at a single observation time is obtained via a generative process dependent on $X \in \mathbb{R}$ and noise variable $N \in \mathbb{R}$ as:

$$Y_i \sim D(\underbrace{w_X^{(i)} X}_{Z_i} + w_N^{(i)} N + \epsilon)$$

The variables X and N are drawn uniformly from $[-1, 1]$. Similarly, the coefficient $w_N^{(i)}$ is drawn from a uniform distribution on $[-1, 1]$, while $\epsilon \sim \mathcal{N}(0, \sigma_\epsilon^2)$ is independent noise. Finally, $w_X^{(i)}$ is drawn from the standard conjugate prior for the distribution D.

Results. We conducted these simulations and measured the error in the estimated Z versus the true Z. Due to inherent variability caused by sampling, the error will not go down to zero. We present below the results on Poisson regression, while other results can be found in the appendix.

Figure 4 shows the estimation error for Poisson regression as a function of the dimension n of Y_{-1}, i.e., the number of auxiliary variables available for denoising. In Fig. 4(a) we plot the mean square error against the true value of Z_1 aggregated across the runs. Clearly increasing n reduces error. This is expected, as the noise

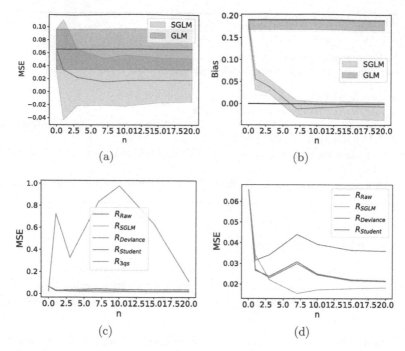

Fig. 4. Results on synthetic data. Mean Squared error (a), bias (b), and residual comparison (c,d) vs dimension of $|Y_2|$ on Poisson regressions

N can be better estimated using more auxiliary variables. This in turn leads to lower error in estimates. Due to the effect of N, the standard GLM estimates are biased. The simulation results bear this out, where we get more than *10%* bias in Poisson regression estimates. On the other hand, by being able to correct for the noise variable on a observation basis, our approach gives bias less than *3%*. We plot in Fig. 4(b) the bias of these estimates for Poisson. Once again as expected, increasing n reduces the bias.

We also experiment with other possible versions of residuals $\mathcal{R}^2$ including residual deviance and student residuals. In Fig. 4(c) we have plotted the results of these methods against the method of Shankar et al. (2019) (by fitting linear models on transformed observations). As is evident from the figure, data transformation, while a common practice, leads to substantially larger errors. Figure 4(d) presents the effect of changing the residual definition. Under our interpretation of sibling regression (Lemma 1), any version of residual would be acceptable. This intuition is borne out it in these results as all the residuals perform reasonably. However from the figure, our proposed residual definition is the most effective.

---

[2] Details in the Appendix.

5.2   Discover Life Moth Observations

Our next set of experiments use moth count surveys conducted under the Discover Life Project[3]. The survey protocol uses screens illuminated by artificial lights to attract moths and then records information of each specimen. This data has been collected over multiple years at different locations on a regular basis.

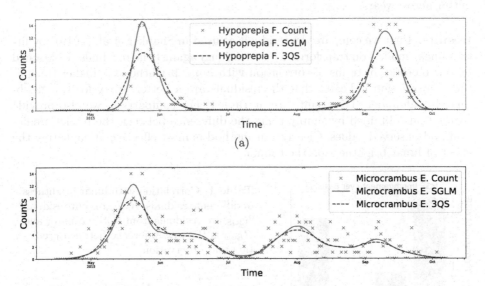

(a)

Fig. 5. Seasonal patterns of a) *Hypoprepia fucosa* and b) *Microcrambus elegans* as estimated by 3QS regression and our method alongside the observed counts. Note the higher peaks and better overall fit of our method.

A common systematic confounder in such studies is moonlight. Moth counts are often low on full moon nights as the light of the moon reduces the number of moths attracted to the observation screens. In their paper, Shankar et al. (2019) present the 'three-quarter sibling' (3QS) estimator and use it to denoise moth population counts. However, to apply the model they used least-squares regression on transformed count variables. Such transformations can potentially induce significant errors in estimation. The more appropriate technique would be to build models via a Poisson generalized additive model (GAM). In this experiment, we use our technique to directly learn the underlying Poisson model.

Description. We follow the methodology and data used by Shankar et al. (2019). We choose moth counts from the Blue Heron Drive site for 2013 through 2018. We then follow the same cross-validation like procedure, holding each year out as a test fold, while using all other years for training. However, instead of

---

[3] https://www.discoverlife.org/moth.

transforming the variables, we directly estimated a Poisson GAM model with the pyGAM (Servén and Brummitt 2018) package. Next, we compute Z_i with our SGLM-algorithm. This procedure is repeated for all folds and all species. We compare the MSE obtained by our estimates against the 3QS estimates. Note here that due to the absence of ground truth, the prediction error is being used as a proxy to assess the quality of the model. The hypothesis is that correcting for systematic errors such as the ones induced by the moon will help to generalize better across years.

Results. First we compare the residuals as used in Shankar et al. (2019) (by fitting linear models on transformed observations) against the residuals as obtained by our method, in terms of correlation with lunar brightness. A higher (magnitude) correlation indicates that the residuals are a better proxy for this unobserved confounder. The results are in Table 1. For comparison, we also provide correlations obtained by simply using the difference between the model prediction and observed values. Clearly, our method is most effective at capturing the effect of lunar brightness on the counts.

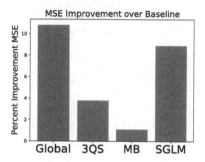

Fig. 6. Average percent improvement in predictive MSE relative to a GAM fitted to the raw counts

Table 1. Correlation with lunar brightness of different residuals. $\hat{R}_{\text{SGLM}}$ is our residual, $\hat{R}_{\text{3QS}}$ are residuals from 3QS estimator and R_{Raw} is the raw difference between prediction and observation.

Species	Moonlight Correlation		
	R_{SGLM}	R_{3QS}	R_{raw}
Melanolophia C	**−0.55**	−0.21	−0.42
Hypagyrtis E	**−0.66**	−0.42	−0.62
Hypoprepia F	**−0.65**	−0.36	−0.59

Next, we compare the decrease in prediction error obtained by our method against the methods tested by Shankar et al. (2019). These results are presented in Fig. 6. The mean squared error (MSE) is computed only on data from the test-year with moon brightness zero. "Global" is an oracle model shown for comparison. It is fit on multiple years to smooth out both sources of year-to-year variability (intrinsic and moon phase). "MB" is a model that includes moon brightness as a feature to model detection variability. From the figure, we can see that our method not only improves substantially over the 3QS estimator (9% vs 4%) but is comparable with the global model fit on multiple years (9% vs 10%).[4] This is partly because the transformed linear model is unsuited for these

---

[4] The global model is included as a rough guide for the best possible generalization performance, even though it does not solve the task of denoising data within each year.

variables. Our technique on the other hand can directly handle a broader class of conditional models and more diverse data types.

Finally, to present the difference in behavior of the two approaches, we plot the estimated moth population curves in Fig. 5. These curves are plotted for two different species with the dotted line representing the 3QS regression model, while the bold line represents our method. The actual observed counts are also plotted as points. One can clearly see the impact of the transformation, which produces flatter curves. For example, the height of the peaks for *Hypoprepia fucosa* end up significantly lower than the observed counts. On the other hand, our method predicts higher and narrower peaks, which better match the observed values.

6   Ethical Impact

Applications. Our method is more focused towards ecological survey data applications. Such surveys provide information useful for setting conservation policies. However there are other potential domains of application. Measurement noise is ubiquituous in experimental studies, and applied scientists often used different schemes to protect against confounding (Genbäck and de Luna 2019) and measurement effects (Zhang et al. 2018). As such our method may be useful applied to domains such as drug reporting (Adams et al. 2019) and epidemiology (Robins and Morgenstern 1987).

Implications. Our method provides an approach to handle the presence of unobserved confounding noise. The unobserved confounder however need not be a nuisance variable. Empirical datasets often exhibit different biases due to non-nuisance confounders (Davidson et al. 2019), which can lead to unfairness with respect race, gender and other protected attributes (Olteanu et al. 2016; Chouldechova et al. 2018). In some cases a protected attribute itself might be a confounder, in which case our approach can have implications for developing fairer models. Since our method partially recovers the unobserved confounder (noise), it can lead to identification or disclosure of protected attributes even when such data has been hidden or unavailable. This could lead to issues regarding privacy and security. The proposed method does not handle such issues, and adequate measures may be warranted for deploying this method.

7   Conclusion

Our paper has two primary contributions: a) reinterpreting sibling regression in residual form which enabled the generalization to GLMs and b) presenting a residual definition which corresponds to the case of noise in natural parameters. Based on these we designed a practical approach and demonstrated its potential on an environmental application.

A future line of work would be to develop goodness-of-fit tests for these models. A second question could be to generalize this chain of reasoning to complex non-linear dependencies. Finally since this method partially recovers the unobserved confounder (noise), it can potentially lead to identification of protected attributes even when such data has been hidden. As such another venue of future research is in the direction of how sibling regression can affect fairness and security of models.

References

Adams, R., Ji, Y., Wang, X., Saria, S.: Learning models from data with measurement error: tackling underreporting. arXiv:1901.09060 (2019)

Athey, S., Imbens, G.: Recursive partitioning for heterogeneous causal effects. Proc. Nat. Acad. Sci. **113**, 7353–7360 (2016)

Bang, H., Robins, J.: Doubly robust estimation in missing data and causal inference models. Biometrics **61**, 962–973 (2005)

Belsley, D.A., Kuh, E., Welsch, R.E.: Regression Diagnostics: Identifying Influential Data and Sources of Collinearity, vol. 571. John Wiley & Sons, Hoboken (2005)

Chouldechova, A., Benavides-Prado, D., Fialko, O., Vaithianathan. R.: A case study of algorithm-assisted decision making in child maltreatment hotline screening decisions. In: Proceedings of the 1st Conference on Fairness, Accountability and Transparency, volume 81 of Proceedings of Machine Learning Research, pp. 134–148. PMLR (2018)

Davidson, T., Bhattacharya, D., Weber, I.: Racial bias in hate speech and abusive language detection datasets. In: Workshop on Abusive Language Online (2019)

Formann, A.K., Kohlmann, T.: Latent class analysis in medical research. Statist. Methods Med. Res. **5**(2), 179–211 (1996)

Genbäck, M., de Luna, X.: Causal inference accounting for unobserved confounding after outcome regression and doubly robust estimation. Biometrics **75**(2), 506–515 (2019)

Horton, N.J., Laird, N.M.: Maximum likelihood analysis of generalized linear models with missing covariates. Statist. Methods Med. Res. **8**(1), 37–50 (1999)

Hutchinson, R.A., He, L., Emerson, S.C.: Species distribution modeling of citizen science data as a classification problem with class-conditional noise. In: AAAI, pp. 4516–4523 (2017)

Ibrahim, J.G., Weisberg, S.: Incomplete data in generalized linear models with continuous covariates. Australia J. Statist. **34**(3), 461–470 (1992)

Ibrahim, J.G., Lipsitz, S.R., Chen, M.-H.: Missing covariates in generalized linear models when the missing data mechanism is non-ignorable. J. R. Statist. Soc. Ser. B (Statist. Methodol.) **61**(1), 173–190 (1999)

Jones, M.P.: Indicator and stratification methods for missing explanatory variables in multiple linear regression. J. Am. statist. Assoc. **91**(433), 222–230 (1996)

Knape, J., Korner-Nievergelt, F.: On assumptions behind estimates of abundance from counts at multiple sites. Methods Ecol. Evol. **7**(2), 206–209 (2016)

Koller, D., Friedman, N.: Probabilistic Graphical Models: Principles and Technique. MIT Press, Cambridge (2009)

Kupperman, M.: Probabilities of hypotheses and information-statistics in sampling from exponential-class populations. Ann. Math. Statist. **29**(2), 571–575 (1958). ISSN 00034851. http://www.jstor.org/stable/2237349

Lele, S.R., Moreno, M., Bayne, E.: Dealing with detection error in site occupancy surveys: what can we do with a single survey? J. Plant Ecol. **5**(1), 22–31 (2012)

Little, R.J.: Regression with missing x's: a review. J. Am. Statist. Assoc. **87**(420), 1227–1237 (1992)

MacKenzie, D.I., Nichols, J.D., Lachman, G.B., Droege, S., Royle, A., Langtimm, C.A.: Estimating site occupancy rates when detection probabilities are less than one. Ecology **83**(8), 2248–2255 (2002)

Menon, A., van Rooyen, B., Ong, C., Williamson, R.: Learning from corrupted binary labels via class-probability estimation. Journal Machine Learning Research, vol. 16 (2015)

Natarajan, N., Dhillon, I., Ravikumar, P., Tewari, A.: Learning with noisy labels. In: Advances in Neural Information Processing Systems (2013)

Nelder, J., Wedderburn, R.: Generalized linear models. J. R. Statist. Soc. **135**(3), 370–384 (1972)

Olteanu, A., Castillo, C., Diaz, F., Kiciman, E.: Social data: biases, methodological pitfalls, and ethical boundaries. CoRR (2016)

Robins, J., Morgenstern, H.: The foundations of confounding in epidemiology. Comput. Math. Appl. **14**, 869–916 (1987)

Royle, J.A.: N-Mixture models for estimating population size from spatially replicated counts. Biometrics **60**(1), 108–115 (2004)

Schölkopf, B., et al.: Removing systematic errors for exoplanet search via latent causes. In: Proceedings of the 32nd International Conference on Machine Learning (ICML) (2015)

Servén, D., Brummitt, C.: pygam: Generalized additive models in python, March 2018. https://doi.org/10.5281/zenodo.1208723

Shankar, S., Sheldon, D., Sun, T., Pickering, J., Dietterich, T.: Three-quarter sibling regression for denoising observational data, pp. 5960–5966 (2019). https://doi.org/10.24963/ijcai.2019/826

Sharma, A.: Necessary and probably sufficient test for finding valid instrumental variables. CoRR, abs/1812.01412 (2018)

Sólymos, P., Lele, S.R.: Revisiting resource selection probability functions and single-visit methods: clarification and extensions. Methods Ecol. Evol. **7**(2), 196–205 (2016)

White, H.: Estimation, Inference and Specification Analysis. Econometric Society Monographs. Cambridge University Press, Cambridge (1994). https://doi.org/10.1017/CCOL0521252806

Yu, J., Hutchinson, R.A., Wong, W.-K.: A latent variable model for discovering bird species commonly misidentified by citizen scientists. In: Twenty-Eighth AAAI Conference on Artificial Intelligence (2014)

Zhang, Y., Jenkins, D., Manimaran, S., Johnson, W.: Alternative empirical bayes models for adjusting for batch effects in genomic studies. BMC Bioinformatics, vol. 19 (2018)

Privacy Amplification via Iteration for Shuffled and Online PNSGD

Matteo Sordello[1][(✉)], Zhiqi Bu[2], and Jinshuo Dong[3]

[1] Department of Statistics, Wharton School, University of Pennsylvania,
Philadelphia, USA
sordello@wharton.upenn.edu
[2] Graduate Group in AMCS, University of Pennsylvania, Philadelphia, USA
zbu@sas.upenn.edu
[3] IDEAL Institute, Northwestern University, Evanston, USA
jinshuo@northwestern.edu

Abstract. In this paper, we consider the framework of privacy amplification via iteration, which is originally proposed by Feldman et al. and subsequently simplified by Asoodeh et al. in their analysis via the contraction coefficient. This line of work focuses on the study of the privacy guarantees obtained by the projected noisy stochastic gradient descent (PNSGD) algorithm with hidden intermediate updates. A limitation in the existing literature is that only the early stopped PNSGD has been studied, while no result has been proved on the more widely-used PNSGD applied on a shuffled dataset. Moreover, no scheme has been yet proposed regarding how to decrease the injected noise when new data are received in an online fashion. In this work, we first prove a privacy guarantee for shuffled PNSGD, which is investigated asymptotically when the noise is fixed for each sample size n but reduced at a predetermined rate when n increases, in order to achieve the convergence of privacy loss. We then analyze the online setting and provide a faster decaying scheme for the magnitude of the injected noise that also guarantees the convergence of privacy loss.

Keywords: Differential privacy · Online learning · Optimization

1 Introduction

Differential privacy (DP) [11,12] is a strong standard to guarantee the privacy for algorithms that have been widely applied to modern machine learning [1]. It characterizes the privacy loss via statistical hypothesis testing, thus allowing the mathematically rigorous analysis of the privacy bounds. When multiple operations on the data are involved and each intermediate step is revealed, composition theorems can be used to keep track of the privacy loss, which combines subadditively [16]. However, because such results are required to be general, their associated privacy bounds are inevitably loose. In contrast, privacy amplification provides a privacy budget for a composition of mechanisms that is less that

© Springer Nature Switzerland AG 2021
N. Oliver et al. (Eds.): ECML PKDD 2021, LNAI 12976, pp. 796–813, 2021.
https://doi.org/10.1007/978-3-030-86520-7_49

the budget of each individual operation, which strengthens the bound the more operations are concatenated. Classic examples of this feature are privacy amplification by subsampling [4,8], by shuffling [14] and by iteration [3,15]. In this paper, we focus on the setting of privacy amplification by iteration, and extend the analysis via contraction coefficient proposed by [3] to prove results that apply to an algorithm commonly used in practice, in which the entire dataset is shuffled before training a model with PNSGD. We emphasize that the shuffling is a fundamental difference compared to previous work, since it is a necessary step in training many machine learning models.

We start by laying out the definitions that are necessary for our analysis. We consider a convex function $f : \mathbb{R}^+ \to \mathbb{R}$ that satisfies $f(1) = 0$. [2] and [9] define the f-divergence between two probability distribution μ and ν is as

$$D_f(\mu\|\nu) = \mathbb{E}_\nu\left[f\left(\frac{\mathrm{d}\mu}{\mathrm{d}\nu}\right)\right] = \int f\left(\frac{\mathrm{d}\mu}{\mathrm{d}\nu}\right)\mathrm{d}\nu$$

For a Markov kernel $K : \mathcal{W} \to \mathcal{P}(\mathcal{W})$, where $\mathcal{P}(\mathcal{W})$ is the space of probability measures over $\mathcal{W}$, we let $\eta_f(K)$ be the contraction coefficient of kernel K under the f-divergence, which is defined as

$$\eta_f(K) = \sup_{\mu,\nu:D_f(\mu\|\nu)\neq 0} \frac{D_f(\mu K\|\nu K)}{D_f(\mu\|\nu)}$$

If we now consider a sequence of Markov kernels $\{K_n\}$ and let the two sequences of measures $\{\mu_n\}$ and $\{\nu_n\}$ be generated starting from μ_0 and ν_0 by applying $\mu_n = \mu_{n-1}K_n$ and $\nu_n = \nu_{n-1}K_n$, then the strong data processing inequality [19] for the f-divergence tells us that

$$D_f(\mu_n\|\nu_n) \leq D_f(\mu_0\|\nu_0) \prod_{t=1}^{n} \eta_f(K_t)$$

Among the f-divergences, we focus on the E_γ-divergence, or hockey-stick divergence, which is the f-divergence associated with $f(t) = (t-\gamma)_+ = \max(0, t-\gamma)$. We do so because of its nice connection with the concept of (ϵ, δ) differential privacy, which is now the state-of-the-art technique to analyze the privacy loss that we incur when releasing information from a dataset. A mechanism $\mathcal{M}$ is said to be (ϵ, δ)-DP if, for every pair of neighboring datasets (datasets that differ only in one entry, for which we write $D \sim D'$) and every event $\mathcal{A}$, one has

$$\mathbb{P}(\mathcal{M}(D) \in \mathcal{A}) \leq e^\epsilon \mathbb{P}(\mathcal{M}(D') \in \mathcal{A}) + \delta \tag{1}$$

It is easy to prove that a mechanism $\mathcal{M}$ is (ϵ, δ)-DP if and only if the distributions that it generates on D and D' are close with respect to the E_γ-divergence. In particular, for $D \sim D'$ and $\mathbb{P}_D$ being the output distribution of mechanism $\mathcal{M}$ on D, then $\mathcal{M}$ is (ϵ, δ)-DP if and only if

$$E_{e^\epsilon}(\mathbb{P}_D\|\mathbb{P}_{D'}) \leq \delta. \tag{2}$$

It has been proved in [3] that the contraction coefficient of a kernel $K : \mathcal{W} \to \mathcal{P}(\mathcal{W})$ under E_γ-divergence, which we refer to as $\eta_\gamma(K)$, satisfies

$$\eta_\gamma(K) = \sup_{w_1, w_2 \in \mathcal{W}} E_\gamma(K(w_1) \| K(w_2))$$

This equality improves on a result proved by [5] and makes it easier to find an explicit form for the contraction coefficient of those distributions for which we can compute the hockey-stick divergence. Two such distributions are the Laplace and Gaussian, and [3] investigate the privacy guarantees generated by this privacy amplification mechanism in the setting of PNSGD with Laplace or Gaussian noise. As the standard stochastic gradient descent (SGD), the PNSGD is defined with respect to a loss function $\ell : \mathcal{W} \times \mathcal{X} \to \mathbb{R}$ that takes as inputs a parameter in the space $\mathbb{K} \subseteq \mathcal{W}$ and an observation $x \in \mathcal{X}$. Common assumptions made on the loss functions are the following: for each $x \in \mathcal{X}$

- $\ell(\cdot, x)$ is L-Lipschitz
- $\ell(\cdot, x)$ is ρ-strongly convex
- $\nabla_w \ell(\cdot, x)$ is β-Lipschitz.

The PNSGD algorithm works by combining three steps: (1) a stochastic gradient descent (SGD) step with learning rate η; (2) an injection of i.i.d. noise sampled from a known distribution to guarantee privacy and (3) a projection $\Pi_\mathbb{K} : \mathcal{W} \to \mathbb{K}$ onto the subspace $\mathbb{K}$. Combined, these steps give the following update rule

PNSGD
$$w_{t+1} = \Pi_\mathbb{K}\left(w_t - \eta(\nabla_w \ell(w_t, x_{t+1}) + Z_{t+1})\right)$$

which can be defined as a Markov kernel by assuming that $w_0 \sim \mu_0$ and $w_t \sim \mu_t = \mu_0 K_{x_1} \ldots K_{x_t}$, where K_x is the kernel associated to a single PNSGD step when observing the data point x. The application of the PNSGD on a dataset D assumes that the entries of the dataset are passed through the algorithm in a fixed order that depends on their index, hence w_1 is defined observing the first entry x_1 and so on. With this definition, one can find an upper bound for δ by bounding the left hand side of (2). The specific bound depends on the index at which the neighboring datasets D and D' differ and the distribution of the noise injected in the PNSGD. [3] investigate the bound for both Laplace and Gaussian noise, which we report in the following theorem.

Theorem 1 (Theorem 3 and 4 in [3]). *Define*

$$Q(t) = \frac{1}{\sqrt{2\pi}} \int_t^\infty e^{-\frac{u^2}{2}} du = 1 - \Phi(t)$$

where Φ is the cumulative density function of the standard normal,

$$\theta_\gamma(r) = Q\left(\frac{\log(\gamma)}{r} - \frac{r}{2}\right) - \gamma Q\left(\frac{\log(\gamma)}{r} + \frac{r}{2}\right) \tag{3}$$

and the constant

$$M = \sqrt{1 - \frac{2\eta\beta\rho}{\beta + \rho}}$$

which depends on the parameters of the loss function and the learning rate of the SGD step. If $\mathbb{K} \subset \mathbb{R}^d$ is compact and convex with diameter $D_{\mathbb{K}}$, the PNSGD algorithm with Gaussian noise $N(0, \sigma^2)$ is (ϵ, δ)-DP for its i-th entry where $\epsilon \geq 0$ and

$$\delta = \theta_{e^\epsilon} \left(\frac{2L}{\sigma} \right) \theta_{e^\epsilon} \left(\frac{MD_{\mathbb{K}}}{\eta\sigma} \right)^{n-i}$$

If instead we consider $\mathbb{K} = [a, b]$ for $a < b$, then the PNSGD algorithm with Laplace noise $\mathcal{L}(0, v)$ is (ϵ, δ)-DP for its i-th entry where $\epsilon \geq 0$ and

$$\delta = \left(1 - e^{\frac{\epsilon}{2} - \frac{L}{v}} \right)_+ \left(1 - e^{\frac{\epsilon}{2} - \frac{M(b-a)}{2\eta v}} \right)_+^{n-i}$$

To slightly simplify the notation, we can present the guarantees in Theorem 1 as $\delta = A \cdot B^{n-i}$ where for the Gaussian case

$$A = \theta_{e^\epsilon} \left(\frac{2L}{\sigma} \right), \quad B = \theta_{e^\epsilon} \left(\frac{MD_{\mathbb{K}}}{\eta\sigma} \right) \tag{4}$$

and for the Laplacian case

$$A = \left(1 - e^{\frac{\epsilon}{2} - \frac{L}{v}} \right)_+, \quad B = \left(1 - e^{\frac{\epsilon}{2} - \frac{M(b-a)}{2\eta v}} \right)_+ \tag{5}$$

To get a bound that does not depend on the index of the entry on which the two datasets differ, the authors later consider the randomly-stopped PNSGD, which simply consist of picking a random stopping time for the PNSGD uniformly from $\{1, ..., n\}$. The bound that they obtain for δ in the Gaussian case is $\delta = A/[n(1-B)]$. Based on their proof, it is clear that the actual bound contains a term $(1 - B^{n-i+1})$ at the numerator and that the same result can be obtained if we consider the Laplace noise.

In Sect. 3 we prove that a better bound than the one obtained via randomly-stopped PNSGD can be obtained by first shuffling the dataset and then applying the simple PNSGD. In Sect. 4 we study the asymptotic behavior of such bound and find the appropriate decay rate for the variability of the noise level that guarantees convergence for δ to a non-zero constant.

2 Related Work

In the DP regime, (ϵ, δ)-DP (see (1)) is arguably the most popular definition, which is oftentimes achieved by an algorithm which contains Gaussian or Laplacian noises. For example, in NoisySGD and NoisyAdam in [1,6], and PNSGD in this paper, a certain level of random noise is injected into the gradient to achieve DP. Notably, as we use more datapoints (or more iterations during the optimization) during the training procedure, the privacy loss accumulates at a rate that depends on the magnitude of the noise.

It is remarkably important to characterize, as tightly as possible, the privacy loss at each iteration. An increasing line of works have proposed to address this difficulty [1,3,4,7,10,13,17,18,20], which bring up many useful notions of DP, such as Rényi DP, Gaussian DP, f-DP and so on. Our paper extends [3] by shuffling the dataset first rather than randomly stopping the PNSGD (see Theorem 5 in [3]), in order to address the non-uniformity of privacy guarantee. As a consequence, we obtain a strictly better privacy bound and better loss than [1,3], and an additional online result of the privacy guarantee.

Furthermore, our results can be easily combined with composition tools in DP [1,10,16,17]. In Theorem 2, Theorem 3 and Theorem 4, the (ϵ, δ) is computed based on a single pass of the entire dataset, or equivalently on one epoch. When using the shuffled PNSGD for multiple epochs, as is usual for modern machine learning, the privacy loss accumulates and is accountable by Moments accountant (using Renyi DP [18]), f-DP (using functional characterization of the type I/II errors trade-off) and other divergence approaches.

3   Shuffled PNSGD

In this section, we prove the bound on δ that we can obtain by first shuffling the dataset and then apply the PNSGD algorithm. The simple underlying idea here is that, when shuffling the dataset, the index at which the two neighboring datasets differ has equal probability to end up in each position. This is a key difference compared to the randomly-stopped PNSGD, and allows us to get a better bound that do not depend on the initial position of that index.

Theorem 2. *Let $D \sim D'$ be of size n. Then the shuffled PNSGD is (ϵ, δ)-DP with*

$$\delta = \frac{A \cdot (1 - B^n)}{n(1 - B)} \tag{6}$$

and the constants A and B are defined in (4) for Gaussian noise and (5) for Laplace noise.

Proof. Let's start by considering the simple case $n = 2$, so that $D = \{x_1, x_2\}$ and $D' = \{x'_1, x'_2\}$ and let $i \in \{1, 2\}$ be the index at which they differ. Let μ be the output distribution of the shuffled PNSGD on D, and ν be the corresponding distribution from D'. If we define $S(D)$ and $S(D')$ to be the two datasets after performing the same shuffling, then we can only have either $S(D) = \{x_1, x_2\}$ or $S(D) = \{x_2, x_1\}$, both with equal probability $1/2$. The outcomes of the shuffled PNSGD on D and D' are then

$$\mu = \frac{1}{2}\mu_0 K_{x_1} K_{x_2} + \frac{1}{2}\mu_0 K_{x_2} K_{x_1}$$

$$\nu = \frac{1}{2}\mu_0 K_{x'_1} K_{x'_2} + \frac{1}{2}\mu_0 K_{x'_2} K_{x'_1}$$

By convexity and Jensen's inequality we have that

$$E_\gamma(\mu\|v) \le \frac{1}{2}E_\gamma\left(\mu_0 K_{x_1} K_{x_2} \| \mu_0 K_{x_1'} K_{x_2'}\right) + \frac{1}{2}E_\gamma\left(\mu_0 K_{x_2} K_{x_1} \| \mu_0 K_{x_2'} K_{x_1'}\right)$$

and now we have two options, based on where the two original datasets differ. If $i = 1$, in the first term the privacy is stronger than in the second one (because x_1 is seen earlier), and we have

$$E_\gamma(\mu\|\nu) \le \frac{1}{2}A \cdot B + \frac{1}{2}A = \frac{1}{2}A(B+1)$$

If $i = 2$, now the privacy is stronger in the second term, and

$$E_\gamma(\mu\|\nu) \le \frac{1}{2}A + \frac{1}{2}A \cdot B = \frac{1}{2}A(B+1)$$

Since in both cases the bound is the same, this means that for any $i \in \{1, 2\}$ the privacy guarantee of the shuffled PNSGD algorithm is equal to $A(B+1)/2$. From here we see that, when $n > 2$, the situation is similar. Instead of just two, we have $n!$ possible permutations for the elements of D, each one happening with the same probability $1/n!$. For each fixed index i on which the two neighboring datasets differ, we have $(n-1)!$ permutations in which element x_i appears in each of the n positions. When, after the permutation, element x_i ends up in last position, the bound on $E_\gamma(\mu\|\nu)$ is the weakest and just equals A. When in ends up in first position, the bound is the strongest and is equal to $A \cdot B^{n-1}$. We then have that, irrespectively of the index i,

$$E_\gamma(\mu\|v) \le \frac{1}{n!}(n-1)!A\sum_{j=0}^{n-1}B^j = \frac{A \cdot (1 - B^n)}{n(1 - B)}$$

This bound is indeed better than the one found in [3] for the randomly stopped PNSGD since it contains an extra term $(1 - B^n)$ at the numerator which does not depend on i and is smaller than 1. If n is large and B is fixed, this difference is negligible because it decays exponentially. However, we will see later that when the injected noise is reduced at the appropriate rate we can guarantee that $B \approx 1 - O(1/n)$, so that the extra term ends up having an impact in the final bound. It is also important to notice that shuffled PNSGD achieves in general better performance than randomly stopped PNSGD and it is much more commonly used in practice. We see in Fig. 1 that this is the case for both linear and logistic regression, and that the variation in the result in shuffled PNSGD is less than for the early stopped case, due to the fact that we always use all the data available for each epoch. In the next section we look at the asymptotic behavior of (6) when n grows and the variance of the injected noise is properly reduced to guarantee convergence.

4   Asymptotic Analysis for δ When Using Shuffling and Fixed Noises

In this Section we investigate the behavior of the differential privacy bound in (6) when the size n of the dataset grows. In Sect. 4.1 we prove a results for the

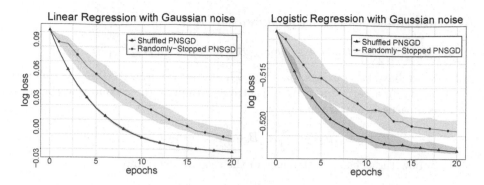

Fig. 1. Comparison between shuffled PNSGD and randomly-stopped PNSGD with Gaussian noise in linear and logistic regression. On the y-axis we report the log loss achieved. The parameters used are $n = 1000, d = 2, \sigma = 0.5, \theta^* = \Pi_{\mathbb{K}}(1,2)$ and $\mathbb{K}$ is a ball of radius 1. The learning rate is 10^{-4} in linear regression and $5 \cdot 10^{-3}$ in logistic regression.

shuffled PNSGD with fixed Laplace noise, while in Sect. 4.2 we prove the same result on the shuffled PNSGD with fixed Gaussian noise.

4.1 Laplace Noise

We present first a result that holds when we consider a fixed Laplace noise $\mathcal{L}(0, v)$ injected into the PNSGD algorithm for each update. In order to get a convergence result for δ as the size n of the dataset grows, the level of noise that we use should be targeted to the quantity n. The decay of v is regulated by two parameters, C_1 and C_2. While C_1 is set to be large, so that δ converges to a small value, the use of C_2 is simply to allow the noise level not to be too large for small n, but does not appear in the asymptotic bound.

Theorem 3. *Consider the shuffled PNSGD with Laplace noise $\mathcal{L}(0, v(n))$ which is fixed for each update, where*

$$v(n) = \frac{M(b-a)}{2\eta \log (n/C_1 + C_2)}. \tag{7}$$

Then, for n sufficiently large the procedure is (ϵ, δ)-DP with $\delta = \delta^* + O(1/n)$ and

$$\delta^* = \frac{1 - e^{-C_1 \exp(\epsilon/2)}}{C_1 e^{\frac{\epsilon}{2}}} \tag{8}$$

Proof. We use the result in Theorem 2 combined with (5), and get that

$$\delta = \frac{\left(1 - e^{\frac{\epsilon}{2} - \frac{L}{v(n)}}\right)_+ \cdot \left[1 - \left(1 - e^{\frac{\epsilon}{2} - \frac{M(b-a)}{2\eta v(n)}}\right)_+^n\right]}{n \cdot e^{\frac{\epsilon}{2} - \frac{M(b-a)}{2\eta v(n)}}}$$

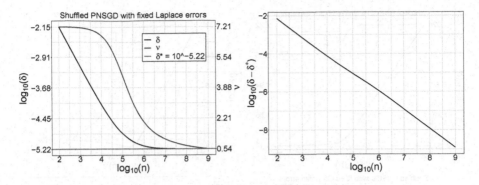

Fig. 2. (left) Convergence of δ to δ^* in (8). We plot in black the behavior of δ as a function of n, and in blue the corresponding behavior of $v(n)$ in (7). (right) We show that the convergence rate is $1/n$. The parameters used are $L = 10, \beta = 0.5, \rho = 0, \eta = 0.1, \epsilon = 1, (a, b) = (0, 1), C_1 = 10^5$ and $C_2 = 2$. (Color figure online)

Once we plug in the $v(n)$ defined in (7) we have that, when n is sufficiently large,

$$
\delta = \frac{\left(1 - e^{\frac{\epsilon}{2} - \frac{2L\eta \log(\frac{n}{C_1} + C_2)}{M(b-a)}}\right)_+ \left[1 - \left(1 - \frac{C_1 e^{\frac{\epsilon}{2}}}{n + C_1 C_2}\right)^n_+\right]}{n \cdot e^{\frac{\epsilon}{2} - \log(\frac{n}{C_1} + C_2)}}
$$

$$
= \frac{\left[1 - \left(1 - \frac{C_1 e^{\frac{\epsilon}{2}}}{n + C_1 C_2}\right)^n\right]}{n \cdot \frac{C_1 e^{\frac{\epsilon}{2}}}{n + C_1 C_2}} \cdot \left(1 + O\left(\frac{1}{n}\right)\right)
$$

$$
= \frac{1 - e^{-C_1 \exp(\epsilon/2)}}{C_1 e^{\frac{\epsilon}{2}}} + O\left(\frac{1}{n}\right)
$$

The convergence result in Theorem 3 is confirmed by Fig. 2. In the left plot we see that δ converges to the δ^* defined in (8), while in the right plot we observe that the convergence rate is indeed $1/n$.

4.2   Gaussian Noise

Similarly to what we just proved in Sect. 4.1 we now discuss a result for the shuffled PNSGD with Gaussian noise $N(0, \sigma^2(n))$.

Theorem 4. *Consider the shuffled PNSGD algorithm with Gaussian noise $N(0, \sigma^2(n))$ which is fixed for each update, where*

$$
\sigma(n) = \frac{M D_{\mathbb{K}}}{2\eta \sqrt{W\left(\frac{n^2}{2C_1^2 \pi} + C_2\right)}} \tag{9}
$$

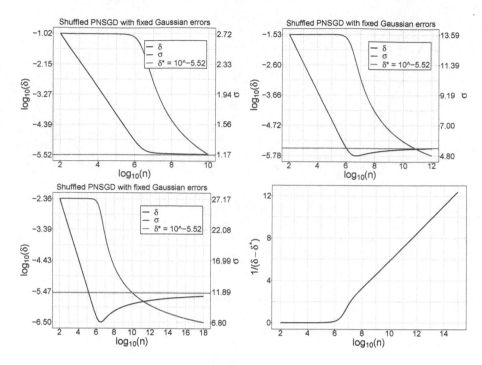

Fig. 3. Convergence of δ to δ^* defined in (10). We report in black the behavior of δ and in blue that of $\sigma(n)$ in (9). We consider $\eta \in \{0.1, 0.02, 0.01\}$ and the other parameters are $L = 10, \beta = 0.5, \rho = 0, \epsilon = 1, D_{\mathbb{K}} = 1, C_1 = 10^5$ and $C_2 = 100$. In the right-bottom panel we show that the convergence rate is $1/\log(n)$. (Color figure online)

and W is the Lambert W function. Then, for n sufficiently large, the procedure is (ϵ, δ)-DP with $\delta = \delta^* + O\left(\frac{1}{\log(n)}\right)$ and

$$\delta^* = \frac{1 - e^{-2C_1 e^{\frac{\epsilon}{2}}}}{2C_1 e^{\frac{\epsilon}{2}}} \tag{10}$$

Just like $v(n)$, the decay of the standard deviation $\sigma(n)$ is regulated by the parameters C_1 and C_2. The difference here is that, instead of a simple logarithmic decay, we now have a decay rate that depends on the Lambert W function, which is slightly harder to study analytically than the logarithm. Even though the Lambert W function is fundamentally equivalent to a logarithm when its argument grows, the difference with the Laplace case is also evident in the fact that the convergence of δ to δ^* happens more slowly, at a rate of $1/\log(n)$. The proof of the theorem is in the Supplementary Material, and makes use of the following Lemma, also proved in the Supplementary Material.

Lemma 5. *For $\theta_\gamma(r)$ defined in (3), a sufficiently small σ and two constants c and ϵ, we have*

$$\theta_{e^\epsilon}\left(\frac{c}{\sigma}\right) = 1 - \frac{1}{\sqrt{2\pi}}e^{\frac{\epsilon}{2}}e^{-\frac{c^2}{8\sigma^2}}\left(\frac{4\sigma}{c} + O(\sigma^3)\right).$$

The behavior described in Theorem 4 is confirmed by what we see in Fig. 3, where we can also observe that there are different patterns of convergence for δ, both from above and from below the δ^* defined in (10). In the right-bottom panel we also see a confirmation that the convergence rate is the one we expected, since $(\delta - \delta^*)^{-1}$ increase linearly with respect to $\log(n)$ when n is sufficiently large (notice that the y-axis is rescaled by a factor 10^6).

5 Multiple Epochs Composition

We now consider a simple yet important extension of the result in Theorem 2, where the shuffled PNSGD is applied for multiple epochs. In real experiments, e.g. when training deep neural networks, usually multiple passes over the data are necessary to learn the model. In such scenario, the updates are not kept secret for the whole duration of the training, but are instead released at the end of each epoch. The result proved in Theorem 2 states that for each epoch the procedure is (ϵ, δ)-DP with $\delta \leq A \cdot (1 - B^n)/[n(1 - B)]$. We can then easily combine these privacy bounds using state-of-the-art composition tools, such as the Moments Accountant [1], f-DP and Gaussian DP [10]. We present some popular ways to compute the privacy loss after E epochs.

At the high level, we migrate from (ϵ, δ) in DP to other regimes, Gaussian DP or Rényi DP, at the first epoch. Then we compose in those specific regimes until the end of training procedure. At last, we map back from the other regimes back to (ϵ, δ)-DP.

f-DP and Gaussian DP: At the first epoch, we compute the initial (ϵ, δ) and derive the four-segment curve $f_{\epsilon,\delta}$ for the type I/II errors trade-off (see Equation (5) and Proposition 2.5 in [10]). Then by Theorem 3.2 in [10], we can numerically compose this trade-off function with Fourier transform for E times, which can be accelerated by repeated squaring. When the noise is Gaussian, we can alternatively use μ in GDP to characterize the trade-off function (i.e. the mechanism is μ-GDP after the first epoch). Next, we apply Corollary 3.3 in [10] to conclude that the mechanism is $\sqrt{E}\mu$-GDP in the end. We can compute the final (ϵ, δ) reversely from GDP by Corollary 2.13 in [10].

Moments Accountant: Moments Accountant is closely related to Rényi DP (RDP), which composes easily: at the first epoch, we compute the (ϵ, δ) of our PNSGD. By Proposition 3 in [18], we can transfer from (ϵ, δ)-DP to $(\alpha, \epsilon + \frac{\log \delta}{\alpha - 1})$ RDP. After the first epoch, the initial RDP can be composed iteratively by Moments Accountant[1]. The final (α', ϵ') RDP is then mapped back to (ϵ, δ)-DP with $\epsilon = \epsilon' - \frac{\log \delta}{\alpha' - 1}$.

---

[1] See https://github.com/tensorflow/privacy/blob/master/tensorflow_privacy/privacy/analysis/rdp_accountant.py.

6   Online Results for Decaying Noises

We now go back to the original framework of [3] and consider the PNSGD algorithm applied to the non-shuffled dataset. This time, however, we want to apply a different level of noise for each update, and see if we can get a convergence result for δ when $n \to \infty$. We then need to consider values of A and B in (4) and (5) that depend on the specific index, and the privacy bound for the PNSGD with non-fixed noises and neighboring datasets that differ on index i becomes

$$\delta = A_i \cdot \prod_{t=i+1}^{n} B_t \tag{11}$$

Here the definition of A_i and B_i is the same as in (4) and (5) but the noise level v and σ is now dependent on the position of each element in the dataset. In this scenario we can actually imagine adding new data to the dataset in an online fashion, without having to restart the procedure to recalibrate the noise level used for the first entries. It is clear that, in order to get convergence, the decay of the injected noise should be faster than in Theorem 3 and Theorem 4, since now the early entries receive an amount of noise that does not vanish as n becomes large. However it is interesting to notice that for both the Laplace and Gaussian noise the only difference needed with the decay rate for $v(n)$ and $\sigma(n)$ defined before is an exponent $\alpha > 1$.

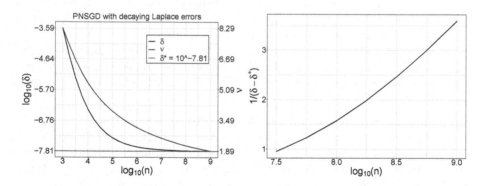

Fig. 4. (left) Convergence of δ to δ^* defined in (13). We report in black the behavior of δ and in blue that of v_n defined in (12). The parameters considered are $L = 10, \beta = 0.5, \rho = 0, \epsilon = 1, \eta = 0.01, \alpha = 1.5, (a, b) = (0, 1), i = 100, C_1 = 100$ and $C_2 = 100$. (right) The convergence rate is approximately $1/\log(n)$. (Color figure online)

6.1   Laplace Noise

We prove here the online result for the PNSGD with Laplace noise that decays for each entry. As anticipated, the decay is no longer the same for all entries and proportional to $1/\log(n)$ but now for the entry with index j we have a decay which is proportional to $1/\log(j^{\alpha})$.

Theorem 6. *Consider the PNSGD where for update j we use Laplace noise $\mathcal{L}(0, v_j)$, and*

$$v_j = \frac{M(b-a)}{2\eta \log\left(j^\alpha / C_1 + C_2\right)} \tag{12}$$

for $\alpha > 1$. Then as $n \to \infty$ the procedure is (ϵ, δ^*)-DP where

$$\delta^* = \left(1 - e^{\frac{\epsilon}{2} - \frac{2L\eta \log(i^\alpha/C_1 + C_2)}{M(b-a)}}\right)_+ e^{\int_{i+1}^\infty \log\left(1 - \frac{C_1 e^{\frac{\epsilon}{2}}}{x^\alpha + C_1 C_2}\right) dx} \tag{13}$$

and i is the index where the neighboring datasets differ.

Proof. We show again that δ converges to a non-zero value as n goes to ∞. In fact, again following the proof of ([3] Theorem 3), we get that,

$$\delta = \left(1 - e^{\frac{\epsilon}{2} - \frac{L}{v_i}}\right)_+ \cdot \prod_{t=i+1}^n \left(1 - e^{\frac{\epsilon}{2} - \frac{M(b-a)}{2\eta v_t}}\right)_+$$

$$= \left(1 - e^{\frac{\epsilon}{2} - \frac{2L\eta \log(\frac{i^\alpha}{C_1} + C_2)}{M(b-a)}}\right)_+ \prod_{t=i+1}^n \left(1 - \frac{C_1 e^{\frac{\epsilon}{2}}}{t^\alpha + C_1 C_2}\right)_+$$

We know that, for a sequence a_t of positive values, $\prod_{t=1}^\infty (1 - a_t)$ converges to a non-zero number if and only if $\sum_{t=1}^\infty a_t$ converges. Here we have that

$$\sum_{t=i+1}^\infty \frac{C_1 e^{\frac{\epsilon}{2}}}{t^\alpha + C_1 C_2} \leq \sum_{t=i+1}^\infty \frac{C_1 e^{\frac{\epsilon}{2}}}{t^\alpha}$$

and, since $\alpha > 1$ the right hand side converges, hence δ converges to a non-zero number. Let now $f(n) = \prod_{t=i+1}^n \left(1 - \frac{C_1 e^{\frac{\epsilon}{2}}}{t^\alpha + C_1 C_2}\right)_+$. To find the limit $f(\infty)$ we can first log-transform this function, and then upper bound the infinite sum with an integral before transforming back. Since $\log\left(1 - \frac{C_1 e^{\frac{\epsilon}{2}}}{t^\alpha + C_1 C_2}\right)$ is monotonically increasing in t, we have

$$\log(f(n)) = \sum_{t=i+1}^n \log\left(1 - \frac{C_1 e^{\frac{\epsilon}{2}}}{t^\alpha + C_1 C_2}\right)$$

$$< \int_{i+1}^n \log\left(1 - \frac{C_1 e^{\frac{\epsilon}{2}}}{t^\alpha + C_1 C_2}\right) dt \to \int_{i+1}^\infty \log\left(1 - \frac{C_1 e^{\frac{\epsilon}{2}}}{t^\alpha + C_1 C_2}\right) dt.$$

This integral can be written in closed form using the hypergeometric function, or approximated numerically.

The convergence result that we get is slightly conservative, since δ^* in Eq. 13 is an upper bound. However, following the previous proof, we can find an easy lower bound by just noticing that $\log(f(\infty)) > \int_i^\infty \log\left(1 - \frac{C_1 e^{\frac{\epsilon}{2}}}{t^\alpha + C_1 C_2}\right) dt$. When i is not too small, the difference between the upper and lower bound is negligible, as it is confirmed by what we see in the left plot of Fig. 4, where the convergence to the upper bound appears to be impeccable. Since the convergence is not exactly to δ^*, we cannot find an explicit convergence rate the same way we did in Sect. 4. However, we see in the right plot of Fig. 4 that the convergence rate empirically appears to be $1/\log(n)$.

6.2   Gaussian Noise

When working with the Gaussian noises, the cumbersome form of the functions in (4) does not prevent us from finding a closed form solution for the limit δ^*. Just as in the Laplace case we can find a conservative upper bound for δ^* which is very close to the true limit, as confirmed by the left plot of Fig. 5. Just as before, we notice again empirically from the right plot of Fig. 5 that the convergence rate is $1/\log(n)$.

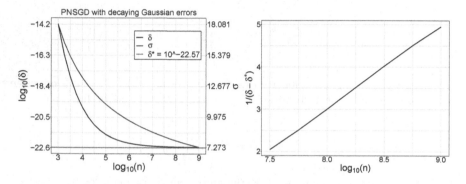

Fig. 5. (left) Convergence of δ to δ^* defined in (15). We report in black the behavior of δ and in blue that of σ_n defined in (14). The parameters considered are $L = 10, \beta = 0.5, \rho = 0, \epsilon = 1, \eta = 0.01, \alpha = 1.5, D_{\mathbb{K}} = 1, i = 100, C_1 = 100$ and $C_2 = 100$. (right) The convergence rate is approximately $1/\log(n)$. (Color figure online)

Theorem 7. *Consider the PNSGD where for update j we use Gaussian noise $N(0, \sigma_j^2)$, and*

$$\sigma_j = \frac{MD_{\mathbb{K}}}{2\eta\sqrt{W\left(\frac{j^{2\alpha}}{2\pi C_1^2} + C_2\right)}} \tag{14}$$

for $\alpha > 1$. Then as $n \to \infty$ the procedure is (ϵ, δ^*)-DP where

$$\delta^* = \theta_{e^\epsilon}\left(\frac{2L}{\sigma_i}\right)e^{\int_{i+1}^{\infty}\log\left(\theta_{e^\epsilon}\left(2\sqrt{W\left(\frac{x^{2\alpha}}{2\pi C_1^2}+C_2\right)}\right)\right)dx} \tag{15}$$

and i is the index where the neighboring datasets differ.

The proof of this result is in the Supplementary Material, and makes use again of Lemma 5 to show that asymptotically the terms B_t in (11) behave approximately as $1 - O(1/t^\alpha)$, so that convergence is guaranteed for the same reason as in Theorem 6.

7 Conclusion

In this work, we have studied the setting of privacy amplification by iteration in the formulation proposed by [3], and proved that their analysis of PNSGD also applies to the case where the data are shuffled first. This is a much more common practice than the randomly-stopped PNSGD, originally proposed, because of a clear advantage in terms of accuracy of the algorithm. We proved two asymptotic results on the decay rate of noises that we can use, either the Laplace or the Gaussian injected noise, in order to have asymptotic convergence to a non-trivial privacy bound when the size of the dataset grows. We then showed that these practical bounds can be combined using standard tools from the composition literature. Finally we also showed two result, again for Laplace or Gaussian noise, that can be obtained in an online setting when the noise does not have to be recalibrated for the whole dataset but just decayed for the new data.

Acknowledgement. The authors would like to thank Weijie Su for his advice and encouragements.

Supplement to: Privacy Amplification via Iteration for Shuffled and Online PNSGD

Proof of Lemma 5

Recall from the definition (3):

$$\theta_{e^\epsilon}\left(\frac{c}{\sigma}\right) = Q\left(\frac{\epsilon\sigma}{c} - \frac{c}{2\sigma}\right) - e^\epsilon Q\left(\frac{\epsilon\sigma}{c} + \frac{c}{2\sigma}\right) \qquad (16)$$

We apply the following approximation of the normal cumulative density function, valid for large positive x,

$$Q(x) := \frac{1}{\sqrt{2\pi}}\int_x^\infty e^{-\frac{u^2}{2}}\,du = \frac{1}{\sqrt{2\pi}}e^{-\frac{x^2}{2}}\left(\frac{1}{x} + O\left(\frac{1}{x^3}\right)\right)$$

and similarly, for large negative values of x

$$Q(x) := \frac{1}{\sqrt{2\pi}}\int_x^\infty e^{-\frac{u^2}{2}}\,du = 1 + \frac{1}{\sqrt{2\pi}}e^{-\frac{x^2}{2}}\left(\frac{1}{x} + O\left(\frac{1}{x^3}\right)\right).$$

Therefore (16) can be reformulated as

$$
\theta_{e^\epsilon}\left(\frac{c}{\sigma}\right) = 1 + \frac{1}{\sqrt{2\pi}}e^{-\frac{1}{2}\left(\frac{\epsilon^2\sigma^2}{c^2}+\frac{c^2}{4\sigma^2}\right)+\frac{\epsilon}{2}}\left(\frac{1}{\frac{\epsilon\sigma}{c}-\frac{c}{2\sigma}} + O\left(\frac{1}{\frac{\epsilon\sigma}{c}-\frac{c}{2\sigma}}\right)^3\right)
$$

$$
- \frac{1}{\sqrt{2\pi}}e^{\epsilon}e^{-\frac{1}{2}\left(\frac{\epsilon^2\sigma^2}{c^2}+\frac{c^2}{4\sigma^2}\right)-\frac{\epsilon}{2}}\left(\frac{1}{\frac{\epsilon\sigma}{c}+\frac{c}{2\sigma}} + O\left(\frac{1}{\frac{\epsilon\sigma}{c}+\frac{c}{2\sigma}}\right)^3\right)
$$

$$
= 1 - \frac{1}{\sqrt{2\pi}}e^{\frac{\epsilon}{2}}e^{-\frac{1}{2}\left(\frac{\epsilon^2\sigma^2}{c^2}+\frac{c^2}{4\sigma^2}\right)}\left(\frac{4\sigma}{c} + O(\sigma^3)\right)
$$

$$
= 1 - \frac{1}{\sqrt{2\pi}}e^{\frac{\epsilon}{2}}e^{-\frac{c^2}{8\sigma^2}}\left(\frac{4\sigma}{c} + O(\sigma^3)\right)
$$

Proof of Theorem 4

From Theorem 2 we know that

$$
\delta = \frac{\theta_{e^\epsilon}\left(\frac{2L}{\sigma(n)}\right)\cdot\left[1-\theta_{e^\epsilon}\left(\frac{MD_\mathbb{K}}{\eta\sigma(n)}\right)^n\right]}{n\cdot\left[1-\theta_{e^\epsilon}\left(\frac{MD_\mathbb{K}}{\eta\sigma(n)}\right)\right]} \tag{17}
$$

We show that with $\sigma(n)$ that decays according to (9) we have that

$$
\theta_{e^\epsilon}\left(\frac{2L}{\sigma(n)}\right) \to 1 \quad \text{and} \quad \theta_{e^\epsilon}\left(\frac{MD_\mathbb{K}}{\eta\sigma(n)}\right) \to 1 - \frac{2C_1e^{\frac{\epsilon}{2}}}{n}.
$$

Let's first focus briefly on the behavior of the Lambert W function. Formally, the Lambert W function is an implicit function defined as the inverse of $f(w) = we^w$, meaning that for any x one has $W(x)e^{W(x)} = x$. As an interesting fact, we note that the Lambert W function's behavior is approximately logarithmic, e.g. $\log(x) > W(x) > \log_4(x)$, where by log we denote the natural logarithm. We also denote the argument of the W Lambert function in $\sigma(n)$ as $x = \frac{n^2}{2C_1^2\pi} + C_2$. Using this fact, an immediate consequence of Lemma 5 is that, when plugging in the $\sigma(n)$ from (9), we get

$$
\theta_{e^\epsilon}\left(\frac{2L}{\sigma(n)}\right) = 1 - o(\sigma^3) = 1 - o\left(\frac{1}{\sqrt{W^3(x)}}\right) = 1 - o\left(\frac{1}{(\log n)^{3/2}}\right)
$$

since $e^{-\frac{c^2}{8\sigma^2}} \cdot \frac{1}{\sigma^2} \to 0$ as the exponential decays faster than the polynomial. Next, we study $\theta_{e^\epsilon}\left(\frac{MD_\mathbb{K}}{\eta\sigma(n)}\right)$. Again by Lemma 5, we have

$$
\begin{aligned}
\theta_{e^\epsilon}\left(\frac{MD_\mathbb{K}}{\eta\sigma(n)}\right) &= 1 - \frac{1}{\sqrt{2\pi}} e^{\frac{\epsilon}{2}} e^{-\frac{M^2 D_\mathbb{K}^2}{8\eta^2 \sigma(n)^2}} \left(\frac{4\eta\sigma(n)}{MD_\mathbb{K}} + O(\sigma(n)^3)\right) \\
&= 1 - \frac{1}{\sqrt{2\pi}} e^{\frac{\epsilon}{2}} e^{-\frac{W(x)}{2}} \left(\frac{2}{\sqrt{W(x)}} + O\left(\frac{1}{W(x)^{3/2}}\right)\right) \\
&= 1 - \frac{2e^{\frac{\epsilon}{2}}}{\sqrt{2\pi}} \frac{1}{\sqrt{e^{W(x)} W(x)}} + O\left(\frac{1}{\sqrt{e^{W(x)} W(x)^3}}\right) \\
&= 1 - \frac{2e^{\frac{\epsilon}{2}}}{\sqrt{2\pi x}} + O\left(\frac{1}{\sqrt{x}\log(x)}\right) \\
&= 1 - \frac{2C_1 e^{\frac{\epsilon}{2}}}{n} + O\left(\frac{1}{n\log(n)}\right)
\end{aligned}
\tag{18}
$$

Going back to the expression in (17) we finally have that

$$
\begin{aligned}
\delta &= \frac{\left(1 - o\left(\frac{1}{(\log(n))^{3/2}}\right)\right)\left[1 - \left(1 - \frac{2C_1 e^{\frac{\epsilon}{2}}}{n} + O\left(\frac{1}{n\log(n)}\right)\right)^n\right]}{n \cdot \left[1 - \left(1 - \frac{2C_1 e^{\frac{\epsilon}{2}}}{n} + O\left(\frac{1}{n\log(n)}\right)\right)\right]} \\
&= \frac{\left(1 - o\left(\frac{1}{(\log(n))^{3/2}}\right)\right)\left[1 - \left(1 - \frac{2C_1 e^{\frac{\epsilon}{2}} + O(1/\log(n))}{n}\right)^n\right]}{2C_1 e^{\frac{\epsilon}{2}} + O\left(\frac{1}{\log(n)}\right)} \\
&= \frac{\left(1 - o\left(\frac{1}{(\log(n))^{3/2}}\right)\right)\left[1 - e^{-2C_1 e^{\frac{\epsilon}{2}} + O(1/\log(n))}\right]}{2C_1 e^{\frac{\epsilon}{2}} + O\left(\frac{1}{\log(n)}\right)} \\
&= \left(1 - o\left(\frac{1}{(\log(n))^{3/2}}\right)\right)\left[1 - e^{-2C_1 e^{\frac{\epsilon}{2}}} + O\left(\frac{1}{\log(n)}\right)\right]\left(\frac{1}{2C_1 e^{\frac{\epsilon}{2}}} - o\left(\frac{1}{\log(n)}\right)\right) \\
&= \frac{1 - e^{-2C_1 e^{\frac{\epsilon}{2}}}}{2C_1 e^{\frac{\epsilon}{2}}} + O\left(\frac{1}{\log(n)}\right).
\end{aligned}
$$

Proof of Theorem 7

This proof combines elements of the proofs of Theorem 4 and Theorem 6. We start by studying the behavior of $\theta_{e^\epsilon}\left(\frac{MD_\mathbb{K}}{\eta\sigma_t}\right)$ as t grows. We define $x = \frac{t^{2\alpha}}{2\pi C_1^2} + C_2$ so that $\sigma_t = \frac{MD_\mathbb{K}}{2\eta\sqrt{W(x)}}$ and get, as in (18),

$$
\begin{aligned}
\theta_{e^\epsilon}\left(\frac{MD_\mathbb{K}}{\eta\sigma_t}\right) &= 1 - \frac{1}{\sqrt{2\pi}} e^{\frac{\epsilon}{2}} e^{-\frac{M^2 D_\mathbb{K}^2}{8\eta^2 \sigma_t^2}} \left(\frac{4\eta\sigma_t}{MD_\mathbb{K}} + O(\sigma_t^3)\right) \\
&= 1 - \frac{2e^{\frac{\epsilon}{2}}}{\sqrt{2\pi x}} + O\left(\frac{1}{\sqrt{x}\log(x)}\right) = 1 - \frac{2C_1 e^{\frac{\epsilon}{2}}}{t^\alpha} + O\left(\frac{1}{t^\alpha \log(t)}\right)
\end{aligned}
$$

This already confirms us that δ^* converges to a finite non zero value, since the asymptotic behavior of each term in the infinite product is the same as in the Laplace case. To express such limit in a more tractable way we follow the proof of Theorem 6 and write $f(n) = \prod_{t=i+1}^{n} \theta_{e^{\epsilon}} \left(\frac{MD_{\mathbb{K}}}{\eta \sigma_t} \right)$ and approximate the infinite sum $\log(f(\infty))$ with an integral.

$$\log(f(n)) = \sum_{t=i+1}^{n} \log \left(\theta_{e^{\epsilon}} \left(\frac{MD_{\mathbb{K}}}{\eta \sigma_t} \right) \right)$$

$$= \sum_{t=i+1}^{n} \log \left(\theta_{e^{\epsilon}} \left(2\sqrt{W \left(\frac{t^{2\alpha}}{2\pi C_1^2} + C_2 \right)} \right) \right)$$

$$< \int_{i+1}^{n} \log \left(\theta_{e^{\epsilon}} \left(2\sqrt{W \left(\frac{x^{2\alpha}}{2\pi C_1^2} + C_2 \right)} \right) \right) dx$$

$$\rightarrow \int_{i+1}^{\infty} \log \left(\theta_{e^{\epsilon}} \left(2\sqrt{W \left(\frac{x^{2\alpha}}{2\pi C_1^2} + C_2 \right)} \right) \right) dx$$

This confirms us that

$$\delta^* = \theta_{e^{\epsilon}} \left(\frac{2L}{\sigma_i} \right) \cdot \exp \left\{ \int_{i+1}^{\infty} \log \left(\theta_{e^{\epsilon}} \left(2\sqrt{W \left(\frac{x^{2\alpha}}{2\pi C_1^2} + C_2 \right)} \right) \right) dx \right\}.$$

References

1. Abadi, M., et al.: Deep learning with differential privacy. In: Proceedings of the 2016 ACM SIGSAC Conference on Computer and Communications Security, pp. 308–318 (2016)
2. Ali, S.M., Silvey, S.D.: A general class of coefficients of divergence of one distribution from another. J. R. Stat. Soc. Ser. B (Methodol.) **28**(1), 131–142 (1966)
3. Asoodeh, S., Diaz, M., Calmon, F.P.: Privacy amplification of iterative algorithms via contraction coefficients. arXiv preprint arXiv:2001.06546 (2020)
4. Balle, B., Barthe, G., Gaboardi, M.: Privacy amplification by subsampling: tight analyses via couplings and divergences. In: Advances in Neural Information Processing Systems, pp. 6277–6287 (2018)
5. Balle, B., Barthe, G., Gaboardi, M., Geumlek, J.: Privacy amplification by mixing and diffusion mechanisms. In: Advances in Neural Information Processing Systems, pp. 13298–13308 (2019)
6. Bu, Z., Dong, J., Long, Q., Su, W.J.: Deep learning with gaussian differential privacy. Harvard Data Science Review, vol. 2020, no. 23 (2020)
7. Bun, M., Steinke, T.: Concentrated differential privacy: simplifications, extensions, and lower bounds. In: Hirt, M., Smith, A. (eds.) TCC 2016. LNCS, vol. 9985, pp. 635–658. Springer, Heidelberg (2016). https://doi.org/10.1007/978-3-662-53641-4_24
8. Chaudhuri, K., Mishra, N.: When random sampling preserves privacy. In: Dwork, C. (ed.) CRYPTO 2006. LNCS, vol. 4117, pp. 198–213. Springer, Heidelberg (2006). https://doi.org/10.1007/11818175_12

9. Csiszár, I., Shields, P.C.: Information theory and statistics: A tutorial. Now Publishers Inc (2004)
10. Dong, J., Roth, A., Su, W.J.: Gaussian differential privacy. arXiv preprint arXiv:1905.02383 (2019)
11. Dwork, C., Kenthapadi, K., McSherry, F., Mironov, I., Naor, M.: Our data, ourselves: privacy via distributed noise generation. In: Vaudenay, S. (ed.) EUROCRYPT 2006. LNCS, vol. 4004, pp. 486–503. Springer, Heidelberg (2006). https://doi.org/10.1007/11761679_29
12. Dwork, C., McSherry, F., Nissim, K., Smith, A.: Calibrating noise to sensitivity in private data analysis. In: Halevi, S., Rabin, T. (eds.) TCC 2006. LNCS, vol. 3876, pp. 265–284. Springer, Heidelberg (2006). https://doi.org/10.1007/11681878_14
13. Dwork, C., Rothblum, G.N.: Concentrated differential privacy. arXiv preprint arXiv:1603.01887 (2016)
14. Erlingsson, Ú., Feldman, V., Mironov, I., Raghunathan, A., Talwar, K., Thakurta, A.: Amplification by shuffling: from local to central differential privacy via anonymity. In: Proceedings of the Thirtieth Annual ACM-SIAM Symposium on Discrete Algorithms, pp. 2468–2479. SIAM (2019)
15. Feldman, V., Mironov, I., Talwar, K., Thakurta, A.: Privacy amplification by iteration. In: 2018 IEEE 59th Annual Symposium on Foundations of Computer Science (FOCS), pp. 521–532. IEEE (2018)
16. Kairouz, P., Oh, S., Viswanath, P.: The composition theorem for differential privacy. In: International Conference on Machine Learning, pp. 1376–1385. PMLR (2015)
17. Koskela, A., Jälkö, J., Honkela, A.: Computing tight differential privacy guarantees using fft. In: International Conference on Artificial Intelligence and Statistics, pp. 2560–2569. PMLR (2020)
18. Mironov, I.: Rényi differential privacy. In: 2017 IEEE 30th Computer Security Foundations Symposium (CSF), pp. 263–275. IEEE (2017)
19. Raginsky, M.: Strong data processing inequalities and ϕ-sobolev inequalities for discrete channels. IEEE Trans. Inform. Theor. **62**(6), 3355–3389 (2016)
20. Wang, Y.X., Balle, B., Kasiviswanathan, S.P.: Subsampled rényi differential privacy and analytical moments accountant. In: The 22nd International Conference on Artificial Intelligence and Statistics, pp. 1226–1235. PMLR (2019)

Author Index

Printed in the United States
by Baker & Taylor Publisher Services

Printed in the United States
by Baker & Taylor Publisher Services